Lecture Notes in Electrical Engineering

Volume 912

The book series *Lecture Notes in Electrical Engineering* (LNEE) publishes the latest developments in Electrical Engineering - quickly, informally and in high quality. While original research reported in proceedings and monographs has traditionally formed the core of LNEE, we also encourage authors to submit books devoted to supporting student education and professional training in the various fields and applications areas of electrical engineering. The series cover classical and emerging topics concerning:

- Communication Engineering, Information Theory and Networks
- Electronics Engineering and Microelectronics
- Signal, Image and Speech Processing
- Wireless and Mobile Communication
- Circuits and Systems
- Energy Systems, Power Electronics and Electrical Machines
- Electro-optical Engineering
- Instrumentation Engineering
- Avionics Engineering
- Control Systems
- Internet-of-Things and Cybersecurity
- Biomedical Devices, MEMS and NEMS

For general information about this book series, comments or suggestions, please contact leontina.dicecco@springer.com.

To submit a proposal or request further information, please contact the Publishing Editor in your country:

China

Jasmine Dou, Editor (jasmine.dou@springer.com)

India, Japan, Rest of Asia

Swati Meherishi, Editorial Director (Swati.Meherishi@springer.com)

Southeast Asia, Australia, New Zealand

Ramesh Nath Premnath, Editor (ramesh.premnath@springernature.com)

USA, Canada:

Michael Luby, Senior Editor (michael.luby@springer.com)

All other Countries:

Leontina Di Cecco, Senior Editor (leontina.dicecco@springer.com)

**** This series is indexed by EI Compendex and Scopus databases. ****

More information about this series at https://link.springer.com/bookseries/7818

Sangchul Lee · Cheolheui Han · Jeong-Yeol Choi ·
Seungkeun Kim · Jeong Ho Kim
Editors

The Proceedings of the 2021 Asia-Pacific International Symposium on Aerospace Technology (APISAT 2021), Volume 1

APISAT 2021
November 15-17, 2021
Ramada Plaza Jeju Hotel, South Korea

Springer

Editors
Sangchul Lee
School of Aerospace and Mechanical Engineering
Korea Aerospace University
Gyeonggi-do, Korea (Republic of)

Cheolheui Han
Department of Mechanical Design and Aeronautical Engineering
Korea National University of Transportation
Chungju, Korea (Republic of)

Jeong-Yeol Choi
Department of Aerospace Engineering
Pusan National University
Busan, Korea (Republic of)

Seungkeun Kim
Department of Aerospace Engineering
Chungnam National University
Daejeon, Korea (Republic of)

Jeong Ho Kim
Department of Aerospace Engineering
Inha University
Incheon, Korea (Republic of)

ISSN 1876-1100 ISSN 1876-1119 (electronic)
Lecture Notes in Electrical Engineering
ISBN 978-981-19-2688-4 ISBN 978-981-19-2689-1 (eBook)
https://doi.org/10.1007/978-981-19-2689-1

This Springer imprint is published by the registered company Springer Nature Singapore Pte Ltd.
The registered company address is: 152 Beach Road, #21-01/04 Gateway East, Singapore 189721, Singapore

About APISAT 2021

APISAT has been the primer symposium and network for advancing the aerospace sciences and technologies in the Asia-Pacific region since its first inception in 2008 by the national aerospace societies of Korea (KSAS), Japan (JSASS), China (CSAA), and Australia (RAeS Australian Division).

Hosted by the Korean Society for Aeronautical and Space Sciences (KSAS), the APISAT 2021 was held at the Jeju Ramada Jeju Hotel in Korea in a hybrid format from November 15–17, 2021, due to the influence of COVID-19.

Nearly 430 authors from a total of five countries, including Korea, Japan, China, Australia, and the United States, attended online and offline.

Starting with the introduction of Professor Sangchul Lee, who is the chairman of the International Program Committee, KSAS President Hyochoong Bang's welcome greeting and five lectures were included as representatives of the four countries as follows.

Flight Testing a 1910 Design Aircraft – My Experience

AVM (retd) Mark Skidmore AM (Executive Chairman, Skykraft, Australia)

Smart Composite Structures and Their Aerospace Applications

Prof. Jinsong Leng (Centre for Composite Materials and Structures of Harbin Institute of Technology (HIT), China)

Opportunities and Challenges of UAM

Dr. Jaiwon Shin (President, UAM Division of Hyundai Motor Group, Korea)

Computational Laminar Flow Technology

Prof. Shigeru Obayashi (Institute of Fluid Science of Tohoku University, Japan)

Current Status and Future Plans of KARI

Dr. Sang-Ryool Lee (President, Korea Aerospace Research Institute (KARI), Korea)

A total of 310 papers, 67 sessions with four topics of [1]Aerodynamics and Design, [2]Dynamics/Control/Avionics, [3]Structures and Materials, [4]Combustion and Propulsion were presented online and offline. Among them, 179 high-quality papers that agreed to publish were published in this book.

Organization

The Asia-Pacific International Symposium on Aerospace Technology is Co-organized by

The Korean Society for Aeronautical and Space Sciences (KSAS)
The Japan Society for Aeronautical and Space Sciences (JSASS)
The Chinese Society of Aeronautics and Astronautics (CSAA)
The Australian Division of the Royal Aeronautical Society (RAeS)

Aim and Scope

The Asia-Pacific International Symposium on Aerospace Technology 2021 (APISAT2021) is the collaborative endeavor among the Korean, Japanese, Chinese, and Australian Societies for aerospace technology.

The symposium, hosted by Korean Society for Aeronautical and Space Sciences, will be held from November 15 to 17, 2021, Ramada Plaza Jeju Hotel, Jeju, Korea.

The symposium provides the opportunity to Asia-Pacific nations not only for the researchers of universities and academic institutes, but also for the industry engineers, to discuss the current and future advanced topics in aerospace technology. The official language of the symposium is English.

Topics of Interest

Aerodynamics and Design

- Computational Fluid Dynamics
- Wind Tunnel Testing
- Flow Visualization
- Unsteady Aerodynamics
- Acoustics
- Aircraft, Helicopter, and UAV Design

Dynamics/Control/Avionics

- Flight Simulation
- Navigation
- Guidance and Control
- ATM/CNS/UAM
- Sensors and Actuators
- Satellite Attitude Control

Structures and Materials

- Structural Analysis
- Structural Testing
- Smart Structures
- Composite Structures
- Structural Dynamics and Control
- Aeroelasticity and Control

Combustion and Propulsion

- Combustion Analysis
- Fuels/Fuel Injection
- Engines
- Turbines
- Cooling Systems
- Spacecraft Propulsion

Organization

Organizing Committee

Executive Committee

Hyochoon Bang, President, KSAS
Isao Kawano, President, JSASS
Zuoming Lin, President, CSAA
Geoff Wilkinson, President, RAeS Australian Division

International Program Committee

Sangchul Lee, Chairman, KSAS
Nam Seo Goo, Vice Chairman, KSAS
Dongyoun Kwak, Co-Chairman, JSASS
Hidetoshi Iijima, Secretariat, JSASS
Xinguo Zhang, Co-Chairman, CSAA
Song Fu, Co-Chairman, CSAA
Cees Bil, Co-Chairman, RAeS Australian Division
Mark Skidmore, Past-President, RAeS Australian Division

KSAS National Organizing Committee

Sangchul Lee, Chairman
Nam Seo Goo, Vice Chairman
Cheolheui Han, Secretariat
Kybeom Kwon

Deog-Kwan Kim
Seungkeun Kim
Jeong-Ho Kim
Sangkyung Sung
Jeong-Yeol Choi
Yoo-Jin Choi
Jae-Hung Han
Kwanjung Yee
Jae-Sung Bae
Wongi Kang

JSASS National Organizing Committee

Dongyoun Kwak, Chairman
Hidetoshi Iijima, Secretariat
Yoshinori Matsuno
Shinsuke Nishimura
Kazuho Setoguchi
Hirotomo Kimata
Zhong Lei
Masato Taguchi
Tatsunori Yuhara

CSAA National Organizing Committee

Junchen Yao, Chairman
Ce Yu, Vice Chairman
Xue Zhang, Secretariat
Xiasheng Sun
Yingchen Chen
Zhenghong Gao
Yongling Fu
Jianping Wang
Pinqi Xia
Qin Huang
Gao Xiaowei
Yaoxing Shang
Jian Fu
Xiangfan Nie
Yufei Zhang

RAeS Australian National Organizing Committee

Cees Bil, Chairman
Douglas Nancarrow
Murray Scott
Richard Morgan
John Vincent
Noel Schmidt

Contents

Editor Bios

Sangchul Lee received his B.S. and M.S. degrees in Aeronautical Engineering from Seoul National University, Korea, in 1986 and 1988, respectively. He got his Ph.D. degree from Texas A&M University in 1994. Then he worked for Samsung Aerospace Industries, Ltd. and Korea Aerospace Industries, Ltd. from 1994 to 2006. He joined Korea Aerospace University in 2006, currently serving as a professor of the School of Aerospace and Mechanical Engineering. His research interests include flight dynamics and aircraft control, dynamics and control of space structures, avionics system design, system engineering and management, and software engineering. Currently, he is serving as a BK21FOUR program director of Smart Drone Convergence E&R center of Korea Aerospace University.

Cheolheui Han graduated from the Department of Mechanical Engineering of Hanyang University in 1993 and got master and Ph.D. degrees in the same department in 1998 and 2003, respectively. His Ph.D. degree was on the aerodynamic analysis and design of a flying train inside a track utilizing the wing-in-ground effect. While he was a Ph.D. student, in 2000, we worked at the Ocean Engineering Group, Dept. of Civil Engineering, University of Texas at Austin (USA), as a visiting scholar. After graduation, we worked at the Dept. of Aerospace and Ocean Engineering, Virginia Tech (USA) as a visiting postdoctoral researcher. In 2004, he worked at Kengwon Tech as a CFD engineer. After working as a lecture professor at Hanyang University for a year and doing research at Artificial Muscle Center at Konkuk as a research professor, he joined the Aeronautical and Mechanical Design Department of Korea National University of Transportation, Chungju, Korea, in 2006. He is now serving as a professor at the same department. He was included in Marqui's Who's who in the World in 2007. He spent his sabbatical year at the Cavitation Laboratory in the University of Texas at Austin in 2012. His current research interests are unsteady aerodynamics of insect wings and flying birds. He is developing a hybrid method between the boundary integral, vortex, and stochastic methods.

Jeong-Yeol Choi is a professor of Aerospace Engineering, Pusan National University, Republic of Korea. He received his Ph.D. from the Seoul National University in 1997. His research interests include dynamics of energetics flows, high-pressure combustion, detonation, reaction kinetics, multi-phase/particle flows, and turbulence–chemistry interaction in propulsion systems. He has co-authored more than 140 archival papers and more than 400 conference papers with international colleagues. He is an associate fellow of the AIAA, and his other professional activities include The Combustion Institute, IDERS, KSAS, KSPE, KOSCO, and KSCFE. He served as the program manager of Space R&D of National Research Foundation (NRF) of Korea, planning and evaluation of Space R&D programs including the development of space launch vehicle, micro- to GEO satellites, space exploration, and basic space researches sponsored by the Ministry of Science, ICT, Republic of Korea Government. Currently, he is serving as a BK21FOUR program director of DRONE E&R center of Pusan National University.

Seungkeun Kim received his B.Sc. degree in Mechanical and Aerospace Engineering from Seoul National University (SNU), Seoul, Korea, in 2002, and then acquired his Ph.D. degree from SNU in 2008. He is currently a professor at the Department of Aerospace Engineering, Chungnam National University, Korea. He was an associate professor and an assistant professor at the same university from 2012 to 2020. He was a visiting scholar at George Washington University at Washington D.C. in 2019. Previously, he was a research fellow and a lecturer at Cranfield University, UK, from 2008 to 2012. He is interested in micro-aerospace systems, aircraft guidance and control, estimation, sensor fusion, fault diagnosis, fault-tolerant control, and decision-making for autonomous systems.

Jeong Ho Kim graduated from the Department of Aerospace Engineering of Seoul National University in 1992, and he got master and Ph.D. degrees in department of Aerospace Engineering at the same university in 1994 and 1999, respectively. His Ph.D. degree was on the parallel computation technique of finite element method for large-scale structural analysis. As soon as he got a Ph.D. degree, he entered the national supercomputing center of Korea Institute of Science and Technology Information (KISTI) as a senior researcher. After nine years' service, he moved to MIDAS IT as a main developer of the commercial finite element analysis software package, midas NFX. He joined Inha University, Incheon, Korea, in 2012, currently serving as a professor of Department of Aerospace Engineering. He was nominated as one of the Finalists of the Gordon Bell Prize in 2003 for his high-performance parallel sparse direct solver for efficient finite element analysis. This sparse solver has been used as the main linear solver for many research and commercial finite element analysis codes. His current research interests are high-performance computing for computational structural mechanics, additive manufacturing for aerospace applications, and machine learning and data mining for structural mechanics.

Aerodynamics and Design

Asia–Pacific Region's Aircraft Design and Development Strategy for Single-Aisle Aircraft in the Oligopoly Market of Boeing and Airbus

Myung-kwan Ahn, Jong-soo Lee, Heung-jae Kim, and Hyeong-yu Jang

Abstract There have been severe supply and demand imbalances in the global aircraft operation situation, but various factors make it difficult to address this market imbalance. Considering the number of commercial jet flights in December 2019, Asia–Pacific (AP) countries operate 34%, but supply is 0%. North America operates 25%, while Boeing supplies 55%. Europe operates 21%, while Airbus supplies 44%. Why this imbalance happened, how to address it, and how to develop aircraft will be suggested.

In the 1960s, European countries had tried to correct the severe mismatch between aircraft supply and demand. France and Germany agreed to design and develop aircraft for the development of the European aerospace industry by international joint development and established Airbus in 1970. The A320 was designed and developed, and was finally launched in 1988. Airbus outperformed Boeing by taking a 52% market share with 305 units in 2003, which contributed significantly to the European economy. Europe established EADS in 2000 and integrated the European aeronautical defence industry, which contributed greatly to European peace. In 2020, Airbus delivered 566 units, occupying a 78% market share (Boeing 162), remaining the world's No. 1 aircraft supplier and maintaining Europe's economy and peace. Regional Comprehensive Economic Partnership (RCEP) countries should benchmark the Airbus A320 success stories to correct AP aircraft demand–supply distortions and lay the groundwork for Asia–Pacific aerospace industry development. To achieve long-term goals, it is effective to divide international joint development into three stages. In the first stage, individual companies (company-company) promote

M. Ahn (✉)
Changwon Moonsung University, Changwon, Korea
e-mail: mkahn@cmu.ac.kr

J. Lee
Department of Aircraft Part Production, ASTK, Sachon, Korea

H. Kim
Korea Aircraft Development Center (KADC), Changwon, Korea

H. Jang
Department of Business Administration, Gyeongsang National University, Ginju, Korea

S. Lee et al. (eds.), *The Proceedings of the 2021 Asia-Pacific International Symposium on Aerospace Technology (APISAT 2021), Volume 1*, Lecture Notes in Electrical Engineering 912, https://doi.org/10.1007/978-981-19-2689-1_1

international co-development, and it is similar to the current environment of AP countries. The second stage is the case of Japan-promoted international joint development (JADC-Boeing) (companies consortium-company), and the third is international joint development between countries, as in the case of Airbus. This study is based on a quantitative approach to provide clear findings. AP aircraft manufacturers should secure the basis for the RSP international joint development project.

Keywords Electric aircraft · RCEP · AAMC · RSP IJD · Quantitative approach

1 Introduction

Piketty [4] studied financial and labor income in France. If financial income is higher than labor income, the gap between the rich and the poor widens and the social conflict index is high. As a result, the partnership between members of society breaks down, the overall organizational efficiency is lowered in the long term, and it is difficult to expect a virtuous cycle of social development. This phenomenon is not limited to one country but occurs across continents. Therefore, we can prove Piketty's theory (Capital in the Twenty-First Century) by analyzing commercial aircraft manufacturing in the Americas, Europe, and Asia.

RCEP countries account for 34% of the world's fleet, but their supply is 0%. Achieving a reasonable aircraft supply is one way to resolve conflict in the RCEP region [5].

Cho and Porter [2] argued that intercontinental movement is caused by rising wages. In the past, the shipping industry moved from the European continent to the RCEP through the US continent. There are core competencies for industrial development, but over time, these core competencies have changed, and intercontinental movements have occurred in terms of competitiveness.

Aircraft manufacturing also moved from the United States to Europe, and now it is time for it to move to RCEP countries [1]. Nishiguchi [3] analyzed the competitiveness of the automotive industry in Japan and the United States. The supply chain of automobile manufacturers indicates that cooperative relationships between parts suppliers and assembly suppliers maximizes the efficiency of the overall supply chain.

2 Commercial Aircraft International Joint Development (IJD)

This study aimed to predict the sales volume of the AP aircraft manufacturing industry in 2039 and derive an achievement plan. Therefore, many survey findings should be cited and assumptions should be introduced. To improve the reliability of forecast data, we will cite the JADC and Boeing annual reports to support our findings and

Table 1 RSP I.J.D. characteristics

Category	Stage 1	Stage 2	Stage 3
Host	Individual company	Japan, JADC	Europe, Airbus
Major models	Embraer E2, etc.	.B767/777/787/777X, etc.	.A320/A350/A380, etc.
Type of production	Business of companies	Offset trading	Agreement among countries
Major participants	Korea and other AP countries	USA, Europe, Japan	France, Germany, Spain, UK
Government role	Company support	Support system establishment	Law amendments for support
Motivation	Profit creation	Secure the manufacturing volume	Cope with USA. Monopoly

will introduce the management science (mathematic method) concepts to support the objective assumptions.

First, we surveyed the volumes of aircraft parts manufacturing in AP countries, such as Korea, China, Japan, and so forth. In addition, we analyzed the success of Japanese commercial aircraft parts production and export and then analyzed the success cases of Airbus's commercial aircraft production and delivery.

Second, by combining the status of the first-stage AP countries and the success cases of the second and third stages, the process of improving the commercial aircraft manufacturing industry output in AP countries is summarized according to three stage as shown in Table 1.

2.1 *Demand and Supply Imbalance by Region*

Table 2 shows the demand for major jet transport (operation) in each region of the world and the number of aircraft supplied (delivery) by producers in 2019. JADC, Boeing, and others have reported that the aircraft demand of the AP region is expected to soar in the next 20 years. The number of aircraft supplied (delivery) by producers in 2019 is shown in Table 2. A review of the major jet supply regions (completed aircraft supplier) for each region reveals a significant supply and demand imbalance.

2.2 *Aircraft Production of AP Countries*

Boeing (2019) forecasted the 20-year (2019–2039) demand in the aircraft market. Table 3 summarizes aircraft demand and market value. Table 3 divides the commercial aircraft market into four categories: regional jet, single-aisle, widebody, and freighter. In terms of deliveries, single-aisle is the biggest with 74%; widebody is the

Table 2 Demand and supply status of major jet aircraft in 2019

Classification		Demand (operation)														
		N. A	L. A	W. E	E. E	CIS	Africa	M. E	Oceania	China	Japan	N.E. A	S.E.A	S.W.A	Sum	%
Supply (delivery)	Airbus	1827	712	2380	220	375	211	685	172	2013	105	225	1017	469	10,411	46.5
	Boeing	3568	558	1929	164	508	453	606	284	1761	465	375	573	206	11,450	51.2
	M.D	237	41	10	11	3	10	43				0	4	1	360	1.6
	etc.		2			118	9	7				9			145	0.7
	Sum	5632	1313	4319	395	1004	683	1341	456	3774	570	609	1594	676	22,366	100
	%	25	5.9	19	1.8	4.5	3.1	6.0	2.0	17	2.5	2.7	7.1	3.0	100	

Source: JADC, Commercial Aviation Data Collection (2020) II-19

Table 3 World major jet aircraft delivery and market value

Classification		Regional jet	Single-aisle	Widebody	Freighter	Total
Deliveries (A)	Ship set	2,240	32,420	8,340	1,040	44,040
	%	5	74	19	2	100
Market value (B)	Billion dollar	105	3,775	2,630	300	6,810
	%	2	55	39	4	100
(B)/(A)	Million dollar/fleet	47	116	315	288	155

Source: Boeing, Commercial Market Outlook 2019–2039 (2019), p. 1

second with 19%, regional jet is the third with 5%, and freighter is the fourth with 2%.

The total AP commercial aircraft manufacturing output of 2019 shown in Fig. 1 can be expressed by Eq. 1. The production results of commercial aircraft parts are shown as (A), and the improvement of production amount is represented as (B) by securing the production volume by conducting a two-phase RSP (commercial aircraft offset trade). The production increase caused by China's independent development of the C919 is classified as (C). In the case of promoting RSP international joint development after 2020, when the C919-300 is released to the market after 2029, the increased amount of production in RCEP countries is expressed as (D).

(A): To calculate (A), we assumed that the production performance in 2019 can be expressed by Eqs. 1 and 2, and if there is no improvement in the current method, the production performance will be

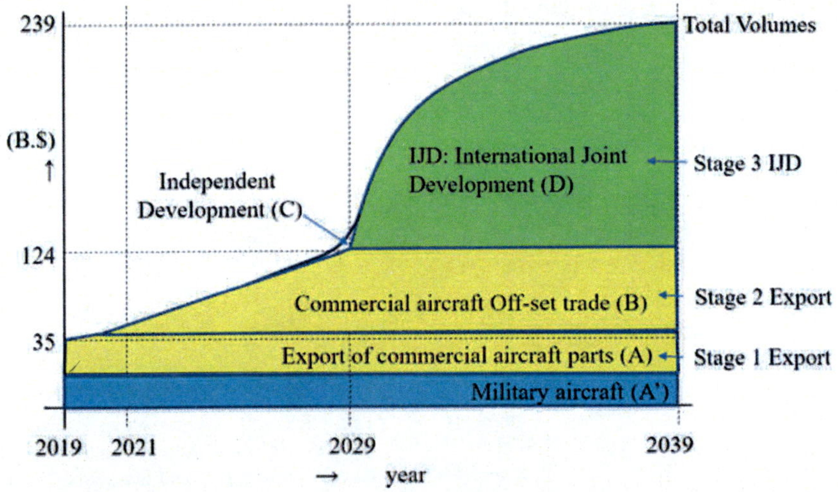

Fig. 1 Expected production volumes in AP countries

$$AP_{(2019)} = AA'_{(2019)} \times AL''_{(2019)} \tag{1}$$

$$AP_{(2029)} = AA'_{(2029)} \times AL''_{(2029)} \tag{2}$$

$$TP_{(2039)} = A_{(2039)} + B_{(2039)} + D_{(2039)} = \$238,831 \text{ million} \tag{3}$$

(C): It shows the effect of increasing the production value of C919, which is currently being developed in China. This effect is limited to China. Therefore, it is appropriate to manage it separately because it does not have any effect in other AP countries. It is a project involving only china. The market share will be limited like YS-11.

(D): RCEP countries need a third stage RSP international joint development. In the absence of international co-development among AP countries, the negotiation power of off-set trade will disappear, and aerospace technology development will be limited.

3 RSP Stage 2: Japan Success

In 1974, the Commercial Transport Development Corporation (CTDC) was established; the second stage RSP was undertaken, and in 1982 it was changed to Japan Aircraft Development Corporation (JADC). Since 1974, CTDC and JADC jointly developed the B767, B777, B787, and B777X in RSP to secure large quantities of commercial aircraft parts manufacturing. As a result, the volume of Japanese commercial production and repair amounted to 11,259 million dollars in 2018 (see Table 4).

The number of Japanese aircraft operating is 1.4 times of the number of Korea aircraft, but the parts exports of commercial aircraft are 6.4 times greater, and the

Table 4 JADC off-set trade (international joint development)

Classification	B767(181–375)	B777 (312–550)	B787 (242–330)	B777X (350–425)
Development	1974	1988	2003	2015
Investor	5 companies (MHI, KHI, SUBARU, NIPPI, Shinmaywa)			
Invest. ratio	15%	21%	35%	21%
Supplier	19 companies including KYB	13 companies including KYB	12 companies including KYB	Procurement of composite materials, fuselage structure parts, LRU
Remark	–	6 suppliers reduce 6% increase in ivest	1 supplier decrease. 14% increase in invest	Production in Japan: $36 billion over the next five years

Source: JADC, (2019), VIII-25-26, 2019

efficiency of manufacturing per operation is 4.4 times higher. For AP countries to improve their sales by 4.4 times in 2018, each country established an organization to process, promote, and negotiate the offset trade in commercial aircraft, such as Japan's JADC. Thus, part and part assembly workshares have increased.

3.1 Target Achievement Time of AP RSP Stage 2

To improve the performance of parts production due to the offset trade of commercial aircraft, the relevant organizations of the pertinent countries should work together to shorten the time to achieve the goal. In the case of a target achievement index of 1 for the relevant country, the will of each sector in 2020 was the same as it was in 2019; if the index was 1 or more, the will was relatively higher, and if it was less than 1, the will was relatively weak. In Eq. 4, a country's aircraft manufacturing business environment is difficult to adjust, and the time required to achieve production targets depends on the desire of aircraft manufacturing industry people and related organizations. {Ministry of Industry: MOI $D_{(n)}$, Ministry of Transportation: MOT $D_{(n)}$, Aircraft Manufacturer: AM $D_{(n)}$, Airline: AL $D_{(n)}$, and other supporting organizations: SO $D_{(n)}$}.

$$\mathrm{BY}_{(n)} = \mathrm{BY}_{(n-1)} \times \left\{ \begin{array}{l} \mathrm{AMS}_{(n)} \times \mathrm{MOI\ D}_{(n)} \times \mathrm{MOT\ D}_{(n)} \\ \times \mathrm{AM\ D}_{(n)} \times \mathrm{AL\ D}_{(n)} \times \mathrm{SO\ D}_{(n)} \end{array} \right\} \tag{4}$$

$$\mathrm{BY}_{(n)} = 2\mathrm{BY}_{(2019)} \times \left\{ \begin{array}{l} \mathrm{AMS}_{(n)} \times \mathrm{MOI\ D}_{(n)} \times \mathrm{MOT\ D}_{(n)} \times \\ \mathrm{AM\ D}_{(n)} \times \mathrm{AL\ D}_{(n)} \times \mathrm{SO\ D}_{(n)} \end{array} \right\}$$

$$2 = \sum_{c=1}^{16} \mathrm{Country\ (c)} \sum_{n=2019}^{2028} \left\{ \begin{array}{l} \mathrm{AMS}_{(n)} \times \mathrm{MOID}_{(n)} \times \mathrm{MOTD}_{(n)} \\ \times \mathrm{AMD}_{(n)} \times \mathrm{ALD}_{(n)} \times \mathrm{SOD}_{(n)} \end{array} \right\} \times \% \tag{5}$$

3.2 Japan Success Factors

Table 5 summarizes the operating status of major jet aircraft in AP countries. There have been some peculiarities in the purchase status of Japanese airline companies. Japan Airlines owned 165 Boeing aircraft (100%). JAL was a state-owned company until 1987 (since its establishment in 1951). JADC has participated in B767, B777, B787, and B777X international joint development since 1974. It is a motivation for offset trading. SRP international co-development was promoted due to offset trading, and SRP co-development contributed to the growth 4.4 times greater than that of South Korea as seen in Table 6. Although China and other AP countries have a high proportion of Airbus, the proportion of Boeing's aircraft is higher overall

Table 5 Major jet operations in AP airlines

Country		Japan				Korea				China	Others	A.P total
Airline		Japan air lines	JAL group others	Other	Sum	KAL	Asiana	Other	Sum			
Fleet number	Airbus	0	24	75	99	39	53	32	124	1895	1597	3715
	Boeing	165	16	275	456	127	27	111	265	1839	1239	3799
	Others	0	0	0	0	0	0	0	0	0	24	24
	Total	165	40	350	555	166	80	143	389	3734	2860	7538
%	Airbus	0.0	0.3	1.0	1.3	0.5	0.7	0.4	1.6	25.1	21.2	49.3
	Boeing	2.2	0.2	3.6	6.0	1.7	0.4	1.5	3.5	24.4	16.4	50.4
	Others	0.0	0.0	0.0	0.0	0.0	0.0	0.0	0.0	0.0	0.3	0.3
	Total	2.2	0.5	4.6	7.4	2.2	1.1	1.9	5.2	49.5	37.9	100

Table 6 Correlation between operation and production

Classification	No. of operation (aircraft)	Commercial production and repair (M. $)	Remark
Japan (A)	548	10,629	Commercial aircraft manufacturing and repair
Korea (B)	389	1,836	Exports of such as aircraft, engines, parts, etc.
(A)/(B) (%)	140%	579%	Korea efficiency is 24% that of Japan because of aircraft buying strategy

Source: Aerospace Industry Statistics 2020, Aerospace Industry Promotion Association, 13, 24 P, JADC, March 2018, II-53, VIII-14 P

in the AP region due to Japan's lopsided in Boeing. The strategic management of the Japanese government has resulted in successful cases in which Japanese aircraft manufacturing output is higher than that of other AP countries.

Therefore, the second stage RSP success case of this study was set as the Japanese aircraft manufacturing industry. The Korean government founded the 'Korean National Airlines' in 1946. It was renamed as Korean Air Corporation in 1962, and privatized as Korean Air in 1969. As seen in Table 5, Korea owns 1.7% of Boeing aircraft, which is lower than the 6.0% in Japan. We can infer that The Korean government has insufficient buyer bargaining power. AP countries should successfully pursue the second stage (commercial aircraft offset trade) by learning from the Japanese success stories shown in Table 6.

3.3 *Government, Airlines, and Manufacturers*

According to the UK DTI [5] aircraft industry supply chain, the relationship between airline service and aviation manufacturers is shown on the right side of Fig. 2. AP aircraft parts manufacturing companies (F1, F2, and F3) have weak bargaining power to supply parts to commercial aircraft manufacturers (E) (Airbus, Boeing). Airbus and Boeing apply a cost-competitive market structure, and the AP aircraft manufacturing companies are becoming less competitive. In Korea and other AP countries, the competitiveness of the aircraft manufacturing industry is insufficient compared to government support because strong nonprofit organizations (left side in Fig. 2) do not exist in AP countries except Japan. The commercial aircraft manufacturing department (C) of each government should adopt a policy to improve the commercial aircraft manufacturing industry to improve the competitiveness of the AP aircraft parts manufacturing industry. (C) They should request cooperation with the commercial aircraft management department (B) of other governments. Japan has developed

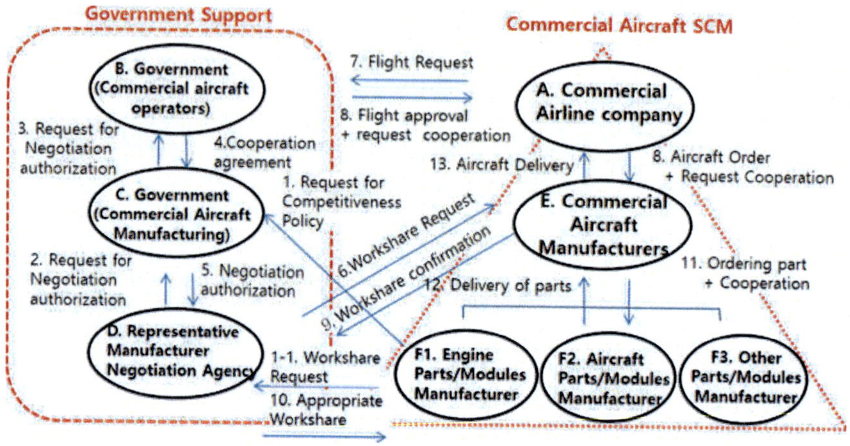

Fig. 2 Cooperation between airlines and aircraft manufacturers

a system similar to the supply chain shown in Fig. 2, so that (F) had occupied the core competitiveness, such as structure section design and market analysis, rather than just manufacturing. Military aircraft off-set trade has generated the aircraft manufacturing industry in AP countries, and commercial aircraft off-set trade should bring about the advancement of the AP aircraft manufacturing industry. The development of the aircraft parts industry was delayed due to the little off-set trading conception of commercial aircraft in Korea. R&D of Thermally Assisted Machining of Titanium (TAM) developed by Lockheed in 2012 is underway in 2021.

4 RSP Stage 3: Europe Success Case

4.1 Airbus Success Factors

Airbus is an example of successfully maximizing the benefits of RSP international co-development as seen in Fig. 3, Tables 7, and 8. It is one of the greatest business success stories that should be studied in the aerospace industry. When Boeing delivered 95% of its major jet aircraft in 1968, a dramatic supply structure was created, others aircraft manufacturers throughout the world delivered only 5%, including Europe. In 1970, Airbus' innovative success story began with the approval of the French government's special law (Société par actions simplifiée).

This study omits the various innovation activities (enactment of laws, A320 technology innovation, the establishment of special corporations, government support, etc.) in the development process of Airbus, and we would provide guidelines for the development of RCEP aircraft based on the success case of Airbus.

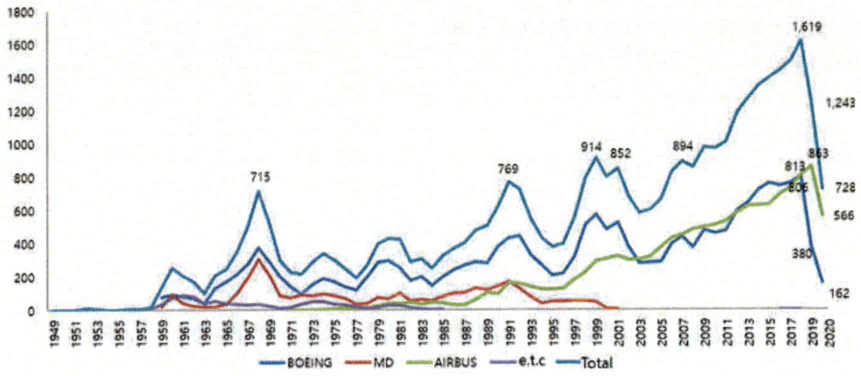

Fig. 3 Major jet aircraft delivery performance

Table 7 Comparison of SAA delivery efficiency (IJD vs. ID)

Manufacturer	Aircraft name	Delivery period	Number of deliveries	Average del./year
Airbus (IJD)	A318, A319, A320, A321	30 Years (1988–2018)	9,647 fleet	322 units/year
Boeing (ID)	B707, B717, B720, B737	60 years (1958–2018)	13,440 fleet	224 units/year

Table 8 Relationship of number of participants and efficiency in IJD (airbus)

Classification	Delivery period	Del. no	Delivery/year	Number of participant
A350	2014~(4y.)	235	58.5	6 (France, UK, Germany, Spain, China, Russia)
A340	1993~2012(19y.)	377	17.7	4 (France, England, Germany, Spain)
A380	2007~(11y.)	242	22.0	4 (France, England, Germany, Spain)

According to the results of the international joint development (IJD) investment effect, IJD has a greater effect on investment than independent development (ID) (Table 7). Investment effectiveness is greater if several countries participate rather than just a few countries in IJD (Table 8).

Therefore, AP countries should start to newly develop electric single-aisle aircraft by IJD. The more partners involved in IJD), the greater the efficiency will be.

4.2 Aerospace Production Value

The segmentation, targeting, and positioning (STP) strategy related to aircraft manufacturing is needed to calculate the effect of RSP in 2039. The results are summarized in Table 9. First, in the case of fighter/attack in Table 9, fighters account for 6% of the world's aerospace industry at $24.8 billion. Other military T. aircraft ($10.4 billion), missiles ($15.3 billion)), C. & M.S. ($9.8 billion), the total military demand amounted to $60.3 billion (16%), less than commercial aircraft demand, counted for 84% of the market demand. Commercial aircraft is expected to account for 78% ($302 billion) of the total demand in the aircraft manufacturing industry. We should review the commercial aircraft sector to improve the value of AP aerospace production.

Moreover, if we break down the aircraft market according to Table 9, we can take advantage of the development of electric single-aisle aircraft (SAA) which had the largest market value as shown in Table 4.

Third, we can summarize the operation and development status of the world SAA to select SAA for RSP as Table 10. AADC can participate in the electric B737Max, electric A320neo. electric C919-300, and so forth. It is reasonable to participate on the new electric aircraft development of SAA.

4.3 The International Joint Development Effect of SAA

Due to the COVID-19 pandemic, we use data from 2019 because would be difficult for aircraft operators and manufacturers to provide exact data and accurately forecast demand after 2019. Boeing (2019) predicted demand for 44,040 jet aircraft over the next 20 years (2019–2039). Boeing (2019) estimated demand for 32,420 SAAs. According to JADC (2019), the number of major jet aircraft was 22,204 in 2019. Table 11 shows the status of the world's single-aisle jets. In 2018, 15,373 SAA were in operation, and it is estimated that 32,420 SAA will be required in the next 20 years, and 3,345 SAA could be used among 15,373 SAAs in 2038. According to JADC (2019), the number of delivery orders for SAAs was 10,819 aircraft set in 2018. The data are summarized in Fig. 4.

4.4 RSP Revenue Forecast in AP

We calculated the expected sales volume of AP countries' RSP joint international development by applying the criteria shown in Table 12. The number of deliverables was calculated by reflecting the market share of Airbus by year. The AP national RSP organization is referred to as the Asia Aircraft Development Consortium (AADC).

With reference to the delivery data of Boeing's Commercial Market Outlook 2019–2039 (2019) and Market Value, the number of delivered products, market

Table 9 Annual average production value over the next five years

Classification	Military demand								Commercial demand						Sum	
	Fighter/attack		Military T. Aircraft		Missiles		C. & M. S		Commercial Aircraft		Business Aircraft		Commercial Rotorcraft			
	$B	%	$B	%	$B	%	$B	%	$B	%	$B	%	$B	%	$B	%
Boeing	3.9	1	3.0	1	–	–	1.2	–	153.0	40	–	–	–	–	161.0	42
Airbus	–	–	3.6	1	–	–	0.7	–	136.6	35	–	–	1.1	–	142.0	37
Lockheed M	14.6	4	1.9	1	1.0	–	1.4	–	0.0	0	–	–	0.0	–	18.9	5
Etc.	6.3	2	1.9	0	14.3	4	6.5	2	12.5	3	20.6	5	1.0	–	64.9	17
Sum	24.8	6	10.4	3	15.3	4	9.8	3	302.0	78	20.6	5	2.1	1	386.8	100

Source: Aviation Week and Space Technology (2019), December 24~January 13, 55~96p

Table 10 Development and operation status of SAA

Classification	Operation		Development			Planning
Model name	B737	A320	C-series	MS-21	Easy Jet	C919-300
Development country	USA	Multinational	Canada	Russia	USA	China
Manufacturer	Boeing	Airbus	Bombardier	United AC	–	COMAC
Status	Operation	Operation	Test flight	Development	First article	Concept plan
Production year	1966~	1987~	–	–	2027~	–
Project cost	–	–	$4.4 billion	$4.6 billion	–	–
Remarks	B737MAX Crash, 2 times	A320NEO delivered	–	–	Electric aircraft	Independent develop

value, and global market share are in Table 13. According to JADC (2019), 10,819 fleet delivery orders had not been completed for Boeing B737 and Airbus A320 in 2018. According to Boeing (2019), because a fleet of 32,420 SAAs will be needed during the next 20 years, delivery orders for a fleet of an additional 12,166 SAAs will be issued. In Table 13, the cumulative orders for AADC are 8,030, so that Boeing and Airbus can additionally take orders for 13,571 SAAs. The three kinds of orders are shown in Fig. 5.

As seen in Fig. 4, the actual aggregate SAA total delivery numbers do not appear as a single equation, but aircraft market analysis agencies, such as JADC, Boeing, and Airbus, are shown as the single equation when they forecast aircraft demand. In the future, when a new theory of prediction techniques is established, it will be more accurately represented. If 20,028 SAAs are delivered during the period from 2019 to 2039 and Eq. 6 is applied, *a* will be 1,002, as seen in Fig. 5.

$$Y = a X + b \tag{6}$$

4.5 Long-Term Outlook on AADC

If the AADC was established, it could account for 49% of the SAA market by 2039 as seen in Table 13. Changes are predictable in the single-aisle market for the long-term, as shown in Fig. 6. Then, Boeing and Airbus can each account for 26% in 2039, as shown on the left. In this case, due to economies of scale after 2039, AADC's market share will increase, while that of Boeing and Airbus will gradually decrease.

In addition, according to Boeing (2019), when comparing the market value to the number of aircraft, that of widebody aircraft is 2.71 times higher than that of SAAs.

Table 11 Demand and Supply Status of Single-Aisle Aircraft in 2018

Classification		Demand (operation)														
		NA	LA	WE	EE	CIS	Africa	ME	Oceania	China	Japan	NEA	SEA	SWA	Sum	%
Supply (delivery)	Airbus	1,416	629	1,805	163	346	120	349	112	1,503	102	92	670	395	7,702	49
	Boeing	2,419	492	1,323	164	320	359	209	211	1,510	176	173	444	188	7,988	51
	etc.	–	–	–	–	41	–	–	–	–	–	6	–	–	47	0.3
	Sum	3,835	1,121	3,128	327	707	479	558	323	3,013	278	271	1,114	583	15,737	100
	%	24	7	20	2	4	3	4	2	19	2	2	7	4	100	

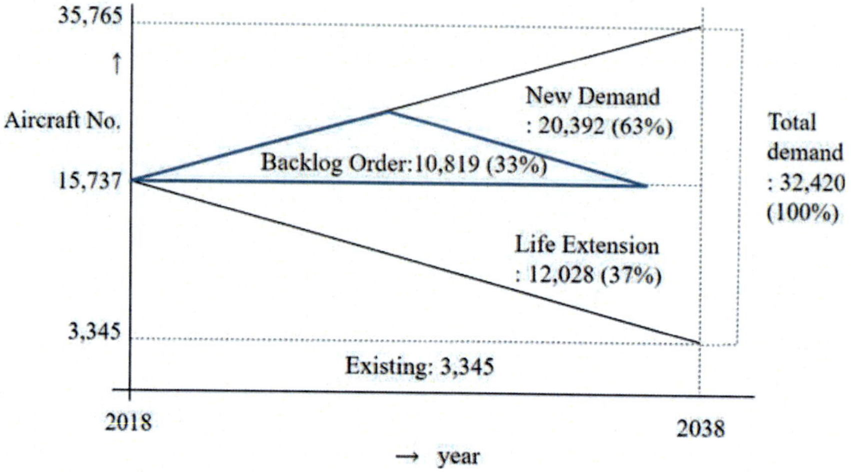

Fig. 4 Outlook of SAA demand in 2038

Table 12 Estimated criteria of AADC SAAs

Category		Contents
Market share	Base	– Application of Airbus A320 market share (%)
	Yearly Market Share	– AADC country · If some country is a participant in RSP, it will be an active purchaser · After participation in AADC, the 200% A320 market share since 1988 · Coverage Market Share by Year (Y (%)): YAADC, 2021 + n (%) = YA320, 1988 + n' × 2 ($0 \leq n \leq 10$), (n = subsequent years after 2021, n' = subsequent years after 1988) – Non AADC countries · It will be passive in purchasing aircraft · Before participation in AADC, 50% of A320 market share since 1988 · Market Share by Year (Y (%)): Y no AADC, 2021 + n (%) = YA320,1988 + n ' × 0.5 ($0 \leq n \leq 10$), (n = subsequent year after 2021, n' = subsequent year after 1988)
Delivery	Base	– YDP (Year Delivery Possibility) =Number of aircraft demand × market share (by country by year) – YDP number of deliveries by country and year =D country, yearly aircraft demand by country × Y country, market share by year (%)
	Total	– Total Aircraft Delivery Number (TADN) = Total Delivery Number from 2018 to 2039

Table 13 Dlivery comparison of AADC and Chinese independent development

Category	Delivery (aircarft/year)			Market value ($ million)			World market share (%)		
	2029	2039	Cum	2029	2039	Cum	2029	2039	Cum
AADC (A)	0	986	8,030	292	114,789	934,998	0	49	25
Chinese indep. (B)	3(2022)	352	3,525	292	41,028	410,481	0	18	11
%, (A/B*100)	100	280	228	100	280	228	100	280	228

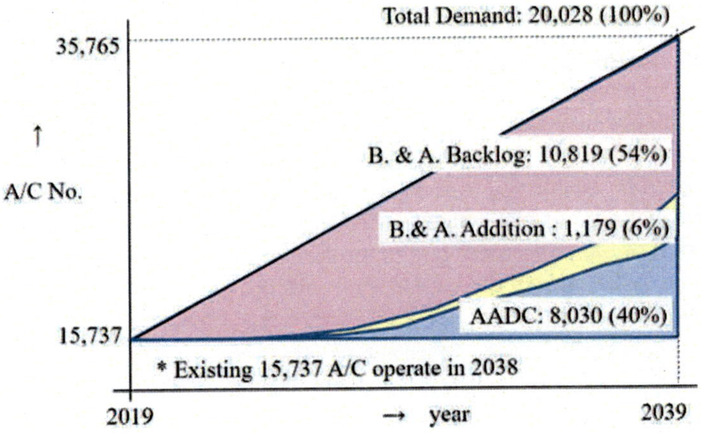

Fig. 5 Delivery order estimation in the global SAA market

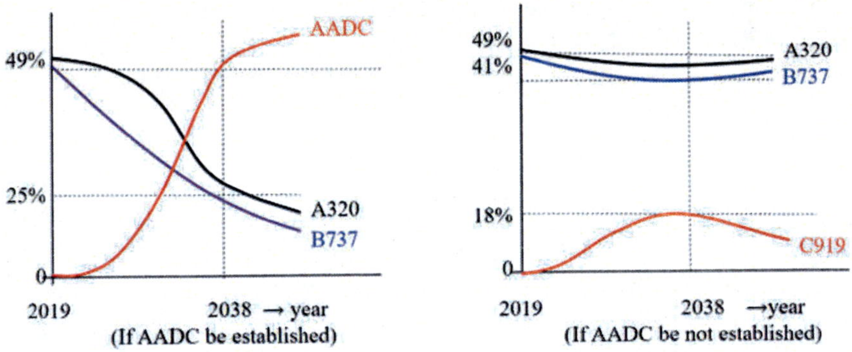

Fig. 6 Long-term outlook for global SAA market trend

Thus, Boeing and Airbus will focus more on high value-added widebody and have less interest in that SAA market share. Therefore, after 2039, AADC's market share of SAAs will increase.

Table 14 Comparison of AP and European aerospace industries

Category	(Future) A.P	(Current) Europe
Commercial aircraft	AADC	Airbus
Aeronautic defense and space	AADS	EADS
Union	Asia- pacific union	Europe union
Major country	China, Japan, Korea, etc.	France, Germany, Spain, etc.

On the other hand, if AADC is not established, COMAC's SAA market share will be 18% in 2039, while that of Boeing and Airbus will be 41%. In this case, the economies of scale will cause COMAC's market share to gradually decline, and it will eventually face difficulties like Japan's YS11, and AP countries will have to depend of the aircraft components suppliers of Boeing and Airbus. As a result, AADC should be established as soon as possible to successfully enter and develop the commercial aircraft manufacturing industry of AP countries. Table 14 shows a comparison of the projected future of the AP aerospace industry and the current European aerospace industry. First start, first develop and became the major country in the aerospce industry.

5 Conclusions

To address the imbalances in supply and demand, RSP should be pursued. RSP international co-development can be divided into three stages. After analyzing the success cases of Japan JADC (Stage 2) and Europe Airbus (Stage 3), we proposed to establish AADC to promote the RSP co-development of AP countries. The four main points to establish AADC are summarized as follows.

First, if AADC promotes the cooperation of AP countries, it can calculate the 89 billion dollars seen Fig. 1 as the result of the second stage in 2029 and 115 billion dollars in Fig. 1 as the result of the third stage in 2039. Through AADC establishment, the AP aerospace industry could achieve the production value of more than 239 billion $/year.

Second, AP countries could occupy 49% of the SAA market in 2039, and the SAA market will be further expanded by the economics of scale. Then we generate additional delivery in the widebody, freighter, and regional jet markets, and AADC will become the center of the global aerospace industry in the long run.

Third, AADC will be developed into Asia Aerospace and Defense Systems (AADS), and AADS will contribute to maintaining RCEP peace. Eventually, AADS will contribute to the establishment of the Asian Union (AU). The AU will create an ideal continent to secure economic development and peacekeeping in the AP region.

Fourth, to establish AADC, we should explain the justifications of the AADC to many organizations, including aircraft manufacturers, the Ministry of Land, Infrastructure and Transportation (MLIT), the Ministry of Economy, Trade and Industry (METI), the Trilateral Cooperation Secretariat (TCS), RCEP, the Asian-ROK Summit, and so forth. We just do it new, the sooner, the better.

Funding This work was supported by the Korea Institute for Advancement of Technology (KIAT) grant funded by the Korean government (MOTIE) (No. 2021-228-005).

References

1. Ahn MK, Jang HY (2020) The investment effect analysis of commercial aircraft development-focusing on international joint development cases. J Aviat Manag Soc Korea 18(4)
2. Cho DS, Porter ME (1987) Changing global industry leadership: the case oh shipbuilding. Competition in Global Industries. Harvard Business School, pp 539–567
3. Nishiguchi T (1989) Strategic dualism: an alternative in industrial societies. Oxford University Press, England
4. Piketty T (2015) Capital in the twenty-first century. Harvard University Press. ISBN 978-0674430006
5. UK DTI (2000) A study on the impact of e-business on the UK aerospace sector. Department of Trade and Industry Ref: URN00/1309: Appendix G-1

Study on Wind Resistance Characteristics of Multi-rotor UAV

Zhan-ke Li, Shi-gao Su, Jin-shuo Cao, and Si-jie Luo

Abstract Based on the wind field characteristics of high-rise buildings, this paper studies the wind resistance stability of multi-rotor UAV flight platform, conducts research on the design technology of high-wind resistance flight platform, improves the safety and reliability of fire fighting and extinguishing UAV, and provides technical support for the design and optimization of fire fighting and extinguishing UAV. Based on various aerodynamic data under no-load and on-load conditions, the wind resistance of the UAV is analyzed, and a set of wind resistance evaluation methods for the multi-rotor fire fighting and extinguishing UAV is proposed. The wind static model of the multi-rotor fire fighting UAV is established to evaluate the wind resistance of the UAV. The results show that: the designed fire-fighting UAV can theoretically resist the wind of level 7 (16 m/s), and the UAV can achieve the balance of force and torque by adjusting its attitude under the horizontal action of level 7 wind.

Keyword Aerodynamic performance · Multi-rotor UAV · Wind resistance

1 Fire Fighting UAV Flight Platform

The fire-fighting UAV designed in this paper is used to mount fire-fighting bombs to perform fire-fighting tasks in high-rise buildings. Due to the large load capacity, strong wind resistance and long endurance of multi-rotor UAVs with large wheelbases, eight rotors are unmanned Compared with the six-rotor and four-rotor UAVs, the aircraft has better stability. An eight-rotor UAV flight platform is designed, and a fire extinguisher is mounted under the flight platform. In order to reduce the overall size and structural weight of the UAV, the flight platform adopts a layout with overlapping blades, and the adjacent propellers are staggered up and down during installation to prevent the tip of adjacent propellers from colliding. The multi-rotor layout is shown in Fig. 1a. In Fig. 1b, the black paddle is on the top and the blue paddle is on the bottom.

Z. Li (✉) · S. Su · J. Cao · S. Luo
School of Aeronautics, Northwestern Polytechnical University, Xi'an, People's Republic of China
e-mail: lzk@nwpu.edu.cn

S. Lee et al. (eds.), *The Proceedings of the 2021 Asia-Pacific International Symposium on Aerospace Technology (APISAT 2021), Volume 1*, Lecture Notes in Electrical Engineering 912, https://doi.org/10.1007/978-981-19-2689-1_2

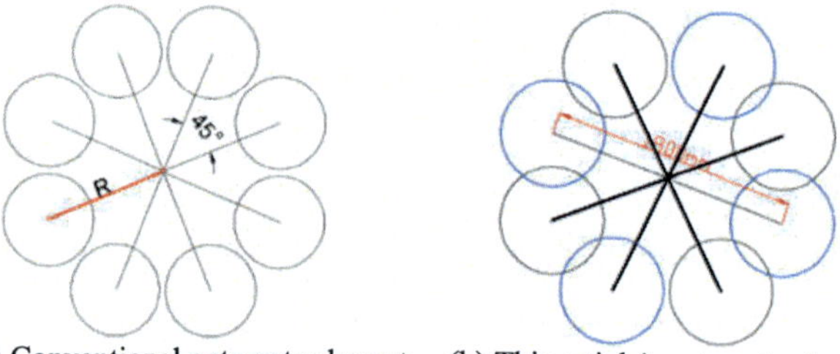

(a) Conventional octo-rotor layout (b) This article's octo-rotor layout

Fig. 1 Multi-rotor layout

The three-dimensional model of the fire fighting UAV is shown in Fig. 2.

According to the propeller pulling force test and considering the safety margin of the propeller, the propeller motor is sufficient to meet the flight requirements at a speed of 4000 r/min, so the calculated boundary pulling force selects the pulling force at 4000 r/min. When the propeller speed is 4000 r/min, the characteristic curve of its maximum pulling force changes with the axial flow speed. As shown in Fig. 3 the abscissa in the figure is the axial flow velocity, and the ordinate is the value of the maximum pulling force of the rotor. It can be seen from the figure that the maximum pulling force of the rotor decreases with the increase of the axial flow speed, showing an inverse relationship. The axial flow velocity of the UAV during wind resistance is calculated from the incoming wind speed and the wind resistance angle of attack. In order for the multi-rotor UAV to still adjust its attitude when

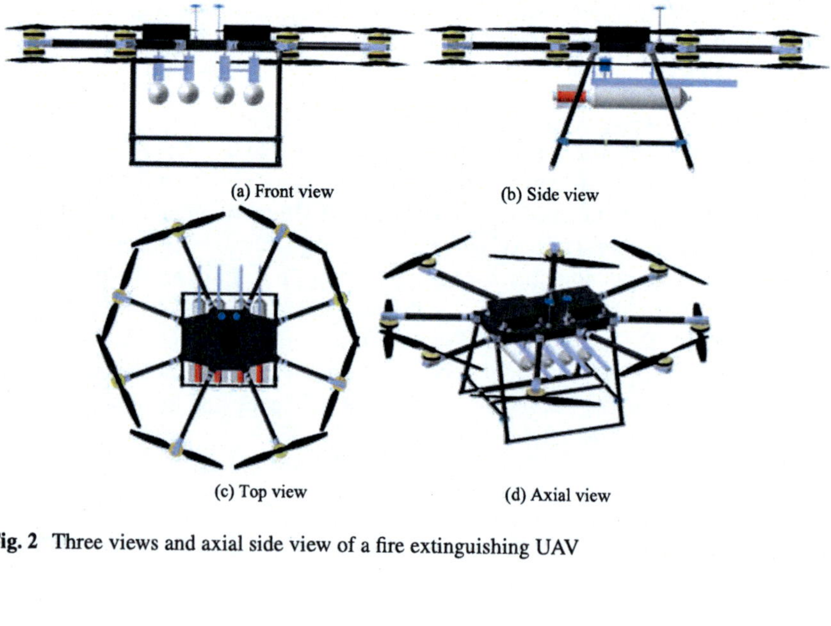

(a) Front view (b) Side view

(c) Top view (d) Axial view

Fig. 2 Three views and axial side view of a fire extinguishing UAV

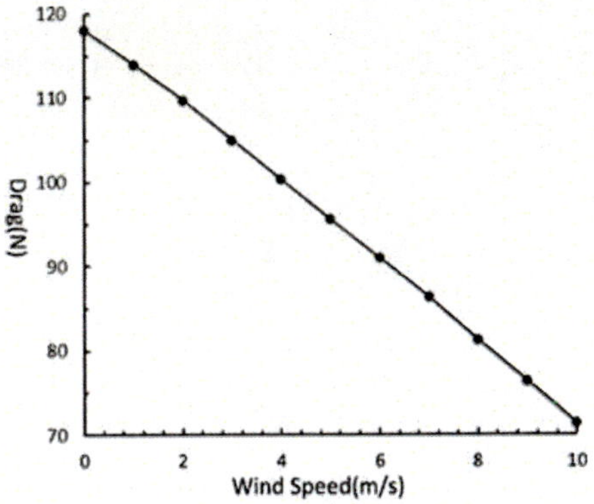

Fig. 3 The maximum pull of the rotor at different axial flow speeds

encountering emergencies, the motor pull force should be left with a certain safety margin.

2 Analysis of Anti-wind Characteristics

The definition of wind resistance: In the design process of the UAV flight platform, the wind resistance level is an important indicator [1]. When the UAV is disturbed by the external wind, whether the UAV can maintain a fixed attitude and initial position, the resistance of the UAV Wind performance can be understood as the ability to adjust the attitude of the UAV through the rotor power system so that it can still maintain a stable hover under the action of wind. The maximum wind speed at which the UAV can maintain a stable hover is used to measure the strength of its wind resistance. The greater the wind speed, the stronger its wind resistance.

Assuming that the actual wind farm environment is steady wind, the premise of the wind resistance analysis in this paper is that the dynamic changes of the wind farm environment and the dynamic control characteristics of the flight control system are not considered. Ignoring the interference between the rotors, the simplified model only focuses on the influence of power, and the rotor system is simplified to be the source of force and moment [2]. Taking the load condition as an example, the statics model of UAV wind resistance is established based on the balance of forces and moments in various directions. The motor number and direction of rotation are shown in Fig. 4.

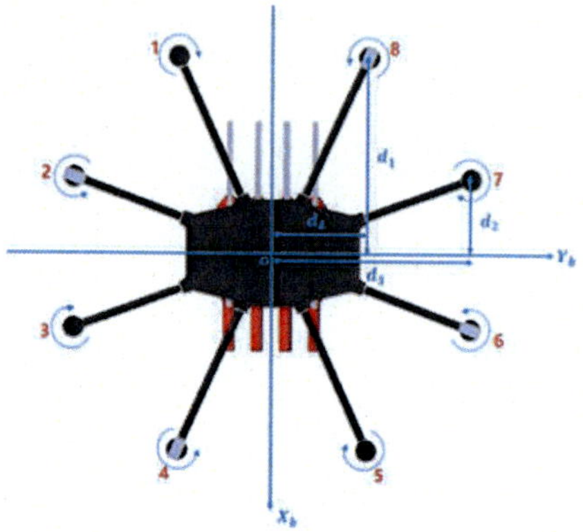

Fig. 4 Motor number and steering

2.1 Forward and Incoming Flow Analysis

The schematic diagram of the UAV hovering at any angle of attack and forward and incoming wind speed is shown in Fig. 5.

Project the UAV's own gravity and the aerodynamic force received on the body axis system:

$$\left[G_x\ G_y\ G_z\right]^T = A_g\left[0\ 0\ mg\right]^T \tag{1}$$

$$\left[F_x\ F_y\ F_z\right]^T = A_w\left[-D\ Y\ -L\right]^T \tag{2}$$

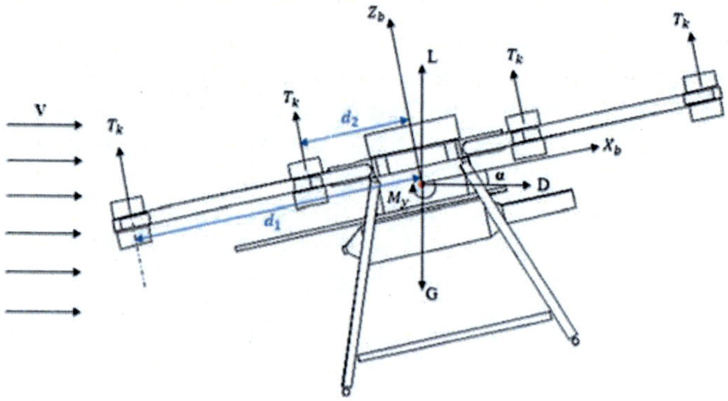

Fig. 5 Schematic diagram of the UAV hovering stationary at any angle of attack and forward and incoming wind speed

D, L, Y are the resistance, lift, and side force of the UAV, G_x, G_y, G_z respectively are the projections of gravity on the OX_b axis, OY_b axis, and OZ_b axis of the airframe coordinate system, F_x, F_y and F_z respectively are the aerodynamic forces on the OX_b axis, OY_b axis, and OZ_b axis of the airframe coordinate system, respectively The projection on the axis. A_g is the conversion matrix from the earth axis system to the airframe coordinate system, and A_w is the conversion matrix from the wind axis system to the airframe coordinate system. The sideslip angle is zero under the flow condition of the nose, the conversion matrix:

$$A_g = A_w \begin{bmatrix} \cos\theta\ 0\ \sin\theta \\ 0 \quad\ 1\ 0 \\ \sin\theta\ 0 - \cos\theta \end{bmatrix} \tag{3}$$

The aerodynamic formula is:

$$L = \frac{1}{2}\rho v^2 C_L S \tag{4}$$

$$D = \frac{1}{2}\rho v^2 C_D S \tag{5}$$

$$Y = \frac{1}{2}\rho v^2 C_Y S \tag{6}$$

L, D, Y are lift, drag and side force respectively, v are the incoming wind speed, ρ is the air density, C_L is the lift coefficient, C_D is the drag coefficient, C_Y is the side force coefficient, and S is the reference area.

Assuming that the UAV has the ability to resist wind, it can still be in a static equilibrium state by adjusting the position and attitude after being disturbed by the incoming flow. In the case of the UAV in the forward and incoming flow, the pulling force generated by the rotor is set as T_k, Then at any angle of attack α and the incoming wind speed ν is to be able to hover at rest, the following equations must be satisfied:

In the OX_b axis direction:

$$F_x(\alpha, \nu) + G_x = 0 \tag{7}$$

In the OZ_b axis direction:

$$F_z(\alpha, \nu) + \Sigma T_k + G_z = 0 \tag{8}$$

Establish the moment balance equation around the OY_b axis:

$$\left(\sum_{i=1,8} T_i - \sum_{j=4,5} T_j\right) \times d_1 + \left(\sum_{m=2,7} T_m - \sum_{n=3,6} T_n\right) \times d_2 + M_y(\alpha, \nu) = 0 \quad (9)$$

$M_y(\alpha, \nu)$ is the pitching moment of the whole machine calculated according to the moment coefficient:

$$M_y(\alpha, \nu) = \frac{1}{2}\rho\nu^2 C_m Sc \quad (10)$$

It is the distance from the central axis of the motor No. 1, 8, 4, and 5 to the OY_b axis, and d_2 is the distance from the central axis of the motor No. 2, 7, 3, and 6 to the OY_b axis.

According to the geometric relationship of the flight platform:

$$d_1 = 813\,\text{mm}$$

$$d_2 = 315\,\text{mm}$$

Assuming that the motor tension has the following relationship:

$$T_1 = T_8,\ T_4 = T_5,\ T_2 = T_7,\ T_3 = T_6 \quad (11)$$

The pulling force of the motor cannot exceed the limit of the maximum pulling force of the motor, and the maximum pulling force of the motor is related to the axial flow speed, and the pulling force of the motor is related to the torque, namely:

$$T_k \le T_{\max},\ T_{\max} = f(V_T) \quad (12)$$

V_T is the axial flow speed of the motor:

$$V_T = \nu \sin\alpha \quad (13)$$

The relationship between the incoming wind speed and the pulling force of the motor can be obtained.

2.2 *Analysis of Lateral Incoming Flow*

The schematic diagram of the UAV hovering at any angle of attack and lateral incoming wind speed is shown in Fig. 6.

When there is a load, the UAV is asymmetric about the axis, and theoretically generates a moment coupled with pitch, yaw, and roll. For a multi-rotor UAV, the

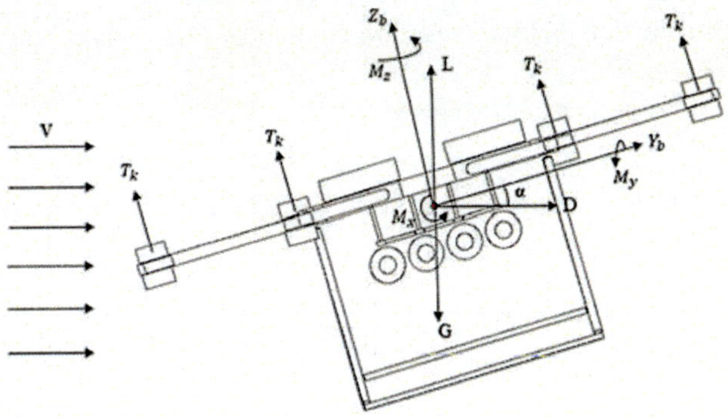

Fig. 6 Schematic diagram of UAV hovering at any angle of attack and lateral incoming wind speed

pitch and roll operations are mainly realized by adjusting the propeller tension [3]. The yaw control is achieved through the propeller anti-torque. Since the yaw moment is balanced by the propeller anti-torque, aerodynamic analysis cannot be calculated, and through the analysis of the aerodynamic data, it is found that the yaw moment is very small, compared to other forces and moments. Even smaller, after discussion, in the case of lateral flow, the generated yaw moment is ignored, and only the roll moment and pitch moment are considered.

In order to facilitate the analysis, use Eqs. (1) and (2) to project the gravity and aerodynamic forces received by the UAV to the body axis system. The conversion matrix:

$$A_g = A_w = \begin{bmatrix} 0 & 1 & 0 \\ -\cos\theta & 0 & -\sin\theta \\ \sin\theta & 0 & -\cos\theta \end{bmatrix} \tag{14}$$

The modeling method is the same as that of the forward and incoming flow. After the UAV is disturbed by the incoming flow, the UAV can still be in a static equilibrium state by adjusting the position and attitude, and the UAV has the ability to resist wind. In the case of side flow, the pulling force generated by the rotor is T_k, and the following equations must be satisfied to be able to hover at any angle of attack α and the flow wind speed ν.

From the analysis data, the side force received by the UAV is very small, approximately zero, in the OX_b axis direction.

In the OY_b axis direction:

$$F_y(\alpha, \nu) + G_y = 0 \tag{15}$$

In the OZ_b axis direction:

$$F_z(\alpha, \nu) + \Sigma T_k + G_z = 0 \tag{16}$$

Establish the moment balance equation around the OX_b axis:

$$M_T + M_x(\alpha, \nu) = 0 \tag{17}$$

$M_x(\alpha, \nu)$ is the rolling moment of the whole machine calculated according to the moment coefficient:

$$M_x(\alpha, \nu) = \frac{1}{2}\rho \nu^2 C_m Sc \tag{18}$$

M_T is Moment generated for propeller pull:

$$M_T = \left(\sum_{p=2,3} T_p - \sum_{q=6,7} T_q\right) \times d_3 + \left(\sum_{v=1,4} T_v - \sum_{w=5,8} T_w\right) \times d_4 \tag{19}$$

It is the distance from the central axis of the motor No. 2, 3, 6, and 7 to the OX_b axis, and d_4 is the distance from the central axis of the motor No. 1, 4, 5, and 8 to the axis.

According to the geometric relationship of the flight platform:
$d_3 = 848\,\text{mm}$.
$d_4 = 400\,\text{mm}$.

Establish the moment balance equation around the $\left(\sum_{i=1,8} T_i - \sum_{j=4,5} T_j\right) \times d_1 + \left(\sum_{m=2,7} T_m - \sum_{n=3,6} T_n\right) \times d_2 + M_y(\alpha, \nu) = 0$ axis:

$$\left(\sum_{i=1,8} T_i - \sum_{j=4,5} T_j\right) \times d_1 + \left(\sum_{m=2,7} T_m - \sum_{n=3,6} T_n\right) \times d_2 + M_y(\alpha, \nu) = 0 \tag{20}$$

$M_y(\alpha, \nu) = \frac{1}{2}\rho \nu^2 C_m Sc$ is the pitching moment of the whole machine calculated according to the moment coefficient:

$$M_y(\alpha, \nu) = \frac{1}{2}\rho \nu^2 C_m Sc \tag{21}$$

3 Wind Resistance Evaluation Results

According to the calculation and analysis, the change trend of the Aerodynamic Coefficient is the same at different inlet velocity, and the corresponding aerodynamic coefficient is very small at the same angle of attack, therefore, the aerodynamic force and moment of UAV can be calculated by aerodynamic force and moment formula at any velocity, and the wind resistance of UAV at each velocity can be analyzed. In order to obtain more conservative results, the aerodynamic coefficients of 8 m/s flow at wind speed are calculated theoretically [4].

3.1 *Forward and Incoming Flow*

Solve Eqs. (1)–(13) simultaneously, and find the anti-wind attack angle of the UAV under different incoming wind speeds, as shown in Fig. 7. That is, the UAV needs to be balanced at a certain incoming flow speed [5]. In the figure, the abscissa is the incoming wind speed, and the ordinate is the UAV's anti-wind angle of attack. It can be seen from the figure that the anti-wind angle of attack that satisfies the balance condition increases with the increase of the incoming flow speed. Under the same wind speed, the anti-wind angle of attack with load is smaller than that of no load. When the wind speed reaches 16 m/s. When there is a load, the anti-wind attack angle is about −8° under load, and the anti-wind attack angle is about −14° under no load, both of which have not reached the UAV's ultimate pitch angle of 30°, which is within the range that the UAV can control. Inside.

Bring the calculated wind speed and wind resistance angle of attack into the force and moment balance equation to find the maximum and minimum pull forces of the eight propellers under different incoming wind speeds, as shown in Fig. 8, and Fig. 8a shows that for the load condition, Fig. 8b is the no-load condition. The horizontal

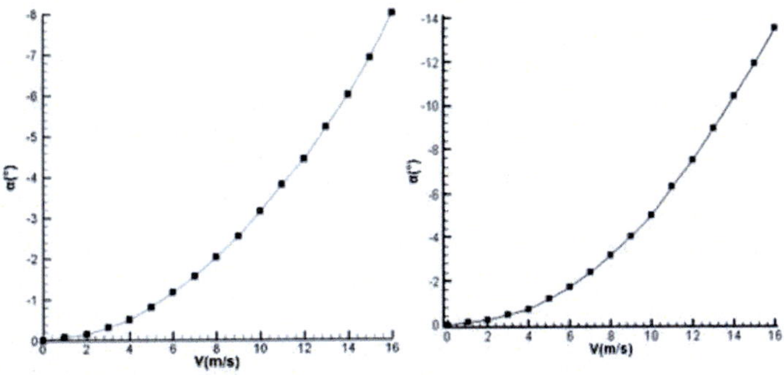

(a)with load (b)without load

Fig. 7 Anti-wind angle of attack under different forward and incoming wind speeds

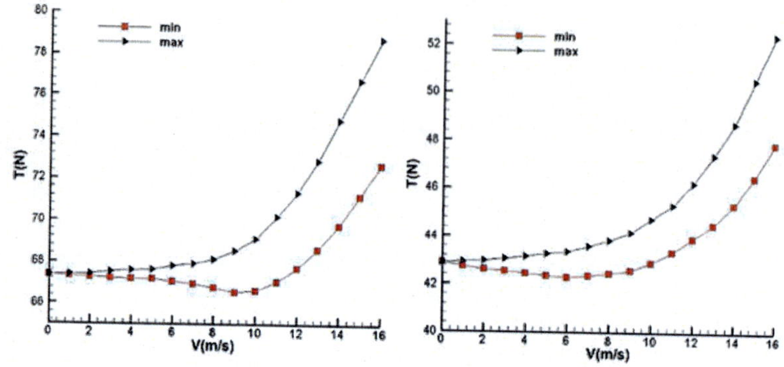

(a) with load (b)without load

Fig. 8 The maximum and minimum pull forces of the propeller under different incoming wind speeds

axis in the figure is the incoming wind speed, and the vertical axis is the propeller pull.

From the analysis of the data, it can be seen that in the case of a forward and incoming load, the maximum motor tension corresponding to a wind speed of 16 m/s is about 79 N, which meets the maximum motor tension limit, and there is still a 26% safety margin at this time; in the case of no load, The maximum motor pull force corresponding to a wind speed of 16 m/s is about 52 N, which meets the maximum motor pull force limit. At this time, there is still a safety margin of 46%.

3.2 *Lateral Incoming Flow*

Solve the above Eqs. (15)–(21) together, and find the anti-wind attack angle of the UAV under different lateral incoming wind speeds, as shown in Fig. 9, that is, the

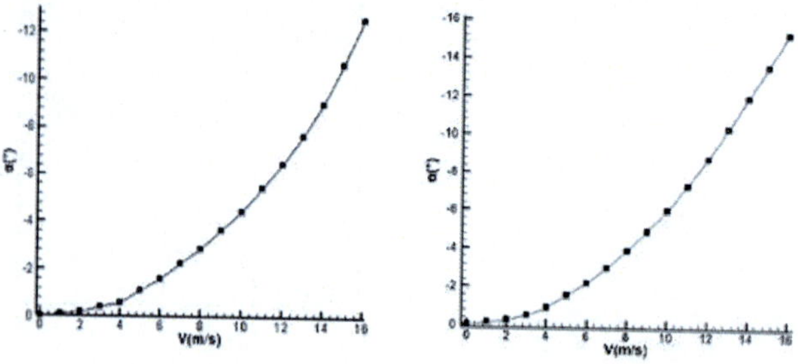

(a)with load (b)without load

Fig. 9 Anti-wind angle of attack under different lateral incoming wind speeds

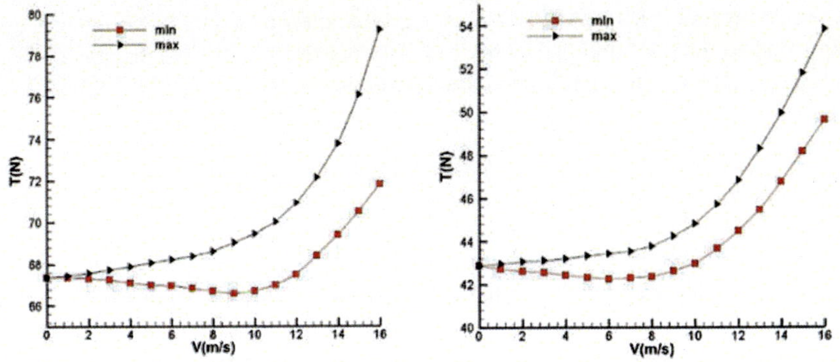

(a)with load (b)without load

Fig. 10 The maximum and minimum pull forces of the propeller under different incoming wind speeds

UAV needs to be balanced at a certain incoming flow speed. In the figure, the abscissa is the incoming wind speed, and the ordinate is the UAV's anti-wind angle of attack. It can be seen from the figure that the anti-wind angle of attack that satisfies the balance condition increases with the increase of the incoming flow speed. Under the same wind speed, the anti-wind angle of attack with load is smaller than that of no load. When the wind speed reaches 16 m/s. When there is a load, the wind resistance angle of attack is about $-13°$ under load, and the wind resistance angle of attack is about $-16°$ under no load, both of which have not reached the UAV's limit roll angle of 30°, which is within the range that the UAV can control. Inside.

Bring the calculated wind speed and wind resistance angle of attack into the force and moment balance equations to find the maximum and minimum pull forces of the eight propellers under different incoming wind speeds, as shown in Fig. 10. The horizontal axis in the figure is the incoming flow. Wind speed, the longitudinal axis is the propeller pull.

When there is a load in the lateral direction, the maximum motor tension corresponding to a wind speed of 16 m/s is about 80 N, which meets the maximum motor tension limit, and there is still a safety margin of 20% at this time; in the case of no load, the wind speed of 16 m/s corresponds to The maximum motor pull force of is about 54 N, which meets the maximum motor pull force limit. At this time, there is still a safety margin of 44%.

4 Analysis Process of Wind Resistance Characteristics

This article proposes a set of analysis procedures for wind resistance evaluation of multi-rotor UAVs. This process is not only for octo-rotors, but can also be applied to any multi-rotor UAVs so as to initially evaluate the wind resistance of UAVs during the design process [6]. The wind resistance evaluation is added to the design process of the multi-rotor UAV flight platform, and it is checked and iterated repeatedly. The analysis process of wind resistance characteristics of multi-rotor UAV is summarized as follows:

(1) Establish the aerodynamic model of the multi-rotor UAV, and calculate the aerodynamic force and torque of the UAV under a certain incoming wind speed and different angles of attack through CFD software.
(2) Analyze the force in a stable hovering state, and establish the force balance and moment balance equations of the multi-rotor UAV with six degrees of freedom, that is, the statics model of the multi-rotor UAV against wind and stability.
(3) Solve the statics model, decoupling by programming, and obtain the numerical solution of the wind speed that makes the model valid under the limitation of the maximum pulling force of the rotor power system. The result obtained is that the multi-rotor UAV can resist wind speed.

5 Summary

(1) Under the forward and incoming wind, the UAV can withstand 7-level (16 m/s) wind by pitching 8° under load. At this time, the maximum motor tension required is 79 N, and the remaining motor tension is still 26%. The remaining motor tension it can effectively realize the attitude control. The UAV can withstand seven-level (16 m/s) wind when the UAV rolls 13° under load in the lateral flow wind. The maximum motor pull force required at this time is 80 N, and the remaining pull force of the motor is still 20%.
(2) Under the forward and incoming wind, the UAV can withstand seven-level (16 m/s) wind at a pitch of 14° without load. At this time, the maximum motor tension required is 52 N, the remaining motor tension is still 46%, and the remaining motor tension it can effectively realize the attitude control. Under the side flow wind, the UAV can resist the seven-level (16 m/s) wind when it rolls 16° without load. At this time, the maximum motor pulling force required is 54 N, and the remaining motor pulling force is still 44%.
(3) In summary, the fire-fighting UAV designed in this paper is theoretically able to withstand the seventh wind (16 m/s). Under the horizontal action of the seventh wind, the UAV can achieve force and moment balance by adjusting its attitude.

References

1. Wei Chen (2017) Overall design of multi-rotorcraft with large payload. Northwestern Polytechnical University, Xi'an

2. Changqing F, Bing C (2007) Fundamentals of multi-rotor UAV technology. Tsinghua University Press, Beijing
3. Xiaogang L, Guoshou S (2011) Research on wind resistance of UAV. In: Proceedings of the 13th China annual conference on system simulation technology and its applications, pp 856–859
4. Hang Z, Bifeng S, Haifeng W (2019) Wind-resistant characteristics and influencing factors of fixed-rotor UAV during vertical take-off and landing. Syst Eng 41(7):1525–1535
5. Yao L, Yuxia J (2018) Analysis of anti-wind disturbance aerodynamic characteristics of small coaxial rotor in natural flow. Acta Armamentarii 39(6):1225–1232
6. Yao L, Rongzhao L, Zhiquan W (2019) Hover efficiency analysis of small multi-rotor aircraft. Acta Armamentarii 40(6):1323–1328

Inverse Design Method of Transonic Airfoil Based on Deep Neural Network

Zhiliang Bai, Wei Zhang, and Ruyue Wei

Abstract This research introduces a deep neural network-based transonic airfoil inverse design approach. A deep neural network is established using the predicted pressure coefficient distribution, angle of attack, and Mach number around the airfoil as inputs and parameterized airfoil parameters as outputs. The network solves the transonic flow field of an airfoil using the Euler equation. The neural network is trained to grasp the impacts of geometric changes on the pressure coefficient and shock wave at various places on the airfoil surface. Using the finite volume approach and multigrid acceleration method, the Euler equation is employed to batch compute the airfoil database and build the training set. As the training set's output vector, the CST parameterization technique is utilized to parameterize each synchronously generated airfoil into 10 parameters. The results demonstrate that the deep learning approach developed in this research is capable of achieving the reverse design of a transonic airfoil, and that the trained neural network can generate the airfoil with the requisite aerodynamic properties rapidly and directly. The generalization error of the anticipated airfoil geometry information is less than 1.5 percent when compared to the accurate results. The results reveal a high level of precision. The impact of the deep neural network's shape and the activation function of neurons on the subjects under investigation is also examined. The results reveal that the activation function used is critical to the outcome, and the design of the neural network also has an impact on the outcome.

Keywords Inverse design · Transonic airfoil · Parameterization method · Deep neural network

1 Introduction

The optimization design of wing airfoil under cruise condition is a problem of great concern to modern aircraft designers, especially for the design of commercial aircraft. Finding the best airfoil in cruise state can greatly improve the economy of the aircraft.

Z. Bai · W. Zhang (✉) · R. Wei
School of Aeronautics, Northwestern Polytechnical University, Xi'an 710072, China

S. Lee et al. (eds.), *The Proceedings of the 2021 Asia-Pacific International Symposium on Aerospace Technology (APISAT 2021), Volume 1*, Lecture Notes in Electrical Engineering 912, https://doi.org/10.1007/978-981-19-2689-1_3

At present, the CFD method of computing airfoils is well-established and frequently utilized in the design and calculation of airfoils [1]. The typical design strategy for airfoil design is to iteratively adjust the local geometry based on the initial airfoil calculation results, run CFD calculations again for the updated calculation results, and repeat the process. Finally, when the calculation results reach a value that relatively meets the design index, stop the iteration, and take this result as the final design. This method is very dependent on the designer's experience, and usually can not get the optimal results in theory. At the same time, it usually consumes a lot of iteration time. Therefore, researchers have proposed a series of more efficient optimization methods. The employment of the adjoint optimization approach, for example, can enhance the efficiency of gradient optimization with several design variables, but it still requires many calls to the CFD simulation in certain optimization situations [2, 3]. Surrogate method can greatly reduce the use frequency of CFD in the optimization process and greatly improve the optimization efficiency [4, 5]. For example, Han Shaoqiang and others used gradient enhanced Kriging model to realize aerodynamic inverse design of high-dimensional design variables [6]. Kriging and other Surrogate methods, on the other hand, are often only suited for single input single output issues, making them inefficient for multi-dimensional design optimization problems. Machine learning technologies have shown increasing promise in aerodynamic design and prediction in recent years [7–9]. This research introduces a deep neural network-based transonic airfoil inverse design approach. The predicted pressure coefficient distribution, angle of attack, and Mach number of the airfoil are sent into the deep neural network, which outputs the parameterized airfoil parameters. The network solves the transonic flow field of an airfoil using the Euler equation. The neural network can grasp the impact of geometric modifications at different points on the airfoil on the change in pressure coefficient and shock wave after sufficient training. After the neural network has been trained, the airfoil form with a particular goal pressure coefficient distribution can be predicted immediately and at a high rate.

2 Modeling Method

2.1 Numerical Calculation Method

The Euler equation, which ignores viscosity, is employed as the governing equation of transonic airfoil computation in order to swiftly acquire a large number of trustworthy calculation results of airfoil pressure distribution. The multigrid acceleration approach and the upwind finite volume method are employed. The solver software is designed in Fortran to assure high performance of the computation process and to make calling the Python neural network program easier.

The calculation grid adopts structured O-shaped grid. The grid creation software is designed in Fortran to greatly automate the grid production process. It can read in two-dimensional airfoil coordinate points and produce the grid rapidly and automatically.

To assure the grid's orthogonality, the conformal transformation method is utilized to produce it. Figure 1 shows an example grid created by the grid generating tool.

The mature open source solver Cfl3d is utilized to compare with the calculation results of the solver in this work in order to verify the reliability of the aerodynamic calculation results in this study. For comparison, the RAE2822 airfoil was calculated at 0.73ma and 2.79° angle of attack. Figure 2 shows the results.

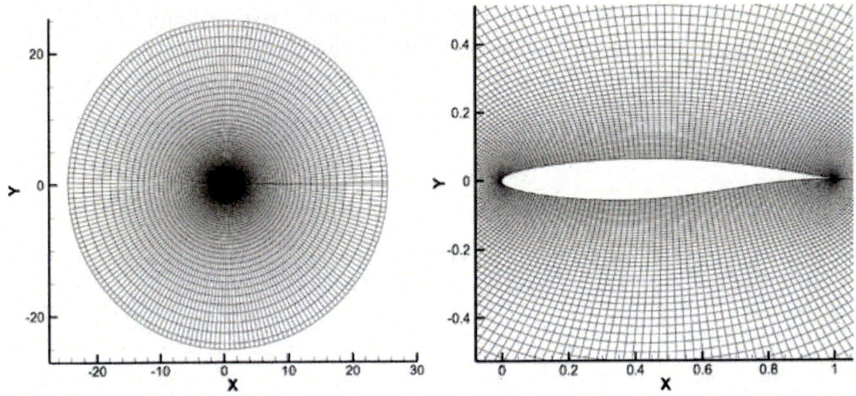

Fig. 1 2D airfoil grid

Fig. 2 The solver in this paper is compared with Cfl3d

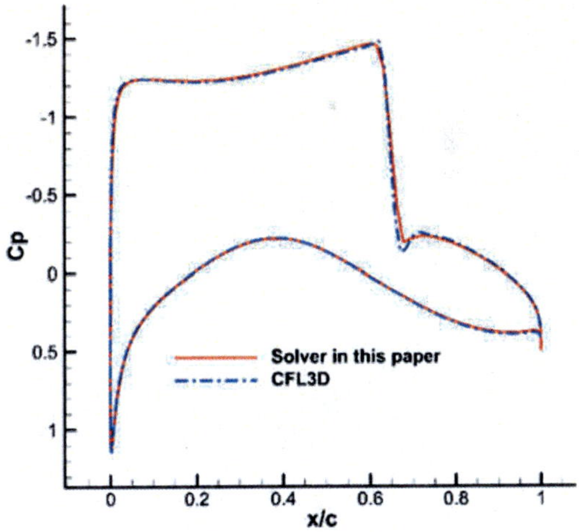

2.2 Airfoil Parameterization Method

In order to reduce the design variables in the inverse design of airfoil and make the airfoil as smooth as possible, CST parameterization method is used to parameterize the airfoil. (Class Function/Shape Function Transformation, CST) is an airfoil parameterization method proposed and developed by Kulfan et al. [10–12]. It has good robustness, few design parameters and high precision. It is widely used in airfoil design research. This method uses a combination of a class function to define the geometric shape and a shape function to modify the class function to express the airfoil. The airfoils used in this paper are closed at the trailing edge. For airfoils with closed trailing edges, CST parameterization can be expressed as:

$$\zeta(\psi) = C_{N_2}^{N_1}(\psi)S(\psi) \tag{1}$$

where $\zeta = z/c$, $\psi = x/c$. Where C is the chord length of the wing, x is the x-axis coordinate of the airfoil, z is the z-axis coordinate, and the category function is expressed as:

$$C_{N_2}^{N_1}(\psi) = \psi^{N_1}(1-\psi)^{N2} \tag{2}$$

For the round-headed pointed-tailed airfoil, the category function has $N_1 = 0.5$, $N_2 = 1.0$. The shape function $S(\psi)$ can be defined using a variety of methods; the method used in this paper is a weighted sum of nth-order Bernstein polynomials as an expression subset of $S(\psi)$, which is expressed as:

$$\begin{cases} S(\psi) = \sum\limits_{i=0}^{n} b_i B_n^i(\psi) = \sum\limits_{i=0}^{n} b_i\left[K_n^i \psi^i (1-\psi)^{n-i}\right] \\ K_n^i = \frac{n!}{i!(n-i)!} \end{cases} \tag{3}$$

where B_n^i is the Bernstein polynomial, i is the exponent in the polynomial, n is the order of the polynomial, K_n^i is the combination number, b_i is the weight factor. In this paper, the airfoil is parameterized by 10th order polynomial, that is, $n = 10$, which can reduce the design parameters of the upper and lower surfaces of the airfoil to 10.

2.3 Neural Network Model

The neural network model used in this paper is BPNN (back propagation neural network), which is a fully connected neural network through forward propagation calculation output and back propagation calculation error. It is characterized by multi-input and multi-output modeling, with strong nonlinear fitting ability, and is widely

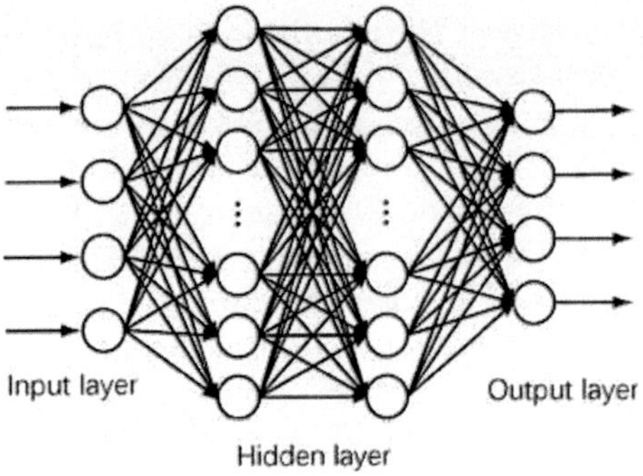

Fig. 3 Neural network model

used in classification or regression problems. A typical BPNN is composed of three parts: input layer, hidden layer and output layer, as shown in the Fig. 3.

There are m neurons in the input layer, n neurons in the output layer, and one or more layers in the hidden layer. In general, the more layers, the lower the overall network error, but it also increases network complexity and training time. The number of neurons in the hidden layer is usually selected according to experience. The referential experience obtained in engineering practice is as follows: the number of hidden neurons should be between the input layer and the output layer. The number of hidden neurons should be 2/3 of the size of the input layer plus the size of the output layer. The number of hidden neurons should be less than twice the size of the input layer. Passing weight of neurons in each layer ω Connected with the activation function, the training steps of the neural network are as follows:

1. Select the activation function and initialize the weight ω;
2. Forward propagation calculates the output values of each layer;
3. Calculate the error between the output value and the real value;
4. Calculate the neuron error of each layer by back propagation;
5. Update weight ω;
6. Repeat steps 2 to 5 until the model meets the training requirements.

3 Numerical Examples and Analysis

3.1 Training Set Establishment

Neural network training requires sufficient training sets. In this paper, about 600 different airfoils are selected, including common transonic airfoils and low-speed airfoils. The airfoil data are taken from NACA, NASA-SC and RAE. RAE-2822, NASA-SC-0412 and NASA-SC-0010 are removed from the airfoil library for subsequent verification. The Euler equation solver described above is used for batch calculation of the airfoil library. The incoming Mach number is 0.73. Seven different angles of attack are calculated for each airfoil, which are 3°, 2.5°, 2°, 1.5°, 1°, 0.5° and 0°

respectively. The airfoil and its corresponding pressure coefficient distribution are calculated.

Generally, the calculation results of the pressure coefficient distribution of the airfoil are given in the solver in the coordinates of the chord line of the point airfoil and the corresponding values at its coordinates. If the training set is directly established in this format, the dimension of input parameters will be too high. Therefore, a fixed method for dividing the chord line of the airfoil is used in this paper, Thus, the coordinate information of the pressure coefficient distribution along the airfoil chord can be omitted.

As shown in Fig. 4, take the midpoint of the airfoil chord line as the center of the circle and $C/2$ as the radius to make a circle. According to an equal angle θ divide the circle, make a vertical line from the point on the circumference to the X axis and intersect with the airfoil surface. The solver takes the Cp value at the intersection and outputs it as a column vector clockwise around the airfoil from the trailing edge of the airfoil. In this paper, the airfoil is divided into 96 parts, so the coordinate information of pressure distribution is omitted.

The training set is divided into two parts: X and Y. X contains the pressure coefficient distribution vector calculated by the solver and the corresponding incoming attack angle information. Y contains the CST parameter vector parameterized by CST for the airfoil corresponding to the pressure coefficient distribution. The dimension of each X vector is 97 and the dimension of each Y vector is 10. The generation of the training set is shown in Fig. 5. The training set calculation program reads the airfoil file, calculates the CST parameters, writes them into the training set Y, generates the calculation grid, solves the flow field around the airfoil, obtains the pressure coefficient distribution and writes them into the training set X. Circularly calculates until all airfoils in the complete airfoil library are calculated and the results are written into the training set.

Fig. 4 Airfoil segmentation strategy

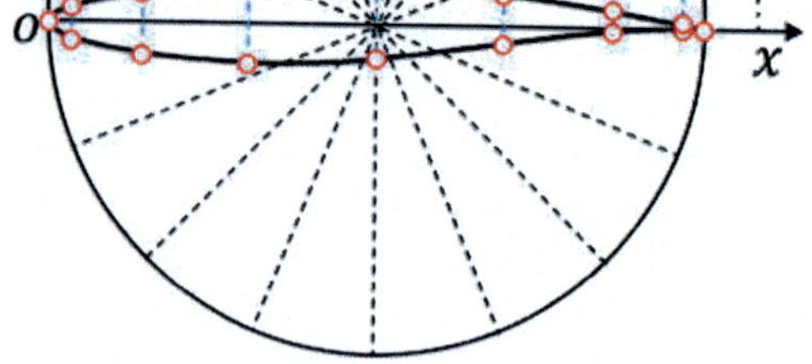

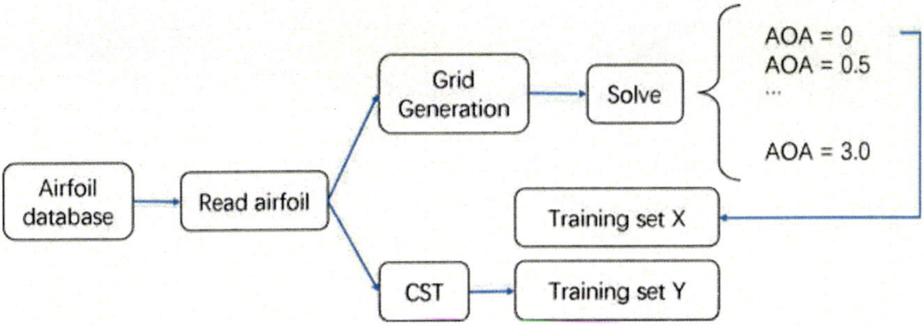

Fig. 5 Training set calculation process

3.2 *Model Training*

The neural network may then be used to train the training set when it has been built. The neural network's structure is developed first. The input layer has the same number of neurons as each X vector, which has a size of 97. The output layer's number of neurons is compatible with the Y vector, which is 10, while the hidden layer's number of neurons is 128. For training, neural network models with varied numbers of layers are created. The optimization method of neural network back propagation adopts Adam optimization method. Before the training set is sent to the neural network, the sorting of the training set is randomized, and then 5% of the training set is selected as the test set in the training process. The training process lasts 500 epochs in total, and each batch is 32 groups of data. An example of convergence in the training process is shown in Fig. 6.

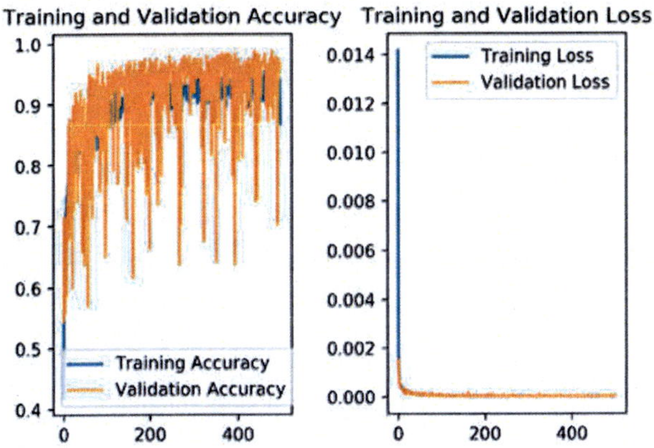

Fig. 6 Convergence in the training process

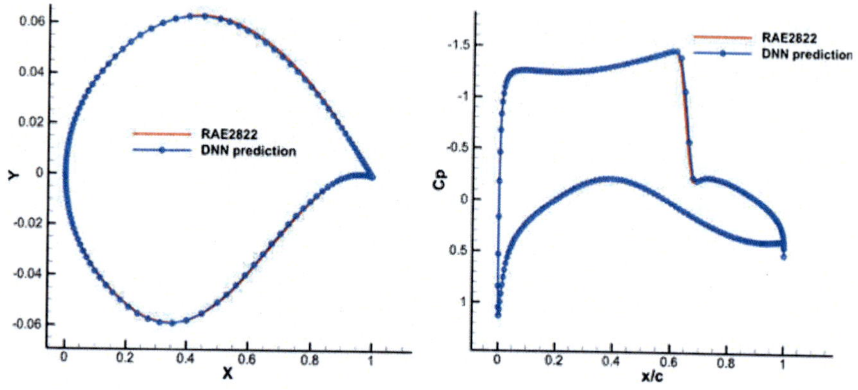

Fig. 7 Example of neural network prediction results

3.3 Analysis of Training Results

The training results are verified by RAE2822 airfoil. Before training, RAE2822 airfoil data were removed from the training set. The test input is the airfoil pressure coefficient distribution of RAE2822 airfoil calculated by CFD under the incoming flow condition of 2.79° angle of attack, Ma = 0.73. The neural network model uses two hidden layers, and the prediction time is only 0.007 s, which is much less than CFD calculation and surrogate method. The predicted airfoil and RAE2822 airfoil are drawn in the same coordinate system, as shown in Fig. 7. The pressure coefficient distribution of the airfoil predicted by neural network under the same flow conditions is compared with the target pressure distribution input by neural network. As shown in Fig. 7, it can be seen that the airfoil predicted by neural network is very close to the target result. The error of airfoil parameters predicted by neural network is defined as:

$$error = \sqrt{\frac{\sum_1^n (y_{dnni} - y_i)^2}{n}} \tag{4}$$

where y_{dnni} is the ith y coordinate point of airfoil predicted by neural network, y_i is the Y coordinate point at the same position of RAE2822, and the error calculation value is only 1.2%.

3.4 Neural Network Hyperparametric Analysis

Examine the impact of the neural network's design settings on the airfoil prediction findings presented in this research. To begin, investigate the impact of the neural network's number of hidden layers on prediction outcomes and training duration. The number of layers picked ranges from 1 to 6. The following are the outcomes:

- 1 hidden layer, training time 57.59 s, error = 1.35%.
- 2 hidden layers, training time 62.74 s, error = 1.2%.
- 3 hidden layers, training time 68.35 s, error = 1.18%.
- 4 hidden layers, training time 73.41 s, error = 1.21%.
- 5 hidden layers, training time 79.52 s, error = 1.32%.
- 6 hidden layers, training time 88.20 s, error = 1.54%.

The training time required by the neural network grows as the number of layers increases, while the verification error between the training results and the real scenario increases. Overfitting can be regarded as an issue when there are too many layers. Figure 8 depicts the relationship between training time, result error, and the number of hidden layers in neural network.

In the previous calculation results, the neuron activation function adopts Tanh function. Replace the activation function in the neural network with the more generally used ReLU function, train again, and forecast the airfoil. Figure 9a shows the images of Tanh function and ReLU function. The results are presented in Fig. 9b.

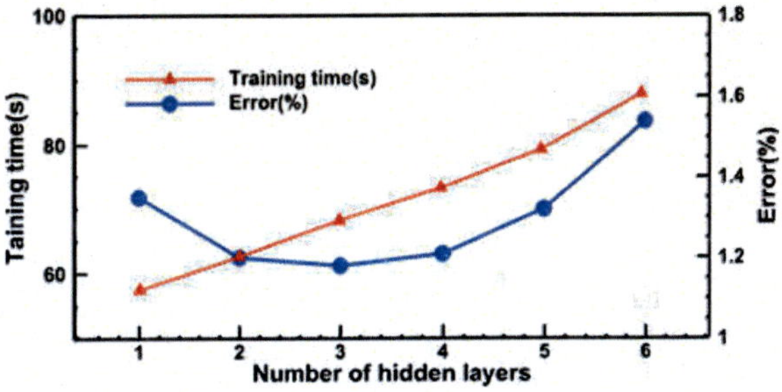

Fig. 8 Training time and result error with different number of hidden layers

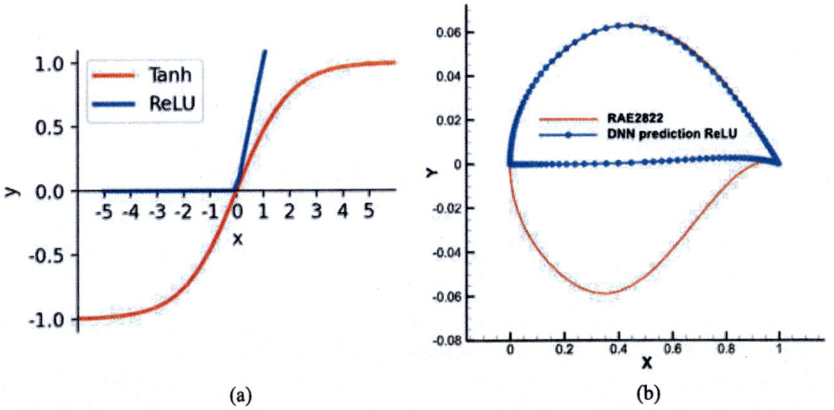

Fig. 9 **a** The images of Tanh function and ReLU function. **b** Example of neural network prediction results using the ReLU function

It can be seen that the neuron using the ReLU function as the activation function completely loses the characteristics of the lower surface of the airfoil. Because the slope of the ReLU function is 0 in the range of $x < 0$, it will smooth out the parameters with the airfoil CST parameters less than 0, which is not suitable to be used as the activation function in the airfoil inverse design problem.

4 Conclusion

In this study, a neural network-based airfoil inverse design approach is created. The neural network can directly predict the airfoil based on the input of target pressure distribution after training the airfoil database. The findings reveal that the airfoil inverse design approach described in this study has high speed and precision, demonstrating the artificial intelligence method's potential in solving aerodynamic design challenges.

The selection of activation function of neural network is very important for the results. Inappropriate activation function may get completely wrong prediction results. The number of hidden layers of neural network should not be too many. Overfitting the results and increasing the training duration will be an issue if there are too many hidden layers.

References

1. Slotnick J, Khodadoust A, Alonso J et al (2014) CFD vision 2030 study—a path to revolutionary computational aerosciences: NASA/CR-2014-218178. NASA, USA
2. Lyu Z, Kenway G, Martins J (2015) Aerodynamic shape optimization investigations of the common research model wing benchmark[J]. AIAA J 53(4):968–985
3. Martins J, Hwang J (2013) Review and unification of methods for computing derivatives of multidisciplinary systems. AIAA J 51(11):2582–2599
4. Zhonghua HAN, Chenzhou XU, Jianling QIAO et al (2020) Recent progress of efficient global aerodynamic shape optimization using surrogate-based approach. Acta Aeronautica et Astronautica Sinica 41(5):30–70 (in Chinese)
5. Vianaf, ST, Balabanov V et al (2014) Metamodeling in multidisciplinary design optimization: how far have we really come? AIAA J 52(4):670–690
6. Shaoqiang HAN, Wenping SONG, Zhonghua HAN et al (2017) Aerodynamic inverse design method based on gradient enhanced Kriging model. Acta Aeronautica et Astronautica Sinica 38(7):138–152 (in Chinese)
7. Saleem W, Kharal A, Ahmad R et al (2015) Comparison of ACO and GA techniques to generate neural network-based Bezier-PARSEC parameterized airfoil. In: The 2015 11th international conference on natural computation. IEEE, USA, pp 1127–1133
8. I K, Kou J, Zhang W (2019) Deep neural network for unsteady aerodynamic and aeroelastic modeling across multiple Mach numbers. Nonlinear Dyn 132:1–21

9. Haixin CHEN, Kaiwen DENG, Runze LI (2019) Utilization of machine learning technology in aerodynamic optimization. Acta Aeronautica et Astronautica Sinica 40(1):52–68 (in Chinese)
10. Kulfan BM, Bussoletti JE (2006) "Fundamental" parametric geometry representations for aircraft component shapes. AIAA-2006-6948
11. Kulfan BM (2007) Recent extensions and applications of the "CST" universal parametric geometry representation method. AIAA-2007-7709
12. Kulfan BM (2007) A universal parametric geometry representation method "CST". AIAA-2007-62

High-Rise Building Wind Field Simulation

Zhan-ke Li, Si-jie Luo, Shi-gao Su, Jin-shuo Cao, and Xiao-min Zhang

Abstract High-rise building fire extinguishing UAV is a new equipment used in high-rise building fire extinguishing and rescue. However, the turbulent environmental airflow in the high-rise building area and the complex wind field pose a threat to the stability of fire fighting and fire fighting drones, which puts forward higher requirements for the stability and wind resistance of multi-rotor UAV flight platforms. Therefore, aiming at multi-rotor drones that perform fire fighting tasks in high-rise buildings, the characteristics of wind fields between buildings in high-rise buildings are studied to provide technical references for the design and optimization of fire-fighting drones, which not only has important theoretical significance, but also has high practical value. With multi-rotor firefighting drones as the background, based on the computational fluid dynamics (CFD) theory and Fluent software, the wind field characteristics around high-rise buildings are studied, carrying out numerical simulation analysis of the wind field around single buildings and multiple buildings to provide a reference for the actual flight of fire extinguishing drones between buildings. The analysis results show that: Compared to the single building, the wind field around the building has changed a lot when the two buildings are arranged, and the aerodynamic interference between the two buildings is significantly weakened as the distance between the buildings increases; Layout and the surrounding buildings play a shielding role. The wind and shadow areas are superimposed on each other, and the air flow is complicated. Drones should avoid flying in these complex wind fields as for as possible.

Keywords Multi-rotor fire-fighting drone · Wind field characteristics · Aerodynamic interference

---

Z. Li (✉) · S. Luo · S. Su · J. Cao · X. Zhang
School of Aeronautics, Northwestern Polytechnical University, Xi'an, People's Republic of China
e-mail: lzk@nwpu.edu.cn

S. Lee et al. (eds.), *The Proceedings of the 2021 Asia-Pacific International Symposium on Aerospace Technology (APISAT 2021), Volume 1*, Lecture Notes in Electrical Engineering 912, https://doi.org/10.1007/978-981-19-2689-1_4

1 Introduction

The turbulent air flow in the high-rise building area and the complex wind field pose a threat to the stability of the fire-fighting UAV. It also puts forward higher requirements for the stability and wind resistance of the multi-rotor UAV flight platform. Therefore, for multi-rotor drones that perform high-rise building fire extinguishing tasks, the characteristics of wind fields between high-rise buildings are studied, the wind resistance stability of multi-rotor drones flying platforms, and the design technology of high wind resistance and stable flying platforms are studied. It not only has important theoretical significance, but also has high practical value for improving the safety and reliability of fire fighting drones and providing technical reference for the design and optimization of fire fighting drones [1].

2 Computational Fluid Dynamics Theory

2.1 Wind Field Theory

The obstruction of the building causes obvious changes in the wind field flowing through the building. Complex physical phenomena such as air impact, trapped vortex, separation, backflow, reattachment, and surrounding flow appear around the building. The wind field surrounding the building is shown in Fig. 1 shown.

2.2 Fluid Mechanics Control Equation

The three basic governing equations of fluid mechanics are continuity equation, momentum equation and energy equation. At the same time, these three basic governing equations are also the most direct mathematical explanations of the three conservation laws of physics [2].

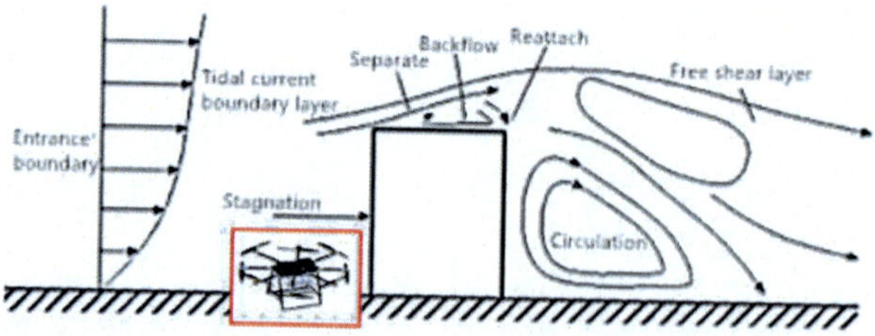

Fig. 1 Schematic diagram of the wind field around the building

Continuous Equation

The continuum equation is one of the basic equations of fluid mechanics. It is a mathematical expression used by the law of conservation of mass in fluid mechanics. In the calculation area, any control body can be selected, and the rate of increase of fluid mass in the body per unit time is equal to the flow rate of inflow and outflow. The differential form of the difference is as follows:

$$\frac{\partial \rho}{\partial t}+\frac{\partial(\rho u_x)}{\partial x}+\frac{\partial(\rho u_y)}{\partial y}+\frac{\partial(\rho u_z)}{\partial z}=0 \quad (1)$$

If the fluid is incompressible and its density is constant, then formula (1) can be simplified to:

$$\frac{\partial(\rho u_x)}{\partial x}+\frac{\partial(\rho u_y)}{\partial y}+\frac{\partial(\rho u_z)}{\partial z}=0 \quad (2)$$

Momentum Equation

According to Newton's second law, during the movement, the amount of change in the momentum of the fluid element per unit time is equal to the sum of the mass force of the element body and the surface force. The momentum equations in the three directions are as follows:

$$\frac{\partial(\rho u_x)}{\partial t}+\nabla\cdot\left(\rho u_x\vec{u}\right)=-\frac{\partial p}{\partial x}+\frac{\partial \tau_{xx}}{\partial x}+\frac{\partial \tau_{yx}}{\partial y}+\frac{\partial \tau_{zx}}{\partial z}+\rho f_x \quad (3)$$

$$\frac{\partial(\rho u_y)}{\partial t}+\nabla\cdot\left(\rho u_y\vec{u}\right)=-\frac{\partial p}{\partial y}+\frac{\partial \tau_{xy}}{\partial x}+\frac{\partial \tau_{yy}}{\partial y}+\frac{\partial \tau_{zy}}{\partial z}+\rho f_y \quad (4)$$

$$\frac{\partial(\rho u_z)}{\partial t}+\nabla\cdot\left(\rho u_z\vec{u}\right)=-\frac{\partial p}{\partial z}+\frac{\partial \tau_{xz}}{\partial x}+\frac{\partial \tau_{yz}}{\partial y}+\frac{\partial \tau_{zz}}{\partial z}+\rho f_z \quad (5)$$

Energy Equation

The essence of the energy equation is the first law of thermodynamics. It is usually expressed as: the internal energy increment of a fluid cell is equal to the sum of the heat that enters the cell through heat conduction, the heat generated in the cell, and the work done by the surrounding fluid on the cell. The expression is as follows:

$$\frac{\partial(\rho T)}{\partial t}+\frac{\partial(\rho u T)}{\partial x}+\frac{\partial(\rho v T)}{\partial y}+\frac{\partial(\rho w T)}{\partial z}=\frac{\partial}{\partial x}\left(\frac{k}{C_p}\frac{\partial T}{\partial x}\right)+\frac{\partial}{\partial y}\left(\frac{k}{C_p}\frac{\partial T}{\partial y}\right)+\frac{\partial}{\partial z}\left(\frac{k}{C_p}\frac{\partial T}{\partial z}\right)+S_T \quad (6)$$

Generally, the energy equation is not considered for incompressible fluids.

3 CFD Simulation of High-Rise Building Wind Field

Based on the ANSYS Fluent 18.0 software platform, ICEM is used as the pre-processing software to provide high-quality grids, and the Tecplot 360 software is used to coordinate post-processing and sorting of images. The wind field around single and multiple buildings is numerically simulated and analyzed to obtain the wind speed and wind around high-rise buildings. The pressure distribution provides a reference for the anti-wind flight of the fire-fighting drone between buildings [3].

3.1 Computational Domain Setting and Meshing

In order to facilitate modeling and calculation and analysis, the geometric model of this paper selects a rectangular column with a full-scale model size of 30 m × 45 m × 100 m (length × width × height) with a flat surface without any attachments. The geometric model is shown in Fig. 2.The positive direction of the x-axis in the Fig is the flow direction of the incoming wind.

The setting of the calculation domain has an important influence on the calculation results. In order to minimize the influence of the calculation domain on the flow field near the model and enable the full development of the flow wind field around the building, the size of the calculation domain should be set reasonably. Generally speaking, the setting of the calculation domain cannot be too small, otherwise it will affect the full development of the flow, but the calculation domain should not be too large, which will increase the amount of calculation and reduce the efficiency of the solution. Whether the setting of the calculation domain is reasonable can be judged by calculating the blocking rate, A_b is used to represent the maximum windward area of the building, and A_d is used to represent the cross-sectional area of the watershed. The formula for the blocking rate is as follows:

$$Blocking\,rate = \frac{A_b}{A_d} \tag{7}$$

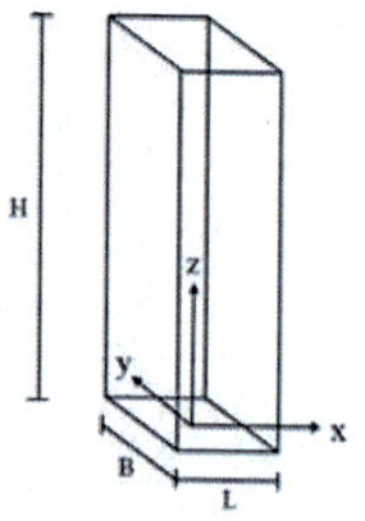

Fig. 2 Schematic diagram of the geometric model

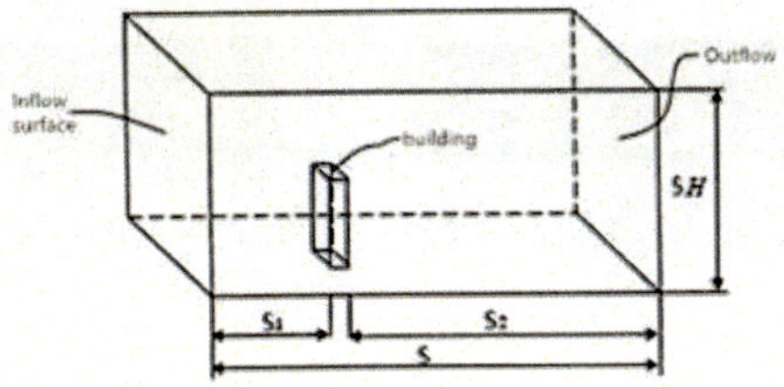

Fig. 3 Schematic diagram of computational domain

When the blocking rate meets the requirements, it is also necessary to consider whether the turbulence in the calculation domain is fully developed to set up the longitudinal calculation domain. Generally, the building is placed in the 1/3 point of the watershed along the flow. In order to fully develop the flow around the building, assuming that the height of the building is H, the general distance between the outflow surface and the leeward surface of the building should be greater than 5H. In order to avoid flow congestion, the width of the calculation domain should be above 6H, and the total height of the calculation domain Must be greater than 4H. The schematic diagram of the computational domain is shown in Fig. 3.

In this paper, the size of the calculation domain of a single building is set to 1200 m × 600 m × 500 m (length × width × height). Since the calculation domain is large enough, the size of the calculation domain does not change when the direction of different incoming winds is changed. By calculating the blocking rate:

$$(45\,\text{m} \times 100\,\text{m}) \big/ (600\,\text{m} \times 500\,\text{m}) = 1.5\% < 3\%$$

The calculated blocking rate is 1.5%, which is less than 3%, which meets the blocking rate requirement.

In this paper, structured grids are selected when dividing the grid. Structured grids are suitable for problems with simple boundary conditions. They have the advantages of good grid quality, fast generation speed, simple data structure, and stable calculation results. After creating the O-Block, define the grid node distribution, generate a grid, the number of grids is 800,000, and the grid division is shown in Fig. 4.

Fig. 4 Mesh division

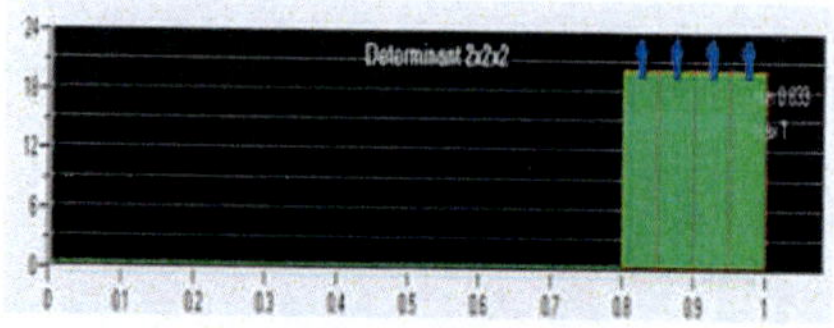

(a) Determinant2×2×2

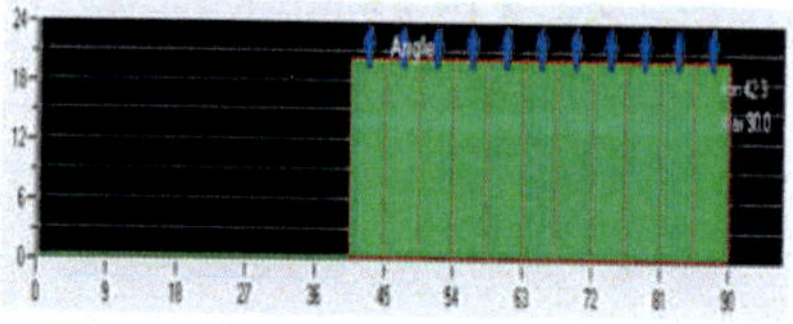

(b) Angle (°)

Fig. 5 Grid quality judgment

Grid quality has a great influence on calculation accuracy and stability. Determine $2 \times 2 \times 2$ (Jacobian determinant criterion) and Angle (angle) are respectively selected as the criterion of grid quality. Figure 5 shows the result of grid quality determination. Figure 5a shows that the judgment value is greater than 0.8, and Fig. 5b shows that the judgment value is greater than 40°, and the mesh quality meets the calculation requirements.

3.2 Result Analysis and Processing

The calculation uses a 3D double-precision, implicit separation pressure solver. Since this article simulates an incompressible flow, the separation solver is selected; an incompressible air model with a constant density is selected; the simulated flow is a steady flow around a turbulent flow model Use Realizable model; use non-equilibrium wall function; use SIMPLEC algorithm to solve the velocity–pressure coupling equation; use second-order upside style discrete velocity, turbulent kinetic energy and turbulent dissipation rate and other convection terms. Set the iterative residual error limit value is all, by checking that the residual error value of all control equations is less than the limit value, the iteration result is considered to be convergent; monitor the lift coefficient or average pressure coefficient on the surface of the model to ensure that it is basically unchanged [4].

Wind Field Simulation of Two High-Rise Buildings

In this section, the calculation results are analyzed and processed. In order to facilitate the analysis of the distribution law of wind pressure and wind speed around the building, velocity cloud diagrams and streamline diagrams at different sections are selected for illustration. The XZ section and the XY section at the height of 2/3H are respectively selected for analysis [5].

XZ (Y = 0) Plane

Figure 6 and Fig. 7 are the wind speed diagram and streamline diagram in the XZ plane, respectively. Through observation, it can be seen that as the height of the building increases, the wind speed increases, which is in line with the characteristics of the exponential wind profile. When the air flow is obstructed by the building, the

Fig. 6 Y = 0 plane wind speed diagram

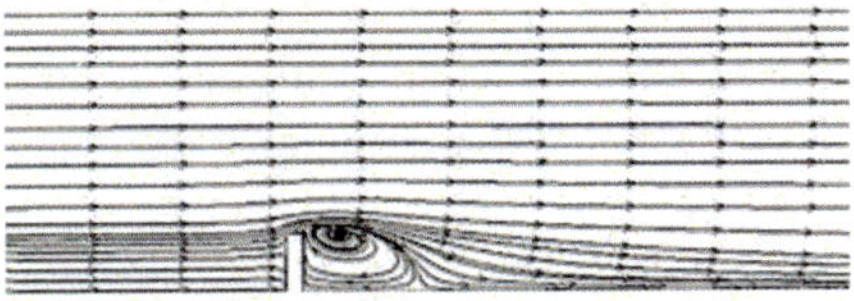

Fig. 7 Y = 0 plane streamline diagram

flow separates. The stagnation point of the airflow appears at about 2/3 of the height of the windward side of the building. The stagnation point rises, and the airflow rises and circles to the top of the building to merge with the mainstream, and the stagnation point moves toward Down, the incoming flow flows towards the ground, returning to form a small vortex at the bottom of the building. A large vortex was formed on the leeward side of the building, which caused suction. A separation point also appeared in the wake area behind the bottom of the building, where the airflow was turbulent. The vortex in the streamline diagram corresponds to the minimum wind speed in the velocity cloud diagram.

XY Plane at 2/3H Height

Figure 8 and Fig. 9 are respectively the wind speed diagram and streamline diagram in the XY plane at a height of 2/3H. It can be clearly seen from the Fig that the incoming flow from a distance will circulate when it encounters the building. The incoming flow circulates on both sides of the building, and the boundary layer separates at the corner. The speed at the corner is the highest, and the side airflow accelerates toward When it flows downstream, the velocity of the airflow to the back side decreases

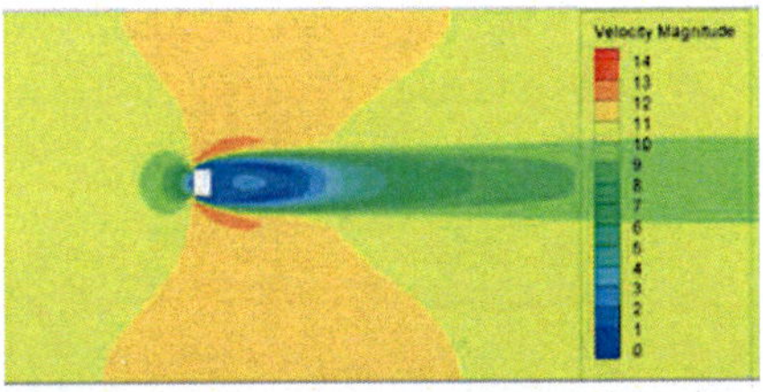

Fig. 8 Wind speed diagram at 2/3H height

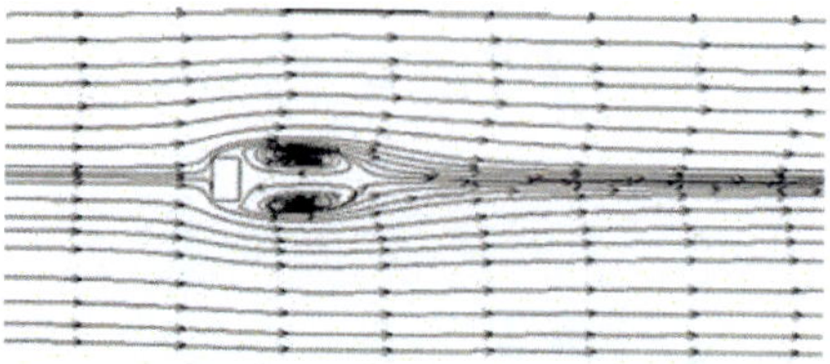

Fig. 9 Streamline diagram at 2/3H height

and a backflow occurs. It flows from the high-pressure area to the low-pressure area, thus forming two symmetrical vortices on the side and back of the building, causing suction. Similarly, the vortex in the streamline diagram corresponds to the minimum wind speed in the velocity cloud diagram.

Wind Speed Distribution Curve around the Building

The calculation results are processed with Tecplot software, and the wind speeds in the front, back, left, and right directions at different heights of the building are obtained as shown in Fig. 10.

In Fig. 10, the horizontal axis is the normal distance from the building wall, and the vertical axis is the wind speed. Five curves with different colors show the velocity distribution at heights of 20, 40, 60, 80, and 100 m.

It can be seen from the Fig that the speed gradient increases as the height increases, which conforms to the exponential rate; at the same height, as the flow field gradually moves away from the building, the flow is less and less disturbed by the building, and the speed gradually increases. The speed on the back side of the building is significantly reduced, and the wind speed is between 0.5 and 3 m/s. This is caused by the shielding effect of the building on the airflow, and the existence of the vortex on the back side of the building makes the wind speed curve turbulent. Since the flow field is symmetrical, the wind speed distribution curves on the left side of the building and the right side of the building are the same.

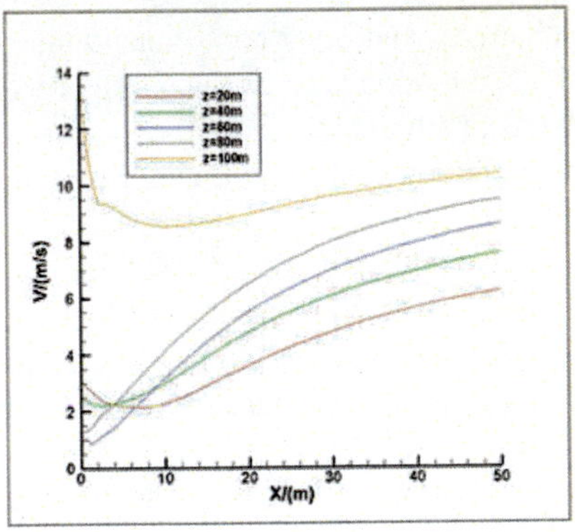

(a) The front side of the building

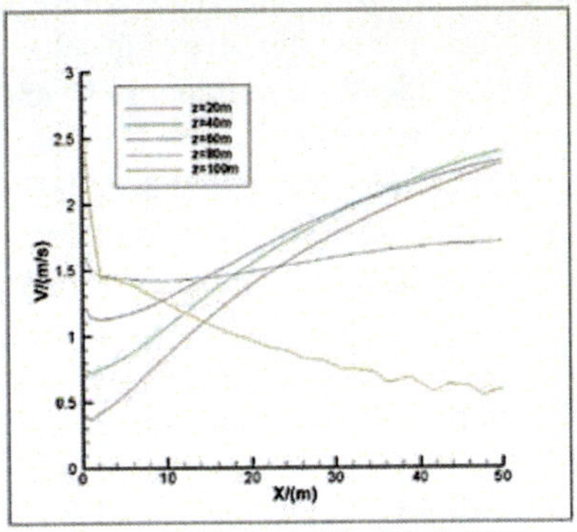

(b) The back side of the building

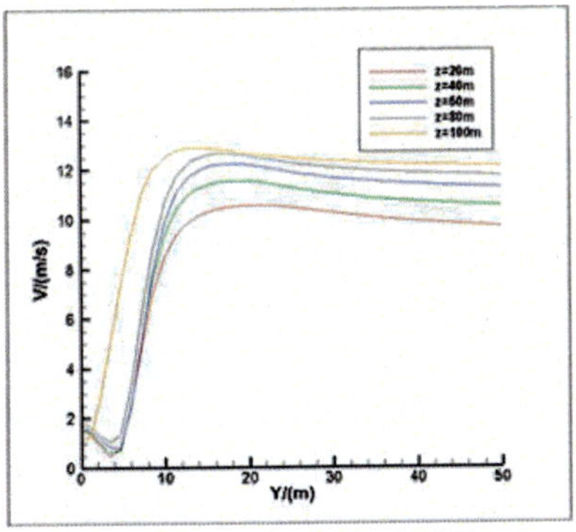

(c) The left side of the building

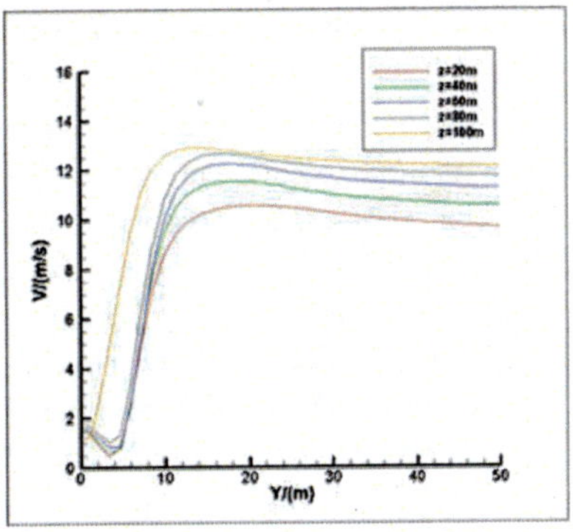

(d) The right side of the building

Fig. 10 Wind speed curve

3.3 *Study on the Interference Effect of Two Buildings*

The characteristics of the wind field around the two buildings are more complicated than that of a single high-rise building. When the arrangement of the two high-rise buildings is different, even if the incoming wind speed is the same, their wind field circumfluence characteristics are also significantly different. In this section, the aerodynamic interference effects between two layout buildings are studied by changing the spacing, and the flow field characteristics of a single building are compared and analyzed [6].

Two buildings with a height of 100 m are selected as the research objects. According to the 《Code for Fire Protection Design of High-Rise Civil Buildings 》. When high-rise buildings are arranged in parallel, the building spacing should not be less than 0.4 times the height of the higher building and not less than 20 m. Therefore, according to the standard specification, the building spacing should be selected at about 45 m, and then the spacing should be changed to study the interference effects between buildings under different spacings. Assuming that only the net distance of the building in the X direction is changed, and other conditions remain unchanged, the net distance of the change in the X direction is set as the width of the windward side of the building as D, and three sets of simulation experiments are set for comparison. The distance between the two buildings is 45 m, 90 m and 135 m respectively. The layout is shown in Fig. 11.

Due to the existence of two buildings, the corresponding computational domain will also expand. In order to ensure that the computational domain meets the blocking rate requirements, and at the same time to ensure that the two sides of the building to the boundary of the computational domain are long enough to enable the flow field to develop fully, the computational domain size is taken as 2400 m × 800 m × 500 m (length × width × height), and the building is placed in the calculation The domain is 1/3 forward along the flow, and the blocking rate at this time:

$$(45\,\text{m} \times 100\,\text{m}) / (800\,\text{m} \times 500\,\text{m}) = 1.13\% < 3\%$$

The calculated blocking rate is 1.13%, which is less than 3%, so the setting of the calculation domain meets the blocking rate requirement. The setting of other calculation conditions is the same as that of a single building. Using a structured grid, define the grid node distribution after creating the O-Block, divide the grid, the number of grids is 1.2 million, the grid division is shown in Fig. 12, here is the grid when the net distance is 45 m Take for example.

When the two buildings are arranged, the flow field is much more complicated than that of a single building. Figures 13 and 14 are the velocity flow diagrams and speed cloud graph at 2/3H height of the two buildings arranged in two buildings.

From Figs. 13 and 14, it can be seen that compared to a single building, the flow field around the two buildings has undergone some changes, and as the distance between the buildings increases, the impact of this change gradually Decrease. Same as the case of a single building, the boundary layer separation occurs at the inflection point in front of the building after the incoming flow passes through the disturbing building, and backflow appears on the leeward side of the building, forming a stationary symmetrical vortex pair in the gap between the two buildings, which is on the leeward side of the disturbing building All the disturbed buildings afterwards were in the negative pressure zone. The disturbed building also has boundary layer

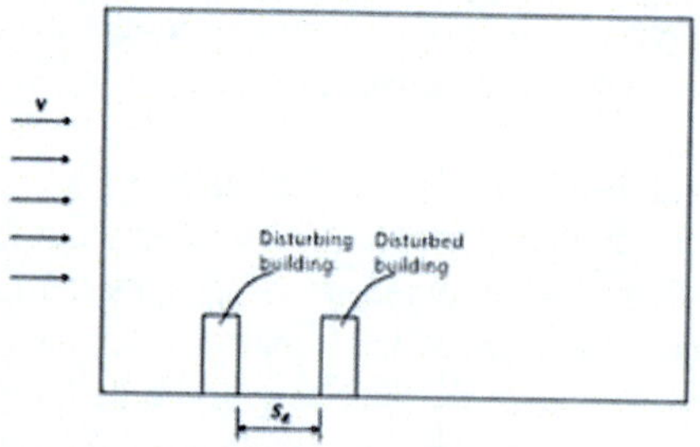

Fig. 11 Relative location map of building

separation, but the separation position is relatively behind the disturbing building, and the lateral flow of the building is relatively stable. Like the disturbing building, a recirculation zone formed by a pair of stationary symmetrical vortex pairs is also generated on the leeward side.

As the distance between the two buildings increases, the degree of squeezing of the vortex between the buildings is reduced. The vortex that disturbs the leeward side of the building gradually becomes oblate, and the vortex on the leeward side of the disturbed building gradually becomes smaller. As the spacing increases, the aerodynamic interference between buildings is weakened, and the flow can be more fully developed at this time. When the spacing between buildings reaches 135 m, the flow condition is very close to that without interference, and the downstream flow is basically the same as that under no interference. The leeward side of the disturbed building has a large vortex, while the leeward side of the disturbed building has only a small vortex backflow. The separated airflow gradually returns to the same flow direction as the original flow near the leeward side of the disturbed building, as shown in Fig. 13c As shown, this shows that the aerodynamic interference between the two buildings becomes weaker and weaker as the distance between the buildings increases.

Figure 15 is the velocity streamline diagram of the Y = 0 section. When the building spacing is 45 m, the boundary layer separates near the top of the building to form a small vortex. Because the vortex is squeezed by the building spacing, there is a strong direction between the buildings. Downwind flows, forming a small backflow at the bottom of the building. Because the disturbed building is in the negative pressure

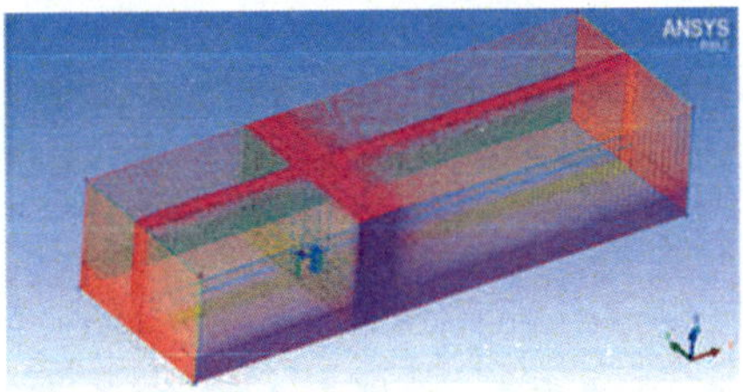

Fig. 12 Schematic diagram of grid division when the spacing is 45 m

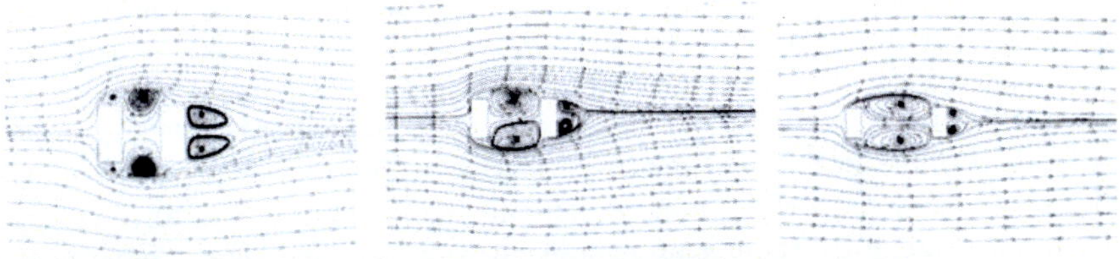

(a) The building spacing is 45 meters (b) The building spacing is 90 meters (c) The building spacing is 135 meters

Fig. 13 Streamline diagram at 2/3H height

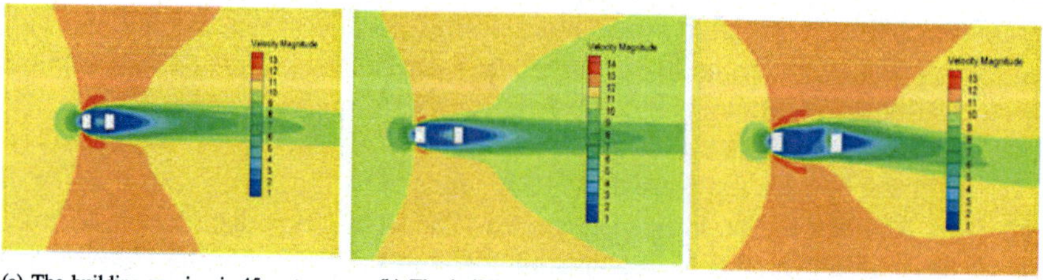

(a) The building spacing is 45 meters meters (b) The building spacing is 90 meters (c) The building spacing is 135

Fig. 14 Velocity cloud diagram at 2/3H height

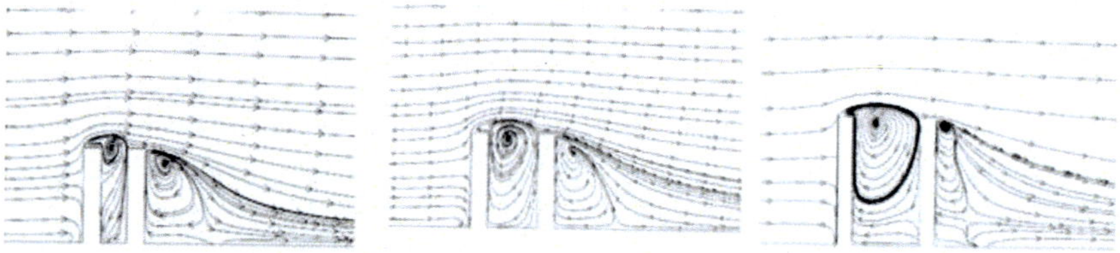

(a) The building spacing is 45 meters meters (b) The building spacing is 90 meters (c) The building spacing is 135

Fig. 15 Velocity streamline diagram of Y = 0 section

area, a larger vortex appears on the leeward side, and the back separation surface is far from the building. As the distance between the buildings increases, the flow gradually develops fully. The vortex between the two buildings gradually returns to the size of the vortex on the leeward side of a single building without interference, and the vortex on the leeward side of the disturbed building becomes smaller. The separating surface is close to the building. In short, as the distance between buildings increases, the aerodynamic interference between the two buildings is significantly reduced.

(a)–(d) in Fig. 16 are the curves of the wind speed at different heights behind the disturbed building of a single building and two buildings with the distance from the building wall.

Figure 16a shows that the speed gradient increases with the height of a single building, which conforms to the exponential rate. At the same height, as the flow field gradually moves away from the building, the flow is hindered by the building and the wake vortex interference becomes smaller and smaller. Figure 16b, c show that when the distance between buildings is close, the interference effect is more obvious due to the vortex squeezing between buildings, and the velocity distribution is disordered; when the distance between buildings increases to 135 m, As shown in Fig. 16d, the velocity distribution has returned to the velocity distribution of a single building—that is, the velocity distribution without interference.

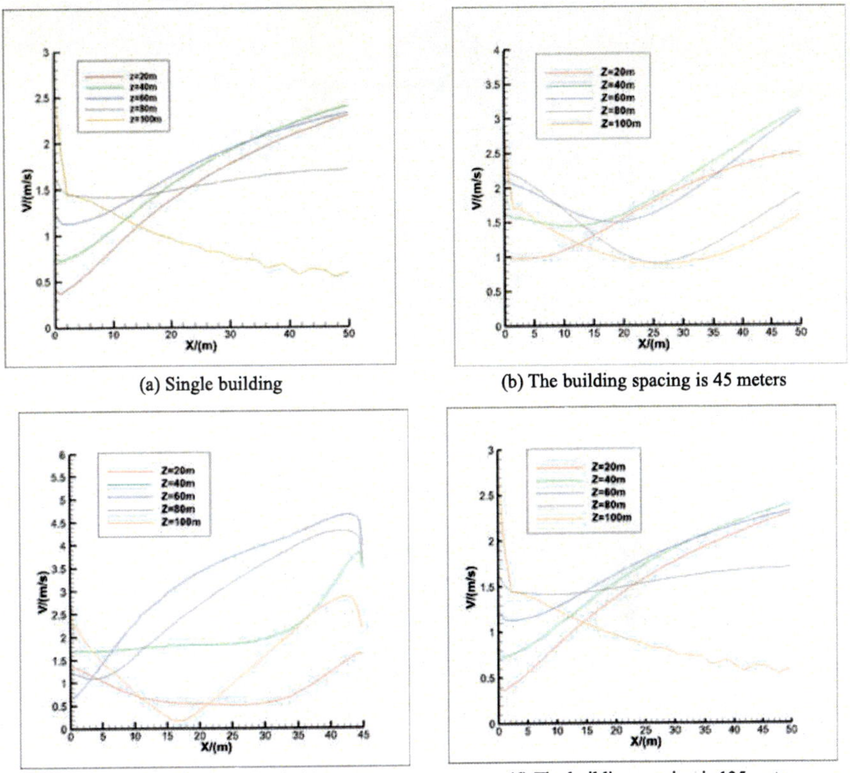

(a) Single building

(b) The building spacing is 45 meters

(c) The building spacing is 90 meters

(d) The building spacing is 135 meters

Fig. 16 The wind speed curve behind the disturbed building

3.4 Enclosed Layout

Enclosed layout refers to a layout formed by multiple buildings enclosed on all sides. This layout can form a nearly closed and relatively large public space in the middle. The geometric model of the enclosed layout building group is also established. As shown in Fig. 17, the cross-sectional size of the building remains unchanged at 30 m × 45 m, and the building height is still 100 m. The eight buildings form a circle and are adjacent to each other. The distance between the left and right of the building is 20 m, and the distance between the front and back is 45 m. The incoming wind speed is still 12 m/s at the reference height of 100 m.

Figure 18 and Fig. 19 are the velocity cloud diagram and velocity streamline diagram at the height of 2/3H of the enclosed building group respectively. It can be seen that under the action of the incoming wind, the wind speed in the slit between the left and right of the building is the highest. It can reach 13–14 m/s, and the wind speed at the corners of the building becomes larger, and the wind speed can reach 11–12 m/s. The wind speed in the internal enclosed area is significantly reduced.

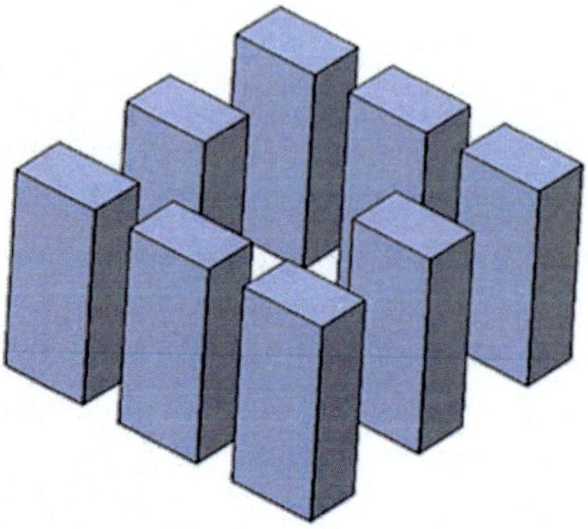

Fig. 17 Geometric model of enclosed layout building group

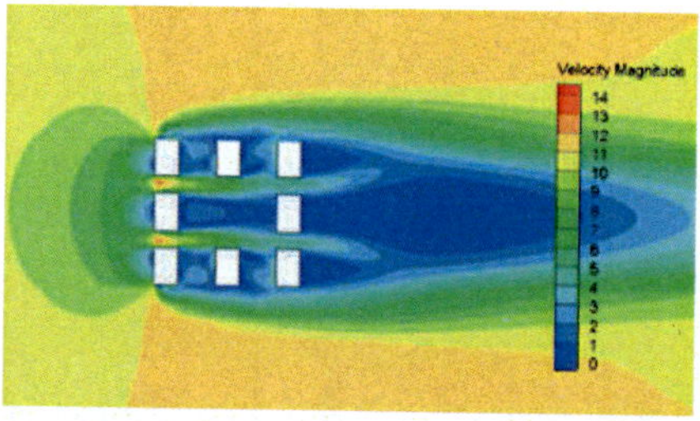

Fig. 18 Speed cloud diagram

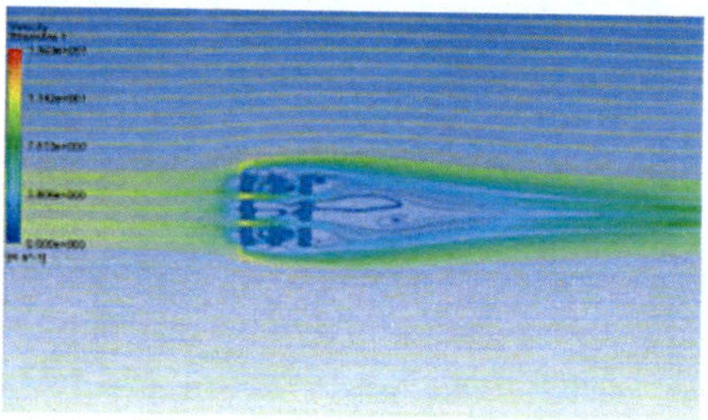

Fig. 19 Speed streamline diagram

This is because the building complex is enclosed on all sides, and the surrounding buildings play a role of shielding; in addition, the wind shadow area overlaps with each other and the air flow is complicated, which is generated on the leeward side of the building. The complex vortex is not conducive to the flight of multi-rotor drones.

4 Conclusion

Studying the distribution of wind fields around high-rise buildings is to provide a reference for the actual wind-resistant flight of fire-fighting drones between buildings. The wind field between high-rise buildings is complex and changeable. Multi-rotor drones are very susceptible to interference from incoming wind during actual flight. Therefore, areas with high wind speeds should be avoided, such as corners of buildings and slits between buildings. Excessive wind speed will cause serious yaw of the drone and damage the stability of the drone; also avoid areas with large vortexes, where there is serious backflow, which will affect the flight safety of the drone. It can be seen from the pressure cloud chart that the wind pressure on the leeward side of the building is negative, which produces suction. The fire-fighting drone should maintain a distance of at least 30 m from the building wall during actual flight; the maximum wind speed can be seen in the speed cloud chart. The wind speed value is about 16 m/s.

The three-dimensional numerical simulation analysis of the wind field of high-rise buildings was carried out using the CFD software platform. First, the numerical simulation of the wind field of a single high-rise building was carried out; then the height of the super high-rise building was changed to study the influence of the height of the building on the wind field around the flow; The interference effect between two buildings was simulated, the distance between the buildings was changed to study the interference situation, and the results were compared with the results of a single building. The study found that the wind field around the two buildings was much larger than that of a single building. With the change, the aerodynamic interference between the two buildings is obviously weakened with the increase of the building spacing; finally, the wind field simulation of the high-rise building group is carried out, and the wind field distribution of the high-rise building group under the two enclosed layouts is studied. Finally, the significance of the simulation study of the wind field around the high-rise building for the actual flight of the fire-fighting UAV is explained.

References

1. Bairagi AK, Dalui SK (2014) Optimization of interference effects on high-rise buildings for different wind angle using CFD simulation. Electron J Struct Eng 14:557–569
2. Braun AL, Awruch AM (2009) Aerodynamic and aeroelastic analyses on the CAARC standard tall building model using numerical simulation. Comput Struct 87(9–10):564–581
3. Davis RW, Moore EF (1982) A numerical study of vortex shedding from rectangles. J Fluid Mech 116:475–506
4. Hui Y, Yoshida A, Tamura Y (2013) Interference effects between two rectangular-section high-rise buildings on local peak pressure coefficients. J Fluids Struct 37:120–133
5. Murakami S, Mochida A (1988) 3-D numerical simulation of airflow around a cubic model by means of the model. J Wind Eng Ind Aerodyn 31(2):283–303
6. Yoshikawa M, Tamura T () LES of fluctuating pressures on a high-rise building influenced by windward building. APCWE8, pp 1–10

Study on Aerodynamic Characteristics of "Propeller/Wing" System with Overset Mesh Method

Zhang Zhitao, Xie Changchuan, and Yang Chao

Abstract In this paper, the propeller unsteady aerodynamic and "propeller/wing" aerodynamic interaction are investigated using the Computational Fluid Dynamics (CFD) method. Based on the true three-dimensional (3D) propeller simulation, a steady RANS solution for ANSYS FLUENT was presented using an actuator-disk approach, with a user-defined function that was proposed for the commercial solver. The full 3D unsteady RANS numerical simulation work was performed using four models: (1) a clean wing with span length of 1 m; (2) single propeller model composed of four blades; (3) the "propeller/wing" model containing propeller and 1-m wing; (4) a clean wing with span length of 3 m. On the other hand, the actuator-disk model contains three models: one disk model, one disk/wing model, and two disk/wing models. The unsteady aerodynamic results revealed that propeller slipstreams affect the wing characteristics by changing the lift and drag. Meanwhile, when the freestream angle of attack is not equal to 0°, the distribution of the pressure coefficient of the blade is very different at various phase positions. Actuator-disk method can greatly reduce the calculation cost and provide satisfied prediction of the trends of the distributed propeller aerodynamic performance.

Keywords "Propeller/wing" system · CFD · Unsteady aerodynamic · Actuator-disk · Distributed propeller

1 Introduction

Due to the high oil cost, the propeller aircraft has received increasing attention from the aerospace community. Propeller wakes can significantly change the flow field of the wing surfaces and further influences the aerodynamic characteristics. Therefore, it is essential to investigate propeller and enhance prediction capabilities for the highly complex flow around the propeller/wing. As a result, several ways, including experiment [1–3], flighting test [4] and Computational Fluid Dynamics

Z. Zhitao · X. Changchuan (✉) · Y. Chao
School of Aeronautic Science and Engineering, Beihang University, Beijing 100191, China
e-mail: xiechangc@buaa.edu.cn

S. Lee et al. (eds.), *The Proceedings of the 2021 Asia-Pacific International Symposium on Aerospace Technology (APISAT 2021), Volume 1*, Lecture Notes in Electrical Engineering 912, https://doi.org/10.1007/978-981-19-2689-1_5

(CFD) [5, 6] were adopted to explore the problem. However, experiment and flighting test are expensive in investigating the details of this phenomenon. Furthermore, recent advances in CFD method could rapidly and accurately predict the flow field surrounding the propeller and wing.

The utilization of CFD was generally divided into two types to investigate the propeller phenomena. One was to establish a true 3D propeller/wing simulation model and solve Euler or (u)RANS equations [7], which can not only accurately predict the aerodynamic performance, but also capture the flow details of mutual interference between propeller and wing. However, this approach spends a lot of time and computing resources. Therefore, some scholars proposed a simplified model that simplifies the true propeller as a "0" thickness disk or a disk of certain thickness, and then imposes boundary conditions or source terms on it to simulate the propeller, so it is called actuator-disk or virtual blade method [8–11]. Related calculations showed that the actuator-disk or virtual blade method can balance the accurate prediction of the aerodynamic performance and the acceptable calculation cost.

In this present paper, two methods mentioned above were used to study the aerodynamics of the propeller and "propeller/wing" system. The distributed propeller has more complicated aerodynamic phenomena, such as nonlinear aerodynamic interference [12–14], a preliminary discussion on the propeller was also carried out. The later work will be focused on the optimization of the research method and deep exploration of the distributed propeller aerodynamic performance.

2 Computational Strategy

2.1 3D Models

2.1.1 Single Propeller Model

In this work, the overset technique that has been proved to be very suitable for simulating rotating parts was used to simulate the true 3D propeller, especially in the condition of unsteady aerodynamic. The overset mesh includes background mesh and component mesh, where the propeller is unstructured component mesh and the background mesh is structured, as shown in Fig. 1. Grids are generated by the meshing software ICEM and the total number of all grids is about 310 millions. In particular, the grid of the leading and trailing edges of blades are sophisticated. The standard k-epsilon turbulence model and Transient Reynolds-Averaged-Navier–Stokes (RANS) viscous solver were chosen to calculate the model.

(a) Hybrid grids for whole computational domains

(b) Unstructured grids for rotational domain

(c) Grids on propeller surface

Fig. 1 Single propeller computational mesh

(a) Hybrid grids for whole computational domains

(b) Unstructured grids for "propeller/wing"

Fig. 2 "Propeller/wing" system computational mesh

2.1.2 "Propeller/Wing" System Model

Standard k-epsilon turbulence model and RANS viscous solver were selected to calculate the "propeller/wing" mutual interference model. In order to capture the details of the flow, the surface of wing directly behind propeller was densely meshed and the total number of the grids was 400 millions, as displayed in Fig. 2.

2.1.3 Wing Models

This part contains two wing models: one has a span length of 1 m and another has a span length of 3 m. Similarly, grids is elaborately disposed in the crucial area of the wing to ensure the accuracy of calculation, as illustrated in Fig. 3 and the S-A turbulence model was applied in simulation.

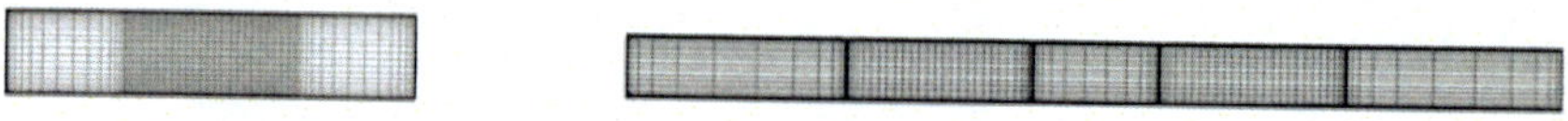

(a) 1 meter wing (b) 3 meters wing

Fig. 3 Wing models

2.2 *Actuator-Disk Models*

In this part, body force was added to the RANS equation as a source term, as demonstrated in formula (1).

$$\frac{\partial}{\partial t}\int_{\Omega}\vec{W}d\Omega+\oint_{\Omega}\left(\vec{F}_c-\vec{F}_v\right)dS=\vec{S}_{\Phi}, \tag{1}$$

Where $\vec{W}$ is conservation term, $\vec{F}_c$ refers to advection term, $\vec{F}_v$ means viscosity term, and S_Φ marks source term.

Based on the calculation results of the single propeller, the propeller blade was divided into a series of blade elements along the radial direction, and then the axial force, f_t and tangential force, f_q were solved on each element separately. Among them, axial force is related to the propeller thrust and the tangential force is expressed as the torque that causes the slipstream to rotate. The equations of the body forces can be written as [9],

$$f_t=\frac{1}{2}\rho V_\infty^2\sum_0^n A_n r^n, \tag{2}$$

$$f_q=\frac{1}{2}\rho V_\infty^2\sum_0^n B_n r^n, \tag{3}$$

where A_n and B_n represent the polynomial coefficient, n stands for the order of the polynomial, and r means the radial distance from the axis of rotation. The axial and tangential forces of per unit volume are provided by (4) and (5).

$$\frac{f_t}{2\pi rh}, \tag{4}$$

$$\frac{f_q}{2\pi rh}, \tag{5}$$

Where h is the thickness of the propeller disk, and the thickness marks 0.027 m in this paper. Supposing that the number of propeller blades is N, and the total axial and tangential force use formulas (6) and (7), which are computed for each cell of the actuator-disk.

$$F_t = \frac{Nf_t}{2\pi rh}, \quad (6)$$

$$F_q = \frac{Nf_q}{2\pi rh}, \quad (7)$$

After projecting the force to the cartesian coordinate, the momentum source term S_Φ is described as

$$S_\Phi = -\left[\left(\vec{F_t} + \vec{F_q}\right)_x \left(\vec{F_t} + \vec{F_q}\right)_y \left(\vec{F_t} + \vec{F_q}\right)_z\right]^T, \quad (8)$$

There are three actuator disk models in this paper, as shown in Fig. 4.

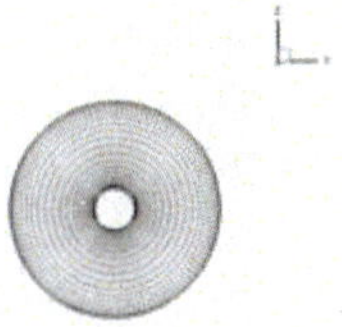

(a) Single actuator disk model

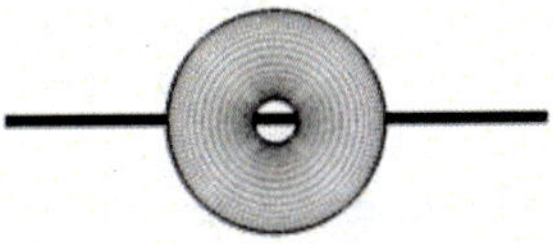

(b) "One actuator disk/wing" model

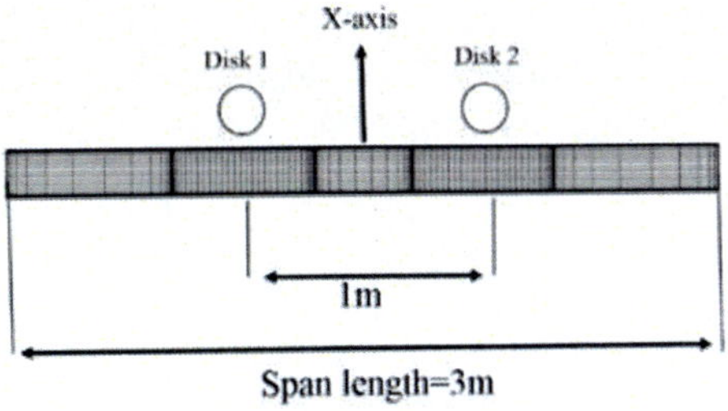

(c) "Two actuator disk/wing" models

Fig. 4 Actuator disk model

3 Results

3.1 Single Propeller

3.1.1 True 3D Models

By appling the above calculation strategy, the aerodynamic characteristics of the propeller at different advance ratios were calculated. The propeller advance ratio J is defined as

$$J = \frac{60V_\infty}{Ns \cdot D}, \tag{9}$$

Where D is propeller diameter, and other parameters and the overset mesh calculation conditions are listed in Table 1.

When the incoming flow angle of attack is equal to 0°, the propeller thrust curve, torque curve, power curve and the efficiency curve are illustrated in the Fig. 5. From the result, in the case of a given rotated speed, the thrust, torque and the power decrease along with the advance ratio inceases. It is noteworthy that the efficiency of the propeller increases at first and then decreases, and reaches the peak when $J = 0.9$. It can be observed from the Fig. 5 that when $J > 1.4$, propeller enters the wind-milling state. The trend of these parameters with the advance ratio is consistent with the experimental results of propeller in concerning literature [2], which proves that the overset mesh model established in this work is effective. In addition, propeller related experiments are also in the schedule of our follow-up works. Propeller thrust coefficient C_T, torque coefficient C_M and the aerodynamic efficiency η are defined by the following formulas.

$$C_T = \frac{T}{\rho n^2 D^4}, \tag{10}$$

$$C_M = \frac{M}{\rho n^2 D^5}, \tag{11}$$

Table 1 Overview of the single propeller calculation conditions

Parameter	Symbol	Value
Freestream velocity	V_∞	8.12–24.36 m/s
Reynolds number	Re	27,000
Propeller rotated speed	Ns	2400 rpm
Propeller angle of attack	α_P	0°
Propeller advance ratio	J	0.5–1.5

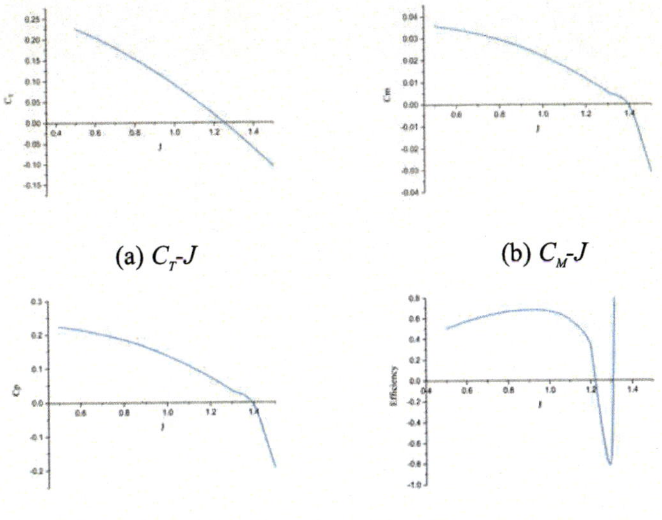

(a) C_T-J (b) C_M-J

(c) C_P-J (d) η-J

Fig. 5 Propeller aerodynamic characteristic

$$C_P = \frac{P}{\rho n^3 D^5}, \tag{12}$$

$$P = 2\pi n \cdot M, \tag{13}$$

$$\eta = J \cdot \frac{C_T}{C_P}, \tag{14}$$

Where T is propeller thrust, M represents propeller torque, P stands for propeller power, η marks propulsion efficiency, n means rotation speed.

Along with the variation of α_P, separately analyses have been conducted about the aerodynamic characteristics of the propeller when the $J = 0.9$. Related parameter setting are exhibited in Table 2.

Table 2 Overview of the single propeller ($J = 0.9$) calculation conditions

Parameter	Symbol	Value
Freestream velocity	V_∞	14.616 m/s
Reynolds number	Re	27,000
Propeller rotated speed	Ns	2400 rpm
Propeller angle of attack	α_P	0–10°
Propeller advance ratio	J	0.9

(a) Windward side (b) Leeward side

Fig. 6 Blade C_P distribution at $J = 0.9, \alpha_P = 0°$

When $\alpha_P = 0°$, the pressure coefficient (C_P) distribution of the propeller blades is demonstrated in Fig. 6, and the C_P distribution on the blades is almost the same. Since the propeller is in "steady" state when $\alpha_P = 0°$ and the flow around four blades will not change during the rotation period. However, in the actual flight of a propeller aircraft, it is quite rare that the angle of attack $\alpha_P = 0°$. Therefore, further study and discussion on the aerodynamic performance of the propeller will be implemented.

It should be noted that the case for $\alpha_P \neq 0°$ and $\alpha_P = 10°$ were selected as a situation to be illustrated. When $\alpha_P = 10°$, the C_P distribution on the blades is obviously different, as shown in Fig. 7, because there is an angle between the propeller axis and the airflow, and causes the airflow through the blade to be different at every moment. Under other conditions, such as $\alpha_P = 4°, 6°$ and $8°$, due to the space limitations, the calculation results are omitted, while the C_P distribution propeller blades also changes with time. Thus, it could be concluded that when α_P is not equal to 0°, the propeller shows unsteady aerodynamic characteristics.

In order to understand the unsteady characteristics more clearly, the propeller was equally divided into 20 parts along the radial direction, which is displayed in Fig. 8. The root part is numbered 1, and the tip is 20. For the case of $\alpha_P = 10°$, the part numbered 1, 5, 16 and 18 were selected to study the changes of the axial force and the tangential force during the rotation period. Besides, for the case of $\alpha_P = 0°$, only

(a) Windward side (b) Leeward side

Fig. 7 Blade C_P distribution at $J = 0.9, \alpha_P = 10°J = 0.9, \alpha = 10°J = 0.9, \alpha = 0°$ (physical time = 0.08 s)

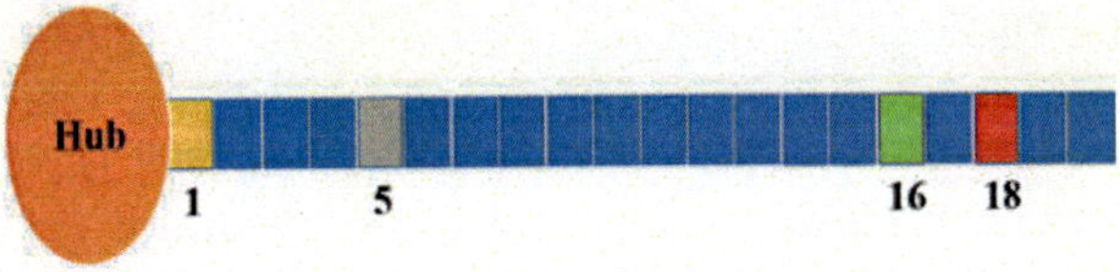

Fig. 8 Propeller division diagram

part 1 and 5 were selected as the research objects. It was concluded that when $\alpha_P = 0°$, the axial force and the tangential force are almost constant with the calculation time, but when $\alpha_P = 10°$, each part shows remarkable periodicity, which also proves that the C_P distribution law in Fig. 6 and Fig. 7 is reliable. According to the curve in Fig. 10, when the advance ratio is 0.6, 0.9 and 1.0, the axial forces and tangential forces reach the maximum at about 70%-80% radius, which is in good agreement with Fig. 9 and other literature results [11].

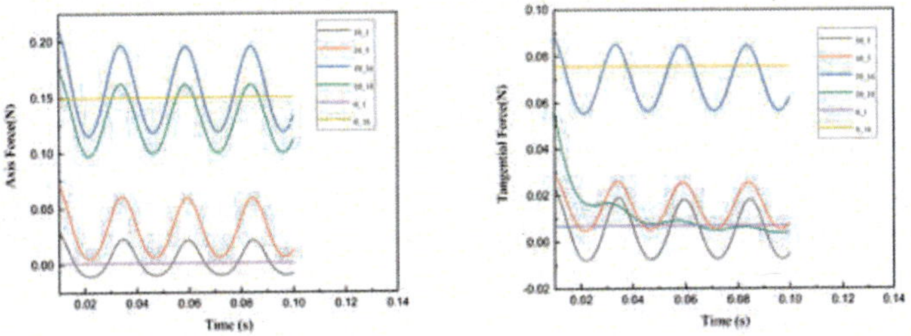

(a) Txial force (b) Tangential force

Fig. 9 Body force during the rotation period

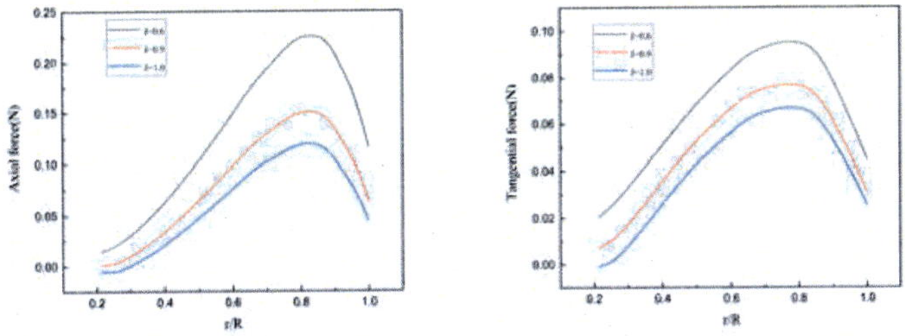

(a) Txial force (b) Tangential force

Fig. 10 Relationship between body force and advance ratio

3.1.2 Actuator Disk

To simulate the aerodynamic of the propeller more realistically, this paper adopted a 3D actuator-disk model. It was assumed that it is evenly distributed, and the max thickness of the true 3D model is selected as the thickness of the actuator-disk. In this work, the calculation results of single propeller at $\alpha_P = 0°$ were used. After post-processing, the axial force and tangential force on each blade are fitted as

$$f_t = \frac{\rho V_\infty^2}{2}\left(-1244.69955r^4 + 378.71862r^3 - 31.81888r^2 + 1.66888r - 0.03897\right), \tag{15}$$

$$f_q = \frac{\rho V_\infty^2}{2}\left(-116.18254r^4 - 19.66914r^3 + 9.93698r^2 - 0.09627r - 0.0062\right), \tag{16}$$

The axial force (15) and tangential force (16) were substituted into Eqs. (6)–(8), and the source term can be obtained. After being solved, the pressure distribution before and after the disk is exhibited in Fig. 11. The static pressure distribution on the disk is circular, and the pressure distribution at the same radius is completely consistent. As shown in Fig. 12, the airflow flowing through the disk surface shrinks and rotates, and the velocity of airflow through the disk is accelerated, which shows the slipstream rotation effects and the acceleration effects of the true propeller. Compared with the calculation cost of a true 3D propeller, the calculation time of the actuator disk is shortened by 95%.

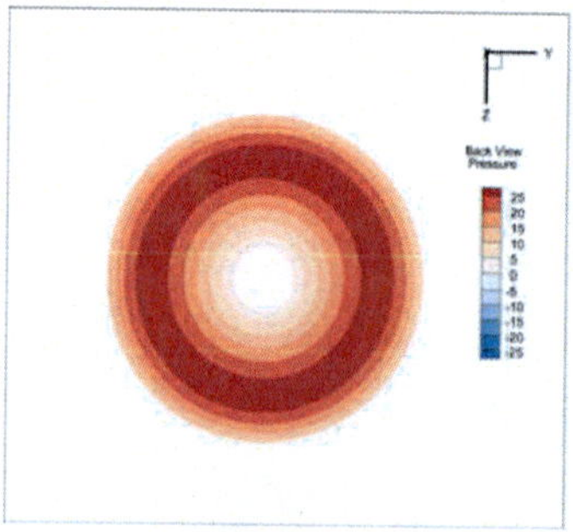

(a) Back view

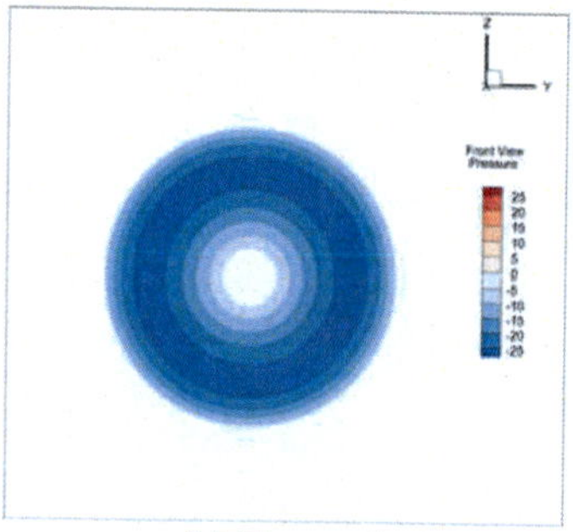

(b) Front view

Fig. 11 3D actuator disk pressure contours

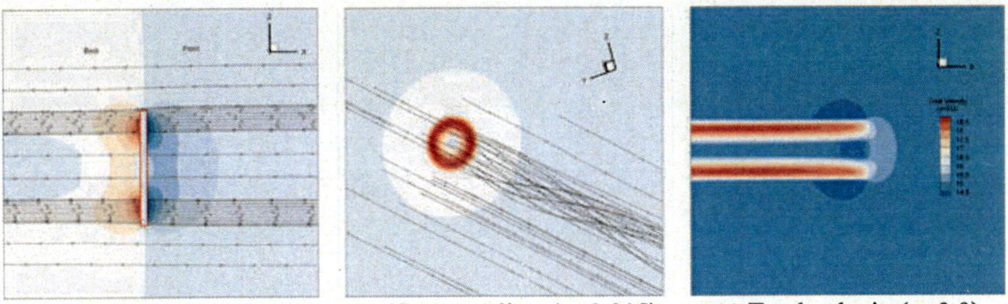

(a) Surface streamlines(y=0.0) (b) 3D streamlines(x=0.015) (c) Total velocity(y=0.0)

Fig. 12 Slipstream simulation

3.2 Propeller/Wing

3.2.1 True 3D Model

The influences of the propeller on the aerodynamic characteristics of the wing mainly come from the slipstream, and accordingly, the wing also has effects on the vortex structure of the slipstream. "Propeller/wing" model calculation conditions are reported in Table 3.

In the light of the parameters in Table.3, the lift, drag, pitch, yaw and roll moment characteristics of the wing with or without propeller were calculated. When there is no slipstram and $\alpha = 0°$, the C_P distribution on upper surface of the clean wing is illustrated in Fig. 13a. However, when there is slipstream, the C_P distribution on upper surface of the wing changes as shown in Fig. 13b.

Figure 14 indicates the influences of slipstream existence on the lift, drag and moment characteristics of the wing during a rotation period when $\alpha = 0°$. When there is no slipstream, the flow exhibits steady characteristics, and the lift, drag and moment characteristics of the wing hardly change as shown in Fig. 14a. On the contrary, the flow on the wing surface shows strong unsteadiness, and the lift drag and moment characteristics appear obviously changes, as shown in Fig. 14b, when the slipstream exists.

Table 3 Overview of the "Propeller/wing" model calculation conditions

Parameter	Symbol	Value
Freestream velocity	V_∞	14.616 m/s
Reynolds number	Re	200,000
Propeller rotated speed	Ns	2400 rpm
Propeller/wing angle of attack	α	0–10°
Propeller advance ratio	J	0.9

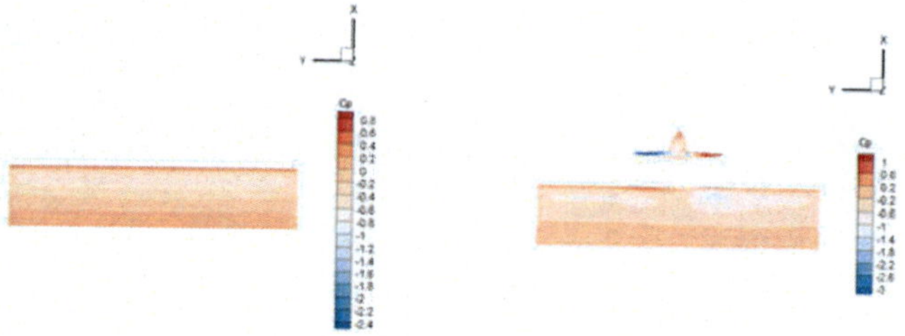

(a) Clean wing (b) Propeller/Wing

Fig. 13 Wing C_P distribution

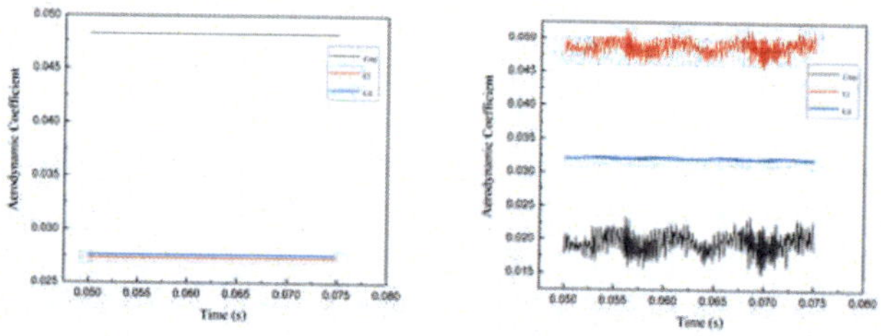

(a) Clean wing (b) Propeller/Wing

Fig. 14 Aerodynamic coefficient

3.2.2 Actuator Disk/Wing

In this part, two models shown in Fig. 4b, c were calculated separately: one is the one disk/wing model, and another is the two disk/wing models. As demonstrated in Fig. 15, compared with the clean wing, the pressure distribution on the wing surface behind the disk has changed significantly, which is consistent with the calculation results of the real 3D model. In particular, the disk calculation time is greatly shortened by 96% and the calculation resource is saved. Accordingly, it is very suitable for preliminary estimation of the engineering projects.

The distributed electric propulsion (DEP) concept considered in this paper is composed of two rotated directions. The direction of rotation displayed in Figs. 16 and 17 is specified as two types, namely co-rotation and counter-rotation. The influences of the rotating direction on the aerodynamic of the wing were explored. From Fig. 16b and Fig. 17b, when it passes the wing, the structure of slipstream is destroyed and divided into upper part and lower part, and it also varies the flow filed on the wing. Later, the influences of the propeller installation position, number, rotation speed and other parameters on the aerodynamic of the wing will be carried out.

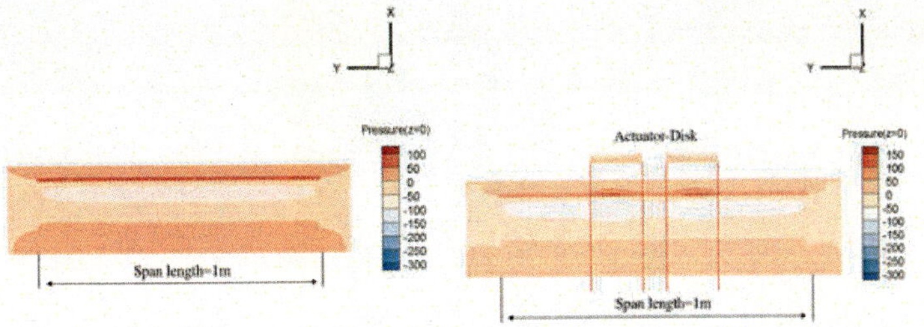

(a) Clean wing (b) One disk/Wing

Fig. 15 Pressure distribution

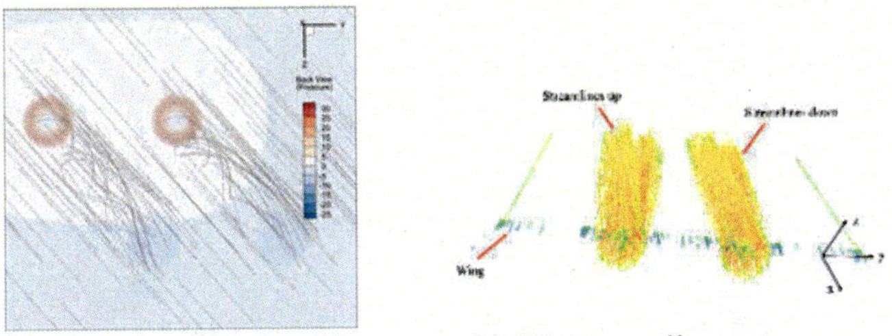

(a) Surface streamlines(y=0.0) (b) 3D streamlines

Fig. 16 Co-rotation

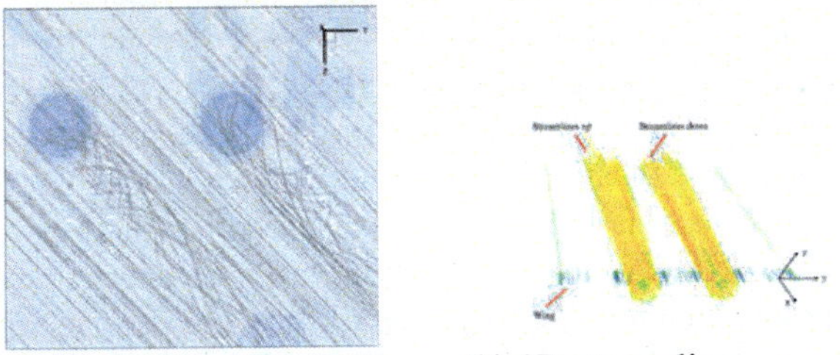

(a) Surface streamlines(y=0.0) (b) 3D streamlines

Fig. 17 Counter-rotation

Different from the clean wing, due to the presence of disk, there has new low-pressure areas appear on the surface of the DEP, which leads to higher lift to some extent. Owing to the difference of rotating direction, the positions of the low-pressure areas are also different. No matter what the rotating direction of the twin propellers installed in one wing is, the wing drags show little difference, while the lifts generated by the counter-rotation are slightly larger than that of co-rotation. When taking the

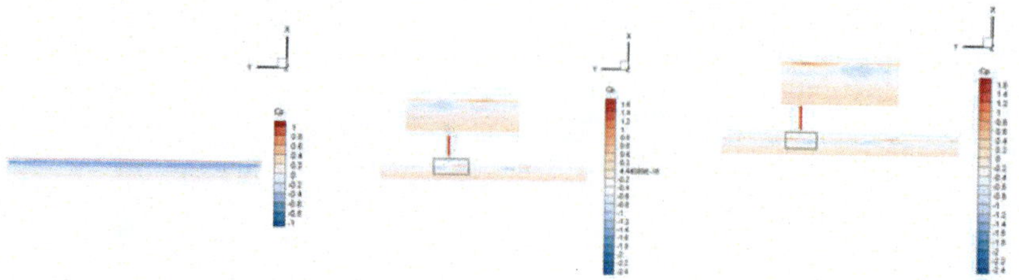

(a) 3m clean wing (b) Co-rotation of two disks (c) Counter-rotation of two disks

Fig. 18 C_P distribution

twin propellers working under the conditions set in Table 3 into consideration, the results of 0° angle of attack indicate a significant increase in wing lift to 22%, and a slight increase in drag to 5.4%. Since the results above are quite acceptable, reasonable distributed propeller parameters setting can achieve the increasing lift and lift-drag ratio, which is very important to guide aircraft design (see Fig. 18).

4 Conclusions

In this paper, the overset mesh approach was applied to fluid modeling, and the RANS equation was used to solve the flow field. Two models, namely true 3D model and actuator-disk model, were used to study the aerodynamic characteristics of a single propeller and propeller/wing interaction, which manifests good agreements of related predictions between true 3D model and actuator-disk model. The calculation results proved that the 3D actuator-disk model can not only simulate the true propeller well, but also greatly reduce the calculation cost. This method is prospective in the engineering field.

It should be noticed that, when the incoming attack angle is not equal to 0°, the distribution of the pressure coefficient of blades is significantly distinguished at different phase positions, with distinct periodic variation. In addition, when the propellers have the same advance ratio, the pressure coefficient distribution of blades is remarkably different between inflow attack angle 0° and 10°. Moreover, the mutual aerodynamic interference between propeller and wing manifests that the propeller varies the characteristics of the surface flow field of the wing, and, in turn, the wing changes the flow field structure of the slipstream, thus making the flow direction of the slipstream shift.

Finally, this paper preliminarily studied the influences of the actuator-disk model with different rotation directions on the aerodynamic of the DEP when the incoming flowing is 0°. The results revealed that the distributed propeller can effectively improve the lift of the wing. However, it should be noticed that this paper only discussed the layout of distributed propellers and obtained richer conclusions. Tremendous amount of calculations and analysis will be performed next. The future work will be concentrated on the aerodynamic of DEP with different incoming flow angles of attack, propeller installation positions, propeller numbers, rotation speed and other parameters.

References

1. Fratello G, Favier D, Marescaf C (1991) Experimental and numerical study of the propeller/fixed wing interaction. J Aircraft. https://doi.org/10.2514/3.46036
2. Britcher CP, Landman D, Duvall B et al (2017) An Aeroacoustic study of propellers for small electric propulsion aircraft. In: 23rd AIAA/CEAS aeroacoustics conference. Denver Colorado
3. Catalano FM(2004) On the effects of an installed propeller slipstream on wing aerodynamic characteristics. J Acta Polytechnica.https://doi.org/10.14311/562
4. Jones JR (2017) Development of a very flexible testbed aircraft for the validation of nonlinear aeroelastic codes. PhD thesis, Michigan University
5. Ruiz-Calavera LP, Perdones-Diaz D (2012) CFD based aeroelastic calculation of propeller loads. In: 28th international congress of the aeronautical sciences. Brisbane, Australia
6. Ruiz-Calavera LP, Perdones-Diaz D (2013) CFD computation of in-plane propeller shaft loads. In: 49th AIAA/ASME/SAE/ASEE Joint Propulsion Conference. San Jose, CA
7. Xue C, Zhou Z (2019) Propeller-wing coupled aerodynamic design based on desired propeller slipstream. J Aerospace Sci Technolhttps://doi.org/10.1016/j.ast.2019.105556
8. Stajuda M, Obidowski D, Karczewski M, Jozwik K (2018) Modified virtual blade method for propeller modelling. Mech Mech Eng. https://doi.org/10.2478/mme-2018-0048
9. Yifei Z (2020) Aerodynamic optimization design of propeller aircraft based on the actuator disc method. PhD thesis, Tsinghua University (基于激励盘的螺旋桨飞机气动优化设计)
10. Lombardi N, Ordonez-Sanchez S, Zanforlin S, Johnstone C (2020) A hybrid BEM-CFD virtual blade model to predict interactions between tidal stream turbines under wave conditions. J Marine Sci Eng. https://doi.org/10.3390/jmse8120969
11. Chauhan SS, Martins JRRA (2020) RANS-based aerodynamic shape optimization of a wing considering propeller-wing interaction. In: AIAA scitech 2020 forum, Orlando, FL
12. Hoover CB, Shen J, Kreshock AR (2017) Whirl flutter stability and its influence on the design of the distributed electric propeller aircraft X-57. In: 17th AIAA aviation technology, integration, and operations conference, Denver, Colorado
13. Deere KA, Viken SA, Carter MB et al (2017) Comparison of high-fidelity computational tools for wing desing of a distributed electric propeller aircraft. In: 35th AIAA applied aerodynamic conference, Denver, Colorado
14. Bohari B, Borlon Q, Santos PBM (2018) Conceptual design of distributed propellers aircraft: non-linear aerodynamic model verification of propeller-wing interaction in high-lift configuration. In: AIAA aerospace sciences meeting. Kissimmee, Florida

An Ontology Based Single Source of Truth (SSOT) Construction Approach for Aircraft Modeling and Simulation

Yuchen Zhang, Chuangye Chang, Weijia Wang, and Gang Xiao

Abstract Single source of truth (SSOT) is a key element in the realization of digital engineering or model-based system engineering. During today's aircraft development, huge number of models are built and simulated for different design purposes at different development stages. This brings a crucial challenge for achieving and maintaining model consistency. The single source of truth is a solution to meet model consistency challenge. It emphasizes on building an authoritative data source and all other models shall refer back to it, so that no inconsistent duplicate exists and model consistency is ensured.

In this paper, an ontology based single source of truth construction approach is proposed. Ontology is brought out to provide a generic and language-independent description framework for aircraft and SSOT can be built based on this ontology. Then, with the model transformation mechanism introduced, which defines the matching relations between ontology and models, the reference ontology can be transformed into different kinds of aircraft design models and also support the transformation between different models. To confirm the usability of this approach, a practice of this approach on a fix wing unmanned aircraft system (UAS) development is introduced According to the practice, the ontology based single source of truth construction approach initially confirms its feasibility and usability.

Keywords Single source of truth · SSOT · Ontology · Aircraft · Modeling and simulation

Y. Zhang
Chinese Aeronautical Establishment, Beijing, China
e-mail: zhangyc082@avic.com

C. Chang (✉)
Institute of Unmanned System, Beihang University, Beijing, China
e-mail: changchy@buaa.edu.cn

W. Wang · G. Xiao
National Key Laboratory for Complex Systems Simulation, Beijing, China

S. Lee et al. (eds.), *The Proceedings of the 2021 Asia-Pacific International Symposium on Aerospace Technology (APISAT 2021), Volume 1*, Lecture Notes in Electrical Engineering 912, https://doi.org/10.1007/978-981-19-2689-1_6

1 Introduction

Single source of truth (SSOT) is a key element in the realization of digital engineering or model-based system engineering. The goal of digital engineering is completing the paradigm shift from document-centric development to model-centric development, it raises model and data to an unprecedented high level in research and development process. In order to continuously transfer model and data across disciplines and through the whole life cycle, SSOT is proposed in digital engineering to support the management and usage of model and data.

SSOT is a collection of unified and authorized data. The idea of single source of truth is building an authoritative data source and all other usages of data shall refer back to the primary data source, so that no inconsistent duplicate exists and data consistency is ensured. Single source of truth is the single and authoritative data source for the analysis, design, implementation, verification and validation process in systems engineering, and is the key to guarantee the model and data consistency through all these processes.

Especially, during today's aircraft development, huge number of models are built and simulated for different design purposes at different development stages. From different levels, there are mission level models, engagement level models, engineering level models, and from different types, there are entity/structure models and behaviour models. According to our practice, these too many models bring a crucial challenge for achieving and maintaining model consistency. In this situation, the single source of truth comes up to our mind as a solution to meet model consistency challenge.

In this paper, an ontology based single source of truth construction approach is proposed. Ontology is brought out to provide a generic and language-independent description framework for aircraft modelling, defining and describing the basic information of aircraft. Ontology supports the construction of single source of truth and serves as the primary data source for all other aircraft models.

This paper firstly introduces how a SSOT is constructed based on a common and language-independent ontology description of aircraft. Then, it introduces how this ontology based SSOT solves our model consistency problem with model transformation mechanism. According to our practice, the ontology based single source of truth construction is feasible and several types of models in aircraft development are proved to be transformed and generated from this ontology based SSOT.

2 An Ontology Based SSOT Construction Approach

2.1 A Reference Ontology for Aircraft

Originating from philosophy, ontology is the study of being and existence. Ontology studies the formal description of existence and formal representation of knowledge.

An ontology describes the basic concepts in a certain domain, endeavoring to capture the essential meaning of concepts in a language-independent way [1].

In aircraft development, engineers from different disciplines build models from different aspects of aircraft to support their design and implementation. All these models describe the same aircraft, but from different aspects and in different modelling languages. This brings inconsistencies among models.

Since ontology is the study of essential concepts and describing concepts regardless of languages. It can be used to identify and describe model element which represent the same thing in different model representations. Thus, ontology naturally comes up to our mind as a solution to model inconsistency problem.

Referring to the models used in our aircraft development practices, mainly focusing on the mission analysis and concept design phase, we propose a reference ontology for aircraft. This ontology contains two parts: structural aspect shown in Fig. 1 and behaviour aspect shown in Fig. 2.

The figures below demonstrate the ontology in a UML way. Concepts are shown as rectangular annotations, and relationships are shown as different lines. Solid line means *Association*, line with filled diamond means *Composition*, line with hollow diamond means *Aggregation*, and line with hollow triangle means *Generalization*.

The core concept of this ontology is *Object*. An object is an active entity of system, such as aircraft or sub-system of aircraft. Object executes in its operational *Environment*. Each object owns *Properties* which represent its characteristics and owns *Interfaces* through which object interacts with outside.

For most ontology studies of aircraft [2–4], they only considered how to represent the structural aspect of aircraft, such as has-part and has-property relationships, while

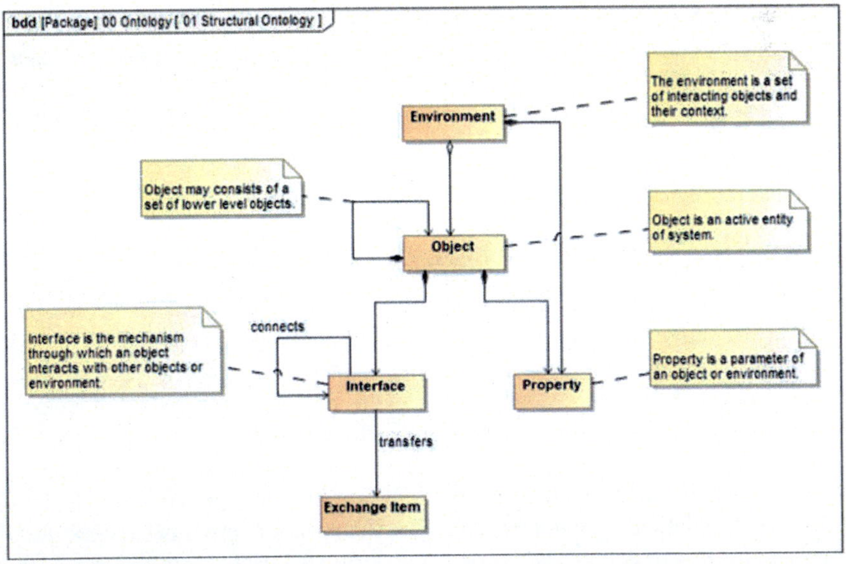

Fig. 1 Ontology from structural aspect

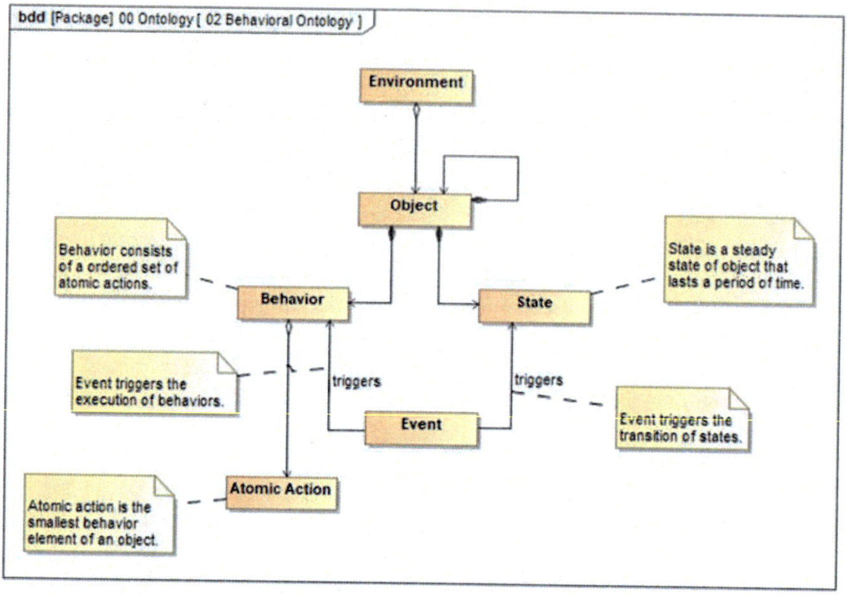

Fig. 2 Ontology from behavioral aspect

ignored the behavior aspect. In our case, because we need to ensure consistency of several behavior models, so we tried to propose an ontology also including behaviour aspect representation.

In this ontology, each object has its *Behaviors*. The behavior consists of an ordered set of atomic actions, and represent a process that object executes [5]. Besides of behaviors, each object has *States* as its dynamic properties. State describes the steady condition of object that lasts a period of time and can transit to other states. Each behavior executes in a state of object. The transition of states and the execution of behaviors are triggered by *Event*.

2.2 Construct SSOT Based on Ontology

Based on the reference ontology proposed above, single source of truth (SSOT) of aircrafts can be constructed. In our practice, a SSOT of a fix wing unmanned aircraft system (UAS) is built. Core concepts of this fix wing UAS, including system, subsystem, functions, measures, are all matched to the reference ontology.

In brief, this fix wing UAS consists of air vehicle and control station two parts. This three system and subsystem concepts are matched to *Object* in ontology and the whole-part relationship is represented with *Composition*. From behavior aspect, the behavior characteristics of aircraft are all represented using *State* and *Behavior*. While from structural aspect, the parameters and measures of aircraft are matched

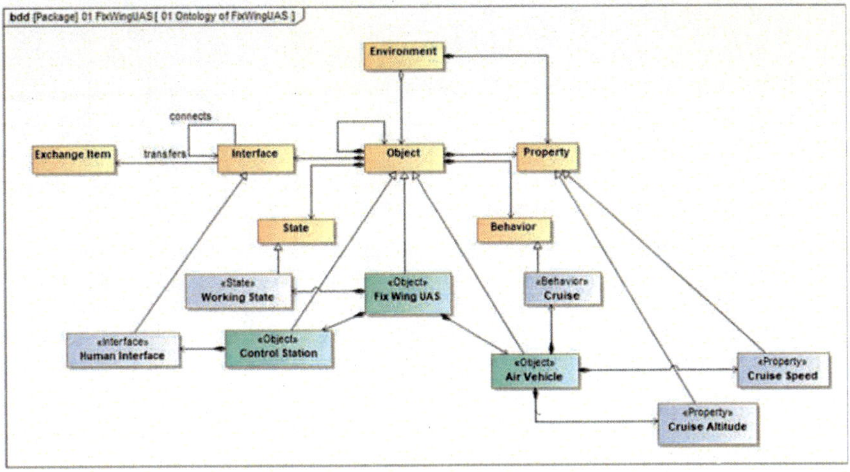

Fig. 3 Construct SSOT of fix wing UAS based on ontology

to *Property*, and the interaction mechanism of aircraft including ports and interface are mapped to *Interface* (Fig. 3).

With the approach above, most of the important concept of aircraft can be described using the reference ontology. And this unified description can serve as a SSOT for aircraft. It captures semantic from modelling languages, identifies core concepts of aircraft and represents these concepts by formal semantics. With this SSOT, all other representations of aircraft, namely other models, can be translated and transformed from this ontology based SSOT.

3 Model Transformation with Ontology Based SSOT

3.1 Behavior Model Transformation Example

With the ontology based SSOT, consistency of different models can be improved. Models can firstly transform to the ontology representation, then other kinds of models can be generated from the ontology. This process is automated or semi-automated, thus it avoids the repeated and redundant modeling and avoids the inconsistency occurring in the modeling process. Blow two examples of the model transformation is introduced. The transformation mechanism is mapping the semantics in different modeling languages to reference ontology, with the ontology based SSOT as medium, different models can be translated to each other.

The first example is a mission-level behavior model transformation example. Two models are DoDAF architecture description model and Maxsim wargame model. In our aircraft development process, for the fix wing UAS, its operational process is

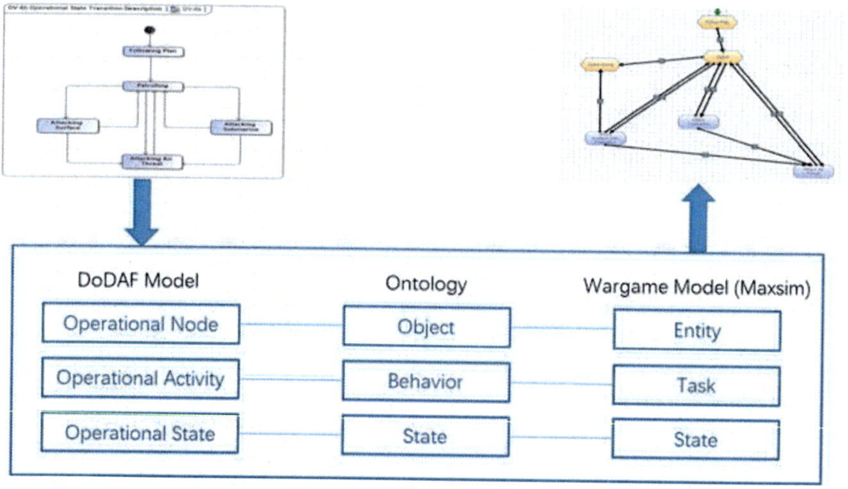

Fig. 4 Behavior model transformation example between DoDAF and Maxsim

firstly described with DoDAF model in operational view, then the operational process is simulated and evaluated with wargame model in Maxsim software. In the past, these two kinds of model were built respectively and the consistency was ensured by engineers. With the SSOT, the wargame model now can be transformed from DoDAF model automatically and the consistency is ensured by transformation mechanism (Fig. 4).

The semantics in two modeling languages are mapping to the semantics in ontology. As shown in the figure above, *Operational Node* in DoDAF and *Entity* in Maxsim are both represent entities, representing different kinds of organizations, systems, and materials. They are matched to *Object* in reference ontology. In a similar way, *Operational Activity* and *Task* both represent entity's behavior, thus matched to *Behavior* in ontology. With the ontology as medium, semantics in two modeling languages are mapped together, then the wargame model can be transformed from DoDAF model. There is no longer need to build model twice in different software and the less inconsistency mistakes will occur.

3.2 Structural Model Transformation Example

The second example is an engineering-level structural model transformation example. Two models are System concept model and Modelica analysis model. In our development practice, an aircraft is firstly described in SysML, analysing its requirements, functions and architecture, then takes multi-discipline simulation with Modelica model, analysing its design and performance.

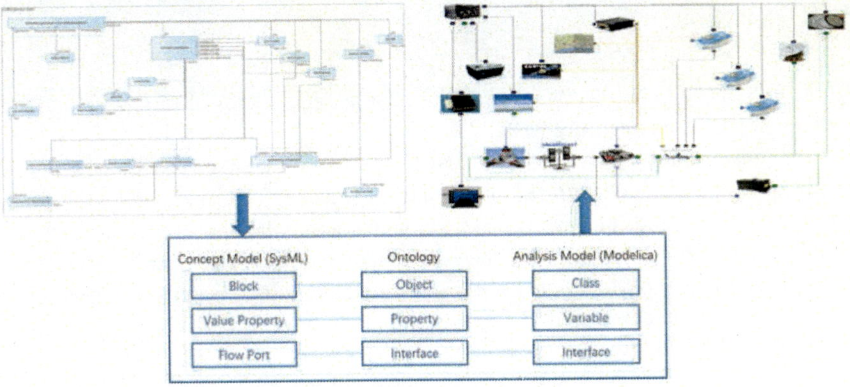

Fig. 5 Structural model transformation example between SysML and Modelica

Similar with above, semantics in SysML and Modelica are also mapped to the ontology [6, 7]. For example, in this time, *Block* in SysML and *Class* in Modelica both represent entity, thus these two semantics are matched to the *Object* in ontology. After the semantic mapping, model transformation can be realized (Fig. 5).

One question may be asked is that why don't we map the semantics of different languages directly and without the ontology as medium [8]. In fact, for two languages, directly mapping is more convenient. However, if there are many modeling languages, the directly mapping will be complex and redundant. For example, in our practice, *Operational Node* in DoDAF, *Entity* in Maxsim, *Block* in SysML, *Class* in Modelica are all represent entity. If we make pairs respectively, there will be six mappings. However, with a SSOT in the central, topological structure of mappings will be simple and clear.

4 Conclusion

In this paper, an ontology based SSOT construction approach is proposed. Firstly, a reference ontology for aircraft is introduced, capturing essential concepts of aircraft and providing generic description framework. A SSOT of fix wing UAS is constructed based on this ontology. Secondly, the model transformation mechanism is studied, demonstrating how different aircraft models are transformed by matching to the reference ontology. With the ontology based SSOT, several kinds of models of fix wing UAS are successfully transformed.

According to the practical results on fix wing UAS, the ontology based SSOT construction approach initially confirms its feasibility and usability. Ontology is an appropriate solution for SSOT construction and model transformation.

5 Discussion

At the end of the paper, we want to discuss some deficiencies of this work and put forward expectation of future work.

First, our primary motivation and goal of this work are ensuring the consistency of the models used in our aircraft development practice. The reference ontology in this paper is proposed for this purpose. So, when define the concepts of aircraft, we mainly refer to the modeling language involved in our practice, such as DoDAF, Maxsim, SysML, and Modelica. For this reason, the ontology proposed in this paper may be not complete and shall be further improved or extended. Our ontology can serve as a foundation for later work.

Second, the representation of the reference ontology is in a UML way. However, since ontology pursues to describe concepts language-independently, so there is actually no "right" language to represent ontology. So, the representation of ontology shall be more seriously considered, maybe in OWL or maybe in other ways.

References

1. Ni Y, Fan Y (2008) Ontology based cross-domain enterprises integration and interoperability. In 2008 IEEE congress on services part II (services-2 2008). IEEE, pp 133–140
2. Ast M, Glas M, Roehm T, Luftfahrt VB (2014) Creating an ontology for aircraft design. Deutsche Gesellschaft für Luft-und Raumfahrt-Lilienthal-Oberth eV
3. Glas M (2013) Ontology-based model integration for the conceptual design of aircraft. Doctoral dissertation, Technische Universität München
4. Reiss M, Moal M, Barnard Y, Ramu JP, Froger A (2006) Using ontologies to conceptualize the aeronautic domain. In: Proceedings of the international conference on human-computer interaction in Aeronautics. Cépaduès-Editions, Toulouse, France, pp 56–63
5. Gorodetsky VI, Samoylov VV, Trotskii DV (2015) The reference ontology of collective behavior of autonomous agents and its extensions. J Comput Syst Sci Int 54(5):765–782
6. Johnson T, Kerzhner A, Paredis CJ, Burkhart R (2012) Integrating models and simulations of continuous dynamics into SysML. J Comput Inf Sci Eng 12(1)
7. Paredis CJ, Bernard Y, Burkhart RM, de Koning HP, Friedenthal S, Fritzson P, Schamai W (2010). 5.5. 1 An overview of the SysML-Modelica transformation specification. In: INCOSE international symposium, vol 20, No 1, pp 709–722. (2010, July)
8. Kappel G, Kargl H, Kramler G, Schauerhuber A, Seidl M, Strommer M, Wimmer M (2007) Matching metamodels with semantic systems-an experience report. In: BTW workshops, pp 38–52

Experimental Investigations on Aerodynamic and Psychoacoustic Characteristics of Loop-Type Propeller

Jianwei Sun, Koichi Yonezawa, Eiji Shima, and Hao Liu

Abstract This study proposes a low noise loop-type propeller that provides lift by loops rather than blades, aiming to reduce the noise sound pressure level (SPL) and improve sound quality. Noise data were acquired via a hover stand test in an anechoic chamber. We then made a comparison of the SPL and psychoacoustic metrics (loudness, sharpness, roughness, and fluctuation strength) of the DJI Phantom 3 propeller and the loop-type propeller using acoustic analysis. Furthermore, we discussed the tonal noise and broadband noise components of both propellers. The aerodynamic force and torque results show that the two propeller types provide very similar figure of merit values, while the DJI Phantom 3 propeller shows higher lift and torque coefficients. In terms of aeroacoustics, the loop-type propeller can reduce the noise SPL at different positions under 5400 rpm. With Fastl and Zwicker's psychoacoustic annoyance model, we predict and demonstrate that the loop-type propeller can improve a human's psychological response to the sound via higher-level broadband noise and lower-level tonal noise components.

Keywords Aeroacoustics · Psychoacoustic · Loop-type propeller · Noise experiment

J. Sun · H. Liu (✉)
Graduate School of Engineering, Chiba University, Chiba 263-8522, Japan
e-mail: hliu@faculty.chiba-u.jp

K. Yonezawa
Central Research Institute of Electrical Power Industry, Abiko, Japan

K. Yonezawa · H. Liu
Center for Aerial Intelligent Vehicles, Chiba University, Chiba, Japan

E. Shima
Japan Aerospace Exploration Agency, Tokyo, Japan

S. Lee et al. (eds.), *The Proceedings of the 2021 Asia-Pacific International Symposium on Aerospace Technology (APISAT 2021), Volume 1*, Lecture Notes in Electrical Engineering 912, https://doi.org/10.1007/978-981-19-2689-1_7

1 Introduction

In recent years, UAVs (unmanned aerial vehicles), particularly rotorcrafts, have played a crucial role in impacting civilian tasks, including urban logistics, airborne imagery, and even transportation. UAVs have emerged as useful platforms for flying over or hovering in cities where urban populations live. Amazon, one of the largest online retailers in the world, developed its fleet of Prime Air delivery rotorcrafts to achieve unmanned package delivery, which received Federal Aviation Administration (FAA) approval in August 2020. Alphabet and UPS have also previously received the approval of FAA [1]. Thus, drone-based package delivery and other civilian tasks may soon be achieved. However, a problem emerging given the rapidly increasing needs and applications of various drones is rotor noise, which seriously disturbs the environment and community. Previous studies have attempted to reduce noise via bio-inspired structures [2] or add-in structures [3].

A survey by Christian and Cabell [4] reported that the rotorcraft-generated noise could become a severe issue causing human beings even more annoyance even than car noise; thus, the psychoacoustics of rotor noise has recently attracted significant interest [5]. In this regard, objective psychoacoustic metrics are becoming a vital method to evaluate the annoyance of products' noise [6].

In this study, we propose a low noise loop-type propeller that provides lift by loops rather than blades, aiming to reduce the noise sound pressure level (SPL) and improve sound quality [7]. Here, we show the results of a hover stand test performed in an anechoic chamber under different rotation speeds to understand the aerodynamic and psychoacoustic characteristics of the new type of propeller compared with a DJI Phantom 3 propeller.

2 Experiment Setup

Aerodynamic (thrust and torque) and acoustic measurements in the hover stand experiment were performed simultaneously in an anechoic chamber at Nishi Chiba Campus, Chiba University. The room has internal dimensions of 3.86 m long × 1.56 m wide × 2.4 m tall.

2.1 Thrust and Torque Measurement

A high-resolution six-degree-of-freedom force-torque cell (Leptrino PFS 030YA151) was attached between the stand and motor adapter. A Kyowa EDX-100A DAQ and a10 Hz low-pass analog filter were used to reduce the noise of voltage signal, with a sampling rate (f_s) of 20 kHz used. The thrust (F_z) range of the force-torque cell is ± 150 N with a precision of 0.075 N; the corresponding torque (M_z) range is ±

1 N·m with a precision of 0.0005 N·m. An aluminum frame structure of 800 mm height was used to make the hover stand. The thrust vector was set to the bottom to eliminate the ground effect.

2.2 *Sound Signal Measurement*

All noise signal measurements were performed with RION CO. UC-59 microphones, using a Kyowa EDX-100A data acquisition system at a sampling rate (f_s) of 20 kHz. An arc array was used in this study, which positions four microphones at a distance of 520 mm (i.e., 2.18*d*, where *d* is the diameter of the propellers). The polar angles are between the propeller disk plane, the line of the monitor point, and the propeller center. The four locations' polar angles (θ) are 0, 25, 50, and 75°. In addition, a line microphone array was set at a polar angle (θ) of 45° at distances of 476 mm (2*d*), 714 mm (3*d*), 952 mm (4*d*), and 1190 mm (5*d*).

2.3 *Model Setup*

The loop-type propeller was 3D printed with laser-curing resin. Figure 1a presents the four-blade loop-type propeller, where the tips of two adjacent blades are connected to become two circle loops. As shown in Fig. 1b, the experiment's 3D printed Phantom 3 propeller is used as a baseline. The diameter of both rotors is 238.8 mm (*d*). A HIOKI FT3406 tachometer was used to monitor the propeller rotation speed, which was input to a closed-loop rotation speed control program, achieved via NI USB-6229 and Labview. The propulsion system is composed of an EMAX GT2820/06 motor and Hobbywing Platinum 40A Electronic Speed Controllers (ESC). Figure 2 shows the experimental setup in the anechoic chamber.

Fig. 1 Photographs of **a** the Loop-type propeller and **b** the DJI Phantom 3 propeller

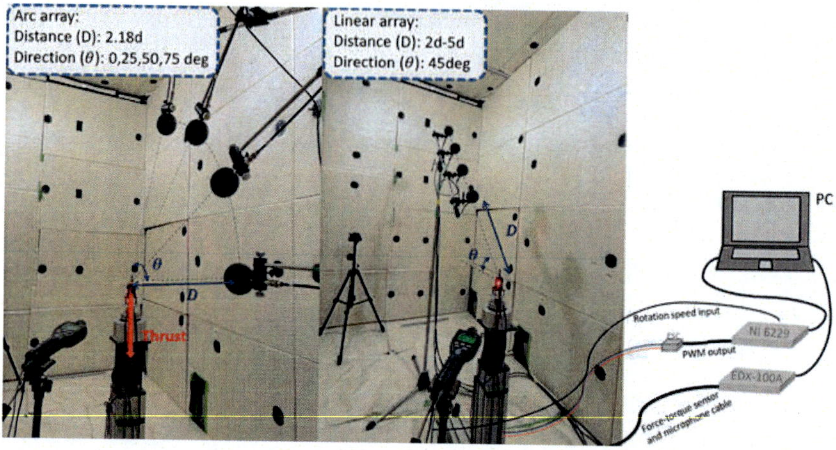

Fig. 2 Arrangement of the instrumentation in the arc and line microphone arrays

3 Result

3.1 Aerodynamic Performance

The aerodynamic performance of the loop-type propeller and DJI Phantom 3 propeller are shown in Fig. 3, where the thrust and torque were obtained using a force-torque sensor at different rotation speeds. Under the same rotation speed, the DJI Phantom 3 propeller produces higher thrust and torque. To compare the aerodynamic performance, the thrust and axial torque were normalized as thrust coefficient (C_T), torque coefficient (C_Q), and figure of merit (FM) using the well-known definitions given in Eqs. (1–3) below:

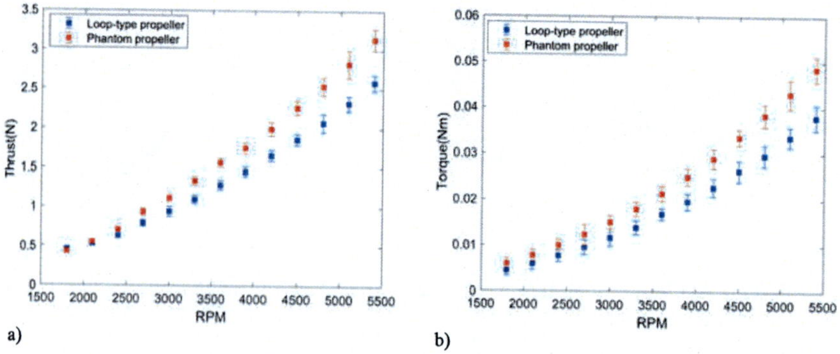

Fig. 3 Representations of the loop-type and DJI phantom propellers' **a** thrust and **b** torque

$$C_T = \frac{F_z}{\rho (r\Omega)^2 A} \tag{1}$$

$$C_Q = \frac{M_z}{\rho (r\Omega)^2 R A} \tag{2}$$

$$FM = \frac{C_T^{3/2}}{\sqrt{2} C_Q} \tag{3}$$

where the r is the propeller radius, Ω is the rotation speed in rad/s, ρ is the density of air, A is the area of propeller disk, and F_z and M_z are the measured thrust and torque, respectively. As shown in Fig. 4, the Phantom 3 propeller showed significantly higher values of C_T and C_Q, while the FM value is similar (<5% difference) between the two kinds of propellers at 3,600 and 5,400 RPM.

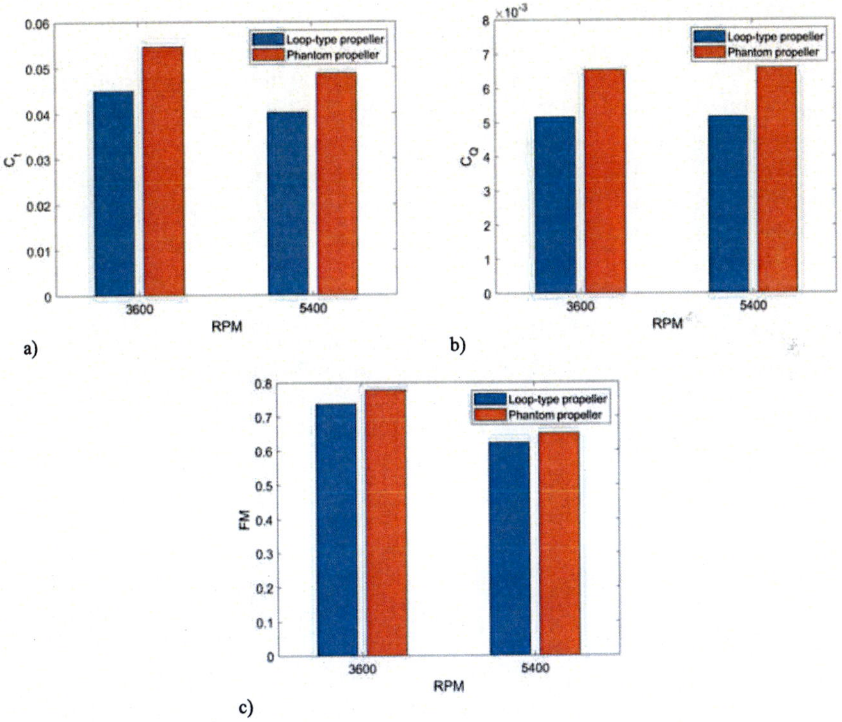

Fig. 4 Representations of the loop-type and DJI phantom propellers' **a** thrust coefficient, **b** torque coefficient, and **c** figure of merit

3.2 Acoustic Performance

For acoustic performance, 10-s sound signals were acquired by a data logger at a sampling rate of 20 kHz. The spectrum presented in this report was transferred by Welch's method of power spectrum estimation. The data were each 8,192 samples in length with a rectangular window and 50% overlap between records. The sound pressure level is calculated by Eq. (4), where the reference sound pressure (P_{ref}) of air is 20 μPa/Hz.

$$SPL = 10\log_{10}\frac{PSD}{P_{ref}{}^2} \tag{4}$$

Figure 5 compares the power spectral density (PSD) of the Phantom 3 propeller and loop-type propeller at 5,400 RPM, as measured at location 3. As there are seven pairs of magnets in the motor, the 7th harmonic (1,260 Hz) appears to be the sound source of the motor. The spectrum shows that the most significant component of motor noise is single-frequency noise at the 7th harmonic (f_7). To avoid the motor's influence, we did not consider the 7th harmonic (1,260 Hz) in the comparison. Notably, the spectrum indicates that the loop-type propeller produces more high-frequency broadband noise, above 6,000 Hz, and decreases the tonal noise at blade passing frequency (BPF; 1st BPF (f_1) = 180 Hz) and the harmonics (f_i). Also, only the Phantom propeller has obvious higher-order harmonic (i > 30) peaks in the PSD spectrum. To understand the acoustic characteristics of the loop-type and Phantom 3 propellers, Fig. 6 illustrates the SPL variance dependency of direction and RPM. The colors in this figure represent the change in sound pressure level, ΔSPL. At high frequencies of 6,000–10,000 Hz, the broadband noise clearly increases at different rotation speeds and polar angles. In the line array, as shown in Fig. 7, this characteristic of noise variation does not depend on distance, which reduces the noise at BPF and the harmonics while broadband noise in the high-frequency range is increased.

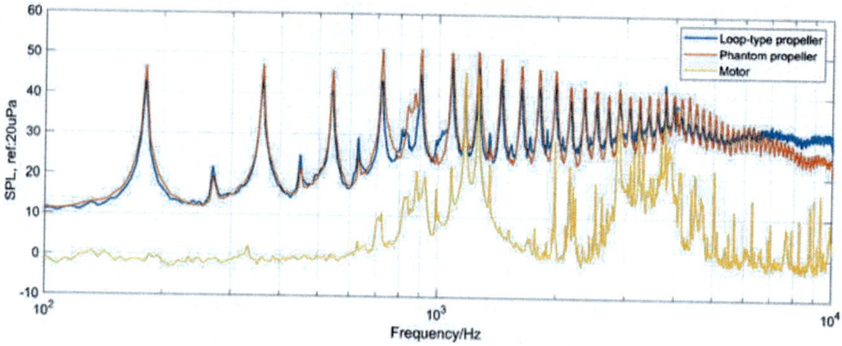

Fig. 5 PSD spectra at location 3 ($\theta = 50°$, $D = 2.18\,d$) under 5400 RPM of the loop-type propeller, DJI phantom propeller, and motor

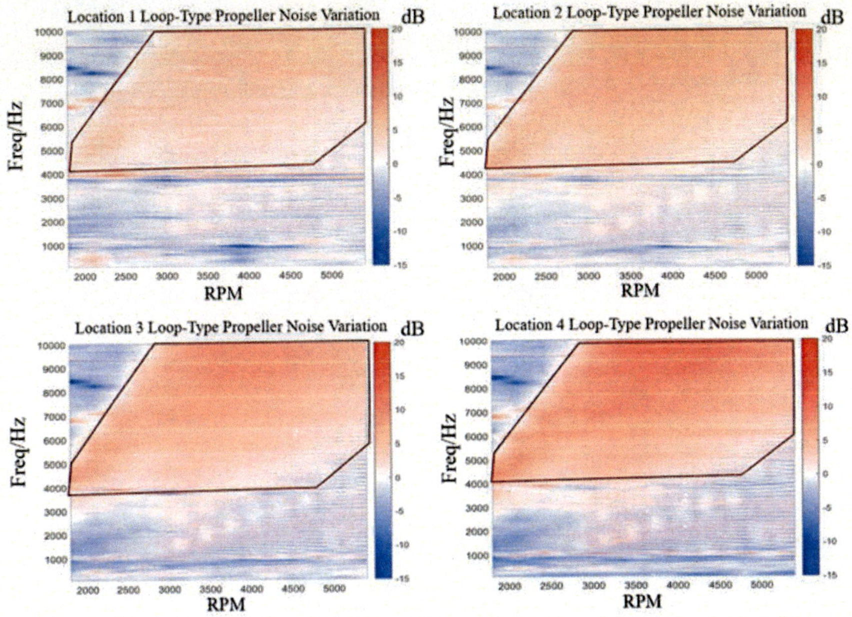

Fig. 6 Sound pressure level (SPL) change as a function of frequency and rotation speed at different locations

The overall sound pressure level (OASPL) values at different locations at 5,400 RPM and 3,600 RPM are compared in Fig. 9f and Fig. 10f. The loop-type propeller reduces noise OASPL by as much as 0.2–1.9 dB at the locations of Microphones 1–4 at 5,400 RPM in the arc array configuration. However, noise reduction at 3,600 RPM is not significant, with noise increasing at the location of Microphone 4.

3.3 *Objective Psychoacoustic Metrics*

The objective psychoacoustic metrics include loudness, sharpness, roughness, and fluctuation strength. Psychoacoustic annoyance (PA) is calculated from the objective psychoacoustic metrics using Eq. (5) and is used to evaluate people's feelings of annoyance in response to different sounds [8, 9].

$$PA = N_5\left(1 + \sqrt{w_s^2 + w_{FR}^2}\right) \tag{5}$$

$$w_s = \begin{cases} (S - 1.75) \bullet (0.25\log_{10}(N_5 + 10)), \, S > 0 \\ 0, \, S \le 1.75 \end{cases}$$

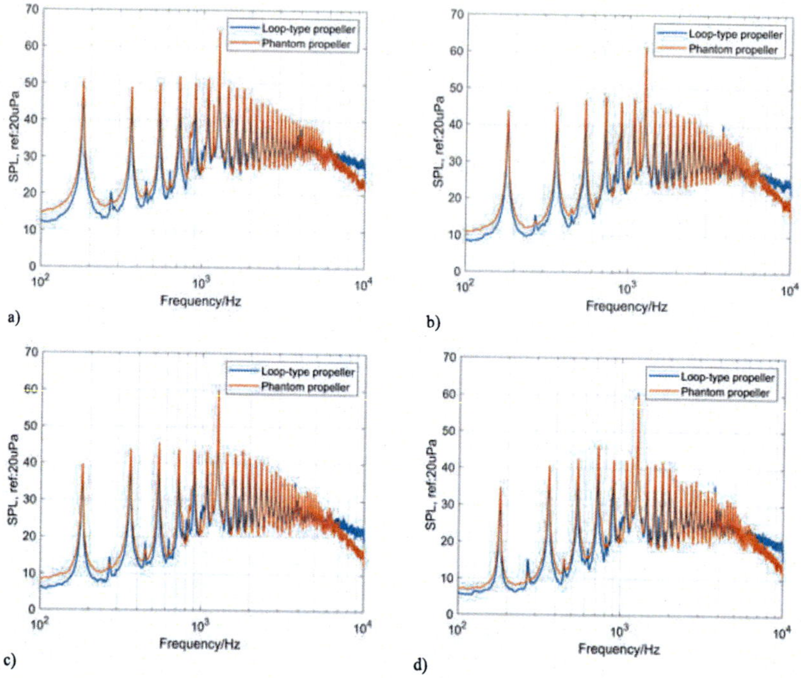

Fig. 7 PSD spectra at **a** location 1 ($\theta = 45$ deg, $D = 2\,d$), **b** location 2 ($\theta = 45$ deg, $D = 3\,d$), **c** location 3 ($\theta = 45$ deg, $D = 4\,d$), and **d** location 4 ($\theta = 45$ deg, $D = 5\,d$) for the loop-type and DJI phantom propellers at 5,400 RPM

$$w_{FR} = \frac{2.18}{(N_5)^{0.4}}(0.4F + 0.6R)$$

N_5 is the percentile loudness in sones (i.e., the level that is only exceeded 5% of the time). w_s and w_{FR} are standardized measures of sharpness, roughness, and fluctuation strength, where S, R, and F are the sharpness, roughness, and fluctuation strength, respectively.

The 7th BPF harmonic (i.e., 1,260 Hz), a peak in the PSD spectrum, is mainly produced by the motor with its seven-pair magnets. This peak in the frequency domain may influence the loudness due to the simultaneous masking effect. A notch filter of 1,260 Hz was applied during data processing of the sound signal time series to reduce the influence of motor noise. The loudness and PSD spectra of the filtered signal at location 3 are compared in Fig. 8. Because the tonal noise is decreased via the loop-type structure, the loudness of the Phantom 3 propeller is greater at frequencies below 6000 Hz. In the high-frequency range, the loop-type propeller structure increases the loudness.

The objective psychoacoustic metrics of the two kinds of propellers at 3,600 and 5,400 RPM are shown in Figs. 9 and 10, respectively. Loudness reflects people's

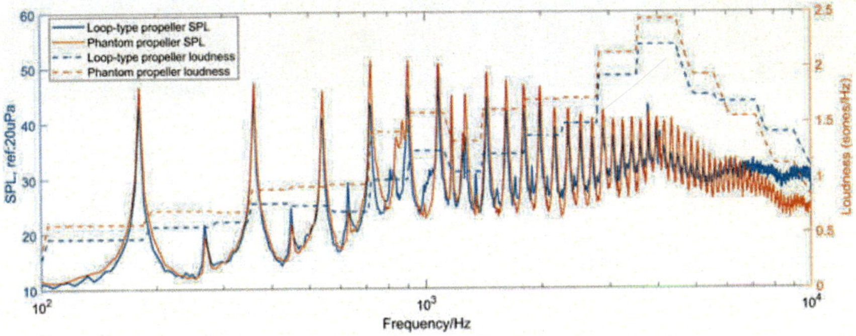

Fig. 8 Sound pressure level (SPL) and specific loudness of loop-type propeller and DJI phantom propeller

subjective perception of sound pressure. At location 4 at 5,400 RPM, the loudness values of the loop-type propeller and Phantom 3 propeller were 19.2 sone and 22.1 sone, respectively, while the corresponding SPL values were 66.9 and 67.9 dB. Sharpness reflects the high-frequency component of the sound signal, which is a comparison of the amount of high-frequency energy to the total energy. At all locations for rotation speeds of 3,600 and 5,400 RPM, the sharpness value of the loop-type propeller is higher than that of the Phantom 3 propeller. This can be explained by the observation that the loop-type propeller increases high-frequency broadband noise. Roughness reflects noticeable or annoying sounds as heard by the human ear at the modulation frequency, which is the BPF in the case of propeller noise. At both rotation speeds, the roughness of the loop-type propeller shows a clear decrease, particularly at locations at higher polar angles. This outcome is potentially due to the loop-type propeller reducing the signal frequency noise at BPF. Fluctuation strength is heard at values below 20 Hz modulation frequency. In this study, the value of fluctuation strength is too small to affect the PA significantly overall. Using these data, the value of PA was calculated using Eq. (5). Interestingly, the trends of PA and OASPL are not entirely the same. At location 4 at 3,600 RPM, the loop-type propeller reduces the PA, despite producing a higher SPL noise value. This indicates that loop-type propellers can improve people's perceived annoyance even though the SPL does not change.

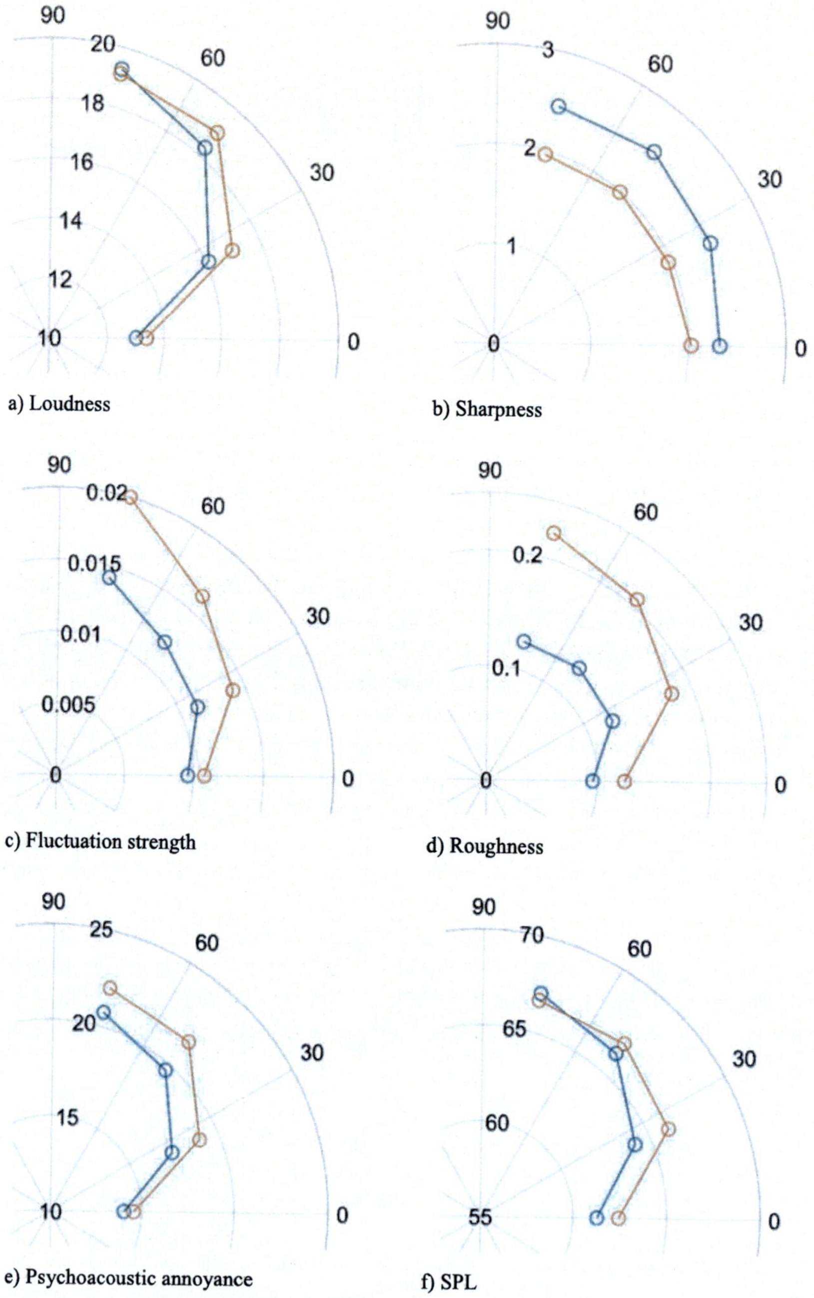

Fig. 9 Plots showing **a** loudness, **b** sharpness, **c** fluctuation strength, **d** roughness, **e** psychoacoustic annoyance, and **f** overall sound pressure level (OASPL) at 3,600 RPM

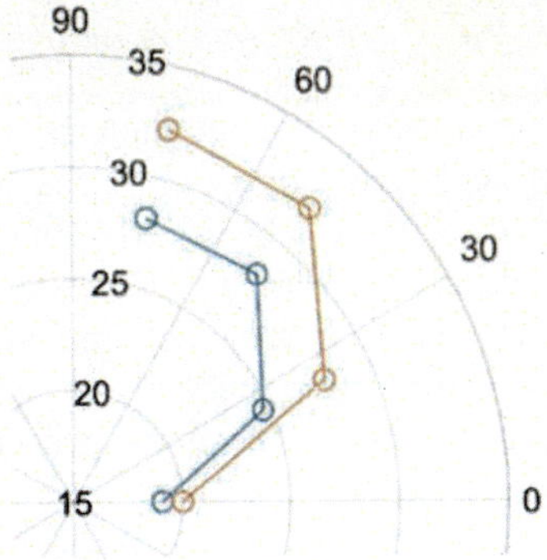

a) Loudness

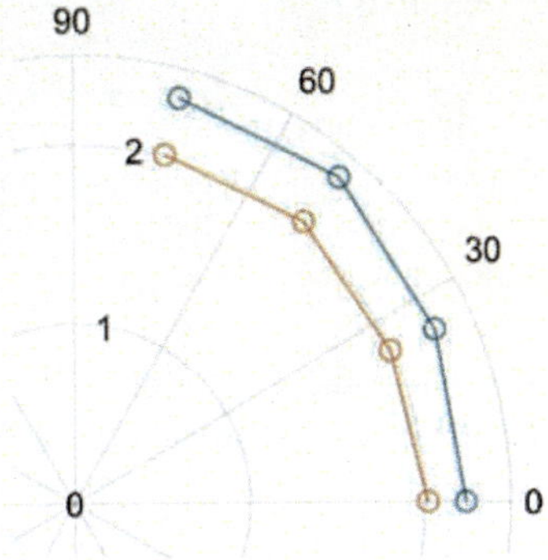

b) Sharpness

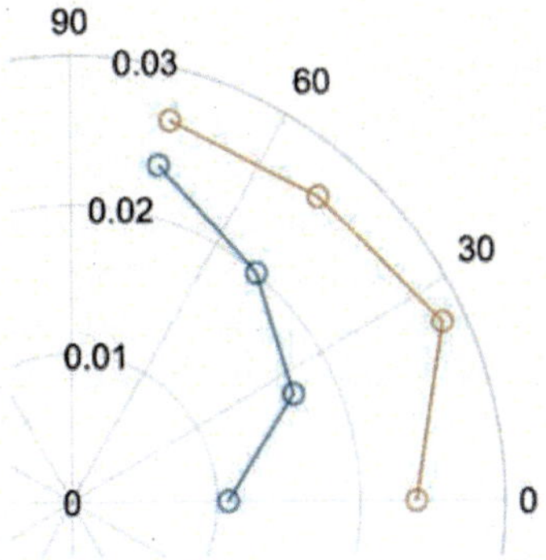

c) Fluctuation strength

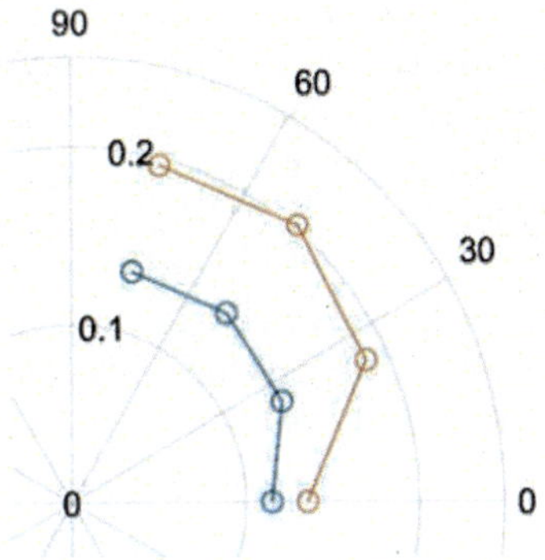

d) Roughness

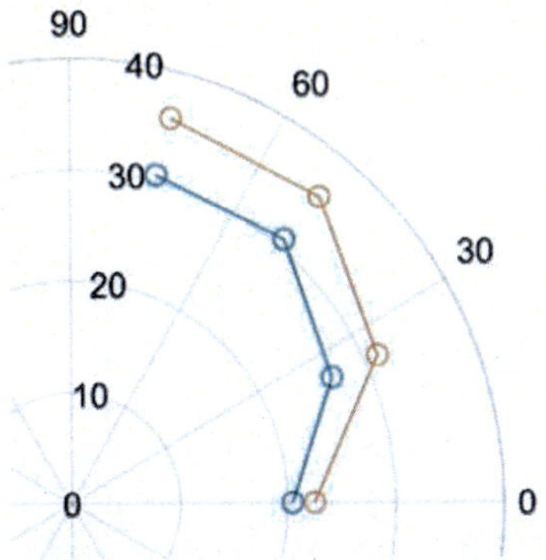

e) Psychoacoustic annoyance

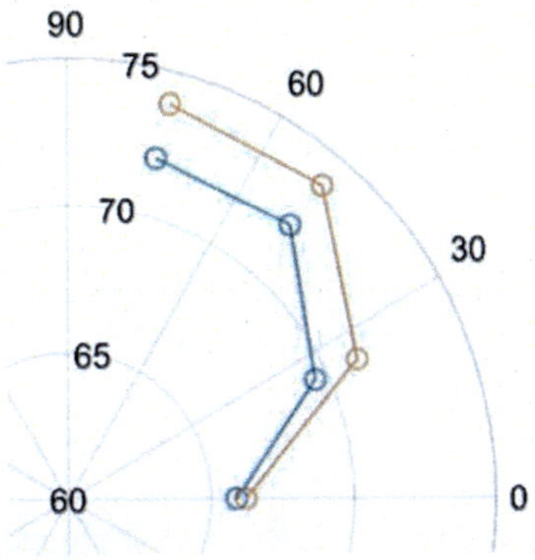

f) SPL

Fig. 10 Plots showing **a** loudness, **b** sharpness, **c** fluctuation strength, **d** roughness, **e** psychoacoustic annoyance, and **f** overall sound pressure level (OASPL) at 5,400 RPM

4 Discussion

Previous studies have noted the importance of psychoacoustics and the amount of tonal noise or broadband noise components in the design of propellers. The aerodynamic noise of a propeller includes periodic noise, induced by periodic rotating, and broadband noise, which includes vortex and turbulence noise [9]. Periodic or tonal noise describes the sound identified at a discrete frequency occurring at the BPF and the harmonics. In contrast, broadband noise describes sound whose energy is distributed over a broad section of the frequency domain. Human annoyance in response to tonal noise is generally at higher level. Surprisingly, the acoustic characteristics of the loop-type propeller, which are composed of lower-level tonal noise and higher-level broadband noise, improve sound quality in terms of psychoacoustic metrics.

This interesting acoustic characteristic can be explained by the distinctive structure of the loop-type propeller. This propeller reduces the structural symmetry, which provides thrust from four blades, of which the tips of two adjacent blades are connected to become two circle loops, rather than from two completely symmetrical blades. The annoying tonal noise, or periodic noise, which is usually produced by rotation of a symmetrical blade, will decrease in a loop-type structure. However, the complex structure and higher wetted area tend to increase the broadband noise composed of vortex noise and turbulence noise. This finding, while preliminary, suggests that changing the propeller structure can help to improve the acoustic characteristics and sound quality. Thus, this study provides support for multi-blade propellers or fans' acoustic designs.

5 Conclusion and Future Work

This paper presents a study focused on investigating the aerodynamic and psychoacoustic characteristics of a loop-type propeller that provides lift by loops rather than blades. Thrust and torque measurements show that this loop-type propeller provides lower C_T and C_Q values and similar FM compared to a DJI Phantom 3 propeller. The loop-type propeller, which increases broadband noise from 6,000 Hz to 10,000 Hz and decreases tonal noise at BPF and the harmonics, overall improves the sound quality.

There are still some unanswered questions regarding psychoacoustic evaluation. For example, the broadband noise at high frequencies was calculated to have high sharpness. However, this high-frequency broadband component of propeller aerodynamic noise is usually white noise which does not tend to cause significant annoyance. Further work is required to establish the viability of Fastl and Zwicker's psychoacoustic annoyance (PA) model for propeller noise and the correlation between the objective psychoacoustic model and subjective jury test results. Another potential disadvantage of this study is that the sound signal of outdoor flight tests is not considered. Given the high decay rate of the high-frequency component of noise signals outdoor and the interactions with other propellers, the SPL and the psychoacoustic results may be different. Thus, the psychoacoustic jury test and flight noise test will be the next steps in our future research.

Acknowledgements This work is partially supported by the Grant-in-Aid for Scientific Research of KAKENHI No. 19H00750, JSPS, and the Centre for Aerial Intelligent Vehicles (CAIV), Graduate School of Engineering, Chiba University.

References

1. Palmer A (2020) Amazon wins FAA approval for Prime Air drone delivery fleet. CNBC
2. Rao C, Liu H (2020) Effects of reynolds number and distribution on passive flow control in owl-inspired leading-edge serrations. Integr Comp Biol 60(5): https://doi.org/10.1093/icb/icaa119
3. Yonezawa K (2020) Experimental investigation of noise characteristics of rotors. In: QUIET DRONES international e-symposium on UAV/UAS noise
4. Christian AW, Cabell R (2017) Initial Investigation into the psychoacoustic properties of small unmanned aerial system noise. In: 23rd AIAA/CEAS aeroacoustics conference. https://doi.org/10.2514/6.2017-4051. (5 June 2017)
5. Gwak DY, Han D, Lee S (2020) Sound quality factors influencing annoyance from hovering UAV. J Sound Vib 489: https://doi.org/10.1016/j.jsv.2020.115651
6. Novaković T, Ogris M, Prezelj J (2020) Validating impeller geometry optimization for sound quality based on psychoacoustics metrics. Appl Acoust 157: https://doi.org/10.1016/j.apacoust.2019.107013
7. Shima E (2019) Preliminary study on innovative loop propellers for quiet eVTOL. In: 8th Asian/Australian Rotorcraft Forum
8. Zwicker E, Fastl H (2013) Psychoacoustics: v. 22: facts and models. 2nd edn. Springer
9. Made JE, Kurtz DW (1970) A review of aerodynamic noise from propellers, rotors, and lift fans. https://ntrs.nasa.gov/search.jsp?R=19700005920

Effect of Dynamic Micro Vortex Generator on Corner Shock Wave Boundary Layer Interactions Based on DES

Yong-sheng Zhao, Jun-fei Wu, and Jian Zhou

Abstract Based on the OpenFOAM numerical simulation platform, the dynamic mesh technology and Detached-eddy simulation (DES) are used to study the flow field characteristics of corner shock wave boundary layer interactions (SWBLI) when the micro vortex generator (MVG) moves downstream at a certain speed. The inlet Mach number is 2.5, and the speed of the MVG moving downstream is 0 m / s, 10 m / s and 20 m / s. The results show that when MVG moves downstream, the high pressure area in SWBLI region will gradually weaken or even disappear. The boundary layer thickness of SWBLI will gradually decrease; this shows that dynamic MVG has a good control effect on SWBLI. As the movement speed increases, the control effect becomes more obvious. MVG controls SWBLI through the wave-system structure generated by MVG and the streamwise vortex pair of wakes.

Keywords Dynamic micro vortex generator · SWBLI · Detached-eddy simulation

1 Introduction

Shock wave boundary layer interactions (SWBLI) are widely present in the internal and external fields of supersonic/hypersonic aircraft [1–3]. SWBLI will have a great impact on the flight stability control and structural safety of the aircraft. Therefore, the importance of SWBLI control need not be repeated [4–7]. In the research on SWBLI, the compressed flow at the corner has always been a hot issue that has attracted much attention. It is essentially the movement of supersonic airflow around the corner surface. In the supersonic inlet or on the surface of the aircraft, if the wall slope undergoes a sudden compressive change, this interference flow field will be generated [8–10]. The control of this type of SWBLI is crucial to the design of supersonic/hypersonic aircraft.

The anti-vortex generated by the micro vortex generator (MVG) [11–15] can exist for a long distance in the boundary layer. MVG has excellent control effect

Y. Zhao (✉) · J. Wu · J. Zhou
China Academy of Aerospace Aerodynamics, Beijing 100074, China
e-mail: zhaoyongshengying@163.com

S. Lee et al. (eds.), *The Proceedings of the 2021 Asia-Pacific International Symposium on Aerospace Technology (APISAT 2021), Volume 1*, Lecture Notes in Electrical Engineering 912, https://doi.org/10.1007/978-981-19-2689-1_8

on SWBLI. At the same time, MVG has less interference and higher robustness. Therefore, the research on MVG has always been a hot spot in the field of SWBLI control.

The published research on MVG has a common feature: it does not have dynamic characteristics [16–19], that is, the disturbance frequency and amplitude of the wake vortex, streamwise vortex and vortex ring structure cannot be changed. This largely limits its performance and applicable working conditions. However, supersonic/hypersonic aircraft operate in unsteady pressure environments caused by a variety of significant turbulence and compression effects. It also includes the pulsating flow field caused by SWBLI. A strong shear layer is formed between the internal flow and the external flow in the separation zone. The large-scale vortex structure in the shear layer evolves downstream with the flow. These large-scale vortex structures are closely related to the unsteady characteristics of SWBLI [20–22]. Wu et al. generate animations of DNS results. It is found that when the size (length and height) of the separation bubble changes, it has obvious low-frequency characteristics, which is similar to the low-frequency motion of the shock wave [23]. Dolling summarizes the research on SWBLI, and proposes that the low-frequency motion in the interference zone has extremely important significance for SWBLI [24]. The research shows: (1) Low-frequency unstable motion originates from the inherent instability of the separated flow [25]; (2) The low-frequency unstable motion is driven by the upstream turbulent boundary layer [26]; (3) The time scale of the low-frequency oscillation is about ($10\delta/U_\infty \sim 100\delta/U_\infty$), δ is the thickness of the upstream boundary layer, and U_∞ is the free flow velocity [24]. The above research shows that SWBLI has dynamic characteristics.

Kim and Sung [27] use spanwise slits to blow air regularly and periodically into the boundary layer. Comparing the results, it is found that periodic high-frequency blowing can induce small-scale vortex structures, which increase the pulsation of the boundary layer spanwise and normal vortices. Hao [28] imposes acoustic disturbance on the boundary layer. It is found that there is a perceptual connection between the inner and outer layers of the boundary layer, and the closer the disturbance frequency is to the natural frequency of the turbulence of the boundary layer, the stronger the interaction and the faster the attenuation of the disturbance amplitude. Liu [29] uses synthetic jets to study the effects of disturbance frequency and disturbance amplitude on the stability of the boundary layer. It is found that relative to low frequency, high frequency disturbance is more conducive to the flow stability of the supersonic boundary layer. When the disturbance frequency exceeds 400 Hz, the time growth rate of the second mode disturbance wave decreases, and the second mode is more stable. Based on the above research results, the following conclusions can be easily inferred: Injecting dynamic disturbances into the boundary layer will be more conducive to the control of SWBLI.

Therefore, based on the OpenFOAM numerical simulation platform, this paper uses the dynamic mesh technology and Detached-eddy simulation (DES) to study the control effect of the dynamic micro-vortex generator on the corner SWBLI. The changes of pressure and boundary layer in the SWBLI region are emphasized. At the same time, the control mechanism of dynamic MVG is analyzed from the wave system structure and the vortex structure.

2 Numerical Simulations

2.1 Numerical Method

Based on the open source object-oriented computational fluid dynamics (CFD) software OpenFOAM, the density-based dynamic solver rhoCentralDymFoam is used to solve the following three-dimensional compressible NS equation [30–32].

$$\frac{\partial \rho}{\partial t} + \nabla \cdot (\rho U) = 0 \tag{1}$$

$$\frac{\partial \rho U}{\partial t} + \nabla \cdot (\rho U U) + \nabla \cdot p = \nabla \cdot (\tau \nabla U) \tag{2}$$

$$\frac{\partial \rho E}{\partial t} + \nabla \cdot (U \rho E) - \alpha_{eff} \nabla e = -\nabla \cdot (U p) + \nabla \cdot (\tau \cdot U) \tag{3}$$

where ρ is density, t is time, U is velocity, p is pressure, E is internal energy, α_{eff} is heat transfer coefficient, e unit mass internal energy, and τ is viscosity coefficient. Assuming that the gas is an ideal gas, the following equation of state is given: p = ρ R T.

The turbulence model adopts k-omega-SST DES. The turbulent kinetic energy and the turbulent dissipation rate are solved by the following equation [33].

$$\frac{\mathrm{D}}{\mathrm{D}t}(\rho\omega) = \nabla \cdot (\rho D\omega \nabla \omega) + \rho\gamma\frac{G}{\tau} - \frac{2}{3}\rho\gamma\omega(\nabla \cdot U) - \rho\beta^{*}\cdot\omega^{2} - \rho(F_1 - 1)CD\omega k + S_\omega \tag{4}$$

$$\frac{\mathrm{D}}{\mathrm{D}t}(\rho k) = \nabla \cdot (\rho Dk \nabla k) + min(\rho G, (c_1\beta^{*})\rho k\omega) - \frac{2}{3}\rho k(\nabla \cdot U) - \rho\frac{k^{1.5}}{\tilde{d}} + S_k \tag{5}$$

$$\tilde{d} = min(C_{DES}\Delta, \frac{\sqrt{k}}{\beta^{*}\omega}) \tag{6}$$

where w is the turbulent dissipation rate, γ is the specific heat ratio, G is the vorticity term, ν is the viscosity coefficient, k is the turbulent kinetic energy, c_1, β^*, C_{DES} is a constant, Sω, Sk is the turbulence generation, F1 is the switching function, and $\tilde{d}$ is the length scale. Δ is the maximum side length of the grid cell. Therefore, the selection of the turbulence model is determined, that is, when the turbulence scale is smaller than Δ, the SST two-equation model is adopted; when

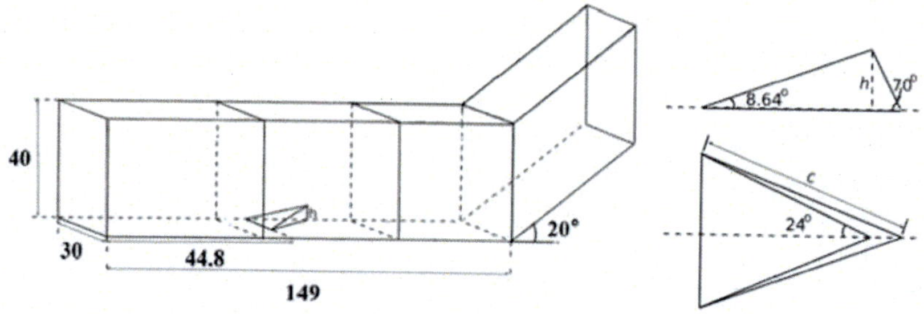

Fig. 1 Computational domain model (unit / mm)

the turbulence scale is larger than Δ, the LES model is adopted. Among them, the sub-grid of LES adopts the dynamic Smagorinsky sub-grid model.

Kurganov and Tadmor (KT/KNP) format is used to discretize the above mathematical model, and the time advancement adopts the second-order time format (C-N format), At the same time, the transmission speed of the wave is considered, and a smaller dissipation coefficient is determined according to the wave speed to obtain higher accuracy.

2.2 Computational Domain Model

The computational domain model is shown in Fig. 1. Through the slope, a shock wave is generated at the corner, and the slope angle is 20°. MVG is installed on the lower wall, 44.8 mm away from the inlet, to control SWBLI. MVG adopts a wedge structure with a half-opening angle of 24°, where h = 5 mm. The bottom is an isosceles triangle, where c = 7.2 h.

2.3 Mesh and Boundary Conditions

SnappyHexMesh in OpenFOAM is adopted for mesh division. The geometry is divided into three areas, namely the stretching zone 1, the moving zone 2 and the compression zone 3, as shown in Fig. 2a. When dynamic simulation is carried out, the mesh in zone 1 is stretched along the speed direction, the mesh in zone 2 moves at a set speed, but the grid size remains unchanged, and the mesh in zone 3 is compressed along the speed direction. In order to ensure that the stretched mesh size of zone 1 will not affect the numerical calculation results, the mesh number of zone 1 is determined by mesh independence verification. At the same time, in order to ensure the accuracy of the calculation results, local encryption is performed on the MVG, SWBLI area and the boundary layer. The turbulence model is k-omega-SST DES.

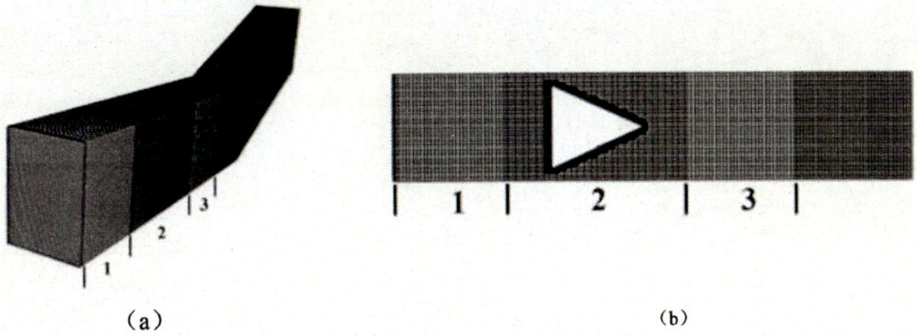

Fig. 2 Mesh division methods

The speed inlet and pressure outlet are adopted. The wall surface is a non-slip wall surface.

The inlet Mach number (Ma) is 2.5, the total pressure is 0.276 atm, the total temperature is 300 K, and the MVG moving speed is 0 m/s, 10 m/s and 20 m/s.

3 Verification of Numerical Methods

3.1 Mesh Independence

Three sets of meshes are divided by the above method. The number of nodes in the base mesh (lev1) is 3.21 million, the maximum cell size is $0.04\delta * \times 0.25\delta * \times 0.25\delta *$, ($\delta *$ is the displacement thickness of the boundary layer at the MVG), and the y + range of the grid is from 60 to 90. The base mesh is encrypted, and the number of nodes after encryption is 4.25 million (lev2) and 5.23 million (lev3). Three sets of meshes are used to calculate the flow field characteristics when the MVG velocity is 0 m / s. The pressure and Mach number distributions of SWBLI near the left wall in three directions (flow direction (x), normal direction (y) and spanwise direction (z)) simulated by three sets of meshes are compared, as shown in Fig. 3. It can be seen from the results that when the number of nodes is 4.25 million, the increase in the number of nodes has little effect on the calculation results. Therefore, lev2 can be used for dynamic calculation. The maximum cell size of the mesh is $0.02\delta^* \times 0.12\delta^* \times 0.12\delta^*$, and the y + of the first layer of mesh is less than 20.

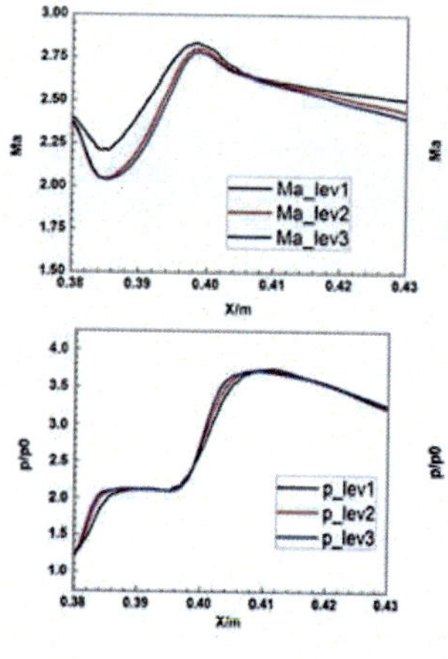

Fig. 3 Verification of mesh independence

3.2 Numerical Simulation Accuracy

Using the above methods, the uninstalled MVG model studied by Lee [34] is numerically simulated. The simulated conditions are inlet Mach number Ma = 3.0, inlet pressure P = 70760 Pa, and temperature T = 582.3 k. The velocity and pressure distribution at the same location (x / δ * = 86.2) are extracted. The current simulation results are compared with Lee's LES simulation results and WPAFB experiment results, as shown in Fig. 4. It can be seen that the mesh division and numerical simulation method can obtain reliable results.

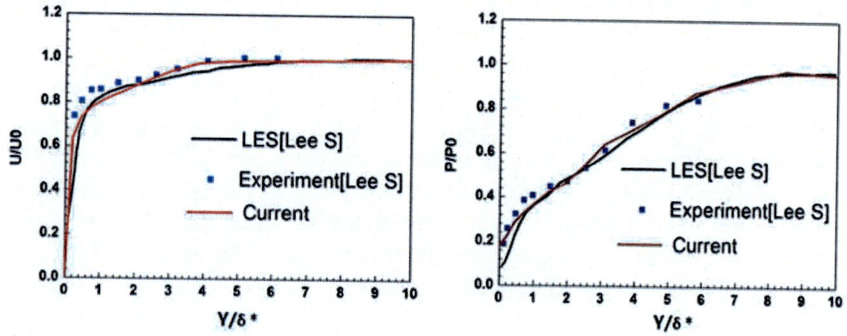

Fig. 4 Velocity and pressure distribution at x / δ * = 86.2

4 Results and Analysis

4.1 The Impact of Dynamic MVG on SWBLI Area Pressure

Figure 5 shows the effect of static MVG on the pressure in the SWBLI area. It can be seen that (1) when the supersonic air flows through the upward corner, a shock wave is formed, and an expansion wave is formed on the upper wall. Due to the interaction between the expansion wave and the shock wave, a high pressure area is formed on the lower wall, as shown in Fig. 5a; As the shock wave interacts with the boundary layer at the corner, the front edge of the shock wave moves upstream and a high pressure area is formed, as shown in Fig. 5b; (2) Due to MVG, two compression waves are formed. The compression wave interacts with the shock wave on the lower wall, which changes the shape of the shock wave on the lower wall, as shown in Fig. 5c. A diamond-shaped wake appears downstream of the MVG. The wake interacts with SWBLI, thereby changing the shape of the bottom high pressure zone as shown in Fig. 5d. The high-pressure zone moves downstream, indicating that MVG has achieved control over SWBLI.

When the MVG moves downstream at a certain speed, the cloud diagram of the bottom pressure is shown in Table 1. In order to verify the influence of the moving speed on the flow field control, the bottom pressure change when the MVG moves to the same position is analyzed. The calculation results show that: (1) when the MVG moves downstream, the shape of the bottom high pressure zone will change. When the moving speed is 10 m/s, the bottom high-pressure zone maintains its shape, but the area decreases. When the MVG moves 4 mm downstream, the central high-pressure zone disappears, and high-pressure zones are formed near the left and right walls. (2) When the MVG moves 1 mm at a moving speed of 20 m/s, the area of the high-pressure zone decreases; after moving 2 mm, the shape of the high-pressure zone

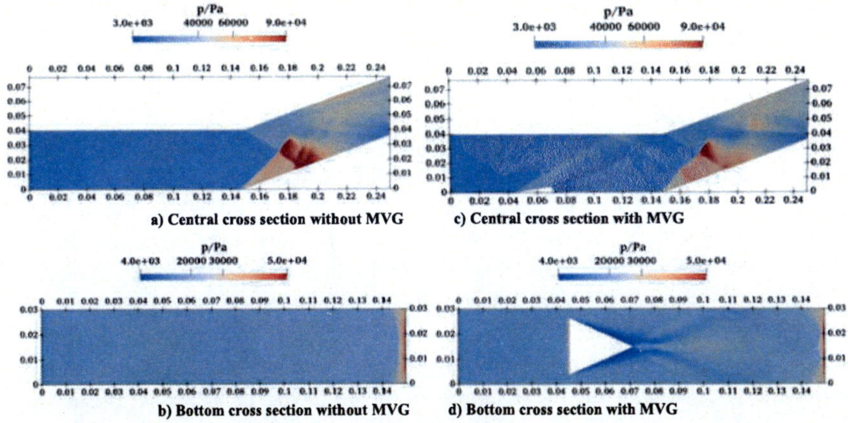

Fig. 5 Influence of static MVG on flow field (unit / m)

Table 1 Influence of dynamic MVG on SWBLI

Motion distance /mm	Uvg = 10 m/s	Uvg = 20 m/s
1	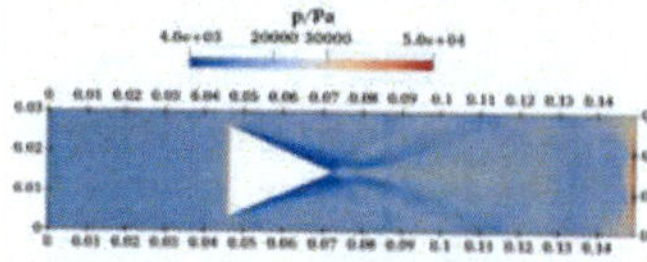	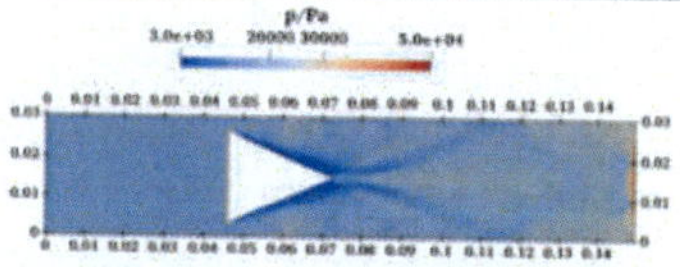
2	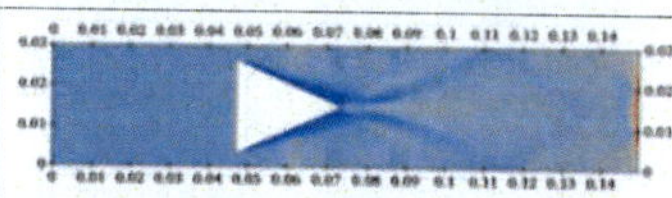	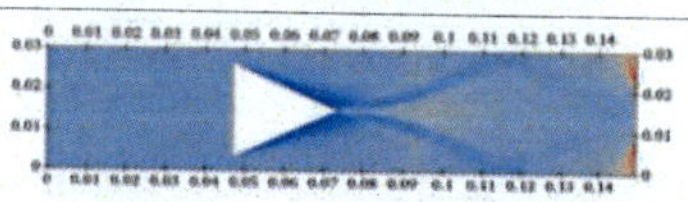
3	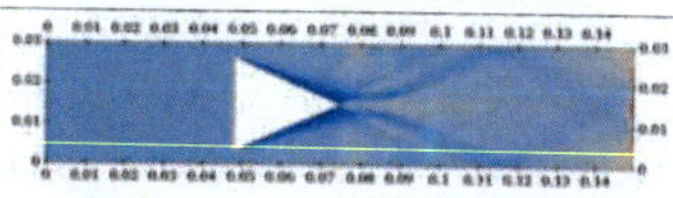	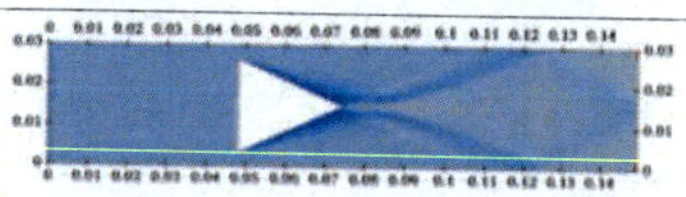
4		
5		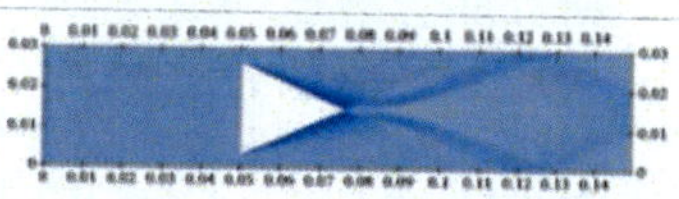

changes; When the MVG continues to move downstream, the high-pressure zone disappears. The following conclusions can be drawn from the above phenomenon: (1) Dynamic MVG can better control the bottom high pressure area, that is, reduce the influence of SWBLI; (2) With the increase of moving speed, the control effect will be more obvious.

4.2 *The Impact of Dynamic MVG on the Boundary Layer of the SWBLI Region*

The thickness of the boundary layer can further characterize the strength of SWBLI. Figure 6 shows the velocity distribution on the center line at X = 0.149 m (i.e.: SWBLI center area). It can be seen that: (1) when there is no MVG, the velocity of the bottom boundary layer is low, and the ability to resist the adverse pressure gradient is weak. At the same time, due to the influence of SWBLI, the thickness of the boundary layer is relatively large. When there is a static MVG, the velocity at the bottom of the boundary layer increases, while the thickness of the boundary layer decreases; (2) When the MVG moves, the bottom velocity of the boundary layer further increases, while the thickness of the boundary layer further decreases,

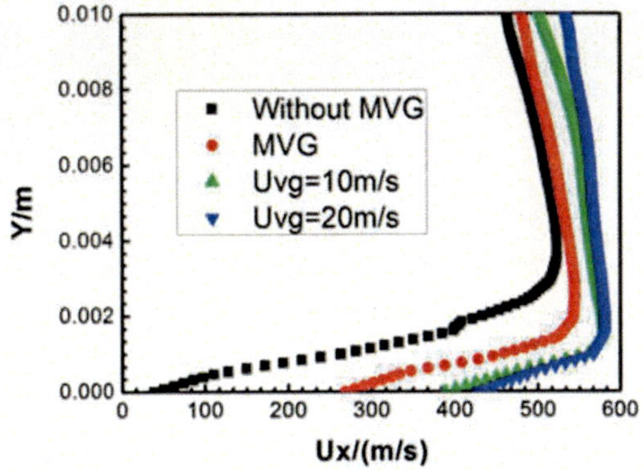

Fig. 6 The velocity distribution of the boundary layer (X = 0.149 m)

and the greater the velocity, the more obvious the above phenomenon. The above research shows that, compared with static MVG, dynamic MVG has better control effect on the boundary layer, and the higher the speed, the more obvious the control effect.

4.3 Control Mechanism of Dynamic MVG on SWBLI

Due to the effect of MVG in the supersonic flow field, a complex wave system structure and vortex structure are produced. In order to analyze the control mechanism of dynamic MVG on SWBLI, the numerical schlieren and vortex structure in the flow field are extracted. Figure 7 shows the wave system structure of the flow field under the action of MVG. It can be seen that (1) two compression waves are formed at the leading edge and trailing edge of the MVG, and the wedge-shaped surface forms an expansion wave, as shown in Fig. 7a. Two compression waves can penetrate the shock wave at the corner and affect SWBLI. (2) It can be seen from the numerical schlieren at the bottom that a complex wave system structure is formed at the trailing

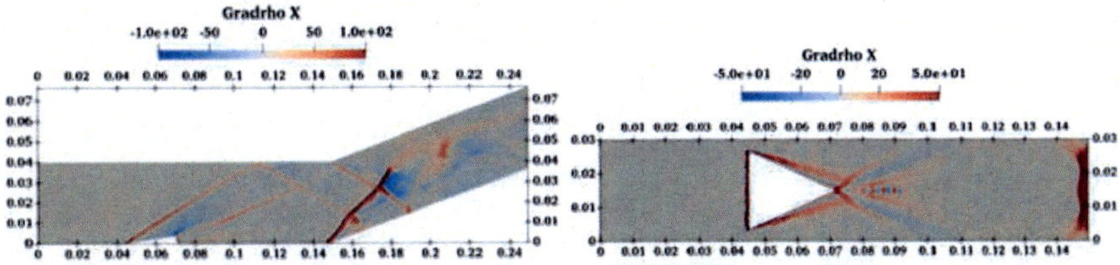

(a) (b)

Fig. 7 Wave structure of MVG

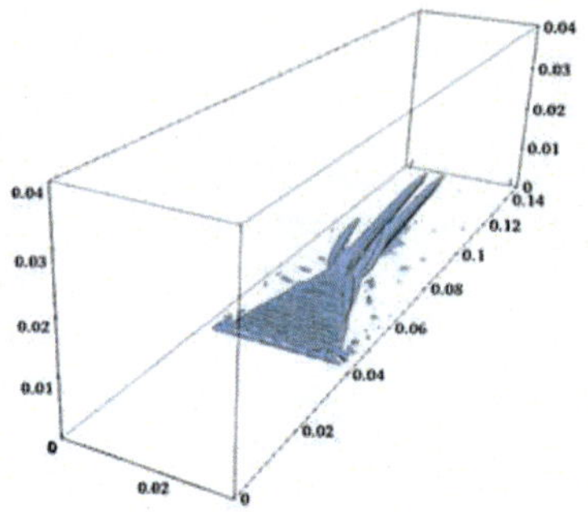

Fig. 8 Vortex structure of MVG

edge of the MVG. Compression waves and shock waves alternate with each other, as shown in Fig. 7b. The wave system structure at the tail interacts with SWBLI to change the shape of the bottom high pressure zone.

The vortex structure is characterized by the second order invariants of the velocity gradient (Q). Figure 8 shows the vortex structure near the MVG. It can be seen that there are two pairs of relatively strong streamwise vortex pairs at the trailing edge of MVG, of which one pair is longer and the other pair is higher. The streamwise vortex pair can promote the energy transport between the mainstream and the interior of the boundary layer, thus increasing the internal energy of the boundary layer and inhibiting flow separation. When the streamwise vortex pair interacts with SWBLI, the velocity of the bottom boundary layer will be increased and the thickness of the boundary layer will be reduced, so as to achieve the control effect.

Figure 9 shows the comparison of the vortex intensity in the SWBLI region when the MVG moves to the same position at different speeds. It can be seen that when the moving speed of the MVG increases, the vortex intensity in the SWBLI region increases. It shows that as the moving speed increases, the energy transport between the boundary layer and the mainstream becomes stronger, and SWBLI is better controlled.

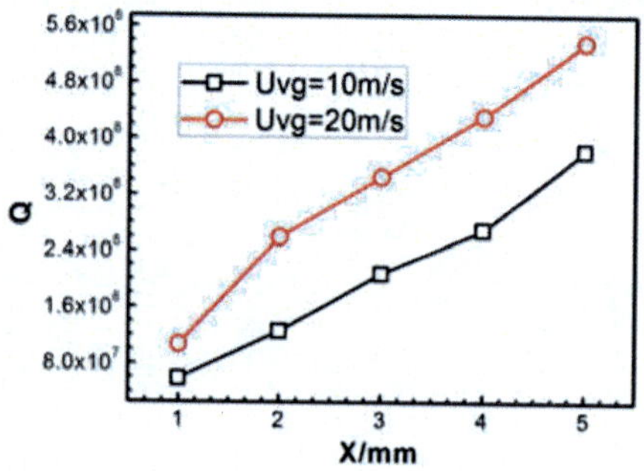

Fig. 9 Q varies with MVG moving speed

5 Conclusions

Based on dynamic mesh technology and DES model, this paper studies the influence of dynamic MVG on corner SWBLI. The focus is on the changes in the high-pressure zone and boundary layer in the SWBLI region. Combining the wave structure and vortex structure, the control mechanism of dynamic MVG on SWBLI is analyzed. The following conclusions are drawn:

(1) When the MVG moves downstream, the shape of the high-pressure zone in the SWBLI region will be changed, and will gradually weaken or even disappear. The thickness of the SWBLI boundary layer will gradually decrease. It shows that dynamic MVG has a good control effect on SWBLI.
(2) As the moving speed increases, the strength of the vortex structure formed by MVG will increase, and the control effect will be more obvious.
(3) MVG controls SWBLI through wave system structure and vortex structure.

References

1. Babinsky H, Harvey JK (2011) Shock wave-boundary layer interaction. Cambridge University Press, New York, pp 5–6
2. Gaitonde DV (2015). Progress in shock wave/boundary layer interactions. Prog Aerosp Sci 72(1):80–99
3. Délery J, Dussauge JP (2009) Some physical aspects of shock wave/boundary layer interactions. Shock Waves 2009(19):453–468
4. Gad-el-Hak M, Bushnell DM (1991) Separation control: review. J Fluids Eng 1991(113):5–30
5. Gad-el_Hak M (1994) Interactive control of turbulent boundary layers: a futuristic overview. AIAA J 32:1753–1765
6. Ashill PR, Fulker JL, Hackett KC (2005) A review of recent developemts in flow control. Aeronaut J 2005(109):205–232
7. Gursul I, Wang Z, Vardaki E (2007) Review of flow control mechanisms of leading-edge vortices. Prog Aerosp Sci 2007(43):246–270
8. Wu Y, Yi S-H, Chen Z et al (2013). Experimental investigations on structures of supersonic laminar/turbulent flow over a compression ramp. Acta Physica Sinica 62(18):308–319
9. Ganapathisubramani B, Clemens N, Dolling D (2007) Effects of upstream coherent structures on low-frequency motion of shock-induced turbulent separation. AIAA Paper 2007–1141:2007
10. Jian Z, Weiwei Z, Chuanqiang G et al (2014) Side force control on slender body by vortex generators. Chin J Theor Appl Mech 2014(46):308–312
11. Priebe S, Tu JH, Rowley CW et al (2016) Low- frequency dynamics in a shock-induced separated flow. J Fluid Mech 2016(807):441–477
12. Ghosh S, Choi J, Edwards JR (2008) RANS and hybrid LES/RANS simulations of the effects of micro vortex generators using immersed boundary methods. AIAA Paper 2008–3726:2008
13. Li Q, Li CQ (2010). LES for supersonic ramp control flow using MVG at M=2.5 and Re=1440. AIAA Paper 2010-592
14. Da-Wen XUE, Zhi-Hua C, Xiao-Hui SUN et al (2013) Investigations on the flow characteristics of supersonic flow past a micro-ramp. Eng Mech 30(4):455–460
15. Da-Wen XUE, Zhi-Hua C, Xiao-Hui SUN et al (2014) Micro-ramp control of the boundary separation induced by the flow past an airfoil. Eng Mech 31(8):217–222

16. Lin JC (2002) Review of research on low-profile vortex generators to control boundary-layer separation. Prog Aerosp Sci 2002(38):389–420
17. Babinsky H, Li Y, Pitt Ford CW (2009) Microramp control of supersonic oblique shock-wave boundary-layer interactions. AIAA J 2009(47):668–675
18. Hamstra JW, Miller DN, Truax PP, Anderson BH, Wendt BJ (2000) Active inlet flow control technology demonstration. ICAS Paper 2000-6.11.2
19. Lu FK, Li Q, Liu CQ (2012) Micro vortex generators in high-speed flow. Prog Aerosp Sci 2012(53):30–45
20. Mohammed-Taifour A, Weiss J (2016) Unsteadiness in a large turbulent separation bubble. J Fluid Mech 2016(799):383–412
21. Clemens NT, Narayanaswamy V (2014) Low- frequency unsteadiness of shock wave/turbulent boundary layer interactions. Annu Rev Fluid Mech 2014(46):469–492
22. Fulin T, Xinliang L, Zhigong T (2017) Numerical analysis of unsteady motion in shock wave/transitionalboundary layer interaction. Chin J Theor Appl Mech 2017(49):93–104
23. Wu M, Martn M P (2008). Analysis of shock motion in shockwave and turbulent boundary layer interaction using direct numerical simulation data. J Fluid Mech 594(1):71–83
24. Dolling DS (2001) Fifty years of shock-wave/boundary-layer interaction research: What Next?. AIAA J 39(8):1517–1531
25. Erengil ME (1993). Physical causes of separation shock unsteadiness in shock wave/turbulent boundary layer interactions. Texas University Press, Austin, pp 15–40
26. Dussauge JP, Dupont P, Debiève JF (2006). Unsteadiness in shock wave boundary layer interactions with separation. Aerosp Sci Technol 10(2):85–91
27. Kim K, Sung HJ (2003) Effect of cavitation on transverse injection into subsonic crossflows. AIAA Paper 2003–4788:2003
28. Hao G, Jiang N (2013) Evolution of periodic wall blow/suction disturbance in turbulent boundary layer. J Mech Strength 35(6):730–736
29. Qiang L, Zhen-Bing L, Xiong D et al (2017) Linear stability of supersonic boundary layer with synthetic cold/hot jet control. Acta Physica Sinica 66(23):234701–1-11
30. Greenshields CJ, Weller HG, Gasparini L, Reese JM (2010) Implementation of semi-discrete non-staggered central schemes in a colocated polyhedral finite volume framework for high-speed viscous flows. Int J Numer Methods Fluids 6(3):1–21
31. Kurganov A, Tadmor E (2000) New high-resolution central schemes for nonlinear conservation laws and convection–diffusion equations. J Comput Phys 2000(160):241–282
32. Johnsen E, Johan L, Ankit VB, William HC et al (2010) Assessment of high-resolution methods for numerical simulations of compressible turbulence with shock waves. J Comput Phys 2010(229):1213–1237
33. Strelets M (2001) Detached eddy simulation of massively separated flows. In: 39th aerospace sciences meeting and exhibit, Reno, NV, USA
34. Lee S, Loth E, Wang C (2007) LES of supersonic turbulent boundary layers with [mu] VG's. AIAA Paper 2007–3916:2007

Aerodynamic Characteristics of A Compound Deflected LEF/TEF Rotor

Hualong Wang, Xiayang Zhang, Guoqing Zhao, and Qijun Zhao

Abstract In order to analyze the comprehensive influence of a compound leading edge flap (LEF) and trailing edge flap (TEF) on the rotor aerodynamic characteristics of a 2-D airfoil and a 3-D rotor, a grid generation method including the LEF/TEF is developed. Based on the Navier–Stokes equation, a high-precision analysis method of aerodynamic characteristics under compound deflected LEF/TEF is established. Through a numerical analysis, variations of the aerodynamic characteristics under different LEF/TEF deflection states are studied and the effects of compound flap deflections on the leading-edge vortex and shock wave in 2-D and 3-D cases are investigated thoroughly. The results indicate that the LEF deflects downward and the range of the leading-edge vortex decreases; the pressure distribution on the leading edge of airfoil with differently deflected TEF is nearly identical at a large Mach number. The effect of the compound deflected LEF/TEF on the aerodynamic characteristics is approximately equivalent to the superposition of the separate deflection effects of the LEF/TEF.

Keywords Deflected LEF/TEF · Aerodynamic characteristics · Shock wave · Leading-edge vortex

1 Introduction

The rotor is the main vibration source in a helicopter, and reducing its aerodynamic excitation level has been one of the core and most difficult points in helicopter design. At present, smart rotor technology is a research hotspot in the field of helicopter vibration reduction [1]. Among these components, the leading edge flap (LEF) and the trailing edge flap (TEF) are two smart rotor technologies with great application potential [2].

Flaps were initially used on fixed-wing aircraft and later used on wind turbines and helicopter rotors [3]. Vladimir [4] points out that the TEF can change the local

---

H. Wang · X. Zhang · G. Zhao · Q. Zhao (✉)
National Key Laboratory of Rotorcraft Aeromechanics, Nanjing University of Aeronautics and Astronautics, Nanjing 210016, China
e-mail: zhaoqijun@nuaa.edu.cn

S. Lee et al. (eds.), *The Proceedings of the 2021 Asia-Pacific International Symposium on Aerospace Technology (APISAT 2021), Volume 1*, Lecture Notes in Electrical Engineering 912, https://doi.org/10.1007/978-981-19-2689-1_9

(a) wind turbine (b) rotor blade (c) fixed wing

Fig. 1 Examples of how flaps are used in different fields

aerodynamic load on the blade, and the LEF can offset the additional pitching moment caused by the deflection of the TEF. Junaid et al. [5] posit that the TEF can suppress gust-induced wing bending moment, while the LEF can compensate for wing torque caused by the deflection of the TEF (Fig. 1).

In the field of smart rotor technology, there have been many studies of the TEF. Chan [6] verifies that the TEF plays an important role in suppressing retreating blade stall and improving the flight envelope. Friedmann [7] holds that using a vTEF to reduce the 4/rev vibration load is effective for a helicopter with four blades. Shen [8] illustrates that excessive flap aerodynamic over-balancing and mass unbalancing are the main causes of flap instability, while Viswamurthy [9] and Dalli [10] propose that under an identical damping effect, using multiple TEFs is better than a single flap configuration. Lim [11] claims that with an increase in the rotational speed, active control of the TEF has a significant effect on the vibration reduction of the rotor hub. Dehaeze [12] finds that the TEF is beneficial for improving the aerodynamic performance of a rotor under the condition of high thrust during forward flight, while Ma [13] proposes that the TEF can better suppress the airfoil dynamic stall phenomenon under certain operating states. Further research [14] indicates that the TEF can give full play to the lift potential of the rotor on the advancing side and can reduce the drag and torque of the blade during a dynamic stall on the retreating side.

In contrast, there are few studies of the LEF. Jain [15] finds that aerodynamic performance improvements of the LEF are more effective than those of the TEF and active twist under the condition of high thrust. Zhao [16] illustrates that the variable-droop leading edge can significantly suppress the dynamic stall vortex, thereby alleviating the rotor dynamic stall phenomenon.

The TEF has a significant effect on vibration and noise reduction of a helicopter rotor, and the LEF can significantly suppress the dynamic stall. It has been demonstrated that in the case of a non-rotating wing, the shortcomings of using the LEF or TEF alone can be effectively resolved by the combined LEF/TEF technology; meanwhile, the advantages of the LEF and TEF are retained. In contrast, there is a lack of research achievements on LEF/TEF combination technology in relation to these rotors. Given this background, the present paper aims to study the aerodynamic characteristics of a rotor with a compound deflected LEF/TEF configuration and to reveal the coupling effects on the aerodynamic performance and vortex flow field. The conclusions also establish a theoretical basis for practical rotor deflection LEF/TEF technologies.

2 Numerical Method

2.1 Grid Generation Method

To achieve the deflection of the LEF/TEF, the CST method [17] is used to construct the surface points of the airfoil. First, the position of the pivot point is fixed, and the points in the deflection part are deflected around the pivot point to a given angle. After the deflection process, the CST method is used to update and reconstruct the surface points of the airfoil so as to avoid dense or sparse local points in the deflection part. Finally, the appropriate distribution of the surface points of the airfoil is obtained, as shown in Fig. 2. At the same time, when the LEF deflects upward, the deflection angle of the LEF is defined as negative; when the TEF deflects upward, the deflection angle of the TEF is defined as positive.

After obtaining the distribution of the 2-D airfoil surface points, the initial mesh generation method [18] is used to generate the initial mesh of the airfoil with the LEF/TEF, after which the Poisson equation mesh generation method is used to correct the mesh quality. Based on the two-dimensional airfoil mesh generation process, the Lagrange interpolation method is used to interpolate the grid of the blade section. At

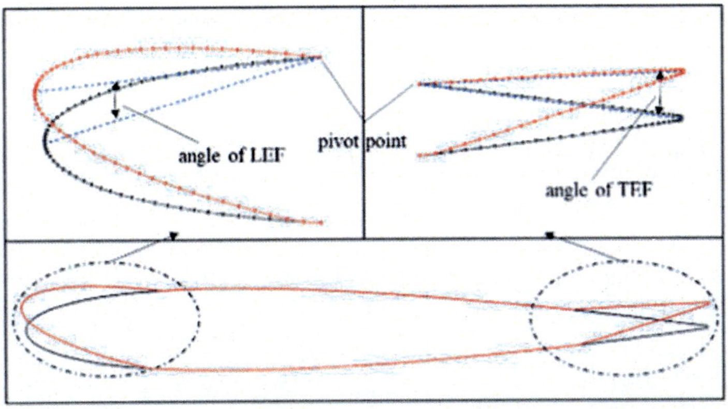

Fig. 2 Surface point reconstruction of an airfoil

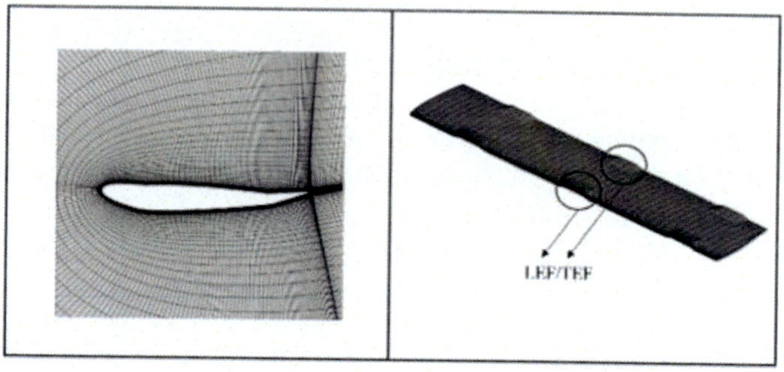

Fig. 3 Body-fitted grid around an airfoil and rotor blade

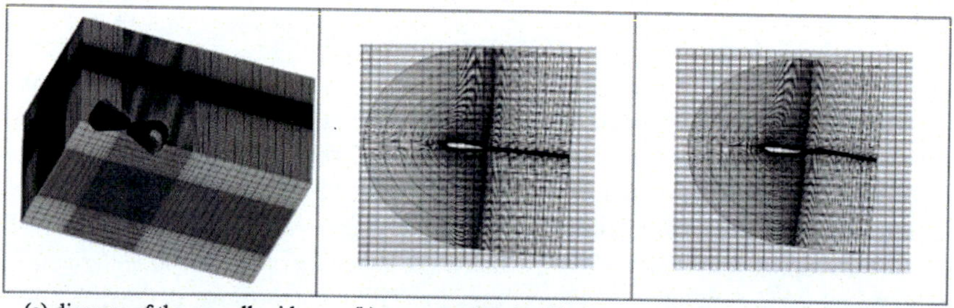

(a) diagram of the overall grid (b) cross-sectional grid of the normal part (c) cross-sectional grid of the deflected part

Fig. 4 Moving-embedded grid system

the blade tip and blade root, the blade spanwise O-type grid folding method is used. Figure 3 presents the airfoil surface grid and blade surface grid with the LEF/TEF.

The background grid is a rectangular Cartesian grid. For an accurate simulation of the flow field and vortex near the blade, the background grid near the blade is densified. The final moving-embedded grid system is shown in Fig. 4.

2.2 Aerodynamic Analysis Method

The CLORNS code for CFD simulation of helicopter rotor [19] is used to calculate the flow field. Compressible Navier–Stokes equations are used to simulate a 3-D rotor in a hovering condition.

$$\frac{\partial}{\partial t}\iiint_V \vec{W}dV + \iint_S \left(\vec{F}_c - \vec{F}_v\right)\cdot\vec{n}dS + \iint_S \vec{\Omega}\cdot\vec{n}WdS = \iiint_V \vec{Q}dV$$

Here, $\vec{W}$ represents the conserved variables, $\vec{F}_c$ and $\vec{F}_v$ are the convective and viscous fluxes, $\vec{n}$ is the unit vector of the control volume, $\vec{\Omega} = u_\Omega \vec{i}_x + v_\Omega \vec{i}_y + w_\Omega \vec{i}_z$ is the rotor speed, $\vec{Q} = \left[0 \; -\rho\Omega w \; 0 \; \rho\Omega u \; 0\right]^T$, V is the volume of the grid cell, S is the boundary of the grid cell, and (u, v, w) and $(u_\Omega, v_\Omega, w_\Omega)$ are correspondingly the absolute speed and rotating speed of rotor blade in the rotating reference frame.

The Roe-MUSCL scheme is used to discretize the convective terms and the Spalart–Allmaras turbulence model is used for all simulations. Time marching is performed using an implicit LU-SGS scheme.

3 Results and Discussion

In order to verify the effectiveness of the 2-D aerodynamic analysis method, numerical simulations of an NACA0012 airfoil at Ma = 0.7, $\alpha = 6.0°$ (α is the angle of attack of the airfoil), and Re $= 1.6 \times 10^7$ are carried out, with the results compared with the experimental results, as shown in Fig. 5 [20]. It can be seen that the calculated surface pressure distribution of the airfoil is in good agreement with the experimental value and that the calculation method can effectively simulate the shock wave of an airfoil.

In order to verify the effectiveness of the 3-D aerodynamic analysis method, the calculated surface pressure distribution of the C-T rotor (for a model consisting of two rectangular blades configured as an NACA0012 airfoil with no pretwist, no precone, and an aspect ratio of 6) at 0.80R (R is the radius of the blade) and 0.96R is compared with the experimental value [21] under $Ma_{tip} = 0.439$ and $\theta = 8°$(θ is the collective pitch of the rotor), the results are found to be in good agreement (Fig. 6), indicating that the numerical method in this study can effectively simulate the aerodynamic characteristics of the blade.

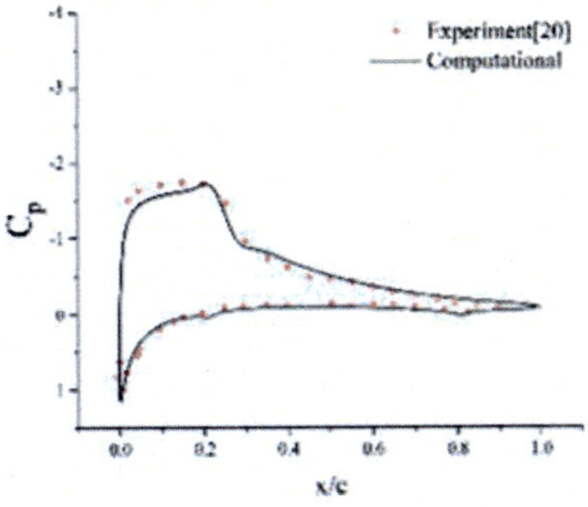

Fig. 5 Verifications of pressure coefficients of the NACA0012 airfoil

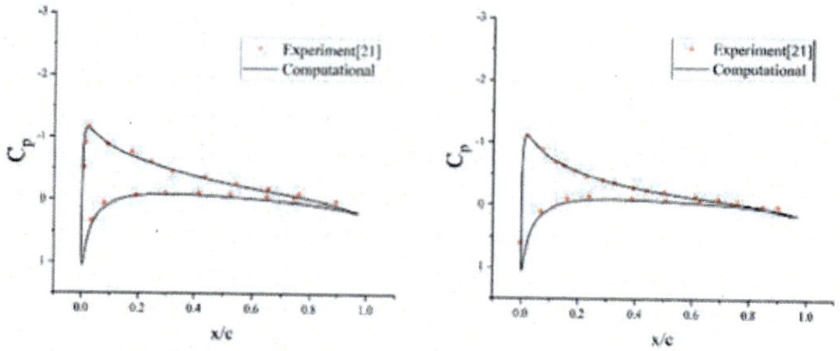

(a) r=0.80R (b) r=0.96R

Fig. 6 Verifications of pressure coefficients in different C-T rotor blade profiles

3.1 Aerodynamic Characteristics of an Airfoil with the LEF/TEF

The airfoil used is the NACA0012 airfoil, and the LEF/TEF is a plain flap with a length of 20% c (c is the chord length of the airfoil). To explore the aerodynamic characteristics of the 2-D airfoil, this part simulates the aerodynamic characteristics of the airfoil under different conditions of $Ma_\infty = 0.3$, angle of attack of $-5° \sim 17.5°$, and LEF/TEF deflection angles of 0° and ± 10°. These results are shown in Fig. 7.

According to the different deflection states of the LEF, it is found that when the LEF deflects downward, C_L changes slightly and C_m decreases; when the LEF deflects upward, C_L changes slightly and C_m increases. These outcomes occur because when the LEF deflects downward, although the increase in the airfoil camber increases C_L, the decrease in the angle of attack on the leading edge of the airfoil will effectively reduce C_L such the comprehensive change of C_L is minor. At the same time, the LEF deflects downward, and the pressure difference between the upper and lower surfaces of the leading edge of the airfoil decreases (Fig. 8a), equivalent to providing nose-down pitching moment for the airfoil, after which the C_m will decrease.

According to the different deflection states of the TEF, it is found that when the TEF deflects downward, C_L increases and C_m decreases; when the TEF deflects

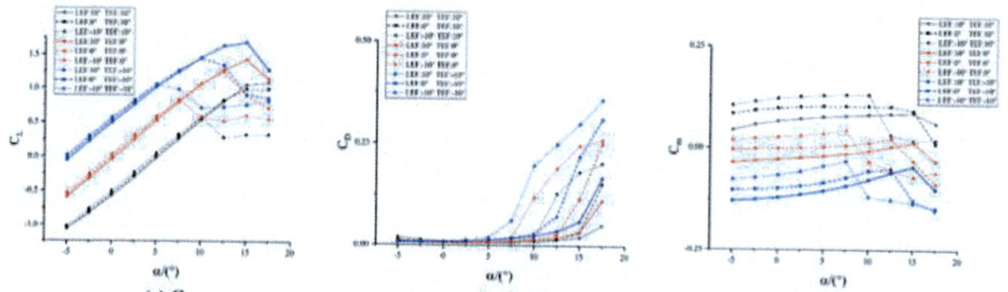

(a) C_L (b) C_D (c) C_m

Fig. 7 Aerodynamic coefficients of an airfoil with a deflected LEF/TEF at $Ma_\infty = 0.3$

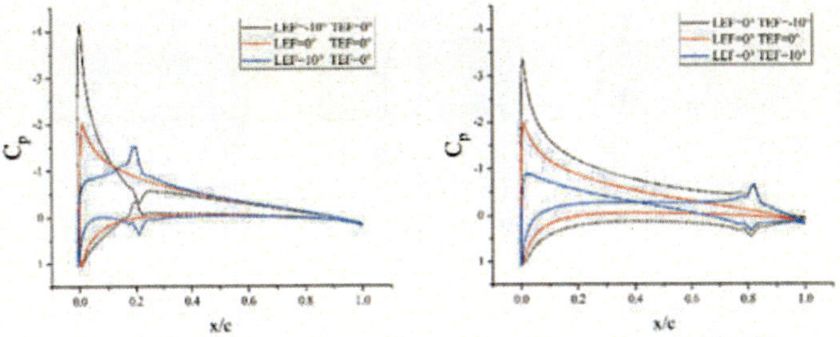

(a) deflected LEF (b) deflected TEF

Fig. 8 Pressure coefficient distributions of an airfoil with a deflected LEF/TEF at $Ma_{\infty} = 0.3$ and $\alpha = 5°$

upward, C_L decreases and C_m increases. This occurs because when the TEF deflects downward, the increase in the camber of the airfoil increases C_L. At the same time, when the TEF deflects downward, the pressure difference between the upper and lower surfaces of the airfoil increases. Although the pressure difference from the aerodynamic center to the leading edge provides the nose-up pitching moment, the pressure difference from the aerodynamic center to the trailing edge provides a larger nose-down pitching moment, which is equivalent to providing nose-down pitching moment for the airfoil. Subsequently, C_m will decrease (Fig. 8b).

As shown in Fig. 8, some pressure fluctuation occurs at the connection position between the deflected flap and the airfoil. This arises because when the flap is deflected, a local convex (concave) surface will be formed at the connection position. When the airflow flows through the convex (concave) surface, the velocity will increase (decrease) and the pressure coefficient will decrease (increase).

Figure 9 presents the influences of flap deflection on the airfoil stall characteristics when $\alpha = 15°$.

According to the different deflection states of the LEF, it is found that when the LEF deflects downward, the stall angle increases, C_{Lmax} increases and C_D decreases; when the LEF deflects upward, the stall angle decreases, C_{Lmax} decreases and C_D increases (Fig. 7). These outcomes occur because when the LEF deflects downward, the detached vortex that will be generated at the leading edge of the airfoil disappears (Fig. 9a–c). At the same time, when the LEF deflects downward, the camber of the airfoil increases. At a large angle of attack, the face area of the airfoil decreases and the pressure drag decreases.

According to the different deflection states of the TEF, it is found that when the TEF deflects downward, the stall angle decreases, C_{Lmax} increases and C_D increases; when the TEF deflects upward, the stall angle increases, C_{Lmax} decreases and C_D decreases (Fig. 7). This occurs because when the TEF deflects downward, the range of the detached vortex generated at the leading edge of the airfoil increases (Fig. 9d, b, e). At the same time, when the TEF deflects downward, the camber of the airfoil increases, the face area of the airfoil increases, and the pressure drag increases.

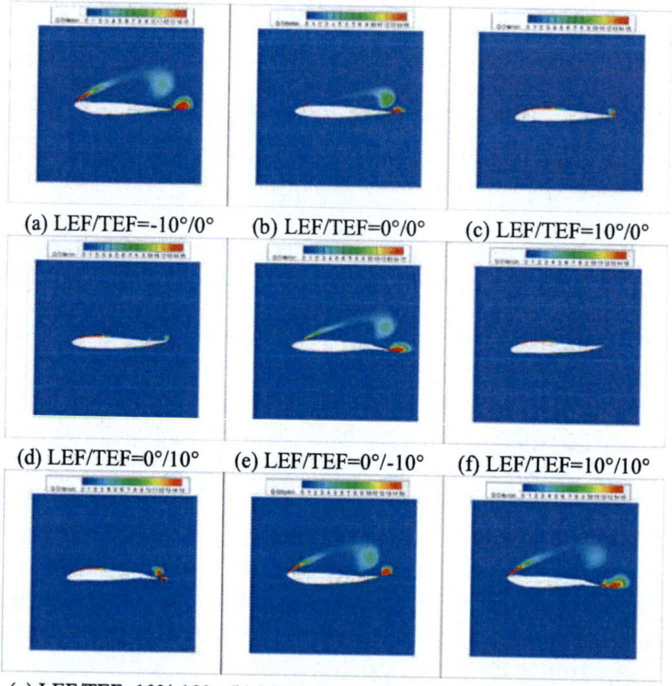
(a) LEF/TEF=-10°/0° (b) LEF/TEF=0°/0° (c) LEF/TEF=10°/0°
(d) LEF/TEF=0°/10° (e) LEF/TEF=0°/-10° (f) LEF/TEF=10°/10°
(g) LEF/TEF=10°/-10° (h) LEF/TEF=-10°/10° (i) LEF/TEF=-10°/-10°

Fig. 9 Vorticity contours of airfoil with a deflected LEF/TEF at $Ma_\infty = 0.3$ and $\alpha = 15°$

According to the different combined deflection states of the LEF/TEF, it is found that the effect of the combined deflection of the LEF/TEF on the stall characteristics of the airfoil in the 2-D state is approximately equivalent to the superposition of the separate deflection effects of the LEF/TEF: the LEF deflects downward, the stall angle increases, C_{Lmax} increases, C_D decreases and the range of the detached vortex decreases. In addition, the TEF deflects downward, the stall angle decreases, C_{Lmax} increases, C_D increases and the range of the detached vortex increases.

Figure 10 presents the influences of flap deflection on the shock wave when $Ma_\infty = 0.8$.

According to the different deflection states of the LEF, it is found that when the LEF deflects downward, the shock wave moves slightly to the leading edge of the airfoil and its intensity is nearly identical to the original configuration; when the LEF deflects upward, the shock wave moves to the leading edge of the airfoil and its intensity increases greatly.

According to the different deflection states of the TEF, it is found that the pressure distributions on the leading edges of airfoils with differently deflected TEFs are nearly identical at a large Mach number. The difference is that when the TEF deflects downward, the shock wave moves to the trailing edge of the airfoil.

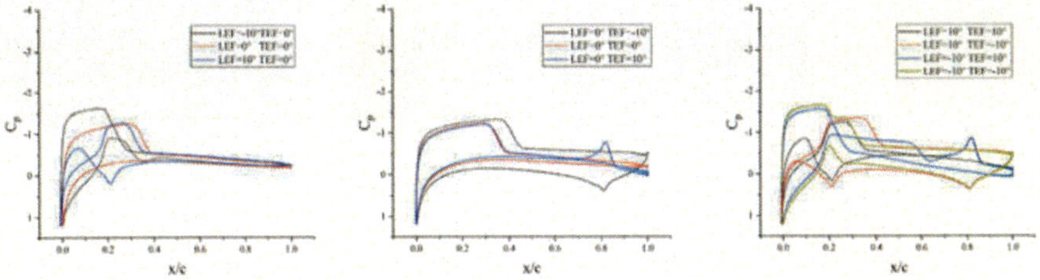

(a) deflected LEF (b) deflected TEF (c) deflected LEF/TEF

Fig. 10 Pressure coefficient distribution of an airfoil with a deflected LEF/TEF at $Ma_\infty = 0.8$ and $\alpha = 5°$

According to the different combined deflection states of the LEF/TEF, it is found that the effect of the combined deflection of the LEF/TEF on the shock wave characteristics of the airfoil in the 2-D state is nearly equivalent to the superposition of the separate deflection effects of the LEF/TEF: the LEF deflects upward, the shock wave moves to the leading edge of the airfoil and its intensity increases; the TEF deflects downward, the shock wave moves to the trailing edge of the airfoil, and the pressure distributions on the leading edges of airfoils with differently deflected TEF are nearly identical at a large Mach number.

3.2 *Aerodynamic Characteristics of a Rotor with a LEF/TEF When Hovering*

In this study, a C-T rotor with a rotor tip speed of $Ma_{tip} = 0.877$ is used, and a LEF/TEF with a radial length of 0.07R is set such that r = 0.34R, 0.63R and 0.91R [9, 10] (corresponding to relative inflow velocities of Ma = 0.3, Ma = 0.55 and Ma = 0.8, respectively). The chordwise distribution is identical to that of the 2-D part.

In order to determine the aerodynamic characteristics of a rotor with a deflected LEF/TEF in hovering motion, the aerodynamic characteristics of the rotor are simulated at collective pitch angles of 5° and 15°, with the deflection angles of the LEF/TEF set to 0° and ± 10°. These results are shown in Table 1.

At a smaller collective pitch angle, the TEF deflects downward, C_T increases, and m_K increases; the TEF deflects upward, C_T decreases, and m_K decreases. The LEF deflects downward, C_T decreases slightly, and m_K increases; the LEF deflects upward, C_T increases slightly, and m_K increases. These outcomes arise because at a smaller collective pitch angle, nearly all sections of the blade are in an unstall state, and the unstall aerodynamic law dominates the entire blade. When the TEF deflects downward, the camber of the airfoil increases, the lift increases, and C_T increases. At the same time, the face area of the airfoil increases, the pressure drag increases, and m_K then increases. The LEF deflects downward, although the reduced angle of attack reduces the airfoil lift coefficient, the increased camber will weaken the effect of the

Table 1 Rotor aerodynamic data under different deflected LEF/TEF combinations in a hovering state

	$\theta = 5°$				$\theta = 15°$			
	C_T	m_K	$\Delta C_T/C_{T0}$ (%)	$\Delta m_K/m_{K0}$ (%)	C_T	m_K	$\Delta C_T/C_{T0}$ (%)	$\Delta m_K/m_{K0}$ (%)
LEF = 10° TEF = 10°	0.00194	3.68E-04	−30.22	−1.42	0.01136	0.00201	−9.48	−13.73
LEF = 0° TEF = 10°	0.00204	3.35E-04	−26.62	−10.14	0.01150	0.00209	−8.37	−10.30
LEF = -10° TEF = 10°	0.00212	3.88E-04	−23.74	3.91	0.01131	0.00223	−9.88	−4.29
LEF = 10° TEF = 0°	0.00266	4.06E-04	−4.32	8.79	0.01236	0.00225	−1.51	−3.43
LEF = 0° TEF = 0°	0.00278	3.73E-04	0.00	0.00	0.01255	0.00233	0.00	0.00
LEF = -10° TEF = 0°	0.00282	4.23E-04	1.44	13.48	0.01235	0.00248	−1.59	6.44
LEF = 10° TEF = -10°	0.00408	5.47E-04	46.76	46.51	0.01384	0.00257	10.28	10.30
LEF = 0° TEF = -10°	0.00419	5.16E-04	50.72	38.21	0.01402	0.00265	11.71	13.73
LEF = -10° TEF = -10°	0.00424	5.65E-04	52.52	51.36	0.01383	0.00280	10.20	20.17

decreases in the airfoil lift coefficient such that C_T overall shows a small decrease. At the same time, with a small collective pitch angle, the deflected LEF will cause the face area to increase; the pressure drag increases and m_K then increases as well.

At a larger collective pitch angle, the TEF deflects downward, C_T increases, and m_K increases; the TEF deflects upward, C_T decreases, and m_K decreases. The LEF deflects downward, C_T decreases slightly, and m_K decreases; the LEF deflects upward, C_T decreases slightly, and m_K increases. This occurs because when the TEF deflects downward, the camber of the airfoil increases, the lift continues to increase, and C_T increases. At the same time, the face area of the airfoil increases, the pressure drag increases, and m_K then increases. The LEF deflects upward, the angle of attack near the blade tip increases, and the airflow separation becomes more serious, thus reducing the aerodynamic force of the profile. When the LEF deflects downward, the angle of attack of the flap near the blade root decreases, and a negative angle of attack can even occur, with the aerodynamic force of the section also decreasing. Therefore, the deflection of the LEF will reduce C_T. At the same time, when the collective pitch angle is large, the deflection of the LEF will reduce the face area of the airfoil, reduce the pressure drag, and then reduce m_K.

In order to ascertain the influence of flap deflection on the aerodynamic characteristics, the previous results at $\theta = 15°$ are selected and comparisons of the aerodynamic characteristics of rotors with different control strategies are shown in Figs. 11, 12 and 13.

In Fig. 11, the LEF deflects downward and the pressure difference at the leading edge of the profile decreases. In addition, the TEF deflects downward and the pressure difference at the trailing edge of the profile increases.

As shown in Figs. 12 and 13, according to the different deflection states of the LEF, it is found that when the LEF deflects downward, C_n undergoes minor fluctuation and C_m decreases slightly; when the LEF deflects upward, C_n also undergoes minor fluctuation and C_m increases slightly. These outcomes occur because when the LEF deflects downward, despite the increasing camber of the airfoil increasing the lift coefficient of the airfoil, the decreased angle of attack will weaken the effect of this increase, causing C_n to undergo a small amount of fluctuation. At the same time, the

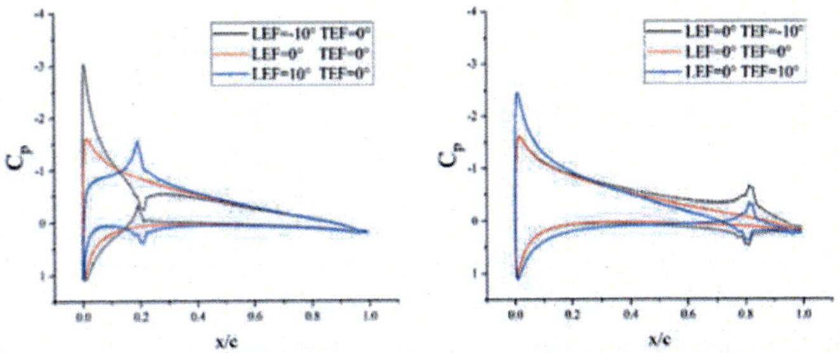

(a) deflected LEF (b) deflected TEF

Fig. 11 Pressure coefficient distribution of a blade with a deflected LEF/TEF at r = 0.34 R

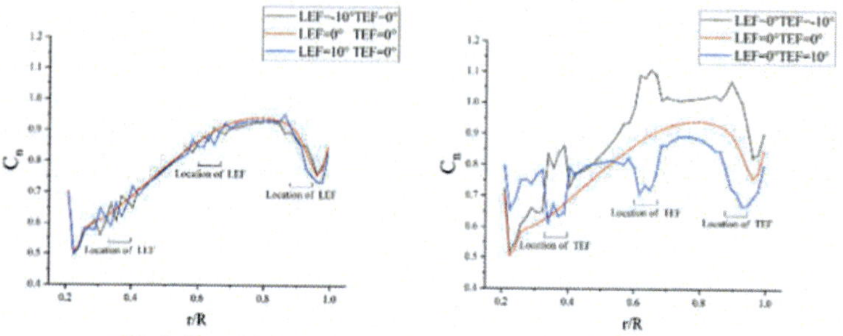

(a) deflected LEF (b) deflected TEF

Fig. 12 Normal force coefficient at r = 0.34 R of a blade

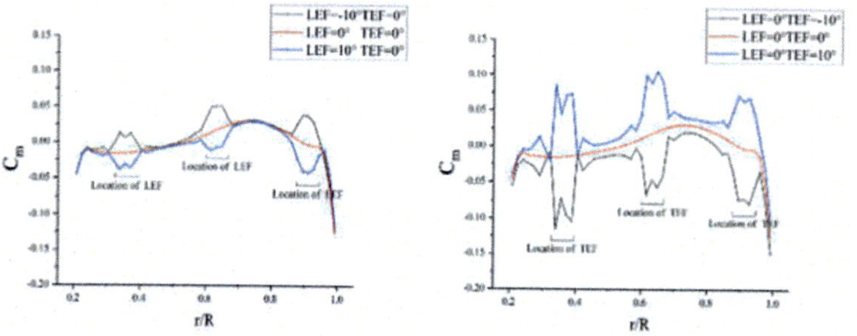

(a) deflected LEF (b) deflected TEF

Fig. 13 Moment coefficient at r = 0.34 R for a blade

pressure difference at the leading edge of the profile decreases, which is equivalent to providing nose-down pitching moment for the blade profile, causing C_m to decrease slightly.

According to the different deflection states of the TEF, it is found that when the TEF deflects downward, C_n increases greatly and C_m decreases greatly; when the TEF deflects upward, C_n decreases greatly and C_m increases greatly. These outcomes arise because the downward deflection of the TEF increases the camber of the airfoil, with the lift of the profile increasing strongly such that C_n increases greatly. At the same time, the downward deflection of the TEF will increase the pressure difference between the upper and lower surfaces near the trailing edge of the airfoil, which is equivalent to providing strong nose-down pitching moment so that C_m is greatly reduced. (C_n denotes the normal force coefficient and C_m is the moment coefficient).

Figure 14 presents the influences of flap deflection on the vorticity of the blade section when the collective pitch of the rotor is $\theta = 20°$:

According to the different deflection states of the LEF, it is found that when the LEF deflects downward, the range of the leading-edge vortex decreases; when the LEF deflects upward, the range of the leading-edge vortex increases.

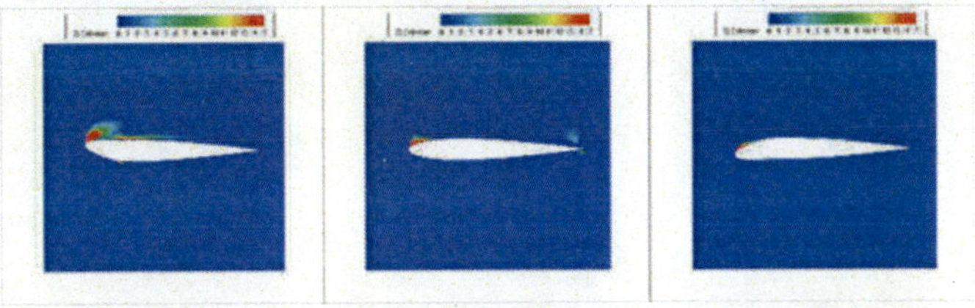

(a) LEF/TEF=-10°/0° (b) LEF/TEF=0°/0° (c) LEF/TEF=10°/0°

(d) LEF/TEF=0°/10° (e) LEF/TEF=0°/-10° (f) LEF/TEF=10°/10°

(g) LEF/TEF=10°/-10° (h) LEF/TEF=-10°/10° (i) LEF/TEF=-10°/-10°

Fig. 14 Vorticity contours of a blade with a deflected LEF/TEF at r = 0.91R and $\theta = 20°$

According to the different deflection states of the TEF, it is found that the deflection of the TEF has little effect on the leading-edge vortex.

According to the different combined deflection states of the LEF/TEF, it is found that the effect of the combined deflection of the LEF/TEF on the leading-edge vortex of the blade section in the 3-D state is equivalent to the superposition of the separate deflection effects of the LEF/TEF: the LEF deflects downward and the range of the leading-edge vortex decreases; the LEF deflects upward and the range of the leading-edge vortex increases. The deflection of the TEF has little effect on the leading-edge vortex.

Figure 15 presents the effect of flap deflection on the shock wave when the collective pitch of the rotor is $\theta = 5°$:

According to the different deflection states of the LEF, it is found that when the LEF deflects downward, the shock wave moves to the trailing edge and its intensity is slightly improved; when the LEF deflects upward, the shock wave moves to the leading edge and its intensity is greatly improved.

According to the different deflection states of the TEF, it is found that the pressure distributions on the leading edges of airfoils with differently deflected TEFs are approximately the same at a large Mach number. The difference arises when the TEF deflects downward, in which case the shock wave moves to the trailing edge.

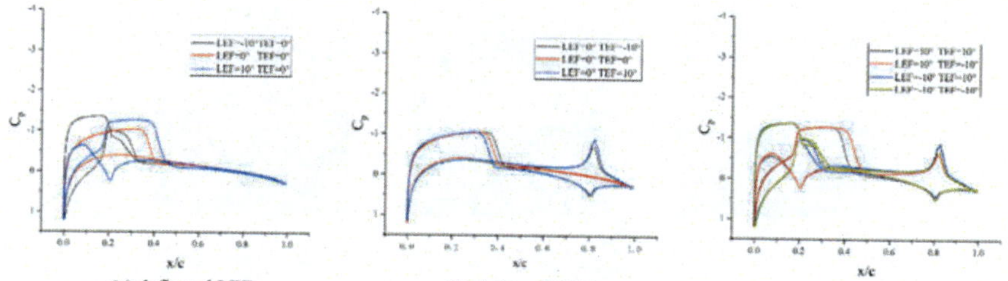

(a) deflected LEF (b) deflected TEF (c) deflected LEF/TEF

Fig. 15 Pressure coefficient distribution of a blade with a deflected LEF/TEF at r = 0.91R and $\theta = 5°$

According to the different combined deflection states of the LEF/TEF, it is found that the effect of combined deflections of the LEF/TEF on the shock wave in the 3-D state is approximately equivalent to the superposition of the separate deflection effects of the LEF/TEF: the LEF deflects upward, the shock wave moves to the leading edge of the airfoil and its intensity increases greatly; the TEF deflects downward, the shock wave moves to trailing edge of airfoil, and the pressure distributions on the leading edges of airfoils with different deflected TEFs are approximately the same at a large Mach number.

4 Conclusion

In this study, the effects of the deflected LEF/TEF on the aerodynamic performance of a rotor and a rotor airfoil are calculated under static conditions. The final conclusions are as follows:

(1) Aerodynamic characteristics: The effect of the compound deflected LEF/TEF on the aerodynamic characteristics is approximately equivalent to the superposition of the separate deflection effects of the LEF/TEF. In the 2-D state, the deflection of the LEF has little effect on C_L, and downward deflection of the TEF can increase C_L. In the 3-D state, the deflection of the LEF has little effect on C_T, and downward deflection of the TEF can increase C_T.

(2) Leading-edge vortex: The effect of the compound deflected LEF/TEF on the leading-edge vortex is approximately equivalent to the superposition of the separate deflection effects of the LEF/TEF: In the 2-D state, the LEF deflects downward and the range of the leading-edge vortex decreases; the TEF deflects upward and the range of the leading-edge vortex decreases. In the 3-D state, the LEF deflects downward and the range of leading-edge vortex decreases, but the deflection of the TEF has little effect on the leading-edge vortex.

(3) Shock wave: The effect of the compound deflected LEF/TEF on the shock wave is approximately equivalent to the superposition of the separate deflection effects of the LEF/TEF. In the 2-D state, the LEF deflects downward and

the shock wave moves slightly to the leading edge and its intensity becomes nearly identical to that in the original configuration; moreover, the pressure distributions on the leading edges of airfoils with different deflected TEFs are approximately the same at a large Mach number. In the 3-D state, the LEF deflects downward and the shock wave moves to the trailing edge and its intensity is slightly improved, with the deflection effect of the TEF identical to that in the 2-D state.

Acknowledgements This work is supported by the National Natural Science Foundation of China (Nos. 12072156, 12032012 and 12102186), the National Key Laboratory Foundation of China (No. 61422200101), the Natural Science Foundation of Jiangsu Province (No. BK20200433), and by the Priority Academic Program Development of Jiangsu Higher Education Institutions.

References

1. Li CM (2001) Research on vibration control for smart rotor. NUAA
2. Millott TA, Friedmann PP (1994) Vibration reduction in helicopter rotors using an actively controlled partial span trailing edge flap located on the blade. NASA CR-1994-4611
3. Ni XP (2008) Development prospect of future helicopter technology. Aeronaut Manuf Technol 3:32–37
4. Leble V, Barakos GN (2016) Trailing and leading edge flaps for load alleviation and structure control. Springer
5. Ullah J, Lutz T, Klug L et al (2021) Active gust load alleviation by combined actuation of trailing edge and leading edge flap at transonic speeds. AIAA-2021-1831
6. Chan W, Brocklehurst A (2001) Performance enhancement evaluation of an actuated trailing edge flap. Aeronaut J 105(1049):391–399
7. Friedmann PP, Terlizzi MD, Myrtle TF (2001) New developments in vibration reduction with actively controlled trailing edge flaps. Math Comput Model 33(10–11):1055–1083
8. Shen J, Chopra I (2003) Aeroelastic stability of trailing-edge flap helicopter rotors. J Am Helicopter Soc 48(4):236–243(8)
9. Viswamurthy SR, Ganguli R (2004) An optimization approach to vibration reduction in helicopter rotors with multiple active trailing edge flaps. Aerosp Sci Technol 8(3):185–194
10. Dalli U, Yiiksel S (2011) Identification of flap motion parameters for vibration reduction in helicopter rotors with multiple active trailing edge flaps. Shock Vib 18(5):727–745
11. Lim IG, Lee I (2009) Aeroelastic analysis of rotor systems using trailing edge flaps. J Sound Vib 321(3–5):525–536
12. Dehaeze F, Baverstock K, Barakos G (2015) CFD simulation of flapped rotors. Aeronaut J 119(1222):1561–1583
13. Ma YY, Zhao QJ, Zhao GQ (2017) Dynamic stall control of rotor airfoil via trailing-edge flap. Acta Aeronautica et Astromautica Sinica 38(3):127–137 (in Chinese)
14. Ma YY, ZHAO Q J, (2018) Control mechanism and parameter analyses of aerodynamic characteristics of rotor via trailing-edge flap. Acta Aeronautica et Astromautica Sinica 39(5):14–27 (in Chinese)
15. Jain R, Yeo H, Chopra I (2010) Computational fluid dynamics—computational structural dynamics analysis of active control of helicopter rotor for performance improvement. J Am Helicopter Soc 55(4):42004-1–42004-14
16. Zhao GQ, Zhao QJ (2015) Dynamic stall control optimization of rotor airfoil via variable droop leading-edge. Aerosp Sci Technol 43:406–414

17. Guan XH, Li ZK, Song BF (2012) A study on CST aerodynamic shape parameterization method. Acta Aeronautica et Astromautica Sinica 33(4):625–633 (in Chinese)
18. Zhao QJ, Xu GH (2016) Foundations of helicopter computational fluid dynamics. Science Press
19. Zhao QJ, Zhao GQ, Wang B et al (2018) Robust Navier-Stokes method for predicting unsteady flowfield and aerodynamic characteristics of helicopter rotor. Chin J Aeronaut 31(2):214–224
20. Han BZ, Huang YY, Zhang QW et al (1987) An experiment of pressure measurement for NACA0012 airfoil in a transonic wind tunnel. J Nanjing Aeronaut Inst 19(2):97–107
21. Caradonna FX, Tung C (1981) Experimental and analytical studies of a model helicopter rotor in hover. NASA TM-1981-81232

Analysis of Aerodynamic Characteristics and Influence of Parameters of the Quad Tiltrotor Aircraft

Jinshuai Shi, Muyang Lin, Xiayang Zhang, Qijun Zhao, and Guoqing Zhao

Abstract In this work, computational fluid dynamics (CFD) and momentum source methods are combined to perform aerodynamic characteristic analysis and to determine the influence of aerodynamic layout parameters on the design of a quad tiltrotor aircraft. First, a numerical method is established to simulate the whole-aircraft interference flow field of a quad tiltrotor aircraft. Then, the aerodynamic characteristics of the rear rotor under various interferences are analyzed based on the established CFD method, and the interference mechanism of the quad tiltrotor aircraft explored. Finally, an influence analysis of parameters under different rotor rotation direction combinations, wing installation angle, rotor transverse distance, and rotor longitudinal distance are carried out. The numerical results indicate that the aerodynamic characteristics of the rear rotors are greatly affected by whole aircraft interference in forward flight, and that the thrust of the rear rotor is reduced by 36.7% relative to an isolated rotor. The left rotors and right rotors rotate in opposite directions, which is conducive to greater wing lift. Increasing the longitudinal and transverse distance between the front and rear rotors can reduce interference of the front wings with the rear wings and improve the aerodynamic characteristics of the rear wings.

Keywords Quad tiltrotor aircraft · CFD method · Momentum source method · Aerodynamic characteristics · Parameter influence

1 Introduction

A quad tiltrotor aircraft can transition between the functions of a helicopter and a fixed-wing aircraft [1]. The conversion from helicopter mode to fixed-wing aircraft mode is realized by tilting rotors that not only can take off and land vertically but also can cruise at high speed. Given the same power used by a dual tiltrotor aircraft, a quad tiltrotor aircraft (shown in Fig. 1) has higher hovering efficiency [2], greater take-off weight, and longer travel distance. Due to the particular configuration of a

J. Shi · M. Lin · X. Zhang (✉) · Q. Zhao · G. Zhao
National Key Laboratory of Rotorcraft Aeromechanics, College of Aerospace Engineering, Nanjing University of Aeronautics and Astronautics, Nanjing 21006, China
e-mail: zhangxiayang@nuaa.edu.cn

S. Lee et al. (eds.), *The Proceedings of the 2021 Asia-Pacific International Symposium on Aerospace Technology (APISAT 2021), Volume 1*, Lecture Notes in Electrical Engineering 912, https://doi.org/10.1007/978-981-19-2689-1_10

Fig. 1 Quad tiltrotor aircraft

quad tiltrotor aircraft, there are complex mutual phenomena involving interference of wings with rotors, between front and rear rotors, and of rotors with the fuselage [3–6]. In this paper, we present the aerodynamic characteristics of whole aircraft interference and the influence of aerodynamic layout parameters under forward flight, which has significance for specific guidance in the design of quad tiltrotor aircraft.

At present, some research has been carried out on the aerodynamic characteristics of dual tiltrotor aircraft at home and abroad, but investigations on quad tiltrotor aircraft are relatively few.

In 2015, Mylapore et al. [7] carried out an experimental study on the influence of ground effect under hover and low-speed forward flight conditions for a 0.031 scale model of a quad tiltrotor aircraft. The test results showed that when there is no ground effect, a download of about 9% of the total thrust is caused by the presence of the wings in the rotor downwash, which also causes fountain flows on both the front and rear wings. In 2016, Zhou et al. [8] established an overall parameter optimization design method for quad tiltrotor aircraft based on a performance index and weight constraint, and carried out multi-objective optimization of the overall parameters. The results indicated that reduction of the wing area could reduce the downwash load of a wing in helicopter mode and enhance the vertical flight ability. In 2018, Wang [9] established a set of numerical calculation methods to study the aerodynamic interference involving the front rotor, fuselage, rear rotor, and wings of a quad tiltrotor aircraft in both vertical and forward flight states. It was concluded that the presence of a front rotor reduces fuselage drag when the forward speed is high, and that aerodynamic interference between rotors and wings is small in forward flight. In 2020, Wu et al. [10] studied the lift characteristics of an X-type UAV quad tiltrotor aircraft during vertical flight and tilting transition process based on the computational fluid dynamics (CFD) method. The numerical results indicated that in transition mode, the influence of the rear rotor on the rear wing is stronger due to the existence of the front rotor, and that the influence of the rotor on the wing is gradually reduced with increase of the forward speed.

The CFD method has the advantages of high precision and effective simulation of flow details, which makes it an important method for studying the aerodynamic characteristics of quad tiltrotor aircraft. Therefore, in the work reported in this paper, the momentum source method [11] was used to simulate the influence of the rotor on the flow field, and a set of numerical methods suitable for the whole aircraft interference flow field simulation of the quad tiltrotor aircraft were established. The aerodynamic characteristics of the rear rotor under different interferences in forward flight mode of a quad tiltrotor aircraft were analyzed, and a preliminary exploration of the interference mechanism performed. The influences of different combinations of rotor rotation directions, wing installation angles, and the transverse and longitudinal distances between front and rear rotors on the interference characteristics of the whole aircraft were investigated. A preliminary summarization of the influences is presented, which provides a reference for the next stages of improvement of the aerodynamic characteristics of quad tiltrotor aircraft.

2 Numerical Method

2.1 Geometric Model and Grid Generation

The main parameters of quad tiltrotor aircraft are shown in Table 1. The pitching angle and sideslip angle of the fuselage are both 0°. Some parameters are explained in Fig. 2. When the influence of a certain parameter on the quad tiltrotor aircraft was studied, only that parameter was changed.

The calculation domain of the flow field is set as a cuboid with the size $-13L \times 13L \times 7L$, with L being the length of the fuselage. The unstructured grid generation method was used to generate the whole aircraft computing grid. The grid is a polyhedron type and the grid number is about 2.2 million. In order to simulate the flow of the fuselage boundary layer effectively, the boundary layer grid is generated on the fuselage surface. The height of the first layer grid is 0.09 mm, which satisfies $y+ < 1$. A total of 15 layers of the boundary layer grid are generated, and the grid growth rate is 1.3. At the same time, the grid of the rotor region is encrypted to add

Table 1 Main parameters of a quad tiltrotor aircraft

Length of fuselage	7.0 m	Wing installation angle	6°
Height of fuselage	2.2 m	Airfoils	Boe106m/VR7/VR8
Span of wing	6.0 m	Forward speed	102 m/s
Rotor tip speed	200 m/s	Rotor longitudinal distance	2.8 m
Rotor nacelle length	0.75 m	Rotor vertical distance	0 m
Rotor radius	1 m	Rotor transverse distance	0 m
Rotor root cutting	20%	Number of blades per rotor	3
Left front rotor	Right-handed	Right front rotor	Left-handed
Left rear rotor	Left-handed	Right rear rotor	Right-handed
Distance between rotor center and fuselage symmetry plane			3.2 m

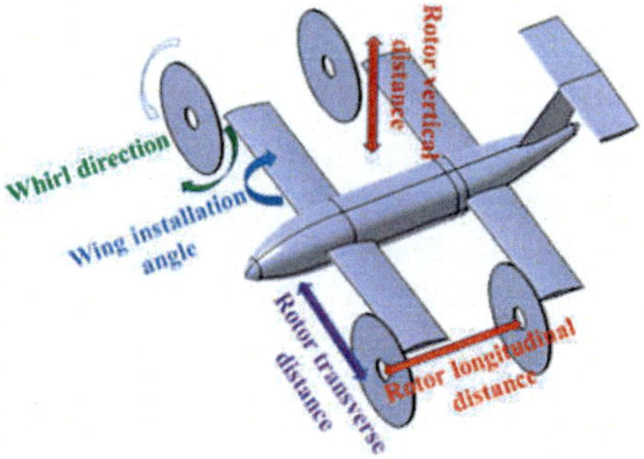

Fig. 2 Some parameter explanation

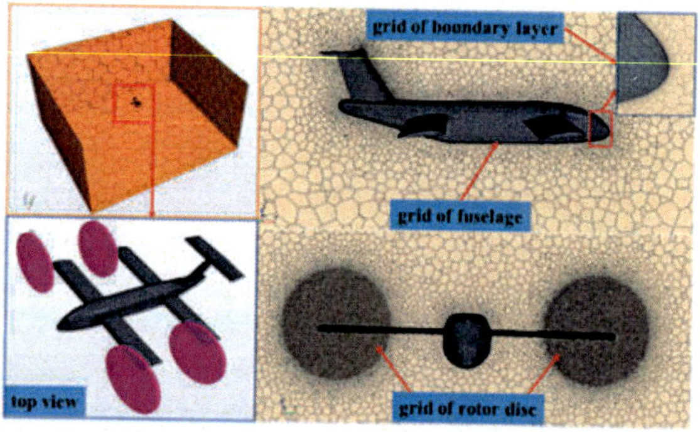

Fig. 3 Calculation domain and fuselage grid

an effective momentum source term. The calculation domain and fuselage grid are shown in Fig. 3.

2.2 Calculation Method

The momentum source method is an engineering method that can effectively simulate the effect of a rotor on the flow field. The basic thought is that the effect of a blade on the airflow is equivalent to the time-averaged momentum source term that is added to the control equation (Euler or Navier–Stokes). The most important feature of the momentum source method is that the effect of a blade on the airflow is reflected in the flow-field simulation by adding a momentum source term, ignoring the detailed flow near the blade, and transforming the periodic flow into "quasi-steady" flow using the time averaging method. Thus, the momentum source method can ensure a certain level of engineering accuracy, reduce the grid quantity, and improve the computational efficiency.

In this work, numerical simulation of the whole aircraft aerodynamic characteristics of the quad tiltrotor aircraft was carried out by solving the Reynolds average N-S equation. The implicit format was used to solve the N-S equation. The flow field medium characteristic selected was "ideal gas." The boundary condition was "pressure far-field boundary," and the wall boundary condition was "non-slip."

2.3 *Verification of the Method*

The validity of the CFD method was verified by calculation of the aerodynamic characteristics of the ROBIN fuselage. The fuselage length was 2 m, and the angle of attack −5°. The far-field airflow velocity was 0.3 Ma, the temperature 300 K, and the pressure 101,325 Pa. The established CFD method was used to simulate numerically the flow field of the fuselage in forward flight mode, and the calculation data was compared with the experimental data [12].

The pressure coefficient distribution on the cross-sectional surface of the ROBIN fuselage at the middle and end sections of the convex platform was compared with the experimental value. The comparison results are shown in Fig. 4. It can be seen that the calculated value at the middle of the convex platform is in good agreement with the experimental value, indicating that the CFD method established in this paper is effective for numerical simulation of the surface pressure of the fuselage and the flow field near the fuselage.

3 Results and Analysis

The rotor collective pitch was 40° in all calculated conditions. Because the quad tiltrotor aircraft in this paper is symmetrical on the left and right side of the layout, the left part of the quad tiltrotor aircraft was selected for the analysis below. To simplify communication in this paper, the parts of the aircraft are labeled: front wing (FW), rear wing (RW), front rotor (FR), and rear rotor (RR).

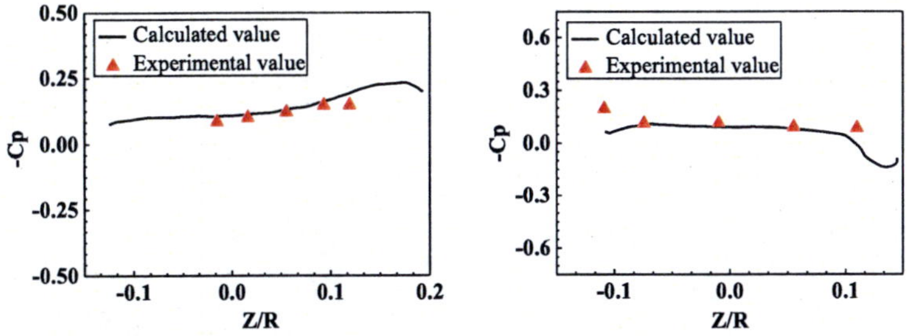

(a) X/L = 0.6003 (b) X/L = 1.0008

Fig. 4 Verification example comparison results

3.1 Interferences on the Rear Rotor

In forward flight mode, the rear rotor of the quad tiltrotor aircraft is between the front rotor and the wings, and the interference is more complex than for the front rotor. In this section, analyses of the influence of different interferences on the performance of the rear rotor of a quad tiltrotor aircraft in forward flight mode provides a preliminary exploration of the mechanism of whole aircraft interference. Different interferences: the isolated rotor (1), the rotor combination (2), no front rotor (3), and whole aircraft interference (4), as shown in Fig. 5.

The rear rotor thrust and cruise efficiency under different interferences are shown in Fig. 6. In this paper, the momentum source method is used for the rotor part, so the rotor in Fig. 5 is virtual, as indicated. Figure 6 shows that the interference between the front rotor and wing will cause the rear rotor thrust to decrease. Under interference from the front rotor, the rear rotor thrust is reduced by 4.3% compared with that of an isolated rotor. When subjected to wing and fuselage interference, the rear rotor thrust is reduced by 32.2%, compared with an isolated rotor. With whole aircraft interference, the rear rotor thrust is reduced by 36.7% compared with an isolated rotor. Therefore, the influence of wing interference on the aerodynamic performance

Fig. 5 Different interferences

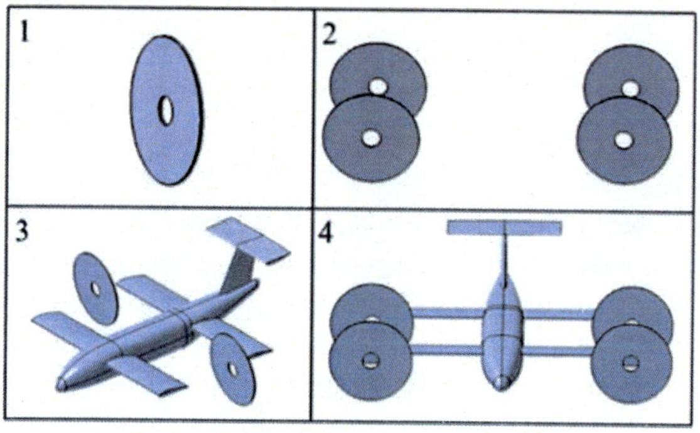

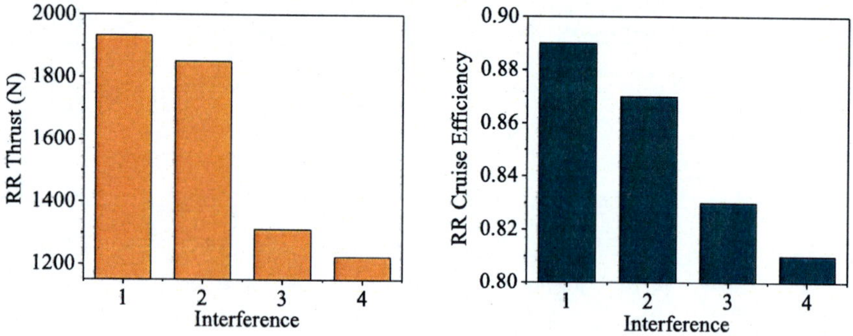

Fig. 6 Rear rotor thrust and cruise efficiency under different interferences

of the rear rotor is stronger, which should be considered in the design of a quad tiltrotor aircraft.

It can be seen from Fig. 6 that the interference between the front rotor and the wing are unfavorable for the cruise efficiency of the rear rotor. The influence of wing interference is stronger than that of the front rotor interference. The cruise efficiency of the rear rotor under whole aircraft interference is 9% lower than that of an isolated rotor. To reduce fuel consumption of the quad tiltrotor aircraft during cruise mode, the interference of the wing with the rear rotor should be reduced.

The pressure contours of the rotor central cross-section under different interferences are shown in Fig. 7. It can be seen from Fig. 7a and b that, due to the interference of the front rotor, the high-pressure area behind the rear rotor increases and the area of low pressure in front of the rear rotor decreases more significantly. Finally, the rear rotor thrust decreases. This is mainly due to the acceleration effect of the front rotor on the airflow in front of the rear rotor, which leads to increase of the inflow velocity of the rear rotor. For the blade element micro-segment, increase of the inflow velocity increases the local inflow angle, so that the aerodynamic angle of attack of the blade element decreases, and finally the whole rear rotor thrust decreases. It can be seen from Fig. 7a and c that, due to the interference of the front and rear wings, the flow near the rear rotor is more complex. There is a large area of low pressure behind the rear rotor, and a high-pressure area in front of the rear rotor. Compared with the front rotor interference, the pressure distribution near the rear rotor caused by wing interference changes more intensely. In Fig. 7c and d, the pressure distribution near the rear rotor is similar, so a preliminary inference could be that wing interference is the main reason for the decrease of rear rotor thrust caused by the whole aircraft interference.

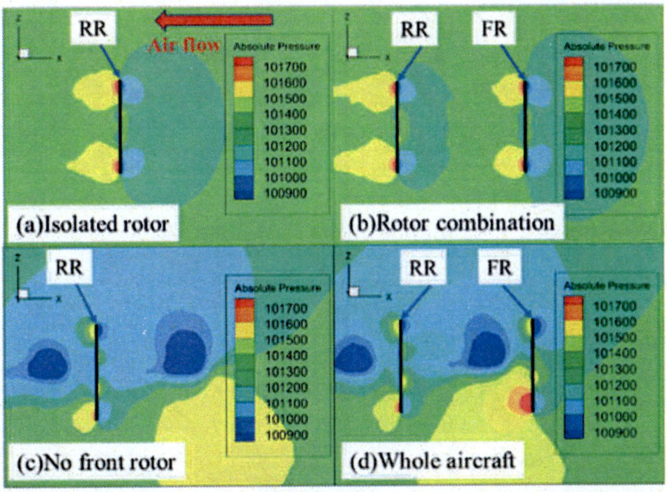

Fig. 7 Pressure contours of rotor central cross-section under different conditions

3.2 *Influence of the Rotor Rotation Direction*

The wing will be affected by the rotor wake when the quad tiltrotor aircraft flies forward, and the rotor rotation direction will directly affect the flow direction and velocity of the rotor wake, thus also affecting the pressure and aerodynamic load distribution of the wing. Therefore, it was necessary to research the influence of different rotor rotation direction combinations (RDCs) on the aerodynamic characteristics of the whole aircraft.

To balance the anti-torque of the rotor, the four rotors of a quad tiltrotor aircraft should meet the requirement that the rotation directions of the front two rotors are opposite, and the rotation directions of the rear two rotors are opposite. Thus, there are four types of rotor RDCs. The definition of rotor rotation direction is as follows: a rotor that rotates counter-clockwise is right-handed and a rotor that rotates clockwise is left-handed when facing the nose of the aircraft. The four rotor RDCs of the quad tiltrotor aircraft are shown in Fig. 8.

The influences of different rotor RDCs are shown in Fig. 9. It can be seen from Fig. 9a that the lift of the front wing is greater under RDC1 and RDC2. Under these

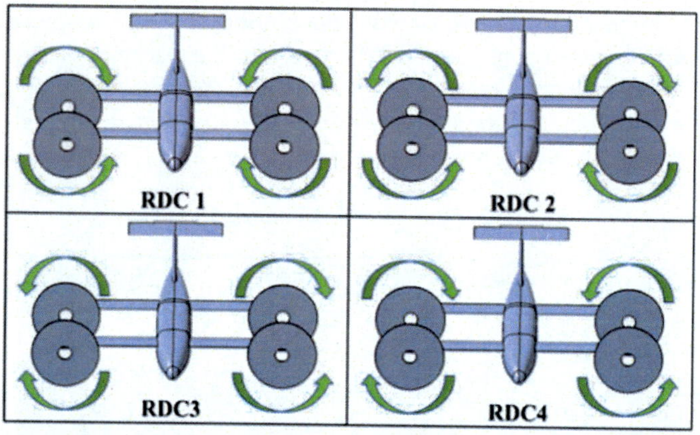

Fig. 8 Rotor rotation direction combinations

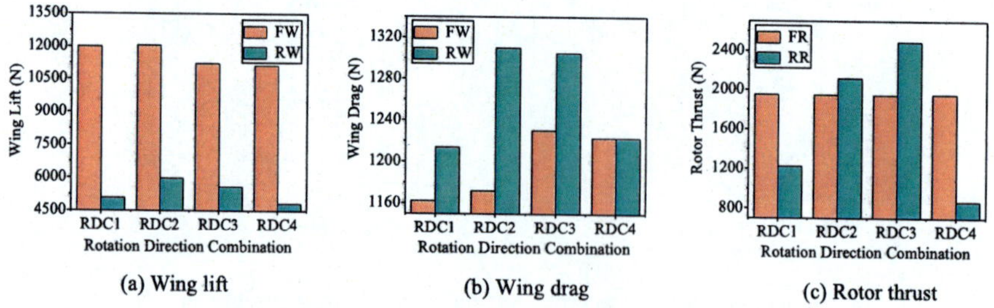

Fig. 9 Influence of the rotor rotation direction combination

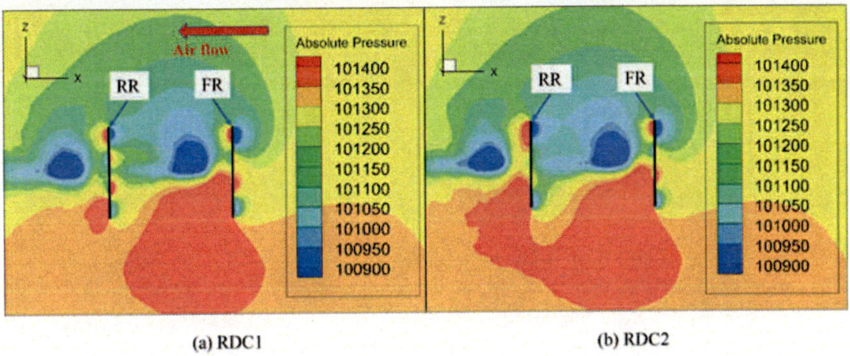

Fig. 10 Pressure contours of the rotor central cross-section

two rotation direction layouts, the rotation direction of the front rotor are all such that the left rotor is right-handed and the right rotor is left-handed. The lift of the rear wing of the RDC2 and RDC3 is larger. Under these two rotation direction layouts, the rotation directions of the rear rotor are all such that the left rotor is right-handed and the right rotor is left-handed. It can be seen that the wing lift can be improved to a certain extent when the left rotor is right-handed and the right rotor is left-handed. As can be seen from Fig. 9b, the front wing drag is small with RDC1 and RDC2, and the rear wing drag is small with RDC1 and RDC4. It can be seen from Fig. 9c that the RDC has a great influence on the rear rotor thrust. The rear rotor thrust is larger with RDC2 and RDC3. Under these two rotation direction layouts, the rotation directions of the rear rotor are such that all the left rotors are right-handed and the right rotors are left-handed, so this rotation layout is conducive to increasing the rear rotor thrust. The rear rotor thrust of RDC2 is greater than with RDC3, indicating that the front rotor rotation direction also has a certain impact on the rear rotor thrust.

The pressure contours of the rotor central cross-section of RDC1 and RDC2 are shown in Fig. 10. It can be seen from Fig. 10 that the high-pressure area behind the rear rotor and the low-pressure area ahead of the rear rotor in RDC2 is larger than in RDC1, so the rear rotor thrust of RDC2 is larger than that in RDC1. This may be due to the circumferential airflow caused by the rotation of the rear rotor backward flows hitting the lower surface of the rear wing and forming the high-pressure area.

3.3 Influence of the Wing Installation Angle

The influence of the wing installation angle is shown in Fig. 11. It can be seen from Fig. 11a that, with increase of the installation angle of the wing, the lift of the front and rear wings increases linearly, and lift of the front wing increases faster than that of the rear wing. In addition, increase of the installation angle of the wing increases interference of the front wing with the rear wing. It can be seen from Fig. 11b that

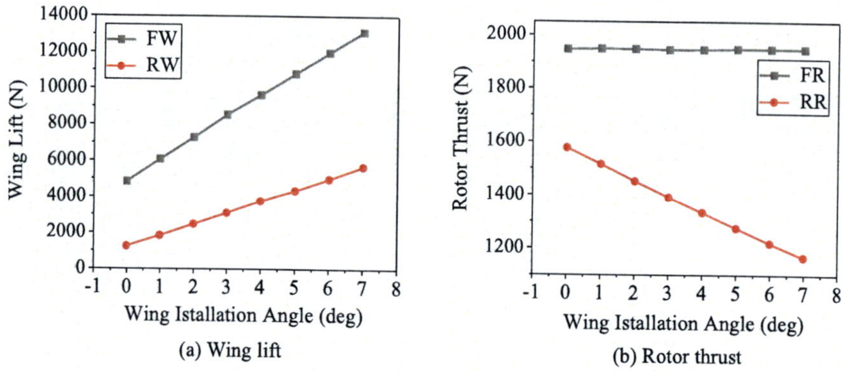

Fig. 11 Influence of the wing installation angle

the wing installation angle has little effect on the front rotor thrust, while the rear rotor thrust decreases linearly with increase of the wing installation angle.

The velocity contours and pressure contours of the Y = 2.6 m cross-section are shown in Fig. 12. It can be seen from Fig. 12a and b that when the wing installa-

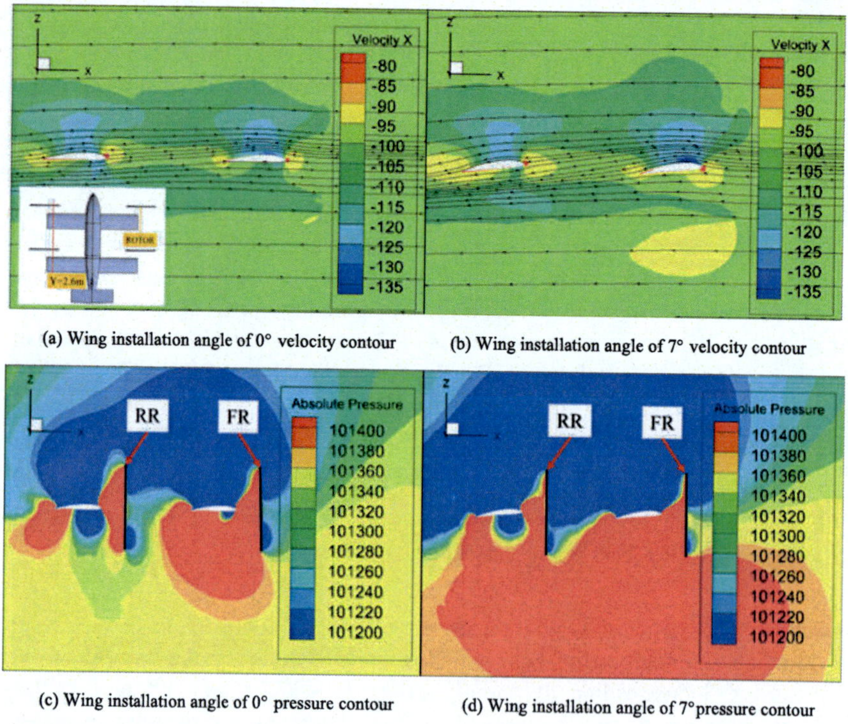

Fig. 12 Velocity and pressure contours of the Y = 2.6 m cross-section

tion angle is 7°, the airflow velocity difference between the front and rear wings is larger, and the aerodynamic angle of attack at the rear wing is smaller. Therefore, with increase of the wing installation angle, the lift difference between front wing and rear wing tends to increase. It can be seen from Fig. 12c and d that when the wing installation angle is 7°, the high-pressure area in front of the rear rotor increases, and the low-pressure area behind the rear rotor also increases, resulting in decrease of the rear rotor thrust. When the wing installation angle increases, the airflow velocity declines more substantially and the static pressure increases below the wing. Compared with a wing installation angle of 0°, when the wing installation angle is 7°, the high-pressure area below the front wing increases more obviously than that of the rear wing, so the lift difference between the front and rear wing increases.

3.4 Influence of the Rotor Transverse Distance

Change the transverse distance of the front and rear rotors on the same side is realized by shortening the front wing. The front rotor approaches the plane of fuselage symmetry by the same distance as the shortening of the front wing. The influence of the rotor transverse distance is shown in Fig. 13. It can be seen from Fig. 13a and b that, as the rotor transverse distance increases, the lift and drag of the front wing decreases due to decrease of the front wing area. The shielding area of the front wing to the rear wing is reduced, the aerodynamic angle of attack of the unshielded part of the rear wing increases, and the lift and drag of the rear wing increases. It can be seen from Fig. 13c that, as the rotor transverse distance increases, the interference of the front rotor with the rear rotor decreases, influence of the front rotor acceleration effect on the rear rotors weakens, the aerodynamic angle of attack of the rear rotor increases, and the thrust of the rear rotor increases. When the rotor transverse distance is 0.7 m, the rear rotor thrust is increased by 47% compared with 0 m. Reducing the shielding area of the front rotor and the front wing to the rear rotor can increase the rear rotor thrust.

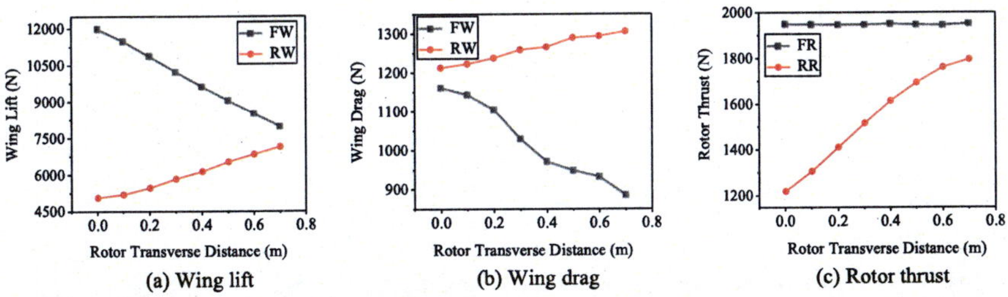

Fig. 13 Influence of the rotor transverse distance

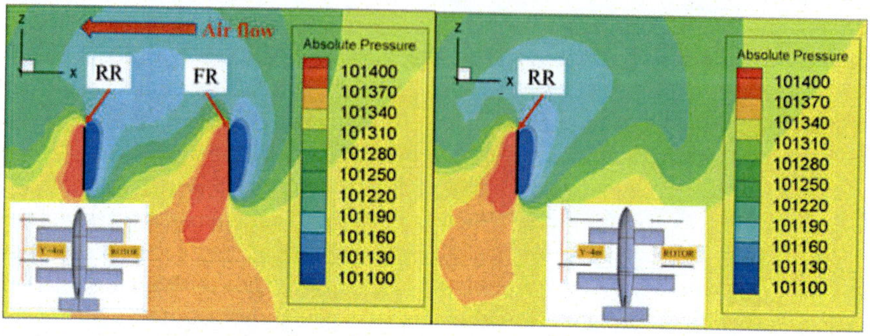

(a) Rotor transverse distance 0 m (b) Rotor transverse distance 0.7 m

Fig. 14 Pressure contours of the Y = 4 m cross-section

The pressure contours of the Y = 4 m cross-section, when the rotor transverse distance is 0 m or 0.7 m, is shown in Fig. 14. It can be seen from the figure that, with transverse distance of 0.7 m, the pressure in front of the rear rotor is higher than that with a transverse distance of 0 m. That is, the airflow velocity there decreases, which is consistent with the analysis result of the velocity above. The interference of the front wing and front rotor with the rear rotor is reduced, the pressure behind the rear rotor increases significantly, the high-pressure area formed at the front of the rear rotor due to the interference of the front wing disappears, and the rear rotor thrust increases.

3.5 *Influence of the Rotor Longitudinal Distance*

Change of the longitudinal distance between the front and rear rotor is realized by changing the longitudinal position of the rear wing in the model. The relative positions of the rear rotor and the rear wing are unchanged. The influence of the rotor longitudinal distance is shown in Fig. 15. It can be seen from Fig. 15a that the lift of the rear wing increases with increase of the rotor longitudinal distance. Compared with the rotor longitudinal distance of 2.4 m, the lift of the rear wing at the rotor longitudinal distance of 3.2 m increases by 16%. It can be seen from Fig. 15b that the rear wing drag is smallest near the rotor longitudinal distance of 2.5 m, and largest near the rotor longitudinal distance of 3.1 m, showing an overall increasing trend.

The velocity contours at the Y = 2.6 m cross section are shown in Fig. 16. It can be seen from Fig. 16a and b that, when the longitudinal distance of the rotor is 2.4 m and 3.2 m, respectively, the airflow velocity near the rear wing shows little difference; however, the aerodynamic angle of attack at the rotor longitudinal distance of 2.4 m is smaller than that at 3.2 m. It can be seen from Fig. 16a that, after the airflow passes through the front wing, there is a downward deflection that results in decrease of the

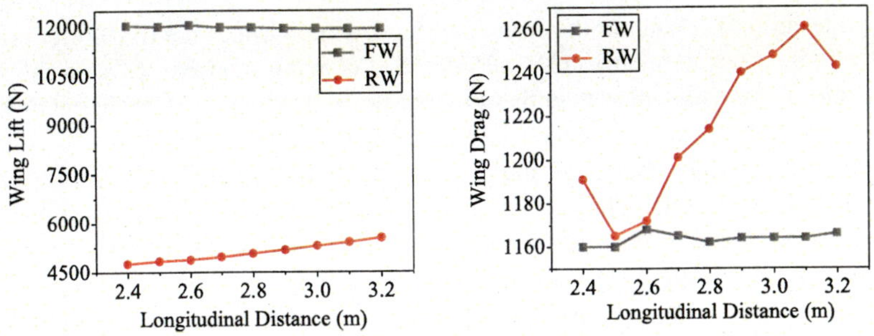

(a) Wing lift (b) Wing drag

Fig. 15 Influence of the rotor longitudinal distance

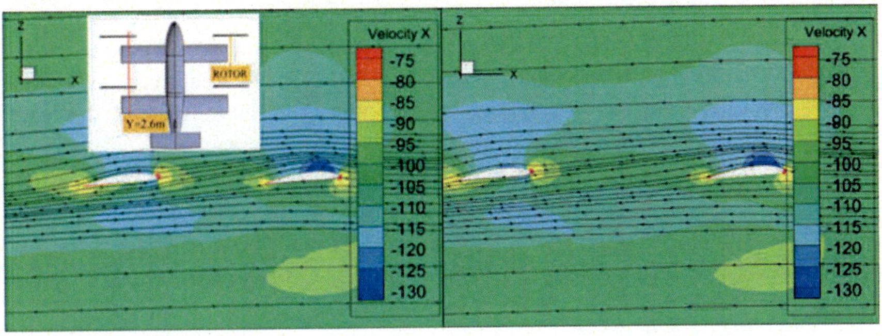

(a) Rotor longitudinal distance 2.4 m (b) Rotor longitudinal distance 3.2 m

Fig. 16 Velocity contours at the Y = 2.6 m cross section

airflow angle of attack at the rear wing. The closer the rear wing is to the front wing, the greater the decrease of the airflow angle of attack is. Therefore, the lift of the rear wing will increase with increase of the longitudinal distance between the front and rear rotors.

4 Conclusions

In this paper, a set of numerical methods suitable for a whole aircraft interference flow field simulation of quad tiltrotor aircraft was used to analyze the interference characteristics in forward flight. The influences of different interferences, rotor rotation direction combination, wing installation angle, rotor transverse distance, and rotor longitudinal distance on the aerodynamic characteristics of a quad tiltrotor aircraft in forward flight were studied. The following conclusions were obtained.

(1) Interference involving the front rotor and wing reduces the thrust and cruise efficiency of the rear rotor. The influence of wing interference with the rear rotor is much stronger than that of front rotor interference with the rear rotor.
(2) When the left rotor of the wing is right-handed and the right rotor is left-handed, lift of the wing is enhanced and along with it, the aerodynamic performance of the whole aircraft. This is because the wing is affected by the circumferential rotation speed of the rotor wake, which increases the aerodynamic angle of attack of the wing behind the rotor.
(3) With increase of the wing installation angle, the interference of the front wing with the rear wing is stronger, and the difference in lift between the front and rear wings is larger. The interference of the wing with the rear rotor is also enhanced and the thrust of the rear rotor is reduced.
(4) With increase of the transverse distance of the rotor, the interference of the front wing with the rear wing is weakened, the interference of the front rotor and front wing with the rear rotor is weakened, and the lift of the rear wing and thrust of the rear rotor are increased. The shorter the longitudinal distance between the front and rear rotor is, the stronger the interference of the front wing with the rear wing, and the smaller the lift of the rear wing.
(5) For future study, two aspects of this study will be improved. The body-fitted grid method is considered to simulate the rotor motion, and its use should improve the calculation accuracy and obtaining the flow field details. Considering the influence of additional parameters should provide a more comprehensive grasp of the whole aircraft interference characteristics of a quad tiltrotor aircraft.

Acknowledgements This work is supported by the National Key Laboratory Foundation of China (no. 61422200101), the National Natural Science Foundation of China (nos. 11872211 and 12102186), the Natural Science Foundation of Jiangsu Province (no. BK20200433), and also the Priority Academic Program Development of Jiangsu Higher Education Institutions.

References

1. Zhao G (2019) Aerodynamic interaction analysis and wing optimization of tiltrotor aircraft. Nanchang Hangkong University, Nanchang, China
2. Wu XM, Mou XW (2021) The key technology of helicopter and its future development and vision. Acta Aerodyn 39(3):6–8
3. Sheng C, Narramore JC (2009) Computational simulation and analysis of Bell Boeing quad tiltrotor aircraft aero interaction. J Am Helicopter Soc 54(4):042002
4. Gupta V, Baeder J (2002) Investigation of quad tiltrotor aircraft aerodynamics in forward flight using CFD. AIAA 2002-2812, June 2002
5. Han D, Barakos GN (2020) Aerodynamic interference model for multi-rotors in forward flight. J Aircr 57(6):1220–1223
6. Sitaraman J, Baeder J (2013) Analysis of quad-tiltrotor blade aerodynamic loads using coupled CFD/free wake analysis. AIAA 2002-2813, June 2002
7. Mylapore AR, Schmitz FH (2015) An experimental investigation of ground effect on a quad tilt rotor in hover and low speed forward flight. J Bacteriol 60(1):9–13

8. Zhou QC, Li CH (2016) Research on optimization design of overall parameters of high-speed quad-tiltrotor. Helicopter Technol 01:3–5
9. Wang C (2018) Aerodynamic interference analysis of tilt quad-rotors aircraft. NUAA, Nanjing, China
10. Wu YC, Chen ZL, Ji JJ (2020) Research on aerodynamic characteristics of a quad-tiltrotor UAV. J Ordnance Equip Eng 41(3):74–75
11. Zhao QJ, Xu GH (2016) Foundations of helicopter computational fluid dynamics. Science Press, Beijing, China
12. Freeman CE, Mineck RE (1979) Fuselage surface pressure measurements of a helicopter wind-tunnel model with a 3.15-meter diameter single rotor. NASA-TM-1979-80051

A Two-Step Optimization Method Using POD-Based Geometric Parameterization for Aerodynamic Shape Optimization

Chenliang Zhang, Yanhui Duan, Guangxue Wang, and Hongbo Chen

Abstract A two-step optimization method is proposed for aerodynamic shape optimization (ASO), and the efficiency of the 2nd-step optimization is improved by proper orthogonal decomposition-based (POD-based) geometric parameterization. The two-step optimization method is constituted by particle swarm optimization (PSO) combined with Hicks-Henne shape functions and steepest descent algorithm (SDA) bonded with POD-based parameterization method. After the 1st-step optimization, superior data (SD) given with better values of objective functions which can satisfy all the constraints are filtered to extract POD bases. The POD bases only cover design space near the 1st-step optimized solution and can parameterize the geometric shape with fewer design variables (DVs). Fewer DVs and smaller design space can improve the efficiency of the 2nd-step optimization,. The two-step optimization method is validated by a case of NASA Rotor 37 aiming to increase peak adiabatic efficiency. The efficiency of the 2nd-step optimization is improved by using POD-based geometric parameterization, which is proved by comparing with SDA using the conventional parameterization method.

Keyword ASO · POD · Hybrid optimization · Rotor 37

1 Introduction

When solving problems of aerodynamic shape optimization (ASO), the effect and efficiency of optimization methods are two critical performances to determine the methods. The effect denotes whether the optimized solutions are global or not, and the efficiency denotes how many objective functions should be evaluated to obtain the solutions. There are two kinds of methods to solve optimization problems, which are local and global optimization methods. The merits of the local methods are their

C. Zhang · Y. Duan (✉) · H. Chen
School of Systems Science and Engineering, Sun Yat-Sen University, Guangzhou, China
e-mail: duanyh7@mail.sysu.edu.cn

G. Wang
School of Aeronautics and Astronautics, Sun Yat-Sen University, Shenzhen, China

S. Lee et al. (eds.), *The Proceedings of the 2021 Asia-Pacific International Symposium on Aerospace Technology (APISAT 2021), Volume 1*, Lecture Notes in Electrical Engineering 912, https://doi.org/10.1007/978-981-19-2689-1_11

high efficiency, and their drawbacks are the local optimums.,On the contrary, the global methods can obtain global solutions but with quite low efficiency.

Since hybrid methods can overcome drawbacks of each optimization method, they were widely researched [1–5]. And the efficiency can be further improved by replacing NS solver with surrogate models during the global optimization [2, 5, 6] and solving adjoint equations to obtain gradient during the local optimization [3, 5]. It can be found that the purposes of previous researches always focus on how to calculate objective functions and gradients to improve the efficiency of the hybrid methods. Few people research the effect of geometric parameterization, which is also the core part of ASO. In this paper, geometric parameterization is considered to improve the efficiency of two-step optimization methods which play important roles in the hybrid optimization methods.

In two-step optimization methods, the 1st-step optimized solutions will be provided to the 2nd-step optimization as the start point, which is quite close to the global optimum and further refined by the 2nd-step optimization. It can be known that the design space of the 2nd-step optimization is much smaller than the 1st-step optimization, so conventional geometric parameterization methods just considering requirements of the 1st-step optimization that design space is as large as possible are not suitable for the 2nd-step optimization. It is necessary to find an appropriate geometric parameterization method to explore a contractible design space of the 2nd-step optimization.

POD-based geometric parameterization can reduce the order of the problem of ASO, meanwhile keeping the characteristics of the database or snapshots [7–11]. All these researches focused on optimization of airfoils, except Ref. [8] researched more on optimization of wing and aircraft shapes. What should mention above is that in Ref. [11], adaptive DVs bounds were deeply studied, which was very important for POD-based geometric parameterization. In addition, POD was also used in the active subspace method, based on gradients, to discover and exploit low-dimensional and monotonic trends in the design space as a function of the DVs [12]. Besides POD, other methods were also used to reduce the order of the problem of shape optimization. Ref. [13] obtained orthogonal basis by least-squares fitting to the smoothed airfoils; a recent work [14] compare linear principal component analysis (PCA) with deep autoencoder (DAE, a non-linear extension of PCA) by optimizing the hull form and the solutions showed that DAE is better than PCA. There are two different points between the paper and previous researches, and one is that the training database of POD is from the data generated at each iteration of the 1st-step optimization rather than samples obtained from design space by certain sampling technique; another is that optimized geometric shape is not 2D airfoil but 3D compressor blade which can be considered a typical case of industrial application.

To optimize the compressor blades need more DVs than wings for their more complex geometric shape. Although the adjoint method can provide gradients to optimize these problems, for it is not sensitive to the number of DVs, it is impossible to overcome the drawback that the optimum of gradient-based optimization could not find the global optimum. The POD-based parameterization, which can reduce the order of the problem of ASO, provides an opportunity to find the global optimum when optimizing the compressor blades with affordable computational cost. In this study, a single compressor rotor was optimized by two-step optimization. And POD-based parameterization is only used for the 2nd-step optimization with training data base generated during the 1st-step optimization, which is also the different from previous researches.

In this paper, POD-based geometric parameterization and its characteristics are introduced firstly, then two-step optimization method including conventional geometric parameterization and grid deformation are described. Finally, the case of Rotor 37 is proceeded to validate the two-step optimization and prove the effectiveness of the 2nd-step optimization using POD-based geometric parameterization.

2 POD-Based Geometric Parameterization and Its Characteristics

2.1 *Principle of POD*

POD is a procedure that provides a set of optimal bases for the construction of multidimensional data, which allows a reduction in the order of the original system under consideration from very large numbers to very small ones. Basic POD [15] can provide a series of bases by solving an eigenvalue problem of an autocorrelation matrix, and the order of the matrix is generally very large, which is hard and sometimes impossible to solve. Snapshot POD [16] is a more efficient way to determine POD bases, since the order of the autocorrelation matrix is equal to the number of snapshots, which is much smaller than the basic POD. Here only snapshot POD is introduced shortly, and to make the description simple, 'Snapshot POD' is replaced by POD. Assume that the original system is described by a set of vectors: $\boldsymbol{s}_1, \boldsymbol{s}_2, \cdots, \boldsymbol{s}_{\mathrm{m}}$, which are also named as snapshots. To obtain their POD bases: $\boldsymbol{\phi}_1, \boldsymbol{\phi}_2, \cdots, \boldsymbol{\phi}_{\mathrm{m}}$, the following eigenvalue problem Eq. (1) should be solved.

$$\boldsymbol{R}\mathbf{a} = \lambda \mathbf{a}, \tag{1}$$

where $\boldsymbol{R}$ is the autocorrelation matrix defined as Eq. (2).

$$\boldsymbol{R}_{ij} = \frac{1}{m}(\boldsymbol{s}_i, \boldsymbol{s}_j), \tag{2}$$

After solving Eq. (1), POD bases can be calculated by Eq. (3).

$$\boldsymbol{\phi}_k = \sum\nolimits_{i=1}^{m} a_i^k \boldsymbol{s}_i, \tag{3}$$

where $\boldsymbol{\phi}_k$ denotes the k-th POD basis, a_i^k denotes the i-th element of the corresponding eigenvector of the k-th eigenvalue.

When using POD bases to describe the original system, the corresponding eigenvalue denote the "energy" of the basis in the system. The "energy" of the first h bases can be calculated as follows,

$$E_h = \frac{\sum_{i=1}^{n} \lambda_i}{\sum_{i=1}^{m} \lambda_i} \times 100, \tag{4}$$

It is believed that the system can be described well when E_{h} is greater than a certain value. Furthermore, h is always much less than m to make E_{h} satisfy the requirement.

2.2 POD-Based Geometric Parameterization

Parameterization based on POD includes two steps: generating POD bases and parameterizing geometry shapes by the bases. According to Ref. [8], snapshot data consists of either implicit or explicit shape DVs, and the explicit shape DVs consisting of the geometric coordinates proved more practical.

As mentioned above, the snapshots consist of the coordinates of the geometric shape of superior data (SD) of 1st-step optimization, which are with better values of objective functions and satisfy all constraints. All the geometric shapes are composed of surface grid points. To describe the geometry shapes, define a set of vectors: $\mathbf{g}_1$, $\mathbf{g}_2, \cdots, \mathbf{g}_{\mathrm{m}}$, each of which consists of the coordinates of grid points arranged at certain sequence, such as $\mathbf{g}_{\mathrm{i}}$ can be denoted as Eq. (5).

$$\mathbf{g}_{\mathrm{i}} = \begin{bmatrix} \mathbf{X} \\ \mathbf{Y} \\ \mathbf{Z} \end{bmatrix}_{\mathrm{i}}, \tag{5}$$

where **X**, **Y**, **Z** are the coordinates of the **X**, **Y**, **Z**-axis separately, which are arranged at the sequence according to the index of the grid points. If the number of grid points of each geometry shape is n, the dimension of each vector is 3n.

Then snapshot of POD is calculated as follow,

$$s_i = \mathbf{g}_i - \mathbf{g}_0, \tag{6}$$

where $\mathbf{g}_0$ is the vector of the original geometry shape. Equation (6) means that the snapshots only contain variations of each geometry shape. Then POD bases can be calculated by Eqs. (1), (2) and (3).

Then an arbitrary geometric shape (denoted by **g**) in the design space covered by the snapshots can be represented by the POD bases,

$$\mathbf{g} = \sum\nolimits_{j=1}^{m} a_j \boldsymbol{\phi}_j + \mathbf{g}_0, \tag{7}$$

Using Eq. (7) can parameterize the geometry shape, but not all POD bases are necessary to parameterize the geometric shape, which will provide too many DVs. Only those with certain energy can describe the original system exactly enough. In most cases, the number of necessary POD bases is always much less than the number of original DVs.

2.3 Merits and Drawbacks

Merits:

- Reduce the order of the optimization problem, since the number of used POD bases (equal to the number of DVs) is much less than that of the original problem.
- Narrow the design space, since the POD bases only contains the characteristics of SD.

A drawback mainly comes from the second merit. If the characteristics contained in SD are not wanted, such as that of local optimum, the following optimization must not find the global optimum. It is assumed that the drawback can be overcome when the solutions of the 1st-step optimization are good enough because no theory can prove the solutions are global for objective functions in a black box.

3 Two-Step Optimization Method for ASO

As shown in Fig. 1, the two-step optimization are composed of four parts.

Part 1, the 1st-step optimization. This part optimizes the geometric shapes by PSO and conventional parameterization methods.

Part 2, POD-based parameterization. Firstly, SD given with better values of objective functions which can satisfy all the constraints are filtered from all samples generated at each iteration of the 1st-step optimization. Then geometric POD basis will be extracted from SD, by which the 1st-optimized solution will be represented. Coefficients of the POD basis are parameters or DVs.

Part 3, the 2nd-step optimization. This part optimizes the 1st-optimized solution by SDA and POD-based parameterization.

Part 4, Common module. This part provides functions of updating the grid and calculating objective functions, which are necessary for both the 1st and 2nd-step optimization.

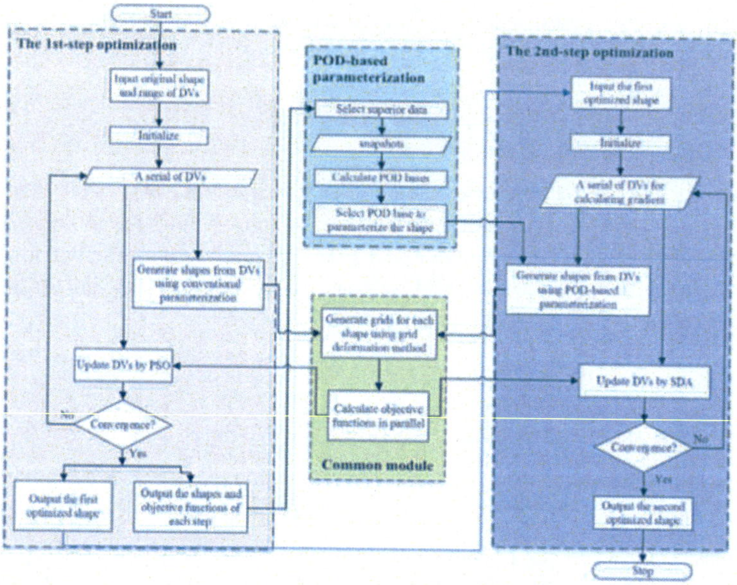

Fig. 1 Flowchart of the two-step optimization using POD-based geometric parameterization

Details of PSO, conventional parameterization method based on Hicks-Henne functions and grid deformation method based on algebraic interpolation can be found in Ref. [21]. More information about PSO can be found in Ref. [17–19]. Details of SDA can be found in Ref. [20], an improvement calculating gradient parallel of which is made to increase the efficiency.

4 Validation Case of Rotor 37

4.1 The 1st-Step Optimization

The paper is intended to maximize the peak adiabatic efficiency of Rotor 37 under the design rotational speed, meanwhile limiting changes of both total pressure ratio and mass flow rate at the design point. During the 1st-step optimization, the cost function is defined as Eq. (8).

$$\text{Find: } \boldsymbol{x}*, \tag{8}$$

$$\begin{aligned} \text{Min}: f &= \begin{cases} \frac{1}{\eta} & \text{satisfy constraints} \\ \frac{5}{\eta} & \text{not satisfy constrains} \end{cases} \\ \text{st.}: \quad & |(\dot{m} - m_0)/m_0| \le 0.5\%, \\ & |(\pi - \pi_0)/\pi_0| \le 0.5\%, \end{aligned} \tag{9}$$

Table 1 Distribution of DVs

Index	Surface	Spanwise (%)	Chord wise	Range (cm)
1–10	Suction	25	Average distributed along curve length; from leading edge to trailing edge	(–0.01,0.01)
11–20	Suction	50	The same as above	(–0.01,0.01)
21–30	Suction	75	The same as above	(–0.01,0.01)

Table 2 Solutions comparisons between the original and the 1st-step optimized blade

	Adiabatic efficiency	Total pressure ratio	Mass flow (kg/s)
Original	0.8588	1.9971	20.81
1st-step optimized	0.8699	1.9887	20.91
Changed	1.29%	–0.42%	0.48%

where η, $\dot{m}$ and π denote the peak adiabatic efficiency, the mass flow rate and the total pressure ratio, subscript "0" denotes the variable is from Rotor 37. There are a total of 30 DVs to change the geometry of Rotor 37, the distribution of which is shown in Table 1. The objective function is calculated by an in-house and structure-grid CFD code, which is validated in Ref. [17]. The case of the 1st-step optimization is the same as that of our previous work [21]. So only important solutions are listed below, details of the case can be found in the published paper [21].

The optimization was evolved 25 generations with 80 particles, because a bad solution (the blade with zero thickness) was obtained at 26 generations. Table 2 compares the performance between the original (marked by original) and 1st-step optimized solutions (marked by 1st-step optimized). The optimized blade increases the adiabatic efficiency by 1.29%, with a 0.42% reduction in the total pressure ratio and a 0.48% increase in the mass flow rate. The original and the optimized sections at 25, 50 and 75% span are depicted in Fig. 2. Maximum thickness isincreased at 75 and 50% span, but decreased at 25% span, positions of which are all moved toward trailing edge.

4.2 SD Selection and POD-Based Geometric Parameterization

The number of particles generated during 25 generations is 2000. Adding 80 particles of initial generation, there are a totally of 2080 particles. SD include 1734 particles which are 83% of the total particles. Be different from our previous work [21], and POD bases are chosen by analyzing the energy.

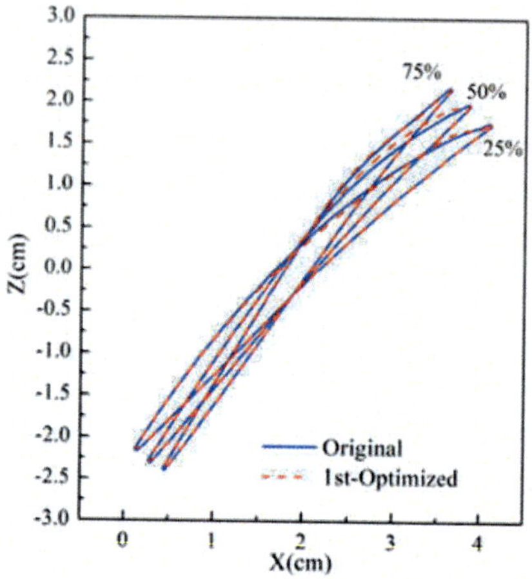

Fig. 2 Geometric comparisons between original and the 1st-step optimized blade

The coordinates of the surface grid points on the suction surface of each SD are selected to be transferred to snapshots according to Eqs. (5) and (6), and then the POD bases are extracted. To gain a better understanding, each basis is transferred by Eq. (10).

$$\boldsymbol{\phi}_{\iota}^{\mathrm{T}} = \sqrt{\boldsymbol{\phi}_{\iota}^{2} + \boldsymbol{\phi}_{n+i}^{2} + \boldsymbol{\phi}_{2n+i}^{2}} \quad I = 1, 2, \ldots, n, \tag{10}$$

As mentioned above, n denotes the number of the surface grid points. $\boldsymbol{\phi}_{l}$, $\boldsymbol{\phi}_{n+l}$ and $\boldsymbol{\phi}_{2n+l}$ are respectively the part of X, Y, Z coordinates of POD basis. In fact, ϕ_{l}^{T} is the geometric variation represented by the POD basis at the corresponding grid point. The $\boldsymbol{\phi}_{l}^{T}$ contours of the first three bases on the suction surface of Rotor 37 are depicted in Fig. 3, coordinates of which are normalized respectively by local chord and local span. By using the same contour level, it can be concluded the first basis contributes primarily to the most part of the suction surface change, and contributions of the second and the third basis are smaller in proper order. Another characteristic is that almost all variations are located between mid-chord and trailing edge for all three bases. The maximum $\boldsymbol{\phi}^{T}$ is located at about 50% span and 85% chord of the first POD basis. Variations of $\boldsymbol{\phi}^{T}$ at about 25% span and 65% chord of the first POD basis, and at about 75% span and 95% chord of the second POD basis are also obvious. All variations mentioned above are consistent with the geometric variations shown in Fig. 2, which proves that the POD basis includes 'excellent' geometric characteristics.

When using POD bases to parameterize the geometric shape as shown in Eq. (7), not all bases are needed, but only those contain 99.99% system energy. The energy of the each first ten bases are depicted in Fig. 4, and the first eight bases are selected to parameterize the blade, sum of the energy of which is larger than 99.99%. Comparisons of three sections between exact and POD-based blade are shown in Fig. 5, that the three sections are very closed illustrate eight bases are sufficient to parameterize

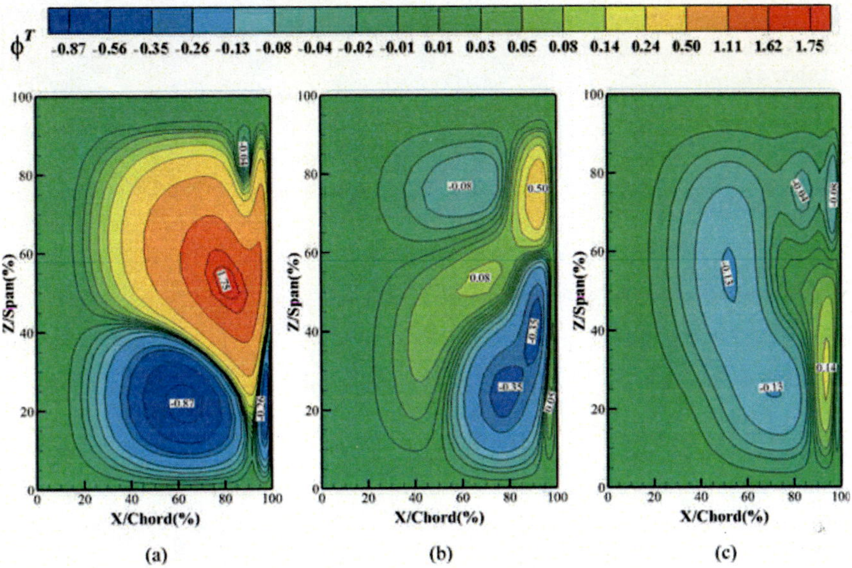

Fig. 3 The first three POD basis (Unit of VG is cm) **a** The first basis function, **b** The second basis function, **c** The third basis function

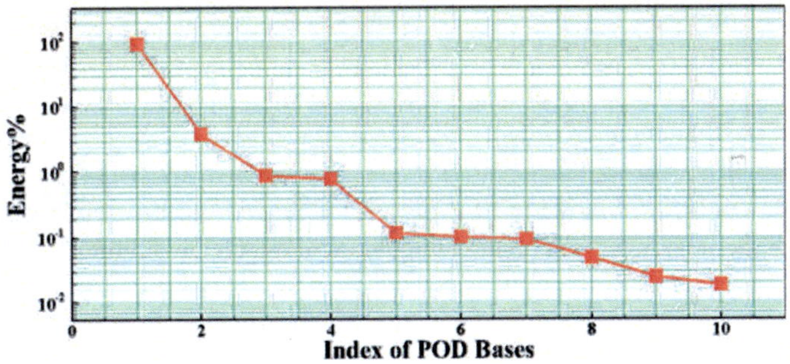

Fig. 4 Energy of each POD basis

the blade. It is a coincidence that the number of selected POD bases is consistent with the number judged by errors in previous studies [21]. The performance of both exact and POD-based shape have been calculated and compared in previous work [21]. The results show that all points are closed except that two points near stalling are with a little deviation implies the reliability of POD-based geometric parameterization.

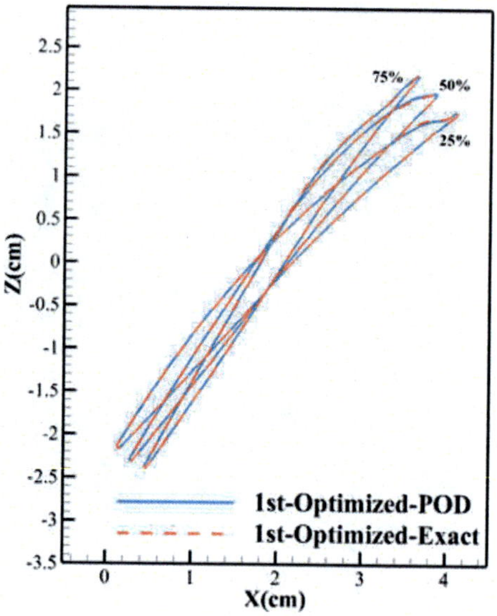

Fig. 5 Comparisons of three sections between exact and POD-based blade

4.3 The 2nd-Step Optimization

The 2nd-step optimization evolving by SDA starts from the 1st-optimized blade. To prove the optimization efficiency of POD-based geometric parameterization, two cases are proceeded. One parameterizes the geometric shape in the same way with the 1st-step optimization as shown in Table 1, the other parameterized by 8 POD basis described in Sec.4.2. SDA can not use the objective function of the 1st-step optimization defined in Eq. (8) for it is discontinuous, so another objective function defined in Ref. [22], which is equal to Eq. (8), is used as shown in Eq. (11).

$$\begin{aligned} &\text{Find}:\quad \boldsymbol{x}^*,\\ &\text{Min: } f = \frac{s}{s_0} + \delta_1\left(\frac{\dot{m}}{\dot{m}_0} - 1\right)^2 + \delta_2\left(\frac{\pi}{\pi_0} - 1\right)^2, \end{aligned} \tag{11}$$

where δ_1 and δ_2 are the weighting factors taking following values: $\delta_1 = \delta_2 = 100$, s denotes the entropy generation which is given by,

$$s = \frac{\int_e \rho u \frac{\rho}{\rho'} ds_x}{\int_e \rho u ds_x} - \frac{\int_i \rho u \frac{\rho}{\rho'} ds_x}{\int_i \rho u ds_x}, \tag{12}$$

where subscript e denotes the exit plane of a computational domain, and i denotes the inlet plane of a computational domain.

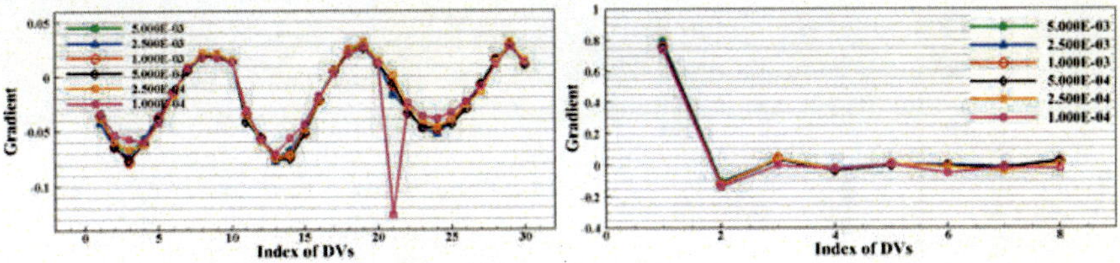

(a) (b)

Fig. 6 Comparison of several gradients with 6 steps **a** 30 DVs based on Hicks-Henne, **b** 8 DVs based on POD bases

SDA will convergence when the normal of the gradient is less than 1.0×10^{-6} or the value of searching step α in SDA is less than 1.0×10^{-4}. And the initial value of α is set to be 1.0×10^{-3}, and attenuation coefficient c is set to be 0.5. The step of calculating gradient is determined by comparing several gradients with 6 steps from 5.0×10^{-3} to 1.0×10^{-4}. The comparisons are shown in Fig. 6, for both 30 Hicks-Henne functions and 8 POD basis, the gradients give bad tendency when the step is small to 1.0×10^{-4}. To avoid truncate error, 1.0×10^{-3} is selected for the two cases.

Table 3 compares the performance between the two 2nd-step optimized blades parameterized respectively by conventional (marked by 2nd-step optimized-HH) and POD-based (marked by 2nd-optimized-POD) method. The former increases the adiabatic efficiency by 1.72%, which is a little better than the latter increases the adiabatic efficiency by 1.67%. Figure 7 shows the history of the objective function of the two 2nd-step optimization designs versus function evolutions. It can be seen that the POD-based 2nd-step optimization convergences after calculating 193 functions is much more efficient than that based on conventional parameterization, which convergences after calculating 655 functions. After calculating the 193 functions, the value of objective function for the HH-based optimization is about 0.8925, which is much higher than the result of the POD-based optimization.

As shown in Fig. 8, the 2nd-step optimized blade parameterized by POD basis further increases the maximum thickness of all three sections, and the blade parameterized by Hicks-Henne functions makes a little less maximum thickness at 50% span. Figure 9 shows comparisons of relative Mach number contours between the

Table 3 Performance comparisons between the two 2nd-step optimized blades parameterized respectively by conventional (marked by 2nd-step optimized-HH) and POD-based (marked by 2nd-optimized-POD) method

	Adiabatic efficiency	Total pressure ratio	Mass flow (kg/s)
2nd-step optimized-8POD	0.8731	1.9849	20.91
Changed	1.67%	–0.61%	0.48%
2nd-step optimized-30HH	0.8736	1.9907	20.96
Changed	1.72%	–0.32%	0.72%

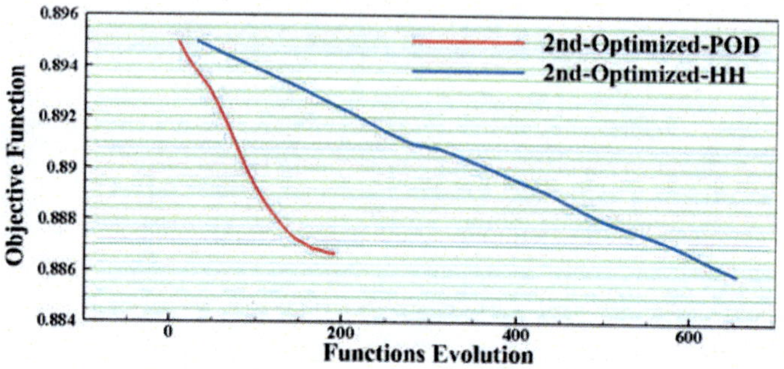

Fig. 7 The objective function history of the two 2nd-step optimization design versus function evolutions

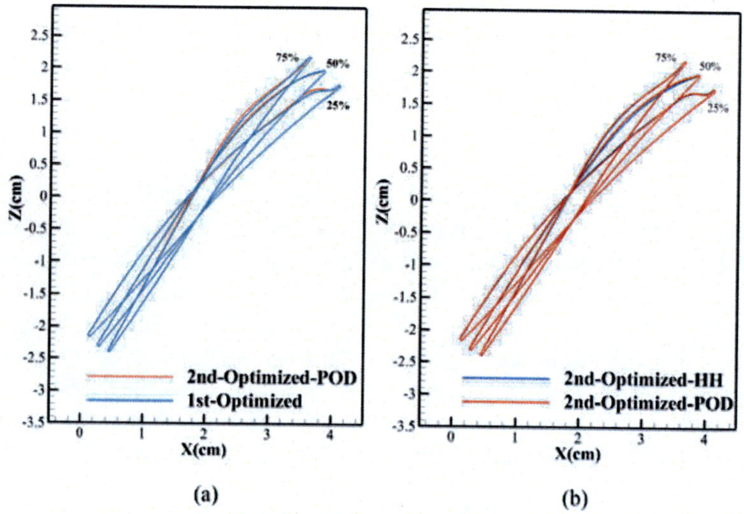

Fig. 8 Comparisons of sections at three spans of three optimized blades **a** 1st-step optimized and 2nd-step optimized-POD, **b** 2nd-step optimized-POD and 2nd-step optimized-HH

two 2nd-step optimized blades at 25, 50, 75% span. As can be seen that, shock losses of Rotor 37 are generated by two shocks, a bow shock and a weak passage normal shock (marked by S_3), for one passage the former can be divided into two parts, its oblique extension ahead of the leading edge (marked by S_1) and the first passage shock (marked by S_2). The patterns of the shock wave of both the 2nd-step optimized blade at three spans are almost the same, and compared with that of the 1st-step optimized blade [21], S_2^{rs}(reflection shock of S_2) at 50% span are eliminated completely. By the analysis above, it can be concluded that the 2nd-optimization using POD-based geometric parameterization is much more efficient than that using Hicks-Henne functions (marked by HH), and the adiabatic efficiency is a little less than that using POD-based geometric parameterization.

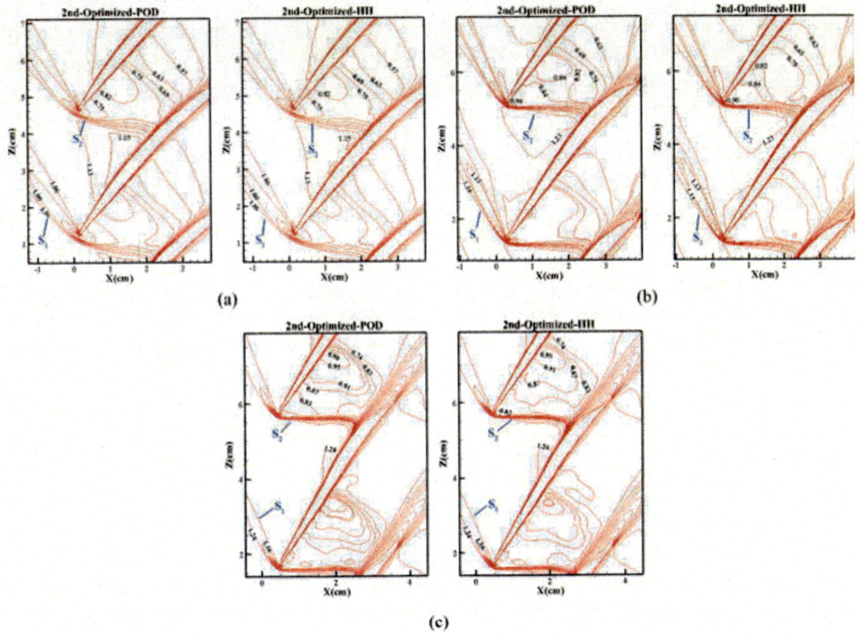

Fig. 9 Comparisons of relative mach number contours between two 2nd-step optimized blades at three spans **a** 25% span, **b** 50% span, **c** 75% span

5 Conclusions

1. The efficiency of the 2nd-step optimization can be improved significantly by using POD-based geometric parameterization, with a little sacrifice of the performance of the optimized solution.
2. Filtering SD by better values of objective functions and satisfied constraints and determining the number of POD basis used for geometric parameterization by system energy larger than 99.99% are both reliable to POD-based geometric parameterization.
3. Comparisons of the flow field and geometric shapes show that values and positions of maximum thickness of the sections of Rotor 37, which narrows the separation region included by the shock wave and weakens the strength of the shock wave, increase the adiabatic efficiency.

References

1. Vicini A, Quagliarella D (1999) Airfoil and wing design through hybrid optimization strategies. AIAA J 37(5):634–641

2. Duan YH, Cai JS, Li YZ (2012) Gappy proper orthogonal decomposition-based two-step optimization for airfoil design. AIAA J 50(4):968–971
3. Chernukhin O, Zingg DW (2013) Multimodality and global optimization in aerodynamic design. AIAA J 51(6):1342–1354
4. Cao CQ, Cai JS, Qu K et al (2017) An efficient multistep ROM method for prediction of flows over airfoils. In: 55th AIAA aerospace sciences meeting
5. Li JC, Cai JS, Qu K (2018) Adjoint-based two-step optimization method using proper orthogonal decomposition and domain decomposition. AIAA J 56(3):1133–1145
6. Han ZH, Zhang Y, Song CX et al (2017) Weighted gradient-enhanced kriging for high-dimensional surrogate modeling and design optimization. AIAA J 55(12):4330–4346
7. Toal DJJ, Bressloff NW, Keane AJ (2010) Geometric filtration using proper orthogonal decomposition for aerodynamic design optimization. AIAA J 48(5):916–928
8. Ghoman S, Wang Z, Chen P et al (2012) A POD-based reduced order design scheme for shape optimization of air vehicles. In: 53rd AIAA/ASME/ASCE/AHS/ASC structures, structural dynamics and materials conference 20th AIAA/ASME/AHS adaptive structures conference 14th AIAA.
9. Poole DJ, Allen CB, Rendall TCS (2015) Metric-based mathematical derivation of efficient airfoil design variables. AIAA J 53(5):1349–1361
10. Masters DA, Taylor NJ, Rendall TCS et al (2017) Geometric comparison of aerofoil shape parameterization methods. AIAA J 55(5):1575–1589
11. Cinquegrana D, Iuliano E (2018) Investigation of adaptive design variables bounds in dimensionality reduction for aerodynamic shape optimization. Comput Fluids 174:89–109
12. Lukaczyk TW, Constantine P, Palacios F et al (2014) Active subspaces for shape optimization. In: 10th AIAA multidisciplinary design optimization conference
13. Robinson GM, Keane AJ (2001) Concise orthogonal representation of supercritical airfoils. J Aircraft 38(3):580–583
14. D'Agostino D, Serani A, Campana EF et al (2018) Deep autoencoder for off-line design-space dimensionality reduction in shape optimization. In: 2018 AIAA/ASCE/AHS/ASC structures, structural dynamics, and materials conference
15. Holmes P, Lumley JL, Berkooz G et al (2012) Turbulence, coherent structures, dynamical systems and symmetry. Cambridge University Press
16. Sirovich L, Kirby M (1987) Turbulence and the dynamics of coherent structures. part 1: coherent structures. Q Appl Math 45(3):561–571
17. Duan YH, Wu WH, Fan ZL et al (2016) An introduction of aerodynamic shape optimization platform for compressor blade. Turbo expo: power for land, sea, and air. american society of mechanical engineers
18. Shi Y, Eberhart RC (1998) Parameter selection in particle swarm optimization. International conference on evolutionary programming. Springer, Berlin, Heidelberg.
19. Venter G, Sobieszczanski-Sobieski J (2003) Particle swarm optimization. AIAA J 41(8):1583–1589
20. Reuther J, Jameson A (1995) Aerodynamic shape optimization of wing and wing-body configurations using control theory. In: 33rd aerospace sciences meeting and exhibit
21. Duan YH, Wu WH, Zhang PH et al (2019) Performance improvement of optimization solutions by POD-based data mining. Chinese J Aeronaut 32(4):826–838
22. Wang DX, He L, Li YS et al (2010) Adjoint aerodynamic design optimization for blades in multistage turbomachines—part ii: validation and application. J Turbomach 132(2)

Anti-icing Performance of Engine Inlet Cone by Hot Air Film

Xinwei Jiang, Guochao Liu, Yundan Li, Qi Jia, and Jianjun Zhou

Abstract Numerical simulation and icing wind tunnel test were carried out to understand the anti-icing performance of engine inlet cone by hot air film. The cones with different slots and holes were investigated under the typical icing conditions. For slot film structures, as the inclination angle decreases from 90° to 30°, the heating efficiency increases about 17% and the amount of ice decreases about 26% under the same hot air mass flow rate. Compared with inclination angle, the effect of slot width on heating and accretion is much weaker. For circular hole, as the diameter of the holes decrease from 6 to 3 mm, the averaged wall temperature increases by about 2–5 °C, and the amount of ice decreases more than 50%. The results of icing wind tunnel tests agree well with numerical results. Increasing the temperature and flow rate of hot air can obviously reduce the amount of ice on the cone. Overall, reducing the slot inclination angle and diameter of circular hole can both improve anti-icing performance.

Keywords Anti-icing · Ice accretion · Hot air film · Engine cone

1 Introduction

When an aircraft passes through a cloud containing supercooled water droplets, icing will occur on the surface of engine inlet components. If appropriate measures are not taken in time, it will bring serious threat to flight. At present, the most widely used icing protection method of aero-engine is hot air anti-icing system, that is, hot air is bleed from compressor through the pipeline to the anti-icing parts of engine for heating and anti-icing. For engine inlet cone, the hot air tends to be discharged from the discrete holes or slots at suitable axial location, forming a film on the cone surface to prevent or reduce icing. In recent years, numerical simulations and experimental studies have been carried out on the thermal loads [1]and the heating effects on the surface of anti-icing parts [2]. Through the research on the small aero-engine cone

X. Jiang (✉) · G. Liu · Y. Li · Q. Jia · J. Zhou
AECC Shenyang Engine Research Institute, Shenyang, China
e-mail: jiangxinwei_nwpu@163.com

S. Lee et al. (eds.), *The Proceedings of the 2021 Asia-Pacific International Symposium on Aerospace Technology (APISAT 2021), Volume 1*, Lecture Notes in Electrical Engineering 912, https://doi.org/10.1007/978-981-19-2689-1_12

anti-icing, the results show that the air film anti-icing system is effective on the leading edge of the cone [3]. Further, the boundary layer shape of the exit area on the aero-engine inlet components was investigated [4], and the jet impingement heat transfer characteristics of the internal flow and the heating characteristics of the external air film were calculated and analyed [5]. A hot air film anti-icing system suitable for the structure of the cone was proposed [6]. The effect of different geometric parameters of outlet slots on anti-icing at the aero-engine inlet was obtained [7]. The results of the water droplets impingement characteristics show that the hot air film has an obvious sweep effect on the droplets [8].

Based on the analysis of above literatures, this paper studied the icing and anti-icing characteristics of a typical engine inlet cone with hot air film structures. The influences of geometric and aerodynamic air film parameters on anti-icing performance of cone surface were investigated under the typical icing conditions. The results can help researchers understand the mechanisms and laws of anti-icing by hot air film more deeply.

2 Numerical Investigations

2.1 *Model and Methods*

When modeling, the complex structures of hot air channel inside the cone are simplified. In order to model the mainstream, an area consistent with the engine inlet is established outside the cone, and a steady flow section with a length of 500 mm and 300 mm is set in the inflow area and outflow area respectively. As shown in Fig. 1, Take 1/9 fan section in the circumferential direction of the cone for modeling.

The local water collection efficiency β is defined as:

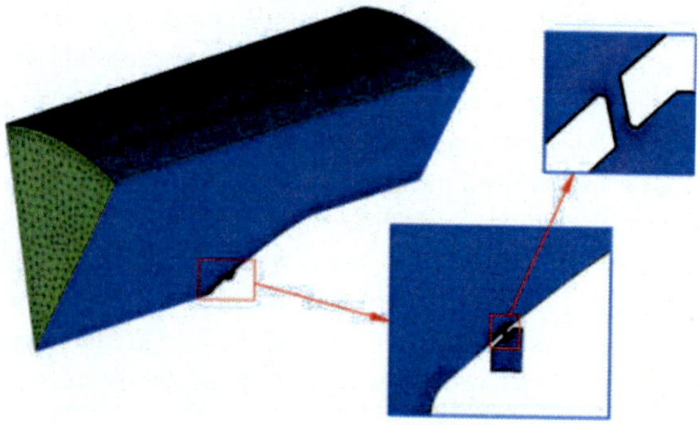

Fig. 1 Meshes of the computational domain

$$\beta = -\frac{\partial \vec{V}_d . \vec{n}}{(LWC)V_\infty} \tag{1}$$

LWC is the amount of liquid water in the mainstream. $\vec{n}$ is the normal vector of the impingement surface, V_∞ is the velocity of free flow, and V_d is the velocity of droplets.

By solving the water film velocity on the wall surface, the ice calculation continuity equation is obtained.

$$\rho_w \left[\frac{\partial h_f}{\partial t} + \nabla \cdot \left(V_f h_f \right) \right] = V_\infty LWC\beta - \dot{m}_{evap} - \dot{m}_{ice} \tag{2}$$

The energy equation is:

$$\rho_w \left[\frac{\partial h_f c_f \tilde{T}_f}{\partial t} + \nabla \cdot \left(V_f h_f c_f \tilde{T}_f \right) \right]$$

$$= \left[c_w \left(\tilde{T}_\infty - \tilde{T}_f \right) + \frac{V_d^2}{2} \right] V_\infty LWC\beta + \left(L_{fusion} - c_{ice}\tilde{T} \right) \dot{m}_{ice} \tag{3}$$

$$-L_{evap}\dot{m}_{evap} + \varepsilon\sigma \left(T_\infty^4 - T_f^4 \right) - c_h \left(\tilde{T}_f - \tilde{T}_{ice,rec} \right) + Q_{anti-icing} \tag{4}$$

h_f is the thickness of the water film, and V_f is the velocity of the water film on the wall. ρ_w is the density of water, c_w is the specific heat at constant pressure of water, ε is the emissivity of water, σ as the radiation constant, L_{evap} as the latent heat of evaporation of water, and L_{fusion} is the latent heat of melting of ice. c_{ice} is the specific heat at constant pressure of ice, β is the local water collection efficiency for droplets on the wall, $\tilde{T}$ as the Celsius temperature and the others are kelvin temperature, $Q_{anti-icing}$ is the anti-icing heat flow rate.

2.2 *Boundary Conditions*

During the calculation, boundary conditions are set as follows: the temperature of water droplets in the mainstream is minus 10 °C, the Liquid Water Content (LWC) in the state is 2 g/m^3, the Medium Volume Diameter (MVD) of water droplets is 20 μm, the mainstream velocities are 45 and 180 m/s, the hot air flow temperatures for anti-icing are 70 and 140 °C, the heat flow rates are 20, 40, 60 g/s and 80 g/s, the inclination angles of the slot are 30°, 60° and 90°, and the ice accretion time is 100 s.

2.3 *Results and Discussions*

2.3.1 Influence of Film Slot Structure Parameters on Icing Characteristics

The ice thickness of the cone surface under different slot angles (from 30 to 90) and hot air mass flow (20 and 40 g/s) is given in Fig. 2. From the graph, the largest ice developed at the leading edge, and the film slot has formed a protection for its rear area. As the film slot inclination angle increases, the ice thickness of this area decreases. When the inclination angle is 30, there is no ice developed behind the film slot. Figure 3 shows the accretion mass curves of each structure. Through comparison, the amount of ice on 30° angle slot is about 26% less than 90° slot.

Figure 4 shows the influences of different slot inclination angles on the cone surface temperatures. Results show that 30° inclination angle structures have highest

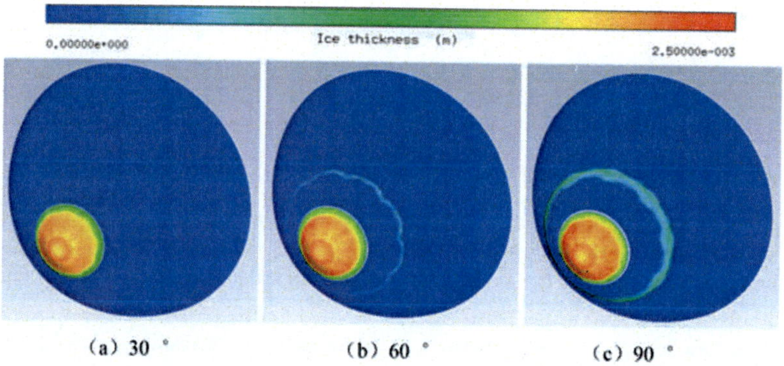

Fig. 2 Ice thickness of cone surface with different inclination angle slots (mainstream = 45 m/s)

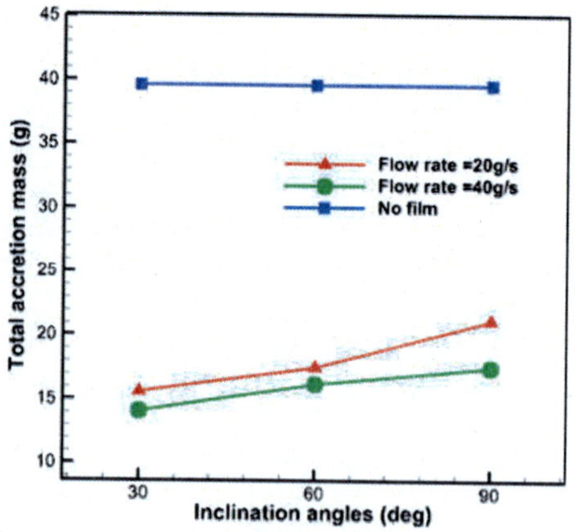

Fig. 3 Accretion mass of different inclination angle slots

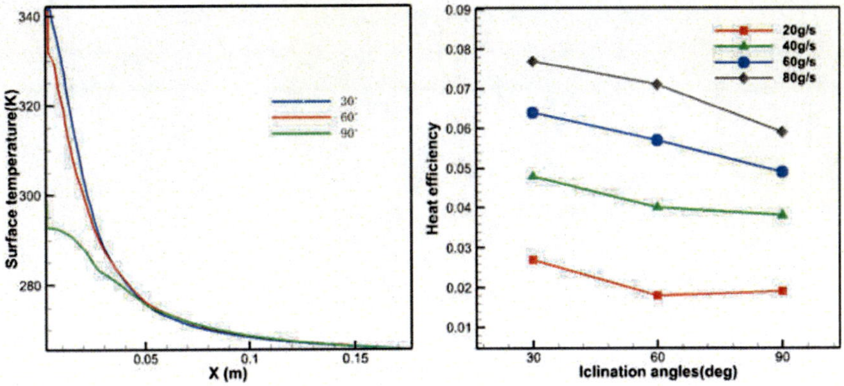

(a) Surface temperature (b) Heating efficiency

Fig. 4 Influence of different inclination angle slots on heating effect of cone surface

average wall temperature, the 90° angle structures are the lowest, and the temperature behind the slot are much lower than the other two structures. The decrease of average temperature is more obviously as the inclination angles increase. This trend indicates that with the increase of the slot inclination angle, more hot air will mix with mainstream cold air, which reduces the heating flow temperature near the wall, so the wall temperature decreases accordingly. As shown in the heat efficiency curves, 30° inclination slots have the best heating efficiency, which is about 17% higher than 90°.

Figure 5 shows the accretion mass curves on the cone surface under different widths of film slots. It can be seen that under the given conditions of three slot widths, there is no significant influence on the icing thickness over the cone surface.

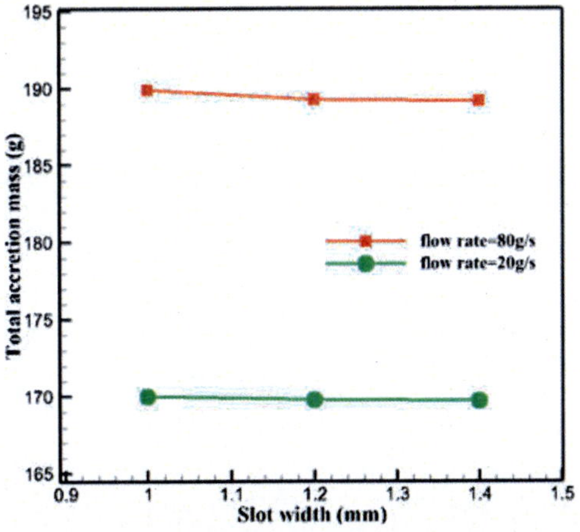

Fig. 5 Influence of different slot widths on the amount of ice over the cone

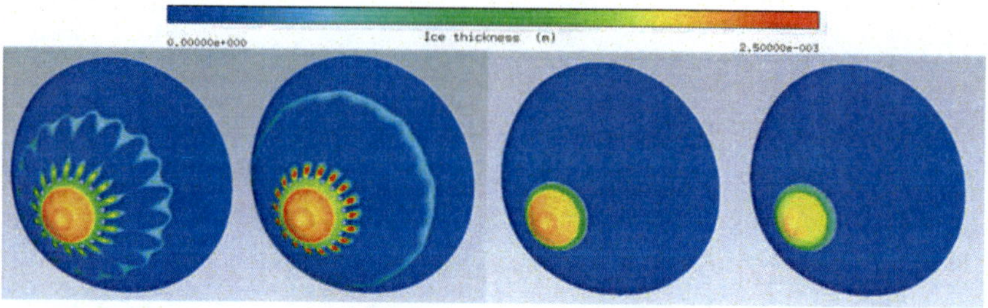

a. d=6mm, g=40g/s b. d=6mm, g=80g/s c. d=3mm, g=40g/s d .d=3mm, g=80g/s

Fig. 6 Contour plots of anti-icing characteristics on the cone surface with different film hole diameters (d as diameter g as the heat mass flow rate)

In general, ice accretion on the cone surface is not sensitive to the width of the film slot.

2.3.2 Influence of Film Holes Diameter on Icing Characteristics

Figure 6 shows the contour plots of anti-icing characteristics on the cone surface of 3 mm and 6 mm diameter film holes. For the 6 mm diameter hole structures, the area between the two film holes on the cone surface will produce a certain amount of ice, and then a small amount of circumferential ice will be formed near the downstream. With the increase of the hot air flow rate, due to the sweeping and heating effects, the icing amount in the middle area of the cone surface decreases and the icing coverage area moves to downstream of the cone. For the 3 mm diameter hole structure, when the mainstream flow rate is 45 m/s, there is almost no ice on the cone surface behind the film hole.

Figure 7 shows the comparison curves of accretion mass for different calculation conditions. Under the condition of low mainstream velocity, the big holes structure of 6 mm have more ice accretion significantly than the small holes of 3 mm, which is about 50% more. The influence of the hole diameters on the ice amount over the cone is not obvious under high velocity mainstream conditions. But the accretion mass of the small holes structure is still less than that of the big holes structure.

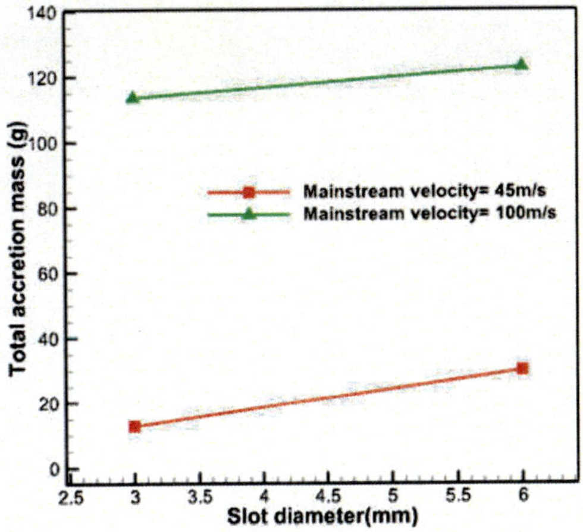

Fig. 7 Accretion mass of different film hole diameters

3 Experimental Investigations

3.1 Experiment Systems

The verification of experimental technology was carried out on the icing wind tunnel of AECC Shenyang Engine Research Institute. The facility is composed of eleven parts, including wind tunnel, air dynamic system, cooling system, atomization water spray system, vacuum system, secondary flow system, circulating water system, softening water system, corner heat system, control system, viewing system and corresponding auxiliary systems. It is mainly used for engine inlet cone, strut and other inlet components tests and verifications. The structure of the experimental system is shown in Fig. 8. The specified water droplet parameters will be achieved by spraying water through the atomization system. The icing and anti-icing process were recorded by cameras, and the wall temperatures of cone were measured by thermocouples. Figure 9 shows the experimental products of different film structure, the big film hole diameter is 6 mm, the small film hole diameter is 3 mm, and the film slot inclination angles are 30°, 60°,90°.

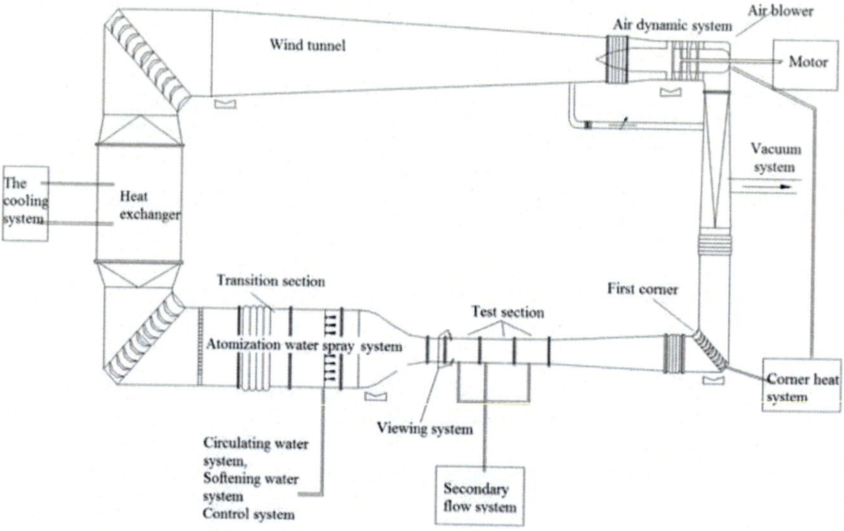

Fig. 8 Experimental systems

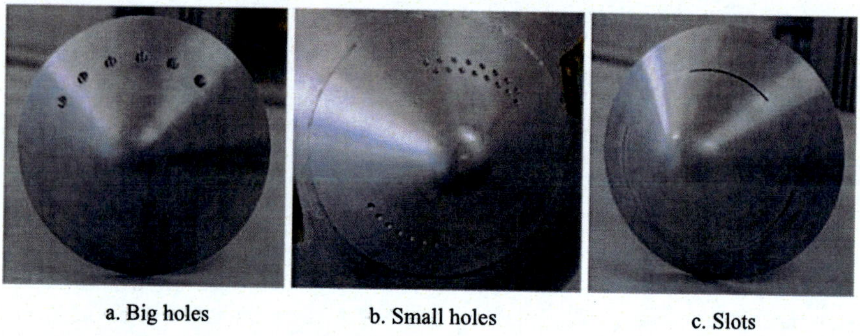

Fig. 9 Experimental products of different film structure

3.2 Experimental Results and Discussions

3.2.1 Influence of Hot Film Parameters on Anti-icing

The influences of hot air flow rate and temperature on anti-icing were obtained through the experimental investigation. Figure 10 shows the ice accretion on the cone surface under different heat flow rates and temperatures. It can be seen from the figure that at a small heat flow rate, a thin layer of ice was formed on the cone surface, covering the area from the middle of the cone to the trailing edge. When the hot air flow increases, the ice behind the film hole had a certain extent melting, but the overall ice amount did not change much. Furthermore, when the temperature of

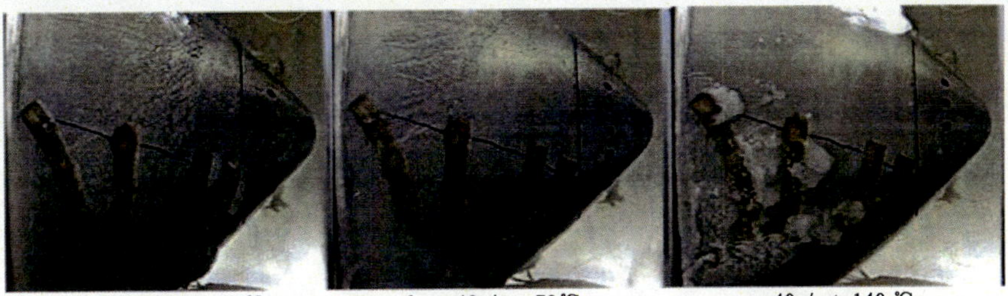

Fig. 10 Ice accretion on the cone surface of different heat flow rates and temperatures

the hot air increases from 70 to 140 °C, the amount of ice in the rear area behind the film hole decreased significantly, and there is almost no ice in the middle of the cone, with only a very small amount of ice in the tail. As a whole, increasing the temperature and flow rate of hot air can obviously reduce the amount of ice on the cone.

3.2.2 Influence of Film Structure Parameters on Anti-icing

Figure 11 shows the comparison of icing on the cone surface with 6 and 3 mm diameter slot under the same flow rate conditions. It can be seen that icing occurs on the surface of the 6 mm large holes structure. For the small holes structure, there is basically no icing behind the film hole except for a slight icing at the thermocouple points. This indicates that within the range of working conditions studied in this paper, the anti-icing effect of 3 mm small hole structures is better than that of 6 mm large hole structures.

Figure 12 shows the comparison curve of wall temperatures for the two diameter structures in icing environment. The average wall temperature of small hole structure was about 2–5 °C higher than the large hole structure over the whole surface of cone. It can be confirm that small hole film structure can heat the temperature of the rear wall to a high level than big hole under the same condition of mainstream and secondary

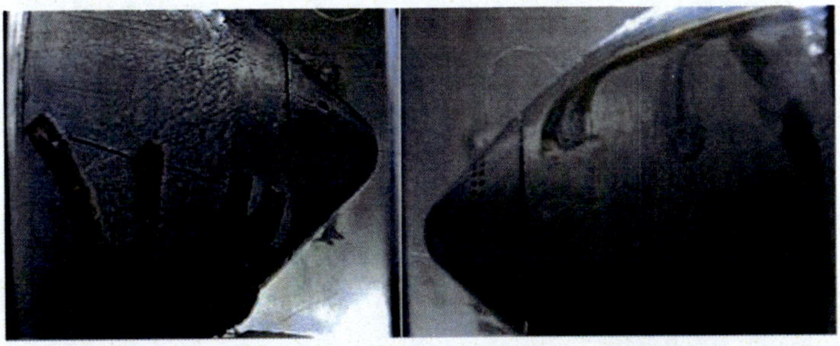

Fig. 11 Comparison of icing on the cone surface with 6 and 3 mm diameter slot

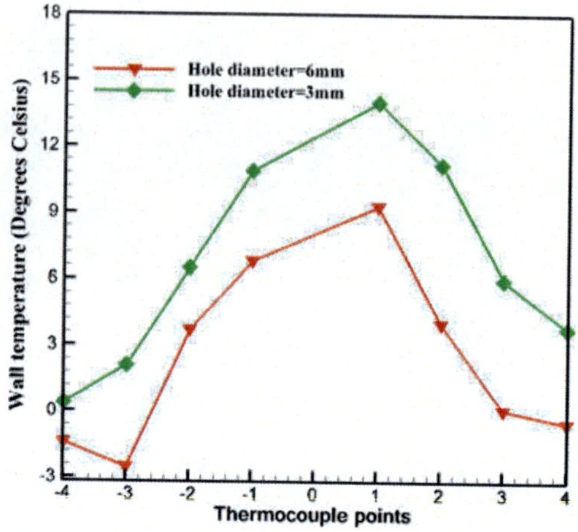

Fig. 12 Comparison curve of the wall temperature

flow, and it is easier to provide a protection when the supercooled water droplets impinge on the surface.

3.2.3 Influence of Film Slots Inclination Angles on Anti-icing

Figure 13 shows the influence of different film slot angles on the anti-icing over the cone surface. On the left, the film slot angles are 30° and 90°. And the film angles on the right are 60°. By comparing the accretions on the cone, the icing amount behind 30° slot was the smallest. With the increased of the angle, the icing amount behind slot increased gradually. And the accretion behind 90 was the heaviest. This is because the smaller the angle of the slot, the stronger the sweeping and heating effects of the hot air on the surface. When the angle increases, more hot air will mix with the mainstream, leading to the reduction of the heating effect of the hot air near the wall.

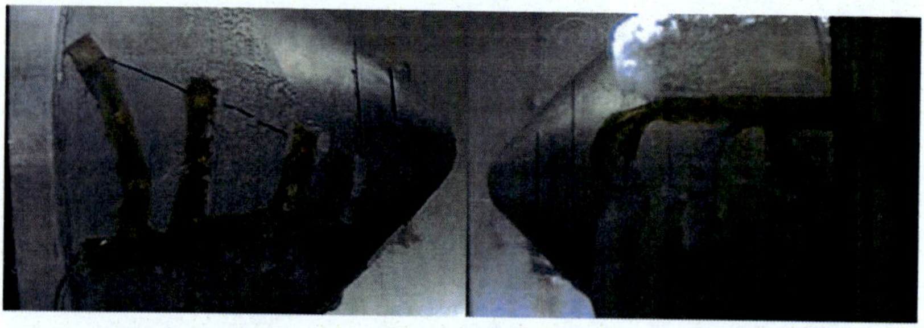

Fig. 13 Influence of different film slot angles on the anti-icing

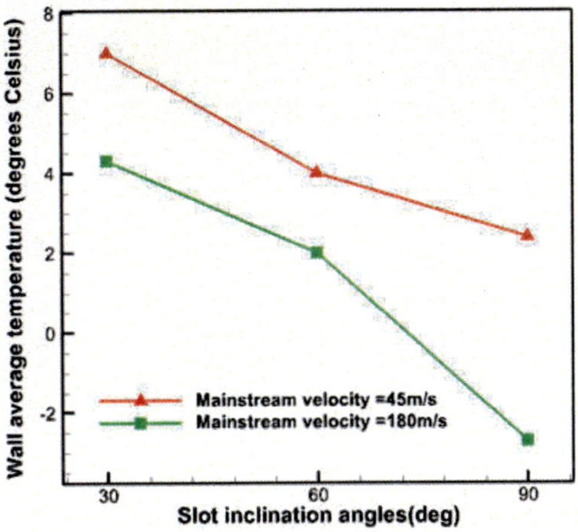

Fig. 14 Comparison curve of the wall temperature with different film slot angles

Figure 14 shows the wall temperature curves with different film slot angles and mainstream velocities. It can be seen from the figure that with the increase of the film slot angle, the wall temperature on the surface keeps decreasing, with a declination of 7 °C, and its anti-icing effect in consequence weakens gradually.

4 Conclusions

In this paper, the hot air film anti-icing performances of engine inlet cone under typical icing conditions were investigated by experiments and numerical simulations. The following conclusions can be obtained.

1. For film slot structures, as the inclination angle decreases, the heating efficiency of the cone surface increases and the amount of ice decreases accordingly, and the accretion is not sensitive to the width of the film slots.
2. For film hole structures, the anti-icing effect of small hole structures is better than that of large hole structures. The film with small hole structure can heat the temperature to a higher level than large one.
3. Increasing the temperature of hot air film can effectively raise the wall temperature of the cone, and the flow rate of hot air is important factor affecting icing as well.
4. The results can help researchers understand the mechanisms and laws of anti-icing by hot air film more deeply.

References

1. Hua L, Wen G, Jun Y (2012) Numerical investigation of anti-icing thermal loads for engine inlet cowl. Gas Turbine Exper Res (01)
2. Planquart P, Vanden Borre G, Buchlin JM (2005) Experimental and numerical optimization of a wing leading edge hot air anti-icing system. In: AIAA 2005–1277
3. Dong W, Zhu J, Zheng M et al (2015) Thermal analysis and testing of nonrotating cone with hot-air anti-icing system. J Propul Power 31(3):1–8
4. Downs SJ, James EH (1987) Boundary layer profile investigations downstream of an aero-engine intake anti-icing air exhaust slot. In: ASME gas turbine conference and exhibition
5. Peng KE, Huiyun YANG, Junkai WANG (2018) Heating characteristics of aero-engine nose cone with film-heating anti-icing system. J Aerosp Power 33(3):530–539
6. Yang H, Ke P, Yang C (2016) Analysis of droplet impingement characteristics of aero-engine nose cone with hot air film. In: Asme turbo expo: turbomachinery technical conference and exposition
7. Riley SJ, James EH (1994) Influence of various exhaust geometrical parameters on the effectiveness of an aero-engine intake thermal anti-icing system. In: ASME proceedings of the international gas turbine and aero-engine congress and exposition
8. Peng K, Yun Z; Guang-Feng Y (2017) Influence of exterior hot-film on droplet impingement characteristics over aero-engine inlet strut. J Aerosp Power 32(3):621–629

Aerodynamic Characteristics of Close UAV Formation

Ziyu Li, Zhengping Wang, and Zhou Zhou

Abstract The aim of this work is to study the aerodynamic characteristics of solar unmanned aerial vehicle (UAV) wake surfing. First, the application of the following methods to this kind of aircraft was compared: rapid method (i.e., horseshoe vortex model, lifting line theory and nonlinear lifting line theory) and computational fluid dynamics (CFD). The results showed that nonlinear lifting line theory (NLLT) is more accurate in the calculation of the aerodynamic force. The flow field of wake surfing was evaluated with CFD and the local angle of attack was calculated with a local lift coefficient from the CFD results. It was found that the linear method that added induced angle of attack directly to the rear wing overestimated the influence of wing-tip vortices on the rear aircraft. The classical vortex model was introduced to explain this error in a way of vortex decay, which means the rear wing can influence trailing vortices of the lead wing. The error is quantitatively studied by introducing vortex core radius into the calculation, and it was found that the closer trailing vortices act on the rear plane, the stronger the decay shows. After analysis of the characteristics, the vortex core radius was finally introduced into the fast method to modify errors caused by vortex decay. The modified method can reduce the amount of computation in engineering evaluation and give more accurate predictions.

Keywords Aerodynamic characteristics · Formation flight · Wake surfing · Vortex model

1 Introduction

Many birds, such as Canada goose in North America and ibis in Europe, perform long-distance migration in specific seasons. During migration, they often fly in a V-shaped formation, which can provide a positive induced angle of attack for individuals in the formation, thus reducing the overall flight energy consumption. This phenomenon has sparked research interest in the aviation community because of its potential for

Z. Li (✉) · Z. Wang · Z. Zhou
Northwestern Polytechnical University, Xian, China
e-mail: npu_lzy@foxmail.com

S. Lee et al. (eds.), *The Proceedings of the 2021 Asia-Pacific International Symposium on Aerospace Technology (APISAT 2021), Volume 1*, Lecture Notes in Electrical Engineering 912, https://doi.org/10.1007/978-981-19-2689-1_13

Fig. 1 Helios solar UAV

energy conservation, with researchers focusing on fixed-wing drone formation wake surfing.

The aim of this work is to study the aerodynamic characteristics of wake surfing of a solar-powered unmanned aerial vehicle (UAV). Solar UAV is a kind of UAV that pursues long endurance. Its goal is to achieve sustainable flight in the sky. They are often used in reconnaissance and signal base stations. Formation flight of several solar UAVs with relatively small aspect ratio can further increase their flying time and improve the aeroelastic problems caused by their very high aspect ratio. Helios is a typical solar UAV as shown in Fig. 1. In 2003, it broke down due to aeroelastic problems.

1.1 Previous Work

The study of bird formation aerodynamics began in the early twentieth century. In 1914, Wieselsberger, a researcher in Gottingen, first correctly studied the aerodynamics of this formation [1]. When birds are in formation, the trailing vortices of the lead bird produce an upwash and downwash area, which changed the local angle of attack of the rear wings. This alters lift and deflects the direction of the local lift at the same time, resulting in the change of total lift and drag. Therefore, in the study of formation flight, the most important aspect is to study the local angle of attack along the spanwise direction of the rear wing. In 1942, Schlichting, one of Prandtl's students, used the horseshoe vortex model (HS), instead of the elliptical wing, to achieve energy saving of a two aircraft formation [2]. In 1970, Lissaman and Shollenberger made a comprehensive discussion on formation shape and energy consumption based on the lifting line theory (LLT). According to their calculation, the average maximum lift-drag ratio of 25 birds in a V-shaped formation can be increased by a maximum of 71%, compared with a single bird [3]. In 1974, zoologists Gould and Heppner took pictures of Canada geese with three cameras and restored the formation shape of birds through geometric calculations. They discovered that the distance between wingtips of birds in formation was often negative, which means that wingtips overlapped in the spanwise direction [4]. Later, other biologists observed and studied various bird formations by means of photography, radar, GPS and inertial navigation systems [5, 6].

With the prospect of reducing drag in formation flight, several literature articles exist on formation flight, most of which are based on potential flow theory. Hummel extended the HS model to more general wings, studied with Dock how to balance the roll and yaw moment caused by the formation with the rudder surface, and conducted flight tests with two DO-28 s. According to his experiments, the energy saving of the rear aircraft in the formation of two aircraft was about 15% [8]. In 2004, Blake and Gingras conducted wind tunnel tests on two delta-wing aircraft, compared them with the results predicted by the vortex lattice method, and found that the induced drag could be reduced by a maximum of 25% [7]. In 2015, Portugal and Hubel [9] studied the formation of ibis with high-precision inertial navigation and GPS, and believed that the flapping of the wings would have an impact on the structure of the wing-tip vortex, with the birds behind also capturing upstream by adjusting the frequency and phase of the flapping wing. NASA conducted an experiment in 2017 in which two C-20A transport aircraft flew in formation [10]. Wei Liu and Minghua Zhang used the Reynolds-averaged Navier–Stokes equation (RANS) method to simulate wing-tip vortices, and compared them with the wind tunnel test results. The results showed that the Realizable k-ε(RKE) turbulence model was better at simulating the wing-tip vortex [11, 12]. Yuan Yuan conducted computational fluid dynamics (CFD) simulations on the aerodynamic characteristics of two X-47B aircraft formation, and compared the results with those based on the HS aerodynamic model [13]. Based on Kriging model, Yang Tao established a response surface model of aerodynamic force of rear aircraft [14].

1.2 Our Work

Fast methods to simulate the wake of the aircraft were evaluated and compared with the CFD simulation. The horseshoe vortex model (HS) predicted an inaccurate position of wake vortex core, while lifting line theory (LLT) produced higher accuracy in wake calculation. The nonlinear lifting line theory (NLLT) was also evaluated in the calculation of the aerodynamic force with the local angle of attack, and showed good accuracy for most areas of wings. Nevertheless, for the area close to wingtip, the accuracy of this method decreased.

To study the influence of the rear wing to the trailing vortices of lead wing, the wake flow field of the single aircraft was first simulated, and the local induced angle of attack at the rear wing position was calculated from the previous CFD simulation. With the calculated local induced angle of attack, the aerodynamic force distribution of the rear wing was also calculated. Because the above calculation assumes that the existence of the rear wing has no effect on the trailing vortices from the front, the difference between the CFD numerical simulation can be regarded as the result of the influence we aim to study. The results showed that NLLT predicted a larger aerodynamic interference than the CFD simulation. By introducing the classical vortex model, the influence of the rear aircraft is explained by the decay of vortex. A parameter of decay radius was introduced to study the decay quantitatively, and we discovered that the decay was small and indistinguishable when the vortex core of the lead aircraft was far from the following aircraft. When the lead wing tip vortex hits the back aircraft, this decay happens. The more trailing vortices get closer to the fuselage of the rear wing, the stronger the decay becomes.

Finally, the aerodynamic characteristics of the rear aircraft were determined by calculating the decay radius of other relative positions. Our modified fast method fit well with CFD results and can be extended to more general type aircraft, which means CFD calculations can be reduced by using methods.

2 Basic Method and Validation

In this section, the classical vortex model, wake model and NLLT are introduced and tested by numerical simulation results in the flight state of solar UAV.

2.1 Vortex Model

A flow field section of a single wing-tip vortex can be regarded as a special two-dimensional flow, after some distance; its internal flow is driven by a circle of flow for forced motion and a free vortex is generated outside, which can be approximated as the flow field generated by a rotating cylinder with radius r_c. This kind of model is called Rankine vortex. Its velocity profile in polar coordinates is:

$$\begin{cases} v_\theta(r) = \frac{\Gamma}{2\pi r}\frac{r^2}{r_c^2}, r \leq r_c \\ v_\theta(r) = \frac{\Gamma}{2\pi r}, r > r_c \end{cases} \quad (1)$$

The Rankine vortex model gives a relatively graphic analogy, but the model has a sudden change in the slope of the velocity, which is impossible in the actual flow, and the model also overestimates the tangential velocity. Aiming to address these shortcomings, Vatistas proposed a family of smooth algebraic velocity models approximating the Rankine vortex. [17]

$$v_\theta = \frac{\Gamma}{2\pi r}\frac{r^2}{\left(r_c^{2n} + r^{2n}\right)^{1/n}}, \quad (2)$$

where the model when $n = 1$ is also called Scully vortex, as shown in Eq. (3). Burnham and Hallock also proposed this equation. When $n = 2$, this velocity model also simulates the wing-tip vortex well, and when n approaches infinity, this algebraic model approximates the Rankine vortex.

$$v_\theta = \frac{\Gamma}{2\pi r}\frac{r^2}{r^2 + r_c^2} \tag{3}$$

The solution is obtained by Oseen and Lamb, also known as Oseen–Lamb vortex. The velocity type of Oseen–Lamb vortex is derivative, and when it is equal to zero, the vortex core radius $r_c(T)$ (i.e., the radius at the maximum tangential velocity) is obtained by the iterative method. The constant is the kinetic viscosity coefficient, which can be solved for as:

$$r_c(t) = \sqrt{4\alpha \nu t} \tag{4}$$

The decay of the wing-tip vortex is similar to that of line vortex. However, since the wing-tip vortex has stronger three dimensional and viscous effects than the two-dimensional line vortex, the wing-tip vortex decays faster. Therefore, the decay equation of the line vortex cannot be directly applied, but the equation for the radius change of wing-tip vortex can be written in a similar form:

$$r_c = c_{mod}\sqrt{t}, t = \frac{x_{vor}}{V_\infty}, \tag{5}$$

where x_{vor} is the flow direction distance from the trailing edge of the wing to the wing-tip vortex, and the empirical equation given by the literature is $r_C = 0.52\sqrt{t}$. [22]

A straight wing with a chord length of 1 m, aspect ratio of 16 and airfoil of e387 as was used an example to verify this equation. The angle of attack of incoming flow was 0°, and the finite volume method was used to solve the Reynolds average Navier–Stokes equation, by referring to the calculation setting of reference [11, 12]. The RKE turbulence model was adopted, the incoming flow velocity was 30 m/s, and the altitude was 500 m in the standard atmosphere. The Calculated Reynolds number was 2 mil and the incoming flow turbulence was 5%. The half-mode symmetry plane was used in the calculation. The mesh quantity was 4.5 mil, with denser mesh in the wing-tip and wing-tip vortex regions. Mesh fit the condition of $y+ < 1$. The calculated CFD flow field was extracted, the wing-tip vortex core radius distribution was obtained from the wake flow field, and the parameter was obtained according to the least square estimation (Fig. 2):

$$r_c = 0.58\sqrt{t} \tag{6}$$

The origin of the x-coordinate in the figure above was at the front edge of the wing symmetry, and the direction was along the flow direction. This equation can fit the variation of wing-tip vortex core size well, but there are certain differences between this example and the parameter in the literature, which may be caused by

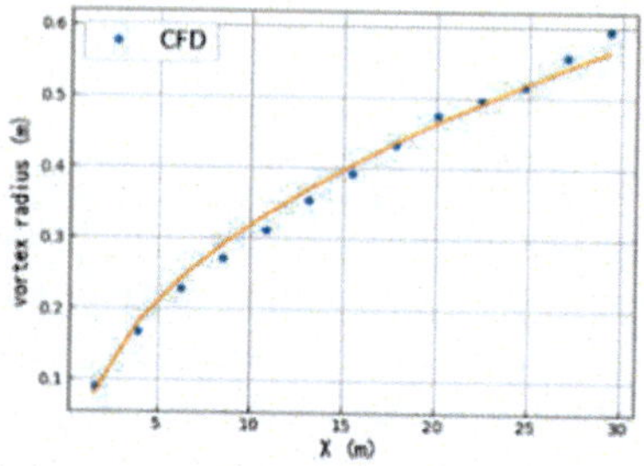

Fig. 2 Vortex core radius variation with distance to wing

the different flight altitudes. In the example in this section, the kinetic viscosity coefficient was $\nu = 1.53e - 5$. Since r_c is proportional to $\sqrt{\nu}$, the parameters c_{mod} for other flight altitudes and temperatures can be modified according to the kinetic viscosity coefficient.

2.2 *Nonlinear Lifting Line Theory*

Classical LLT relates the geometric angle of attack, induced angle of attack and twist chord length distribution. However, the slope of the lift—angle of attack line is assumed to be linear, and the accuracy of the calculated results is low, when applied to UAV with low flight speed.

The NLLT based on the iterative method is able to calculate the lift distribution of the aircraft more accurately. The method was used in this paper to establish the relationship between the angle of upwash and the lift distribution.

For a wing with a general lift distribution, the spanwise lift distribution and induced angle of attack can be written as follows:

$$\frac{c_l c}{b} = \sum A_n sin(n\theta), \alpha_i = \frac{180}{4\pi sin(\theta)} \sum nA_n sin(n\theta) \quad (7)$$

With the order of Fourier series r-1, and θ divided evenly into r parts, $\theta = \frac{m\pi}{r}, m = 1, 2, \ldots, r-1$, the equation can be obtained (Fig. 4):

$$\alpha_{i_k} = \sum_{m=1}^{r-1} \left(\frac{c_l c}{b}\right)_m \cdot \frac{180}{4\pi r sin(k\pi/r)} \sum_{n=1}^{r-1} n\left[cos\left(n\frac{(k-m)\pi}{r}\right) - cos\left(n\frac{(k+m)\pi}{r}\right)\right] \quad (8)$$

The lift distribution of the wing can be obtained through the iteration of (8). Steps are as follows:

1. Give an initial lift distribution.
2. Induced angle of attack distribution is obtained by Eq. (8).

3. The actual angle of attack distribution is calculated by twist, upwash (from the forward wing-tip vortex), the incoming flow angle of attack and induced angle of attack with equation $\alpha_{eff} = \alpha_{twist} - \alpha_i + \alpha_{free} + \alpha_{UP}$.
4. According to the airfoil of the corresponding section, the actual angle of attack distribution and flow conditions, the two-dimensional airfoil aerodynamic characteristics are calculated by CFD.
5. If the lift distribution does not converge, go back to step 2, otherwise get the lift distribution of the wing.

The algorithm was verified by the example in Sect. 1.1, and the boundary conditions, turbulence model, calculation settings and grid structure of the two-dimensional airfoil were consistent with those of the three-dimensional example. In the calculation of NLLT in this paper, the order of sine was 30 (Fig. 6).

The airfoil pressure distribution corresponding to the local angle of attack calculated by NLLT are presented in Fig. 3, and the points obtained by CFD simulation. The corresponding local angle of attack was -0.2°, -1°, and -1.8°.

In most areas of the wing, the theoretical results match the two-dimensional aerodynamic force with the three-dimensional aerodynamic forces. However, at the tip of the wing, results of the NLLT deviate from the results of the CFD simulation, because the flow no longer conforms to the hypothesis that the spanwise flow is small (Fig. 5).

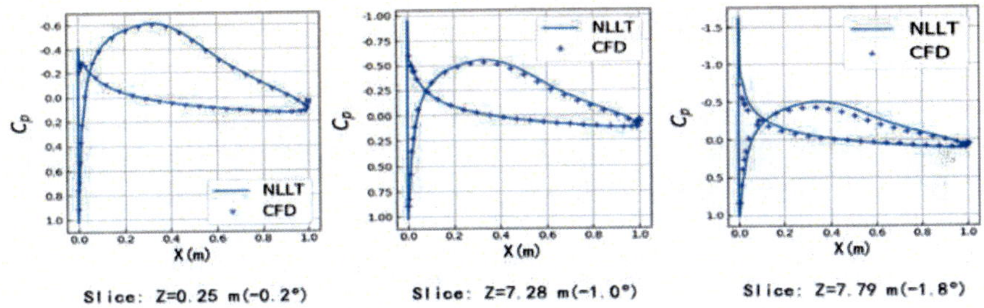

Fig. 3 C_p calculated by CFD and NLLT

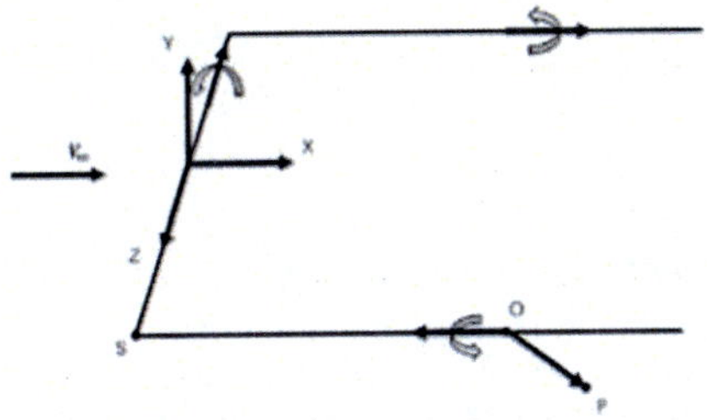

Fig. 4 Horse shoe vortex model

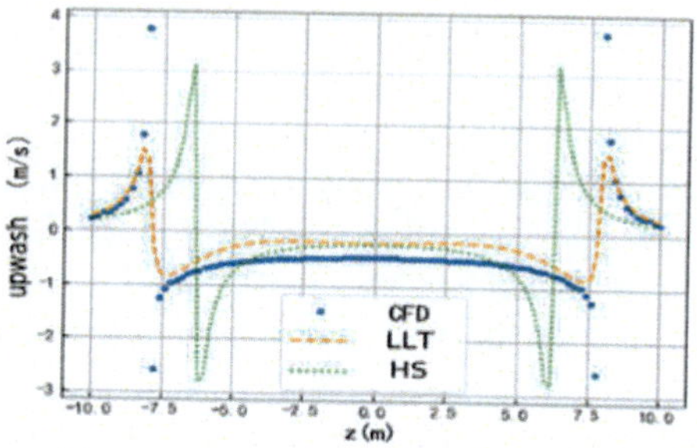

Fig. 5 Wing-tip vortex velocity distribution calculated by different methods

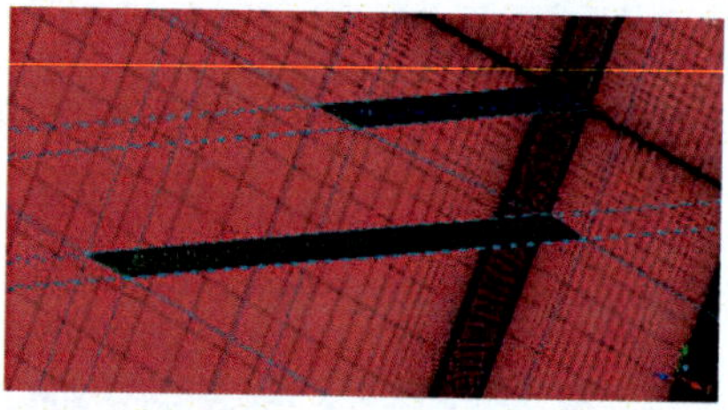

Fig. 6 Mesh structure of formation

2.3 Trailing Vortex Model

The HS model and lift line model are common methods to simulate the upwash and downwash caused by wing-tip vortex. For a general wing, the spanwise distribution of circulation can be written as Fourier series [21]:

$$\Gamma(\theta) = 2bV_\infty \sum_1^N A_n sin(n\theta) \tag{9}$$

Using transformation $z = (b/2)cos(\theta)$, integrate along the spanwise direction to obtain the lift force:

$$L = \rho_\infty V_\infty \int_{-\frac{b}{2}}^{\frac{b}{2}} \Gamma(\theta)dz = A_1 \frac{\pi}{2} b^2 \rho_\infty V_\infty{}^2 = q_\infty S A_1 \pi AR \tag{10}$$

For a wing with a given shape and incoming flow, the total lift force is related only to the first-order component of the sine series of the given circular distribution. In particular, the lift of an elliptical wing can be written as:

$$L = \rho_\infty V_\infty \Gamma_0 \frac{b}{2} \int_0^\pi sin^2\theta d\theta = \rho_\infty V_\infty \Gamma_0 \left(\frac{\pi}{4} b\right) \tag{11}$$

In other words, the HS with $b' = (\pi/4)b$, $\Gamma = \Gamma_0$ and the wing with elliptical lift distribution are equivalent in lift force, and the wing with general lift distribution can also be equivalent to the elliptical lift distribution through the first-order components of Fourier series. Therefore, some literature works about formation studies use the HS with $b' = (\pi/4)b$, $\Gamma = \Gamma_0$ to simulate the wake flow field of the front aircraft in formation flight. This method also simply characterizes the shrinking effect of the wing-tip vortex.

According to Biot-Savart theorem, the induced velocity of the small vector $\vec{dl}$ at point O in Fig. 3 to a certain point P in space is:

$$dW = \frac{\Gamma}{4\pi}\left|\frac{\vec{dl} \times \vec{r}}{||r||^3}\right| = \frac{\Gamma}{4\pi}\frac{sin\theta dl}{r^2} = \frac{\Gamma}{4\pi d}sin\theta d\,\theta \tag{12}$$

Then the induced velocity of the free vortex extending from point S to point P is:

$$||W|| = \frac{\Gamma_0}{4\pi\sqrt{y_p^2 + (z_o - z_p)^2}}\left(\frac{x_p}{SP} + 1\right) \tag{13}$$

Equation (13) has a singularity problem. It can be avoided by adding vortex core radius R_c to the vortex line. R_c is calculated by Eq. (2). For the decay structure of the vortex, Eq. (3) is mathematically convenient. The first-order Vatistas vortex model was selected, and the induced velocity after adding the vortex core is:

$$\vec{W} = \frac{\Gamma_0}{4\pi}\frac{\sqrt{y_p^2 + (z_o - z_p)^2}}{y_p^2 + (z_o - z_p)^2 + r_c^2}\left(\frac{x_p}{SP} + 1\right) \cdot \frac{sign\left(\vec{dl}\right)}{\sqrt{y_p^2 + (z_o - z_p)^2}}(0, z_o - z_p, y_p) \tag{14}$$

The LLT model is composed of numerous HS vortices. The strength of the vortex line from the attachment vortex is equal to the change rate of the circular along the span direction. For the general wing, similar to the derivation of the HS model, the calculation results of the LLT model can be obtained as follows:

$$W_y = \int_{-b/2}^{b/2} \frac{\partial\Gamma/\partial z_s(z_s - z_p)}{4\pi\left(y_p^2 + (z_s - z_p)^2 + r_c^2\right)}\delta dz_s \tag{15}$$

The accuracy of several methods was verified by the example in Sect. 2.1, and the velocity distribution was calculated as follows:

It is shown that the HS method predicts an inaccurate position of the vortex core, while LLT fit the CFD simulation better.

3 Nonlinear Aerodynamic Characteristics of Wake Surfing

The aerodynamics of a two aircraft formation in different relative positions are discussed in this section. Since the influence of the presence of the rear aircraft on the wing-tip vortex is illustrated in this section, the wake field of the front aircraft was obtained directly by CFD simulation to eliminate the change caused by the modeling of the wake field.

In the formation calculation example, the two wings were of the same shape, with aspect ratio of 16, chord length of 1 m and airfoil of E387. The NLLT method first calculated the local additional angle of attack α_{up}, with the wake field of the front aircraft (i.e., 180 points were taken along the spanwise direction, where the quarter of the string of the rear wing was located), and then the lift distribution of the rear aircraft was obtained according to the steps described in Sect. 2.1.

The CFD method used to verify the example was modeled by the semi-mode plane of symmetry method, and the mesh quantity was 10.5 million. The RKE model was used for turbulence model, pressure far field was used for boundary condition, and other boundary conditions and calculation settings were consistent with the example in 1.1.

The coordinate system was defined as: looking forward from the rear of the wing, with the Y-axis facing up, the Z-axis facing left, and the X-axis heading forward. The front plane position refers to the leading edge of the front plane of symmetry, and the rear plane position refers to the leading edge of the rear plane of symmetry. The local lift coefficient C_l is the lift coefficient of each airfoil section along the spanwise direction of the wing and is defined as Eq. 16.

$$C_l = L'/(0.5\rho_\infty V_\infty^2 c) \tag{16}$$

where L' is the local lift (obtained by stress integral), ρ the incoming flow density, V the incoming flow velocity, and c the local chord length. The local lift coefficient distribution of the rear aircraft obtained by NLLT and CFD simulation is shown in Fig. 7.

An error between the lift distribution calculated by NLLT and the results obtained by CFD simulation is present, which can be divided into two parts: The first is the error generated by the method itself when calculating the lift force of the wing,

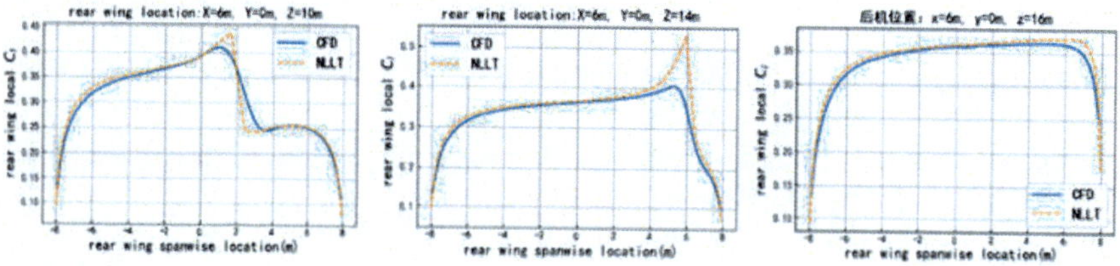

Fig. 7 Spanwise lift distribution of rear wing at different positions

including the error caused by the spanwise flow at the wing tip, etc. The second one is caused by the existence of the rear wing, which cannot be simply simulated by a linear method. In order to study the second type of influence, the first type of error was simply corrected by the CFD results and NLLT results of the single front aircraft. In Eq. (17), C_l is the local lift coefficient of the station with different position, with subscripts 1 representing the front aircraft and 2 representing the rear aircraft.

$$C_{l,2} = C_{l2,CFD} + (C_{l1,NLLT} - C_{l1,CFD}) \tag{17}$$

To study the influence of the presence of the rear aircraft on the wing-tip vortex, the lift distribution obtained by the CFD simulation was used to iterate and get the local angle of attack.

The process to calculate the local angle of attack of the rear wing is as follows:

1. Calculate the difference between the lift distribution obtained by NLLT and CFD and get ΔC_l. (Note: the spanwise stations are different between the two data sources, so interpolation or a fitting function should be used first before subtraction.)
2. Update the wash angle of attack with $\alpha_{upwash}^{n+1} = \alpha_{upwash}^{n} + \Delta C_l/(2\pi)$.
3. NLLT was used to calculate the lift distribution according to the new upwash angle of attack, and the difference between it and the lift distribution obtained by CFD was calculated. If the difference tends to zero, the angle of attack is obtained; otherwise, return to the first step.

The local angle of attack of the rear wing is shown in Fig. 8. The solid line represents the distribution of the upwash angle of attack with the existence of the rear wing, and the dotted line represents the distribution without the rear wing.

When the wing-tip vortex is far away from the wing-tip of rear wing, the presence of the rear aircraft has no great influence on the wing-tip vortex, and an accurate result can be obtained directly by linear superposition (Fig. 8). However, when the wingtip vortices of the front aircraft hit the rear aircraft, the presence of the rear aircraft will decay the front wingtip vortex.

This decay is similar to the decay law of Oseen–Lamb vortex decay model, and the impact position will affect the level of decay. The closer the hit position to the plane of symmetry of rear wing is, the stronger the decay that develops. First-order Vatistas vortex model (Scully vortex) was used to modify this situation. The modified

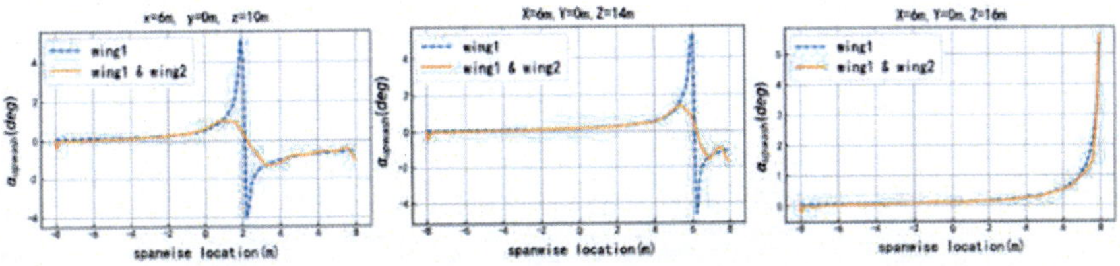

Fig. 8 Local angle of attack distribution with or without rear aircraft

function is as follows:

$$\alpha_{up,mod} = \alpha_{up} * \frac{(z - z_{vortex})^2}{r_{decay}^2 + (z - z_{vortex})^2}, \tag{18}$$

where z_{vortex} is the position of the vortex center, which can be directly obtained from the calculation results of the wake field. In this example, $z_{vortex} = 7.89$ m. The only parameter of the correction method is the decay radius r_{decay}. The correction parameters at different positions were obtained with the aim of minimizing the correction error. The correction error is defined as $\sqrt{\Sigma_1^N (\alpha - \alpha_{modf})^2 / N}$, and then the relationship between the decay radius and the spanwise positions of the two wings can be quantified.

Figure 7 shows the relationship between the decay radius and the overlap length of the two wing tips in the spanwise direction, when the rear wing position is x = 6 m, y = 0 m and the z coordinate of the rear wing position is 10 m, 12 m, 14 m, 15 m and 16 m.

The modified parameter decay radius of the front wing tip vortex increases with the increase of the overlap length of the two wings, following an almost proportional trend. To ensure engineering applicability, a simple method was used. With linear fitting we can get following equation:

$$r_c = 2.1 \frac{\Delta z}{b} \tag{19}$$

The fitting decay radius and the actual decay radius (point in Fig. 9) fitted with Eq. (19) were used to calculate lift distribution of the aircraft. In the fitting, the average maximum absolute error and the average maximum relative error of the lift coefficient in five examples were 0.006 and 2%, respectively, which are in the acceptable range.

The modified equation was verified. The wing-tip vortex decay effect brought by the rear aircraft at 8, 0 and 14 m was calculated by Eq. (19), and verified by

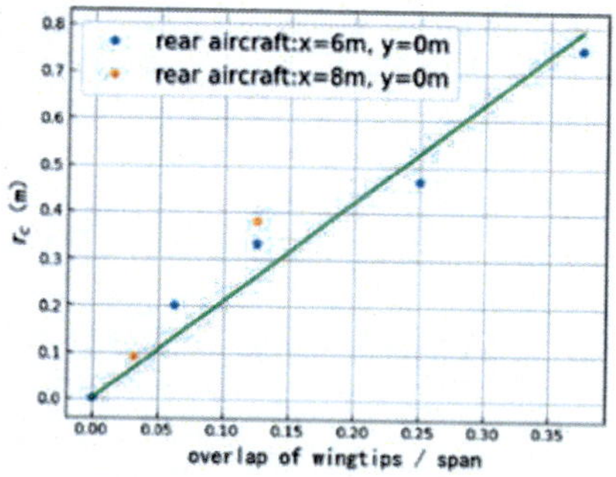

Fig. 9 Relationship between spanwise overlap length and decay radius

CFD calculation. The results obtained after the modification considering decay and unmodified results are shown in Fig. 8:

The established aerodynamic prediction model can better predict the aerodynamic characteristics of the formation of straight wings (Fig. 8).

To further validate the proposed model, the rear wing was changed and the model was evaluated. The root twist was 3°, the wingtip twist was 0, and airfoil was e387. The platform shape is shown in Fig. 9.

In this case, the downwash was modified with Eq. (21) and the lift distribution was solved with NLLT. Results of the modified method and CFD are shown in Fig. 10, where it can be seen that our methods give a good prediction (Figs. 11 and 12).

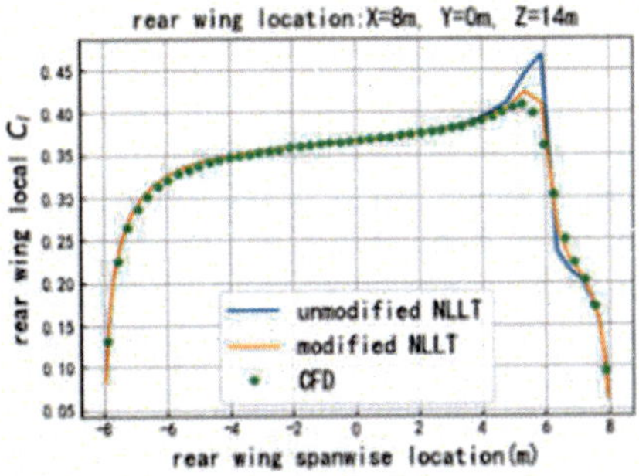

Fig. 10 Lift distribution when vortex decay is considered

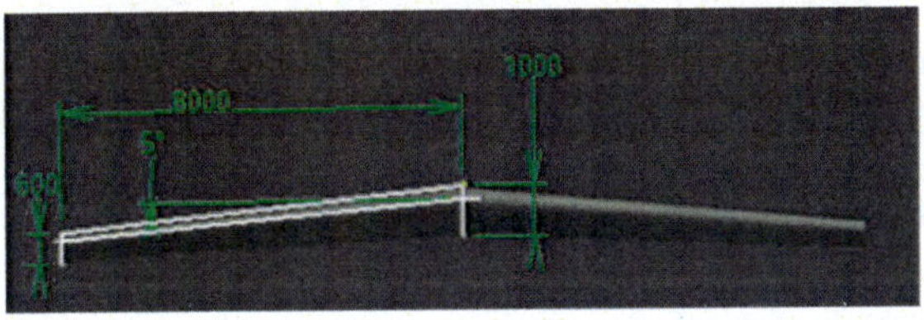

Fig. 11 Shape of the wing

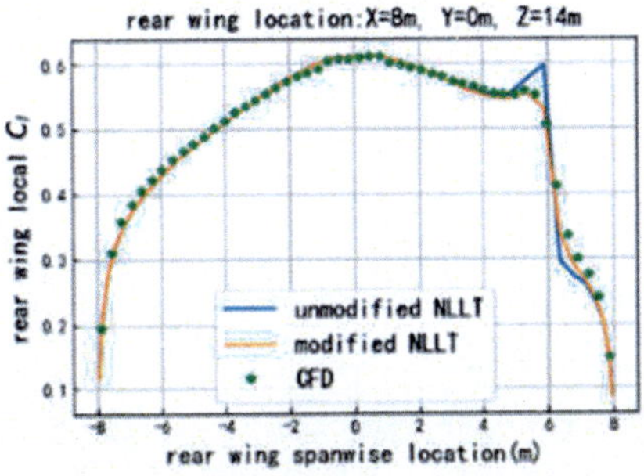

Fig. 12 Special case test

4 Conclusion

First, aiming at a kind of solar unmanned aerial vehicle (UAV), several calculation modeling methods of wing-tip vortices were compared. Studies have shown that the lifting line model is better than the horseshoe vortex (HS) model to predict the velocity distribution of the wingtip vortex. The commonly used method of equating the lifting line to a HS cannot accurately predict the position of the wake vortex core. The vortex core equations given in existing papers were evaluated, and we discovered that although the coefficients of the equations were different, the equations can still reflect the decay law of the wing-tip vortex core. Then, the accuracy of the nonlinear lifting line theory (NLLT) for aerodynamic prediction of this type of aircraft was verified. We found that the NLLT can accurately predict the pressure distribution on most aeras of the wing, but the accuracy drops near the wing tip, because of the strong spanwise flow.

Formation aerodynamic characteristics in different relative positions were studied. To eliminate the error of the wake flow field model, computational fluid dynamics (CFD) wake field was adopted to obtain the local angle of attack on the rear wing position. The aerodynamic force of the rear wing was calculated with NLLT and compared with CFD results. According to the calculation, the aerodynamic force of the rear wing can be predicted by using the linear superposition with high accuracy, when wing-tip vortices did not hit the rear wing; in other words, the existence of the rear wing has little influence on the front wing-tip vortex in this situation. However, when the core of the wing-tip vortex hit the rear aircraft, the aerodynamic interference of the rear aircraft was more affected by the linear superposition method than that of CFD. By referring to the classical vortex model, the decay of the vortex was used to explain this nonlinear effect, and the decay radius was introduced to quantitatively study this effect. We found that the closer the span distance between two aircraft was, the higher the wing-tip vortex decay became, which was manifested as the larger vortex core in our model.

By calculating the formation flow field and the size of the vortex core in a few states, the aerodynamic data of other states were obtained by simple function fitting of the vortex core size. The method was verified and found that it is more accurate than CFD. The modification algorithm also can be generalized to more complex aircraft.

References

1. Bajec IL, Heppner FH (2009) Organized flight in birds. Anim Behav 78(4):777–789
2. Hummel D (1983) Aerodynamic aspects of formation flight in birds. J Theor Biol 104(3):321–347
3. Lissaman PBS, Shollenberger CA (1970) Formation flight of birds. Science 168(3934):1003–1005
4. Gould L L, Heppner F. The vee formation of Canada geese. The Auk, 1974: 494–506.

5. Williams TC, Klonowski TJ, Berkeley P (1976) Angle of Canada goose V flight formation measured by radar. Auk 554–559
6. Badgerow JP, Hainsworth FR (1981) Energy savings through formation flight? A re-examination of the vee formation. J Theor Biol 93(1):41–52
7. Blake WB, Gingras DR (2004) Comparison of predicted and measured formation flight interference effects. J Aircr 41(2):201–207
8. Hummel D (1995) Formation flight as an energy-saving mechanism. Israel J Ecol Evol 41(3):261–278
9. Portugal SJ, Hubel TY, Fritz J et al (2014) Upwash exploitation and downwash avoidance by flap phasing in ibis formation flight. Nature 505(7483):399–402
10. Reynolds JR, Pahle J, Hanson C (2018) Design, development, and flight evaluation of pilot displays and long-track control for wake surfing applications. In: 2018 atmospheric flight mechanics conference, vol 3562
11. Liu W, Song GP, Numerical simulation and analysis of a wingtip vortex in the near field based on ansys fluent. Flight Dyn 2015(2):111–115. (in Chinese)
12. Zhang MH, Li K, Wang J, Numerical simulations of reynolds stress in wake vortex field. Aeronaut Comput Tech 2017(06):69–72
13. Yuan Y, Yan JG, Qu YH (2013) The aerodynamic coupling simulation in formation flight of multi-JAV. Flight Dyn 01:29–32
14. Tao Y, Xiong N, Wang X et al (2021) Experimental and computational investigation of hybrid formation flight for aerodynamic gain at transonic speed. Chinese J Aeronaut 34(1):32–43
15. Zhang J, Yan J, Zhang P (2020) Multi-UAV formation forming control based on adaptive method under wind field disturbances. Acta Aeronaut Astronaut Sin 41(1):323385–323385
16. Xu B, Zhang D (2020) Tight formation flight control of UAVs based on pigeon inspired algorithm optimization by quantum behavior. Acta Aeronaut Astronaut Sin 41(8):323722–323722.
17. Vatistas GH, Kozel V et al (1991) A simpler model for concentrated vortices. Exp Fluids 11(1):73–76
18. Saffman PG (1995) Vortex dynamics. Cambridge University Press
19. Schlichting H, Gersten K (2016) Boundary-layer theory. Springer
20. Dogan A, Venkataramanan S, Blake W (2005) Modeling of aerodynamic coupling between aircraft in close proximity. J Aircr 42(4):941–955
21. Anderson Jr JD (2010) Fundamentals of aerodynamics. Tata McGraw-Hill Education
22. Sivells JC, Neely RH (1947) Method for calculating wing characteristics by lifting-line theory using nonlinear section life data. Natl Advis Commit Aeronaut

Conceptual Design and System Level Analysis of Tilt-Duct eVTOL Aircraft

Jiechao Zhang, Yaolong Liu, Tianhong Jiang, and Yao Zheng

Abstract To take the potential aerodynamic advantages of ducted fan, a novel aircraft configuration utilizing tilt-duct eVTOL concept is developed in this study. The current paper focuses on the overall performance estimation of the tilt-duct aircraft via conceptual design and system level analysis for a typical eVTOL design mission with maximal take-off weight of 2500 kg, 4 passengers, 241 km/h cruise speed, 120 km flight range including reserves. For state-of-the-art battery and electric motor technologies, technically feasible solutions have been found to satisfy the design requirements. For design passenger payload and flight range mission, the maximal take-off weight is 2443 kg. The work in this paper lays the foundation of more detailed disciplinary studies and optimization for tilt-duct eVTOL aircraft, which could further help the real-world implementation of environmentally friendly, cost-effective and safe urban air mobility.

Keywords Electric vertical takeoff and landing · Tilt-duct · Urban air mobility · Conceptual aircraft design

1 Introduction

Recently, electric vertical take-off and landing (eVTOL) aircraft have attracted both academic and industrial interests. According to "The Electric VTOL News™" by Vertical Flight Society, worldwide almost 500 eVTOL concepts have been introduced since 2016 [1, 2]. As one key element for urban air mobility (UAM) that could revolutionize the commuting transportation, eVTOL aircraft have the advantages in zero-in-flight emissions, low noise levels, 3-D transportation, etc. [3]

Similar to conventional vertical take-off and landing (VTOL) aircraft concepts, eVTOL aircraft can be divided into the following possible concepts: multi-rotor, helicopter, lift plus cruise [4], tilt-rotor, tilt-duct, etc. Reference [5] has compared

J. Zhang · Y. Liu (✉) · T. Jiang · Y. Zheng
School of Aeronautics and Astronautics, Zhejiang University, Hangzhou 310027, China
e-mail: liuyaolong@zju.edu.cn

S. Lee et al. (eds.), *The Proceedings of the 2021 Asia-Pacific International Symposium on Aerospace Technology (APISAT 2021), Volume 1*, Lecture Notes in Electrical Engineering 912, https://doi.org/10.1007/978-981-19-2689-1_14

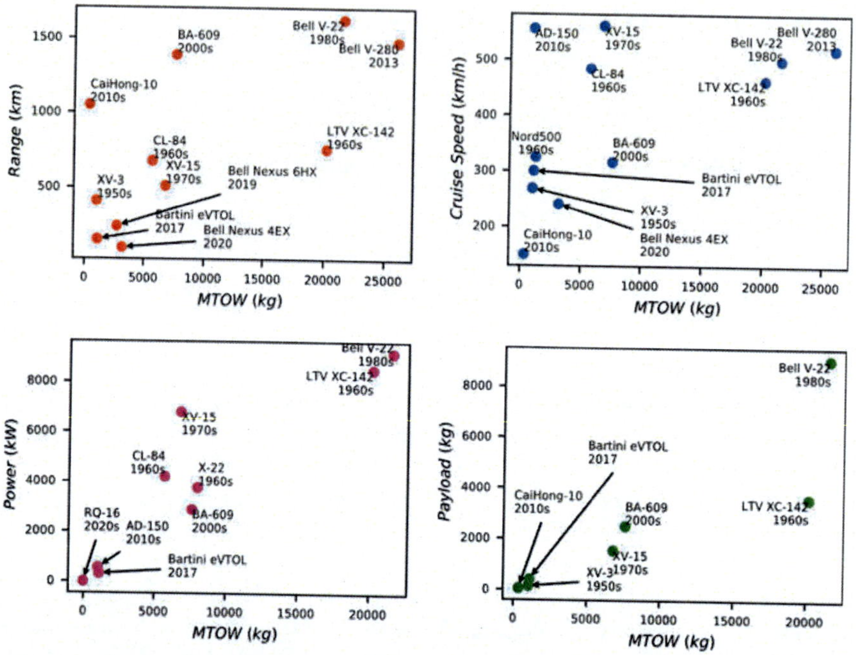

Fig. 1 Tilting class aircraft research

different VTOL configuration and further pointed out the advantages and disadvantages for different flight missions of each configurations.

Due to the high power and energy demand of eVTOL and relatively low specific energy of batteries, it is very important to improve the overall efficiency of eVTOL aircraft. As such, we select the tilt concepts for the studies. Figure 1 summarizes the tilting class aircraft: tiltwing [6], tiltrotor and tilt-duct, which includes the corresponding maximum take-off weight and other top-level aircraft parameters.

Further reason for tilt-duct concept is its potentially higher aerodynamic performance and noise shielding effects due to ducted propellers. The development of tilt-ducted fan can be traced to 1950s with quite some the theoretical research related to open propellers and ducted fans. Between 1960 and 1980s, some prototypes of tilt-duct aircraft like France's Nord500 and the U.S. Navy's X-22 were produced to study the related aerodynamics, structure and flight control issues.

Recently, a lot of tilt-duct aircraft concept have been introduced especially for UAVs and eVTOLs. Figure 2 shows a short development history of tilt-duct aircraft. As can be seen in the figure, the tilt-duct aircraft vary in number of ducted fans, fixed-wing layout, ducted fan positions, etc. It has to be noted that most work in literature either focuses on the other types of tilting concepts such as tilt-rotor concepts or only single disciplinary study of tilt-duct such as flight controls. In this paper, the conceptual design and system level analysis for the novel tilt-duct eVTOL configuration is based on multi-fidelity multi-disciplinary studies, thus quite

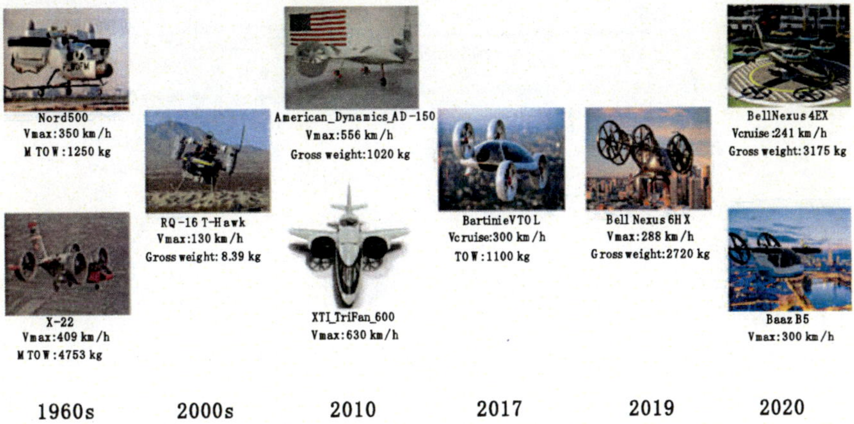

Fig. 2 Tilt ducted fan development history

comprehensive. The modelling approach includes both detailed airframe parameters and flight condition parameters, which makes the overall study more reliable especially for off-design performance. First, the feasibility and design space exploration study in this paper lays the foundation of more detailed disciplinary studies for tilt-duct eVTOL aircraft. Second, the multi-fidelity multi-disciplinary methods developed within this study can also be used for future operation-oriented tilt-duct eVTOL aircraft design and optimization. Lastly, the system level results give some insight into future eVTOL research and development work, which could further help the real-world implementation of environmentally friendly, cost-effective and safe urban air mobility.

2 Design Requirements and Assumptions

2.1 Design Requirements

According to Uber in 2016 [7], a typical design mission for eVTOL aircraft in UAM scenario can be summarized as follows [8]:

- 2 min of takeoff and climb to cruise altitude
- Cruise at optimal altitude, e.g. 1000 m
- Cruise at optimum speed, e.g. 241 km/h
- 2 min of landing and descent from cruise altitude
- Fuel/energy reserves of at least 10% of the mission, or 20 min at optimal endurance
- 4 passengers plus 1 pilot with a flight range of 120 km including 24 km for reserves

According to the typical design missions eVTOL abovementioned, the basic mission profiles are designed as shown in Fig. 3. The MTOW constraint is set as

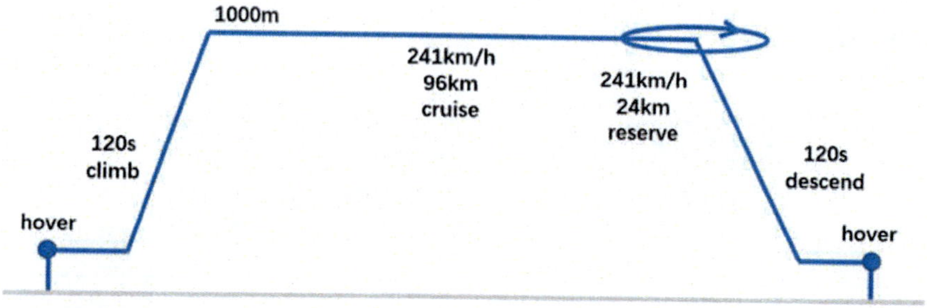

Fig. 3 Flight mission profile

Table 1 Aircraft design requirements

Parameter	Value
MTOW	2500 kg
Payload	500 kg to fulfill one pilot and 4 passengers with luggage
Hover climb time	120 s
Cruise speed	241 km/h
Cruise altitude	1000 m
Range	120 km including reserves
Hover landing time	120 s

2500 kg for required design payload and design flight range. Table 1 lists the top-level aircraft requirements for the tilt-duct eVTOL design.

2.2 *Design Assumptions*

The optimization model also considers advanced technology assumptions, as the eVTOL UAM is primarily a future concept, so an installed battery energy density of 300 wh/kg is assumed. The SOC (state of charge) = 0.2 is assumed, which means 80% of all energy stored in the battery is available. The efficiency of the motor is assumed to be 90% and does not vary with the flight altitude and the degree of battery energy consumption, and the motor power density is assumed to 4.3 kw/kg.

To ensure a certain level of cruise and hover performance, proper values need to be selected for the wing loading and the disc loading. The number of ducted fan is selected as 2 and the number of blades of per ducted fan is selected as 3.

3 Methods

The overall analysis process follows a conventional aircraft design process (Lukaczyk et al. 2015), with some additional considerations listed as follows.

- **Mass estimation**: The weight of wing, tail, ducted fan and fuselage is calculated through the weight module in SUAVE [9] and Kadhiresan [5]. The wing weight is related to the basic parameters of the wing, such as: aspect ratio, quarter chord sweep angle, taper and wing span. The ducted fan mainly consists of the rotor and the duct, so the weight calculation of the ducted fan is performed in a similar way to the weight calculation of the propeller. Interpolating the continuous power of existing aircraft motors and calculating a power density of the motor as a whole to calculate the weight of the motor [10], and the weight of the motor controller is also related to the power of the motor [11]. Other components such as: Landing Gear, Flight Controls, Avionics, Furnishing are modelled using empirical equations from [12, 13].
- **Aerodynamics:** The aerodynamic performance of the tilt-duct concept are modelled using NASA VSPaero tool and empirical equations from Ref. [14]. The zero-lift drag C_{D_0} can be calculated by VSP. The induced drag is calculated using VSPaero tool [15] with corrections of duct effects from literature [16].
- **Cruise and hover performance:** eVTOL typically have thrust-to-weight ratios of 1.15 or greater to provide additional climb power and control power margins. Turbine and piston engines are typically capable of providing 10–20% short-term emergency power ratings, but electric motors are typically capable of generating more than 50% additional power in 1–2 min until they overheat, and these peak ratings are not accounted for in the thrust-to-weight ratio, but are reserved for emergency operations (such as motor failure) [7]. The rotor efficiency reference method in the paper [17]. Since a ducted fan has one more duct than an open propeller, a correction to this equation is required. For the power-thrust ratio, a simple analysis based on momentum theory is performed as follows [16]:

$$\frac{P}{T} = \frac{1}{4}\left[3v + \sqrt{v^2 + 4\phi\left(\frac{T}{\rho A}\right)}\right], \tag{1}$$

where P is the power, T is the thrust, V is the axial air velocity into the duct and $\varnothing$ is the ratio of the propeller disc area to the exhaust area. For open propellers, $\varnothing = 2$, for equal area ducted fans, $\varnothing = 1$. ρ is the air density, A is the disc area.

4 Results

Based on the design requirements and design methods introduced in Sects. 2 and 3, the conceptual design of the eVTOL has been carried out, with a quite conventional layout similar to tiltrotor aircraft designed by Kadhiresan [5] and the side-by-side helicopter designed by Johnson [9]. The three views of the tilt ducted fan were shown in Fig. 4. The main parameters of the lifting surfaces are shown in Table 2.

Figure 5 shows the VSPaero calculated drag polar and the corrected drag polar for the tilt-duct eVTOL at cruise flight conditions. With a calculated cruise lift coefficient of0.425, the calculated total drag coefficient and lift-to-drag ratio are 0.060 and 7.1, respectively. Figure 6 presents the operating empty mass breakdown of the tilt-duct aircraft, which is comparable to results in literature. In our study, we also take a look at the mission behavior of the tilt-duct aircraft. As the aircraft flies with full electric

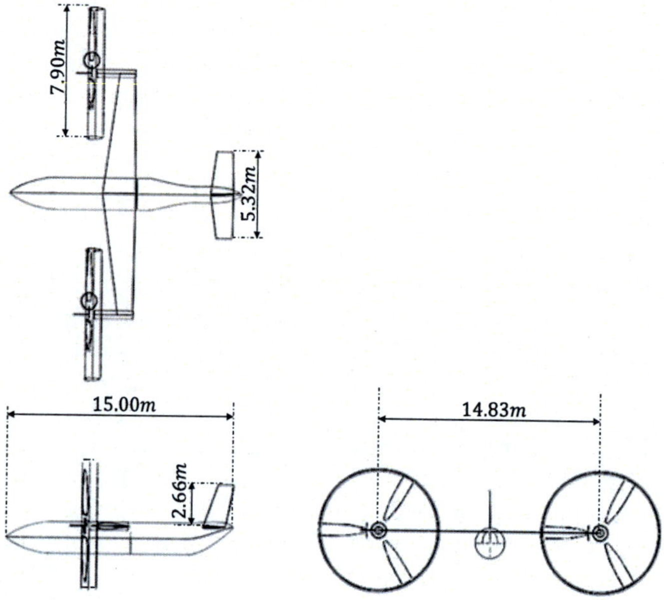

Fig. 4 Aircraft model three views

Table 2 Design assumptions

Parameter	Value
Battery energy density	300 wh/kg
SOC (state of charge)	0.2
Motor efficiency	90%
Motor power density	4.3 kw/kg
Wing loading	106.11 kg/m^2
Disc loading	24.87 kg/m^2
Number of ducted fan	2
Number of blades	3

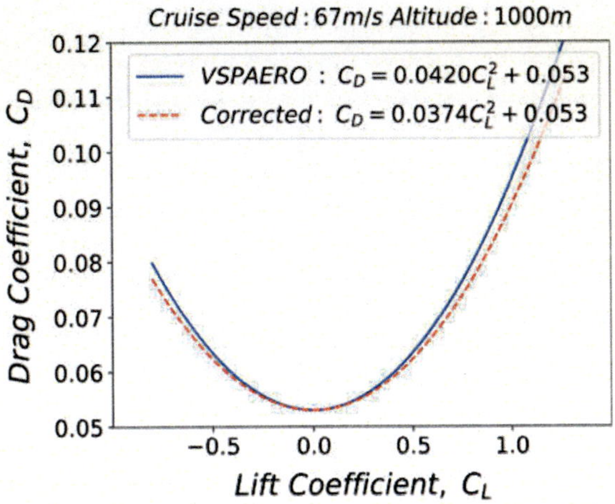

Fig. 5 Aerodynamic polar of the tilt-duct eVTOL

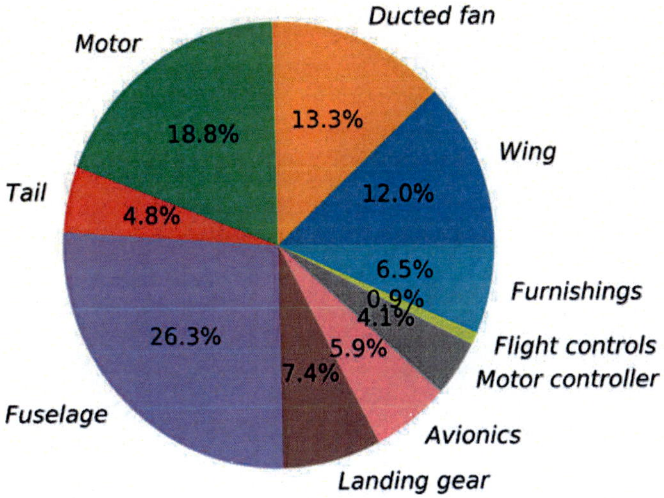

Fig. 6 Weight ratio of each component of the aircraft

propulsion and the mass remains constant during the whole flight, only the battery energy with flight time is shown (see Fig. 7).

Table 3 summarizes the system level design results of our study. Note that it is a work in progress and further sensitivity on technology levels as well as airframe and operation parameter optimization studies will be carried out to see the design potentials for tilt-duct eVTOL aircraft.

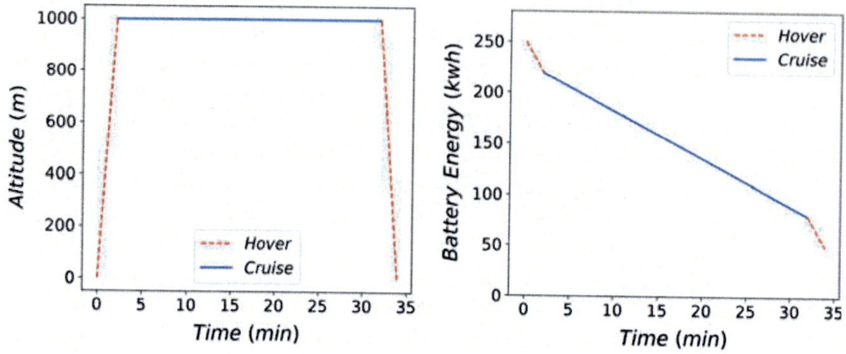

Fig. 7 Altitude and battery energy change with flight time

Table 3 Lift surface parameters

Parameter	Wing	Horizontal tail	Vertical tail
Areas (m^2)	23.02	6.65	3.33
Span (m)	14.83	5.32	2.66
Aspect ratio	9.55	4.26	2.13
Tip chord length (m)	0.97	1.04	1.04
Root chord length (m)	2.14	1.46	1.46
Taper	0.45	0.71	0.71
Average chord (m)	1.55	1.25	1.25
Leading edge sweep (degrees)	8	8	20

Table 4 Design results

Parameter	Value
Takeoff weight	2443 kg
Empty weight	1109 kg
Battery mass	834 kg
Payload	500 kg
Lift to drag ratio	7
Cruise drag	3417 N
Cruise power	277 kw
Cruise rotor efficiency	91.6%
Hover climb drag	27,532 N
Hover climb power	932 kw
Hover climb rotor efficiency	47.3%
Hover efficiency	82.8%

5 Conclusion

In this study, we have reviewed existing tilting especially tilt-duct concepts for an eVTOL aircraft potentially used in UAM context. Based on methods in literature and the existing conceptual aircraft design framework SUAVE and NASA VSP tool, we have conceptually designed a tilt-duct eVTOL aircraft with design passenger of 4 and design flight range of 120 km. The estimated performance of the designed eVTOL satisfied the design requirements under current technology level assumptions. The work is still in progress and the future plan is to carry out airframe parameter, flight altitude and flight speed parameter, energy system technology sensitivity studies and concurrent optimization to extend the design space, and to improve the tilt-duct eVTOL aircraft performance.

References

1. Brown A, Harris WL (2018) A vehicle design and optimization model for on-demand aviation. In: AIAA/ASCE/AHS/ASC structures, structural dynamics, and materials conference, vol 210049. https://doi.org/10.2514/6.2018-0105.
2. Brown A, Harris WL (2020) Vehicle design and optimization model for urban air mobility. J Aircr 57(6):1003–1013. https://doi.org/10.2514/1.C035756
3. Felix Finger D, Braun C, Bil C (2019) Impact of engine failure constraints on the initial sizing of hybrid-electric GA aircraft. In: AIAA Scitech 2019 forum, pp 1–18. https://doi.org/10.2514/6.2019-1812
4. Bacchini A, Cestino E, Van Magill B, Verstraete D (2021) Impact of lift propeller drag on the performance of EVTOL lift+cruise aircraft. Aerosp Sci Technol 109:106429. https://doi.org/10.1016/j.ast.2020.106429.
5. Kadhiresan AR, Duffy MJ (2019) Conceptual design and mission analysis for evtol urban air mobility flight vehicle configurations. In: AIAA aviation 2019 forum, pp 1–19. https://doi.org/10.2514/6.2019-2873
6. Chauhan SS, Martins JRRA (2020) Tilt-wing EVTOL takeoff trajectory optimization. J Aircr 57(1):93–112. https://doi.org/10.2514/1.C035476
7. Holden J, Goel N (2016) Fast-forwarding to a future of on-demand urban air transportation, pp 1–98
8. Johnson W, Silva C, Solis E (2018) Concept vehicles for VTOL air taxi operations. In: Proceedings of the AHS international technical meeting on aeromechanics design for transformative vertical flight 2018
9. Lukaczyk TW, Wendorff AD, Colonno M, Economon TD, Alonso JJ, Orra TH, Ilario C (2015): SUAVE: an open-source environment for multi-fidelity conceptual vehicle design. In: 16th AIAA/ISSMO multidisciplinary analysis and optimization conference: American institute of aeronautics and astronautics (AIAA AVIATION Forum)
10. Hascaryo RW, Merret JM (2020) Configuration-independent initial sizing method for Uam/Evtol vehicles. In: AIAA aviation 2020 forum, 2020, p 16. https://doi.org/10.2514/6.2020-2630
11. Pornet C, Gologan C, Vratny PC, Seitz A, Schmitz O, Isikveren AT, Hornung M (2015) Methodology for sizing and performance assessment of hybrid energy aircraft. J Aircr 52(1):341–352. https://doi.org/10.2514/1.C032716
12. Roskam J (1999) Airplane design part v: component weight estimation. DARcorporation
13. Prouty RW (2002) Helicopter performance, stability, and control. Krieger

14. Raymer DP (1992) Aircraft design: a conceptual approach, 2nd edn. American Institute of Aeronautics and Astronautics, Washington, DC
15. VSPAERO, "Software Package, Version 3.22.0," NASA, http://openvsp.org/, 2021
16. Armutcuoğlu Ö, Kavsaoğlu MŞ, Tekinalp O (2004) Tilt duct vertical takeoff and landing uninhabited aerial vehicle concept design study. J Aircr 41(2):215–223. https://doi.org/10.2514/1.271
17. Leishman JG (2016) Principles of helicopter aerodynamics. Cambridge University Press, Cambridge

Research on Wind Tunnel Test of the Total Pressure Probe Layout for Civil Aircraft

Xing Zhou, Yuan Zhong, and Jiaqi Song

Abstract The total pressure probe of civil aircraft plays an important role in the atmospheric data system. It is used to measure the total pressure of the incoming flow during flights. The total pressure probe should be installed at a correct position on the nose which local angle can be covered by local angle requirement of the total pressure probe in all flight envelope. According to the above wind tunnel tests of the total pressure probe with part model and full model were designed. The characteristics of total pressure loss of total pressure probe were obtained in the full model wind tunnel test. In the full model wind tunnel test a simplified fuselage with two types of miniature probe was designed to carry out the aerodynamic layout test of total pressure probe. Two types of miniature probes include a similar shaped probe and a seven-hole probe which measure the local total pressure and the local angle of the probe respectively. The variation characteristics of the local angle and the total pressure loss coefficient with the altering fuselage angle of attack and sideslip angle were obtained in the full model test. From the two aspects above, the satisfaction of the local angle and the total pressure loss coefficient for the probe installation position on the aircraft nose can be determined. The design of the miniature probe fills the technical blank of the total pressure probe wind tunnel test model. This type of wind tunnel test can accurately examine the reasonability of the layout of the total pressure probe for civil aircraft.

Keywords Total pressure probe · Local angle · Wind tunnel test · Miniature probe · Total pressure loss coefficient

X. Zhou (✉) · Y. Zhong · J. Song
Shanghai Aircraft Design and Research Institute, Commercial Aircraft Corporation of China, Ltd, Shanghai, China
e-mail: jowing1982@163.com

Y. Zhong
e-mail: zhongyuan@comac.cc

J. Song
e-mail: songjiaqi@comac.cc

S. Lee et al. (eds.), *The Proceedings of the 2021 Asia-Pacific International Symposium on Aerospace Technology (APISAT 2021), Volume 1*, Lecture Notes in Electrical Engineering 912, https://doi.org/10.1007/978-981-19-2689-1_15

1 Introduction

Civil aircraft need to obtain various airspeeds during flight, such as indicated airspeed, calibrated airspeed, vacuum speed, and etc. These airspeeds are not directly obtained. They are measured by atmospheric data sensors and passed into ADM (air data model) for calculation. Most modern civil aircrafts use multi-channel air data sensor signals for voting mechanisms to enhance the security of the system architecture [1]. Airspeed is very important to the safety of aircraft. If the air data sensor itself is not accurate enough, the layout is unreasonable, or a certain signal is lost, then the flight parameters obtained by the air sensor deviate greatly from the real one, which is catastrophic for the aircraft. For example, in February 1992, Bergen Airlines Flight 603 suffered from flight data disturbance 9 min after taking off in Dominia, and finally lost control and fell into the Atlantic Ocean, killing 189 people on the plane. On October 2, 1996, Air Peru flight 603 crashed after taking off in Peru [2]. The investigation found that the pitot tube sleeve was not removed after the aircraft was cleaned, resulting in unreliable airspeed. On June 1, 2009, Air France Flight 447 crashed [3]. The investigation report pointed out that the plane crash was caused by the crew who were not able to deal with it effectively after the airspeed was unreliable.

The air data sensors related to airspeed include total pressure probe, total static pressure probe, static pressure source and total temperature probe. Multi-channels are coupled together in the aircraft's avionics architecture [4]. For example, channel one is left total pressure probe and right static pressure source, and channel two is right total pressure probe and left static pressure source. In the airworthiness regulation CCAR25.1323 [5], it is stipulated that the airspeed error shall not exceed 3% or 5 knots within the range of most speed envelopes, whichever is the maximum. In order to be able to fly in RVSM [6] airspace, the airspeed error of advanced civil aircraft usually does not exceed 2 knots. The total pressure probe is one of the important sensors of the airspeed system. The lip of the total pressure probe is facing forward to measure the total pressure. A bracket keeps the probe 100 mm away from the fuselage to reduce the disturbance to the airflow. There are 1 group of drainage holes on the side of the probe. It is generally required that the angle between the drainage hole and the direction of gravity be less than 7°. The anti-icing heater inside the probe is used to prevent the probe from freezing. The lip of the total pressure probe expands outwards in a tapered shape to ensure a small total pressure loss at a large angle. It is very important and frequent to check the lip of the total pressure probe in many civil aircraft maintenance manuals. If the wear or loss reaches about 1.5 mm, the probe must be replaced.

The total pressure probe is arranged on the nose to feel the total pressure in different flow directions. Once the direction of the local airflow exceeds the angle of use of the probe, the total pressure loss coefficient will increase sharply and the airspeed error will increase. If the error exceeds the threshold, the signal of a certain channel will be judged to be invalid, which will affect the airspeed voting. Therefore, the installation position of the total pressure probe on the nose requires a design to meet the requirements of product equipment. Product equipment usually requires local airflow angle less than the threshold. The application range of the total pressure probe not only considers the flight envelope of the angle of attack, but also considers the flight envelope of the sideslip angle. In the layout design of total pressure probes, most of them use the CFD [7] method for verification, but the calculation of the high angle of attack and large sideslip angle is not accurate enough. This article developed a miniature probe to simulate a total pressure probe to measure the local angle and local total pressure. The method of wind tunnel test [8] verifies the reasonability of the aerodynamic layout of the total pressure probe and makes up for the insufficiency of CFD calculation.

2 Test Model

Wind tunnel test model need to meet the requirements of wind tunnel width ratio and blockage. If you use a complete plane to scale, the ratio is usually 1:20 or less. The diameter of the total pressure probe is about 10 mm and the height is about 100 mm. According to the ratio of 1:20, the size of the total pressure probe is very small, and it cannot simulate the function when it falls into the surface layer of the nose. Therefore, it is necessary to design a simplified aircraft shape that can not only meet the requirements of the wind tunnel but also simulate the functional requirements of the total pressure probe. The local angle of the nose is an important factor that affects the result of the total pressure measurement. In the shape design process, three shapes were selected to compare the direction of the local flow lines. They are the whole plane, the outer truncated wing with nacelle, and the outer truncated wing without nacelle. The number of grids is about 20 million, as shown in Fig. 1. The solver is ANSYS CFX. The turbulence model is SST model. The calculation condition of Mach number is 0.78. The AOA is 4 degrees. In the calculation results, the front and rear points on the fuselage were selected for speed comparison, as shown in Fig. 2.

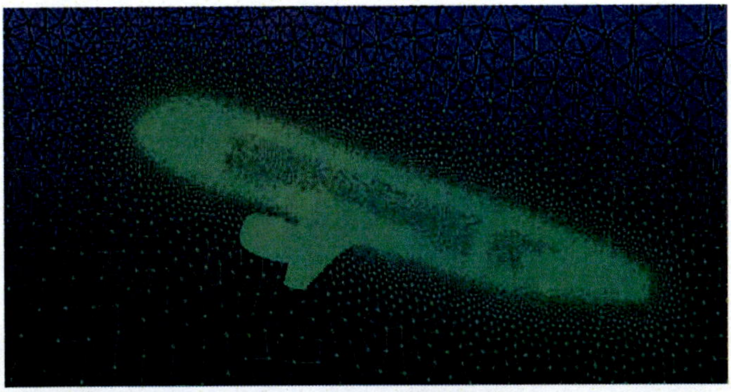

Fig. 1 An aircraft grid

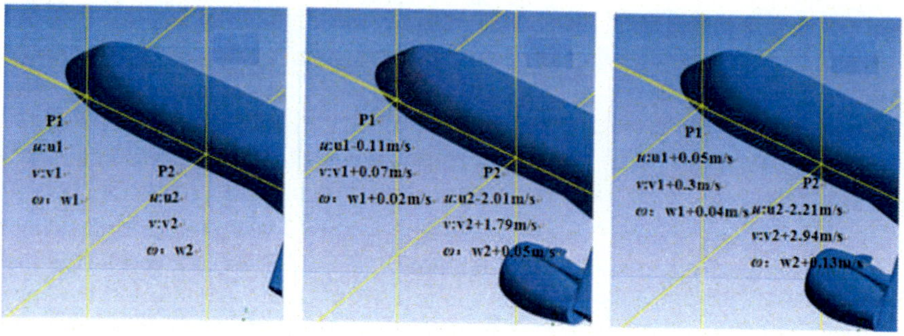

Fig. 2 Speed calculation results

The results show that the three shapes at point P1 have speed differences in the v direction, and the three shapes at point P2 have differences in the three directions of u, v, and ω, and there is a big difference between fuselage with or without nacelles. Therefore, if you verify the position of the total pressure probe in the nose area, you can use the simplified shape without the nacelle. If you verify the position of the total pressure probe in the middle fuselage or later, it is best to use the simplified shape with the nacelle for model design.

This article mainly studies the layout of the total pressure probe in the nose area, so a simplified model without nacelle is used. The model scale is about 1:10, which can coordinate the development of miniature probes, including its size and function. The diameter of fuselage is about 350 mm and the length of fuselage is about 3 m. The distance between the nose and the leading edge of wing root is about 1.3 m. The model has a blockage of 2% and is made of all metal. It is mainly composed of a nose section, a connecting section, a simplified wing section, a rear body fairing, and a miniature probe. Low speed test uses a ventral sting, see Fig. 3. High speed test uses a rear sting. The low-speed wind tunnel test was carried out in a continuous wind tunnel of 4 m × 3 m.

Fig. 3 Wind tunnel test installation diagram

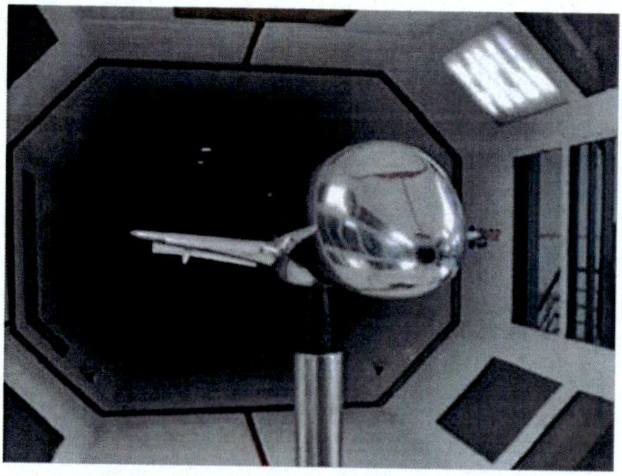

3 Miniature Probe

Miniature probes are smaller probes than the full-size. It is necessary to simulate the shape of the full- size pressure probe as much as possible, especially the shape of the lip. A miniature total pressure probe can be designed to directly measure the total pressure or use a seven-hole probe to measure the local streamline angle. The miniature total pressure probe is small in size, and the use range is generally smaller than the full-size. The seven-hole probe is generally used for low-speed tests but be used for high-speed tests less because it is expensive to calibrate in high-speed wind tunnel. The two measuring devices can be complementary to each other.

3.1 *Seven-Hole Probe*

The current probes used to measure the angle of the flow field have five holes and seven holes. The measuring range of the five-hole probe is only ± 45°, which cannot meet the test requirements. Therefore, the seven-hole probe is selected for the test [9]. The measuring range of the seven-hole probe is ± 70°, and the measuring accuracy is 0.5°—1°. The design and processing of the seven-hole probe are very difficult. Seven holes whose diameter is 0.6 mm parallel to the axis of the column are to be machined on a cone column with a diameter of 4 mm. The basic principle of the seven-hole probe is as follows [10]. When the flow angle is small, the pressure in the center hole is maximum. But as the flow angle increases, the maximum pressure hole will gradually move from the center to the pressure hole facing the lateral airflow. Therefore, according to the pressure of the seven pressure holes, the flow space is divided into seven areas. Name the number of a hole based on the highest pressure. The flow space facing the seventh hole is called the inner zone, the other six zones are called the outer zone. Each district occupies 60 degree fan-shaped space. The θ and Φ angles of the incoming flow relative to the probe coordinate system are obtained by the pressure coefficients of the seven holes.

For the outer zone ($\theta > 24$ degree), because the probe has three pressure holes in the separation zone under a large flow angle, the pressure of the four pressure holes facing the wind must be used to determine the angle of attack and the azimuth of the airflow from the Fig. 4.

The test probe is mainly composed of a bracket and a seven-hole probe. The shape of the bracket is similar to the real shape of the bracket of the total pressure probe. The seven-hole probe is embedded in the cylinder of the bracket. The probe coordinate system coincides with the bracket coordinate system, and there is no roll angle and pitch angle, as shown in Fig. 5. The measuring point of the seven-hole probe coincides with the measuring point of the total pressure probe.

Fig. 4 Seven-hole probe coordinate system

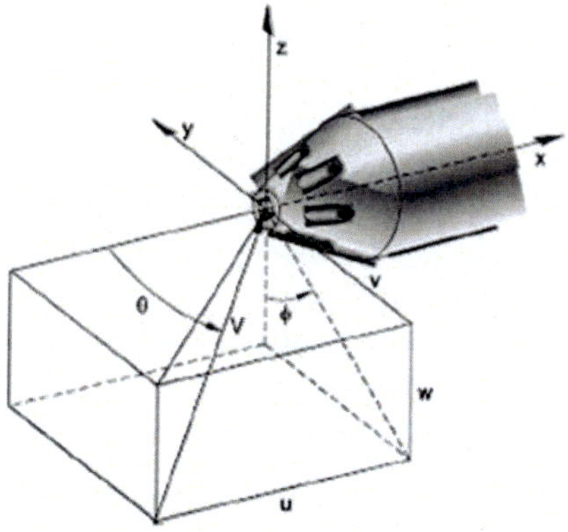

Fig. 5 Test probe

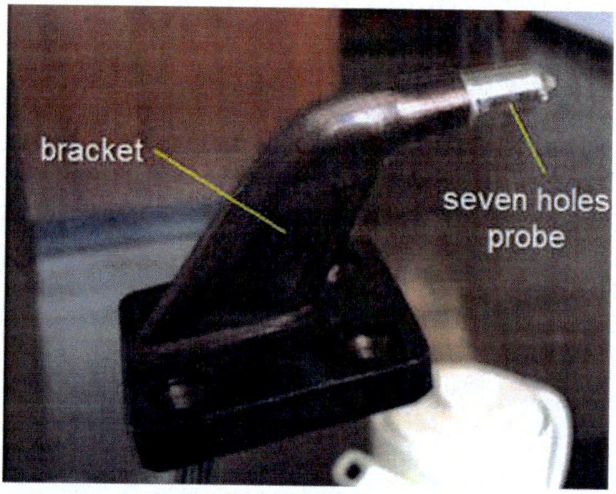

3.2 *Miniature Total Pressure Probe*

The miniature total pressure probe is created referring to the pressure measurement pipeline parameter design and the internal structure of the full-size total pressure probe. It takes into account the model scale meanwhile. The shape of the key area of the sensor is designed with reference to the shape of the full-size probe, which truly restores the lip profile of the total pressure probe. The inside of the sensor is composed of a single total pressure measurement pipeline. The pipe diameter is 0.8 mm, which meets the requirements of wind tunnel pressure test. The internal pipeline is smoothly designed to ensure the air path smooth and measurement accuracy. The sensor is designed with a special installation base, which can be connected with the nose of the wind tunnel test model by using traditional screws to ensure the installation strength of the sensor in the wind tunnel test, as shown in Fig. 6. The miniature total pressure probe adopts 3D printing technology to enhance the simulation of the lip surface.

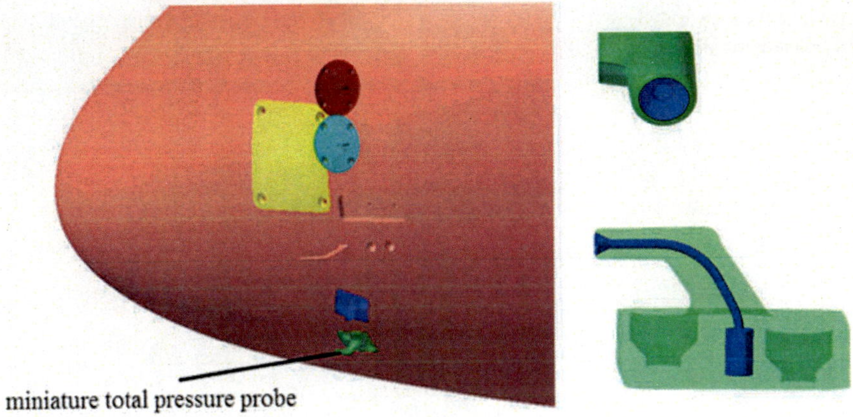

Fig. 6 Miniature total pressure probe

4 Analysis of Test Results

The test results include two aspects. One is the local angle measured by the seven-hole probe, and the other is the local total pressure measured by the miniature total pressure probe.

4.1 Results of the Seven-Hole Probe

Considering the drainage function and the range of the longitudinal angle of attack, the total pressure probe will be assembled by an angle. At high angles of attack, the local angle of the total pressure probe is less than the permitted, so that the total pressure loss coefficient is less than threshold of dynamic pressure. Total pressure probe is arranged to consider not only the longitudinal angle of attack, but also consider the large slip angle. In this paper, two positions of the nose are selected to study local angle along the sideslip angle of the fuselage. Position 1 is at the upper side, which is close to the maximum width line. Position 2 is below the bottom, which is close to the bottom line of the fuselage, as shown in Fig. 7. In the test, seven-hole probes were installed on the left and right sides of the nose at the same position. The test results are shown in Fig. 8. As the angle of attack of the fuselage increases, the local angle of the probe also increases. When the angle of attack is small and the angle of sideslip is large, the local angle of the probe is easier to exceed the threshold. The local angle at the P1 position does not exceed the threshold in most sideslip angle ranges. When the sideslip angle of fuselage reaches the medium, the local angle at the P2 exceeds the threshold of the probe. Therefore, for the position of the nose, the lower the probe is, the easier the local angle will exceed the threshold. That means

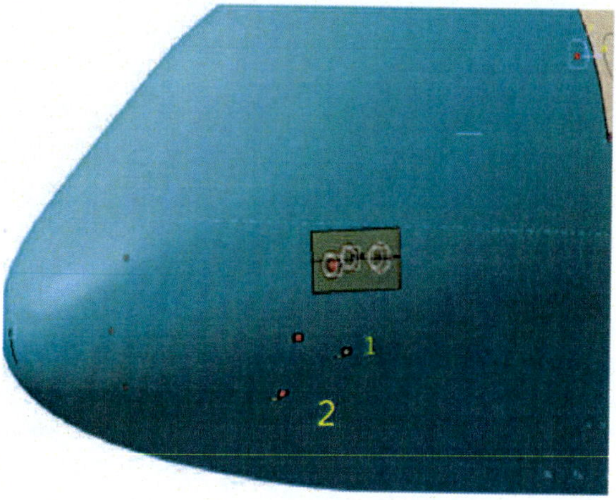

Fig. 7 Different positions of total pressure probe

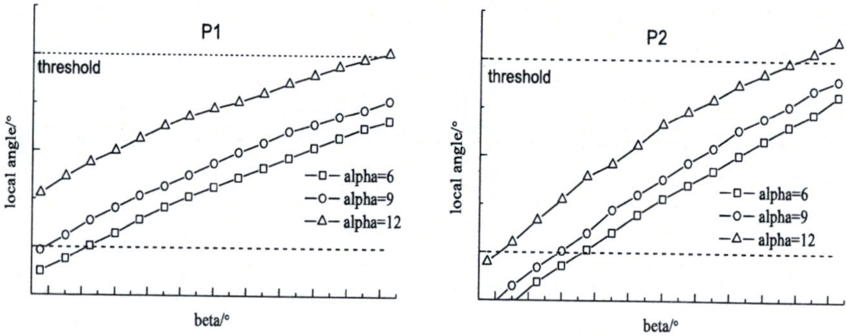

Fig. 8 Comparison of test results between position 1 and position 2

if the probe is close to the maximum width line, the local angle will conform to the use range of the total pressure probe.

4.2 Results of Miniature Total Pressure Probe

The miniature total pressure probe directly measures the pressure. Compared with the seven-hole probe measuring angle, it is more intuitive and directly judge the loss of total pressure. Due to the small size of the miniature total pressure probe, the shape of the lip is little different from that of the full-size total pressure probe because of reduced-scale. The measurement range is often smaller than a full-size probe. So before the wind tunnel test, the miniature total pressure probe needs to be calibrated in the small wind tunnel alone. Fix the miniature probe on the wall of the wind tunnel or on the flat plate to avoid the influence of the surface layer on the wall. The local angle of attack of the probe is changed through the angle mechanism, and the curve of the total pressure loss coefficient of the miniature probe with the local angle of attack is obtained. If the angle at which the total pressure loss increases sharply is smaller than the full-size probe, then the angle difference needs to be corrected in the wind tunnel test data. The installation diagram of the miniature probe in the wind tunnel test is shown in Fig. 9. The position of the total pressure probe in the wind tunnel test is close to the bottom line, and the results of the wind tunnel test are shown in Fig. 10.

In Fig. 10, Pt0 is the total pressure of the incoming flow, which is a constant value in the figure. In Fig. 10a the total pressure loss of the probe with the angle of attack changes is less than threshold of dynamic pressure, which meets the requirements of airspeed calibration. In Fig. 10b, there is basically no loss of total pressure when the probe is on the leeward side of the fuselage, but there is a loss of total pressure on the windward side. The greater the angle of attack, the smaller the total pressure loss. As the sideslip angle of the fuselage increases, the total pressure loss increases. The total pressure loss in Fig. 10b exceeds the airspeed calibration requirement. From the test results in Figs. 8 and 10, the lower and closer to the nose the total pressure probe is installed, the greater the local angle and the side slip angle change. The shape of the

Fig. 9 Wind tunnel test of miniature total pressure probe

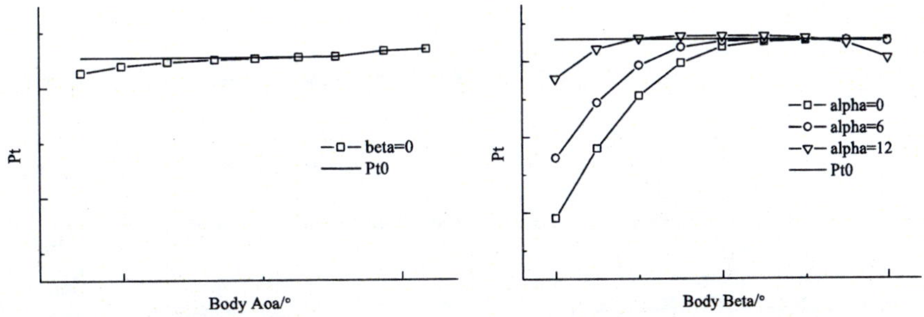

(a) Longitudinal test results (b) Horizontal test results

Fig. 10 Test results of miniature total pressure probe

nose has a weaker guiding effect on the airflow, so the total pressure probe is better to place close to the maximum width line and farther away from the front of nose.

5 Conclusion

This paper conducts a layout wind tunnel test study on the total pressure probe of a civil aircraft. The research includes the design of the main model, the design of the miniature probe, the wind tunnel test and the analysis of the test results, and the following conclusions are obtained.

1. The main model design needs to consider the size requirements of the wind tunnel, and the size of the miniature probe should be coordinated after scaling down. For the total pressure probe at the nose position, a simplified model with truncated wing and without a nacelle can be selected. For the total pressure probe at the position of the fuselage, it is recommended to use a simplified model with the truncated wing and a nacelle.
2. The design of the miniature probe includes a seven-hole probe and a miniature total pressure probe. The seven-hole probe observes whether the local angle meets the requirements of the total pressure probe and the miniature pressure probe measures intuitively the local total pressure. Both test devices can meet the test needs of total pressure probes.
3. In the wind tunnel test, we should not only pay attention to the characteristics of the total pressure probe with the angle of attack, but also the characteristics of the sideslip angle. The selected position of the total pressure probe must satisfy that the total pressure loss coefficient is less than the threshold of dynamic pressure or the local angle in the range of attack and sideslip is less than the threshold of the angle.

References

1. Tao J, Zhang Z (2012) Air data sensors location on aircraft analysis design. Microcomput Appl 28(9):6–8
2. Gu S (2003) Baptism of blood: exploring the global air disaster. Beihang University Press
3. Yang M, Luo Y (2020) Analysis of air france flight 447 crash and safety countermeasures. Electron Technol 49(06):29–31
4. Li J, Chen C (2021) Functional analysis of air data modules for civil aircraft. Adv Aeronaut Sci Eng 12(2):151–156
5. Dai W (2014) Application of trailing cone in calibration of airspeed system. Meas Control Technol 33(12):151–153
6. Zhang X, Pan W (2009) Safety assessment of vertical separation in airspace of RVSM. Aeronaut Comput Tech 39(5):5–8
7. Du Z, Shi Y et al (2017) Aerodynamic design, simulation and experiment of high-speed aircraft total static pressure probes. Metrol Meas Technol

8. Zhou X, Yang S et al (2013) Study on low-speed verification test of atmospheric sensors of civil aircraft. Adv Aeronaut Sci Eng 4(1):43–48
9. Ma X, Ming X (2012) Experimental research and error analysis on seven-hole probe measurements of shearing flow field. Acta Aerodyn Sin 30(1):57–62
10. Zou Y, Lin J et al (2020) Experimental calibration of seven-hole probe at the large flow angle. Gas Turbine Technol 33(127):11–16

Study on the Dynamic Aerodynamic Performance of Airfoil with Direct Force Measurement

Yuqin Jiao, Chunsheng Xiao, and Dengke Wu

Abstract An in-depth study on the dynamic aerodynamic performance of the pitching oscillating airfoil is carried out in the two-dimensional test section of the NF-3 low-speed wind tunnel of Northwestern Polytechnical University. The experimental model is a span-wise three-section force measuring model, and the force measure is only performed in the middle section of the model to reduce the influence of the sidewall interference of the wind tunnel. In the experiment, the transient angle of attack of the model is collected, the inertial force and pitching moment on the middle section of the model are calculated, and the data collected from the balance is subtracted to correct the influence of the model's inertia on the results. The results show that the angle of attack exceeding the positive or negative static stall angles of attack is a necessary condition for the lift and pitch moment coefficients to produce a large hysteresis loops. As the oscillation reduced frequency increases, the dynamic stall is delayed, the lift coefficient hysteresis loop increase, the drag coefficient increase, and the pitch moment coefficient near the maximum angle of attack decrease. When the angle of attack is less than the static angle of attack of stall or exceeds a small range, with the increase of the reduced frequency of the airfoil oscillation, the pitch moment coefficient of the airfoil decreases when it goes up and increases when it goes down. With the increase of the oscillation amplitude, the hysteresis loops of both dynamic lift coefficient and pitching moment coefficient of the oscillating airfoil increase. As the average angle of attack increases, the angle of attack of airfoil enters the positive stall zone more, the lift coefficient hysteresis loop increases, and the minimum pitch moment coefficient decreases. The Reynolds number has no obvious effect on the hysteresis loop of lift, drag and pitch moment coefficients; however, in the downward process, as the Reynolds number increases, the lift recovery advances, and the hysteresis loop decreases.

Keywords Airfoil · Wind tunnel test · Dynamic aerodynamic performance · Direct force measurement · Inertia correction

Y. Jiao (✉) · C. Xiao · D. Wu
National Key Laboratory of Science and Technology on Aerodynamic Design and Research, Northwestern Polytechnical University, No.127 Youyixi Road, Beilin, Xi'an, Shaanxi, China
e-mail: jiaoyuqin@nwpu.edu.cn

S. Lee et al. (eds.), *The Proceedings of the 2021 Asia-Pacific International Symposium on Aerospace Technology (APISAT 2021), Volume 1*, Lecture Notes in Electrical Engineering 912, https://doi.org/10.1007/978-981-19-2689-1_16

1 Introduction

The maneuverability requirements of fighter jets, the study of the dynamic characteristics of helicopter rotors and wind turbines in the field of wind energy all require more in-depth research on the dynamic aerodynamic performance and dynamic stall characteristics of the airfoil [1–3]. In the 1940s, Himmelskamp first discovered the existence of dynamic stalls in experiments, but it was not until the study of helicopter rotor experiments in the 1960s that the study of dynamic stalls received attention [4–9]. A variety of test methods have been applied to dynamic stall research to measure lift, drag, pitching moment and related flow parameters and flow characteristics during the stall. The commonly used measuring methods for dynamic stall research include: (1) using dynamic pressure sensors to measure airfoil surface pressure [6, 7, 9–11], (2) using Hot-Wire probes [6, 7, 12] or Particle Image Velocimetry (PIV) measure the velocity field around the airfoil [13–15], and (3) using the direct force measurement to study the influence of various airfoil motion parameters on its dynamic aerodynamic performance [12, 16–19]. The surface pressure measurement can only obtain the differential pressure drag, and the results of lift and pitch moment also have larger errors. Although the direct force measurement cannot obtain the detailed information of the flow field, it can better obtain the dynamic aerodynamic performance data of the airfoil.

The measurement results of the aerodynamic balance in the dynamic force measuring test include a certain amount of inertial force and pitching moment. In order to eliminate this error, in the above-mentioned direct force measuring study, a test with the pitching oscillation of airfoil is usually carried out without running of the wind tunnel (i.e., wind-off) before the formal test, and then the test with the same pitching oscillation of airfoil is performed under the required test wind speed in the wind tunnel (i.e., wind-on). And then, the wind-on data subtract the wind-off data, which is considered to eliminate the influence of the inertial force and pitching moment under the required test wind speed. Since the airfoil model is still subjected to aerodynamic forces even though the wind tunnel is not being driven, this method of eliminating the influence of inertia has errors. The National Key Laboratory of Science and Technology on Aerodynamic Design and Research of Northwestern Polytechnical University has carried out research on this issue. By accurately measuring the transient angle of attack of the model movement during the wind tunnel test, calculating the inertial force and pitching moment and correcting the test data, it has been verified that the inertial influence has been successfully corrected. In this paper, the direct force measurement is used to study the dynamic aerodynamic performance of a wind turbine airfoil with a relative thickness of 18% with both Reynolds number of millions and sinusoidal pitch oscillation.

2 Descriptions of Experiments

2.1 *Experimental Equipment and Instruments*

The experiment was carried out in the two-dimensional test section of the NF-3 low-speed wind tunnel of the National Key Laboratory of Science and Technology on Aerodynamic Design and Research of Northwestern Polytechnical University. This wind tunnel is low-speed open circuit wind tunnel of the all-steel structure and of length of 80 m, which power is 1120 kw. The two-dimensional test section has a rectangular cross-section of 8 m long, 3 m wide and 1.6 m high. The maximum wind speed is 130 m/s, the minimum steady wind speed is 10 m/s for the empty wind tunnel, the axial static pressure gradient is $dCp/dx = 0.0011(1/\text{m})$ and airflow turbulence is $\varepsilon = 0.045\%$.

The model oscillating motion driving mechanism is composed of a DC motor, a large flywheel, a reducer, a "-|"-shaped rod, an adjustable connecting rod and a swing lever. The DC motor with tachometer drives the "-|"-shaped rod, adjustable connecting rod and swing lever through a large flywheel and reducer to make the model to reciprocate in the sinusoidal law in the air flow of the test section. The motor is the Z4-132–3 type DC speed-adjusting motor with a rated power of 18.5 kW, a rated rotation speed of 1540 rpm, and a maximum rotation speed of 3000 rpm. The cylindrical worm reducer model is WHC18, which reduction ratio is 8, and the oscillation frequency range of airfoil model is 0–5 Hz. It is shown through theoretical analysis and experimental verification that the maximum deviation between the mechanism and the theoretical sinusoïdal pitching oscillation motion angles is less than 0.146% (Fig. 1).

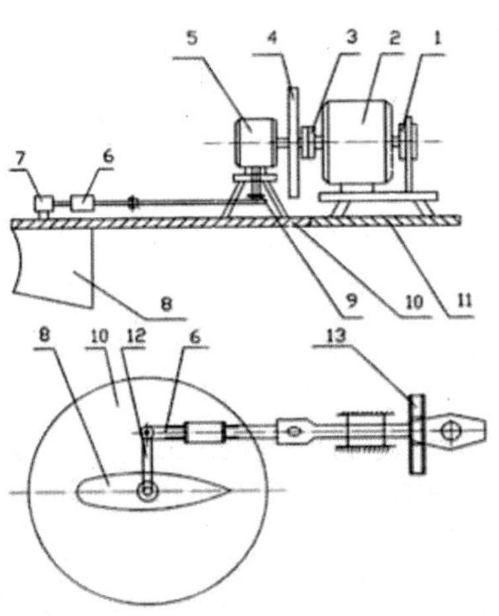

1—speed measuring machine
2—motor
3—flexible coupling
4—flywheel
5—cylindrical worm reducer
6—adjustable rod
7—angle sensor
8—model
9—rotating arm
10—turntable door
11—wind tunnel wall
12—swing lever
13—"-|"-shaped rod

Fig. 1 Schematic diagram of model oscillating motion driving mechanism

Table 1 Load capacity and static calibration performance index of balance

Item	Y	X	Z	M_z
Design load (N, N·m)	2200	600	500	200
Calibration load (N, N·m)	2000	500	500	200
Absolute error (N, N·m)	8	1.2	1	0. 7
Static calibration (%)	0.3	0.2	0.2	0.35
Limit error (%)	0.9	0.6	0.6	1.05
Comprehensive accuracy (%)	0.1	0.06	0.04	0.1

The aerodynamic force measurement uses a four-component box-type airfoil aerodynamic balance, which laboratory number is YXTP-4. The measuring components are lift Y, drag X, lateral force Z, and pitching moment Mz. The overall dimensions of the balance are 150 mm in length, 65 mm in width and 77 mm in height. The design center of the balance is 36.5 mm from the bottom. The balance material is high-strength aging steel 00Ni18Co8Mo5TiAl (Table 1).

An angle sensor is installed on the model shaft rotation. In the experiment, the signal of angle sensor and the four-component signals of the balance are collected by the dynamic measurement module of the Agilent VXI E8401A data acquisition system. The dynamic measuring performance index are that there are 32 measurement channels, each channel can be independently sampled in parallel, the range can be set independently, the acquisition speed of each channel is 100 kHz, there is a 16-bit independent A/D converter and the input signal voltage range is ± 12.5 mV ~ ± 10.0 V. The dynamic measuring accuracy of the system is better than 0.1% FS.

2.2 Experimental Model

The experimental model uses a 18% thick wind turbine airfoil with a span of 1.6 m and a chord length of 0.6 m. The overall frame span-wise three-section structure is used, and the upper and lower sections of the airfoil model are directly processed on the overall frame [20]. The upper and lower sections of the model adopt the structure with the support plates and outer skin, the interior is hollow, the outer skin is made of glass fiber reinforced plastic, the thickness is 5 mm, and the support plates are wing-shaped metal or fiberboard (Fig. 2). The middle section of the model is made up of upper and lower parts, which are made of high-strength, low-density 7075 aluminum alloy, and suspended on the overall frame of the model through the balance (Fig. 3). This not only ensures the overall strength and stiffness of the model, but also minimizes the mass and moment of inertia of the model in order to increases the oscillation frequency and reduced frequency during the test. The weight of the middle section of the processed model is 13.5 kg, the position of the center of gravity of the middle section of the airfoil model is $x_G = 216.41$ mm, $y_G = 13.71$ mm, and $z_G = -9.74$ mm.

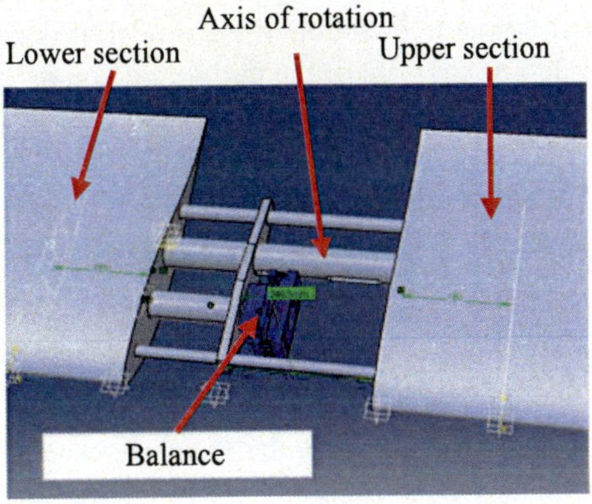

Fig. 2 Structure diagram of the airfoil dynamic force measuring model

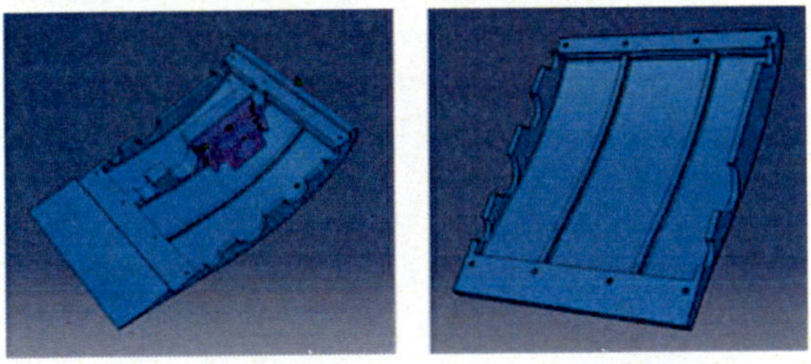

Fig. 3 The upper and lower cover of the middle section of the airfoil model

The experimental model is vertically suspended between the upper and lower turntables of the two dimensional test section of the wind tunnel. The upper and lower rotation shafts of the model are supported in the center holes of the turntables with bearings and can rotate flexibly. The upper rotation shaft reaches out through the center hole of the upper turntable and connects with the model pitching oscillating motion driving mechanism.

3 Elimination Method of Inertial Force and Pitching Moment

In the airfoil dynamic force measuring experiment, the airfoil model performs sinusoidal pitch oscillation driven by the driving mechanism, and the middle section of the model is driven direct by the aerodynamic balance. The force and pitching moment measured by the balance include both the aerodynamic force and pitching moment on the middle section of the model by the flow field and the inertial force and pitching moment of the middle section of the model. It is necessary to evaluate the inertial force and pitching moment of the middle section of the airfoil model and correct the direct force measuring results.

3.1 Elimination of Inertial Pitching Moment

When the center of gravity of the middle section of the airfoil model is on the model's axis of rotation, as shown in Fig. 4, because the middle section of the model only produces a inertial pitching moment on the balance while model move as pitching oscillation about the model's axis of rotation, and no inertial forces are generated in the x-axis direction and the y-axis direction. Of course, the moment reference center of the aerodynamic balance often does not coincide with the axis of rotation of the model and when the inertial force and pitching moment is deducted, the measured value is first converted to the coordinate system with origin at the axis of rotation of the model. For the force balance installed in the middle of the model in this article, the moments measured and transmitted by the balance are similarly

$$M_{z,b} = M_{z,d} = M_z + I_z\ddot{\alpha} \tag{1}$$

where $M_{z,d}$ and $M_{z,b}$ are the driving pitching moment and the pitching moment measured by the balance, which have been translated to a coordinate system centered on the axis of rotation according to the principle of coordinate system translation, α is the pitching oscillation transient angle of attack, t is the time, and $\ddot{\alpha}$ is the second derivative of the transient angle of attack α with respect to time t, and I_z is the moment of inertia of the middle section of the model around the axis of rotation parallel to the z-axis. Therefore, the aerodynamic pitching moment acting on the model is

$$M_z = M_{z,b} - I_z\ddot{\alpha} \tag{2}$$

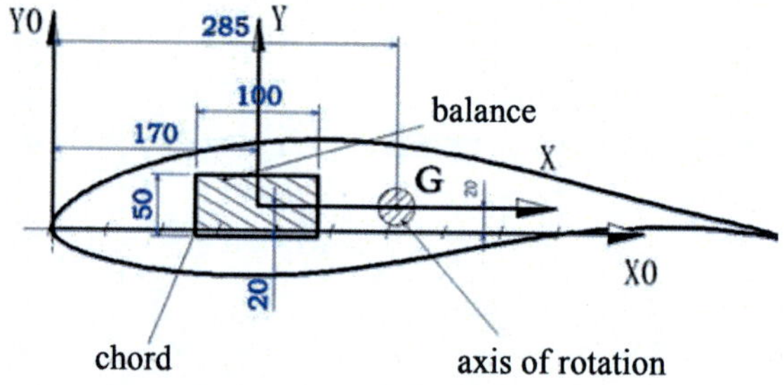

Fig. 4 Sectional diagram of the middle section of the model (the center of gravity G of the model on the axis of rotation of model)

The moment of inertia I_z is obtained by the Compound Pendulum Method and $Iz \approx 0.3505$ kg·m^2.

3.2 Elimination of Inertial Force

As shown in Fig. 5, when the center of gravity of the middle section of the model G is not on the central axis of rotation and the model rotates around the axis of rotation, it is subjected to centrifugal force. This centrifugal force is included in the measuring values of the built-in aerodynamic balance in this article. Of course, as above-mentioned in Sect. 3.1, the measured value is first converted to the axis of rotation of the model. Assuming that the angle between the line connecting the center of gravity of the middle section of the model G with the model's axis of rotation (actually a point in two dimensions) and the x-axis direction is β, and the distance from the center of gravity of the middle section of the model to the axis of rotation is r, the inertial force of the middle section of the airfoil model can be obtained,

$$|\boldsymbol{F}_i| = mr\dot{\alpha}^2 \tag{3}$$

Then the aerodynamic force of the middle section is $F = F_b - F_i$, where F_b is the measured value of the aerodynamic balance. Under the condition of two-dimensional motion, the components of the aerodynamic forces in the x- and y-axis directions, i.e., the normal force Y and the tangential force X, are represented as

$$Y = Y_b - Y_i = Y_b + |F_i| \sin\beta \tag{4}$$

$$X = X_b - X_i = X_b + |F|_i \cos\beta \tag{5}$$

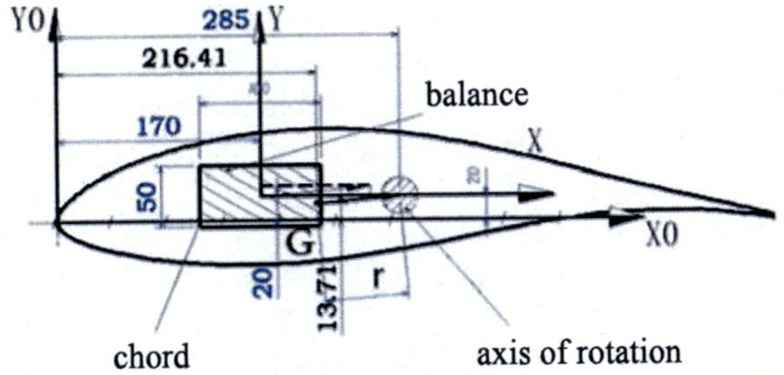

Fig. 5 Sectional structure diagram of the middle section of the model (the center of gravity G of the model not on the axis of rotation of model)

Finally, the calculated normal force Y and tangential force X are converted to the model wind axis system, and the final results of the dynamic lift L and drag D can be obtained as

$$\begin{aligned} L &= Y\cos\alpha - X\sin\alpha \\ D &= Y\sin\alpha + X\cos\alpha \end{aligned} \tag{6}$$

where L, D, α, $\dot{\alpha}$ are the model's lift, drag, and transient angle of attack of pitching motion and its first derivative with respect to time t respectively, and the subscripts b and i represent the balance measuring results and the amount of inertia.

The lift, drag and pitching moment calculated by formulae (2) and (6) are translated to the 1/4 chord point of the airfoil model, and nondimensionalized by the flow dynamic pressure q, the reference area s of the middle section of the model and the chord length c of model to obtain the lift, drag and pitching moment coefficients.

4 Data Processing and Analyses

4.1 Data Filtering and Averaging Processing

The cutoff frequency of the physical low-pass filter used in this study is 5 kHz. Generally, in large-amplitude ($\alpha \geq 10°$) oscillation motion experiments, the 5 or 6 times harmonic components of the oscillation frequency must be taken into account. The amplitudes of pitching oscillation of airfoil studied in this paper are all greater than 10°, and the maximum frequency of pitch oscillation that can be achieved in the experiments is 2.188 Hz. Therefore, the cut-off frequency of data digital low-pass filtering should be set to greater than 13 Hz.

The analysis of the measuring data shows that the force measuring signal contains a larger signal value with a frequency in the range of 16–60 Hz. By analyzing and comparing the data with different cutoff frequencies, it is determined that a digital low-pass filter with cutoff frequencies of 16 Hz for normal and tangential forces and 8 Hz for pitching moment is used for the force measurement. A Chebyshev filter with zero phase displacement is designed to ensure the phase consistency of the forces and pitching moments before and after filtering. Unless otherwise specified, the result data is the result of filtering with this Chebyshev filter. The data of multiple periods are averaged as the final result of a complete period.

4.2 Calculation of the First and Second Derivatives of the Angle of Attack with Respect to Time

A high-precision angle sensor is installed on the model, and the dynamic data acquisition system is used to collect the transient changes of the angle of attack over time in the experiment. In order to eliminate the error caused by the gap of the model motion driving mechanism, a designed model motion law formula is adopted as follows

$$\alpha = \alpha_0 + A\sin(2\pi f t + \varphi) \tag{7}$$

to fit the collected angle of attack to get the actual value of α_0, A, f, φ, and substitute these value into formula (7) as the actual model motion law formula. And adopt

$$\begin{aligned} \dot{\alpha} &= 2\pi f A\cos(2\pi f t + \varphi) \\ \ddot{\alpha} &= -(2\pi f)^2 A\sin(2\pi f t + \varphi) \end{aligned} \tag{8}$$

to calculate the first and second derivatives of angle of attack α with respect to time, $\dot{\alpha}$ and $\ddot{\alpha}$, where α_0, A, f, φ are the average angle of attack, the amplitude of the angle of attack, the frequency, and the initial phase of the airfoil pitching oscillating motion respectively.

5 Result Analyses

5.1 The Impact of Reduced Frequency

Figure 6 shows the dynamic aerodynamic characteristics of the oscillating airfoil at different reduced frequencies for average angle of attack $\alpha_0 = 15°$, amplitude $A = 15°$ and wind speed $V = 25$ m/s with natural transition of airfoil surface flow. The frequencies in the figure is $f = 0.531$, 0.995, 1.326, 1.989 respectively, which correspond to the reduced frequency $k = 0.04$, 0.075, 0.1, 0.15, respectively. It can be clearly seen from Fig. 6a that as the reduced frequency increases, the airfoil stall is delayed, the reattachment process is also delayed when descending, and the area of the hysteresis loop increases, which is in line with the general law of the dynamic experimental results. It can be seen from Fig. 6b, c that the drag coefficient increases, the pitch moment coefficient near the maximum angle of attack decreases with the increase of the reduced frequency, and the absolute value of its negative value is larger. In the range indicated by the dash line in the figure, the pitching moment coefficient decreases with the increase of the reduced frequency when going up, and increases with the increase of the reduced frequency when going down.

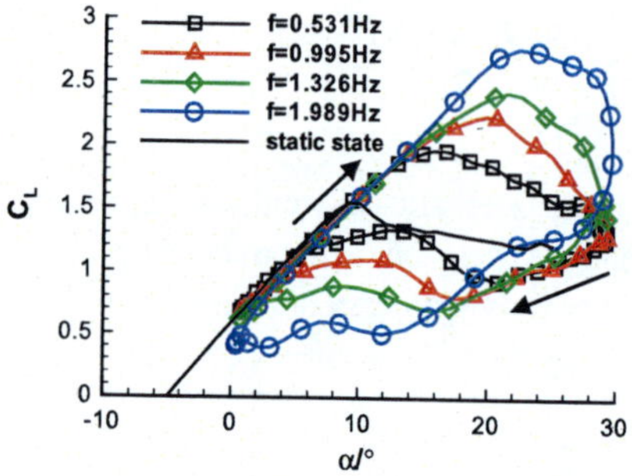

(a) Lift coefficient

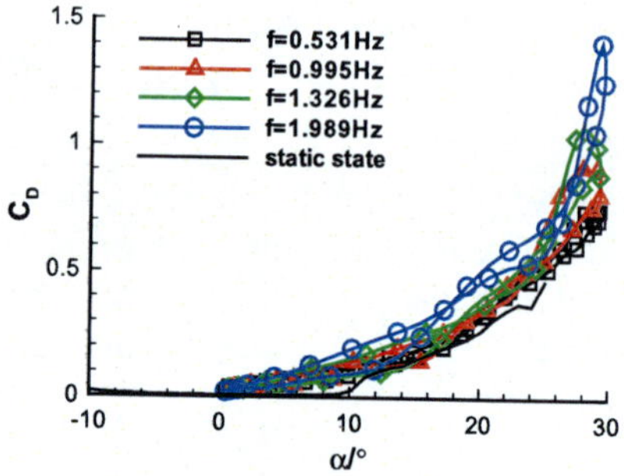

(b) Drag coefficient

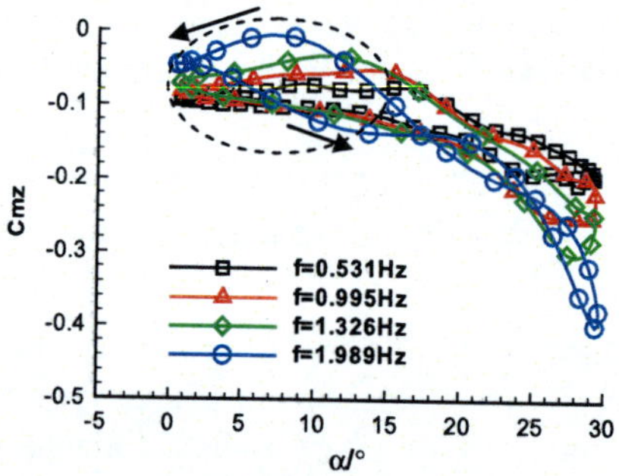

(c) Pitching moment

Fig. 6 Dynamic aerodynamic characteristics at different frequencies ($\alpha_0 = 15°$, $A = 15°$, $V =$ 25 m/s)

5.2 *Influence of Amplitude*

Figure 7 gives the experimental results of the pitching oscillation of airfoil model with the parameters of average angle of attack $\alpha_0 = 15°$, wind speed $V = 25$ m/s, frequency $f = 0.531$, reduced frequency $k = 0.04$ which the amplitudes are $A =$ 10, 12, 15° respectively. It is shown in Fig. 7 that as the amplitude of the oscillating motion increases, the hysteresis loops of the lift coefficient and the pitch moment coefficient increase.

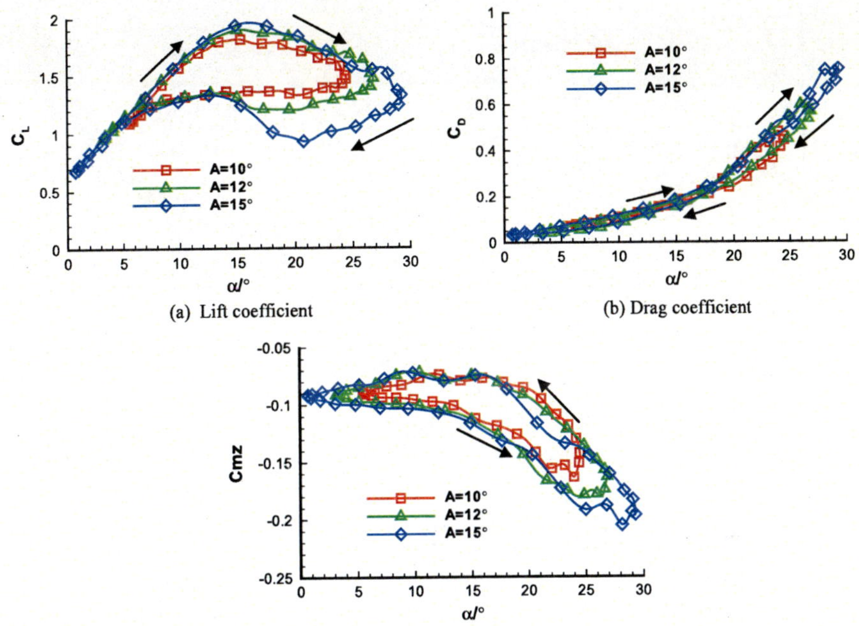

(a) Lift coefficient

(b) Drag coefficient

(c) Pitching moment

Fig. 7 Dynamic aerodynamic characteristics of different amplitudes ($\alpha_0 = 15°$, $V = 25$ m/s, $k = 0.04$)

5.3 *Influence of Reynolds Number*

In order to investigate the influence of Reynolds number on the dynamic aerodynamic performance of airfoil, Fig. 8 gives the experimental results of the pitching oscillation of airfoil model with the parameters of average angle of attack $\alpha_0 = 15°$, wind speed $V = 25$ m/s, reduced frequency $k = 0.075$, which the Reynolds number are $Re = 1.0, 1.6, 1.8, 2.0, 2.2 \times 10^6$ respectively. In Fig. 8a, it can be clearly seen from the area indicated by the elliptical dashed line that in the downward process, the lift recovery advances, and the hysteresis loop decreases with the increase of the Reynolds number. Reynolds number has no obvious influence on drag coefficient and pitch moment coefficient hysteresis loops. These conclusions are the same as these obtained by Xia YS (10) in the dynamic pressure test of NACA0012 airfoil in 1991 on the whole, but there are the differences somewhat on the shape of lift hysteresis loops.

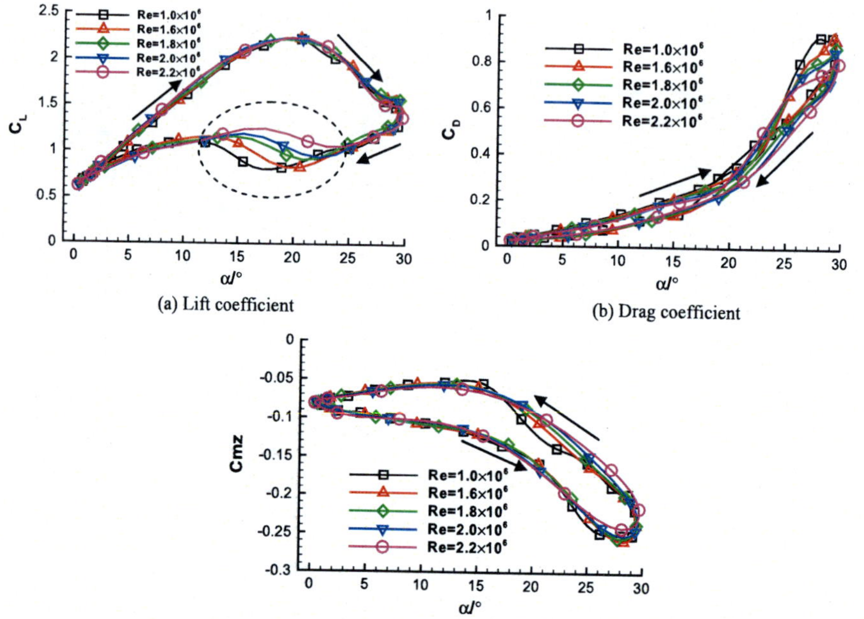

(a) Lift coefficient (b) Drag coefficient (c) Pitching moment

Fig. 8 Dynamic aerodynamic characteristics of different Reynolds numbers ($\alpha_0 = 15°, A = 15°$, $k = 0.075$)

5.4 *Effect of Average Angle of Attack*

Figure 9 gives the experimental results of the pitching oscillation of airfoil model with the parameters of amplitude $A = 15°$, wind speed $V = 25$ m/s, frequency $f = 0.531$ and reduced frequency $k = 0.04$, which the average angles of attack are $\alpha_0 = 0°, 5°, 10°, 15°$ respectively. It can be clearly seen from Fig. 9 that when the average angle of attack $\alpha_0 = 0°, 5°, 10°$ gradually increases, the stall angle of attack gradually increases, and the maximum lift coefficient also increases. But, when the average angle of attack further increases to $\alpha_0 = 15°$, the stall angle of attack remains unchanged and is 16°, the maximum lift coefficient is also unchanged. As the average angle of attack increases, the lift hysteresis loop increases. When the angle of attack approaches the lower limit, the hysteresis loop also appears, and the drag coefficient has a more obvious hysteresis loop near the lower limit of the angle of attack. When the average angle of attack increases successively, $\alpha 0 = 0°, 5°, 10°, 15°$, the pitch moment coefficient hysteresis loop moves in the direction of increasing angle of attack, and the value of the minimum pitch moment coefficient decreases. At the average angle of attack of $\alpha_0 = 0°$, since the minimum angle of attack is already smaller than the negative static stall angle of attack, it enters a negative stall, forming the "8"-shaped hysteresis loop of lift and pitching moment coefficients. Since the average angle of attack controls the range of angle of attack

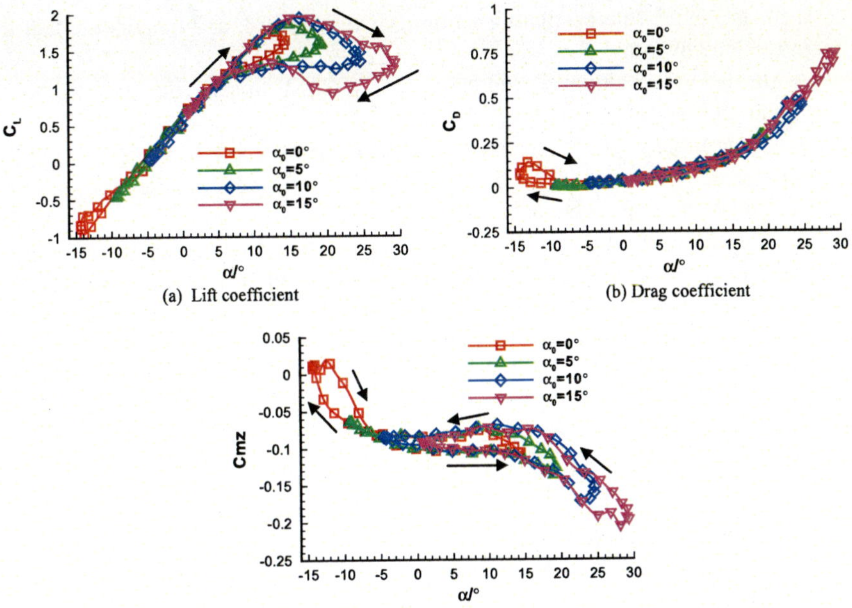

(a) Lift coefficient (b) Drag coefficient

(c) Pitching moment

Fig. 9 Dynamic aerodynamic characteristics of the different average angle of attack ($A = 15°, f = 0.531$, $V = 25$ m/s)

of the oscillating airfoil, when the average angle of attack increases, the oscillation range more enters the positive deep stall zone, and the hysteresis loops of lift and pitching moment coefficient moves and increases in the direction of the increasing angle of attack. Therefore, the angle of attack exceeding the positive or negative stall angle of attack is a necessary condition for a large hysteresis loop to form.

6 Conclusions

The main research conclusions are:

1. The angle of attack exceeding the positive or negative stall angle of attack is a necessary condition for the lift coefficient and pitching moment coefficient to produce a large hysteresis loops.
2. As the oscillation reduced frequency increases, the dynamic stall is delayed, the lift coefficient hysteresis loop increases, the drag coefficient increases, and the pitch moment coefficient near the maximum angle of attack decreases, and the absolute value of its negative value is greater. When the angle of attack is less than the static stall angle of attack or exceeds a small range of angle of attack, with the increase of the reduced frequency, the pitching moment coefficient

decreases when the oscillating motion moves upward, and increases when it moves downward.
3. As the oscillation amplitude increases, the hysteresis loops of the lift coefficient and pitch moment coefficient increase.
4. As the average angle of attack increases, the airfoil angle of attack enters the positive stall zone more, the lift coefficient hysteresis loop increases, and the minimum pitch moment coefficient decreases.
5. The Reynolds number has no obvious effect on the drag coefficient and pitch moment coefficient hysteresis loop, and has a small effect on the lift coefficient. In the downward process, as the Reynolds number increases, the lift recovery advance and then the hysteresis loop decreases.

References

1. Carr LW (1988) Progress in the analysis prediction of dynamic stall. J Aircr 25(1):6–17
2. Leishman JG (2000) Principles of helicopter aerodynamics. Cambridge University Press, Cambridge, pp 302–318
3. Conlisk AT (2001) Modern helicopter rotor aerodynamics. Prog Aerosp Sci 37(5):419–476
4. Ham ND (1968) Aerodynamic loading in a two-dimensional airfoil during dynamic stall. AIAA J 6(10):1927–1934
5. McCroskey WJ, Pucci SL (1982) Viscous-inviscid interaction on oscillating airfoil in subsonic flow. AIAA J 20(2):167–174
6. Carr LW, McAlister KW, McCroskey WJ (1977) Analysis of the development of dynamic stall based on oscillating airfoil experiments. National Aeronautics and Space Administration, Washington D. C. NASA TN D-8382
7. McAlister KW, Carr LW, McCroskey WJ (1978) Dynamic stall Experiments on the NACA0012 Airfoil. National Aeronautics and Space Administration, Washington D. C. NACA TP 1100
8. Conger RN, Ramaprian BR (1994) Pressure measurements on a pitching airfoil in a water channel. AAIA J 32(1):108–115
9. Piziali RA (1994) 2-D and 3-D oscillating wing aerodynamics for a range of angles of attack including stall. National Aeronautics and Space Administration, Washington D. C. NASA TM 4632
10. Xia YS et al (1996) Dynamic of stall characters of NACA0012 airfoil and investigations of dynamic pressure measure methods. Acta Aeronautica et Astronautica Sinica. 17(7):25–30
11. Hui ZH, Wang L, Xu Q (2012) Dynamic pressure measurement techniques on wind turbine air-foil. J Experiments Fluid Mech 26(4):6–10
12. Panda J, Zaman KBMQ (1994) Experimental investigation of the flow field of an oscillating airfoil and estimation of lift from wake surveys. J Fluid Mech 265:65–95
13. Werner P et al (1996) Experimental and numerical investigations of dynamic stall on a pitching airfoil. AIAA J 34(5):982–989
14. Oshima H, Ramapian BR (1997) Velocity measurements over a pitching airfoil. AIAA J 35(1):119–126
15. Wang Q, Zhao QJ, Zhao GQ (2014) PIV experiments on flow field characteristic of rotor airfoil dynamic stall and modifications of L-B model. Chinese J Theoret Appl Mech 46(4):631–635
16. Tang RY, Hua XM, Wu SJ (1992) Experimental study of dynamic stall on an oscillating airfoil. J Nanjing Aeronaut Inst 24(5):506–512
17. Tsang KYK (2006) Direct force measurements of a two-dimensional airfoil undergoing dynamic stall. Hong Kong Polytechnic University, Hong Kong, pp 1–125

18. Rival D, Tropea C (2010) Characteristics of pitching and plunging airfoils under dynamic-stall conditions. J Aircraft 47(1)
19. Lin YF et al (2012) Experimental investigation of rotor airfoil dynamic stall characteristics. Aeronaut Sci Technol Special Helicopter Technol 2012(4):25–28
20. Jiao YQ et al (2020) Three-section airfoil experimental model with integral frame for dynamic force measurement in wind tunnel. China National Intellectual Property Administration, Peking. 2020–6–9. CN201921940956.4[P]

The Research of Correlationship and Critical Ice Shape Acquisition in CIRA-IWT Icing Wind Tunnel

Bai Feng, Yan Wei, and Li Haixing

Abstract Icing wind tunnel is a mature way to obtain ice shape, and the SAE has conducted ice shape correlation work in the icing wind tunnels as IRT, BRAIT, COX, Goodrich, etc. On the basis of this, the NACA0012 standard model was adopted to implement series ice shapes test in CIRA-IWT, meanwhile, the ice shape was also calculated by CFD for comparison, then the icing parameters sensitivity was analysed. All this work confirmed the correlationship of CIRA-IWT with other icing wind tunnel around the world and was the basis of the following critical ice shape research. According to the FAR25 App.C, the 45 min holding ice shape acquired under continuous cloud icing condition can result in the most severe aerodynamic degradation of aircraft, which called the critical ice shape. A method based on the total water collection (TWC) was proposed to determine the critical icing parameters such as velocity, altitude, temperature, LWC and MVD. Furthermore, the method was verified in CIRA-IWT by observing the ice shedding at the end of 45 min duration and by the comparison of ice shape calculated by CFD. Finally, the critical ice shape prediction method based on the combination of CFD and icing wind tunnel test was the future research direction of this paper.

Keywords Critical ice shape · Mean volume diameter (MVD) · Liquid water content (LWC) · Total water content (TWC) · Ice shedding

---

B. Feng (✉) · Y. Wei · L. Haixing
Flight Physics Department, Shanghai Aircraft Design and Research Institute of COMAC, Shanghai, China
e-mail: baifeng@comac.cc

Y. Wei
e-mail: yanwei@comac.cc

L. Haixing
e-mail: Lihaixing@comac.cc

B. Feng
Shanghai Aircraft Design and Research Institute of COMAC, No.5188, Jin Ke Road, Pu Dong, Shang Hai, China

S. Lee et al. (eds.), *The Proceedings of the 2021 Asia-Pacific International Symposium on Aerospace Technology (APISAT 2021), Volume 1*, Lecture Notes in Electrical Engineering 912, https://doi.org/10.1007/978-981-19-2689-1_17

1 Correlation Research Program

The SAE conducted the icing wind tunnel correlation research work based on NACA0012 standard aerofile in the mature industrial icing wind tunnel in Europe and the United States, as show in Table 1. The correlation was studied by comparing the standard model ice shapes in these wind tunnels calibrated respectively [1].

The CIRA-IWT icing wind tunnel was not included because of the corresponding tests in CIRA-IWT was not finish in time. CIRA-IWT is a single-circuit, multi-test section icing wind tunnel, as shown in Fig. 1.

This article is based on the second test section of CIRA-IWT to carry out correlation comparison work as shown in Table 2 [2], and the corresponding test tasks are shown in Table 3 [1].

Table 1 Mature industrial icing wind tunnels

No	Wind tunnel	Test section height (mm)	Test section length (mm)
H	AIWT, NRC	571.5	3962.4
P	GKN ATS, UK	762	4876.8
F	BRAIT, Boeing	1828.8	6096
A	COX	1168.4	5791.2
E	Goodrich	1117.6	6400.8
M	IRT, NASA	2743.2	13,411.2

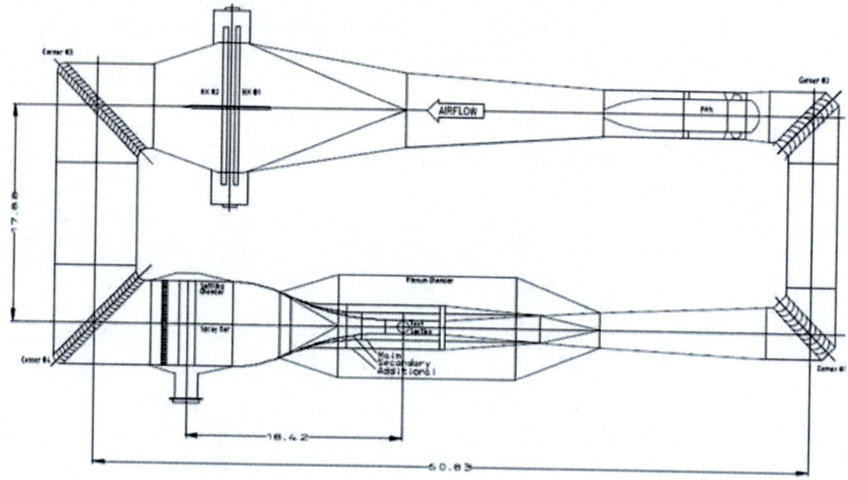

Fig. 1 CIRA-IWT wind tunnel

Table 2 The second test section of IWT

Height (m)	Width (m)	Length (m)	Mach	SAT (°C)	Static pressure (Pa)	MVD (μm)	LWC (g/m^3)
2.35	1.15	5	Up to 0.7	−40	39,000 to 145,000	10 to 50	0.1 to 4.0

Table 3 2 D standard model test case

CASE	Altitude (ft) (ft)	AOA (°)	V (m/s) (m/s)	SAT (°) (°C)	MVD (μm) (μm)	LWC (g/m^3) (g/m3)	Duration (min) (min)
1	1500	3	67.0	−7	20	0.5	25
2	1500	3	90.0	−7	20	0.5	20
3	1500	3	67.0	−7	40	1.0	20
4	1500	3	67.0	−23	20	0.5	25
5	1500	3	90.0	−30	20	0.5	20

2 Standard Model Test

2.1 *Test Model*

The test mode was NACA0012 aerofoil with chord length 1410 mm and an extension length 1130 mm. It was installed horizontally in the test section, and the two sides are connected with the turntable of the side walls to adjust the angle of attack, as shown in Fig. 2. A total of 173 pressure measuring points were arranged on the upper and lower surface at the model central cross section. The pressure distribution was

Fig. 2 Ice measurement bracket

Fig. 3 Pressure match

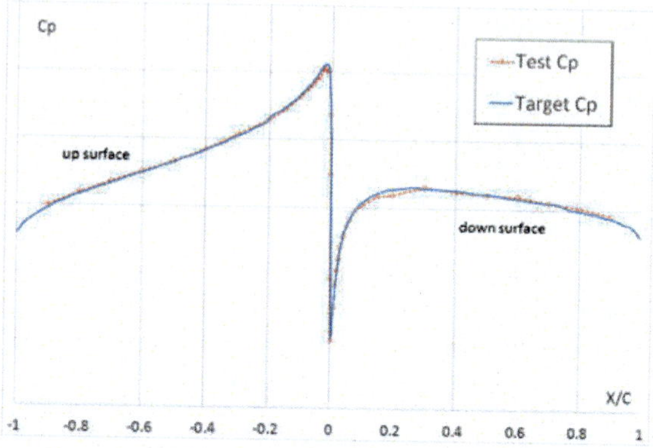

Fig. 4 Ice shape achievement

matched to the target value by adjusting the model angle to eliminate the influence of the wind tunnel wall to the airflow angle, then ensure that the water droplet impact characteristics near leading edge of the aerofoil are similar to that of the theoretical one, as shown in Fig. 3. The ice shape measuring bracket installation interfaces were set at 400 mm aside of the central line. After ice accretion process, a hot knife was used to cut the ice, then the ice shape was traced on the graph paper, as shown in Fig. 4.

2.2 *Icing Parameters Sensitivity*

Five differerent icing test cases were shown in Table 3, and the variation icing parementers contained the velocity, static temperature, MVD and LWC, and the test results compared with CFD are shown in Figs. 5, 6, 7, 8 and 9. By comparing CASE1 and CASE4, when the static temperatures were −7 °C and −23 °C respectively, the ice texture characteristics of glaze ice and rime ice can be distinguished. According

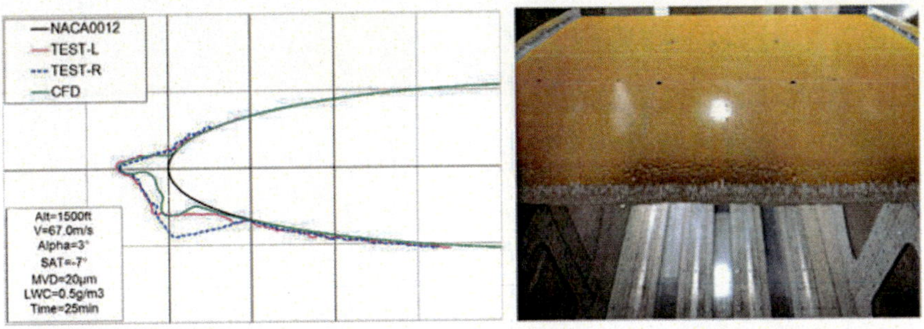

Fig. 5 Ice shape of CASE1

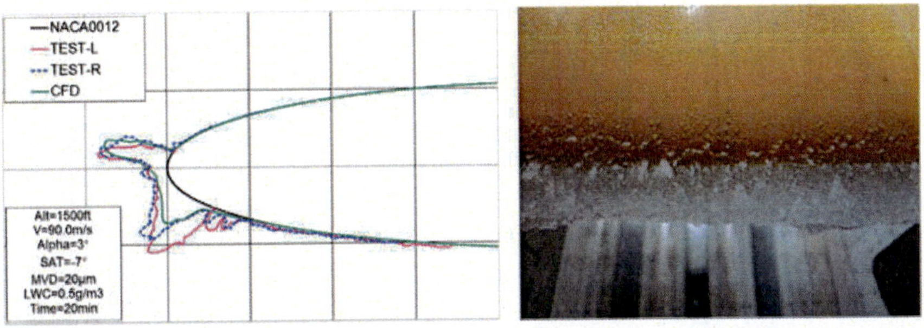

Fig. 6 Ice shape of CASE2

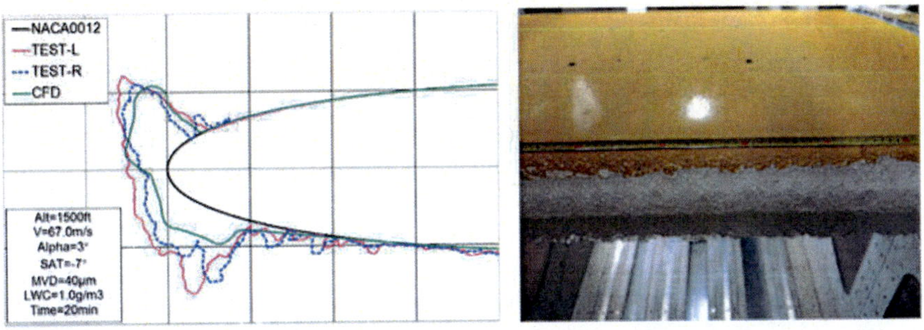

Fig. 7 Ice shape of CASE3

to the ice shape, it shows that the horn ice shape is more likely to be formed under higher temperature. The reason is that, at higher temperature, the water droplets will not freeze immediately, then the initial freezing position is located farther downstream of the stagnation point. Thus, the following coming water droplets will flow backword which affected by the aerodynamic force, result in bigger ice horn angle and higher ice height [13].

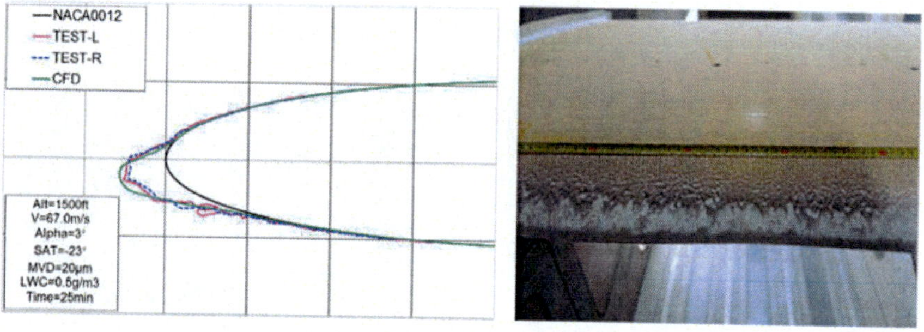

Fig. 8 Ice shape of CASE4

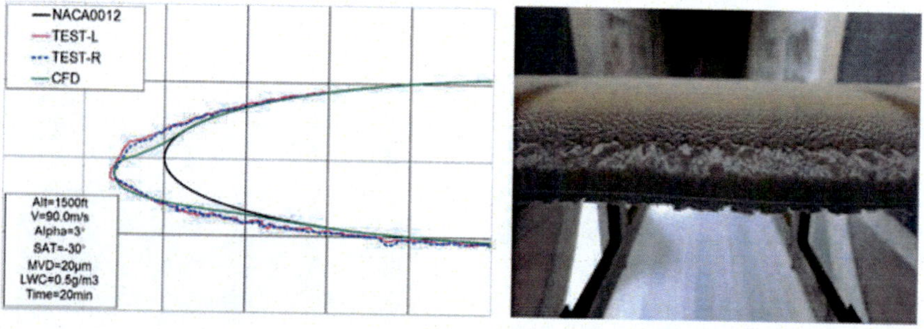

Fig. 9 Ice shape of CASE5

Comparing CASE1 and CASE2, when the velocities were 67.1 m/s and 89.4 m/s (duration is 25 min and 20 min respectively) while the static temperature was both −7 °C, the direction of the ice horn and the cross-sectional area are obviously different. CASE2 is more closer to a typical double horn ice with higher test velocity. The mainly reason is that the increase in velocity will cause higher total temperature, then postpone the freezing time of the water droplets, and the inertia of the water droplets is larger, so it is easier to flow backwards [14].

Comparing CASE1 and CASE3, when the MVD is 20 μm and 40 μm while the LWC is 0.5 g/m^3 and 1.0 g/m^3 (the duration is 25 min and 20 min respectively). The larger MVD of CASE3 causes the greater inertia of the water droplets, which increases the water collection rate. And the larger LWC causes the greater total water droplet collection volume and the more amount of ice.

For the whole rime ice and the upper horn of the glaze ice shape, the wind tunnel restuls shown good agreement with that ot the CFD results. The typical ice shape parameters (ice horn angle and height) were defined as shown in Fig. 10, and compared in Table 4.

By comparing the parameters of ice shape, the results of CFD calculation of rime ice are consistent with the test results. For the glaze ice, the upper horn angle are similar while the lower horn angle and the total amount of ice shape area were smaller [15].

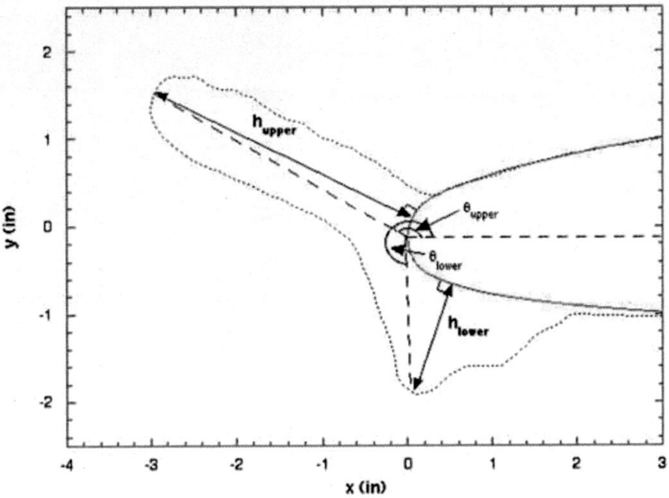

Fig. 10 Ice shape parameters definition

Table 4 Comparison of ice shape parameters

	Ice shape	hu (h/c)	φu (°)	hl (h/c)	φl (°)
CASE1	Test left cut	0.034	180	0.027	248
	Test right cut	0.033	180	0.022	275
	CFD	0.033	180	0.019	270
CASE2	Test left cut	0.049	169	0.045	265
	Test right cut	0.047	165	0.038	252
	CFD	0.045	160	0.022	267
CASE3	Test left cut	0.059	118	0.055	283
	Test right cut	0.050	113	0.044	295
	CFD	0.044	104	0.020	290
CASE4	Test left cut	0.031	104	\	\
	Test right cut	0.028	104	\	\
	CFD	0.034	104	\	\
CASE5	Test left cut	0.038	100	\	\
	Test right cut	0.036	100	\	\
	CFD	0.037	100	\	\

2.3 *Ice Shape Correlationship*

In order to compare the correlations with the SAE wind tunnels more clearly, the ice shape with good consistency in the SAE tests were averaged firstly, for example, the AFMP shown in Fig. 11 means the averaged ice shape of COX (A), BRAIT (F),

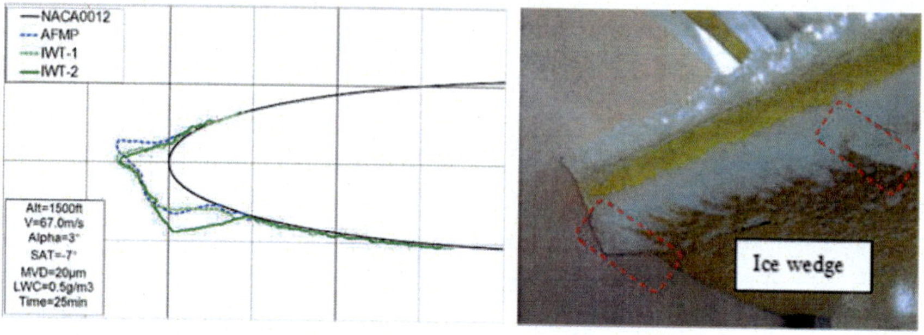

Fig. 11 Comparison of CASE1

NASA-IRT (M) and GKN ATS (P), and then the comparison with the IWT (CIRA-IWT) were shown in Figs. 11 , 12, 13, 14 and 15. As shown in Figs. 11 and 12, the IWT-1 and IWT-2 mean the ice shape obtained at two different cross sections, the consistency of the upper ice angle is excellent, and the IWT-2 lower ice horn is larger due to the small ice wedge as show in right photo of Fig. 11.

Fig. 12 Comparison of CASE2

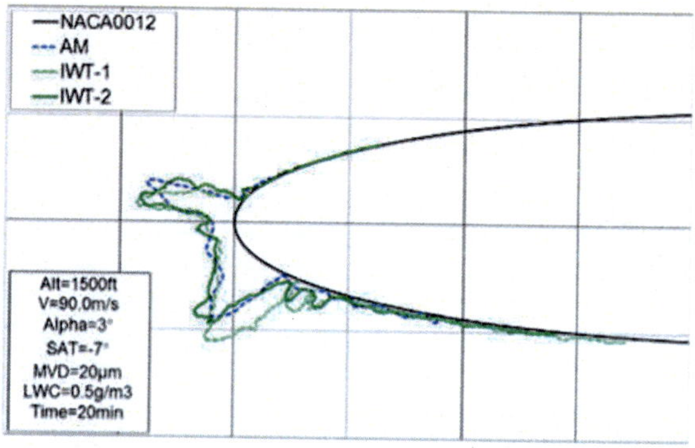

Fig. 13 Comparison of CASE3

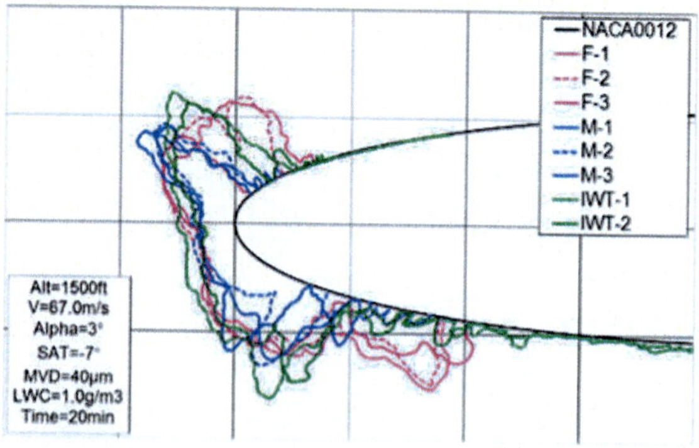

Fig. 14 Comparison of CASE4

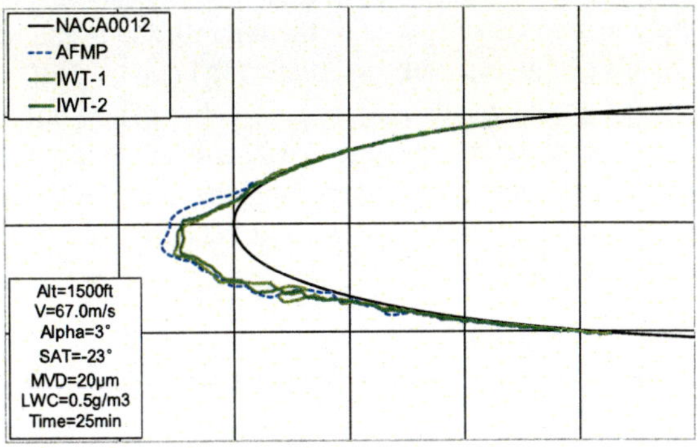

Fig. 15 Comparison of CASE5

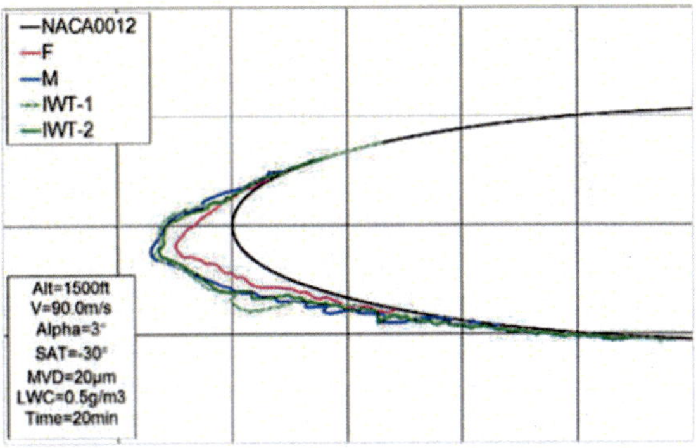

Comparing the results of the BRAIT(F) and the IRT(M) under the glaze icing condition in CASE3, it shows that the repeatability of the ice shape of same wind tunnel is very good, however, the ice shape difference between different wind tunnel is obviously. Contrasting with IRT, the upper ice horn of IWT is slightly thicker and with higher angle. But the BRAIT result has the shortest upper horn and the biggest upper angel of them. Judging from the difference between the amount of ice accumulation and the range of icing, it can be inferred that the three wind tunnels have differences in the control precision of droplet diameter and water content.

By comparing to that of the glaze icing condition, the consistency of each wind tunnel under rime icing conditions is better, as shown in CASE4 and CASE5.

As the comparison above, the uniformity of rime ice is great due to the lower temperature, while that of the difference of double horn ice (glaze ice) is obviously. According to the results of the SAE research, it was believed that although each wind tunnel satisfied the icing wind tunnel's flow field and cloud field calibration indicators [8], the difference of the calibration equipment and calibration procedures will also affect the result. In addition, the difference in model's surface quality and wind tunnel structure will also lead to the difference in the final ice shape. In general, the test results of the IWT wind tunnel and IRT wind tunnel were in good agreement, and the aerodynamic loss of the aircraft due to the ice shape from these two wind tunnels can be estimated as same [11, 12].

On the other hand, from the perspective of airworthiness certification, if the important ice parameters (upper ice angle and height, etc.) of the icing shape were in good agreement, it can be considered that the impact on aerodynamic force is equally, and slightly conservative ice shape was encouraged.

3 Critical Ice Shape Research

3.1 Critical Icing Conditions

The parameters that affect icing during aircraft flight include flight status (velocity, altitude), meteorological conditions (temperature, LWC, MVD), etc. Therefore, when determining critical icing conditions, these parameters need to be combinatory analysis. According to the results of a numerous of cloud detections, the MVD in continuous cloud is mostly around 20 μ, and this is most critical parameter of MVD through sensitivity analysis. In addition, MVD is calculated as a secondary parameter that affects the ice shape, so MVD 20 μ is used for most critical ice shape analysis [16].

Furthermore, ice accretion may only occur when the total temperature is below zero in flight. When the holding speed of the aircraft is change, the icing altitude-temperature envelope is different. The relationship between total temperature (TAT) and static temperature (SAT) is shown in the following formula (4), and the relationship between Ma and the corrected airspeed (CAS) is shown in the following formula (5).

$$TAT = SAT \cdot \left(1 + 0.2Ma^2\right) \tag{4}$$

$$Ma = \sqrt{5\left[\left(\frac{1}{\Delta}\left\{\left[1 + 0.2\left(\frac{CAS}{661.4786}\right)^2\right]^{3.5} - 1\right\} + 1\right)^{\frac{1}{3.5}} - 1\right]} \tag{5}$$

where Δ is the static pressure ratio, which can be expressed as:

$$\Delta = [(288.15 - 0.0019812 \times h_p)/288.15]^{5.25588} \tag{6}$$

where h_p is the pressure altitude.

According to formula (4) and formula (5), for different flight speeds and air pressure altitudes, the boundary line of TAT equal to zero can be draw out according to the altitude-temperature envelope of Appendix C, as shown in Fig. 16. The freeze

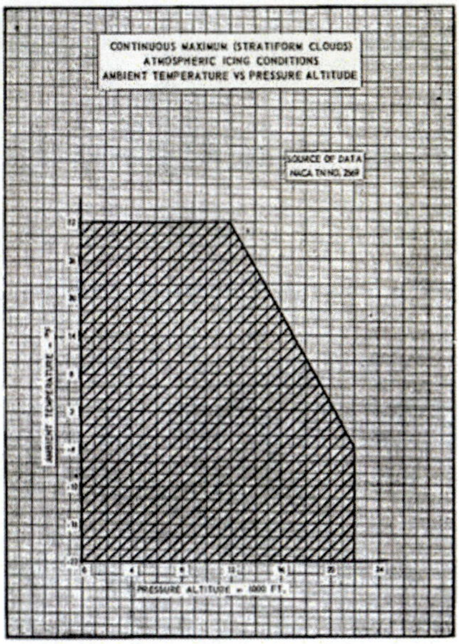

Fig. 16 Ice Alt-T envelope of App.C

in flight can only happen below this boundary line, so this boundary line is called the speed-altitude-temperature envelope of icing shown in Fig. 17.

According to the static temperature corresponding to the zero-degree line in Fig. 17, combined with the MVD-LWC envelope in Appendix C, as Fig. 18, the altitude-LWC curve of specific flight speed can be obtained, see Fig. 19.

Assume that the aerofoil projected area along the airflow direction is S and the water droplet collection coefficient is Ac. Then the total water collection can be

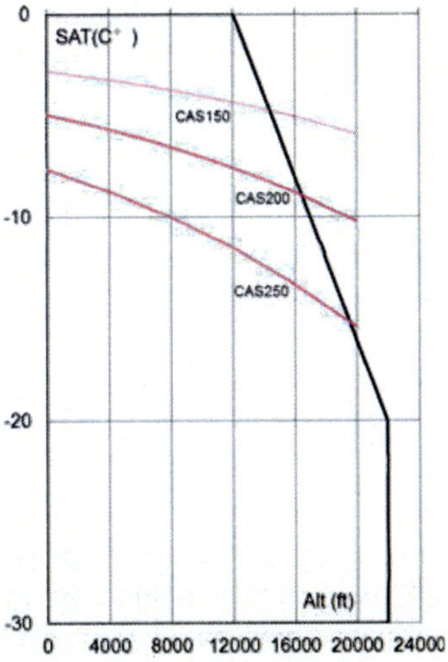

Fig. 17 Zero TAT borderline

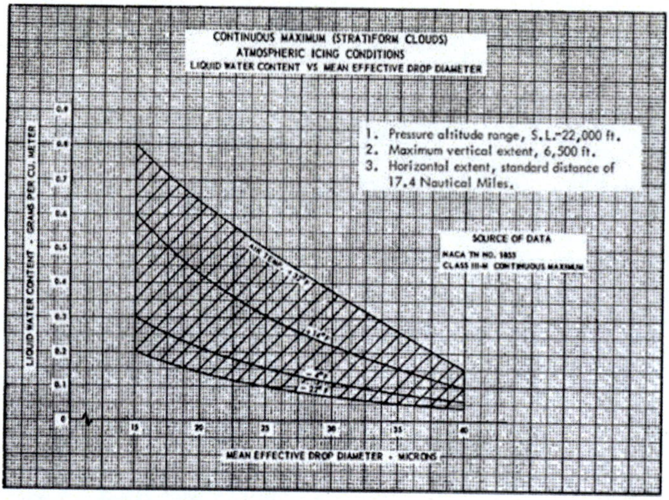

Fig. 18 Ice MVD-LWC envelope of App.C

Fig. 19 Zero TAT max TWC point

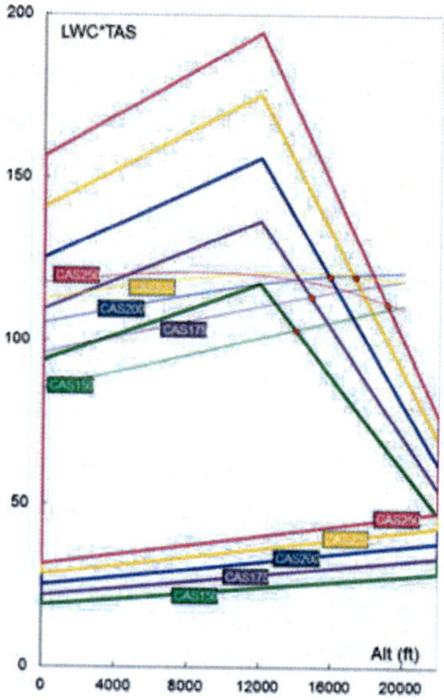

expressed as formula (7) [17]:

$$TWC = Time * TAS * S * LWC * Ac \tag{7}$$

In Fig. 19, for different flight speeds, the cross point (red dot) of the dimensionless total water collection curve (dashed line) and the zero-total temperature boundary line (quadrangle line) is the critical freezing point of the accordingly flight speed.

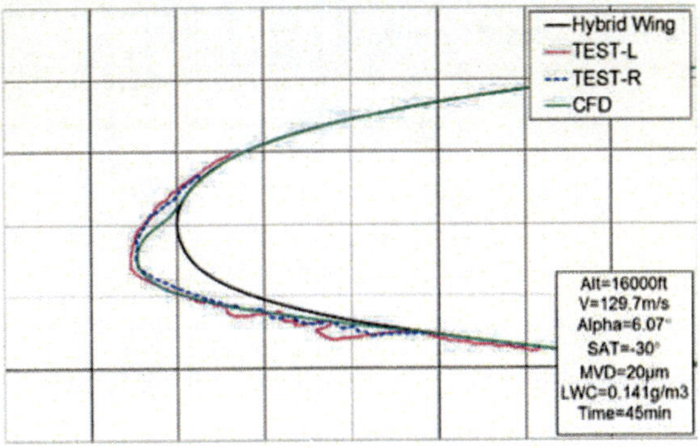

Fig. 20 Ice shape of CASE1

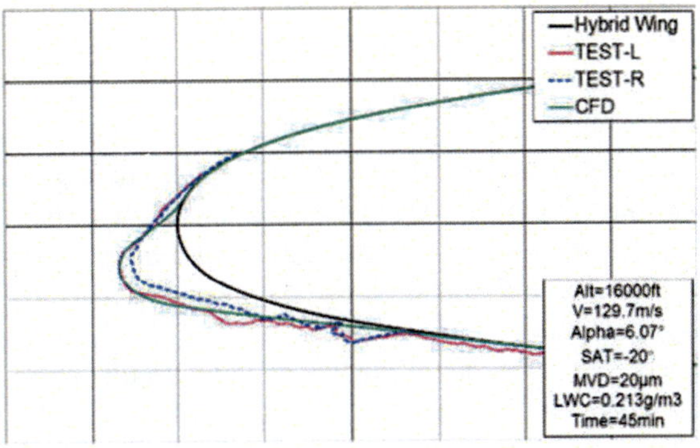

Fig. 21 Ice shape of CASE2

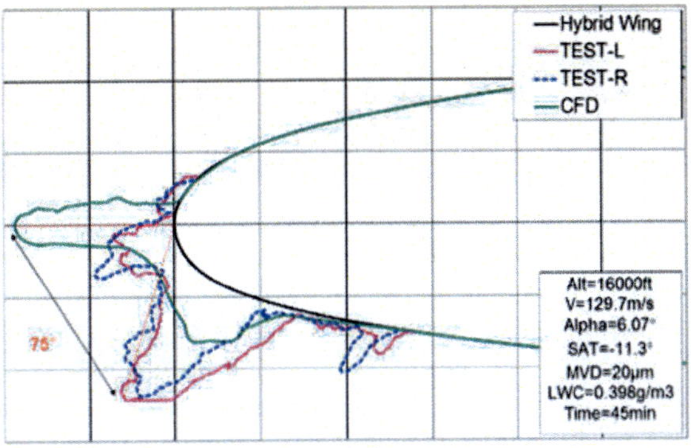

Fig. 22 Ice shape of CASE3

Then, the most critical icing conditions can be estimated by pick the largest water collection volume (vertical axis).

Fig. 23 Ice shape of CASE4

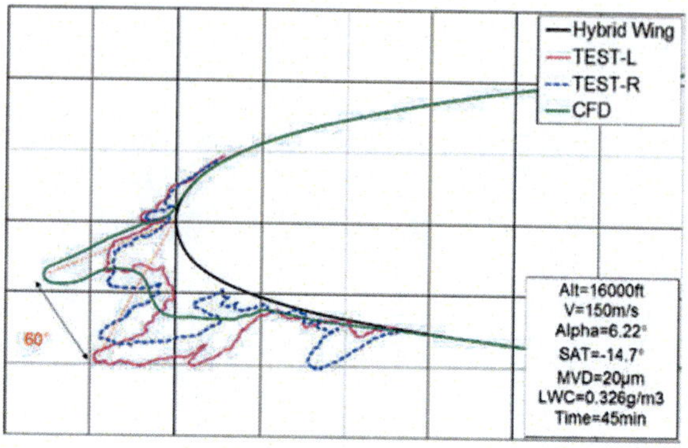

3.2 *Critical Ice Shape Test*

To verify the critical icing parameters, icing test was carried out in the IWT wind tunnel by 2D supercritical aerofoil model. The test process is same with that of the previous NACA0012 standard model test. The critical ice state is determined to be 16000ft in height, 200 knots (129.7 m/s) in speed, SAT −11.3 °C, MVD 20 μm, LWC 0.398 g/m^3, and icing test duration set as 45 min. The test status is arranged as shown in Table 5.

The test result of ice shape and CFD calculation results were shown in Figs. 20, 21, 22, 23, 24, 25, 26 and 27. For the CFD results, ice shape of CASE3 is the most critical one with maximum horn angle and height. But the ice shape of CASE3 and CASE4 is shedding during the test, which leads to a large difference in ice shape compared with CFD, the ice horn angel difference is 75°, even the ice section area almost same.

The ice shedding process is shown in Figs. 24 and 25, and the two ice shapes cutting position is marked by red line. The shedding of the ice shape is due to the aerodynamic force that greater than the freeze adhesion force, since the aerodynamic force is mainly determined by the size of the ice horn, and the freeze adhesion force is determined by the ice texture. For the critical ice shape of glaze ice, the ice horn is the largest and thus reaches the extreme aerodynamic force; while the glaze ice has a slow freezing process (small air bubble being extrude out) cause solid texture and

Table 5 Icing test case of wing

CASE	Altitude (ft)	AOA (°)	V (m/s)	SAT (°C)	MVD (μm)	LWC (g/m3)
1	16,000	6.07	129.7	−30	20	0.141
2	16,000	6.07	129.7	−20	20	0.213
3*	16,000	6.07	129.7	−11.3	20	0.398
4	16,000	6.22	150	−14.7	20	0.326

Note Case3 is the expected most critical condition

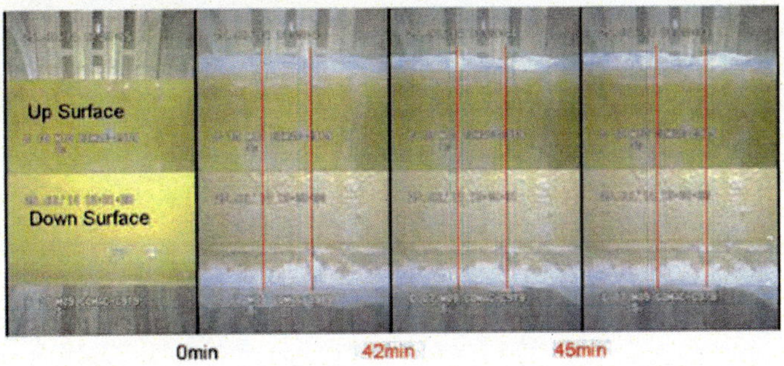

Fig. 24 Ice shedding of CASE3

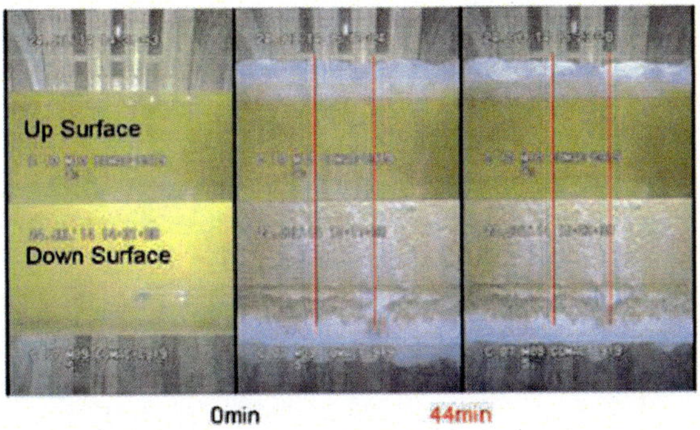

Fig. 25 Ice shedding of CASE4

strong adhesion. Therefore, it can be inferred that ice horn has reached critical size before it falls off but can't be measured finally. And according to the video record, the ice horn shedding occurred nearly 45 min, so it is mean that the ice type reached the critical size just before shedding. At the same time, according to the CFD calculation results of CASE3, the ice angle height has reached about 3 inches at 45 min, which is also the critical ice type size recognized by the FAA [18, 19].

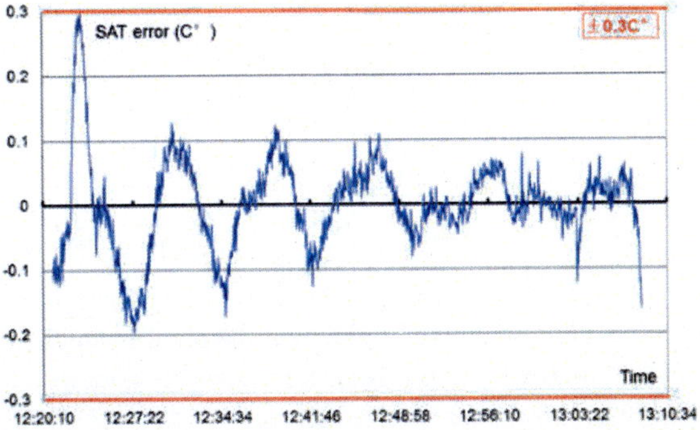

Fig. 26 Temperature stability of CASE4

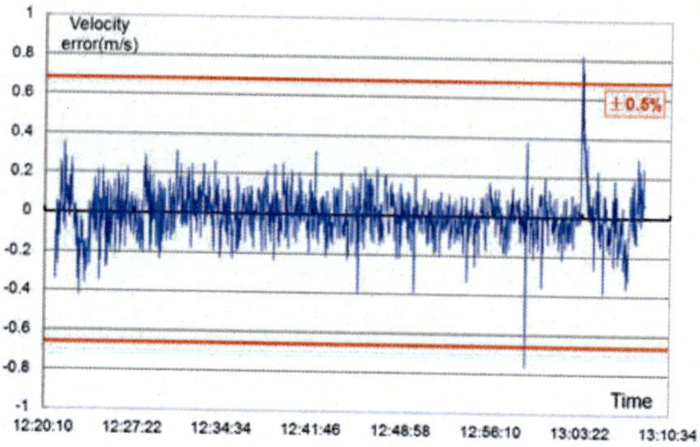

Fig. 27 Velocity stability of CASE4

In addition, the speed and temperature parameter stability during the test were analysed, and they were all within the error control range [5], which eliminates the possibility of ice shedding caused by the fluctuation of air flow.

According the above analysis, the critical icing condition estimation method based on the total water collection (TWC) can determine the critical icing parameters, and the ice shape shedding just before the end of 45 min ice accretion is probably due to the limitation of freeze adhesion in the icing wind tunnel test. For this critical ice type, it will be more difficult to get in the actual atmospheric flight because of the parameters fluctuated more severely and uncertainly. Therefore, the method combining CFD and wind tunnel test was raised as:

(1) Obtain an ice shape of short period in the icing wind tunnel (for example in 10 min);
(2) Base on this geometry of aerofoil and ice shape, conducting CFD calculation for time remaining (35min), then acquire the complete 45 min ice shape;
(3) Change the time combination of wind tunnel test and CFD to verify the ice shape uniformity;
(4) Verify the smaller ice shape (for example 25 min) obtained in step (3) by comparing with the 25 min ice shape (without ice sheeding) obtained entirely in the icing wind tunnel.

This method combining CFD and icing wind tunnel test for obtaining critical ice shape is a future research direction of the paper.

4 Conclusion

This article uses a 2D NACA0012 standard aerofoil to confirmed the correlationship between the CIRA-IWT icing wind tunnel with the major icing wind tunnels in Europe and the United States. Furthermore, a method for determining critical icing conditions based on total water collection (TWC) was proposed and verified by

2D supercritical aerofoil model icing test in CIRA-IWT. Finally got the following conclusions:

(1) The IWT wind tunnel has good agreement with the SAE verified icing wind tunnels, especially with the IRT wind tunnel in terms of upper ice shape angle characteristics;
(2) A method determining critical icing condition based on TWC was proposed, and this method was verified by the dramatic ice horn size based on the CFD calculation and the ice shedding phenomenon at the end of ice accretion process in the icing wind tunnel.
(3) Since the completely critical ice shape can't directly be obtained in icing wind tunnel because of the ice shedding at critical moment, the method combining CFD and icing wind tunnel test was raised to obtain theoretical critical ice shape.

References

1. Icing Wind Tunnel Interfacility Comparison Tests. AIR5666–2012–10–03
2. Vecchione L, De Matteis PP, Leone G An overview of the CIRA icing wind tunenl. AIAA 2003–900
3. Irvine TB, Oldenburg JR, Sheldon DW (1997) The new icing cloud simulation system at NASA LEWIS'Icing research tunnel. AIAA
4. Bhargava C, Loth E (2005) Simulating the aerodynamics of the NASA John H. Glenn icing research tunnel. J Aircraft 42(3)
5. Herman E (2006) Goodrich icing wind tunnel overview improvements and capabilities. AIAA 9–12
6. Bellucci M, Esposito BM, Marrazzo M, Fatigati G, Ferrigno F Calibration of the CIRA IWT in the low speed configuration. AIAA 2007–1092
7. Esposito BM1, Marrazzo M Application of PDPA system with different opticalconfiguration to the IWT Calibration. AIAA 2007–1094
8. Calibration and acceptance of icing wind tunnels[S]. SAE ARP 5905
9. Olsen W, Takeuchi D, Adams K (1983) Experimental comparison of icing cloud instruments. AIAA 83–0026
10. Jeck RK (2007) Calibration and use of goodrich model 0871FA ice detectors in icing wind tunnels. J Aircraft 44(1)
11. Broeren AP, Lee S, Shah GH, Murphy PC Aerodynamic effects of simulated ice accretion on a generic transport model. SAE International 2011–38–0065
12. Yuping W, Wenbiao M, Yanhong T (2008) Aerofoil icing influence to the flight performance. Civil Aircraft Design Res 1674–9804, 01, 003
13. Yanxia D, Yewei G, Chunhua X, Xian Y, Jun L (2009) Investigation on liquid/ solid phase change heat transfer of aircraft icing. J Aerospace Power 24(8)
14. Miller* DR, Potapczuk MG () Additional investigations of ice shape sensitivity to parameter variations. AIAA 2006–469
15. Shinan C, Qiuming Y, Yan L (2011) Quasi-steady numerical simulation of ice accretion on aerofile. Acta Aerodyn Sinica 29(3)
16. Xian Y, Guolin Z, Kaichun W, Yewei G (2010) Selection of water droplet diameter in icing wind tunnel test. Acta Aeronautica Et Astronaut Ica Sinica 31(5)
17. Jeck R FAA Technical Center Atlantic City, NJ. Other ways to characterize the icing atmosphere. AIAA-94–0482

18. Yanxia D, Yewei G, Peng K, Xunnian W (2011) Investigation on the ice type microstructure characteristics of aircraft icing based on the fractal theories. J Aerospace Power 26(5)
19. Xian Y, Guolin Z, Yewe G (2008) Modification and evaluation of an icing scaling law. J Exp Fluid Mech 22(2)

Parametric Studies of the Fuselage Geometry of a Double-Bubble Aircraft Configuration

Kenya Takahashi, Ryutaro Furuya, Toshiyuki Nomura, Dongyoun Kwak, and Katsuyoshi Fukiba

Abstract The double-bubble fuselage aircraft configuration is a suitable aircraft concept for achieving higher environmental performance. This study aims to improve the aerodynamic performance of the fuselage geometry design of a double-bubble aircraft configuration (JAXA's technology reference aircraft, TRA2035A). In this regard, to understand the aerodynamics of the fuselage geometry, CFD analyses were conducted considering several geometric parameters, such as the height and length of the nose and tail parts. The parametric studies regarding only the fuselage geometry showed that higher nose and lower tail positions improved the lift-to-drag ratio and reduced the trim drag by nose-up pitching moment. Therefore, the fuselage geometry of a wing-body configuration was re-designed. The re-designed geometry showed a higher aerodynamic performance than a baseline geometry. Moreover, the aerodynamics of the wing-body configuration exhibited the same trend as that of the fuselage-only configuration.

Keywords Aerodynamics · CFD · Double-Bubble fuselage · Lifting-body

1 Introduction

Recent global concerns on environmental problems have encouraged aviation industries to improve the environmental performance of aircraft. In addition to economic efficiency, reducing CO_2 emissions and airport noise are required for commercial aircraft. The ICAO community noise standards have been revised from Chaps. 4 to 14, and restrictions on CO_2 emissions are also becoming stricter [1, 2]. To satisfy

K. Takahashi (✉) · K. Fukiba
Department of Engineering, Shizuoka University, Hamamatsu, Shizuoka, Japan
e-mail: takahashi.kenya.16@shizuoka.ac.jp

R. Furuya
Aerospace Engineering Solution Division, Ryoyu Systems Co., Ltd, Nagoya, Aichi, Japan

T. Nomura · D. Kwak
Aeronautical Technology Directorate, Japan Aerospace Exploration Agency, Mitaka, Tokyo, Japan

S. Lee et al. (eds.), *The Proceedings of the 2021 Asia-Pacific International Symposium on Aerospace Technology (APISAT 2021), Volume 1*, Lecture Notes in Electrical Engineering 912, https://doi.org/10.1007/978-981-19-2689-1_18

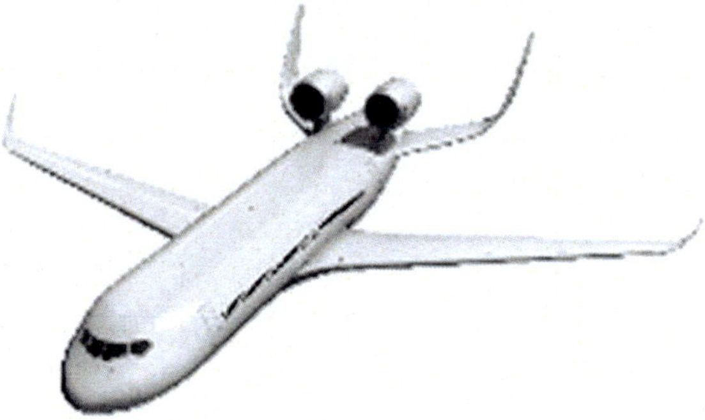

Fig. 1 Double-bubble fuselage aircraft (JAXA TRA2035A)

these requirements, researchers have been focusing on using propulsion and airframe technologies [3–5]. The Japan Aerospace Exploration Agency (JAXA) conducted research and development on technologies for drag reduction and structural weight reduction to improve environmental performance in the Eco-Wing Project. Aircraft configurations, including unconventional concepts, have been studied for environmental compatibility. A double-bubble fuselage aircraft concept [6, 7] was considered a suitable aircraft configuration. In particular, in this design, the engine noise was shielded by its wide fuselage and tailplanes (Fig. 1). However, the double-bubble fuselage increases structural weight compared with a cylindrical fuselage. Nevertheless, the double-bubble fuselage geometry improves the aerodynamic performance [8].

This study aims to improve the aerodynamic performance of the aerodynamic design of the double-bubble fuselage geometry. In this regard, numerical analyses are conducted to understand the flow field and evaluate the aerodynamic performance.

2 Design Process

The design of the double-bubble fuselage aircraft was as follows: First, the aerodynamic characteristics were studied considering fuselage configurations with different geometries regarding the nose and tail parts. Subsequently, a wing-body configuration was developed. After that, the fuselage geometry was modified considering the entire aircraft configuration, including tails, engine nacelles, and pylons. Some design requirements and restrictions, such as cabin layout and tip back angle, were also considered in the design process. Finally, the optimal fuselage geometry with the highest aerodynamic and acoustic performance was obtained for the double-bubble fuselage aircraft configuration.

Fig. 2 TRA2035A design geometry

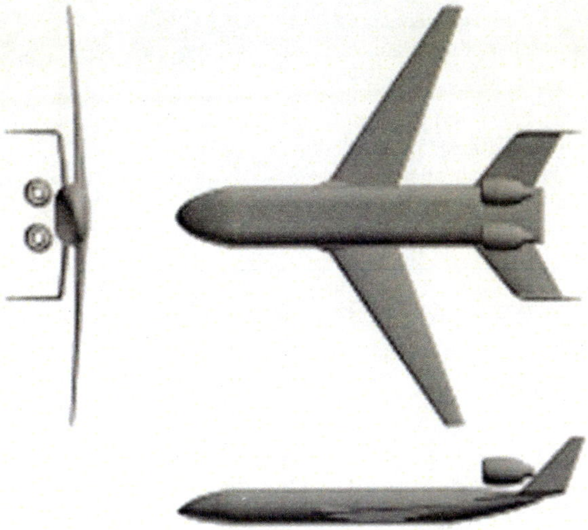

Step1. Only the fuselage geometry is studied.
Step2. The wing is added (wing-body geometry) to the model.
Step3. Constraints of the aircraft feasibility are considered.
Step4. Tailplanes and propulsion systems are added to the model.
The results obtained at design steps 1 and 2 are reported in this paper.

3 Numerical Analysis

3.1 Baseline Aircraft Geometry

In this study, TRA2035A, a double-bubble fuselage aircraft designed by JAXA, was considered. TRA2035A is a technology reference aircraft that is assumed to enter service in the 2040s [6]. Figure 2 shows the geometry of TRA2035A, designed as a twin-aisle aircraft with the specifications of 220 passengers, a cruise Mach number of $M = 0.8$, and a range of 12,000 km. These specifications are similar to those of B767.

3.2 Grid Generation

The fuselage model of the TRA2035A was created in CATIA software (Fig. 3). Table 1 lists the dimensional specifications of the fuselage model.

Fig. 3 Fuselage geometry of the baseline configuration

Table 1 TRA2035A fuselage specifications

Length	
Total	42.955 m
Nose	7.209 m
Middle	23.875 m
Tail	11.871 m
Height	
Height	3.999 m
Nose position	1.497 m
Tail position	2.831 m
Width	
Width	6.958 m

The grid was generated on the fuselage model using MEGG3D, a 3-dimensional mixed-element grid generator [9]. The grid is hemispherical, as shown in Fig. 4a. Moreover, an unstructured grid was adopted because of its high adaptability to complex shapes. A hybrid grid was applied to generate meshes with high accuracy. The grid comprised triangular and square elements for the surface grid and tetrahedron, prism, and pyramid elements for the space grid. The grid point density near the fuselage was increased by refining the grid near the center of the hemisphere, as shown in Fig. 4b. Moreover, considering the boundary layer, prism elements are used around the fuselage, as shown in Fig. 4c. The total number of grid points was approximately 2,130,000.

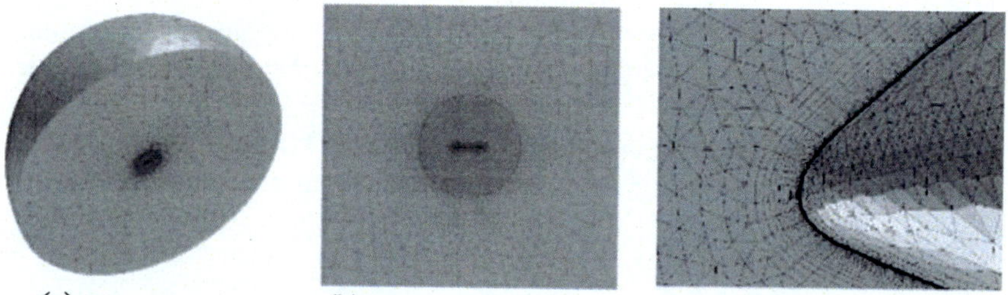

(a) Grid overview (b) Refined area (in the circle) (c) Gird near the leading-edge surface

Fig. 4 CFD grid

Table 2 Details of the CFD model

CFD settings	
Governing equation	3D Navier–Stokes
Time integration	LU-SGS
Gradient reconstruction	GLSQ
Reconstruction	UMUSCL
Inviscid flux	HLLEW
Turbulence model	SA-noft2-R-QCR2000
Discretization	cell vertex
Iteration	60,000
Flow conditions	
Mach number	0.8
Altitude	10.668 km
Re (*MAC* based)	4.17×10^7
Angles of attack	$a = -2, 0, 2, 4, 6$ deg

3.3 CFD Analysis

Reynolds Averaged Navier–Stokes (RANS) CFD analyses were conducted using JAXA's in-house CFD solver FaSTAR (FAST Aerodynamic Routines) [10]. The calculation conditions are presented in Table 2. The design point was the cruise flight conditions of TRA2035A. Therefore, the atmospheric properties were set to an altitude of 10.7 km. The mean aerodynamic chord length (MAC) and the wing area of TRA2035A were used to calculate the Reynolds number *Re* and the nondimensional aerodynamic coefficients. The boundary conditions were set as follows: (1) non-slip adiabatic wall condition on the fuselage surface, (2) uniform flow condition on the outer boundary surface, and (3) symmetric boundary condition on the circular cross section.

4 Results

4.1 Parametric Study on Fuselage Configuration

A parametric study on the fuselage geometry was conducted, considering only the fuselage configuration (see Fig. 3). The aerodynamic characteristics, such as the lift, drag, and pitching moment coefficients C_L, C_D, *and* C_m, were investigated for different nose and tail geometries. The fuselage geometry was divided into the nose, middle, and tail (Fig. 5). The geometry of the middle part of the cabin was fixed. Four geometric parameters were selected for the fuselage: nose height (Case 1), nose length (Case 2), tail height (Case 3), and tail length (Case 4). Table 3 and Fig. 6

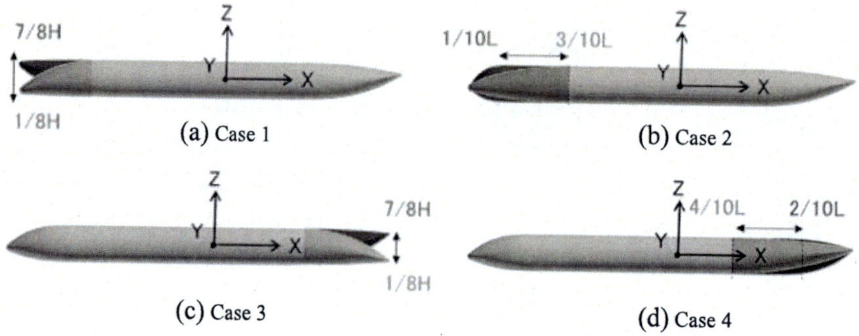

Fig. 5 Typical geometries for each parameter case

Table 3 Details of the geometric parameters

	Nose		Tail	
	Height	Length	Height	Length
Case 1	1/8*H*, 3/8*H*, 4/8*H*, 5/8*H*, 7/8*H*	Baseline	Baseline	Baseline
Case 2	Baseline	1/10*L*, 2/10*L*, 3/10*L*	Baseline	Baseline
Case 3	Baseline	Baseline	1/8*H*, 3/8*H*, 4/8*H*, 5/8*H*, 7/8*H*	Baseline
Case 4	Baseline	Baseline	Baseline	2/10*L*, 3/10*L*, 4/10*L*

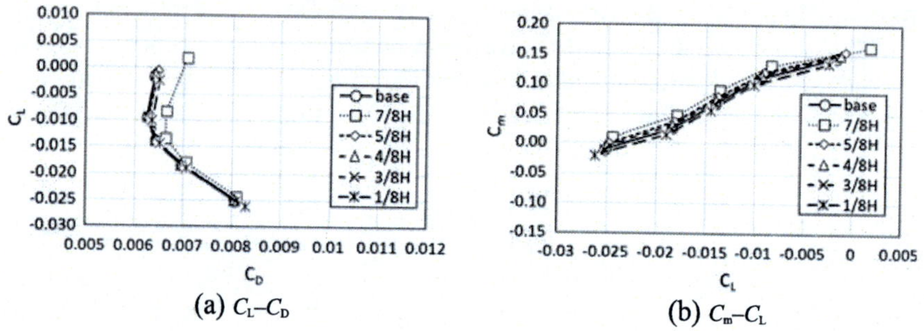

Fig. 6 Aerodynamic characteristics (Case 1)

show the geometric parameters and typical configurations considered in this study. In Table 3, "H" indicates the fuselage height, and "L" is the total fuselage length. The nose height (Case 1) and tail height (Case 3) indicate the vertical position of the apex.

(1) Case 1 (nose height)

The nose height was varied from 1/8*H* to 7/8*H*. Figure 6 shows the C_L–C_D and C_m–C_L curves. Except for the case of 7/8*H*, the C_L–C_D is almost the same.

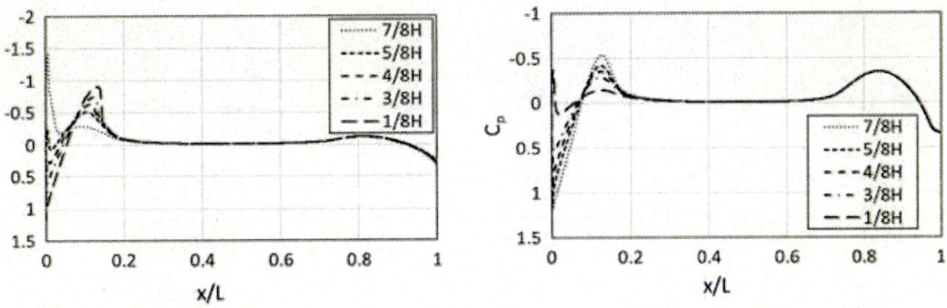

(a) Upper surface (b) Lower surface

Fig. 7 C_p distribution (case 1, y = 0, a = 6 degree)

The C_D increases significantly at a high angle of attack for 7/8*H*. Figure 7a shows the distribution of the static pressure coefficient C_p on the centerline of the upper surface. The distribution shows a large negative pressure around $x/L = 0.1$ for 7/8*H*. In contrast, the C_p distribution on the lower surface has a large area of positive C_p near the nose apex (Fig. 7b), which increases the C_L and nose-up pitching moment. At high angles of attack, C_p contributes to the increase in both C_L and C_D. In addition, the C_m–C_L curve shows that C_m increases with a high nose position. Thus, the nose height slightly affects C_L and C_D, except for 7/8*H*. However, it significantly affects C_m.

(2) Case 2 (nose length)

The nose length was varied from 1/10 L to 3/10 L. Figure 8 shows the C_L–C_D and C_m–C_L curves. Note that C_D significantly increases for 1/10*H*. The reason is the generation of shock waves (Fig. 9). The short nose induces a large wave drag downstream of the high-curvature surface. In the case of no shocks, the nose length slightly affects the aerodynamic characteristics.

(3) Case 3 (tail height)

The tail height was varied from 1/8*H* to 7/8*H* (see Fig. 5c). Figure 10 shows the C_L–C_D and C_m–C_L curves. C_L increases as the tail height decreases. This

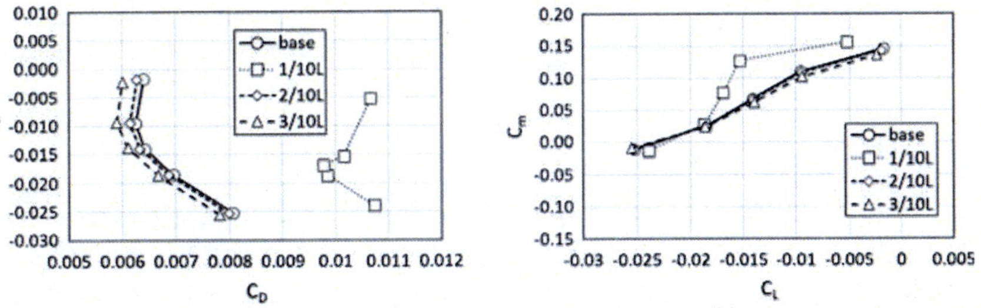

(a) C_L–C_D (b) C_m–C_L

Fig. 8 Aerodynamic characteristics (Case 2)

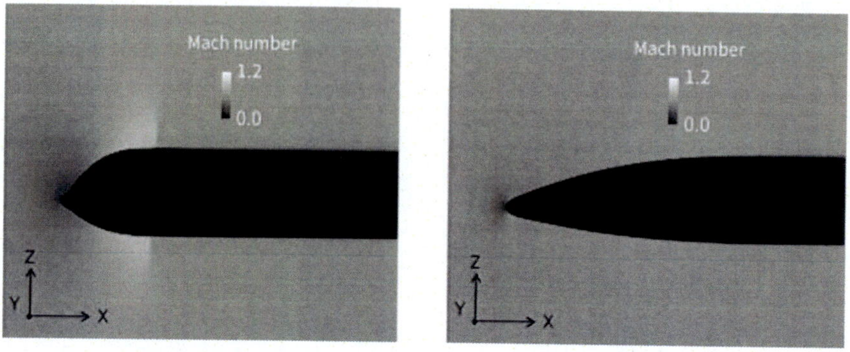

(a) Shortest (1/10*L*) (b) Longest (3/10*L*)

Fig. 9 Shortest and longest noses (Mach distribution at $a = 0$ degree)

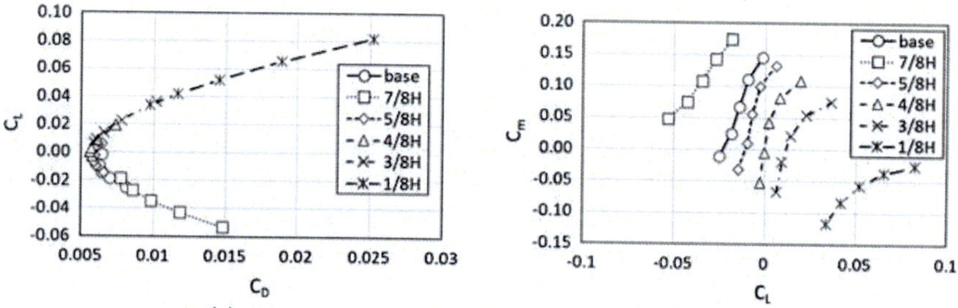

(a) C_L–C_D (b) C_m–C_L

Fig. 10 Aerodynamic characteristics (Case 3)

is induced by an increase in the curvature of the upper surface. Although the fuselage configuration has a low aspect ratio compared with the wing, the geometry of the lower tail position increases the camber effect, as mentioned in the wing geometry. The increase in C_L at the tail part increases the pitch-down moment. This result is confirmed from the C_p distribution on the upper surface shown in Fig. 11). The figure shows a large suction peak on the upper surface for 1/8*H* compared with that for 7/8*H*. Notably, the effect of the tail height on C_L and C_m is much higher than that of the two other parameters in Cases 1 and 2.

(4) Case 4 (tail length)

The tail length was varied from 2/10 L to 4/10 L. Figure 12 shows the C_L–C_D and C_m–C_L curves. When the tail is extremely short (2/10 L), C_L increases by 0.007 with respect to the baseline geometry. The abrupt increase in C_D for 2/10*L* is due to the flow separation around the tail part (see Fig. 13). The minimum C_D is obtained for 4/10 L. Without flow separation, the tail length slightly affects the aerodynamic characteristics.

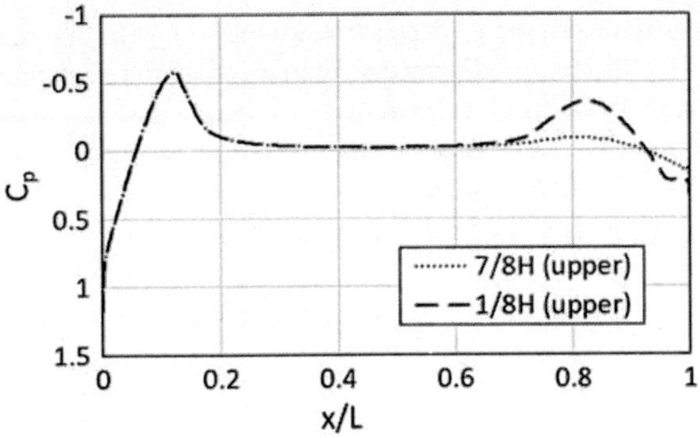

Fig. 11 C_p distribution on the upper surface (y = 0, a = 0 degree)

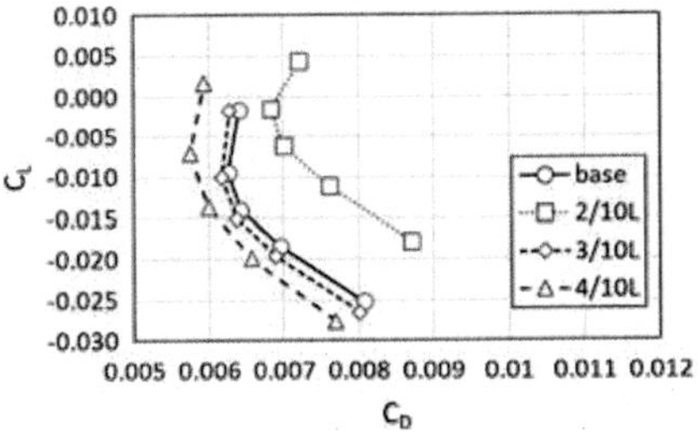

(a) C_L–C_D

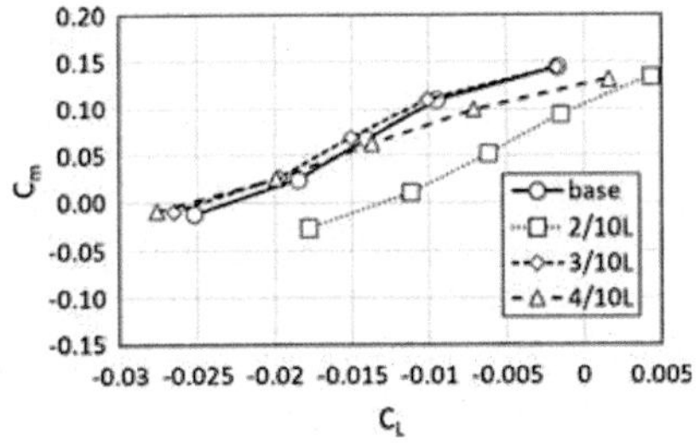

(b) C_m–C_L

Fig. 12 Aerodynamic characteristics (case 4)

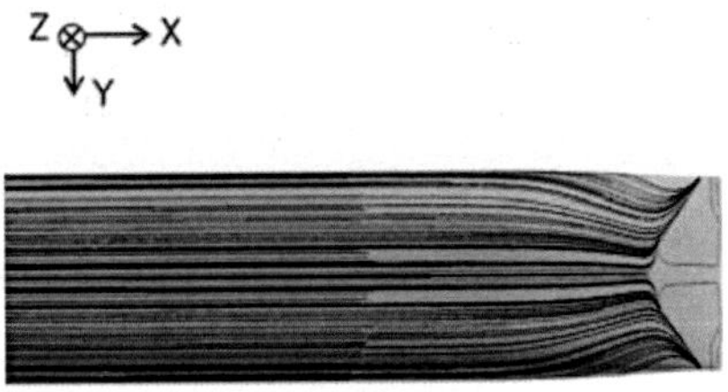

(a) Shortest (2/10*L*)

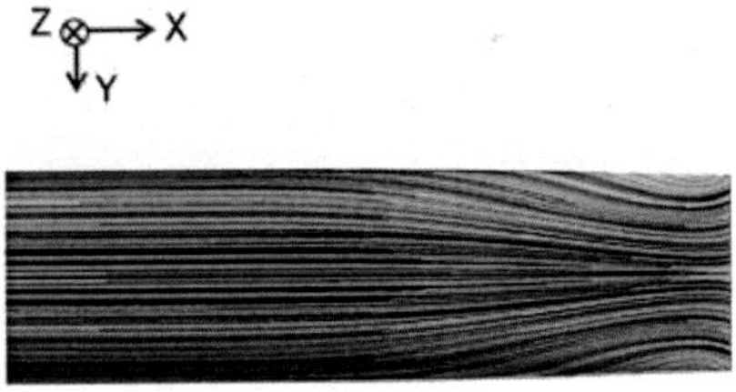

(b) Longest (4/10*L*)

Fig. 13 Streamlines on the lower surface of the tail (a = 0 degree)

(5) Summary of the parametric study

The aerodynamic performance of the fuselage geometry can be improved by increasing the lift-to-drag ratio and C_L. Moreover, increasing the value of C_m is suitable for pitch-up because the pitch-up moment reduces the downward lift required for the tails. The reduction of the trim drag generated by the tails reduces the total drag of the entire aircraft configuration. From these parametric studies, C_L was found to be highly dependent on tail height. Furthermore, lowering the tail position increases C_L. In this case, the value of C_m decreases; therefore, it is necessary to increase the nose height for the pitch-up moment. In addition, an extremely short nose and tail are not adequate for the fuselage geometry because they induce strong shock waves and flow separation that drastically increases the drag.

4.2 *Design of Wing-Body Configuration*

Geometry modifications based on the parametric study were applied to the wing-body configuration of TRA2035A. The wing and fairing geometries were fixed. A nose height of 7/8*H* was selected to obtain a large pitch-up effect. A tail height of 3/8*H* was selected to obtain a large C_L without increasing C_D with respect to the baseline configuration. The nose and tail lengths were fixed as the drag reduction due to increasing the lengths was small, and the cabin volume must be maintained. Figure 14 shows the baseline and modified configurations of the wing-body. The CFD analysis results for the two configurations are shown in Fig. 15. The C_L on the modified geometry increases by $\Delta C_L = 0.07$ from the baseline geometry over a wide range of angles of attack. The increment in C_L is mainly attributed to tail height. However, an increment of $\Delta C_L = 0.03$ is obtained for 3/8*H* (see Fig. 10a).

Fig. 14 Wing-body configuration

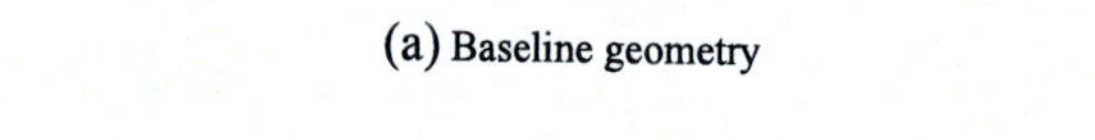

(a) Baseline geometry

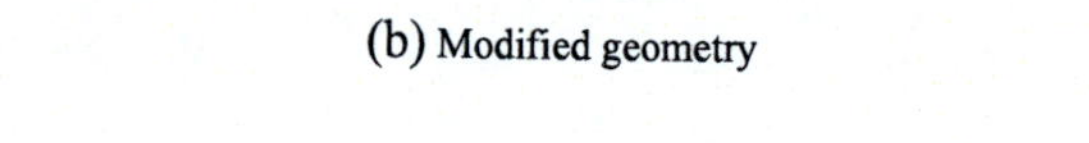

(b) Modified geometry

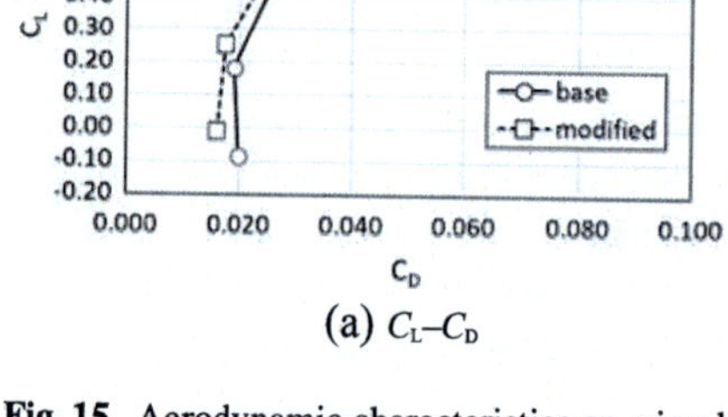

(a) C_L–C_D

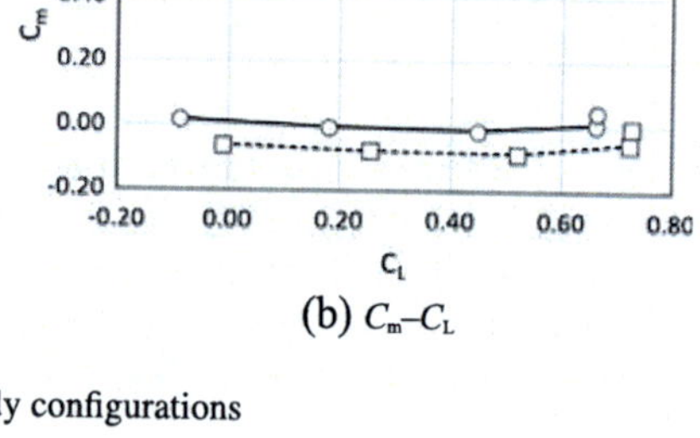

(b) C_m–C_L

Fig. 15 Aerodynamic characteristics on wing-body configurations

This result indicates that the tail height affects the lift of the wing-body configuration more than the fuselage-only configuration. Moreover, a reduction in C_D with respect to the baseline geometry was observed around the cruise $C_L = 0.45$. This implies that the improvement in cruise performance is obtained by lowering the position of the fuselage tail. However, C_m decreased by approximately 0.07, implying that the pitch-down effect of the lower fuselage tail surpasses the pitch-up effect owing to the higher fuselage nose.

The analysis of the wing-body configuration relied on the same geometric parameters used in the above parametric studies. Therefore, some requirements for the feasibility of the aircraft should be considered in the design phase.

5 Conclusions

The effects of the height and length of the nose and tail on the aerodynamic characteristics of the double-bubble aircraft, TRA2035A, were investigated by analyzing its fuselage configuration using CFD. The following results were obtained from geometric-parameter studies:

- A high nose increases C_m.
- When the nose is extremely short, C_D increases. However, in the cases with nose length, no significant effect on aerodynamics was observed.
- Lowering the tail height improves C_L; however, C_m decreases simultaneously.
- Shortening the tail length increases C_L, but it separates the flow, followed by an increase in C_D.

From the above results, a fuselage geometry with a higher nose and lower tail positions was designed and applied to the TRA2035A wing-body configuration. The CFD results indicated an improvement in the lift-to-drag ratio.

Acknowledgements The numerical work was conducted using the computational resources of JAXA Supercomputer System Generation 2 and 3 (JSS2 and JSS3).

References

1. ICAO Reduction of Noise at Source. https://www.icao.int/environmental-protection/pages/reduction-of-noise-at-source.aspx
2. ICAO ICAO Council adopts new CO2 emissions standard for aircraft. https://www.icao.int/Newsroom/Pages/ICAO-Council-adopts-new-CO2-emissions-standard-for-aircraft.aspx
3. Clean sky (2017) Clean Sky2 joint undertaking, development plan. https://www.cleansky.eu/sites/default/files/inline-files/51.%20CS2DP%20December%202017.pdf
4. NASA Aeronautics research mission directorate, strategic implementation plan. https://www.hq.nasa.gov/office/aero/pdf/armd-strategic-implementation-plan.pdf

5. Reneaux J (2004) Overview on drag reduction technologies for civil transport aircraft. In: Proceedings of European congress on computational methods in applied sciences and engineering
6. Nomura T (2017) Redesign of a passenger aircraft with a double-bubble fuselage in consideration of its feasibility. Proc. of JSASS annual meeting, 1E06. (in Japanese)
7. Uranga A, Drela M, Greitzer EM, Hall DK, Titchener NA, Lieu MK, Siu NM, Casses C, Huang AC, Gatlin GM, Hannon JA (2017) Boundary layer ingestion benefit of the D8 transport aircraft. AIAA J 55(11)
8. Kwak D, Nomura T, Tokugawa N, Kurita M, Murayama M (2016) Introduction of research project for environmental conscious aircraft technology in JAXA. Greener Aviation 2016 Symposium, no.180, Brussels
9. Ito Y (2013) Efficient hybrid surface/volume mesh generation using suppressed marching-direction method. AIAA J 51(6)
10. Hashimoto A, Murakami K, Ishiko K (2012) Toward the fastest unstructured CFD code "FaSTAR", AIAA 2012–1075

Wind Tunnel Free Flight Model Attitude Determination Based on Non-contact IR Optical Measurement

Litao Fan, Bowen Nie, and Yong Jiang

Abstract Model position and attitude measurement is one of the key techniques for wind tunnel free flight test. The state-of-the-art model attitude is determined by a wind vane sensor mounted near the nose of the model, angular sensors or attitude and heading reference system located in the model. However, the large wind vane sensor will influence the flow distribution of the model surface, which increases the difficulty of system modelling and identification. The inner sensors occupy the limited space in the model, making it more difficult to design model and inertia similarity match. This paper proposes a solution using OptiTrack system mainly applied in motion capture for model position and attitude based on wind tunnel free flight reality. Static test and Φ3.2 m wind tunnel dynamic test results show that the static position measurement accuracy is better than 0.07 mm and angular accuracy is better than 0.05°. Dynamic tests indicate the good following performance of the system, meeting the no less than 100 Hz requirement of free flight test data update.

Keywords Low speed wind tunnel · Non-contact measurement · Free flight test · Optical measurement · Attitude determination · Dynamic test

1 Introduction

Wind tunnel free flight is a significant platform, on which a scale model can be controlled remotely with six DOF flight without support in a wind tunnel test section, to achieve aerodynamics/flight/control integration design and develop new generation and high performance aircraft through advanced flight control algorithms study, high angle of attack stall and deviation investigation, system identification and aircraft special mode tests [1]. Wind tunnel free flight can offer a test environment to simulate full size aircraft flight. Model attitude determination is one of the key techniques because of the necessary attitude angle and angular velocity data for control law design and system identification [2].

L. Fan (✉) · B. Nie · Y. Jiang
Aerodynamics Research and Development Center, No 6, South Section, Second Ring Road, Mianyang, China
e-mail: litaofan_cardc@outlook.com

S. Lee et al. (eds.), *The Proceedings of the 2021 Asia-Pacific International Symposium on Aerospace Technology (APISAT 2021), Volume 1*, Lecture Notes in Electrical Engineering 912, https://doi.org/10.1007/978-981-19-2689-1_19

Normally a wind vane sensor mounted (Fig. 1) near the nose of a model is used for alpha and beta angle measurement, resulting in flow disturbance to the model surface which will affect modelling accuracy. Angular velocity sensor and AHRS, mounted in the model for angular velocity, pitch, yaw and roll angle measurement between model body axes and Earth axes, enhance the difficulty of model design and manufacture. Since the limited inner space need to be allocated to control surface actuators (Fig. 2), engines, inlet, cables and other necessary accessories. Meanwhile, a free flight model needs to keep mass and inertia similarity [3].

To overcome these deficits, CARDC always places the non-contact measurement technique as a priority for development. Developed were the OptoTrak system for continuous sweep technique to enhance test efficiency and data quality in FL-12 Wind Tunnel with 4 m × 3 m closed test section [4], and a full field view measurement system for vertical wind tunnel spin model attitude determination in FL-15 Wind Tunnel with Φ5 m open test section [5]. Despite the strong real-time performance and high precision, the narrow field view, thick markers fixed on the model surface and keeping hole for signal cables limit the application of the OptoTrak system. The developed full field view measurement system can only provide poor real-time

Fig. 1 Wind vane sensor

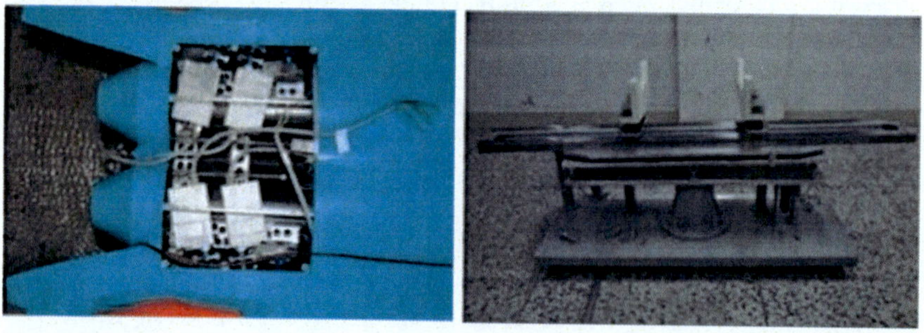

Fig. 2 Control surface actuators (left) and inertia measurement instrument (right)

performance, up to 20 Hz, which cannot meet the requirement of minimum 100 Hz for attitude data update frequency in control law design.

The non-contact measurement system, OptiTrack system, based on motion capture overcomes these shortcomings very well. The technology employed by this system belongs to passive infrared optical positioning technique. Several cameras emit 850 nm infrared light to the thin and small markers (thickness of 0.12 mm and diameter of 6 mm), the material on the markers can enhance the reflective ability of infrared light and distinguish the markers from the surrounding environment and improve real-time performance.

The data processing flow can be described briefly as follows: cameras receive the reflected infrared light by a marker and store the maker information as gray images. The marker position relative to the 2D image can be solved through image gradation information [6]. A marker can be positioned in 3D space through photogrammetry principle if it can be "seen" by at least two cameras at the same time. After position acquisition, choose the line connecting the most distant two makers as baseline, then the model attitude can be transformed from the position vector in the reference frame at different time.

2 OptiTrack System

2.1 System Composition

OptiTrack system is generally composed of cameras, cables, hubs, motive software, markers, esync, calibration tools, hardware key etc. (Fig. 3).

Fig. 3 OptiTrack sytem

A set of OptiTrack system often includes 6 to 24 infrared Ethernet cameras (USB cameras are beyond the scope of this paper) connected through CAT-6 Ethernet cables, supporting PoE and net communication. The cameras need GigE PoE/PoE + switches for power supply and data transmission. Each port power of the switch is determined by camera specification, 30 W for Prime41 and Prime17 camera or 15.4 W for other camera. Motive software controls the OptiTrack system, calibrates the system, processes and captures 3D data, and transmits the data to other software platform through off-line or real-time mode. There are two kinds of markers, adhesive and sphere. The latter has better measurement precision. Each kind of markers include active luminescence and passive reflective makers. Esync synchronizes the time code between the OptiTrack system and other equipment. The calibration wand kit determines the relative position among different cameras and the calibration square constructs the world coordinate system.

OptiTrack camera time is synchronized through IEEE1588 precise time synchronization protocol [7]. This protocol acquires time delay and clock bias among master clock and slave clocks through delay measurement method and provides sub-microsecond precision.

2.2 *Specification*

There are total 18 Prime41 cameras which can be freely combined according to actual necessity. Table 1 lists the camera specification.

Table 1 Prime 41 camera specification

Image sensor	Resolution: 2048 × 2048 Frame rate: 30-180FPS(adjustable) Latency: 5.5 ms Shutter type: global
Capture distance	30 m
LED	170 LEDs, 850 nm IR
Image processing types	Object, segment, raw grayscale, mjpeg grayscale
Precision	Postion: 0.1 mm
Field of view	51° × 51°
Input/output & power	Data:GigE, camera sync: ethernet, power:PoE or PoE +

2.3 Camera Layout

18 cameras will be mounted on the top of two side walls of FL-13 Wind Tunnel during a free flight test, covering the field of 8 m (length) × 6 m (width) × 4 m (height), according to measurement precision, field of view and tunnel environment requirement. Figure 4 presents the sketch of camera layout.

3 Attitude Determination

The world axes o_g-$x_g y_g z_g$ (Fig. 5) is constructed through the calibration square, and can be transformed to the body axes o_b-$x_b y_b z_b$ through a transformation matrix defined by three Euler angles, ψ(about y axis)、θ(about z axis)and ϕ(about x axis). The transformation matrix from the world axes to the body axes can be written as follows:

$$R_{bg} = R_x(\phi)R_z(\theta)R_y(\psi) \tag{1}$$

And this equation can be expanded as

$$R_{bg} = \begin{bmatrix} \cos\theta\cos\psi & \sin\theta & -\cos\theta\sin\psi \\ R_{21} & \cos\phi\cos\theta & R_{23} \\ R_{31} & -\sin\phi\cos\theta & R_{33} \end{bmatrix} \tag{2}$$

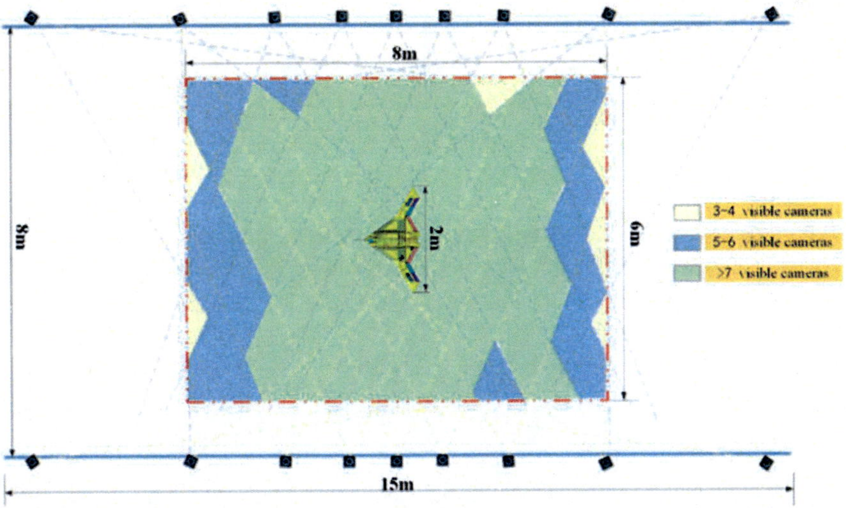

Fig. 4 Camera layout

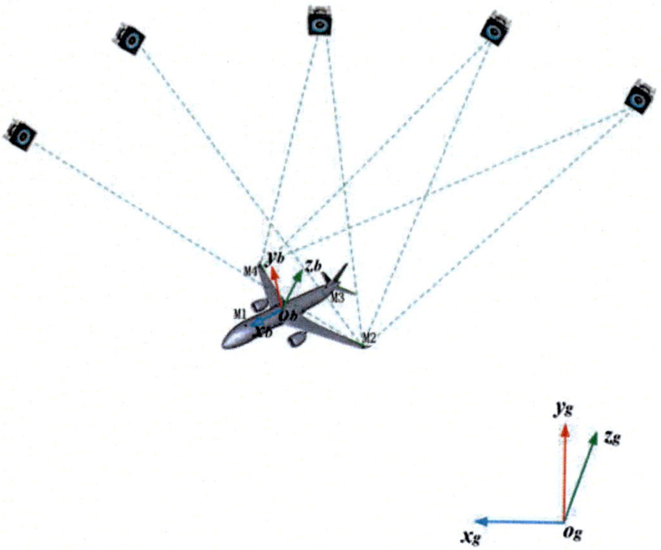

Fig. 5 World axes and body axes

and

$$R_{21} = -\cos\phi\sin\theta\cos\psi + \sin\phi\sin\psi \quad (3)$$

$$R_{23} = \cos\phi\sin\theta\sin\psi + \sin\phi\cos\psi \quad (4)$$

$$R_{31} = \sin\phi\sin\theta\cos\psi + \cos\phi\sin\psi \quad (5)$$

$$R_{33} = -\sin\phi\sin\theta\sin\psi + \cos\phi\cos\psi \quad (6)$$

Assume that the body axes are aligned with the world axes at the original time, the coordinate of a baseline vector from one marker to another marker in world axes is $[x_1, y_1, z_1]^T$ at this time, and at time t the coordinate of the baseline vector in world axes is $[x_2, y_2, z_2]^T$, then

$$\begin{bmatrix} x_1 \\ y_1 \\ z_1 \end{bmatrix} = R_{bg} \begin{bmatrix} x_2 \\ y_2 \\ z_2 \end{bmatrix} \quad (7)$$

If there are at least three baselines, the transformation matrix can be solved through the least square method. Then the three Euler angles can be calculated through the following equations:

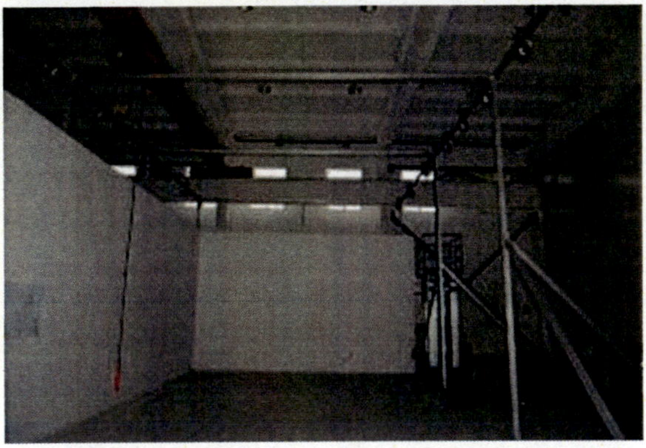

Fig. 6 Ground test environment

$$\begin{aligned}\theta &= \arcsin\left(R_{bg12}\right)\\ \psi &= -\arcsin(R_{bg13}/\cos\theta)\\ \phi &= -\arcsin(R_{bg32}/\cos\theta)\end{aligned} \tag{8}$$

If position measurement precision is better than 0.1 mm, the Euler angle precision will be better than 0.01° for baselines larger than 1 m theoretically.

4 Test Scheme

Before application for wind tunnel free flight test, the OptiTrack system needs to be tested on ground and in a smaller tunnel (FL-14 Wind Tunnel). The static ground tests will validate the static position and attitude precision. The dynamic tests in FL-14 Wind Tunnel will verify the system latency.

For the static ground tests, a frame coving the space of 10 m × 6 m × 6 m is built to simulate the test section of FL-13 Wind Tunnel (Fig. 6).

5 Test Results

5.1 Ground Tests

Ground tests include position tests and attitude tests.

For position tests, a calibrated ruler was utilized adhered with markers, comparing the move distance of the ruler with the distance measured by the system to evaluate the measurement accuracy of the OptiTrack system (Fig. 7). Test results indicate that the static position measurement accuracy is better than 0.07 mm.

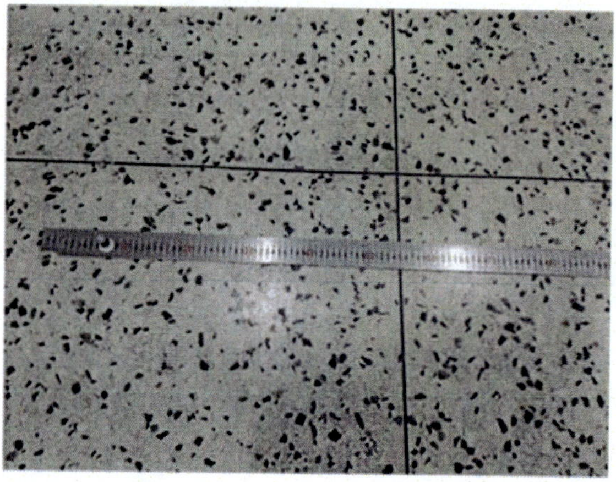

Fig. 7 Position test

Attitude tests were carried out through a 3-DOF turntable (Fig. 8) with precision of 1′. Angle ranges are: $\psi \in [0°, 360°]$, $\theta \in [-45°, 45°]$, $\phi \in [-90°, 90°]$。

Turntable tests show that pitch angle, from −10° to 30°, measurement bias mean value is 0.06°, and RMS is 0.002°. Yaw angle, from −30° to 30°, measurement bias mean value is 0.05°, and RMS is 0.002°. Roll angle, from −60° to 60°, measurement bias mean value is 0.04°, and RMS is 0.006°. Figure 9 presents the repeatable results of pitch angle.

Fig. 8 Turntable test

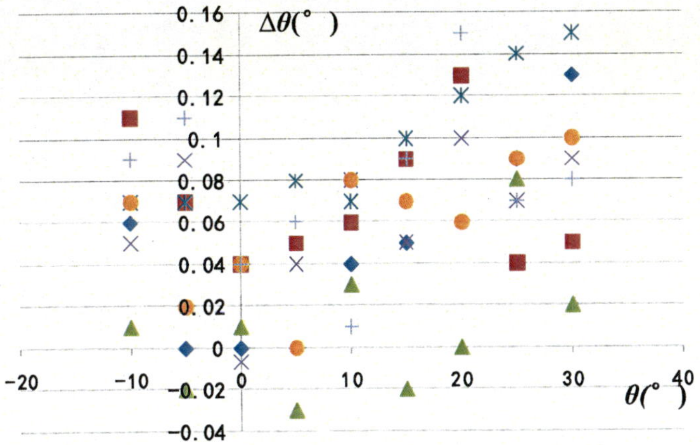

Fig. 9 Pitch angle repeatable tests

5.2 *Dynamic Tests*

Dynamic tests were performed in FL-14 Wind Tunnel. 8 Prime41 cameras were mounted on the top platform of the test section. A light carbon fiber transport model adhered with 8 markers (Fig. 10) was driven by a 2 DOF (pitch and roll) mechanics to realize harmonic motion. The movement angle of the mechanics was measured by a potentiometer.

Figures 11, 12 and 13 show the comparison results between the OptiTrack system and the potentiometer under pitch oscillation. The nominal amplitude of the mechanics is 10°, 15° and 25°. It can be seen that the oscillation period measured by two instruments are almost the same, and the OptiTrack system shows the good following performance. The amplitude measured by the OptiTrack system is larger

Fig. 10 Dynamic tests

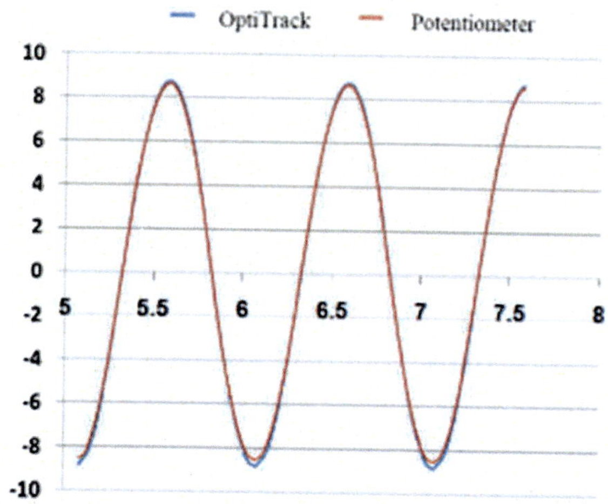

Fig. 11 10° amplitude

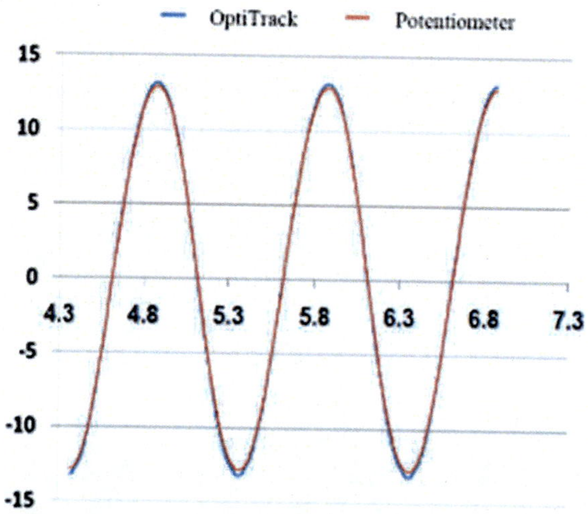

Fig. 12 15° amplitude

than that determined by the potentiometer. This may be due to model and mechanics aeroelastic deformation. The model acceleration reaches maximum at wave crest and wave trough points, resulting in large aeroelastic deformation. Table 2 shows that larger amplitude leads to larger deformation.

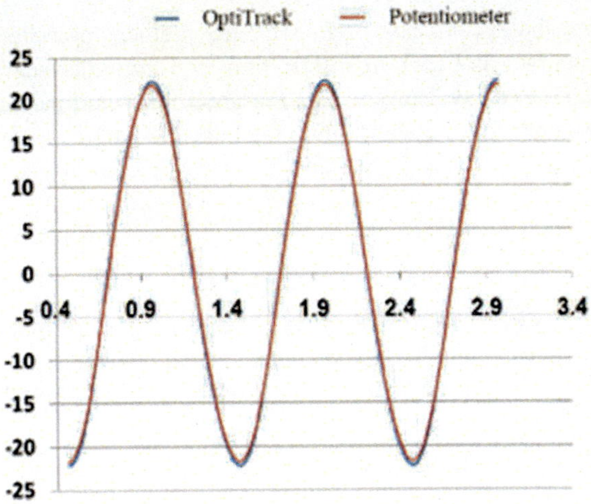

Fig. 13 20° amplitude

Table 2 Amplitude differences measured by OptiTrack and potentiometer

Nominal amplitude (°)	OptiTrack (°)	Potentiometer (°)	Difference (°)
10	−8.72	−8.54	−0.18
	8.73	8.63	0.10
15	−13.26	−12.91	−0.35
	13.16	12.91	0.25
25	−22.30	−21.76	−0.54
	22.20	21.70	0.50

6 Future Work

In the near future, a free flight test will be carried out in FL-13 Wind Tunnel. We will record the model attitude, angular rate, acceleration solved by the OptiTrack system and compare them with the corresponding information measured by the wind vane sensor, angular rate sensor and AHRS to evaluate the application feasibility of the system.

7 Conclusions

This paper analyzes the necessity of developing the non-contact measurement system for a wind tunnel free flight model, and proposes a solution based on the motion capture OptiTrack system. Groud static tests and dynamic tests in the smaller wind tunnel show that the system has high static mesurement precision, the position measurement accuracy is better than 0.07 mm, and atttitude measurement accuracy is about 0.05°. The system presents the good followind performance, which will meet the requirement of a free flight model data update.

References

1. Haisheng S, Fei C, Bowen N (2011) Liuzhitao, "Present research status and prospective application of wind-tunnel free-flight test technique." J Experiments Fluid Mech 25(4):103–108
2. Nowack J, Alles W (2008) Free flight wind tunnel tests for parameter identification. RWTH Aachen University
3. Zhitao L, Fei C, Bowen N, Litao F, Haisheng S (2017) Similarity criteria and simulation method for flight control system of free-flight test in low speed wind tunnel. Acta Aerodynamics Sinica 35(5):693–699
4. Gu Y, Fei C, Yuchang W, Zhitao L, Binghui C (2013) Research about continuous scanning test technique based on non-contact measurement technique in low speed wind tunnel. J Experiments Fluid Mech 27(5):98–104
5. Ma J, Jin S, Bei L, Sanchun Q, Jianjun X, Min J (2016) Design and implementation for full field of view measurement scheme in vertical wind tunnel. J Experiments Fluid Mech 30(6):66–70
6. Junwei W, Xi Q, Qiqiang F (2017) Precision testing of OptiTrack system for point measurement. Beijing CeHui S1:294–300
7. Li Z, Jingyu Z, Junhao N (2012) Research on clock synchronization technology based on IEEE1588. Instru Techniq Sensor 12:99–102

Design and Performance Analysis of a Fuel Cell Powered Heavy-Lift Multirotor Drone

Allie Foong Yi Chia and Kyoung Moo Min

Abstract The growing use of unmanned aerial vehicles (UAVs) in various sectors has created market demands for drones which can carry heavier payloads and have longer flight endurance. As the endurance of a UAV is highly dependent on the weight of the payload that it carries, drone makers are exploring new and more efficient power sources to produce UAVs that can fly longer while carrying a heavy payload. Previous studies have mainly presented the potentials of fuel cell (FC) as a power source solution for mini to small sized fixed-wing UAVs. The design considerations and performance when implementing the FC system in larger sized UAVs for the purpose of carrying heavier payload and have extended flight endurance has yet been evaluated. This paper describes the design and construction of a 39 kg fuel cell powered heavy-lift medium sized hexacopter and presents the results of its performance, using an analysis of real data from flight tests. It also discusses the possibility of extending flight endurance by increasing the amount of hydrogen fuel using a larger hydrogen tank. In order to provide enough power for the 39 kg medium-sized hexacopter, a 5.2 kw hydrogen fuel cell system was produced by linking two 2.6 kW class power modules. Experiments were conducted to check the performance of the 5.2 kW hydrogen fuel cell system before mounting the system into the hexacopter. Results showed that hydrogen discharge pressure of the hydrogen tanks distributed in Korea cannot cope with the required pressure of 1 bar when loaded with the 5.2 kW power module, so a regulator was used to control discharge pressure. The hexacopter was specially designed to be loaded with the linked FC systems and optimized in terms of weight and structural strength. Structural strength analysis was performed using NASTRAN and PASTRAN to check the design safety margin of the composite skin and the internal structures during structural optimization. Special design considerations were also given to the airflow intake to ensure that the FC system was provided with enough airflow to reach high performance. The flight control system communication and power platform were also modified to adapt to the linked power modules. Flight tests were performed to check the performance and endurance of the UAV. The flight test results demonstrated the feasibility of the designed fuel cell powered heavy-lift hexacopter to perform stable flights. The

A. F. Y. Chia (✉) · K. M. Min
R&D Center, Earth and Aerospace, Sacheon-si, Gyeongsangnam-do, South Korea
e-mail: allie@samcokorea.com

S. Lee et al. (eds.), *The Proceedings of the 2021 Asia-Pacific International Symposium on Aerospace Technology (APISAT 2021), Volume 1*, Lecture Notes in Electrical Engineering 912, https://doi.org/10.1007/978-981-19-2689-1_20

consumption rate of hydrogen mass was evaluated from the flight test results and the flight time was estimated for the case of a larger hydrogen fuel amount. The consumption rate of hydrogen mass was calculated to be 6 g/min and it was estimated that the flight time could be increased by 6 min by using a larger hydrogen tank to store an additional 36 g of hydrogen fuel.

Keywords Fuel cell · Drone · Design · Multicopter · Unmanned aerial vehicle

1 Introduction

As Unmanned Air Vehicle (UAV) technology advances, there has been growing demand for UAVs for various applications including surveying, mapping, surveillance, agriculture, and military attack. For a drone to perform a mission in any of these applications, it requires various sensors, such as a gimbal camera, Light Detection and Ranging (LiDAR) sensor or a multispectral camera. Since the endurance of a UAV is highly dependent on the weight of the payload that it carries, drone makers are exploring new and more efficient power sources to produce UAVs that can fly longer while carrying a heavy payload. LiPo batteries can provide a typical flight endurance of 20–50 min for multirotor drones, but this can be drastically reduced when the drones are carrying a heavy payload and this reduces operating time to a level that is insufficient to cover the entire mission area.

The implementation of fuel cell (FC) systems in UAVs has gained attention as a possible solution for extending flight endurance because of their high energy densities over weight. Previous studies [1, 2, 5, 8, 9, 15] have focused on mini to small sized fixed-wing UAVs (up to the take-off weight of 25 kg [13]). And although the fuel cell systems have been implemented on larger sized fixed-wing UAVs [4, 6, 12], there is still very little literature on these implementations in order for researches to fully understand the capabilities and limitations of fuel cell applications on higher power consuming platforms.

Ref. [14] demonstrated that although the fuel cell system weighs more than the LiPo battery system for a low power system, things take a different turn when the energy stored is more than 333 Wh. The fuel cell systems start to weigh less than the LiPo battery system and the weight gap becomes larger as the energy storage becomes larger, as shown in 500 and 1000 W systems. They also demonstrated that batteries have much higher power densities than fuel cells although they have lower energy densities [14]. Since drones do not require maximum power during the entire operation, a hybrid system consisting of the fuel cell and batteries could be an effective combination, as the fuel cell could provide constant power while the batteries could respond to dynamic power peaks during flight. The hybrid system could be a good solution for UAVs with larger take-off weights, to achieve higher energy at a lower weight and ultimately expand flight time. As multirotor drones have higher power intake and more dynamic load profiles than fixed wing UAVs [1], it is significant to look at the performance of hybrid fuel cell systems, especially when implemented on heavy-lift multirotor drones.

The growing technology of fuel cell systems has led to a number of commercially available power modules that can be implemented on drones. To date, a 2.6 kW system, the DP30 [3] developed by Doosan Mobility is the world's largest power fuel cell power pack module [7]. The power modules are based on Polymer Electrolyte Membrane (PEM) fuel cell that uses compressed hydrogen. This lightweight fuel cell type is the most developed and widely available [1].

This paper presents a 39 kg heavy-lift hexacopter drone model powered by a 5.2 kW hydrogen fuel cell system that produced by linking two 2.6 kW class power modules in parallel. The system is supplied with a 314 bar compressed hydrogen from a 10.8 L hydrogen tank. This research presents the design and construction of the hexacopter and presents the analysis of flight test results to evaluate the performance of the fuel cell systems when implemented on a heavy-lift medium sized multirotor drone. The design considerations, and performance when implementing the FC system in a medium sized UAV are evaluated in this study.

1.1 kW Fuel Cell System

The 5.2 kW hydrogen fuel cell system was built by connecting two 2.6 kW class power modules. Each of the 2.6 kW fuel cell modules has two fuel cell power stacks which are controlled by a power board each. The power boards have an inbuilt LTE communication module which enables wireless communication with the ground control station. The power modules are connected in parallel and the connected system is supplied with 314 bar compressed hydrogen from a 10.8 L hydrogen tank. LiPo batteries are also connected to the system in order to manage power peaks. The bridge circuit block diagram configuration and drive control logic block diagram configuration for two 2.6 kW power modules to output 5.2 kW was constructed. The hardware configuration was also designed such that the current 2.6 kW DC-DC 2 channel can output 5.2 kW. The system control features a 2 in 1 system communication platform. The control logic configuration was set to less than 500 W deviation when outputting 5 kW from the two 2.6 kW power modules. Figure 1 shows the layout of the fuel cell stacks, hydrogen cylinder tank, power boards and LiPo support batteries for the 5.2 kW system.

Experiments were conducted in order to check the performance and efficiency of the 5.2 kW system. In the first experiment, two 2.6 kW power modules were connected to a 6 kW electronic loader. Hydrogen was supplied to the two power packs at the same time and a LiPo battery was connected to each of the power modules. The experiment results showed that when the input load was 40 A or below, only one power module provided power output and when the input load was above 40 A, both the power modules provided power output. This was also confirmed by the cooling fan duty values, which showed that the cooling fan for power module 1 was maintained according to the temperature in power module 1 while the cooling fan

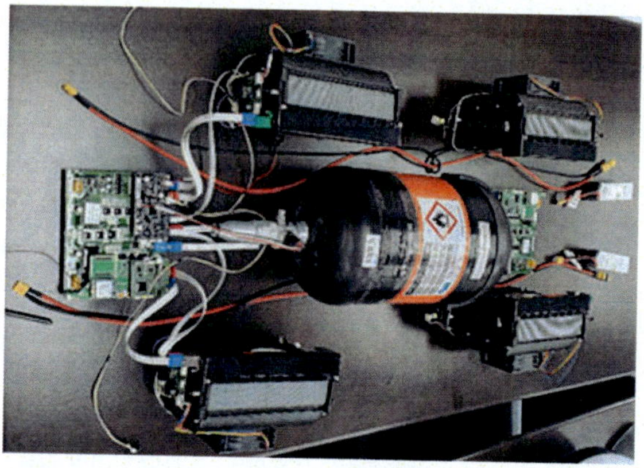

Fig. 1 Fuel cell power stacks, hydrogen cylinder tank, power boards and LiPo batteries

duty value of power module 2 decreased as the temperature dropped during the low load input of less than 20 A, when power module 2 did not provide power output.

In the second experiment, an adjustable regulator was connected to the hydrogen cylinder in additional to the setup used in experiment 1. The purpose of the experiment was to check the stack hydrogen supply dynamic pressure and flow rate. Experiment results showed that the maximum hydrogen discharge pressure required was 1.5 bar and the flow rate was 60 to 80 NLPM when 5.2 kW output was applied. However, the hydrogen cylinder distributed in Korea has a dynamic discharge pressure of only 0.7 bar, which is lower than the required pressure for the load of 5 kW. The solution to this problem was to apply a regulator to control the discharge pressure of the hydrogen so that the required hydrogen supply pressure of above 1 bar could be achieved for the 5 kW system.

2 Multicopter Design

2.1 Sizing and Weight Estimation

The take-off mass of a multirotor drone can be calculated using Eq. 1 below [11].

$$W_0 = W_{empty} + W_{energy} + W_{payload} \tag{1}$$

where the empty weight, W_{empty} includes the mass of the airframe, propulsion system and flight control system. W_{energy} is the weight of the energy system needed to power the drone, which is the fuel cell system and all the components related to it. $W_{payload}$ applies to the mass of the mission payload that the drone is designed to carry. As the weight of the W_{energy} and $W_{payload}$ is known, the appropriate empty weight portion against the total weight can be estimated using the empty weight fraction (W_e/W_o)

relationship in Eq. 2 below, which is derived from statistical data of commercial heavy-lift drones [10].

$$\frac{W_e}{W_0} = 0.4666{W_0}^{-0.02} \quad (2)$$

Considering the total weight, W_0 of 39 kg, the appropriate $W_e = 16.91$ kg, which is 43% of the total weight. The fuselage sizing was performed by taking into account the area and volume needed to mount the 5.2 kW fuel cell system. After the sizing of the fuselage, the weight estimation of the composite fuselage skin was performed based on the total area and the composite layout design of the skin. The composite skin was constructed from carbon fiber and glass fiber layers using the wet layup method. Composite skins were used to optimize the weight of the hexacopter as they have a high strength to weight ratio. An initial take-off weight estimation was performed and the motor and propeller were selected for the configuration based on a target maximum take-off weight, MTOW. The wheelbase sizing was performed based on the number of motors selected and the size of the propellers. The estimated take-off weight of the hexacopter was 39 kg as shown in Table 1. The empty weight, W_e was 16.53 kg, which is 42.4% of the total weight. This is slightly lower than the estimation of 43% from the statistical data as calculated above, due to weight optimization.

Figure 2 shows the weight breakdown of the components. Considering the empirical correlation of the battery fraction trend [10] from the statistical data of commercial drones in Eq. 3, if batteries were used as the power source for a 39 kg drone, it

Table 1 Component weight table

Classification	Components	Weight (kg)
Frame	Fuselage + Motor arms	9.50
	Landing pad	0.80
Propulsion	Motor	5.27
	Propeller	
	ESC	
5.2 kW FC system (2 × 2.6 kW System)	10.8L fuel tank	16.47
	Fuel Cell Stack + Air Ducts	
	Power board	
	Support battery	
Flight control system	Flight computer battery	0.96
	Flight computer, antenna	
Payload	Payload	6.00
Total take-off weight		39.00

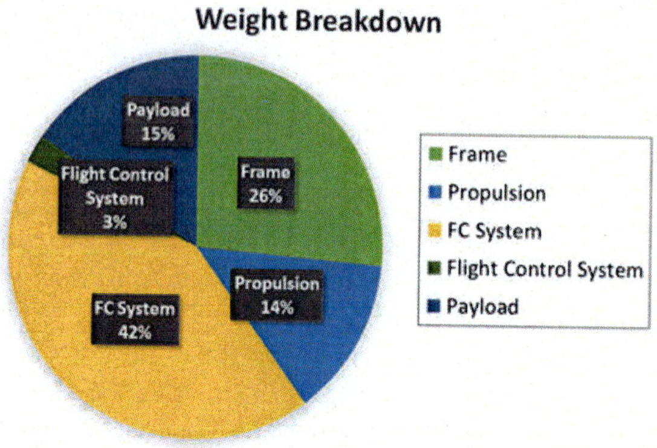

Fig. 2 Component weight breakdown chart

would only account for 11.56% of the total aircraft weight. On the other hand, the FC system accounted for 42% of the total aircraft weight. Although the weight of the FC system is higher than the weight of batteries required to power a similar sized drone, the weight disadvantages are outweighed by the high energy density advantages of the FC system over the battery system.

$$\frac{W_b}{W_0} = 195.27{W_0}^{-0.703} \tag{3}$$

2.2 *Propulsion System*

After determining the target MTOW of the UAV, the propulsion system was designed. The fuel cell system provides 5.2 kWh of power, therefore the current (A) of the system was calculated to be 5.2 kW/50 V = 104 A and the current supplied to each motor was 104A/6 = 17 A. The motor and propeller were selected based on the results of the motor thrust test. Considering a safety margin, the thrust at 50% throttle was selected. The selected motor and propeller combination provide a thrust of 6.77 kg at 50% throttle. This will provide a total of 40.62 kg of thrust for the hexacopter. A total power consumption of 4.32 kW is required and the FC system can provide 5.2 kW of power, which is more than the estimated required power. The specifications are shown in Table 2.

Table 2 Drone specifications

Classification	Specifications
Drone type	Hexacopter
Size (mm)	1,770 × 1,770 × 630
Wheelbase	1770 mm
Airframe weight	16.5 kg
Payload weight	6 kg
Take-Off weight	39 kg

2.3 *System Controller*

A flight control system was designed for the 5.2 kW fuel cell system powered hexacopter. The mission boards from two 2.6 kW modules send data through a UART communication to the flight control computer (FCC) to provide the status of each fuel cell. The FCC has an inbuilt gyro sensor, accelerometer and barometric sensor which can measure the angular velocity, direction, vibration and atmospheric pressure, to determine and maintain flight stability. Figure 3 shows a schematic diagram of the drone system controller.

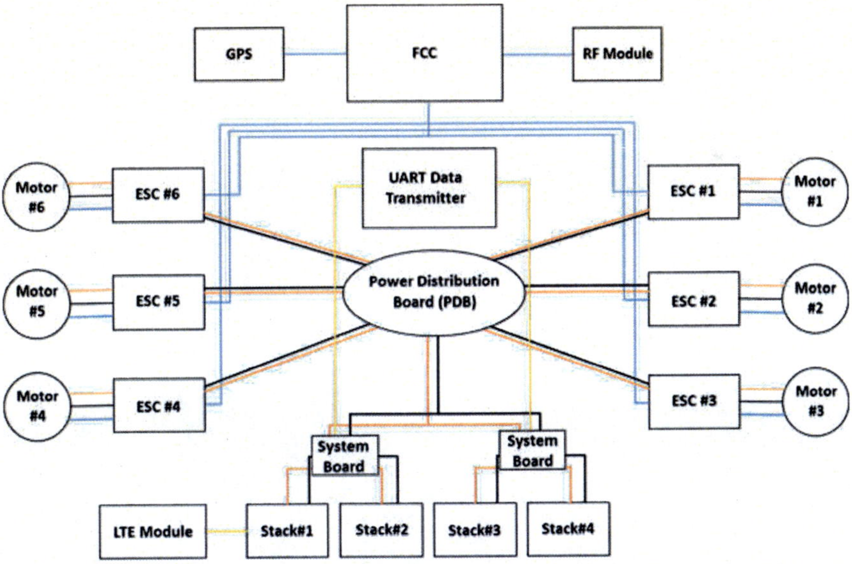

Fig. 3 Schematic diagram of the system controller

2.4 *Structure Optimization*

To evaluate the structural integrity of the hexacopter, the margin of safety (MOS) for each part was calculated. The safety margin is a coefficient indicating how much additional load the component receiving the maximum load can withstand before failure. If it has a value of 0 or more, it is judged to be structurally safe. In the case of parts made of isotropic materials such as aluminum and ABS, the safety margin was calculated by dividing the maximum stress applied to the member for both the yield strength and ultimate strength. For parts which have no elasticity, such as carbon plates, the safety margin was calculated only for the ultimate strength. For parts made of laminated composite materials of glass and carbon fabric, unlike isotropic materials, the safety margin was calculated with the failure index according to the failure theory. In this way, the failure index was derived using the commonly used Tsai-Wu failure theory, and the safety margin was calculated from these values. When the Tsai-Wu failure index is less than 1, it indicates that no failure occurs.

Figures 4 show a comparison of Tsai-Wu index distribution before and after design optimization when the value was decreased to below 1 after design changes. Design optimization was been performed by changing the internal structure design to improve load distribution to the structures and to fix weak areas. Table 3 shows the stress distribution before and after design optimization, which indicates higher stress values on the skin and inner structures after design optimization. This means that the loads are better distributed. Table 4 shows the margins of safety before and after design optimization, which indicate a higher margin of safety or structural strength of the fuselage skin which had some weak points before design optimization. All the margins of safety values were checked to be above or close to 1 and all failure index values were checked to be below 1 after design optimization.

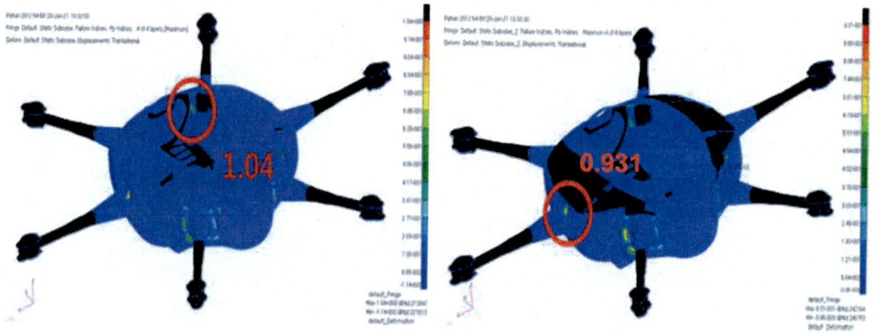

Fig. 4 Tsai-Wu failure index distribution before (left) and after (right) design optimization

Table 3 Stress values before and after design optimization

Type	Before design optimization		After design optimization	
	von Mises stress [MPa]	Maximum principal stress [MPa]	von Mises stress [MPa]	Maximum principal stress [MPa]
Fuselage skin part 1	–	126	–	140
Fuselage skin part 2	–	122	–	114
Fuselage skin part 3	–	99.1	–	119
Motor mount	–	14.8	–	15.5
Carbon folding motor arm	–	9.93	–	9.89
Aluminum folding mtor arm	8.03	–	9.80	–
Folding motor arm holder	9.31	–	9.69	–
Fuselage inner structure	25.8	–	37.8	–

Table 4 Margin of safety before and after design optimization

Type	Margin of safety before design optimization	Margin of safety after design optimization
Fuselage skin part 1	1.673	0.953
Fuselage skin part 2	0.703	1.652
Fuselage skin part 3	1.538	2.690
Motor mount	46.73	44.58
Carbon folding motor arm	70.14	70.43
Aluminum folding motor arm	34.94(S_Y) / 53.4(S_{VT})	32.06(S_Y)/ 46.85 (S_{VT})
Folding motor arm holder	1.148(S_Y) / 2.179(S_{VT})	1.063(S_Y) / 2.054 (S_{VT})
Fuselage inner structure	2.255	1.245

2.5 *Airflow Design*

The airflow intake is important in order to ensure that the FC system achieves high performance. To ensure good airflow to the airframe, air inlet and outlet areas were added to the top skin of the airframe. The air inlets were designed at the front side of the airframe and air outlets were designed at the back side of the airframe so that air could flow in from the front side and flow out through the back side. There were also openings at the side areas to allow air from the stack fans to escape so that the hot air from the stacks could flow out and not overheat the stacks. The total surface area of the air inlet/outlet was estimated to be 90 cm^2. To check whether the airflow in the airframe was sufficient for the FC system, ground tests were conducted by connecting the 5.2 kW system to a 6 kW electronic loader to check the performance of the FC

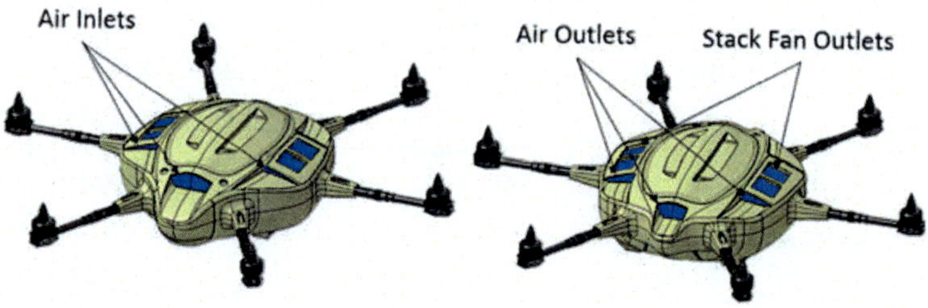

Fig. 5 Air Inlet/Outlet Areas (marked in blue) on the front side (left) and back side (right) of the airframe

system. Test results showed the system operated normally, where the required power output was achieved and the temperature of the stacks was well controlled by the cooling fans. Figure 5 below shows the air inlet areas on the airframe design.

3 Flight Test and Analysis

3.1 Flight Test Conditions

The flight test is performed inside a factory in order to have a better controlled environment, as it was conducted in winter. The take-off weight of the hexacopter was at 39 kg while carrying a 6 kg payload. As humidity affects the performance of the fuel cells, the humidity was measured and controlled at 25%. The flight test was first performed using safety wires connected to the hexacopter body and after checking that the hexacopter was stable enough to hover on its own, the safety wires were removed and the endurance test was performed. For the endurance test, the hydrogen cylinder was filled at 314 bar which is slightly below the maximum storage allowance of 350 bar. The test was conducted by hovering the hexacopter in the same position until the low fuel warning alarms takes off and landed in 3 min with 48.5 bar of remaining hydrogen. Figure 6 shows the hexacopter during the flight test.

3.2 Flight Test Results

The endurance of the FC system hexacopter was 36 min with a consumption of 265.5 bar of hydrogen. The hydrogen consumption was estimated to be 6 g/min. The X, Y, Z acceleration graph of the accelerometer mounted in the FCC is shown in Fig. 7 below. The acceleration graph indicates a stable flight as there are no large variations in the values.

Fig. 6 Hexacopter during flight test

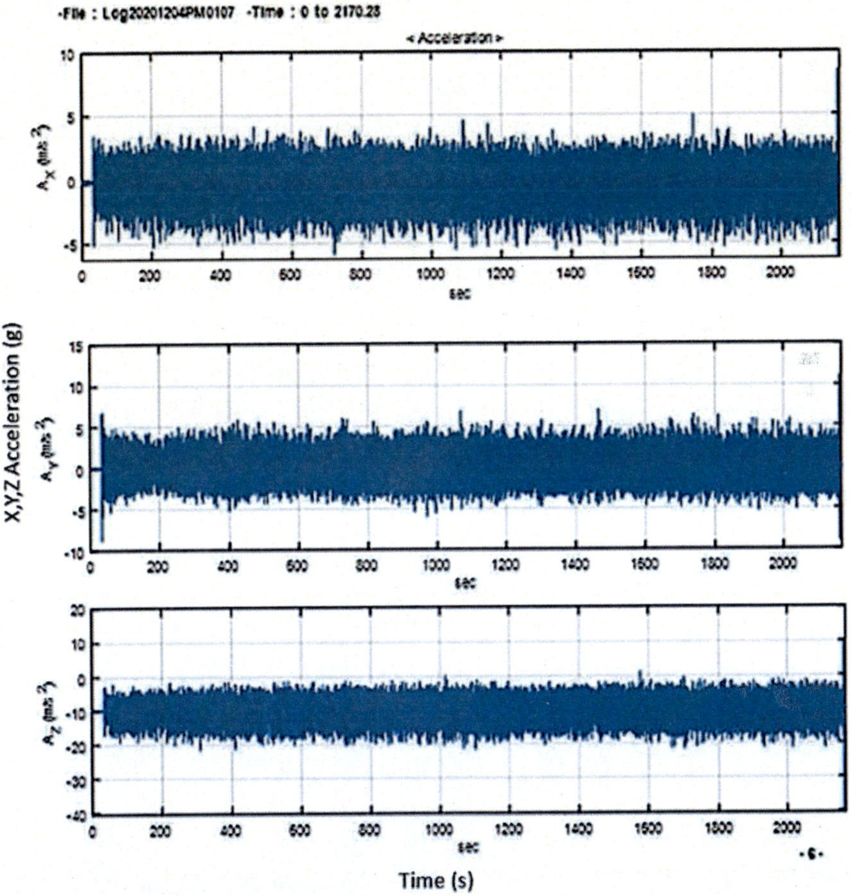

Fig. 7 Vibration (Acceleration (g) versus Time(s)) graph

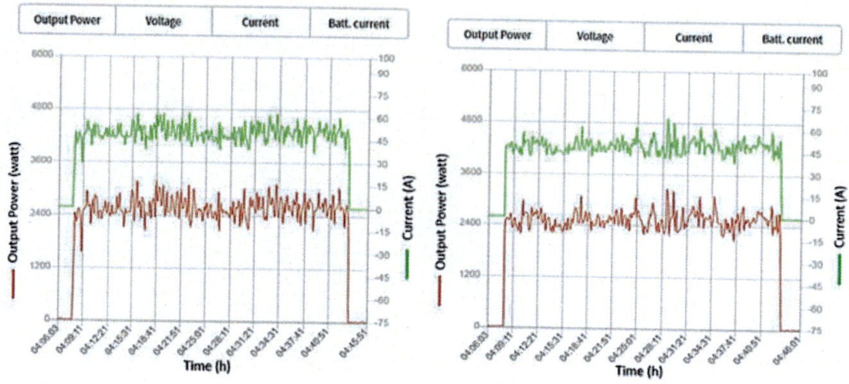

Fig. 8 Output power (watt) and current (A) versus time (h) of the left side and right side of the fuel cell system

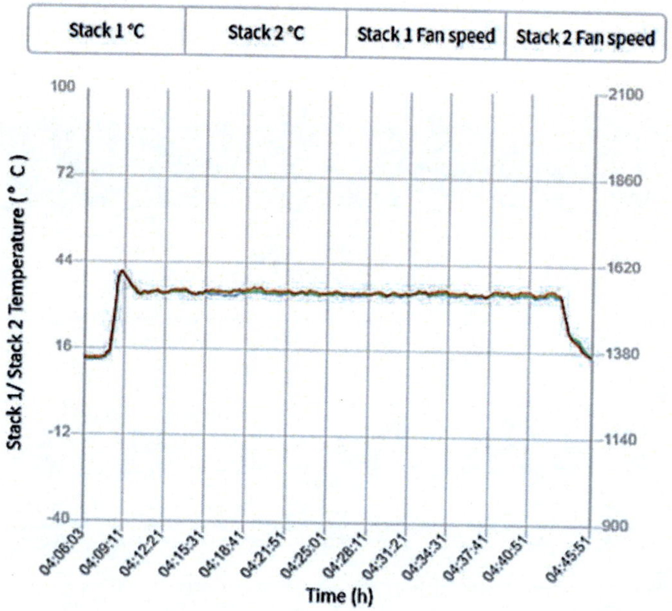

Fig. 9 Temperature of the right side and left side of the fuel cell system (°C) versus time (h)

Figure 8 shows the output power and current of each side of the fuel cell system. The combined average value of ampere draw is 98 A, the maximum value of momentary peak in current is 161 A and net power output is 4507 Wh. The momentary current peaks were resolved by the 7000 mAh support batteries which are rated 60 C, which means that a maximum discharge of 420 Ah can be drawn from the battery.

Figure 9 show the temperature graph of stack 1 and 2 of the left and right fuel cell systems. The temperature graphs show almost identical temperatures in stack 1 and stack 2 for both systems, indicating that both the stacks were functioning at the same time.

4 Flight Time Estimation

Based on the flight test, the total mass of hydrogen consumed was 265.5 bar and the flight time was 36 min. This amounts to 216 g of hydrogen which means the hydrogen consumption was 6 g/min for the take-off weight of 39 kg. In order to increase flight time, a larger hydrogen tank can be used. If the 10.8 L hydrogen tank is replaced with a 12 L hydrogen tank of similar weight, it is estimated that an additional 36 g of hydrogen mass can be stored and this will result in an additional 6 min of flight assuming that the take-off weight is approximately the same.

5 Conclusion

This research presents the design and development of a 39 kg fuel cell powered heavy-lift medium-sized hexacopter and the analysis of data obtained from a flight test performed using the hexacopter developed in the research. The linking of two 2.6 kW class power modules to produce a 5.2 kW hydrogen fuel cell system was successfully accomplished, using a regulator to control hydrogen discharge pressure. Special design considerations were taken when designing the hexacopter in order to achieve optimized weight and structural strength. Airflow intake was also taken into considerations to ensure that there will be enough airflow to the FC system to achieve high performance. The flight test results proved the feasibility of the 5.2 kW fuel cell system medium-sized hexacopter to perform stable flights. Based on the results, a hydrogen consumption of 6 g/min was estimated and the possibility of extending the flight time by 6 min is presented by using a 12 L hydrogen tank. It can be concluded that fuel cell systems can be a possible solution for extending the flight endurance of heavy-lift drones. This could give drone manufacturers more design freedom when designing drones to carry heavier payloads, while providing enough endurance to perform its mission.

Acknowledgements This paper has been supported by Korea Agency for Infrastructure Technology Advancement (KAIA) in 2021.

References

1. Apeland J, Pavlou D, Hemmingsen T Suitability analysis of implementing a fuel cell on a multirotor drone. J Aerosp Tecnol Manag, 12 e3220
2. Bradley TH, Moffitt BA, Mavris DN, Parekh DE (2007) Development and experimental characterization of a fuel cell powered aircraft. J Power Sources 171(2):793–801
3. DP30 Powerpack.). [Cited 2021 3 June]. www.doosanmobility.com/en/technology/tech_01
4. DS30 from Doosan Mobility: [Cited 2021 3 June]. https://www.doosanmobility.com/en/products/drone-ds30
5. Dudek M et al (2013) Hybrid fuel cell–battery system as a main power unit for small unmanned aerial vehicles (UAV). Int J Electrochem Sci 8(6):8442–8463
6. Global Observer from AeroVironment. [Cited 2021 3 June]. https://www.airforce-technology.com/projects/globalobserverunmann
7. Jørgen A, Pavlou DG, Tor H (2021) Sensitivity study of design parameters for a fuel cell powered multirotor drone. J Intell Robot Syst 102.1
8. Kim T, Kwon S (2012) Design and development of a fuel cell-powered small unmanned aircraft. Int J Hydrogen Energy 37(1):615–622
9. Lapeña-Rey N et al (2017) A fuel cell powered unmanned aerial vehicle for low altitude surveillance missions. Int J Hydrogen Energy 42(10):6926–6940
10. Ong W, Srigrarom S, Hesse H (2019) Design methodology for heavy-lift unmanned aerial vehicles with coaxial rotors. AIAA Scitech 2019 Forum
11. Raymer P (1992) Aircraft design: a conceptual approach, 2nd ed. American Institute of Aeronautics and Astronautics, Washington, DC, USA
12. "Thunderbird" (LN60F). [Cited 2021 3 June]. https://www.suasnews.com/2012/08/thunderbirds-are-go-chinese-fuel-cell-powered-uav-first-flight/. Accessed on 3 June 2021
13. Types of Drones. [Cited 2021 3 June]. https://www.casa.gov.au/drones/rules/drone-types
14. Ustolin F, Taccani R (2018) Fuel cells for airborne usage: energy storage comparison. Int J Hydrogen Energy 43(26):11853–11861
15. Yacoubi M et al (2012) Study of the propulsion system of a VTOL MUAV using fuel cells. 9 th National congress on theoretical and applied mechanics, Brussels

Multi-objective System Optimization of Suborbital Spaceplane by Multi-fidelity Aerodynamic Analysis

Shintaro Tejika, Takahiro Fujikawa, and Koichi Yonemoto

Abstract An effective approach has not been established for the reusable launch vehicle, particularly for spaceplanes, whereas the conceptual design method has been established to some extent for aircraft. Spaceplanes that have wing like airplanes flight various environment, and the airframe and trajectory design problem are closely linked. Therefore, the multidisciplinary optimization method is required to optimize the airframe and the flight trajectory design at the same time in the spaceplane conceptual design. Due to computational cost constraints, a low-accuracy Computational Fluid Dynamics (CFD) method was used in a previous study to evaluate the aerodynamic characteristics of the airframe. This has been a source of concern regarding the accuracy of optimization calculations. In this study, a multi-fidelity approach is applied where the low fidelity and high-fidelity CFD are used in a complementary way. This surrogate model was connected into the manned spaceplane's Multidisciplinary Design Optimization (MDO) framework. As the load limit was reduced, the wing area grew larger and the initial mass increased.

Keywords Multidisciplinary design optimization · MOEA/D · Surrogate model · Multi-fidelity · Suborbital spaceplane

1 Introduction

It is well understood in the design process of aerospace vehicles that the quality of the conceptual study and conceptual design at the start of the process has a significant impact on the overall design, manufacturing, and operation costs. While the conceptual design method has been established to some extent for aircraft, an effective approach has not been established for reusable launch vehicles, especially for spaceplanes. Spaceplanes flight various environment, and the airframe and trajectory design problems are closely linked. Furthermore, due to the strict weight feasibility of space transportation systems, optimization techniques must be used in their design.

S. Tejika (✉) · T. Fujikawa · K. Yonemoto
Tokyo University of Science, Yamazaki 2641, Noda, Japan
e-mail: 7521536@ed.tus.ac.jp

S. Lee et al. (eds.), *The Proceedings of the 2021 Asia-Pacific International Symposium on Aerospace Technology (APISAT 2021), Volume 1*, Lecture Notes in Electrical Engineering 912, https://doi.org/10.1007/978-981-19-2689-1_21

Therefore, the MDO method is required to optimize the airframe and the flight trajectory design at the same time in the spaceplane design. Because the spaceplane MDO problem is multi-objective and multi-modal, stochastic optimization is promising. Due to computational cost constraints, a low-accuracy CFD method was used in a previous study [1] to evaluate the aerodynamic characteristics of the airframe.

Considering this issue, we focused on a multi-objective optimization algorithm using a surrogate model, which replaces a high-cost evaluation of the objective function with a surrogate model to search for good individuals. The surrogate model typically requires training data representing several input–output relationships, and after training, the estimated value of the objective function can be calculated using simple algebraic computation, enabling a much lower computational cost than CFD analysis. A multi-objective optimization algorithm that proceeds with the search while creating a surrogate model in the optimization loop is generally called Efficient Global Optimization (EGO) [2]. Since Jones et al. reported in 1998, research on EGO algorithms has been actively reported, and they have attracted attention for the ability to reduce the computational cost of optimization calculations. Among them, MOEA/D-EGO [3], a method that applies the concept of EGO to MOEA/D [4] has shown promise as a multi-objective optimization method.

In this paper, as a preliminary step to implementing EGO in MDO, outside of the optimization loop, a surrogate model is created, and a multi-objective optimization using MOEA/D is applied to a suborbital spaceplane problem and its performance is validated.

2 Aerodynamic Calculation Methods

In this study, a surrogate model is used for estimating aerodynamic characteristics, which will be discussed in Sect. 3, and various types of training data can be introduced. It is explained how to calculate high-accuracy and low-accuracy aerodynamic data to be used as training data.

2.1 *Low-Accuracy Aerodynamic Calculation Method*

The panel method was used for low-accuracy CFD. The panel method obtains information on velocity and pressure distribution primarily from panel data on the aircraft surface, without the use of a spatial grid, and aerodynamic force is obtained by integrating the pressure distribution. In this paper, a newly developed unstructured panel code is used for the analysis. The code is a significant extension of CPanel [5] and is based on linear potential flow theory with the Prandtl–Glauert modification [6] for subsonic speeds, and the modified Newtonian method [7] for supersonic and hypersonic speeds. The modified Newtonian method [7] and the Prandtl–Meyer expansion flow theory are used for supersonic and hypersonic flows. The van Driest II method

Table 1 Analysis conditions of FaSTAR

Velocity	Subsonic	Supersonic	Hypersonic
Domination equation	Full Navier Stokes equation		
Turbulence model	SA-noft2-R	SA-noft2-R	SA-noft2-R
Scheme for advection term	SLAU	HLLE	HLLE
Time integration method	LU-SGS	LU-SGS	LU-SGS
Slope limiter factor of N.S. equation	Quadratic	Quadratic	Linear
Slope limiter factor of turbulence model	Quadratic	Quadratic	Linear
Reconstruction method	MUSCL	MUSCL	MUSCL

for calculating frictional drag [8] and an empirical model for calculating base drag [9] have also been implemented.

2.2 *High-Accuracy Aerodynamic Calculation Method*

HexaGrid, an automatic grid generation tool, was used as a computational grid generation tool. We used this tool because it can generate hexahedron-based unstructured grids using only STL data from the target geometry and input parameters, and it is scalable to automatic processing. In this study, I use Fast Aerodynamic Routines (FaSTAR), a fast fluid analysis tool developed by Japan Aerospace Exploration Agency (JAXA), as a high-accuracy aerodynamic calculation method. FaSTAR is a three-dimensional compressible fluid analysis solver for unstructured meshes that is especially useful for aircraft and spacecraft aerodynamic analysis.

Table 1 shows the calculation method for each velocity. For the advection term, SLAU, in the subsonic region, a scheme corresponding to full velocity is used. The turbulence model SA-noft2-R eliminates the ft2 term involved with the trip in the Spalart–Allmaras model [10] and replaces it with the rotation correction [11].

3 Surrogate Model

3.1 *Multi-fidelity*

In multi-fidelity, not only conventional data with high computational cost and high accuracy can be used as training data, but also data with low computational cost and low accuracy, lowering the computational cost even further. The Co-Kriging model [12] is used in this paper.

The Co-Kriging model is an extension of the Kriging model to multiple input variables; it allows estimation using multiple training data that are correlated with each other, rather than with one type of training data as in the Kriging model. For more information on kriging, the reader may wish to consult [13] or [14]. Two kinds of sample points must be prepared: one for the high-accuracy function and one for the low-accuracy function. It is preferable to set the sample points to cover the design variable space to prevent convergence to wrong local solutions. Because Co-Kriging model is based on the difference between high and low accuracy, the high-accuracy sample points should be included in the low-accuracy sample points so that the difference can be calculated.

3.2 *Surrogate Model for Aerodynamic Analysis*

This section discusses the surrogate model for estimating aerodynamic characteristics in a MDO, including the method of construction and the model's accuracy. Co-Kriging is used as a stand-in model.

3.2.1 Surrogate Model Building Method

As training data, I prepared 150 low-accuracy data and 2 high-accuracy data. For each individual, only static design variables related to the airframe shape described in Fig. 1 were generated using the Latin Hypercube method, and these data were trained to construct the surrogate model. Each individual's aerodynamic coefficients were calculated for a total of 27 cases in which the Mach number was 0.3, 0.6, 0.7, 1.5, 2.0, 3.0, 4.0, 5.0, and 10.0 [−] and the angle of attack was varied in three patterns of 0, 10, and 20 [deg.], respectively. Required time to prepare the 150 low-fidelity and 2 high-fidelity training data is 90 h. However, once the data is prepared, the time required to calculate the aerodynamic characteristics of one individual was approximately 4 min using a desktop computer (CPU: Intel Core i7-9700) for the low-accuracy CFD and

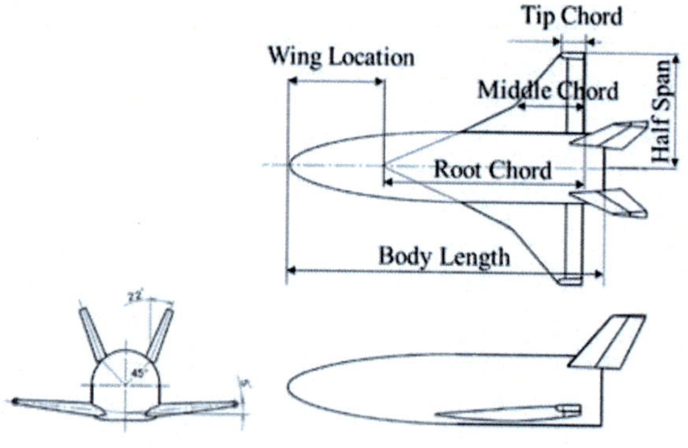

Fig. 1 Airframe shape of HIMES and variables

approximately 40 h (not including the waiting time for processing) using the JAXA supercomputer system for the high-accuracy CFD.

In this paper, the shape of the airframe is based on the WSV C-5 of the Highly Maneuverable Experimental Space (HIMES) flight vehicle [15], which was studied at ISAS/JAXA, and the variables are the body length, wing root-chord length, wing middle-chord length, wingtip chord length, and half span length. The tail wing's shape was modified from rectangular to NACA0012. Rectangular airfoils have typically been used for supersonic flight because of their good steering effect. However, there is almost no air at the altitude at which the spaceplane enters the supersonic region, and it is expected that the airfoil will be ineffective in cutting the steering surface. The front and rear positions of the wing are adjusted to match the aerodynamic center position of HIMES.

The HIMES shape was preferred as a starting point because it is designed for stability and maneuverability from subsonic to hypersonic speeds and is intended to be a practical spaceplane in the future. It is also possible to use data of wind tunnel test results obtained in the past as aerodynamic characteristics.

In this paper, OpenVehicle Sketch Pad (OpenVSP) is adopted as the software for modeling the 3D shape and generating the surface grid necessary for CFD analysis.

3.2.2 Evaluation

To determine whether the accuracy of the surrogate model has improved, the aerodynamic characteristics of the HIMES geometry with a body length of 11.5 [m] were compared. The panel method, a low-accuracy aerodynamic calculation method, and FaSTAR, a high-accuracy aerodynamic calculation method, were used to compare the aerodynamic characteristics of the HIMES shape. In Figs. 2 and 3, the shapes of high-accuracy data are represented. Figures 4 show the results of evaluating the aerodynamic data of the HIMES geometry using the low-accuracy and high-accuracy aerodynamic calculation methods and the multi-fidelity surrogate model. Despite the fact that the error is large in the subsonic region with a high angle of attack, there is an improvement in accuracy in both the subsonic and supersonic regions when

Fig. 2 The airframe of high-accuracy data 01

Fig. 3 The airframe of high-accuracy data 02

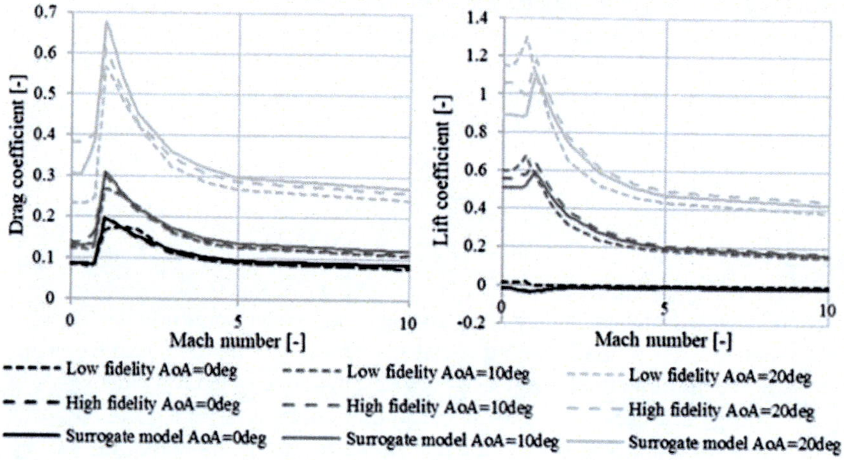

Fig. 4 Surrogate model vs. High-accuracy CFD vs. Low-accuracy CFD for drag and lift coefficient

compared to the low-accuracy aerodynamic calculation method, regardless of the fact that only two high-accuracy individuals were trained. Furthermore, the lift coefficient's peaks were consistent. Although accuracy verification was done only at one point, it was considered sufficient as a stage before EGO implementation and integrated into the analysis model.

4 MDO of Suborbital Spaceplane with the Surrogate Model

The mission is to return to the launch site after reaching an altitude of 120 km with six crew members, including passengers. The rough flight sequence is shown in Fig. 5. The following is the aircraft's assumed concept:

Fig. 5 Flight sequence

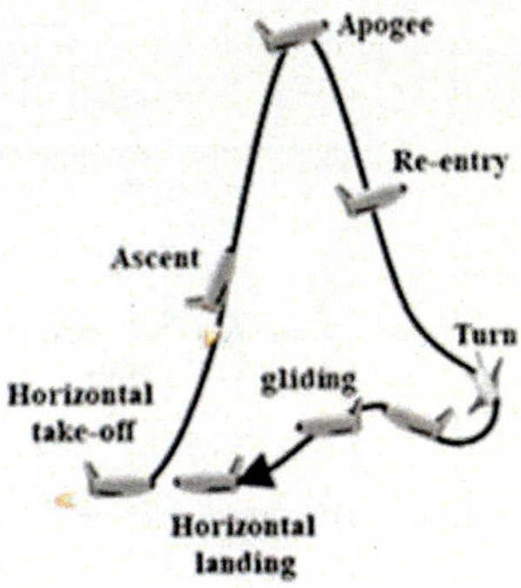

- It will be a completely reusable single-stage vehicle with a horizontal launch and landing.
- Multiple LOX/LNG engines are installed in a vehicle.

4.1 Design Optimization Methodology

Using the evolutionary computation method and the gradient method, I generated an optimization problem for a manned suborbital spaceplane based on the previous study [1]. The objective function of the optimization is to minimize the total mass of the aircraft at launch and the maximum load factor during flight. The variables to be optimized are shown in Table 2. The optimization problem is formulated in

Table 2 Design variables of suborbital spaceplane optimization

Category		Variable	Unit
Vehicle design	Airframe	Body length	[m]
		Root-chord length	[m]
		Middle-chord length	[m]
		Tip-chord length	[m]
		Half span length	[m]
		Main wing location	[m]
		Maximum operating load factor	[G]
		Maximum operating dynamic pressure	[Pa]
		Initial thrust-to-weight ratio desired	[–]
	Engine	The nozzle expansion ratio of engines	[–]
Trajectory	Powered ascent	State variables, Control variables, Thrust-cutoff time	
	Coasting ascent	State variables, Control variables, Apogee time	
	Nose-first return	State variables, Control variables, Return time	

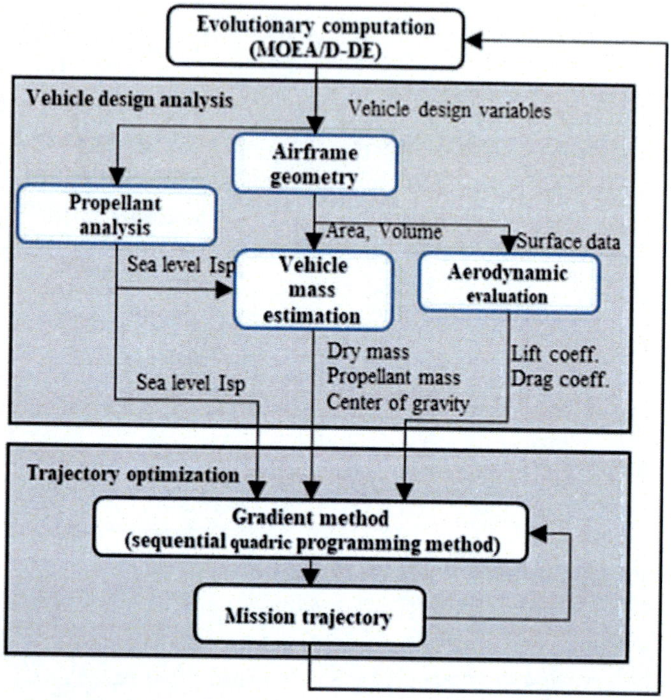

Fig. 6 Flow of optimization

this section using the framework shown in Fig. 6, and the analysis procedure is as follows.

1. In the external optimization loop, the airframe design variables in Table 2 are optimized by MOEA/D-DE [16]. The airframe design variables are input into numerical models and an airframe design analysis is performed in the individual evaluation of MOEA/D-DE. The masses and center of gravity of the airframe components are output by inputting the airframe data and the flight load conditions. These masses are added together to obtain the dry mass of the aircraft. In addition, the following constraints on the airframe design are evaluated, and when all of them are satisfied, it proceed to the next process. If any conditions are not satisfied, the individual will be regenerated.

   Constraint 1: The engine must fit in the fuselage base plane.

   Constraint 2: The trailing edge of the main wing fits into the fuselage base plane.

   Constraint 3: The thrust-to-weight ratio at lift-off is greater than 1.

   Constraint 4: The thrust-to-weight ratio at lift-off is less than 2.5.

   The result of airframe design analysis is passed to the trajectory optimization. The flight trajectory design variables in Table 2 are optimized using the gradient method in the internal optimization loop. The objective function is to maximize the amount of propellant remaining at the time of return to the launch site. However, until the 10th generation, the evaluation in the internal loop didn't take into consideration the calculation's stability. This section's trajectory optimization problem is divided into three phases: (1) powerd ascent, (2) coasting ascent, and (3) return. The state variables include altitude, longitude, latitude, velocity, residual propellant mass, path angle, and flight heading angle. The control variables consist of the angle of attack, bank angle, and throttle opening.

2. The residuals of the following flight trajectory constraints are returned to MOEA/D-DE as the results of individuals evaluation based on the results of the flight trajectory optimization. Constraint 5: The residual propellant level at the end of Phase 3 is greater than or equal to zero. The initial and end conditions and constraints for the trajectory in-flight trajectory optimizations are as follows.
3. The initial path angle, initial speed, and initial altitude of Phase 1 are set to 89.9°, 10 m/s, and 0 m, respectively. At the Taiki-cho multipurpose park, the initial longitude and latitude are set to 143.44° east longitude and 42.50° north latitude, respectively, with the initial flight azimuth set to free.
4. The initial propellant mass is free, but a restriction is imposed on the maximum available propellant mass.
5. The engine is assumed to function only during Phase 1, and the throttle is set to between 0.5 and 1. Following that, the engine is turned off and the throttle is set to zero.
6. The angle of attack shall be limited to between − 20° and 30°.
7. At the end of Phase 2, the altitude must reach the target value, i.e., 120 km, as a constraint.
8. The return to the launch site is subject to specific constraints at the end of Phase 3. The terminal altitude must be at least 1000 m, the terminal path angle must be at least − 3°, and the terminal longitude and latitude must be the same as the launch site. The amount of propellant consumed should not be less than maximum amount of propellant that can be carried.

Since the overall design objectives explained in Sect. 4.1 are to minimize m_{lf} and to maximize h_{apogee} under the constraint that $m_{prop.f} \geq 0$, system-level objectives, F_1 and F_2, are written using a penalty-based constraint handling approach as follows:

$$\min. F_1 = m_{lf} - 1000 \min\left(m_{prop.f}, 0\right) \tag{1}$$

$$\min. F_2 = -0.1 h_{apogee} \tag{2}$$

The settings of MOEA/D are shown in Table 3. A desktop computer was used for optimization calculations.

Table 3 Setting of MOEA/D

Generation	30
Popular size	50
Niche size	5
Crossover method	Differential evolution (Scaling factor is 0.5)
Crossover ratio	1
Mutation method	Polynomial mutation ($\eta = 20$)
Mutation ratio	0.1

4.2 Design Analysis Models

100kN-class Liquid Oxygen/methane rocket engines, which are being studied by aerospace corporations, are installed in the engine performance model.

Hypersonic Aerospace Sizing Analysis (HASA) [17], developed by NASA, is used for vehicle mass analysis. HASA is a statistical estimation model that is based on existing data on space transportation. Inputting the airframe data and flight load conditions yields the masses of the airframe components. These masses are added together to obtain the dry mass of the aircraft.

Flight simulations are carried out using the aircraft's equations of motion in a three-degree-of-freedom mass point model. The trim condition is ignored in this optimization. The atmospheric model is based on the United States Standard Atmosphere 1976 [18], which specifies the atmospheric density, pressure, temperature, and sound velocity for each altitude. In the trajectory optimization calculation, a smooth atmospheric model is implemented by spline interpolation with each value. The equation of motion does not take into account the effect of the earth's rotation, but it does take into account the spherical shape of the earth.

5 Result and Discussions

5.1 Outline

It took 10 days to finish the computations by using a desktop computer with Intel Core i7 9700 CPU and 32 GB RAM. Figure 7 shows the Hypervolume values of the archived individuals in the optimization calculation. Figure 8 depicts the individuals' objective function values as determined by the optimization calculation. The Hypervolume value is almost coverging, and the evolution is assumed to have progressed sufficiently. The computation took three days up to 7 generations, and the computational cost was caused by the unstable initial computation. It has been identified that the load factor and total mass have an inverse relation. Solutions are searched towards the optimal lower-left direction while maintaining a high level of diversity, indicating that the optimization is effective.

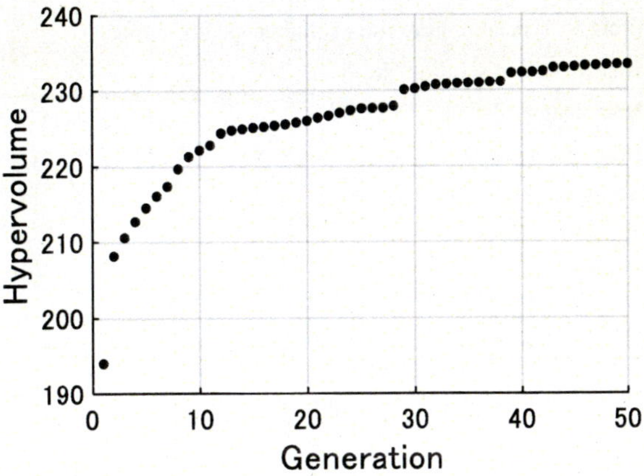

Fig. 7 Hypervolume

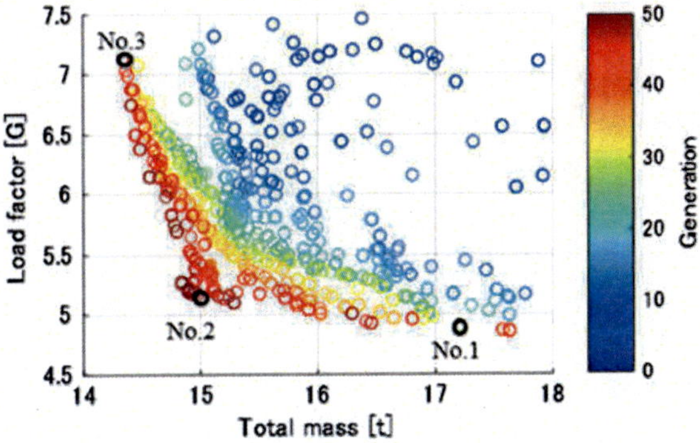

Fig. 8 All optimization solutions

5.2 *Airframe and Trajectory*

Table 4 shows the characteristics of typical solutions obtained in the optimization calculations.

Figures 9, 10, 11 depict typical optimization solutions' aircraft shape and flight trajectory. It was discovered that as the wing area grew larger and the initial mass increased, the load limit was reduced. Around 400 s into Solution No. 1's flight trajectory, the angle of attack becomes unstable. This output is because the continuity constraint is not strict, but it is not classified as an infeasible solution because it is considered to have little effect on the entire trajectory. When the wing area is reduced to near zero, the load factor rises and the propellant mass falls as the initial thrust-to-weight ratio rises. No.3's trajectory is omitted because its load factor is too large to be practical.

Table 4 Typical optimization solution specifications

Representative solution No		1	2	3
Body length	[m]	13.9	13.4	13.3
Root-chord length	[m]	6.35	4.67	4.61
Middle-chord length	[m]	4.42	1.46	1.14
Tip-chord length	[m]	0.96	1.31	1.08
Half span length	[m]	5.34	4.26	4.19
Main wing location	[m]	6.53	6.04	5.91
Maximum operating load factor	[G]	4.86	5.17	7.13
Maximum operating dynamic pressure	[kPa]	29,351	43,524	21,250
Initial thrust-to-weight ratio desired	[-]	1.27	1.21	1.60
The nozzle expansion ratio of engines	[-]	29.9	25.0	25.1
Sea level I_{sp}	[s]	341	338	338
Dry mass	[kg]	6230	5730	5870
Propellant mass	[kg]	11,400	9250	8470
Total mass	[kg]	17,600	15,000	14,300
Number of engines	[-]	3	2	3

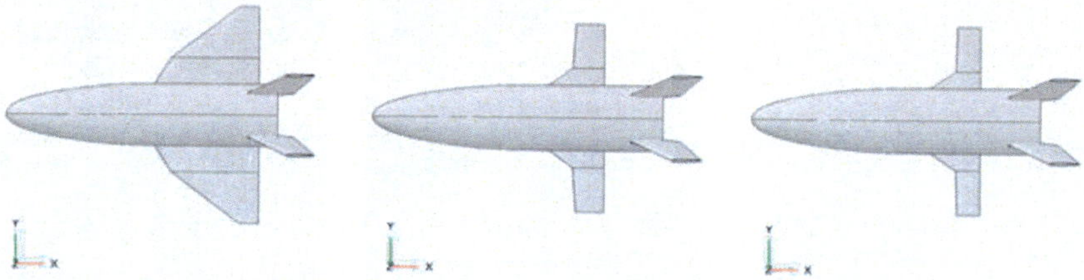

A) Solution No.1 B) Solution No.2 C) Solution No.3

Fig. 9 The airframe of typical optimization solutions

6 Conclusion

It was confirmed that the multi-fidelity surrogate model can evaluate aerodynamics with higher accuracy than the conventional low-accuracy aerodynamic evaluation method. Furthermore, it was confirmed that the model could be implemented in the MDO of suborbital spaceplanes and that optimization calculations could be performed successfully. The future work identified in this study are as follows.

- It is necessary to take into account of static stability and trim conditions in the trajectory calculation. This is because the trim condition is ignored in this optimization. In addition, it is necessary to estimate the aerodynamic characteristics of the shape with the elevon steering.

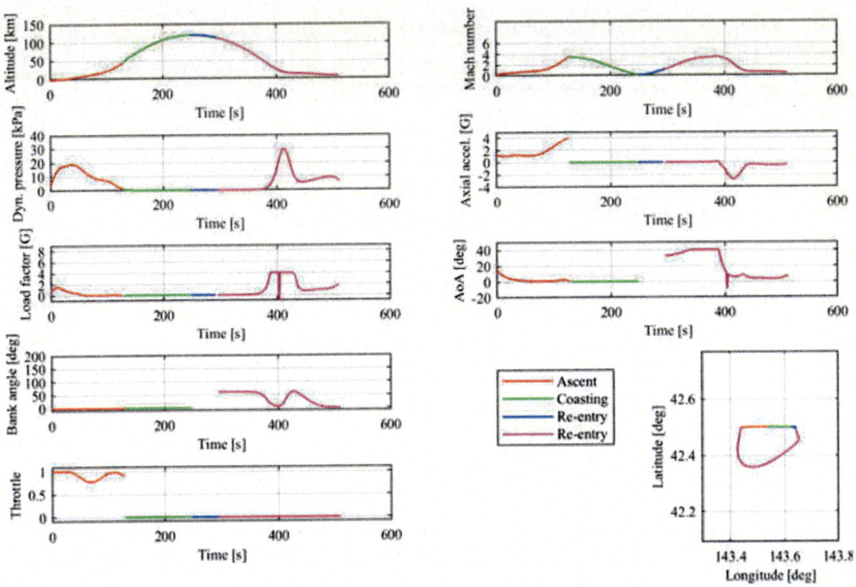

Fig. 10 Flight trajectory of Solution No.1

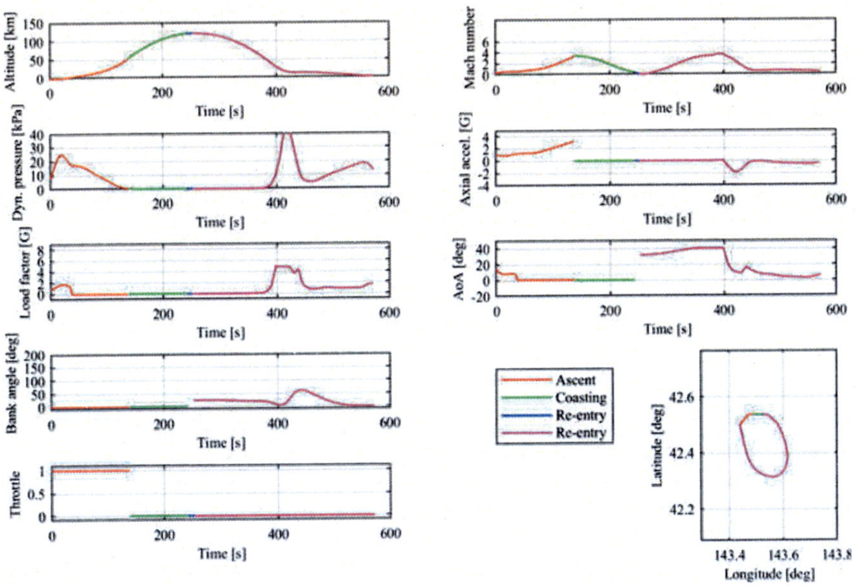

Fig. 11 Flight trajectory of Solution No.2

- The accuracy of the surrogate model for aerodynamics has been confirmed only for the HIMES geometry, and comparisons with other geometries are necessary.
- It is necessary to consider a trajectory calculation method that is stable in the initial generation of optimization.
- In this paper, the surrogate model of the aerodynamic characteristic evaluation method was constructed outside the optimization loop. However, when adding high-accuracy sample points, it is inefficient to do so randomly, and conducting CFD analysis in the loop can improve the solution's reliability. As a result, it is preferable to use MOEA/D-EGO.

References

1. Fujikawa T, Tsuchiya T, Tomioka S (2017) Multidisciplinary design optimization of a two-stage-to orbit reusable launch vehicle with ethanol-fueled rocket-based combined cycle engines. Trans Japan Soc Aeronaut Space Sci 60(5):265–275
2. Jones DR, Schonlau M, Welch WJ (1998) Efficient global optimization of expensive black-box functions. J Global Optim 13:455–492
3. Zhang Q, Liu W, Virginas B (2010) Expensive multiobjective optimization by MOEA/D with Gaussian process model. IEEE Trans Evolut Comput 14(3)
4. Li H, Zhang Q (2007) MOEA/D: a multiobjective evolutionary algorithm based on decomposition. IEEE Trans Evol Comput 11(6):712–731
5. Satterwhite CR (2015) Development of CPanel, an unstructured panel code, using a modified TLS velocity formulation. Master Thesis, California Polytechnic State University, San Luis Obispo
6. Hess RV, Gardner CS (1949) Study by the Prandtl-Glauert method of compressibility effects and critical mach number for ellipsoids of various aspect ratios and thickness ratios, NACATN-1792
7. Anderson JD (2006) Hypersonic and high-temperature gas dynamics (AIAA Education Series)
8. White FM (1974) Viscous fluid flow. McGraw-Hill, Inc., New York, NY, USA, p 639
9. Bonner E, Cloever W, Dunn K (1991) Aerodynamic preliminary analysis system II Part I—Theory. NASA CR-182076, p 58
10. Spalart PR (1992) A one-equation turbulence model for aero-dynamic flows. AIAA Paper 1992–439
11. Lei Z (2005) Effect of RANS turbulence models on computation of vertical flow over wing-body configuration. Trans. JSASS,48 pp 152–160
12. Forrester AIJ, Sóbester A, Keane AJ (2010) Multi-fidelity optimization via surrogate modelling. IEEE Transactions on Evol Comput 14(3)
13. Jones DR (2001) A taxonomy of global optimization methods based on response surfaces. J Global Optim 21:345–383
14. Forrester AIJ, Keane AJ, Bressloff NW (2006) Design and analysis of 'noisy' computer experiments. AIAA J 44:2331–2339
15. Kawaguchi J, Inatani Y, Yonemoto K, Hosokawa S (1987) On the flight control system of a winged space vehicle and physical simulation test. Institute of Space and Astronautical Science report 64:3–199
16. Li H, Zhang Q (2009) Multiobjective optimization problems with complicated pareto sets, MOEA/D and NSGA-II. IEEE Trans Evol Comput 13(2):284–302
17. Harloff GJ, Berkowitz BM (1988) HASA—hypersonic aerospace sizing analysis for the preliminary design of aerospace vehicles. NASA CR–182226
18. NOAA, NASA and U. A. Force (1976) U. S. Standard atmosphere. NASA TM-X-74335

Experimental Validation of Nonlinear Coupled Constitutive Relations in Continuum Flows

Shuhua Zeng, Yunlong Qiu, Zhongzheng Jiang, and Weifang Chen

Abstract The nonlinear coupled constitutive relations (NCCR) model has recently gained great success in stable numerical predictions of high-speed flows over three-dimensional complex flight configurations. However, there is an urgent demand for performing essential experimental measurements to validate the new theoretical model before further promotion to engineering applications. In this paper, the numerical simulations of the NCCR model with Jiang's undecomposed algorithm and Navier–Stokes-Fourier (NSF) equations by in-house code are in comparison with experimental measurements to validate the applicability and accuracy of the NCCR theory in hypersonic flows. The wind tunnel experiments of HB-2 model were conducted at Ma 5 with five different angles of attack ($\alpha = \pm 4°, \pm 2°, 0$) in the Φ120 mm hypersonic wind tunnel. It was found that the flow field simulation results by NSF equations and the NCCR model, including aerodynamic force and shock wave structure position, are consistent with the experimental results. Most importantly, the NCCR model could completely recover the flow field solutions of NSF equations in hypersonic continuum flows. Thus, the accuracy and physical consistency of the NCCR model is critically examined and validated for hypersonic flow in continuum region.

Keywords Hypersonic wind tunnel · Nonlinear coupled constitutive relations · Experimental validation · Undecomposed algorithm

1 Introduction

Hypersonic vehicles, a key field in aerospace engineering, have gained widespread attention in recent years. Developing hypersonic vehicles is and will continue being one of the most crucial technological ambitions as the accurate and fast evaluation of their aerodynamic performance is of great significance [1]. For the design and performance assessment of hypersonic aircraft, hypersonic wind tunnel tests

S. Zeng · Y. Qiu · Z. Jiang · W. Chen (✉)
School of Aeronautics and Astronautics, Zhejiang University, Hangzhou 310027, China
e-mail: chenwfnudt@163.com

S. Lee et al. (eds.), *The Proceedings of the 2021 Asia-Pacific International Symposium on Aerospace Technology (APISAT 2021), Volume 1*, Lecture Notes in Electrical Engineering 912, https://doi.org/10.1007/978-981-19-2689-1_22

and new numerical calculation models and methods applied to computational fluid dynamics (CFD) are emerging rapidly with the development of computers and wind tunnel experimental equipment. The wind tunnel tests and CFD technology can be complementary as the former provides a reference for the examination and verification of the latter, and CFD plays an irreplaceable role in the initial stage of wind tunnel construction and operation as well.

To extend the Navier–Stokes-Fourier Hydrodynamics developed over many years into more widespread areas, a new set of generalized hydrodynamic equations (GHE) proposed by Eu in 1980 from the viewpoint of generalized hydrodynamics formulated via the non-equilibrium ensemble method, provide an alternative to the Boltzmann equation approach to the gas flow modelling [2]. However, these equations are extremely difficult to be solved numerically. Subsequently, Myong developed a set of nonlinear coupled algebraic equations based on GHE, called the non-linear coupled constitutive relations (NCCR) model in 1999 [3]. The main innovation of the NCCR model lies in the transport equations of viscous stress and thermal conduction directly derived from the Boltzmann equations by considering the entropy condition, instead of using the traditional linear Newton's viscosity law and Fourier's conduction law. To date, this new model has gained appreciable results in stable numerical predictions of one-dimensional shock wave structure and some multi-dimensional flows over three-dimensional complex flight configurations [4]. Moreover, many previous research results have already pointed out that the NCCR model could also simulate the flow phenomenon in transition regime [5].

However, the verification and validation of the NCCR model are merely subject to theoretical derivation and numerical comparisons with other kinetic methods and have less experimental proof. Before being applied to engineering applications, some experimental measurements are urgently needed to validate this new theoretical model. To achieve this demand, a hypersonic wind tunnel experiment platform has been built up in the Near Space Vehicle Research Centre of Zhejiang University in recent years.

In the present study, the wind tunnel experiments of some classical hypersonic flow cases are designed and conducted. The numerical simulations of the NCCR model with Jiang's undecomposed algorithm and Navier–Stokes-Fourier (NSF) equations by in-house code were conducted for high-speed flows around the HB-2, which have been already tested in Ma = 5 with five different angles of attack ($\alpha = \pm 4°, \pm 2°, 0$) by the Φ120 mm hypersonic wind tunnel. This aims at validating the applicability and accuracy of the theory in hypersonic continuum flows by comparing with our experimental measurements and the flow field simulation results from the NSF equations.

The rest of this paper is structured as follows. A brief introduction of experimental details is presented in Sect. 2. The NSF equations and NCCR model with corresponding numerical solver are explained in Sect. 3. Section 4 will discuss the general flow-field and aerodynamic force computed by the new model and present a detailed comparison with the linear NSF and our experimental results to assess the accuracy and physical consistency of the model in continuum regime, which are followed by the conclusions in Sect. 5.

2 Experimental Details

Hypersonic wind tunnels are the most essential hypersonic ground test equipment. Generally, wind tunnel with flow Mach number greater than five is called hypersonic wind tunnels, which can work for more than ten seconds. Furthermore, its working medium is limited to dry and pure air as well. The experimental details conducted in the present study are organized as follows.

2.1 Facility

The experiment of hypersonic flows around HB-2 model with varying angles of attack was tested in the Φ120 mm hypersonic wind tunnel, which was installed in the Near Space Vehicle Research Centre of Zhejiang University in 2018. It can achieve the measurement of aircraft aerodynamic/aerothermodynamics and the distribution of surface pressure precisely, as well as effectively implement the research of turbulence structure, shock boundary layer interference and rarefied gas effects. A schematic view of the nozzle and test section of this hypersonic wind tunnel is shown in Fig. 1. The conventional "front blow and back suction" temporary and direct connection layout is adopted in this wind tunnel. Its performance parameters are presented in Table 1.

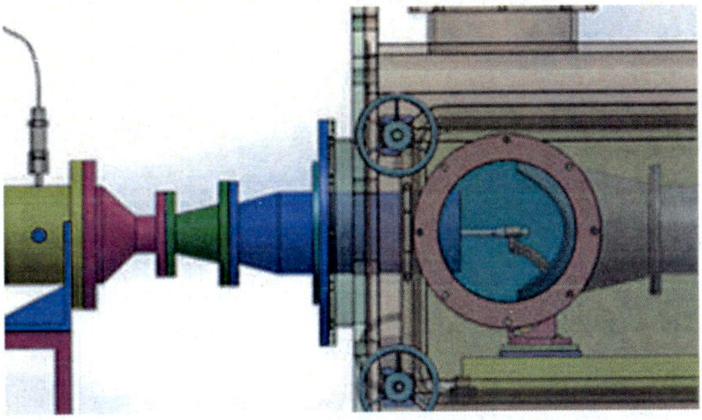

Fig. 1 Nozzle and test section of the Φ120 mm hypersonic wind tunnel in Zhejiang University

Table 1 The performance parameters of Φ120 mm hypersonic wind tunnel

Property of this wind tunnel	
The range of Mach number	5/6/7
The diameter of nozzle outlet	120 mm
The maximum total temperature	500 K
The scope of total pressure	0.1–1.0 MPa
The working time	$t >= 15$ s

2.2 Model Configuration

In the present test, the HB-2 model demonstrated in Fig. 2 is a classic configuration proposed by AGARD and Supersonic Tunnel Association (STA) in the 1950s, and it is often used for verifying the performance of hypersonic wind tunnels [6]. The installation drawing of model is also presented in Fig. 2. The model configuration is axis-symmetric and consists of four parts, a sphere, a cone, a cylinder, and a flare skirt. The overall length is 122.5 mm and the diameter at the rear edge is 40 mm. A six-component balance is set up inside the model, and the balance centre is located at the centre of model, 61.4 mm away from the top. According to Table 2, five experimental conditions for the HB-2 configuration have been considered.

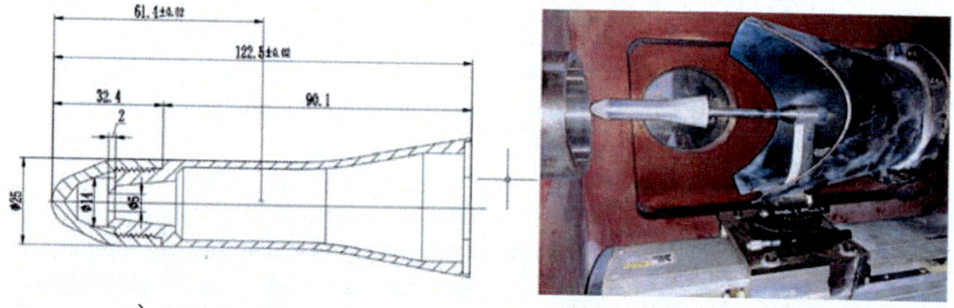

a) HB-2 standard model b) Installation drawing of the model

Fig. 2 The HB-2 standard model and its installation drawing

Table 2 Experimental conditions for the HB-2 model

Case	1	2	3	4	5
Mach number	5	5	5	5	5
Angle of attack (°)	−4	−2	0	2	4
Total pressure (kPa)	451	401	494	346	401
Total temperature (K)	415	417	410	402	417
Static pressure (Pa)	852.4	757.9	933.7	653.95	757.9
Static temperature (K)	69.2	69.5	68.3	67.0	69.5
Velocity (m/s)	833.53	835.54	828.50	820.37	835.54

3 Numerical Method

3.1 Navier–Stokes-Fourier Equations

Conventionally, gas flows in near-equilibrium regime have been theoretically investigated in computational fluid dynamics based on the Navier–Stokes-Fourier (NSF) theory framework, which provides excellent and reliable results for the continuum flows. The NSF equations are made up of three conservation equations (including mass, momentum, and total energy) and other evolution equations of physical variables that we are interested in.

The conservation laws of mass, momentum and total energy for diatomic gases can be given as follows

$$\frac{\partial}{\partial t}\begin{bmatrix}\rho \\ \rho\mathbf{u} \\ \rho E\end{bmatrix} + \nabla\cdot\begin{bmatrix}\rho\mathbf{u} \\ \rho\mathbf{uu}+p\mathbf{I} \\ (\rho E+p)\mathbf{u}\end{bmatrix} + \nabla\cdot\begin{bmatrix}0 \\ \mathbf{\Pi}+\Delta\mathbf{I} \\ (\mathbf{\Pi}+\Delta\mathbf{I})\cdot\mathbf{u}+\mathbf{Q}\end{bmatrix} = 0 \tag{1}$$

in which, t denotes the time and $\nabla\cdot()$ is the divergence.p, $\mathbf{u}$, E and $\mathbf{I}$ denote the pressure, the velocity vector of the fluid, total energy per unit mass and the unit second-rank tensor, respectively. The densityρ, momentum density$\rho\mathbf{u}$, and total energy density ρE are defined as conserved variables; The shear stress$\mathbf{\Pi}$, excess normal stress Δ and heat flux $\mathbf{Q}$ are called non-conserved variables.

However, the Eq. (1) are a multidimensional open system of PDE. For the sake of closing the system so that it can be solved numerically, the expression forms of these non-conserved variables (including $\mathbf{\Pi}$, Δ and $\mathbf{Q}$) need to be given in more detail. Generally, under the linear constitutive framework system of traditional NSF equations,$\mathbf{\Pi}_{NSF}$,Δ_{NSF} and $\mathbf{Q}_{NSF}$ are determined by Newton's law of shear and bulk viscosity and Fourier's law of heat conduction, as given below

$$\begin{aligned}\mathbf{\Pi}_{NSF} &= -2\eta[\nabla\mathbf{u}]^{(2)} \\ \Delta_{NSF} &= -\eta_b\nabla\cdot\mathbf{u} \\ \mathbf{Q}_{NSF} &= -\lambda\nabla T\end{aligned} \tag{2}$$

where T denotes the temperature; η, η_b, λ represent the coefficients of shear viscosity, bulk viscosity and heat conduction, respectively. $[\mathbf{A}]^{(2)}$ is the traceless symmetric part of the second-rank tensor $\mathbf{A}$, and it is defined as

$$[\mathbf{A}]^{(2)} = \frac{1}{2}\left[\mathbf{A}+\mathbf{A}^T\right] - \frac{1}{3}Tr(\mathbf{A})\mathbf{I} \tag{3}$$

According to Stokes' hypothesis in conventional NSF equations, it is necessary to mention that $\Delta_{NSF} = 0$, namely the bulk viscosity is set to zero, is appropriate for both flows in monatomic gases and diatomic gases.

3.2 Nonlinear Coupled Constitutive Relations

To generalize the Navier–Stokes-Fourier Hydrodynamics into more widespread areas, Eu proposed Generalized Hydrodynamic Equations (GHE) based on the Boltzmann equation or Boltzmann-Curtiss kinetic equation [7], which are strictly consistent with the second law of thermodynamics. Using the adiabatic assumption, Eu's closure theory and Myong's simplifications, the 3-D NCCR model for diatomic gases can be summarized below as

$$\begin{aligned} \mathbf{\Pi}_{NCCR} q(\kappa) &= -2(p+\Delta_{NCCR})\tfrac{\eta}{p}[\nabla \mathbf{u}]^{(2)} - 2\tfrac{\eta}{p}[\mathbf{\Pi}_{NCCR}\cdot\nabla\mathbf{u}]^{(2)} \\ \Delta_{NCCR} q(\kappa) &= -\eta_b \nabla\cdot\mathbf{u} - 3\tfrac{\eta_b}{p}(\mathbf{\Pi}_{NCCR}+\Delta_{NCCR}\mathbf{I}):\nabla\mathbf{u} \\ \mathbf{Q}_{NCCR} q(\kappa) &= -\tfrac{\lambda}{p}\mathbf{\Pi}_{NCCR}\cdot\nabla T - \tfrac{\lambda}{p}(p+\Delta_{NCCR})\nabla T \end{aligned} \tag{4}$$

Here,

$$q(\kappa) = \frac{\sinh\kappa}{\kappa}, \kappa = \frac{(mk_B)^{1/4}}{\sqrt{2}d}\frac{T^{1/4}}{p}\left[\frac{\mathbf{\Pi}:\mathbf{\Pi}}{2\eta} + \gamma'\frac{\Delta^2}{\eta_b} + \frac{\mathbf{Q}\cdot\mathbf{Q}}{\lambda T}\right]^{1/2} \tag{5}$$

In this expression, $q(\kappa)$ in Eq. (5) is the nonlinear factor related to entropy production and κ is the first-order cumulant of the cumulant approximation for dissipation terms. m is the mass of a molecule, k_B represents the Boltzmann constant, and d is the mean diameter of the molecule.γ' is equal to $(5-3\gamma)/2$, where γ means the specific heat ratio. The bulk viscosity and rotational effect are not necessarily taken into consideration for monatomic gases, thus the NCCR model for a monatomic gas can be obtained by removing the equation evolution of Δ_{NCCR} and setting the excess normal stress related to bulk viscosity $\Delta_{NCCR} = 0$ in other equations.

By substituting the linear constitutive relations (2) as aforementioned into Eq. (4), the format of NCCR model can be rewritten as [8]

$$\begin{aligned} \mathbf{\Pi}_{NCCR} q(\kappa) &= \mathbf{\Pi}_{NSF} - 2\eta\tfrac{\Delta_{NCCR}}{p}[\nabla\mathbf{u}]^{(2)} - 2\tfrac{\eta}{p}[\mathbf{\Pi}_{NCCR}\cdot\nabla\mathbf{u}]^{(2)} \\ \Delta_{NCCR} q(\kappa) &= \Delta_{NSF} - 3\tfrac{\eta_b}{p}(\mathbf{\Pi}_{NCCR}+\Delta_{NCCR}\mathbf{I}):\nabla\mathbf{u} \\ \mathbf{Q}_{NCCR} q(\kappa) &= \mathbf{Q}_{NSF} - \lambda\tfrac{\Delta_{NCCR}}{p}\nabla T - \tfrac{\lambda}{p}\mathbf{\Pi}_{NCCR}\cdot\nabla T \end{aligned} \tag{6}$$

By contrasting Eq. (2) with the format above, there are two apparently different distinctions between them. One distinction is that the shear stress $\mathbf{\Pi}_{NCCR}$ and excess normal stress Δ_{NCCR} in the NCCR model are nonlinearly relevant to the velocity gradient. Meanwhile, $\mathbf{Q}_{NCCR}$ is also not linearly related to the space derivatives of temperature. And the other is that the nonlinear dissipation factor $q(\kappa)$, as shown in Eq. (5), is a pivotal parameter coupling the gas physical properties (mass, molecular diameter, shear viscosity, bulk viscosity, and thermal conductivity), macroscopic physical variables (temperature and pressure), and non-conserved variables (shear stress, excess normal stress, and heat flux) together. Nevertheless, in conventional Navier–Stokes-Fourier constitutive relations, the non-conserved variables are irrelevant to each other.

The non-conserved variables can be computed explicitly from the first-order derivatives of conserved variables in NSF relations (2) but implicitly in NCCR model (6). Therefore, this paper focuses on the solution of nonlinear coupled constitutive equations by Jiang's undecomposed algorithm [9], where three equations given above are merged into one scalar formulation by using the Rayleigh–Onsager dissipation function as

$$\begin{aligned}\widehat{R}^2 q\left(c\widehat{R}\right) &= \left(1+\widehat{\Delta}\right)\widehat{\Pi}:\widehat{\Pi}_0+\widehat{\Pi}:\left[\widehat{\Pi}\cdot\nabla\widehat{\mathbf{u}}\right]^{(2)}+\frac{2\gamma'}{f_b}\widehat{\Delta}\widehat{\Delta}_0 \\ &+3\gamma'\widehat{\Delta}\left(\widehat{\Delta}\mathbf{I}+\widehat{\Pi}\right):\nabla\widehat{\mathbf{u}}+\left(1+\widehat{\Delta}\right)\widehat{\mathbf{Q}}\cdot\widehat{\mathbf{Q}}_0+\widehat{\Pi}:\widehat{\mathbf{Q}}_0\widehat{\mathbf{Q}}\end{aligned} \tag{7}$$

where,

$$\begin{gathered} f_b=\frac{\eta_b}{\eta},\ c=\frac{(mk_B)^{1/4}}{2d_\infty}\frac{T_\infty^{1/4}}{\eta_\infty^{1/2}},\ \widehat{R}=\left[\widehat{\Pi}:\widehat{\Pi}+2f_b\gamma'\widehat{\Delta}^2+\widehat{\mathbf{Q}}\cdot\widehat{\mathbf{Q}}\right]^{1/2}, \\ \frac{\eta_\infty}{\rho_\infty a_\infty L_0}=N_\delta=\frac{\mathrm{Ma}}{\mathrm{Re}},\ \widehat{\Pi}=\frac{N_\delta}{p}\Pi,\ \widehat{\Delta}=\frac{N_\delta}{p}\Delta,\ \widehat{\mathbf{Q}}=\frac{N_\delta}{p}\frac{\mathbf{Q}}{\sqrt{T/(2\varepsilon)}}, \\ \nabla\widehat{\mathbf{u}}=-2\eta\frac{N_\delta}{p}\nabla\mathbf{u},\ \varepsilon=\frac{1}{\mathrm{Pr}(\gamma-1)} \end{gathered} \tag{8}$$

And $\widehat{\Pi}_0$, $\widehat{\Delta}_0$ and $\widehat{\mathbf{Q}}_0$ in the above equation are calculated by NSF relations (2). The fixed-point iterative method sometimes makes it hard to construct a convergent iterative expression for Eq. (7), thus we coupled the fixed-point iterative formulations and Newton iterative formulations together to overcome this deficiency in this work.

4 Results and Discussion

In this section, the hypersonic flows around the type HB-2 model as aforementioned with five different angles of attack at Ma 5 using the nonlinear coupled constitutive relations (NCCR) are presented and then compared with experiments and NSF validation data to assess and validate the capability and physical consistency of the NCCR model in hypersonic continuum flows. The working gas is assumed to be pure air, and the specific inputs for the free-stream conditions are listed in Table 2, which is the same as the experimental conditions conducted in the Near Space Vehicle Research Centre of Zhejiang University.

The comparisons of the lift and drag coefficients of HB-2 model obtained by experimental measurements and numerical methods, consisting of the NSF equations and NCCR model, are made in Fig. 3, respectively. It is easily observed that the lift and drag coefficients by NCCR model are both in good agreement with NSF equations and experimental results. Because of the various unpredictable factors including the

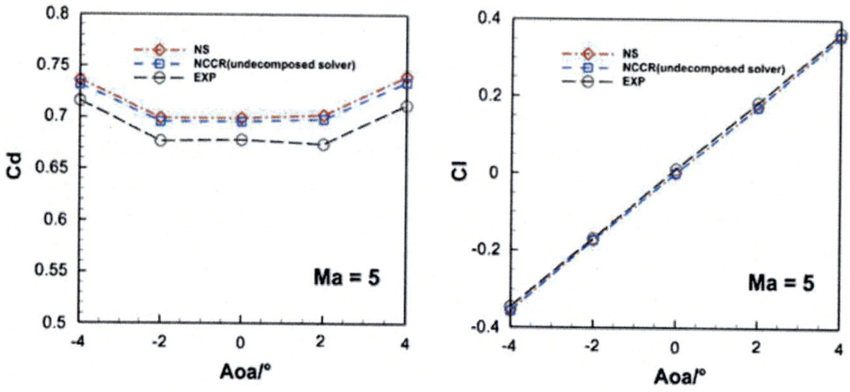

Fig. 3 Drag (left) and lift (right) coefficients of HB-2 model versus angle of attack at Ma 5

balance and test instrument accuracy, model scale and manufacturing deviation and so on, the aerodynamic measurement results might be affected in a wind tunnel. The maximum deviation of the drag coefficients between CFD and experimental results is within three percent, and it is known as an acceptable fact.

From the experimental schlieren photographs shown in Fig. 4, there is a strong detached shock wave forming at the head of HB-2 model and an attached shock wave taking place in the flare skirt. Meanwhile, two expansion waves appear at the near-wake flow and the junction of the head and cylinder, respectively. The positions of these shock waves and expansion waves are both accurately captured by the previously described numerical methods, as shown in Fig. 5. Moreover, the detailed contours of Mach number, density and streamline computed by the NSF equations and NCCR model are nearly consistent, except for some weak discrepancies of Mach number contours and streamline arising in near-wake region.

To discuss these hypersonic flows in detail, Boyd et al. [10] proposed a continuum breakdown parameter, namely the maximum gradient length local Knudsen number,

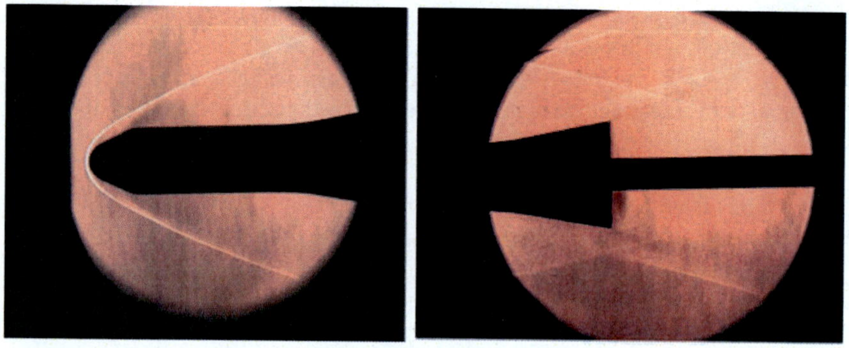

Fig. 4 Experimental schlieren photographs at Ma 5

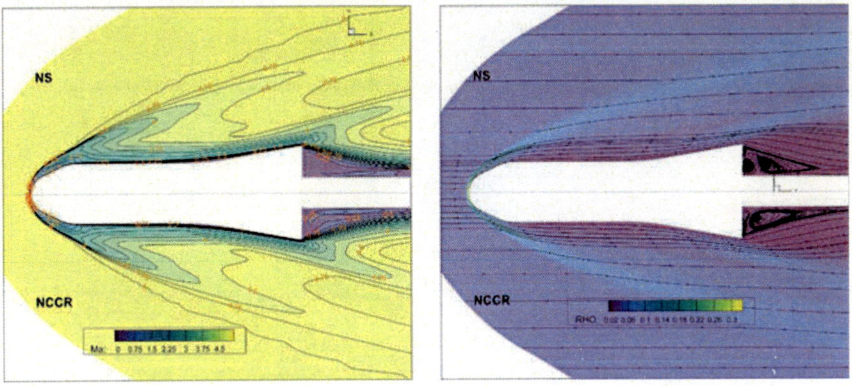

Fig. 5 The detailed contour comparisons of Mach number (left), density and streamline (right) between NSF and NCCR model (Ma = 5, $\alpha = 0$, $T_w = 300\,K$, symmetrical plane Z = 0)

which is given as below

$$\mathrm{Kn}_{GLL} = \frac{\lambda}{Q}\left|\frac{dQ}{dl}\right| \tag{9}$$

where the derivative is taken in the direction of the maximum gradient, and Q is a variable of interest such as density, pressure or temperature. It has been suggested that a value of Kn_{GLL} above 0.05 is identified as a continuum breakdown point [11], hence the flows around the model can be easily determined as continuum flows in Fig. 6a. The flow variables including Mach number, density and temperature along the stagnation line are also presented in Fig. 6. These curve profiles depict a strong shock computed by the NSF equations and NCCR model in hypersonic region, and the NCCR solution computed by undecomposed solver yields results in great agreement with the NSF solution.

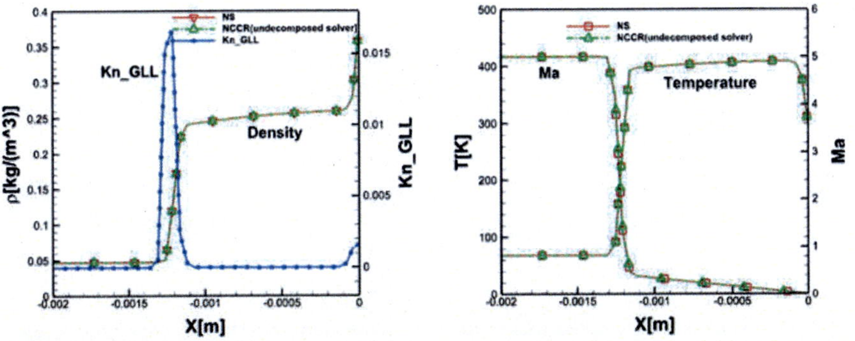

a) density and gradient length local Knudsen

b) Mach number and temperature

Fig. 6 The profiles along the stagnation line (Ma = 5, $\alpha = 0$)

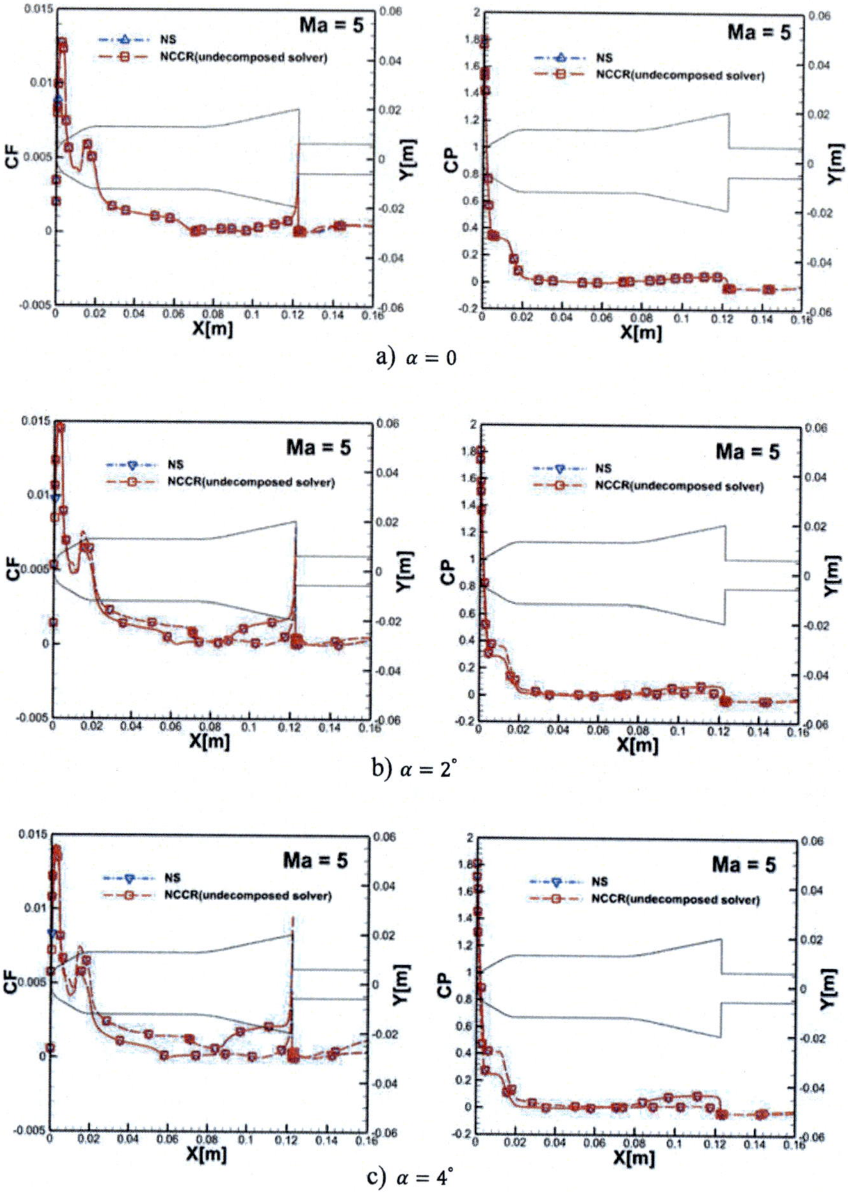

Fig. 7 Comparisons of surface friction (left) and surface pressure (right) coefficients with the NSF and NCCR model (Ma = 5, $T_w = 300\ K$, symmetrical plane Z = 0)

Figure 7 shows the comparisons of the surface properties (skin friction coefficient $CF = \tau/(0.5\rho_{\infty}\mathbf{u}_{\infty}^2)$, and pressure coefficient $CP = (p - p_{\infty})/(0.5\rho_{\infty}\mathbf{u}_{\infty}^2)$ along the model symmetrical plane Z = 0 with three different angles of attack. Note that the discrepancies of the flow profiles along the HB-2 model surface between NSF and NCCR are totally non-existent, which indicates that the NCCR model gives excellent results in continuum region just as the NSF model. We can also clearly find that the difference of the surface aerodynamic properties in the windward and leeward gradually rises as the angle of attack increases, especially in the flare skirt of HB-2 configuration.

5 Conclusions

The main emphasis of this paper is the experimental validation of nonlinear coupled constitutive relations (NCCR) in hypersonic continuum flows. The numerical simulations of NCCR model with undecomposed solver and Navier–Stokes-Fourier (NSF) equations by in-house code were conducted for high-speed flows around the HB-2 model, which have been already tested in Ma = 5 with five different angles of attack ($\alpha = \pm 4^\circ, \pm 2^\circ, 0$) by Φ120 mm hypersonic wind tunnel built up in recent years in the Near Space Vehicle Research Centre of Zhejiang University. By comparing with our experimental measurements and the flow field simulation results from NSF equations, the applicability and accuracy of NCCR theory in hypersonic continuum flow is thoroughly examined and validated.

Specifically, the lift and drag coefficients calculated by NSF and NCCR are in good agreement with the experimental measurements. Besides, the surface properties and distribution of variables in the flows around the HB-2 model indicate an excellent agreement between NSF and NCCR model. These results show that NCCR model can obtain excellent performance in continuum region just like NSF model. On the other hand, the hypersonic wind tunnel developed by the Near Space Vehicle Research Centre of Zhejiang University has been able to provide accurate and effective aircraft aerodynamic data. Moreover, many previous studies have already highlighted that NCCR model could also simulate the flow phenomenon in transition regime. Therefore, the NCCR model undoubtedly has great application value and potential for the accurate and rapid calculation of aerodynamics/heat of hypersonic vehicles in hypersonic and continuum-rarefaction coupled flows.

Acknowledgements This research was supported by the National Natural Science Foundation of China (Grant No. 12002306).

References

1. Bertin JJ, Cummings RM (2003) Fifty years of hypersonics: where we've been, where we're going. Prog Aerosp Sci 39(6–7):511–536
2. Eu BC (1980) A modified moment method and irreversible thermodynamics. J Chem Phys 73(6):2958–2969
3. Myong RS (1999) Thermodynamically consistent hydrodynamic computational models for high-Knudsen-number gas flows. Phys Fluids 11(9):2788–2802
4. Jiang Z, Zhao W, Yuan Z, Chen W, Myong RS (2019) Computation of hypersonic flows over flying configurations using a nonlinear constitutive model. AIAA J 57(12):5252–5268
5. Xiao H, Tang K (2017) A unified framework for modeling continuum and rarefied gas flows. Sci Rep 7(1):13108
6. Kodera M, Tanno H, Takahashi M, Komuro T, Satoh K, Itoh K (2009) Comparison of re-entry flow computations with high enthalpy shock tunnel experiments. In: 16th AIAA/DLR/dglr international space planes and hypersonic systems and technologies conference
7. Curtiss CF (1981) The classical Boltzmann equation of a gas of diatomic molecules. J Chem Phys 75(1):376–378
8. Yuan Z, Zhao W, Jiang Z, Chen W (2021) Numerical simulation of hypersonic reaction flows with nonlinear coupled constitutive relations. Aerospace Sci Technol 112
9. Jiang Z (2019) An undecomposed hybrid algorithm for nonlinear coupled constitutive relations of rarefied gas dynamics. Commun Comp Phys 26(3):880–912
10. Boyd ID, Chen G, Candler GV (1995) Predicting failure of the continuum fluid equations in transitional hypersonic flows. Phys Fluids 7(1):210–219
11. Holman T, Boyd I (2008) Numerical investigation of the effects of continuum breakdown on hypersonic vehicle surface properties. In: 40th Thermophysics Conference

A Propeller Evaluation and Selection Tool for Multicopter and VTOL Design

Franco Maurice Staub, Yuji Shimizu, Dai Tsukada, Shosuke Inoue, Emery Premeaux, Chris Raabe, and Takeshi Tsuchiya

Abstract Along with the emergence of electric vertical takeoff and landing (eVTOL) aircraft arose the need to predict propeller performance, namely thrust and power, at unconventional free stream angles. For cases where this angle is either 0° (hover) or 90° (forward flight), it is not difficult to estimate the performance. However, in the case of long-range eVTOLs, where the free stream angle is usually in between these angles, the propeller behaves differently. Blade element theory (BET) models can account for this but are impractical due to their large number of geometric parameters. As a substitute, we present an empirical model built on wind tunnel data that uses only propeller diameter and pitch to define the geometry, while capturing the effects of free stream angle on thrust and power. Herein, we discuss the experiment used to gather data for our model, its mathematical derivation, validation against an in-house BET model, and limitations. We end with an example demonstrating how to apply the model to solve the non-trivial problem of selecting an appropriate propeller for a tiltrotor aircraft with maximum range.

Keywords Propeller · Model · Empirical · eVTOL · Multicopter · Design

1 Introduction

Over the past two decades, electric vertical takeoff and landing (eVTOL) aircraft have emerged in the marketplace for a plethora of applications. Most of these vehicles are of the multicopter type, which—as the name suggests—attain flight and control using multiple propellers. These propellers are almost always fixed-pitch, making them mechanically simple and lightweight. Insofar as the operating flight conditions of the aircraft do not vary too much, there is usually some *pitch ratio* (r)—which is the blade's pitch divided by the blade diameter—that optimizes performance.

F. M. Staub (✉) · Y. Shimizu · T. Tsuchiya
The University of Tokyo, Tokyo, Japan
e-mail: francostaub@g.ecc.u-tokyo.ac.jp

D. Tsukada · S. Inoue · E. Premeaux · C. Raabe
Autonomous Control Systems Laboratory, Tokyo, Japan

S. Lee et al. (eds.), *The Proceedings of the 2021 Asia-Pacific International Symposium on Aerospace Technology (APISAT 2021), Volume 1*, Lecture Notes in Electrical Engineering 912, https://doi.org/10.1007/978-981-19-2689-1_23

For hovering operations, where the propeller experiences steady, axial, and mostly attached flow, the ideal pitch ratio is about 0.25–0.33, with the lower value usually being reserved for "heavy-lift" drones. Thanks to the well-behaved flow, it is not difficult to estimate a propeller's performance, namely its thrust and power, for these operations.

One option is to use a simplified blade-element or simplified momentum model [1]. If this approach is too complicated, we can also use online tools such as [2], which estimates static thrust within 26%, eCalc™, which requires more vehicle information, or propeller manufacturer datasheets like those found on the T-motor™ website. Alternatively, The University of Illinois Urbana-Champaign (UIUC) provides three volumes of propeller thrust and power data [3–5]. All these approaches are practical because they do not demand much technical knowledge about propeller aerodynamics. They also happen to work for forward flight propellers, such as those on fixed wing aircraft. However, their drawback is that they are unreliable for situations where the angle between the free stream velocity and the propeller disc plane, herein referred to as the *free stream angle* (θ), is neither 0° nor 90°. In recent developments, especially for long-range eVTOLs, we are witnessing more aircraft which operate in this intermediate range.

Delivery multicopters often tilt forward about 30° during cruise. Tiltrotors and tailsitters fly through a flight mode transition, in which the propellers experience high free stream angles. In some cases, an aircraft may even cruise at such flight conditions [6–8]. As will be shown by our wind tunnel data, thrust and power for a given RPM and diameter will vary significantly with free stream angles and free stream velocity. This fact is corroborated by [9, 10], and cannot be neglected when selecting an appropriate propeller for long-range eVTOL aircraft. Unfortunately, to evaluate the thrust and power of propellers in these conditions, we need to increase the model fidelity. Usually this entails using blade element theory (BET) based models as discussed in [1], which may be unnecessarily detailed because the user effectively needs to design a propeller geometry from scratch; and impractical because that propeller geometry as designed probably does not exist in the market.

As a solution, we created an empirical model from extensive wind tunnel data that captures propeller thrust and power dependencies on the free stream angle and free stream velocity. Specifically, it defines thrust and power coefficients as functions of advance ratio, which is the ratio of the free stream velocity to the propeller tip speed, and free stream angle, which was defined above. The user only needs to define six parameters to attain thrust and power: propeller diameter, propeller pitch, rotational velocity, air density, free stream velocity and free stream angle (Fig. 1).

Diameter and pitch are the only parameters used to define propeller geometry, and are considered highly practical for designers because this is how propellers are defined as products. Although this is a simplification, it is known that propellers' thrust and power coefficients can be approximated according to the twist angle at the three-quarter radial position, which we relate to the propeller's pitch ratio. This relation seems to hold true for propellers of interest in multicopter and eVTOL design—namely fixed pitch propellers with a pitch ratio between 0.25 and 0.65. Beyond a pitch ratio of 0.65, the trends observed for this model diverge; and hovering becomes

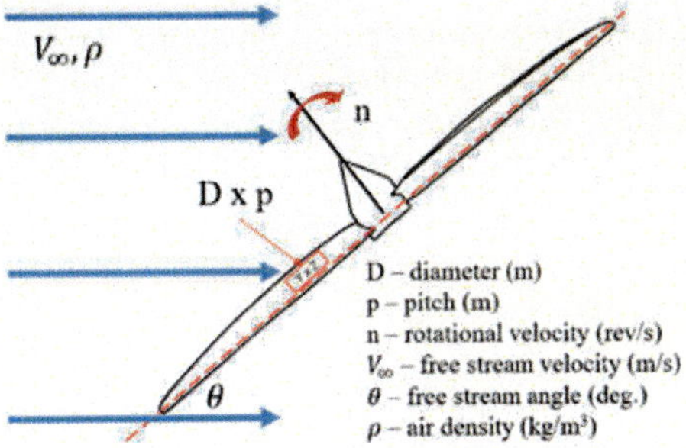

Fig. 1 Schematic of six parameters required to solve for the thrust and power of a propeller using the model in this study

highly uneconomical from an energy consumption standpoint. We emphasize that this model is not intended for propeller design, but for initial full-vehicle design. It should be used as a first estimate of which propeller is suitable for a given aircraft, taking into consideration its expected operating conditions.

In the Methodology section, we will discuss the wind tunnel procedure used to gather data for the model and gloss over the theory of the BET model used for our validation. Following that, we will share some data from the wind tunnel experiment, as well as the mathematical derivation of our model in the Results section. In the Discussion section, we compare the results of this model against the BET model, discuss its limitations, and demonstrate how it can be utilized to optimize the propeller pitch of a tiltrotor delivery drone. We end with a summary, conclusions, and suggestions for future work.

2 Methodology

2.1 Wind Tunnel Procedure

The first step towards developing our model was gathering the necessary empirical data. This was accomplished by testing five propellers of varying pitch and diameter (Table 1) at the Japan Aerospace Exploration Agency's (JAXA) low-speed wind tunnel facility (LWT 1). While we managed to test a variety of diameters, time and hardware constraints prevented us from testing a large propeller with a pitch ratio

Table 1 Propellers used for wind tunnel tests. In this paper, we share the results of the 15 × 5 and 15 × 8 propellers

Size	Brand	Material	Pitch ratio
15 × 5	T-Motor	Carbon fiber + epoxy	0.33
18 × 6.1	T-Motor	Carbon fiber + epoxy	0.34
15 × 8	Falcon	Carbon fiber + epoxy	0.53
28 × 9.2	T-Motor	Carbon fiber + epoxy	0.33
30 × 10	T-Motor	Carbon fiber + epoxy	0.33

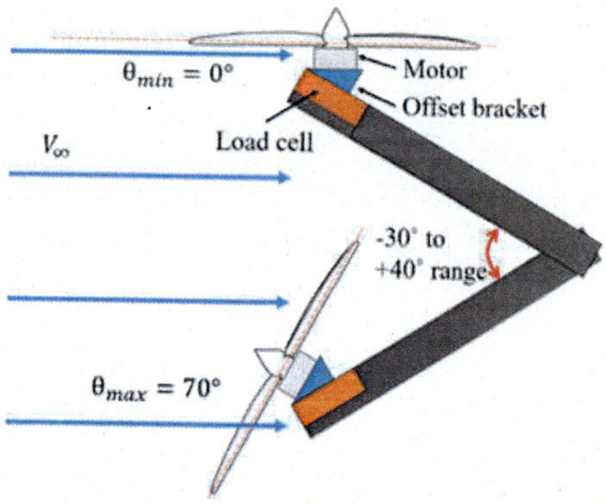

Fig. 2 Schematic of the experiment set up used to gather data for various free stream angles and advance ratios

greater than 0.33. All propellers had two blades and were constructed out of carbon fiber.

The test procedure for each propeller was as follows. We mounted each propeller to a 40° offset bracket because JAXA's LWT 1 sting sweep is limited to 70° (−30° to +40°). This bracket was attached to a 6-axis load cell at the end of the sting. Then we swept the sting so the freestream angle varied from 0° to 70° in increments of 10° (Fig. 2). At each increment, we ran an automated, PWM step input program to increment the propeller rotational speed. There were nine steps to the step input, and at each step the rotational speed was measured using a light sensor and reflection tape attached to the motor. The current, voltage, load cell forces and moments were also measured at each step. This process was repeated for airspeeds 0, 10, 15, and 20 m/s.

2.2 *Blade Element Theory Model*

Our empirical model was crafted not just from our experimental data, but also by referencing data from [3–5, 9–11]. While it was important that our model agreed with these experimental data, we wanted to ensure that the extrapolated regions and general trends of our model were accurate as well. To this end, we compared our *data*-based model against a *physics*-based, in-house BET model.

Many physics-based, aerodynamic models of rotors have been developed and their validity has been proven to a great extent by comparison against experimental results [12, 13] Among these, we used a combination of momentum theory and BET, which is a popular way to estimate the thrust, torque and power of a rotor at both low and high frequencies [14]. Momentum theory estimates the aerodynamic force acting on a rotor from the change in momentum between upstream and downstream air flow. Thrust is generated by accelerating a large mass of air through a relatively small change in velocity. In momentum theory, this thrust can be calculated as the product of inflow mass and downstream velocity or as the product of the rotor disc area and the change in air pressure across that disc. BET is a way to calculate forces and moments of the propeller by discretizing it into small, wing-like elements and integrating the forces and moments along the entire blade and over one rotation. BET needs information about the induced velocity because the induced velocity relates to the angle of attack faced by each element. Therefore, momentum theory and BET work in harmony to solve the total propeller thrust, power, and torque.

In our BET implementation, we made the following assumptions. The propeller had a linear twist angle. The airfoils for each blade element had a constant drag coefficient and a constant lift-curve slope, hence stall and induced drag are not captured. Lastly, the magnitude of the vertical free stream velocity at the rotation plane was much smaller than the that of the horizontal component. Unfortunately, a mathematical description of our BET implementation is beyond the scope of this paper, so we refer the reader to [15], which was the main source for developing our BET model.

3 Results

3.1 Wind Tunnel Data

For the sake of brevity, we have decided to share only the results for the 15 × 5 and the 15 × 8 propellers, as the data from these propellers most clearly illustrates the trends that were observed for the other propellers. To standardize our data, we non-dimensionalized it using the following equations and units for thrust coefficient, power coefficient, and advance ratio:

$$C_T = \frac{\mathrm{T}}{\rho n^2 D^4} \tag{1}$$

$$C_P = \frac{\mathrm{P}}{\rho n^3 D^5} \tag{2}$$

$$\mu = \frac{V_\infty}{nD} \tag{3}$$

T is thrust in Newtons, P is power in Watts, ρ is density in $\frac{kg}{m^3}$, n is angular velocity in $\frac{rev}{s}$, D is the propeller diameter in meters, and V_∞ is the freestream velocity in $\frac{m}{s}$.

Figure 3 shows the thrust and power coefficients of the 15 × 5 propeller as functions of advance ratio for various free stream angles. The trends agree with results found in [10], showing that the slopes of C_T-μ curves decrease with increasing free stream angle. The y-intercept of all these curves represents the static thrust. We can observe that the disparity in thrust between free stream angles increases dramatically with advance ratio, especially for advance ratios above 0.2.

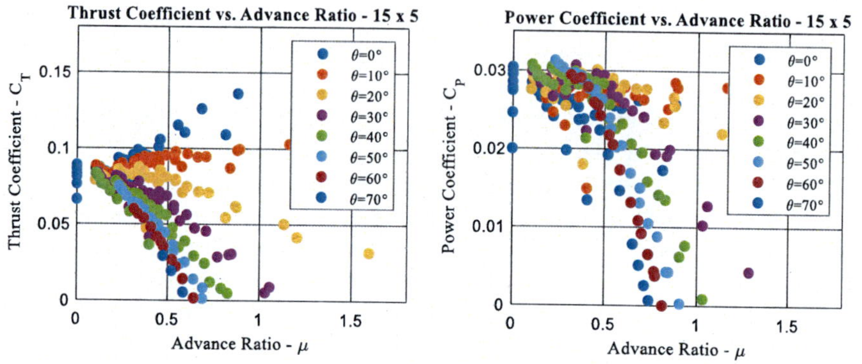

Fig. 3 Thrust and power coefficients of the 15 × 5 propeller as a function of advance ratio for various free stream angles

The power coefficient also shows the decreasing slopes with increasing free stream angle. Unfortunately, the data is noisier in comparison. This is because power was calculated from the load cell moment measurements, which were smaller and noisier than the axial force measurements used for the thrust coefficient. We can also see that, unlike for the thrust coefficient, there is no positive slope for low free stream angles.

Figure 4 shows the same results, but for the 15 × 8 propeller. As expected, the steeper pitch can create thrust at higher advance ratios for high free stream angles. When we increase the propeller pitch, we also expect that the power consumption and thrust will increase across all advance ratios. We can observe this phenomenon when comparing the 15 × 8 data with the 15 × 5 data, though it is a relatively small increase. The increase in static thrust is particularly hard to discern due to the large range of static thrusts for the 15 × 5 propeller. Another contrast is that we can see positive-slope regions of the C_P-μ data where the free stream angle is close to 0°.

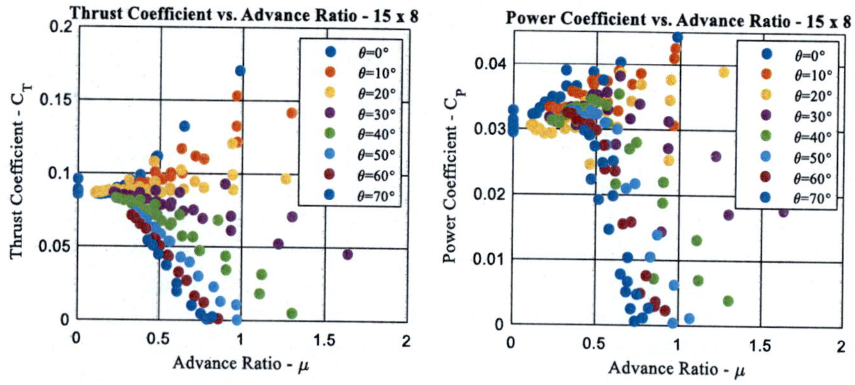

Fig. 4 Thrust and power coefficients of the 15 × 8 propeller as a function of advance ratio for various free stream angles

3.2 Mathematical Formulation

Our model formulation is intuitive, as it revolves around identifying physically meaningful data points and then performing second order polynomial fits to those points. These second order polynomials describe the relationship between advance ratio and thrust/power coefficient for a given free stream angle. We assume the polynomial is symmetric about the y-axis, which means we only needed two points to define a 2nd order polynomial. The first point is where advance ratio is zero (static thrust), and the second point is one out of the series of points at or directly above where the propeller reaches the windmill (zero thrust) state at a 90° free stream angle. As will be shown, this point is a function of free stream angle.

The power coefficient is obtained in an identical manner, but with different coefficients for the polynomials. To avoid confusing the polynomial coefficients with the thrust and power coefficients, we will refer to the former as parameters instead of coefficients from here onward. First, we will derive the general form of the "X coefficient" vs. advance ratio. Then, we share the parameters specific to the thrust coefficient, and those specific to the power coefficient in a table at the end. To convert the general "X coefficient" to the thrust and power coefficient equations, replace the X subscript in the following equations with either power (P) or thrust (T) and plug in the corresponding parameters from the table at the end of the section.

First, we define the first point of interest, the static thrust/power, as a function of pitch ratio. A higher pitch ratio yields a higher static thrust/power. This relationship seems to be quadratic up to a pitch ratio of about 0.65, hence our model upper limit. Our equation for this relationship is as follows:

$$C_{X0} = Ar^2 + Br + C \tag{4}$$

where C_{X0} is the static thrust/power coefficient, and r is the propeller pitch ratio.

The next point of interest is the advance ratio at which the propeller begins to windmill (where thrust goes to zero), specifically when the free stream angle is 90°. This value will be denoted by μ_{w90}, where the w subscript indicates windmill, and the *90* indicates the free stream angle. Like Eq. (4), this value is determined explicitly by the propeller's pitch ratio. The relation is as follows:

$$\mu_{w90} = Dr^2 + Er + F \tag{5}$$

Using these two equations alone, we can actually derive the thrust/power coefficient polynomial for the case for a fixed wing aircraft (where $\theta = 90°$). However, we wish to add to these results so we can determine the thrust/power coefficient polynomials at non-ninety-degree free stream angles.

We do this by adding one more function that describes the movement of the thrust/power coefficient at the advance ratio μ_{w90} as a function of free stream angle, θ, written as $C_X|_{\mu=\mu_{w90}}(\theta)$. This is probably the most complicated function due to its relatively high nonlinearity. However, the strategy for deriving this function is the same as before: first, we find points of physical meaning from the data, and then we fit a curve to them. In this case, the curve is a 3rd order polynomial. We chose four points to uniquely define this 3rd order polynomial; where $\theta = 0°$, θ where $\frac{\partial C_T}{\partial \mu} = 0$, where $\theta = 70°$, and where $\theta = 90°$. For $\theta = 0°$, we defined $C_X|_{\mu=\mu_{w90}}$ as follows:

$$C_X|_{\mu=\mu_{w90}}(0) = Gr + H \tag{6}$$

The free stream angle for which the thrust/power coefficient curve vs. advance ratio slope is 0, $\theta|_{\frac{\partial C_X}{\partial \mu}=0}$, also seems to be a function of the propeller pitch ratio.

$$\theta|_{\frac{\partial C_X}{\partial \mu}=0} = Ir + J \tag{7}$$

Equations (6) and (7) may be nonlinear in reality, but we only have two points from the wind tunnel experiment to describe this relation, so we assume linearity for now.

By definition,

$$C_X|_{\mu=\mu_{w90}}\left(\theta|_{\frac{\partial C_X}{\partial \mu}=0}\right) = C_X|_{\mu=0} = C_{X0} \tag{8}$$

which was defined in Eq. (4). At $\theta = 70°$,

$$C_X|_{\mu=\mu_{w90}}(70) = \mathrm{K} \cdot C_X|_{\mu=\mu_{w90}}(0) \tag{9}$$

and at $\theta = 90°$,

$$C_X|_{\mu=\mu_{w90}}(90) = L \tag{10}$$

Now that we have defined 4 points, we can use MATLAB to create 3rd order polynomials that describe $C_X|_{\mu=\mu_{w90}}$ as a function of θ, which also defines how the 2nd order polynomials for the thrust/power coefficient vs. advance ratio vary with θ. Thus, our model is complete. The last step is to define the parameters of the above functions for the case of the thrust coefficient and for the case of the power coefficient. Please refer to the Table 2 for these values.

4 Discussion

4.1 Comparison of Models

Using this working model, we can plot and compare the results for our model against those of the BET model. The top two plots of Fig. 5 show $C_T(\mu)$ and $C_P(\mu)$ of a

Table 2 Summary of parameters corresponding to the generalized form of the power and thrust coefficient functions defined above

Parameter	C_T	C_P	Parameter	C_T	C_P
A	−0.0617	0.1353	G	0.01032	0.081
B	0.1451	−0.0555	H	0.077	−0.0009
C	0.0459	0.0289	I	28.571	23.810
D	0.0747	0.0747	J	5.857	22.381
E	0.827	0.827	K	0.075	0.0075
F	0.2776	0.2776	L	0	0.005

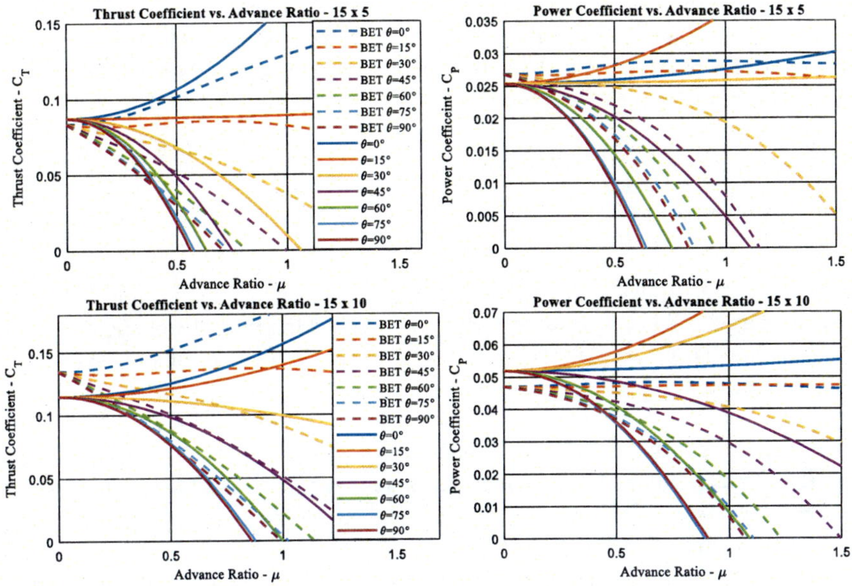

Fig. 5 Comparison of BET model and model derived in the present study. The general trends of the two models agree, but outputs differ in some regions

propeller with a pitch ratio of 0.33 for a variety of free stream angles, θ. The BET model had a root pitch of 19° and a twist angle of −12.5°. The bottom two plots show the same comparison, but for a pitch ratio of 0.67. In this case, the BET model had a root pitch of 26° and a twist of −16.5°. Note that we did not have data for the exact twist distributions of these propellers, so we referred to [10], which shows this data for propellers of similar pitch ratio as our two examples here. The solid lines show our model, and the dashed lines show the BET model results for similar propellers.

The general trends of the two models agree fairly well except where the free stream angle is low and the advance ratio is high, where our model diverges to unrealistically large values. This was considered unimportant as it is very rare for propellers to experience this free stream condition—especially as a steady-state flight condition. If an aircraft is flying fast (high advance ratio), it will have high drag, and therefore the propellers will need to point more aggressively into the wind (high free stream angle) to counteract this drag.

Another point of disagreement is in the static thrust/power values. Because the BET model does not account for stalled propellers, we believe it may overestimate the static thrust for propellers with high pitch ratios. Also, we observe that either our model is conservative or the BET model is generous in predicting the windmill advance ratio. As discussed earlier, we had a limited data base from which we built up the model; and because a specific combination of diameter and pitch can consist of different twist and chord distributions, they can produce a range of thrust and power coefficients despite identical flow conditions. It is possible that our selection of propellers used to create the model lie in the more conservative end of that range. A deeper cross-reference with the UIUC data set, or additional data would reveal which model is more accurate. In general, it is good to be conservative during early design stages.

4.2 *Limitations*

From the plots above, it should be clear that we cannot trust this model for free stream conditions in which both the advance ratio is high and the free stream angle is low. Also, this model is not valid for negative advance ratios, where the rotational velocity or the free stream velocity is negative. It may be possible to rotate the coordinate system to create a quasi-3D model, but we have not attempted this. In general, we also discourage users from applying this model for transient analyses. As will be shown later, rapid changes in model parameters can induce oscillations in the results. Lastly, we remind the user that this model is meant to represent the average propeller of the specified diameter and pitch. In reality, a specific diameter and pitch can consist of a variety of materials, chord and/or twist distributions. Results from this model will likely differ from real propellers of the same pitch and diameter, but they can be helpful for identifying design trends and for pointing the user towards an appropriate first guess.

In summary, our recommendation is to apply this model for 2D, steady-state analyses—which are usually most relevant during the full-vehicle design stage. We demonstrate this with an example next.

4.3 *Example Use Case*

Let us look at an example that demonstrates the importance of selecting an appropriate propeller for a tiltrotor delivery drone. The vehicle in question was designed in NASA's OpenVSP, and is shown in Fig. 6 with the rotors at a free stream angle of 90° on the left and 0° on the right. The rotors can also rotate to any angle in between.

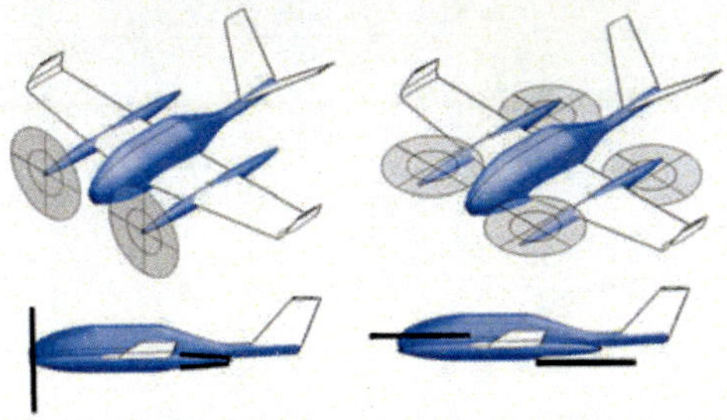

Weight (5 kg payload)	16 kg
Propeller diameter	61 cm (24 ") x4
Battery	10,000 mAh 6S x2
Wing area (S_{ref})	0.858 m^2
C_{Lmax} (at α=10°)	1.0

Fig. 6 Overview of tiltrotor delivery drone configuration with basic specifications. All four propellers can tilt 90°, and the rear propellers shut down during forward flight for better efficiency

The propeller diameter is constrained to 24 inches (61 cm) by the airframe geometry, so the task is to optimize the propeller pitch for best range, given a 5-min hovering requirement.

For speeds less than the stall speed, the aircraft flies at 10° angle of attack, which yields the maximum lift coefficient. In this regime, the propeller angle, theta, is determined by:

$$\theta = \tan^{-1}\left(\frac{D}{W - L}\right) \tag{11}$$

where D is the drag, W is the vehicle weight, and L is the lift at that speed. Lift and drag coefficients were calculated using NASA's VSPAERO. The magnitude of the thrust is then solved for so that horizontal and vertical forces are balanced. Note that pitching moments are not considered in this analysis, but can be if so desired.

$$\mathrm{T} = \frac{D}{\sin(\theta)} \tag{12}$$

When flying faster than the stall speed, the angle of attack is adjusted so the lift cancels out the weight; and the thrust, angled at a free stream angle of 90°, cancels out the drag at that angle of attack. Since it is more efficient to fly with only two rotors in this operating regime, we assume the back rotors are folded away (Fig. 6).

The equilibrium forces, as well as the free stream angle of the propellers, are plotted as a function of cruise airspeed in Fig. 7. Using the thrust, free stream angle (θ_{prop}), free stream velocity, propeller diameter, and sea level air density as inputs; and setting propeller pitch as a variable, we use our model implicitly to calculate the power as a function of airspeed. We plotted these power-airspeed curves in Fig. 7 for three different propellers with pitch ratios 0.25, 0.45, and 0.65. In one additional step, we can calculate the range, assuming two 10 Ah 6S LiPo batteries and a 5-min hovering period. The range, as well as other metrics of interest like the hover time and max speed, are summarized for various pitch ratios in Table 3.

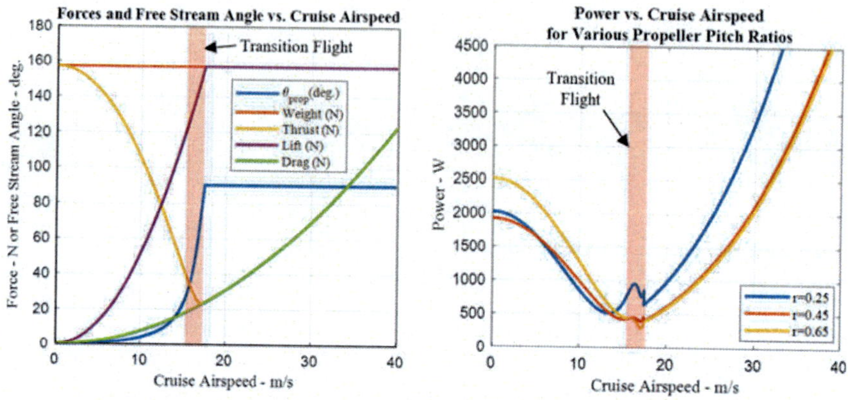

Fig. 7 The left figure shows the equilibrium forces as a function of cruise airspeed, as well as the free stream angle of the propellers at sub-stall speeds. These are the same no matter what propeller is used. The right shows how much power is required from propellers of three pitch ratios (r) to achieve the thrust specified on the left

Table 3 Results from design study in which we vary propeller pitch and evaluate performance metrics of interest. We assumed a 5-min hover time for this analysis. Optimal range is achieved for a pitch ratio of 0.55

Pitch ratio (D = 24″)	$V_{best\ range}$ (m/s)	$\theta_{best\ range}$ (deg.)	Range (km)	Hover Time (min.)	V_{max} (m/s)
0.25	14 or 17.5	16 or 90	27.0	13.2	33.0
0.35	17.5	90	38.3	14.4	36.4
0.45	17.5	90	44.6	14.0	38.3
0.55	17.5	90	44.7	12.6	38.8
0.65	17.5	90	39.8	10.9	39.1

The results teach us several interesting lessons. First, we observe that when the model parameters, in this case θ, vary too rapidly as they do during the transition flight indicated in Fig. 7, it will lead to oscillations. In this example, we ignore results for the transition region as defined by where the free stream angle is between 30° and 90°. As discussed in the previous section, we urge the user to be aware of rapidly varying parameters in their analyses.

With that considered, we can look at the more meaningful results. We can see from Table 3 that if we use a standard quadcopter pitch ratio (0.25–0.33), the best range can be obtained without even rotating the propellers forward the full 90°. In fact, with a pitch ratio of 0.25, you will get the same range with the propellers rotated forward only 16° and flying at 14 m/s as you will with the propellers rotated forward 90° and flying at 17 m/s.

Additionally, we can see that while a higher pitch ratio gives better high-speed efficiency and faster top speeds, it comes at the cost of hovering time. This tradeoff comes at diminishing returns when approaching the pitch ratio 0.55. The results indicate that the optimal propeller pitch ratio will depend on the mission requirement. Generally speaking, the longer the hovering time, the lower the optimal pitch ratio, and vice versa. In the case of a 5-min hovering time, the pitch ratio for the best range appears to be about 0.55, yielding a range of 44.7 km.

5 Conclusions

We have shown that the appropriate selection of propellers for eVTOL aircraft is not always trivial. In cases where the aircraft tilts forward during flight like they often do for long range eVTOLs, propeller performance will behave differently from how it does during hover ($\theta = 0°$) or forward flight at ($\theta = 90°$). BET-based models can handle this problem effectively, but introduce the complication of designing a propeller geometry from scratch. This geometry is impractical because when we purchase propellers, we usually make a decision based on diameter and pitch, not chord and twist distribution. Our empirical model utilizes these practical parameters to estimate propeller thrust and power, while accounting for how tilting the propeller forward affects the power and thrust.

In this paper, we discussed the experiment used to gather data for our model, showed the mathematical derivation for it, and compared its results with an in-house BET-based model. The two showed similar trends, with our model giving more conservative estimates for the windmill advance ratio, and showing unrealistic results for conditions where both the free stream angle is low and the advance ratio is high. Our recommendation is to use this model for 2D, steady-state, full-vehicle performance analysis. A rapid change in model parameters will lead to instabilities, so the user should be aware of this when running analyses.

At the end, we shared an example of how this model can be used to evaluate and maximize the range, hover time, or maximum speed of a tiltrotor delivery drone. The results showed that propeller selection for such vehicles is indeed non-trivial, and requires analysis tools that capture the effects of free stream angle and advance ratio on propeller thrust and power.

There are a number of ways in which this model can be improved upon. The most obvious is to integrate more data, preferably from a greater variety of propellers. Extending the range of pitch ratios to include those typical of fixed-wing aircraft would be helpful as well. Additionally, we would like to better define the relationship between pitch ratio and parameters like windmill advance ratio and static thrust/power, or between the 90° windmill advance thrust/power coefficient and free stream angle. This would also fix the inaccurate divergence discussed in the limitations section.

Acknowledgements We would like to acknowledge Japan Aerospace Exploration Agency for allowing us to use their Low Speed Wind Tunnel 1 facilities for the purpose of gathering data for this propeller model. Also, we would like to thank the University of Tokyo Doctoral Students Special Incentives Program for their financial support.

References

1. Gur O, Rosen A (2008) Comparison between blade-element models of propellers. Aeronaut J 112(1138):689–704
2. Staples G (2014) electricaircraftguy.com. Retrieved August 16, 2021, from https://www.electricrcaircraftguy.com/2013/09/propeller-static-dynamic-thrust-equation.html
3. Brandt JB, Selig MS (2011) Propeller performance data at low reynolds numbers. American Institute of Aeronautics and Astronautics, Orlando
4. Deters RW, Ananda GK, Selig MS (2014) Reyolds number effects on the performance of small-scale propellers. American Institute of Aeronautics and Astronautics, Atlanta
5. Dantsker OD, Caccamo M, Deters RW, Selig MS (2020) Performance testing of aero-naut CAM folding propellers. Virtual Event, American Institute of Aeronautics and Astronautics
6. Staub FM, et al (2021b) Wind tunnel testing of a lift-augmented quadcopter. Virtual Event, American Institute of Aeronautics and Astronautics
7. Staub FM, Tsukada D, Inoue S, Tsuchiya T (2021a) Modeling and design of a lift-augmented quadcopter. Virtual Event, American Institute of Aeronautics and Astronautics
8. Theys B, Vos GD, Schutter JD (2016) A control approach for transitioning VTOL UAVs with continuously varying transition angle and controlled by differential thrust. In: International Conference on Unmanned Aircraft Systems (ICUAS), pp 118–125
9. Theys B, et al (2014) Wind tunnel testing of a VTOL MAV propeller in tilted operating mode. International Conference on Unmanned Aircraft Systems (ICUAS), Orlando
10. Kolaei A, Barcelos D, Bramesfeld G (2018) Experimental analysis of a small-scale rotor at various inflow angles. Int J Aerospace Eng 2018(2560370):1–14
11. McCormick BW (1979) Aerodynamics, aeronautics and flight mechanics. Wiley, Michigan
12. Dreier ME (2018) Introduction to helicopter and tiltrotor flight simulation. AIAA Education Series
13. Rotatu C, Todorov M (2017) Helicopter flight physics. In: Volkov K, (ed.) Flight physics–models, techniques and technologies. IntechOpen
14. Rubin RL, Zhao D (2021) New development of classical actuator disk model for propellers at incidence. AIAA J 59(3)
15. 河内啓二 (1999) ロータの空気力学. 日本航空宇宙学会誌 47(550):246–251

Design and Development Status of Experimental Winged Rocket WIRES#015

Akira Watanabe, Koichi Yonemoto, Takahiro Fujikawa, Takahiro Matsukami, Tsuyoshi Otsuki, Yuto Kitazono, Yasuhiro Koshida, Masaaki Murakami, Tomataka Watanabe, Hina Atarashi, Hidetoshi Takeyama, Shintaro Tejika, Yuito Fujii, Raizo Matsuda, Yusuke Mine, Ayaka Yamazaki, Sho Yoshida, and Toshiki Morito

Abstract Research on winged reusable space transportation systems and flight demonstration by experimental winged rockets called WInged REusable Sounding rocket (WIRES) have been conducted by researchers and laboratory graduates since 2005. WIRES#015 has been developed by Tokyo University of Science, the university start-up SPACE WALKER Inc., and JAXA in collaboration with the German Aerospace Center (DLR) and Swedish Space Corporation (SSC). It has a total length of 4.6 m, launch mass of 1000 kg, and maximum flight altitude of 5.5 km, which is a comprehensive technology demonstrator for technical issues such as the liquid oxygen/liquefied natural gas engine, reaction control system, airframe structure, and propellant tanks made of carbon fiber reinforced plastics and autonomous guidance, navigation, and control systems. Currently, WIRES#015 is in the sustaining design phase. The airframe structure will be manufactured by March 2022 in cooperation with Toray Carbon Magic Co., Ltd. and Mooncraft Co., Ltd. and subjected to structural tests. In 2023, the captive firing test (CFT) will be conducted. The first flight demonstration will be performed in early 2024. Three additional flight demonstrations are planned from 2025 to 2026. In this paper, we describe the development status of WIRES#015 and its development schedule.

Keywords Winged rocket · Flight demonstration · LOX/LNG engine · Guidance and control

A. Watanabe (✉) · K. Yonemoto · T. Fujikawa · T. Matsukami · T. Otsuki · Y. Kitazono · Y. Koshida · M. Murakami · T. Watanabe · H. Atarashi · H. Takeyama · S. Tejika · Y. Fujii · R. Matsuda · Y. Mine · A. Yamazaki · S. Yoshida
Tokyo University of Science, Yamazaki, Noda 2641, Japan
e-mail: 7521565@ed.tus.ac.jp

T. Morito
Japan Aerospace Exploration Agenccy, Tsukuba Space Center, 2-1-1 Sengen, Tsukuba, Ibaraki, Japan

S. Lee et al. (eds.), *The Proceedings of the 2021 Asia-Pacific International Symposium on Aerospace Technology (APISAT 2021), Volume 1*, Lecture Notes in Electrical Engineering 912, https://doi.org/10.1007/978-981-19-2689-1_24

1 Introduction

In the United States, many private companies are developing Reusable Launch Vehicles (RLVs), such as Falcon 9 of SpaceX, SpaceShip Two of Virgin Galactic, and New Shepard of Blue Origin. RLVs can significantly reduce transportation costs compared with Expendable Launch Vehicles (ELVs). There are two types RLVs: winged and wingless RLVs. Winged RLVs can fly back to the aiming runway with high range performance.

Tokyo University of Science has been developing experimental winged rockets called WInged REusable Sounding rocket (WIRES) since 2005. Small WIRES#011 had its flight tests from 2008 to 2009 five times to evaluate attitude controllability based on H_∞ theory. From 2010 to 2011, WIRES#012 had its flight test to validate the two-stage parachute system, which comprises deceleration and main chutes. From 2012 to 2019, WIRES#014 had its flight tests to validate basic guidance and control technologies. WIRES#014–1 equipped with CAscaded MUltistage Impinging-jet (CAMUI) hybrid rocket engine of Hokkaido University reached the maximum altitude of 600 m, which was about half of the planned altitude to fail the recovery. According to the post-flight analysis, the causes of failure were not only insufficient engine performance but also the malfunction of Air Data Sensing (ADS) and attitude control systems. WIRES#014–2 employed a hybrid rocket engine called HyperTEK M1000 manufactured by Cesaroni Technology Inc. to improve the performance. Unfortunately, during the preflight ground combustion test, the nozzle of the combustor broke, resulting in airframe burning. The engine combustor of the hybrid rocket engine HyperTEK M1000 was reinforced with carbon fiber for WIRES#014–3; the flight test was conducted in November 2015. Figure 1 shows the flight test of WIRES#014–3. It was safely recovered using the two-stage parachute and airbag systems but was unable to guide the flight path as expected. The flight test of WIRES#014-3A, an improved model of WIRES#014–3, was conducted in March 2020. Figure 2 shows the flight test of WIRES#014-3A. Since the combustor

Fig. 1 Flight test of WIRES#014–3

Fig. 2 Flight test of WIRES#014-3A

continued to burn after the engine thrust was decreased, the deceleration chute burned and could not deploy the main chute to recover WIRES#014-3A safely on the ground. There was an error of the coordinate system in the guidance program to cause an unexpected trajectory [1].

In this paper, we present the design and development status of the experimental winged rocket WIRES#015 under development by Tokyo University of Science, SPACE WALKER, and JAXA. Further, the flight demonstration will be planned in Esrange in Kiruna of Sweden [2].

2 Details of Experimental Winged Rocket WIRES#015

2.1 Major Dimensions and Specifications

The aerodynamic shape of WIRES#015 is based on the Highly Maneuverable Experimental Space vehicle (HIMES) [3, 4], which is a suborbital winged rocket researched at the Institute of Space and Astronautical Science (ISAS) of the Ministry of Education in the 1980s. The major dimensions and specifications of WIRES#015 are shown in Fig. 3 and Table 1. Table 2 shows the mission objectives.

Fig. 3 Two-views drawing

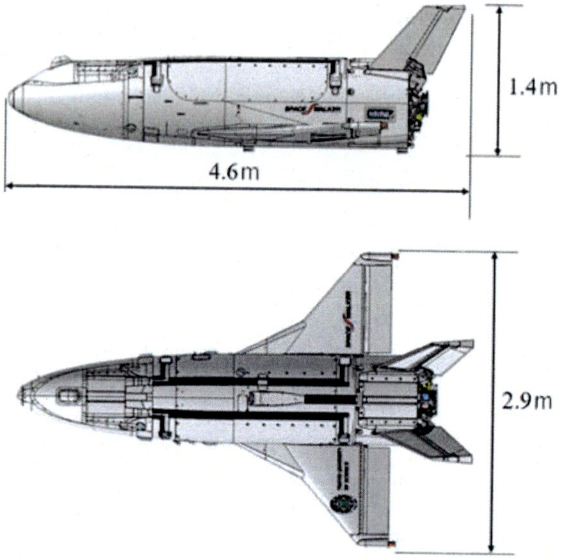

Table 1 Major dimensions and specifications

Initial mass	[kg]	1000
Total length	[m]	4.6
Body length	[m]	4.0
Wing span	[m]	2.9
Body diameter	[m]	0.91
Height	[m]	1.4
Max. thrust at sea level	[kN]	17.8
Combustion time	[s]	30
Apogee altitude	[km]	5.1
Maximum speed	[Mach]	0.67

2.2 *Nominal Flight Profile*

WIRES#015 makes a powered ascent for 30 s. After the engine cutoff, it will enter a coasting ascent. It controls the nose down by RCS at the apogee. Then, it makes reentry flight and equilibrium gliding to the aiming recovery point. Figure 4 shows the nominal flight profile.

Table 2 Mission objectives

Navigation, guidance, and control (NGC) system
• Flush-type Air Data Sensing (FADS) system for estimating aerodynamic attitude and speed of dynamic pressure and Mach number with two failure tolerance capability
• Real-time optimal guidance system using Dynamically Distributed Genetic Algorithm (DynDGA)
• Adaptable attitude control system using Dynamic Inversion (DI) theory
• Reaction Control System (RCS) for attitude control at apogee
Recovery system
• Recovery by two-stage parachute and airbags
• Airbag ventilation system
Propulsion system
• Full expander cycle LOX/LNG engines
• Propellant feed system

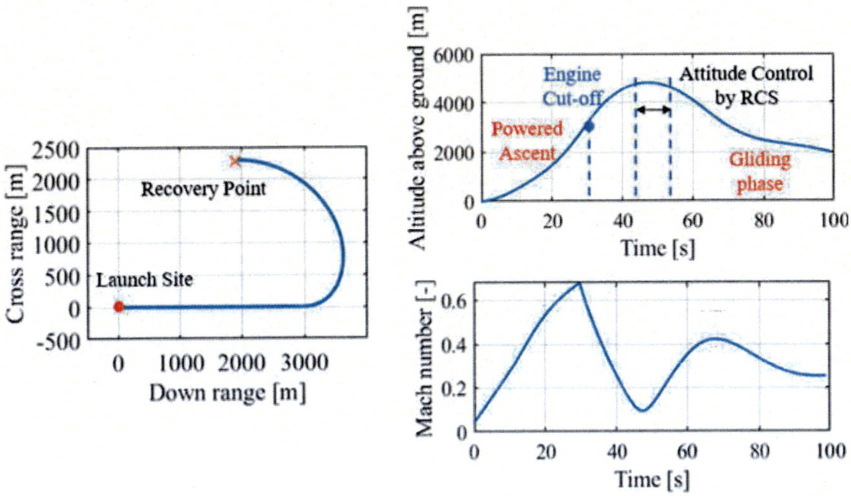

Fig. 4 Nominal flight profile

2.3 *Autonomous Flight System*

2.3.1 Navigation

The FADS of WIRES#015 [5], which measures aerodynamic attitude, dynamic pressure, and Mach number, has nine measurement static pressure ports and eight reference static pressure ports for improving measurement accuracy and realizing two failure tolerance capabilities in Fig. 5.

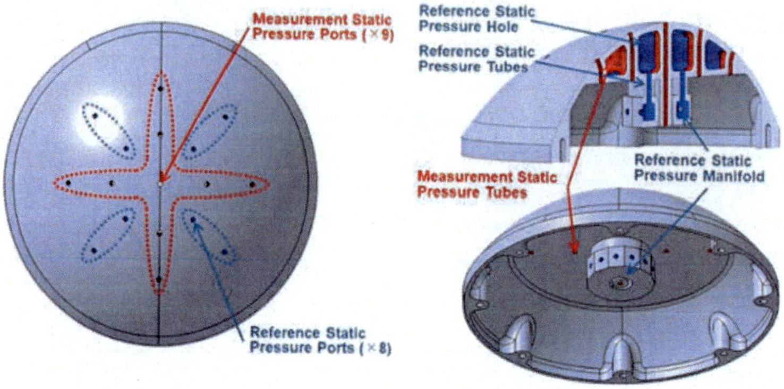

Fig. 5 FADS system

The commercial-off-the-shelf Inertial Navigation System (INS)/Global Navigation Satellite System (GNSS) is used as a hybrid navigation system for guidance and control.

2.3.2 Guidance

RLVs need onboard trajectory optimization and flexible guidance methods to cope with unexpected flight conditions and abort flight due to the system malfunction and failure, which differs from the ELVs guided by predesigned trajectory. WIRES#015 employs an onboard real-time trajectory optimization and guidance designed by Bezier curve using Particle Swarm Optimization (PSO) [6] and evolutionary algorithm, DynDGA [7] in Fig. 6, along the flight phases in Fig. 7 and Table 3.

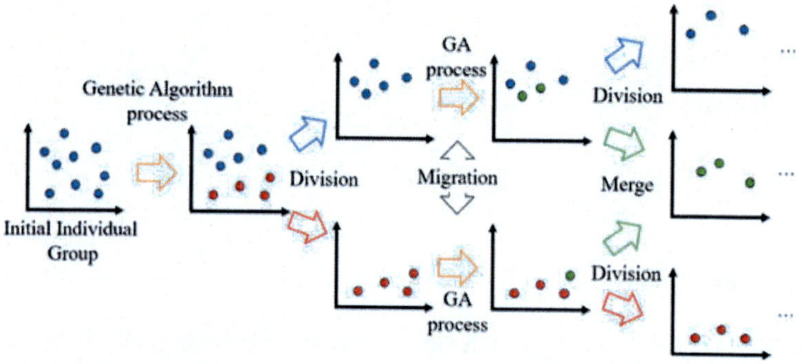

Fig. 6 Schematics of DynDGA

Fig. 7 Flight phases

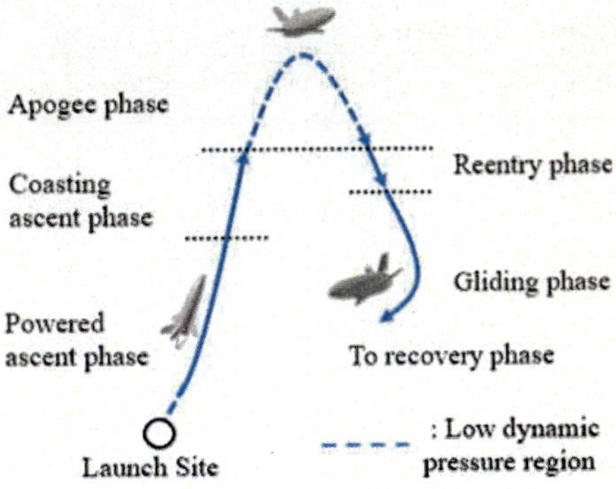

Table 3 Trajectory optimization and control commands

Phase	Optimization algorithm	Optimization variables and control commands
Powered Ascent	PSO	Angle of attack and bank angle
Coasting Ascent		
Apogee	–	–
Reentry	DynDGA	Angle of attack and bank angle
Gliding	PSO	

2.3.3 Control

The flight conditions of winged rockets, such as high dynamic pressure during the powered flight, low dynamic pressure at the apogee of trajectory, and subsonic to hypersonic Mach number at the ascent coasting and reentry flight, change significantly.

WIRES#015 employs an adaptive control law using DI [8, 9] in Fig. 8 and Table 4.

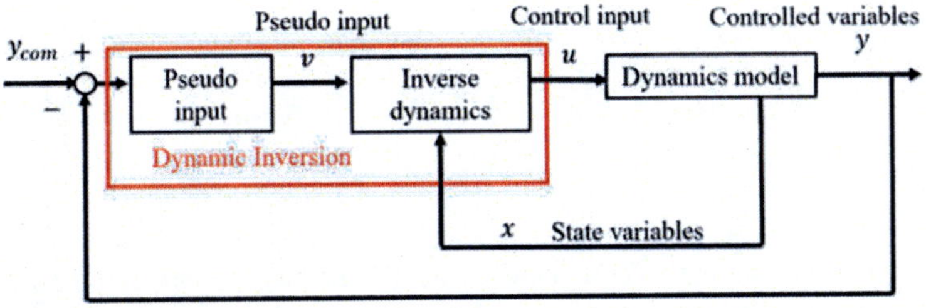

Fig. 8 Block diagram of DI method

Table 4 Control law

Phase	Control devices	Control law	Control variables	Control commands
Powered ascent	Aerodynamic surface, thrust vector control (TVC)	DI	Angle of Attack, Sideslip Angle, Roll Angle	Elevator Angle, Aileron Angle, Rudder Angle, Gimbal Angle
Coasting ascent	Aerodynamic surface			Elevator Angle, Aileron Angle, Rudder Angle
Apogee	RCS	Phase plane	Roll angle	Torque around Each Body-axis
Reentry	Aerodynamic surface	DI	Roll angle, Pitch Angle, Yaw Angle	Elevator Angle, Aileron Angle, Rudder Angle
Gliding			Angle of Attack, Sideslip Angle, Roll Angle	

2.4 *Structure and Onboard Equipment*

The airframe structure of WIRES#015 is made of carbon fiber reinforced plastic (CFRP) and partially Glass Fibber Reinforced Plastic (GFRP) with honeycomb to reduce weight and maintain high rigidity.

The airframe structure comprises forward fuselage, mid-fuselage, wings, and rear fuselage for installing propellant tanks, engine and propulsion system, and transportability.

The avionics and main parachute system are located in the forward fuselage covered by a nose cone made of GFRP. FADS locates at the nose top. Helium gas (GHe) tank, nitrogen gas (GN2) tank, LNG tank, LOX tank, and battery package are mounted in the mid-fuselage. In the rear fuselage LOX/LNG engine, the deceleration chute system and RCS are mounted.

Figure 9 shows the airframe structure's components and onboard equipment.

2.5 *Avionics*

Avionics comprises a control computer, guidance computer, navigation computer, communication computer, air data computer, data computer, and engine computer, which communicate with each other via an ARINC429 data bus. Figure 10 shows the onboard computers and data bus.

The navigation computer sends the INS/GNSS and GNSS data to each computer via the ARINC429 data bus. The guidance computer calculates optimal trajectories

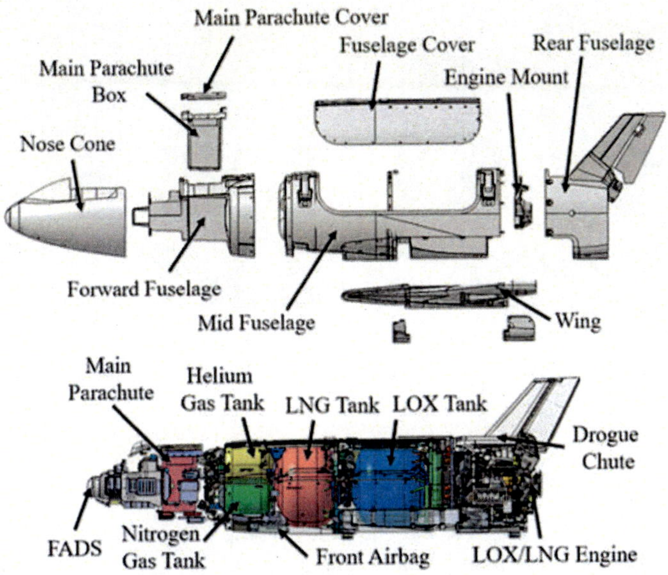

Fig. 9 Airframe structure components and onboard equipment

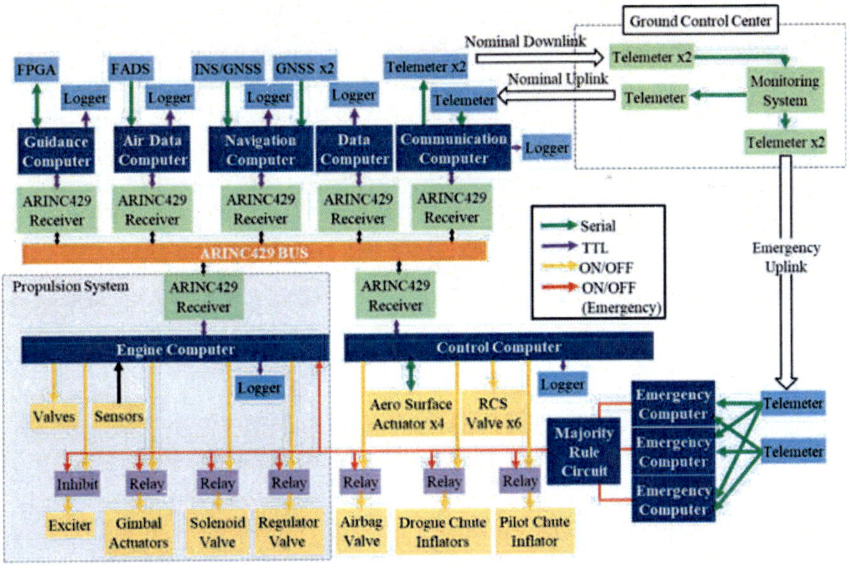

Fig. 10 Onboard computers and data bus

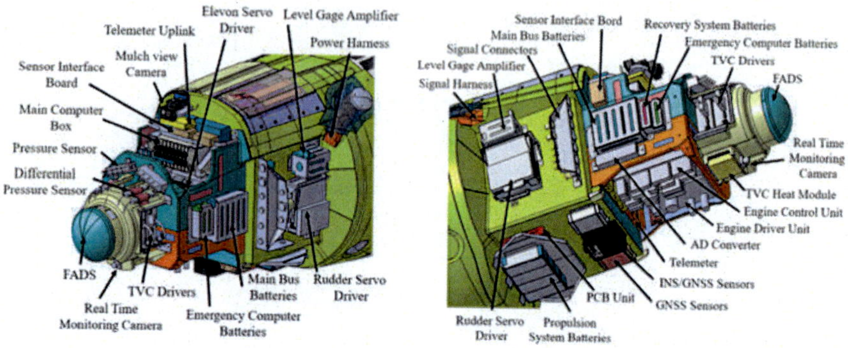

a) Front View b) Bottom View

Fig. 11 Avionics bay arrangement

with assistance from a field-programmable gate array (FPGA) to send the control command to the ARINC429 data bus. The control computer controls the aerodynamic surface actuators. The communication computer is connected with two uplink and one-downlink telemetry units. The air data computer calculates the dynamic pressure, Mach number, and aerodynamic attitudes such as angle of attack and sideslip angle from the FADS data. The data computer records the flight data via the ARINC429 data bus in the data logger. The engine computer controls the engine and propulsion system, such as intuition and shut-off, and performs the health monitoring of the engine and propulsion system. There are three emergency computers to activate flight termination, which are independent of the ARINC429 data bus. The arrangement of the avionics bay is shown in Fig. 11.

2.6 *Propulsion System*

2.6.1 Engine

The engine of WIRES#015 is a full expander cycle LOX/LNG engine being developed by JAXA. Figure 12 shows the LOX/LNG engine. During the powered ascent, the attitude is controlled by TVC deflecting thrust vector with gimbal actuators with the assistance of aerodynamic control surfaces of elevon and rudder.

2.6.2 Tank of Propulsion System

The propellant tanks comprise LOX and LNG tanks. High-pressure GHe and GN2 are used for the propulsion system, RCS, and gas seal of the propellant pump and purge propulsion system. The LOX and LNG tanks are first made of aluminum but will

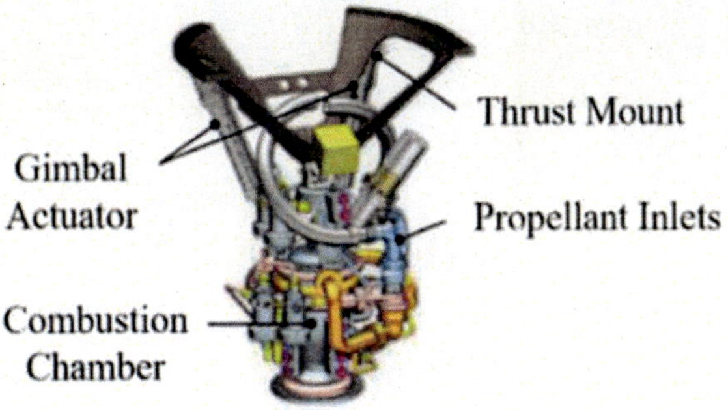

Fig. 12 Full expander cycle LOX/LNG engine

be replaced with those of CFRP. The GHe and GN2 tanks are made of polyethylene liner reinforced with carbon fiber filament winding.

2.6.3 Reaction Control System

Six thrusters are mounted at the rear fuselage to control the attitude control during the ballistic flight under low dynamic pressure conditions near the apogee of trajectory. Each thruster has 50 N using GN2 in Fig. 13. The pitch and roll attitude are controlled by the upper and lower four thrusters. Yawing is controlled by the left and right thrusters [10].

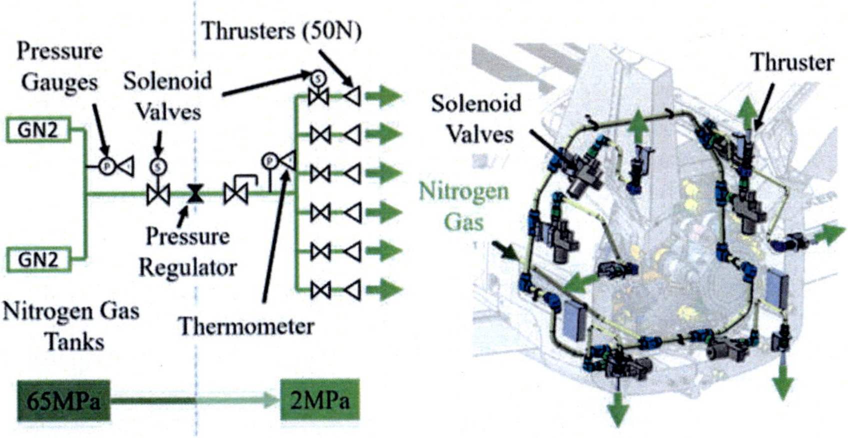

Fig. 13 Block diagram and arrangement of RCS

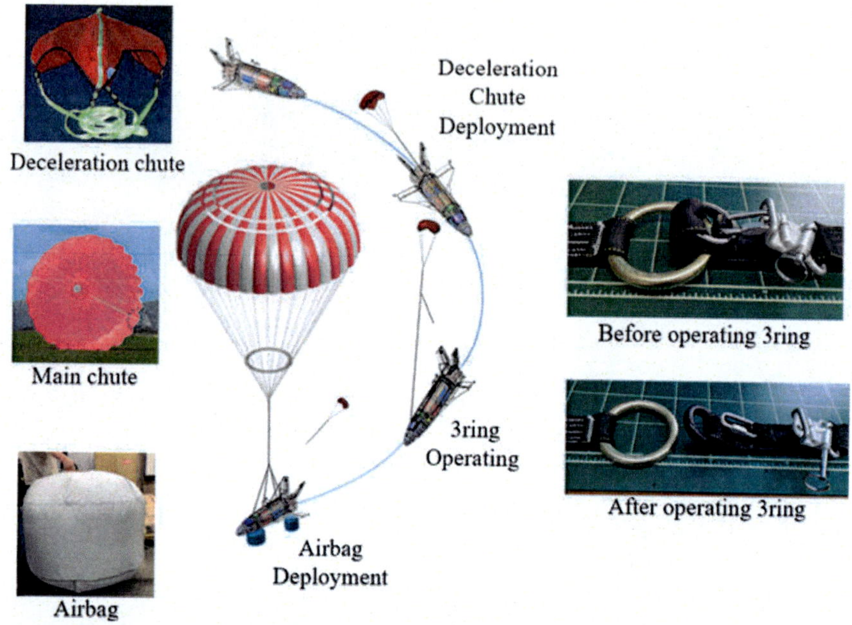

Fig. 14 Two-stage parachute system

2.7 *Recovery System*

2.7.1 Parachute

The parachute of WIRES#015 uses a two-stage parachute system in Fig. 14. First, the deceleration chute is released from the rear fuselage to reduce the flight speed to pull out the main chute for deployment by a 3ring system. Three airbags are deployed when descending with the main chute.

2.7.2 Airbag

After deploying the main chute, the three airbags are deployed to reduce the impact load. One airbag is installed in front of the mid-fuselage, and the other two are installed inside of the main wing. The airbag ventilation system mitigates the impact load in Fig. 15 [11].

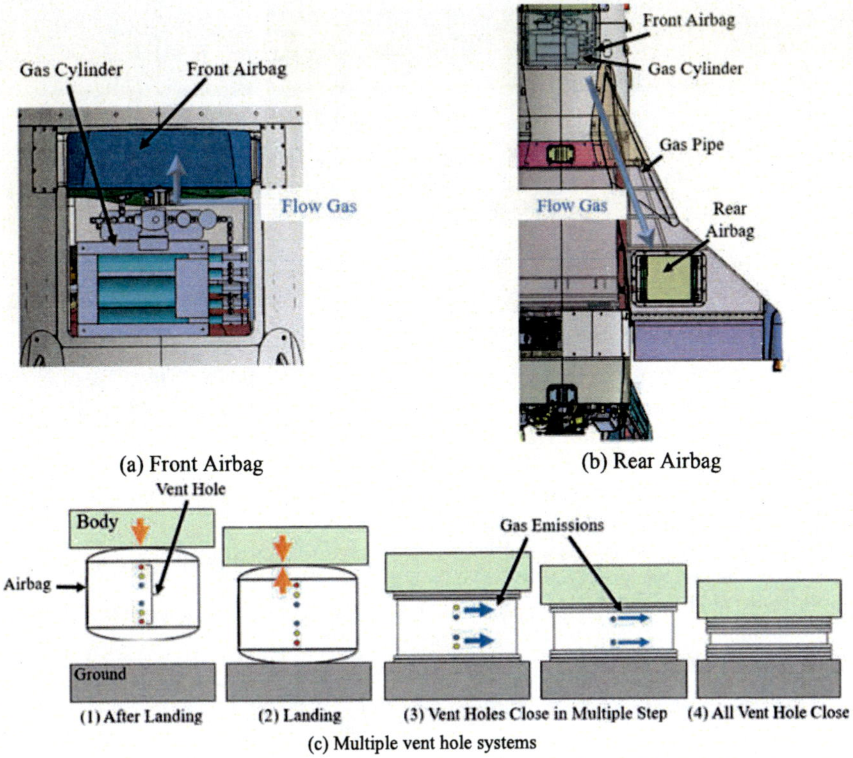

Fig. 15 Airbag system

2.8 *Ground Support System*

WIRES#015 will be launched by Mobile MAN-2 Launcher and ground support systems of the German Aerospace Center (DLR) in Fig. 16. The communication system of Swedish Space Corporation (SSC), including telemetry, emergency uplink, and ground communication between the launch site and control center, will be used. The propellant supply system will be prepared by DLR. The monitoring system of flight status and command transmission system in the control station will be prepared by Tokyo University of Science and SPACE WALKER.

3 Development and Flight Demonstration Plan of WIRES#015

The development and flight demonstration schedule of WIRES#015 is shown in Fig. 17. Currently, WIRES#015 is in the sustaining design phase. The airframe

Fig. 16 Grand support system of DLR

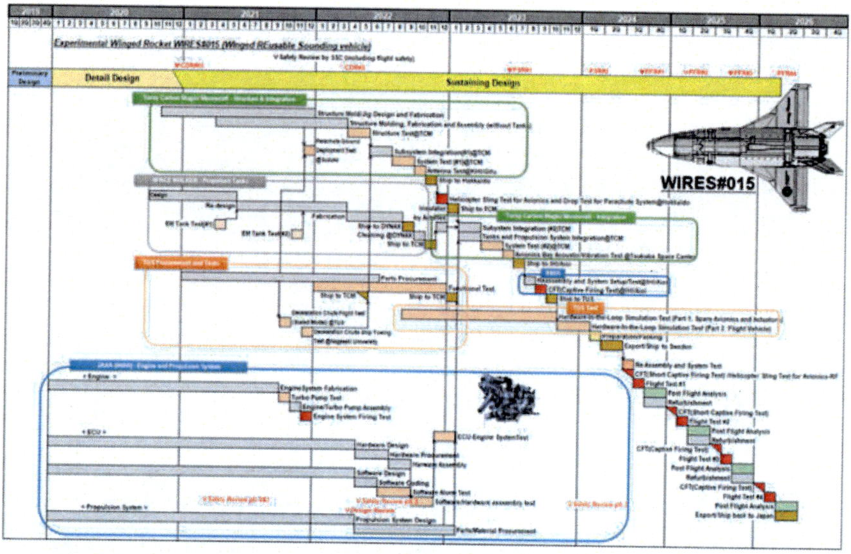

Fig. 17 Development and flight demonstration schedule

structure will be manufactured by March 2022 in cooperation with Toray Carbon Magic and Mooncraft and subjected to structural tests.

The onboard avionics will be installed from July 2020, followed by the first system test. A helicopter sling test will be conducted in December 2022 to verify the NGC and two-stage parachute systems. In April 2023, the engine, propulsion system, and tanks will be installed to conduct a second system test. In September 2023, the Captive Firing Test (CFT) will be conducted. After completing the final hardware-in-the-loop test in the first quarter of 2024, WIRES#015 will be shipped to Esrange in Kiruna of Sweden. The first flight demonstration will be conducted in the third quarter of 2024. Three additional flight demonstrations are planned from 2025 to 2026.

4 Conclusions

The design and development status of experimental winged rocket WIRES#015 is presented. Tokyo University of Science and SPACE WALKER have started hardware manufacturing to get into a sustaining design. The flight demonstration will be conducted in cooperation with DLR and SSC.

References

1. Kunitsu Y, Yonemoto K, Fujikawa T (2019) Design and development status of experimental winged rocket WIRES#015 and WIRES#013. In: 32nd International Symposium on Space Technology and Science, Japan
2. Yonemoto K, Fujikawa T (2021) Development status of experimental winged rocket WIRES#015 and its flight demonstration plan at esrange of sweden in cooperation with university, agency, industries and international partners. Glob Space Explor Conf 7:62871
3. HIMES Research Group (1987) Conceptual design of HIMES (Winged Test Rocket). Institute of Space and Astronautical Science, Japan (in Japanese)
4. Inatani Y, Kawaguchi J, Yonemoto K (1990) Status of "HIMES" reentry flight test project, AIAA 2nd International Aerospace Planes Conference. Florida, USA
5. Gossamsetti GS, Yonemoto K, Fujikawa T (2017) Design and enhancement of flush air data system (FADS) for winged rocket. In: Asia-Pacific International Symposium on Aerospace Technology, Seoul, Korea
6. Fujikawa T, Yonemoto K (2020) Real-Time optimization of guidance trajectories and commands using evolutionary computation with convex quadratic programming. In: Proceedings of 63rd Japan Joint Automatic Control Conference, IC3–5(in Japanese)
7. Ichige M, Yonemoto K, Fujikawa T (2017) Preliminary design of dynamically distributed genetic algorithm onboard guidance system for experimental winged rocket. Asia-Pacific International Symposium on Aerospace Technology, Seoul, Korea
8. Shirakata K, Yonemoto K, Fujikawa T (2018) Flight attitude controller using dynamic inversion theory with stability margin guarantee. J Eng Res Appl 8:16–27
9. Kitazono Y, Yonemoto K, Fujikawa T Control system design for the powered ascent of winged rocket based on dynamic inversion theory. In: 60th annual conference of the society of instrument and control engineers of Japan. pp 767–772
10. Yamazaki A, Yonemoto K, Fujikawa T, Seki D (2020) Detail design status of experimental winged rocket WIRES#013 and WIRES#015. Proc Space Sci Technol Conf 64:1106 (in Japanese)
11. Fujii Y, Yonemoto K, Fujikawa T (2019) Numerical analysis and development of airbag system for experimental winged rocket WIRES. In: 32nd International Symposium on Space Technology and Science, Japan

Multiobjective Optimization of a Highly Maneuverable Supersonic Airfoil Using Multifidelity EGO

Tomotaka Watanabe, Koichi Yonemoto, Takahiro Fujikawa, and Ayaka Yamazaki

Abstract In supersonic flight regimes, maneuverability reduces owing to the retreat of the aerodynamic center. This paper describes an optimization that objective is the shift forward of the aerodynamic center to solve the maneuverability problem. For this optimization, multifidelity efficient global optimization (EGO) was used, which is an extension of EGO that uses a surrogate model for an efficient search and allows using two different evaluation functions with different fidelities. B-spline curves, which allow for a flexible airfoil shape, were used to generate the airfoil shape. We could determine the type of the airfoil shape that would move the aerodynamic center forward from the pressure distribution and control points of the B-spline curve based on the optimization results. Therefore, we succeeded in finding the geometric features that determine the tradeoff between the aerodynamic center position and lift–drag ratio. The contribution of the geometry of the airfoil to the aerodynamic center is discussed herein.

Keywords Supersonic airfoil · Aerodynamic optimization · Multiobjective optimization · Multifidelity optimization · Evolutionary algorithm

1 Introduction

The aerodynamic characteristics of aircraft vary greatly depending on the flight Mach regime. The position of the aerodynamic center retreats significantly in the supersonic region compared to the subsonic region. This causes static stability to become too strong, and maneuverability, which is a characteristic relative to static stability, remarkably deteriorates. Flying in the supersonic regime makes it difficult for airplanes to change their attitude. This maneuverability problem characteristic disappears completely as planes enter the hypersonic regime. However, it is an unavoidable problem when an airplane accelerates to the hypersonic or decelerates from the hypersonic regime even if it is cruising in the hypersonic regime.

T. Watanabe (✉) · K. Yonemoto · T. Fujikawa · A. Yamazaki
Tokyo University of Science, 2641 Yamazaki, Noda, Japan
e-mail: 7520580@ed.tus.ac.jp

S. Lee et al. (eds.), *The Proceedings of the 2021 Asia-Pacific International Symposium on Aerospace Technology (APISAT 2021), Volume 1*, Lecture Notes in Electrical Engineering 912, https://doi.org/10.1007/978-981-19-2689-1_25

To improve maneuverability, an airfoil with forward-located aerodynamic centers was investigated in this study. However, because the aerodynamic center alone is inadequate to maintain a practical airfoil, the lift–to–drag ratio was also included in the objective function.

Although various effective methods have been proposed for genelating airfoils [1–3], only a few use an optimization focusing on the aerodynamic center as proposed herein. Because there is no prerequisite knowledge on whether the existing methods can represent the optimal airfoil shape for the objective function of this optimization, a more flexible airfoil expression method is required. Moreover, airfoils with shape edge, such as double-wedge airfoils, are effective in the supersonic regime. To utilize the existing knowledge of aerodynamic design, representing and optimizing such airfoils are necessary.

The rest of this paper is structured as follows: The optimization framework is presented in Sect. 2. Section 3 explains how to express an airfoil, and Sect. 4 provides an evaluation technique of aerodynamic characteristics. The optimization results are then presented and discussed in Sect. 5. Finally, Sect. 6 concludes this paper.

2 Optimization Framework

Airfoil optimization adopts an evolutionary computation, which does not require differential information and is characterized by high robustness. However, the evolutionary computation has a disadvantage in that the objective function is evaluated frequently. When using the Cmputation Fluid Dynamics (CFD) analysis, for example, the calculation cost keeps rising. Therefore, herein, the optimization is conducted using the efficient global optimization (EGO) algorithm, which reduces the computational cost by employing a surrogate model [4, 5]. Furthermore, Ordinary EGO can only input single fidelity inputs. However, we use Multi-fidelity EGO algorithm, which is extended to support multiple fidelity inputs, for more efficient exploration [6].

2.1 Optimization Algorithm

Multiobjective Evolutional Algorithm based on Decomposition using Differential Evolution (MOEA/D-DE) is an evolutionary computation algorithm that adopts a non-Pareto approach [4], which is often used in combination with EGO, and this algorithm is also used in this paper's optimization. By dividing the search direction and explorering every divided cluster, the global search which maintains the diversity is possible.

The EGO algorithm is a method for objective function estimate by using a surrogate model. The individuals to be evaluated can prioritize regions that have not yet been explored or regions that have been explored to some extent and are known to

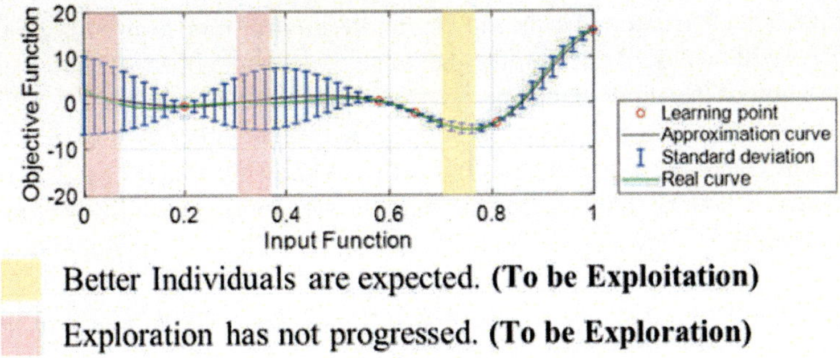

Fig. 1 Example of kriging estimation objective function and variance

have good individuals, thus reducing the evaluation of the objective function. This algorithm is suitable for optimization where the computational cost of evaluating the objective function is high, such as in this optimization. In addition, this optimization uses the Mlti-Fidelity-EGO algorithm, which is an extension of EGO. This is a more efficient way to evaluate the objective function by using different evaluation methods with different fidelity.

2.2 *Surrogate Model*

The surrogate model estimates the objective function of the new individual based on the information of the individual evaluated until that generation. Kriging, a surrogate model, is used in the proposed optimization. Kriging implements a Gaussian process, which allows the expected value and variance of the objective function to be calculated simultaneously. Kriging reveals the range where excellent individuals are expected to be found and the range where the search has not progressed due to a large variance, as shown in 1. In EGO in the optimization, the renewal of the surrogate model and search of the good individual are conducted simultaneously, and this exploration and exploitation are compatible. Therefore, the search can be effective. Furthermore, in this study, a more efficient search is enabled using a surrogate model (co-kriging) in which two types of data with different accuracies are made to correspond to the input (Fig. 1).

2.3 *Objective Function*

The aerodynamic coefficients are obtained by the analysis when the angle of attack is 4° steps at 0–20°. The objective function is calculated after interpolating each aerodynamic coefficient for the angle of attack. To avoid overshoot, the interpolation technique using PCHIP[7].

The first objective function is the maximum lift–to–drag ratio at an angle of attack of 4–16°. It will be described later that the range of the angle of attack is limited for the calculation of the aerodynamic center.

The second objective function is to improve maneuverability by minimizing the aerodynamic center position at the maximum lift–to–drag ratio angle of attack. When L denotes the lift, D the drag, and x_{ac} the aerodynamic center, the objective function can be expressed as follows.

$$Maximize\ f_1 = max\ L/D, Minimize\ f_2 = x_{ac}@max\ L/D \tag{1}$$

A method for determining the aerodynamic center position from the aerodynamic coefficients is explained here.

First, the pitching moment coefficient around the leading edge, C_{m0}, can be calculated as follows using the aerodynamic center, x_{ac}, pitching moment around the aerodynamic center, C_{mac}, and normal direction force coefficient, C_n:

$$C_{m0} = C_{mac} - \frac{x_{ac}}{c} C_n \tag{2}$$

when differentiated by angle of attack α,

$$\frac{\partial C_{m0}}{\partial \alpha} = \frac{\partial C_{mac}}{\partial \alpha} - \frac{\partial x_{ac}}{\partial \alpha} \frac{C_n}{c} - \frac{x_{ac}}{c} \frac{\partial C_n}{\partial \alpha}, \tag{3}$$

Here, the first and second terms of the right side become zero from the definition of the aerodynamic center.

The relation between the leading edge and pitching moment around $x = 0.25$ is given by the following equation:

$$C_{m0.25} = 0.25 C_n + C_{m0}, \tag{4}$$

when differentiated by α and deform,

$$\frac{\partial C_{m0}}{\partial \alpha} = -\frac{\partial C_{m0.25}}{\partial \alpha} + 0.25 \frac{\partial C_n}{\partial \alpha}. \tag{5}$$

Equation 3 and Eq. 5 are combined and arranged by x_{ac}. Finally, x_{ac} can be obtained as follows:

$$\frac{x_{ac}}{c} = 0.25 - \frac{dC_{m0.25}}{dC_n}. \tag{6}$$

2.4 *Flight Conditions in Optimization*

The flight conditions in optimization is based on the flight environment of Overture, a supersonic transportation (SST) of Boom Technology. The Mach number was changed to 1.5, which is thought to show the strongest characteristics in the supersonic regime. Table 1 is the conditions.

3 Airfoil Expression

There are only a few examples of optimizing the position of the aerodynamic center as an objective function in the supersonic regime, so it is an unknown problem. In such a case, it is considered that the optimized shape differs significantly from the existing airfoil. Therefore, an airfoil expression method with high freedom, which is not caught in the existing airfoil shape, is suitable.

Therefore, the airfoil is expression using B-spline curves, which connect the upper and lower surfaces separately. As design variables, the control points of the B-spline curve are used. Furthermore, to expression airfoils with sharp angles, such as double-wedge airfoils, division points on the B-spline curves of the upper and lower surfaces are set separately, and their positions are also optimized.

An example of airfoil expression is shown in Fig. 2. The lower surface is not divided by the B-spline curve, and the upper surface is divided by the first control point from the leading edge.

4 Aerodynamic Characteristic Estimation Technique

Because a Multi-Fidelity-EGO capable of inputting two types of fidelity is used, two types of analyses with different accuracy are required for the aerodynamic characteristics estimation technique. In this study, computation fluid dynamics (CFD) analysis is used for the high-accuracy evaluation, and the panel technique is used for the low-accuracy evaluation. In this section, these two techniques are explained.

Table 1 Flight conditions of the optimization

Mach number [-]	1.5
Altitude [km]	18
Reynolds number [-]	3.8×10^6

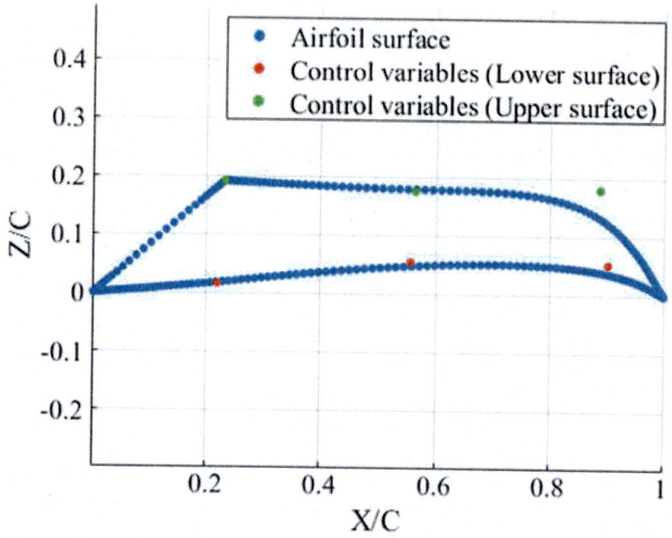

Fig. 2 Example of airfoil expression

4.1 CFD for the High-Fidelity Estimation

As a high-fidelity aerodynamic characteristic estimation technique, the FAST aerodynamic routines (FaSTAR) of compressible fluid solver developed by the Japan Aerospace Exploration Agency was adopted [8]. FaSTAR is a high-speed CFD code. The model parameters for the CFD analysis are shown in the (Table 2).

The mesh necessary for the CFD analysis uses a free software, i.e., two-dimensional structure grid generator or construct2D. An example of the calculation grid is shown in Fig. 3. The calculation region is set to be approximately 30 times the chord length, with a total of 250 points on the airfoil surface and 120 points in the radial direction. The first layer of the airfoil's surface has a minimum lattice width of approximately 3.56E-6 [1/Chord]. The boundary conditions are the upstream semicircle of the O-grid as the inlet boundary and the downstream semicircle as the extrapolation boundary.

Table 2 CFD analysis parameters

Software	FaSTAR
Meshing	Construct 2D O-Grid
Governing equation	Navier–Stokes Equation
Turbulence model	SA-noft2
Angle of attack	0–20° 4 steps

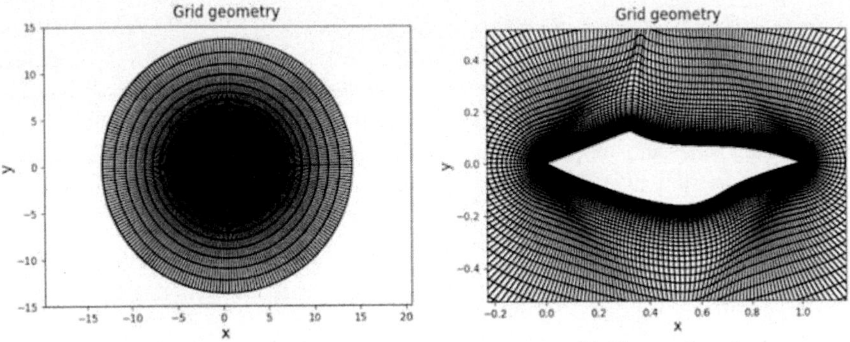

(a) Far-field view (b) Close-surface view

Fig. 3 CFD mesh example

4.2 Panel Method for the Low-Fidelity Estimation

The local surface inclination method is adopted for the low-fidelity estimation technique. In the supersonic regime, this is a useful technique. The airfoil surface is divided into panels, and the pressure coefficient is calculated from the angle formed by the main flow and panel, ignoring the panels' aerodynamic mutual interference. The tangent-wedge method is applied to the compression region, and the Prandtl–Meyer expansion flow theory is applied to the expansion region.

- Tangent-wedge method

The tangent-wedge method is a technique for estimating the flow-field state from the relational expression of a two-dimensional oblique shock wave. The solution of the following cubic equation for the shock angle θ_s is obtained.

$$[sin^2(\theta_s)]^3 + b[sin^2(\theta_s)]^2 + c[sin^2(\theta_s)] + d = 0 \tag{7}$$

Equation 7 is arranged as follows, where $R = sin^2\theta_s$

$$R^3 + bR^2 + cR + d = 0 \tag{8}$$

each coefficient becomes as follows. δ indicates the angle of inclination of the panel relative to the freestream.

$$b = -\frac{(M^2 + 2)}{M^2} - \gamma \sin^2(\delta), \tag{9}$$

$$c = \frac{(2M^2 + 1)}{M^4} - \left[\frac{(\gamma + 1)^2}{4} + \frac{(\gamma - 1)}{M^2}\right] \sin^2(\delta) \tag{10}$$

$$d = \frac{cos^2\delta}{M^4}, \tag{11}$$

The three solutions of Eqs. 9–11 are expressed as R_1, R_2, andR_3, respectively, in an ascending order and are physically fit solutions. When the shock angle is known, the pressure coefficient is calculated using the equation for oblique shock waves:

$$C_p = \frac{4M_\infty^2 sin^2\theta_s - 1}{(\gamma + 1)M_\infty^2} \tag{12}$$

where γ denotes the specific heat ratio. The relationship between the shock angle and angle between the freestream and panel is obtained using the following equation:

$$sin\theta_s = \frac{1}{2}(\gamma + 1)sin\delta. \tag{13}$$

- Prandtl–Meyer expansion wave theory

When the angle between the freestream and panel is greater than 90°, the Prandtl–Meyer expansion wave theory is applied to the expansion region. The Prandtl–Meyer flow is a two-dimensional steady supersonic isentropic flow. The Prandtl–Meyer function is given by the Mach number.

$$\nu(M) = \frac{1}{\lambda}tan^{-1}\lambda\beta - tan^{-1}\beta \tag{14}$$

Here, $\beta = \sqrt{M^2 - 1}$ and $\lambda = \sqrt{(\gamma - 1)/(\gamma + 1)}$. Determine the Mach number after expansion. First, Eq. 14 is modified as follows:

$$F(\beta) = \frac{1}{\lambda}tan\left[\lambda(\nu + tan^{-1}\beta)\right] - \beta = 0 \tag{15}$$

If β_e satisfies, then Eq. 15 is calculated numerically using the Newton–Raphson method. The sequential update formula, β_e, is given below:

$$\beta_{n+1} = \beta_n + \frac{(1 + \beta_n^2)F(\beta_n)}{(1 - \lambda^2)\beta_n^2}. \tag{17}$$

The Mach number after the expansion can be calculated using Eq. 16, and the static temperature, static pressure, and pressure coefficient after the expansion can be calculated using the relational expression of isentropic flow.

$$T_e = T_\infty \frac{2 + (\gamma + 1)M_\infty^2}{2 + (\gamma + 1)M_e^2}, \tag{17}$$

$$p_e = p_\infty \left(\frac{T_e}{T_\infty} \right)^{\frac{\gamma}{\gamma - 1}} \tag{18}$$

$$C_p = \frac{2}{\gamma M_\infty^2} \left(\frac{p_e}{p_\infty} - 1 \right) \tag{19}$$

5 Optimization Result

5.1 Pareto Front

The Pareto front obtained via the optimization is shown in Fig. 4. Only individuals evaluated with a high accuracy are plotted. The " + " sign in the upper left of the figure is a randomly initialized indivisuals, and the lower right direction is the optimal direction. The color bar represents the number of generation, and the evaluation point near the Pareto front increases as the generation number increases, proving that the explorer is conducted properly. It is possible to generate a sufficient number of airfoils on the Pareto front, and it succeeds in the generation of various airfoils from the airfoil with a large lift-to-drag ratio airfoil in which the aerodynamic center position shift forward. The Pareto front can be used to observe the tradeoff relationship between the

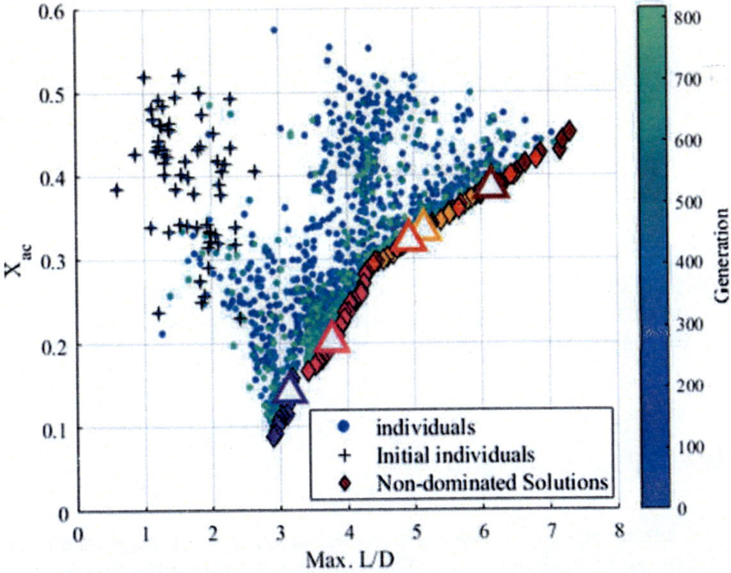

Fig. 4 All the evaluated individuals with Pareto front

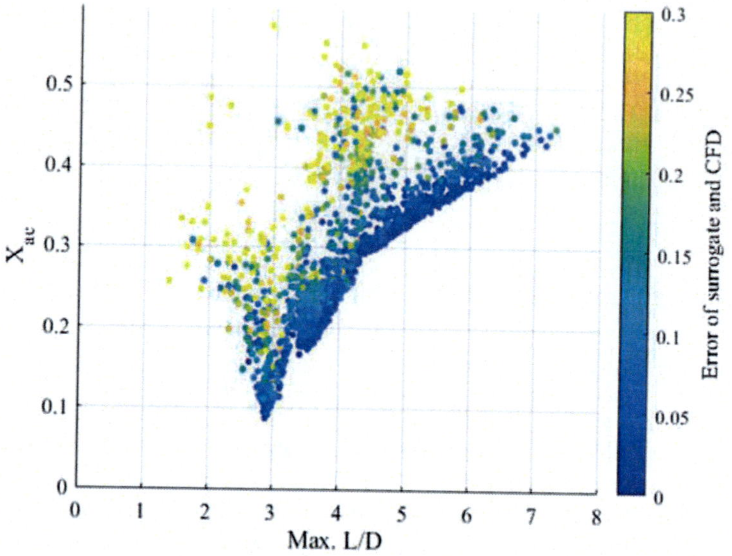

Fig. 5 Accuracy of the surrogate model estimation

lift–to–drag ratio and aerodynamic center position. Nondominanted solution airfoils are color coded and clustered into five clusters. The center of cluster is indicated by "▲,".

5.2 *Surrogate Model Accuracy*

The surrogate model estimation error is shown in Fig. 5. The error is defined here as the difference between the surrogate model estimate just before the evaluation of the individual evaluated through the high-fidelity evaluation during the optimization. Individuals that are far from the Pareto front tend to have larger errors, and those who are closer to the Pareto front tend to have smaller errors. The optimization can efficiently search for the location of a promising individual, and the error can be reduced in the vicinity of that individual.

5.3 *Clustering*

In order to examine the obtained pareto-front airfoil shape, we grouped similar individuals into clusters. Clustering is not done by design variables but by features that represent the shape of airfoils. The features used for clustering are shown in Fig. 6. The definitions of the variables shown in Fig. 6 are shown in Table 3. The represen-

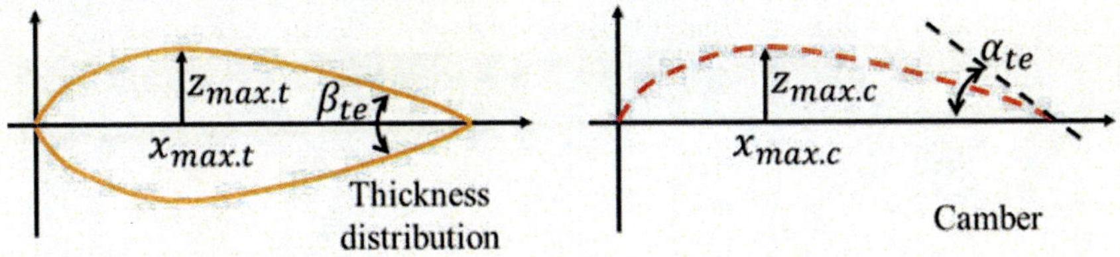

Fig. 6 Geometric feature of clustering

Table 3 Geometric Feature Definition

	Variables	Definition
Thickness distribution	$z_{max.t}$	Max. thickness
	$x_{max.t}$	Max. thickness location
	$r_{max.t}$	Curvature at Max. thickness
	β_{te}	Opening angle at T.E
Camber line	$z_{max.c}$	Max. thickness
	$x_{max.c}$	Max. thickness location
	$r_{max.c}$	Curvature at Max. thickness
	α_{te}	Angle of T.E

tative quantities of the thickness distribution and camber line that define the airfoil shape are selected as the features. Clustering is done using the k-means method. The center of cluster j as m_j and the feature of wing type i as x_i. k-means method is the combination of $m_1 \cdots m_M$ that minimizes the Eq. 20 [9].

$$E(m_1 \cdots m_M) = \sum_{i=1}^{N} \sum_{j=1}^{M} |x_i - m_k|^2 \tag{20}$$

The airfoil closest to the cluster center calculated by the k-means method is called the center of the airfoil cluster. Among the obtained airfoils, the lift–to–drag ratio at the center of the cluster is called 1 to 5 in order of lift–to–drag ratio.

5.4 *Geometric Features of Each Cluster*

Figure 7 depicts the pressure field generated via the CFD analysis around the airfoils. The angle of attack is the angle closest to the objective function's maximum lift–to–drag ratio.

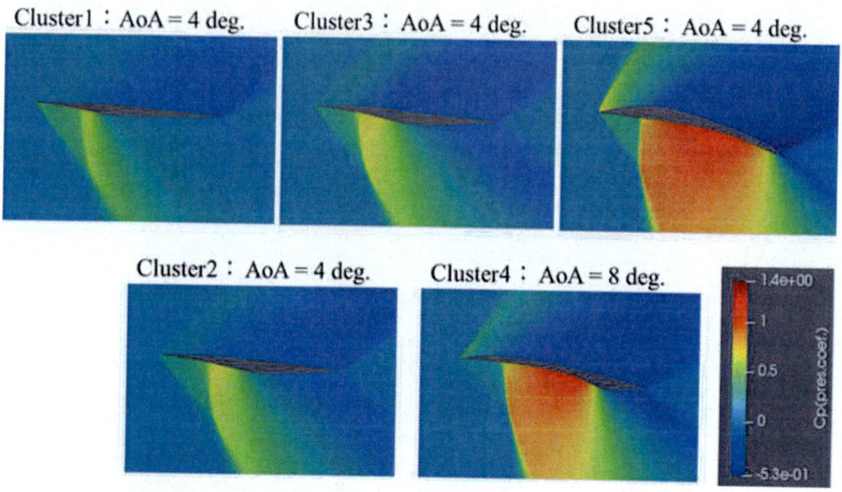

Fig. 7 Pressure Distribution of the Cluster Center Airfoil

First, cluster 1, which has the largest lift-to-drag ratio, is small in thickness, just like a typical supersonic airfoil. However, there is a convexity near the leading edge of the lower surface. The pressure field demonstrates that, shock waves are generated from this convexity. And at the upstream side of the convexity, the lower surface is in the high-pressure region, and at the downstream side, it is in the low-pressure region. The cluster 1–3 has a similar shape and pressure field although the convexity shift rearward as the lift–to–drag ratio decreases. It appears that the convex shape of the lower surface is an important parameter for achieving both the lift–to–drag ratio and the aerodynamic center.

Clusters 4 is significantly different in shape from clusters 1–3, and the convexity that are the characteristic of clusters 1–3 shift rearward. Most of the region on the lower surface is concave except for the convexity near the trailing edge. The shock wave is generated from concave area; and the lower surfaces becomes the high-pressure region in the shock wave downstream. In the convexity near the trailing edge, the transition from the compression wave to the expansion wave occurs in the same way as in the cluster 1–3, and the airfoil surface becomes a low pressure region in the downstream. In other words, the cluster 4 is the same shape as the airfoil of the cluster 1–3, and can be considered as the airfoil in which the convexity remarkably receded.

The airfoil of cluster 5 has no convexity and has a concave shape on the whole lower surface. Because the entire lower surface is concave and has no convexity, all areas of the lower surface form a high pressure area, and the expansion waves only occur at the trailing edge.

The aerodynamic center can be shift forward by making a convexity on the lower surface and transition the compression wave to the expansion wave. And, the tradeoff between aerodynamic center and lift–to–drag ratio is carried out by the position of transition, so it is an important parameter.

On the upper surface, there was no large difference in every cluster unlike the lower surface. The leading edge and the trailing edge are smoothly connected according to the shape of the lower surface.

6 Conclusion

In the supersonic regime, the lift-drag ratio and the aerodynamic center position were optimized as objective functions. To ensure the manueuvability. The airfoil is expressed by B-Spline curve with high degree of freedom, and the improvement to express the edge is added. Because the computational cost is a problem, we use the EGO algorithm, which enables efficient explorer.

As shown in Sect. 5, sufficient number of noninferior solutions to capture a clear tradeoff relationship between the lift–drag ratio and position of the aerodynamic center was obtained. Then, the pressure distribution around the airfoil was examined. Characteristic convexity is generated in the lower surface of airfoil, and this convexity of the lower surface is an important parameter which shifts the aerodynamic center forward. The airframe shape optimization in three dimensions is carried out as future works.

References

1. Balu R, Selvakumar U (2009) Optimum hierarchical Bezier parameterization of arbitrary curves and surfaces. In: 11th Annual CFD Symposium, Indian Institute of Science, Bangalore, India, pp 46–48
2. Sobieczky H (1998) Parametric airfoils and wings. In: Fujii K, Dulikravich GS (ed) Notes on numerical fluid mechanics, vol 68. Vieweg Verlag, pp 71–88
3. Mukesh PR, Selvakumar U (2010) Aerodynamic shape optimization using computer mapping of natural evolution process. In: Proceedings of the 2010 International Conference on Mechanical and Aerospace Engineering April 16–18, vol 5, pp 367–371
4. Zhang Q, Li H (2007) MOEA/D: a multiobjective evolutionary algorithm based on decomposition. IEEE Trans Evol Comput 11(6):712–731
5. Jones DR, Schonlau M, Welch WJ (1998) Efficient global optimization of expensive black-box functions. J Glob Optim 13:455–492
6. Ding F, Kareem A (2018) A multi-fidelity shape optimization via surrogate modeling for civil structures. J Wind Eng Ind Aerodyn 178:49–56
7. Fritsch FN, Carlson RE (1980) Monotone piecewise cubic interpolation. SIAM J Numer Anal 17:238–246
8. Hashimoto A, Murakami K, Aoyama T, et al (2012) Toward the fastest unstructed CFD code "FaSTAR." In: 50th AIAA aerospace sciences meeting including the new horizons forum and aerospace exposition, January 09–12, Nashville, Tennessee
9. Aristidis L, Nikos V, Jakob V (2001) The global k-means clustering algorithm. In: IAS-UVA-01–02, 2001, p 12

A Computational Study on Unsteady Aerodynamic Forces Around a Pitching Airfoil with Shock and Shock-Induced Separation

Noah D. Oyeniran and Hiroshi Terashima

Abstract The present study computationally investigates the characteristics of unsteady aerodynamic forces around an oscillating airfoil in the transonic flow regime. Particular attention is paid to the role of shock wave and shock-induced boundary layer separation in unsteady aerodynamics. The Reynolds-averaged compressible Navier–Stokes equations are solved with SA and SST turbulence models. A well-known forced-pitching NACA64A010 airfoil experiment (Sanford and Gerald in AIAA J 11:1306–1312, 1980, [6]) is simulated, and the freestream Mach number, Reynolds number, and reduced frequency are set to 0.8, 1.2×10^7, and 0.2, respectively. The pitching airfoil with mean angles of attack of 0°, 2°, 4°, and 6° having the amplitude of 1° is parametrically simulated. It is observed that at the mean angle of attack of 0°, there is a phase delay of the lift coefficient against the angle of attack due to the delay of a shock wave movement over the airfoil surface. In contrast, a phase-advanced feature of unsteady aerodynamics appears in increasing the mean angle of attack (e.g., 4° and 6°). There is a phase transition of unsteady aerodynamics from the delay to advance, significantly caused by the appearance of shock-induced boundary-layer separation. The mean angle of attack around 3° may correspond to a transition condition between the phase-delayed and phase-advanced features. The present study demonstrated that the trend of the unsteady aerodynamic characteristics around the transonic oscillating airfoils largely changes with the mean angle of attack. The shock wave, the shock-induced separation, and their interaction play a crucial role in determining the unsteady aerodynamics such as the phase-delayed or the phase-advanced features.

Keywords Unsteady aerodynamics · Pitching airfoil · Shock wave · Shockwave-induced separation · Transonic flow

---

N. D. Oyeniran (✉) · H. Terashima
Division of Mechanical and Aerospace Engineering, Hokkaido University, N13 W8, Kita-ku, Sapporo, Hokkaido, Japan
e-mail: oyenirannoah@gmail.com

H. Terashima
e-mail: htera@eng.hokudai.ac.jp

S. Lee et al. (eds.), *The Proceedings of the 2021 Asia-Pacific International Symposium on Aerospace Technology (APISAT 2021), Volume 1*, Lecture Notes in Electrical Engineering 912, https://doi.org/10.1007/978-981-19-2689-1_26

1 Introduction

The advent of modern aircraft with high maneuver capabilities and high-speed requirements calls for more aircraft flexibilities [1]. Aircraft with these properties tend to be affected by a phenomenon referred to as aeroelasticity. This is due to the flexibility of the aircraft structure and merging interaction between three forces of aerodynamics, elastics, and inertia. The coupling of these forces often leads to aerodynamic instabilities, which are often characterized by a sustained violent vibration that may lead to the destruction of the wings of an aircraft [6, 9]. For example, the limit cycle oscillation is obtained when there is a continuous vibration at a large amplitude until it leads to the destruction of the wings [1, 2]. It can be said that there is a kind of exchange of energy between the fluid and the structure of an aircraft. Thus, this phenomenon involves two main forces, which are the fluid force and structural force of an aircraft. Whenever there is a change in the fluid environment, the change is felt by the structure which then responded to the changes by a considerate amount of displacement. This interaction continues and on. A state is reached when there is no commensuration between these two forces. This often happens when the flow speed becomes so high, and frequency increases due to external disturbance leading to a more aggressive vibration.

An aircraft moving at transonic speed is said to operate in a complicated non-linear flow. This non-linearity is accounted for by the presence of shock waves and their union interaction with boundary layer. The shock wave is characterized by adverse pressure gradient and thus interact with boundary layer to cause separation of the airflow from the wing surface. This phenomenon together with detachment and reattachment of flow and formation of vortices along the trailing edge and wake of the wing bring about an increment in the level of complexities of instabilities that occur at transonic speeds [9]. Some levels of discrepancies are caused by shock-boundary layer interaction which introduces aerodynamic instabilities [9].

Therefore, in order to accurately capture this phenomenon in a way of preventing catastrophic failure that is associated with aerodynamics instabilities leading to flutter and the fact that modern commercial vehicles operate under the transonic flow condition made it a necessity to investigate the unsteady aerodynamics around a forced-oscillating airfoil in the transonic flow regime. The present study computationally investigates the characteristics of unsteady aerodynamic forces around an oscillating airfoil. Particular attention is paid to the role of shockwave and shock-induced boundary layer separation in unsteady aerodynamics.

2 Methodology

The Reynolds-averaged compressible Navier–Stokes equations are solved with the SA [8] and SST [5] turbulence models. SHUS [7] was used for evaluating the inviscid numerical flux. SHUS is one of the Advection Upstream Splitting Method (AUSM) family schemes [4]. The lower–upper symmetric Gauss–Seidel (LU-SGS) implicit scheme [3] was used for time integration. The SA model is a one-equation model that directly deals with turbulent viscosity. It has a good capability in predicting shock waves [8]. The SST model is a two-equation model that hybridizes the two classical turbulence models of k-ε and k-ω. It is a good vortex viscosity predicting model [5].

2.1 Computational Setup

A well-known forced-pitching NACA64A010 airfoil experiment carried out in the NASA Ames 11 by 11 ft Transonic Wind Tunnel [6] was simulated, and the freestream Mach number, Reynolds number, and reduced frequency were set to 0.8, 1×10^7, and 0.2, respectively. The airfoil oscillates around the quarter-chord point. The reduced frequency is given in Eq. (1) as follows:

$$k = \frac{\omega c}{2U_\infty} \tag{1}$$

where ω, c, U_∞ are angular speed, airfoil chord length and free stream velocity, respectively. The pitching airfoil with mean angles of attack of 0, 2, 4, and 6° having the amplitude of 1° was parametrically simulated. It is noted that the conditions with 2 and 6° have little been addressed in past studies. Equation (2) gives the equation for the angle of attack (AoA) with respect to time during oscillation.

$$\alpha(t) = \alpha_m + a_0 \sin(\omega t) \tag{2}$$

where α_m and a_0 are the mean angle of attack and amplitude, respectively.

An overview diagram of the computational grid is shown in Fig. 1. A fixed boundary condition was applied to the outer boundary located 20 times the chord length. At the wall, the density was extrapolated, and the velocity corresponds to the

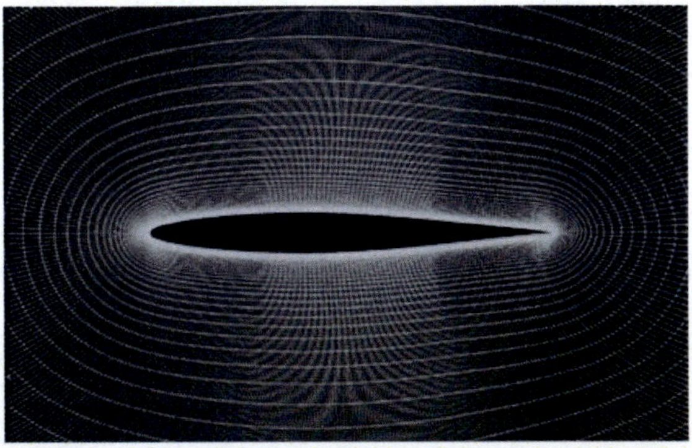

Fig. 1 A computational grid around an NACA64A010 airfoil (O-type grid)

grid speed due to airfoil oscillation given in Eq. (2). The normal momentum equation, which includes the acceleration effect of airfoil surface on the pressure gradients was used to define the pressure at the wall. The grid points are not evenly distributed over the entire airfoil with respect to the airfoil circumference direction, but attention was paid to the airfoil leading edge which has a large rate of changes in curvature and the grid at $x/c = 0.3$–0.7 which can be predicted from experiments or preliminarily performed computations to be a range of shock wave movement. This range between $x/c = 0.3$ and 0.7 was made finer in order to accurately capture shock-boundary layer interaction phenomena. For the vertical direction of the airfoil surface, y^+ equal to or less than one was used as the minimum grid width. A grid convergence study was preliminarily performed, and the effect of time step size was evaluated. As a result, the grid resolution used was 399×100 and the (non-dimensional) time step size used was 2×10^{-3}.

3 Result and Discussion

3.1 Validation

For a validation purpose of the present method and grid resolution, a comparison of results with experimental data [6] using DFT analyses was first carried out using a condition of $\alpha_m = 4°$. For this condition, a shock wave and the shock-induced separation appear on the upper surface (see Fig. 5c). By performing DFT analysis, the changes in pressure coefficients on the airfoil surface are divided into real and imaginary parts based on the changes of angle of attack. The real part represents the phase of the oscillation of the airfoil, while the imaginary part is the phase perpendicular (at 90°) to the oscillation of the airfoil. The graphical plots showing a comparison between the present computational study (using SA and SST models) and experimental work by NASA are provided in Fig. 2. The result demonstrates that the present result is overall in good agreement with the experimental result. The shock wave movement due to the airfoil oscillation is captured well with the peaks having the positive real and imaginary values, while the shock wave position and the range of shock wave movement on the upper surface have some deviations from the experiment. The two turbulence models offer similar prediction accuracy for unsteady aerodynamics and there are no significant differences in the distributions of the real and imaginary parts between them. Thus, the SA model is used during the course of this study.

Furthermore, a comparison was made with a past numerical study [9] by plotting the lift and moment coefficients against AoA at $\alpha_m = 0$ and $a_0 = 1°$ case. Figure 3 shows the degree of closeness between the present study and the past study [9]. The past study was chosen for comparison because it provides a high level of closeness with the experimental data by NASA [6] and also with a computational study using SA model and a high-order scheme [10]. The comparison in Fig. 3 demonstrated a

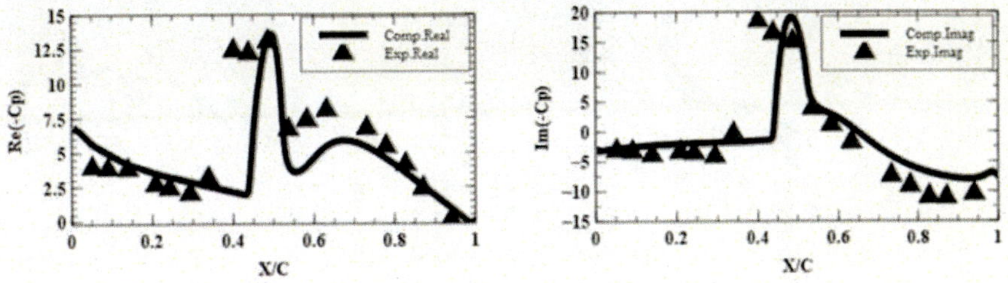

(a) SA model (left: real part, right: imaginary part)

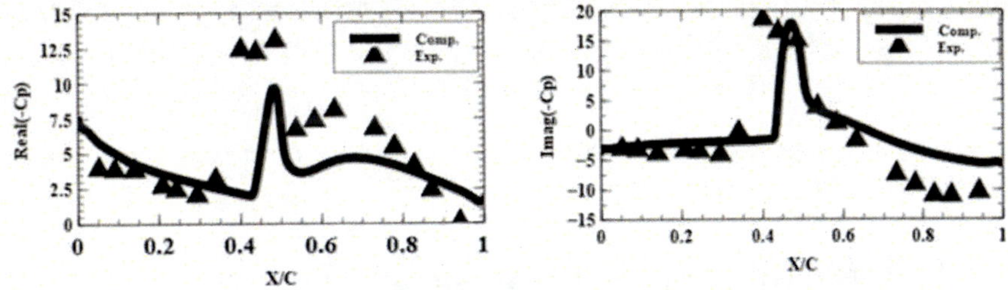

(b) SST model (left: real part, right: imaginary part)

Fig. 2 Comparison of pressure coefficients on the upper surface in terms of real and imaginary parts

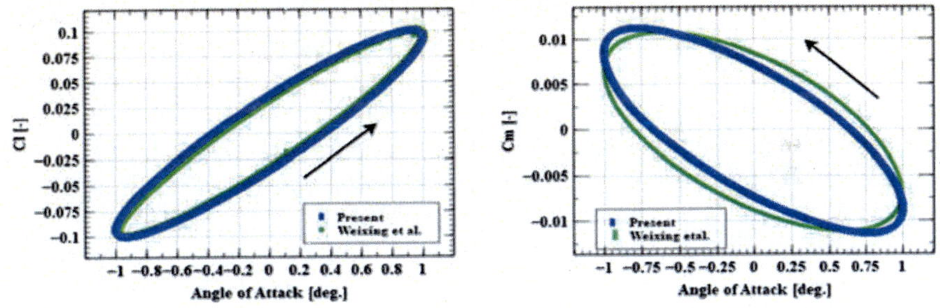

(a) lift coefficient

(b) moment coefficient

Fig. 3 Comparison of lift and pitching moment coefficients of the NACA64A010 airfoil in pitching oscillations under flow condition; $M_\infty = 0.8$, $\alpha_m = 0°$, $Re = 1.2 \times 10^7$

high level of closeness with the result obtained by the past study but with a slight difference in the plot of the moment coefficient against AoA. This discrepancy may be attributed to the fact that the aerodynamic non-linearities are more evident in the pitching moment than in the lift force.

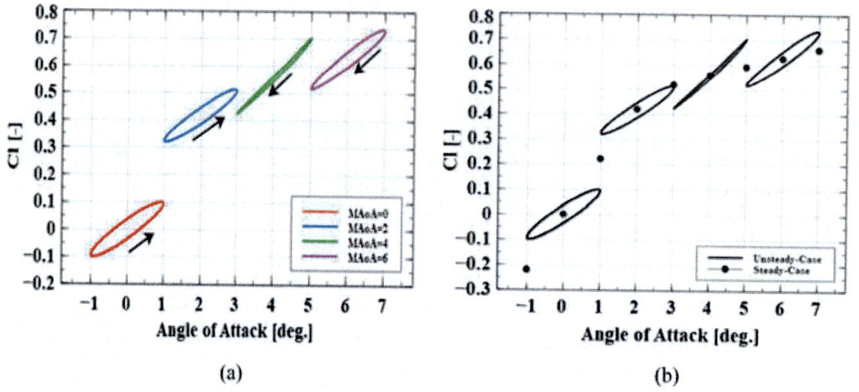

Fig. 4 C_l variations against the angle of attack, showing the result for: **a** each mean angle of attack with direction of loop profile; and **b** both static and oscillating case

3.2 *Unsteady Aerodynamics on Various Mean Angles of Attack*

Figure 4 shows the variation of the lift coefficient (C_l) against AoA for the cases using the mean AoA of 0, 2, 4, and 6° with 1° amplitude of oscillation. The C_l loop at the mean AoA of 0 and 2° follows an anticlockwise direction while the loop direction was reversed in the clockwise direction at the mean AoA equal 4 and 6°. For the mean AoA of 0° case, a typical phase-delayed (anticlockwise loop) profile against AoA is confirmed in the C_l variation. This phase delay may lead to unstable airfoil motion (e.g., in flutter simulations). In contrast, a phase-advanced (clockwise loop) profile of C_l against AoA is observed at the mean AoA of 4 and 6°. Especially, the mean AoA of 4° shows a narrower loop width. This phase-advanced profile is caused by the appearance of massive separation on the upper surface in an event of shock-boundary layer interaction. Figure 5 shows the Mach number distributions for each mean AoA case. The result confirmed that, in the relatively large AoA cases, the massive shock-induced boundary layer separation occurs behind the shock wave on the upper surface. A continuous detachment and reattachment of flows can be observed during the airfoil oscillation and thus the behavior of the separation region significantly changes the shock wave movement. It is noted that, in static (non-oscillating) cases, while a shock wave generally moves aftward in increasing AoA. On the other hand, a shock wave tends to move forward in increasing AoA for higher AoA conditions due to the separation. Thus, the trend of unsteady aerodynamic (lift) forces in the pitching airfoil essentially vary in terms of the mean AoA due to the presence of shock-induced boundary layer separation. In addition, the clockwise profile in the C_l variation may indicate that the unsteady aerodynamics caused by shock-induced boundary layer separation turn out to work for stable airfoil motion.

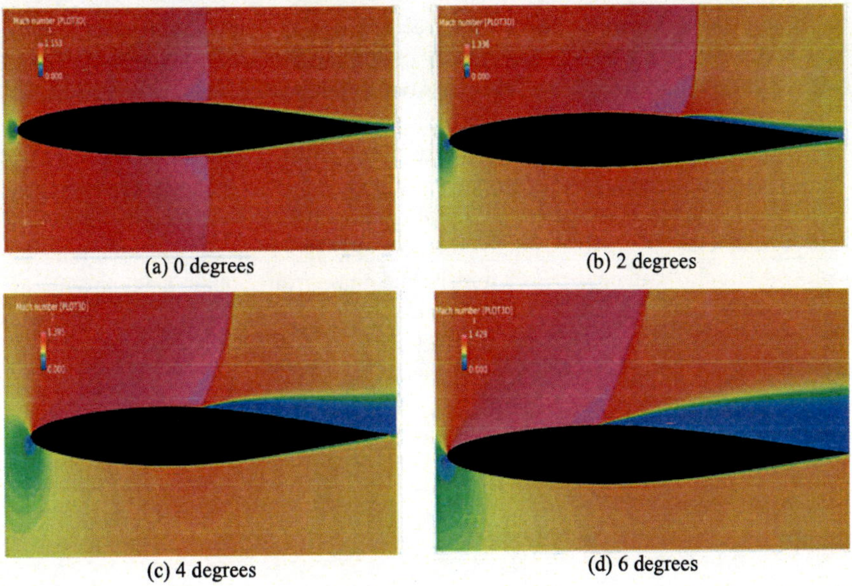

(a) 0 degrees (b) 2 degrees

(c) 4 degrees (d) 6 degrees

Fig. 5 Mach number distributions obtained from static (non-oscillating) airfoil calculations

Furthermore, other effects of shock-induced boundary layer separation on unsteady aerodynamics can be seen, for example, from the result at AoA of 5°. Generally, the maximum and minimum C_l in a pitching airfoil take smaller values than the corresponding static (non-oscillating) airfoil case due to the dynamic behavior (inertia effects). In fact, C_l at AoA of 1° in a static airfoil takes approximately 0.22, while, for the pitching airfoil, C_l at AoA of 1° takes approximately 0.1 in the $\alpha_m = 0$ case and 0.32 in the $\alpha_m = 2$ case, respectively. On the other hand, C_l at AoA of 5° in the pitching airfoil takes 0.7 for in the $\alpha_m = 4$ case, which is larger than that in a static airfoil case (0.56). This unusual behavior can be also observed at AoA of 7° in the case of $\alpha_m = 6$. Thus, the unsteady aerodynamic characteristics significantly vary from low AoA conditions to that of higher AoA condition. The present result indicates that the shock-induced boundary layer separation plays a significant role in determining the unsteady aerodynamic characteristics and thus, the stability of airfoil motion.

4 Conclusion

A computational study on unsteady aerodynamic forces around a pitching airfoil in a transonic flow was performed. The Reynolds-Averaged Navier–Stokes simulation was carried out using the SA and SST turbulence models. The mean angle of attack for the oscillating airfoil was varied from 0 to 6° to investigate the role of shock waves and shock-induced separation in unsteady aerodynamics. The validation study was firstly performed and a good agreement with the experimental data was obtained. The lift coefficient (C_l) variation against AoA formed a hysteresis loop at various mean AoA under consideration. It was shown that, at mean AoA equal 4°, the C_l loop changes direction from anti-clockwise direction to clockwise direction. This event took place due to the changes in the direction of shock movement and the action of shock induced separation. Different phase characteristics can be observed in the profile. The mean AoA equal 0° can be seen as a typical phase delay while mean AoA equal 2° can be seen as a transitional phase from a phase delay to a phase advance. Mean AoA equal 4 and 6° are seen as a critical angle which can be marked with full starts of phase advance embodied with changes in the direction of loop (from anti-clockwise direction to clockwise direction) to mark a new phase tagged "advance phase". This phase is characterized by the activities of shock movement and interaction with boundary layer resulting into separation.

References

1. Chinmaya P (2009) Aeroelasticity—in general and flutter phenomenon. In: Second international conference on emerging trends in engineering and technology, ICETET-09
2. Gao C, Zhang W, Ye Z (2016) A new viewpoint on the mechanism of transonic single-degree-of-freedom flutter. Aerosp Sci Technol 52:144–156
3. Jameson A, Yoon S (1987) Lower-upper implicit schemes with multiple grids for the Euler equations. AIAA J 25:929–935
4. Liou MS, Steffen CJ (1993) A new flux splitting scheme. J Comput Phys 107:23–39
5. Menter FR, Kuntz M, Langtry R (2003) Ten years of industrial experience with the SST turbulence model. In: Hanjalic K, Nagano Y, Tummers M (eds) Turbulence heat mass transfer 4. Begell House, Inc., pp 625–632
6. Sanford SD, Gerald NM (1980) Transonic shockwave/boundary-layer interactions on an oscillating airfoil. AIAA J 11:1306–1312
7. Shima E, Jounouchi T (1997) Role of CFD in aeronautical engineering (No. 14)—AUSM type upwind schemes—. In: NAL-SP34, Proceedings of 14th NAL symposium on aircraft computational aerodynamics, pp 7–12
8. Spalart PR and Allmaras SR (1994) A one-equation turbulence model for aerodynamic flows. AIAA paper 92-0439
9. Weixing Y, Rimple S, Dominique P (2021) Fully coupled aeroelastic analyses of wing flutter towards application to complex aircraft configurations. J Aerosp Eng 34:04020117
10. Wang B, Zha G (2010) Numerical simulation of transonic limit cycle oscillations using high-order low diffusion schemes. J Fluids Struct 26(4):579–601

A Multi-fidelity Method-Based Aerodynamic Design Strategy for Preliminary Prop-Rotors

Hang Zhang, Qijun Zhao, and Guoqing Zhao

Abstract A design strategy to provide suitable baseline prop-rotors for sophisticated applications is presented. Different from those aiming at improving aerodynamic performance of a matured baseline blade fully based on high-fidelity methods, the new strategy is capable of efficiently working out proper layouts with no baseline, which is more beneficial for customizing prop-rotors in industrial applications. The proposed design strategy is introduced through the design process for the initial blades for a self-developed, unmanned tilt-rotor aircraft. The influences of thrust-weighted solidity and thrust requirements in cruise on the aerodynamic performance of the resulting blades are also discussed. Under pre-set design requirements for hover and cruise flight modes, a representative prop-rotor can reach the figure-of-merit of 0.7 in the design hover state and reach cruise efficiency of 0.62 in cruise state due to a relatively lower thrust demand. Using this strategy, part of the design results can be quite close to the requirements, vastly reducing the design space and supporting more advanced optimizations.

Keywords Prop-rotor · Preliminary aerodynamic design · BEMT · Surrogate model · Multi-objective optimization

1 Introduction

Advanced prop-rotors tailored for specific missions may help achieve higher efficiencies, faster cruising speeds, and longer flight ranges for tiltrotor aircraft. Yet

H. Zhang · Q. Zhao (✉) · G. Zhao
College of Aerospace Engineering, Nanjing University of Aeronautics and Astronautics, Nanjing 210016, PR China
e-mail: zhaoqijun@nuaa.edu.cn

H. Zhang
e-mail: hangzhang@nuaa.edu.cn

G. Zhao
e-mail: zgq198495@nuaa.edu.cn

S. Lee et al. (eds.), *The Proceedings of the 2021 Asia-Pacific International Symposium on Aerospace Technology (APISAT 2021), Volume 1*, Lecture Notes in Electrical Engineering 912, https://doi.org/10.1007/978-981-19-2689-1_27

coming up with a highly efficient prop-rotor for specific requirements is still a challenge [1]. To acquire highly efficient blades, researchers have turned to optimization techniques. Many earlier workers employed some optimization methods available in their time and applied them to improve the aerodynamic performance of some matured blades. Michael et al. [2] detailed the development of the Advanced Technology Blade (ATB) used to replace the original metal blade of the XV-15. Using Nash Games Theory and a surrogate model technique, Leon et al. [3] developed a multi-objective strategy to optimize the model rotor, ERATO, in expectation of increasing the figure-of-merit (FM) and reducing the power required in forward flight. For a similar purpose, Manfred [4] established a set of CFD-CSD–based high-fidelity optimization processes for the 7A rotor. But the time-consuming property of CFD required many sacrifices in other aspects. The study revealed that reducing the computational cost was the key to industrial applications of optimization research. Debbie et al. [5] built a multi-fidelity multi-objective optimization (MOP) loop for the 7A rotor to improve both its FM and lift-drag ratio (L/D) in forward flight. Garcia et al. [6, 7] carried out gradient-based aerodynamic optimizations of the metal blade of the XV-15 using a high-fidelity CFD method. Although the optimization loop could cut calculation cost utilizing guidance from gradient information, it relies heavily on choices for its initial design space.

All the research mentioned above provided good insights into implementing optimization of aerodynamic blade designs, but most of them focused on improving the performance of matured blades. This often involved a simple platform for applying certain operating conditions (defined in current work as advanced design or the second period of blade design). However, there are still many problems that occur when applying optimization processes to specific engineering tasks. Multiple rounds of performance calculations may be demanded to select proper conditions and many design parameters may be required to actually perform optimizations. If the condition of the selected reference blade is far from the design points, a larger design space is required, thus making the design loop inevitably more time-consuming and inefficient, particularly when there are multiple objectives to meet. Moreover, directly coupling with CFD makes it hard for optimization algorithms to provide advantages. The dimensions of the design variables and the introduction of CFD to problems where rotational motions are involved, add further challenges for optimization. On the other hand, a few researchers tried to come up with blade layouts that were close to, or even met, design requirements without a baseline (defined in current work as preliminary design, or the first period of blade design). Droandi et al. [8] built a two-level procedure for aerodynamic-shape design of a new prop-rotor blade corresponding, respectively, to the two periods of blade design. As for the first level, however, the approach used to identify chord and twist variable bounds may be short of universality.

To acquire promising aerodynamic layouts efficiently, of prop-rotor blades that are close to specific engineering design goals or that could even be used directly in engineering projects, we propose a new method for the initial design of prop-rotor blades. To that end, an efficient low-fidelity simulation tool based on BEMT was employed to build surrogate models for higher FM and cruise efficiency, respectively.

These models are used to run multi-objective optimizations based on NSGA-II. And then a high-fidelity simulation tool was introduced to check the performance of potential blade layouts on Pareto fronts rather than directly engage in optimization loops in the preliminary design. Because no baseline prop-rotor will be introduced, the research goal is thought to be achieved by the appearance of the blade layouts in the final Pareto front, of which the FM reaches 0.7 and cruise efficiency exceeds 0.6 at the design points, as verified by the high-fidelity simulation tool.

2 Design Strategy and Methods

Most concerns were focused on boosting the two crucial aerodynamic performance factors, *FM* and η, of the preliminary rotor at the design points. The design procedure follows the flow chart depicted in Fig. 1. The process consists of four main parts, including airfoil selection and deployment, calculation of the ideal distribution of rotor twist and chord, building surrogate models, and searching for optima that meet the multi-objective requirements. Except for the first part, which is left to introduce later along with a practical design example, the rest are detailed in order below. The infilling time in the flowchart was designated as 300 to leave the process enough iterations to converge [9].

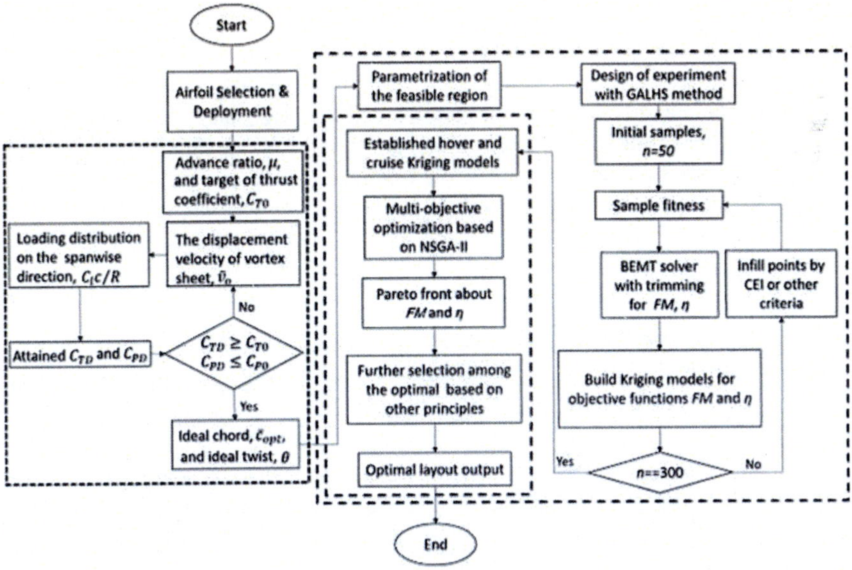

Fig. 1 Flow chat of the design and optimization process

2.1 Identification of Ideal Chord and Twist Distribution

Shown in the left part of Fig. 1, an important process of the method is to acquire the ideal chord and twist distribution along the span-wise direction by iterated calculation of the displacement velocity of the vortex wake from a tiny value. That velocity is assumed to be constant in the radial direction, leading the freestream crossing the propeller disk to form a rigid vortex-sheet moving backward [10]. Formulas contained here are modified in accordance with the velocity triangle depicted in Fig. 2.

Because the circulation around an airfoil Γ can be presented differently as in Eq. (1) from blade element and momentum theory, the local blade loading can be expressed as in Eq. (2).

$$\Gamma = 4\pi R\bar{r}Fw\sin\phi_1 / N = C_L cU/2 \tag{1}$$

$$(C_L c/R) = 8\pi F\bar{r}\bar{w}\sin\phi_1 \Big/ \left(N\sqrt{\mu^2 + \bar{r}^2 - \bar{w}^2}\right) \tag{2}$$

$$\overline{U_1} = \sqrt{\mu^2 + \bar{r}^2 - \bar{w}^2} = U/\Omega R \tag{3}$$

$$F = \frac{2}{\pi}\cos^{-1}\left[\exp\left(-\frac{N(1-\bar{r})}{2\bar{r}\sin\phi_1}\right)\right] \tag{4}$$

in which Ω represents rotational speed and the R blade radius. The w refers to the velocity induced by the vortex weak and $\bar{w} = w/\Omega R$. The terms ϕ_1 and μ refer to the inflow angle and advancing ratio, respectively, and V_∞ represents the speed of forward flight. The term $\overline{U_1}$ represents the non-dimensional local resultant profile velocity at the radial location $\bar{r}$. The term c is the dimensional local chord and N is the blade number. Here, F is Prandtl's tip-loss model. From the principle of minimum energy loss [2, 10] $v_o = w/\cos\phi_1$, Eq. (3) can be further converted into a function of v_o as in Eq. (6).

$$\phi_1 = \tan^{-1}[(\bar{V}_o + \mu)/\bar{r}]; \quad \mu = V_\infty/\Omega R \tag{5}$$

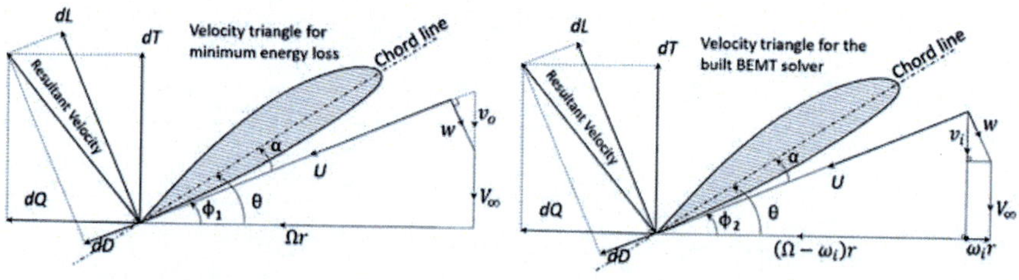

Fig. 2 Velocity triangle on a certain blade element at the relative location of r/R

$$(C_L c/R) = 4\pi \bar{r} \bar{v}_o \sin 2\phi_1 \Big/ \left(N\sqrt{\mu^2 + \bar{r}^2 - v_o^2 \cos^2 \phi_1} \right) \quad (6)$$

Then, as functions of the local blade loading expression $C_L c/R$, the design thrust coefficient C_{TD} and the design power coefficient C_{PD} can be represented below based on the blade element theory:

$$C_{TD} = \frac{N}{2\pi} \int_{\bar{r}_0}^{1} \overline{U_1}^2 \frac{C_L c}{R} (1 - \varepsilon \tan \phi_1) \cos \phi_1 d\bar{r} \quad (7)$$

$$C_{PD} = \frac{N}{2\pi} \int_{\bar{r}_0}^{1} \bar{r} \overline{U}_1^2 \frac{C_L c}{R} (\varepsilon + \tan \phi_1) \cos \phi_1 d\bar{r} \quad (8)$$

Prescribed beforehand, ε is the reciprocal of the lift-drag ratio of the local profile under certain conditions, such as the designated optimal conditions of this paper, corresponding with the optimal angles of attack α_{opt} in Eq. (9) and the root-cut position $\bar{r}_0$. Once the design coefficients, C_{TD} and C_{PD}, meet the design requirements, the ideal chord distribution $(c/R)_{opt}$ and local fixed angle θ can be obtained as:

$$(c/R)_{opt} = (C_L c/R)_{opt}/C_{Lopt}; \quad \theta = \tan^{-1}[(\overline{v}_o + \mu)/\bar{r}] + \alpha_{opt} \quad (9)$$

Showed in the left part of Fig. 1, the resolution process starts with the non-dimensional displacement velocity of a vortex sheet initialized as in Eq. (10). Then, the velocity gets revised by increasing a small amount (of 0.0005) until the target coefficients are attained.

$$\overline{v_o} = \frac{1}{2}\left(\sqrt{\mu^2 + 2C_{To}} - \mu\right) \quad (10)$$

2.2 *Multi-fidelity Aerodynamic Simulation Methods*

For aerodynamic optimization problems of rotors where rotational motions are involved and trimming is requested (which indicates a demand for numerous performance calculations during the process), an efficient aerodynamic solver with adequate accuracy is necessary. BEMT was employed as a performance evaluator of blade candidates to build surrogate models. The small inflow angle assumption in the classical theory was replaced by a more general consideration of large angle approach [1, 11]. This is necessary for applications involving propellers, prop-rotors, and wind turbines in which large inflow angles are common. The non-dimensional formulations upon which the solver is based, are provided as follows:

$$dC_T^{MT} = 4\bar{r}(\mu + \lambda)\lambda F d\bar{r}; \quad dC_Q^{MT} = 4\bar{r}^3(\mu + \lambda)\xi F d\bar{r} \tag{11}$$

$$dC_T^{BE} = \frac{1}{2\pi}\overline{U_2}^2 N(C_L \cos\phi_2 - C_D \sin\phi_2)\bar{c}d\bar{r} \tag{12}$$

$$dC_Q^{BE} = \frac{1}{2\pi}\overline{U_2}^2 N(C_L \sin\phi_2 + C_D \cos\phi_2)\overline{cr}d\bar{r} \tag{13}$$

$$\overline{U_2}^2 = (\mu + \lambda)^2 + (1 - \xi)^2\bar{r}^2; \quad \phi_2 = \tan^{-1}\left[(\mu + \lambda)/(1 - \xi)\bar{r}\right] \tag{14}$$

where dC_T and dC_Q, respectively, refer to the thrust coefficient and the torque coefficient of a blade element with a span-wise width of $d\bar{r}d\bar{r}$ at the radial location of $\bar{r}$, C_L lift coefficient, and C_D drag coefficient. Here, $\overline{U}_2$ represents the non-dimensional local resultant profile velocity and $\bar{c}$ the non-dimensional local chord. The up-index "*MT*" indicates that the expression is derived from momentum theory and "*BE*" indicates derivation from blade element theory. The terms λ and ξ refer to the axial and the azimuthal induced factor, respectively, and they can be expressed as:

$$\lambda = v_i/\Omega R; \quad \xi = \omega_i/\Omega \tag{15}$$

where v_i and ω_i refer to the axial and azimuthal velocity, respectively. All the variables above form a local velocity triangle, as depicted in Fig. 2.

The aerodynamic performance of any candidate layout can be obtained by equalizing different expressions of the same thrust and torque coefficients from the two theories to establish a coupled non-lineal system, and then solving it by numerical methods (like Newton-iteration and fixed-point iteration). Note that the velocity triangle formed for the BEMT solver differs from the one for minimum energy loss (and also for identifying the ideal distributions) which omits the influence of azimuthal induced velocity [10]. The subscript numbers in terms $\overline{U}_1$, $\overline{U}_2$, ϕ_1, and ϕ_2 were set so as to correspond to different theories or the different velocity triangles above.

As for the part with high-fidelity numerical simulations, it was carried out using a home-made CFD code CLORNS [12], which is capable of performing steady and unsteady calculations. Because most calculation states involved in the current work were hovering and cruising flights of a single prop-rotor, the periodic boundary condition was employed and steady-state solutions were carried out.

The finite-volume approach was employed to discretize steady-state flow fields spatially. During simulation, the Jameson scheme was tried in order to compute the convective flux. The implicit Lower–Upper Symmetric Gauss–Seidel (LU-SGS) approach was used to perform temporal discretization. The eddy viscosity was modeled using the *Spalart-Allamas* turbulence model. The grid nodes for a blade mesh were set using 468 × 95 × 179, and that of the background was set as 133 × 302 × 181. The grid height of the first layer on the blade surface was set to 4.0 × 10^{-6}, which means that $y+$ will be <1.0 all over the blade.

2.3 Optimization Methods

The aerodynamic design of the prop-rotor to be discussed in this paper involves more than single objective functions to be considered. The nondominated sorting genetic algorithm II (NSGA-II [13]), an advanced version of NSGA, was introduced to deal with multi-objective optimization. Instead of the established BEMT solver, two Kriging models [14, 15] were built to combine with NSGA-II for robustness considerations. These models were built under the performance values of blade layouts from the BEMT solver according to point-infilling schemes, like constraint-free Expected Improvement (EI) and Constrained Expected Improvement (CEI).

Because constraints imposed on the current work are mainly about blade geometry, and because the aerodynamic performances were trimmed before inputting into the Kriging models, there was no particular need to build surrogate models for the constraint functions. For a blade layout **x**, the CEI scheme can be expressed as

$$\mathrm{CEI}(\mathbf{x}) = \mathrm{EI}(\mathbf{x}) \times \prod_{j=1}^{c} F_j(\mathbf{x});\; F_j(\mathbf{x}) = \begin{cases} 1, & \textit{(if the jth constrain is met)} \\ 0, & \textit{(or not)} \end{cases} \tag{16}$$

where EI(**x**) is the original constraint-free expected improvement value of **x** and $F_j(\mathbf{x})$, the fitness value of the constraints. This scheme is simpler than those based on surrogate models but also stricter and more effective as a result of many practical operations.

All of the introduced optimization algorithms came from two python-based open-source code packages, pyKriging 0.2.0 [16] and Inspyred 1.0 [17], and were customized according to design requirements.

2.4 Validation of the Methods

The first experimental object employed was a full-scale, three-bladed, NACA high-speed propeller [18]. Figure 3a shows comparisons of the numerical predictions from the BEMT solver with the experimental measurements and Ref. [11]. The inflow speed and the collective pitch were valuable under all experimental conditions where the rotational speed was constantly maintained at 1600 rpm. For each set of conditions, or each set of points forming an oblique line in the figures below, the experiments were carried out by augmenting inflow speed or the forward speed ratio within a certain range, while keeping the collective pitch the same. Measurements of the full-scale metal-blade rotor of the XV-15 [19] were also employed to perform validations. Shown in Fig. 3b are test values of the XV-15 at the tip Mach number of 0.69 (the predicted result from Ref. [20]) and that from the newly built BEMT solver. The comparison shows that the predictions from the BEMT solver match well the measurements within the main operating range.

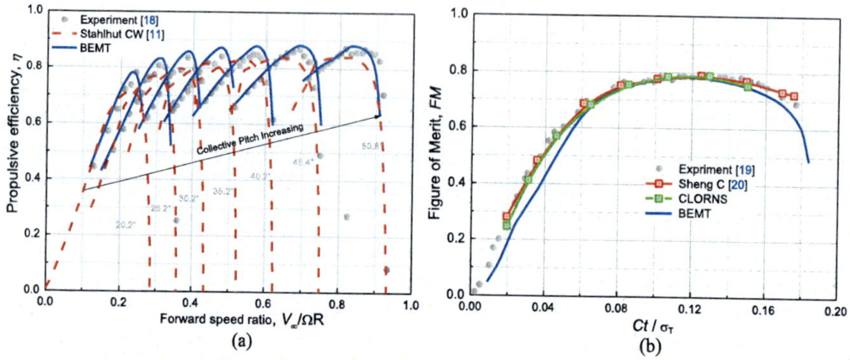

Fig. 3 Validations of the BEMT solver against relative predicted results and the measurements of **a** a three-bladed NACA high-speed propeller; **b** FM of the XV-15 full-scale metal-blade rotor at $M_{tip} = 0.69$

3 Design of a Preliminary Prop-Rotor Blade

The preliminary design for a homemade unmanned tiltrotor aircraft, is considered in this paper. The aircraft has a gross weight of 100 kg, two nacelles mounted on its wing tips, and a prescribed cruise speed of 147 km/h. The aim is to design an efficient initial layout based on the provided design requirements listed in Table 1. An extra download of 7% of gross weight (due to the rotor's downwash), and 10% for consideration of the thrust margin, was added to the thrust demand during the process. In addition, the rotational speed in both flight models were specifically designed to be the same to avoid extra weight associated with a transmission mechanism catering for multi-speed driving.

Airfoil distribution is determined by comparing the lift coefficients at each airfoil's maximum lift-drag ratio in an attempt to make good use of their lift advantages and simultaneously minimize profile loss as much as possible. Two airfoil candidates were the Boeing VR-7 and VR-8, both used as outboard parts of the advanced technology blades of the XV-15. Also included was a modified version of the Boeing 106 (denoted in this paper as Boe106m), of which the thickness was increased from 13.06% of chord to 18.06%. As the predicted lift coefficient showed in Fig. 4a, the

Table 1 Design requirements for the rotor of the unmanned tiltrotor aircraft

Requirements	Hover	Cruise	Requirements	Hover	Cruise
Blade number	3	3	Rotational speed	2150 rpm	2150 rpm
Radius	0.8 m	0.8 m	Gross weight	100 kg	100 kg
Cruise altitude	–	1500 m	Total thrust per rotor	60 kg	6 kg
Cruise speed	–	147 km/h	Power per rotor	≤15 kW	≤15 kW
Advance ratio	–	0.228			

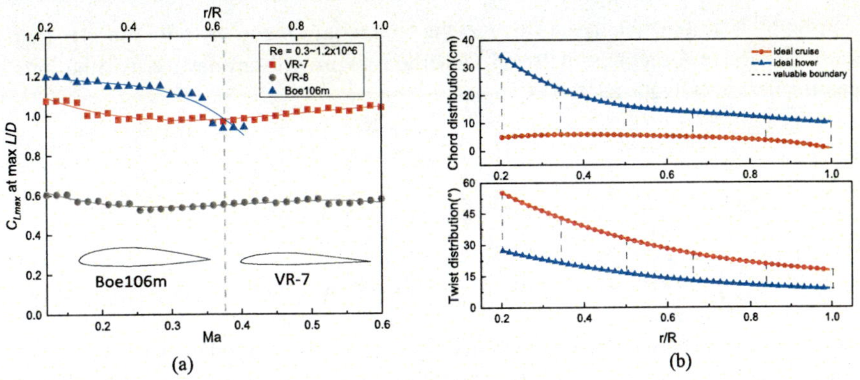

Fig. 4 **a** Comparison of lift coefficient at the range of design tip Mach number and Reynolds number among the airfoil candidates; **b** Ideal twist and chord distribution based on the airfoil deployment

VR-7 was employed within the main part of the rotor, beginning from the radial position of 0.62 to the rotor tip, and the Boe106m served as the root airfoil at 0.2R.

Based on the previously mentioned methods, ideal distribution regions of both twist and chord lines along the blade span were formed within the corresponding ideal lines contained in Fig. 4b. The curves are quite different in the two ideal operation models. As shown in the pictures, under the minimum energy loss assumption, highly efficient hovering requires a rotor with relatively high solidity and a slightly twisted shape, which is just opposite the qualities demanded for the desired cruise flight criteria. The huge distribution gap between the two vastly differing conditions posts a challenge to present the same blade shape to satisfy the highly efficient requirements simultaneously. For all prospective design layouts contained in the feasible regions, those that first meet the thrust demand while hovering, and meet the thrust margin for other maneuvering abilities, will be chosen and used to make further trade-offs to realize efficient cruising.

A parametrization approach was created to represent numerically the feasible regions and layout individuals within the regions. Each individual was represented by a vector $\mathbf{x} = [x_1, x_2, ..., x_{12}]$. The first six variables denoted the local chord value at six radial positions from root to tip, as depicted in Fig. 4b and collectively determined the chord distributions relative to the 1/4 chord line. So did the next six ones for twist distributions. All the curves were formed by approximating these variables with polynomials as an effort to partly prevent shape distortion. In this case, the optimization problem can be expressed as follows:

$$
\begin{aligned}
&find\text{:}\ \mathbf{x} = [x_1, x_2, \ldots, x_{12}] \\
&max\text{:}\ \ FM = C_T^{3/2}/\sqrt{2}C_P;\ \eta_P = TV_\infty/P = \mu C_T/C_P \\
&s.t.\text{:}\ \ \ (1)\ \ |T - T(\mathbf{x})| \le 0.2\% \times T \text{ both in cruse and hover} \\
&\qquad\quad (2)\ \ P(\mathbf{x}) \le 15\ \text{kW}; \\
&\qquad\quad (3)\ \ 0.077\ \text{m} \le c_{0.2R} \le 0.158\ \text{m};\ \ 0.070\ \text{m} \le c_{1.0R} \le 0.101\ \text{m}
\end{aligned}
$$

It should be noted that layouts to meet the final requirements are not limited to that mentioned above. Other parametrization methods to provide any desired distributions need further investigation.

4 Results

The evolution processes of the kriging models began from the same initial sample of 50 individuals, generated by the GA-LHS method, and these were updated 250 times to ensure evolutionary convergence. The built kriging models were then used as the fitness functions during the process of the multi-objective optimization based on NSGA-II, which started with an initial sample of 100 randomly generated points and ended as the evolution process reached 350 generations.

Figure 5 shows the typical candidates on the final Pareto front in terms of their chord and twist distributions. Based on the design requirements and restrictions of the current problem, most of these chord lines are closer to that of an ideal cruise mode and represent a quasi-linear feature. Other planforms (like propellers) can also be obtained by adding numbers of span-wise variables or control points and readjusting their relative radial positions. More sophisticated 3-D characteristics of blades, such as an anhedral tip, dihedral tip, or forward or backward sweep, should be left for

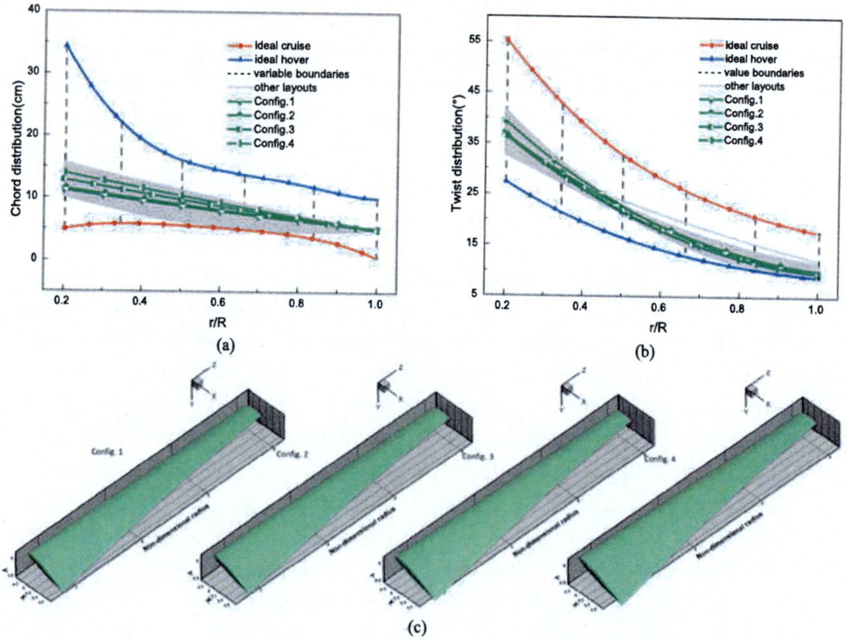

Fig. 5 **a** Chord distributions; **b** Twist distributions; **c** 3-d profiles of typical blade layouts

the advanced design period, where high-fidelity simulation tools are commonly used and can give more accurate reflections of complex blade shapes during simulations. Thus, the chord lines shown in the figure are acceptable as initial points. The twist carves in the picture generally express a non-linear property and most of them have a greater twist angle (up to 30°) than normal helicopter blades have. Local twist angles at the root and mid-span region (up to 0.5R) more approach the middle value of locally available design space, and then gradually decrease to values required by the ideal hover flight at the tip region.

Contained in Table 2 are geometry parameters and aerodynamic performances of the four blades, the associated thrust-weighted solidity σ_T, overall twist angle θ, achieved lift coefficients in different flight models after trimming, and aerodynamic efficiencies in hover and cruise, as predicted by CLORNS. The table shows that the thrust-weight solidity increases in turn, while the twist distributions of blades with Config. 2, 3, and 4 are close. All four pairs of blades could reach the required design load after trimming. Among them, the FM of the Config. 4-blade with the maximum thrust-weight solidity and moderate twist is the highest, reaching 0.724, but its cruise efficiency is the lowest among the four blades, only 0.56. The FM of the Config. 1 blade is the lowest (with the minimum thrust-weighted solidity and the maximum twist), but it obtains the highest cruise efficiency (0.681) among the four blades.

Furthermore, the span-wise load distributions of these blades in the design states were compared. As shown in Fig. 6, by visually comparing the load peaks under the two design flight states, it can be seen that the load of the tiltrotor aircraft in cruise

Table 2 Aerodynamic performances of typical configurations after trimming

Config	σ_T	$\Delta\sigma_T$	θ	Ct_{cruise}	Ct_{hover}	FM	η
1	0.0783	–	29.23°	0.00150	0.0148	0.659	0.681
2	0.0793	+1.3%	26.69°	0.00151	0.0149	0.658	0.651
3	0.0820	+4.7%	27.41°	0.00147	0.0148	0.699	0.623
4	0.0853	+8.9%	26.81°	0.00148	0.0148	0.724	0.562

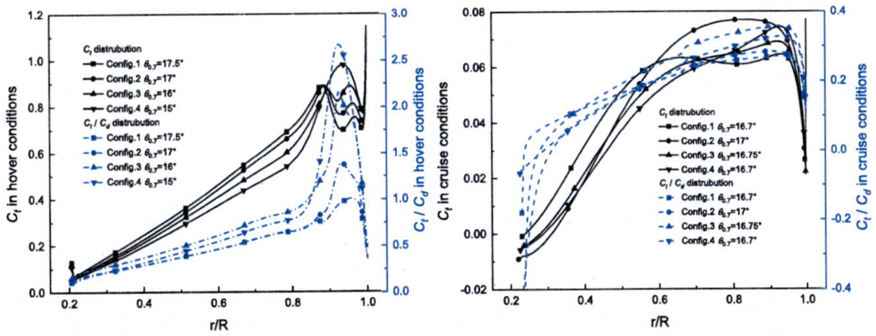

Fig. 6 Radial loading distributions of the typical blade layouts in designed hover (left) and cruise (right) conditions

state is much smaller than that in hover, and it has generated a gap of 10 times under the design states preset in this paper. The load peak of the four pairs of blades in the hovering state is concentrated in the tip area around 0.9R, and the peak value is about 0.9–1.0. However, their load distribution in the design cruise state is relatively uniform and the load peak value is less than one tenth that when in hover. From the span-wise load distributions, influences of a blade's thrust-weighted thrust on the actual aerodynamic performances under each design state can be observed. It can be seen through all the load distributions in the hovering state from Fig. 6 that the lower σ_T is, the higher the collective pitch of a blade must reach in order to meet the design weight, resulting in gradual reduction of the load peak and rapid reduction of the lift-drag ratio in the tip area. It shows that, although each layout can reach the design weight, those with small σ_T have stalled. In the load distributions under the cruise state, there is no stall. The main reason is that even the layout with the smallest σ_T is fully capable of meeting the force demand in the designed airplane mode given the huge gap between the design loads. It also reflects, on the other hand, that σ_T in a specific design mission has a great impact on FM. If there is no appropriate design method, it is difficult to determine the suited σ_T. This is why, in many advanced optimization studies, fixing σ_T is often selected to avoid many problems. Therefore, finding an appropriate σ_T in the initial design stage is also a key parameter for advanced optimization.

Figure 7 shows the hover performance of the four initial configurations by CLORNS. In general, as mentioned earlier, appropriate σ_T can help avoid stalls when the rotor reaches the design weight. On the other hand, the rotor blades with small σ_T, such as Config. 1 and 2, can achieve higher FM under the same Ct/σ_T in the non-stall range, indicating that their compact shape may be suitable for a lower takeoff weight. From this point of view, the initial blade design strategy proposed in this paper can provide potential prop-rotor profiles with acceptable aerodynamic performance and proper σ_T.

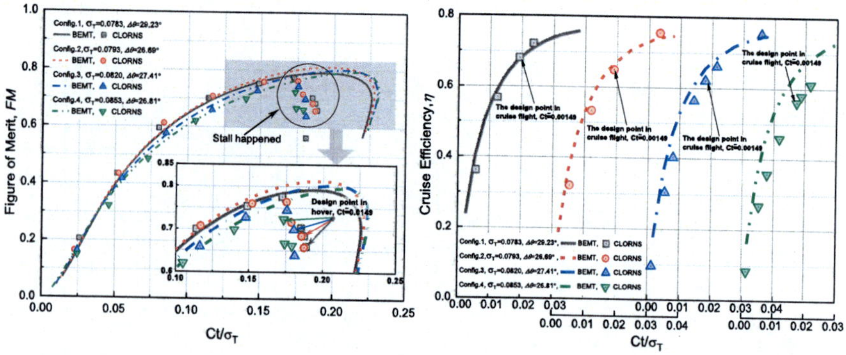

Fig. 7 Aerodynamic efficiencies of the typical blade layouts in designed hover (left) and cruise (right) conditions

From the cruise performance lines in Fig. 7, the highest efficiency of these blades at the design point is about 0.68, which is lower than that in some references. With further increase in disk loading, cruise efficiency of these prop-rotors can continue to increase in the available power range, indicating the possibility of realizing higher speeds. Due to the conservative design for low-speed cruising, the preset cruise load in this paper is only one tenth of the hovering load, which is much lower than the values of 1/3 to even 1/2 [8] in other studies. Therefore, this should be carefully considered when proposing the load demand for cruise states.

5 Conclusions

The proposed design strategy was introduced in detail through the preliminary blade design for a self-developed, unmanned, tilt-rotor aircraft. Under the conservative design requirements, hover performance of the representative Config. 3 at the hover design state reaches 0.7, while its cruise performance reaches 0.62 due to the relatively low disk loading required in the design cruise condition. It is acceptable as a blade baseline. Most of the design results from the strategy can be quite close to the design requirements, vastly reducing design space, and benefitting advanced optimizations.

Moreover, through the current work, the influences of some important design parameters on design of prop-rotors were explained. Namely, a blade layout with adequate thrust-weight solidity can achieve a desired FM without stall, but it may sacrifice cruise efficiency. And, properly set requirements for airplane mode can help obtain higher cruise efficiency.

Acknowledgements This research was supported by the National Natural Science Foundation of China (11872211, 12032012), the National Key Laboratory Foundation of China (no. 61422010401), and as a project funded by the Priority Academic Program Development of Jiangsu Higher Education Institutions (PAPD).

References

1. Leishman JG, Rosen KM (2009) Challenges in the aerodynamic optimization of high-efficiency proprotors. In: 65th annual forum of the American helicopter society, Grapevine, TX, 27–29 May 2009
2. Michael AM, Harold JR, Francis JM (1983) Aerodynamic design of the XV-15 advanced composite tilt rotor blade. In: 39th annual forum of the American helicopter society, St. Louis, Missouri, May 1983
3. Enric RL, Arnaud LP, Jean AD, David A, Michel C (2013) Concurrent aerodynamic optimization of rotor blades using a Nash Game method. In: 69th annual forum of American helicopter society, May 2013, Phoenix, Arizona, USA
4. Manfred I (2012) High-fidelity optimization framework for helicopter rotors. Aerosp Sci Technol 23:2–16. https://doi.org/10.1016/j.ast.2011.12.011

5. Debbie L, David A, Paola C, Jean CR (2013) Aerodynamic rotor blade optimization at Eurocopter—a new way of industrial rotor blade design. In: 51st AIAa aerospace sciences meeting including the new horizons forum and aerospace exposition, 7–10 January, Grapevine (Dallas/Ft. Worth Region), Texas
6. Garcia AJ, Biava M, Barakos GN, Baverstock KD, Gates S, Mullen P (2017) Tiltrotor CFD part II: aerodynamic optimisation of tiltrotor blades. Aeronaut J 121(1239):611–636. https://doi.org/10.1017/aer.2017.21
7. Garcia AJ, Barakos GN, Gates S (2017) Tiltrotor CFD part I: validation. Aeronaut J 121(1239):577–610. https://doi.org/10.1017/aer.2017.17
8. Droandi G, Gibertini G (2015) Aerodynamic shape optimization of a proprotor and its validation by means of CFD and experiments. Aeronaut J 119(1120):1223–1251
9. Han ZH (2016) Kriging surrogate model and its application to design optimization: a review of recent process. Acta Aeronauticaet Astronautica Sinica 37(11):3197–3225 (in Chinese). https://doi.org/10.7527/S1000-6893.2016.0083
10. Glauert H (1963) Airplane propellers. Aerodynamic Theory, Dover, New York, pp 251–254
11. Stahlhut CW, Leishman JG (2012) Aerodynamic design optimization of proprotors for convertible-rotor concepts. In: 68th annual forum of the American helicopter society, Fort Worth, TX, 1–3 May 2012
12. Zhao QJ, Zhao GQ, Wang B, Wang Q, Shi YJ, Xu GH (2017) Robust Navier-Stokes method for predicting unsteady flowfield and aerodynamic characteristics of helicopter rotor. Chin J Aeronaut 31(2):214–224. https://doi.org/10.1016/j.cja.2017.10.005
13. Deb K, Pratap A, Agrawal S, Meyarivan T (2002) A fast elitist non-dominated sorting algorithm for multi-objective optimization: NSGA-II. 6th Parallel Problem Solving from Nature Conference, pp. 182–197, Paris, France
14. Krige DG (1951) A statistical approach to some basic mine valuations problems on the Witwatersrand. J Chem Metallur Min Eng Soc South Africa 52(6):119–139
15. Sacks J, Welch WJ, Mitchell TJ, Wynn HP (1989) Design and analysis of computer experiments. Stat Sci 4(4):409–423. https://doi.org/10.1214/ss/1177012413
16. pyKrigin. https://www.pykriging.com/
17. Inspired. https://pypi.org/project/inspyred/
18. Evans AJ, Liner G (1958) Wind tunnel investigation of the aerodynamic characteristics of a full-scale supersonic-type three-blade propeller at Mach numbers to 0.96. NACA Report 1375, May 1958
19. Felker FF, Betzina MD, Signor DB (1985) Performance and loads data from a hover test of a full-scale XV-15 rotor. NASA TM-86833, November 1985
20. Sheng C, Zhao Q, Hill M (2016) Investigations of XV-15 rotor hover performance and flow field using U^2NCLE and HELIOS codes. In: 54th AIAA aerospace sciences meeting, AIAA SciTech, 4–8 January 2016, San Diego, California

Aerodynamics Characteristics and Flight Dynamics Analysis of Multi-body Aircraft

Chao An, Linpu Wang, Changchuan Xie, and Chao Yang

Abstract As a new aircraft concept composed of independent aircrafts, multi-body aircraft (MBA) is connected together by hinge in wing tip to wing tip configuration. The dynamic characteristics of this kind of aircrafts are different from traditional aircrafts. If the independent aircraft is stable, MBA will be unstable in this configuration. In this paper, a dynamic modeling method of MBA was proposed. The state-space vortex lattice method (VLM) was used for the calculation of aerodynamics coefficients. Newton–Euler equation was used to obtain the flight dynamic model with absolute coordinates. In addition, trim and stability analysis were presented based on the dynamic model and the particular characteristics of this aircrafts were illustrated.

Keywords Multi-body · Aerodynamics · Vortex lattice method · Dynamics · Stability

1 Introduction

As the representative of the very flexible airplane, high-altitude long-endurance (HALE) aircrafts have attracted extensive attentions for decades. Unfortunately, because of the light weight and large flexibility, the wing of HALE produces large deformation during the flight, which leads to the geometrical nonlinear aeroelastic problem [1, 2]. With another perspective, a new concept of HALE, namely multi-body aircraft (MBA), was investigated to solve the aircraft with a large-aspect-ratio with superiorities. This concept is based on the combination of several independent rigid body aircraft to a single HALE aircraft in the mission altitude. Large-aspect-ratio wing can be accomplished without the consideration of structural geometric nonlinearities, so that the accurate nonlinear aeroelastic framework can be omitted temporary. Lower bending moment on the aircraft structure ensures the increase of lifetime due to the joints. Meanwhile, a deacrease in wing-tip vortices ensure the

C. An (✉) · L. Wang · C. Xie · C. Yang
School of Aeronautics Science and Engineering, Beihang University, Beijing 100080, China
e-mail: ac@buaa.edu.cn

S. Lee et al. (eds.), *The Proceedings of the 2021 Asia-Pacific International Symposium on Aerospace Technology (APISAT 2021), Volume 1*, Lecture Notes in Electrical Engineering 912, https://doi.org/10.1007/978-981-19-2689-1_28

similar boost in range and endurance as HALE. Moreover, each single aircraft can fly to the MBA for combing or return to ground for changing the flight mission with different module, aiming to maintain or other reasons.

Experimentally, MBA was first done by the United States Air Force and code-named "Project TipTow". This project consisted of two F-84s connected to the wing tips of a B-29. One flight test experienced unstable oscillations and the aircrafts crashed. Since the end of Project TipTow, some researches have been conducted on flight dynamics and flight control. Magill [3] analyzed the best position of the MBA docking for the best aerodynamics benefit by wind tunnel test. Cooper et al. looked at the dynamics and control of fixed wing aircraft with the capability to link together. With the simulation and modeling, the aerodynamic loads and linkage froces were calculated [4]. Montalvo introduced meta aircraft conception and the overall flight dynamic changes in meta aircraft were investegated as well. The simulation revealed that fundamental rigid body flight dynamic modes changed as a function of the number of aircrafts, and the experienced modes changed, with the characterisitcs of the joint mechanism [5]. The flight test of multiple UAVs connected has been investigated recently, and the control system of take-off and land-off were investigated [6, 7]. An et al. used Newton–Euler dynamics equation adding constranhts to single aircraft. The dynamic modes changes were also presented [8]. Alexander et al. adopted the lifting line method and Kane method to model aerodynamic loads and multi-body dynamic system [9–11]. Two configurations were examined in their researches. The first configuration permited a pitch motion between the independent aircraft and the second configuration allows a pitch and roll motion. Unstable phenomenon occured in both two configurations.

For a wingtip-jointed multi-body aircraft system, the aerodynamics force and moment were highly coupled between each other. Lifitng line method and 2-D aerodynamics model were applied for aerodynamics modeling. The accuracy was limited and precised aerodynamics modeling was reqiured for aerodyanmic load simulation. The unsteady vortex lattice method (VLM) is the time domain aerodynamics calculation method that allows for the simulation of the flow around the wing undergoing arbitrary motions. VLM is especially popular in geometrically nonlinear aeroelastic analysis of very flexible aircraft with large structural deformations. In this paper, state-space VLM was used for the calculation of aerodynamic derivatives. Newton–Euler equation was applied for modeling of multi-body dynamic system with absolute coordinates. Under the trim configuration with linearization of state-space model, trim and stability analyses were carried out with difference with traditional independent aircraft.

2 Theory

A flight mechanic model was proposed to investigate the steady and stable flight mechanical characteristics of the MBA. Newton–Euler method was used to model the multi-body dynamics [8]. State-space VLM and derivation model were proposed to simulate the aerodynamic characteristics. Trim and stable analyses can be proposed based on the flight dynamics model.

2.1 Flight Dynamics

Each aircraft has six coordinates to describe state, including three Cartesian coordinates $r_i = (x_i, y_i, z_i)$ corresponding to Earth coordinate system and three Euler angles $\varphi_i = (\phi_i, \theta_i, \psi_i)$. The equations of motion for a rigid body are derived from the equations of motion for a general system of particles with Euler's law:

$$m_i r_i = A^{(0,i)} F_i^{(i)}, (i = 1, 2, 3, \ldots, n) \tag{1}$$

$$J_i^{(i)} \dot{\omega}_i^{(i)} + S(\omega_i^{(i)}) J_i^{(i)} \omega_i^{(i)} = M_i^{(i)}, (i = 1, 2, 3 \ldots, n) \tag{2}$$

where m_i is the mass of aircraft, $F_i^{(i)}$ represents the active forces, $M_i^{(i)}$ means the active moments, $J_i^{(i)}$ marks the inertia of aircraft, $\omega_i^{(i)} = (p_i, q_i, r_i)$ stands for the angle velocity, and $A^{(0,i)}$ is the transfer matrix from body coordinates to inertial coordinates. Standard shorthand notation was used for trigonometric functions:$cos\alpha \equiv c\alpha$, $sin\alpha \equiv s\alpha$. The superscript (i) denotes the item in equation and is corresponding to respective body coordinate system. $S()$ is the skew symmetric operator. Multiplying this matrix by a vector is equivalent to a cross product:

$$S(\omega_i^{(i)}) = \begin{pmatrix} 0 & -r_i & q_i \\ r_i & 0 & -p_i \\ -q_i & p_i & 0 \end{pmatrix} \tag{3}$$

Equation (1) governs the motion of the centre of mass of the rigid body and Eq. (2) governs its angular motion. The relationship in angular motion is added as:

$$\omega_i^{(i)} = D_i \dot{\varphi}_i = \begin{pmatrix} 1 & 0 & -s\theta \\ 0 & c\phi & s\phi c\theta \\ 0 & -s\phi & c\phi c\theta \end{pmatrix} \dot{\varphi}_i \tag{4}$$

As a result, the dynamic equations of each single aircraft are expressed as:

$$A\ddot{u} = B \tag{5}$$

$$\ddot{u} = [u_1^T, u_2^T, u_3^T, \ldots, u_n^T]^T \tag{6}$$

$$A = diag(A_1, A_2, A_3, \ldots, A_n) \tag{7}$$

$$B = diag(B_1, B_2, B_3, \ldots, B_n) \tag{8}$$

With the hinge joint restricting relative roll motion between ith and jth aircraft, the constraint equations are formed as:

$$r_i + A^{(0,i)} c_{ij}^{(i)} - r_j - A^{(0,j)} c_{jj}^{(j)} = 0 \tag{9}$$

$$A^{(0,j)} p_j^{(i)} - A^{(0,j)} p_j^{(j)} = 0 \tag{10}$$

where c_{ij}, c_{jj} is the direction vector from center of mass of aircraft O_{ci}, $O_{cj}$to connection point O_j and p_j is the direction vector of roll motion. It should be noted that the joint is assumed as ideal without friction, damping or spring forces. The constraint equations can be expressed in a brief form:

$$\Phi(u) = 0 \tag{11}$$

Considering that n single aircraft is connected, dynamic equations with the constraint of MBA are expressed as:

$$\begin{pmatrix} A & {\Phi_u}^T \\ \Phi_u & 0 \end{pmatrix} \begin{pmatrix} \ddot{u} \\ \lambda \end{pmatrix} = \begin{pmatrix} B \\ \zeta \end{pmatrix} \tag{12}$$

$$\zeta = -[(\Phi_u \dot{u})_u \dot{u} + 2\Phi_{ut} \dot{u} + \Phi_{tt}] \tag{13}$$

where $\lambda$is Lagrange multipliers conjugated to constraints, t marks time and Φ_u, Φ_{ut}, Φ_{tt} refers to Jocobi matrix.

For a stability analysis, the nonlinear has to be linearized. Rewrite Eq. (12) as:

$$F(\ddot{u}, \dot{u}, u, \lambda) = 0 \tag{14}$$

when introducing a variable $z = \left(u^T, \eta^T\right)^T$, $\ddot{\eta} = \lambda$, the linearized equation can be expressed as:

$$\tilde{M}\Delta\ddot{z} + \tilde{C}\Delta\dot{z} + \tilde{K}\Delta z = 0 \tag{15}$$

where the $\tilde{M}$, $\tilde{C}$, $\tilde{K}$ is the Jocobi matrix of Eq. (14) with respect to $\ddot{z}$, $\dot{z}$, z, respectively. The stability analysis is based on the state-space model of Eq. (15).

2.2 Aerodynamics Model

Aerodynamic forces and moments were calculated with the state-space VLM. The continuous-time state-space UVLM was a time-domain aerodynamic solver based on potential theory. Different from the frequency domain aeroelastic methods, such as doublet lattice method (DLM), the transient aerodynamic response under arbitrary motions can be obtained directly without harmonic vibration assumptions. According to the theory of VLM, first, the lifting surface is discretized to several panels. Here, the lifting surface was selected as the mean camber surface of the wing, which can be non-planar in geometry due to the airfoil camber, twist, sweep and structural deflections, as illustrated in Fig. 1.

With no penetration boundary condition stated as [12]:

$$\mathbf{K}_1\boldsymbol{\Gamma}_b(t) + \mathbf{K}_2\boldsymbol{\Gamma}_{w0}(t) + \mathbf{K}_3\boldsymbol{\Gamma}_{wl}(t) = -\mathbf{V}\cdot\mathbf{n} \tag{16}$$

where $\mathbf{K}_1$, $\mathbf{K}_2$, and $\mathbf{K}_3$ are influence coefficient matrices and define the normal induced velocities to the wing surface induced by the bound vortex $\boldsymbol{\Gamma}_b$, the first row of wake $\boldsymbol{\Gamma}_{w0}$, and the other wake elements $\boldsymbol{\Gamma}_{wl}$ on the collocation points. Each element of the influence coefficient matrices represents the normal induced velocity computed by the Biot-Savart Law. $\mathbf{V}\cdot\mathbf{n}$ represents the contribution of the freestream velocity and arbitrary motion of the wing. The aerodynamics force and moments can be obtained by:

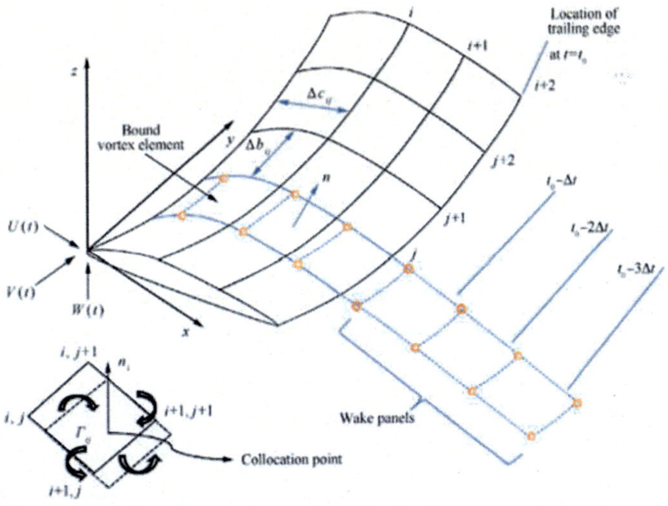

Fig. 1 Aeromodel of the unsteady vortex lattice method

$$\begin{bmatrix} \mathbf{F} \\ \mathbf{M} \end{bmatrix} = \mathbf{L}_9\mathbf{\Gamma}_w + \mathbf{L}_{10}\boldsymbol{\alpha} + \mathbf{L}_7\dot{\boldsymbol{\alpha}} + \mathbf{L}_{11} \tag{17}$$

where $\boldsymbol{\alpha}$ is the perturbation angle of attack on the wing surface, $\mathbf{L}_9$ represents the contribution of the wing and wake, $\mathbf{L}_7$ and $\mathbf{L}_{10}$ refer to the contribution of perturbing the freestream flow, and $\mathbf{L}_{11}$ marks the contribution of the motion of the wing surface. The accurate presentation of matrix can be obtained in Ref. [12].

Aerodynamics coefficients C_i can be obtained by solving Jacobi matrix of Eq. (17) with $\boldsymbol{\alpha}$ and $\mathbf{L}_{11}$, because the contribution of the motion of the wing surface and perturbing the freestream flow are included in $\boldsymbol{\alpha}$ and $\mathbf{L}_{11}$,

$$C_i = \partial \begin{bmatrix} \mathbf{F} \\ \mathbf{M} \end{bmatrix} / \partial[\boldsymbol{\alpha}, \mathbf{L}_{11}]. \tag{18}$$

3 Numerical Studies

3.1 Aircrafts Model

As described before, the multi-body aircraft is based on the configuration, and several independent aircrafts are connected to assmeble a HALE aircraft. With original assumption, each aircraft can fly individually with different missions or maintain demand. The wing tip to wing tip configuration in this paper permits a roll motion between two independent aircrafts with a hinge joint [8]. All degrees of freedom between two coupled aircraft are constrained. The schematic configuration has been presented in Fig. 2.

The single aircraft in wing tip to wing tip configuration has a straight wing, a horizontal tail and two vertical tails. The wing of single aircraft has three sections and the spanwise length of each section is 1000 mm and the chordwise length is 270 mm. Design parameters have been presented in Table 1. It should be illustrated that the design of fuselage refers to the X-HALE of University of Michigan [13].

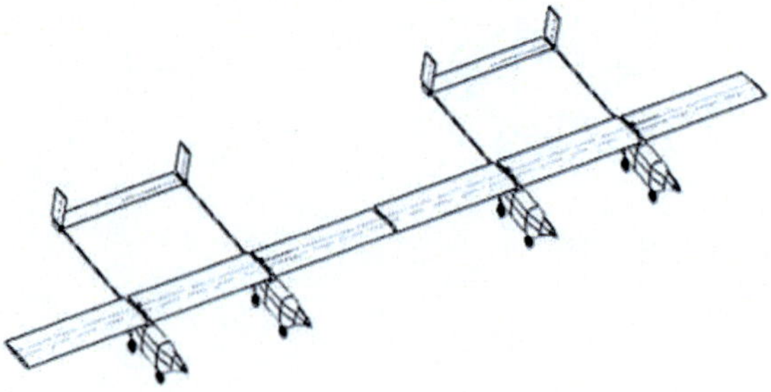

Fig. 2 Configuration of multi-body aircraft

Table 1 Design parameters of single aircraft

Parameter	Value (mm)	Parameter	Value
Span length of aileron	1000	Height of vertical tail	240 mm
Chord length of aileron	90	Chord of vertical tail	120 mm
Span of horizontal tail	960	Distance of tail	1020 mm
Chord of horizontal tail	120	Weight	8 kg

3.2 *Aerodynamics Analysis Results*

The left aircraft A and right aircraft B have both roll angle and angle of attack with zero, the aerodynamics coefficients calculation results comparison about angle of attack with software ZONAIR are illustrated in Fig. 3. Different locations of moment centre are different. The lift coefficients stay a constant value and the pitch moment coefficients change. It is obvious that the state-space VLM proposed here can get agreement with ZONAIR.

The left aircraft A has positive roll angle ϕ_A and right aircraft B has positive roll angle $\phi_B = -\phi_A$. The aerodynamics pressure distributions on wing surface and lift distribution in spanwise are illustrated in Figs. 4 and 5. State-space VLM is efficient for multi-body aircraft aerodynamics modelling. Roll angle ϕ_A are given 8° and 12°, respectively. The angle of attack in this two conditions are both 5°.

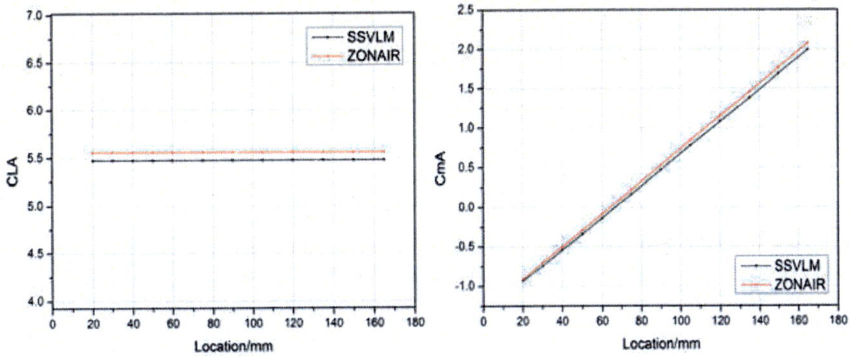

(a) Lift coefficients (b) Pitch moment coefficients

Fig. 3 Comparison of aerodynamics coeffcients. **a** Lift coefficients. **b** Pitch moment coefficients

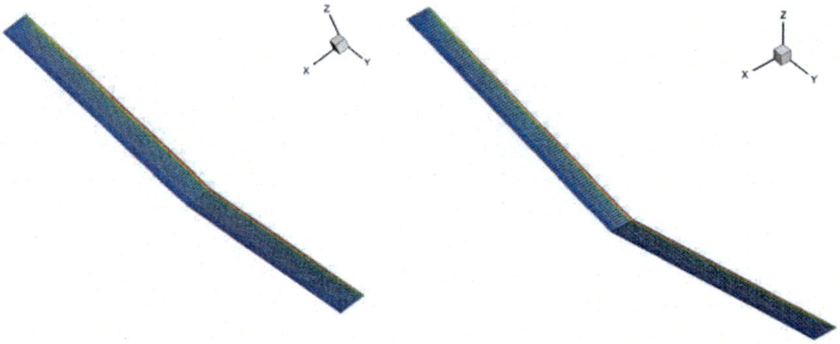

(a) roll angle 8 deg (b) roll angle 12 deg

Fig. 4 Aerodynamics pressure distributions. **a** Roll angle 8°. **b** Roll angle 12°

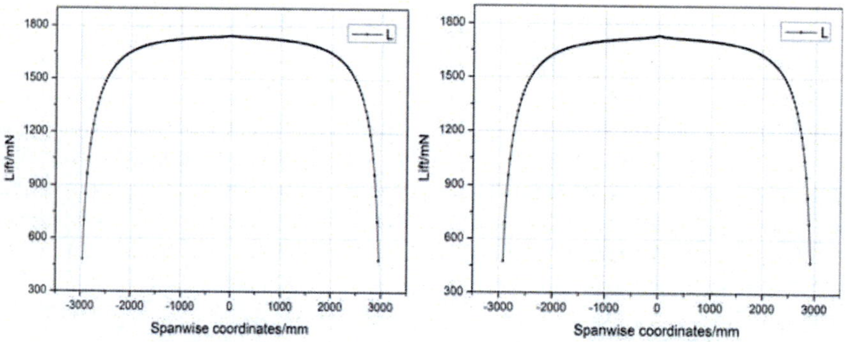

(a) roll angle 8 deg (b) roll angle 12 deg

Fig. 5 Lift distributions in spanwise. **a** Roll angle 8°. **b** Roll angle 12°

3.3 *Trim Analysis and Stability Analysis Results*

The MBA system is trimmed under a given flight condition. With two connected aircraft, the MBA contains seven degrees of freedom. Equation (12) has twelve motion equations and five constraint equations. It is assumed that two rudders have the same deflection angle and two ailerons deflect in the equal number angle but opposite direction. As a result, for two single aircrafts in level flight, seven trim variables are specified as:

$$\tilde{z} = [\alpha_A, \alpha_B, \delta_e, \phi_A, \phi_B, \delta_{aA}, \delta_{aB}] \tag{19}$$

Under the trim condition with 20 m/s velocity and relative air density 1.225, the trim results of two connected single aircraft in level flight have been displayed in Table 2.

Table 2 Trim results of two single aircraft

Trim variable	Value/rad
AOA of aircraft A	0.1218
AOA of aircraft B	0.1218
Elevator	−0.2262
Roll angle of aircraft A	0.031
Roll angle of aircraft B	−0.031
Aileron deflect of aircraft A	0.209
Aileron deflect of aircraft B	−0.209

Table 3 Eigenvalues analysis results

No.	Mode	Eigenvalues	State
1–2	Short-period	0.9087 ± 1.19i	Stable
3	Integral roll	−4.471	Stable
4–5	Dutch roll	0.1177 ± 0.56i	Stable
6	Relative roll	−0.8754	Stable
7	Relative roll	1.1542	Unstable
8–9	Phugoid	0.0326 ± 0.04i	Unstable

After trim analysis, the stability analysis can be carried out under the steady equilibrium [8]. The eigenvalues of the state matrix in Eq. (15) determined the dynamic behavior of the multi-body aircraft. The eigenvalues analysis results are exhibited in Table 3. Obviously, mode 7 is an unstable mode with an aperiodic relative roll eigenform. The characteristics time is very short, so the MBA system could not flight without control system. It is different to the traditional aircraft and need to be concerned.

4 Conclusion

The work presented here investigated the dynamics modeling and analysis of multi-body aircrafts system. The state-space VLM was used for aerodynamic modeling and coefficients calculations. The Newton–Euler equation with absolute coordinates was applied for dynamic system modeling. Aerodynamcis modeling approach was validtaed from ZONAIR. Trim analysis and stability analysis were presetned based on dynamics equation of motion. Unstable flight modes were detected and different from traditional aircrafts. Most importantly, stability control system should be used for this kind of aircraft, and it will be the focuse in the future.

References

1. Patil MJ, Hodges DH, Cesnik CES (2001) Nonlinear aeroelasticity and flight dynamics of high-altitude long-endurance aircraft. J Aircr 38(1):88–94
2. Patil MJ, Hodges DH (2000) On the importance of aerodynamics and structural geometrical nonlinearities in aeroelastic behavior of high-aspect-ratio wings. In: 41st AIAA/ASME/ASCE/AHS/ASC structures, structural dynamics, and material conference and exhibit, Atlanta, GA, USA
3. Magill S, Schetz J, Mason W (2003) Compound aircraft transport: a comparison of wing tip-docked and close-formation flight. In: 41st aerospace sciences meeting and exhibit
4. Cooper JR, Rothhaar PM (2018) Dynamics and control of in-flight wing tip docking. J Guid Control Dyn 41(11):2327–2337
5. Montalvo C, Costello M (2015) Meta aircraft flight dynamics. J Aircraft 52(1)
6. Cobar M, Montalvo C (2021) Takeoff and landing of wing-tip-connected meta aircraft with feedback control. J Aircraft 58(4)
7. Cobar M, Montalvo C (2018) Experimental control of two connected fixed wing aircraft. Aerospace 5(113)
8. An C, Xie CC, Meng Y, Yang C (2019) International forum on aeroelasticity and structural dynamics, Savannah, Georgia, USA
9. Alexander K, Robert L (2015) Flight mechanical modeling and analysis of multi-body aircraft. In: 16th international forum on aeroelasticity and structural dynamics, Saint Petersburg, Russia
10. Alexander K, Alexander B, Alexander H, Phillip N, Danny N, Robert L (2017) Closed-loop flight tests with an unmanned experimental multi-body aircraft. In: 17th international forum on aeroelasticity and structural dynamics, Como, Italy
11. Alexander K, Robert L (2017) Flight path control for a multi-body HALE aircraft. In: 4th CEAS specialist conference on guidance, navigation & control
12. Werter NPM, De Breuker R, Abdalla MM (2018) Continuous-time state-space unsteady aerodynamic modeling for efficient loads analysis. AIAA J 56:905–916
13. Cesnik CES, Su WH (2011) Nonlinear aeroelastic simulation of X-HALE: a very flexible UAV. In: 49th AIAA aerospace sciences meeting including the new horizons forum and aerospace exposition, Orlando, Florida

Hydrophilic Coatings for Natural Laminar Flow

Hidetoshi Iijima, Yoshimi Iijima, Takeo Suga, Hidehiko Minami, and Kazumi Takao

Abstract The presence of surface contaminants on the wing leading edge of aircraft can expedite boundary-layer transition from laminar to turbulent. For aircraft wings to maintain natural laminar flow and reduce skin-friction drag, it is important to ensure their surfaces remain free from contaminants and/or insects. Japan Aerospace Exploration Agency has recently developed a method for maintaining laminar flow over wings. It involves the application of hydrophilic coatings on wing surfaces, which can be subsequently washed clean with water. These coatings prevent insects from adhering to the wing surface. The results of the coupon and ground tests performed in this study reveal that compared to conventional aircraft coatings, the proposed hydrophilic coatings demonstrate washabilities improved by up to 20%.

Keywords Natural laminar flow · Insect contamination · Hydrophilic coating · Photocatalyst · Polymer brush

1 Introduction

Aircraft drag reduction is a high-priority task in the development of commercial, military, and general-aviation aircraft owing to its impact on the overall aircraft fuel consumption and resulting CO_2 emissions. Because skin-friction drag accounts for more than 50% of the total drag of a transport aircraft under cruising conditions [1], several studies have been performed concerning the development of its reduction techniques, such as the design of the riblet surface and laminar flow wings

H. Iijima (✉) · Y. Iijima
Aeronautical Technology Directorate, Japan Aerospace Exploration Agency, Tokyo, Japan
e-mail: iijima.hidetoshi@jaxa.jp

T. Suga
Department of Applied Chemistry, Waseda University, Tokyo, Japan

H. Minami
Technical Engineering Department, Hoei Sangyo Co., Ltd., Osaka, Japan

K. Takao
OPTIMUS Inc., Osaka, Japan

S. Lee et al. (eds.), *The Proceedings of the 2021 Asia-Pacific International Symposium on Aerospace Technology (APISAT 2021), Volume 1*, Lecture Notes in Electrical Engineering 912, https://doi.org/10.1007/978-981-19-2689-1_29

[2, 3]. In addition, Japan Aerospace Exploration Agency (JAXA) has performed research concerning the design of natural laminar flow (NLF) airfoils that demonstrate reduced skin-friction drags [4]. In general, laminar flow is characterized by a lower skin-friction drag than turbulent flow. Therefore, a method to reduce the aircraft skin-friction drag is to design the wing airfoil shape such that the flow over its surface remains laminar over the maximum possible extent in the streamwise direction. However, their effectiveness can be limited by the presence of insects or surface contaminants on the aircraft wing leading edges. To address this concern, many researchers have investigated the use of surface-coating techniques to prevent insect adherence [5–8]. A joint team of National Aeronautics and Space Administration and Boeing applied hydrophobic coatings to wing surfaces, and their flight-test results demonstrated 37% suppression in insect adhesion compared to the uncoated surface [9]. However, the use of hydrophilic coatings for the inhibition of insect contamination has been seldom investigated. JAXA has recently developed a new method for maintaining laminar flow over wings. It involves the application of hydrophilic coatings on the wing surfaces. An advantage of this approach is that the coatings can be easily washed off the wing surfaces using water along with any insects that may have adhered to the surface. This paper presents an overview of the coatings used in the proposed approach, including their structures, manufacturing methods, and functions. Moreover, the results of the coating-evaluation, coupon, and ground tests are reported herein.

2 Insect- and Contaminant-Adherence Mitigation

2.1 Self-cleaning Properties of Hydrophilic Coatings

Although several researchers have focused on the use of hydrophobic coatings to prevent wing-surface contamination, these coatings are not durable compared to their hydrophilic counterparts. Moreover, they demonstrate the tendency to be damaged by the lens effect of water droplets. Meanwhile, surfaces coated with hydrophilic coatings are characterized by a small water-contact angles and self-cleaning properties. Therefore, surface contaminants, such as insects, dust, and oil, can be easily washed off using water. Figure 1 presents a schematic of the hydrophilic and hydrophobic

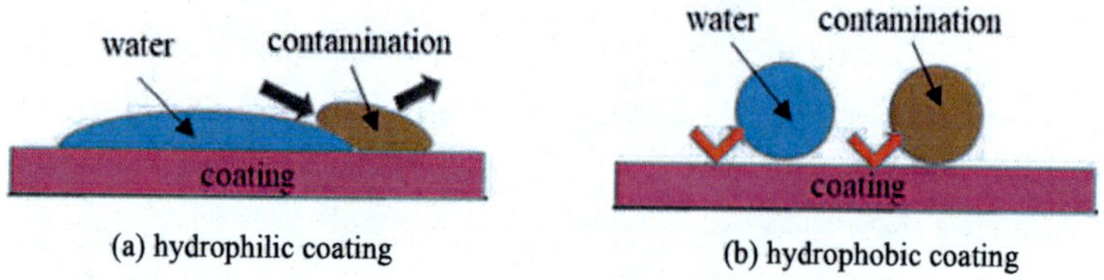

Fig. 1 Schematic of hydrophilic and hydrophobic surface coatings

coatings. However, owing to the time and effort required to wash with water, the surfaces coated with hydrophilic coatings have been seldom investigated to date. By applying hydrophilic coatings to wing surfaces and simple water washing after flights, insects adhering to wing surfaces can be efficiently removed in a short time compared to conventional contaminant-mitigation methods. To promote the maintenance of laminar flow over wing surfaces, JAXA has developed several hydrophilic coatings with high washabilities.

2.2 *Photocatalytic Coating with Meshed-Polymer Brush*

JAXA has developed a hydrophilic coating that combines a photocatalytic coating with a meshed-polymer brush. The structure of this coating can be divided into two layers—the photocatalyst layer and polymer-brush—as showed in Fig. 2. The photocatalyst layer not only performs a hydrophilic function but also decomposes organic substances, such as insects. While the photocatalytic coating sufficiently demonstrates its effectiveness during flight, while the effect of ultraviolet radiation is strong, it is not as potent in the absence of ultraviolet radiation. In the latter case, the hydrophilic function is performed by the mesh-like polymer-brush layer atop the photocatalyst coating. The chemically bonded photocatalyst and polymer-brush layers exhibit excellent adhesion. In addition, polymer brushes can be applied over a large area and demonstrate potential for use in practical applications. Insects adhering to the coated aircraft surfaces can easily be separated by cavities in the mesh polymer brush. Because the hollow capsules attached to the photocatalysts make them float on the coating-film surface layer (water-based coating (C3 Optimus), the photocatalyst itself does not erode an aircraft coating made by organic substances. Figure 3 shows the hollow capsule structure. Subsequently, a polymer brush layer, which measures several hundred nanometers thick, is grown atop the polymerization

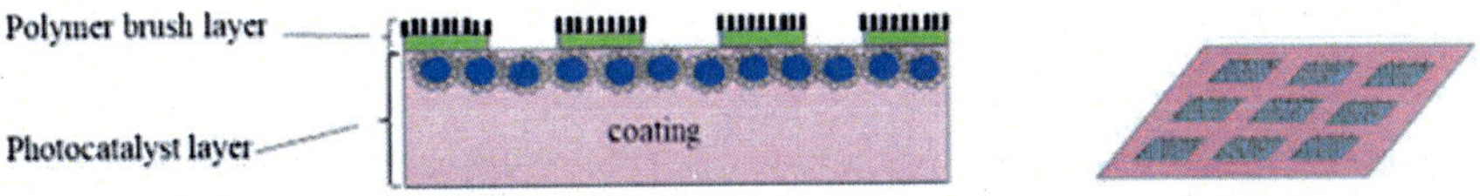

Fig. 2 Proposed photocatalytic coating with meshed-polymer brush

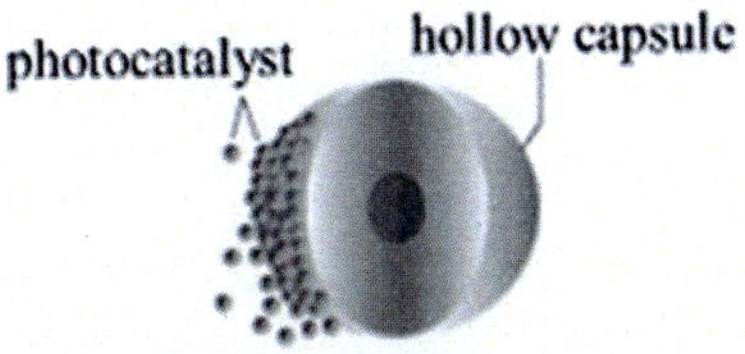

Fig. 3 Hollow capsules with attached photocatalysts

initiator layer [10] via the surface-initiated atom transfer radical polymerization, as showed in Fig. 4. The thin initiator layer, which starts the polymerization reaction, comprises (p-Chloromethyl) phenyl trim ethoxy silane (CMPTMS) and Tetra ethoxy silane (TEOS), and the corresponding chemical scheme is showed in Fig. 5. The polymer-brush layer is fabricated using 2-acrylamido-2-methyl-1-propanesulfonic acid sodium salt (SPMK) as a monomer, and $CuCl_2$, N, N, N′, N″, N″-pentamethyl diethylene triamine (PMDETA), and ascorbic acid as catalysts. The corresponding chemical scheme is showed in Fig. 6. Because polymer brushes can be fabricated on large-scale real-life substrates under ambient conditions without using complex equipment or harsh reaction conditions, they demonstrate significant potential for use in aerospace applications. The mesh pattern is formed by printing through mesh or screen sheets.

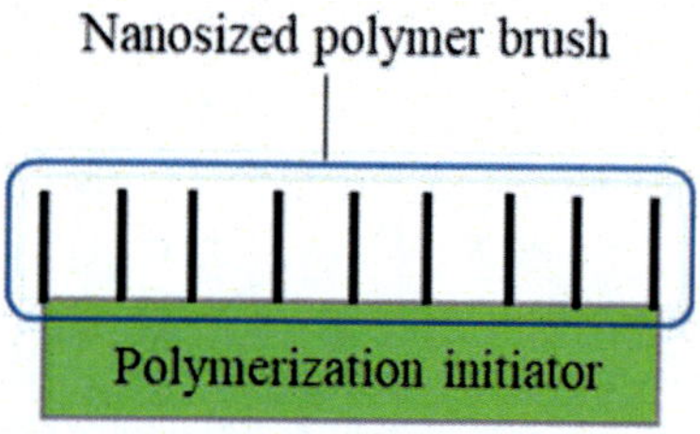

Fig. 4 Schematic of polymer brush

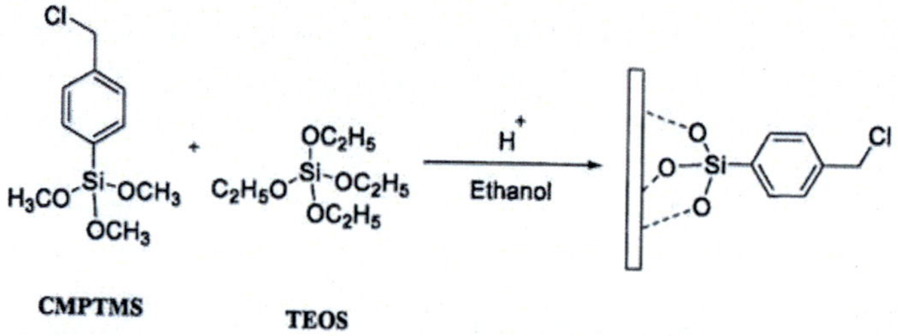

Fig. 5 Chemical scheme of polymerization initiator

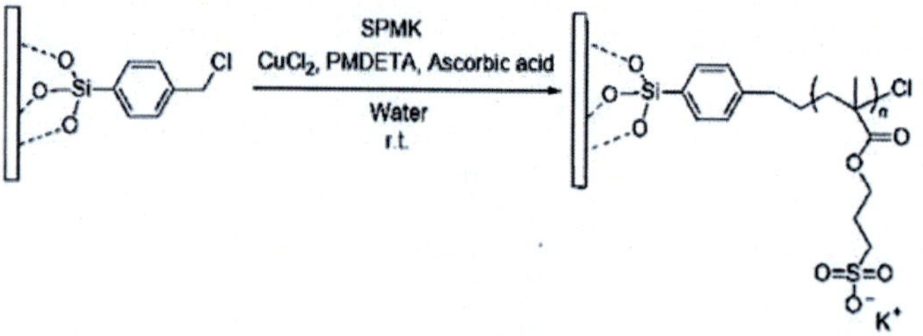

Fig. 6 Chemical scheme for polymer brush

3 Evaluation of Coating Washability

In this study, we performed two types of washability tests—ground and coupon tests. In each test, the coating washability was evaluated based on the number of insect marks observed before and after cleaning. The washing rate (%) was defined as the percentage of washed insects per total insects attached.

A conventional coating for the comparison was used the coating manufactured by Nihon Tokushu Toryo Co., Ltd.

3.1 Ground Test

The ground tests were performed at the race circuit located in North Tsukuba in the Ibaraki Prefecture, Japan and the National Institute of Advanced Industrial Science and Technology (AIST). The racecourse measures 3200 m in circumference and comprises a 750 m long straight section. During the tests, the adherence of insects to a coated wing model was realized by driving a car installed with the model at approximately 100 km/h. Subsequently, the coated model was washed with water, and the coating washability was evaluated by counting the number of washed insects. The tests were performed for three coatings—photocatalytic coating, photocatalytic coating with the meshed-polymer brush, and conventional coating. The washing for each coating was completed at the same time in approximately 20 s. The tests were performed with rotation of the position to eliminate the location dependence in the insect adhesion. Figure 7 depicts the wing model with adhering insects at the leading edge (magnified view), its installation on the car, and the washing process. The average number of attached insects before washing is approximately 30. Since nearly all insects were found to be bursting, insect blood increases adhesive strength on the surface.

Fig. 7 Ground tests

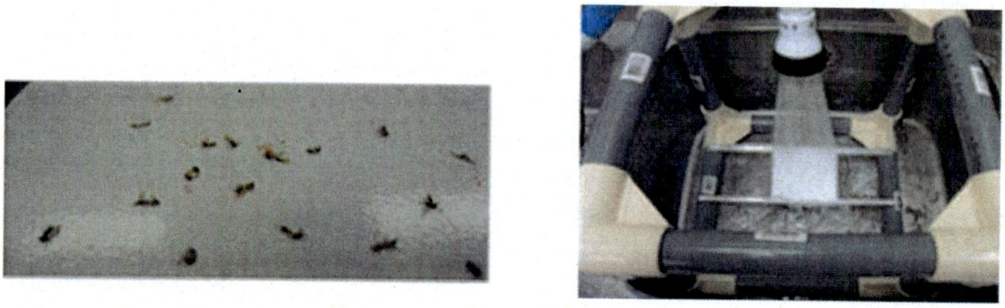

Fig. 8 Coupon tests

3.2 Coupon Tests

The effectiveness of the polymer-brush coating, photocatalytic coating with the meshed-polymer brush was evaluated with a conventional coating in the coupon tests. During these tests, approximately 20 insects were manually attached to the surfaces of 10 × 10 cm aluminum test plates. Thereafter, the test plates were placed under running water set at a constant flow rate (11.8 L/min) for approximately 30 s. The insects used in these tests were *Drosophila melanogaster* (fruit flies) and measured 1–2 mm in size. To eliminate the bias caused by the insect-adhesion pressure, the tests were repeated 10 times for each coating type. Figure 8 illustrates the coupon-test setup used in this study. Prior to the washability evaluation, the hydrophilicity of each coating type was evaluated by measuring the water contact angle using a drop-shape analyzer (KRUSS).

4 Results

4.1 Ground Tests

Figure 9 presents the washability results obtained for the ground test. The average washing rates of the conventional, photocatalytic only, and photocatalytic with meshed-polymer brush coatings are 62.5%, 61.3%, and 69.6%, respectively. The washing rate of the photocatalytic coating is approximately equal to that of the conventional coating. This can be attributed to the short duration of ultraviolet irradiation, owing to which a sufficient photocatalytic effect and hydrophilicity could not be achieved. Moreover, more repetitions of the test involving the photocatalytic coating alone should be considered. Furthermore, the results confirm that the proposed photocatalytic coating with the meshed-polymer brush improves the washability by approximately 7.1% compared to the conventional coating. This washability can be further improved by in-flight photocatalytic effects.

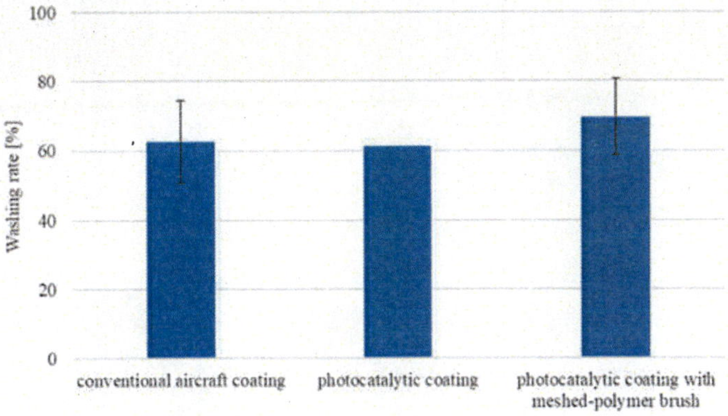

Fig. 9 Washability results obtained during ground tests

4.2 *Coupon Tests*

Figure 10 shows the washability obtained for three cases considered during the coupon tests. The average washing rates observed for the conventional, polymer-brush, and photocatalyst with the meshed-polymer brush coatings are 58.7%, 76.5%, and 79%, respectively. The washing rate of the proposed photocatalytic coating with the meshed-polymer brush exceeds that of the conventional coating by 20.3%. In addition, the results obtained using the polymer-brush coating and photocatalytic coating with the meshed-polymer brush confirm the slightly high effectiveness of the rough meshed-polymer brush in increasing the washability. Moreover, because the roughness height of the meshed-polymer brush is below 1 μm, the height is small enough not to be effective in boundary layer transition. Figure 11 shows the surface wetting properties (water contact angles) of the three coating types considered

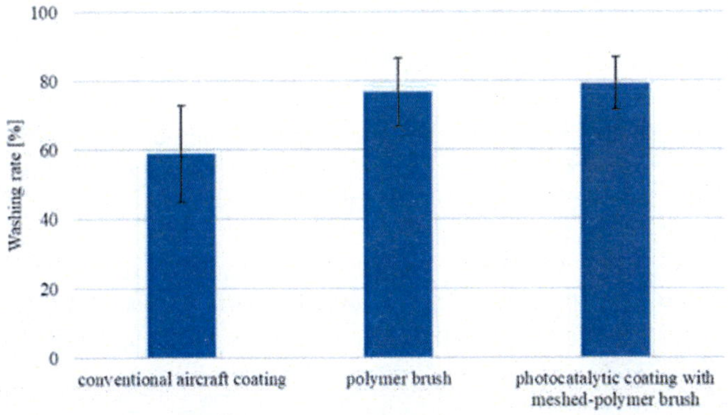

Fig. 10 Washability results obtained during coupon tests

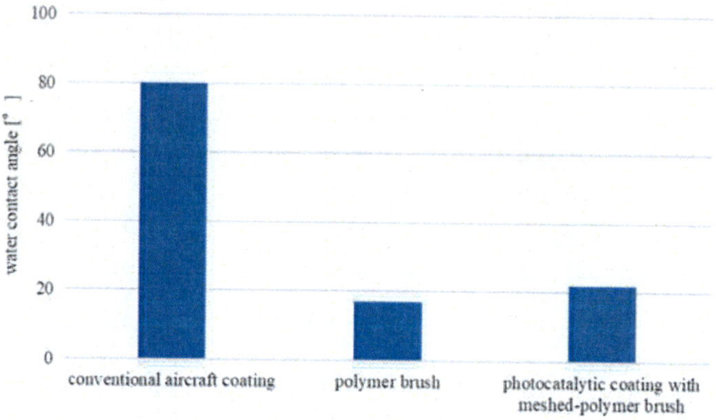

Fig. 11 Comparison between water contact angles of coatings considered during coupon tests

during the coupon tests. The water contact angles of the conventional, polymer-brush, and photocatalytic with meshed-polymer brush coatings are 80°, 17°, and 22°, respectively. The latter two coatings are significantly more hydrophilic compared to the conventional coating. This finding is consistent with the washability-test results.

5 Conclusions

This paper presents the concept of hydrophilic coatings developed by JAXA. The proposed coating mitigates the effects of the early transition from laminar to turbulent flow over aircraft surfaces caused by the adherence of insects near the aircraft wing leading edge. The structure of the proposed coating is described in this paper. Moreover, the results of the ground and coupon washability tests performed using the proposed and conventional coatings are presented. These results confirm that the washability of the proposed hydrophilic coating exceeds that of conventional coatings by approximately 20%. In future studies, we intend to evaluate and improve the proposed coating design to meet the prescribed aerospace application specifications, such as low-temperature brittleness, oil resistance, and long-term durability, while maintaining a low surface roughness to sustain laminar flow. The proposed coating technology is expected to be used in not only aircraft aerodynamics but also that pertaining to trains, cars, and buildings.

Acknowledgements The authors thank Mr. Muneyoshi Nakagawa, Dr. Naoko Tokugawa, and Dr. Dongyoun Kwak for extending their support during the evaluation of the coatings used in this study and providing helpful suggestions.

References

1. Szodruch J (1991) Viscous drag reduction on transport aircraft. In: 29th aerospace sciences meeting, AIAA Paper-91-0685
2. Joslin RD (1998) Aircraft laminar flow control. Ann Rev Fluid Mech 30:1–29
3. Viswanath PR (2002) Aircraft viscous drag reduction using riblets. Prog Aerosp Sci 38:571–600
4. Miyazaki M, Kuroda F, Ueda Y, Kimura T, Tokugawa N (2016) Natural laminar flow wing design for civil aircraft. In: 2016 Asia-Pacific international symposium on aerospace technology, Toyama, F8-2
5. Carter D, Loth E, Wohl C (2018) Effect of insect impact conditions and surface coating composition on the residue height. In: AIAA aerospace sciences meeting
6. Siochi E, Smith J, Wohl C, Gardner J, Penner R, Connell J (2013) Engineered surfaces for mitigation of insect residue adhesion. In: 51st AIAA aerospace sciences meeting including the new horizons forum and aerospace exposition
7. Wohl C, Smith J, Gardner J, Penner R, Connell J, Siochi E (2014) Novel epoxy particulate composites for mitigation of insect residue adhesion on. future aircraft surfaces. In: 37th annual meeting of the adhesion society
8. Young T, Tobin E, Kok M (2012) Laboratory testing of insect contamination for laminar flow Ap plications using an insect-impact test facility. In: 28th international congress of the aeronautical science
9. Michael G, Alexander F, Harris K (2016) Active Flow Control (AFC) and Insect Accretion and Mitigation (IAM) system design and integration on the boeing 757 ecoDemonstrator. In: AIAA AVIATION Forum, Washington, DC, 16th AIAA aviation technology, integration, and operations conference, 13–17 June 2016
10. Sato T, Dunderdale GJ, Urata C, Hozumi A (2018) Sol–Gel Preparation of initiator layers for surface-initiated ATRP: large-scale formation of polymer brushes is not a dream. Macromolecules 51:10065–10073

Modeling of Crossflow-Induced Boundary Layer Transition

Makoto Hirota, Yuki Ide, and Yuji Hattori

Abstract Analytical models for the primary and secondary instabilities of three-dimensional boundary layer are developed by assuming locally parallel shear flow and scale similarity at each position. More specifically, both linear and nonlinear growth rates of the stationary crossflow modes are estimated as functions of a normalized wavenumber and a Reynolds number, which are locally defined by the characteristic velocity and length scales of the inflectional profile of shear flow. By choosing one of them as the primary mode, the maximum linear growth rate of the secondary mode is also estimated by assuming scale similarity. The quantitative expressions of these scaling laws are determined so that they agree with several results of direct numerical simulation and numerical stability analysis. Transition to turbulence is triggered when a secondary mode grows nonlinearly. In other words, if a nonlinear primary mode is less unstable to secondary instability, it may not cause transition but laminarize the boundary layer. For various three-dimensional boundary layers, our models are expected to be useful for quickly predicting whether a certain crossflow mode leads to transition or laminarization under a certain disturbance level.

Keywords Boundary layer transition · Crossflow instability · Secondary instability

1 Introduction

On the widely used swept wings of aircrafts, friction drag is significantly increased by laminar-turbulent transition of three-dimensional boundary layer, which is dominantly caused by the crossflow instability (see [1, 6] for reviews). Reduction of friction drag is therefore one of the major issues in the development of more fuel-efficient aircrafts. When trying to laminarize a certain boundary layer by some control devices, it is necessary to predict the transition location and prove how far it is

M. Hirota (✉) · Y. Hattori
Institute of Fluid Science, Tohoku University, Sendai, Japan
e-mail: makoto.hirota.d5@tohoku.ac.jp

Y. Ide
Japan Aerospace Exploration Agency, Tokyo, Japan

S. Lee et al. (eds.), *The Proceedings of the 2021 Asia-Pacific International Symposium on Aerospace Technology (APISAT 2021), Volume 1*, Lecture Notes in Electrical Engineering 912, https://doi.org/10.1007/978-981-19-2689-1_30

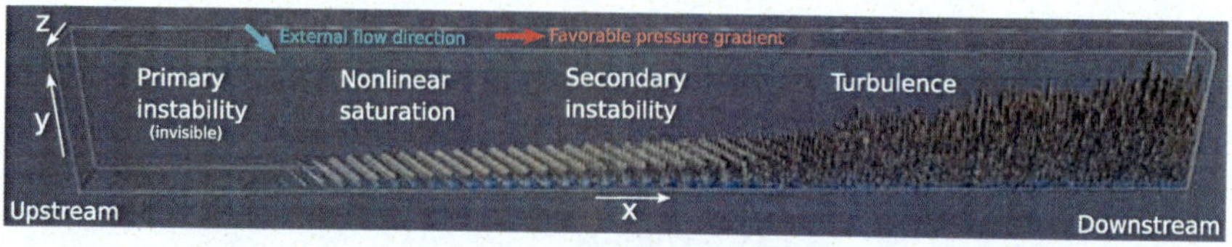

Fig. 1 Crossflow-induced transition process demonstrated by DNS for the boundary layer specified in Appendix, where vortices are visualized by the lambda2 method

delayed by means of the devices. However, the standard e^N method for the primary crossflow instability does not always correctly predict the transition location, since three-dimensional boundary layer often becomes turbulent only when a secondary instability grows enough after the primary mode is saturated nonlinearly. In Fig. 1, this typical transition scenario is demonstrated by our direct numerical simulation in a cuboid domain on a flat plate which is periodic in the spanwise (z) direction. The Fig. 1 shows that the crossflow vortex street breaks down into turbulence via secondary instability. On the other hand, if any secondary instability does not occur notably, such a primary mode is not harmful but rather useful as an agent to control the flow artificially. The laminarization technique by the discrete roughness elements (DREs) utilize this subcritical mode to stabilize the mean flow profile [5, 8]. This technique is, however, still under development and more computational studies are required to understand the influence of DREs [7].

We can perform direct numerical simulation (as in Fig. 1) and various stability analyses (as reported by [2]). Using these numerical tools, the primary and secondary instabilities are investigated in detail for a given laminar boundary layer and a given disturbance level. For example, depending on the spanwise wavenumber of the primary mode, its nonlinear saturation level and the growth rate of the subsequent secondary instability can be found to differ in a consistent manner. However, such the tendency and knowledge obtained from one case are not directly applied to other cases. Moreover, these numerical approaches are computationally high cost and cannot be quickly carried out for various boundary layers and various disturbances. Thus, the goal of our study is to develop a simplified model that can predict the primary and secondary instabilities almost analytically.

For simplicity, we will assume locally parallel shear flow and its scale similarity with respect to the normalized wavenumber and Reynolds number at each position. First, we will model the linear and nonlinear growth rates of the primary mode, which can estimate the saturation level of the modal amplitude (that is the scale of crossflow vortices). Next, we further model the maximum linear growth rate of the secondary instability, which occurs due to the velocity shear of low-speed streaks. The estimated N-factor of the most unstable secondary mode is expected to predict the transition location more plausibly than the N-factor of the primary mode. The nonlinearly saturated primary modes at relatively high wavenumbers and low Reynolds numbers are shown to be less unstable to secondary instability, which indicates that they do not cause transition under a sufficiently low disturbance level. We remark that our models are by no means rigorous and do not replace the numerical analyses. Nevertheless, our models will be useful for predicting transition location approximately, and for finding a suitable control mode which should be excited by the DREs to laminarize arbitrary boundary layer in question.

2 Characterization of Three-Dimensional Boundary Layer

Around a swept wing, let (x, y, z) be the chordwise, wall-normal and spanwise directions, respectively, and (U, V, W) be the corresponding velocity components of boundary layer, which are assumed to be uniform in z. When developing our models, we always focus on a fixed position x and approximate the problem by a *parallel shear flow*$(U(y), 0, W(y))$. Because of the swept angle, the external flow direction $(U_e, 0, W_e)$ at the upper edge of boundary layer $y \to \infty$ is not parallel to the chordwise direction, and the flow direction $(U(y), 0, W(y))$ inside the boundary layer further veers in the presence of crossflow. If we introduce a rotation transformation by angle θ,

$$U_\xi = -U\sin\theta + W\cos\theta, \tag{1}$$

$$U_\zeta = U\cos\theta + W\sin\theta, \tag{2}$$

the typical profile of three-dimensional boundary layer possesses a unique value of θ such that $U_\xi(y_I) = d^2U_\xi/dy^2(y_I) = 0$ holds at a position $y = y_I$. This angle θ is known to be the wave front of the stationary crossflow mode and, namely, corresponds to the direction of the crossflow vortex street, which slightly differs from the external flow direction (by a few degrees like Fig. 11). Examples of U_ξ and U_ζ are shown in Fig. 2, for the case of the FSC boundary layer in Appendix. The velocity component U_ξ is unstable to the primary crossflow instability, whereas the other component U_ζ

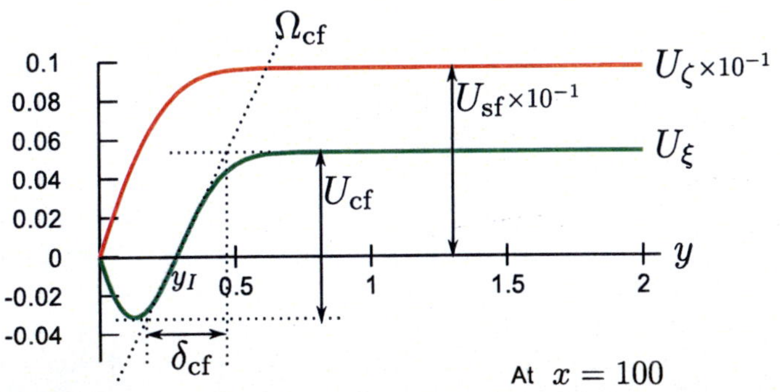

Fig. 2 Typical flow profile of three-dimensional boundary layer and its scale parameters

becomes unstable to the secondary instability only after the primary mode grows nonlinearly.

Given the base flow profiles of U_ξ and U_ζ at a position, we define the following scale parameters.

$$U_{\mathrm{cf}} := \max\ U_\xi(y) - \min\ U_\xi(y), \tag{3}$$

$$U_{\mathrm{sf}} := \max\ U_\zeta(y) - \min\ U_\zeta(y), \tag{4}$$

$$\delta_{\mathrm{cf}} := U_{\mathrm{cf}} / \max \frac{dU_\xi}{dy}(y). \tag{5}$$

These are also shown in Fig. 2. Note that δ_{cf} is in the same order of the boundary layer thickness and we will use δ_{cf} as the representative length scale. By neglecting minor differences, three-dimensional boundary layer is locally characterized by these three parameters. Finally, we introduce the local Raynolds numbers,

$$Re_{\mathrm{cf}} = \frac{U_{\mathrm{cf}}\delta_{\mathrm{cf}}}{\nu},\ Re_{\mathrm{sf}} = \frac{U_{\mathrm{sf}}\delta_{\mathrm{cf}}}{\nu}, \tag{6}$$

for each velocity components, where ν is the kinematic viscosity at the same position.

3 Linear Growth Rate of Primary Instability

The crossflow instability is nothing but the Kelvin–Helmholtz instability due to the inflection point y_I of $U_\xi(y)$. Since the dominant disturbance source of this instability is natural roughness on the wing surface, we only consider stationary modes $\propto \exp(i\alpha x + i\beta z)$ where $\alpha = \alpha_r + i\alpha_i \in \mathbb{C}$ and $\beta \in \mathbb{R}$ are the wavenumbers. The spatial growth rate $-\alpha_i$ can be estimated by using the local scale parameters in a normalized form,

$$-\alpha_i = \frac{U_{\mathrm{cf}}}{U_e\delta_{\mathrm{cf}}} f\left(\hat{k}, Re_{\mathrm{cf}}\right), \tag{7}$$

where the normalized wavenumber $\hat{k}$ is defined by

$$\hat{k} = \delta_{\mathrm{cf}}\sqrt{\alpha_r^2 + \beta^2} = \frac{\delta_{\mathrm{cf}}\beta}{\cos\theta}. \tag{8}$$

This parameter $\hat{k}$ specifies the ratio of the wavelength to the boundary layer thickness δ_{cf}. To estimate the spatial growth rate from the temporal growth rate, one needs

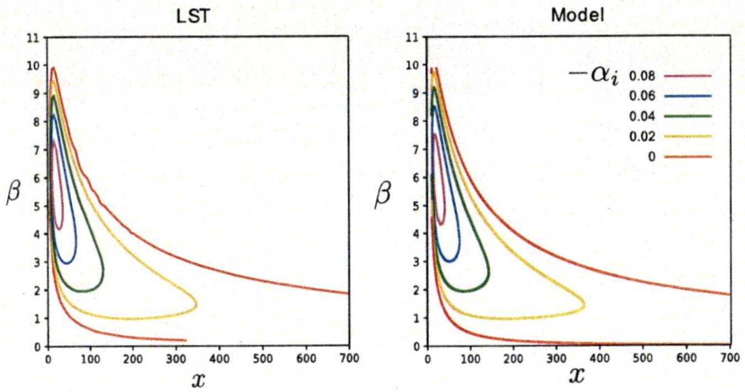

Fig. 3 Spatial linear growth rate $-\alpha_i$ of primary mode (steady crossflow mode) calculated by LST analysis (left) and analytical model (right), in the case of the reference FSC boundary layer in Appendix

to know the group velocity of the wave in general (called Gaster's transformation). In Eq. (7), its scale is represented by the external velocity U_e in the x direction. More plausibly, one can choose $U(y_I)$ instead of U_e, but their difference is not essential here because $U(y_I) \sim U_e$.

In any events, we must determine the above nondimensional function f. By solving the Orr-Sommerfeld equation for the explicit profile $U_\xi(y)$ in Fig. 2, we can numerically calculate the growth rate; this procedure is called the LST (Linear Stability Theory) analysis. The result is not so complicated and well approximated by

$$f\left(\hat{k}, Re_{\text{cf}}\right) = 0.24\hat{k}\left(1.83 - \hat{k}\right) - \frac{12}{Re_{\text{cf}}}, \tag{9}$$

as shown in Fig. 3. In this example, the effect of viscosity, $-12/Re_{\text{cf}}$, is negligible except for the neighbourhood of the leading edge $x \lesssim 100$. Then, the neutral curve is simply predicted by $\hat{k} = 1.83$ and the growth rate is maximum at β satisfying $\hat{k} = 0.92$. Note that β is constant for one crossflow mode, but all U_{cf}, δ_{cf}, θ and U_e are gently varying functions of x as shown in Figs. 11 and 12.

It should be remarked here that we have not allowed for the effects of nonparallel flow and wall curvature. Therefore, it is not surprising that the error of the estimated growth rate amounts to 10–15% for general boundary layers on swept wings.

4 Nonlinear Saturation of Primary Instability

We assume that the scale similarity holds even for the nonlinear saturation phase of primary mode. At a position x, suppose that a crossflow mode with a single spanwise wavenumber β grows nonlinearly and the base flow profile $U_\xi(y)$ turns into $u_\xi(y,z)$. Its spanwise average, denoted by $U_\xi^{(1)}(y) = \langle u_\xi(y,z)\rangle$, represents the mean flow which is nonlinearly distorted by the primary mode. We similarly define $U_{\rm cf}^{(1)}$ and $\delta_{\rm cf}^{(1)}$ for this profile $U_\xi^{(1)}(y)$. As shown in Fig. 4, the vortical motion of the crossflow mode induces momentum exchange between the lower and upper parts of boundary layer, which tends to decrease $U_{\rm cf}^{(1)}$ and increase $\delta_{\rm cf}^{(1)}$ by mean flow distortion. Obviously, $U_\xi^{(1)}(y)$ becomes less unstable to the crossflow instability than $U_\xi(y)$; that is why the primary mode is saturated and the mean flow distortion contributes laminarization (unless the secondary instability occurs).

Now, we *define* the amplitude of the primary mode as

$$A^{(1)} := \sqrt{\frac{\delta_{\rm cf}^{(1)}}{\delta_{\rm cf}} - 1}. \tag{10}$$

Since the mean flow distortion is caused by a second order nonlinearity in the linear (and weakly nonlinear) phase of the primary mode, this $A^{(1)}$ indeed grows exponentially $\propto e^{-\alpha_i x}$ in the linear phase with the growth rate (7). Although it is difficult to predict the strongly nonlinear phase, we can expect that saturation occurs when β approaches the neutral stability curve of $U_\xi^{(1)}(y)$, namely,

$$\frac{\delta_{\rm cf}^{(1)}\beta}{\cos\theta} \to 1.83. \tag{11}$$

Based on this ansatz, the saturation amplitude is roughly predicted by

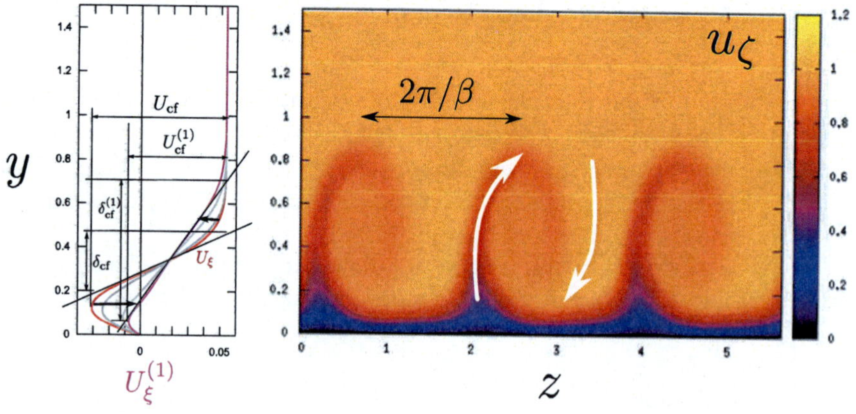

Fig. 4 Mean flow distortion of the crossflow component $U_\xi(y)$ into $U_\xi^{(1)}(y) = \langle u_\xi(y,z)\rangle$ (left) and the streak-flow component $U_\zeta(y)$ into $u_\zeta(y,z)$ (right)

$$A_{\text{sat}}^{(1)} = \sqrt{\frac{1.83}{\widehat{k}} - 1}, \tag{12}$$

which depends on how much the normalized wavenumber $\widehat{k}$ is close to the neutral curve.

For the purpose of validating this nonlinear model, the actual boundary layer is rather complicated because the normalized wavenumber $\widehat{k}$ and the Reynolds number Re_{cf} vary along the chordwise (x) direction. Therefore, we have performed DNS by fixing the base flow *artificially* to a parallel shear flow $(U(y), 0, W(y))$ that extends over a sufficiently wide range in the x direction. In fact, we extract this base flow profile from the reference boundary layer at positions $x = 100, 200, 300, 400$ and make them uniform everywhere in x. Using these parallel shear flows, we can fix the Reynolds numbers to $Re_{\text{cf}} = 223, 334, 452, 552$ and $Re_{\text{sf}} = 2630, 4590, 6500, 8530$, respectively. Moreover, in each case, we choose β so that it triggers either the most unstable mode ($\widehat{k} \simeq 1.0$) or the moderately unstable mode ($\widehat{k} \simeq 1.5$). In total, we have carried out DNS for eight cases designated by the circle symbols in Fig. 5. In this way, we can fix $\widehat{k}$ and Re_{cf} everywhere in the computational domain.

As a result, the nonlinear growth rate $(dA^{(1)}/dx)/A^{(1)}$ of the amplitude $A^{(1)}$ is shown in Fig. 6 for the above eight cases. We note from Fig. 6 that the most unstable mode ($\widehat{k} \simeq 1.0$) leads to a large saturation level $A^{(1)} \simeq 1$. On the other hand, the moderately unstable mode ($\widehat{k} \simeq 1.5$) results in a moderate saturation level $A^{(1)} \simeq 0.5$. These trends agree with our heuristic prediction $A_{\text{sat}}^{(1)}$. The nonlinear evolution of $A^{(1)}$ is therefore modelled by a Stuart-Landau equation,

$$\frac{dA^{(1)}}{dx} = -\alpha_i \left[1 - \left(\frac{A^{(1)}}{1.2A_{\text{sat}}^{(1)}} \right)^2 \right] A^{(1)}, \tag{13}$$

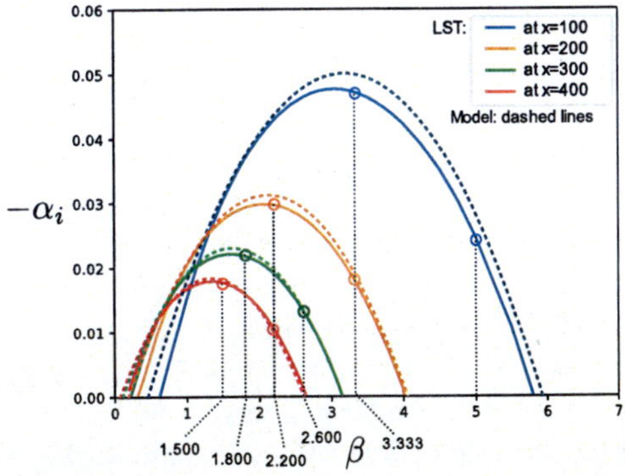

Fig. 5 Spatial growth rate of primary mode at $x = 100, 200, 300, 400$ (see also Fig. 3)

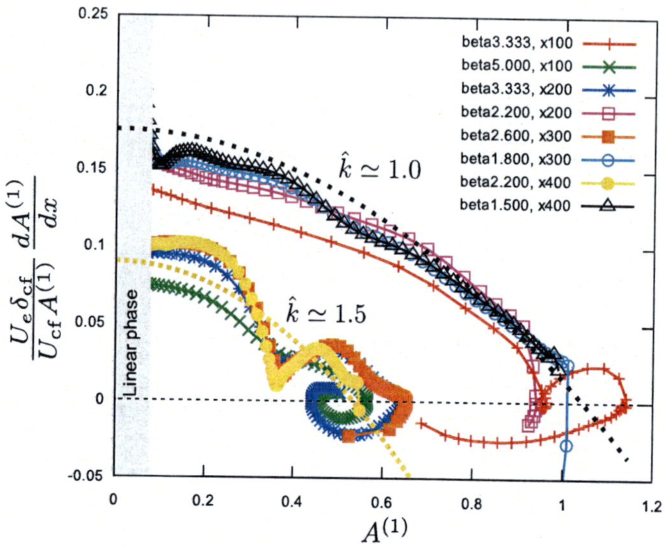

Fig. 6 Normalized spatial growth rates of primary modes versus the amplitude $A^{(1)}$

which is shown as dotted lines in Fig. 6 for comparison. This is the simplest model that starts with the exponential growth $e^{-\alpha_i x}$ and ends up with the saturation at $A^{(1)} \simeq 1.2 A^{(1)}_{\mathrm{sat}}$.

5 Linear Growth Rate of Secondary Instability

When the primary crossflow mode is saturated, the crossflow vortex street (i.e., Kelvin–Helmholtz billows) is generated as shown in Fig. 4. The associated vortical motion also distorts the originally monotonic velocity profile $U_\zeta(y)$ into a two-dimensional one $u_\zeta(y, z)$ whose velocity distribution is so called low-speed streaks. Since $u_\zeta(y, z)$ is again an inflectional shear flow, some secondary instability occurs due to it and finally the two-dimensional flow $u_\zeta(y, z)$ breaks down into fully three-dimensional turbulence. The two-dimensional LST (2dLST) analysis can calculate the linear growth rate of the secondary modes numerically [2, 3]. This analysis is performed extensively for the four cases of parallel shear flows discussed in the previous section. In Fig. 7, the temporal growth rates of the secondary modes are shown as functions of wave number $k_r^{(2)}$ along the vortical axis, where each case is calculated at a specific position where the primary mode is saturated, and the fastest growth of secondary instability is observed. The eigenfunctions of the most unstable modes, indicated by (i)–(iv) in Fig. 7, are shown in Fig. 8. All these modes are localized around the shoulders of the Kelvin–Helmholtz billows, at which $u_\zeta(y, z)$ is the most inflectional. They seem to be the so-called type-I modes (or 'z'-modes).

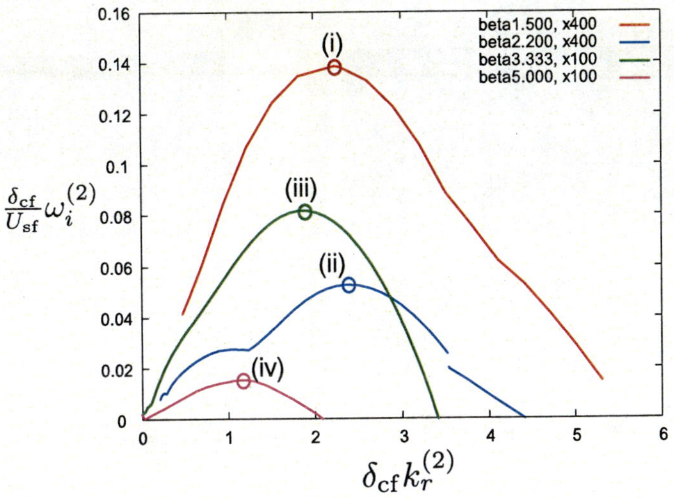

Fig. 7 Temporal growth rate $\omega_i^{(2)}$ of secondary instability calculated by 2dLST analysis

We confirm that the phase velocity $\omega_r^{(2)}/k_r^{(2)}$ of the secondary instability is almost constant [3] and $\omega_r^{(2)}/k_r^{(2)} \simeq U_{sf}$. Therefore, we can estimate the spatial growth rate along the vortical axis by $-k_i^{(2)} = \omega_i^{(2)}k_r^{(2)}/\omega_r^{(2)}$. In Fig. 9, only the maximum growth rate at each position is plotted versus the primary mode amplitude $A^{(1)}$ at the same position. The positions denoted by the square symbols in Fig. 9 are, in fact, chosen to exhibit in Fig. 7. We can find that the most unstable primary mode ($\hat{k} \simeq 1.0$) leads to a large growth rate of the secondary instability when it is saturated at $A^{(1)} \simeq 1.0$, especially when the Reynolds number Re_{sf} is large. The moderately unstable primary mode ($\hat{k} \simeq 1.5$) results in a moderate growth rate of the secondary instability since its saturation level is low, $A^{(1)} \simeq 0.5$. In either case, the viscous damping seems to be strong even at the value of $Re_{sf} = 2630$. This fact indicates that secondary instability (or transition) does not tend to occur near the leading edge even when a primary mode is nonlinearly saturated there.

By invoking the scale similarity again, the maximum spatial growth rate must be written in the form of using some nondimensional function g. Then, the amplitude $A^{(2)}$ of the secondary instability is expected to grow in the x direction as follows.

$$-k_i^{(2)} = \frac{1}{\delta_{cf}} g\left(A^{(1)}, \hat{k}, Re_{sf}\right) \tag{14}$$

$$\frac{dA^{(2)}}{dx} = \frac{1}{\delta_{cf}\cos\theta} g\left(A^{(1)}, \hat{k}, Re_{sf}\right) A^{(2)}. \tag{15}$$

We can predict the N-factor of the secondary instability by integrating this equation along x. Although the numerically studied cases are very limited at present, we propose a model

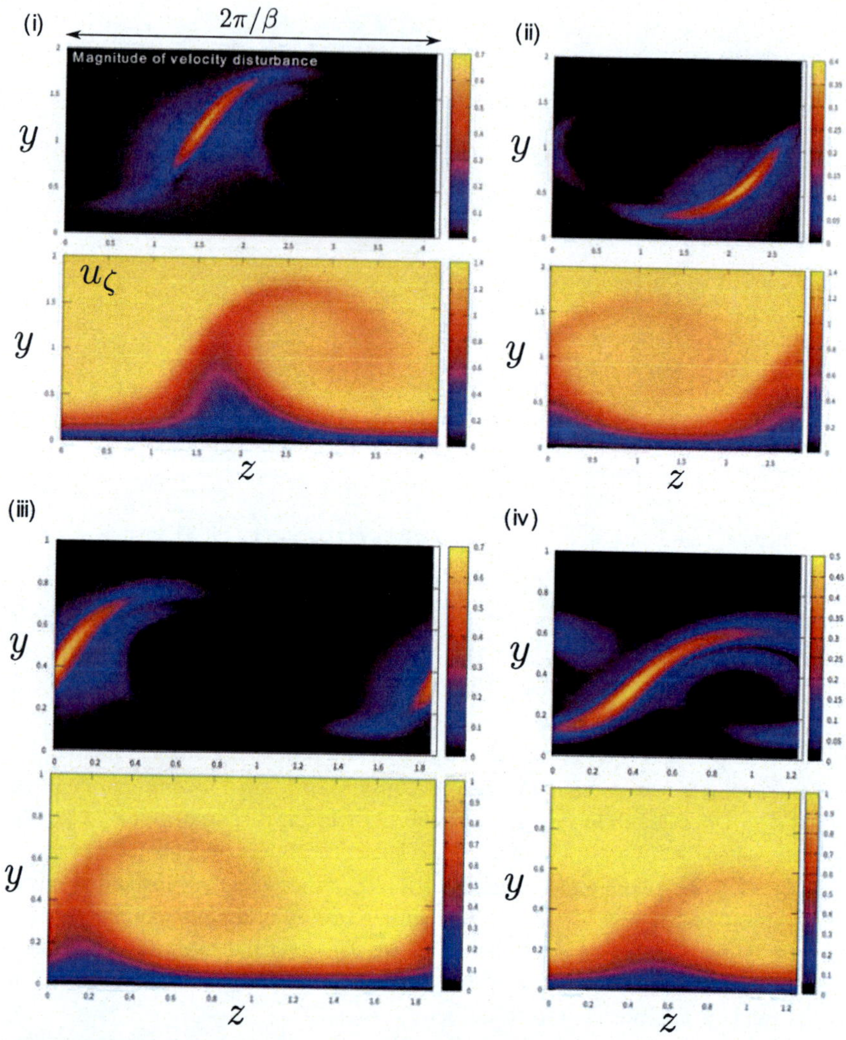

Fig. 8 Magnitude of velocity disturbance indicating mode structure of secondary instability and distribution of two-dimensional flow $u_\zeta(y, z)$ at the same position. These four modes (i)–(iv) are respectively the most unstable ones in Fig. 7

$$g\left(A^{(1)}, \hat{k}, Re_{\rm sf}\right) = 0.41A^{(1)}_{\rm sat}\left(A^{(1)} - 0.3A^{(1)}_{\rm sat}\right) - 480\frac{\left(A^{(1)}\right)^2}{Re_{\rm sf}}, \tag{16}$$

as a choice of g. For comparison, this function is plotted with dotted lines in Fig. 9 by substituting the corresponding values of $\hat{k}$ and $Re_{\rm sf}$.

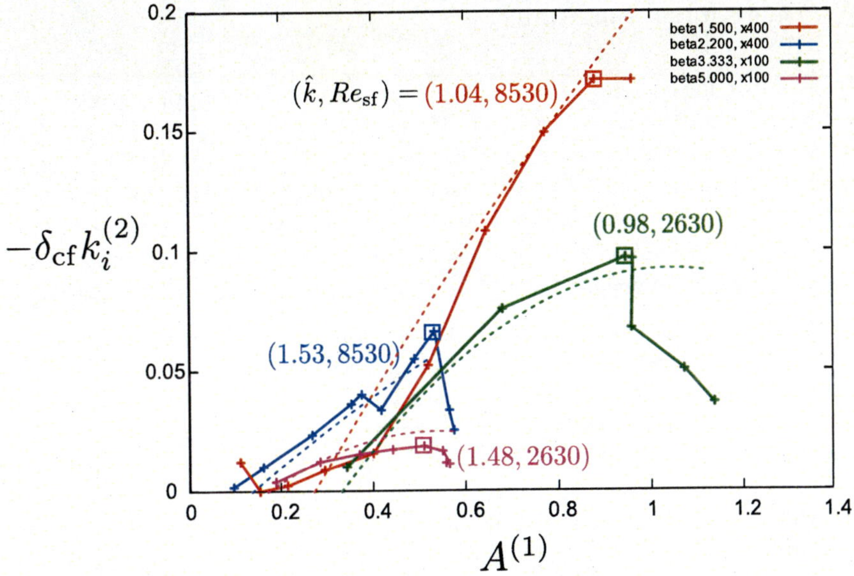

Fig. 9 Maximum spatial growth rate $-k_i^{(2)}$ of secondary instability versus nonlinear amplitude of primary mode $A^{(1)}$

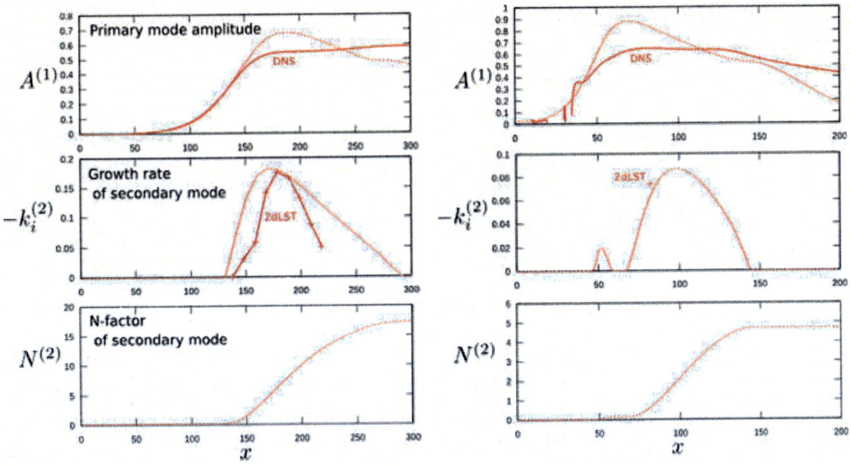

Fig. 10 Predictions by the model (dotted lines) for the cases of $\beta = 3.333$ and $A_{\text{ini}}^{(1)} = 0.0005$ (left), $\beta = 5$ and $A_{\text{ini}}^{(1)} = 0.032$ (right)

6 Validation of the Model

In this paper, the analytical models for the linear and nonlinear primary mode (13) and the most unstable secondary mode (15) are proposed. The Eq. (13) can be solved for a given wavenumber β and an initial amplitude $A_{\rm ini}^{(1)}$ (i.e., ambient disturbance level). Using the solution $A^{(1)}$, we can estimate the growth rate of $A^{(2)}$ by (14) and the N-factor $N^{(2)}$ of the secondary instability, $A^{(2)} = A_{\rm ini}^{(2)} \exp N^{(2)}$, by integrating (15). In the case of the FSC boundary layer in Appendix, this calculation is demonstrated as shown in Fig. 10 when $\left(\beta, A_{\rm ini}^{(1)}\right) = (3.333, 0.0005)$ and $(5, 0.032)$. Although the linear growth rate of $A^{(1)}$ already shows a good agreement in Fig. 3, there are noticeable differences between the model and DNS in the saturation phase of $A^{(1)}$, which indicates that the Stuart-Landau equation (13) is too simple to reproduce the nonlinear behaviour accurately. In Fig. 10, both primary modes are excited strongly enough that they are saturated. The model can estimate $N^{(2)}$ in no time and show that the $\beta = 5$ mode is less likely to cause transition than the $\beta = 3.333$ mode, even though the former is excited more strongly than the latter. This tendency is exactly what we wanted to reproduce by using the model. Namely, it is appropriate to drive this $\beta = 5$ mode artificially by DREs to laminarize the present FSC boundary layer.

7 Concluding Remarks

In terms of local scale parameters of three-dimensional boundary layer, the growth rates of primary and secondary unstable modes are estimated by assuming locally parallel shear flow and scale similarity. Non-dimensional functions, such as f and g, are empirically determined by reference to numerical results for a certain FSC boundary layer (given in Appendix). Of course, our models do not rigorously apply to other three-dimensional boundary layers. However, the prediction of our models cannot be wrong by orders of magnitude, for it is based on scale analysis of minimal physics. Our purpose is to explain the qualitative tendency of crossflow-induced transition efficiently by developing simple models. For example, dependencies on the wavenumber and Raynolds number are already elucidated in this work.

At present, more rigorous calculations such as DNS and 2dLST analysis are computationally high cost and hence limited to case study on a small area of wing surface. It also takes a very long time to do a parametric study with respect to even one or two parameters. Therefore, the simplified analytical models derived in this work would be useful as another complementary approach for quickly assessing what happens to a given boundary layer and how it changes depending on some parameter.

Acknowledgements This work is based on results obtained from a project commissioned by the New Energy and Industrial Technology Development Organization (NEDO) in Japan.

Appendix: Numerically Studied Boundary Layer Flow

In this work, our models (or scaling laws) are finally determined such that they agree with the numerical results for a certain three-dimensional boundary layer. The specifications of this boundary layer are summarized as follows. We adopt the Falkner-Skan-Cooke (FSC) boundary layer that is generalized to compressible fluid by Reshotko and Beckwith [4]. Distributions of the external flow (U_e, 0, W_e) and the displacement thickness δ are plotted in Fig. 11, where velocity and length are already normalized by the sound speed $a_0^* = 311$ m/s and $L^* = 1$ mm, respectively. The kinematic viscosity at $x = 0$ is $\nu_0^* = 3.24 \times 10^{-5}\text{m}^2/\text{s}$ and hence $Re = a_0^* L^* / \nu_0^* = 9610$. The Prandtl number is Pr = 1 and the specific heat ratio is $\gamma = 1.4$. The angle θ defined in Sect. 2 is calculated for this boundary layer and plotted in Fig. 12.

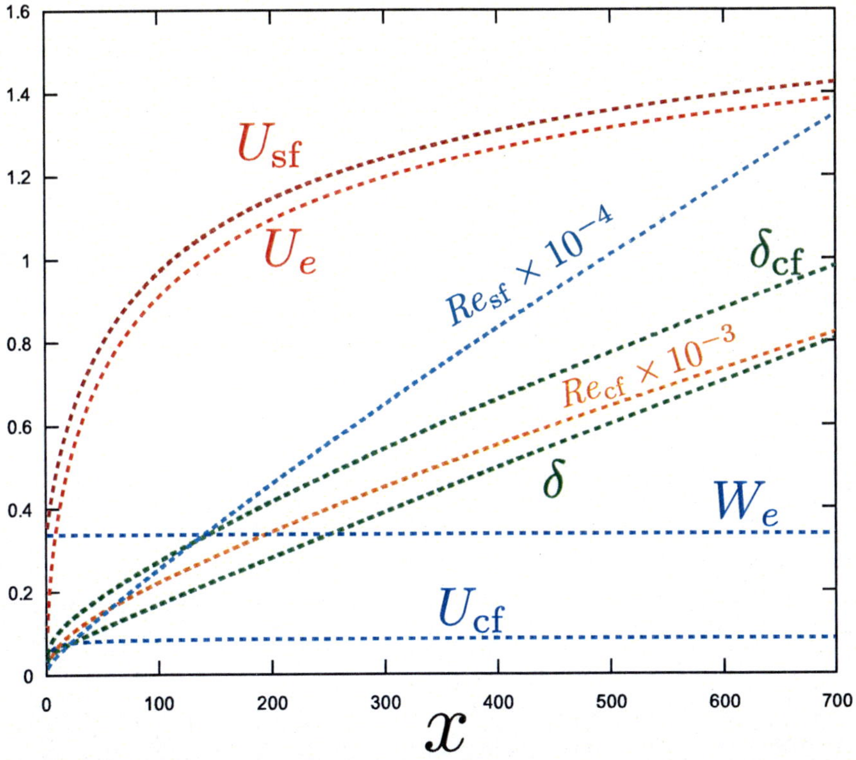

Fig. 11 External velocity (U_e, 0, W_e) and displacement thickness δ of the FSC boundary layer studied in this paper. The other quantities defined in (3)–(6) are also plotted

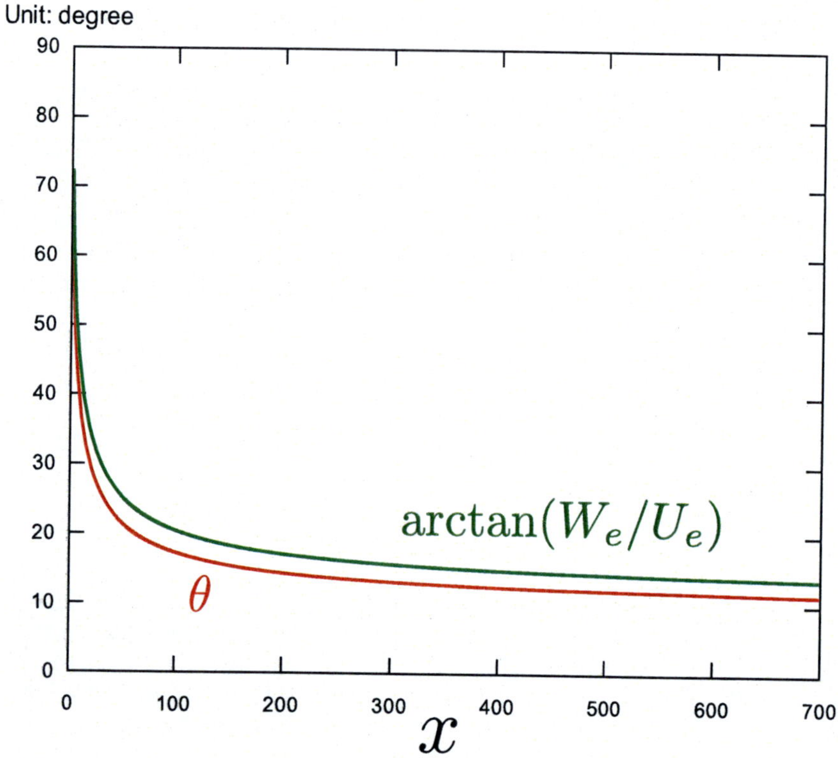

Fig. 12 Angle of external velocity and angle θ of crossflow vortical axis in the FSC boundary layer studied in this paper

References

1. Bippes H (1999) Prog Aero Sci 35:363–412
2. Ide Y, Hirota M, Tokugawa N (2021) Phys Fluids 33:034112
3. Malik MR, Li F, Choudhari MM, Chang C-L (1999) J Fluid Mech 85–115
4. Reshotko E, Beckwith IE (1958) NACA Rep 1379:1–49
5. Saric WS, Carrillo RB, Reibert MS (1998) AIAA Paper No. 1998-0781
6. Saric WS, Reed HL, White EB (2003) Ann Rev Fluid Mech 35:413–440
7. Saric WS, West DE, Tufts MW, Reed HL (2019) AIAA J 57:641–654
8. Wassermann P, Kloker M (2002) J Fluid Mech 456:49–84

On the Characteristics of Anti-contamination Devices on Attachment-Line of Subsonic Transport Aircraft

Keisuke Ohira and Naoko Tokugawa

Abstract The natural laminar flow (NLF) wing design approach effectively reduces the fuel consumption and environmental impact by reducing the friction drag for a subsonic civil transport aircraft. This study conducts a numerical analysis based on the Reynolds-averaged Navier–Stokes equations applied to a technology reference aircraft, TRA2022 (120 pax) developed by Japan Aerospace Exploration Agency (JAXA), to analyze the characteristics of a passive flow control device known as the anti-contamination device (ACD), on an attachment line. It is introduced to prevent the propagation of contaminated flow from the fuselage to the wing. Three different types of ACDs are considered: split ACD, chevron ACD, and streamwise groove. The split ACD can provide a wider area to apply the NLF design because it is placed at the wing-body junction. However, it significantly affects the aerodynamic characteristics. Conversely, the chevron ACD and streamwise groove do not significantly affect the characteristics. The split and chevron ACD do not cause flow separation or additional disturbance to isolate the contaminated flow. The streamwise groove includes a turbulent vortex along the groove along with the mechanism used to prevent the contamination flow, which corresponds to this vortex. The effect of the angle of attack between the end and beginning of the cruise flight is analyzed. The change in the boundary layer thickness immediately upstream of the apex of the chevron ACD is approximately 14% of the ACD height, and it can be increased locally based on the ACD shape. The span of the split ACD and the height of the chevron ACD must be considered this effect to ensure their decontamination characteristics.

Keywords Natural laminar flow · Laminar flow control · Attachment-line transition · Aerodynamic device

K. Ohira (✉)
Ryoyu Systems Co. Ltd., 6-19 Oe, Minato-ku, Nagoya, Aichi 455-0024, Japan
e-mail: ohirak@chofu.jaxa.jp

N. Tokugawa
Japan Aerospace Exploration Agency, 6-13-1, Osawa, Mitaka, Tokyo 181-0015, Japan

S. Lee et al. (eds.), *The Proceedings of the 2021 Asia-Pacific International Symposium on Aerospace Technology (APISAT 2021), Volume 1*, Lecture Notes in Electrical Engineering 912, https://doi.org/10.1007/978-981-19-2689-1_31

1 Introduction

The natural laminar flow (NLF) wing design approach effectively reduces the fuel consumption and environmental impact by reducing the friction drag for a subsonic civil transport aircraft. The swept wing faces the risk of flow contamination owing to the turbulent boundary layer on the fuselage, which propagates through the wing attachment-line [1–3]. Attachment-line contamination is widely known with the Poll criterion, $\overline{R}$. The turbulent flow from the fuselage becomes sustainable along with the attachment line if this criterion exceeds a threshold of approximately 250 and propagates towards the outboard of the wing, due to which turbulent flow can develop over the wing. The value of the $\overline{R}$ for subsonic civil transport aircraft is typically higher than the critical threshold [4]. Therefore, the flow over the NLF wing must be protected against contamination.

An anti-contamination device (ACD) is a passive device used to prevent the propagation of contaminated flow; it presents considerable potential since it can coexist with other devices such as active flow control [4, 5] and usually does not require complicated mechanisms [5]. Various types of ACDs have been proposed and described in previous studies along with their characteristics.

Gaster [6] designed the widely known passive aerodynamic device called the "Gaster bump" to overcome the issue of attachment-line contamination. This device can effectively stagnate the propagation of the contaminated flow along the attachment line on the bump shape, and the laminar boundary layer is the newly formed outboard side of the stagnation point. However, the device is not sufficiently capable to apply it to the high $\overline{R}$ conditions such as subsonic civil transport aircraft [4].

Seyfang [7] experimentally analyzed several types of ACDs installed on a simple swept-wing model and confirmed the effectiveness of a few of the ACDs, such as step-down and groove types. For a step-down type ACD, the contamination flow is trapped by the vortex formed behind the step, and non-turbulent free stream is attached at the outboard of the step. For the groove case, Seyfang stated that the contamination flow was transported downstream by the strong pressure gradient near the attachment line along the groove. Both the devices characterized by flow separation at the attachment line and the flow mechanism appear to vary from those of the other ACDs such as the Gaster bump.

Krier and Scipto [5] proposed unique ACD types, such as the positive and the negative arrow type ACD which, along with the surface streamline and the split ACD, isolates the fuselage boundary layer from the attachment line by placing the ACD at the wing-body junction. They also presented guidelines to optimize the ACD shape with practical considerations, such as the incidence range during cruise flight and variation of the boundary layer thickness.

Fiore et al. [8] demonstrated through both numerical and experimental approaches that their optimized chevron-shaped ACD can successfully maintain up to $\overline{R} \simeq 580$. Based on the stability analysis, this level of $\overline{R}$ indicated that it reached the critical threshold required to cause the natural transition of the newly generated boundary layer along the attachment line behind the ACD.

The adverse effects on the aerodynamic characteristics of the aircraft caused by the ACD must be minimized and the drag increment due to the ACD must not exceed the gain obtained by the NLF wing design. Therefore, a practical perspective must determine the effect of an ACD on the overall performance of the aircraft in addition to analyzing the characteristics of preventing the propagation of contamination.

This study presents a numerical analysis based on the Reynolds-averaged Navier–Stokes equations applied to the technology reference aircraft TRA2022 (120pax) [9], developed by the Japan Aerospace Exploration Agency (JAXA) to observe the characteristics of the ACDs. Three types of ACDs with characteristic shapes are selected: split ACD [5], chevron ACD [8], and streamwise groove [7]. Following the ACD implementation confirmation in terms of the basic characteristics, such as isolation of contaminated flow from the fuselage and no obvious disturbance generation from the ACD, it is analyzed for the characteristics of individual ACDs and the effect of the ACDs on the aerodynamic characteristics of the aircraft.

2 Geometry Model and Flow Condition

2.1 Baseline

The baseline geometry includes the tertiary design geometry of the TRA2022 [9] for a civil subsonic transport aircraft within the same class as the Airbus A319 with a cruising Mach number of 0.78, a capacity of 120 passengers, and a range of 3,600 nm. Figure 1 depicts the baseline configuration applied to the computations, which includes the primary components during cruise flight, such as the wing, fuselage, vertical/horizontal tails, pylons, and engine-nacelles. The nominal flow conditions corresponding to cruise flight considered in this study are as follows:

Fig. 1 JAXA's technology reference aircraft (TRA2022)

Mach number, $M = 0.78$, Reynolds number based on the mean aerodynamic chord, $Re_{\mathrm{MAC}} = 22 \times 10^6$, and angle of attack, $a = 0.5563°$. Additionally, two values of a are considered to determine the effect of a during cruise flight. One is $a = 0°$, which corresponds to the end of the cruise, and the other is $a = 0.9°$, which corresponds to the beginning of the cruise.

2.2 ACD Type

Figure 2 presents the ACD types used in this study. Each ACD is applied to TRA2022 at the wing root leading edge (LE). Although all the ACDs are defined through some parametric studies, they are not optimized for specific conditions. The first is the split ACD shown in Fig. 2a, which was proposed by Krier and Sucipto [5]. This device can isolate the flow over the wing from the turbulent flow over the fuselage by the cut-out part with no swept sharp LE, placing the wing-body junction. The span size was adjusted to ensure that it is higher than the boundary layer thickness of the fuselage.

The second is the chevron ACD [8], which is shown in Fig. 2b. This device can smoothly divide the flow along the streamline based on the chevron shape, and a new laminar boundary layer is developed on the top side of the ACD. The height of the device is adjusted to ensure that it is higher than the local boundary layer thickness at the apex. The top surface of the device was smoothly connected to the main wing geometry at the back end of the device. The remaining geometrical parameters such as the chordwise width, apex half-angle, and top front edge radius, follow the optimized values normalized by the total length [8]. However, the lateral part of the current geometry shown in Fig. 2b is not smoothly closed as that of the original ACD obtained by Fiore et al. [8] owing to the parameterization of the geometric model.

The last ACD type is the streamwise groove proposed by Seyfang, shown in Fig. 2c [7]. The groove is installed parallel to the aircraft axis and is smoothly connected to the main wing at the end of the device. The depth of the groove at the attachment line, which is normalized by the local LE radius, is set to the same order as that presented in reference [7].

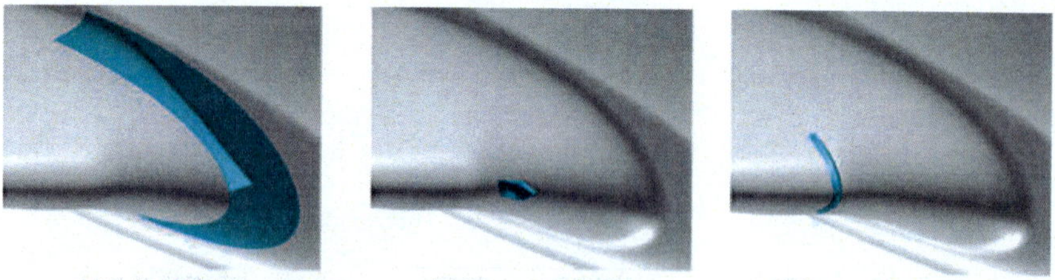

(a) Split ACD [5] (b) Chevron ACD [8] (c) Streamwise Groove [7]

Fig. 2 ACD types implemented on TRA2022. Cyan indicates the parts where the geometry is changed from the baseline (no ACD) geometry

3 Computational Method

3.1 Solver

The unstructured computational fluid dynamics (CFD) code, FaSTAR [10], developed by JAXA, is used for the flow simulations. A cell-vertex finite volume method is used to discretize the Reynolds averaged Navier–Stokes equations (RANS). The Harten-Lax-van Leer-Einfeldt-Wada (HLLEW) [11] method is applied for the inviscid flux, the Green-Gauss-based weighted least square (GLSQ) [12] method is used for gradient calculation, Hishida's (van Leer type) limiter [13] is used as the slope limiter, and LU-SGS is used for time integration in this study. The Spalart–Allmaras one equation model [14] with rotation correction [15, 16] is used for the turbulence model. According to the website, *Turbulence Modeling Resources*, provided by NASA [17], the specific model version used in the study is denoted as SA-noft2-R (Crot = 1). The turbulence model is activated over the entire wall surface.

3.2 Computational Mesh and Boundary Conditions

The commercial software, Pointwise (Cadence Design Systems, Inc.), is used to generate the computational mesh. Figure 3 presents an overview of the computational mesh, which is constructed as an unstructured mesh. Prismatic layers (a maximum of 50 cells) are inserted at the boundaries of the wall to accurately represent the boundary layer (Fig. 3c). The mesh of the baseline geometry comprises approximately five million nodes. The mesh size at the nearest wall along the wall-normal direction is 0.005 mm, and the corresponding mesh size in a wall unit (y^+) is such that $y^+ < 1$. The far-field boundary is located at a point 500 times the fuselage length away from the fuselage apex along the streamwise, spanwise, and vertical directions, as shown in Fig. 3b. The wall boundary is assumed to be an adiabatic viscous wall and the far-field boundary is obtained by using the Riemann invariants. A symmetric condition is applied to the symmetrical plane.

Re-generating the entire mesh for a geometric change in an area as small as the ACD is inefficient and introduces unnecessary grid dependencies into the result [18, 19], which can hinder a physical effect. Therefore, the meshes used for all the cases in the study are generated from the baseline mesh by re-meshing only in the vicinity of the ACD area. The remaining region retains the same mesh as the baseline. Consequently, the changes in the solution due to re-meshing in other areas are eliminated, and the aerodynamic changes in the ACD can be captured accurately. Acheson et al. [19] employed a similar method to predict the effect of small devices, such as vents and drains, on the aerodynamic performance of a commercial aircraft. Figure 3d illustrates the mesh region, including the ACD area that targets local re-meshing. When comparing the drag coefficient (C_D) between the streamwise groove

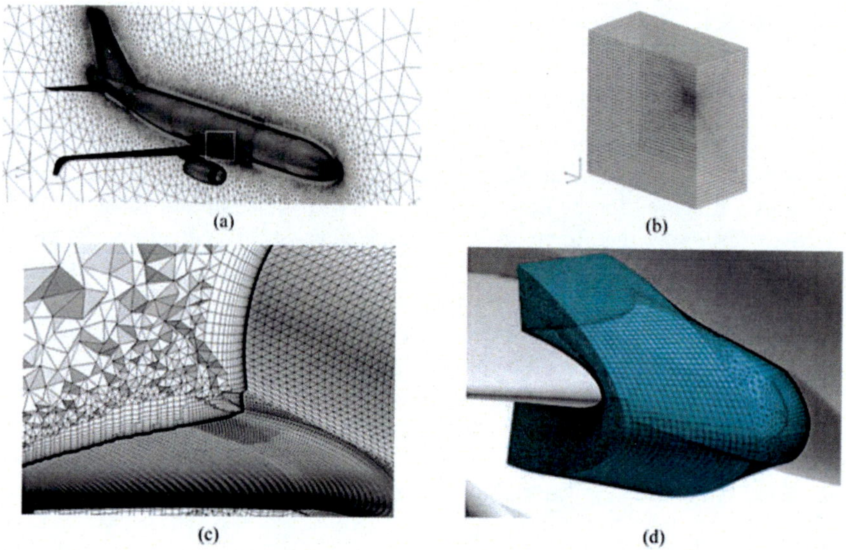

Fig. 3 Overview of the computational mesh **a** overall aircraft view with the walls and the symmetrical plane. The white rectangle at the wing root represents the area displayed in (**c**), **b** far-field boundary, and **c** enlarged view near the wing root with walls and spatial mesh at a plane perpendicular to the aircraft axis at the middle point of the root-chord, **d** local mesh region (cyan) including ACD area for the local re-meshing

and the baseline geometry (no ACD) by regenerating the entire mesh, a difference is observed in each component even in the areas far from the ACD, such as the outboard wing (order of 0.1 drag count), while such a difference is not observed in the local-remeshing method (a drag count below 0.001). It is assumed that the difference in the drag in the former approach is produced by a numerical error owing to the grid dependencies. The degree of numerical error in the drag difference, i.e., the drag count, in the local re-meshing approach is estimated to be approximately 0.02 based on the ratio of the wetted area local re-meshing region to the overall aircraft.

4 Results and Discussions

4.1 Overall Flow Field Without ACD

Figure 4 depicts the flow field with the baseline geometry without an ACD. Figure 4a shows the overall flow field of the baseline model is shown in and Fig. 4b presents an enlarged view of the LE at the wing-body junction. In Fig. 4b, the surface streamline along the LE represent the attachment line, which is connected from the fuselage to the wing via the wing-body fairing. The total pressure distribution over the fuselage

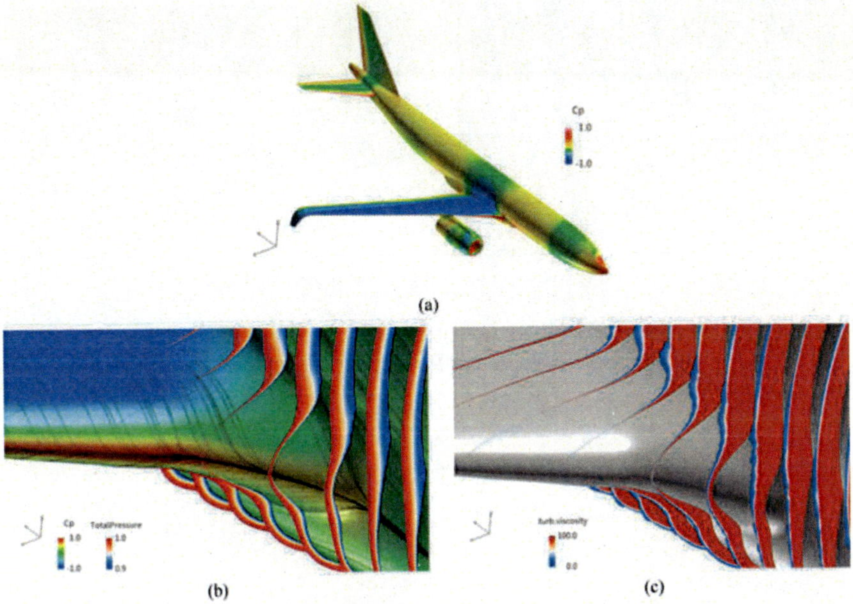

Fig. 4 Flow field with the baseline geometry without ACD. **a** Overall aircraft with pressure coefficient distribution, **b** LE at the wing-body junction with total pressure distributions in space, pressure coefficient distribution on the wall surface and the surface streamlines (black solid line), **c** LE at wing-body junction with turbulent kinematic viscosity

represents the existence of the boundary layer, and its thickness reduces significantly along the attachment line towards the outboard of the wing. Figure 4c presents the same perspective as shown in Fig. 4b, with the turbulent kinematic viscosity distributions, which represent the contribution of the turbulence in the RANS equations. The turbulent kinematic viscosity distribution also displays a fuselage turbulent boundary layer entering the main wing.

4.2 Flow Field Around ACD

Split ACD

Figure 5a and b illustrate the flow field around the split ACD where the total pressure and turbulent kinematic viscosity distributions indicate that the fuselage boundary layer cannot enter the wing owing to the sharp LE, and the boundary layer on the wing-body fairing becomes significantly thinner when compared to that of the baseline (see Fig. 4b and c). There is no evident separation or disturbance induced by the ACD on the wing. Figure 6 illustrates the C_P distributions between the split ACD and the baseline geometry sliced by a plane perpendicular to the lateral axis at the outer end

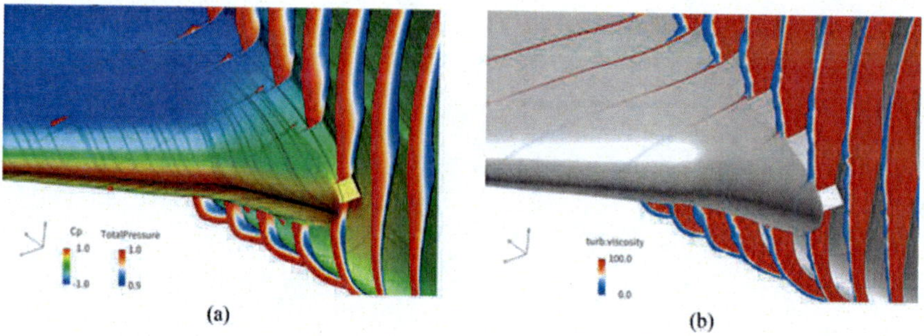

Fig. 5 Flow field around the split ACD. **a** total pressure distributions in space, pressure coefficient distribution on the wall surface and the surface streamlines (black solid line) **b** turbulent kinematic viscosity distributions

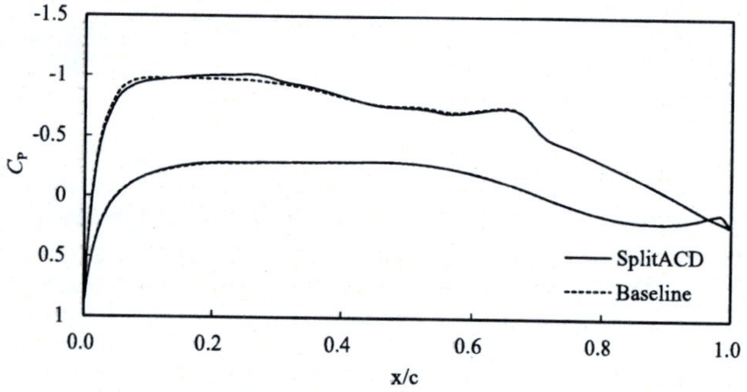

Fig. 6 Comparison of C_P distribution between Split ACD and baseline geometry at the outer end of the wing-body fairing

of the wing-body fairing. In Fig. 6, a slight difference can be observed in the C_P distribution on the forward and upper sides near the area where the ACD is placed, while there is no obvious difference in the rest of the area. This indicates that the change in the flow field due to the introduction of the split ACD is negligible and is limited to its proximity.

Chevron ACD

Figure 7 depicts the flow field around the chevron ACD model, which is placed at the end of the wing-body fairing. The ACD placement is adjusted to maintain the attachment line through the center of the ACD at a specific angle of attack (a = 0.5563°). This is confirmed by the surface streamlines shown in Fig. 7a. The contaminated flow is smoothly divided by the chevron shape as shown in the total pressure (Fig. 7a) and the turbulent kinematic viscosity (Fig. 7b) distributions, and a thin boundary layer with a very low level of turbulent kinematic viscosity is formed on the top side of the device. Figure 7 also shows the flow separation on the lateral sides of the device, which induces a thicker boundary layer in the wake. However, the effect

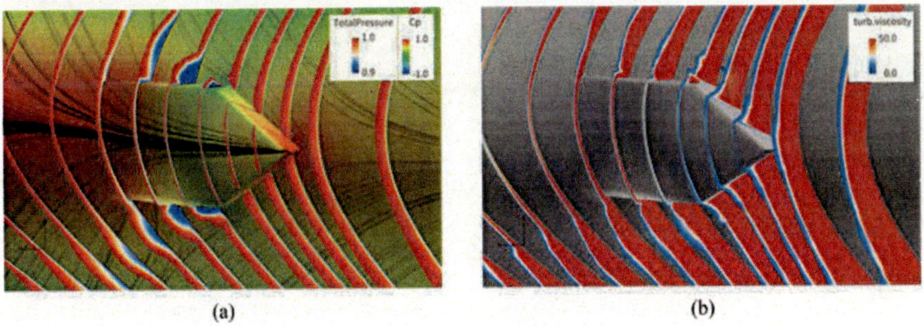

Fig. 7 Flow field around the chevron ACD. **a** Total pressure distributions in space, pressure coefficient distribution on the wall surface and the surface streamlines (black solid line) **b** Turbulent kinematic viscosity distributions

is assumed to be limited to a narrow area since the boundary layer propagates towards the TE along with the contaminated flow. This separation may be less pronounced or suppressed in the original chevron ACD defined by Fiore et al. [8], as explained in Sect. 2.2.

Streamwise Groove

Figure 8 illustrates the flow field around the streamwise groove installed at the outboard end of the wing-body fairing. There is no obvious difference when compared to the baseline geometry (Fig. 4b), apart from a longitudinal vortex at the groove wake formed inside the groove. Three different geometries with varying span (b_{swg}) and depth of the groove at LE (d_{swg}) are compared to observe the isolation of the contaminated flow, as shown in Fig. 9. Both the design parameters are normalized by the LE radius at the outer edge of the wing-body fairing (r_{LE}). In Fig. 9, the total pressure distributions depict a single vortex formed inside the groove, which is strengthened

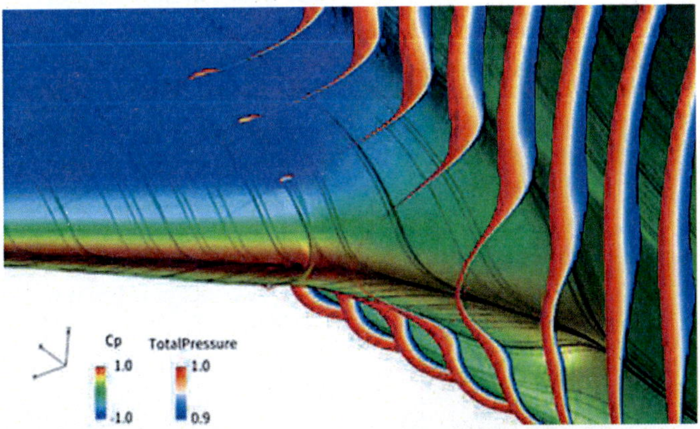

Fig. 8 Flow field with the streamwise groove with total pressure distributions in space, pressure coefficient distribution on the wall surface, and the surface streamlines (black solid line)

(a) $b_{swg} = 0.1$ and $d_{swg} = 0.1$ (b) $b_{swg} = 0.2$ and $d_{swg} = 0.1$ (c) $b_{swg} = 0.2$ and $d_{swg} = 0.2$

Fig. 9 Total pressure distributions and surface streamlines around the streamwise groove

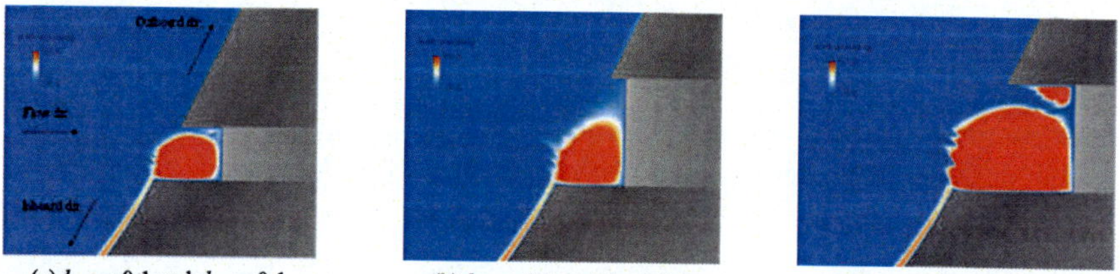

(a) $b_{swg} = 0.1$ and $d_{swg} = 0.1$ (b) $b_{swg} = 0.2$ and $d_{swg} = 0.1$ (c) $b_{swg} = 0.2$ and $d_{swg} = 0.2$

Fig. 10 Turbulent kinematic viscosity distribution on a plane perpendicular to the vertical axis at LE of the groove

by increasing the span, b_{swg}, and depth, d_{swg}. Figure 10 presents the turbulent kinematic viscosity distribution sliced in a plane perpendicular to the vertical axis at the LE of the groove. The turbulent kinematic viscosity is considerable inside the groove corresponding to the vortex and is significantly reduced behind the groove along the attachment line. This indicates that the propagation of contaminated flow from the upstream is prevented by merging into the vortex, and a boundary layer is newly formed on the outboard side of the groove. This decontamination process appears to be similar to that of the step-down type ACD explained earlier in the literature [7], which also involves a vortex along the step. This process varies from the other ACDs since the other ACDs are not obviously related to flow separation. There is no obvious difference between the cases corresponding to the decontamination process.

The vortex formed inside the groove is fully turbulent, as shown in Fig. 10. It is necessary to understand the behavior of the disturbance inside the streamwise groove, and to suppress the disturbances to prevent them from being contaminated by the outer wing in order to follow the basic strategy required to minimize the disturbance caused by the ACD introduced in the literature [4]. A direct numerical simulation or experimental approach is required for further analysis.

4.3 Effect of the Angle of Attack

The dependency of the flow around the ACDs on the angle of attack is determined for the chevron ACD and the streamwise groove. Figure 11 presents a comparison of the turbulent kinematic viscosity distributions with the chevron ACD for $a = 0°$ and 0.9°, which correspond to the end and the beginning of the cruise, respectively. The boundary layer of the contaminated flow propagated from the upstream at $a = 0°$, is slightly thicker than that of $a = 0.9°$, as shown in Fig. 11. Figure 11 also depicts the effect of the change in the boundary layer thickness on the flow on the top side of the ACD. The difference in the boundary layer thickness immediately upstream of the apex of the chevron is approximately 14% of the height of the ACD, based on the turbulent kinematic viscosity distribution. The difference can be further amplified locally near the device, based on the ACD type or geometry. Therefore, the margin must be considered carefully for the parameters such as the height of the chevron ACD and the span of the split ACD based on the boundary layer thickness, to ensure their decontamination characteristics for any changes in the boundary layer thickness during the cruise flight.

Figure 12a and b present a comparison of the turbulent kinematic viscosity distributions with the streamwise groove for $a = 0°$ and 0.9°. The figures demonstrate a similar flow field with a nominal value of $a = 0.5563°$ (Fig. 10b). Additionally, the flow on the outboard side of the groove is still maintained at a very low level when the range is expanded from −3° to +3°, and is quite similar to the cruise conditions even though the boundary layer thickness on the inboard side is changed significantly. Remarkably, the decontamination process remains uncorrelated with *a*. This feature contrasts with that of the other ACDs.

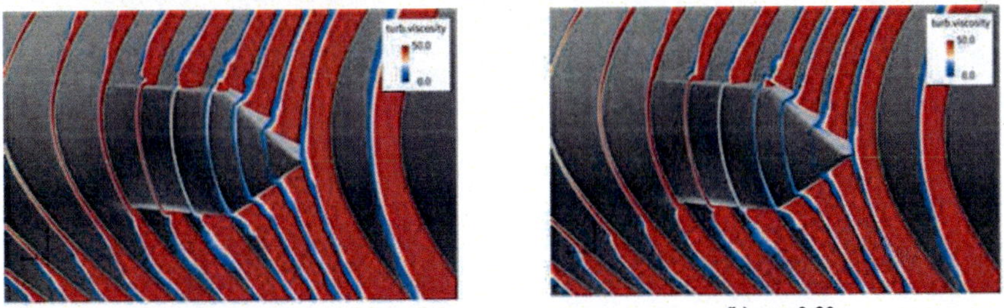

(a) $\alpha = 0°$ (b) $\alpha = 0.9°$

Fig. 11 Comparison of turbulent kinematic viscosity distributions around the chevron ACD between **a** $a = 0°$ corresponds the end of cruise, and **b** $a = 0.9°$ corresponds the beginning of cruise

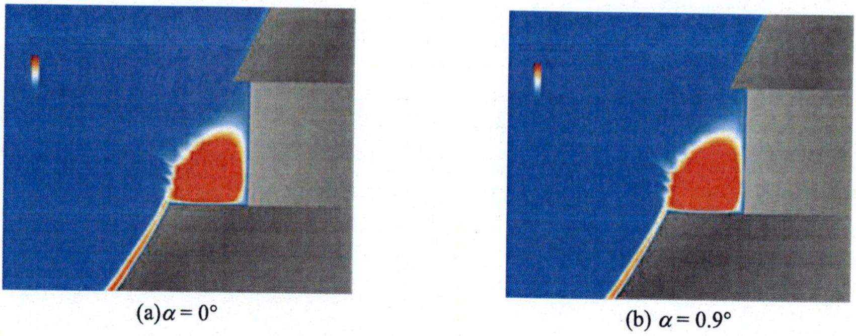

Fig. 12 Comparison of turbulent kinematic viscosity distributions around the streamwise groove between **a** $a = 0°$ and (b) 0.9° corresponds to the end and the beginning of the cruise, respectively

4.4 Aerodynamic Force

This section presents the changes in the aerodynamic characteristics due to the introduction of the ACDs. It introduces a simple correction procedure using the result of a single CFD with ACD and static aerodynamic derivatives in terms of a and the deflection angle of the horizontal tail plane, δ_{HTP}, with the baseline geometry. The correction values of α and δ_{HTP} are first evaluated to match the trimmed state at the design point and the corrected drag coefficient is then estimated. The correction procedure can be written as:

$$\begin{pmatrix} \alpha_{cor} \\ \delta_{HTP,cor} \end{pmatrix} = \frac{1}{\frac{\partial C_L}{\partial \alpha}\frac{\partial C_m}{\partial \delta_{HTP}} - \frac{\partial C_L}{\partial \delta_{HTP}}\frac{\partial C_m}{\partial \alpha}} \begin{pmatrix} \frac{\partial C_m}{\partial \delta_{HTP}} & \frac{\partial C_L}{\partial \delta_{HTP}} \\ \frac{\partial C_m}{\partial \alpha} & \frac{\partial C_L}{\partial \alpha} \end{pmatrix} \begin{pmatrix} C_{L,design} - C_{L,CFD} \\ C_{m,design} - C_{m,CFD} \end{pmatrix} + \begin{pmatrix} \alpha_{CFD} \\ \delta_{HTP,CFD} \end{pmatrix}, \quad (1)$$

$$C_{D,cor} = C_{D,CFD} + (\alpha_{cor} - \alpha_{CFD})\frac{\partial C_D}{\partial \alpha} + (\delta_{HTP,cor} - \delta_{HTP,CFD})\frac{\partial C_D}{\partial \delta_{HTP}}, \quad (2)$$

where the subscript, *design*, indicates the design point of the cruise flight, *cor* represents the corrected value, and *CFD* represents the CFD result with the ACD. The variations in the aerodynamic characteristics evaluated by the difference in C_D, α, and δ_{HTP} are expressed as:

$$\Delta C_D = C_{D,cor} - C_{D,base}, \quad (3)$$

$$\Delta \alpha = \alpha_{cor} - \alpha_{base}, \quad (4)$$

$$\Delta \delta_{HTP} = \delta_{HTP,cor} - \delta_{HTP,base} \quad (5)$$

Table 1 drag coefficient increment and amount of correction $D\alpha$ and $D\delta_{HTP}$

ACD type	$\Delta C_D \times 10^4$	$D\alpha(°)$	$D\delta_{HTP}(°)$
Split ACD	0.65	−0.008	−0.025
Chevron ACD	0.00	0.001	0.003
Streamwise Groove	0.04	0.000	0.001

where the subscript, *base*, indicates the value of the baseline geometry at the design point. Table 1 presents the results of the correction procedure. The split ACD indicates the highest drag increase, while the drag of chevron ACD does not vary from the baseline, although it includes the grid dependency in the local remeshing region as explained in Sect. 3.2. The split ACD and streamwise groove are the highest and lowest values in terms of $\Delta\alpha$ and $\Delta\delta_{HTP}$, respectively. As shown in Fig. 6, the split ACD affects the flow field only around the device. However, the size of the device is relatively large when compared to the others since it is defined based on the boundary layer thickness of the fuselage, which is significantly larger than that of the other ACDs. Additionally, the ACD placed on the inboard side, such as the split ACD, expands the target area of the NLF design. Therefore, a trade-off between the amount of the area where the NLF design is applied and the amount of change in the aerodynamic characteristics due to the introduction of the ACD may have to be considered.

5 Conclusion

Numerical simulations were performed to determine the characteristics of the Anti-Contamination Device (ACD) on an attachment line on a subsonic civil transport aircraft with cruise flight conditions based on the Reynolds-averaged Navier–Stokes equations. Three different types of ACD were installed on JAXA's technology reference aircraft, TRA2022.

All the ACDs considered in this study can effectively isolate the contaminated flow from the fuselage. When using the split ACD, the sharp-LE part can isolate the contaminated flow field before the flow enters the wing without additional disturbance. When using the chevron ACD, the chevron shape can smoothly divide the contaminated flow into the upper and lower wing surfaces and a thin boundary layer is formed on the top side of the device. There is no obvious disturbance generated from the device, except for the separation on the lateral side of the device, which may not by caused by the original chevron ACD in the literature [8]. When using the streamwise groove, the contaminated flow on the attachment line was prevented by merging a longitudinal vortex formed along the groove and a boundary layer formed behind the device along the attachment line. However, further analysis is required to determine the risk of contamination on the outer wing since vortex generation leads to full turbulence, which may be a source of contamination.

The change in the angle of attack from 0 to 0.9° corresponds to the end and beginning of the cruise flight, respectively, inducing a change in the boundary layer thickness by approximately 14% of the chevron ACD height immediately upstream of its apex. The span of the split ACD and the height of the chevron ACD must be considered to ensure their decontamination characteristics. Conversely, the decontamination function of the streamwise groove does not depend on the angle of attack, which ranges from −3 to 3°.

A wider area can be provided for the NLF over the wing using the split ACD, owing to its placement at the wing-body junction. However, it also presents a relatively large impact on the aerodynamic characteristics of the overall aircraft as it increases the drag count by 0.65, and the size of the device is relatively large since it is determined based on the fuselage boundary layer thickness, which is much greater than that of the other ACDs.

Acknowledgements The authors thank Dr. T. Ishida for valuable comments on the paper and Dr. K. Yoshida, Dr. D. Kwak, and Mr. N. Hayabe for fruitful discussions on ACD. This work used the computational resources of JAXA Supercomputer System Generation 3 (JSS3).

References

1. Pfenninger W (1965) Flow phenomena at the leading edge of swept wings. Recent developments in boundary layer research. AGARDograph 97(Part IV)
2. Poll DIA (1978) Some aspects of the flow near a swept attachment line with particular reference to boundary layer transition. Cranfield Institute of Technology, CoA report no. 7805
3. Arnal D (1992) Boundary layer transition: prediction, application to drag reduction. In: AGARD FDP/VKI course on skin friction drag reduction. Brussels, Belgium
4. Arnal D, Juillen JC, Reneaux J, Gasparian G (1997) Effect of wall suction on leading edge contamination. Aerosp Sci Technol 8:505–517
5. Krier JV, Sucipto T (2009) Highly optimizable laminar flow control devices. In: 19th AIMETA congress. Ancona, Italy
6. Gaster M (1967) On the flow along swept leading edges. Aero Q XVIII(Part 2)
7. Seyfang GR (1987) Turbulence reduction on swept leading edges. In: Proceedings of the international conference, turbulent drag reduction by passive means. London
8. Fiore M, Vermeersch O, Forte M, Casalis G, Francois C (2016) Characterization of a highly efficient chevron-shaped anti-contamination device. Exp Fluids 57:59
9. Kwak D, Nomura T, Tokugawa N, Kurita M, Murayama M (2016) Introduction of research project for environmental conscious aircraft technology in JAXA, greener aviation 2016 symposium. Belgium, Brussels
10. Hashimoto A, Murakami K, Aoyama T, Ishiko K, Hishida M, Sakashita M, Lahur P (2012) Toward the fastest unstructured CFD code 'FaSTAR'. AIAA-2012-1075
11. Obayashi S, Guruswamy GP (1995) Convergence acceleration of a navier-stokes solver for efficient static aeroelastic computation. AIAA-J 33(6):1134–1141
12. Shima E, Kitamura K, Haga T (2013) Green–gauss/weighted-least-squares hybrid gradient reconstruction for arbitrary polyhedra unstructured grids. AIAA-J 51(11):2740–2747
13. Hishida M, Hashimoto A, Murakami K, Aoyama T (2011) A new slope limiter for fast unstructured CFD solver FaSTAR. In: JAXA-SP-10-012, pp 85–90. (in Japanese)
14. Spalart PR, Allmaras SR (1992) A one-equation turbulence model for aerodynamic flows. Rech Aero 1:5–21

15. Dacles-Mariani J, Zilliac GG, Chow JS, Bradshaw P (1995) Numerical/experimental study of a wingtip vortex in the near field. AIAA-J 33(9):1561–1568
16. Lei Z (2005) Effect of RANS turbulence models on computation of vortical flow over wing body configuration. Trans Jpn Soc Aero Space Sci 48(161):152–160
17. NASA langley research center turbulence modeling resource, https://turbmodels.larc.nasa.gov/. Cited 16 Aug 2021
18. Ito Y, Murayama M, Yamamoto K, Shih AM, Soni BK (2009) Efficient computational fluid dynamics evaluation of small-device locations with automatic local remeshing. AIAA-J 47(5):1270–1276
19. Acheson KE, Kornegay BS, Lau H-FM (2011) Influence and control of unstructured volume meshes for enhanced fidelity performance predictions. AIAA-Pap 2011–3984

Preliminary Investigation on Flyover Noise of Propeller-Powered Aircraft Using a Phased Microphone Array

Zhang Yingzhe, Zhang Weiguang, Lin Dakai, and Liu Peiqing

Abstract With people's heightened awareness of environmental protection, the noise level of large civil aircraft during take-off and landing are monitored with strict restrictions by the ICAO (International Civil Aviation Organization). This has triggered people's interest in learning about the noise characteristics of aircraft components. Hence, researchers are developing phased microphone array test technology to analyze the characteristics of external noise sources of civil aircraft and verify the effect of noise reduction design. At present, most aviation research institutions around the world, have developed their own phased microphone array test technology. They have also carried out the corresponding flight noise test verification, widely used in aircraft low noise design and verification. To better understand the functioning of the external noise of civil aircraft and development the flight noise test capability of civil aircraft, COMAC Beijing Aircraft Technology Research Institute (BATRI) has also carried out the research on the flyover noise test technology. This research, based on a phased microphone array, has designed microphone array and built a set of phased microphone array test systems. For the research, a multi-arm spiral array with a diameter of 15 m and 84 channels were arranged, at the end of the runway of Zhanghe airport in Hubei Province. The domestic single-engine propeller-powered aircraft is used for multiple flights, such as different flight altitudes, different flight speeds, landing gear put-down, and retraction configurations. The test results provide effective test data. Based on this test data, Doppler effect correction is carried out on the flight test data, the test data of different working conditions are compared and analyzed by using the conventional Beamforming method and the noise source identification results are analyzed. The test data processing results show that the main noise source of single-engine propeller aircraft is the blade passing frequency noise of the propeller, and the pure tone noise has obvious characteristics, which are conducive

Z. Yingzhe · L. Peiqing (✉)
School of Aeronautic Science and Engineering, Beihang University, Beijing, China
e-mail: lpq@buaa.edu.cn

Z. Yingzhe · Z. Weiguang · L. Dakai
Aerodynamic and Aeroacoustic Research Department, COMAC Beijing Aircraft Technology Research Institute, Beijing, China

Beijing Key Laboratory of Simulation Technology for Civil Aircraft Design, Beijing, China

S. Lee et al. (eds.), *The Proceedings of the 2021 Asia-Pacific International Symposium on Aerospace Technology (APISAT 2021), Volume 1*, Lecture Notes in Electrical Engineering 912, https://doi.org/10.1007/978-981-19-2689-1_32

to data correction. The results also indicate that the conventional beamforming noise source inversion algorithm can accurately identify the noise location. However, the sound source location accuracy has a certain deviation due to the influence of array design and aircraft flight condition control. This test has verified the microphone array test system, test process, data post-processing process, and accumulated field flight noise test experience. It also lays a foundation for later large-scale array and large-scale civil aircraft flight external noise source positioning tests.

Keywords Phased microphone array · Flyover test · Noise source location · Beamforming · Propeller noise

1 Introduction

With people's heightened awareness of environmental protection and their own health, their focus has increasingly shifted to noise pollution. This has led to an increase in ICAO's requirements for civil aircraft noise emission. The increasing attention to noise emission has prompted representatives of the Federal Aviation Administration, the Joint Aviation Administration, the International Civil Aviation Administration, the European Union, aircraft manufacturers, and aircraft engine manufacturers, to consider developing more stringent noise emission control measures. In addition to the basic requirement of airworthiness noise, airports have also set up their own airport noise standards in response to people's attention, such as Heathrow Airport, Gatwick Airport, and Stansted Airport.

For aircraft companies, increasingly stringent noise emission requirements and airport noise regulations make the noise design of civil aircraft increasingly important, forcing aircraft companies to invest more energy in the noise reduction design of civil aircraft. To identify the complex noise sources outside the aircraft, people designed and developed the sound source identification test technology of microphone array. This is based on the phased array [1] and studies the characteristics of noise sources outside the aircraft. The microphone array test technology is different from the test of measuring the overall noise of a sound source with a single microphone. It can display the noise source of the aircraft by image and obtain the domain of each source of the aircraft. With the development of hardware and software, the use of microphone array technology has become a standard means of acoustic tests.

European researchers have carried out prior research on microphone array test technology. Howell et al. [2] first studied the noise of aircraft engine fans by using a line array composed of four microphones in 1986. Michel et al. [3–6] continued to carry out test research on flight noise source identification. Using a microphone array, he conducted research on external engine noise and airframe noise of fighter and civil aircraft, to improve the performance of microphone array noise source identification. Piet et al. [7–10], since 1997, have carried out experimental research on the external noise of Airbus aircraft from a wind tunnel to flight. He carried out configuration changes of noise reduction measures on A340 aircraft, including landing gear, and successfully verified the noise reduction effect of noise reduction configuration by using a microphone array, landing gear, lifting device, and power system. He also evaluated the noise reduction amount of different configuration changes. Siller et al. [11–16] continued to use microphone array to carry out many flight tests on A319, MD11, A340, A320, and other aircraft. He also worked on flight tests, data correction, sound source imaging post-processing, and noise source decomposition.

Stoker et al. [16] used microphone array to study the airframe noise of the B-777 model in a wind tunnel in 2001. He compared the results with the flight test [17, 18] and successfully confirmed the anti-de-icing hole noise, and carried out noise reduction design and verification. From this study, Boeing recognized the great value of microphone array and subsequently carried out a series of research work [19–24]. In the QTD series of studies, the microphone array is used to carry out flight tests to verify the noise reduction effect of external low-noise design technology, improve the maturity of the technology and apply it to products. At present, Boeing's microphone array has reached 1000 channels. JAXA [25–27] conducted research on noise reduction measures for landing gear and high-lift devices on its flight verification platform "HISHO" by using a microphone array and verifying the noise reduction design.

BATRI has been carrying out research on microphone array test technology since 2013. From the initial 24 channels small principle verification array, the design and verification of 320 channels microphone array have now been completed. In 2018, the microphone array test was conducted on the domestic single-engine propeller aircraft -Eagle 500- at Zhanghe airport in Hubei. This paper presents the preliminary results of this test.

2 Flight Test Setup

This test was carried out in 2018 to verify the phased microphone array test process, test methods, data correction, noise source identification process, and noise decomposition technology. The phased microphone array system is built on the NI PXIe system, and 84 PCB quarter array microphones are used for noise data acquisition.

2.1 Microphone Array Design

The microphone array test of static sound source has been carried out in the early stage, which can accurately reflect the location of the noise source [28]. This test mainly verifies the ability of the phased microphone array test system to identify and locate the moving noise source. In this test, no specialized GPS system is installed in the aircraft, and it is found that the flight path of small aircraft is greatly affected by the environment. Therefore, during the flight test, the aircraft cannot pass directly above the array when flying over the microphone array. Hence the microphone array

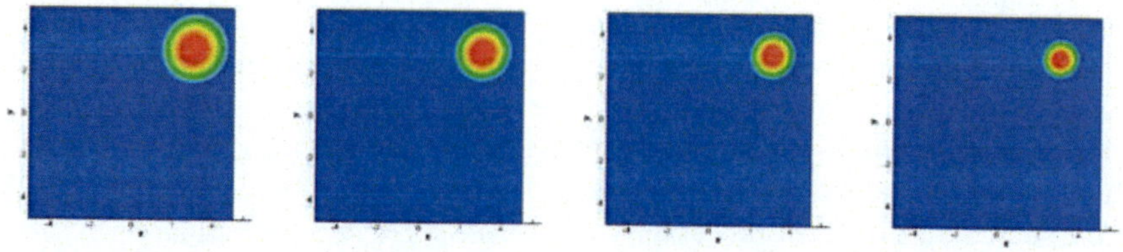

Fig. 1 Noise source simulation results at 400 Hz (from left to right, the diameter is 10, 12, 15 and 20 m respectively)

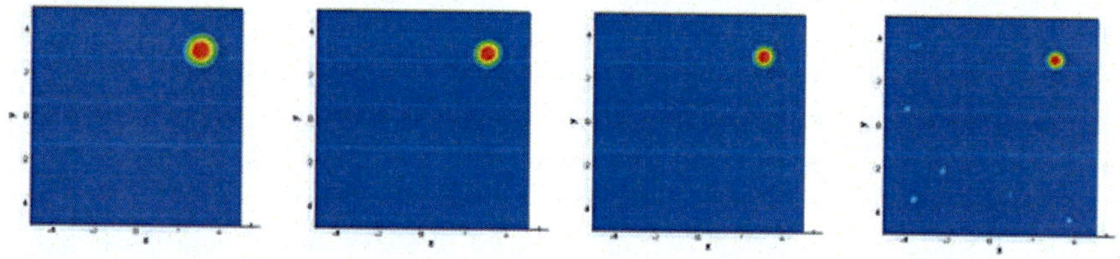

Fig. 2 Noise source simulation results at 800 Hz (from left to right, the diameter is 10, 12, 15 and 20 m respectively)

is designed accordingly and the performance of the array for noise source position inversion when the sound source deviates from the center of the array is mainly considered. At the same time, the resolution of an array to noise source with different array diameters is compared and analyzed.

The object of this test is a small, single-engine propeller aircraft. The noise is mainly tone at Blade Passing Frequency (BPF), and the frequency is relatively low. 84 microphones are planned on being used, and the array is designed in the form of a multi-arm spiral. Figures 1 and 2 show the noise source position and spatial resolution inverted by the microphone array at the simulated noise source coordinates (3, 3 m), where the frequencies are 400 Hz and 800 Hz respectively under four array diameters (10, 12, 15 and 20 m). As shown in Figs. 1 and 2, the arrays with four diameters can accurately reflect the position of the simulated noise source. It was found that the larger the diameter of the microphone array, the higher was the spatial resolution of the noise source. As seen in Fig. 2, with the increase of the microphone array, for higher frequency noise, the side-lobe (dummy noise source) decreases. Finally, the microphone array design consists of 7 spiral arms, each spiral arm has 5 microphones, and the diameter of the array is 15 m, as shown in Fig. 3.

2.2 *Flight Test*

The test aircraft used in this test is the domestic Eagle-500 aircraft, which is a single-engine propeller aircraft with two blades. The test site is Zhanghe airport in the Hubei Province. The microphone array is arranged at the end of the runway, as shown in

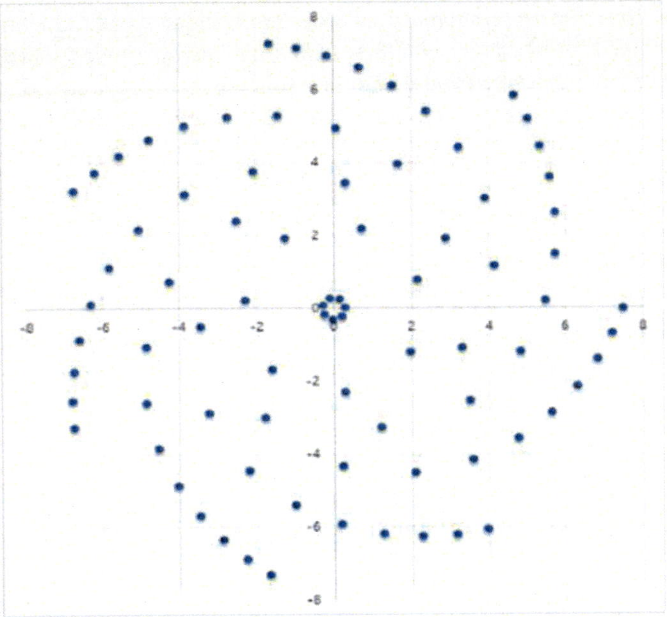

Fig. 3 Microphone array geometry

Fig. 4. All microphones are arranged horizontally and perpendicular to the aircraft route. The acquisition and control equipments are arranged at the end of the runway, away from the array. The system power supply is from the battery.

After the system connection is completed, each channel is tested to prevent channel damage. Background noise is collected before the test. When the aircraft enters the view, data acquisition software is run to record data. When the aircraft disappears from the view, data acquisition is stopped. During data acquisition, the test personnel

Fig. 4 Field test of Eagle-500

stays away from the test equipment and does not walk or speak. After single data acquisition, the test data at the zenith crossing time for rapid sound source identification is quickly intercepted and post-processing analysis is conducted to confirm the effectiveness of the test data. The test conditions are horizontal overflight at 50, 100 and 150 feet and 80, 100, and 120 knots.

3 Data Post-process

In the test, 84 channels of sound pressure signals were synchronously collected at high speed, and the sampling frequency was set to 65536 Hz. The 8.0 s time-length data near the aircraft overhead time was intercepted for post-processing analysis. Because the noise source frequency and flight altitude are relatively low, the influence of atmospheric absorption effect is ignored in this study. Doppler effect correction is mainly carried out on the test data, and the differences in array noise source identification results at different altitudes and speeds are compared and analyzed.

3.1 Doppler Effect Correction

When measuring a moving sound source with a microphone, when the sound source moves close to the microphone, the sound signal is compressed, and the frequency of the sound source is increased. When the sound source moves away from the microphone, the sound signal is stretched and the frequency of the sound source is reduced, which is the Doppler effect, as shown in Fig. 5. If a microphone received a noise signal at time T_R, that noise signal was actually emitted when the airplane was at a position corresponding to time T_S, not T_R. Consequently, there is a time delay

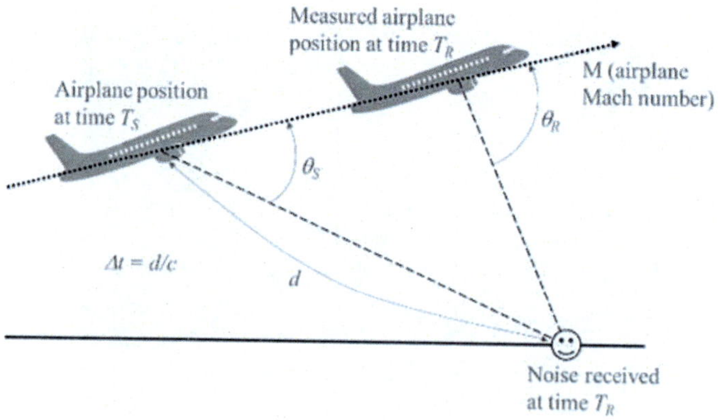

Fig. 5 Basis of Doppler effect

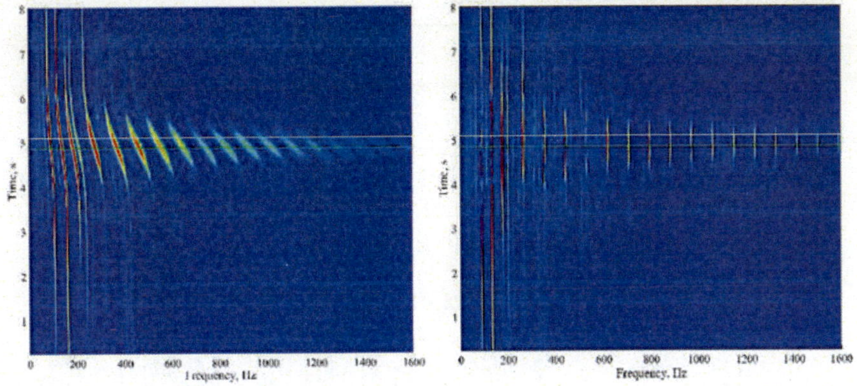

Fig. 6 Spectrum contour of signal (Left: Original; Right: Dedopplerised)

between source and receiver time.

$$T_R = T_S + \Delta t \quad (1)$$

In order to get the correct sound source position, the measured sound signal needs to be corrected. The relation between measured time and dedopplerized time is:

$$(T_{R2}^* - T_{R1}^*) = \frac{1}{(1 - M \cos \theta_s)}(T_{R2} - T_{R1}) \quad (2)$$

Here M is Mach number and θ_s is Doppler angle. The expression $\frac{1}{(1-M\cos\theta_s)}$ is the definition of the Doppler factor and it is important to use the angle at the source time (θ_s) rather than the angle at the time received (θ_R) to account for the Doppler factor. Since the source time and airplane position at each station are known, it is straightforward to calculate the received time corresponding to the source time through interpolation between stations. It is possible to find the source time corresponding to received time equals to T_R. If we know the source time T_S then it is possible to know the position of the airplane and calculate the proper Doppler angle corresponding to received time equals to T_R. Figure 6 shows the spectrum contour of original signal and dedopplerized signal. It is obvious from Fig. 6 that the eagle 500 aircraft noise is mainly pure tone, and the frequency of the original noise signal changes obviously with time. After Doppler correction, the pure audio rate does not change with time.

3.2 *Noise Source Location Results at Different Altitude*

Figure 7 shows the noise signal spectrum of the microphone at the center of the array when the aircraft flies over the microphone array at a speed of 80 knots with three

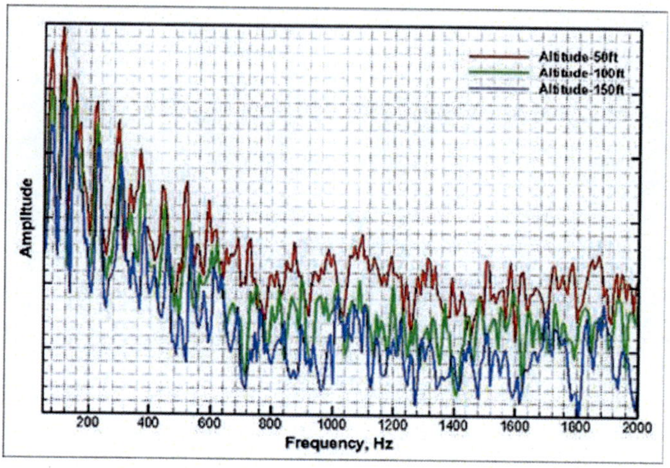

Fig. 7 Spectrum of three different altitudes of flight

different altitudes (50, 100, and 150 ft). It can be seen from Fig. 7 that the main noise of the single-engine propeller aircraft is the tone noise of the propeller blade passing frequency, and the blade noise is significantly higher than the other aircraft noises at low frequencies. As the frequency increases, the propeller noise mixes with other aircraft noises, such as engine noise et al., and no longer presents a clear tone. Figures 8 and 9 show the beamforming results under three different altitudes with narrow-band center frequencies of 292 and 902 Hz. From Fig. 8, it can be seen that the noise source is at the engine blades. Also, as the height of the aircraft increases, the spatial resolution of the beamforming results decreases. Figure 9 shows that since the noise signal is no longer a strong tone, a side-lobe noise is primarily observed in the result, which is especially most obvious at low altitudes. If we want to have a high resolution of high-frequency noise source location, one way is to increase the microphone numbers.

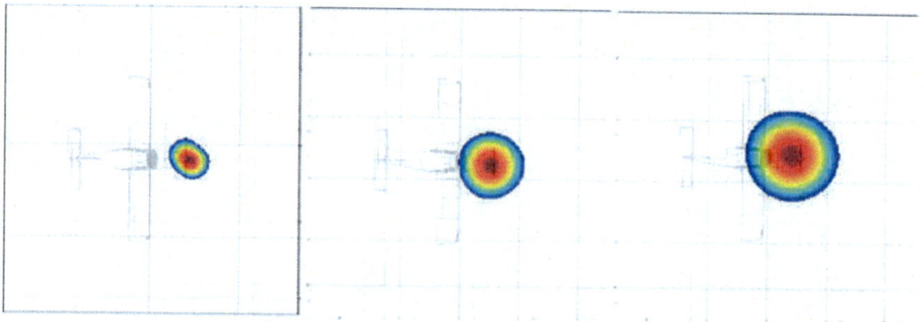

Fig. 8 Beamforming results of three different altitudes (50, 100 and 150 ft; 292 Hz)

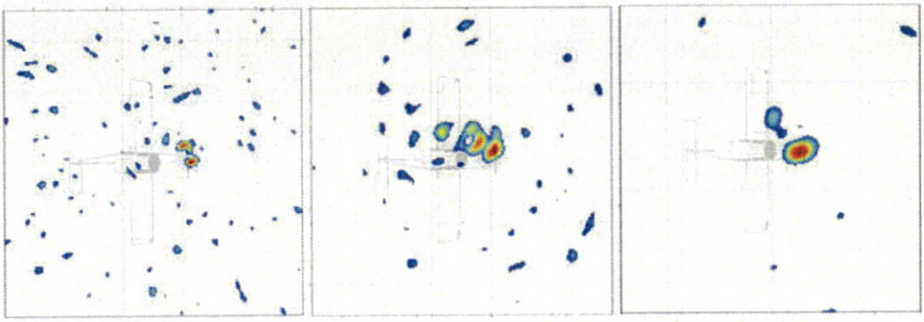

Fig. 9 Beamforming results of three different altitudes (50, 100 and 150 ft; 902 Hz)

3.3 Noise Source Location Results at Different Speeds

Figure 10 shows the noise signal spectrum of the microphone at the center of the array, when the aircraft flies over the microphone array at three different flight speeds (80, 100, and 120 knots), at an altitude of 100 feet. It can be seen that at the lowest speed, the propeller noise is much lower than the other two flights. This is mainly because of the low engine power setting and a lower propeller speed which generates low noise. When the speed is 80 knots, the aircraft propeller noise is only noticeable below 600 Hz. At 100 and 120 knots, the propeller noise dominates even over 1000 Hz. It can be seen from Fig. 11 that at low frequencies, the array results clearly locate the propeller noise at the three speeds. However, as the frequency increases, at about 1000 Hz, the position of the microphone array is no longer located at the spiral under low-speed conditions, but for the two speeds of 100 and 120 knots, the result of the microphone array is still located at the propeller, as shown in Fig. 12.

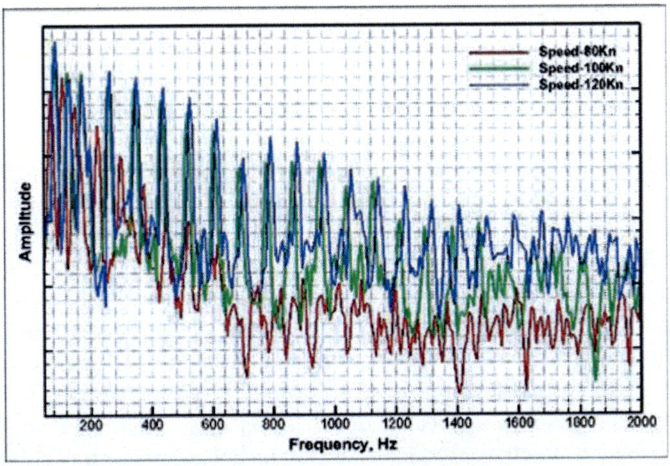

Fig. 10 Spectrum of three different speeds flight

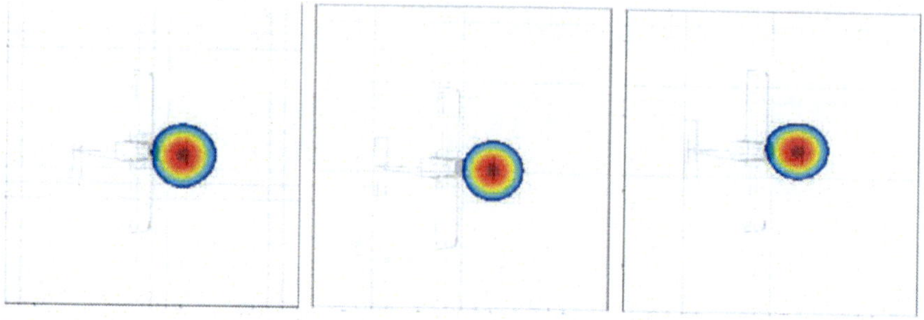

Fig. 11 Beamforming results of three different speeds (100 ft; 80, 100, 120 Kn; around 300 Hz)

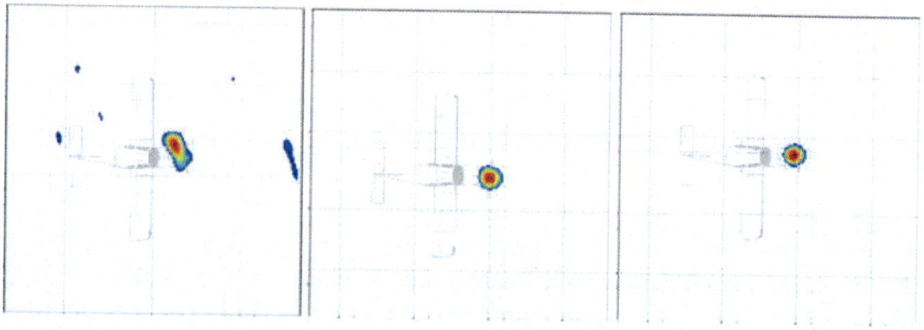

Fig. 12 Beamforming results of three different speeds (100 ft; 80, 100, 120 Kn; around 1000 Hz)

4 Conclusion

This paper introduces the microphone array flight noise test and sound source identification for single-engine propeller aircraft. When designing an array with a given number of channels, it is necessary to fully consider the target noise frequency range and flight path uncertainty. It is also necessary to comprehensively consider the resolution and dynamic range to ensure that the sound source identification results are optimal in all the noise target ranges.

The test data is post-processed and analyzed, which includes Doppler correction, noise source identification, and location. The noise source location results at three heights at 80 knots are compared and analyzed. For the results of the microphone array noise source identification at the same frequency, as the height of the sound source increases, the spatial resolution decreases. The results of the noise source identification at three different speeds at a height of 100 feet are compared and analyzed. At the three speeds, the designed microphone array can accurately obtain the location of the noise source.

Based on the previous basic research, this test has achieved relatively excellent test results and obtained effective test data. It verifies the microphone array noise source identification test system, test process, and test data post-processing algorithm, and lays the foundation for the external noise test of civil aircraft in the future.

Acknowledgements The authors express gratitude and appreciation to the staff of Zhanghe airport for permitting the use of the airport as a test site and providing on-site support for the successful execution of the flight test.

References

1. Michel U (2006) History of acoustic beamforming. In: 1st Berlin beamforming conference
2. Howell GP, Bradley J, McCormick MA et al (1986) De-dopplerization and acoustic imaging of aircraft flyover noise measurements. J Sound Vib 105:151–167
3. Michel U, Barsikow B, Haverich B, et al (1997) Investigation of airframe and jet noise in high-speed flight with a microphone array. In: 3rd AIAA/CEAS aeroacoustics conference, AIAA paper 1997-1596
4. Michel U, Barsikow B, Helbig J, et al (1998) Flyover noise measurements on landing aircraft with a microphone array. In: 4th AIAA/CEAS aeroacoustics conference, AIAA paper 1998-2336
5. Michel U, Bohning P (2002) Investigation of aircraft wake vortices with phased microphone arrays. In: 8th AIAA/CEAS aeroacoustics conference, AIAA paper 2002-2501
6. Michel U, Barsikow B, Böhning P, et al (2004) Localisation of moving sound sources with phased microphone arrays. Inter-Noise 2004
7. Piet JF, Elias G (1997) Airframe noise source localization using a microphone array. In: 3rd AIAA/CEAS aeroacoustics conference, AIAA paper 1997-1643
8. Piet JF, Elias G, Lebigot P (1999) Localization of acoustic source from a landing aircraft with a microphone array. In: 5th AIAA/CEAS aeroacoustics conference, AIAA paper 1999-1811
9. Piet JF, Michel U, Bohning P (2002) Localization of the acoustic sources of the A340 with a large phased microphone array during flight tests. In: 8th AIAA/CEAS aeroacoustics conference, AIAA paper 2002-2506
10. Piet JF, Davy R, Elias G, et al (2005) Flight test investigation of add-on treatments to reduce aircraft airframe noise. In: 11th AIAA/CEAS aeroacoustics conference, AIAA paper 2005-3007
11. Siller HA, Michel U (2002) Buzz-saw noise spectra and directivity from flyover tests. In: 8th AIAA/CEAS aeroacoustics conference, AIAA paper 2002-2562
12. Siller HA, Michel U (2006) Reduction of Approach Noise of the MD-11. 12th AIAA/CEAS Aeroacoustics Conference, AIAA Paper 2006–2464
13. Siller HA, Drescher M, Saueressig G, et al, Fly-over source localisation on a Boeing 747–400. In: Berlin beamforming conference, BeBeC-2010-13
14. Siller HA (2012) Localisation of sound sources on aircraft in flight. In: Proceedings of the internoise 2012/ASME NCAD meeting, NCAD2012-0575
15. Siller HA, Wolfram H, Timo S (2018) Source localisation on aircraft in flight—new measurements with the DLR research aircraft Airbus 320 ATRA. In: 7th Berlin beamforming conference, BeBeC-2018-D01
16. Stoker R, Sen R (2001) An experimental investigation of airframe noise using a model-scale Boeing 777. In: 39th Aerospace sciences meeting and exhibit, AIAA paper 2001-0987
17. Stoker R (2003) Using microphone phased arrays to enable low airframe noise design. In: 41st Aerospace sciences meeting and exhibit, AIAA paper 2003-707
18. Stoker R, Guo Y, Streett C, et al (2003) Airframe noise source locations of a 777 aircraft in flight and comparisons with past model–scale tests. In: 9th AIAA/CEAS aeroacoustics conference, AIAA paper 2003-3232
19. Brusniak L, Underbrink J, Stoker R (2006) Acoustic imaging of aircraft noise sources using large aperture phased arrays. In: 12th AIAA/CEAS aeroacoustics conference, AIAA paper 2006-2715

20. Herkes W, Olsen R, Uellenberg S (2006) The quiet technology demonstrator program: flight validation of airplane noise-reduction concepts. In: 12th AIAA/CEAS aeroacoustics conference, AIAA paper 2006-2720
21. Elkoby R, Brusniak L, Stoker R, et al (2007) Airframe noise test results from the QTD II flight test program. In: 13th AIAA/CEAS aeroacoustics conference, AIAA paper 2007-3457
22. Khorrami MR, Lockard D, Humphreys W, et al (2008) Preliminary analysis of acoustic measurements from the NASA-gulfstream airframe noise flight test. In: 14th AIAA/CEAS aeroacoustics conference, AIAA paper 2008-2814
23. Wong J, Nesbitt E, Jones M, et al (2019) Flight test methodology for NASA advanced inlet liner on 737MAX-7 test bed (quiet technology demonstrator 3). In: 25th AIAA/CEAS aeroacoustics conference, AIAA paper 2019-2763
24. Nark D, Jones M (2019) Design of an advanced inlet liner for the quiet technology demonstrator 3. In: 25th AIAA/CEAS aeroacoustics conference, AIAA paper 2019-2764
25. Yamamoto K, Hayama K, Kumada T, et al (2016) FQUROH: a flight demonstration project for airframe noise reduction technology—concept and current status. In: 22nd AIAA/CEAS aeroacoustics conference, AIAA paper 2016-2709
26. Takaishi T, Ura H, Nagai K, et al (2016) Flyover array measurements with JAXA jet flying test bed 'Hisho'. IN: 22nd AIAA/CEAS aeroacoustics conference, AIAA paper 2016-2710
27. Takaishi T, Ura H, Nagai K et al (2017) Airframe noise measurements on JAXA jet flying test bed "Hisho" using a phased microphone array. Int J Aero 16(4–5):255–273
28. Zhang YZ, Lee I, Lin DK (2017) Measurement of noise from a moving drone using a phased array microphone system. In: 2017 Asia-Pacific international symposium on aerospace technology. Seoul, Korea

Natural Laminar Flow Design for Highly Swept Wings

Naoko Tokugawa, Kosuke Toyoda, Fumitake Kuroda, Yoshine Ueda, and Takahiro Ishida

Abstract Natural laminar flow design has been applied to the next generation of subsonic aircraft, which are expected entry into service in the 2020s, with the aim of reducing greenhouse gas emissions, which is one of the Sustainable Development Goals. The natural laminar flow wing design was applied to highly swept wings, such as the main wing and the vertical tail plane of TRA2022, which is defined as a 120-pax technical reference aircraft in subsonic aircraft research of the Japan Aerospace Exploration Agency. An automatic inverse design system with inverse problem design tools and flow solvers outside of the system was developed. A natural laminar flow design was carried out more efficiently with the help of various innovations. A natural laminar flow design was successfully achieved for not only the main wing but also the vertical tail plane. In both cases, the laminar flow region was extended by more than 20% in the area. This result shows that the natural laminar flow wing design technology of the Japan Aerospace Exploration Agency is effective for highly swept wings. The objective of the natural laminar flow wing design of the Japan Aerospace Exploration Agency has been achieved.

Keywords Inverse design · Drag reduction · Boundary layer transition · Transition prediction

N. Tokugawa (✉) · K. Toyoda · T. Ishida
Japan Aerospace Exploration Agency, 6-13-1 Osawa, Mitaka-shi, Tokyo 181-0015, Japan
e-mail: tokugawa.naoko@jaxa.jp

F. Kuroda
Ryoyu Systems Co., LTD., 6-19 Oe, Minato-ku, Nagoya-shi, Aichi 455-0024, Japan

Y. Ueda
Advantec Co., Ltd., 2-7-1 Nishi-Shinjyuku, Shinjyu-ku, Tokyo 163-0703, Japan

S. Lee et al. (eds.), *The Proceedings of the 2021 Asia-Pacific International Symposium on Aerospace Technology (APISAT 2021), Volume 1*, Lecture Notes in Electrical Engineering 912, https://doi.org/10.1007/978-981-19-2689-1_33

1 Introduction

The number of natural disasters caused by global warming and many climatic changes, such as flooding, wildfires, and heat waves have been increasing in the last half a century. The United Nations Intergovernmental Panel on Climate Change (IPCC) has reported that there is no doubt that the Earth is warming due to human activities, and that the climate crisis will continue, causing further heat waves, heavy rains, and droughts [19]. "Taking concrete actions against this climate change" is one of the Sustainable Development Goals (SDGs) adopted at the UN Summit in 2015. Reducing greenhouse gas emissions is a global challenge, and the aviation industry remains a key area for improvement. CO_2 emitted by aircraft accounts for as much as 2% of the total global emissions [20].

There is a wide range of methods of reducing the CO_2 emissions of aircraft, such as the use of sustainable aviation fuel instead of fossil fuels and the development of electric or hydrogen aircraft. However, even with these technologies, the reduction of aerodynamic drag is a fundamental and essential component. Therefore, various attempts to reduce the aerodynamic drag of aircraft are being made around the world [3, 4]. Friction drag, which develops on the surface of an aircraft, accounts for the largest percentage of the aerodynamic drag of subsonic aircraft [21]. Therefore, if the friction drag can be reduced, the total aerodynamic drag can also be significantly reduced. Laminarization can be used to maintain a laminar boundary layer, and therefore reduce friction drag, which is more effective than turbulent boundary layer control. For example, in the case of an A319 class aircraft, if laminar boundary layer is maintained up to 50% of the upper surface area on the main wing, the friction drag is reduced by approximately 8.9 cts. This corresponds to 3% of the total drag. And this drag reduction is approximately 2.8 times greater than when riblets are applied to the turbulent boundary layer over the entire upper surface of the wing. Natural laminar flow (NLF) design is the most effective method of passively laminarizing boundary layer flow. NLF design is a technique that suppresses the laminar-to-turbulent transition of the boundary layer by changing the cross-sectional shape (typically, airfoil). Although NLF airfoils design has been studied extensively a well-known "traditional" concept. However, the degree of difficulty increases significantly when it is applied to a swept airfoil. It has prevented its application to commercial airplanes. The main source of difficulty stems from the instability that dominates the boundary layer transition. As the sweep angle increases, the dominant instability shifts from the Tollmien-Schlichting instability to the crossflow instability, which enhances the turbulent transition. Therefore, the relationship between the sweep angle and the Reynolds number generally determines whether to adopt an NLF airfoil or to control the laminar flow using boundary layer suction [12]. In Europe, an NLF wing was applied to the outboard of the A340 main wing however the sweep angle was reduced, as compared to the original. Research into the application of NLF design to highly swept wings is being performed in the United States [13].

The Japan Aerospace Exploration Agency (JAXA) has been focusing on the design of NLF to the highly swept wings. In particular, the NLF wing design was applied to the wing of a supersonic experimental airplane, the NEXST-1, with a sweep angle of over 60°, which is the first known attempt at a supersonic wing with a subsonic leading edge. The effects of the design were successfully demonstrated in a test flight conducted in Australia in October 2005 [17, 22, 26]. NLF was achieved through an appropriate combination of the suppression of the crossflow instability utilizing a steep rise in the pressure coefficient near the leading edge and the suppression of the Tollmien-Schlichting instability using a gradual acceleration near the mid-chord area [24]. A similar concept has been successfully applied in the numerical design of several large supersonic aircraft [7, 8, 25]. In recent years, the NLF design has been conceptually extended to the next generation of subsonic aircraft, which are expected entry into service in the 2030s. In this study, a natural laminar flow wing design was applied to a highly swept wing to achieve approximately 4 cts. reduction in the drag. The design was also extended to the vertical tail plane to achieve further reduction in overall drag.

2 Natural Laminar Flow Design

2.1 Design Objects

The main wing and vertical tail plane of a single-aisle subsonic commercial aircraft were the targets for applying NLF. In JAXA, the technical reference aircraft, TRA2022, is defined as the final target of subsonic aircraft research (Fig. 1) [1, 10, 11, 15, 16]. The TRA2022 is a 120-seat class of future passenger aircraft designed by JAXA based on the TRA2012A, with the expectation of entry into service in the 2030s. The TRA2012A was also designed for comparison with the TRA2022. It simulates aircraft operating in 2012 and is equivalent to the A319. To achieve superior environmental performance, material and aerodynamic technologies to reduce fuel consumption have been developed and applied in the design of the TRA2022. A coupled aerodynamic-structural design loop was developed to apply lightweight composite structural design and drag reduction technologies. With lightweight composite structure technology, sizing of the airframe was carried out using an original tool to reduce the weight of the structure. Drag reduction technologies include the formulation of a high-aspect-ratio wing, optimization of load distribution, investigation of winglets, riblets, and NLF design. The specifications of the TRA2012A and TRA2022 are shown in Fig. 1.

(a)

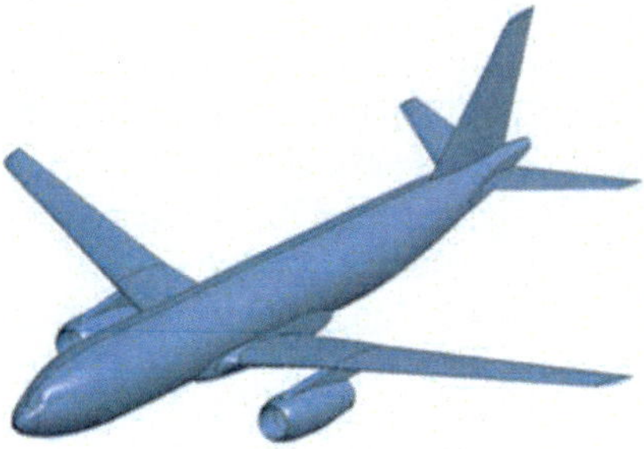

Cruise Mach Number	0.78
Range [nm]	3600
Passengers	126
Overall Length [m]	37.6
Wing Span [m]	34.1
Aspect ratio (main wing)	9.5
Maximum Take-off Weight [kgf]	71182

(b)

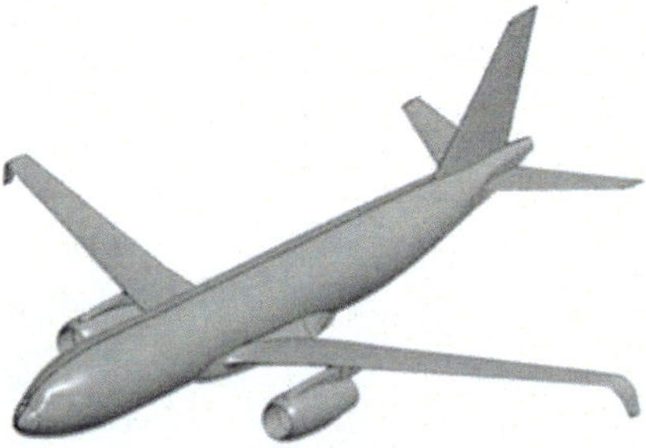

Cruise Mach Number	0.78
Range [nm]	3600
Passengers	126
Overall Length [m]	37.6
Wing Span [m]	34.8
Aspect ratio (main wing)	11
Maximum Take-off Weight [kgf]	60223

Fig. 1 The technical reference aircraft of JAXA's subsonic aircraft research, **a** the TRA2012A and **b** the final shape of the TRA2022 (the fourth shape of TRA2022)

First, NLF design was applied to the main wing of the TRA2012A.Because the coupled aerodynamic-structural design loop of the TRA2022 was carried out in parallel with the natural laminar wing design, the geometry of the TRA2022 was not defined at the beginning of the natural laminar wing design. The baseline shape is called the "TRA2012A-MW," while the wing that is applied NLF is called "TRA2012A-MW-NLF." Similarly, the "TRA2022-2-MW-NLF" wing shape was obtained by applying the NLF design to the main wing of the second shape of the TRA2022, called the "TRA2022-2-MW" (Fig. 2a). The TRA2022-2 wing shape was designed based on the TRA2012A wing, not only by applying a high aspect ratio planform and optimizing the load distribution, but also by taking advantage of the TRA2012A-MW-NLF. However, the TRA2022-2-MW wing shape improves upon the TRA2012A-MW-NLF wing by reducing the shock wave at the inboard and pylon of the lower surface. Both NLF designs were applied to the wing-body configuration to improve the design efficiency. The sweep angle of the leading edge is 26.95°, where the mean aerodynamic chord (MAC) and span length vary depending on the design generation; the TRA2012A-MW wing has an MAC of 3.97 m and semi-span length of 17.05 m, while the TRA2022-2-MW wing has an MAC of 3.68 m and semi-span of 18.32 m. Because the planform is not changed by the natural laminar flow design, these dimensions are maintained in the NLF versions.

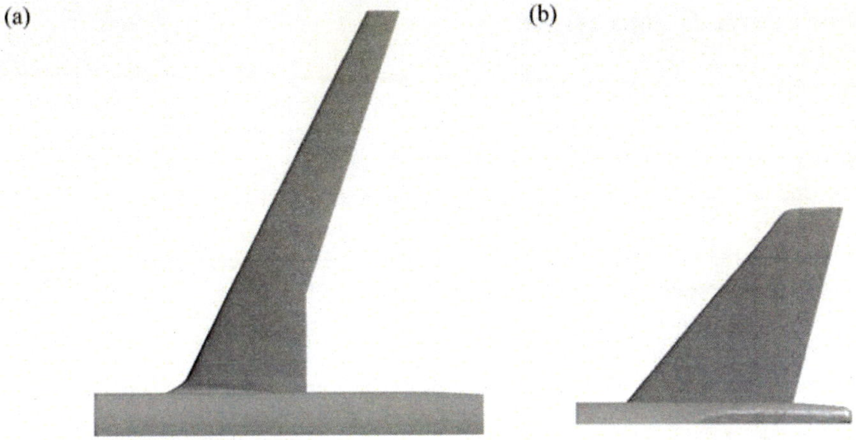

Fig. 2 Design objects. **a** Top view of the main wing "TRA2022-2-MW", and **b** side view of the vertical tail plane "TRA2022-VTP"

To further reduce the friction drag, NLF design was also applied to the vertical tail plane (Fig. 2b). Since the vertical tail plane was applied not only to natural laminar flow design but also to other aerodynamical designs. Therefore, only the side shape mimicked A319 and the airfoil shape was NACA64A-010 modified with a maximum wing thickness ratio of 9.13%. This initial shape is referred to as the "TRA2022-VTP," and the NLF vertical tail plane is named the "TRA2022-VTP-NLF." The sweep angle of the leading edge is 40.18°, MAC was 3.89 m, and the span length is 6.08 m, which was defined as the distance from the wing-body junction to the wing tip. NLF design was applied to the initial shape. An "unusual wing-body configuration," where only the vertical tail plane and fuselage were used to improve the design efficiency of the vertical tail plane.

2.2 Design Concept

The crossflow instability promotes turbulent transition in the three-dimensional boundary layer compared to the Tollmien-Schlichting instability observed in the two-dimensional boundary layer. Therefore, to achieve NLF, the crossflow instability and the Tollmien-Schlichting instability should be suppressed near the leading edge and after the mid-chord, respectively, for both subsonic and supersonic aircraft [24]. JAXA has several types of original concepts.

This concept is applied to the target pressure distribution at the time of design.

2.3 Inverse Design System

The most important feature of JAXA's NLF design technology is its utilization of inverse problem design. Inverse problem design is different from the usual forward problem design, in which the shape is designed by hand, the performance is evaluated, and the shape is modified according to the evaluation. The performance in this experiment (in this case, defined by the pressure distribution) is defined in advance, and the shape is automatically modified to achieve performance. Figure 3 shows the flowchart of the automatic design system.

An inverse problem design system was developed by JAXA and named the "Automatic inverse design System Utilizing Shape-Smoothing (ASUS2)". It is an extremely versatile system because the ASUS2 has interfaces with inverse problem design tools and computational fluid dynamics (CFD) solvers, which are outside of the ASUS2 system. Additionally, it has various modules that not only smooth the designed airfoil shape but also generate surface and spatial grids, thus eliminating the need for manual modification based on experience. Furthermore, the system is able to automatically launch external tools to perform multiple generations of design in succession, making it an automatic inverse problem design system. Three types of inverse problem design tools (SuperWINV [14], WINDES [6], and CDISC [2]), were selected as default settings, depending on the application. In this study, CDISC was used.

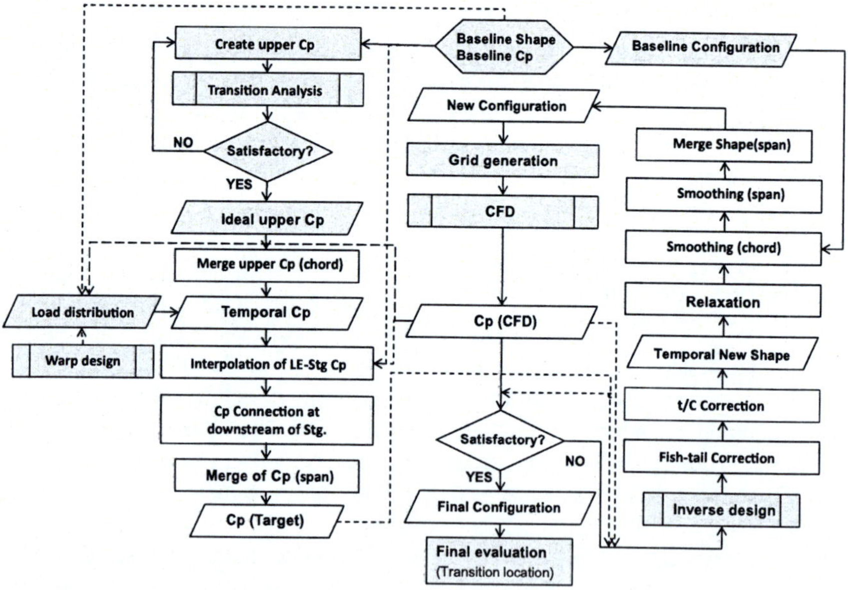

Fig. 3 Simplified flow chart of JAXA's inverse design system "Automatic inverse design System Utilizing Shape -Smoothing (ASUS2)". Gray hatching parts are out of the ASUS2

Airfoils at each cross-section designed for the inverse problem design tool are connected in a spanwise direction and then smoothed in both the chordwise and spanwise directions to obtain a three-dimensional airfoil shape. The grid was automatically generated according to the designed three-dimensional shape in order to automatically perform CFD, as described below. The grid on the surface of the wing was arranged along the 1% chord line in the spanwise direction, and in the chordwise direction. The grid was arranged such that it was denser at the leading edge and the trailing edge and coarser at the mid-chord region since the shape deformation at the leading edge and the trailing edge is delicate. As a result, the surface grid appears similar to a structural grid, even when the spatial grid is unstructured. In the direction perpendicular to the wall, over 20 layers were extruded beyond the boundary layer. In this extruded layer, the grid is structural. The total grid was comprised of approximately 5.6 million points.

For the flow solver, which is another important tool, used from ASUS2, either UPACS [18] corresponding to the structured grid or FaSTAR [5] corresponding to the unstructured grid can be selected. Both solvers were developed by JAXA and solve the three-dimensional full Navier–Stokes equations. In this study, FaSTAR was selected and used on the JAXA Supercomputer System Generation 2 and 3 (JSS2 and JSS3). The number of iterations was set to 30,000.

By using this inverse problem design system, it takes approximately 2 h to design one generation, including design, grid generation, and CFD, which is a reduction to 1/80th of the time required by the NEXST project. Additionally, considering the human resources required for inverse problem design and grid generation, the efficiency of the proposed system is much higher.

2.4 New Innovation

Several innovations have been made in the design of NLF wings through this development and investigation.

The first of these innovations was created to deal with the shock waves which form on the upper surface of the main wing. Unlike supersonic aircraft, supercritical wings, such as the wings of the TRA2022, produce weak shock waves on the upper surface. Because it is difficult to maintain NLF downstream of the shock wave, the target pressure distribution for NLF was defined as the pressure distribution leading up to the shock wave, after which it is deemed uneconomical to manage the pressure distribution of the wing for NLF design. However, because the position and strength of the shock wave of the designed shape varied from design to design, the difference from the target pressure distribution might become large, locally. However, it was challenging to make the target pressure distribution follow the change in the shock wave due to the design shape. Therefore, it was decided not to modify the airfoil after the shock wave.

The second innovation was related to the constraints on the load distribution and wing thickness. The target pressure distribution is optimized to delay the transition from laminar to turbulent flow, but it does not guarantee the other constraints, namely, the load distribution and wing thickness. Therefore, the load distribution was constrained by adjusting the twist angle distribution of the main wing. The wing thickness was artificially modified during the design process to achieve the desired initial thickness. Further innovations were brought to optimize the pressure distribution on the vertical tail plane. Because it is a symmetrical wing, and no lift is generated, load distribution is not a significant concern. However, this symmetry makes it difficult to constrain the thickness of the vertical tail plane. In the case of an asymmetrical main wing, the balance between the natural laminar flow wing design and the constraints imposed by the lift distribution or wing thickness on the lower surface poses a significant challenge. However, in the case of a symmetrical wing, both the natural laminar flow effect and the thickness constraint are imposed on the same surface. Therefore, instead of insisting on a single target pressure coefficient, Cp, based on NLF, the design was carried out while frequently modifying the target Cp. The initial target Cp was considered to be the beginning of the design process. Then, the NLF effect and wing thickness constraint become complementary.

Additionally, an artificial forward problem design process was introduced prior to the inverse problem design process. By starting the design from the initial airfoil with a pressure distribution close to the target Cp, which was obtained through the forward problem design, the design proceeded efficiently and prevented the convergence to an unacceptable shape other than the final exemplary solution.

3 Results

3.1 Transition Prediction

The results from the application of NLF design to the main wing are presented in Fig. 4. This figure shows the transition position predicted by the compressible three-dimensional boundary layer code [24], based on Iyer's method [9] which was developed by JAXA, and the transition prediction tool LSTAB [23] based on linear stability theory. The location corresponds to the contour of the transition index, the so-called N-factor at $N = 12$. The transition location is clearly delayed by the NLF design, as can be seen in Fig. 4. The laminar region extends 24% further for the TRA2012A-MW-NLF, which is eight times further than that of the TRA2012A-MW. The laminar region was extended by 20% for the TRA2022-2-MW-NLF, even though the difference from the performance of the TRA2022-2-MW was only approximately 1%. The degree of delay is larger for the TRA2012A-MW-NLF shape as compared to the TRA2022-2-NLF shape. This difference is due to the shape of the TRA2022-2-MW, which is the basis for the TRA2022-2-MW-NLF and is based on the TRA2012A-MW-NLF shape, to which was designed to utilize NLF. The delay of the location of

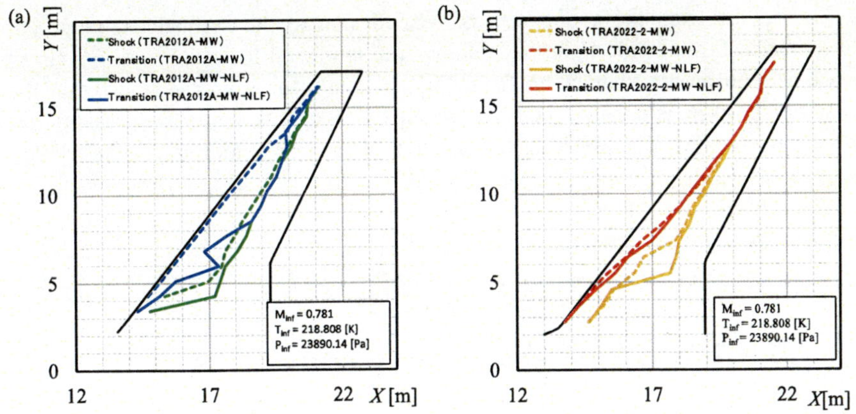

Fig. 4 The numerically predicted transition location of main wing. **a** comparison between the TRA2012A-MW-NLF wing shape and the TRA2012A-MW wing shape, and **b** comparison between the TRA2022-2-MW-NLF wing shape and the TRA2022-MW-2 wing shape. The abscissa is the fuselage station from the aircraft nose. The ordinate is the spanwise location from the aircraft axis

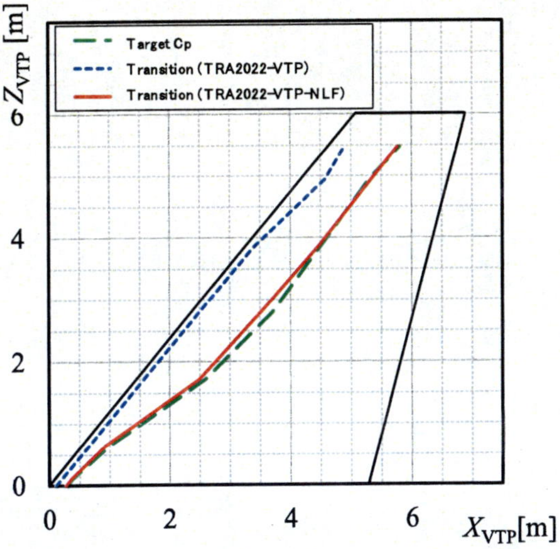

Fig. 5 The numerically predicted transition location of the vertical tail plane. The abscissa is airframe-direction location from the leading edge of vertical tail plane—fuselage junction. The ordinate is the vertical-direction location from the same point

the transition reduced the friction drag by approximately 3.9 cts. for the TRA2012A-MW-NLF shape and by approximately 0.2 cts. for the TRA2022-2-MW-NLF shape. Furthermore, the pressure drag was significantly reduced through the application of the proposed NLF design.

The results from the application of NLF design to the vertical tail plane are presented in Fig. 5. This figure also shows the predicted transition position for N = 12 as well as the main wing from Fig. 4. It is clear from the figure that the transition position is delayed owing to the NLF design. The laminar region extends to 27% of the area. This delay in the location of the transition reduces the friction drag by 1.8 cts.

These results show that JAXA's natural laminar airfoil design technology is also effective for highly swept airfoils.

3.2 *Wind Tunnel Verification*

The effect of the application of NLF wing design on the main wing was verified using a half model at the JAXA 2 × 2 m transonic wind tunnel (TWT1) (Fig. 6).

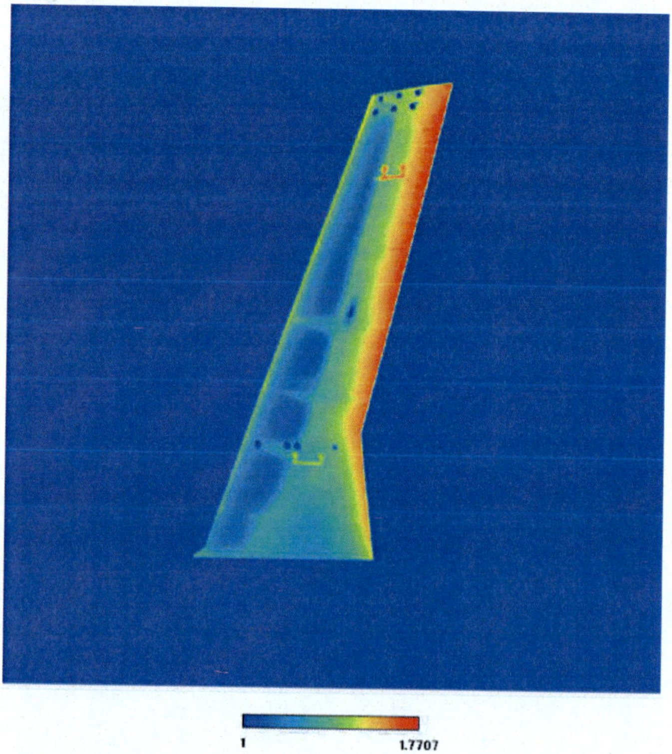

Fig. 6 Surface temperature distribution measured by temperature sensitive paint (TSP) on the main wing of TRA2022-2-MW-NLF at the JAXA TWT1. Here, M = 0.78, $Re_c = 9.5 \times 10^6$, and AoA = 2°. Blue "cool" areas corresponds to laminar flow, and green and red "hot" areas correspond to turbulent flow. A few turbulent wedges, due to pressure taps or dust, were observed

The transition position was detected using temperature sensitive paint (TSP), where it was confirmed that the flow ahead of the shockwave was laminar. However, the Reynolds number based on MAC was $Re_c = 9.5 \times 106$ at $M = 0.78$ in the testing environment of the TWT1. This Reynolds number corresponds to only 40% of the flight Reynolds number of TRA2022. Therefore, this result is reasonable but is not sufficient. To properly verify the design, it is necessary to conduct the verification tests under the flight Reynolds number.

4 Conclusion

In this study, NLF design was applied to the next generation of subsonic aircraft, which are expected to be entry service in the 2030s. The NLF wing design was applied to a highly swept main wing and a vertical tail plane.

A natural laminar flow design was successfully achieved for both the main wing and vertical tail plane. In both cases, the laminar flow region was extended by over 20% in the total area. This result shows that JAXA's wing design technology is effective for highly swept wings.

Acknowledgements The authors would like to thank Dr. K. Yoshida, Dr. D. Kwak, Dr. T. Yuhara, Mr. H. Ishikawa, and Mr. R. Shimada from the Japan Aerospace Exploration Agency for technical discussions. We would also like to thank the technical trainees and interns who have helped us, Dr. M. Sasamori, Mr. M. Miyazaki, Mr. T. Kimura, Mr. H. Adachi, Mr. Y. Yamasaki, Mr. K. Ishizuki, Mr. N. Yuri, Mr. M. Hirata, and Mr. T. Mizumoto. The wind test was conducted with the cooperation of the JAXA Aerodynamics Research Unit. The numerical work was conducted using the computational resources of JAXA Supercomputer System Generation 2 and 3. Part of this study was conducted in collaboration with Mitsubishi Heavy Industries, Ltd.

References

1. Arizono H, Kwak D-Y, Nomura T (2017) Aerodynamics and structural design studies for fuel consumption reduction on subsonic aircraft. IFASD-2017-008
2. Campbell RL (1992) An approach to constrained aerodynamic design with application to airfoils. NASA TP-3260
3. Clean Sky Joint Undertaking, CleanSky2. https://www.cleansky.eu/large-passenger-aircraft
4. Collier FS, Thomas R, Nickol CA, Lee C-M, Tong M (2010) Environmentally responsible aviation—real solutions for environmental challenges facing aviation. Proc Int Cong Aero Sci 1(6):1. (Invited)
5. Hashimoto A, Murakami K, Aoyama T, Ishiko K, Hishida M, Sakashita M, Lahur P (2012) Toward the fastest unstructured CFD code 'FaSTAR'. AIAA-paper-2012-1075
6. Hirose N, Takanashi S, Kawai N (1985) Transonic airfoil design based on Navier–Stokes equation to attain arbitrarily specified pressure distribution—an iterative procedure. AIAA-paper-85-1592
7. Ishikawa H, Tokugawa N, Ueda Y, Ito K, Yoshida K (2016) Development of inverse design system for supersonic natural laminar flow wing on high reynolds number condition. J Jpn Soc Aero Space Sci 64:113–122 (in Japanese)

8. Ishikawa H, Ueda Y, Tokugawa N (2017) Natural laminar flow wing design for a low boom supersonic aircraft. AIAA-2017-1860
9. Iyer V (1993) Three-dimensional boundary layer program (BL3D) for swept subsonic or supersonic wings with application to laminar flow control. NASA CR-4531
10. Kwak D, Arizono H, Nomura T (2016a) Studies on the lift distribution and aspect ratio for the fuel consumption reduction on subsonic aircraft. In: 2016 Asia-Pacific international symposium on aerospace technology
11. Kwak D, Nomura T, Tokugawa N, Kurita M, Murayama M (2016b) Introduction of research project for environmental conscious aircraft technology in JAXA. In: Proceedings of greener aviation 2016, paper no.180
12. Lehner S, Crossley W (2008) Combinatorial optimization to include greener technologies in a short-to-medium range commercial aircraft. Proc Int Cong Aero Sci 2008–4(10):2
13. Lynde MN, Campbell RL, Rivers MB, Viken SA, Chan DT, Watkins AN, Goodliff SL (2019) Preliminary results from an experimental assessment of a natural laminar flow design method. AIAA-2019-2298
14. Matsushima K, Iwamiya T, Nakahashi K (2004) Wing design for supersonic transports using integral equation method. Eng Ana Bound Ele 28:247–255
15. Nomura T (2013) Conceptual design of future passenger aircraft aimed at reducing fuel consumption. JAXA-RR-13-007. (in Japanese)
16. Nomura T (2019) Aerodynamic design of a 120-seat class passenger aircraft at cruise. JAXA-CR-18-001. (in Japanese)
17. Ohnuki T, Hirako K, Sakata K (2006) National experimental supersonic transport project. Proc Int Cong Aero Sci 2006–1(4):1
18. Takaki R, Yamamoto K, Yamane T, Enomoto S, Mukai J (2003) The development of the UPACS CFD environment, high performance computing. Springer, Proc. ISHPC, pp 307–319
19. The Intergovernmental Panel on Climate Change (2021) AR6 climate change 2021: the physical science basis. https://www.ipcc.ch/report/ar6/wg1/#FullReport
20. The International Air Transport Association (2021) Aviation & climate change fact sheet. https://www.iata.org/en/iata-repository/pressroom/fact-sheets/fact-sheet-climate-change
21. Thomas ASW (1985) Aircraft drag reduction technology—a summary. AGARD Rep 723:1–1–1–20
22. Tokugawa N, Kwak D, Yoshida K, Ueda Y (2008) Transition measurement of natural laminar flow wing on supersonic experimental airplane (NEXST-1). J Aircraft 45:1495–1504
23. Ueda Y, Ishikawa H, Yoshida K (2004) Three dimensional boundary layer transition analysis in supersonic flow using a Navier–Stokes code. Proc Int Cong Aero Sci 2004–2(8):2
24. Ueda Y, Yoshida K, Matsushima K, Ishikawa H (2014) Supersonic natural-laminar-flow wing-design concept at high-reynolds-number conditions. AIAA J 52:1294–1306
25. Ushiyama K, Ishikawa H, Tokugawa N, Koike K (2016) Natural laminar flow wing design for a small supersonic transport QSST. JAXA-RR-16-001. (in Japanese)
26. Yoshida K, Makino Y (2004) Aerodynamic design of unmanned and scaled supersonic experimental airplane in Japan. Proc ECCOMAS 2004

Research on Aerodynamic Characteristics of General Long Endurance Vertical Take-Off and Landing Fixed Wing UAV

Zhan-ke Li, Liang-yang Zhang, Si-jia Zhang, Hai-bo Wei, and Hai-yang Han

Abstract This paper proposes a new type of pure electric vertical take-off and landing fixed wing general long endurance UAV, and carries out a series of design, calculation and analysis on it. Firstly, use CATIA to perform three-dimensional modeling of the UAV, determine the aerodynamic shape of the UAV, and analyze the calculation method. Secondly, use CFD to divide the mesh, set the boundary conditions, and verify the independence of the mesh. Finally, Fluent is used to calculate the aerodynamic characteristics, the lift-drag characteristic curve, the pitching torque curve, the corresponding pressure cloud graph and the aerodynamic characteristics of the whole UAV under different lateral wind speeds at different attack angles are obtained. The calculation results show that the aerodynamic characteristics of the UAV are good and meet the design index requirements.

Keywords Vertical take-off and landing fixed wing UAV · Aerodynamic analysis · CFD · Fluent

1 Introduction

At present, due to the high cost of wind tunnel tests and the relatively limited scope of application, it is not used under many special calculation requirements, but only used for verification [1]. CFD analysis has become an indispensable part of the calculation and analysis of aircraft aerodynamic performance. Compared with wind tunnel tests, the time and economic benefits of CFD technology, and CFD is becoming more mature and widely used. In this paper, ICEM is used to divide the UAV mesh, and ANSYS Fluent is used to calculate and analyze the aerodynamic characteristics of the model. Considering that the main mission stage of the composite layout UAV is cruising under the fixed wing, this article only calculates and analyzes its aerodynamic characteristics in the level flight state, and does not perform the calculation

Funded by the Key R & D Projects of Shaanxi Province.

Z. Li (✉) · L. Zhang · S. Zhang · H. Wei · H. Han
School of Aeronautics, Northwestern Polytechnical University, Xian, China
e-mail: lzk@nwpu.edu.cn

S. Lee et al. (eds.), *The Proceedings of the 2021 Asia-Pacific International Symposium on Aerospace Technology (APISAT 2021), Volume 1*, Lecture Notes in Electrical Engineering 912, https://doi.org/10.1007/978-981-19-2689-1_34

and analysis of the aerodynamic characteristics of the multi-rotor state and the study of the vertical take-off and landing system.

Part of the basic parameters of the model are given by the overall parameter design of the aircraft to facilitate subsequent research and analysis. The cruising speed is 22 m/s, the cruising height is 1000 m, the take-off weight is 15 kg, the LA203A airfoil, the fuselage length is 1.26 m, the wing area is 0.71 square meters, the wing span is 2.8 m, and the static stability margin is 0.16.

2 Calculation Model, Calculation Method and Setting of Boundary Conditions

2.1 Calculation Model

This article uses CATIA to construct a three-dimensional model, initially determines the aerodynamic shape, and then builds each component model separately, and assembles the model in the assembly according to the preliminary design plan. In order to ensure the accuracy of the model aerodynamic analysis and appropriately reduce the amount of calculation, the model is partially simplified, the parts that have less impact on the aerodynamic performance in the model are simplified, and the internal structure is deleted. The main mission stage of the composite layout UAV is to cruise under the fixed wing, so this article only calculates and analyzes its aerodynamic characteristics in level flight. In order to ensure the accuracy of the model, Boolean operations are performed on the model and then imported into the meshing software. Figure 1 shows the aerodynamic model of the UAV.

The aerodynamic shape of this calculation model is basically the same as that of the solid model. The fuselage length is 1.26 m, the wing span is 2.8 m, the wing chord length is 0.26 m, and the wing area is 0.71 m^2. The three views of the whole UAV are shown in Fig. 2.

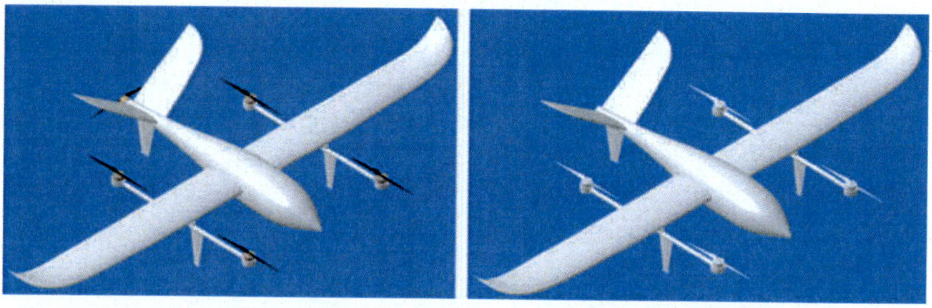

(a) Unsimplified model (b) Simplified model

Fig. 1 Aerodynamic model of UAV

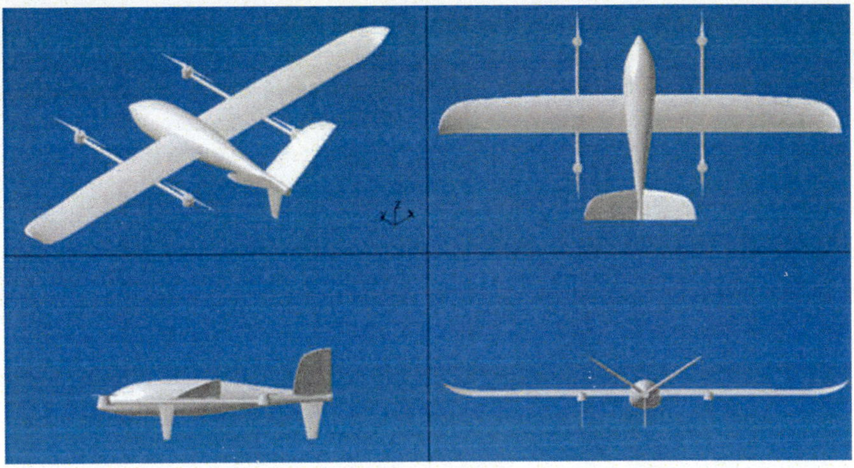

Fig. 2 Three views of UAV model

2.2 Calculation Method

In order to meet the design requirements and meet the actual flight conditions, since the actual flight speed of the vertical take-off and landing fixed wing UAV is relatively low, the incompressible fluid flow continuity equation is selected, which can be expressed as:

$$\frac{\partial v_x}{\partial x} + \frac{\partial v_y}{\partial y} + \frac{\partial v_z}{\partial z} = 0 \tag{1}$$

The equation of motion of the incompressible fluid is:

$$\rho \frac{Dv}{Dt} = \rho g + \nabla \cdot \sigma \tag{2}$$

The energy equation of the incompressible fluid is:

$$\rho c_v \frac{DT}{Dt} = -\nabla \cdot q + \nabla \cdot (\sigma \cdot v) \tag{3}$$

Taking into account the problems of convergence and calculation accuracy, this paper selects the $SST\ k - \omega$ turbulence model, and its control equation is:

$$\frac{\partial(\rho k)}{\partial t} + \frac{\partial(\rho k \mu_i)}{\partial x_i} = \frac{\partial}{\partial x_j}\left[\Gamma_k \frac{\partial k}{\partial x_j}\right] + \overline{G_k} - Y_k + S_k \tag{4}$$

$$\frac{\partial(\rho \omega)}{\partial t} + \frac{\partial(\rho \omega \mu_i)}{\partial x_i} = \frac{\partial}{\partial x_j}\left[\Gamma_\omega \frac{\partial \omega}{\partial x_j}\right] + G_\omega - Y_\omega + D_\omega + S_\omega \tag{5}$$

SST $k-\omega$ turbulence model can accurately simulate and calculate the problems of reverse pressure gradient flow and split flow [2], and is more suitable for calculating the flow around bluff bodies, so it has a wide range of applications. At the same time, the turbulence model is mostly used in the calculation of the external flow field of the aircraft, and its simulation accuracy and realism in the aerodynamic performance of the aircraft are higher than most turbulence models.

2.3 Boundary Conditions and Mesh Settings

In order to reduce the influence of the outer flow field wall on the calculation of the aerodynamic characteristics of the UAV, the length of the outer flow field in this paper is set to 20 times the wingspan before and after the fluid flow direction, and the other directions are set to 10 times the wingspan (The characteristic length of this model is wingspan length). The computational domain meshing method uses unstructured tetrahedral mesh [3]. The number of volume meshes is 3 million, and the number of surface meshes is 520,000. Due to calculation accuracy and turbulence model requirements, the boundary layer mesh y+ value is set to 1, and the number of boundary layer layers is 15 (Fig. 3).

Because the model is more complicated, there are many local details, and the local still has a great influence on the aerodynamic characteristics after being simplified. In order to improve the accuracy of calculation, the local mesh of the model is refined (The trailing edge of the wing and the propeller part). The deformation of the mesh after the encryption process is small. Figure 4 shows the overall mesh of the UAV, and Figs. 5 and 6 show the local mesh of the model.

The calculation of the environmental conditions of the external flow field follows the real atmospheric conditions [4]. According to the design requirements and the previous calculation data, the flight altitude is selected as 1000 m, the atmospheric temperature is 281.65 K, the atmospheric pressure is 89875.8pa, the atmospheric density is 1.11168 kg/m^3, the atmospheric viscosity is $1.7578 \times 10^{-5} \times \text{N} \times \text{s/m}^2$,

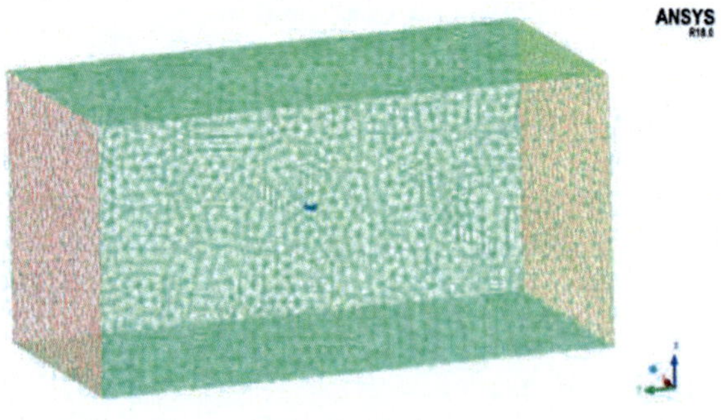

Fig. 3 Model computational domain mesh diagram

Fig. 4 Whole UAV model surface mesh

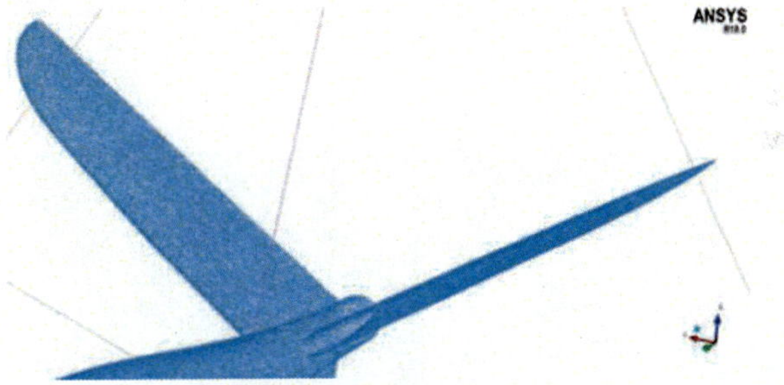

Fig. 5 Tail surface mesh after local refinement

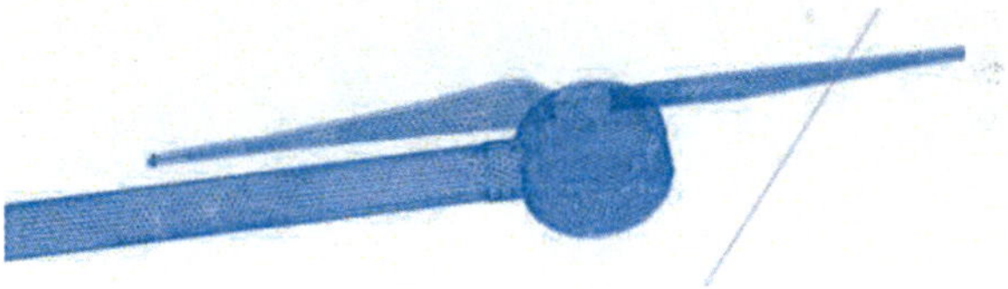

Fig. 6 Propeller surface mesh after local refinement

and the Ma is 336.43 m/s. The inlet of the external flow field is set as the velocity inlet boundary condition, the outlet boundary condition is set as the pressure outlet boundary condition, and the outlet pressure is set as the ambient pressure.

3 Feasibility Verification and Mesh Independence Verification

3.1 Feasibility Verification

In order to verify that the analysis method in this paper is effective, compare the airfoil database data with the airfoil calculation data, and set the environmental data and flight data according to the same settings in this paper, so as to carry out the two-dimensional pressure parameter analysis of the LA203A airfoil. Use the two-dimensional unstructured tetrahedral meshing method generated by ICEM to mesh the UAV. The length of the external flow field is set to 100 times the airfoil chord length before and after the flow direction, and 50 times the airfoil chord length for the upper and lower sides. The following Figs. 7 and 8 show the pressure cloud diagram and velocity streamline diagram near the airfoil.

Fig. 7 Relative pressure cloud map near the airfoil

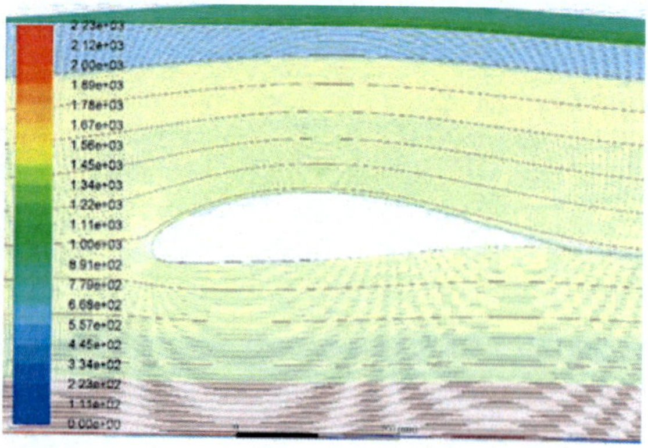

Fig. 8 Velocity streamlines near the airfoil

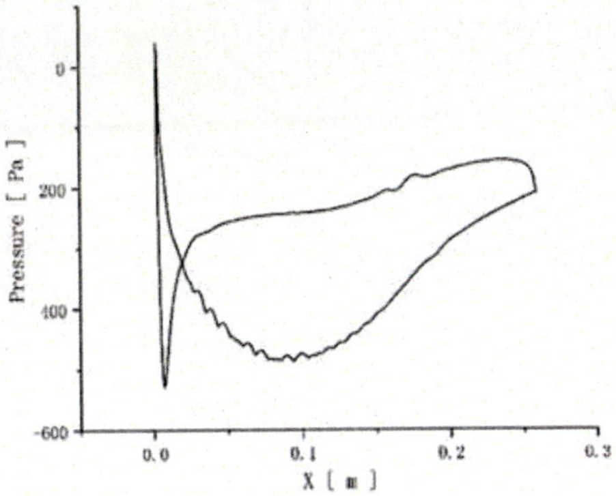

Fig. 9 Relative pressure distribution curve of airfoil

The airfoil of the UAV is selected as LA203A airfoil. Because the airfoil is a low-speed airfoil, the leading edge pressure of the airfoil is relatively high, and the velocity cloud diagram also reflects the pressure difference between the upper and lower surfaces of the airfoil, which is the reason for the lift of the airfoil. Figure 9 shows the calculated airfoil pressure distribution curve, which is basically consistent with the pressure distribution of LA203A in the airfoil database, which shows that the algorithm is reliable.

3.2 *Mesh Independence Verification*

In order to ensure the calculation accuracy and appropriately reduce the calculation amount and calculation cycle, the mesh independence verification is carried out. After it is proved that the number of meshes is encrypted to the number of calculations, the accuracy of the calculation is not much related to the number of meshes [5], and there is no need to continue to refine the meshes. The results of the mesh convergence verification are shown in Fig. 10.

It can be seen from the figure above that there are still big changes in the calculation results of the lift coefficient when the number of body meshes is 2 million and 2.53 million. When the volume meshes are encrypted to 3.01 million, 3.45 million and 3.91 million, the lift characteristics of the UAV calculated by the three have not changed much, and the numerical results have an error within 1.5%, which tends to be stable. Considering the amount of comprehensive calculations, this paper uses a mesh of 3 million volume meshes to calculate and analyze the aerodynamic characteristics of the UAV.

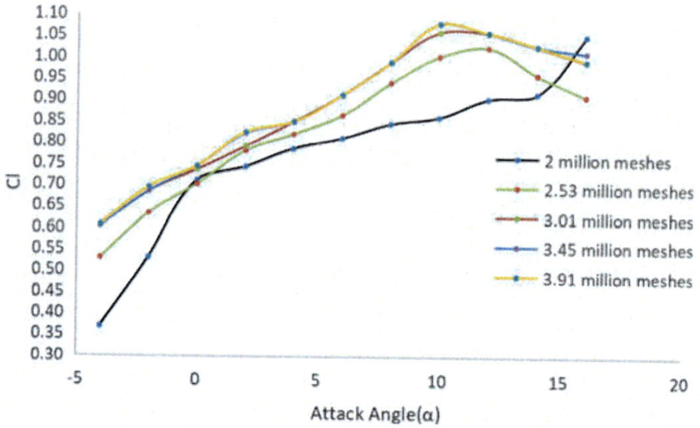

Fig. 10 Curves of lift coefficients under different numbers of meshes

4 Analysis of UAV Aerodynamic Characteristics

4.1 Analysis of UAV Lift and Drag Characteristics

Analyze the changes of the aerodynamic parameters of the UAV body with the attack angle when the incoming flow velocity, flight altitude and atmospheric conditions are constant. The calculated speed and altitude are the cruise speed and cruise altitude of the UAV: the cruise speed is 22 m/s, the cruise altitude is 1000 m, and the gravity center coordinates of the UAV are selected as the reference point of the pitching moment. Using ANSYS Fluent to calculate the lift-drag characteristics and pitching moment characteristics of the UAV at an attack angle of −2 to 16° with an interval of 2° gradient, the aerodynamic parameter data is shown in Table 1.

From the data on the lift and drag characteristics of the UAV in the above table, it can be seen that the lift coefficient of the flight attack angle increases with the increase of the attack angle when the attack angle is from −4 to 10°. When the attack angle is 10°, the UAV's lift coefficient reaches the maximum $C_{LMAX} = 1.06$, and when the attack angle is 10–16°, the lift continues to decrease. This is due to the separation of the airflow at the trailing edge of the upper wing surface of the UAV wing, which causes the lift to drop sharply; The drag of the UAV has been increasing. As the drag increased relatively rapidly, the lift decreased after 10° attack angle, and after reaching the stall angle of 10°, the drag increased sharply; The lift-to-drag ratio of the UAV first rises and then decreases. The maximum lift-to-drag ratio appears at −2° attack angle, with a maximum of 16.81. After −2° attack angle, the lift-to-drag ratio continues to decrease.

Perform curve fitting on the above data to obtain the curve of the aerodynamic parameters of the whole UAV with the attack angle shown in Fig. 11.

From the above aerodynamic characteristic curve, the change of the lift and drag characteristics of the UAV can be seen more intuitively, and it can be concluded that

Table 1 Calculated data of UAV aerodynamic parameters

Attack angle (°)	Lift coefficient	Drag coefficient	Lift-to-drag ratio	Pitching moment coefficient
−4	0.603783	0.037645	16.0388	−0.093703
−2	0.683971	0.040679	16.8138	−0.107010
0	0.735145	0.047250	15.5586	−0.113493
2	0.790642	0.065378	12.0934	−0.121512
4	0.848494	0.092822	9.1411	−0.131388
6	0.911325	0.123120	7.4019	−0.143347
8	0.989102	0.155654	6.3545	−0.157832
10	1.059803	0.190435	5.5652	−0.169724
12	1.058142	0.237969	4.4466	−0.181126
14	1.026369	0.284327	3.6098	−0.196593
16	0.991654	0.330456	3.0009	−0.212066

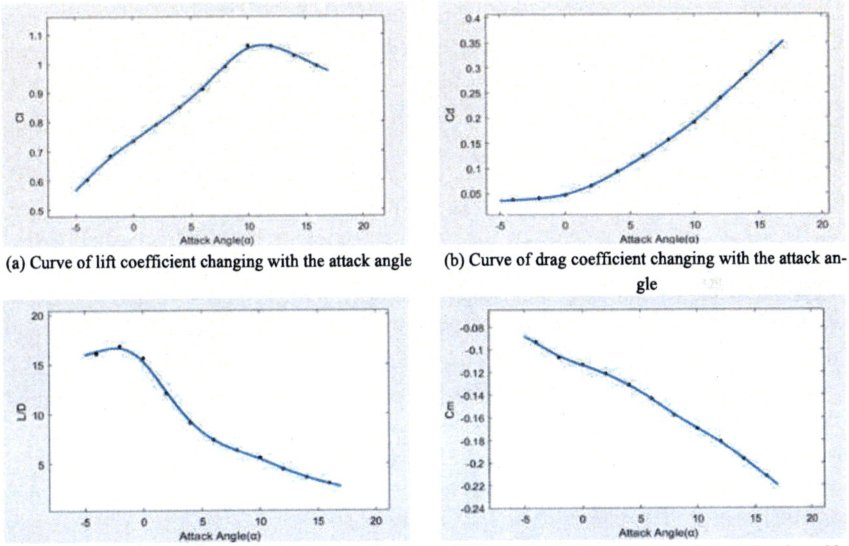

(a) Curve of lift coefficient changing with the attack angle

(b) Curve of drag coefficient changing with the attack angle

(c) Curve of lift-to-drag ratio changing with the attack angle

(d) Curve of pitching moment coefficient changing with the attack angle

Fig. 11 UAV aerodynamic characteristic curve

the UAV has better aerodynamic characteristics at an attack angle of −4 to 10°. The UAV's lift coefficient increases rapidly from −4 to 0°, and reaches its maximum at 10° attack angle; The drag coefficient is increasing rapidly, and the drag coefficient has a larger increase after 10° attack angle; The lift-to-drag ratio decreases rapidly from $-2°$ attack angle to 6° attack angle, and the downward trend slows down after 6° attack angle, but it is still a continuous downward trend.

From the above analysis, it can be obtained that the cruise attack angle of the UAV is $\alpha_c = 1.8°$, the lift coefficient during cruise is $C_{LC} = 0.785$, and the lift-drag ratio during cruise is $L/D = 12$.

From the data in Table 1 and curve 4-1(d), it can be seen that the pitch moment coefficient of the whole aircraft is negative, that is $C_m < 0$, indicating that the UAV is longitudinally static and stable [6]. The UAV pitching moment changes little with the attack angle, and as the attack angle increases, the pitching moment continues to decrease, and the pitching moment reaches the minimum at an attack angle of 16°, which is -0.212066. The static stability margin of the UAV calculated from the aerodynamic data is $K_n = 16.64\%$, which is consistent with the previous estimate.

4.2 UAV Surface Pressure Data Analysis

This section analyzes the UAV's surface pressure distribution at an attack angle from -4 to 12°, analyzes the cause of lift, and further analyzes the optimization plan of the UAV aerodynamic characteristics. Figure 12 shows the relative pressure cloud diagram on the surface of the UAV under different attack angles, and Fig. 13 shows the cloud diagram of the flow field under the stall attack angle.

From the pressure cloud diagrams at different attack angles, it can be seen that the low pressure area of the whole UAV body is mainly concentrated on the upper surface of the wing. The greater the flight attack angle, the lower the pressure on the upper surface of the wing. At an attack angle of 12°, the pressure on the upper surface of the wing is the lowest; The high-pressure area of the whole UAV body is mainly distributed at the tip of the nose, the leading edge of the wing and the leading edge of the tail, and the steering gear. This is caused by the accumulation of airflow and the low flow rate [7]; The pressure distribution in the remaining parts is relatively uniform, especially in the fuselage part, the pressure distribution is relatively reasonable, and a conclusion can be drawn that the aerodynamic performance is relatively good. Therefore, the aerodynamic data from -4 to 10° attack angle is selected for subsequent research and analysis.

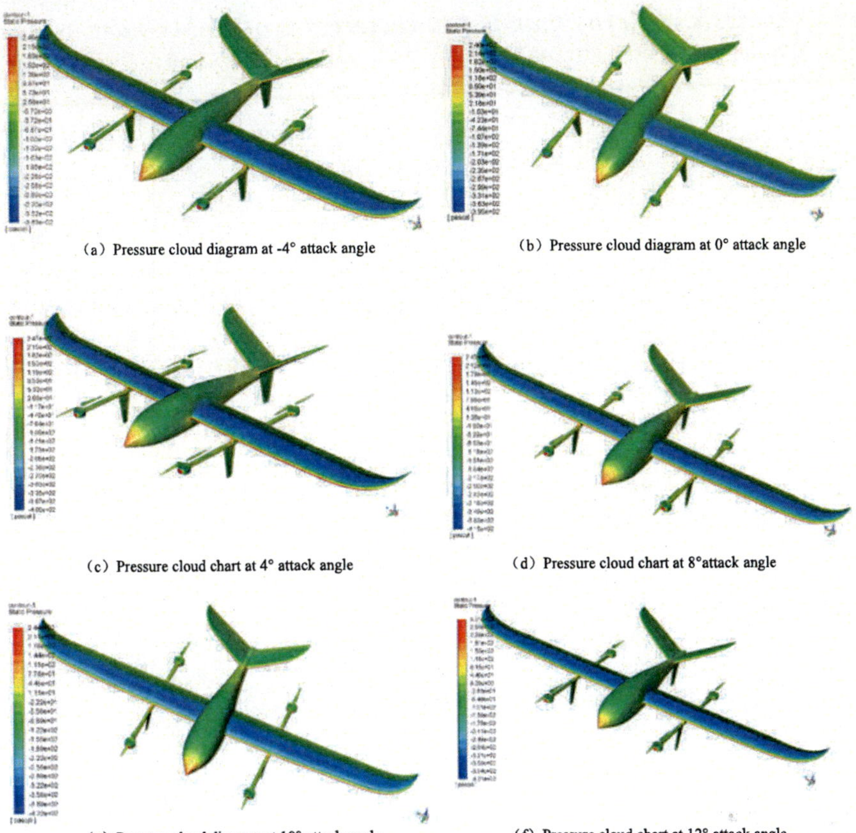

(a) Pressure cloud diagram at -4° attack angle

(b) Pressure cloud diagram at 0° attack angle

(c) Pressure cloud chart at 4° attack angle

(d) Pressure cloud chart at 8°attack angle

(e) Pressure cloud diagram at 10° attack angle

(f) Pressure cloud chart at 12° attack angle

Fig. 12 Surface pressure cloud map of UAV at different attack angle

Fig. 13 Cloud image of flow field at 10° attack angle

4.3 Analysis of Aerodynamic Characteristics of UAV Under Crosswind

Considering the engineering practicability, ANSYS Fluent is used to calculate the aerodynamic performance of the UAV under the lateral wind speed. The lateral wind speeds of 3, 5, and 7 m/s are selected for calculation and analysis of aerodynamic characteristics at an attack angle of −4 to 10°. As shown in Table 2, Table 3, and Table 4, the aerodynamic characteristic data of the UAV body subjected to side wind speeds of 3, 5, and 7 m/s, respectively.

According to the data in the above table, use PYTHON to draw the curve of aerodynamic parameters with the angle of attack at 3, 5, and 7 m/s lateral wind speed, as shown in Fig. 14.

It can be seen from the above table and curve that when affected by the side wind speed, the lift, drag, lift-to-drag ratio, pitching moment curve and the curve without crosswind tend to be consistent; It can be seen from the lift Fig. 14a and drag Fig. 14b that the crosswinds at different attack angles have different effects on

Table 2 Aerodynamic data under 3 m/s crosswind wind speed

Attack angle (°)	Lift	Drag	Side force	Lift-to-drag ratio	Pitch moment	Rolling moment	Yaw moment
−4	117.2182	7.3603	0.8934	15.9258	−4.8002	−6.1224	1.8051
−2	132.5498	8.1997	0.8714	16.1651	−5.4793	−5.4374	1.6592
0	143.5396	9.6902	0.8330	14.8128	−5.9297	−4.8171	1.4531
2	155.0970	13.0565	0.7786	11.8789	−6.2893	−4.1854	1.2622
4	167.6266	18.5430	0.9040	9.0399	−6.8622	−3.2643	1.4851
6	180.1087	24.7119	0.8961	7.2884	−7.4208	−2.4957	1.3679
8	196.8757	31.3274	1.0191	6.2845	−8.2616	−1.4135	1.5512
10	211.8106	38.5928	1.3171	5.4883	−8.9340	−0.5856	2.0983

Table 3 Aerodynamic data under 5 m/s crosswind wind speed

Attack angle (°)	Lift	Drag	Side force	Lift-to-drag ratio	Pitch moment	Rolling moment	Yaw moment
−4	117.7067	7.6757	1.5123	15.3350	−4.8159	−8.1044	2.8945
−2	133.8268	8.6773	1.4638	15.4226	−5.5497	−7.3300	2.6221
0	145.0114	10.0909	1.5210	14.3705	−6.0580	−6.3239	2.6163
2	157.8462	13.5792	1.4717	11.6241	−6.4899	−5.4611	2.3212
4	171.1814	19.3254	1.4377	8.8578	−7.1465	−4.5285	2.0598
6	185.9291	25.9398	1.7310	7.1677	−7.8579	−2.7433	2.5716
8	203.9365	33.4775	1.5783	6.0917	−8.7206	−1.8821	2.1329
10	219.1969	41.8288	1.4215	5.2403	−9.3241	−0.9310	1.8128

Table 4 Aerodynamic data under 7 m/s crosswind wind speed

Attack angle (°)	Lift	Drag	Side force	Lift-to-drag ratio	Pitch moment	Rolling moment	Yaw moment
−4	121.9005	8.8156	1.9010	13.8278	−4.9560	−11.8852	3.4529
−2	138.1821	9.8513	2.5143	14.0268	−5.6877	−9.9161	4.8293
0	152.8363	11.5704	2.5086	13.2093	−6.3811	−8.1743	4.4492
2	170.2078	15.4698	2.4737	11.0026	−7.0243	−6.2515	4.0071
4	187.6742	22.2691	2.3787	8.4276	−7.8901	−4.8710	3.5068
6	201.4652	29.9834	2.0992	6.7192	−8.5275	−4.2269	2.6742
8	222.4993	38.6443	2.3294	5.7576	−9.6064	−2.5604	2.8107
10	242.0243	48.4427	2.5888	4.9961	−10.5216	−1.0138	2.8525

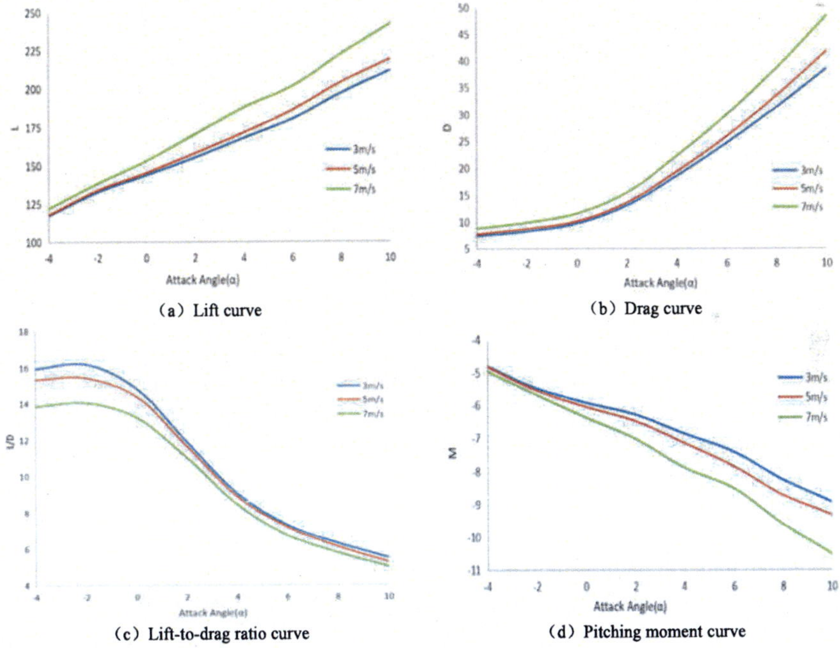

Fig. 14 Aerodynamic characteristic curves under different crosswind wind speeds

the UAV. However, in general, the lift resistance increases to varying degrees under crosswinds, and with the increase of the lateral incoming wind speed, the higher the lift and the resistance, and the increase more rapidly; It can be seen from the lift-to-drag ratio Fig. 14c that the lift-to-drag ratio has increased or decreased to different degrees, which is mainly caused by the different increase in lift and drag, and as the crosswind wind speed increases, the lift-to-drag ratio decreases; From the pitching moment Fig. 14d, it can be seen that the pitching moment under the lateral incoming wind speed is still negative, and the absolute value increases with the increase of the lateral incoming flow speed.

From the above analysis, it can be obtained that in the range of −4 to 10° attack angle, the increase and decrease of the lift-drag characteristic data at 3 m/s, 5 m/s, and 7 m/s is not very large. Therefore, it can be judged that the UAV is less affected by the crosswind, the flight is relatively stable, and it can effectively resist the influence of the lower speed of the crosswind on the normal flight state of the UAV. Based on the above analysis, the design attack angle in this case is 1.8°.

5 Conclusion

This article first uses ANSYS Fluent to calculate the aerodynamic parameters of the UAV, and obtains the pressure cloud diagram on the surface of the whole UAV; Then the aerodynamic characteristics of the UAV under crosswind wind speed were calculated and analyzed, and it was concluded that the lift and drag characteristics of the UAV were less affected by the crosswind, had good aerodynamic characteristics, and the UAV was reasonably designed; Finally, it is proved that the UAV can meet the design index requirements.

References

1. Wang S-A, Yin G-L (2002) Application of CFD in aircraft design. Aviat Sci Technol 6:33–36
2. Loureiro JBR, Alho ATP, Silva Freire AP (2008) The numerical computation of near-wall turbulent flow over a steep hill. J Wind Eng Ind Aerodyn 96(5):540–561
3. Ji B-B, Chen J-P (2012) ANSYS ICEM CFD meshing technology example detailed explanation. China Water Conservancy and Hydropower Press, Beijing
4. Zeng M, Liu W, Zou J-J (2016) Basics of aerodynamics. Science Press, Beijing, p 2
5. Feng J-A, Tang X-Q, Wang W-B et al (2017) Reliability verification method of numerical simulation based on mesh independence and time independence. J Shihezi Univ 35(01):52–56
6. Yechout TR, Morris SL, Bossert DE, et al (2003) Introduction to aircraft flight mechanics: performance, static stability, dynamic stability, and classical feedback control. AIAA
7. Bao R-J (2019) Design and research of small vertical take-off and landing fixed wing UAV for forestry. Northeast Forestry University

Flux Limiting Schemes for Fine Resolution Detonation Wave Cell Structure

Jaehoon Ryu, Mohammed Niyasdeen, and Jeong-Yeol Choi

Abstract The dynamic feature of the detonation wave is characterized by the cell structure formed by the traces of triple points in the detonation wave-front and transverse shock wave behind the detonation wave. In this paper, detonation wave is numerically traced by the local maximum pressure and the fully resolved results for the dynamic structure of the detonation wave is studied. A comparative study is carried out for the selection of different numerical schemes which can provide fine resolution without sacrificing the overall accuracy and robustness. For this study, a weak condition is purposefully chosen in such a way that making a single detonation cell to propagate in a two-dimensional (2D) channel of unit width. Grid resolution study has also been carried out for a wide range of radial grid spacing (Δy) By considering the finest resolution results in 256 million grid points in 2D simulation, a moving computational window technique is adopted to calculate only the vicinity of detonation wave front where important physics happens. For low resolution cases, numerical fluxes are evaluated by Roe, RoeM, AUSMDV and AUSMPW+ schemes and high resolution was achieved by 3rd-oder MUSCL, 3rd-order WENO, 5th-order WENO and 5th-order oMLP schemes. This paper shows the full details about the comparative performance of each scheme to capture dynamics structure of detonation cell structure including the dynamic maximum value on Von Neumann spike.

Keywords Flux limiting schemes · Von Neumann peak · Detonation wave · Cell structure

1 Introduction

Detonation wave is a dynamic shock wave physics driven by heat addition behind the shock wave. The dynamic feature of the detonation wave is characterized by the cell structure formed by the traces of triple points in the detonation wave-front

J. Ryu · M. Niyasdeen · J.-Y. Choi (✉)
Department of Aerospace Engineering, Pusan National University, Busan 46241, Korea
e-mail: aerochoi@pusan.ac.kr

S. Lee et al. (eds.), *The Proceedings of the 2021 Asia-Pacific International Symposium on Aerospace Technology (APISAT 2021), Volume 1*, Lecture Notes in Electrical Engineering 912, https://doi.org/10.1007/978-981-19-2689-1_35

and transverse shock wave behind the detonation wave. Detonation wave propagation mechanism and the characteristics of its dynamic structure can be resolved only from the detailed understanding of the physical process involved in the unsteady detonation wave front with Zel'dovich-von Neumann-Döring (ZND) structure. Detonation wave-front instability was first observed experimentally in 1961 [1]. In the experiments, detonation wave propagation left the traces of the structure so-called detonation cell on a wall coated with a thin layer of soot or smoke foil. Even though a few extended numerical studies have been conducted to investigate detonation wave structure [2–8], only few studies focus on numerical requirements on detonation phenomenon with relatively low grid resolution.

In this paper, a series of numerical investigation has been carried out to find a numerical requirement for capturing detonation cell structure, which can explain its characteristics efficiently. Grid resolution is studied for Δy = 1/100, 1/200, 1/400, 1/800, 1/1,600. Numerical fluxes are evaluated by Roe, RoeM, AUSMDV and AUSMPW+ schemes those compared with each other for low resolution cases. High resolution was achieved by 3rd-oder MUSCL, 3rd-order WENO, 5th-order WENO and 5th-order oMLP schemes. This fundamental numerical investigation and its detailed numerical schemes comparison on detonation wave behavior will give an insight and deep understanding on numerical approach to investigating detonation phenomenon.

2 Governing Equations & Numerical Methodology

2.1 Governing Equations

The governing equations used are two-dimensional compressible conservation equations for mass, momentum, energy, and reaction progress variable, with ideal gas equation of state for the reactive gas.

$$\frac{\partial}{\partial t}\begin{bmatrix}\rho\\ \rho u\\ \rho v\\ \rho e\\ \rho Z\end{bmatrix}+\frac{\partial}{\partial x}\begin{bmatrix}\rho u\\ \rho u^2+p\\ \rho uv\\ (\rho e+p)u\\ \rho uZ\end{bmatrix}+\frac{\partial}{\partial y}\begin{bmatrix}\rho v\\ \rho uv\\ \rho v^2+p\\ (\rho e+p)v\\ \rho vZ\end{bmatrix}=\begin{bmatrix}0\\ 0\\ 0\\ 0\\ \dot{\omega}\end{bmatrix}$$

$$\dot{\omega}=\rho(1-Z)\cdot A\cdot\exp\left(\frac{-E_a}{p/\rho}\right)$$

$$p=(\gamma-1)\cdot\rho\cdot\left\{e-\frac{1}{2}\left(u^2+v^2\right)+Zq\right\}$$

where r is the density, u and v are the velocity components, and e is the specific total energy. The reaction progress variable Z accounts for the mass fraction of product species which varies from 0 to 1 for unburned (reactant) to burned (product) state respectively. The chemical kinetics is modeled by a 1-step irreversible Arrhenius reaction model. Where $\dot{w}$ is the mass rate of product formation, T_a is the activation temperature, and A is the pre-exponential factor. The pressure and temperature are obtained through the equations of state.

2.2 Numerical Methodology

The governing equations are numerically solved using a cell-vertex finite volume method. The convective fluxes in the cell boundaries are calculated by using Roe, RoeM, AUSMDV, AUSMPW+ scheme [9–12]. As flux splitting schemes, 3rd-order MUSCL, 3rd-order WENO, 5th-order WENO and 5th-order oMLP schemes are used and compared each other [13–16]. The discretized equations are integrated in time using a 4th-order accurate 4-step Runge–Kutta method scheme. To maximize the computational efficiency and the parallelization of the code, OpenMP is used.

2.3 Initial ZND Condition

For initiation of detonation wave, 1-Dimensional (1D) steady ZND structure solution is used. A particular ZND conditions are chosen in such a way that a single detonation cell structure is obtained in the channel of unit width. Its specific heat ratio of burned to unburned gases is assumed as 1.3, its dimensionless heat addition (Q) is 25, pre-exponential factor (A) is 60, and activation energy (E_a) is 10, which results in detonation Mach number, so called Chapman-Jouguet (C-J) Mach condition of 6.0393. The minimum grid points in the $L_{1/2}/\Delta x \geq 5$ which appears to be sufficient to capture the detonation cell structure [17] (Figs. 1 and 2).

2.4 Geometry & Grid System

In this paper, straight channel with aspect ratio 100 is used with equal grid spacing in both x and y directions. For grid refinement test, grid system varies as shown in the Table 1.

Since the finest resolution results in 256 million grid points, moving computational window technique is used to capture the important physics happening in the vicinity of the detonation wave front. Incoming boundary is selected by checking the protruding detonation wavefront and the exit boundary is set to 1.5 times the channel width where characteristic boundary conditions is applied with C-J condition. The moving

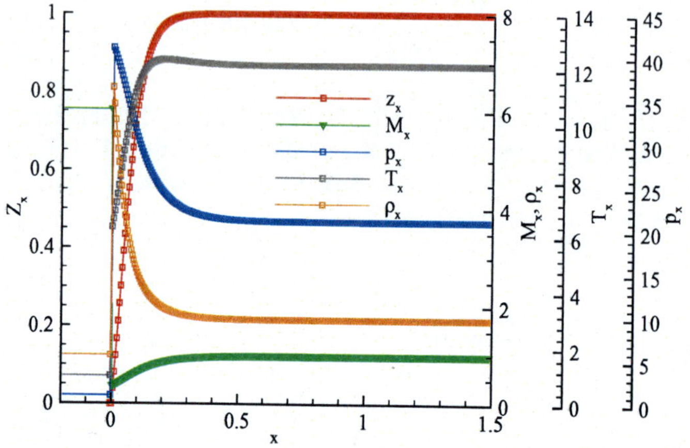

Fig. 1 ZND structure for initial condition

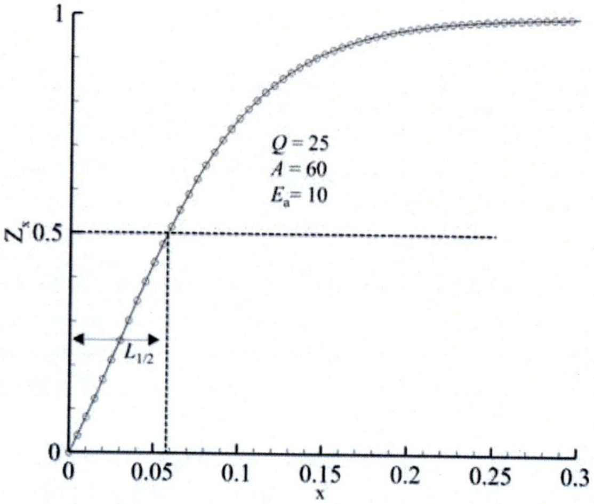

Fig. 2 Half reaction length in ZND structure

Table 1 Grid system configuration

Number of cells in unit channel width, *n*	Grid system for entire computational domain	Size of moving computational window	Minimum spacing (Δy)	$L_{1/2}/\Delta x$ (number of grid points within $L_{1/2}$)
100	10,001 × 101	151 × 101	1/100	6
200	20,001 × 201	301 × 201	1/200	12
400	40,001 × 401	601 × 401	1/400	24
800	80,001 × 801	1,201 × 801	1/800	47
1,600	160,001 × 1,601	2,401 × 1,601	1/1600	93

computational window technique practically reduced the computational resolution to 3.84 million grid points for finest case.

3 Results & Discussion

3.1 Grid Refinement Test

Figure 4 shows the maximum pressure history at bottom wall for various grid resolutions. It is plotted in inverted direction for convenience. Figure 5 is the close-up view at the end of computation. The averaged values of peak pressure and the cell length are plotted in Fig. 6. The averaged values are taken for the stabilized range x/H = 20–100, because the early part is the transient period from the initial condition. Length of detonation cell is calculated by measuring the distance between two consecutive peak pres-sure points (Fig. 3).

Figure 6 shows the converging solution of the peak pressure and cell length as the grid resolution is refined. It is found that variation of call length is less affected by the grid resolution. Largest case is n = 400, but the deviation is not so big. On the other hand, the average peak max pressure value converges as the number of grid points along the width increases. Both 1600 grid points and 800 grid points along the width seems reasonably good enough to get a fine resolution considering the computational cost. Present result implies that very fine resolution, up 47 grid points within half reaction length, is needed to have fully converged value of Von Neuman peak pressure.

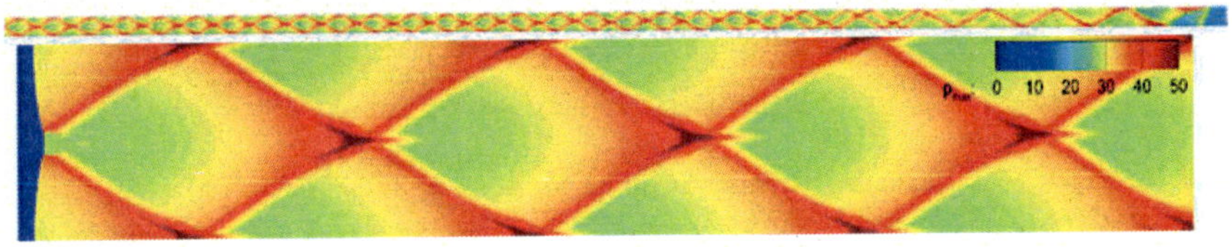

Fig. 3 Entire numerical smoked-foil record for the channel with aspect ratio 100 showing the evolution process and stabilized detonation (upper), Magnified plot at the end of computation (lower)

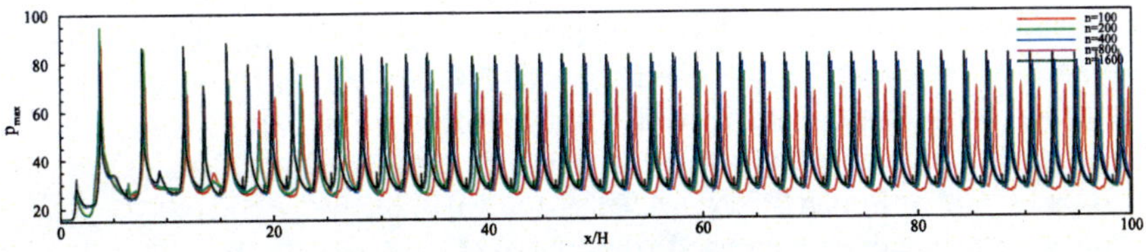

Fig. 4 Maximum pressure history at bottom wall for various grid resolutions

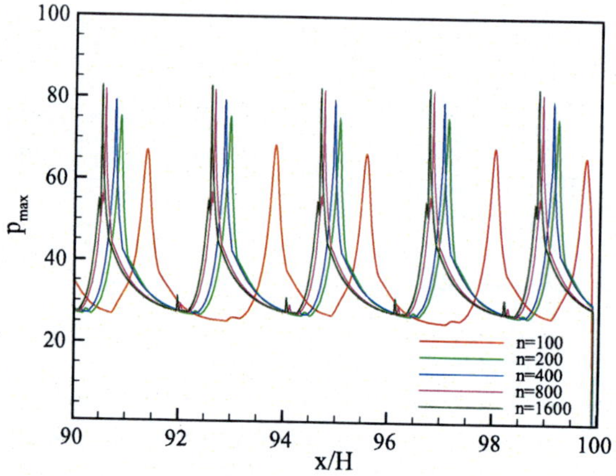

Fig. 5 Close-up view of maximum pressure history for various grid resolutions

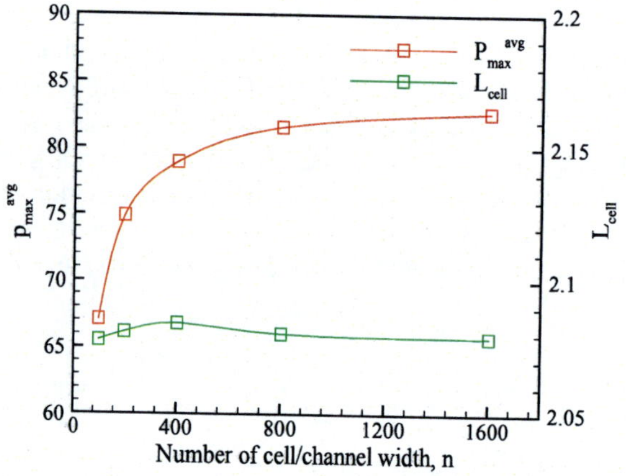

Fig. 6 Averaged peak maximum pressure & cell length along grid numbers within unit channel width, n

3.2 Dependency on High Resolution Scheme

Dependency on high resolution scheme has been investigated for 3rd-order MUSCL, 3rd-order WENO, 5th-order WENO, and 5th-order oMLP with RoeM flux scheme fixed with 80,001 × 801 grid system. Figure 7 shows the max pressure history comparison for various schemes with close-up view at the end of computation. From Fig. 8, 5th-order WENO scheme shows the finest values followed by 5th-order oMLP and 3rd-order WENO scheme. Even though 5th-order oMLP scheme is multi-dimensionally extrapolated from physical properties, it doesn't give fine

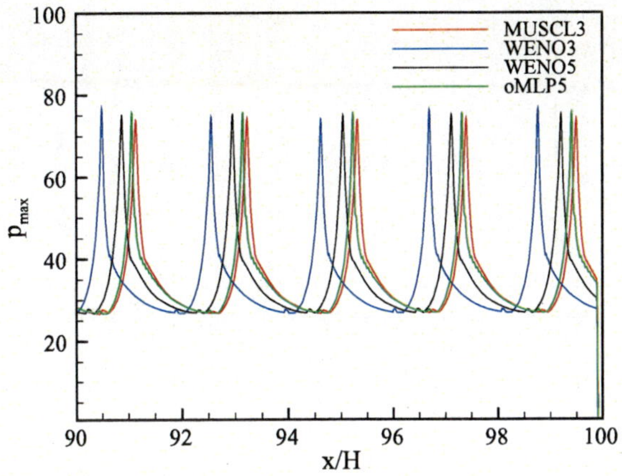

Fig. 7 Close-up view of maximum pressure history for various high resolution schemes

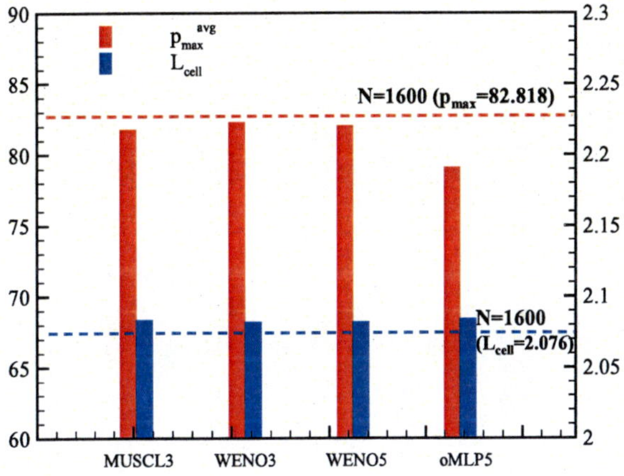

Fig. 8 Averaged peak maximum pressure & cell length for various high resolution schemes

results compared to the others which are extrapolated from one-dimensional physical properties. When it comes to cell length, oMLP scheme shows the worst solution.

3.3 Dependency on Flux Splitting Scheme

Dependency on flux splitting scheme has been investigated for Roe, RoeM, AUSMDV, and AUSMPW + with 5th-order WENO scheme fixed with 80,001 × 801 grid system. Maximum pressure and cell length are compared with each other and the finest resolution result for n = 1,600 in Fig. 10. As Fig. 10 shows, peak max

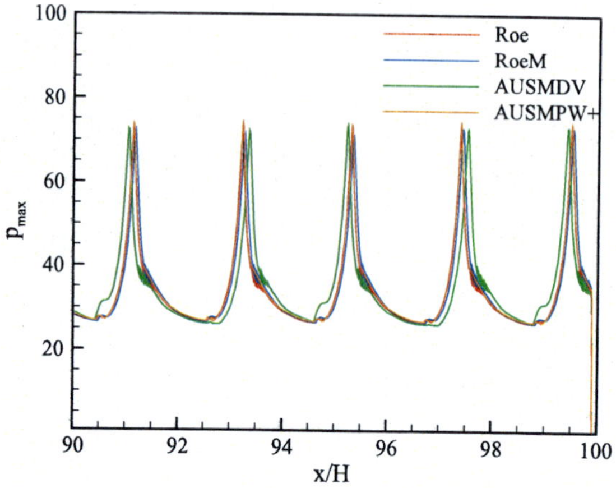

Fig. 9 Close-up view of maximum pressure history for various flux splitting schemes

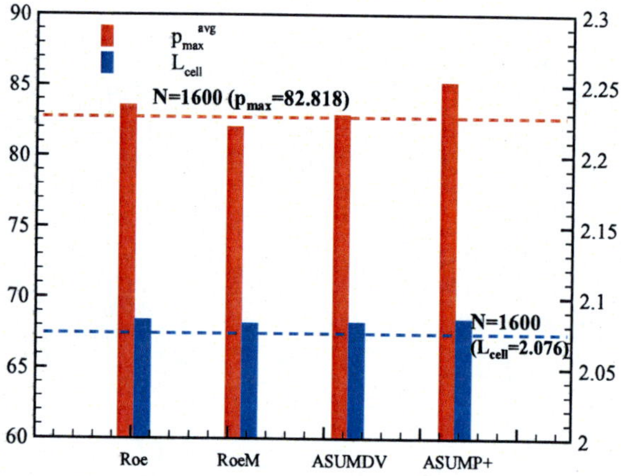

Fig. 10 Averaged peak maximum pressure & cell length for various flux splitting schemes

pressure average values have relatively big difference. Amongst, RoeM shows the finest values compared to the finest resolution result. For cell length, all schemes predicted under 1% erorr (Fig. 9).

4 Conclusion

Series of numerical investigation have been carried out to find out the numerical requirements for capturing fine detonation cell structure. Averaged peak max pressure and averaged cell length have been investigated for various grid resolutions, higher-resolution schemes and flux splitting schemes. Grid refinement study suggest that very fine resolution is required to have converged solution of Von Neumann peak pressure while the cell length is less affected by grid resolution. In the practical consideration, n $\geq$ 400 ($L_{1/2}/\Delta x \geq 24$) provides the peak pressure 5% less than the converged solution. Investigating dependency on various numerical schemes suggests that 5th-order WENO flux scheme and RoeM flux splitting scheme provides the finest solution in the detonation cell structure study on straight channel.

Acknowledgements This work was supported by the Basic Science Research Program Grant (2019R1A2C1004505) through the National Research Foundation (NRF) of Korea, funded by the Ministry of Science and ICT (MSIT) of the Republic of Korea Government.

References

1. White DR (1961) Turbulent structure of gaseous detonation. Phys Fluids 4(4):465–480
2. Oran ES, Boris JP, Young T, Flanigan M, Burks T, Picone M (1981) Numerical simulations of detonations in hydrogen-air and methane-air mixtures. Symp Int Combust 18(1):1641–1649
3. Kailasanath K, Oran ES, Boris JP, Young TR (1985) Determination of detonation cell-size and the role of transverse-waves in two-dimensional detonations. Combust Flame 61(3):199–209
4. Bourlioux A, Majda AJ (1992) Theoretical and numerical structure for unstable two-dimensional detonations. Combust Flame 90(3):211–229
5. Choi JY, Jeung IS, Yoon Y (2000) Computational fluid dynamics algorithms for unsteady shock-induced combustion. Part 1: validation. AIAA 38(7):1179–1187
6. Taki S, Fujiwara T (1978) Numerical analysis of two-dimensional nonsteady detonations. AIAA J 16(1):73–77
7. Taki S, Fujiwara, T (1981) Numerical simulation of triple shock behavior of gaseous detonation. Symp (Int) Combust 18(1):1671–1681
8. Lee John HS (2008) Detonation phenomenon. Cambridge University Press. https://doi.org/10.1017/CBO9780511754708
9. Roe PL (1981) Approximate Riemann solvers, parameter vectors, and difference schemes. J Comput Phys 43(2):357–372
10. Kim SS, Kim C, Rho OH, Hong SK (2003) Cures for the shock instability: development of a shock-stable Roe scheme. J Comput Phys 185(2):342–374
11. Wada Y, Liou MS (1997) An accurate and robust flux splitting scheme for shock and contact discontinuities. SIAM J Sci Comput 18(3):633–657
12. Kim KH, Kim C, Rho OH (2001) Methods for the accurate computations of hypersonic flows: I. AUSMPW+ scheme. J Comput Phys 174(1):38–80
13. Van Leer B (1979) Towards the ultimate conservative difference scheme, V. A second-order sequel to Godunov's method. J Comput Phys 32(1):101–36
14. Jiang GS, Shu CW (1996) Efficient implementation of weighted ENO schemes. J Comput Phys 126(1):202–228
15. Henrick AK, Aslam TD, Powers JM (2005) Mapped weighted essentially non-oscillatory schemes: achieving optimal order near critical points. J Comput Phys 207(2):542–567
16. Kim S, Lee S, Kim KH (2008) Wavenumber-extended high-order oscillation control finite volume schemes for multi-dimensional aeroacoustic computations. J Comput Phys 227(8):4089–4122
17. Choi JY, Ma FH, Yang V (2008) Some numerical issues on simulation of detonation cell structures. Combust Explos Shock Waves 44(5):560–578

Experimental Investigation of Performance for Korean Electric Unmanned Helicopter on Ground

Sanghyun Chae, Hong-Tae Yeo, Kyuhoon Lee, and Byung Il Yoon

Abstract Korean Electric Unmanned Helicopter(K-EUH) has been developing to achieve the major performance target of 60 km flight distance with 20 kg payload. The first prototype weighing 126 kgf was manufactured and was tested. Objectives of this ground test is to investigate whether the helicopter has sufficient performance for take-off, and to acquire know-how on the operation of the electric power system. The ground test equipment has loadcells for measuring thrust of the main rotor, and the tail rotor. Rotating power of rotors is supplied by the battery system of nominal DC 355 V. As a result of the test, it was confirmed that the maximum thrust of the main rotor was 1.27 times of MTOW, and the tail rotor also confirmed to have the thrust capable of anti-torque of the main rotor. Experiments have also shown that it is advantageous to configure the electric power system at high voltage. If the required current is low by feeding the high voltage, heat generation and wire thickness can be reduced. The other significant fact is that the voltage drop of battery becomes more severe when the required power is high such as landing approach. Since such a sudden voltage drop reduces the rotating speed of the rotor, a dangerous situation may occur.

Keywords Electric unmanned helicopter · BLDC motor · Voltage drop · Hover performance

1 Introduction

Due to environmental regulations to reduce carbon dioxide, electric power is being applied to vertical take-off and landing(VTOL) vehicles. In the case of manned

S. Chae (✉) · B. Il Yoon
Korea Aerospace Research Institute, 34133 Daejoen, Republic of Korea
e-mail: nyugnas@kari.re.kr

H.-T. Yeo
Sungwoo Engineering, 28114 Chungju, Republic of Korea

K. Lee
DBRAIN, 34104 Daejoen, Republic of Korea

S. Lee et al. (eds.), *The Proceedings of the 2021 Asia-Pacific International Symposium on Aerospace Technology (APISAT 2021), Volume 1*, Lecture Notes in Electrical Engineering 912, https://doi.org/10.1007/978-981-19-2689-1_36

helicopters, Sikorsky's firefly has been developed with electric power based on the S-300C as a technology demonstrator [1]. In addition to helicopters, various types of UAMs such as multi-rotor and lift cruises are being developed with electric power for VTOL. In unmanned VTOLs, electric power has been widely used although internal combustion engines are still used in unmanned aerial vehicles above ultra-light VTOLs (115 kgf or less of empty weight in Korea).

Most of ultra-light unmanned VTOLs are multi rotors configurations, especially quad rotors. Quadrotor drones are popular in areas of hobbies, leisure, and aerial photography, because mechanical systems (RPM control) are simple, inexpensive and low cost of maintenance. In order to increase the utility of the multi rotor VTOLs, their size has been increased. However, RPM controlled multi-rotor VTOLs have a disadvantage that the handling quality decreases as the weight increases [2]. Therefore, a request was proposed for the development of an electric powered VTOL in the configuration of a conventional helicopter with of 150 kgf or less of a maximum take-off weight.

The Korean Electric Unmanned Helicopter(K-EUH) is being developed to be used for goods delivery. The major performance target of K-EUH is 60 km flight distance with 20 kg payload. Last year, detailed design and the first prototype production were completed, and the ground test was conducted. Objectives of this ground test is to investigate whether the helicopter has sufficient performance for take-off, and to acquire know-how on the operation of the electric power system.

2 Experimental Setup

The maximum takeoff weight (MTOW) of K-EUH is 126 kgf, that consists of 106 kgf of empty weight and 20 kgf of payload. The main rotor and a tail rotor had two blades, and radius of 1.63 m, 0.66 m respectively. The distance between the main rotor and the tail rotor is 1.96 m. Specific configuration information of K-EUH is shown in Fig. 1.

K-EUH has two batteries, one of which is a main battery of nominal DC 355 V and the other is an avionic battery of nominal DC 25 V. The main battery supplies rotating power of two rotors and is made of lithium polymer with 8.6 kWh (328–400 V, 22 Ah, 54 kgf). The tail rotor is powered by DC 56 V dropped by a DC-DC converter from the main battery. The avionic battery supplies power of the battery management system(BMS), the flight control computer, actuators, etc.

The main motor and the ESC for the main rotor are ROTEX REX 50/8 and MGM Compro HBC-400400-3. The tail motor and the tail ESC for the tail rotor are KDE 8218XF-120 and KDE-UAS125UVC-HE. These motors were selected in consideration of the power required to rotate the rotors. The main rotor RPM is shifted from the main motor by the belt and pulleys, and the tail rotor is directly connected to the motor.

The entire prototype was installed on a ground test equipment as shown in Fig. 2. When measuring the performance of the main rotor, the thrust was measured with

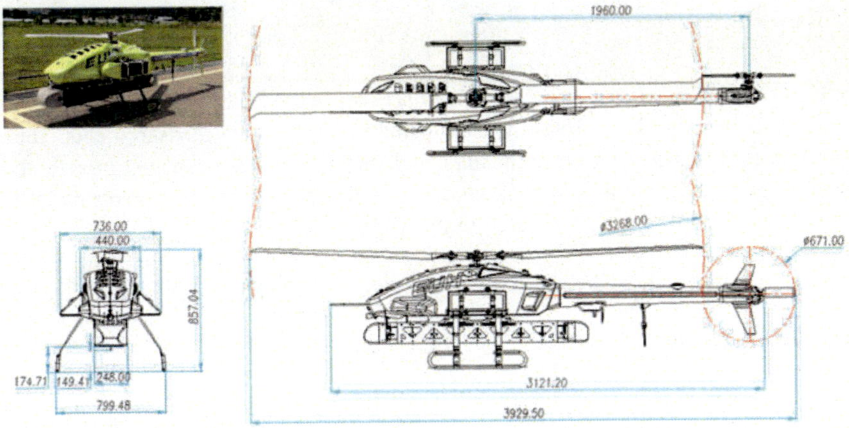

Fig. 1 K-EUH configuration

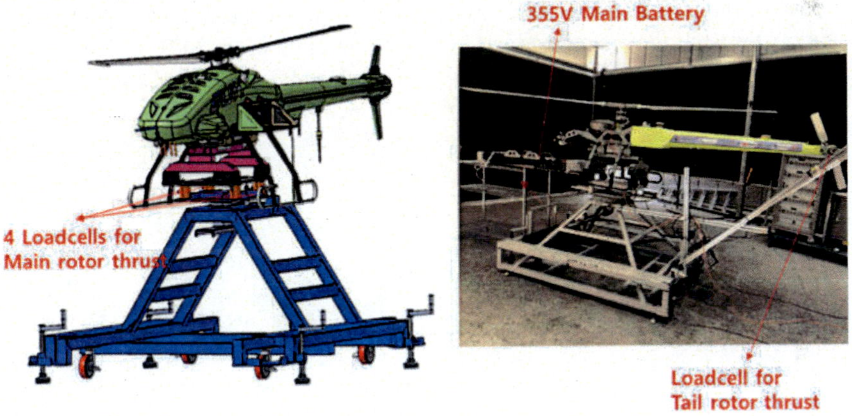

Fig. 2 K-EUH ground test equipment

four load cells installed under the fuselage of K-EUH, the torque was measured with the load cell located on the tail rotor. The main motor temperature and the tail motor temperature were measured with separate temperature sensors. Continuous temperature monitoring was essential on this ground test because electric motors have a risk of burning or melting of coils when operated at a temperature above a limit.

3 Results

The experimental results are presented and discussed in the following order: (1) main rotor measurements, (2) tail rotor measurements, and (3) battery measurements. The performance(thrust and power) of the main and the tail rotor has been measured three or more times, and the measured values were converted into international standard atmospheric conditions and presented in this paper.

3.1 Main Rotor

Thrust and torque of the main rotor according to the change of the collective pitch were measured, and the mechanical output power is calculated by applying the main rotor rotation speed of 860 RPM to the measured torque in Fig. 3. The electrical input power is calculated by measured the voltage and current of the power source (the power supply or the battery system). It is confirmed by the experiments that the maximum thrust of the main rotor was 1.27 times or more of the MTOW. When the main rotor generated the thrust of MTOW, the mechanical output was 14.5 kW and the electric input of 19.5 kW was measured, the electrical efficiency(the mechanical output/the electrical input) was 74.5% as shown in Fig. 4. This means that the effi-

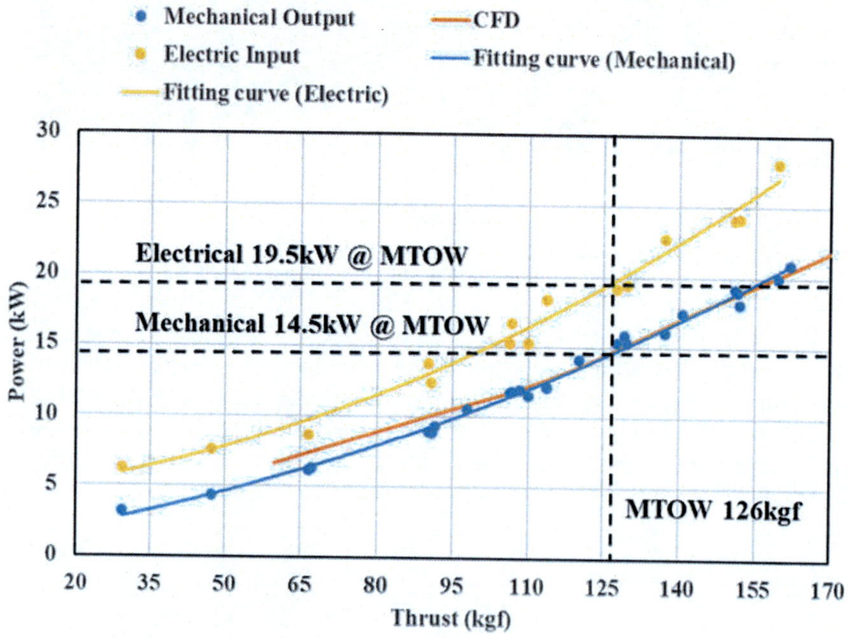

Fig. 3 Measured thrust and power of the main rotor

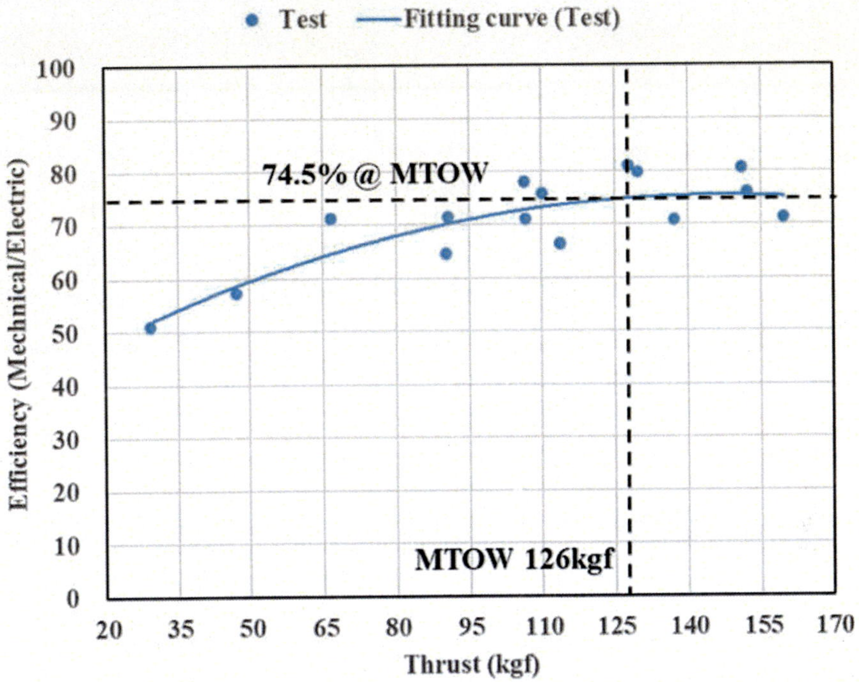

Fig. 4 The efficiency of the main rotor

ciency of the motor and the ESC (about 86%) are multiplied, resulting in an efficiency of 74.5%.

3.2 *Tail Rotor*

The tail rotor was tested at 3600–5500 RPM to determine the operating rotation speed as shown in Fig. 5. The thrust of the tail rotor was measured with the load cell, and the electric power was measured from the tail rotor ESC. In order to investigate the temperature change according to the operating RPM, the ESC board temperature is also shown.

As a result, it was confirmed that the tail rotor generates thrust that can match the torque of the main rotor even under the condition of maximum thrust of the main rotor. It is shown that the ESC temperature decreases as the tail rotor RPM increases in Fig. 5. A higher RPM means a higher voltage of the tail motor, which reduces the amount of current required for a similar level of thrust. Based on the test results, 5100RPM, which requires low power and low temperature under the maximum thrust condition of the main rotor, was selected as the tail rotor operating condition.

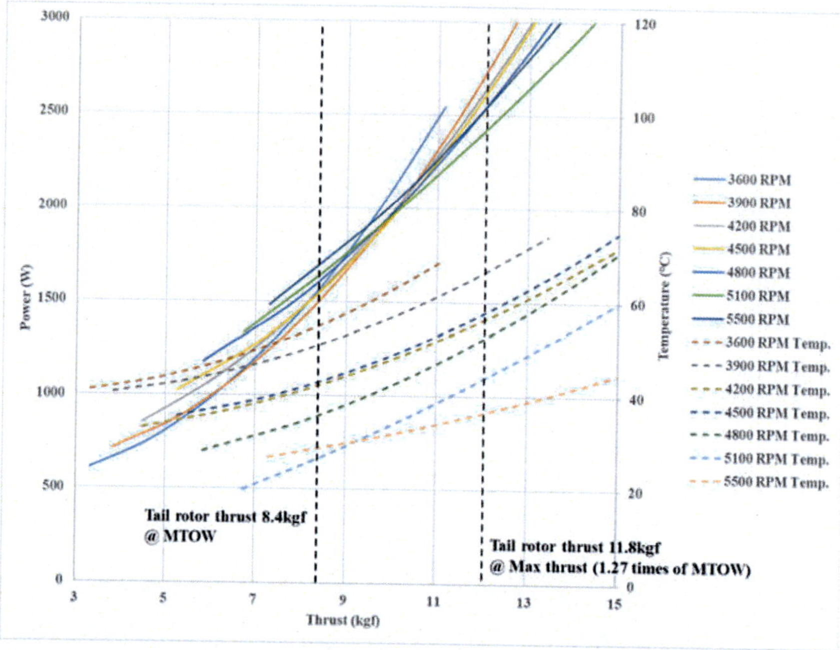

Fig. 5 Thrust, power and ESC temperature of the tail rotor

3.3 *Battery System*

The main battery BMS not only controls charging and discharging for each battery cell, but also records voltage, current, and state of charge (SOC). Figure 6 shows the information recorded in the BMS when the K-EUH was operated 4 times under similar thrust conditions. The graph of the middle section in the figure presents the battery supply power, and it can be seen that all four times are similar values, and in the graph at the bottom, it can be seen that the RPM of the main rotor is continuously 860. The upper graph shows the voltage and current changes of the battery, and the battery voltage is continuously decreasing and the current is increasing.

There are two things to note here. The first is that the difference between the voltage of the ESC (inv. in figure) and the voltage measured from the battery does not change continuously. This indicates that there is no problem in the wiring harness of K-EUH. The second is that the voltage of the battery is 390 V at the beginning but decreases to 330 V at the end. The average voltage is considered at the stages of design and analysis, but in actual operation, the voltage drop is significant at the higher discharge conditions(high C-rate). Since the voltage is proportional to the number of motor rotations, it is important to set the operating RPM in the lowest voltage condition.

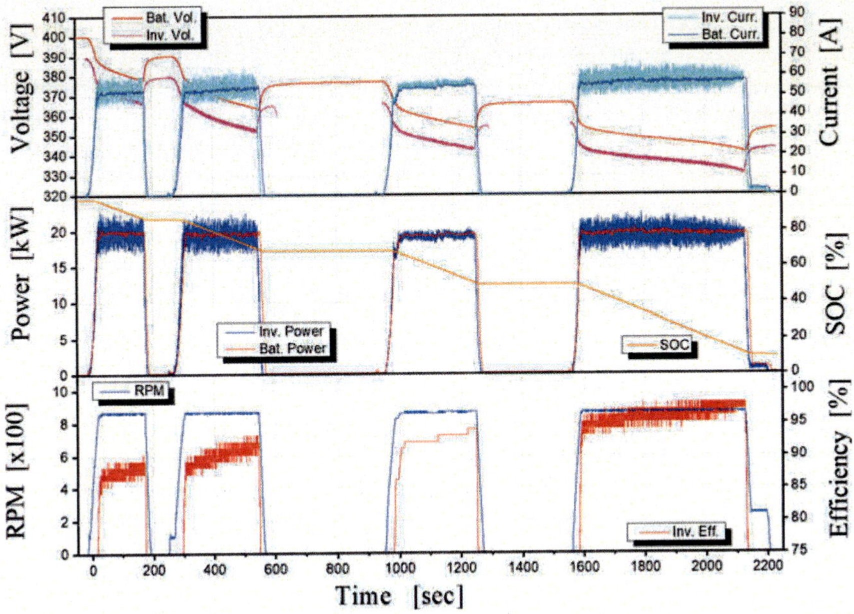

Fig. 6 Voltage, power of the main battery system

4 Conclusions

Korean Electric Unmanned Helicopter(K-EUH) has been developing to achieve the major performance target of 60 km flight distance with 20 kg payload. The first prototype was manufactured and was tested. Objectives of this ground test is to investigate whether the helicopter has sufficient performance for take-off, and to acquire know-how on the operation of the electric power system. As a result of the test, it was confirmed that the maximum thrust of the main rotor was 160 kgf (1.27 times of MTOW), and the tail rotor also confirmed to have the thrust capable of anti-torque of the main rotor. The tail rotor was tested at 3600–5500 RPM to determine the operating rotation speed. As a result of the tail rotor test, it was found that the higher the operating rotational speed, the lower the temperature of the motor and the electric speed controller. A higher RPM means a higher voltage, which reduces the amount of current required for a similar level of thrust. If the required current is low, heat generation and wire thickness (weight) are reduced, so it is advantageous to configure the electric power system with a high voltage. The other significant fact confirmed in this experiment is that the voltage drop of battery be-comes more severe when the required power is high (high C-rate). The voltage drop problem causes serious problems when helicopters land. With the battery voltage reduced while flying, the voltage decreases further as a higher C- rate is required when landing. Since such a sudden voltage drop reduces the rotating speed of the rotor, a dangerous situation may occur. In an electric power system, the reduction ratio should be determined in consideration of this voltage drop condition.

Acknowledgements This work was supported by the Technology Innovation Program (20005054, Development of Electric Powered Unmanned Helicopter Less Than MTOW 150 kg with Capabilities of 20 kg Payload and 60 km Range) funded by the Ministry of Trade, Industry and Energy (MOTIE, Korea). We would like to thank engineering staffs of Sungwoo Engineering who worked hard to support the ground test of K-EUH.

References

1. Burgeson J (2010) Sikorsky to debut first all-electric helicopter. In: Connecticut post of Bridgeport, connecticut, July 25, 2010
2. Niemiec R, Gandhi F, Lopez MJ S et al (2020) System identification and handling qualities predictions of an eVTOL urban air mobility aircraft using modern flight control methods. VFS international 76th annual forum & technology display, Virginia Beach, VA, Oct. 6–8, 2020

Aerothermodynamic Shape Optimization of a Hypersonic Lifting Body

Haoge Li, Chengrui Li, Weifang Chen, and Wenwen Zhao

Abstract Hypersonic lifting vehicles experience severe aerodynamic heating while on hypersonic flight regime during atmospheric reentry or unpowered glide. Complex flow interaction is observed in a certain range of flight conditions that results in high heat flux banded region on the windward surface, which should be taken into consideration for thermal protection designing. To obtain desirable surface heat flux distribution and aerodynamic efficiency of a lifting body, shape optimization to reduce the interaction intensity and increase the spanwise distance of the banded heating region on the windward surface was performed in this study. Both single- and multi-objective designs were developed with multiple constraints, including volume, lift-to-drag ratio, the heating rates in stagnation region, and the base section profile of the body. The free-form deformation method was employed for evaluating surface deformation followed by the transfinite interpolation method to update the computational mesh. The range of Pareto front obtained was consistent with the results of single-objective optimization. Additionally, a contradictory and non-linear correlation between the two objectives was elucidated. The optimal solutions exhibited superior comprehensive aerodynamic performance, and the aerothermodynamic shape optimization method was demonstrated to be efficient and practicable.

Keyword Aerothermodynamic optimization · Flow interaction · Multi-objective optimization · Free-form deformation · Lifting body

1 Introduction

Typical hypersonic vehicles developed in recent years present as lifting bodies, e.g., X-33, X-38, HYFLEX, and HTV-2. Design of a hypersonic lifting vehicle is an important issue for the atmospheric long endurance flight. Specifically, aerodynamic shape design is in the focus of researchers due to its impacts on the flight envelope, flight stability, maneuverability, and thermal protection system (TPS). Based on the

H. Li · C. Li · W. Chen · W. Zhao (✉)
School of Aeronautics and Astronautics, Zhejiang University, Hangzhou 310027, China
e-mail: wwzhao@zju.edu.cn

S. Lee et al. (eds.), *The Proceedings of the 2021 Asia-Pacific International Symposium on Aerospace Technology (APISAT 2021), Volume 1*, Lecture Notes in Electrical Engineering 912, https://doi.org/10.1007/978-981-19-2689-1_37

theory and design experience in engineering applications, aerodynamic configuration of lifting vehicles generally comprises geometrical characteristics, such as flat fuselage, swept-back wings, blended wing body, and blunt leading edge. However, more challenges in designing are encountered due to the growing demand for multidisciplinary design optimization of complex aerodynamic configurations under various flight conditions.

The lift-to-drag ratio (*L/D*) is the most important parameter to describe the aerodynamic efficiency, and therefore, it is a principal objective in the lifting body design. The hypersonic aircraft X-33 has been modeled and optimized for maximizing *L/D* using the Newton panel method [1]. For a typical lifting body defined using the class/shape transformation method, the *L/D* and volume efficiency have been optimized with the constraints of volume and longitudinal stability, and the competitive relationship between the objective functions and effects of constraints were revealed [2]. Besides, in the optimal design of HTV-2 for higher *L/D* [3], the free-form deformation method (FFD) of different basis functions has been studied for parameterization, and the volume of a frustum inside the body was considered as a constraint in the optimization process.

Hypersonic flow is accompanied with high temperature and enthalpy, and thus, complex flow structure and severe aerodynamic thermal environment are inevitable in design optimization. The TPS design is crucial for successful flight; however, the TPS design cycle and flexibility can be seriously restricted once the aerodynamic shape optimization (ASO) has been finalized. Therefore, it is beneficial to improve the aerodynamic efficiency and alleviate aerothermal environment simultaneously during hypersonic shape optimization. For generic lifting hypersonic vehicle configuration in the hypersonic flight regime, there are three important heating regions [4], including the nose, swept wing leading edge attachment line, and windward body centerline. The shape of the nose is closely related to the stagnation heating rate, which is usually added into the optimization as a constraint [5]. Conversely, hypersonic ASO involving heat flux distribution of the whole vehicle is relatively deficient. To decrease the heating load, Xia [6] performed lifting body shape optimization to delay the boundary layer transition onset through incorporating *L/D* and also the volume. Without the fluence of turbulence, an excessive local heating load was observed on/near the windward surface centerline of a blunt-nosed 2:1 elliptic cone in certain range of incidence [7]. The evidently high heating regions were also observed in hypersonic lifting vehicles, which poses a challenge on the TPS design. In this study, aerothermodynamic shape optimization for a lifting vehicle was formulated and performed to alleviate the local exorbitant heating load and maintain the aerodynamic efficiency.

The main novelty of this study is that the TPS-oriented aerothermal and aerodynamic lifting body optimization were performed simultaneously for the first time, and geometry parameterization based on the FFD technique was also applied based on the geometry characteristics and flow analysis.

2 Problem Formulation

The hypersonic lifting vehicle chosen as the baseline exhibits a flat and smooth windward surface, which provides most lift force as the windward surface is configured at an angle of attack. The length and maximum width of the body are 4.0 m and 1.825 m, respectively. Aerothermodynamic characteristics of the baseline were analyzed on Mach 17 with an angle of attack of 8° at 50 km of height.

The three-dimensional N-S equations were solved using the finite volume method on a multi-block structured mesh. Inviscid fluxes were discretized using the AUSMPW + method, and viscid fluxes were centrally discretized to second order. Implicit time integration with multi-grid acceleration was applied to obtain the steady-state solutions. Due to symmetrical flow, a half model was simulated with a symmetry boundary condition at the meridian plane. The computational domain, of which the surface mesh is shown in Fig. 1, consisted of 13 blocks, resulting in a total mesh size of approximately 1.3 million. The mesh was algebraically clustered at the surface to perform viscous calculations, and mesh independence was also guaranteed.

2.1 *Flowfield Analysis*

Stagnation regions of the nose and leading edge enduring severe aerodynamic heating as expected are depicted in Fig. 2. The flow pattern over the nose and leading edge was also sketched, and flow reattachment could be distinctly observed. The reattachment phenomenon located at about x = 0.084 m resulted from the interaction of the nose shock wave and the shock wave caused by the swept wing leading edge. Consequently, as presented in Fig. 3, local high pressure and high heating rates were profound near this region. The high energy flow moved downstream, which formed a high heat flux banded region near the centerline.

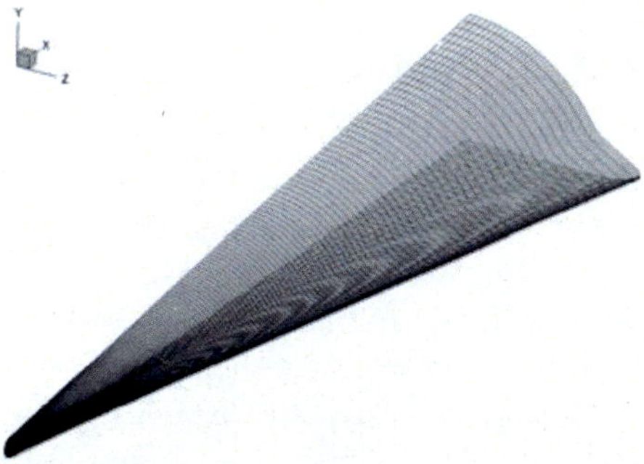

Fig. 1 Computational surface mesh

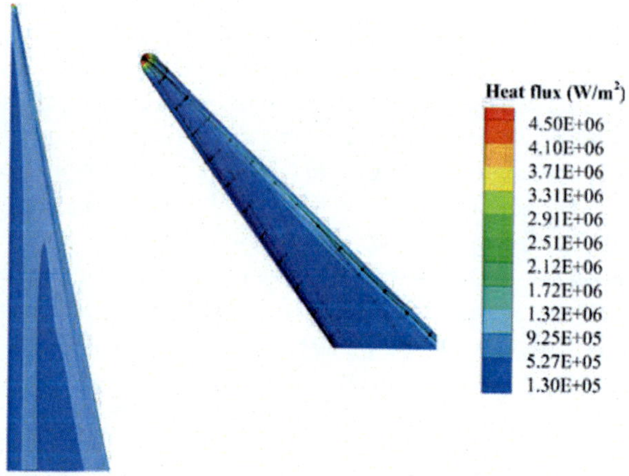

Fig. 2 Heat flux distribution on the windward side and flow reattachment line

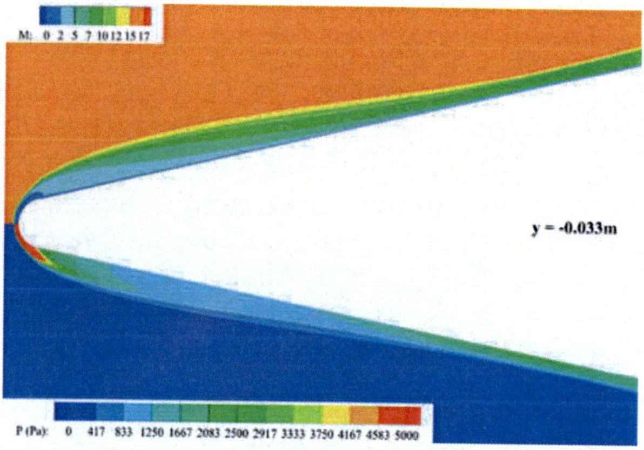

Fig. 3 Zoom-in view of Mach contour and pressure distribution on the section of y = −0.033 m

2.2 *Description of ASO Problem*

In terms of the TPS design of the unpowered glide vehicle, low intensity of aerodynamic heating near the windward surface centerline was desirable, where the flow interaction and the banded heating were required to be weakened, besides, the banded heating zones needed to be as far away from the windward centerline as possible. Improvements in the aerothermal environment should not be at the cost of compromising on aerodynamic efficiency in the hypersonic flight regime. Due to the demand of fuselage structure and inner loading, the volume (V) size was considered as well. In this study, we aimed to optimize the shape of the windward side, and the ASO problem is described as follows.

- The single/multi-objective functions
  - Minimization of the heating flux of the banded heating region
  - Maximization of the spanwise distance of the banded heating region from the windward surface centerline
- The constraints
  - $Q_s \leq 1.1Q_{s,baseline}$
  - $V \geq V_{baseline}$
  - $L/D \geq (L/D)_{baseline} - 0.1$
  - The unchanged cross-section at x = 4.0 m

In the above lists, Q_s represents the heating flux at the stagnation point of the optimal configuration.

2.3 *Optimization Framework*

Surrogate-assisted optimization framework was constructed to reduce the number of precise simulations. Based on the sampling plan, the radial basis function neural network was trained via cross-validation because this model is adept for modeling complex landscapes. For the single-objective optimization of the lifting body, the Multi-island Genetic Algorithm was utilized as the optimizer, with a generation of 100, and a population size and island number of 30. To investigate the relationship between the objective functions and identify the optima that possesses comprehensive aerothermodynamic performance, the non-dominated sorting genetic algorithm II was then employed to perform the optimizations.

3 FFD Modeling of the Lifting Body

Parameterization is a critical step in the optimization process, since the parameterization method correlates with the design space, dimensionality, and success of design optimization. The FFD technique [8] was employed for the lifting body optimization because of its superiority in terms of surface deformation flexibility and surface smoothness. After the surface was deformed in sample plan, the computational domain mesh was updated using the transfinite interpolation method to facilitate the design efficiency.

The FFD technique takes a control lattice to contain the shape to be deformed, and the mapping relationship between the surface mesh points and control points is created. If the tri-variate Bernstein polynomial basis function was used, the surface point coordinates could be calculated using the following equation:

$$X = \sum_{i=0}^{l} C_l^i (1-s)^{l-i} s^i \sum_{j=0}^{m} C_m^j (1-t)^{m-j} t^j \sum_{k=0}^{n} C_n^k (1-u)^{n-k} u^k \boldsymbol{P}_{i,j,k} \quad (1)$$

where, s, t, and u indicate the local coordinates, and $(l + 1)$, $(m + 1)$, and $(n + 1)$ are the number of control points in three coordinate directions.

The windward surface of the lifting body was embedded into the control lattice, as shown in Fig. 4. As an irregular FFD lattice, the control surface was found to be close to the windward surface and swept wing leading edge to guarantee deformation sensitivity. Control point manipulation was constrained to the y axis, such that a negative change in y coordinates would result in an increase in local thickness. The active control points (denoted by blue spots) and fixed ones (denoted by red squares) are presented in Fig. 4 along with the sequences $(i, 0, k)$ of the active lattice. Movements of all active control points can be described by 10 variables y1–y10, as shown in Table 1.

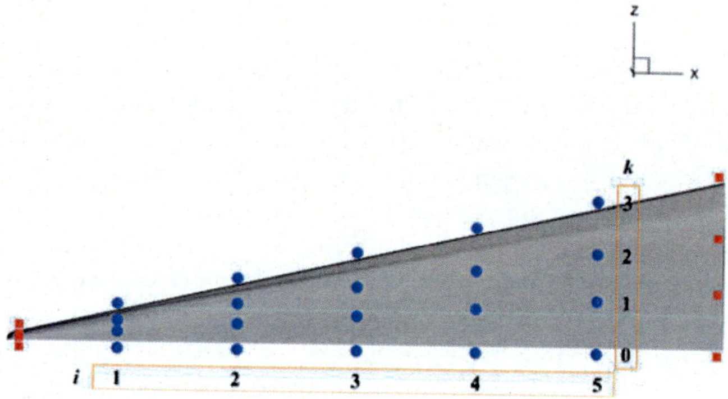

Fig. 4 Control points on the windward surface of the lifting body

Table 1 Sequences of the control points on FFD lattice ($j = 0$)

k	i				
	1	2	3	4	5
0	y1	y2	y3	y4	y5
1	1.5y1	1.5y2	1.5y3	1.5y4	1.5y5
2	– 1.1y1	– 1.1y2	– 1.1y3	– 1.1y4	– 1.1y5
3	y6	y7	y8	y9	y10

4 Results and Discussion

Here, we adopted the averaged value of the maximum heating rates, and averaged location of the banded heating zone on four cross-sections to quantify the objectives described above, which are represented by "*Qa*" and "*za*" respectively. A comparison of the heat flux on the four cross-sections of the lifting body is plotted in Fig. 5.

4.1 *Optimization of* Qa *Minimization*

As presented in Table 2, the surrogate model was accurate for optimization, since the maximum error was less than 0.07. The optimization within the constrained bounds in ten-variable design space significantly weakened the banded heating zone, especially on the posterior segment of the optimal shape, as shown in Fig. 6. It can be observed in Table 3 that the constraints were satisfied, besides, the averaged location of the banded heating zone increased by 8.3%. The *Qa* decreased by about 7.9% albeit the aerodynamic heating near the swept wing leading edge exhibited no perceptible change.

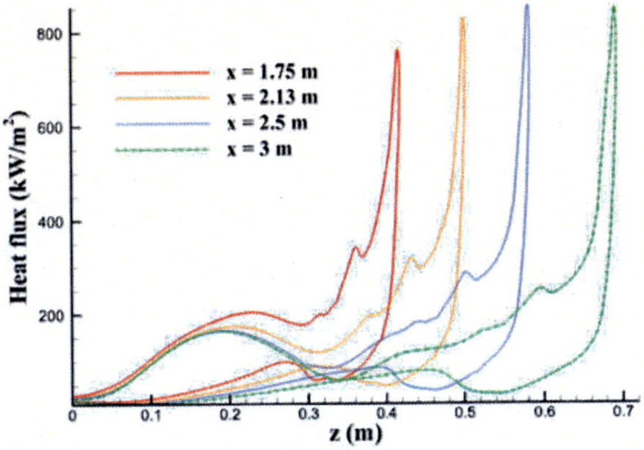

Fig. 5 Comparison of the heat flux on four cross-sections

Table 2 Surrogate estimation of *Qa* minimization

	Average error	Maximum error
L/D	0.0123	0.0394
Q_s	0.0021	0.0070
Qa	0.0080	0.0657

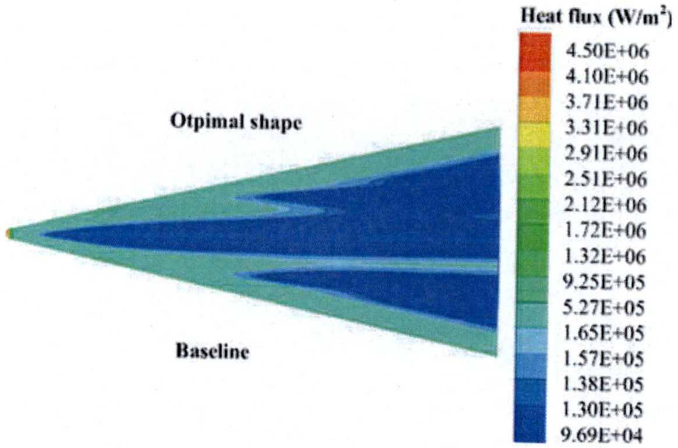

Fig. 6 Heat flux contour on the windward surface of *Qa* minimization

Table 3 Optimization results of *Qa* minimization

	Baseline	Optimization
L/D	2.690	2.685
Q_s (kW/m^2)	594.1	594.8
V (m^3)	0.371	0.377
Qa (kW/m^2)	177.1	163.1
za (m)	0.204	0.221

4.2 Optimization of Za Maximization

Upon comparison, the surrogates of spanwise distance of banded heating zone exhibited relatively larger errors, as presented in Table 4. However, it was still considered accurate enough to perform the optimization, as the maximum error was below 0.3, which is a common criterion for surrogate estimation. The windward aerodynamic heating of the optimal geometry, as depicted in Fig. 7, was similar to that of the optimized shape for *Qa* minimization. Nevertheless, *za* shown in Table 5 increased by nearly 20% with all constraints satisfied. We noted that the windward heat flux, both at stagnation point and banded heating region, increased slightly, which suggested the existence of a contradictory correlation between *za* and *Qa*.

Table 4 Surrogate estimation of *za* maximization

	Average error	Maximum error
L/D	0.0110	0.0404
Q_s	0.0022	0.0084
za	0.0744	0.2284

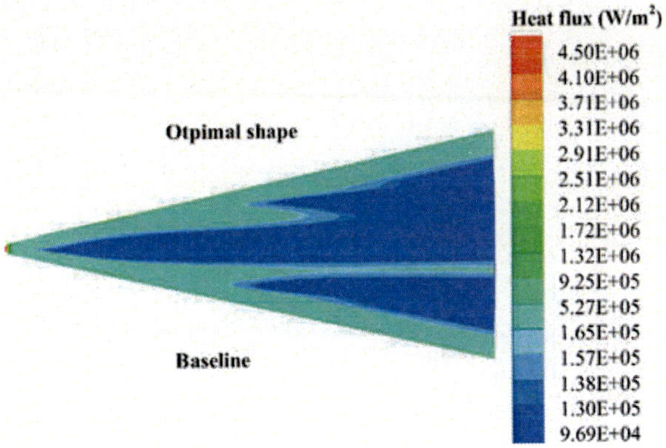

Fig. 7 Heat flux contour on the windward surface of *za* maximization

Table 5 Optimization results of *za* maximization

	Baseline	Optimization
L/D	2.690	2.694
Q_s (kW/m^2)	594.1	599.2
V (m^3)	0.371	0.384
Qa (kW/m^2)	177.1	177.7
za (m)	0.204	0.244

4.3 Multi-Objective Optimization

To investigate the correlation between the intensity and location of the banded heating region on the windward side and further identify the optimal shape to achieve high aerodynamic efficiency and preferable heating distribution, multi-objective shape optimization of the lifting body was performed. Pareto front (PF), as shown in Fig. 8, demonstrated a contradictory relation between the two objectives and, meanwhile, revealed a non-linear trend.

As shown in Fig. 8, if the spanwise location of the banded heating region is not taken into consideration, the heat flux at average can be achieved as low as approximately 163 kW/m^2. Similarly, if the average of the *z* coordinate of the high heating region is considered as the only objective, *za** of 1.223 could be obtained. Here, we introduce *za**, which represents the normalized *za* with the baseline value for clarity of illustration. The range of PF is consistent with the results obtained in the previous single-objective design.

Three typical feasible solutions from the PF, named A, B, and C, were simulated and analyzed. The windward side geometry has been compared in Fig. 9 on the cross-section at x = 1.75 m. Tabulated in Table 6 is the comparison between the baseline and typical solutions. It is evident that the parameterization in this study, which thickens the body to some extent, can push the banded region toward the

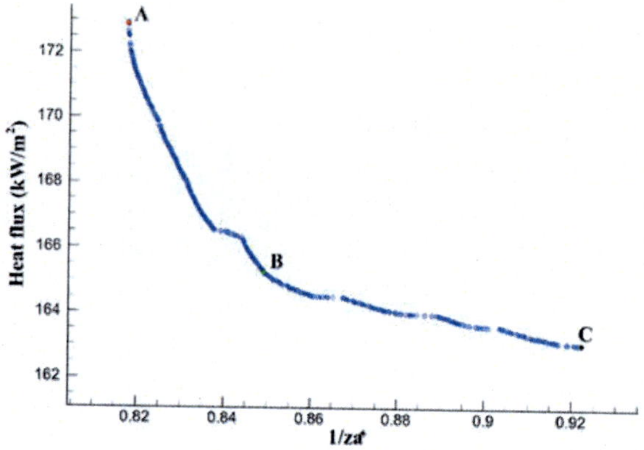

Fig. 8 Pareto front of the heat flux value and normalized *za*

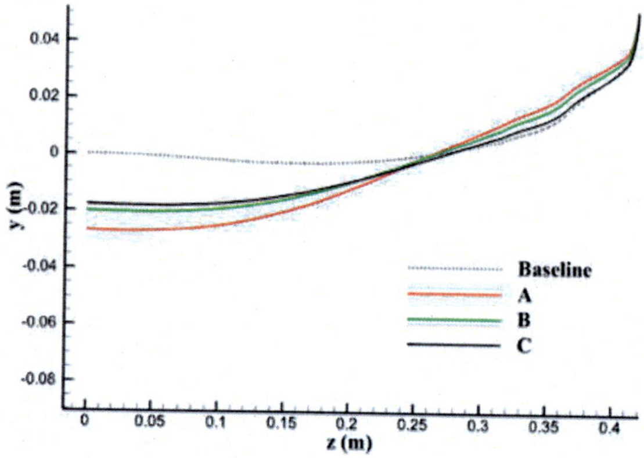

Fig. 9 Comparison of the windward surface geometry on the cross-section at x = 1.75 m

Table 6 Multi-objective optimization results

	Baseline	A	B	C
L/D	2.690	2.715	2.691	2.683
Q_s (kW/m^2)	594.1	599.2	599.9	595.5
V (m^3)	0.371	0.383	0.378	0.377
Qa (kW/m^2)	177.1	177.3	169.5	163.8
za (m)	0.204	0.244	0.240	0.221

wing leading edge effectively. Moreover, the design to reduce the heating rates on the windward surface can be accessible, through slightly increasing the stagnation region heat flux and maintaining L/D. Within the limited ten-variable design space, the optimized results revealed thickened surface and shrinkage near the leading edge.

5 Conclusion

In this study, single- and multi-objective aerothermodynamic shape optimization with multiple constraints of a hypersonic lifting body has been presented. Based on the flow interaction analysis, the TPS-oriented shape design was formulated and resolved to reduce the intensity of the banded heating zone on the windward surface and move the region away from the centerline. The results indicated that the parametrization using the FFD technique and the surrogate-assisted optimization framework are efficient and practicable. Optimal shapes with superior comprehensive performance have been achieved with the constraints of *L*/*D*, volume, heating rate in the stagnation region, and bottom section profile. The aerodynamic simulation methods, which take the high-temperature effects into consideration, could be beneficial to facilitate the reliability of the optimization of hypersonic vehicles. Additionally, the relaxation in design space is expected to further improve the optimization results.

Acknowledgements This work was supported by the National Numerical Wind tunnel project (Grant No. NNW2019-ZT3A08 and No. NNW2018-ZT1B01).

References

1. Su H, Gu L, Gong C (2015) Research on geometry modeling method based on three-dimensional CST parameterization technology. In: 16th AIAA/ISSMO multidisciplinary analysis and optimization conference. American Institute of Aeronautics and Astronautics, Dallas, TX
2. Ma Y, Yang T, Feng Z, Zhang Q (2015) Hypersonic lifting body aerodynamic shape optimization based on the multiobjective evolutionary algorithm based on decomposition. Proc Inst Mech Eng, Part G 229(7):1246–1266
3. Zhang B, Feng Z, Xu B, Yang T (2019) Efficient aerodynamic shape optimization of the hypersonic lifting body based on free form deformation technique. IEEE Access 7:147991–148003
4. Inger GR (1995) Non-equilibrium boundary layer effects on the aerodynamic heating of hypersonic vehicles. Acta Astronaut 36(4):205–216
5. Eyi S, Hanquist KM, Boyd ID (2019) Shape optimization of reentry vehicles to minimize heat loading. J Thermophys Heat Transfer 33(3):785–796
6. Xia C, Tao Y, Jiang T, Chen W (2016) Multiobjective shape optimization of a hypersonic lifting body using a correlation-based transition model. Proc Inst Mech Eng Part G 230(12):2220–2232
7. Atkinson M, Poggie J, Camberos J (2012) Hypersonic flow computations for an elliptic cone at high angle of incidence. J Spacecr Rockets 49(3):496–506
8. Sederberg TW, Parry SR (1986) Free-form deformation of solid geometric models. In: Proceedings of the 13th annual conference on computer graphics and interactive techniques (SIGGRAPH '86). Association for Computing Machinery New York, USA

Experimental Studies on Behavior of Laminar Separation Bubble Formed on the Hofsass Espada Airfoil Near Stall

Masashi Kawakami and Kenichi Rinoie

Abstract For micro air vehicles development, it is necessary to investigate the aerodynamic characteristics at low Reynolds numbers. Hofsass Espada airfoil has been developed, and its aerodynamic characteristics at low Reynolds numbers have been studied. This study uses wind tunnel experiments to understand the flow field around the Hofsass Espada airfoil near the stall with a laminar separation bubble. In this study, simultaneous flow visualization and time-resolved surface pressure measurements were conducted. Based on the chord length, the Reynolds number is 6.5×10^4. At the angle of attack near stall, flow around Hofsass Espada airfoil switches between the formation of a laminar separation bubble and a large separated region over the airfoil after the bubble burst. This behavior is similar to that of a relatively thick airfoil, such as the NACA0012. Owing to the smaller leading-edge radius than the NACA0012 airfoil, quasi-periodic switching behavior of flow does not exist.

Keywords Laminar separation bubble · Airfoil stall · Low reynolds number

1 Introduction

Recently, micro air vehicles (MAVs) have attracted attention because of advances in electronic devices' miniaturization, weight reduction, and performance [1]. Aircraft, which are smaller and fly at slower speeds, fly in the low Reynolds number, around 10^4. Low Reynolds number conditions have distinct aerodynamic characteristics than high Reynolds number conditions found in conventional aircraft [2]. To improve the aerodynamic performance of MAVs, it is important to understand the flow field around the airfoil at low Reynolds number conditions.

In the field of model airplanes, airfoils with high aerodynamic performance at low Reynolds numbers comparable to MAVs have been developed through competitions. In particular, the Hofsass Espada airfoil has a proven track record in model airplane competitions [3]. At low Reynolds numbers, the Hofsass Espada airfoil has a higher

M. Kawakami (✉) · K. Rinoie
Department of Aeronautics and Astronautics, The University of Tokyo, Tokyo, Japan
e-mail: atc2mk12@gmail.com

S. Lee et al. (eds.), *The Proceedings of the 2021 Asia-Pacific International Symposium on Aerospace Technology (APISAT 2021), Volume 1*, Lecture Notes in Electrical Engineering 912, https://doi.org/10.1007/978-981-19-2689-1_38

lift and lower drag than the NACA0012 airfoil and the 4% camber airfoil through surface pressure and force measurements [4, 5]. At a low angle of attack, which corresponds to the MAV flight state, the Hofsass Espada airfoil is considered aerodynamically superior because, unlike other airfoils, it can increase the lift without forming bubbles near the leading edge of the airfoil [4].

A laminar boundary layer is separated from the airfoil surface in a flow field around the airfoil with a low Reynolds number, but it can reattach to the airfoil after a transition to turbulent flow. A laminar separation bubble is a region between separation and reattachment. There are two types of laminar separation bubbles: short bubbles, whose chordwise length decreases as the angle of attack increases, and long bubbles, whose length increases. The bubble's behavior has a significant effect on the airfoil's stall characteristics [6].

The flow around an airfoil with a short bubble shows quasi-periodic behavior at the angle of attack near the stall. When the angle of attack is near stall angle, the flow around NACA0012 airfoil oscillates between an attached flow with the laminar separation bubble created near the leading-edge and a largely separated flow with Reynolds number, $Re_c = 1.3 \times 10^5$ based on the airfoil chord length. A large-scale vortical structure is released from the leading edge when the bubble is broken down and reformed [7–9].

This study uses wind tunnel experiments to better understand the flow field around the Hofsass Espada airfoil with a laminar separation bubble near the airfoil stall. In this study, simultaneous flow visualization and time-resolved surface pressure measurements were conducted on the Hofsass Espada airfoil and the experimental results were compared with those of NACA0012 airfoil, which had previously been researched.

2 Experimental Details

In this chapter, the experimental setups for surface pressure measurement and flow visualization are discussed in detail.

2.1 Wind Tunnel

A low-speed suction-type wind tunnel was used in the measurements. The wind tunnel's test section size was 600 mm in height, 200 mm in width, and 1000 mm in length. The freestream velocity, U_∞ was 10 m/s, and the tunnel's freestream turbulence intensity was less than 0.16%. The experimental setup is shown in Fig. 1.

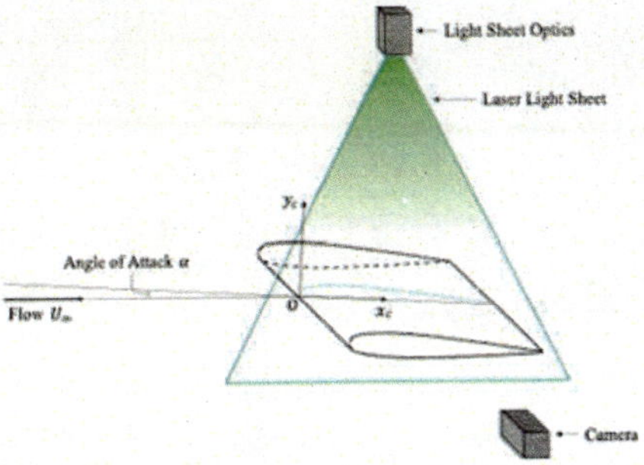

Fig. 1 Schematic diagram of the experimental setup

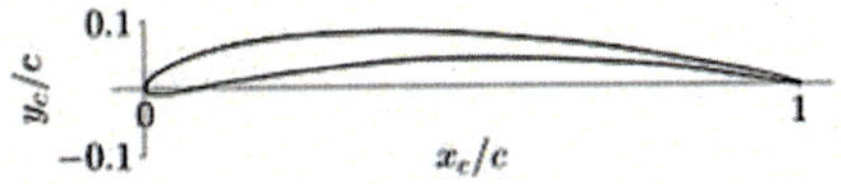

Fig. 2 Hofsass Espada airfoil and coordination system

2.2 *Airfoil Model*

A metallic Hofsass Espada airfoil model, which was mounted on a rotary table in the wind tunnel, was employed. Figure 2 shows the shape of the airfoil. The model's chord length, c was 100 mm, and the Reynolds number, Re_c based on that length was 6.5×10^4. It has a wingspan of 200 mm, a camber ratio of 6%, and a maximum wing thickness ratio of 5.6% at 17% chord length. For pressure distribution measurement, the airfoil model has 44 static pressure holes on the upper and lower surfaces near the center of the span direction.

2.3 *Pressure Measurement System*

Surface pressure measurements were conducted using 21 surface pressure holes on the airfoil's upper surface at a sampling frequency of 1,000 Hz. The pressure transducer (Honeywell pressure transducer DCAL 401 DN, differential pressure type, Range 1 in. H_2O) was connected to the A/D converter and supplied to a PC via a USB serial transducer. This pressure transducer is installed outside the wind tunnel and is connected to the model's surface pressure holes using a pressure tube. It has been confirmed that the surface pressure's characteristics at different positions can be obtained at the same time for the frequency oscillations of several tens of hertz by multipoint simultaneous pressure measurement method using this device [10].

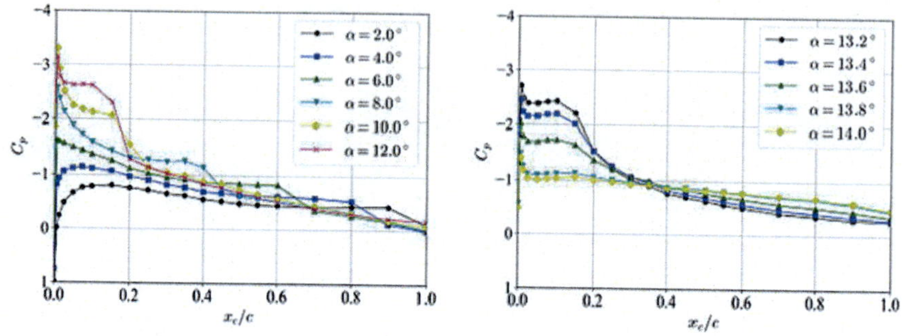

(a) $\alpha = 2.0°- 12.0°$ (b) $\alpha = 13.2°- 14.0°$

Fig. 3 Mean pressure distribution

2.4 *Flow Visualization*

In order to investigate the behavior of the flow field near the stall angle of attack in detail, flow visualization experiments were conducted using a high-speed camera (Dantec Speed-Sense M310) and a Nd:Yag laser light source. A timer box (Dantec 80N77) connects the camera and laser to a PC running Dantec Dynamic Studio, a PIV processing software. Laser irradiation and image recording were performed using Dynamic Studio. For visualization, atomized ondina oil produced by an atomizer was used as tracer particles.

A cylindrical lens was added to enlarge the laser sheet because the laser light sheet lens system alone irradiated a narrower area than required for observation. The image acquisition interval was set to 1000 Hz to synchronize with the pressure measurement, and the measurement lasted for 4 s. The multipoint pressure measurement and visualization were performed simultaneously. However, since it was impossible to automatically synchronize the multipoint pressure measurement and visualization recording, an LED light was attached to the wind tunnel's external wall. While the visualization image was being recorded, a trigger generator output a square wave of 0.5 Hz to the LED light and the pressure measurement system, and the output of the trigger was measured simultaneously with the multipoint pressure measurement in order to synchronize the multipoint pressure measurement result and the visualized image.

3 Results of the Experiments

3.1 Pressure Measurement

Figure 3 shows the mean pressure distribution at each angle of attack. The pressure is expressed as a non-dimensionalized pressure coefficient at freestream dynamic pressure, and the graph's vertical axis shows low pressure upward. In Fig. 3a, the suction peak, which is a strong low-pressure region near the leading edge, becomes stronger as the angle of attack increases. However, a large positive pressure gradient, where the bubble's reattachment point appears to exist, exists at $x_c/c = 1.0$ for $\alpha = 2.0°$, but advances to $x_c/c = 0.2$ for $\alpha = 12.0°$. This indicates the formation of a short bubble [6], whose chordwise length decreases as the angle of attack increases. Figure 3b shows the mean pressure distribution at the angle of attack near the stall of the Hofsass Espada airfoil. The strong suction peak near the leading edge, which is maintained until $\alpha = 13.2°$, is gradually lost from $\alpha = 13.4°$ to $\alpha = 13.8°$. At $\alpha = 14.0°$, the pressure distribution over the entire airfoil is flat, indicating that the airfoil is stalled. This pattern of behavior with increasing angle of attack is also reported in reference [4]. In Ref. [4], the airfoil stalls in the range of $\alpha = 12.5°$ to $\alpha = 13.0°$ at $Re_c = 6.7 \times 10^4$, showing a qualitatively similar mean pressure distribution. The difference in the stall angle of attack between this study and Ref. [4] is primarily due to the difference in the wind tunnel used.

Figure 4 shows the time history of the surface pressure near the stall angle of attack, confirming the flow field's unsteadiness. At $\alpha = 13.4°$, the pressure near the leading edge remains low for almost all the time, indicating that the airfoil is in a pre-stall state. At $\alpha = 13.6°$, the time it takes for the suction peak to be lost near the leading edge increases. Overall, the time when the suction peak is lost is irregular, but there are times when it shows quasi-periodic behavior, such as at $t = 2 - 4$ [s]. The frequency of this quasi-periodic behavior is approximately 3 Hz, as shown in Fig. 4b. At $\alpha = 13.8°$, the loss of the suction peak increases further, and the peak is no longer maintained, but the quasi-periodic behavior can still be observed at $t = 7 - 8$ [s]. In Fig. 4c, the frequency is slightly higher, approximately 3–4 Hz.

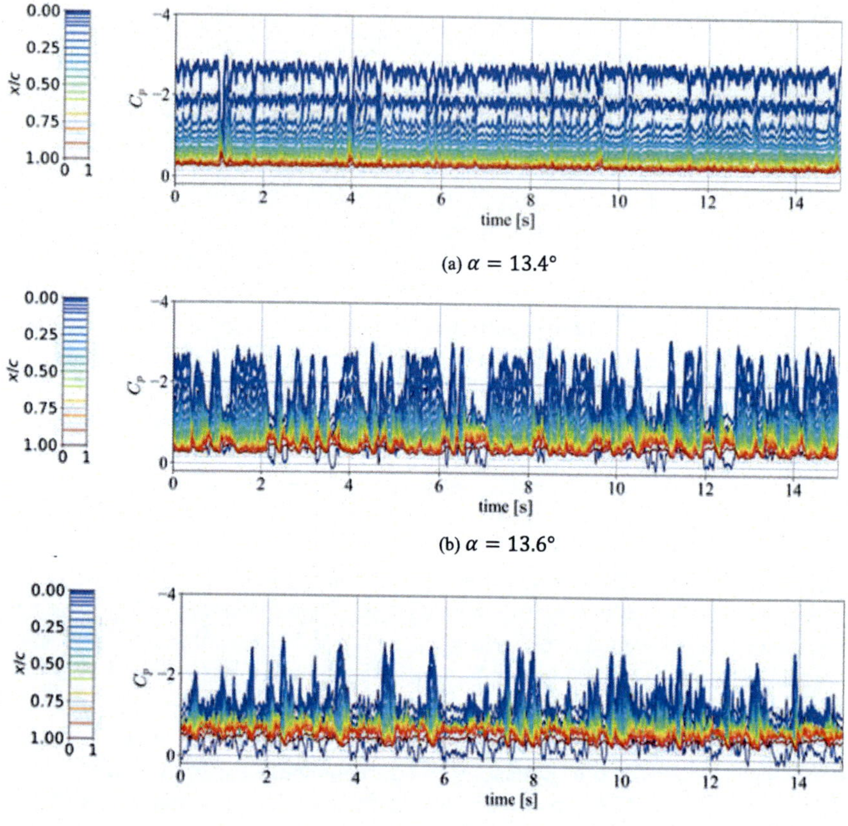

(a) $\alpha = 13.4°$

(b) $\alpha = 13.6°$

(c) $\alpha = 13.8°$

Fig. 4 Time-resolved pressure coefficient

Figure 5 shows the power spectral density (PSD) distribution of the pressure–time history at $\alpha = 13.4°$. The PSD near the leading edge shows a moderate peak at 6–7 Hz. However, the PSD near the trailing edge shows relatively sharp peaks around 25, 50, and 75 Hz.

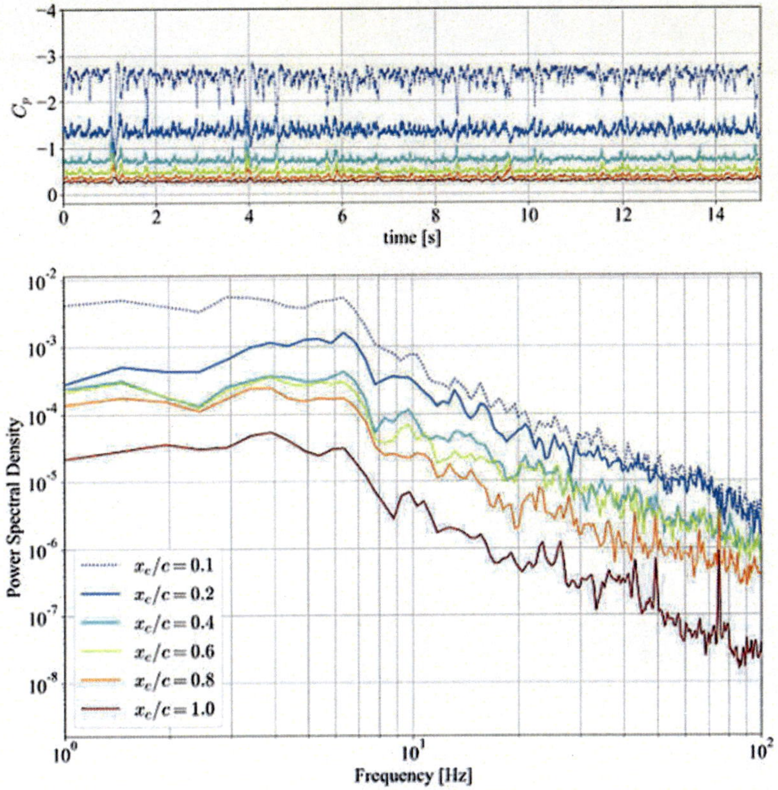

Fig. 5 Power spectral density distribution of pressure fluctuation

3.2 *Flow Visualization*

In this section, the flow visualization is presented. At the angle of attack near the stall, a switching behavior between largely separated flow and attached flow was observed. Two types of visualizations of identical behavior observed at different times during the transition from separated to attached flow are illustrated, one for the entire airfoil (Fig. 6) and the other for the area around the leading edge (Fig. 7).

Figure 6 shows flow visualization around the airfoil at $\alpha = 13.6°$. The flow visualization is on the left, the pressure distribution on the airfoil surface at that time is on the lower right, and the time in the pressure–time history is on the upper right. In Fig. 6a, the flow is separated from the airfoil. In the separated shear layer, small vortical structures come together and a series of relatively large vortical structures emerge near the leading edge. At $x_c/c = 0.3$, a large vortex structure exists near the airfoil surface, and surface pressure fluctuation is observed there. In Fig. 6b, the vortex at $x_c/c = 0.3$ in (a) moves to $x_c/c = 0.6$, and the flow behind the vortex is closer to the airfoil surface. In Fig. 6c, the vortex has moved to $x_c/c = 1.0$. In

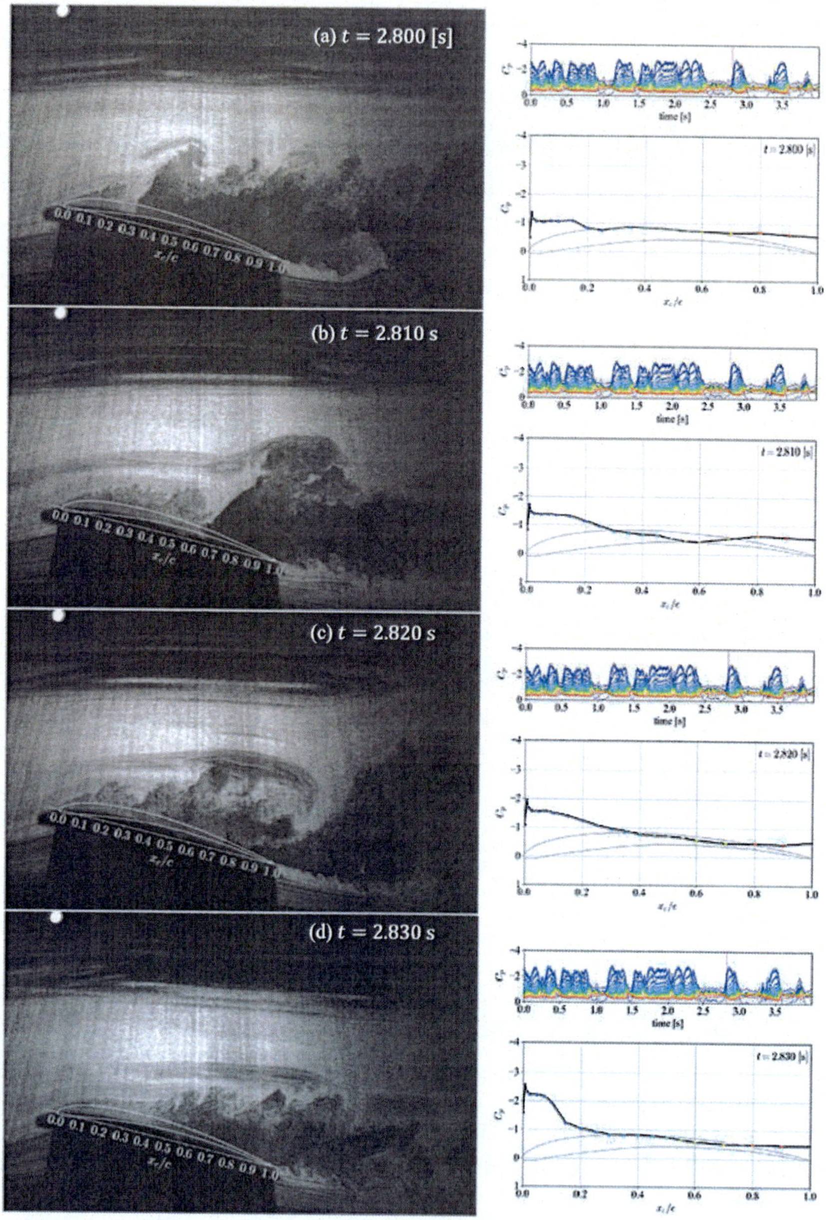

Fig. 6 Flow visualization and surface pressure distribution at $\boldsymbol{\alpha = 13.6^\circ}$

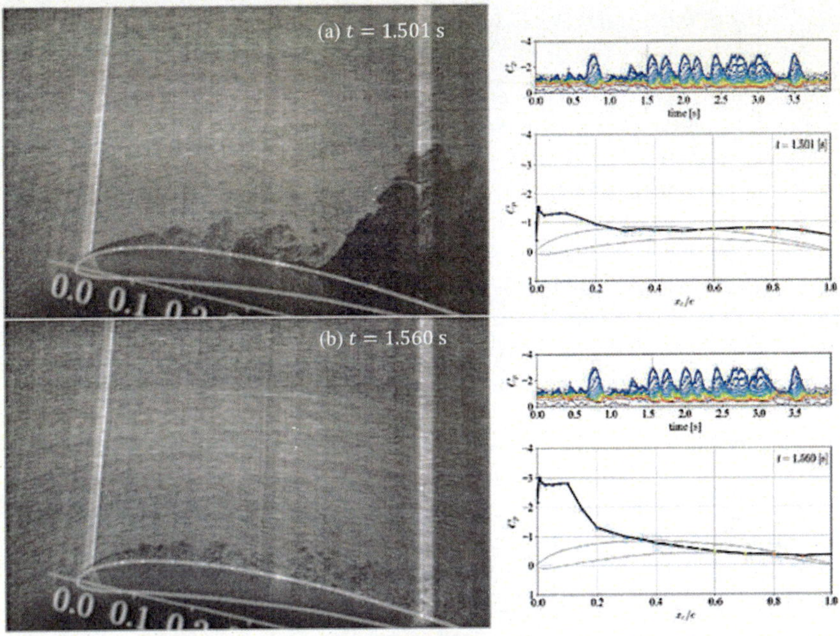

Fig. 7 Flow visualization near the leading edge and surface pressure distribution at $\alpha = 13.6°$

Fig. 6d, the large vortex structure flows across the airfoil, and the flow attached to the surface with a bubble is observed near the leading edge. The pressure distribution shows a suction peak.

Figure 7 shows a similar visualization near the leading edge of a transition from a separated flow to an attached flow at $\alpha = 13.6°$. In Fig. 7a, a large vortex developed in the separated shear layer near the leading edge exists at $x_c/c = 0.3$. After the passage of the vortex, the freestream is directed toward the surface. In addition, in Fig. 7b, a significant suction peak with a bubble is formed around the airfoil. In the visualized image, the separated shear layer transitions to turbulence around $x_c/c = 0.1$, and reattachment is observed around $x_c/c = 0.2$. This corresponds to the flat pressure distribution behind the suction peak and a strong positive pressure gradient from $x_c/c = 0.1$ to $x_c/c = 0.2$ in the pressure distribution at that time.

3.3 Comparison with NACA0012

It has been reported that a quasi-periodic behavior is observed for the NACA0012 airfoil at the angle of attack near the stall, where the flow switches stably between attached and largely separated flow [7]. The frequency of this switching behavior depends on the angle of attack, and this quasi-periodic behavior is said to be a part of it. In addition, when the flow changes, the leading edge emits a large vortical structure [8, 9]. This phenomenon has been replicated in numerical simulations, and in a certain range of angles of attack, quasi-periodic behavior with burst and formation of a laminar separation bubble has been observed [11].

In this study, the switching behavior of the flow with burst and formation of a laminar separation bubble was observed at the angle of attack near the stall for the Hofsass Espada airfoil, and it was discovered that a large vortical structure was emitted from the leading edge. However, at $\alpha = 13.6°$ as shown in Fig. 4, the emergence of switching was unsteady and no stable quasi-periodic behavior was observed.

The presence of a laminar separation bubble is also important in a dynamic stall, and the effect of the leading edge geometry has been investigated [12]. The larger the leading-edge radius, which affects the flow's pressure gradient, the more bubble burst can be suppressed. In this study, the Hofsass Espada airfoil is thinner and has a smaller leading-edge radius than the NACA0012 airfoil. Therefore, the bubble reacts more sensitively to changes in the angle of attack near the stall, and the range of angle of attack where the quasi-periodic behavior appears is narrow. This phenomenon is thought to be related to the leading-edge-type airfoil stall, which shows a sudden decrease in lift.

4 Conclusions

Surface pressure measurements and flow visualization were conducted to investigate the behavior of flow around Hofsass Espada airfoil at the angle of attack near the stall. The Reynolds number based on the airfoil chord length is $Re_c = 6.5 \times 10^4$. The following conclusions were obtained.

1. At the angle of attack near the stall, flow around Hofsass Espada airfoil changes between a pressure distribution with a suction peak accompanied by the generation of a bubble and a flat pressure distribution after the bubble bursts. The frequency of this changing behavior varies with the angle of attack, and the larger the angle of attack, the less probable the bubble will form.
2. When the flow changes, a large vortical structure is emitted from near the leading edge, similar to a relatively thick airfoil, such as NACA0012 airfoil.
3. The quasi-periodic changing behavior of flow around Hofsass Espada airfoil near the stall does not exist stably because the leading edge radius is small and a small change in the angle of attack has a significant effect on the flow.

References

1. Elbanhawi M, Mohamed A, Clothier R (2017) Enabling technologies for autonomous MAV operations. Prog Aerosp Sci 91:27–52
2. Mueller TJ, DeLaurier JD (2001) An overview of micro air vehicle aerodynamics. Prog Astronaut Aeronaut 195:1–10
3. Wakamatsu T, Koike M (2009) Effect of turbulators of wing at low Reynolds numbers. Proceedings of the 41st JSASS fluid dynamics conference, JSASS-2009–0118 (in Japanese)
4. Nakamura Y, Rinoie K, Sunada Y (2010) Experimental studies on low speed aerodynamic characteristics of Hofsass Espada airfoil Part 1 surface pressure distributions. Proceedings of the 42nd JSASS fluid dynamics conference, JSASS-2010–2072 (in Japanese)
5. Tanaka H, Nakamura Y, Sunada Y, Rinoie K (2011) Experimental studies on low speed aerodynamic characteristics of Hofsass Espada airfoil Part 2 force measurements. Proceedings of the 43rd JSASS fluid dynamics conference, JSASS-2011–2103 (in Japanese)
6. Tani I (1964) Low-speed flows involving bubble separations. Prog Aerosp Sci 5:70–103
7. Rinoie K, Takemura N (2004) Oscillating behaviour of laminar separation bubble formed on an aerofoil near stall. Aeronaut J 108:153–163
8. Tanaka H (2004) Flow visualization and PIV measurements of laminar separation bubble oscillating at low frequency on an airfoil near stall. 24th congress of international council of the aeronautical sciences, p ICAS-2004–2.10.4
9. Fujiwara G, Sunada Y, Rinoie K (2018) Vortex shedding from laminar separation bubble undergoing oscillating behavior near airfoil stall. AIAA Sci Tech Forum. AIAA-2018–1084
10. Fujiwara G, Sunada Y, Rinoie K (2017) Instantaneous surface pressure measurements of quasi-periodic behavior of laminar separation bubble in the angles of attack near airfoil stall. J Jpn Soc Aeronaut Space Sci 65:168–175 (in Japanese)
11. Almutairi JH, AlQadi IM (2013) Large-eddy simulation of natural low-frequency oscillations of separating–reattaching flow near stall conditions. AIAA J 51:981–991
12. Benton SI, Visbal MR (2018) Effects of leading-edge geometry on the onset of dynamic stall. AIAA J 56:4195–4198

Influence of Turbulence Intensity on Flow Field Around NACA0012 Wing at Reynolds Number of 20,000

Masataka kase, Makoto Mizoguchi, and Hajime Itoh

Abstract The flow field around an NACA0012 wing is visualized in a towing tank using particle image velocimetry at a low Reynolds number of 2.0×10^4. A semi-span model of the NACA0012 wing with an aspect ratio of 3.0 is used for the experiment. The visualization is performed at the streamwise cross section along the centerline of the wing model. In the experimental investigation of the turbulence effects, a turbulence grid is installed upstream of the wing model. The turbulence is generated by moving the turbulence grid and the wing model at the same velocity. The effect of turbulence on the flow field is investigated by comparing the results obtained without using the grid. The flow field visualization around the wing model is achieved in a turbulent flow. It is found that turbulence leads to the reattachment of the separated boundary layer on the wing surface.

Keywords Low-reynolds-number flow · Towing tank · Flow visualization · Turbulence intensity

1 Introduction

There has been extensive research, development, and utilization of unmanned air vehicles (UAVs). There is growing global demand for UAVs for logistics and observation. UAVs are more susceptible than conventional large aircraft to atmospheric turbulence due to their smaller size, lighter weight, and slower flight speed. Small UAVs fly in environments with very low turbulence, such as indoor areas, or very high turbulence intensity (about 30%) in the lower atmospheric boundary layer [1]. Therefore, it is necessary to clarify the aerodynamic characteristics of wings and associated flow fields in turbulent environments for their safe flight.

M. kase (✉)
Graduate School of Science and Engineering, National Defense Academy of Japan, Yokosuka, Japan

M. Mizoguchi · H. Itoh
Department of Aerospace Engineering, National Defense Academy of Japan, Yokosuka, Japan
e-mail: makoto@nda.ac.jp

S. Lee et al. (eds.), *The Proceedings of the 2021 Asia-Pacific International Symposium on Aerospace Technology (APISAT 2021), Volume 1*, Lecture Notes in Electrical Engineering 912, https://doi.org/10.1007/978-981-19-2689-1_39

Several studies have investigated the effect of airflow turbulence on the aerodynamic characteristics of an airfoil. Tsuchiya et al. [2] investigated the effect of airflow turbulence on the aerodynamics of the NACA0012 airfoil at Reynolds numbers of $Re = 1.1 \times 10^4$–4.7×10^4 and turbulence intensities of $TI = 0.5\%$–2.1%. They reported that the lift coefficient at a low angle of attack and the maximum lift coefficient increased with increasing turbulence intensity. Sytsma et al. [3] investigated the effect of turbulence intensity ($TI = 4.0\%$–21.3%) on the aerodynamics of a rectangular flat wing with rounded leading and trailing edges at $Re = 2.0 \times 10^5$ and an aspect ratio of $AR = 1.0$. In their study, the maximum lift coefficient decreased with increasing turbulence intensity. Moreover, the angle of attack at the maximum lift coefficient was constant regardless of the turbulence intensity. In our laboratory, an active turbulence grid, a turbulence generator developed by Makita et al. [4], was installed in a wind tunnel in front of the test section to generate airflow with strong turbulence. The grid consists of rotating shafts with small blades for airflow agitation. In our laboratory, Fujita et al. [5] investigated the effect of airflow turbulence on the aerodynamics of the NACA0012 airfoil at low Reynolds numbers of 2.0×10^4 and 3.6×10^4. In their study, a decrease in the nonlinearity of the lift coefficient was observed at low angles of attack ($\alpha \leq 4°$) when the turbulence intensity increased. In addition, the maximum lift coefficient increased with increasing turbulence intensity. After a stall, the lift coefficient does not decrease and approaches a certain value when the turbulence intensity is large. Although the effect of turbulence on aerodynamics has been reported in previous studies, even at low Reynolds numbers, few visualization studies have been conducted on the flow field to clarify the flow structure that causes the difference in the aerodynamic characteristics due to turbulence.

In this study, we visualize the flow field around the NACA0012 wing under a turbulent environment. The turbulence intensity is adjusted by towing a turbulent grid and a wing model in a towing tank at a low Reynolds number.

2 Experimental Setup

The visualization experiments in this study were conducted in the towing tank (internal dimensions: 8 m width, 0.9 m depth, and 1.6 m height) installed at the National Defense Academy of Japan. A schematic diagram of the experimental setup is shown in Fig. 1. In the experiment, the turbulence grid and wing model (see below for details) were moved in the width direction of the towing tank. The turbulence grid, the wing model, and a high-speed camera mounted on a rack were moved linearly along linear guides using stepping motors and timing belts. In this study, the rectangular wing of the NACA0012 airfoil was used as the wing model. The wing chord length c was 180 mm, the wingspan b was 540 mm, and the aspect ratio was 3.0. The angle of attack of the wing model (α) was set to 1°, 3°, 5°, 6°, and 10°. The wing model was connected to the axis of rotation via thin struts at three spanwise locations ($0.25b$, $0.50b$, and $0.75b$) on the pressure side of the wing. The struts were connected to the 25% chord of the wing model from the leading edge. An

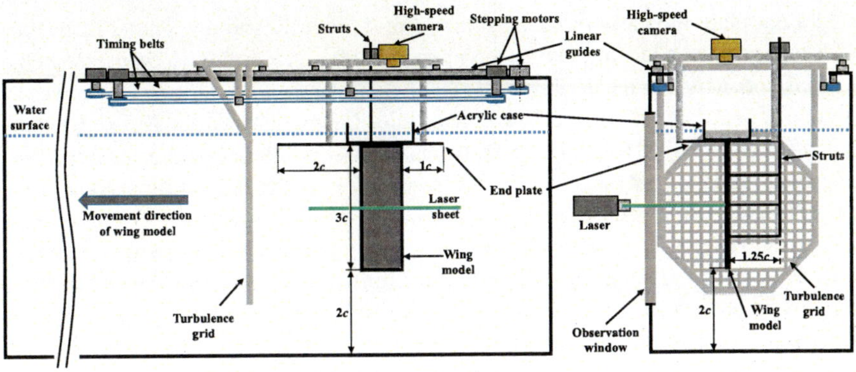

Fig. 1 Schematic diagram of experimental setup

endplate was placed in the towing tank to reduce the influence of the water surface. It also enabled semi-span model experiments to be conducted. The leading edge of the endplate was made of aluminum alloy. The downstream region was made of transparent acrylic for visualization. The leading edge of the endplate was sharpened to make the flow around the endplate laminar. The gap between the wing model and endplate was approximately 2 mm. A transparent case was attached on the endplate to prevent the surface waves generated during towing from affecting the acquired images. The distance from the bottom of the towing tank to the lower wingtip of the wing model was 360 mm and the clearance between the wing model and the sidewall of the towing tank was more than 400 mm. The influence of the towing tank walls could thus be ignored. The mesh size of the turbulence grid (M) was 17.2 mm. The grid consisted of cylinders with a diameter of 1.7 mm. The turbulence intensity generated by the turbulence grid was estimated using data separately measured in a wind tunnel. The distance between the turbulence grid and the 25% chord position of the wing model was set to $X/M = 35.2$ (distance between the model and grid $X = 605$ mm). The turbulence intensity at the 25% wing chord position was estimated to be 1.6%. It was estimated that the turbulence intensity between the leading and trailing edges of the model varied by about 0.3%, which is sufficiently small. The experimental Reynolds number based on the wing chord length was set to 2.0×10^4.

The flow field was visualized using particle image velocimetry (PIV). A laser sheet introduced through the observation window at the sidewall of the tank illuminated the fluorescent particles (SX-117, Shinroishi) in the water. A high-speed camera (K-III, Katokoken Co., Ltd.) mounted on a rack above the tank photographed the scattered light from the particles. A continuous laser (PIV Laser G5000, Katokoken Co., Ltd.) with an output power of 5 W and a wavelength of 532 nm was used. The laser sheet was placed in the spanwise center of the wing model. The specific gravity of the fluorescent particles was 1.2 and their average diameter was 4–6 μm. The particle images were taken with a resolution of 1600 × 1200 pixels at 500 frames per second. Beckwith et al. [6] investigated the transient characteristics of an impulsively moving wing. They conducted aerodynamic measurements and visualization experiments using a flat plate model with an aspect ratio of 4.0 at a Reynolds number of 6.0×10^4 (conditions similar to those in the present study). According to their research, the transient change in the aerodynamic coefficient plateaus after the model moves a distance of 5*c*. In this study, the flow field of the wing model was visualized after the wing model had moved a distance of 6c from its initial position. One thousand frames were used for PIV analysis. The PIV software FlowExpert2D2C was used for the analysis. The recursive correlation method was used. The PIV analysis conditions were as follows: a spatial resolution of about 0.18 mm/pixel, interrogation windows of 20×20, 10×10 and 5×5 pixels, and grid spacings in the chordwise and transverse directions of 2 pixels each. From the results of the PIV analysis, the mean velocity and velocity distribution were obtained.

3 Results and Discussion

Figures 2–4 show the PIV results for angles of attack of 1°, 3°, and 5°, respectively. The results obtained without and with the grid are respectively shown in the top and bottom panels of each figure. The analysis results are shown as contours of the velocity magnitude and velocity vectors. The velocity vectors indicate the velocity in

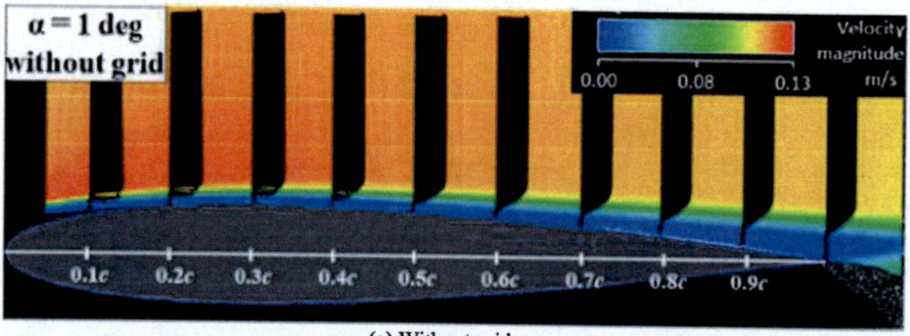

(a) Without grid.

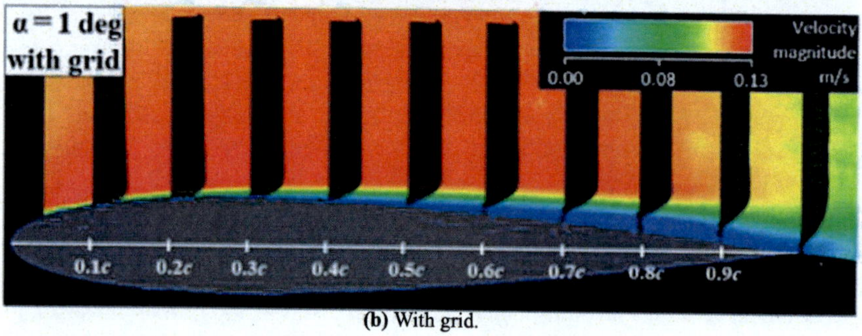

(b) With grid.

Fig. 2 Velocity distribution obtained at $\alpha = 1°$

the cross sections vertical to the wing chord at $0.1c$ intervals from the leading edge. In the contours, regions with relatively low (high) velocity are shown in blue (red).

For an angle of attack of 1° (Fig. 2), a Blasius velocity distribution was observed from the leading edge to $0.4c$, regardless of the turbulence intensity. Hence, the boundary layer of these regions is laminar. Further downstream, the velocity gradient at the surface is close to 0 and the velocity distribution has an inflection point (and is denoted an inflection-type velocity distribution). This suggests the occurrence of laminar separation. However, a difference in the velocity distribution due to the grid was observed around the trailing edge. Without the grid, the low-velocity region is thick. In contrast, with the grid, the low-velocity region is relatively thin. It is thought that the difference in the velocity distribution around the trailing edge is due to boundary layer reattachment, although this reattachment is not obvious in the case with the grid.

For an angle of attack of 3° (Fig. 3), a Blasius velocity distribution was observed from the leading edge to $0.3c$, both with and without the grid. The boundary layer in these regions is laminar. Without the grid, an inflection-type velocity distribution, similar to the case for an angle of attack of 1°, was observed downstream from the $0.4c$ position. No reattachment was observed, also similar to the result for 1°. Both with and without the grid, the velocity gradient near the wing surface is close to 0 and the velocity distribution has an inflection point at $0.5c$. However, with the grid, no inflection point was observed at $0.9c$ or further downstream. Reattachment of the separated

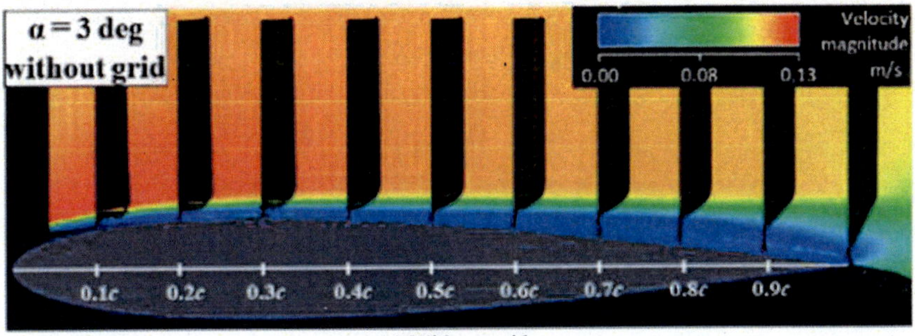

(a) Without grid.

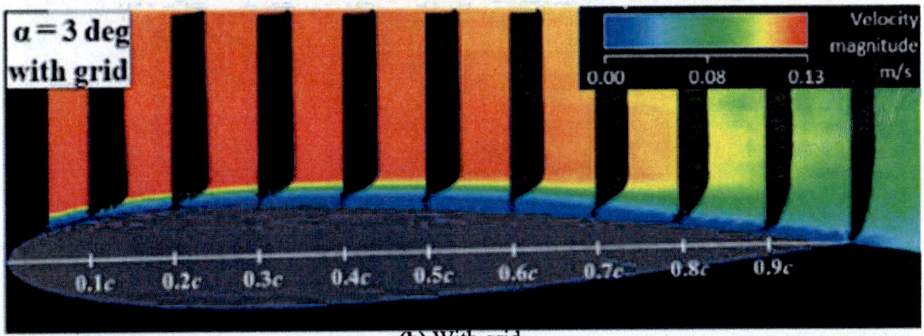

(b) With grid.

Fig. 3 Velocity distribution obtained at $\alpha = 3°$

boundary layer is considered to occur between 0.7*c* and 0.9*c*. Hence, a laminar separation bubble is thought to exist between 0.3*c* and 0.9*c*. In addition, the blue, green, and yellow regions in the contour are thicker after reattachment, indicating that the reattached boundary layer is thick in these regions. Hence, the boundary layer is turbulent after reattachment. The difference in the flow field between the cases with and without the grid is confirmed in the visualization. Furthermore, it is found that the freestream turbulence enhanced the reattachment of the separated boundary layer of the wing surface.

For an angle of attack of 5° without the grid (Fig. 4), a Blasius velocity profile was observed from the leading edge to 0.2*c*. The velocity gradient near the wing surface is close to 0 at 0.3*c* and further downstream. An inflection point was observed in the velocity distribution at these locations. Therefore, laminar separation is considered to occur between 0.2*c* and 0.3*c*. With the grid, a Blasius velocity distribution was observed from the leading edge to 0.2*c*. The velocity gradient near the wing surface is close to 0 at 0.4*c* and an inflection-type velocity distribution appears. Thus, flow separation is considered to occur between 0.2*c* and 0.4*c*. Moreover, no inflection point was observed at 0.6*c* or further downstream. Flow reattachment is thought to occur between 0.4*c* and 0.6*c*. At 0.6*c* and further downstream, the low-velocity region is thick, indicating that the boundary layer is turbulent. Therefore, a laminar separation bubble is considered to exist between 0.2*c* and 0.6*c*. Thus, a laminar boundary layer,

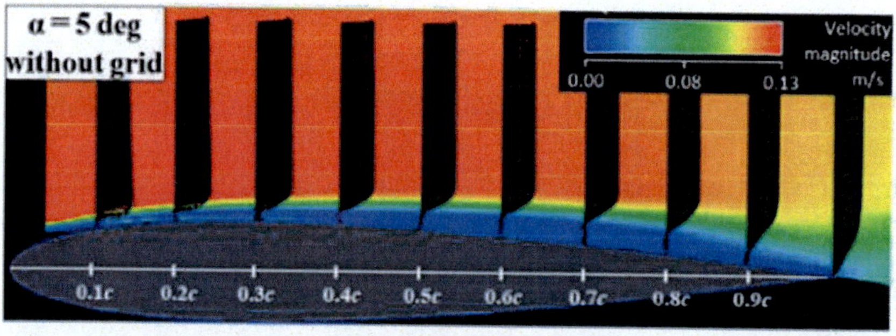

(a) Without grid.

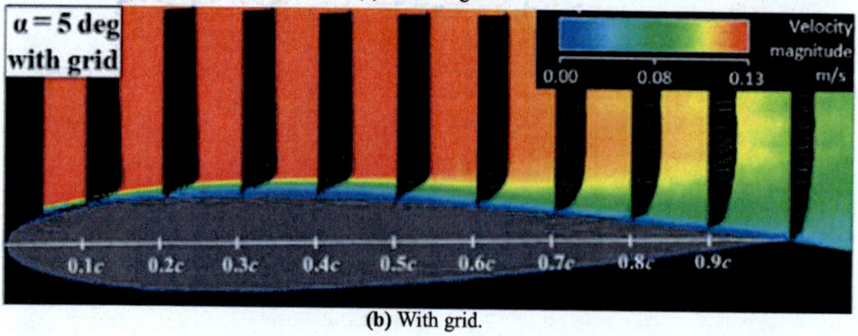

(b) With grid.

Fig. 4 Velocity distribution obtained at $\alpha = 5°$

a reattachment of the separated boundary layer, a laminar separation bubble, and a turbulent boundary layer were observed. This flow structure is similar to that at an angle of attack of 3° with the grid, although the size and position are different. The laminar separation bubble moved toward the leading edge compared to that for 3°. Thus, enhanced reattachment of the separated boundary layer was observed.

4 Conclusions

In this study, the flow field around the NACA0012 wing at a Reynolds number of 2.0×10^4 was visualized in a towing tank. The velocity distribution was analyzed using PIV. In the experiment, a turbulence grid was installed upstream of the wing model and moved with the model at the same velocity, generating turbulence. We investigate the effect of turbulence by comparing the visualization results with and without the grid. Visualization of the difference in the flow field due to turbulence intensity was achieved. The extent of the difference in the flow field depends on the angle of attack. In particular, the difference was smallest at the smallest angle of attack (1°). In addition, it is found that the turbulence intensity leads to the reattachment of the separated boundary layer on the wing surface.

Acknowledgements This work was supported by a grant from the Japan Society for the Promotion of Science KAKENHI (grant number JP20K04933).

References

1. Watkins S, Milbank J, Loxton BJ (2006) Atmospheric Winds and their implications for Microair vehicles. AIAA J 44(11):2591–2600
2. Tsuchiya T, Numata D, Suwa T, Asai K (2013) Influence of turbulence intensity on aerodynamic characteristics of an NACA 0012 at low Reynolds numbers. 51st AIAA aerospace sciences meeting including the new horizons forum and aerospace exposition 2013. AIAA Paper 2013–0065:1029–1043
3. Sytsma MJ, Ukeiley L (2013) Mean loads from wind-tunnel turbulence on low-aspect-ratio flat plates. J Aircraft 50(3):863–870
4. Makita H, Sassa K, Iwasaki T, Iida A (1987) Evalution of the characteristic features of a large-scale turbulence field (1st report, perfomance of the turbulence generator). Trans JSME (B) 53(495):3173–3179 (in Japanese)
5. Fujita K, Mizoguchi M, Itoh H (2020) Effects of strong turbulence on aerodynamic characteristics of NACA0012 airfoil in low Reynolds number flows. Fluid dynamics conference/aerospace numerical simulation symposium 2020 Online JSASS-2020–2023-F+A
6. Beckwith RMH, Babinsky H (2009) Impulsively started flat plate folw. J Aircraft 46(6):2186–2189

Analysis of Rotor Noise During Ramp Collective Pitch Increase by CFD Method

Xi Chen, Kai Zhang, Qijun Zhao, Weiqi Wang, and Siyu Chen

Abstract Based on FW-H equations and the CFD method, aeroacoustic characteristics of rotor during collective pitch aperiodic variation are calculated and analyzed. At first, a set of analysis method for aperiodic rotor aeroacoustic characteristics is developed. The aerodynamic and aeroacoustic characteristics of BO-105 rotor in hover and NACA rotor in a ramp increase of collective pitch are calculated, and the employed numerical analysis method is validated through comparisons with experimental data. Then, the aeroacoustic characteristics of the NACA rotor during an oscillating collective pitch is analyzed, and the sound pressure peak at different collective pitch is discussed in detail, and some conclusions are obtained.

Keywords Aeroacoustic characteristic · Aperiodic · Rotor · Helicopter · Numerical method

1 Introduction

Noise is an important issue of helicopter for both civil and military applications, and much research has been performed to study the noise generation and propagation. Real-world helicopters will maneuver with complex motions, which produce noise different from the current periodic analysis results, called the maneuvering rotorcraft noise [1, 2]. Since the rotor is the main lift and control surface of a helicopter, its aeroacoustic characteristics are the main factors that affect the maneuvering noise of helicopter. In some maneuvering flight missions, such as hover jump, high-speed pull, etc., the control input is a type of fast step input, which brings a step response of aerodynamic and aeroacoustic characteristics. Therefore, it is of great theoretical and practical significance to analyze the aeroacoustic characteristics of rotor blades during collective pitch aperiodic variation.

X. Chen · K. Zhang · Q. Zhao (✉) · W. Wang · S. Chen
Nanjing University of Aeronautics and Astronautics, No 29 Yudao Street, Nanjing 210016, Jiangsu, China
e-mail: zhaoqijun@nuaa.edu.cn

S. Lee et al. (eds.), *The Proceedings of the 2021 Asia-Pacific International Symposium on Aerospace Technology (APISAT 2021), Volume 1*, Lecture Notes in Electrical Engineering 912, https://doi.org/10.1007/978-981-19-2689-1_40

Over the past few decades, researchers have developed many analytical methods for aeroacoustic characteristic of rotor, including the classical Ffowcs Williams-Hawkings (FW-H) equations [3], the Kirchhoff approach [4], and the FW–H_{pds} equations [5]. Recently, Huang [6] presented a frequency domain solution of FW–H_{pds} equation, and studied high speed impulsive noise radiated from rotors with high tip speed. Based on FW–H_{pds} equations, Zhao [7] developed a new permeable adaptive integration surface to evaluate transonic rotor noise. Loiodice [8] developed an accurate and efficient algorithm for the solution of the retarded time equation for rotor noise, and it is faster than the classical Newton and Brent methods. However, these works mainly focus on periodic rotor noise, and the rotor noise characteristic may be different with a collective pitch aperiodic variation.

In maneuvers noise, Brentner [1, 9, 10] developed the maneuver noise prediction code called PSU-WOPWOP, and examines the transient maneuver noise produced by a helicopter rotor during a short-time maneuver. They found that the noise generated during short-duration maneuvers is due to multiple sources: aperiodic blade motions, aircraft attitude changes, and transient aerodynamic loading. Greenwood [11] studied blade-vortex interaction noise generation during maneuvering flight using a semi-empirical noise modeling method. Vouros [12] developed a generalized and efficient methodology for the prediction of the acoustic footprint of realistic rotorcraft operations. In these studies, some simplified analysis methods are applied, such as the compact loading noise formulation, and it is difficult to analyse the relationship between the details of the flowfield around the rotor and the aeroacoustic characteristics in the maneuvering flight.

Considering the typical unsteady characteristics of the helicopter maneuvering flight and the complex of the unsteady vortex flowfield, the quasi-steady rotor aerodynamic model is not suitable for the analysis of rotor maneuvering aeroacoustic problems in detail. Gennaretti [13] investigates a fully unsteady numerical approach and two quasi-steady noise simulation methods, and found that the low level of accuracy of the latter methods when unsteady effects of the maneuvering flight occur. So far, there are some CFD codes developed specifically for helicopter rotor analysis, such as European Rotorcraft Software (EROS) [14], Helicopter Multi-Block (HMB) CFD code [15]. Based on CLORNS [16], Zhang [17] studied the aerodynamic response characteristics of the helicopter rotor during ramp collective pitch increase. The results showed that CFD can provide a reliable time evolution of the blade loads during the transient maneuver, and it is benefit to study the aeroacoustic characteristics of rotor during collective pitch variation in hover.

Based on the previous work [7, 17] and the CLORNS code [16], a set of analysis method for rotor aeroacoustic characteristics during collective pitch aperiodic variation is developed. The main objective of the present work is focused on developing better understanding of the aeroacoustic characteristic of rotor during collective pitch aperiodic variation. The experimental data of NACA Rotor during a ramp increase collective pitch at the Langley helicopter test tower are used to validate the present numerical analysis method. As the main study subject, aeroacoustic characteristics of NACA rotor during collective pitch aperiodic variation are calculated and analyzed, and the variation of the sound pressure peak obtained by single blade are detailed.

2 Calculation Method

2.1 Grid Generation Method and CFD Solver

CLORNS code is employed to predict the complex unsteady rotor flowfield [16], and the governing equations in integral form for predicting the rotor flowfield are described as:

$$\frac{\partial}{\partial t}\iiint \vec{W}\,d\Omega + \iint_S \left(\vec{F} - \vec{F}_v\right)\cdot \vec{n}\,ds = 0 \quad (1)$$

In CLORNS, a disturbance diffraction method for hole-cell identification and a minimum distance scheme for donor cell searches are used with high efficiency and universality. Three alternative spatial discretization schemes are applied in the code, including a second-order central difference scheme, a third-order Roe-MUSCL scheme, and a fifth-order WENO-Roe scheme.

To obtain a reliable rotor flowfield during a collective pitch aperiodic variation in hover, the dual time-stepping approach is adopted, and implicit Lower–Upper Symmetric Gauss–Seidel (LU-SGS) methods [18] is employed in pseudo-temporal iteration.

Grids around rotor blade are generated by interpolating and folding of airfoil section grids constructed by solving Poisson equations. Structured Cartesian grids are used as the background grid of a chimera system.

2.2 Aeroacoustic Analysis Method for Aperiodic Rotor Noise

Based on the sound source information obtained by the CLORNS solver, the Ffowcs-Williams Hawkings (FW-H) method is used for calculating the aeroacoustic characteristic of rotor. The FW-H method is derived by combining the equations of mass and momentum conservation for compressible fluids, to obtain an inhomogeneous wave equation. The Farassat 1A equation to be solved for the FW-H is as follows

$$p(\vec{x}, t) = p_T(\vec{x}, t) + p_L(\vec{x}, t) \quad (2)$$

where,

$$4\pi p_T(\vec{x}, t) = \int_{f=0}\left[\frac{\rho_0 v_n\left(rMa_i\hat{r}_i + c_0 Ma_r - c_0 Ma^2\right)}{r^2(1 - Ma_r)^3}\right]_{ret} ds + \int_{f=0}\left[\frac{\rho_0(\dot{v}_n + v_n)}{r(1 - Ma_r)^2}\right]_{ret} ds \quad (3)$$

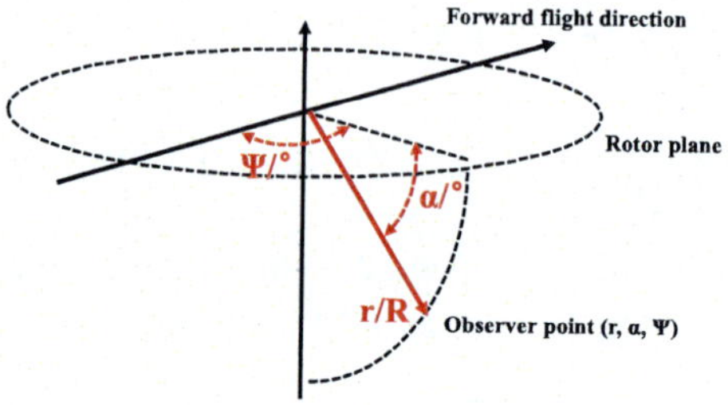

Fig. 1 The relative position between observer points and the rotor plane

and

$$4\pi p_T(\vec{x},t) = \frac{1}{c_0}\int_{f=0}\left[\frac{\dot{l}_i\hat{r}_i}{r(1-Ma_r)^2}\right]_{ret}ds + \int_{f=0}\left[\frac{l_r - l_i Ma_i}{r^2(1-Ma_r)^2}\right]_{ret}ds + \frac{1}{c_0}\int_{f=0}\left[\frac{l_r\left(r\dot{M}a_i\hat{r}_i + c_0 Ma_r - c_0 Ma^2\right)}{r^2(1-Ma_r)^3}\right]_{ret} \tag{4}$$

In addition, the FW-H_{pds} method is used when the quadrupole noise cannot be neglect. In this paper, the location of observer point is defined by the coordinates (**r**, α, Ψ), as shown in Fig. 1. Ψ is azimuthal angle of OBS. **r** is a vector from the center of hub to the location of OBS. α is the included angle between vector **r** and rotor plane.

3 Calculated Results and Analyses

3.1 Validation of Aerodynamic and Aeroacoustic Characteristics on BO-105 Rotor

The test for BO-105 model rotor performed at the China Aerodynamic Research and Development Center were selected for validation [19, 20] The rotor had a radius of 1 m and is equipped with 4 blades. The blade had a cross-sectional shape of NACA23012 airfoil with a chord of C = 60 mm and a negative linear twist of -8°. The collective pitch $\theta_{0.7}$ was set from 0° to 12°. Figure 2 shows pressure coefficient distributions at 0.7R with a rotation speed of 1000 rpm, and Fig. 3 shows sound pressure time histories with a rotation speed of 2000 rpm. As seen, the agreement between calculated results and experimental data is good, indicating that the present method is able to obtain the sound source on the blade and the aeroacoustic characteristic of rotor in hover.

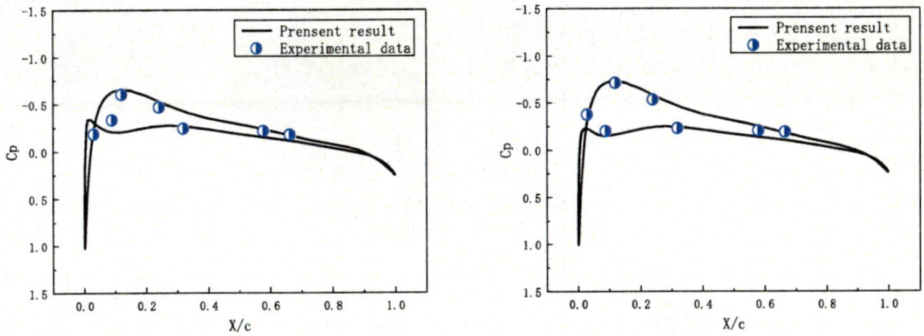

(a) $\theta_{0.7}$=4° (b) $\theta_{0.7}$=6°

Fig. 2 Pressure coefficient distributions of blade section at r = 0.7R

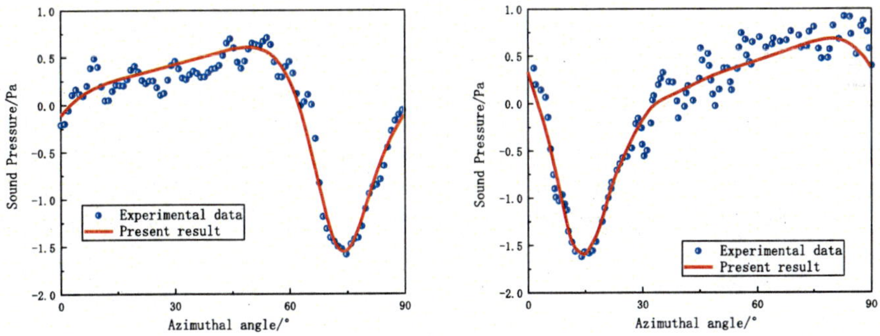

(a) $\theta_{0.7}$=4° (b) $\theta_{0.7}$=6°

Fig. 3 Sound pressure time histories of rotor at OBSs of α = 0° and r = 5R

3.2 Aerodynamic and Aeroacoustic Characteristics on the Ramp Increased Collective Pitch Case

In this section, NACA Rotor during a ramp increase collective pitch at the Langley helicopter test tower [21] were selected, and the transient aerodynamic response and aeroacoustic characteristics of rotor is calculated. The rotor had a rotating radius of 5.79 m and was equipped with 3 blades. The blade had a NACA airfoil with a chord of C = 255 mm, and negative linear twist was 0°. The rotating speed was 220 rpm, and the collective pitch increased from 0° to 12°. The blade surface and the calculated grid are shown in Fig. 4.

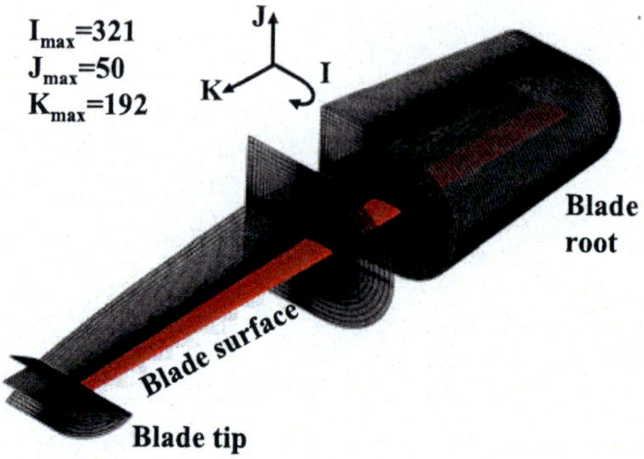

Fig. 4 The blade surface and the calculated blade grid

The step input of the collective pitch was simulated by the linear change of the collective pitch in a short time, with the change rates V_θ of 200°/s, 48°/s and 20°/s, respectively. The calculated and experimental rotor aerodynamic characteristic with collective pitch increase from 0° to 12° is shown in Fig. 5. As seen, the thrust coefficient (CT) of the present method is in particularly better agreement with the test data than those calculated by free wake method [22], indicating that the present method is able to obtain the reliable sound source during collective pitch aperiodic variation in hover.

The case with a V_θ of 48°/s is selected to calculate the aeroacoustic characteristics. The sound pressure time histories at three OBSs are given in the Fig. 6. As a whole, thickness noise is still almost periodic, while loading noise changes a lot in the collective pitch increase process. In addition, the sound pressure value of loading noise has an obvious overshoot with pitch increase, and it is similar to the variation of CT.

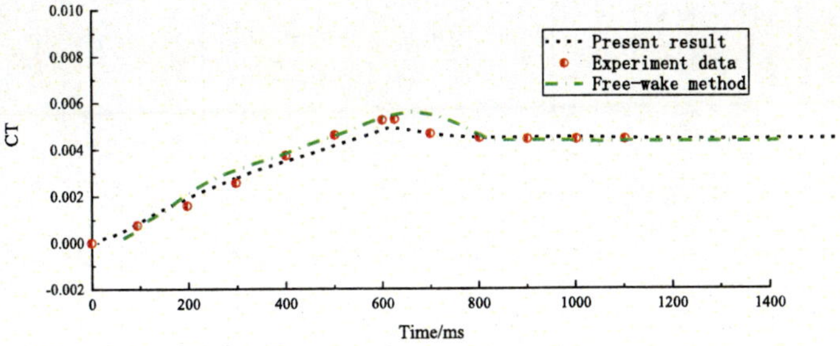

(a) $V_\theta = 20°/s$

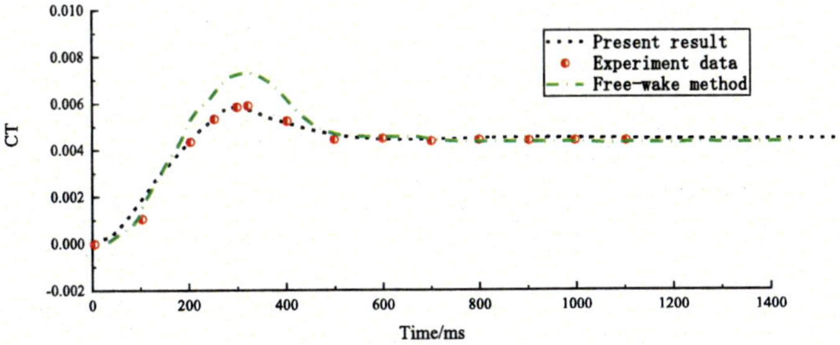

(b) $V_\theta = 48°/s$

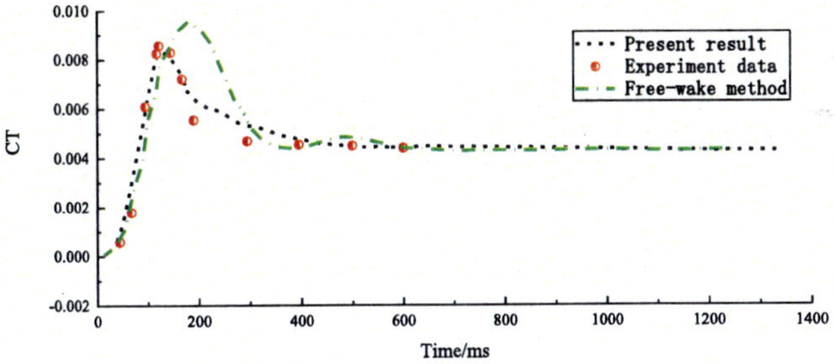

(c) $V_\theta = 200°/s$

Fig. 5 CT variation during the collective pitch increase

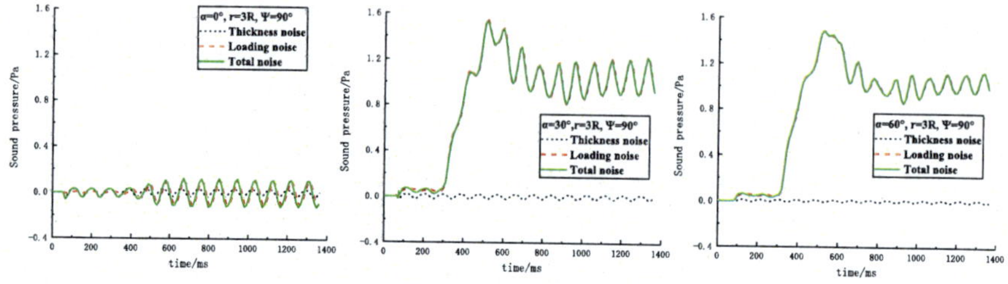

Fig. 6 Sound pressure time histories at different OBSs

3.3 *Aeroacoustic Characteristics Analysis of Rotor with an Oscillation Pitch Movement*

To analysis the influence of collective pitch variation on rotor noise, the aeroacoustic characteristic of rotor with an oscillation pitch movement is calculated in this section. As seen in Fig. 7, the collective pitch increases from 0° to 12° in the first/third cycle, and it decreases from 12° to 0° in the second/fourth cycle. The change rate V_θ is ± 12° per cycle.

Figure 8 shows the variation of CT in the collective pitch oscillation process. As seen, there is a positive relation between CT and collective pitch angle, and CT in the collective pitch increase process is larger than that in the decrease process. In addition, dCT/dt increases firstly and then decreases with the increase of the collective pitch. Predictably, the aerodynamic variation will directly affect aeroacoustic characteristic of rotor in this case.

Figure 9 shows aeroacoustic characteristics of rotor at some typical OBSs. Totally, the thickness noise is almost the same in the collective pitch oscillation process. The amplitude of the loading noise changes a lot, and it increases with the increase of collective pitch.

In order to facilitate analysis, the loading noise of a single blade at $\Psi = 180°$ is given in the Fig. 10a. As seen, there are four sound pressure peaks in the sound pressure time history. Introduction the definition of the doppler amplification factor in the F1-A equations, it is maximum for OBS at $\Psi = 180°$ when the sound source

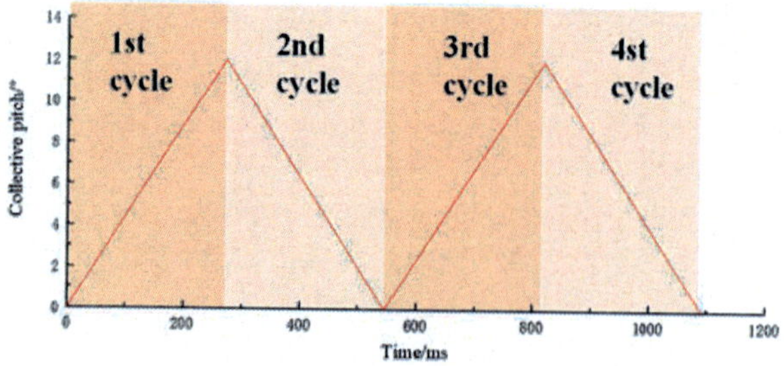

Fig. 7 Variation of collective pitch of rotor in different cycles

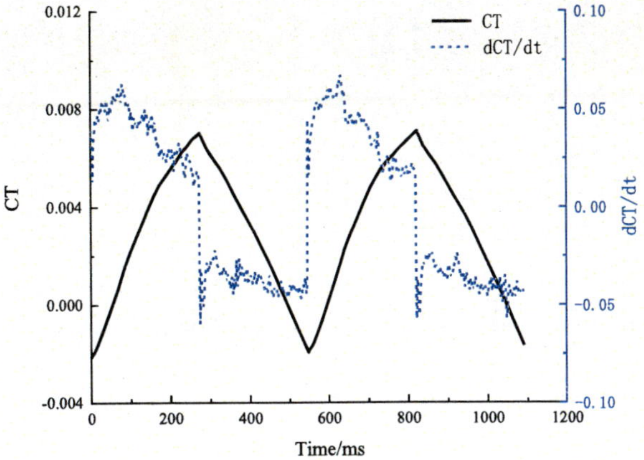

Fig. 8 Variation of CT in different cycles

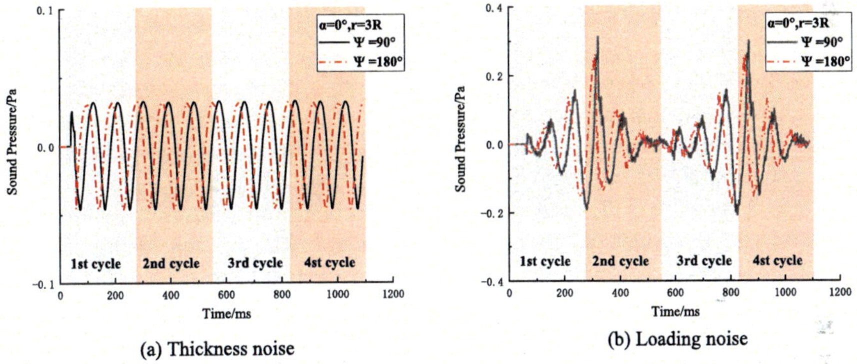

Fig. 9 Sound pressure time histories at some typical OBSs in the rotor plane

of the blade is near to $\Psi = 90°$. As a result, the first sound pressure peak is mainly composed by the blade at $\theta_{0.7} = 3°$, and $\theta_{0.7}$ is increasing (1st cycle). The second sound pressure peak is mainly composed by the blade at $\theta_{0.7} = 9°$, and $\theta_{0.7}$ is decreasing (2nd cycle). Similarly, the sound pressure peak is mainly composed by the sound source near to $\Psi = 270°$ for OBS at $\Psi = 0°$, such as Fig. 10b. Through comparison the sound pressure peak value at $\Psi = 180°$ and $\Psi = 0°$, the sound pressure of blade with an increasing collective pitch is bigger than that with a decreasing collective pitch.

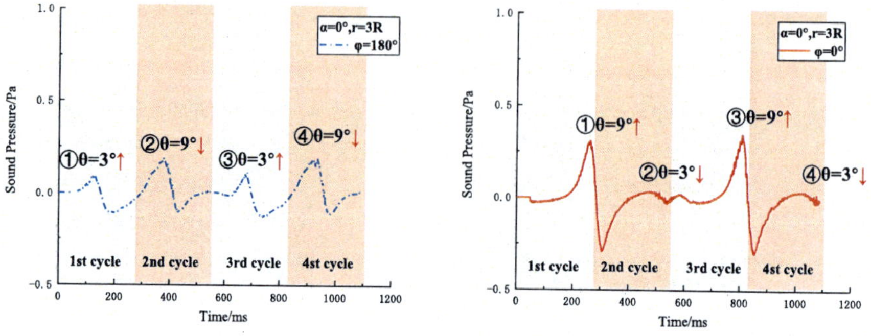

(a) Ψ=180° (b) Ψ=0°

Fig. 10 Sound pressure time histories of single blade at some typical OBSs in the rotor plane

3.4 *Influence of Collective Pitch Change Rate on Aeroacoustic Characteristics of the Rotor*

BO-105 rotor with a Matip = 0.6316 is selected to analysis the aeroacoustic characteristics with a ramp increased collective pitch. As shown in Fig. 11, the step input of collective pitch is simulated by the linear change of the collective pitch in a short time, with the change rates of 24° per cycle, 12° per cycle and 8° per cycle, respectively.

Figure 12 shows the CT variation and sound pressure time histories (a = 30°, r = 3R, Ψ = 90°) in collective pitch increase process. In general, there is a positive relation between CT and collective pitch, and the overshoot of CT for the rapid blade-pitch increase are obviously in each case. Also, *d*CT/*dt* is positive firstly and then negative, and it means that the aerodynamic loading perturbation of blade changes greatly. For the sound pressure time histories, there is a similar overshoot phenomenon. When V_θ is 24° per cycle, the overshoot amplitude of positive sound

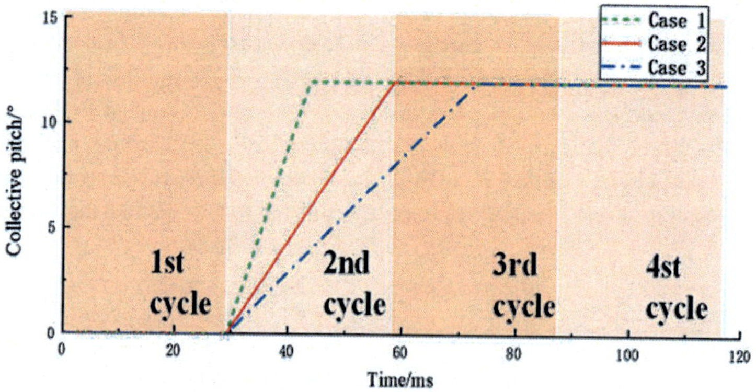

Fig. 11 Variation of collective pitch of rotor in different cycles

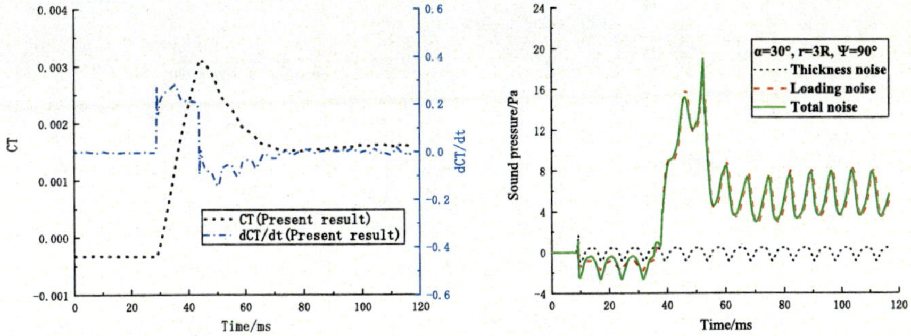

(a) Case 1, $V_\theta = 24°$ per cycle

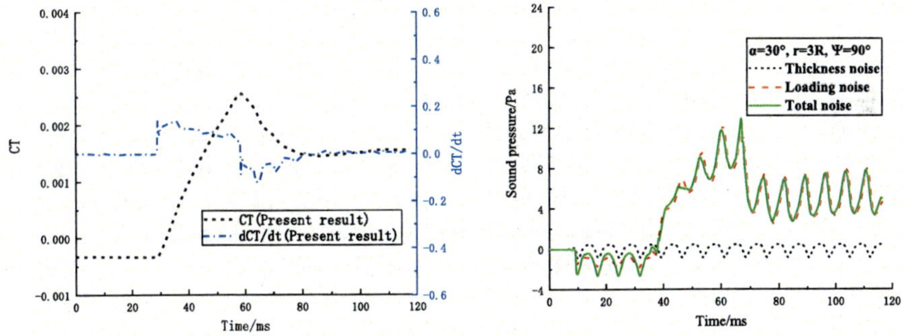

(b) Case 2, $V_\theta = 12°$ per cycle

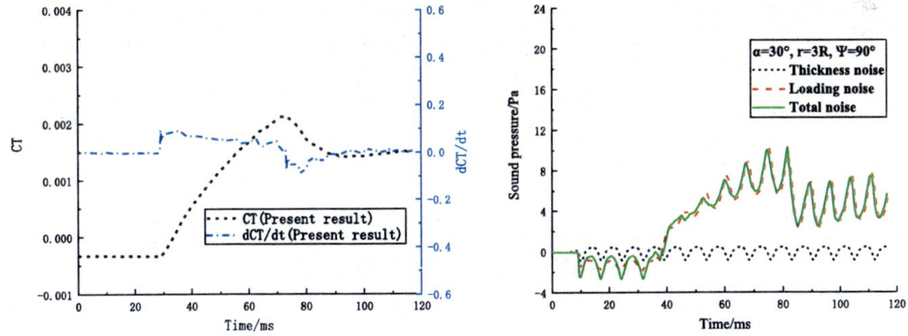

(c) Case 3, $V_\theta = 8°$ per cycle

Fig. 12 Variation of CT (left) and sound pressure time histories (right) in different cases

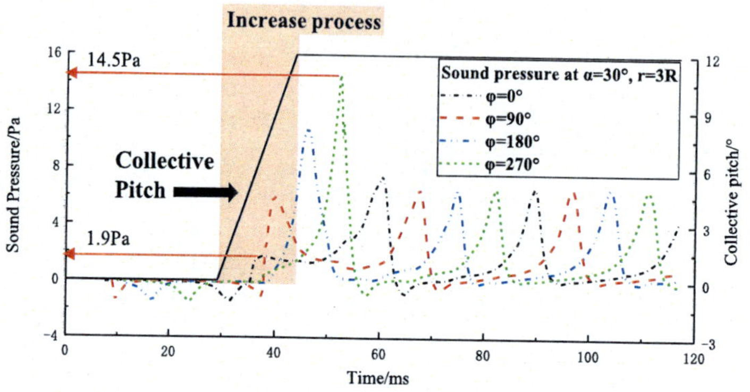

Fig. 13 Sound pressure time histories of loading noise of a single blade

pressure is up to 20 Pa. The collective pitch change rate is larger, the overshoot response amplitude of sound pressure is larger.

To analysis further the variation of loading noise, several sound pressure time histories of a single blade (initiation azimuth angle is 0°) at different OBSs are given in Fig. 13. As seen, there is significant differences among the sound pressure peak values calculated at different OBSs in the collective pitch increase process. The sound pressure peak value increases with the increase of Ψ, such as 1.9 Pa at $\Psi = 0°$ and 14.5 Pa at $\Psi = 270°$.

The overshoot phenomenon in loading noise is largely due to the variation of blade loads. Figure 14 shows the pressure variation on upper surface of blade. The area of negative pressure zone increases with the increase of the collective pitch. After the collective pitch reaches its final value, the area increases further firstly and then decreases for a period time, and remain unchanged at last. Also, there is a similar change trend in the negative pressure value of an upper surface unit at r = 0.8R and c = 0.2C. As seen, CFD method can capture more details of blade loads and flowfield features in the ramp collective pitch state, and can provide a more reliable sound source information compared with the simplified method.

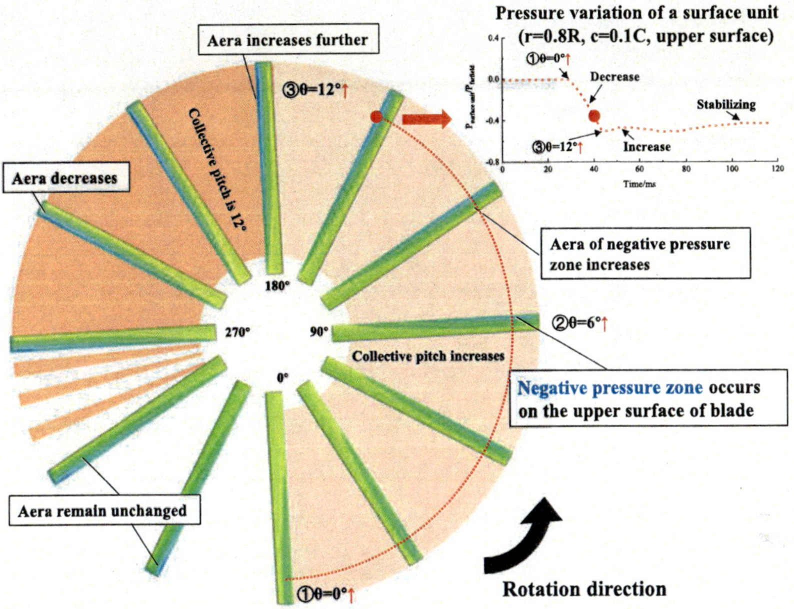

Fig. 14 Pressure variation on upper surface of blade in collective pitch increase process

4 Conclusions

Based on the aperiodic rotor noise analysis method, the influence of the collective pitch change rate on the aeroacoustic characteristics of rotor is calculated and analyzed. The main conclusions are:

1. Compared with the experimental results, the method developed here is effective for the numerical simulation of unsteady flowfield of helicopter rotor, and it can supply a reliable sound source information for aperiodic noise prediction.
2. With the oscillation movement of the collective pitch, loading noise changes a lot, while thickness noise is almost the same. For a certain collective pitch angle, the noise of rotor with an increasing collective pitch is bigger than that with a decreasing collective pitch, and it is due to the trust coefficient overshoot phenomenon in the collective pitch increase.
3. During a ramp increased collective pitch case, collective pitch increases faster, sound pressure of loading noise takes a shorter time to reach the peak value and a longer time to reach the steady state. With the increase of the collective pitch change rate, the overshoot response amplitude of sound pressure is larger.

Acknowledgements This study was co-supported by the National Natural Science Foundation of China (Nos. 12032012).

References

1. Bres GA, Brentner KS, Perez G, Jones HE (2004) Maneuvering rotorcraft noise prediction. J Sound Vib 275(5):719–738
2. Stephenson JH, Tinney CE, Greenwood E, Watts ME (2014) Time frequency analysis of sound from a maneuvering rotorcraft. J Sound Vib 333(21):5324–5339
3. Farassat F, Succi GP (1980) A review of propeller discrete frequency noise prediction technology with emphasis on two current methods for time domain calculations. J Sound Vib 71(3):399–419
4. Farassat F, Myers MK (1988) Extension of Kirchhoff's formula to radiation from moving surfaces. J Sound Vib 123(3):451–460
5. Francescantonio PD (1997) A new boundary integral formulation for the prediction of sound radiation. J Sound Vib 202(4):491–509
6. Huang Z, Siozos-Rousoulis L, Troyer TD, Ghorbaniasl G (2018) Helicopter rotor noise prediction using a convected FW-H equation in the frequency domain. Appl Acoust 140:122–131
7. Chen S, Zhao Q, Ma Y (2019) An adaptive integration surface for predicting transonic rotor noise in hovering and forward flights. Chin J Aeronaut 32(9):2047–2058
8. Loiodice S, Drikakis D, Kokkalis A (2018) An efficient algorithm for the retarded time equation for noise from rotating sources. J Sound Vib 412:336–348
9. Chen H, Brentner KS, Lopes LV, Horn JF (2006) An initial analysis of transient noise in rotorcraft maneuvering flight. International Journal of Aeroacoustics 5(2):109–138
10. Saetti U, Villafana W, Wachspress D, Brentner KS, Horn JF (2016) Helicopter simulations with coupled flight dynamics, free wake and acoustics, AHS 72nd annual forum proceedings, West Palm Beach, FL, USA, May 2016
11. Greenwood E, Schmitz FH, Sickenberger RD (2015) A semiempirical noise modeling method for helicopter maneuvering flight operations. J Amer Helicopter Soc 60(2):1–13
12. Vouros S, Goulos I, Pachidis V (2019) Integrated methodology for the prediction of helicopter rotor noise at mission level. Aerosp Sci Technol 89:136–149
13. Gennaretti M, Bernardini G, Serafini J, Anobile A, Hartjes S (2017) Helicopter noise footprint prediction in unsteady maneuvers. Int J Aeroacoust 16(3):165–180
14. Renzoni P, D'Alascio A, Kroll N (2000) EROS-a common European Euler code for the analysis of the helicopter rotor flowfield. Prog Aerosp Sci 36(5–6):437–485
15. Spentzos A, Barakos G, Badcock K, Richards B (2006) Modelling three-dimensional dynamic stall of helicopter blades using computational fluid dynamics and neural networks. Proc Inst Mech Eng Part G: J Aerosp Eng 220(6):605–618
16. Zhao Q, Zhao G, Wang B, Wang Q, Shi Y, Xu G (2018) Robust Navier-Stokes method for predicting unsteady flowfield and aerodynamic characteristics of helicopter rotor. Chin J Aeronaut 31(2):214–224
17. Zhang K, Zhao Q, Zhang X, Xu G (2021) Analysis of rotor aerodynamic response during ramp collective pitch increase by CFD method. Proc Inst Mech Eng Part G: J Aerosp Eng 235(9):994–1007
18. Yoon S, Jameson A (1988) (1988) Lower–upper symmetric-Gauss–Seidel method for the Euler and Navier-Stokes equations. AIAA J 26(9):1025–1026
19. Jiao L, Liu X, Shi Z, Zhang W et al (2020) A two-dimensional temperature correction method for pressure-sensitive paint measurement on helicopter rotor blades. Exp Fluids 61(104):1–11
20. Zheng X, Wang X, Chen Z, Huang B (2019) Experimental study on aerodynamic noise of BO-105 rotor model, 2nd national conference on aerodynamic noise and control technology, Guiyang, China, August 8–10, 2019
21. Capenter PJ, Fridovich B (1953) Effect of a rapid blade pitch increase on the thrust and induced velocity response of a full-scale helicopter rotor. NACA TN 3044
22. Ananthan S, Leishman JG (2004) Transient helicopter rotor wakes in response to time-dependent blade pitch inputs. J Aircr 41(5):1025–1041

Analyses on Aerodynamic Interactions of Quad-Tiltrotor Aircraft with Variable RPM and Diameter

Hongliang Wang, Qijun Zhao, Guoqing Zhao, Xiayang Zhang, and Jinshuai Shi

Abstract A CFD method for whole aircraft aerodynamic interactions of the quad-tiltrotor is established, where the momentum source method is adopted for the simulation of the rotor flowfield, and the body-fitted grid method is employed for the simulation of the fuselage. Based upon the established CFD method, the aerodynamic interaction characteristics between the front rotor/rear rotor/wing of the quad-tiltrotor under forward and transition flight state are studied, and the effects of rotor diameter, RPM, and forward flight speed on aerodynamic interaction characteristics of quad-tiltrotor are analyzed. The numerical results show that the use of rotors with large diameter may not be able to meet the performance requirements in high-speed cruise, and the aerodynamic efficiency of the rear rotor is slightly higher. As the diameter and RPM decrease, the differences of aerodynamic efficiency for the front and rear rotors decrease. When the forward speed and the tilt angle of the rotors are small, the efficiency of the rear rotor is lower than that of the front rotor, and the interferences between rotors weaken with increasing rotor diameter and RPM.

Keywords Quad-tiltrotor · Variable RPM · Variable diameter · Aerodynamic interactions · Momentum source method

1 Introduction

Compared with helicopter and fixed-wing aircraft, the tiltrotor configuration affords the ability to vertically take off and land and cruise at high speed. In hovering flight, the tiltrotor requires large diameter and high RPM, while a small diameter and low RPM are wanted in forward flight, which restricts the aerodynamic design of the tiltrotor [1, 2]. A rotor with variable RPM and diameter could resolve these two opposing problems [3–5]. Zhu studied the interaction of the rotor and the wing and optimized the wing by the CFD method, improving the lift/drag by 36.78% [6]. Zhao measured aerodynamic interactional characteristics between the rotor and the

H. Wang · Q. Zhao (✉) · G. Zhao · X. Zhang · J. Shi
National Key Laboratory of Rotorcraft Aeromechanics, College of Aerospace Engineering, Nanjing University of Aeronautics and Astronautics, Nanjing 210016, China
e-mail: zhaoqijun@nuaa.edu.cn

S. Lee et al. (eds.), *The Proceedings of the 2021 Asia-Pacific International Symposium on Aerospace Technology (APISAT 2021), Volume 1*, Lecture Notes in Electrical Engineering 912, https://doi.org/10.1007/978-981-19-2689-1_41

wing of a tiltrotor airfoil both in hover and transition modes by use of experimental methods [7]. Li proposed a new highly-effective hybrid CFD method for reducing the calculation load of the CFD simulation on the aerodynamic force of the tiltrotor in transition flight [8].

A quad-tiltrotor has wide range of center of gravity. By reducing the disc loading, the hovering efficiency can be appropriately improved compared with a tiltrotor aircraft [9, 10]. Wu studied aerodynamic interactions of a quad-tiltrotor UAV with the momentum source method, and the numerical results indicated that the influence of the rotor on the wing would decrease with an increase of flight speed under tilting conditions [11]. Zhou established an optimization model of a quad-tiltrotor [12].

Few studies on the aerodynamic interactions of a quad-tiltrotor aircraft with variable RPM and diameter have been reported. In this paper, employing the concept of a tiltrotor with variable RPM and diameter, numerical investigations on aerodynamic interactions between the rotors/wing/fuselage of a quad-tiltrotor aircraft are conducted.

2 Model and Calculation Method

The quad-tiltrotor UAV model used for CFD calculation is shown in Fig. 1. Its specific dimensions is listed in Table 1, which has a tandem wing layout. The front and rear wings have the same installation angle, wingspan, and chord length. The rotors are

Fig. 1 The quad-tiltrotor model

Table 1 Specific dimensions of model

Length of fuselage	7 m
Width of rotors turning (1R)	9.15 m
Wingspan	3 m
Installation angle of wing	5.7°
Chord length	0.95 m
Distance between front and rear wings	3.5 m

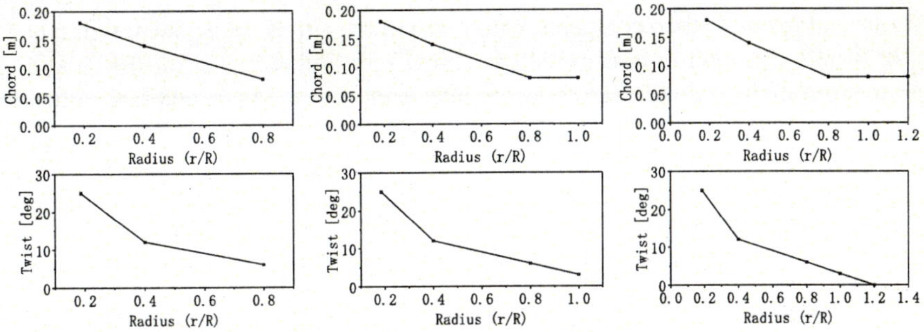

(a) 0.8R rotor (b) 1R rotor (a) 1.2R rotor

Fig. 2 Twist and chord length distribution of the variable diameter rotor

mounted on the tip of the wing and rotate counterclockwise. Front and rear rotors are at the same horizontal height. The zero azimuth angle of the rotors points to the rear of the fuselage when hovering, and points upward when flying forward. The base radius of the rotor is 1375 mm (1R), and the root-cut is 0.185R, The outer section of the rotor can be extended or shortened in the range of 0–0.2R, the twist and chord length distribution of the rotor radius at 0.8R、1R and 1.2R are shown in Fig. 2. The four rotors have the same radius, RPM and tilt angle at the same time.

The CFD method is used to calculate the aerodynamic interference of the quad-tiltrotor, the momentum source method is used to simulate the rotor flowfield, and the body-fitted grid method is used for the fuselage. In order to make the calculation results more reasonable, cyclic pitch trim was performed, and the thrust trim was performed in the forward flight and tilting state, this means that the front and rear rotors have the same thrust at the same time. Given the collective pitch when hovering, and limit the required power of the rotor within a reasonable range.

3 Aerodynamic Interaction Between Rotors in High-Speed Cruise

First, the flowfield of single rotor (including fuselage and wing) and the flowfield of the whole aircraft are calculated respectively. The rotor and flow field parameters are shown in Table 2. Under the same thrust of the rotor, the required power, disc inflow, angle of attack, and normal force distribution of the front and rear rotors of

Table 2 Operation conditions in cruise

Rotate speed	100 rad/s
Rotor radius	1.375 m (1R)
Forward speed	70 m/s
Rotor thrust	0 N, 150 N, 300 N, 450 N

the quad-tiltrotor are then compared with those of the single rotor. The interference of the front and rear rotors is also analyzed. Finally, the influence law of RPM, radius, and forward flight speed on aerodynamic interference of the front and rear rotors is studied.

3.1 Aerodynamic Interaction Between Rotors

In forward flight, the rear rotor of the quad-tiltrotor aircraft is in the wake of the front rotor (Fig. 3, Fig. 4). The required power and collective pitch of the front rotor, rear rotor, and single rotor of a quad-tiltrotor aircraft under different thrust during forward flight are shown in Fig. 5 and Fig. 6. The curve is basically linear, and the required power is large at zero thrust. The data of the front rotor and the single rotor are almost the same, and the required power and collective pitch of the front rotor is larger than that of the rear rotor, which indicates that the rear rotor has little effect on the front rotor, but has beneficial effects under aerodynamic interaction of the front rotor.

When the thrust is 300 N, the inflow velocity, angle of attack and normal force distribution of the front rotor, rear rotor and single rotor are shown in Figs. 7, 8 and 9. The inflow distribution shows that the front rotor increases the inflow at the tip and decreases the inflow at the root of the rear rotor (Fig. 7), so that compared with

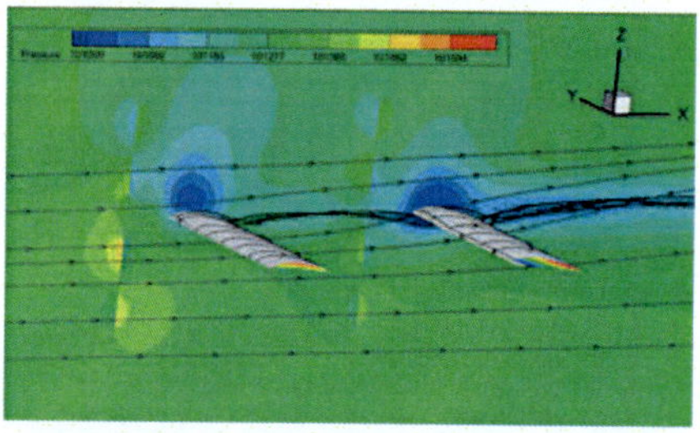

Fig. 3 Pressure distribution on longitudinal plane cross center of disc

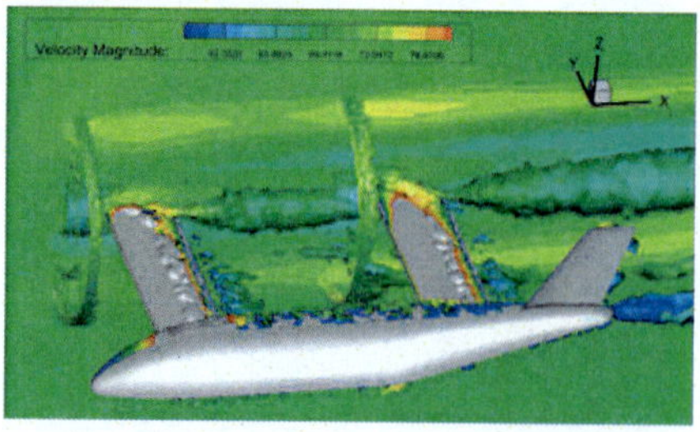

Fig. 4 Velocity distribution on longitudinal plane cross center of disc

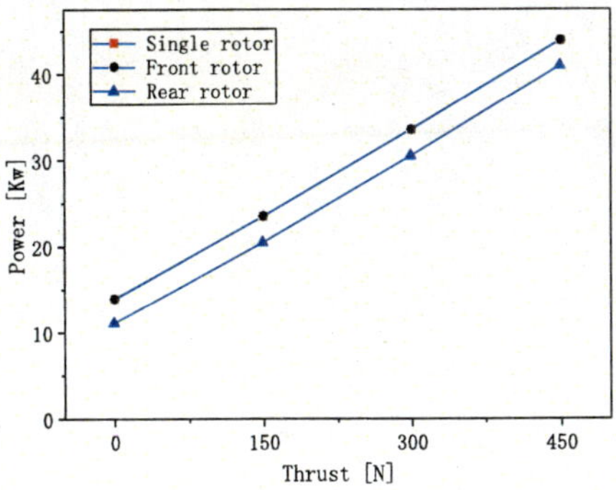

Fig. 5 Required power comparison in forward flight conditions

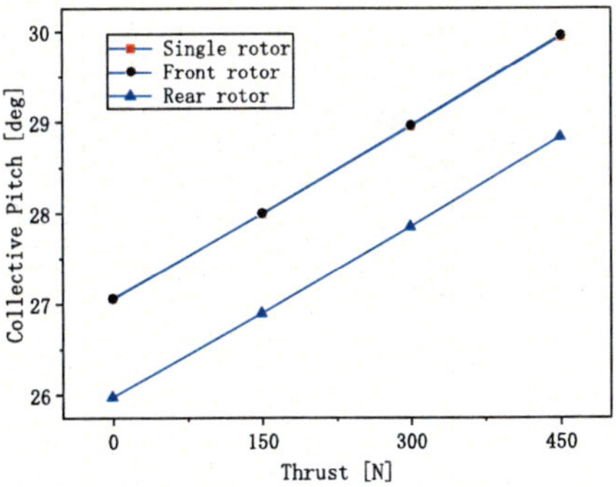

Fig. 6 Collective pitch comparison in forward flight conditions

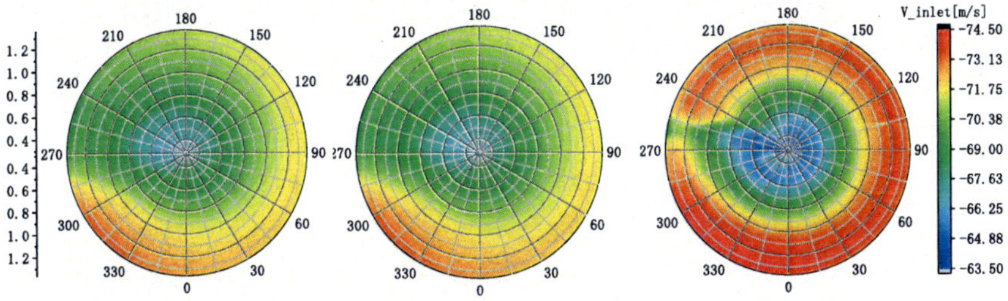

(a) The single rotor (b) Front rotor (c) Rear rotor

Fig. 7 Comparison of inflow velocity distribution on disc

the front rotor, the angle of attack of the rear rotor is slightly closer to the zero angle of attack (Fig. 8), positive thrust of rotor tip and negative thrust of rotor root also decrease (Fig. 9), and the efficiency is improved. The inflow in front of the wing is decelerated, and the airflow above the wing is accelerated, which also affects the inflow distribution of the rotor near the wing side (Fig. 7). Because cyclic pitch trim

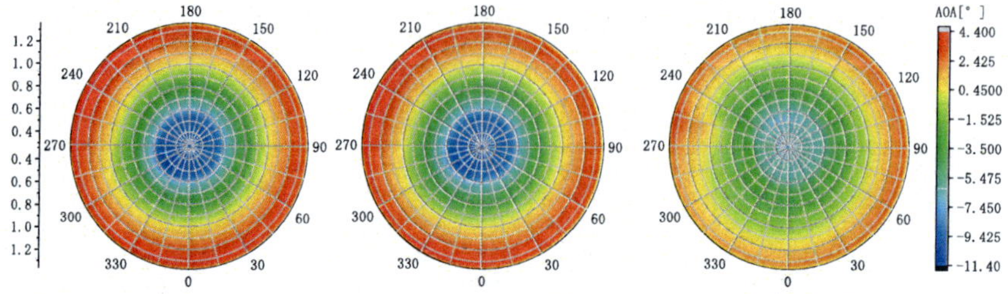

(a) The single rotor (b) Front rotor (c) Rear rotor

Fig. 8 Comparison of angle of attack distribution on disc

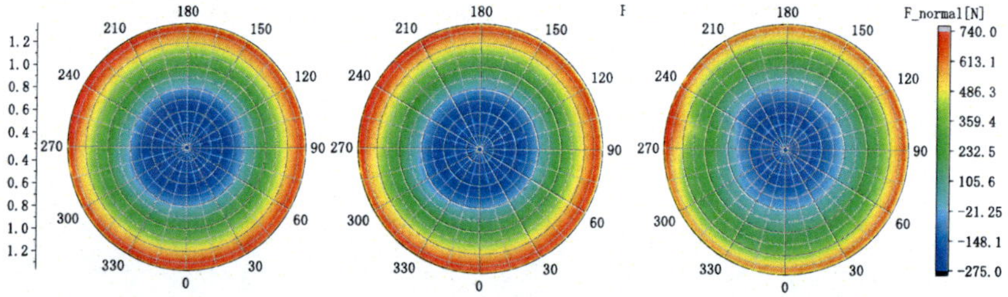

(a) The single rotor (b) Front rotor (c) Rear rotor

Fig. 9 Comparison of normal force distribution on disc

is used, the angle of attack is basically uniform along the azimuth. Since the rear rotor has little influence on the front rotor, this paper uses the required power difference of the front and rear rotors under the same thrust to measure the interference of the front rotor to the rear rotor.

3.2 *The Effect of Rotor Diameter on Aerodynamic Interaction Between Rotors*

Keeping the tip speed, forward speed and angle of attack unchanged (Table 2), the mutual interference of the front and rear rotors in three diameter states (Fig. 2) is calculated and analyzed. When the rotor radius is different, the required power of the front and rear rotors under different thrust is shown in Fig. 10a, and the difference between the required power of the front and rear rotors is shown in Fig. 10b. The results show that the favorable interference of the front rotor on the rear rotor decreases with the decrease of the rotor radius. When the rotor radius is reduced to a certain value, the advantage effect disappears. When the collective pitch increases to a certain value, this advantage disappears. In forward flight, the negative thrust and

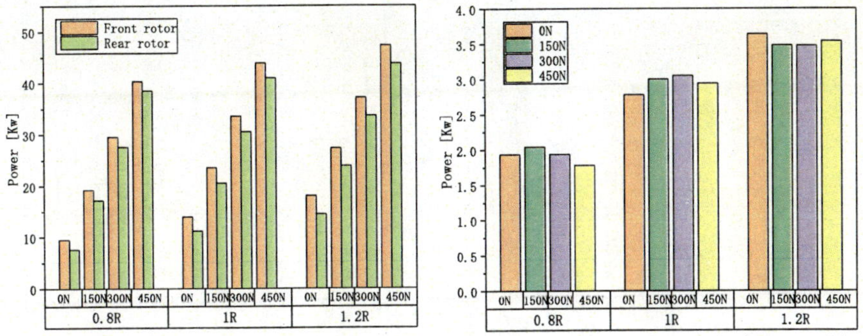

(a) Comparison of the required power (b) Difference between the required power of the front and rear rotors

Fig. 10 The effect of rotor diameter on interaction of the front rotor and the rear rotor

its region of the rotor decreases with the decrease of the rotor radius (Fig. 11), and the efficiency increases with the decrease of the rotor radius.

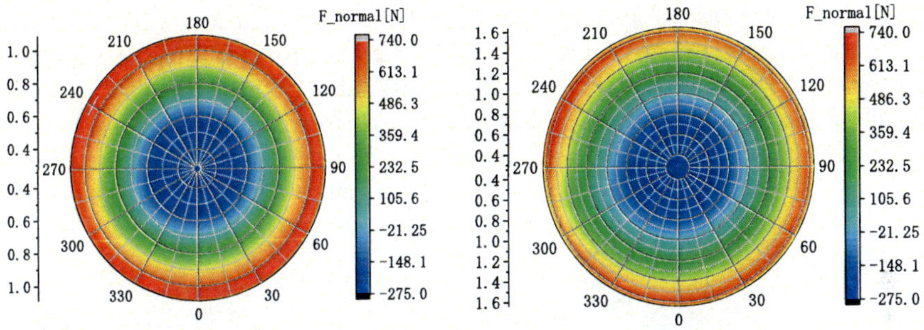

(a) 0.8R (Front rotor) (b) 1.2R (Front rotor)

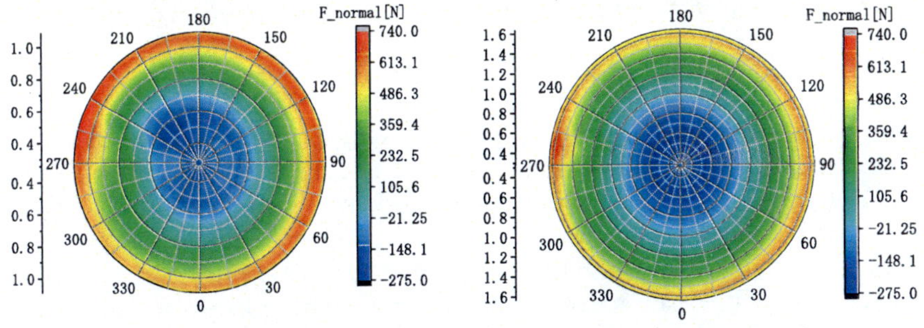

(c) 0.8R (Rear rotor) (d) 1.2R (Rear rotor)

Fig. 11 Comparison of normal force distribution in different rotor diameters (the thrust are all 300 N)

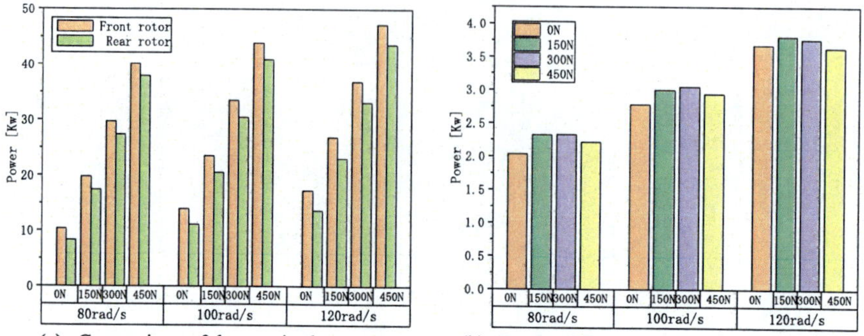

(a) Comparison of the required power (b) Power difference of front and rear rotor under different RPM and thrust

Fig. 12 The effect of RPM on interaction of the front rotor and the rear rotor

3.3 The Effect of RPM on Aerodynamic Interaction Between Rotors

On the basis of the baseline condition shown in Table 2, the influence of RPM on the interference of the front and rear rotors is studied. A comparison of the required power of the front and rear rotors at different RPM of 80 rad/s, 100 rad/s/s, and 120 rad/s depending with different thrust is shown in Fig. 12a. The difference between the required power of the front and rear rotors is shown in Fig. 12b. The decrease of RPM weakens this interference, and it also reduces the required power. When the rotor RPM decreases, the negative thrust area and the negative thrust of the blade root decrease, as shown in Fig. 13.

3.4 The Effect of Advance Ratio on Aerodynamic Interaction Between Rotors

The rotor radius and RPM in Table 2 are used to study the interference of the front and rear rotors when the advance ratio is 0.4, 0.5 and 0.6. The comparison of the required power of the front and rear rotors with different forward ratios is shown in Fig. 14a. The difference between the required power of the front rotor and the rear rotor is shown in Fig. 14b. It also decreases with the decrease of the advance ratio. The advance ratio decreases, and the inflow decreases, which reduces the rotor power consumption. When the advance ratio is reduced, the area of negative thrust and the negative thrust in rotor root are also reduced under the same thrust (Fig. 15).

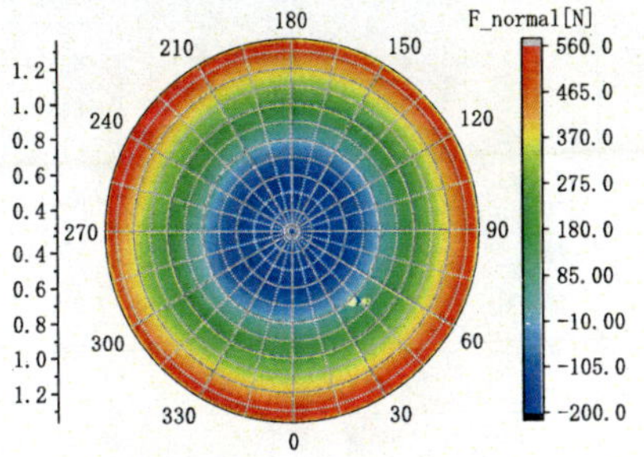

(a) 80 rad/s (Front rotor)

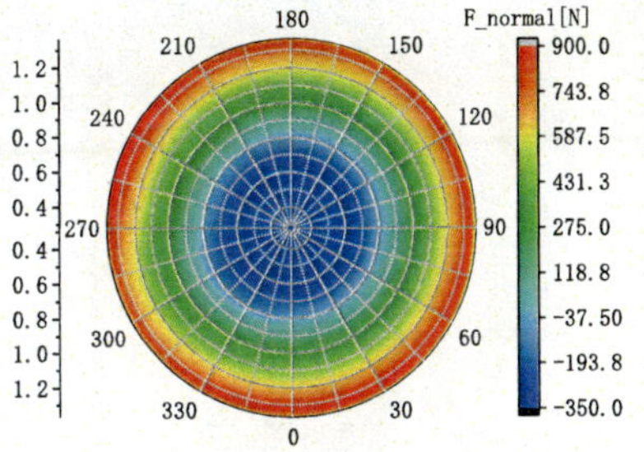

(a) 120 rad/s (Front rotor)

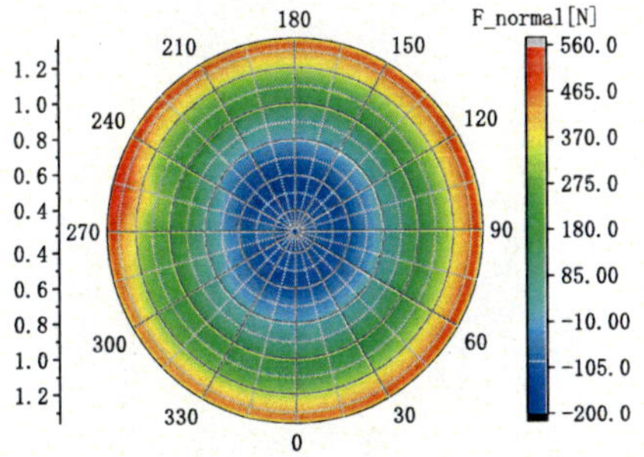

(c) 80 rad/s (Rear rotor)

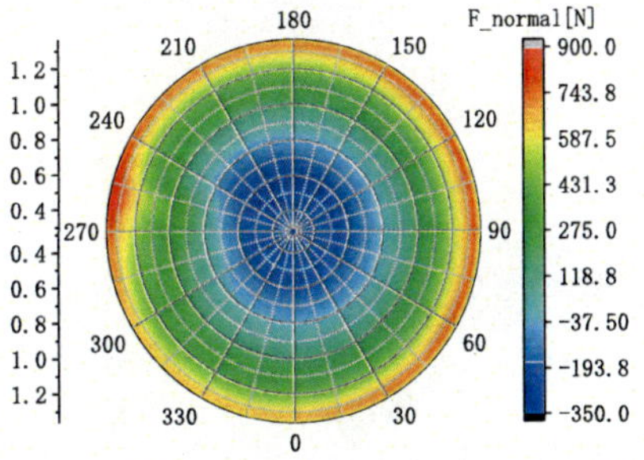

(d) 120 rad/s (Rear rotor)

Fig. 13 Comparison of normal force distribution in different RPM (the thrust are all 300 N)

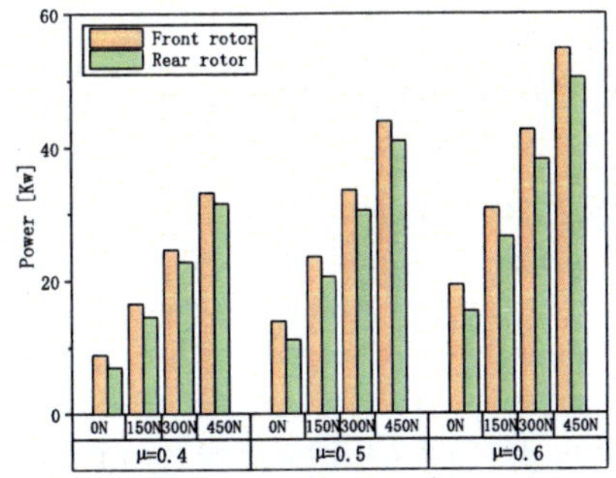

(a) Comparison of the required power

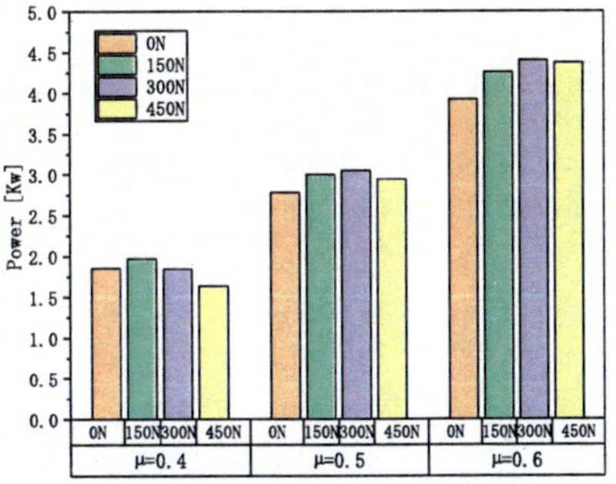

(b) Difference between the required power of the front and rear rotors

Fig. 14 The effect of advance ratio on interaction of the front rotor and the rear rotor

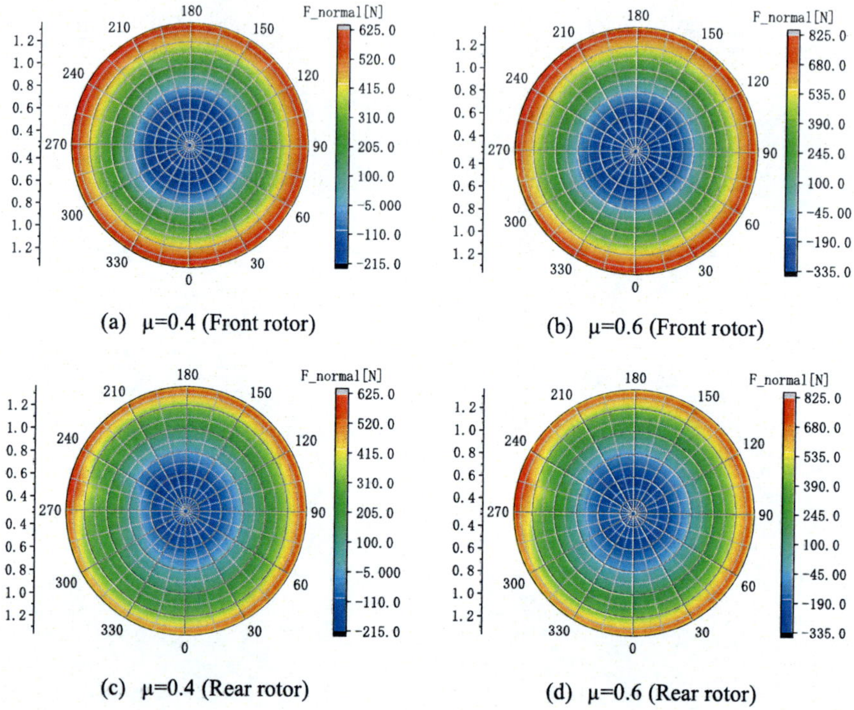

(a) μ=0.4 (Front rotor)

(b) μ=0.6 (Front rotor)

(c) μ=0.4 (Rear rotor)

(d) μ=0.6 (Rear rotor)

Fig. 15 Comparison of normal force distribution with different advance ratios (the thrust is 300 N for all cases)

4 Aerodynamic Interactions Under Tilting Condition

The flowfield of the whole aircraft is calculated under different tilting angles. The rotor and flowfield parameters are shown in Table 3. The interference of the front and rear rotors and its causes are analyzed. The influence law of RPM and radius on the aerodynamic interaction of front and rear rotors is studied under Condition 1.

The required power and cyclic pitch of the front and rear rotors of a quad-tiltrotor aircraft under different tilt angles are shown in Fig. 16. When the forward speed and the tilt angle of the rotors are small, the required thrust of the rotors is large and induced flow of the front rotor results in increment of the inflow velocity of the rear

Table 3 Operation conditions in tilting

	Condition 1	Condition 2	Condition 3
Flight speed	20 m/s	35 m/s	50 m/s
Tilt angle	25°	45°	70°
Attack angle of fuselage	8°	8°	4°

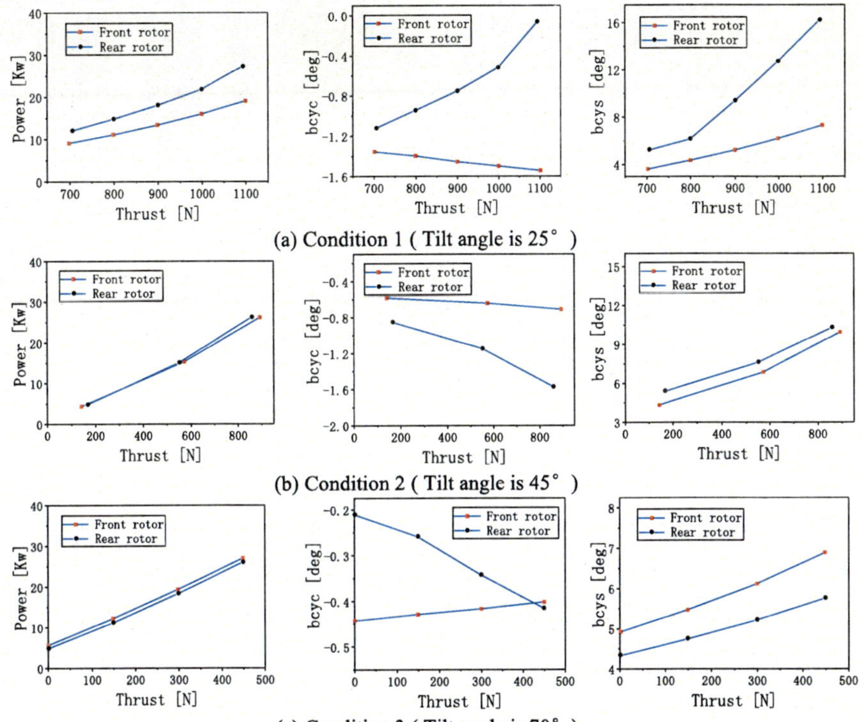

(a) Condition 1 (Tilt angle is 25°)

(b) Condition 2 (Tilt angle is 45°)

(c) Condition 3 (Tilt angle is 70°)

Fig. 16 Comparison of the required powerand collective pitch between front and rear rotors

rotor (Fig. 17), which lowers the efficiency of the rear rotor. When the flight speed and tilt angle increase, the efficiency of the rear rotor becomes higher than that of the front rotor.

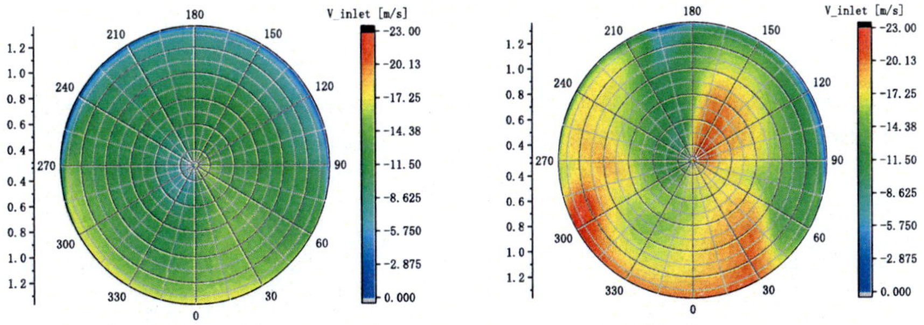

(a) Front rotor (b) Rear rotor

Fig. 17 Comparison of inflow velocity distribution on disc (Condition 1, Thrust are all 1100 N)

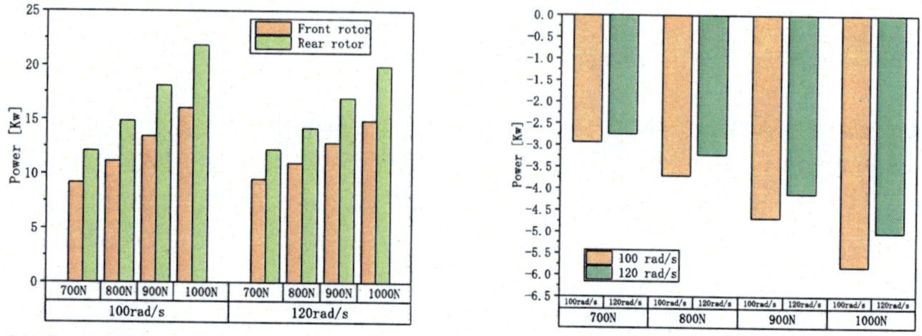

(a) Comparison of the required power

(b) Power difference of front and rear rotor under different RPM

Fig. 18 The effect of RPM on interaction of the front rotor and the rear rotor under tilting condition (Condition 1)

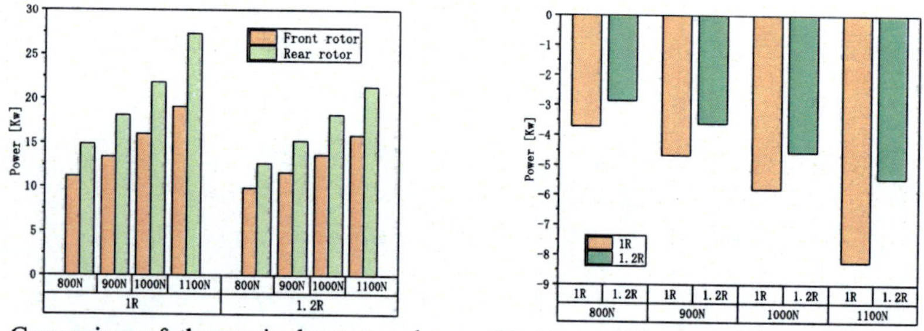

(a) Comparison of the required power under different rotor diameter

(b) Power difference of front and rear rotor under different rotor diameter

Fig. 19 The effect of rotor diameter on interaction of the front rotor and the rear rotor under tilting condition (Condition 1)

Figure 18 shows the effect of RPM on interaction of the front rotor and the rear rotor under a tilting condition, In a reasonable RPM range, the adverse interference is weakened with increasing RPM, which results from the decreased inflow angle.

Figure 19 shows the effect of rotor diameter on the interaction of the front rotor and the rear rotor under a tilting condition. The adverse interference is weakened with increasing rotor diameter, because the velocity of inflow decreases.

5 Conclusions

The aerodynamic interactions between the rotors of a quad-tiltrotor during high-speed cruise and tilting conditions are studied. We arrived at the following conclusions:

(1) When the forward speed and the tilt angle of the rotors are small, the efficiency of the rear rotor is low, this adverse interference is weakened with increasing

rotor diameter and RPM. When the flight speed and tilt angle increase, the efficiency of the rear rotor becomes higher than that of the front rotor.

(2) When the forward speed and the tilt angle are small, the required thrust of the rotors is large, large diameter and proper RPM can improve the efficiency of the rotor.

(3) When cruising at high speed, reducing the rotor RPM and diameter can improve the aerodynamic efficiency of the rotor, and the front rotor has a favorable impact on the rear rotor, which decreases with a decrease of the advance ratio, rotor diameter, and RPM.

Acknowledgements This research is supported by the National Natural Science Foundation of China (11872211), the National Key Laboratory Foundation of China (no. 61422200101) and Project Funded by the Priority Academic Program Development of Jiangsu Higher Education Institutions (PAPD).

References

1. Qijun Z, Guoqing Z, Qing W, Guohua X (2020) Advanced rotor design aerodynamics. Beijing: Science Press
2. Qijun Z, Shuang J, Peng L, Bo W, Hang Z (2017) Aerodynamic optimization analyses of tiltrotor/propeller based on CFD method. Acta Aerodynamica Sinica 35(4):544–553
3. Garavello A, Benini E (2012) Preliminary study on a wide-speed-range helicopter rotor/turboshaft system. J Aircr 49(4):1032–1038
4. Mistry M, Gandhi F (2014) Helicopter performance improvement with variable rotor radius and RPM. J Amer Helicopter Soc 59(4):13–35
5. Zhao G, Zhao Q, Li P et al (2021) Parametric analyses for effects of RPM and diameter on tiltrotor aerodynamic performances in hovering and cruise Mode. Int J Aeronaut Space Sci 22:479–488
6. Qiuxian Z, Qijun Z, Yongfeng L (2017) Analysis and optimizations on aerodynamic interaction of tiltrotor aircraft multi-components on its wing. J Aerosp Power 32(6):1505–1514
7. Qijun Z, Tongbing N, Peng L et al (2018) Experiment on flow mechanism and aerodynamic interaction characteristics od tilt-rotor aircraft. J Aerosp Power 33(12):2900–2912
8. Peng L, Qijun Z, Zhengzhaong W, Bo W (2015) Highly-efficient CFD method for predicting aerodynamic force of tiltrotor in conversion mode. J Nanjing Univ Aeronaut Astronaut,https://doi.org/10.16356/j.1005-2615.2015.02.003
9. Wu X, Mu X (2012) A perspective of the future development of key helicopter technologies. Acta Aerodynamica Sinica 39(3):1–10
10. David S (2000) The quad tiltrotor: it's Beginning and evolution. International powered lift conference
11. Yongcheng W, Zili C, Jinjian J (2020) Quad-tilt rotor UAV aerodynamic characteristics. J Ordnance Equip Eng 41(03):72–75
12. Qichen Z, Chunhua L (2016) Optimization design of primary parameters for high-speed quad tiltrotor. helicopter technique, 1673-1220(2016)01-00-06

Separation Characteristics of a Two-Stage-To-Orbit Winged Rocket by Aerodynamic Interaction Analysis

Tsuyoshi Otsuki, Takahiro Fujikawa, and Koichi Yonemoto

Abstract The booster and orbiter stages of two-stage-to-orbit (TSTO) launch vehicles are separated at supersonic or hypersonic speed regimes under non-negligible dynamic pressure. An aerodynamic interaction exists between the vehicles during the aforementioned separation phase owing to the presence of complicated flow fields with shock wave (SW) interactions. Therefore, the separation operation and attitude control considering such interactions are crucial. In this study, the construction of a database comprising aerodynamic characteristics based on the results obtained using computational fluid dynamics (CFD) analysis is discussed. Based on the results obtained through CFD analysis, the pitch-up and pitch-down moments act on the orbiter and booster stages, respectively when two vehicles are located close. This makes the two vehicles move away from each other. By contrast, the pitch-down and pitch-up moments act on the orbiter and booster stages, respectively, when the two vehicles are located far. The aerodynamic interaction affects the pitching moment may reverse during the separation.

Keywords Two-stage-to-orbit (TSTO) · Separation · Aerodynamic interaction · Hypersonic

1 Introduction

Winged two-stage-to-orbit (TSTO) launch vehicles has been studied as a promising target of future space transportation systems. The booster and orbiter stages of TSTO launch vehicles will be in a state where the engine is cut off; after a sufficient distance, the orbiter engine is ignited when the stages are separated at supersonic or hypersonic

T. Otsuki (✉) · T. Fujikawa (✉) · K. Yonemoto (✉)
Tokyo University of Science, Yamazaki 2641, Noda, Japan
e-mail: 7520510@ed.tus.ac.jp

T. Fujikawa
e-mail: fujikawa.takahiro@rs.tus.ac.jp

K. Yonemoto
e-mail: yonemoto@rs.tus.ac.jp

S. Lee et al. (eds.), *The Proceedings of the 2021 Asia-Pacific International Symposium on Aerospace Technology (APISAT 2021), Volume 1*, Lecture Notes in Electrical Engineering 912, https://doi.org/10.1007/978-981-19-2689-1_42

speed regimes under non-negligible dynamic pressure. An aerodynamic interaction exists between the vehicles during the aforementioned separation phase owing to the presence of complicated flow fields with shock wave (SW) interactions. Therefore, separation operation and attitude control considering such interactions are crucial. Related previous studies include the following: (1) the wind tunnel testing of a booster and orbiter combination with varying relative positions [1] and (2) the numerical simulation of separation flight [2]. These studies are based on the predefined mated position of the vehicles and their separation sequences.

Various vehicle concepts have been envisioned for the TSTO launch vehicles, and it is unclear whether the separation is considered [3] in several rockets. However, separation sequence that involves separating the stages at the launch of TSTO launch vehicles are challenging and has the least reliability. Therefore, establishing a framework to quantitatively evaluate and compare the feasibility of separation is necessary to increase the reliability.

This study aims to clarify the appropriate mated position of TSTO launch vehicles, the separation sequence, and the time history of the angle of attack (AoA) during the separation by optimizing them simultaneously. The objective functions of the optimization are the minimum of the interaction effect during the separation, the maximum of the distance between the vehicles after a given period. This study comprises (1) the construction of a database comprising aerodynamic characteristics as well as interactions and (2) the application of trajectory optimization techniques. In this paper, the construction of an aerodynamic database based on the results obtained using computational fluid dynamics (CFD) analysis is discussed. Furthermore, the deflection of aerodynamic control surfaces required to cancel the interaction effect is shown and the appropriate mated position is discussed.

2 CFD Analysis Conditions

2.1 Numerical Analysis Method

FAST aerodynamic routine (FaSTAR) was used as a three-dimensional compressible fluid analysis solver [4] for unstructured grids developed by the Japan Aerospace Exploration Agency (JAXA). The calculation methods employed in the aforementioned routine are presented in Table 1. We used the three-dimensional compressible Navier–Stokes equation as the governing equation. Reynolds-Averaged Navier–Stokes equations (RANS) Spalart–Allmaras-noft2-R was used as the turbulence model. The cell center method of the finite volume method was employed for spatial discretization. It was evaluated by multiplying the vertical flux on each cell surface by the area. Furthermore, the Harten–Lax–van Leer–Einfeld method, which is a stable method that approximates and solves with two characteristic waves and is effective for flows with strong expansion. This method was employed to determine the non-viscous flow velocity. However, it has the disadvantage of possessing a high

Table 1 Analysis method of FaSTAR

Governing equation	Compressible Navier–Stokes equation
Turbulence model	Spalart–Allmaras-noft2-R
Spatial discretization method	Cell center method and Harten–Lax–van Leer–Einfeld
Time integration method	Lower upper symmetric Gauss–Seidel Implicit method
Grid automatic generation tool	HexaGrid
Number of cells [-]	8.09×10^6–8.52×10^6
Calculator	JAXA Supercomputer System generations 2 and 3

numerical viscosity. The lower upper symmetric Gauss–Seidel implicit method was used as the time integration method. HexaGrid, an unstructured grid automatic generation tool developed by JAXA, was employed as the grid generation tool [5]. As a convergence test, if the difference between the last value of the CFD calculations and the value before it is 0.001 or less, it is considered to have converged. JAXA Supercomputer System generations 2 and 3 were used as the calculators.

2.2 *Analysis Model*

The TSTO launch vehicle considered in this study is based on the conceptual design of a system proposed in a previous study [6]. It is the combination of a reusable winged booster and an expendable wingless orbiter with a separation Mach number of 5.8. It can transport a 100-kg small satellite to a sun-synchronous orbit at an altitude of 700 km as well as perform vertical launch and horizontal landing. The shape of the booster is based on the WSV-C-5 configuration of a highly maneuverable experimental space vehicle [7]. Table 2 and Fig. 1 show the specifications, conceptual design, and flight profile of the target TSTO launch vehicle, respectively. With the tip of the nose of each vehicle as the origin, the centers of gravity of the booster and orbiter were at 61% and 55% of the lengths of the booster and orbiter fuselage, respectively.

2.3 *Analysis Conditions*

Table 3 presents the aerodynamic coefficient nondimensionalization parameters of the target vehicle and the design variables of the proximity between the vehicles during the rise. Figure 2 shows the definition of relative positions of vehicles. When the center of gravity of the booster is the origin, the fixed coordinate system

Table 2 Specifications of TSTO launch vehicle

Target orbit	700		[km]
Payload mass	100		[kg]
Liftoff mass	30.2		[t]
	Booster	Orbiter	
Total length	14.0	8.30	[m]
Body diameter	2.69	1.66	[m]
Representative area	25.4	–	[m^2]
MAC	3.30	3.30	[m]
Wingspan	8.86	–	[m]
Dry mass	4.7	8.2	[t]
Propellant mass	19.6	5.1	[t]
Gross mass	24.3	6.0	[t]
Number of engines	6	1	[-]
Nozzle expansion ratio	25	75	[-]
Total vacuum thrust	582	102	[kN]
Vacuum Isp	338	335	[s]
Center of gravity	7.507	4.565	[m]

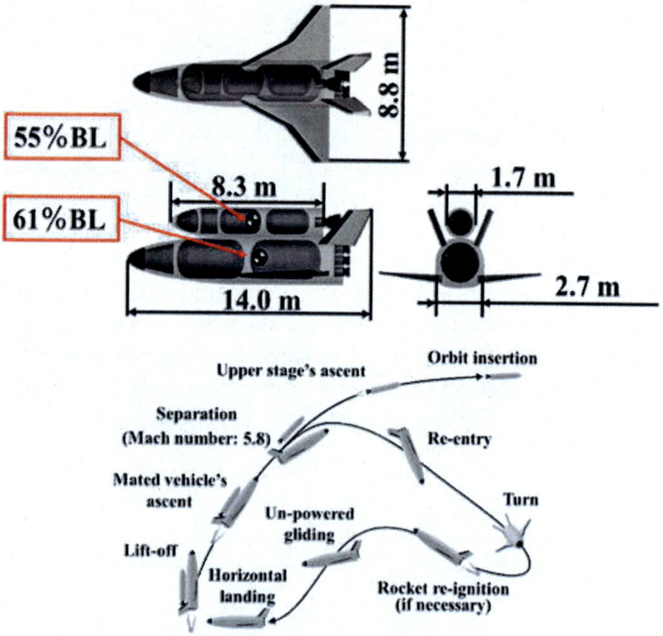

Fig. 1 Schematic and flight profile of TSTO launch vehicle

Table 3 CFD conditions and vehicle relative locations

Distance between bodies $\Delta z/D$	1/20, 1/4, 1/2, 1	[–]
Location along X_b axis $\Delta x/L$	0, 1/20, 1/10, 1/4	[–]
Pitch angle difference $\Delta\theta$	0.0, 5.0, 10.0	[deg.]
AoA of orbiter α_o	−5.0, 0.0, 5.0, 10.0	[deg.]
Booster elevator angle δ_e	0.0, 20.0	[deg.]

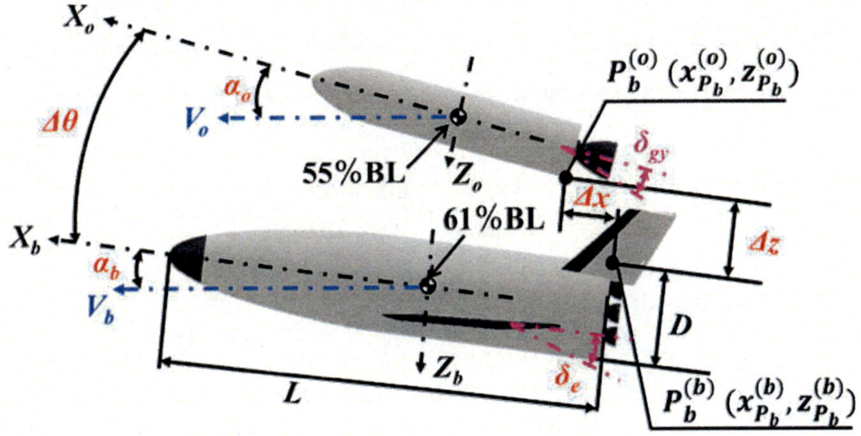

Fig. 2 Definition of the relative position of vehicles

of the booster is (X_b, Z_b) and let the orbiter's fixed coordinate system be (X_o, Z_o). When the uppermost point of the booster's base in the booster fixed coordinate system is $P_b{}^{(b)}(x_{P_b}{}^{(b)}, z_{P_b}{}^{(b)})$ and the lowermost point of the orbiter's base is $P_b{}^{(o)}(x_{P_b}{}^{(o)}, z_{P_b}{}^{(o)})$, the relative position between two vehicles is given by Eqs. 1 and 2. In addition, the angle between the X_b and X_o axes is defined as $\Delta\theta$ and the orbiter gimbal angle is defined as δ_{gy}.

$$\Delta x = x_{P_b}^{(o)} - x_{P_b}^{(b)}, \tag{1}$$

$$\Delta z = z_{P_b}^{(o)} - z_{P_b}^{(b)}. \tag{2}$$

In this study, the orbiter's velocity vector V_o and booster's velocity vector V_b are approximated to be parallel. Equation 3 holds among the pitch angle difference $\Delta\theta$, the orbiter's AoA α_o, and the booster's AoA α_b.

$$\alpha_b = \alpha_o - \Delta\theta. \tag{3}$$

We constructed an aerodynamic database while changing five design variables, the location along X_b axis Δx, the distance between bodies Δz, the Pitch angle difference

Table 4 Flow conditions at separation

Altitude above sea level	35.0	[km]
Mach number	5.8	[–]
Heat capacity ratio	1.4	[–]
Reynolds number	1.33×10^7	[–]
Temperature	237	[K]
Static pressure	5.59×10^2	[Pa]
Atmospheric density	8.21×10^{-3}	[kg/m^3]

$\Delta\theta$, the AoA of orbiter α_o, and the booster elevator angle δ_e, within the range in Table 3.

Table 4 shows the flow conditions at the TSTO launch vehicle separation, which is the value that emerges from the trajectory optimization techniques.

3 Analysis Results and Consideration

3.1 Change of Δz

The result of CFD analysis by changing Δz, the relationship between the distance between body axes ($\Delta z/D$) and pitching-moment coefficient (C_m) around the center of gravity is shown in Fig. 3. When two vehicles are located close ($\Delta z/D < 0.05$), the pitch-up and pitch-down moments acted on the orbiter and booster, respectively. Hence, the two vehicles move away from each other. By contrast, the pitch-down and pitch-up moments acted on the orbiter and booster, respectively, when the two vehicles were far apart ($\Delta z/D > 0.10$). Therefore, the aerodynamic interaction effect on the pitching moment might reverse during the separation. This result can be

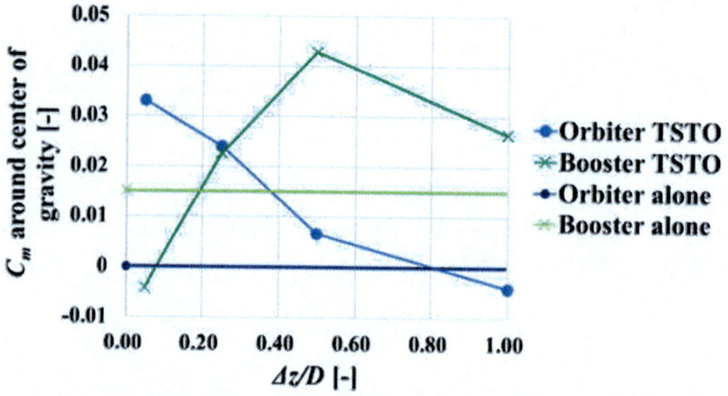

Fig. 3 Pitching-moment coefficient ($\Delta x/L = 0.0$, $\Delta\theta = 0.0$, $\alpha_o = \alpha_b = 0.0$, and $\delta_e = 0.0$)

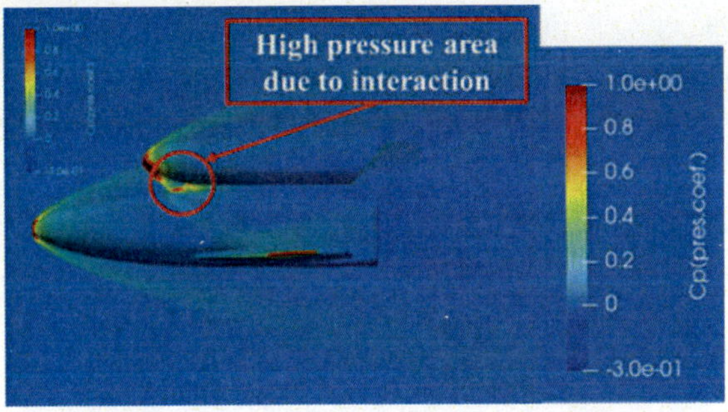

Fig. 4 Pressure coefficient distribution ($\Delta z/D = 0.05$, $\Delta x/L = 0.0$, $\Delta\theta = 0.0$, $\alpha_o = \alpha_b = 0.0$, and $\delta_e = 0.0$)

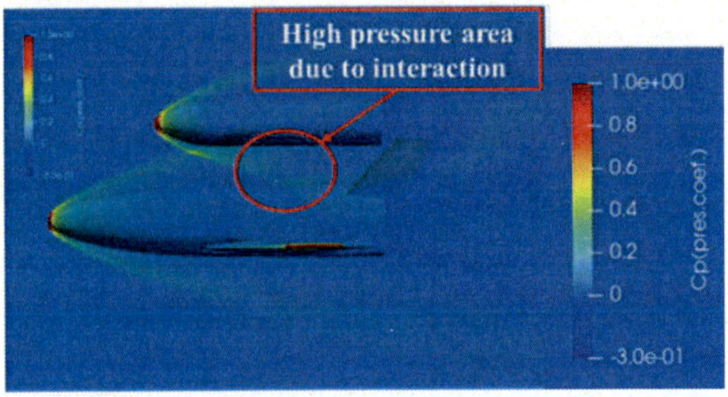

Fig. 5 Pressure coefficient distribution ($\Delta z/D = 0.50$, $\Delta x/L = 0.0$, $\Delta\theta = 0.0$, $\alpha_o = \alpha_b = 0.0$, and $\delta_e = 0.0$)

explained by the pressure coefficient distribution shown in Figs. 4 and 5. When $\Delta z/D = 0.05$, the interactions of the SW generated from noses created a high-pressure region, whereas, when $\Delta z/D = 0.50$, the SW emitted from the orbiter acts on the booster's afterbody. Next, Fig. 6 shows the difference between the C_m of the booster or orbiter in the TSTO launch vehicle form and the C_m of the booster or orbiter alone form ($\Delta C_{m(Booster,\ Orbiter)}$). When $\Delta z/D$ increased, the C_m of the two vehicles approximated the alone form.

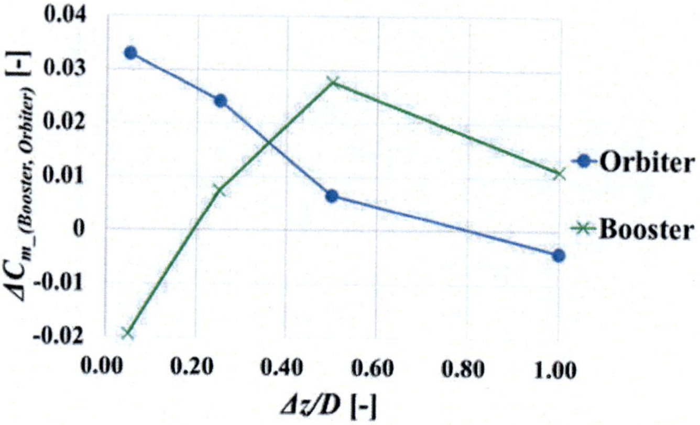

Fig. 6 $\Delta C_{m(Booster,\ Orbiter)}$ ($\Delta x/L = 0.0$, $\Delta\theta = 0.0$, $\alpha_o = \alpha_b = 0.0$, and $\delta_e = 0.0$)

3.2 *Change of* Δx

The result of CFD analysis by changing Δx, the relationship between the $\Delta x/L$ and pitching-moment coefficient around the center of gravity is shown in Fig. 7. Figure 8 shows the C_m difference between the two vehicles. When $\Delta x/L = 0.10$ and $\alpha_o = \alpha_b = -5.0$, the C_m difference was the largest and the pitching- moment acting along the direction, in which the nose of the orbiter and booster separate (the direction in which $\Delta\theta$ increases), was maximized. This result can be explained by the pressure coefficient distribution shown in Fig. 9. Compared with Fig. 4, the pressure around the lower part of the orbiter nose because of the SW interaction from the two vehicles was clearly higher, which was caused by the SW generated at the tip of the booster nose colliding with the orbiter closer. As shown in Fig. 7, the pitching-moment coefficient around the center of gravity decreased as $\Delta x/L$ increased and the nose down was thought to occur more easily because the interfuselage reflected SWs generated by

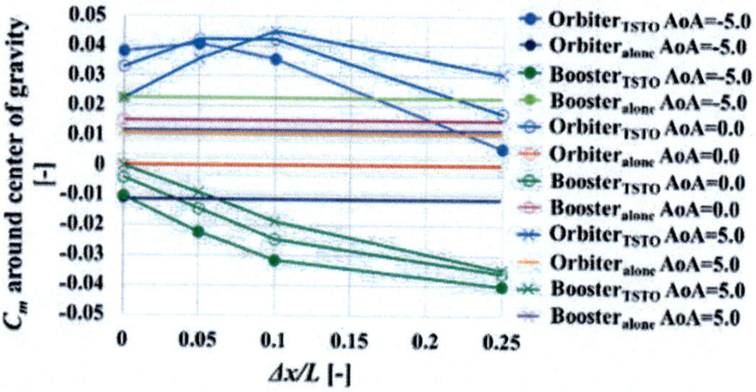

Fig. 7 Pitching-moment coefficient ($\Delta z/D = 0.05$, $\Delta\theta = 0.0$, and $\delta_e = 0.0$)

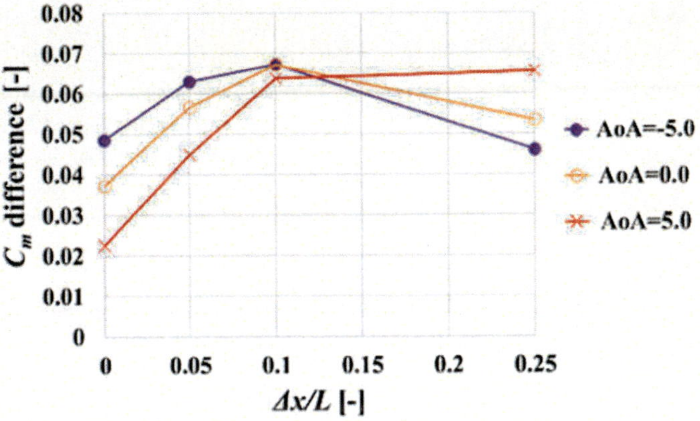

Fig. 8 Difference in C_m ($\Delta z/D = 0.05$, $\Delta\theta = 0.0$, and $\delta_e = 0.0$)

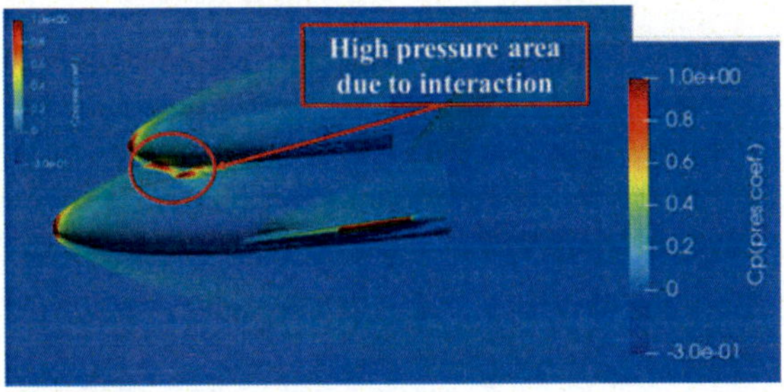

Fig. 9 Pressure coefficient distribution ($\Delta z/D = 0.05$, $\Delta x/L = 0.10$, $\Delta\theta = 0.0$, $\alpha_o = \alpha_b = -5.0$, and $\delta_e = 0.0$)

the orbiter and the SWs emitted from the booster affect. Therefore, the nose is likely to be lowered. However, if the orbiter is located further forward in the X_b axis ($\Delta x/L > 0.10$) compared with the booster, the SW from the booster would act behind the orbiter. Thus, as shown in Fig. 7, the C_m of the orbiter becomes smaller, which is an unsuitable position for separation. Based on the aforementioned results obtained from the proposed TSTO launch vehicle, separation can be easily established by arranging the orbiter appropriately in front of the booster along the X_b axis direction and performing separation at a negative AoA. Next, Fig. 10 shows $\Delta C_{m(Booster,\ Orbiter)}$. When $\Delta x/L$ increased, the amount of interference tended to increase in booster compared to the alone form. In contrast, orbiter approximated the C_m of the alone form.

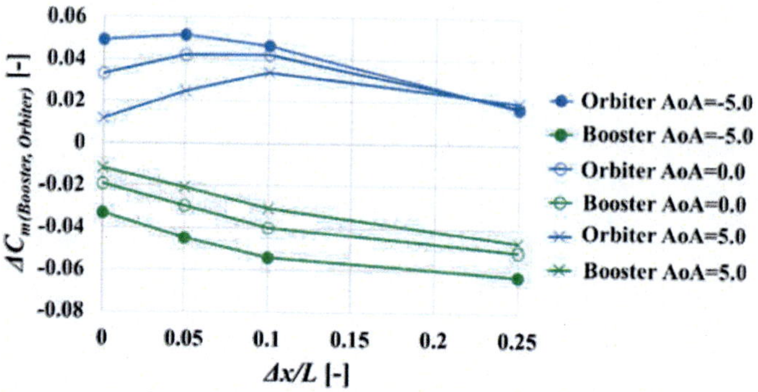

Fig. 10 $\Delta C_{m(Booster,\ Orbiter)}$ ($\Delta z/D\ =\ 0.05$, $\Delta\theta\ =\ 0.0$, and $\delta_e = 0.0$)

3.3 Evaluation of Pitching-Moment Interference with Deflection Angle of Control Devices

When $\Delta x/L\ =\ 0.10$ and other relative positions are changed, $\Delta C_{m\_Booster}$ and $\Delta C_{m\_Orbiter}$, which is the difference between the C_m of the booster in the TSTO launch vehicle form and the C_m of the booster alone form or the C_m of the orbiter alone form, was calculated. Moreover, $C_{m\delta e}$ was calculated from the CFD analysis results for the booster elevator angle of $\delta_e\ =\ 20.0$, $C_{m\delta z}$ was calculated from the calculation for the orbiter gimbal angle $\delta_{gy}\ =\ -\ 1.0$ and the amount of mutual interference in the TSTO launch vehicle form was evaluated to be equivalent to the number of degrees of the steering angle of the booster alone form. This elevator angle is called the equivalent elevator angle $\Delta\delta_e$, the equivalent gimbal angle $\Delta\delta_{gy}$ and is calculated as follows:

$$\Delta\delta_e = \frac{\Delta c_{m}\_Bosster}{c_m\delta_e} \tag{4}$$

$$\Delta\delta_{gy} = \frac{\Delta c_{m}\_Orbiter}{c_m\delta_{gy}} \tag{5}$$

As shown in Fig. 11, when $\Delta z/D$ was small and $\Delta\theta = 0.0$ (for example, $\Delta\delta_e$ when $\Delta z/D = 0.05$ and $\alpha_o =\ -\ 5.0$), $\Delta\delta_e$ tended to be very large. On the other hand, $\Delta\delta_e$ tended to be small when $\Delta z/D$ was between 0.25 and 0.50. Moreover, As shown in Fig. 12, when $\Delta\theta = 0.0$, $\Delta\delta_{gy}$ tended to be small at any relative position. Next, as shown in Fig. 13, when $\Delta z/D$ was small and $\Delta\theta = 5.0$ (for example, $\Delta\delta_e$ when $\Delta z/D = 0.05$ and $\alpha_o = -5.0$), $\Delta\delta_e$ tended to be small. Further, $\Delta\delta_e$ tended to be small even when $\Delta z/D$ was between 0.25 and 0.50. Moreover, as shown in Fig. 14, when $\Delta\theta =$ 5.0, $\Delta\delta_{gy}$ tended to be smaller at any relative position than when $\Delta\theta = 0.0$. These results can be explained by the pressure coefficient distribution shown in Figs. 15 and 16. The interactions of the SW generated from noses created a high-pressure region, which yielded a very large $\Delta\delta_e$ (Fig. 15). Conversely, the pressures in Fig. 16

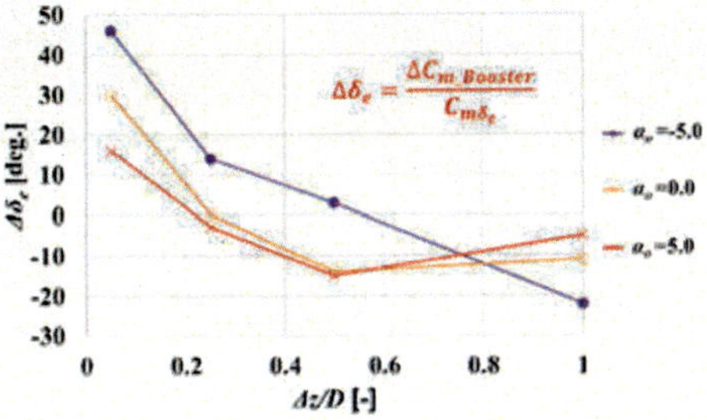

Fig. 11 Equivalent elevator angle $\Delta\delta_e$ ($\Delta x/L = 0.10$, $\Delta\theta = 0.0$)

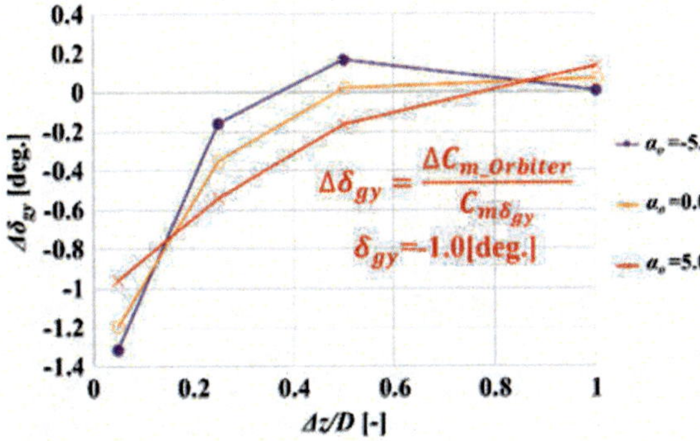

Fig. 12 Equivalent elevator angle $\Delta\delta_{gy}$ ($\Delta z/D = 0.10$, $\Delta\theta = 0.0$)

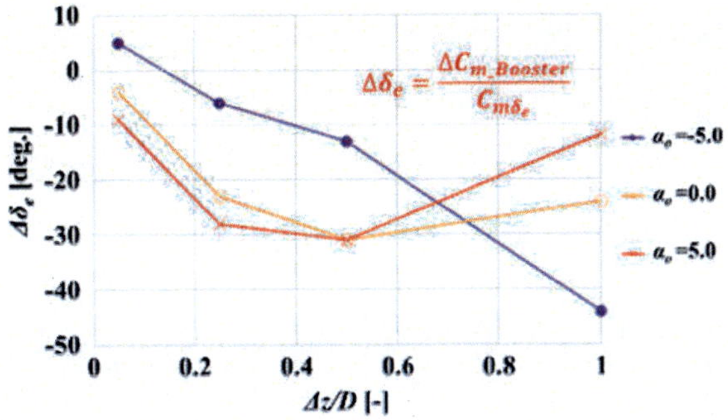

Fig. 13 Equivalent elevator angle $\Delta\delta_e$ ($\Delta x/L = 0.10$, $\Delta\theta = 5.0$)

was lower than that in Fig. 15. Therefore, when $\Delta z/D$ was fitted at a position where it was too small, $\Delta\delta_e$ became large and control was difficult. However, as shown in Fig. 17, when $\Delta\theta$ increased, pitching-moments acted in the direction in which the noses approached each other.

Subsections 3.1–3.3 show that the appropriate mated position should be a position with a negative AoA, a $\Delta z/D$ was between 0.25 and 0.50, a $\Delta x/L = 0.10$ and a $\Delta\theta =$

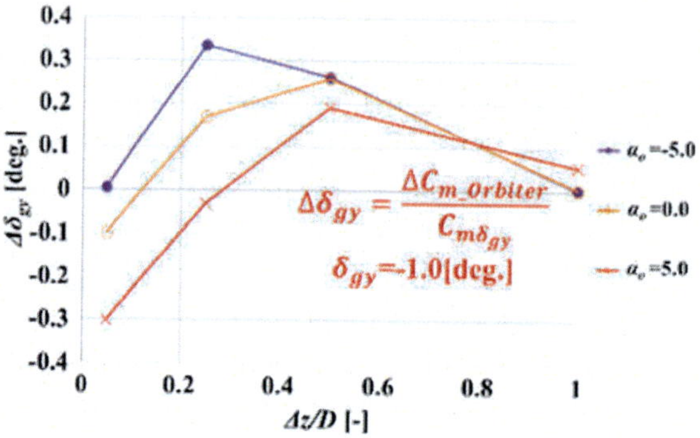

Fig. 14 Equivalent elevator angle $\Delta\delta_{gy}$ ($\Delta z/D = 0.10$, $\Delta\theta = 5.0$)

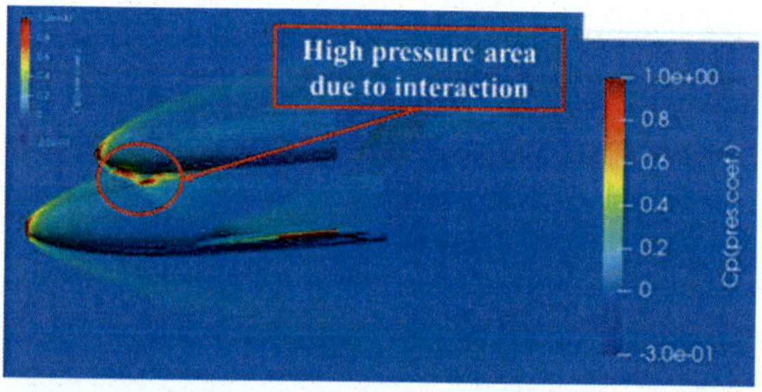

Fig. 15 Pressure coefficient distribution ($\Delta z/D = 0.05$, $\Delta x/L = 0.10$, $\Delta\theta = 0.0$, $\alpha_o = \alpha_b = -5.0$, and $\delta_e = 20.0$)

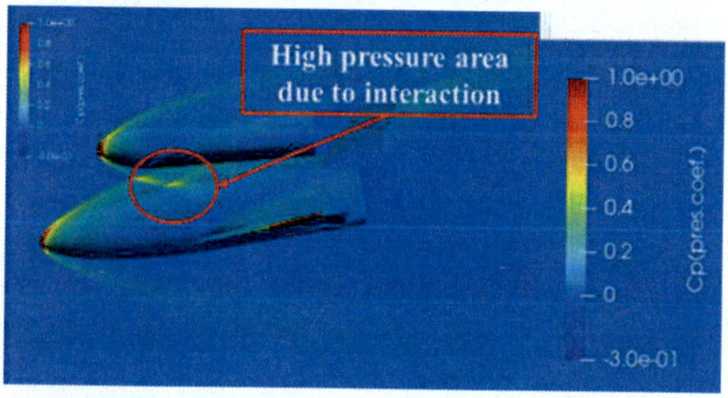

Fig. 16 Pressure coefficient distribution ($\Delta z/D = 0.05$, $\Delta x/L = 0.10$, $\Delta\theta = 5.0$, $\alpha_o = -5.0$, $\alpha_b = -10.0$, and $\delta_e = 20.0$)

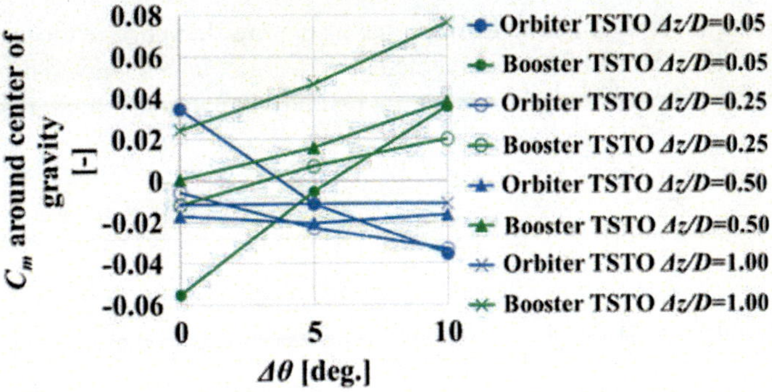

Fig. 17 Pitching-moment coefficient ($\Delta x/L = 0.10$, $\alpha_o = -5.0$, $\alpha_b = -5.0, -10.0, -15.0$ and $\delta_e = 20.0$)

0.0. However, this is just a consideration from the results obtained through the CFD analysis. Therefore, as a future work, performing a separation flight simulation by three-degree-of-freedom vertical motion and examining whether it is an appropriate mated position is necessary.

4 Conclusions and Future Works

CFD analysis was performed in the hypersonic velocity regime under non-negligible dynamic pressure, which is the separation condition of TSTO launch vehicle. Based on the results obtained through the CFD analysis, when two vehicles were close, the pitch-up and pitch-down moments acted on the orbiter and booster, respectively Hence, the two vehicles move away from each other. By contrast, the pitch-down and pitch-up moments acted on the orbiter and booster, respectively, when the two vehicles were far apart.

In addition, arranging the orbiter appropriately in front of the booster along the X_b axis direction and performing separation at a negative AoA facilitates separation in the TSTO launch vehicle.

Furthermore, the deflection of aerodynamic control surfaces required to cancel the interaction effect was investigated and the appropriate mated position should be a position with a negative AoA, a $\Delta z/D$ was between 0.25 and 0.50, a $\Delta x/L = 0.10$ and a $\Delta\theta = 0.0$. Moreover, $\Delta\delta_{gy}$ tended to be small at any relative position.

Future work is as follows.

- To perform simulations including non-stationery and motion calculations and evaluate whether it is possible to prevent collisions during separation, including attitude control.

- To apply trajectory optimization techniques is required to optimize the appropriate mated position of TSTO launch vehicle, the separation sequence, and the time history of the AoA during the separation by optimizing them simultaneously.

References

1. Wayne J, Alonzo L, Darren K (2003) Stage separation wind tunnel tests of a generic two stage-to-orbit launch vehicle. AIAA Paper 2003-4227
2. Cindy W, Paul V, Bandu N (2012) End-to-end simulation of launch vehicle trajectories including stage separation dynamics. AIAA Paper 2012-4836
3. Olds J, Ledsinger L, Bradford J, Charania A, McCormick D, Komar D (1999) Stargazer - A TSTO Bantam-X vehicle concept utilizing rocket-based combined-cycle propulsion. AIAA Paper 1999-4888
4. Hashimoto A, Murakami K, Aoyama T (2012) Lift and drag prediction using automatic hexahedra grid generation method. AIAA Paper 2009-1365
5. Hashimoto A, Murakami K, Aoyama T et al (2012) Toward the fastest unstructed CFD code "FaSTAR." In: 50th AIAA aerospace sciences meeting including the new horizons forum and aerospace exposition, January 09–12, Nashville, Tennessee.
6. Fujikawa T, Yonemoto K (2018) Conceptual study of winged suborbital vehicles as multipurpose space transportation systems.In: 62nd Space sciences and technology conference, 62-2F06, Fukuoka (in Japanese)
7. Kawaguchi J, Inatani Y, Yonemoto K, Hosokawa S (1987) On the flight control system of a winged space vehicle and physical simulation test. Institute of Space and Astronautical Science report, vol 64, pp 3–199. (in Japanese)

Aerodynamic Performance of a Low-Speed Blended-Wing-Body Aircraft with Distributed Ducted Fans

Wenyuan Zhao, Jianghao Wu, and Yanlai Zhang

Abstract The aerodynamic performance of a low-speed, blended-wing-body aircraft with distributed ducted fans was studied. A second-order finite volume flow solver was used to conduct flow simulations around the aircraft. The distributed ducted fans were modeled by applying different boundary conditions at the inlet and outlet of the ducts. Three different geometric models, namely, clean airframe configuration (C1), airframe with a boundary layer ingestion (BLI) fan system (C2), and airframe with a non-BLI fan system (C3), were created to analyze the impact of the BLI fan system on the aerodynamic characteristics of the blended-wing-body aircraft. Then, the aerodynamic performance of these configurations was examined at three design points, namely, climb, cruise, and glide. A comparison of C1 and C3 showed that the installation of the distributed ducted fan system increased the total airframe dissipation in the glide condition. A comparison of the total airframe dissipation of C2 and C3 in the cruise condition revealed that considerable aerodynamic performance improvement was generated by the BLI effects. The aerodynamic performance of the low-speed, blended-wing-body aircraft was sensitive to the installation position of the ducted fan system. The findings of this study provide useful insights into the design of distributed ducted propulsion systems.

Keywords Blended-wing-body · Distributed ducted fan system · Aerodynamic · Installation effects

W. Zhao · J. Wu (✉) · Y. Zhang
School of Transportation Science and Engineering, Beihang University, Beijing, China
e-mail: buaawjh@buaa.edu.cn

W. Zhao
e-mail: zhaowenyuan@buaa.edu.cn

Y. Zhang
e-mail: zhangyanlai@buaa.edu.cn

S. Lee et al. (eds.), *The Proceedings of the 2021 Asia-Pacific International Symposium on Aerospace Technology (APISAT 2021), Volume 1*, Lecture Notes in Electrical Engineering 912, https://doi.org/10.1007/978-981-19-2689-1_43

1 Introduction

Several advanced airframe and propulsion concepts have been proposed to improve the fuel efficiency and reduce the environment impact of future civil aircraft [1]. Among these novel aircraft concept designs, the blended-wing-body (BWB) aircraft features wings merged with the fuselage and has higher aerodynamic efficiency than the conventional tube-and-wing aircraft [2]. With regard to propulsion system design, the distributed propulsion system has the advantage of improved propulsive efficiency over the conventional propulsion system because of its higher bypass ratio [3]. Meanwhile, many aircraft designers have found that the integration of the airframe and propulsor also has a considerable impact on aircraft fuel efficiency and emission [4–7].

This study presents an investigation of the installation effects of a boundary layer ingestion (BLI) fan system on the aerodynamic performance of a BWB aircraft. BLI is an enabling technology that improves the aerodynamic efficiency of the airframe and the propulsive efficiency of the propulsor [8, 9]. The major mechanism of energy saving from BLI effects is the reduction in airframe wake dissipation and propulsive jet dissipation due to wake filling and decrements in propulsion inflow velocity [8, 10]. A reduction of 6–9% in the propulsive power requirement is estimated for the BLI aircraft called D8 [11], which has a double-bubble fuselage to increase lift and a BLI propulsor to improve performance efficiency. For a BWB aircraft, this figure is 3–10% [4, 12] mainly due to the propulsor design and installation method.

BLI effects also lead to a stagnation pressure reduction and distortion of the fan inflow. Thus, other studies have focused on improving the inflow performance of the BLI inlet via flow control technology and inlet shape design [13–15]. In the present study, the inflow performance of the BLI inlet is not discussed in favor of focusing on the aerodynamic performance impact perspective of the BLI propulsor on BWB aircraft. The focus of this study is to quantify the reduction in airframe dissipation brought by BLI effects by using CFD based mechanical energy balance analysis method.

Section 2.1 presents a description of the aircraft configurations used to perform the flow simulation, and Sect. 2.2 provides a detailed description of the flow simulation method. A discussion of the performance metric used to assess the aerodynamic impact of BLI is presented in Sect. 2.3. The main results and related discussions are shown in Sect. 3. The results show that the total airframe dissipation of the BLI configuration is 9% less than that of the non-BLI configuration in the cruise design condition.

2 Flow Simulation Setup

Three geometry configurations were constructed to analyze the installation effects of the distributed ducted fan system on the aerodynamic performance of BWB aircraft.

These configurations were clean airframe configuration (C1), airframe with a BLI fan system (C2), and airframe with a non-BLI fan system (C3). A detailed description of the geometry models is presented in Sect. 2.1. Flow simulations were conducted using ANSYS Fluent, which is a second-order finite volume solver, and the Spalart–Allmaras turbulence model was adopted. A detailed description of the flow simulation method is given in Sect. 2.2. The mechanical energy balance equation was used for the analysis of the aerodynamic performance of the three configurations to account for the interference effects between the airframe and fan system. The definition and calculation expressions of the aircraft aerodynamic performance metric are presented in Sect. 2.3.

2.1 Geometry Configuration Description

We considered a typical BWB aircraft with six electric duct fans installed on the rear part of its upper surface. The wingspan of the aircraft was 5.754 m. The maximum chord length along the spanwise direction was 2.4 m and located at the symmetry plane of the aircraft. Three geometry models were created for the analysis of fan system installation effects on the aerodynamic performance of the BWB aircraft. C1 was used to derive the baseline aerodynamic characteristics of the BWB airframe at climb, cruise, and glide design points. C2 had the same airframe as C1, but the electrical fan system was installed close to the upper surface of the airframe rear part. Part of the fan inlets was immersed in the boundary layer of the airframe, which produced BLI effects. C3 had an identical airframe and fan system as C2; the only difference was the fan system installation height of C3 was larger than that of C2's. Detailed fan system installation information is shown in Fig. 1. The aerodynamic performance of configurations C1 and C3 can be compared to analyze the fan system installation effects, and a comparison of C2 and C3 can be used to analyze the BLI effects.

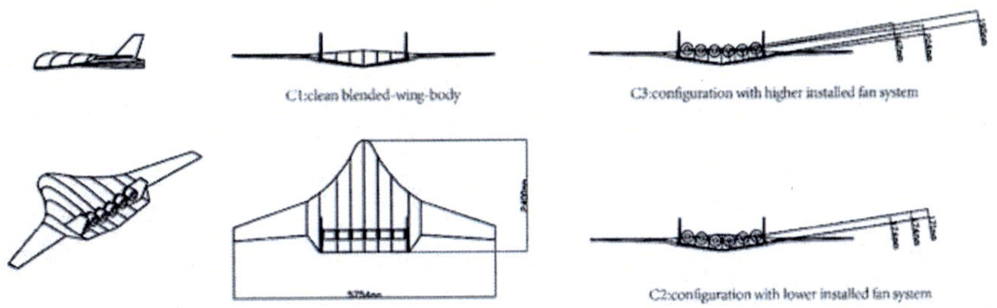

Fig. 1 Geometry configurations

2.2 Flow Simulation Method

Flow field simulations were conducted using ANSYS Fluent, a second-order finite volume solver, and the Spalart–Allmaras turbulence model was adopted. An average wall y^+ of less than 10 was used. Half of the flow field was modeled with a symmetry plane along the centerline. The velocity inlet boundary condition was used at the domain's outer boundary and prescribed the freestream velocity magnitude and direction vector. The wall boundary condition was set at the airframe surfaces and nacelle outer surfaces of the fan system. When performing the flow simulation at the glide design point, no extra boundary conditions were imposed on the fan inlet and outlet planes. Fans were regarded as tubes, and air could flow through them naturally. At the cruise and climb design points, the velocity inlet boundary condition was set at the fan inlet and outlet planes with a specified velocity magnitude and direction. The difference between the glide mesh and the cruise mesh is shown in Fig. 2.

A component-based fan model was built to obtain the flow conditions of the fan inlet and outlet at the cruise and climb design points. It was built using a discrete thermodynamic cycle analysis module and an iterator. The thermodynamic cycle analysis module consisted of three components, namely, diffuser, fan, and nozzle. The discrete Newton method was used in the iterator. The fan station numbers are shown in Fig. 3, where station number 0 represents the freestream condition, 1 denotes the fan inlet, 2 denotes the fan face, 2.1 is for the nozzle inlet, and 7 is for the fan outlet.

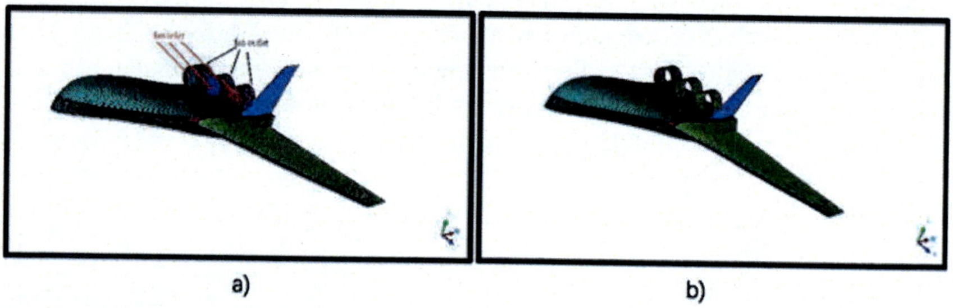

Fig. 2 Mesh at different design points

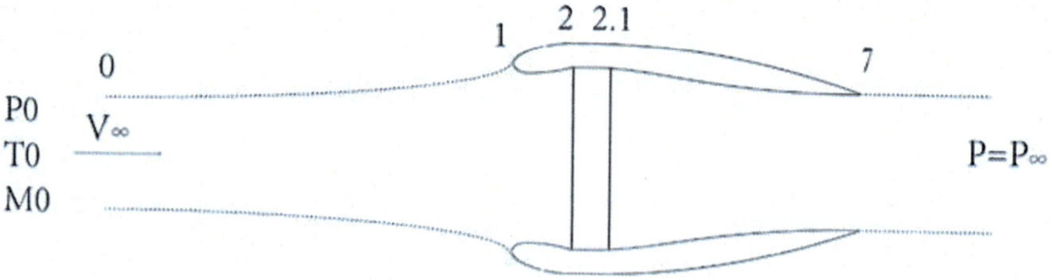

Fig. 3 Fan station numbers

The detailed constitution of the component-based fan model is shown in Fig. 4. From stations 0 to 1, the flow experienced a progression in isentropic compression, and 1D isentropic relations were used to obtain the flow parameters at station 1. The flow path from stations 1–2 was the diffuser before the fan, which can be modeled by total pressure recovery parameter π_d. The flow path from stations 2–2.1 was the fan and modeled by total pressure ratio π_f and polytropic efficiency η_f. The flow path from stations 2.1–7 was the nozzle, which was simulated by total pressure recovery parameter π_{fn}. The flow parameters at the nozzle outlet (station 7) were obtained with the ambient static pressure condition (P7 = P0). The discrete Newton iterator was used to update fan mass flow rate m_f and fan total pressure ratio π_f. The fan operation map was utilized to obtain fan polytropic efficiency η_f with given fan mass flow rate m_f and fan total pressure ratio π_f. A fan operation map was generated from computational fluid dynamics (CFD) calculations of the fan blades under different fan operation conditions. The constraint condition of the component-based fan model was a specified thrust requirement and a mass continuity constraint between the fan and nozzle. The BLI effects were modeled through the averaged total pressure decline before the diffuser. Given that the dynamic pressure of the inflow was lower than the static pressure (about 2–3%), the total pressure decline caused by the BLI effects was small, and the averaged total pressure recovery factor before the diffuser could be set to 1. This study focused on the influence of BLI on aircraft aerodynamic performance, and the influence of BLI on fan efficiency was not considered. π_d and π_{fn} were set to 1 to exclude the inner dissipation generated on the inner surface of the fan nacelle. Consequently, the boundary conditions of all the six fans were identical. Notably, different fan conditions may be set in accordance with the boundary layer conditions before each fan in the future.

The detailed mechanical energy requirement and the resulting fan boundary conditions are summarized in Table 1. The model setup parameters are listed in Table 2.

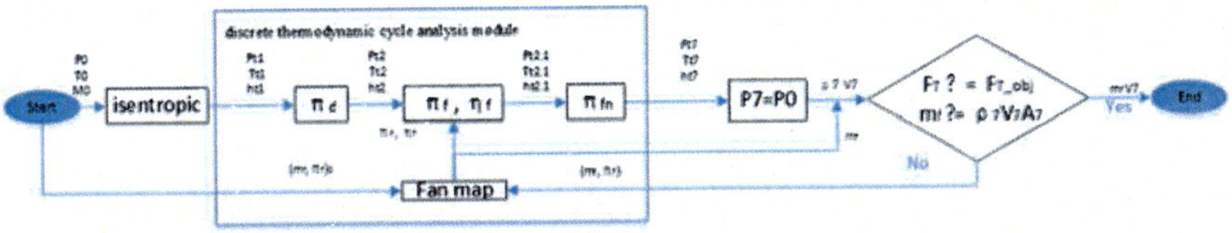

Fig. 4 Component-based fan model

Table 1 Fan boundary conditions

	F_{T_obj}	m_f	V_7	V_1	V_0	P_0	T_0
cruise	44.7	2.6	77.1	59.1	60	70,100	275.15
climb	62.3	2.3	60.4	46.3	33	79,480	268.65

Table 2 Model setup parameters

π_d	π_{fn}	A_1	A_7
1	1	0.0487	0.0374

2.3 *Aerodynamic Performance Metrics*

For a conventional aircraft, aerodynamic efficiency is usually represented by the drag force on the airframe surfaces at a given lift force, and engine propulsive efficiency is represented by the power requirement to produce a required thrust (equal to airframe drag). The drag force of the airframe can be easily obtained through a momentum-conservation-based control volume analysis. The thrust provided by the propulsion system can be calculated based on the produced mass flow rate and velocity difference between the propulsion jet and freestream. The default assumption for conventional aircraft is that the pressure field interference between the airframe and propulsion system can be disregarded; thus, the definitions of thrust and drag do not consider the pressure field interference effect.

The typical thrust–drag definition is unsuitable for advanced aircraft configurations with a tightly integrated propulsion system because the aero-propulsive interactions cannot be neglected. For these advanced aircraft configurations, the only useful streamwise information from near-field force integration is net streamwise force, and decomposition into thrust and drag is difficult and ambiguous. A new performance metric must be introduced to represent the aerodynamic efficiency of such aircraft instead of the thrust–drag terminology. The power balance method introduced by Drela [16] focuses on mechanical power conservation of the flow around the overall airframe–propulsor system and unifies all the power losses and supplements in one mechanical energy balance equation. In the power balance framework, the aerodynamic efficiency of an aircraft is represented by total energy loss Φ of the overall integrated airframe–propulsion system configuration. This metric is convenient for the performance comparison of different aircraft configurations at the same design point (specifying the same lift and net streamwise force). It can also be used to interpret the aerodynamic efficiency variation due to propulsor installation and BLI.

The power balance equation of the aircraft flow field is

$$\Phi = \mathrm{P_K} - \mathrm{F_X V_\infty}, \tag{1}$$

where P_K is the mechanical power inflow provided by the fan system, F_X is the net streamwise force of the airframe–propulsor system, $-F_X V_\infty$ represents the mechanical power used to increase the aircraft's potential energy, and Φ is the total mechanical energy loss of the aircraft and includes the boundary layer loss on the airframe surfaces and the jet loss in the propulsion jet. Given P_K and $-F_X V_\infty$, the total mechanical energy loss Φ can be calculated through the power balance equation, namely, Eq. (1).

The quantification equation of P_K is

$$P_K \equiv \iint_{S_P} -\left[(p - p_\infty) + \frac{1}{2}\rho\left(V^2 - V_\infty^2\right)\right]\mathbf{V} \cdot \hat{n}dS, \tag{2}$$

where S_P is the propulsion surface consisting of the fan inlet and outlet. In a low-speed condition, P_K is the volume flow rate of the total pressure increment between propulsion flow and freestream flow.

F_X can be obtained through near-field force and momentum flux integration as follows:

$$F_X \equiv \oint_{S_B+S_P} -\left[(p - p_\infty)\hat{n} - \boldsymbol{\tau} + \rho\mathbf{V}(\mathbf{V} \cdot \hat{n})\right] \cdot \hat{e_x}dS, \tag{3}$$

where S_B represents all the solid surfaces of the aircraft and S_P represents the propulsion surface.

Given jet velocity V_{jet} and propulsion mass flow rate $\dot{m}$, the jet loss can be presented as

$$\Phi_{jet} = \frac{1}{2}\dot{m}\left(V_{jet} - V_\infty\right)^2. \tag{4}$$

With Φ_{jet} known, the airframe total dissipation $\Phi_{airframe}$ can be separated from aircraft total loss as

$$\Phi_{airframe} = \Phi - \Phi_{jet}. \tag{5}$$

This term is selected as the performance metric to analyze the aerodynamic efficiency of different aircraft configurations.

Notably, the comparison base of BLI and non-BLI configurations contains a similar set of fans and an identical lift coefficient.

3 Results and Discussions

The baseline airframe dissipation of configurations C1, C2, and C3 in the glide condition is analyzed in Sect. 3.1. The results show a significant increment in the total airframe dissipation at the glide design point because of the fan installation. A comparison of the aerodynamic dissipation characteristics of C2 (BLI) and C3 (non-BLI) in the cruise condition is presented in Sect. 3.2. The BLI airframe in C2 consumes 9% less power than the non-BLI airframe in C3 at the cruise design point. A difference of 0.0029 in $C_{\Phi_{airframe}}$ is approximately maintained throughout the range of C_L values simulated.

3.1 Dissipation Characteristic at the Glide Design Point

The airframe dissipation coefficients, $C_{\Phi_{airframe}}$, of configurations C1, C2, and C3 in the glide condition over a range of lift coefficients, C_L, are shown in Fig. 5, where the most bottom blue line represents the aerodynamic performance of C1, the upper yellow line represents that of C2, and the orange line represents that of C3. At the glide design point, the fan system operated at a low rotational speed, and the corresponding jet dissipation was smaller than the dissipations that occurred on the airframe surface. The dissipation characteristic at this design point can be regarded as the baseline aerodynamic performance of each configuration. In the glide condition, the dissipation characteristic of C2 was similar to that of C3. However, the difference in $C_{\Phi_{airframe}}$ between C3 and C1 was 0.021, which was about 30% of C1's value at the glide design point ($C_L = 0.62$). This difference of 0.021 in $C_{\Phi_{airframe}}$ was approximately maintained throughout the range of C_L values simulated.

To explain the remarkable airframe dissipation increase of C2 and C3 relative to C1, we continued the power balance analysis and focused on the viscous dissipation for the freestream settings close to $C_L = 0.62$ (glide design point). We qualitatively examined the total pressure loss on two X cut planes, which were located before the

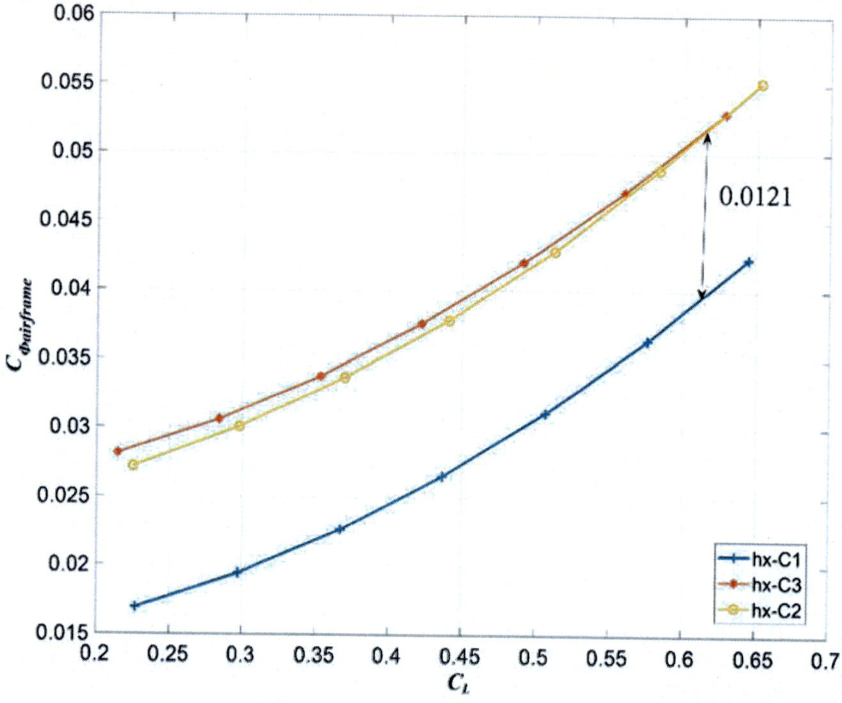

Fig. 5 Non-dimensionalized airframe dissipation $C_{\Phi_{airframe}}$ over a range of lift coefficients C_L of the three configurations at the glide design point

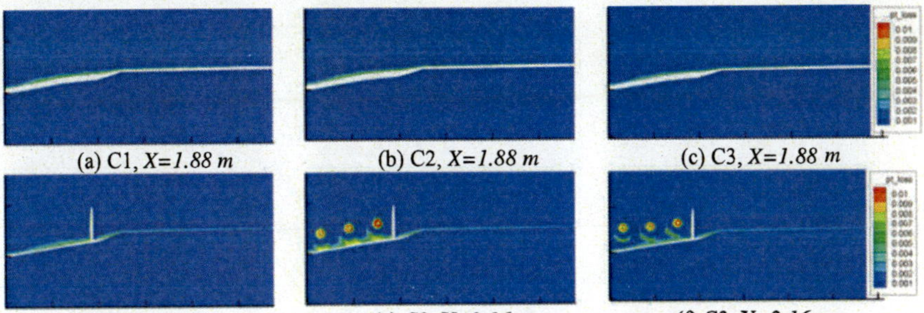

(a) C1, X=1.88 m (b) C2, X=1.88 m (c) C3, X=1.88 m

(d) C1, X=2.16 m (e) C2, X=2.16 m (f) C3, X=2.16 m

Fig. 6 Contour of the total pressure loss coefficient on the planes before the fan inlet ($X = 1.88m$) and behind the fan outlet ($X = 2.16m$) for C1, C2, and C3 in the glide condition

fans' inlet ($X = 1.88m$) and behind the fans' outlet ($X = 2.16m$). The contour of the total pressure loss coefficient on both cut planes for the three configurations (C1, C2, and C3) is plotted in Fig. 6. Sub-figures (a–c) indicate that the dissipation that occurred before the fan system was almost identical in the three configurations, and the installation of the fan system did not affect the fuselage surface dissipation. Sub-figures (d–f) indicate that the only major difference in airframe dissipation occurred in the fan regions and the airframe surface boundary layer regions beneath. With high total pressure loss coefficient regions around the fan system and the expanded airframe surface boundary layer region just under the fan system, the total airframe dissipation of C2 and C3 was much higher than that of C1.

3.2 BLI Benefits at Cruise Design Points

The airframe dissipation benefits of C2 (BLI configuration) over C3 (non-BLI configuration) in the cruise condition are presented in Fig. 5, where $C_{\Phi_{airframe}}$ is plotted against C_L for C2 and C3 for several values of freestream angle of attack. The values of $C_{\Phi_{airframe}}$ plotted in Fig. 5 at $C_L = 0.28$ (cruise design point) could provide a direct comparison of how much power the airframe consumed in the cruise condition. As marked in Fig. 7, the BLI airframe in C2 consumed 9% less power than the non-BLI airframe in C3 at the cruise design point ($C_{\Phi_{airframe}} = 0.0294\ vs.\ 0.0323$). This difference of 0.0029 in $C_{\Phi_{airframe}}$ was approximately maintained throughout the range of C_L values simulated.

To determine the mechanisms that bring performance benefits, we continued to analyze the viscous dissipation distribution of cases near $C_L = 0.28$. We extracted three X cut planes with location $X = 1.88, 1.917, 2.16$ m for each case. In general, two cut planes before and behind the fan are sufficient for a viscous dissipation comparison. However, the airframe and fan system geometry of C2 and C3 are similar; the only difference is the installation height of the fan system. With a lower fan installation height than C3, C2 has an obvious BLI effect that does not exist in

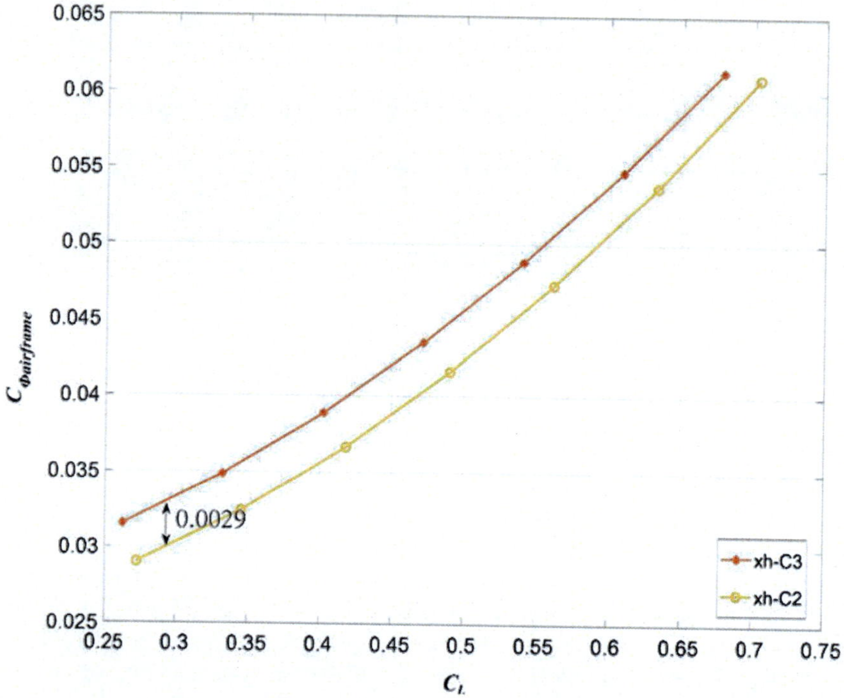

Fig. 7 Non-dimensionalized airframe dissipation $C_{\Phi^*_{airframe}}$ over a range of lift coefficients, C_L, of C2 and C3 at the cruise design point

C3. Thus, we added a cut plane just at the fan inlet location ($X = 1.917m$) to see the BLI effect. The total pressure loss coefficient contours of the extracted planes for C2 and C3 near the cruise design point are plotted in Fig. 8. A comparison of sub-figures (a) and (c) indicated that the dissipation on the airframe surface before the fans was

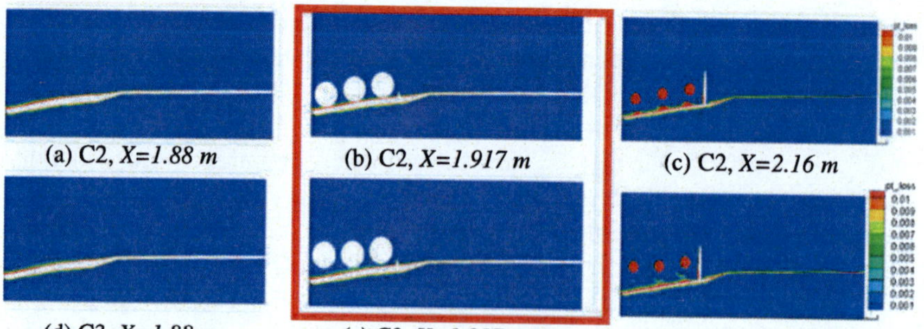

(a) C2, $X=1.88$ m (b) C2, $X=1.917$ m (c) C2, $X=2.16$ m

(d) C3, $X=1.88$ m (e) C3, $X=1.917$ m (f) C3, $X=2.16$ m

Fig. 8 Contour of the total pressure loss coefficient on planes (X = 1.88, 1.917, 1.88 m) for C2 and C3 in the cruise condition

(a) C2, *X=1.917m* (b) C3, *X=1.917m*

Fig. 9 Locally enlarged contour of the total pressure loss coefficient on planes (X = 1.88, 1.917, 1.88 m) for C2 and C3 in the cruise condition

not affected by the installation height of the fan system. A comparison of sub-figures (b) and (e) showed that region with the high total pressure loss in configuration C2 shrank in comparison with that in C3 due to the BLI effect, which is highly obvious in the enlarged total pressure loss coefficient contours in Fig. 9.

In Fig. 9, the region with a high total pressure loss coefficient in C2 looks like a truncated strip (sub-figure (a), colored in red) with several segments ingested by the fans, whereas that in C2 looks like a large continuous strip (subfigure (b), colored in red), which represents high dissipation. We conclude from this result that the 9% benefit in the aerodynamic performance of C2 over C1 can be attributed to the BLI effects. However, as shown in Fig. 8c, the boundary layer flow of the surfaces under the fans in C2 became seperation flow due to the low installation height of the fans, which caused $C_{\Phi_{airframe}}$ to increase. Additional geometry design of the rear part of airframe is underway to reduce the power consumption due to the flow seperation between the fans and the fuselage surface under them.

4 Conclusions

A CFD-based aerodynamic performance analysis of a low-speed BWB aircraft with distributed ducted fans was performed in this study. Three geometry configurations, namely, clean airframe configuration (C1), airframe with a BLI fan system (C2), and airframe with a non-BLI fan system (C3), were constructed to analyze the effect of fan system installation on the aerodynamic performance of the three aircraft configurations. Airframe total dissipation Φ_{airframe} was selected as the aerodynamic efficiency metric of the aircraft.

In the glide condition, an obvious increment in the Φ_{airframe} of C2 and C3 was observed relative to that of C1 because viscous dissipation occurred around the fan system and the airframe surface boundary layer regions beneath it. In the cruise condition, the comparison of the Φ_{airframe} of C2 and C3 showed that the airframe dissipation reduction benefits brought by BLI amounted to 9%, and the absolute difference of 0.0029 in $C_{\Phi_{airframe}}$ between C2 and C3 was approximately maintained throughout the range of C_L values simulated. The analysis of the viscous dissipation distribution of C2 and C3 showed that the region with a high total pressure loss in configuration C2 shrank compared with that in C3 due to the BLI effect.

References

1. Greitzer EM, Bonnefoy PA, Rosa Blanco E (2010) N+3 aircraft concept designs and trade studies, final report. Massachusetts Institute of Technology, Cambridge
2. Liebeck RH (2004) Design of the blended wing body subsonic transport. J Aircr 41(1):10–25
3. Ko A, Schetz JA, Mason WH (eds) (2003) Assessment of the potential advantages of distributed-propulsion for aircraft. In: 16th international symposium on air breathing engines (ISABE)
4. Plas A, Crichton D, Sargeant M, Hynes T, Greitzer E, Hall C et al (2007) Performance of a boundary layer ingesting (BLI) Propulsion System. In: 45th AIAA aerospace sciences meeting and exhibit 2007
5. Kok H, Voskuijl M, van Tooren M (2010) Distributed propulsion featuring boundary layer ingestion engines for the blended wing body subsonic transport. In: 51st AIAA/ASME/ASCE/AHS/ASC structures, structural dynamics, and materials conference 18th AIAA/ASME/AHS adaptive structures conference 12th 2010
6. Valencia EA, Nalianda D, Laskaridis P, Singh R (2014) Methodology to assess the performance of an aircraft concept with distributed propulsion and boundary layer ingestion using a parametric approach. Proc Inst Mech Eng Part G J Aerosp Eng 229(4):682–693
7. Goldberg C, Nalianda D, Laskaridis P, Pilidis P (2018) Installed performance assessment of an array of distributed propulsors ingesting boundary layer flow. J Eng Gas Turbines Power 140(7)
8. Hall DK, Huang AC, Uranga A, Greitzer EM, Drela M, Sato S (2017) Boundary layer ingestion propulsion benefit for transport aircraft. J Propul Power 33(5):1118–1129
9. Uranga A, Drela M, Hall DK, Greitzer EM (2018) Analysis of the aerodynamic benefit from boundary layer ingestion for transport aircraft. AIAA J 56(11):4271–4281
10. Lv P, Rao AG, Ragni D, Veldhuis L (2016) Performance analysis of wake and boundary-layer ingestion for aircraft design. J Aircr 53(5):1517–1526
11. Drela M (2011) Development of the D8 transport configuration. In: 29th AIAA applied aerodynamics conference
12. Hardin L, Tillman G, Sharma O, Berton J, Arend D (2012) Aircraft system study of boundary layer ingesting propulsion. In: 48th AIAA/ASME/SAE/ASEE joint propulsion conference & exhibit. Joint propulsion conferences. American Institute of Aeronautics and Astronautics.
13. Gray JS, Mader CA, Kenway GKW, Martins JRRA (2020) Coupled aeropropulsive optimization of a three-dimensional boundary-layer ingestion propulsor considering inlet distortion. J Aircr 1–12
14. Lei L, Fu C, Kun L, Huachen P, Huawei Q, Fang W et al (2019) Blowing-suction control in S-shaped inlet and its impact on fan-stage performance. AIAA J 57(9):3954–3968
15. Lima LSM, Huebner R, Tobaldini L (2019) Numerical investigations of S-shaped air inlet for embedded engines. J Propul Power 35(2):475–489
16. Drela M (2009) Power balance in aerodynamic flows. AIAA J 47(7):1761–1771

Numerical Study of the Aerodynamic Performance of Two Coaxial Flapping Rotary Wings Under Wake Interaction

Songtao Chu, Chao Zhou, and Jianghao Wu

Abstract Flapping rotary wing (FRW) is a composite flapping wing layout proposed in the last decade for micro air vehicle (MAV) design. FRW flaps actively in the vertical direction coupled with passive horizontal rotation, ensuring that the high-lift mechanisms from wing flapping at low Reynolds numbers and the high aerodynamic efficiency of the rotary wing are applied in this wing layout. In actual MAV designs, multiple FRWs have various arrangements, one of which is the arrangement where FRWs locate coaxially and rotate in the same direction. In this type of arrangement, a complex flow interaction exists between wakes from the upper and lower wings, but it has never been focused on in previous research. Thus, this study investigates the effect of wake interaction on the aerodynamics of two coaxial FRWs by using the lattice Boltzmann method and explores the influence of rotation speed, flapping phase difference ($\Delta\theta$), and wing vertical distance between two wings on the lift and rotating moment, which determine the rotation speed of actual FRW. With inverse flapping (i.e., $\Delta\theta = \pi$), the wings interact strongly, and the MAV can reach the maximum rotating moment. An increase in vertical distance between the two wings and rotation speed can weaken the interaction. Our design allows for high rotating speed and low wing loads, thereby reducing the torque requirements on the motor. This study can enhance our understanding of the complex wake interaction produced by multiple FRWs at low Reynolds numbers and provide theoretical guidance for the design of MAVs, especially FRW MAVs.

Keywords Flapping wing · Micro air vehicle · Lattice Boltzmann method · Unsteady aerodynamics · Wake interaction

S. Chu · C. Zhou (✉) · J. Wu
Beihang University, Beijing, P.R. China
e-mail: zcqh821@163.com

S. Chu
e-mail: st.chu@buaa.edu.cn

J. Wu
e-mail: buaawjh@buaa.edu.cn

S. Lee et al. (eds.), *The Proceedings of the 2021 Asia-Pacific International Symposium on Aerospace Technology (APISAT 2021), Volume 1*, Lecture Notes in Electrical Engineering 912, https://doi.org/10.1007/978-981-19-2689-1_44

1 Introduction

Over the past two decades, the design of micro air vehicles (MAVs) has elicited increased focus due to its civil and military applications in limited spaces. Different from traditional MAV designs, such as those with fixed [1], rotary [2], and flapping wings [3], flapping rotary wing (FRW) was designed to take advantages of traditional designs [4–6]. By coupling active flapping with passive rotating, FRW takes full advantage of the high-lift mechanisms of the flapping wing at a low Reynolds number and uses the high aerodynamic efficiency of the rotary wing.

Researchers have studied the unsteady aerodynamics of FRW in hovering flight by using experimental and numerical methods. They focused on the single rotational plane arrangement. In this arrangement, FRW relies mainly on the downstroke to generate lift, whereas the upstroke contributes minimally to the lift [7]; thus, the load of the MAV wing is high in the case of high load generation, the transient force peak difference between the wing downstroke and upstroke is large, and the aerodynamic force changes dramatically, resulting in increased work and reduced efficiency. To solve this problem, several designers have proposed the concept of multiple wings to reduce the wing load and aerodynamic fluctuations by increasing the number of wings. In a scheme that locates coaxially and rotates reversely [8], the lower wing is completely in the wake of the upper wing, and the lower wing is strongly disturbed by the downwash flow of the upper wing; thus, the aerodynamic efficiency of the lower wing is low.

Previous studies on flow interaction between multiple wings at low Reynolds numbers have revealed that multiple wings can obtain an aerodynamic advantage through wing-to-wing interaction, and one of the effects is the "clap-and-fling" mechanism [9]. This mechanism enhances the production of aerodynamic forces perpendicular to the flapping direction during wing clapping and fling motions. To solve the problem that the lower wing in the previous multiple-wing configuration cannot benefit much from the upper wing wake, other multiple-wing designs must be explored to obtain improved MAVs through the application of multi-wing interaction.

In the present study, a simplified flapping rotary wing model was analyzed to understand the force and moment changes caused by wake interaction. The motion and geometry of the wing were obtained from our previous work. The applied numerical method was introduced and verified, and several of the parameters that affect wake interaction were discussed.

2 Materials and Methods

The wing model and kinematics for the co-axial FRWs were similar to those in our previous research [7], but we added another FRW to the single FRW with different flapping phases and distances of rotational planes. We used the lattice Boltzmann method (LBM) coupled with the immersed boundary method to solve the fluid field with moving wings. This numerical method was verified through algorithm and grid-independence validation.

2.1 Wing Model and Kinematics

The model used in this study is illustrated in Fig. 1. The two FRWs are co-axial with a distance of rotation plane d. A rectangular wing with zero thickness and an aspect ratio λ of 3.0 is used as the wing shape. The kinematics of the wing is decomposed into sinusoidal flapping φ, pitching α, and uniform rotation ψ motions. The wings flap around their root, the pitching axis is 1/4-chord length away from the leading edge, and the axis of rotation motion is perpendicular to the horizontal plane.

We define three coordinate systems, namely, inertial $Oxyz$, rotational $Ox_ry_rz_r$, and body-fixed $Ox_by_bz_b$, with the same origin O. The inertial system $Oxyz$ is on the ground with a horizontal Oxz plane. The body-fixed system $Ox_by_bz_b$ is fixed on the wing, with Oy_b perpendicular to the upper surface and Oz_b pointing to the reverse spanwise direction. For the rotational system, Oy_r is similar to Oy, and plane Ox_rz_r rotates around Oy_r. The three other coordinate systems with origin O' for the lower wing are similar to those for the upper wing. The distance between O and O' is defined as d, which is also the distance between the two rotation planes.

The kinematics of wing motion is defined here in accordance with a previous study [7]. The three coordinate systems coincide initially, and the wing rotates around axis Oz_b at angle α. Next, it rotates around axis Ox_r at angle φ and finally rotates around axis Oy_r at angle ψ. The kinematic function is defined as follows:

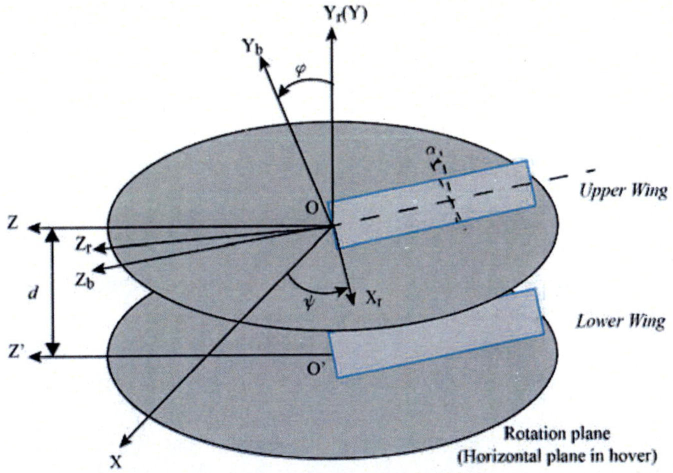

Fig. 1 Coordinate systems for co-axial FRWs

$$\varphi = \Delta\varphi \sin\left(2\pi f t - \frac{\pi}{2} + \theta\right), \alpha = \Delta\alpha \sin(2\pi f t + \theta) + \alpha_0, \psi = \psi_0 t, \quad (1)$$

where $\Delta\varphi$ and $\Delta\alpha$ are the amplitudes of the flapping and pitching motions, respectively, and ψ_0 is the rotational angular velocity. The flapping phase difference between the upper and lower wings is defined as $\Delta\theta = \theta - \theta'$. To simplify, we let $\theta = 0$. The period ratio between rotating and flapping motions is $n = T_f / T_r = \psi_0 / 2\pi f$.

We define reference length and reference velocity as chord length c and average flapping velocity $\overline{U} = 4c\lambda f \Delta\varphi$ at the wingtip, respectively. Thus, the reference time can be derived as $c/\overline{U} = 1/4\lambda f \Delta\varphi$, and the non-dimensional flapping period is $T_f = (1/f)/(c/\overline{U}) = 4\lambda\Delta\varphi$.

Lift C_L and rotating moment C_M are defined as

$$C_L = \frac{L_{y_r}}{0.5\rho(\overline{U})^2 S}, \quad (2a)$$

$$C_M = \frac{M_{y_r}}{0.5\rho(\overline{U})^2 Sc}. \quad (2b)$$

2.2 Numerical Method

To simulate the interaction between wings accurately, we selected LBM coupled with the immersed boundary method (IBM). We developed a solver based on the open-source software Palabos [10] to reduce repetitive basic work. The governing equations of flow are written as follows:

$$f_i(\boldsymbol{x} + \boldsymbol{e}_i, t + 1) - f_i(\boldsymbol{x}, t) = -\frac{1}{\tau}\left[f_i - f_i^{(eq)}\right] + F_i(\boldsymbol{x}, t),$$
$$f_i^{(eq)} = w_p \rho \left\{1 + \frac{\boldsymbol{e}_i \cdot \boldsymbol{u}}{c_s^2} + \frac{1}{2}\left[\frac{(\boldsymbol{e}_i \cdot \boldsymbol{u})^2}{c_s^4} - \frac{u^2}{c_s^2}\right]\right\}, \quad (3)$$

where f_i is the velocity distribution function, $f_i^{(eq)}$ is the equilibrium distribution function, τ is the relaxation time, and F_i is the external force term. To solve the force on the moving wings, we use IBM for LBM proposed by Inamuro [11]. Although it cannot simulate the boundary layer in our program, IBM can compute the force on each lattice, which is enough for this research. The 3D multi-layer mesh refinement algorithm in Palabos is enabled to reduce the amount of computation while maintaining the accuracy of the near-field calculation.

2.3 Validation

The algorithm of the program was validated by comparing it with another algorithm in the study on IB-LBM published by Suzuki and Inamuro [12]. We selected a flapping wing-body model case in which two wings with an aspect ratio of 1 sinusoidally flap and pitch (the case in Appendix C in Suzuki's paper). With all parameters set to be the same ($Re = 500$, $\alpha = \frac{\pi}{4}cos\left(\frac{2\pi}{T}t + \frac{\pi}{2}\right) + \frac{\pi}{4}$, $\theta = \frac{\pi}{4}cos\left(\frac{2\pi}{T}t\right)$, periodic boundary, and same domain size and resolution), the lift and thrust produced by the wing in one flapping period were determined and are shown in Fig. 2. The result of the current work is very close to that of Suzuki's study.

Then, grid-independence validation was performed. We selected a single FRW case with grid resolutions of $c = 20\delta_x, 40\delta_x, 80\delta_x$ and domain sizes of $12.8 \times 19.2 \times 12.8, 25.6 \times 38.4 \times 25.6, 51.2 \times 76.8 \times 51.2$. Owing to the $\delta_t = u_{LB} \bullet \delta_x$ in LBM, the time step was proportional to the grid scale; hence, the time step validation was omitted. The kinematics has already been defined in Sect. 2.1, with $\Delta\varphi = 15°$, $\Delta\alpha = 15°$, $f = 1/\pi$, $and\ \psi_0 = 0.5$. The results are shown in Fig. 3. The results for $c = 40\delta_x$ and $c = 80\delta_x$ almost coincided, whereas the result for $c = 20\delta_x$ was a little different from the others. Thus, we selected the resolution $c = 40\delta_x$. The results of the three domain sizes were very close, and the medium one was used.

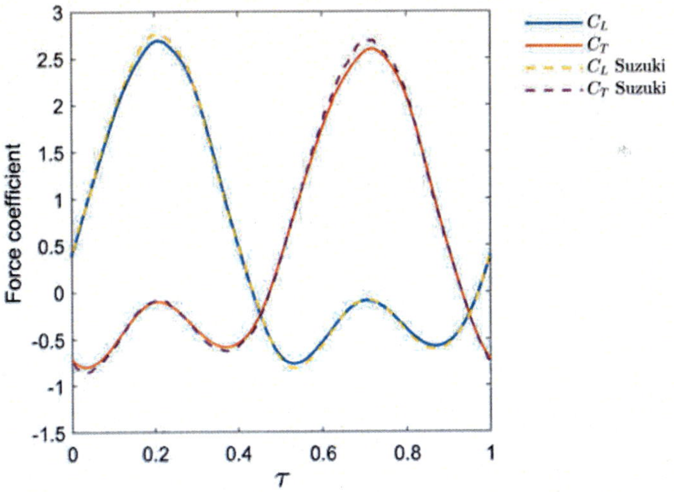

Fig. 2 Force coefficient for algorithm validation

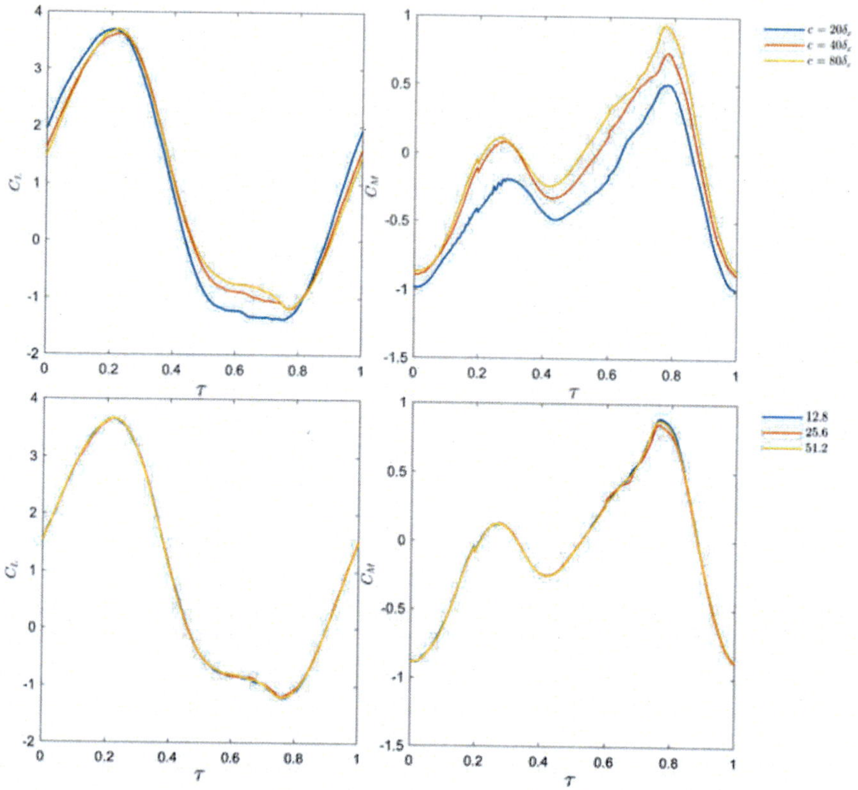

Fig. 3 Force coefficient for grid-independence validation

3 Results and Discussions

The aerodynamic performance of coaxial FRWs, including lift and rotating moment in hovering flight, was addressed. Afterward, the effects of the flapping phase difference ($\Delta\theta$), wing rotating speed (n), and rotational plane distance (d) were examined.

3.1 Aerodynamic Forces and Rotating Moment in Hovering Flight

The aerodynamic performance of coaxial FRWs in hover flights was initially computed based on the selected typical case (i.e., $Re = 3000$, $\Delta\varphi = 15°$, $\Delta\alpha = 15°$, $\alpha_0 = 15°$, $f = 1/\pi$, $\Delta\theta = 180°$, $n = 0.25$, $d = 1.6$). In this case, the two wings had a minimal distance and interacted significantly.

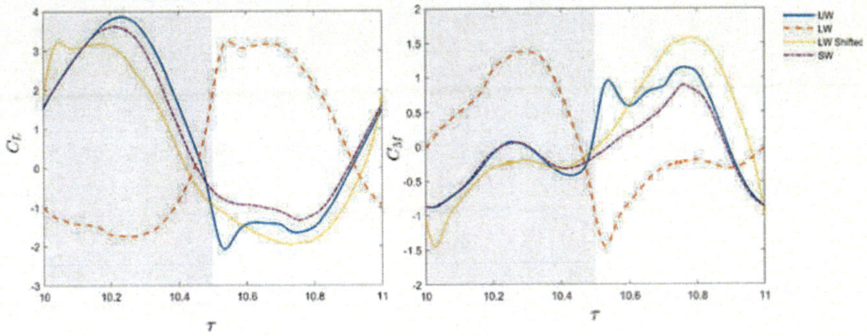

Fig. 4 Lift and rotating moment at $\Delta\theta = 180°, n = 0.25, d = 1.6$.UW = Upper Wing, LW = Lower Wing, SW = Single Wing

The lift and rotating moment in one non-dimensional flapping period are shown in Fig. 4. The lift of the upper wing changed slightly compared with that of the single wing, except at $\tau = 10.5-10.6$ which is the beginning of the upstroke of the upper wing. For the lower wing, the lift was lower than that for the single wing in nearly the entire period, except at the "clap" moment (yellow line with the flapping phase shifted). When the upper wing was at the beginning of the upstroke, the two wings clapped and the "clap-and-fling" mechanism came into play, leading to a low-pressure zone between the wings. As a result, the upper wing lost some lift and increased its thrust, whereas the lower wing increased its lift and thrust during this period. The rotating moment increased significantly due to the increase in thrust. As a result, the lift of each wing was lower than that of the single wing, and the rotating moment was larger than that of the single wing.

3.2 *Effect of Wing Kinematics in Hovering Flight*

3.2.1 Effect of $\Delta\theta$

The effect of $\Delta\theta$ on wing aerodynamic performance was studied based on the given typical case. The aerodynamics of the coaxial FRWs were computed at different $\Delta\theta$ with a step of 45°(i.e., $\Delta\theta = 0°, 45°, 90° \ldots 315°$). The other kinematic parameters were kept the same as those in the typical case.

Figure 5 presents the $\overline{C_L}$ and $\overline{C_M}$ of the coaxial FRWs and the single wing at different $\Delta\theta$ values. The lift of both wings was less than that of a single wing at all $\Delta\theta$ values. For the upper wing, the influence of the interaction was relatively weak, and the lift loss increased at 180°–270°, that is, when the two wings could be relatively close to each other. Given that the lower wing was in the wake of the upper wing, the lift loss of the lower wing was much higher than that of the upper wing due to the influence of downwash. The trend of the rotating moment was opposite to that

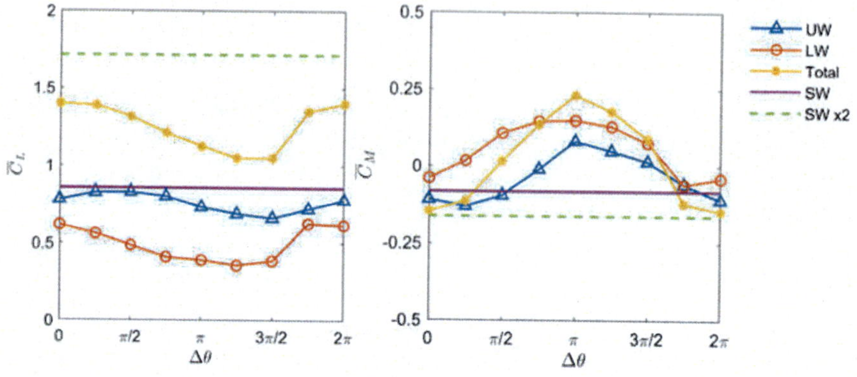

Fig. 5 Lift and rotating moment at different $\Delta\theta$

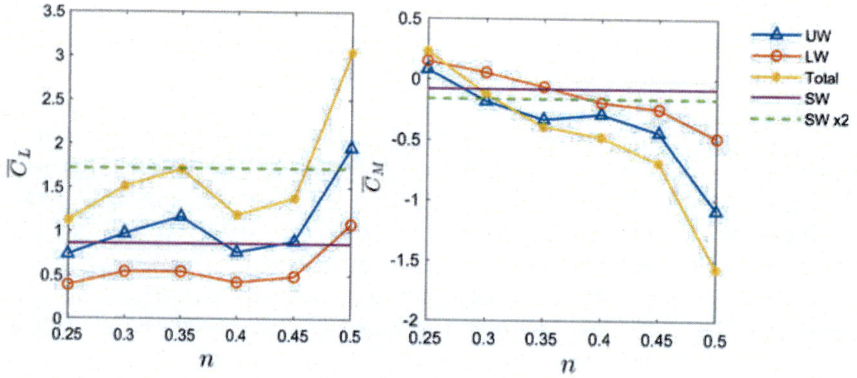

Fig. 6 Finding a comparable case to the single wing

of the lift. We obtained the maximum rotating moment at $\Delta\theta = 180°$, which is the typical case. This result is consistent with the previous analysis of rotation moment in Sect. 3.1.

In actual FRWs, a positive rotating moment causes the rotation to accelerate until the moment is balanced. To compare the above-mentioned results with those of a single wing, we calculated the case where the moment is nearly twice the results of the single wing (the moment is close to balance). The results are shown in Fig. 6. The $n = 0.3$ case (i.e., wake interaction leads to increased rotation and reduced wing load) was selected.

3.2.2 Effect of *n* and *d*

On the basis of the typical case, a set of n values ($n = 0.25, 0.5$) and a set of d values ($d = 1.6, 2.5, 3.5, 4.5$) were selected for analysis. If distance d is increased, then wake interaction will decrease, especially the "clap-and-fling" mechanism. If rotating speed n is increased, then the lower wing may encounter difficulty entering the wake of the upper wing. The results of the computation of the two parameters are shown in Figs. 7 and 8.

With the increase in d, the lift and rotating moment of the upper wing quickly returned to the same level as the single wing. However, when $n = 0.25$, the lift and rotating moment for the lower wing always converged to a single wing until d achieved its maximum value. When $n = 0.5$, the lift and rotating moment tended to be stable at $d = 3.5$.

In summary, when n and d increased, the lift increased, and the moment decreased. The upper wing became stable earlier than the lower wing did.

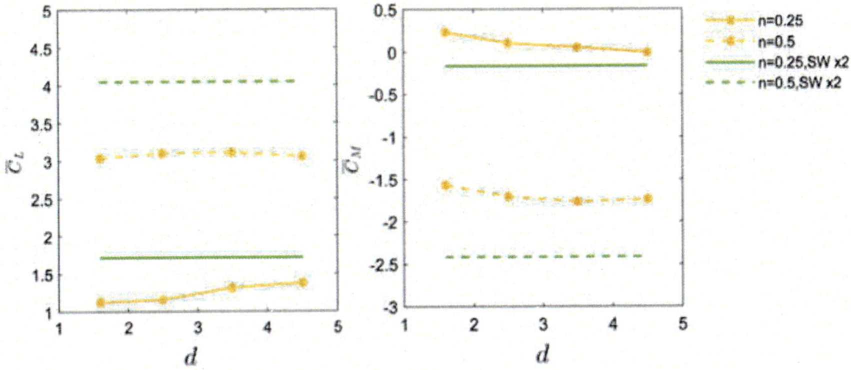

Fig. 7 Total lift and rotating moment at different n and d

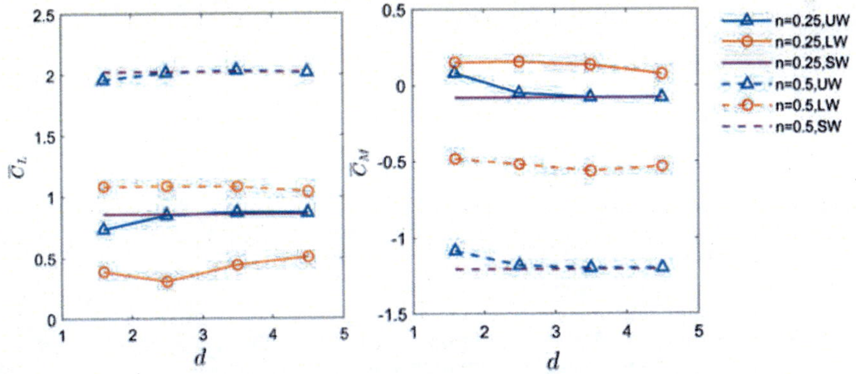

Fig. 8 Lift and rotating moment of separate wings at different n and d

4 Conclusion

We presented a new arrangement for multiple FRWs that locate coaxially and rotate in the same direction. We investigated the effect of wake interaction on the aerodynamics of two coaxial FRWs by using LBM and explored the influence of rotation speed, flapping phase difference, and wing vertical distance between two wings on lift and rotating moment, which determine the rotation speed of actual FRW. The results showed that when the flapping phase difference increased, the total lift initially decreased then increased. The peak of the total lift was reached at $\Delta\theta = 0$, but it was smaller than twice the lift of a single FRW without wake interaction. The reduction in the lift was mainly caused by the lift loss of the lower wing. This result suggests that the wake interaction of coaxial FRWs was not conducive to lift generation. That the rotating moment changed with the phase difference was contrary to the lift scenario, and the maximum was obtained when the two wings flapped inversely, that is, $\Delta\theta = 180°$. In this case, the "clap-and-fling" mechanism in the horizontal direction enhanced the thrust production, leading to a large rotating moment. Overall, the coaxial FRW arrangement was conducive to fast rotating motion and could compensate for the lift loss. Varying either the vertical distance between the two wings or the rotation speed did not change the qualitative law of lift and rotating moment with the flapping phase difference. However, quantitatively, the lift and rotating moment of the upper wing increased and decreased, respectively, as the distance increased. When the distance was more than 0.8 times of the wingspan, the lift and rotating moment recovered to the single wing's state. Meanwhile, the lift of the lower wing was always smaller than the single wing's lift, and the rotating moment was larger than the single wing's moment over a distance range of twice the wingspan. This study can enhance our understanding of the complex wake interaction produced by multiple FRWs at low Reynolds numbers and provide theoretical guidance for the design of MAVs, especially FRW MAVs.

References

1. Combes TP, Malik AS, Bramesfeld G, McQuilling MW (2015) Efficient fluid-structure interaction method for conceptual design of flexible, fixed-wing micro-air-vehicle wings. AIAA J 53(6):1442–1454
2. Ramasamy M, Lee TE, Leishman JG (2007) Flowfield of a rotating-wing micro air vehicle. J Aircr 44(4):1236–1244
3. Ma KY, Chirarattananon P, Fuller SB, Wood RJ (2013) Controlled flight of a biologically inspired, insect-scale robot. Science. 2;340(6132):603 LP–607
4. Fitchett BK (2007) Development and investigation of a flapping rotor for micro air vehicles [Internet]. University of Maryland, College Park. http://hdl.handle.net/1903/7224
5. Guo S, Li D, Wu J (2012) Theoretical and experimental study of a piezoelectric flapping wing rotor for micro aerial vehicle. Aerosp Sci Technol 23(1):429–438
6. Zhou C, Wu J, Guo S, Li D (2014) Experimental study on the lift generated by a flapping rotary wing applied in a micro air vehicle. Proc Inst Mech Eng Part G J Aerosp Eng 228(11):2083–2093

7. Wu J, Wang D, Zhang Y (2015) Aerodynamic analysis of a flapping rotary wing at a low reynolds number. AIAA J 53(10):2951–2966
8. 李道春, 向锦武, 王子瑜. 一种共轴反向双扑旋翼机构. 北京; ZL 201910332059.3, 2021.
9. Weis-Fogh T (1973) Quick estimates of flight fitness in hovering animals, including novel mechanisms for lift production. J Exp Biol 59(1):169 LP–230
10. Latt J, Malaspinas O, Kontaxakis D, Parmigiani A, Lagrava D, Brogi F et al (2021) Palabos: parallel lattice Boltzmann solver. Comput Math Appl 81:334–350
11. Suzuki K, Inamuro T (2011) Effect of internal mass in the simulation of a moving body by the immersed boundary method. Comput Fluids 49(1):173–187
12. Suzuki K, Minami K, Inamuro T (2015) Lift and thrust generation by a butterfly-like flapping wing–body model: immersed boundary–lattice Boltzmann simulations. J Fluid Mech 767:659–695

Twin Support Vector Regression and Its Application on Aerodynamic Design

Pei-Xia Lu, Ke-Shi Zhang, and Peng-Hui Wang

Abstract Surrogate modeling is playing an increasingly important role in multidisciplinary design optimization (MDO) related to different areas of aerospace science and engineering, Support vector regression (SVR), due to its good behavior related to numerical noise filtering and highly nonlinear function modeling, is promising as an alternative modeling method. However, SVR is time-consuming for high dimension large-scale samples problem. Since Twin support vector regression (TSVR) method shows faster modeling speed, this work aims to evaluate the modeling abilities by numerical examples, and introduce the TSVR method into aerodynamic designs to explore its potential in the aerospace science. Through series of numerical examples, including low-dimensional numerical examples, high-dimensional numerical examples and numerical examples with noises, it is shown that TSVR takes much less time for modeling while keeping high modeling accuracy. Then, taking RAE2822 airfoil as an example, TSVR and SVR is compared, in which TSVR still shows higher modeling efficiency and accuracy. It's preliminarily proven of the TSVR has great potential in aerodynamic design.

Keyword TSVR · SVR · Surrogate modelling · MDO

1 Introduction

Surrogate modeling is playing an increasingly important role in multidisciplinary design optimization (MDO) related to different areas of aerospace science and engineering. Surrogate modeling is to build an approximate mathematical model that can replace those complex and time-consuming numerical simulations in optimizations. In the past 30 years, surrogate modeling has attracted more attention in various industries, especially in MDO, in which time-consuming numerical simulations widely exist, due to its high efficiency, highly nonlinear approximation, and global optimization ability.

P.-X. Lu · K.-S. Zhang (✉) · P.-H. Wang
School of Aeronautics, Northwestern Polytechnical University, Xi'an 710072, People's Republic of China
e-mail: zhangkeshi@nwpu.edu.cn

S. Lee et al. (eds.), *The Proceedings of the 2021 Asia-Pacific International Symposium on Aerospace Technology (APISAT 2021), Volume 1*, Lecture Notes in Electrical Engineering 912, https://doi.org/10.1007/978-981-19-2689-1_45

As a surrogate modeling method, support vector regression (SVR) [5, 14] is a special implementation of support vector machine (SVM). Based on the principle of structural risk minimization, SVM was proposed by Vapnik and his colleagues [15]. It surpassed most other systems in a variety of applications in just a few years in 1995, becoming the most popular method in pattern recognition and machine learning. SVR inherits the advantages of SVM and behaves well in high-dimensional space problem [3]. It can either act as an interpolation model or filter numerical noises when setting different ε-tube parameters. In recent years, due to its good ability numerical noise filtering, and highly nonlinear function modeling. SVR was introduced into variety of research fields [1, 2, 9, 10].

However, SVR is time-consuming for the high-dimensional or large-scale samples problems. The researchers tried to proposed the revised version, TSVR is one of them. TSVR solves two smaller quadratic programming problems (QPPs), rather than one large QPP in the classical SVR, which theoretically makes it four times faster than SVR, while keeping good generalization performance. In fact, some studies have shown that TSVR models even much faster. Therefore, in recent years, variety of TSVR methods were proposed in machine learning [7, 11, 13, 16, 18, 19]. In order to overcome shortcomings of the original TSVR, such as lacking sparsity and being affected by outliers. Primal least squares TSVR [13] transforms the inequality constraints into equality constraints by introducing the least squares method, which can improve the learning speed of TSVR. In order to improve the fitting performance of function, ε-TSVR [19] implements the structural risk minimization by introducing the regularization term in primal problems, which yields the dual problems to be stable positive definite quadratic programming problems, and improves generation ability. The outliers existing in the sample points is also a problem we need to consider. Yi and Xu propose weighted TSVR [18], it gives different penalties to samples in the different positions, which can not only weaken the influence of outliers on the model, but also has good anti-interference performance. In addition, the loss function also has a significant effect on reducing sample outliers. Ye [17] et al. transformed ε-insensitive coefficient into L1 insensitive coefficient, which can effectively reduce the impact of outliers on the model. TSVR is widely used different locations [4, 6], However, TSVR method has never been applied in aerospace engineering.

In order to fairly compare the computational efficiency and accuracy of TSVR and SVR, we use the original TSVR and SVR for analysis and comparison in this paper. This work aims to preliminarily evaluate the classical TSVR method via numerical examples and explore its prospect in aerodynamic design problem.

2 Fundamental Theory of TSVR

2.1 Linear Regression Using SVR

Given a training data set $(x_i, y_i)|i = 1, 2, \ldots, n$, where x_i denotes the input vector and y_i is its corresponding response, the regression problem is to find a smooth function close to each point, the function can later be used to predict the response of a new input. SVR is a method of linear regression, as the following expression.

$$f(x) = (w \cdot x) + b \tag{1}$$

where w is a hyperplane vector, and b is a const.

Since ε-SVR method is one of the most popular SVR methods, the basic principle of it is briefly introduced here. ε-SVR sets ε-insensitive tube, which is the extent swept by translating ε up and down the hyperplane.

$$\{x, y|(w \cdot x) + b - \varepsilon \leq (w \cdot x) + b + \varepsilon\} \tag{2}$$

If the tube contains all samples, we can get

$$-\varepsilon \leq y_i - ((w \cdot x) + b) \leq \varepsilon \tag{3}$$

Choosing the flattest function in the feature space leads to a smooth function in the input space. In order to choose the flattest function in the feature space based on the structural risk minimization principle, we can obtain it by minimizing the norm $\|w\|^2$. Then the optimization problem can be written as

$$min\ \frac{1}{2}\|w\|^2$$

$$s.t.\ y_i - (w \cdot x) - b \leq \varepsilon \tag{4}$$

$$(w \cdot x) + b - y_i \leq \varepsilon$$

$$variable : w, b$$

A key assumption of this formulation is that there exists a function $f(x)$ that can approximate all pairs (x_i, y_i) in ε precision. However, sometimes this may not be the case and some error allowance are desired. Thus, the slack variables ξ_i, ξ_i^* are introduced to allow for some error larger than ε and SVR use a loss function to describe how the estimated function deviated from the true one. ε-insensitive loss function is defined in Eq. (5) and graphically depicted in Fig. 1,which enables a sparse set of support vectors to be obtained for regression.

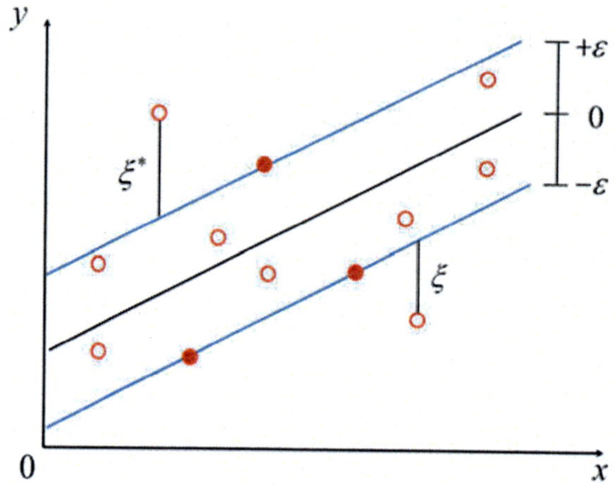

Fig. 1 SVR mathematical model diagram

$$L_\varepsilon(y) = \{ \begin{matrix} 0\ if|f(x) - y| \le \varepsilon \\ |f(x) - y| \le \varepsilon\ otherwise \end{matrix} \tag{5}$$

Since the samples are allowed to fall outside the tube, the penalty parameter C is introduced to express the tolerance of the error. The optimization problem is rewritten as

$$\begin{aligned} &min\ \tfrac{1}{2}\|w\|^2 \\ &s.t.\ y_i - (w \cdot x) - b \le \varepsilon \\ &(w \cdot x) + b - y_i \le \varepsilon \\ &variable : w, b \end{aligned} \tag{6}$$

2.2 *Linear Regression Using TSVR*

In the process of solving the SVR model, a large convex QPP needs to be solved, which usually takes much time, so taking twin support vector machine (TSVM) as reference [8], Peng et al. [12] proposed the TSVR method. TSVR solves two smaller-size bound functions $f_1(x) = w_1x + b_1$, $f_2(x) = w_2x + b_2$ as shown in Fig. 2. In order to better fit the samples, we hope the distance from the estimated function $f_1(x) + \varepsilon_1$ to the samples to be as small as possible. So the first term in the objective

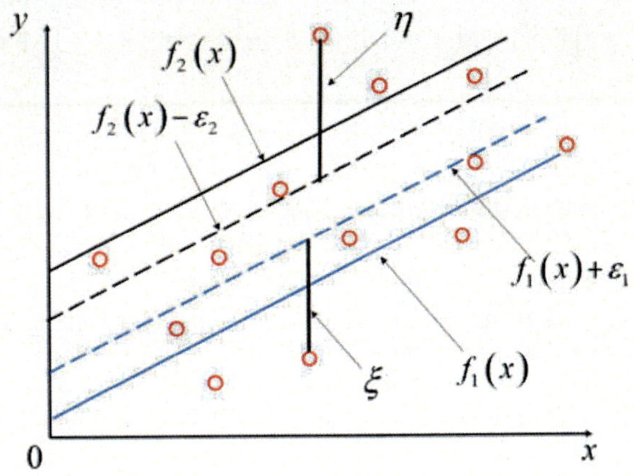

Fig. 2 TSVR mathematical model diagram

function of Eq. (9) aims to minimize the sum of squared distances from samples to the shifted function $f_1(x) + \varepsilon_1$, so do the ε2-insensitive up-bound regressor.

If all samples are located between the $f_1(x) + \varepsilon_1$ and $f_2(x) - \varepsilon_2$ functions, we can get

$$\{(x, y)|(w_1 \cdot x) + b_1 + \varepsilon_1 \le (w_2 \cdot x) + b_2 + \varepsilon_2\}$$

$$\rightarrow y_i - (w_1 \cdot x) + b_1 \ge \varepsilon_1, (w_2 \cdot x) + b_2 - y_i \ge \varepsilon_2 \tag{7}$$

ξ, η is introduced to allow for the samples outside the tube, limited by $f_1(x) + \varepsilon_1$ and $f_2(x) - \varepsilon_2$, the loss function of TSVR is defined in Eq. (8) and as shown in Fig. 2.

$$L_1(y) = \{ \begin{matrix} 0\ if\ y_i - f_1(x) \ge \varepsilon_1 \\ (f_1(x) + \varepsilon_1) - y_i\ otherwise \end{matrix}$$

$$L_2(y) = \{ \begin{matrix} 0\ if\ f_2(x) - y_i \ge \varepsilon_2 \\ y_i - (f_2(x) - \varepsilon_2)\ otherwise \end{matrix} \tag{8}$$

The second term of the objective function introduces a regularization coefficient C to punish the samples outside the tube.

From the above, we can establish the original problem of TSVR.

$$min\ \frac{1}{2}(Y - e\varepsilon_1 - (Aw_1 + eb_1))^T(Y - e\varepsilon_1 - (Aw_1 + eb_1)) + C_1 e^T \xi$$

$$s.t.\ Y - (Aw_1 + eb_1) \ge e\varepsilon_1 - \xi, \xi \ge 0 \tag{9}$$

$$min\ \frac{1}{2}(Y + e\varepsilon_2 - (Aw_2 + eb_2))^T(Y + e\varepsilon_2 - (Aw_2 + eb_2)) + C_2 e^T \eta$$

$$s.t.\ (Aw_2 + eb_2) - Y \geq e\varepsilon_2 - \eta, \eta \geq 0$$

where A is samples matrix, and Y is the response vector, w_1, w_2, b_1, b_2 are parameters to be estimated, e is the unit vector, $C_1, C_2 > 0$ and $\varepsilon_1, \varepsilon_2 > 0$ are hyperparameters, ξ, η are slack vectors.

Usually the optimization problems Eq. (9) will be transformed into the optimization problems that are standard QPP.

$$max\ -\frac{1}{2}\alpha^T G\left(G^T G\right)^{-1} G^T \alpha + f^T G\left(G^T G\right)^{-1} G^T \alpha - f^T \alpha \tag{10}$$

$$s.t.\ 0 \leq \alpha \leq C_1 e$$

$$max\ -\frac{1}{2}\gamma^T G\left(G^T G\right)^{-1} G^T \gamma - h^T G\left(G^T G\right)^{-1} G^T \gamma + h^T \alpha \tag{11}$$

$$s.t.\ 0 \leq \gamma \leq C_2 e$$

where $G = \left[A\ e\right]$,$f = Y - e\varepsilon_1$,$h = Y + e\varepsilon_2$,$u_1 = \begin{bmatrix} w_1 \\ b_1 \end{bmatrix} = \left(G^T G\right)^{-1} G^T (f - \alpha)$, $u_2 = \begin{bmatrix} w_2 \\ b_2 \end{bmatrix} = \left(G^T G\right)^{-1} G^T (h + \gamma)$

Once the vectors u_1 and u_2 are known, we can seek w_1, w_2, b_1, b_2, the upper- and lower-bound functions can be obtained. The approximation function is:

$$f(x) = \frac{1}{2}(f_1(x) + f_2(x)) = \frac{1}{2}(w_1 + w_2)^T x + \frac{1}{2}(b_1 + b_2) \tag{12}$$

2.3 Nonlinear Regression Using TSVR

In order to fit nonlinear functions, TSVR introduces kernel functions into the formulation.

$$f_1(x) = K(x^T, A^T)w_1 + b_1,\ f_2(x) = K(x^T, A^T)w_2 + b_2 \tag{13}$$

Substituting Eq. (13) into Eq. (9), the nonlinear approximation is realized:

$$min\ \frac{1}{2}\left(Y - e\varepsilon_1 - \left(K(A, A^T)w_1 + eb_1\right)\right)^T \left(Y - e\varepsilon_1 - \left(K(A, A^T)w_1 + eb_1\right)\right) + C_1 e^T \xi$$

$$s.t.\ Y - \left(K(A, A^T)w_1 + eb_1\right) \geq e\varepsilon_1 - \xi, \xi \geq 0 \tag{14}$$

$$min\ \tfrac{1}{2}\left(Y + e\varepsilon_2 - \left(K(A, A^T)w_2 + eb_2\right)\right)^T \left(Y + e\varepsilon_2 - \left(K(A, A^T)w_1 + eb_2\right)\right) + C_2 e^T \eta$$

$$s.t.\ \left(K(A, A^T)w_2 + eb_2\right) - Y \geq e\varepsilon_2 - \eta, \eta \geq 0$$

Their dual optimizations are formulated as:

$$max\ -\tfrac{1}{2}\alpha^T H\left(H^T H\right)^{-1} H^T\alpha + f^T H\left(H^T H\right)^{-1} H^T\alpha - f^T\alpha$$

$$s.t.\ Y - \left(K(A, A^T)w_1 + eb_1\right) \geq e\varepsilon_1 - \xi, \xi \geq 0 \tag{15}$$

$$max\ -\tfrac{1}{2}\gamma^T H\left(H^T H\right)^{-1} H^T\gamma - h^T H\left(H^T H\right)^{-1} H^T\gamma + h^T\alpha$$

$$s.t.\ \left(K(A, A^T)w_2 + eb_2\right) - Y \geq e\varepsilon_2 - \eta, \eta \geq 0 \tag{16}$$

where $H = \left[K(A, A^T)\ e\right]$, $f = Y - e\varepsilon_1$, $h = Y + e\varepsilon_2$, $u_1 = \begin{bmatrix} w_1 \\ b_1 \end{bmatrix} = \left(H^T H\right)^{-1} H^T(f - \alpha)$, $u_2 = \begin{bmatrix} w_2 \\ b_2 \end{bmatrix} = \left(G^T G\right)^{-1} G^T(h + \gamma)$

Once the vectors u_1 and u_2 are known, we can seek w_1, w_2, b_1, b_2, the upper- and lower-bound functions can be obtained. The approximation function is:

$$f(x) = \frac{1}{2}(f_1(x) + f_2(x)) = \frac{1}{2}(w_1 + w_2)^T K(A, A^T) + \frac{1}{2}(b_1 + b_2) \tag{17}$$

3 Numerical Examples

In this section, TSVR is compared with SVR via series of examples by modeling accuracy and efficiency, and then preliminarily applied in the airfoil design problems.

3.1 Low-Dimension Function (Dimension ≤ 3)

These functions in Eq. (18), Eq. (19) and Eq. (20) are used to evaluate TSVR for low-dimensional cases.

- 1-D function

Table 1 Results of TSVR and SVR for the low-dimensional functions

Function	D	Samples	Algorithm	MSE	SSE/SST	SSR/SST	Modeling time (s)	QPP time (s)
1-D	1	21	SVR	5.9e-03	2.84e-04	0.9948	0.187	0.123
			TSVR	4.1e-03	1.97e-04	0.9894	*0.023*	*0.017*
Sphere	2	36	SVR	0.3873	0.0032	0.9296	0.192	0.114
			TSVR	1.18e-04	9.64e-06	0.9997	*0.027*	*0.017*
Rosenbrock	2	25	SVR	4.21e + 03	0.0097	1.1043	0.184	0.114
			TSVR	674.1260	0.0016	1.0367	*0.025*	*0.017*

$$f(x) = (6x - 2)^2 \sin(12x - 4), x \in (0, 1) \tag{18}$$

- Sphere function

$$f(x) = \sum_{i=1}^{n} x_i^2, i = 1, \ldots, n \text{ Here } x_i \in [-5, 5], i = 1, \ldots, n \tag{19}$$

- Rosenbrock function

$$f(x) = \sum_{i=1}^{n} (100\left(x_{i+1} - x_i^2\right)^2 + (1 - x_i)^2) \tag{20}$$

Here $x_i \in [-2, 2], i = 1, \ldots, n$

The results are listed in Table 1 and shown in Figs. 3, 4 and 5. It's indicated that the TSVR is about 10 times faster than SVR, while keeping high modeling accuracy.

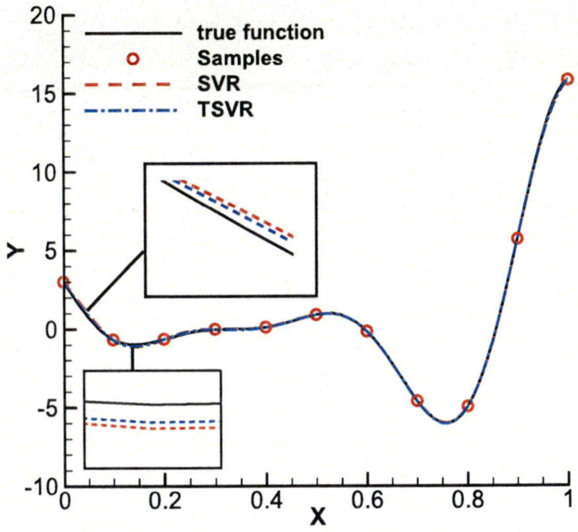

Fig. 3 TSVR and SVR models comparison on 1-Dimension Function

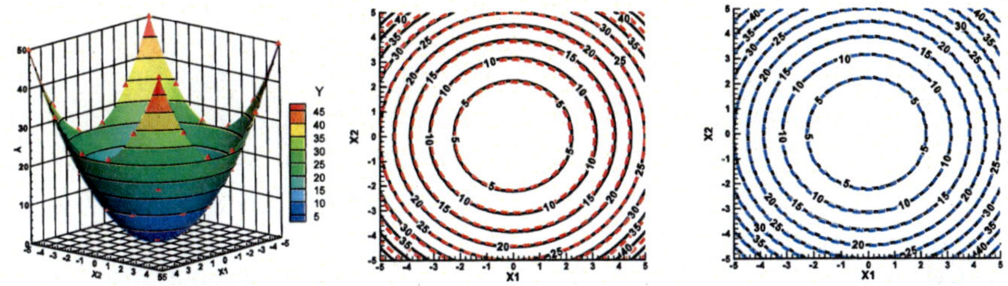

Fig. 4 True Function of Sphere(left), Comparison of fitting results of the SVR (middle) and TSVR (right) (Red triangles: samples; black Line: true Function; red Line: SVR model; blue line: TSVR model)

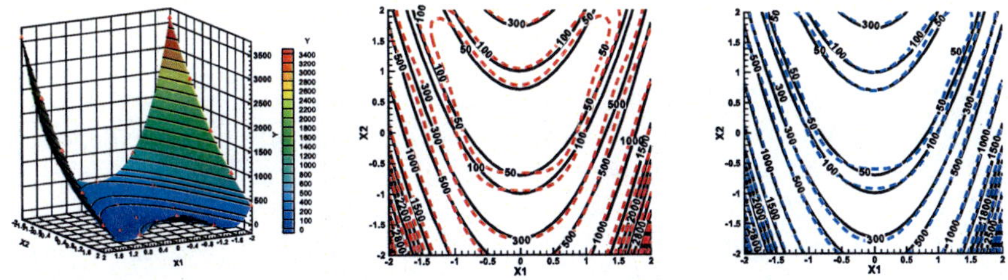

Fig. 5 True Function of Rosenbrock (left), Comparison of fitting results of the SVR (middle) and TSVR (right) (Red triangles: samples; black Line: true Function; red Line: SVR; blue Line: TSVR model)

Table 2 Results of TSVR and SVR for the high-dimensional functions

Function	D	Samples	Algorithm	MSE	SSE/SST	SSR/SST	Modeling time (s)	QPP time (s)
Sphere	10	500	SVR	4.5850	8.3e-03	0.9295	1.149	0.238
			TSVR	0.1455	2.6e-04	0.9823	*0.880*	*0.071*
Rosenbrock	10	500	SVR	9.06e + 05	0.2174	0.7011	1.196	0.292
			TSVR	5.34e + 05	0.1282	0.8288	*0.944*	*0.098*
G9	7	1000	SVR	7.71e + 12	1.3641	0.3641	6.027	2.714
			TSVR	9.74e + 11	0.1723	0.9877	*3.376*	*0.473*

3.2 *High-Dimensional Functions (Dimension > 3)*

The aerospace engineering problems usually include large number of design variables. Therefore, it is very important to evaluate the performance of TSVR based on high-dimensional functions. The sphere function (Eq. (19)), Rosenbrock function (Eq. (20)) and G9 function (Eq. (21)) are used.

- G9 function

$$f(x) = (x_1 - 10)^2 + 5(x_2 - 12)^2 + x_3^4 + 3(x_4 - 11)^2 + 10x_5^6 + 7x_6^2 + x_7^4 - 4x_6x_7 - 10x_6 - 8x_7 \tag{21}$$

where $x_i \in [-10, 10], i = 1, \ldots, 7$

The results are compared in Table 2. It is shown that both TSVR and SVR fit the samples well while TSVR runs faster.

3.3 *Functions with Numerical Noise*

TSVR inherited the advantage of filtering numerical noises of SVR. Numerical noises unavoidably exit in numerical simulations of aircraft design, which brings oscillations to the optimizations. So it is important to see how well it behaves in this case. Two functions are used. One is shown in Eqs. (22) and (23), where Eq. (22) is the smooth function, and an extra function with maximum amplitude 0.02 is introduced in Eq. (23) to model numerical noises. Another function is expressed in Eq. (24), in which case a normal distributed noise of N (0, ± 3) is added. The results are compared in Fig. 6 and Fig. 7 and Table 3. It's shown that the modeling time of TSVR is still much less than that of SVR. Both SVR and TSVR fit data well for the one-dimensional problem, due to not so apparent noises. But they both behave not well in the second function. The reason might be the highly nonlinear shape needs more samples or the hyperparameters need to be optimized.

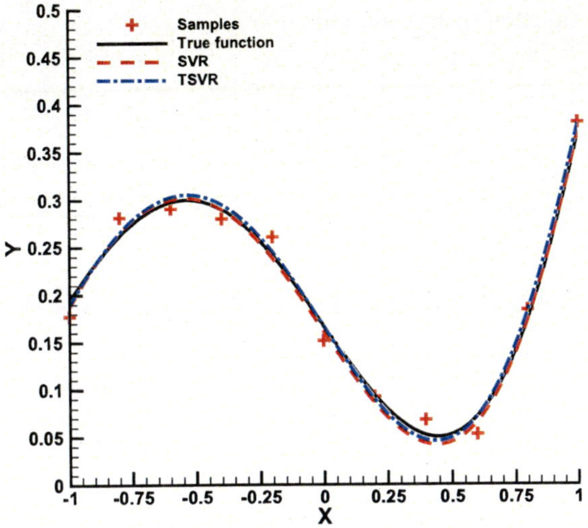

Fig. 6 Fitting moldes of 1-Dimension function with noise levels

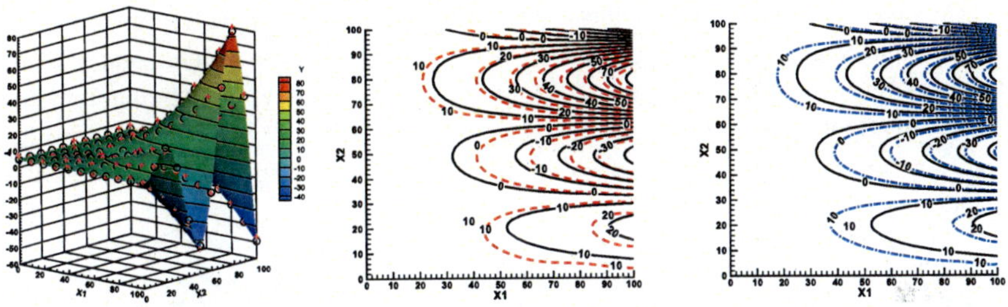

Fig. 7 True Function of 2-Dimension Function(left), Comparison of fitting results of the SVR (middle) and TSVR (right) (' + ': true samples; 'o': samples with numerical noise; black line: true function; red line: SVR model; blue line: TSVR model)

Table 3 Results of SVR, TSVR modeling functions with numerical noise

Function	D	Samples	Algorithm	MSE	SSE/SST	SSR/SST	Modeling Time(s)	QPP Time(s)
1-D (with numerical noise)	1	11	SVR	2.10e-05	2.6e-03	1.0763	0.183	0.120
			TSVR	2.13e-05	2.6e-03	1.0739	*0.024*	*0.018*
Function	D	Samples	Algorithm	MSE	SSE/SST	SSR/SST	Modeling Time(s)	QPP Time(s)
2-D (with numerical noise)	2	121	SVR	9.3799	0.0334	0.9769	0.190	0.118
			TSVR	5.4091	0.0193	1.0354	*0.037*	*0.017*

- 1-Dimension function with numerical noise

$$y = \frac{3}{10} + \sin\frac{16}{15}x - 1 + \sin^2\left(\frac{16}{15}x - 1\right), x \in [-1, 1] \tag{22}$$

$$y = \frac{3}{10} + \sin\frac{16}{15}x - 1 + \sin^2\left(\frac{16}{15}x - 1\right) + \frac{2}{100}\sin(40\left(\frac{16}{15}x - 1\right)), x \in [-1, 1] \tag{23}$$

- 2-Dimension function with numerical noise

$$y = 5 + \frac{x_1^2 x_2 \sin(\frac{x_2}{10})}{10000}, x_1 \in [0, 100], x_2 \in [0, 100] \tag{24}$$

3.4 RAE2822 Airfoil Aerodynamic Modeling

In this section, the TSVR method is preliminarily applied in aerodynamic design of the RAE2822 airfoil. Aerodynamic modeling build up approximation function of C_L, C_d, or C_m w.r.t airfoil geometry, which not only helps understand airfoil aerodynamic performances, but also largely improve efficiency when being applied in aerodynamic optimizations.

- C_d modeling

Grid deformation method is used to generate sample airfoils based on RAE2822 airfoil. The in-house code PMN2D is used to solve the N-S equation at Ma = 0.734 and AOA = 2.79°, S-A turbulence model and the structured meshing are applied. The maximum t/c of the basic airfoil is 0.1207078, It will be varied in the range of [6.719%, 17.641%], which is divided 100 parts in this design space, and calculate the drag coefficient at this 101 sample points.

We take 20 sample points from these 101 samples for SVR and TSVR modelling. The comparisons are made in Fig. 8 and Table 4.

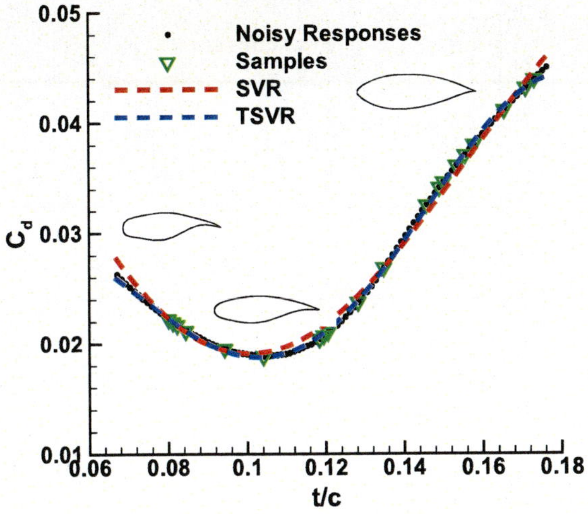

Fig. 8 RAE2822 airfoil drag curve fitting comparisons of TSVR and SVR

Table 4 Fitting comparisons for the RAE2822 airfoil design

Algorithm	Samples	MSE	SSE/SST	SSR/SST	Modeling Time(s)	QPP Time(s)
SVR	20	3.77e-07	0.0016	1.0073	0.238	0.172
TSVR		5.13e-08	1.90e-04	0.9988	*0.033*	*0.018*

It's shown that TSVR almost completely coincides with the original data, overwhelming the SVR model both by accuracy and efficiency.

- C_d modeling with numerical noises

In this case, aerodynamic analysis of the RAE2822 airfoil is performed at Ma = 0.734 and AOA = 2.79°. The inviscid flow is simulated using the commercial CFD software package Fluent. The unstructured mesh of about 15,000 elements (surface grid amount 512) and a second-order accuracy solution scheme are adopted. Numerical noises are induced due to the relatively coarse mesh. The maximum $\overline{f}$ of the basic airfoil is 0.002609, It will be varied in the rage of [−2.854%, 3.245%]. 101 samples are obtained uniformly in the design region.

We take 20 sample points from these 101 samples for SVR and TSVR modeling. The comparisons are made in Fig. 9 and Table 5.

It's shown that although both TSVR and SVR approximate a reasonable trend, however they don't behave well near the minimum, which might be incured by non-uniform distribution of samples in this region. Hyper-parameters adaption or improved sampling might improve it. In this case, TSVR still runs much faster than SVR, which indicates its good prospect in aircraft design.

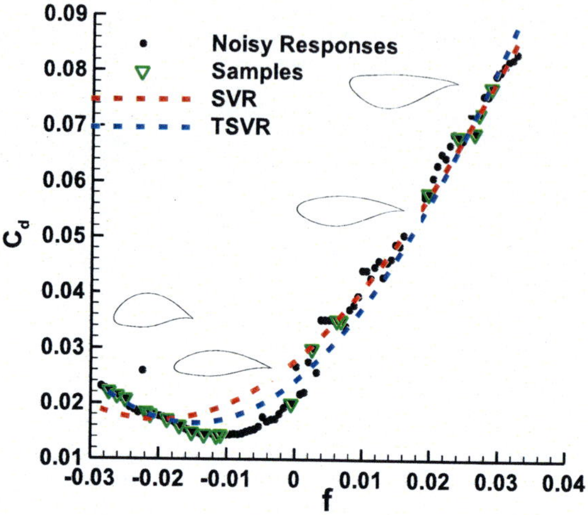

Fig. 9 RAE2822 airfoil drag curve fitting comparisons of TSVR and SVR

Table 5 Fitting comparisons for the RAE2822 airfoil design

Algorithm	Samples	MSE	SSE/SST	SSR/SST	Modeling Time(s)	QPP Time(s)
SVR	20	1.38e-05	0.0077	0.9989	0.238	0.173
TSVR		9.81e-06	0.0054	0.9583	*0.036*	*0.021*

4 Conclusions

This paper evaluates TSVR in its modeling accuracy and efficiency by comparison with SVR. TSVR is preliminarily applied in aerodynamic design. TSVR reveals good modeling ability both in accuracy and efficiency, which indicates that it deserves further study in aerodynamic design.

Acknowledgements The research was sponsored by the Aeronautical Science Foundation(2019ZA053004) and the Natural Science Funding of Shan'Xi Province(2020JM-127).

References

1. Alade (2020) Modeling the viscosity of nanofluids using artificial neural network and Bayesian support vector regression. J Appl Phys

2. Chen Y, Liao Y, Baowen H, Guochang S, Hu Ya han, Z, Yuan (2020) A novel model for electromagnetic properties of complex microstructure composites based on support vector regression. IEEE MTT-S international conference on numerical electromagnetic and multiphysics modeling and optimization
3. Drucker H, Burgers CJC, Kaufmann LSA, Vapnik V (1996) Support vector regression machines, advances in neural information processing systems 779–784
4. Gao C, Shen MG, Liu XP et al (2019) End-point static control of basic oxygen furnace steelmaking based on wavelet transform weighted twin support vector regression. Complexity J 1–6
5. Gunn SR (2000) Support vector machine for classification and regression. Technical Report, image speech and intelligent systems research group, University of Southhampton, UK
6. Houssein EH (2019) Particle swarm optimization-enhanced twin support vector regression for wind speed forecasting. J Intell Syst J 28(5):905–914
7. Hua Juan H, Shi Fei D (2013) Primal least squares twin support vector regression. Zhejiang Univ-Sci C (Comput & Electron) J 14(9):722–732
8. Jayadeva twin support vector machines for pattern classification. IEEE Trans Pattern Anal Mach Intell 29(5):905–910
9. Ke Shi Z, Zhong Hua H (2013) Support vector regression-based multidisciplinary design optimization in aircraft conceptual design. AIAA 1160
10. Li Y, Shuai W, Weng Xiang C, Wei H (2020) Holistic comparison of different kernel functions for support vector regression based on state-of-health prediction of lithium-ion battery. Proceedings—11th international conference on prognostics and system health management, 40–46
11. Peng X (2015) Interval twin support vector regression algorithm for interval input-output data. Int J Mach Learn Cyber 6:719–732
12. Peng X (2010) TSVR: an effificient twin support vector machine for regression. Neural Netw 23:365–372
13. Peng X (2010) Primal twin support vector regression and its sparse approximation. Neurocomputing 73:2846–2858
14. Vapnik V, Golowich SE, A Smola (1996) Support vector method for function approximation, regression estimation, and signal processing. Adv Neural Inform Process Syst, 281–287
15. Vapnik V (1995) The nature of statistical learning theory springer, New York
16. Yanmeng L, Huaijiang S (2020) Multi-output parameter-insensitive kernel twin SVR model. Neural Netw 121:276–293
17. Ye YF, Bai L, Hua XY et al (2016) Weighted Lagrange ε-twin support vector regression. Neurocomputing J 197:53–68
18. Yi tian X (2012) A weighted twin support vector regression, Knowledge-Based Systems 33:92–101
19. Yuan H (2013) An ε-twin support vector machine for regression. Neural Comput Appl 23:175–185

H-Force of Rigid Rotor in Forward Flight of Multi-rotor

Yasushi Morikawa and Takeshi Tsuchiya

Abstract Multi-rotors have certain outstanding features, such as VTOL (Vertical Take-Off and Landing) ability and a simple mechanical system. However, they tend to lack endurance performance. Several types of fixed-wing VTOL have therefore been studied and proposed to minimize this drawback. Our proposal of a novel fixed-wing VTOL is one of these. It has a simple mechanism and has low drag because it is not equipped with control surfaces and stabilizers. Like a multi-rotor, this VTOL is controlled only by the speed of the rotors. In a previous study of our novel VTOL, we conducted a fundamental wind tunnel test of the multi-rotor with and without wings, and acquired data on the drag of the multi-rotor under condition of forward flight. In this paper we describe the results of theoretically examining the drag caused by the rotor. The paper is set out as follows. Section 1 introduces the novel VTOL and the purpose of this research, Sect. 2 describes the wind tunnel tests and the drag measured during forward flight, Sect. 3 describes the theoretical examination of the H-force caused by rotor spin and Sect. 4 verifies the experimental results based on the results of our theoretical study. We draw our conclusions in Sect. 5.

Keywords Multi-rotor · H-force · Forward flight · Blade element theory · Rigid rotor

1 Introduction

Attempts are being made to apply and put into practical use UAVs (Unmanned Aerial Vehicles) such as multi-rotors for the transportation of small goods, aerial photography and infrastructure inspection. Although VTOL (Vertical Take-off and Landing) capability is the key advantage of multi-rotors, they suffer the drawback of

Y. Morikawa (✉)
National Institute of Advanced Industrial Science and Technology, 1-2-1 Namiki, Tsukuba, Ibaraki 305-8564, Japan
e-mail: morikawa.y@aist.go.jp

T. Tsuchiya
The University of Tokyo, 7-3-1 Hongo, Bunkyo-ku, Tokyo 113-8656, Japan

S. Lee et al. (eds.), *The Proceedings of the 2021 Asia-Pacific International Symposium on Aerospace Technology (APISAT 2021), Volume 1*, Lecture Notes in Electrical Engineering 912, https://doi.org/10.1007/978-981-19-2689-1_46

low endurance performance, since their lift comes from rotor thrust only. To solve this problem, some VTOLs with fixed-wing have been researched and developed. Figure 1 shows the Lilium Jet developed by Lilium [1]. The Lilium Jet is a 7-seater whose maximum physical range is designed to be 250 km or more. Figure 2 shows the teTra Mk-5, which is a personal eVTOL [2]. It is being developed with the aim of starting delivery in 2022. We have proposed and studied a novel VTOL with improved endurance performance by adopting a fixed-wing while taking advantage of the simple mechanisms used in multi-rotors. Figure 3 shows one possible variation of the proposed VTOL. The rotors are embedded in the body and wing, and there are no control surfaces and stabilizers. The VTOL is controlled by only rotor spin speed. As a fundamental experiment, we conducted wind tunnel tests of the multi-rotor. The lift and drag of the multi-rotor during forward flight were measured. To develop a novel VTOL, the results of wind tunnel tests need to be understood and applied. The wind tunnel test result is described in detail in our previous paper [3]. However, the theoretical elucidation of the test results was not enough.

On the other hand, research is being actively conducted on flight control of multi-rotors, and studies of the flight dynamics of multi-rotors and their rotors are also

Fig. 1 Lilium Jet [1]

Fig. 2 teTra Mk-5 [2]

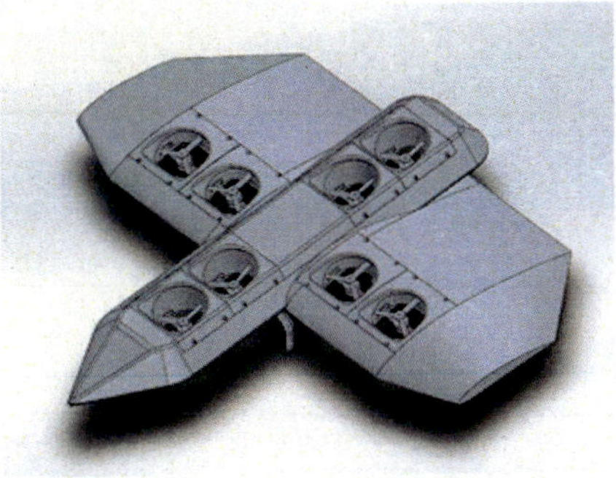

Fig. 3 Image of proposed VTOL

in progress to assist the design of control systems [4, 5]. However, although some studies employ blade element theory and aerodynamic investigations for forward flight, most approximate the thrust and torque of the rotor, and the flight dynamics models are only considered for hovering or low-speed flight. Moreover, these studies do not make a clear distinction between flight speed and rotor speed. Aerodynamic drag acts on the multi-rotor during forward flight, and in a steady state, forward thrust is trimmed with the drag. Our wind tunnel test that simulated forward flight shows that the drag includes a term proportional to velocity in addition to parasite drag which is proportional to the square of the velocity. The term that is proportional to velocity can be explained as the drag generated by the rotor in forward flight. In a helicopter rotor, where blade flapping occurs, the drag generated by the rotor is generally known as the H-force [6, 7]. No flapping of the rotor blades occurs in the multi-rotor because the rotor is rigid. Therefore, in this paper, in which the multi-rotor rotors show no blade flapping, the H-force in forward flight is examined theoretically, and the wind tunnel test result is considered alongside the theoretical study.

2 Wind Tunnel Test Result

In our previous study [3], we conducted a wind tunnel test of a fundamental prototype UAV. The triaxial forces of the UAV in the hovering state and in forward flight state were measured. Figure 4 shows the setup of the wind tunnel test, and Figs. 5 and 6 respectively show the prototype UAV with and without the wing used in the wind tunnel test. The prototype UAV has four 5-inch rotors. The fuselage length is 370 mm and the total weight is 681 g without wings. Figure 7 shows the measured force applied to the x-axis when the free stream velocity is raised from 0 to 20 m/s at 5 m/s intervals while the pitch angle of the UAV is changed from 0 to 30° at 5° intervals. The aerodynamical triaxial forces acting on the UAV were measured using a load cell

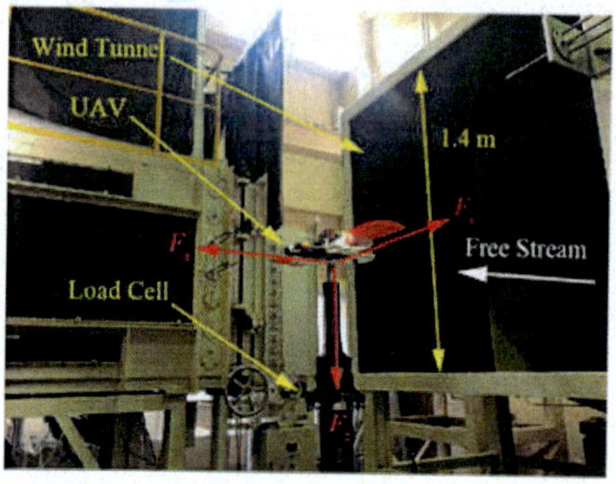

Fig. 4 Low speed wind tunnel

Fig. 5 UAV with wing

Fig. 6 UAV without wing

unit. To simulate the hovering condition, the rotation speed of the rotor was tuned such that the total vertical component of thrust was equal to gravity.

Each x-axis force value at the free stream velocity zero is considered to be a horizontal component of the thrust in Fig. 7. Therefore, Fig. 8 shows the data that are removed the influence of the horizontal component of thrust. It appears to show the drag that acts on the UAV. In Fig. 7, the free stream velocities when the measured

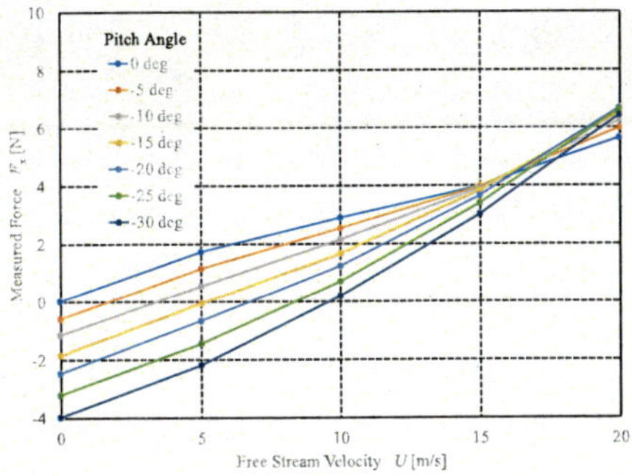

Fig. 7 Measured force in x-axis of UAV without wing (with thrust)

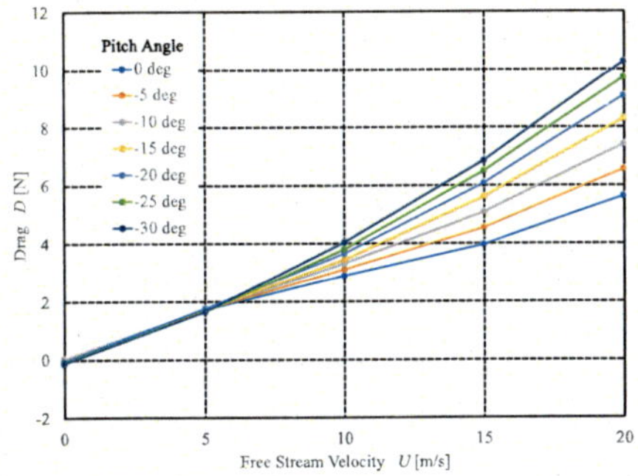

Fig. 8 Drag of UAV without wing (with thrust)

force is equal to zero are the trimmed velocities for each pitch angle. Figure 9 shows the relationship between the pitch angle and the trimmed velocity. In the wind tunnel test, the force of the x-axis when the rotor of the UAV is not rotating was also measured. The force indicates the parasite drag of the UAV. Figure 10 shows the

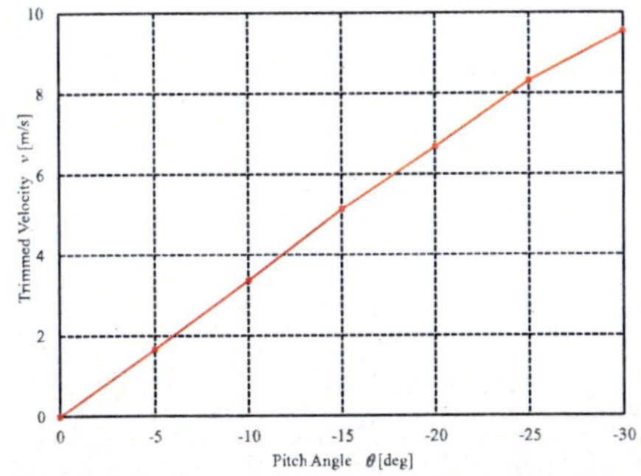

Fig. 9 Trimmed velocity of UAV without wing (with thrust)

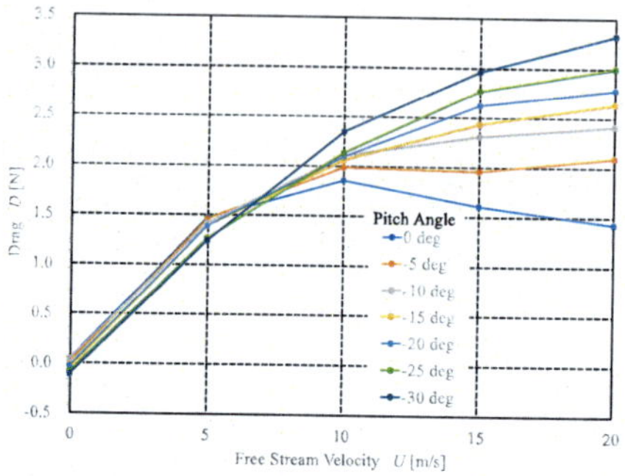

Fig. 10 Additional drag of UAV without wing (with thrust)

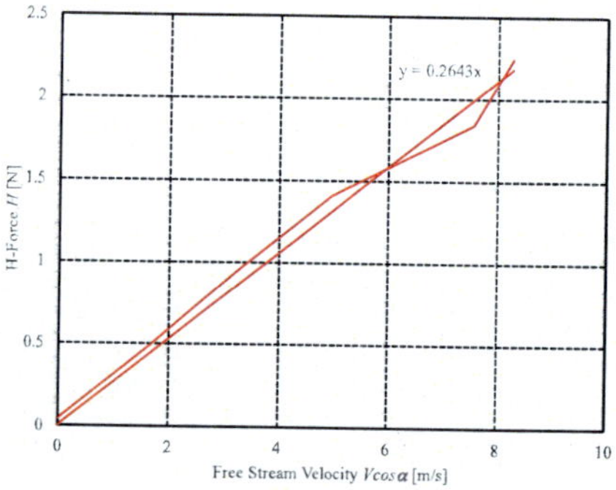

Fig. 11 Drag term proportional to free stream velocity (H-force)

values subtracted the parasite drag from the Fig. 8. It shows the rotor drag that occurs because the rotor is rotating. Figure 11 was created using Figs. 9 and 10 and shows the relationship between the horizontal component of forward flight speed and the drag caused by rotor spin. The relationship seems to be close to proportional. The slope of the approximate curve is 0.2643. The principle for the existence of drag proportional to forward flight speed has not been explained very clearly so far. We will therefore examine it theoretically below.

3 Derivation of H-Force using Blade Element Theory

During flight of a multi-rotor, the airflow velocity of the blade element of the rotor is the sum of the rotation speed of the rotor and the flight speed. It depends on the radial position and azimuth of the rotor. In this section we investigate the thrust and drag of the whole rotor in forward flight using blade element theory. The aerodynamic forces generated in the rotor are broken down into those along the coordinate axes of the control shaft system using a helicopter as an example. They are called the H-force,

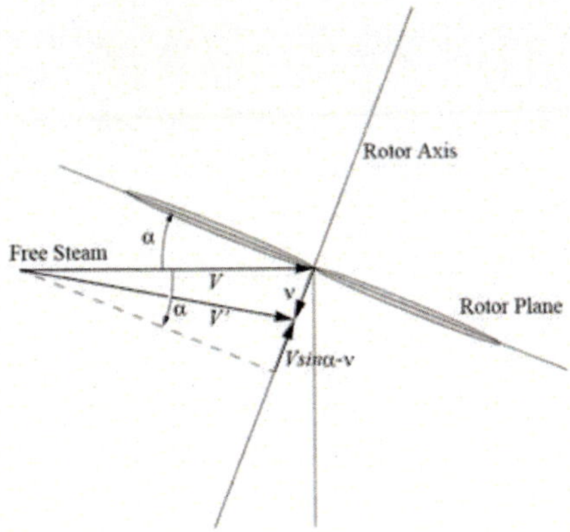

Fig. 12 Side view of rotor plane

the Y-force, and the thrust. However, there is no Y-force in a multi-rotor, because the rigid rotor of the multi-rotor does not experience blade flapping. The aerodynamics of the rigid rotor of a multi-rotor are simpler than those of a helicopter rotor. The H-force of the rigid rotor is examined theoretically below.

Assuming that a rotor is in a free stream as shown in Fig. 12, the horizontal component U_P of relative velocity U to the axis of rotation is expressed by the following equation.

$$U_P = V \sin\alpha - \nu \tag{1}$$

where V is the free stream velocity, α is the angle of attack of the rotor plane, and ν is the inflow to the rotor. However, if the rotor is rotating at a speed Ω as shown in Fig. 13, in the cross-section at a distance r from the axis of rotation, the orthogonal component U_T of relative velocity U is expressed by the following equation

$$U_T = \Omega r + V \cos\alpha \sin\psi \tag{2}$$

where ψ is the azimuth of the rotor. The angle at which the combined velocity V_r of U_T and U_P forms with the rotor plane is ϕ as shown in Fig. 14. The azimuth is ψ as shown in Fig. 15 and the drag dH of a blade element at a distance r from the axis of rotation is expressed by the drag dD and the lift dL as in the equation below.

$$dH = dD \cos\phi \sin\psi - dL \sin\phi \sin\psi \tag{3}$$

Assuming ϕ to be sufficiently small here, the following approximate equation is obtained.

$$dH = dD \sin\psi - dL\phi \sin\psi \tag{4}$$

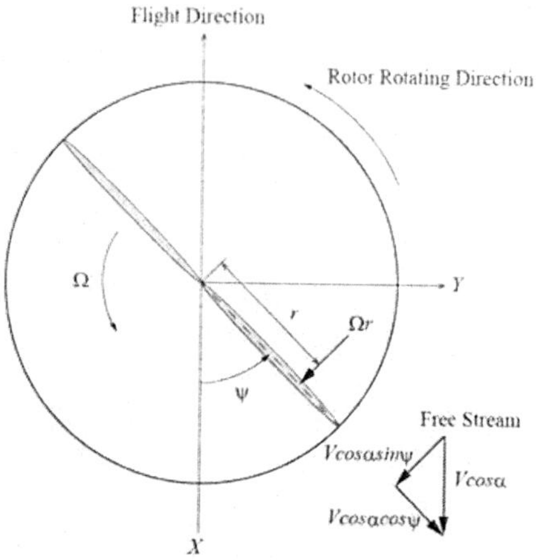

Fig. 13 Overhead view of rotor plane and relative velocity

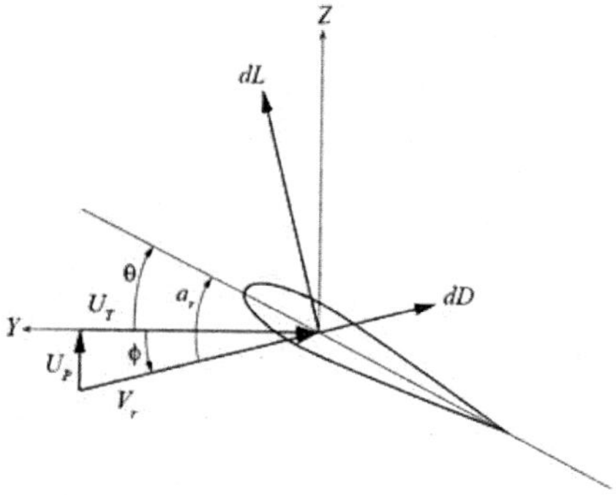

Fig. 14 Blade section showing pitch angle and aerodynamics velocity

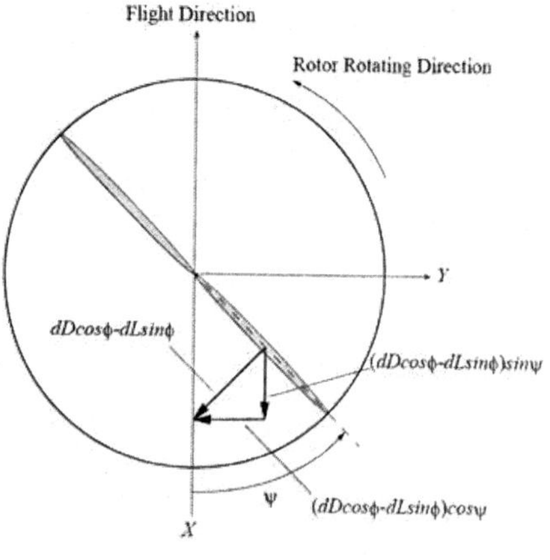

Fig. 15 Elementary H-force

The drag H of the rotor is the sum of the cross-sectional drag H_o and induced drag H_i as expressed by the following equation.

$$H = H_0 + H_i \tag{5}$$

Therefore, from Eqs. 4 and 5, the cross-sectional drag, and induced drag are expressed by the following equations.

$$dH_0 = dD \sin\psi \tag{6}$$

$$dH_i = -dL\phi \sin\psi \tag{7}$$

The lift coefficient of rotor C_l is the angle of attack α_r shown in Fig. 14 multiplied by the lift slope a. The lift coefficient is expressed by the following equation

$$C_l = a\alpha_r = a(\theta + \phi) \cong a\left(\theta + \frac{U_P}{U_T}\right) \tag{8}$$

where θ is the pitch angle of the rotor. The lift dL and the drag dD are therefore expressed by the following equations

$$dL = \frac{1}{2}\rho V_r C_l c dr = \frac{1}{2}\rho a\left(\theta U_T^2 + U_P U_T\right) c dr \tag{9}$$

$$dD = \frac{1}{2}\rho U_T^2 c\delta dr \tag{10}$$

where ρ is the air density and c and δ are the chord of the aerofoil and cross-sectional drag coefficient all of which depend on radial position r. As a result of integrating the cross-sectional drag of the blade element, the whole cross-sectional drag of the entire b rotor blades is expressed by the following equation by substituting Eq. 10 into Eq. 6 and integrating it.

$$H_0 = \frac{b}{2\pi}\int_0^R \int_0^{2\pi} \frac{1}{2}\rho U_T^2 c\delta \sin\psi d\psi dr = \frac{b}{4\pi}\rho \int_0^R \int_0^{2\pi} c\delta U_T^2 \sin\psi d\psi dr \tag{11}$$

The following equation is obtained by using Eq. 2.

$$\begin{aligned} U_T^2 \sin\psi &= (\Omega r + V\cos\alpha \sin\psi)^2 \sin\psi \\ &= \Omega^2 r^2 \sin\psi + \tfrac{1}{2}V^2\cos^2\alpha \sin\psi + \Omega r V \cos\alpha(1 - \cos 2\psi) \\ &\quad - \tfrac{1}{2}V^2\cos^2\alpha(\sin 3\psi - \sin\psi) \end{aligned} \tag{12}$$

Substituting Eq. 12 into Eq. 11, whole rotor drag H_0 is expressed by the following equation.

$$\begin{aligned} H_0 &= \tfrac{b}{4\pi}\rho \int_0^R \int_0^{2\pi} c\delta U_T^2 \sin\psi d\psi dr \\ &= \tfrac{b}{2\pi}\tfrac{1}{2}\rho \int_0^R 2\pi c\delta\Omega r V \cos\alpha dr \\ &= \tfrac{1}{2}\rho b\Omega V \cos\alpha \int_0^R c\delta r dr \end{aligned} \tag{13}$$

On the other hand, similarly, by substituting Eq. 10 into Eq. 7 and integrating it, the induced drag is expressed by the following equation, resulting in zero.

$$\begin{aligned} H_i &= -\tfrac{b}{2\pi}\int_0^R \int_0^{2\pi} \tfrac{1}{2}\rho U_T^2 C_l c\phi \sin\psi d\psi dr \\ &= -\tfrac{b}{2\pi}\tfrac{1}{2}\rho a c \int_0^R \int_0^{2\pi} (\theta U_P U_T + U_P^2) \sin\psi d\psi dr = 0 \end{aligned} \tag{14}$$

The result is that the whole rotor drag depends on only the cross-sectional drag as shown in the following equation.

$$H = H_0 + H_i = \frac{1}{2}\rho b\Omega V \cos\alpha \int_0^R c\delta r dr \tag{15}$$

It is proportional to the flight speed. Strictly speaking, it is proportional to the product of the flight speed and the rotation speed of the rotor.

The fact of the drag of the rotor being proportional to the flight speed is understood as follows. While the rotor is rotating, in the right half, the blade element of the rotor whose azimuth is ψ and whose radical position is r has a relative velocity which is the sum of the rotor spin speed and forward flight speed as shown in Fig. 16. Using Eqs. 2, 6, and 10, the drag of the blade element is expressed by the following equation.

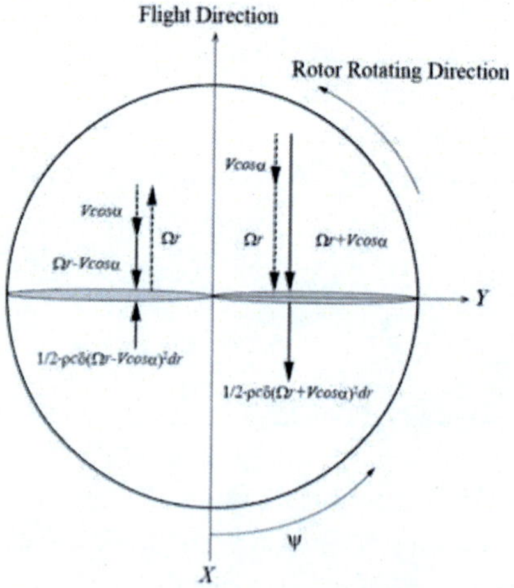

Fig. 16 Relative velocity in the left and right half of rotor plane

$$\begin{aligned} dH_0 &= \tfrac{1}{2}\rho\delta c(r\Omega + V\cos\alpha)^2 dr \sin\psi \\ &= \tfrac{1}{2}\rho\delta c\left(r^2\Omega^2 + 2r\Omega V\cos\alpha + V^2\cos^2\alpha\right)dr \sin\psi \end{aligned} \tag{16}$$

On the other hand, in the left half, the blade element of the rotor has a relative velocity which is different from rotor spin speed to forward flight speed as shown in Fig. 16. The direction of the drag is opposite to that of right half when the rotor is in the azimuth $\psi + \pi$, i.e., rotated 180° from ψ. The drag therefore can be expressed by the following equation.

$$\begin{aligned} dH_0 &= \tfrac{1}{2}\rho\delta c(r\Omega - V\cos\alpha)^2 dr \sin(\psi + \pi) \\ &= -\tfrac{1}{2}\rho\delta c\left(r^2\Omega^2 - 2r\Omega V\cos\alpha + V^2\cos^2\alpha\right)dr \sin\psi \end{aligned} \tag{17}$$

Because the drag of the blade element of the rotor is in opposite directions in the right half and the left half, the square term of the rotor spin speed and the square term of the forward flight speed are offset. The result is that only the first power term remains. Equation 15 is the result of the blade element drag being integrated from 0 to 2π and multiplied by the number of rotor blades.

4 Discussion

Here, the validity of the theoretical examination is verified by comparing the theoretical results and the experimental results. Assuming the chord of the aerofoil and the cross-sectional drag coefficient to be approximately constant, the drag of the rotor is expressed by the following equation.

$$H = \frac{1}{4}\rho bc\delta\Omega R^2 V\cos\alpha \tag{18}$$

The proportionality coefficient of the H-force to the component horizontal of the forward flight speed relative to the rotor is 0.2643 as shown in Fig. 11. Since the prototype UAV used in this wind tunnel experiment is a four-rotor multi-rotor, the proportionality coefficient is about 0.066 when considering one rotor. On the other hand, the diameter of the rotor is 5 in, and the rotor has two blades. Since it rotates at about 50 revolutions per second and has a low Reynolds number, we assume a cross-sectional drag coefficient of 0.5 to confirm the order of the H-force value. Also assuming the average chord of the aerofoil length to be 0.015 mm, and the air density to be about 1.3 kg/m^3, the H-force is calculated as follows using Eq. 18.

$$\begin{aligned} H &= \tfrac{1}{4} \times 1.3 \times 2 \times 0.015 \times 0.5 \times 2\pi \times 50 \times 0.065^2 \times V\cos\alpha \\ &= 0.0065V\cos\alpha \end{aligned} \tag{19}$$

Since the unit of the H-force data in Fig. 11 is Newtons, the proportional coefficient is about 0.063 when the unit is converted. It can be said that the value is on the same order as the value obtained from the wind tunnel test show in Fig. 11. To assess the order, the rotation speed is calculated as a constant at 50 rpm, but strictly speaking, as the flight speed increases, the rotational speed also increases slightly. On the other hand, the wing tip velocity of the rotor reaches about 20 m/s at a rotational speed of 50 rpm. Therefore, it is thought that the reverse flow region increases and becomes less proportional with increasing the forward flight speed. In the left half of the rotor plane in Fig. 16, there is a possibility that the drag against the rotation direction of the rotor increases in the reverse flow area. It appears that this influence appears in the part representing 10 m/s or more in Fig. 11. This needs to be verified in detail in future studies.

5 Conclusion

In the wind tunnel tests conducted in our previous study, the measured drag component included a term proportional to the flight speed during forward flight. In the rotor aerodynamics of a helicopter, this drag generated during forward flight is known as the H-force. In multi-rotors, the H-force can be described more simply because the rotor is rigid. In this paper, the H-force experienced by a multi-rotor during the forward flight was theoretically examined using blade element theory, and the results were compared with the results of the wind tunnel test conducted in our previous study and discussed its validity. The results acquired in this paper to be generally useful for the research and development of multi-rotor type UAVs.

In future works, we will theoretically examine the aerodynamics of a rigid rotor for UAVs in more detail and confirm them experimentally. We will also feed our knowledge into the design of the novel VTOL we are proposing to improve its performance.

References

1. Lilium. https://lilium.com/jet
2. tetra. https://www.tetra-aviation.com/
3. Morikawa Y, Tsuchiya T, Abe H (2019) Fundamental wind tunnel tests of fixed wing VTOL UAV without control surface, APISAT 2019, Gold Coast
4. Gill R, D'Andrea R (2017) Propeller thrust and drag in forward flight. In: 2017 IEEE conference on control technology and applications (CCTA), Hawaii
5. Sartori D, Yu W (2019) Experimental characterization of a propulsion system for multi-rotor UAVs. J Intell Rob Syst 96:529–540
6. Seddon J, Newman S (2011) Basic helicopter aerodynamics, 3rd edn. AIAA Education Series
7. Dreier ME (2018) Introduction to helicopter and tiltrotor flight simulation, 2nd edn. AIAA Education Series

Numerical Investigation of Transonic Flutter Characteristics of a Supercritical Airfoil

Toma Miyake and Hiroshi Terashima

Abstract This study numerically investigates the transonic flutter characteristics of a supercritical airfoil and discusses the similarities and differences compared to a conventional symmetric airfoil. A wing-section model with two-degree-of-freedom was adapted. An SC2-0610 supercritical airfoil was used for the investigation, and the results were compared with those of the NACA64A010 airfoil. Overall, similar flutter characteristics, such as transonic dip appearance, were observed between the two airfoils. The effect of the shape of the supercritical airfoil appears as the angle of attack effect; for example, the supercritical airfoil with an angle of attack of $-2°(0°)$ shows qualitatively similar flutter characteristics to the NACA64A010 airfoil with an angle of attack of $0°(2°)$. Furthermore, in the present study, the supercritical airfoil exhibits unique flutter characteristics under high Mach number conditions because of the development of flow separation near the trailing edge on the lower surface. In particular, when the Mach number was 0.875, the flow reattachment near the trailing edge on the lower surface generated a high-pressure region, which induced a unique flutter oscillation under the second torsion mode.

Keywords Supercritical airfoil · Transonic flutter · Shock oscillation · Flow separation

1 Introduction

Wing flutter is a self-induced oscillation process caused by the coupling of aerodynamic, structural, and inertial forces [1]. When an external force acts on a wing, a phase difference may occur between the structural and aerodynamic forces owing to the deformation of the wing. If energy is continuously supplied to the wing, the oscillation amplitude of the wing may increase with time, which may eventually cause severe damage to the wing structure. In the transonic region, the generation of shock waves has a significant impact on the flutter characteristics [2].

T. Miyake (✉) · H. Terashima
Hokkaido University, Kita 8, Nishi 5, Kita-ku, Sapporo, Hokkaido, Japan
e-mail: toma@eis.hokudai.ac.jp

S. Lee et al. (eds.), *The Proceedings of the 2021 Asia-Pacific International Symposium on Aerospace Technology (APISAT 2021), Volume 1*, Lecture Notes in Electrical Engineering 912, https://doi.org/10.1007/978-981-19-2689-1_47

There have been many experimental and computational studies on the transonic flutter characteristics of conventional symmetric airfoils [3–8]. For example, Isogai [7] conducted a two-dimensional (2-D) transonic flutter simulation using a symmetric airfoil (NACA64A010) and found that the shock wave oscillation on the airfoil surface can induce a unique flutter in the first bending mode. This was the cause of the transonic dip (the flutter speed rapidly decreased in the transonic region). This study focuses on supercritical airfoils that have a flat top surface and a positive camber at the trailing edge. The unique shape of supercritical airfoils may affect the behavior of shock waves and flow separations on the airfoil surface and change the flutter characteristics. For example, Schewe et al. [9] conducted a 2-D flutter experiment using an NLR7301 supercritical airfoil and discussed the effects of shock-induced separation and trailing edge separation on the flutter characteristics. However, there is still a lack of knowledge on the flutter characteristics of supercritical airfoils compared to symmetric airfoils. Therefore, it is important to clarify in detail the flutter characteristics of supercritical airfoils, addressing the effects of the airfoil shape on the flutter characteristics.

In this study, 2-D transonic flutter simulations were conducted using symmetric and supercritical airfoils. A two-degree-of-freedom wing model was used. The similarities and differences in the transonic flutter characteristics between symmetric and supercritical airfoils are discussed.

2 Numerical Methods

2.1 Aerodynamics

The Reynolds-averaged 2-D compressible Navier–Stokes equations were used to compute the transonic, unsteady and aerodynamic characteristics of the airfoil. The inviscid numerical flux was evaluated by SHUS [10], and a higher-order spatial accuracy was obtained using monotone upstream-centered schemes for conservation laws (MUSCL). The lower–upper symmetric Gauss–Seidel (LU-SGS) algorithm [11] was used for time integration. The Spalart–Allmaras (SA) one-equation model [12] was used for the turbulence model.

At the inflow boundary, all variables were specified as the freestream value. At the outflow boundary, the pressure was fixed as the freestream value and the other variables were extrapolated. At the wall, the density was extrapolated and the velocity corresponded to the grid speed due to the airfoil movement. The normal momentum equation, which includes the effects of wall acceleration on the pressure gradients, was used to define the pressure at the wall.

2.2 Structural Dynamics

Figure 1 shows the structural model of an airfoil section of a swept-back wing. The structural model has two-degree-of-freedom: vertical displacement h (downward is positive) and pitching angle α (nose-up is positive). The governing equations of the structural model are as follows [1, 7]:

$$\begin{aligned} &\ddot{h} + S_\alpha \ddot{\alpha} + c_h \dot{h} + k_h h = -L; \\ &S_\alpha \ddot{h} + I_\alpha \ddot{\alpha} + c_\alpha \dot{\alpha} + k_\alpha \alpha = M \end{aligned} \tag{1}$$

where m is the mass per unit length, $S_\alpha = mbx_\alpha$ is the static imbalance, $I_\alpha = mb^2 r_\alpha^2$ is the moment of inertia, c_h and c_α are the structural damping coefficients, $k_h = m\omega_h^2$ and $k_\alpha = m\omega_\alpha^2$ are the spring stiffnesses, L is the lift force, and M is the moment about the elastic axis (nose-up is positive). ω_h and ω_α are the uncoupled circular frequencies.

For Eq. (1), the vertical displacement h and time t are nondimensionalized using half of the wing chord length $b = c/2$ and an uncoupled circular pitching frequency ω_α. Thus, Eq. (1) can be rewritten as follows:

$$[M]\{\ddot{q}\} + [K]\{q\} = \{Q\}, \tag{2}$$

where

$$\begin{aligned} &[M] = \begin{bmatrix} 1 & x_\alpha \\ x_\alpha & r_\alpha^2 \end{bmatrix},\ [K] = \begin{bmatrix} (\omega_h/\omega_\alpha)^2 & 0 \\ 0 & r_\alpha^2 \end{bmatrix}, \\ &\{q\} = \begin{Bmatrix} h \\ \alpha \end{Bmatrix},\ \{Q\} = \frac{V_*^2}{\pi} \begin{Bmatrix} -C_l \\ 2C_m \end{Bmatrix}, \end{aligned} \tag{3}$$

and

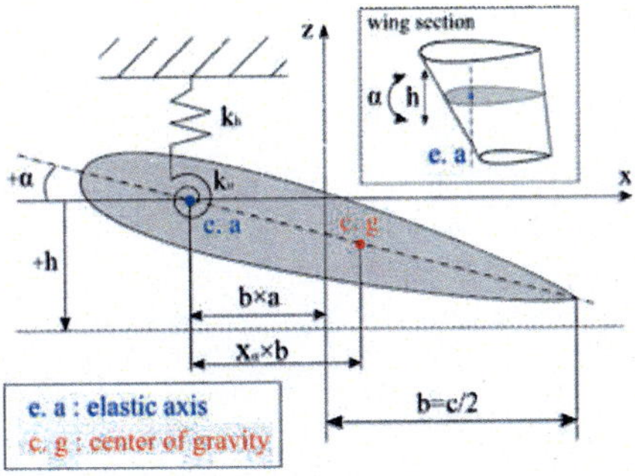

Fig. 1 Typical wing section model with two-degrees-of-freedom

$$V_* = \frac{U_\infty}{b\omega_\alpha\sqrt{\mu}}, \quad \mu = \frac{m}{\rho_\infty b^2 \pi}. \tag{4}$$

In this study, structural damping is not considered, that is, $c_h = c_\alpha = 0$ in Eq. (1). x_α and r_α^2 represent the static imbalance and radius of gyration, respectively.

The fourth-order Runge–Kutta (R–K) method was employed as the time-integration method of the structure equation. The aerodynamic forces C_l and C_m were assumed to be constant at each stage of the R–K method.

2.3 Coupling Method

For the fluid–structure interaction, a loose coupling method was applied, in which each equation was solved separately in the temporal direction, and the variables were exchanged at each time step. The coupled analysis was performed using the following steps:

1. Steady calculations were performed to obtain the flow field around the airfoil.
2. The initial displacement of the airfoil was provided and used to calculate the aerodynamic forces using the Navier–Stokes equations.
3. The obtained aerodynamic forces were substituted into the structural equations and the displacements h and α were obtained.
4. All grids were moved depending on their displacements, after which the aerodynamic forces were calculated on the new grid. Procedures 1 to 3 were then repeated.

Because the dimensionless quantities of the Navier–Stokes (fluid) and structure equations are different, a conversion of the physical quantities is required. Thus, the fluid equations were nondimensionalized using the chord length c and speed of sound a_∞, and the structural equations were nondimensionalized using $b = c/2$ and the ω_α. Thus, the relationships between the length and time scales are given as follows:

$$\begin{aligned} h_{\text{structure}} &= 2H_{\text{fluid}} \\ \Delta\tau_{\text{structure}} &= \tfrac{2M_\infty}{V_*\sqrt{\mu}}\Delta t_{\text{fluid}} \end{aligned} \tag{5}$$

where $h_{\text{structure}}$ and H_{fluid} represent the displacement of the airfoil in the structure and the fluid equation, $\Delta\tau_{\text{structure}}$ and Δt_{fluid} is the time-step size in the structure and the fluid equation, respectively.

2.4 Numerical Condition

NACA64A010 and SC2-0610 airfoils were used in this study. Figure 2 shows the computational grid for the SC2-0610 airfoil. The number of grid points was 803×150

Fig. 2 Computational grid for the SC2-0610 airfoil

for the NACA64A010 airfoil and 903 × 150 for the SC2-0610 airfoil and minimum grid spacing was 10^{-6}. The outer boundary was placed far away from the airfoil surface at a distance of 30c to minimize the boundary effect in the transonic flow simulations [13]. Time-step size was set to $\Delta t_{\mathrm{fluid}} = 10^{-3}$. In the flutter calculations, all grid points were moved at each time step according to the vertical and pitching displacements obtained from the structural equations.

The structural parameters in Eq. (3) and Fig. 1 are $x_\alpha = 1.8, r_\alpha^2 = 3.48, a = -0.2$, $\mu = 60$, and $\omega_h = \omega_\alpha = 100$ rad/s [7]. An eigen value analysis of the structural model indicates that the natural frequencies are 11.35 Hz for the first bending mode and 84.95 Hz for the second torsion mode.

Owing to the asymmetry of the upper and lower surfaces in the supercritical airfoil, there exists a non-zero lift and moment forces at the beginning of the flutter calculations. Therefore, when a flutter calculation is started, the airfoil is displaced to a position where the aerodynamic and structural forces are balanced, and that makes it difficult to control and define the angle of attack in the calculations. Therefore, in this study, the lift force coefficient $C_{l,0}$ and pitching moment coefficient $C_{m,0}$ obtained in the steady state (procedure 0) of the force vector $\{Q\}$ in Eq. (2) and Eq. (3) are used to define $\{Q\}$ as follows:

$$\{Q\} = \frac{V_*^2}{\pi}\left\{\begin{array}{c} -(C_l - C_{l,0}) \\ 2(C_m - C_{m,0}) \end{array}\right\} \tag{6}$$

In this study, three angles of attack, $\alpha_m = -2, 0$, and $2°$, were used for the supercritical airfoil, while $\alpha_m = 0, 2°$ was used for the NACA64A010 owing to the symmetry.

3 Comparison of Flutter Characteristics

Figure 3 shows the speed index (flutter speed) and flutter frequency. In this study, V_* is defined as the flutter speed when the oscillation amplitude is approximately constant and the frequency at that time is defined as the flutter frequency. For the NACA64A010 airfoil, the flutter speed was at a minimum at $M_\infty = 0.825$ for $\alpha_m = 0°$ and the flutter frequency indicated that the airfoil oscillates under the first

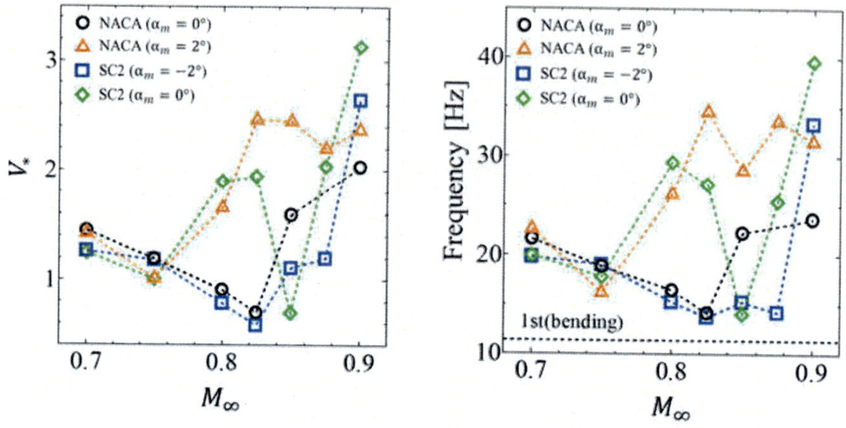

(a) Speed index

(b) Flutter frequency

Fig. 3 Comparisons of the flutter boundaries and frequencies between the NACA64A010 and SC2-0610 airfoils

bending mode. In the cases where the Mach numbers are higher than 0.825, the flutter speed increases rapidly, and the bending-torsion coupling mode becomes dominant. These trends are similar to those of a previous study [13] and represent typical transonic dip characteristics. Conversely, for $\alpha_m = 2°$, the flutter speed is minimum at a lower Mach number, $M_\infty = 0.75$, indicating a shallower dip characteristic than that for $\alpha_m = 0°$. Thus, the comparison of the results of the two angles of attack suggests that as the angle of attack increases, the dip condition of the flutter boundary moves to lower Mach numbers.

For the SC2-0610 supercritical airfoil, the flutter characteristics with $\alpha_m = 2°$ resemble those of the NACA64A010 airfoil obtained at an $\alpha_m = 0°$. Similarly, when the angle of attack increases to $\alpha_m = 0°$, the flutter characteristics become similar to those of the NACA64A010 airfoil with an $\alpha_m = 2°$ (except for $M_\infty = 0.85$). Thus, the results indicate that the overall flutter characteristics of the two airfoils are similar, and the shape effect of the supercritical airfoil appears as an effect of the angle of attack on the transonic flutter characteristics.

The difference in the flutter characteristics between the two airfoils is observed for higher Mach numbers $M_\infty > 0.85$ compared to the range of $M_\infty < 0.825$. This difference may be attributed to the effect of the separation region behind the shock waves. Figure 4 shows the Mach number distributions for the two airfoils under steady-state conditions (the initial flow fields for the flutter calculations). For the NACA64A010 airfoil, the flow field behind the shock wave barely changed with increasing Mach numbers. In contrast, for the supercritical SC2-0610 airfoil case, the separation region, especially on the lower surface, significantly expanded with increasing Mach numbers. The development of the separation region may cause differences in the flutter speed and frequency under higher Mach number conditions. For the SC2-0610 airfoil, a significant drop in the flutter speed is observed at $M_\infty =$

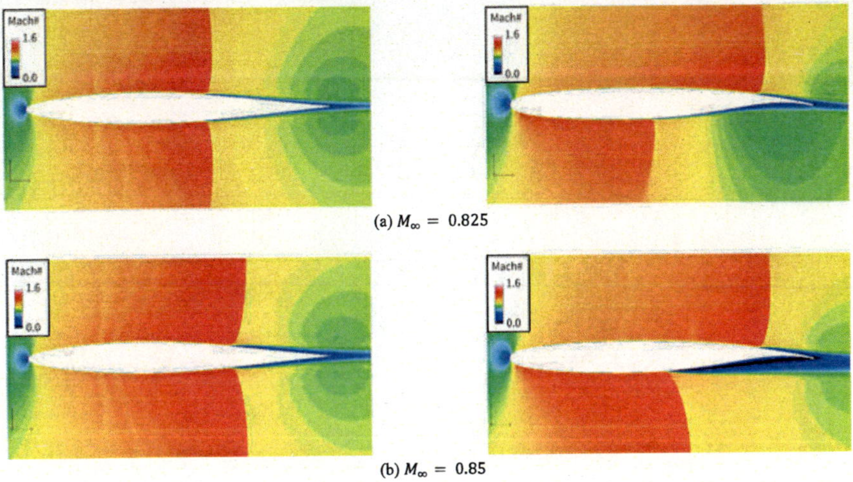

NACA64A010 with $\alpha_m = 0°$ SC2-0610 with $\alpha_m = -2°$

Fig. 4 Comparison of the Mach number distributions under the steady-state (non-oscillating) condition

0.85 with $\alpha_m = 0°$. This may also be attributed to the separation region on the lower surface, which is currently under investigation.

4 Flutter Characteristics Specific to Supercritical Airfoils

Figure 5 shows the flutter boundary and frequency for the SC2-0610 airfoil at $\alpha_m = 2°$. For the $M_{\infty} = 0.9$ case, no flutter was observed even for large V_* conditions. When $\alpha_m = 2°$, the Mach number at which the flutter speed reached a minimum value decreased to $M_{\infty} = 0.7$. This is a similar trend to when the angle of attack was increased from $\alpha_m = -2$ to $\alpha_m = 0°$ as discussed in the previous section. However, the flutter frequency indicated that flutter in the second torsion mode uniquely occurred at $M_{\infty} = 0.875$. This flutter mode only appeared in the SC2-0610 supercritical airfoil and was not present in the NACA64A010 airfoil. The second torsion mode flutter for a supercritical airfoil was presented in a previous study [9], but the detailed mechanism was not discussed. Hence, the flutter mechanism of the second torsion mode is discussed in the following section.

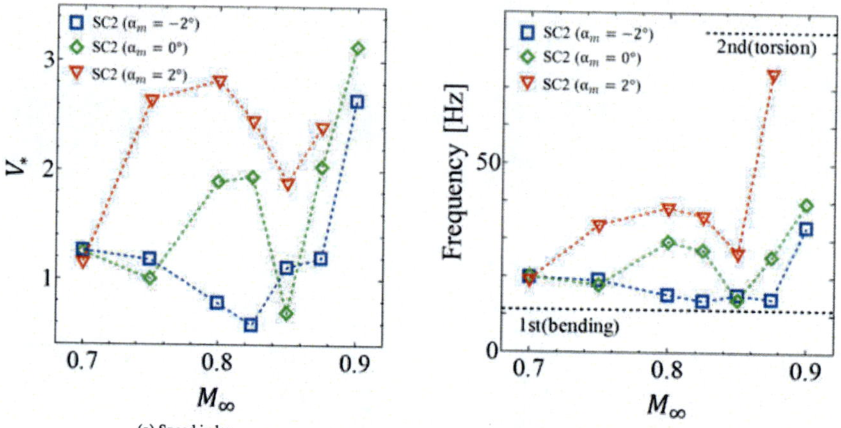

(a) Speed index (b) Flutter frequency

Fig. 5 Effect of the mean angle of attack on the flutter characteristics for the SC2-0610 airfoil

4.1 *Unsteady Aerodynamic Forces in Second Torsion Mode Oscillation*

To investigate the flutter mechanism in the second torsion mode, an analysis using a forced oscillation airfoil was performed. The oscillation is assumed to be a one-degree-of-freedom pitching, and the pitching motion α is given by:

$$\alpha = \alpha_m + \alpha_0 \cos(\omega t), \tag{7}$$

where α_0 is the oscillation amplitude, ω is the circular frequency, and t is the time. The reduced frequency k is defined as follows:

$$k = \frac{\omega c}{2U_\infty}. \tag{8}$$

The parameters of the forced oscillation airfoil calculation are summarized in Table 1.

Figure 6 shows the variation in the pitching moment coefficient with respect to the pitching angle variation at $M_\infty = 0.875$. The arrows in the figure represent the directions of the loop. The C'_m and α' are the variations from the steady-state calculations. Those are defined as:

Table 1 Parameters for the forced oscillating airfoils

Pitching center	0.25 c
Amplitude [°]	1.0
k	0.24 (2nd torsion mode)

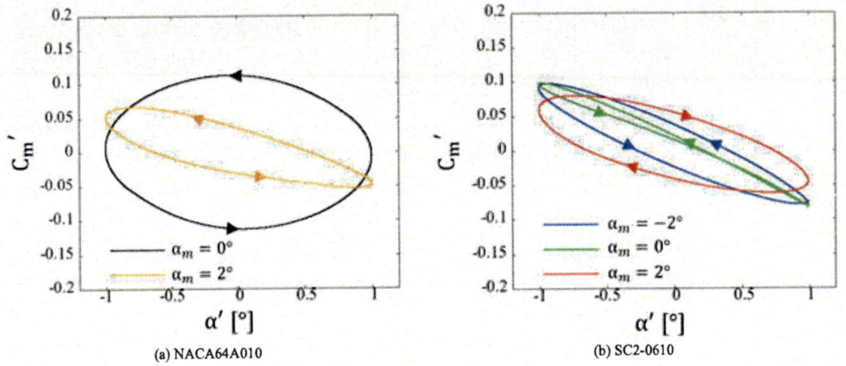

Fig. 6 Hysteresis loop for the pitching moment coefficient at $M_\infty = 0.875$

$$\begin{aligned} C_m' &= C_m - C_{m,0}; \\ \alpha' &= \alpha - \alpha_m. \end{aligned} \tag{9}$$

From Fig. 6, it was confirmed that the variation in the moment exhibits hysteresis characteristics. When the loop was clockwise, energy was supplied from the fluid side to the structure side, and the system became unstable. However, when the loop was counterclockwise, the system became stable. The results in Fig. 6 demonstrated that SC2-0610 with $\alpha_m = 2°$ only exists for a clockwise loop, indicating that the system tends to be unstable. This unstable characteristic of the pitching moment may play an essential role in the occurrence of flutter for the second torsional mode.

Figure 7 shows the distribution of unsteady aerodynamic forces on the airfoil surface. The real part of the pressure coefficient represents the component in phase with the pitching motion, and the imaginary part represents the component out of phase by 90°. The aerodynamic force on the lower surface has an opposite sign in phase to that on the upper surface. The shock movement on the upper surface is confirmed for all the cases with negative real and positive imaginary parts that indicate the phase-advanced behavior of the shock wave. The result of the C_m' variation against

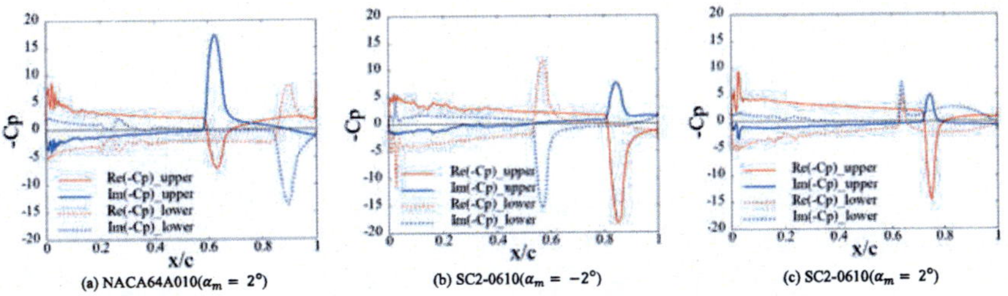

Fig. 7 DFT analysis for the real and imaginary parts of the pressure coefficients at $M_\infty = 0.875$ for the SC2-0610

α' in Fig. 6 shows the negative C'_m in the nose-up motion ($d\alpha/dt > 0$) and the positive C'_m in the nose-down motion ($d\alpha/dt < 0$) at $\alpha' = 0°$ (except for the SC2-0610 supercritical airfoil with $\alpha_m = 2°$). Therefore, the phase-advanced movement of the shock waves on the upper surface (and lower surface) may induce a counterclockwise loop of the pitching moment coefficient, that is, a stabilizing effect. However, the variation in the C_p of the shock wave in the case of the SC2-0610 airfoil with $\alpha_m = 2°$ is smaller than those in the other cases. Thus, the contribution of the shock motion on the pitching moment decreases in the SC2-0610 airfoil at an $\alpha_m = 2°$. Furthermore, on the lower surface, there is a phase delay motion in the aerodynamic force behind the shock wave ($0.7 < x/c < 1$) in the SC2-0610 airfoil with $\alpha_m = 2°$. Thus, the smaller contribution of the shock wave motion on the stabilizing effect and the phase delay in the aerodynamic force near the trailing edge on the lower surface cause the loop of the pitching moment coefficient to turn in a clockwise direction. Therefore, the SC2-0610 airfoil may exhibit instability characteristics against the pitching motion. The following section discusses the cause of the phase delay in the aerodynamic force near the trailing edge on the lower surface.

4.2 Flow Field Near the Trailing Edge on the Lower Side

Figure 8 shows the relationship between the pitching moment coefficient C_m and the angle of attack obtained from the non-oscillatory airfoil (steady-state) calculations. For the SC2-0610 airfoil,C_m is almost constant in the range of $\alpha_m < 0°$, but for $\alpha_m > 0°$, C_m decreases significantly as the angle of attack increases. This variation in the C_m is different from that of the NACA64A010 airfoil. From the flow field shown in Fig. 9, when the angle of attack for the SC2-0610 airfoil increased, the

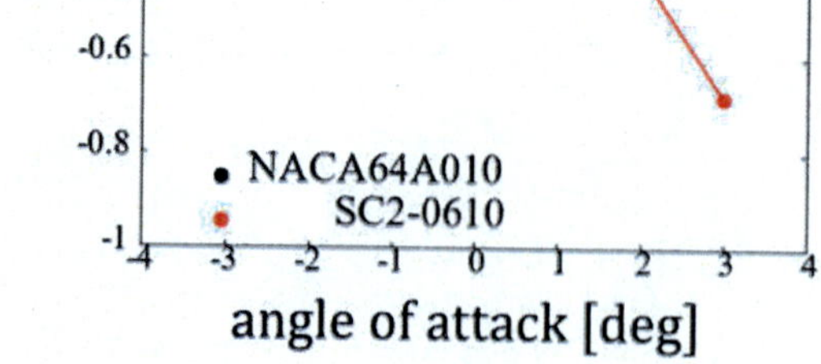

Fig. 8 Pitching moment coefficient of each airfoil in the steady-state condition at $M_\infty = 0.875$

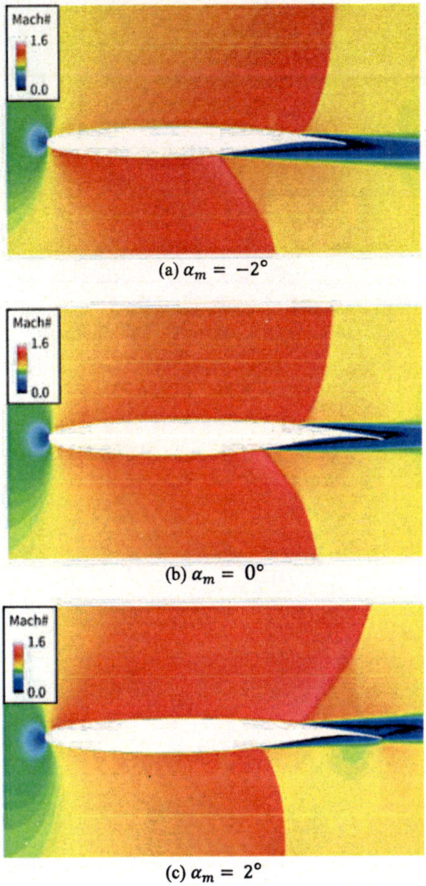

Fig. 9 Mach number distribution for the SC2-0610 airfoil under the steady-state condition at $M_\infty = 0.875$

separation region developed on the upper surface and the shock wave moved forward. In contrast, the separation region on the lower surface shrank, and the shock wave moved backward when the angle of attack increased. Eventually, the variation in the aerodynamic force (C_m) may cancel out such that C_m is almost unchanged when $\alpha_m < 0°$. However, when the angle of attack further increased, a high-pressure region was newly generated owing to the flow reattachment at the trailing edge on the lower surface at $\alpha_m = 2°$ as shown in Figs. 9(c) and 10(c). Hence, the negative pitching moment increased with a significant decrease in C_m. As the substantial decrease in C_m is a specific feature of the SC-0610 airfoil, the appearance of the high-pressure region may cause a phase delay according to the airfoil oscillation (Fig. 7(c)) and may contribute to the flutter mechanism in the second torsion mode.

(a) $\alpha_m = -2°$

(b) $\alpha_m = 0°$

(c) $\alpha_m = 2°$

Fig. 10 Pressure coefficient distribution for the SC2-0610 airfoil under the steady-state condition at $M_\infty = 0.875$

4.3 The Effect of the Airfoil Shape on the Upper Surface

In the previous discussions, it was suggested that the lower surface shape of the supercritical airfoil might be crucial for the generation of flutter in the second torsion mode. To further investigate the effects of a flat upper surface of the supercritical airfoil, a new airfoil shape is considered, as shown in Fig. 11, which does not have a positive camber at the trailing edge on the lower surface. The DFT results for the flutter calculation using the new shape are shown in Fig. 12. The new airfoil does not exhibit the second torsion mode oscillations and offers flutter characteristics similar to those of the NACA64A010 airfoil. In the steady flow field in Fig. 13, the shock wave position on the lower side of the new airfoil is similar to that of the NACA64A010 airfoil. Therefore, it was concluded that the flow separation and

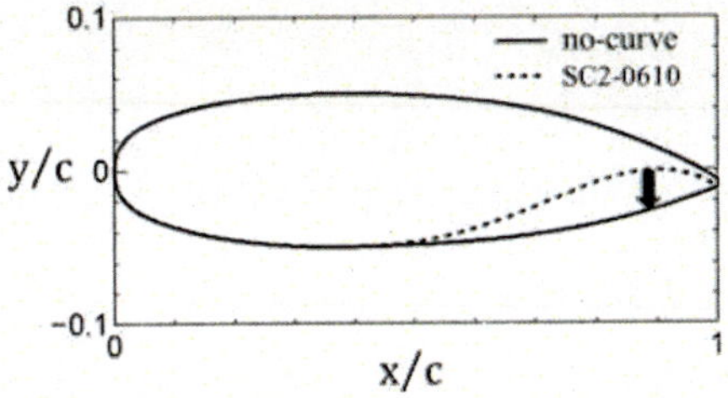

Fig. 11 Geometry of the no-curve airfoil

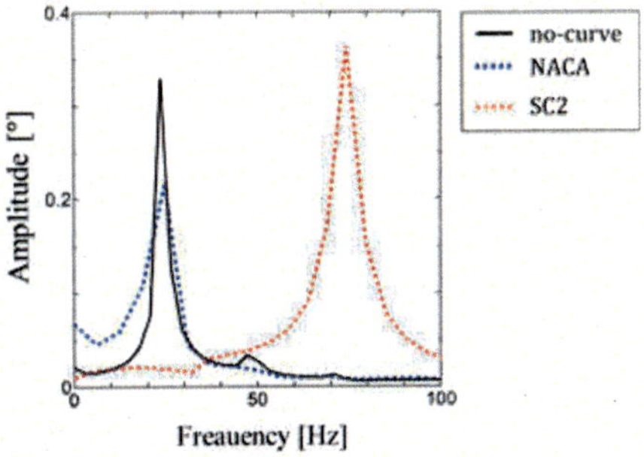

Fig. 12 DFT analysis of the pitching displacement with $\alpha_m = 2°$ at $M_\infty = 0.875$

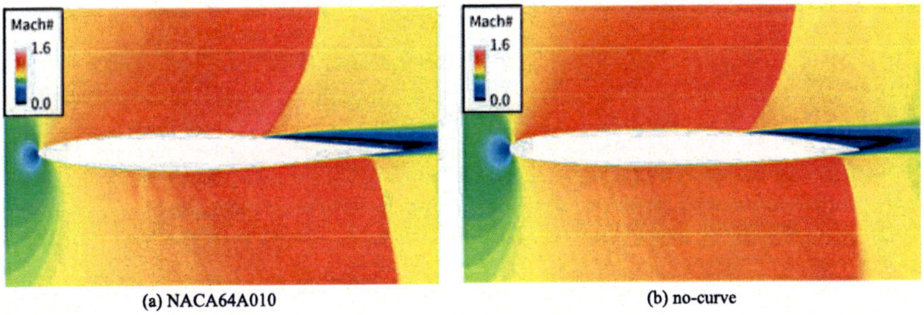

Fig. 13 Comparison of the Mach number distributions between NACA64A010 and no-curve airfoil under the steady-state condition with $\alpha_m = 2°$ at $M_\infty = 0.875$

reattachment on the lower surface caused flutter generation in the second torsion mode.

5 Conclusion

A two-dimensional transonic flutter analysis was performed using a conventional symmetric airfoil (NACA64A010) and a supercritical airfoil (SC2-0610). The effects of the airfoil shape on the transonic flutter characteristics were investigated, and the following conclusions were drawn:

1. Overall, the flutter characteristics in the transonic flow regime were similar between the symmetric and supercritical airfoils.
2. Under low Mach number conditions ($M_\infty < 0.825$), where no significant flow separation occurred, the symmetric and supercritical airfoils exhibited almost identical flutter characteristics in terms of the flutter speed and frequency. The effect of the shape of the supercritical airfoil appears as the angle of attack.
3. Under high Mach number conditions ($M_\infty > 0.825$), the flutter speed and frequency of the supercritical airfoils showed some deviations from that of the symmetric airfoil owing to the development of flow separation.
4. The supercritical airfoil exhibits a unique flutter in the second torsion mode at a high Mach number. The present results indicated that the flow separation and reattachment on the lower side of the airfoil play a crucial role in flutter generation in the second torsion mode.

References

1. Nakamichi J, Tamayama M, Kodama S (2019) Aeroelasticity (Japanese), Maruzen Co
2. Farmer MG, Hanson PW, Wynne CE (1976) Comparison of supercritical and conventional wing flutter characteristics, NASA Technical Memorandum, vol 72827
3. Yates EC Jr (1988) AGARD Standard aeroelastic configurations for dynamic response I-Wing 445.6, NASA Technical Memorandum, vol 100492
4. Doggett RV Jr, Rainey AG, Morgan HG (1955) An experimental investigation of aerodynamic effects of airfoil thickness on transonic flutter characteristics, NASA Technical Memorandum, vol 79
5. Rivera AJ, Dansberry BR, Bennett RM (1992) NACA0012 benchmark model experimental flutter results with unsteady pressure distributions, NASA Technical Memorandum, vol 10751
6. Bendiksen OO (2004) Modern developments in computational aeroelasticity. J Aerosp Eng Proc Inst Mech Eng Part G 218(3):157–178
7. Isogai K (1979) On the transonic-dip mechanism of flutter of a sweptback wing. AIAA J 1:793–795
8. Nakamichi J (1986) Some computations of unsteady Navier-Stoles solutions around oscillating airfoils. NASA Technical Memorandum, vol 88341
9. Schewe G, Mai H, Dietz G (2003) Nonlinear effects in transonic flutter with emphasis on manifestations of limit cycle oscillations. J Fluids Struct 18:3–22
10. Shima E, Jounouchi T (1997) Role of CFD in aeronautical engineering (No. 14)–AUSM type upwind schemes. In: Proceedings of the 14th NAL symposium on aircraft computational aerodynamics, SP-34, pp 7–12
11. Yoon S, Jameson A (1988) Lower-upper symmetric-Gauss-Seidel method for the Euler and Navier-Stokes equations. AIAA J 26:1025–1026

12. Spalart PR, Allamaras SR (1994) A one-equation turbulence model for aerodynamic flows. AIAA Pap 1:5–21
13. Terashima H, Fuji K (2001) Transonic flutter analysis using the fluid/structure coupling method. In: Proceedings of the computational fluid dynamics symposium, vol 15

Control of Wing-Engine-Slat Cut Out Flow Separation Using Nacelle Nozzle Modification

Zheng Cui, Jiangang Wang, Xi Du, Dakai Lin, and Yihua Cao

Abstract A numerical study on active flow control (AFC) to a swept wing in a take-off configuration was conducted. The focus of this study is the feasibility of the innovative active flow control (AFC) method by using the engine jet, and the control of the jet was achieved through the modification of the nacelle nozzle geometry. The modification included a partial cut or extension at the trailing edge of the nacelle. This AFC method is designed mainly for the flow control of wing-engine-slat cut-out position and it is an alternative design to the outboard nacelle strake. The target configuration integrated with this AFC design was a typical twin-engine jetliner with close-coupled High Bypass Ratio (HBR) engine nacelles. The basic take-off configuration was integrated with regular powered nacelles. Two modified configurations with different engine nozzle designs were also investigated and compared. The surface flow visualization and the spatial particle traces illustrated how the different nozzles affected the engine exhaust flow and the further impact on the aerodynamic performance of the wing. The simulation results showed this innovative AFC device can mitigate the aerodynamic performance degradation caused by the installation of large-diameter nacelles, and suppress the flow separation near the wing-engine-slat cut-out position. Meanwhile, the AFC method did not require additional mechanical mechanisms or any special maintenance, nor had any weight penalty. Therefore, this type of AFC device shows a high application potential in commercial transport aircraft.

Keywords Active Flow Control (AFC) · Twin-engine · High-lift configuration · Nozzle

Z. Cui (✉) · Y. Cao
Beihang University, Beijing, Beijing, China
e-mail: zanecui05@hotmail.com

Z. Cui · X. Du · D. Lin
COMAC Beijing Aircraft Technology Research Institute, Beijing, Beijing, China

Beijing Key Laboratory of Simulation Technology for Civil Aircraft Design, Beijing, Beijing, China

J. Wang
DFH Satellite Co, Ltd., Beijing, Beijing, China

S. Lee et al. (eds.), *The Proceedings of the 2021 Asia-Pacific International Symposium on Aerospace Technology (APISAT 2021), Volume 1*, Lecture Notes in Electrical Engineering 912, https://doi.org/10.1007/978-981-19-2689-1_48

1 Introduction

In the field of civil aviation, the improvement of aircraft's economic and environmental performance largely depends on the improvement of engine technology. At present, the wing-mounted twin-engine design has become the mainstream feature form of subsonic civil transport. Such a layout often requires the integration of large-diameter engines and engine nacelles. Although the large size high bypass ratio (HBR) engine can greatly increase the fuel efficiency of the aircraft. But it will also increase the difficulty of engine/airframe integration. First of all, due to the limitation of ground clearance, large nacelles often need to be installed closer to the wing. In this case, the flow interference of the nacelle and engine jet on the wing will be enhanced. The influence of engine integration on the aerodynamic characteristics of the wing needs to be carefully considered. In addition, the increase in the size of the nacelle will further increase the aerodynamic drag of the nacelle itself. Some drag reduction technologies for the nacelle are getting more and more attention. For example, Boeing claims that the natural laminar flow nacelle technology has been applied on their B787 aircraft. Moreover, the 787's nacelle also has integrated chevron nozzles to reduce jet noise. Therefore, integrating more new aerodynamic technologies into the nacelle seems to be a trend in the current aerodynamic design of civil transport.

For high-lift configurations, the nacelle wake can cause a premature wing flow separation. In addition, the integration of larger-diameter nacelle requires a wider cut of leading-edge high-lift devices to avoid it clashing with the nacelle. On the one hand, the wider cut of leading-edge devices will weaken its flow protection ability. On the other hand, the cutting grooves on the wing and end face on the leading-edge high lift devices further increase drag and induce unfavorable flow separations. In response to this problem, civil jets generally use passive flow control methods. One or more nacelle strakes (chine) are installed on each nacelle, each strake can generate a strong detaching vortex heading downstream under appropriate attack angles. These vortices could further mitigate the adverse effects induced by the nacelle wake. In terms of active flow control, Monat and Possti[1] proposed an active flow control device that creates steady suction and pulsed blowing (PB) in the leading-edge region of the wing, and its performance was evaluated by experiments. It was shown that the AFC devices increase the lift by up to 3%. However, similar AFC devices rely on buried ventilation pipes and valves in the wing leading-edge. In fact, this inner space of the wing leading-edge is basically occupied by the actuating mechanism of the high-lift device, and there are greater difficulties in engineering applications.

The focus of this study was the feasibility of the innovative active flow control (AFC) method by using the engine jet, and the control of the jet is achieved through the modification of the nacelle nozzle geometry. This innovative AFC device has the potential to mitigate the aerodynamic performance degradation caused by the installation of large-diameter nacelles, and suppress the flow separation near the wing-engine-slat cut-out position.

2 Methodologies

This chapter introduces the numerical methods and basic aircraft configurations used in this study. Then the basic idea of this AFC method is briefly described. Two different nozzle modifications are proposed and investigated, and the related geometric features and calculation grids are illustrated.

2.1 CFD Method

In this study, all examples are computed using commercial aircraft corporation of China (COMAC) in-house code S-Flow. S-Flow is a finite-volume method solver based on Reynolds average Navier-Stocks (RANS) equation. Roe's scheme is used for space discretization. Lower-upper symmetric Gause-Seidel (LU-SGS) method is used for time advancing. Many turbulent models are available, such as the Spalart–Allmaras (SA) and the k-ω shear stress transport (SST). Its massively parallel computing capability is based on message passing interface (MPI). Multigrid technology is used to accelerate numerical convergence.

The CFD and experiment comparisons in study [2] show that the S-Flow results are in excellent agreement with experiments, showing the grids generation strategy and numerical method are appropriate for this application.

2.2 Configurations

This research focuses on a typical double-aisle twin-engine (BPR 10:1) jetliner with moderate swept wings. The cruising Reynolds number is approximately 5.5×10^7. For high-lift configuration, the droop nose device (DND) is installed at the inboard and the single slot slat is installed at the outboard of the wing leading edge. Two Adaptive Drooped Hinge Flaps (ADHF) are integrated on the wing trailing edge. Two sets of Ailerons are located at the outboard of the flaps. The DND has the advantages of high reliability, lightweight, and low noise [3–5]. The ADHF has the advantage of being lightweight and multifunctionality, which can be used as a high-lift device at low speed, and as an adaptive camber device at cruising [6, 7]. The takeoff configuration and its powered nacelle are shown in Figs. 1 and 2.

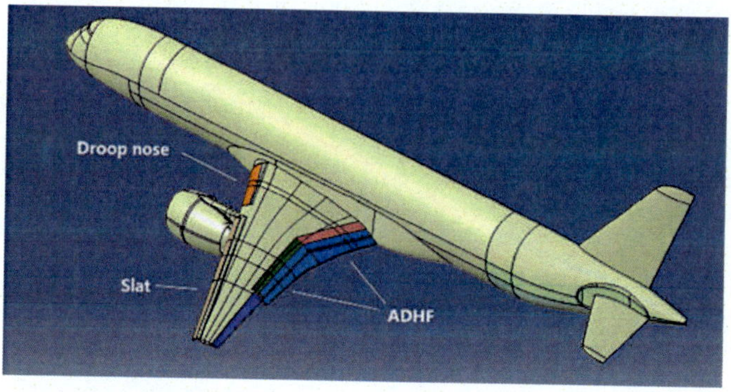

Fig. 1 High-lift configuration

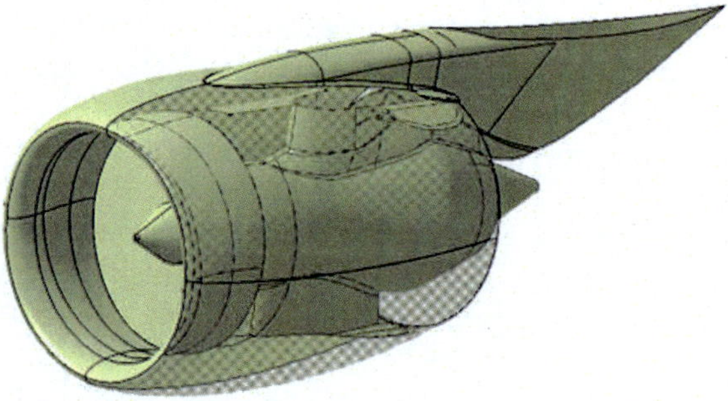

Fig. 2 Powered nacelle (baseline)

2.3 AFC Methods

The power effect of the close-coupled nacelle will have a positive influence on the aerodynamic performance of the wing [8–10]. The power effect can increase lift and suppress the outboard wing flow separation at high angles of attack. The AFC method in this paper uses the nacelle nozzle geometrical modification to slightly affect the total pressure distribution of fan exhaust, but at the same time would not pay too much thrust loss. We have designed two ways to modify the nozzle, one is the partial cut of the nozzle (Type 1 in Fig. 3), and the other is the partial extension of it (Type 2 in Fig. 4). The basic idea is to use nozzle modification to form an ejector, this ejector uses the nacelle's original curvature to form an ejection channel, this channel will guide the high-speed fan exhaust to our desired direction for AFC purposes. Roughly speaking, the Type 1 modification is designed to strengthen the power effect of the close-coupled nacelle, and Type 2 is designed to weaken it for comparisons. In theory, an increase and a decrease in the lift coefficient of the respective configuration should be observed.

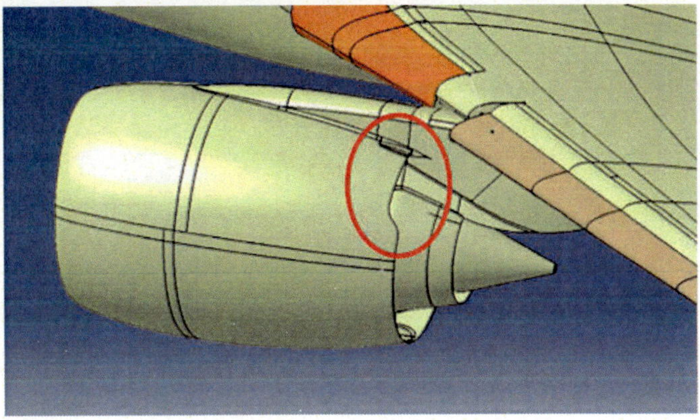

Fig. 3 Type 1 nozzle

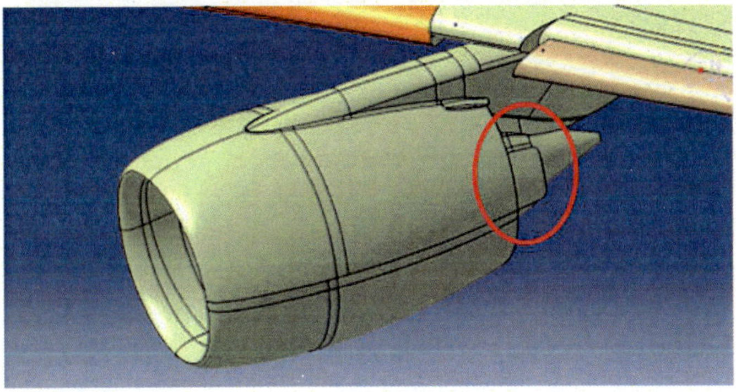

Fig. 4 Type 2 nozzle

2.4 The Grids Generation

The multi-block structured meshes are constructed for this CFD research. O-shape structural grids are applied to wrap the wall of the whole configuration to generate the surface boundary layer. Additional nested O-shape grids (5 layers) are used to generate nacelle and its complex inner ducts and structural surfaces. According to previous studies, this grid strategy could accurately simulate the boundary/vortex interference of the jetliner high lift configurations. The reason why structured grids are used instead of unstructured grids in this research is that we only make very small modifications to the shape of the nozzle. The use of structural grids can ensure that the spatial grid distribution of different configurations can maintain maximum consistency so that we can better distinguish the influences of configuration differences. ANSYS ICEM 11.0 is used to construct the multiblock structured meshes

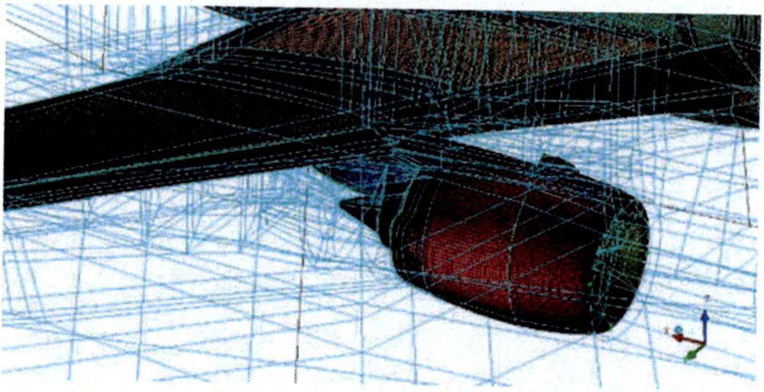

Fig. 5 Grid of Type 1 nozzle

Fig. 6 Grid of Type 2 nozzle

(Figs. 5 and 6). The number of cells of the Type 1 and Type 2 configuration is 58.5 million and 59.6 million, respectively.

3 Result and Discussion

Figure 7 illustrates the lift coefficient of Type 1, Type 2, and baseline configuration. In line with our prediction, Type 1 increased the lift coefficient of the take-off configuration at high angles of attack compared to the baseline, while Type 2 decreased it. The lift coefficient results have been able to show that our AFC method produces the expected flow control effect, even though nozzle geometry modifications are very slight.

Figures 8 and 9 illustrates the pressure coefficient distributions and surface streamlines on the inboard wing at 18° angles of attack. Through the surface streamline of the baseline, we can observe that the inner nacelle strake protects the flow on the inboard wing very well, and there is no obvious flow separation on the inboard wing. However,

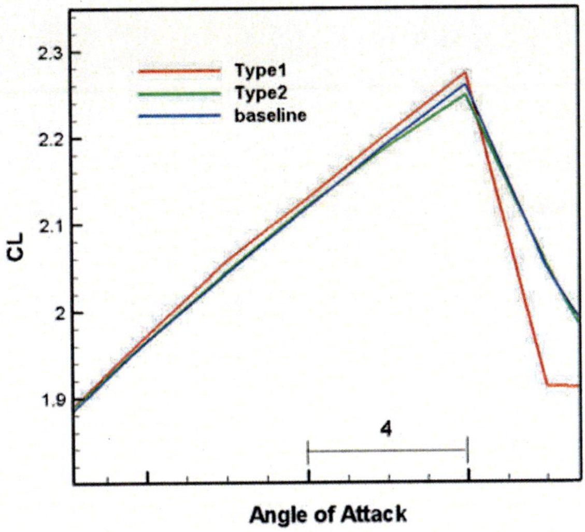

Fig. 7 Lift coefficient comparison

at the outboard wing-engine-slat cut-out position (red circle), the geometric discontinuity of the leading-edge slat leads to a scissor's incision, and a small separation bubble has observed at the incision, and the flow downstream of this separation bubble is affected. The streamline in the red circle has already appeared with an obvious inflection, indicating that a large reverse pressure gradient has appeared here, and the flow here has been in a critical state of separation. Comparing the surface streamline of Type 1 configuration, similarly, the inner nacelle strake protects the flow on the inboard wing very well. After being integrated with the Type 1 nozzle, the inflection trend of the streamline in the red circle has been significantly reduced, indicating that the Type 1 AFC device has played a role in flow protection and suppressing the flow separation which could be induced by the wing-engine-slat cut-out.

Regarding the flow control principle of Type 1 and Type 2 nozzles, we can use the following 3D particle traces and pressure contour (Figs. 10 and 11) to make an explanation. Due to the larger size of the nacelle and its close-coupled integration, the fan exhaust's interference on the wing will be enhanced, this kind of interference is usually manifested in the high angle of attack, and the local incoming flow of the outboard wing can be accelerated by the fan exhaust. As illustrated in Figs. 10 and 11, The streamlines are tracing particles passing through the outboard leading edge. the flow pattern shows these particles passed through the high-pressure zone formed by the engine exhaust, and these particles are dramatically accelerated after passing through the high-pressure zone. Our AFC method has used this principle. From Fig. 10 we can see that the Type 1 nozzle uses a cut at the rear edge of the nacelle to induce more fan exhaust into this high-pressure zone, which further increases the flow pressure and accelerate more income flow heading to the outboard leading edge.

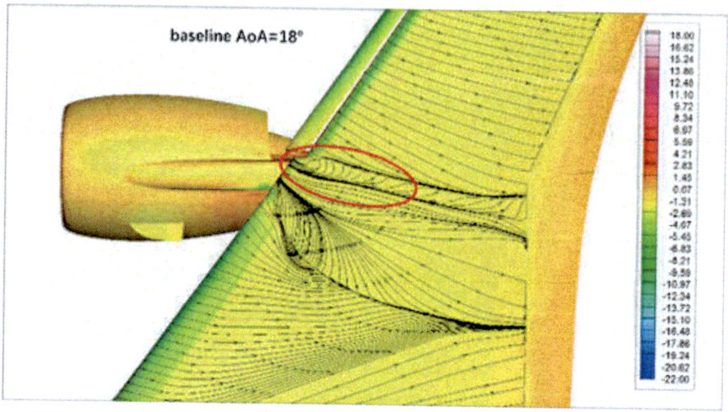

Fig. 8 Baseline at AoA 18°

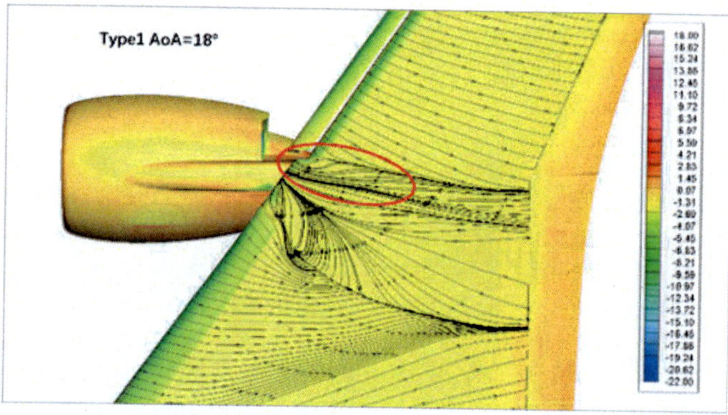

Fig. 9 Type 1 at AoA 18°

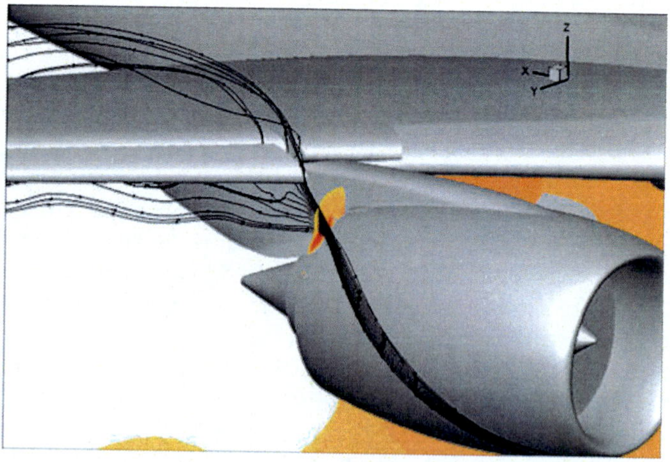

Fig. 10 Particle traces and pressure distribution of Type 1 at AoA 18°

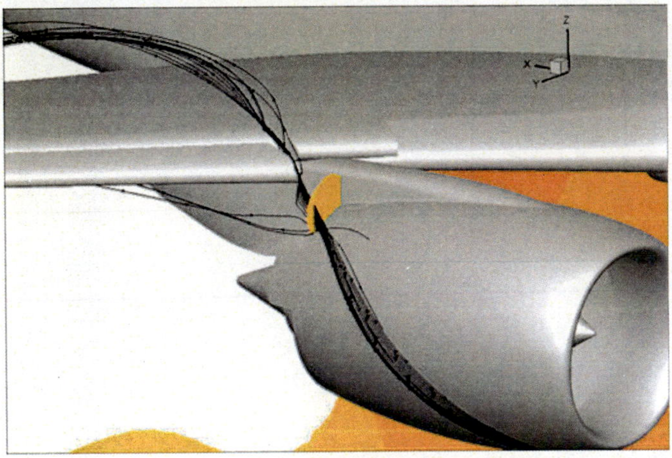

Fig. 11 Particle traces and pressure distribution of Type 2 at AoA 18°

Type 2 produces the opposite control effect, and this indicates fewer partials could obtain momentums from the fan exhaust.

To analyze the adverse effect of the Type 1 device in thrust, The trust-drag book-keeping (TDB) is conducted on the two nacelles of baseline and Type 1 configuration. 13 million cells are applied on each nacelle, and a truncated cone-shape is adopted for the far-field. The length of the far-field is 50 times the length of the nacelle chord. The far-field entrance and outlet are 50 times and 100 times the length of the nacelle diameter, respectively. The thickness of the boundary layer grid is 100 mm. The height of the first layer grid is 0.005 mm, and the growth rate is 1.33. The COMAC in-house drag/thrust analysis code Sflow-TZFJ is used for this thrust-drag book-keeping. The results show that Type 1 has a 0.25% engine thrust loss compared with the baseline. The engine thrust loss induced by this AFC device is acceptable.

Fig. 12 The grid of thrust-drag bookkeeping

4 Conclusion

An innovative active flow control (AFC) method has been proposed and numerical verified, this AFC method by using the engine jet to control the flow on the outboard wing, and the control of the jet is achieved through the modification of the nacelle nozzle geometry. Two different types of nozzle modification are numerically investigated. Type 1 increased the lift coefficient of the take-off configuration at high angles of attack compared to the baseline, while Type 2 decreased it. The lift coefficient results have been able to show that our AFC method produces the expected flow control effect. Surface streamline illustrates this innovative AFC device has the potential to mitigate the aerodynamic performance degradation caused by the installation of large-diameter nacelles and to suppress the flow separation near the wing-engine-slat cut-out position. The AFC method does not require additional mechanical mechanism, has no weight penalty, and requires no special maintenance, the engine thrust loss caused by this AFC device is minor and acceptable. Therefore, this type of AFC device shows a high application potential in commercial transport aircraft.

References

1. Monat S, Maayan P, Bar M, Ariel Y, Avraham S, Ofek D (2020) Suction and pulsed blowing for control of local wing-engine-slat-cut out flow separation: the INAFLOWT project. In: AIAA 2020 Aviation Forum. AIAA 2020-2953
2. Bangmeng X et al (2018) Multi-Objective Wing Optimization of Civil Aircrafts in Engine-Aircraft Integration Configuration. Acta Aerodynamica Sinica 36(6):941–948
3. Adam J, Oliver A (2009) Design of a High-Lift System with Droop Nose Device. Journal of Aircraft 46:731–734
4. Michael PP, Jochen W W, Lothar B (2014) Aerodynamic and Acoustic Design of Silent Leading Edge Devices. In: 20th AIAA/CEAS Aeroacoustics Conference. AIAA 2014-2076
5. Hans M, Markus K et al (2009) Design of a Smart Droop Nose as Leading Edge High Lift System for Transportation Aircrafts. 50th AIAA/ASME/ASCE/AHS/ASC Structures, Structural Dynamics, and Materials Conference. AIAA 2009-2128
6. Thomas S L, Mirko H (2018) An Integrated Design Approach for Advanced Flight Control Systems with Multifunctional Flight Control Devices. In: 2018 Aviation Technology, Integration, and Operations Conference. AIAA 2018-3193
7. James U, Nhan N (2013) A Mission Adaptive Variable Camber Flap Control System to Optimize High Lift and Cruise Lift to Drag Ratios of Future N+3 Transport Aircraft. In: 51st AIAA Aerospace Sciences Meeting including the New Horizons Forum and Aerospace Exposition. AIAA 2013-0214
8. Stefan MW, Roland W, Heiko G (2004) RANS Solutions for a Complex High-Lift Configuration of a Transport Aircraft with Engine Including Improved Resolution of the Nearfield. In: 22nd Applied Aerodynamics Conference and Exhibit. AIAA 2004-5081
9. Zhaoguang T, Yingchun C, Jiangtao S (2014) Study of Power Influences to the Wing-Mounted Civil Aircraft Aerodynamic Characteristics. Journal of Aircraft 51(2):629–636
10. Shaojie G, Peipei Z, Bin W, Xiaotian S (2016) Numerical investigation for influence of power effect on aerodynamic characteristics of civil aircraft take-off configuration. Journal Of Aerospace Power 31(7):1638–1648

Developments of Free-Flight Testing Facility for Aerodynamic Assessment of Martian Entry Capsule

Satoshi Nomura, Kyosuke Itabashi, Masahito Mizuno, and Kazuhisa Fujita

Abstract The evaluation of the aerodynamic characteristics of a Martian entry capsule is crucial for future Mars missions. However, the uncertainty in the dissociation rate of CO_2 makes it difficult to predict them accurately with a CFD analysis. To improve the prediction accuracy of aerodynamic characteristics of hypersonic vehicles, the development of an experimental database is inevitable. In this study, the free-flight tests with a Martian capsule model were performed with using a light gas gun at Japan Aerospace Exploration Agency. The aerodynamic coefficient was determined by reconstructing the trajectory and attitude from high-speed visualization data of the free-flight capsule. The target velocity to reproduce the flight environment was 4.2 km/s for an ambient pressure of 11 kPa. Currently, with a cylindrical projectile, the muzzle velocity of 4.2 km/s was achieved at the pressure of 11 kPa. The shadowgraph image of free-flight capsule model was successfully obtained at the velocity of 3.4 km/s. The template matching method was applied to a capsule shadow, and the pitch angle was determined as 16.0°. With respect to the yaw angle, an improvement in the analysis procedure is necessary.

Keywords Aerodynamic coefficients · Free-flight test · Light gas gun

1 Introduction

For future missions toward Mars, the entry, descent and landing technology are widely and continuously developed and demonstrated globally. In the field of thermophysics of atmospheric entry, understanding real gas effects on aerodynamics at the Martian entry is crucial for precise landing. The flow around an entry vehicle departs from the thermochemical equilibrium state, and the prediction of aerodynamic characteristics

S. Nomura (✉) · M. Mizuno · K. Fujita
Aeronautical Technology Directorate, Japan Aerospace Exploration Agency, Jindaiji Higashimachi 7-44-1, Chofu, Tokyo, Japan
e-mail: nomura.satoshi2@jaxa.jp

K. Itabashi
Department of Advanced Energy, The University of Tokyo, Kashiwa, Japan

S. Lee et al. (eds.), *The Proceedings of the 2021 Asia-Pacific International Symposium on Aerospace Technology (APISAT 2021), Volume 1*, Lecture Notes in Electrical Engineering 912, https://doi.org/10.1007/978-981-19-2689-1_49

is not straightforward. Experimental studies on hypersonic aerodynamics are usually conducted with using a shock tunnel to simulate the Earth reentry condition [1–3]. The careful treatment of a free-fall model enables a free-flight test at the nozzle exit of a shock tunnel. Not only the model visualization, but also the data acquisition of acceleration is achieved by implementing the measurement system inside the model. The real gas aerodynamics at the Earth reentry is well studied grobally. Meanwhile, the experimental approach to Martian entry aerodynamics is limited. In this study, the Martian entry environment is reproduced using a light gas gun, and high-speed observation of a free-flight capsule model is performed. We experimentally determine the aerodynamic coefficients by reconstructing the trajectory and attitude of the capsule model from high-speed visualization images. In this study the current status of the system development is described.

2 Description of Free-Flight Testing Facility

2.1 Light Gas Gun

In this study, hypervelocity light gas gun (HVLGG) at the Japan Aerospace Exploration Agency (JAXA) was used. This facility was transferred from the National Institute for Materials Science (NIMS) in 2012, where the facility was used as 2ST-1 for material studies [4]. The driver section and observation chamber were refurbished for aerodynamic studies; the operation started in 2014. In the beginning, the visualization of a supersonic capsule model was performed for a study of the supersonic parachute. The flight environments at Mach numbers 1.9 and 3.0 were reproduced, and a clear image of the wake region was captured with using a schlieren method [5]. After the refurbishment of the gas driver tank, barrel, and observation area, hypersonic tests were performed in 2019. The schematic view of the current configuration is shown in Fig. 1. The total length was about 20 m, and the inner bore of the barrel was 25 mm. The basic specification is shown in Table 1.

The capsule model used in this work was 0.6% of an aeroshell capsule designed in a MELOS mission [6]. The diameter was 15.6 mm, and the weight was 2.1 g. The target velocity was 4.2 km/s which was based on the previous trajectory analysis [6]. To simulate the flight environment, the binary scaling parameter (rL) was set to the same value as the flight by filling CO_2 in the tank at a pressure of 11 kPa.

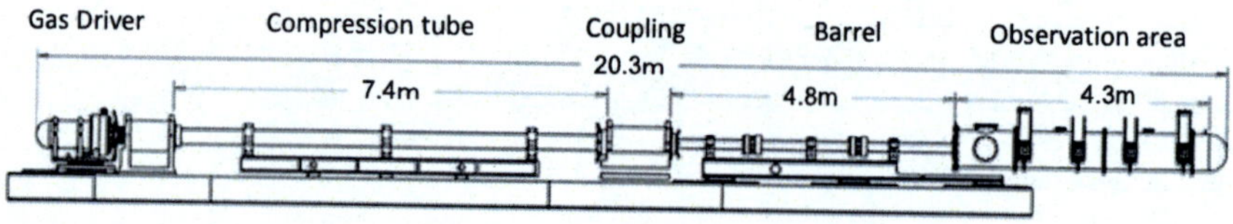

Fig. 1 Schematic of HVLGG

Table 1 Specifications of HVLGG

Gas driver tank	94 L/20 MPa
Piston	6.7 kg
Compression tube	80 mm/7.4 m
Diaphragm	2.0~2.5 mm
Barrel	25 mm/4.8 m
Observation area	4.3 m

Table 2 Operational conditions of HVLGG

	Condition #1	Condition #2
Initial pressure in driver tank	7.1 MPa	13.1 MPa
Initial pressure in compression tube	180 kPa	210 kPa
Diaphragm (depth of groove)	2.0 (0.5) mm	3.2 (0.5) mm
Muzzle velocity	3.4 km/s	4.2 km/s

Four finger sabots were used to cover the capsule model during acceleration in the barrel. A sabot is made of an ultrahigh molecule polyethylene. The diaphragm is made of steel and has a cross groove to achieve a stable rupture. The compression tube was filled with helium gas, and the free piston compresses the helium gas to burst the diaphragm. The gas driver tank was filled with the high-pressure air, and the quick opening valve enabled the rapid acceleration of the free piston. The operational conditions in this experiment are summarized in Table 2. The free-flight visualization test was performed in Condition #1 with the capsule model shown in Fig. 2. As of this moment, the maximum velocity in the free-flight test is 3.4 km/s in Condition #1. The sabot in Fig. 2 did not withstand higher acceleration in the barrel. To realize a flight

Fig. 2 Picture of capsule model in sabots

equivalent environment with a target velocity of 4.2 km/s, a cylindrical projectile was used in Condition #2. As a result, the muzzle velocity of 4.2 km/s was achieved. Using the sabot of a modified design, we expect the achievement of the free-flight test at the target condition in the future.

2.2 *Observation System*

Before the implementation of the observation system, a feasibility study was conducted to determine the appropriate configuration for the aerodynamic assessment [7]. In the feasibility study, the period of pitching motion was estimated. Therefore, to determine the half period of pitching motion, four shadowgraph systems were implemented every 900 mm. An overview of the observation area is shown in Fig. 3. The sabot separation plate was located 800 mm away from the barrel exit. A cross-section of the optical setup is shown in Fig. 4. The parabolic mirrors were used for collimation and focusing. The focus length was 800 mm. To minimize the space for the optics structure, flat mirrors were used, and the parabolic mirrors were mounted above the tank. The pulse diode laser (CAVILUX, Cavitar Ltd) was used as a light source. The pulse duration was short enough to capture the hypersonic flight model, and the high repetition frequency made it possible to capture the multiple shadows of the model during one exposure of a CMOS camera (STC-SPB891PCL, OMRON SENTECH CO., LTD). The laser wavelength was 645 nm, which was suitable to avoid the emission from the shock layer ahead of a model. The coherence was not so high that the image did not show the fringe, and the clear image enabled precise determination of the attitude of a model. The pixel size of the camera was 4096 × 2160, and the focal length of lens (SV-7525H, VS Technology) was 75 mm. The bandpass filter was also used to suppress the emission noise. The trigger signal was generated by the laser cut system which was implemented at the first window just after the barrel and used to burst the laser pulses and start the camera exposure with a delay generator. In this experiment, the repetition frequency was 43 kHz and the pulse duration was 40 ns. The exposure times of the first three cameras were 73.7 ms, whereas the exposure time for the fourth camera was 61.4 ms. There was a shock

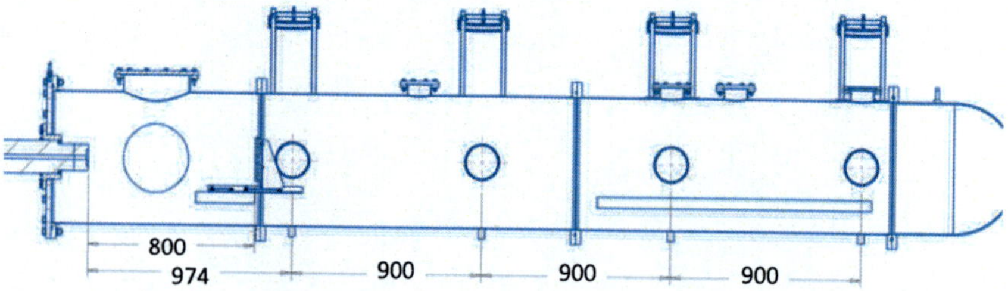

Fig. 3 Overview of the observation area

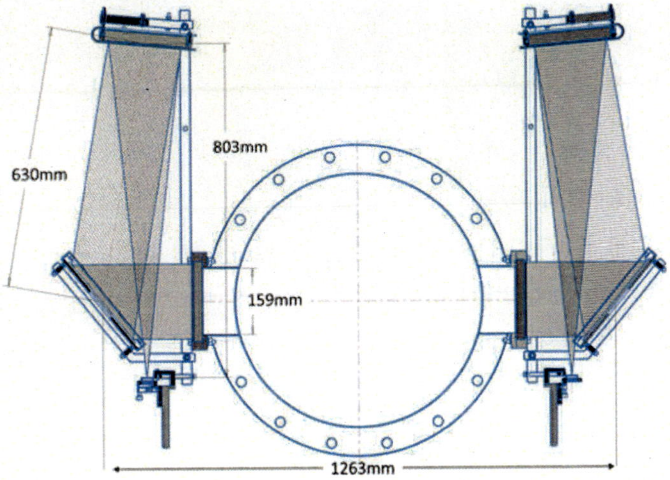

Fig. 4 Cross-section view of the shadowgraph optical setup

absorber just after the fourth observation area, and the impact of model made an emission noise. The exposure time of the fourth camera was shorter than others to avoid the emission noise.

3 Methodology of Attitude Determination

To determine the model attitude by image analysis, the template matching method was introduced in this experiment. The template image of the capsule model at a certain attitude angle (pitch θ and yaw ψ) was generated from the capsule dimensions. The size of the template image was the minimum rectangle, which included the capsule image. The brightness was binary. This template image was scanned in the experiment image data by varying the origin point of the template image (x, y) in the experimental data coordinate. The evaluation function shown in Eq. (1) was used to find the minimum value, $R_{\min}$.

$$R_{\theta,\psi}(x, y) = \frac{\sum_{w,h}\{T(w, h) - I(x + w, y + h)\}^2}{\sum_{w,h} T(w, h)^2 \sum_{w,h} I(x + w, y + h)^2} \quad (1)$$

Here, w and h are the width and height of the template image, respectively. T and I are the brightness intensity at each pixel in the template and experimental image, respectively. By changing the pitch and yaw angle, we could obtain $R_{\min}$ as a function of θ and ψ. Finally, the model attitude was determined by the minimum value of $R_{\min}$ (θ, ψ). To verify this method, the template matching method was applied to the sample shadow image of the capsule model. The model was fixed at

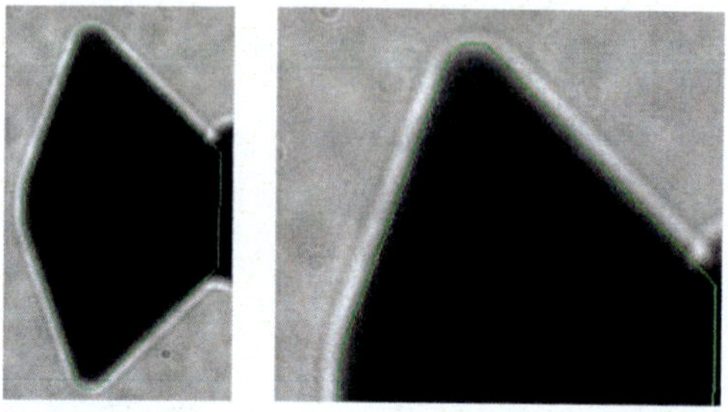

Fig. 5 Template matching for the sample image. The result of matching is shown with a solid line. (Right: close-up view of left image)

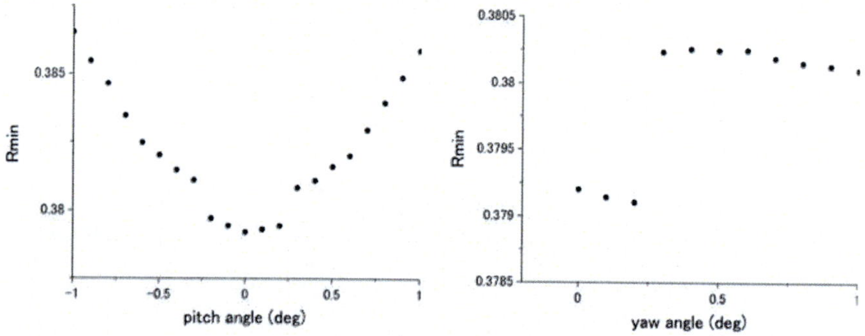

Fig. 6 Plot of R_{min} as a function of pitch angle (left) and yaw angle (right)

pitch and yaw angles of ~0° with a support structure. The solid line in Fig. 5 shows the determination result of the capsule attitude. Behind the base plane, the shadow of the support was projected. The image on the right side in Fig. 5 is the close-up view around the capsule shoulder. Figure 6 depicts the plot of R_{min} as a function of the pitch and yaw angle. In the variation of the pitch angle, R_{min} had a minimum value at 0°. Concerning the yaw angle, R_{min} had a minimum value at 0.2°. However, the variation of R_{min} was relatively small. For accurately determining of the yaw angle, an improvement in the evaluation function was necessary. The change in yaw angle differed in a shadow image at the capsule base plane. The evaluation function, which focused on the base plane, would be important for the yaw angle determination.

4 Results of Free-Flight Test

The results of visualization in the free-flight test are shown in Fig. 7. As of this moment, successful sabot separation was not achieved. In this test, the sabot separation plate was not used, and sabots were also observed in the image. In the first camera image, a bright emission was observed over the capsule shadow. This emission might

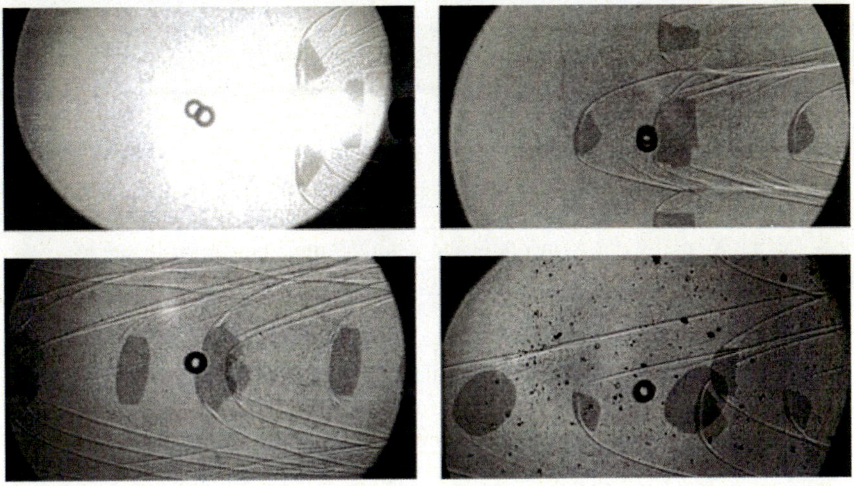

Fig. 7 Shadowgraph images of free-flight testing of Martian capsule. (Top left: first camera image. Top right: second camera image. Bottom left: third camera image. Bottom right: fourth camera image)

come from helium gas behind the model. In the second camera, two capsule shadows were successfully captured. Meanwhile, in the third and fourth camera, the capsule shadows were overlaid by the sabot shadows. The attitude determination method described in Sect. 3 was applied to the image of the second camera image. To avoid the influence of the shadow of the shock wave, the template image was generated only from the back shell configuration. The result is shown in Fig. 8. The pitch angle was determined to be 16.0°, and a good agreement was obtained around the slope region at the back shell. However, the yaw angle was determined as 0°, and the base plane shape was not reproduced correctly, which occurred because the noise in the image is larger than the variation of the evaluation function. For accurate analysis, improvement in the evaluation function was necessary. The capsule attitude was not

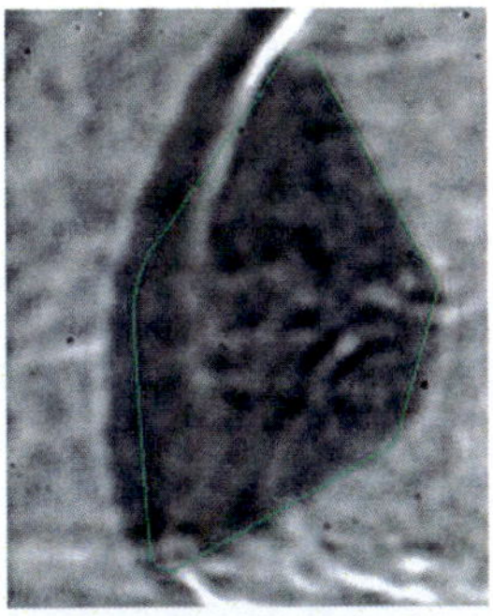

Fig. 8 Attitude determination of free-flight capsule model

derived from the first and third camera images. To obtain the clear capsule image at all four images, the test with a modified sabot is planned in the future.

5 Summary

For the experimental measurements of the aerodynamic coefficient of the Martian capsule, a free-flight testing facility was developed with using the light gas gun. In the current status, the target velocity of 4.2 km/s was not achieved with the capsule model. However, with the cylindrical projectile, the target velocity was reproduced successfully, and we could expect the usage of the more robust sabots enables the experimental simulation of the flight environment. With respect to the high-speed visualization system, multiple shadows of the free-flight capsule model were successfully captured at a velocity of 3.4 km/s. A capsule shadow was analysed using the template matching method, and the pitch and yaw angles were determined to be 16.0° and 0°, respectively. For the accurate determination of yaw angle an improvement in the evaluation function is necessary. The pitching motion reconstruction is not achieved as of this moment. However, the improvement in the sabot design and the image analysis will enable the experimental evaluation of capsule aerodynamics.

Acknowledgements Part of this work is supported by JSPS KAKENHI Grant Number JP 20H02361.

References

1. Tanno H et al (2020) Free-flight measurement technique in the free-piston high-enthalpy shock tunnel. Rev Sci Instrum 85:045112
2. Wey P et al (2012) Determination of aerodynamic coefficients from shock tunnel free flight trajectories. AIAA 2021-3321
3. Kennell C et al (2016) Free flight testing in hypersonic flows: HEXAFLY-INT EFTV. AIAA-2016-1152
4. Sekine T et al (1996) The NIRIM two-stage light-gas gun: performance test results. AIP Conf Proc 370:1201
5. Nomura S et al (2017) Visualization of wake flows of supersonic mars entry capsule with light gas gun. In: International symposium on space technology and science proceedings, e-17
6. Fujita K et al (2013) Design study of EDL demonstrator for MELOS mission. In: 29th International symposium on space technology and science, paper 2013-k-11
7. Fujita K et al (2019) Measurement of real-gas aerodynamics for martian atmospheric entry using a light-gas gun. In: 32rd International symposium on shock waves

Multi-fidelity Modeling via Regression-Based Hierarchical Kriging

Sunwoong Yang, Yu-Eop Kang, and Kwanjung Yee

Abstract Advances in computational technology have enabled engineers to obtain large amounts of data efficiently. However, generating datasets from a single source is still burdensome and inefficient, so there have been many studies on data fusion techniques using multiple sources of data: e.g., low-fidelity data from computational fluid dynamics simulations and high-fidelity data from wind tunnel experiments. The hierarchical Kriging surrogate model is one of these data fusion methods which is known to be simple, robust, and accurate. It is based on the Kriging surrogate model, which predicts quantities of interest using an accumulated database based on the interpolation method. However, serious problems arise when it is built based on the noisy datasets for the following reasons: 1) noise in the dataset induces ill-conditioned correlation matrix; and 2) errors at lower-fidelity levels due to the noise accumulate as the fidelity level increases, adversely affecting the accuracy of the final model. These issues can be mitigated by a method called regression-based Kriging, which incorporates the adaptive regression factor (nugget) to the correlation matrix. This paper extends the corresponding regression technique to the hierarchical Kriging surrogate model, which will be called regression-based hierarchical Kriging. Considering that data fusion methods synthesize data with different fidelity levels to build a surrogate model efficiently, the regression-based hierarchical Kriging model can consider the different degrees of noise at each fidelity level. Herein, this model is demonstrated with several numerical examples with artificial noise and is finally validated with real engineering problem. These examples show that the proposed method has better performance than interpolation-based hierarchical Kriging when trained with the noisy data.

Keywords Hierarchical Kriging · Deterministic noise · Multi-fidelity modeling

---

S. Yang · Y.-E. Kang · K. Yee (✉)
Seoul National University, Seoul 08826, Republic of Korea
e-mail: kjyee@snu.ac.kr

S. Yang
e-mail: sunwoongy@gmail.com

Y.-E. Kang
e-mail: kye72594@snu.ac.kr

S. Lee et al. (eds.), *The Proceedings of the 2021 Asia-Pacific International Symposium on Aerospace Technology (APISAT 2021), Volume 1*, Lecture Notes in Electrical Engineering 912, https://doi.org/10.1007/978-981-19-2689-1_50

1 Introduction

In recent years, computer simulation has become an increasingly important tools in computer aided engineering and advances in computational technology have enabled engineers to obtain large amounts of data efficiently. Nevertheless, it often requires high computational complexity for reasons such as the dense discretization of partial differential equations in computational fluid dynamics (CFD). Limited computational resources and time prevent engineers from increasing the accuracy and number of simulations simultaneously. Especially in aerospace engineering, due to the extraordinary size of the target object compared to other disciplines, computer simulations demand enormous resources. In this regard, the multi-fidelity surrogate modeling is one of the most emerging research areas owing to its potential to ease this problem. Its concept is to fuse the data from several fidelity sources to efficiently predict the highest-fidelity data. Criteria for classifying their fidelities are user-selectable, such as dimensionality, physical model, grid density, and the convergence criterion in the computer simulations. Moreover, any combination of different fidelity types can be used: e.g., the high-fidelity as Reynolds-averaged Navier–Stokes (RANS) results with dense grid, the medium-fidelity as RANS results with coarse grid, and the low-fidelity as Euler solver results. This flexible but efficient concept enables the engineers to predict their quantities of interest (QoI) with reasonable cost and time, and for this reason, various multi-fidelity surrogate models have been proposed. HAN et al. [1] categorized them by correction-based method and cokriging-based method. To briefly introduce, the former uses bridge function to correct the difference between low-fidelity (LF) and high-fidelity (HF) functions [2], whereas the latter takes the LF data as auxiliary data to enhance the prediction accuracy of HF data via covariance matrix [3] (details of these two methods can be found in Ref. [1]). In Ref. [1], Han proposed a novel multi-fidelity surrogate model with their advantages at the same time: it is as simple and robust as the correction-based method, but also accurate and able to provide the model uncertainty as to the cokriging-based method. It is named hierarchical Kriging (HK) owing to its hierarchical construction of the Kriging model towards the highest-fidelity level. The most encouraging feature of the proposed model is that it alleviates the computational cost for the training since it does not require the cross-covariance between the LF and HF data, while providing the reasonable model uncertainty as the trend of the LF is explicitly considered during its estimation. Especially, the reasonable model uncertainty quantification enables this model to be appropriate for efficient global optimization (EGO) based on the expected improvement (EI) [5].

However, as HK is based on the Kriging surrogate model, it has the same drawbacks as of it: the Kriging model trained with noisy data is prone to be overfitted due to its interpolation characteristics. This problem becomes more severe in HK modeling. Since the HK model is trained using the data of each fidelity level hierarchically, in the process of sequentially constructing the Kriging model, inaccuracies due to noises from the lower-fidelity data gradually accumulate, which negatively affects the prediction accuracy of the final HK surrogate model. Accordingly, the accuracy of the multi-fidelity model can be not much different from that of the single-fidelity modeling, which makes the multi-fidelity modeling meaningless. To deal with the issue of erratic behavior due to the noisy data, the regression-based Kriging method in single-fidelity problem was proposed by Forrester et al. [4]. During the training process, they optimized not only correlation parameters but also adaptive regression factor, often called nugget. By adopting the nugget, Kriging loses its interpolation characteristics and shows regression characteristics. This regression-based Kriging filters out the numerical noises and therefore is trained to be robust against them. Moreover, since the model uncertainty is not calculated as zero due to the introduction of the nugget, they also applied re-interpolation process to make the uncertainty zero at the training data. It enables an efficient EGO process by preventing the points used for training from being selected again in the EI process.

In this study, this regression-based Kriging methodology is extended to the HK model, and it is named regression-based hierarchical Kriging (RHK). Its objective is to alleviate the overfitting behavior caused by the deterministic noise in computer simulations while training the HK model. As RHK synthesize data with different fidelity levels consecutively from low to high, it can consider the different degrees of noise at each fidelity level. To verify the performance of this model, several numerical examples are presented and finally it is applied to practical engineering problem. From these examples, when there are noises in the training data, it is confirmed that the proposed method has better performance than interpolation-based hierarchical Kriging (IHK).

2 Methodologies

2.1 Kriging

In the algorithm of HK, ordinary Kriging is generally used as a baseline surrogate model. The mathematical form of ordinary Kriging is defined as follows:

$$y(\boldsymbol{x}) = \beta_0 + Z(\boldsymbol{x}) \tag{1}$$

where β_0 is constant trend term, and $Z(\boldsymbol{x})$ is stationary random process term that locally interpolate the observed points. Suppose we have the computer model y, the n known input $\boldsymbol{X} = \{\boldsymbol{x}_1, \boldsymbol{x}_2, \ldots, \boldsymbol{x}_n\}$, and the resulting output $\boldsymbol{Y} = \{y(\boldsymbol{x}_1), y(\boldsymbol{x}_2), \ldots, y(\boldsymbol{x}_n)\}$. Then, the Kriging predictor $\hat{y}$ at any unknown point $\boldsymbol{x}$ is as follows:

$$\hat{y}(x) = \beta_0 + \boldsymbol{r}^T(\boldsymbol{x})\boldsymbol{R}^{-1}(\boldsymbol{Y} - \beta_0 \mathbf{1}) \tag{2}$$

where $\beta_0 = \left(1R^{-1}1\right)^{-1}1^T R^{-1} Y$; R is the correlation matrix representing the correlation between training points; **r** is the correlation vector representing the correlation between unknown point and training points, and **1** is the vector of ones. The correlation between the points is modeled by a correlation function. Generally, a Gaussian exponential and a cubic spline function are widely used as correlation function. The influence of nearby points is controlled by a correlation parameter, which can be used for model-fitting. Kriging model also provide mean square error (MSE) of estimated value, and it is defined as follows:

$$\text{MSE}[\hat{y}(x)] = \sigma^2\left[1 - r^T R^{-1} r + \left[r^T R^{-1}1 - 1\right]^2 / 1^T R^{-1} 1\right] \tag{3}$$

where $\sigma^2 = (1/n)(Y - 1\beta_0)\mathbf{R}^{-1}(Y - 1\beta_0)$. The MSE can be used as a measure to evaluate the model uncertainty. At training point the MSE is zero because the Kriging predictor returns an accurate estimate, whereas at unknown points the MSE approaches its maximum value of σ^2. This provides useful information in the EGO process since it serves as a basis for identifying where additional training points should be located.

2.2 *Regression-Based Kriging*

The conventional Kriging described in Sect. 2.1 has the interpolation characteristics as the trained model always passes through all training points. Forrester focused on the fact that the erratic behavior of the Kriging model originates from this property [4]: when it is trained with noisy data, interpolation technique can lead to overfitted surrogate model. A numerical example case is presented to illustrate this issue. In a situation where quadratic function is to be modelled via Kriging, it is assumed that the random Gaussian noises are added to the training data: $y = x^2 + N\left(0, 0.1^2\right)$. In Fig. 1, the results of conventional Kriging (interpolation-based Kriging, IK) trained with these noisy data are shown as the black-dashed line. It shows highly wavy trend especially due to the noisy data near $x = 0$. Owing to these overfitting issues when training the Kriging model using the data with deterministic noise, the regression-based Kriging (RK) method was proposed. The deterministic noise herein is referred to "variation in the output due to fluctuations in the error from simulation to simulation as the inputs are varied slightly", as mentioned in Ref. [4]. In other words, when we perform deterministic computer simulations, though the input data are neighboring each other, they can exhibit quite different outputs due to the numerical perturbations (such as discretization error, incomplete convergence, and round-off error during the data processing).

The RK model can control the degree of the regression by considering additional variable to be optimized during the model training process. This additional variable, which activates the regression property in the Kriging, is called the regression

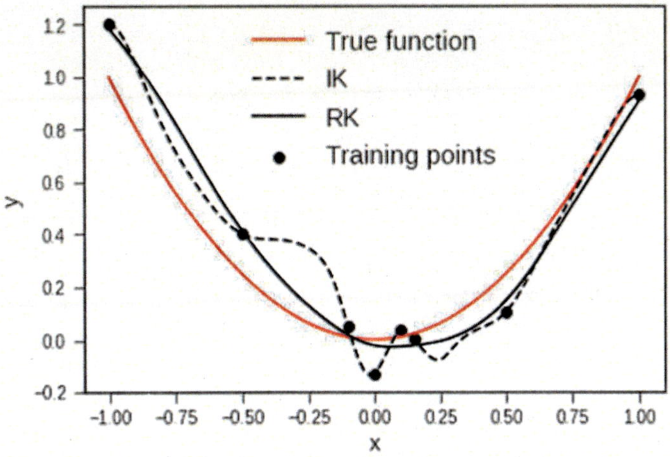

Fig. 1 Results of the IK (interpolation-based Kriging) and RK (regression-based Kriging) models with the noisy data from the function $y = x^2 + \mathrm{N}(0, 0.1^2)$

constant (or nugget) and is added to the diagonal of the correlation matrix $\boldsymbol{R}$. When the nugget is introduced, the Kriging no longer needs to go through the training points. With this flexibility, the RK model can reach higher likelihood during the MLE (maximum likelihood estimation) process. Note that the corresponding regression algorithm requires only one additional variable in the optimization of MLE, indicating there is a negligible increase in terms of the computational cost to realize it. In Fig. 1, the Kriging model with this regression algorithm is trained with the same noisy data as in IK, and its results are shown as the black-solid line. It can be confirmed that RK model has been trained much more realistically than IK model, showing the smooth trend similar to the true function.

2.3 *Hierarchical Kriging*

Using a multi-fidelity training dataset, e.g. $(\boldsymbol{X}_{LF}, \boldsymbol{Y}_{LF})$, $(\boldsymbol{X}_{HF}, \boldsymbol{Y}_{HF})$, the following part will demonstrate how an HK surrogate model is trained. The mathematical form of this model is similar to ordinary Kriging, but the constant trend term is replaced by the LF Kriging $\hat{y}_{LF}$ with scaling factor β (note that the HF model is trained by using the $\hat{y}_{LF}$ multiplied by β as the model trend):

$$\hat{y}_{HF}(\boldsymbol{x}) = \beta \hat{y}_{LF}(\boldsymbol{x}) + Z(\boldsymbol{x}) \tag{4}$$

In this formulation, the LF Kriging model acts as the global trend of HF data, which describes how the fusion of LF and HF data is realized in this multi-fidelity modeling technique. The predictor and MSE of high-fidelity model are defined as follows:

$$\hat{y}_{HF}(\boldsymbol{x}) = \beta \hat{y}_{LF}(\boldsymbol{x}) + \boldsymbol{r}^T(x)\boldsymbol{R}^{-1}(\boldsymbol{Y}_{HF} - \beta_0 \boldsymbol{F}) \quad (5)$$

$$\text{MSE}\left[\hat{y}_{HF}(\boldsymbol{x})\right] = \sigma^2\left[1 - \boldsymbol{r}^T\boldsymbol{R}^{-1}\boldsymbol{r} + \left[\boldsymbol{r}^T\boldsymbol{R}^{-1}\boldsymbol{F} - \hat{y}_{LF}(\boldsymbol{x})\right]^2 / \boldsymbol{F}^T\boldsymbol{R}^{-1}\boldsymbol{F}\right] \quad (6)$$

where $\beta = \left(\boldsymbol{F}\boldsymbol{R}^{-1}\boldsymbol{F}\right)^{-1}\boldsymbol{F}^T\boldsymbol{R}^{-1}\boldsymbol{Y}, \sigma^2 = (1/n)(\boldsymbol{Y} - \boldsymbol{F}\beta)\mathbf{R}^{-1}(\boldsymbol{Y} - \boldsymbol{F}\beta)$, and $\boldsymbol{F}$ is a vector whose elements are the predictions of the LF Kriging at the HF training points. In summary, the LF and HF Kriging models are trained sequentially. This hierarchical feature allows HK to easily extend to multi-fidelity-level problem (more than bi-fidelity) by stacking each fidelity model hierarchically.

In this manuscript, we extended the regression-based Kriging algorithm to the HK. Therefore, the comparison between IHK (interpolation-based hierarchical Kriging) and RHK (regression-based hierarchical Kriging) will be performed in the following results and discussion part. Herein, IHK stands for HK in which all Kriging models at all fidelity levels are trained with the interpolation-based algorithm. On the other hand, RHK indicates regression-based Kriging models are trained at all fidelity levels for HK. Since the nugget is adjusted by MLE at each fidelity level, the RHK model can consider the different degrees of noise at each fidelity level.

3 Results and Discussion

3.1 Numerical Test: Forrester Functions

To compare the IHK and RHK with numerical example, bi-fidelity Forrester functions are used as in Ref. [1]. For the high-fidelity (HF) function, $y_{HF} = (6x-2)^2 \sin(12x - 4)$ is used. For the low-fidelity (LF), $y_{LF}=y_{HF} + 10(x - 1)$ is used. However, for modeling the deterministic noise in the LF function, the random Gaussian noises are additionally considered. Therefore, the LF function used in this numerical test is modified as follows: $y_{noisy\_LF} = \beta^{-1} y_{LF} * N\left(1, \delta_1^2\right) + N(0, \delta_2^2)$. Herein, $N\left(1, \delta_1^2\right)$ is the noise dependent on the scale of the LF function, and $N\left(0, \delta_2^2\right)$ is the noise independent of the LF scale. To control the scale of these noises, the standard deviations δ_1 and δ_2 are set to be adjustable. Similarly, the β is also adjustable. Therefore, it is possible to investigate the effects of noise scale and β (the correlation factor between LF and HF functions) on the accuracy of the HK modeling. Finally, total four cases are classified as in Table 1.

For a quantitative comparison of the two HK models (IHK and RHK), three metrics are adopted: R^2, mean absolute error (MAE), and root mean squared error (RMSE). These metrics are calculated between the real values from the true HF function and predicted values from trained surrogate models with equally spaced

Table 1 Total four cases are classified according to the noise scale and correlation factor

	$\beta = 0.5$ (Small Beta)	$\beta = 2$ (Large Beta)
$\delta_1 = 0.05, \delta_2 = 0.5$ (Small Noise)	SNSB	SNLB
$\delta_1 = 0.1, \delta_2 = 1$ (Large Noise)	LNSB	LNLB

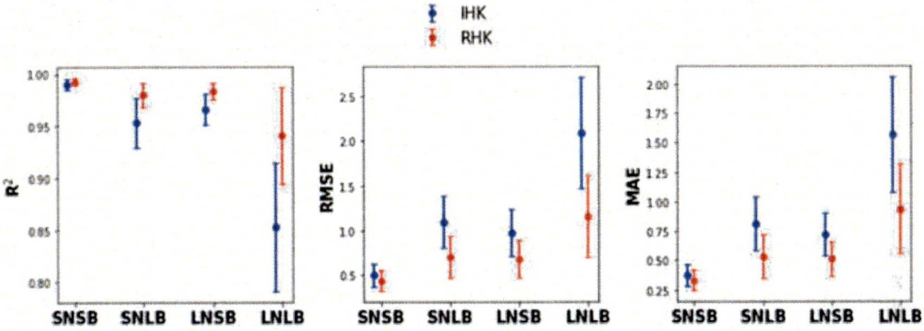

Fig. 2 Results of the three metrics (their averages and standard deviations) comparing IHK and RHK models in the numerical test

101 validation points where $x \in [0, 1]$. Also, due to the randomness in generating Gaussian noises, IHK and RHK are run 50 times for each case and statistics of three metrics are presented. Figure 2 shows the comprehensive results of this numerical test: the average and standard deviation of the metrics in four cases. First, for the cases of small noise (SNSB and SNLB), the differences between IHK and RHK exist depending on the β value. When β is small ($\beta = 0.5$, SNSB), IHK and RHK show almost similar metrics. It can be inferred that the reason for this result is that the erratic behavior in low-fidelity Kriging model due to the noise in LF data is attenuated by half and reflected in the final HK model. This inference is supported by the result of the following case, SNLB. In this case, where the value of β is large ($\beta = 2$), the erratic behavior in low-fidelity Kriging model is doubled and then reflected in the final HK model. Therefore, though it has the same noise scales as the SNSB case, RHK shows relatively better metrics than IHK model owing to its regression property in SNLB case. Similarly, for large noise scales (LNSB and LNLB), the differences between IHK and RHK depend on the β. However, LNSB shows quite different aspects from the SNSB case. As noise scales are larger, the erratic behavior becomes severe than SNSB. Therefore, though beta is small ($\beta = 0.5$), which means the erratic trend of LF Kriging is halved before reflected to the final HK model, the calculated metrics highlight the relative accuracy of RHK over IHK. This phenomenon is further exacerbated in the LNLB case. In this case, the large noises will cause overfitting behavior in the LF Kriging. Furthermore, this behavior will be amplified and reflected in the HK model due to the large β. For these reasons, the differences in metrics between IHK and RHK are the largest in

this case. Additionally, it can be inferred that RHK provides a more robust results when trained with noisy data through the fact that the standard deviation values are smaller than that of IHK in all cases. Please note that when LF function has no noise at all, it is confirmed that IHK and RHK result in exactly the same metrics.

3.2 Engineering Application: RAE 2822 Transonic Airfoil

To show the effectiveness of adopting RHK in real engineering applications, the multi-fidelity modeling of pitching moment coefficients (C_m) with respect to the angle of attack (AoA) is presented. The Mach and Reynolds numbers are set as 0.729 and 6.5×10^6, and the range of AoA is from −2 to 6°. KFLOW flow simulation code with Spalart–Allmaras turbulence model is used. The grid density for the flow analysis is selected as the fidelity criterion (LF data from the structured O-grid with 16,362 nodes and HF data from 64,962 nodes).

Figure 3 shows the results of the engineering application. In Fig. 3a, LF Kriging models which were trained only with LF data (which will be used as the model trends in training HK models) are visualized as dashed lines: IK as blue and RK as red. And final HK models using these LF models as model trends are also visualized as solid lines with the same colors. Figure 3b shows the magnified version of Fig. 3a. It can be confirmed that overfitting behavior in IK (LF only) transmits to the IHK model. On the other hand, the RK (LF only) shows smooth trend and therefore RHK is also trained to be smooth. However, from this figure, as it cannot be judged that RHK is more accurate than IHK, we tried to compare model accuracy through HF validation points, which are 41 equally spaced data where $AoA \in [-2°, 6°]$. Figure 3c compares the real values from these validation data and predicted values from trained HK surrogate models. Herein, three metrics used in previous numerical test are calculated again. Their results are summarized in Table 2 and indicate superior performance of RHK over IHK. The results of the single-fidelity Kriging trained only with HF data are also presented to confirm the effect of the multi-fidelity modeling compared to single-fidelity modeling: we name it "HF only Kriging" and do not compare the regression-based and interpolation-based methods in this case since there was no difference between the results from these methods. For all metrics, they show consistent results: (1) two multi-fidelity modeling techniques (IHK and RHK) show better results than single-fidelity modeling; (2) however, RHK shows significant improvement over interpolation-based multi-fidelity modeling technique (IHK), whereas IHK shows insignificant improvement over single-fidelity modeling technique (HF only Kriging). Finally, the gradients of C_m with respect to AoA are compared in Fig. 3d. Again, RHK shows the most similar gradient trend than HF only Kriging and IHK results. This indicates that based on these trained surrogate models, gradient-based optimization techniques can lead the engineers to totally different optimal solutions (though we cannot guarantee that RHK will always yield an accurate optimal solution from this figure, we can at least assert that it will go through more efficient EGO process than other methods). Note that the beta values

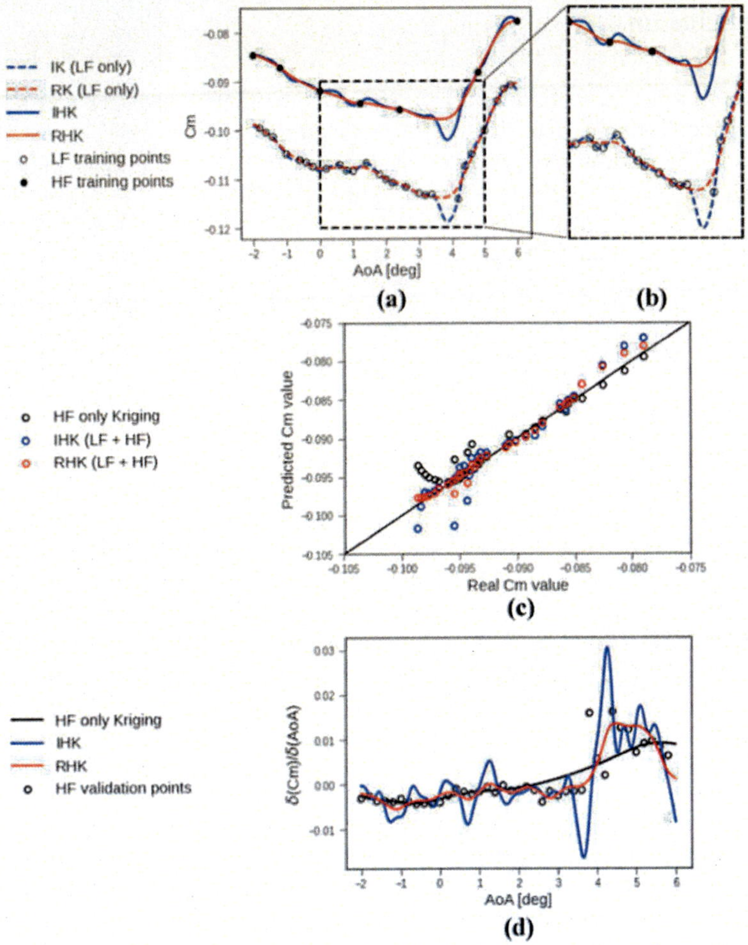

Fig. 3 Results of the engineering problem comparing IHK and RHK models

Table 2 Three metrics are compared between different surrogate models: R^2, RMSE, and MAE

	HF only Kriging	IHK	RHK
R^2	0.9335	0.9390	**0.9866**
RMSE	0.0016	0.0016	**0.0007**
MAE	0.0010	0.0011	**0.0005**

representing the correlation between LF and HF are calculated as 0.8574 and 0.8598 for IHK and RHK, respectively. This indicates that the noise in the LF level significantly affects the final HK result in real engineering problems, even though the correlation between LF and HF is small (near 0.85).

4 Conclusion

This manuscript focuses on the fact that HK has the same drawbacks of Kriging: when trained with noisy data, it is prone to be overfitted due to its interpolation characteristics. This issue is particularly problematic in HK due to its training algorithm: as the Kriging model at each fidelity level is trained hierarchically, inaccuracies due to noise from the lower-fidelity data gradually accumulate, which degrades the prediction accuracy of the final HK surrogate model. This erratic behavior can seriously reduce the usefulness and effectiveness of this multi-fidelity modeling technique.

To ease this problem, the regression-based Kriging algorithm, which controls the degree of the regression by considering nugget parameter during the model training process, is extended to HK algorithm. To verify the performance of this algorithm, several numerical examples and one practical engineering problem were presented. From these examples, we confirmed that the proposed method has better performance than interpolation-based hierarchical Kriging when there are noises in the training data. In particular, in the real engineering case with transonic RAE 2822 airfoil, the RHK model showed superior performance over conventional IHK model in terms of both value prediction and gradient calculation.

Acknowledgements This work was supported by the National Research Foundation of Korea (NRF) Grant funded by the Ministry of Science and ICT (NRF-2017R1A5A1015311).

References

1. Han Z, Görtz S (2012) Hierarchical Kriging model for variable-fidelity surrogate modeling. AIAA J 10(2514/1):J051354
2. Chang K, Haftka R, Giles G, Kao P (1993) Sensitivity-based scaling for approximating structural response. J Aircr 10(2514/3):48278
3. Kennedy M, O'Hagan A (2000) Predicting the output from a complex computer code when fast approximations are available. Biometrika. https://doi.org/10.1093/biomet/87.1.1
4. Forrester A, Keane A, Bressloff N (2006) Design and analysis of "Noisy" computer experiments. AIAA J 10(2514/1):20068
5. Design Rule Extraction Using Multi-Fidelity Surrogate Model for Unmanned Combat Aerial Vehicles
6. Comment on "Novel approach for selecting low-fidelity scale factor in multifidelity metamodeling"

Evaluation of the Use Cases of eVTOLs with High Potential in the Philippines and Thailand

Aki Nakamoto, Patrisha Armie W. Bas, and Masaru Nakano

Abstract This study aims to select and evaluate the promising use cases of multirotor electric vertical takeoff and landing (eVTOL) aircraft in Philippine and Thai markets. With the high potential of eVTOL aircraft for practical use in Southeast Asia, identifying which countries and use cases are promising is necessary. We conducted interviews with doctors of two institutions with helicopter emergency medical services (HEMS) in Thailand to elucidate current challenges in HEMS and airframe requirements for "emergency medical services (EMS)" use case. Next, we conducted interviews with 12 local stakeholders in the Philippines and Thailand to gain information on the challenges surrounding the implementation of eVTOLs for the "Sightseeing/Leisure" and "Urban Air Mobility (UAM)" use cases within the specified locations. Then, we quantitatively evaluated the stakeholder perceptions using the Quantitative Strategic Planning Matrix to select promising use cases. Finally, we performed Fermi estimate to calculate the market size with constraints. The results showed that HEMS in Thailand needs two-seater eVTOLs to dispatch only a doctor to the field as in Japan; however, the feasibility of this use case is lower than other use cases considering the specifications required by doctors. The Sightseeing/Leisure use case in Cebu is the best use case for eVTOL implementation with stakeholder perceptions in mind, whereas the UAM use case in Bangkok is better for sales.

Keywords EVTOL · Demand estimation · Stakeholder perceptions · Quantitative Strategic Planning Matrix (QSPM)

1 Introduction

Electric vertical takeoff and landing (eVTOL) aircraft, the so-called *flying cars*, are being developed worldwide. The aircraft types are largely classified into vectored thrust, lift+cruise, and multirotor (wingless) [1]. In Japan, the multirotor type is

A. Nakamoto (✉) · P. A. W. Bas · M. Nakano
Graduate School of System Design and Management, Keio University, 4-1-1 Hiyoshi, Kohoku-ku, Yokohama, Kanagawa 223-8526, Japan
e-mail: aki.nakamoto@sdm.keio.ac.jp

S. Lee et al. (eds.), *The Proceedings of the 2021 Asia-Pacific International Symposium on Aerospace Technology (APISAT 2021), Volume 1*, Lecture Notes in Electrical Engineering 912, https://doi.org/10.1007/978-981-19-2689-1_51

mainly being developed to be commercialized. Japanese manufacturers include SkyDrive, which is developing electric multirotor airframes with the world's smallest size equivalent to a car and a low price with the goal of launching in 2023 [2].

As for eVTOL markets, the Southeast Asian market probably has a high potential. Regarding the global market of current helicopters, Airbus forecasts that there will be a global demand for 22,000 helicopters for growth and replacement needs over the 2017–2036 period, with Asia–Pacific accounting for 36%, the largest share compared with other regions [3]. Thus, understanding which countries and use cases are promising is necessary to apply multirotors for resolving transportation and social challenges.

Though conventional studies have analyzed demands for eVTOLs, the focus is mainly on those for air taxis, namely, Urban Air Mobility (UAM), in the United States, which has severe traffic congestion challenges [4, 5]. Studies on demands in Europe, including Balac et al. [6], have proposed a methodology for examining the demand for UAM using a multinomial logit model and presented it as a case study in Zurich, Switzerland. However, studies on other cities are needed to grasp the whole picture of the potential of passenger transportation using UAM. Xu and Zhou [7] have shown the possibility of applying eVTOLs to air ambulances in China. However, the aircraft type studied was a battery-powered eVTOL with tandem tilt wings to cover a large service radius (i.e., 100 km), which is close to a normal light helicopter, and a small multirotor was not considered.

Additionally, social acceptance is key to develop eVTOL markets. Social acceptance will be an operational constraint [8], affecting the realization time and market size of eVTOLs. In conventional studies, Eker et al. [9] have analyzed data collected from an online survey toward 584 individuals from the United States and showed that individual-specific sociodemographic, behavioral, and driving attributes influence the perception of eVTOL safety. In terms of ethnicity, Asian respondents were favorable toward the safety benefits of eVTOLs and the suggested security measures. However, studies on the perceptions of eVTOLs among people in countries other than the United States are limited.

Thus, the demands in unstudied countries with high potential should be estimated, while considering stakeholder perceptions. This study aims to select and evaluate the promising use cases of multirotor eVTOLs in Philippine and Thai markets.

This study comprised five sections. Section 2 will explain the flow and methods of evaluating emergency medical service (EMS) use case in Thailand and Sightseeing/Leisure and UAM use cases in the Philippines and Thailand, respectively. Section 3 will present the results, Sect. 4 will discuss the results, and Sect. 5 will conclude the paper with future research topics.

2 Methods

The flow of this study is shown in Fig. 1. To examine the potential of EMS, Sightseeing/Leisure, and UAM use cases, this study focused on the Philippines and Thailand and conducted interviews with stakeholders asking their needs and perceptions on challenges surrounding the implementation of eVTOLs. Then, this study estimated the demands, considering stakeholder perceptions.

2.1 Examination of EMS Use Case

This study examined the EMS use case, which could enjoy a high level of social acceptance. Interviews were conducted with personnels of two institutions with helicopter emergency medical services (HEMS)—Samitivej Sukhumvit Hospital and National Institute for Emergency Medicine (NIEM)—regarding the airframe requirements for eVTOLs to elucidate the airframe requirements in Thailand. Samitivej Sukhumvit Hospital is part of Bangkok Dusit Medical Services (BDMS), the largest private hospital network in Thailand, consisting of 49 hospitals, which was the first HEMS operator in the country [10]. Meanwhile, NIEM is a national institute responsible for administrative management and coordination between agencies related to EMS of both public and private hospitals [11].

The assumption of this use case is that only a flight doctor without a flight nurse is dispatched by a two-seater eVTOL to a patient to assess the clinical state and provide initial treatment in the field, and the patient is transported to a hospital by ground ambulance. This use case has been proposed in Japan, where *doctor helicopters* (DHs) are used in HEMS. A helicopter to transport a physician to the

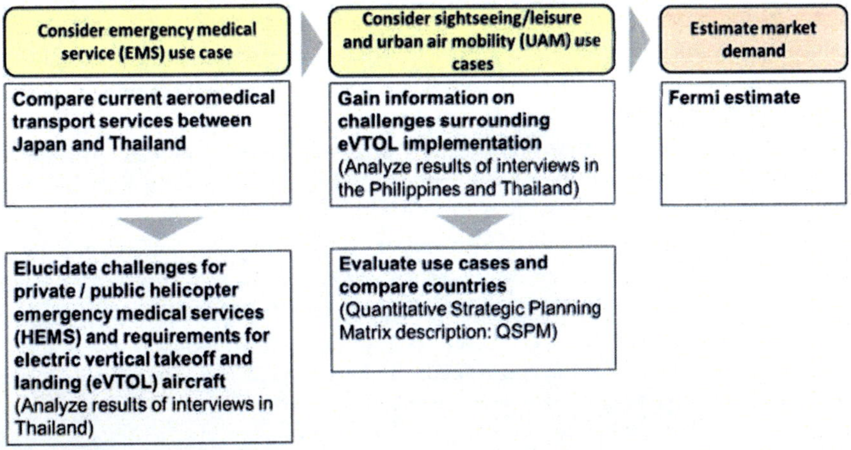

Fig. 1 Flow of this study

intended location of the patient/s (usually a designated rendezvous point closest to the patient) has been used since 2001, with such helicopters being known as *doctor heli*, in Japan. eVTOLs may be additionally introduced to enhance the EMS, addressing challenges of DHs, which are expected by relevant people, including flight doctors and helicopter operators [12].

First, this study compared the current aeromedical transport services between Japan and Thailand because eVTOLs with the aforementioned concept could be added to existing programs. Subsequently, this study elaborated on the interview results on current HEMS's challenges and airframe requirements for eVTOLs to consider the applicability of multirotor eVTOLs.

2.2 Examination of Sightseeing/Leisure and UAM Use Cases

This study examined the Sightseeing/Leisure and UAM use cases, which are expected to be profitable. For the Sightseeing/Leisure use case, Cebu was chosen in the Philippines, and Chiang Mai was chosen in Thailand, among the major tourist destinations of each country. For the UAM use case, capitals were chosen: Manila in the Philippines and Bangkok in Thailand. Interviews were conducted with 12 local stakeholders from the government, aviation business, and tourism in the Philippines and Thailand, where the services are already provided using helicopters, to gain information on challenges surrounding the implementation of eVTOLs within the aforementioned locations.

This study quantitatively evaluated the stakeholder perceptions of the challenges surrounding the implementation of eVTOLs using the Quantitative Strategic Planning Matrix (QSPM) [13] to select promising use cases. This matrix, developed by F.R. David, helps organizations provide an objective approach in evaluating and choosing alternative strategies. This is performed by first listing and assigning weights to the organization's internal strengths and weaknesses and its external opportunities and threats. These factors are used for the vertical axis. Next, alternative strategies are listed as the horizontal axis. Then, each strategy and factor are given an "attractiveness score" according to how attractive the strategy is in addressing the factor. Finally, the scores are multiplied by their weights and summed up. Then, the total attractiveness scores can be compared with one another to determine which strategy is the best for implementation.

Although this method is mainly used in the context of an organization or company evaluating its own alternative strategies considering internal and external factors affecting the organization, this has been modified in this study to assess alternative use cases in the context of an eVTOL service provider evaluating use cases in the Philippines and Thailand considering a speculated vehicle. The modifications to the QSPM steps performed for this study are listed in Table 1.

Table 1 Comparison of the original Quantitative Strategic Planning Matrix (QSPM) procedure with the modified procedure in this study

Original steps for the QSPM	Modified
1. Make a list of an *organization's internal strengths/weaknesses and external opportunities and threats* 2. *Assign ratings* to each internal and external factor (i.e., 1–4, 1 being the most major weakness and 4 being the most major strength) 3. *Determine alternative strategies* that the organization is considering on implementing 4. Determine *attractiveness scores* (ATS) for each factor and strategy. *(1, the strategy is not acceptable; 2, the strategy is possibly acceptable; 3, the strategy is probably acceptable; and 4, the strategy is most acceptable.)* 5. Compute the total attractiveness scores (ATS × factor rating) and compute the sum total attractiveness score for the strategy	1. List down the *external factors and eVTOL specifications* that are to be considered 2. *Assign ratings* to each factor and specification according to how much influence it may have 3. *Determine the use cases and their respective locations* that will be considered for implementation 4. Determine *applicability scores* (APS) for each factor and strategy. *(0, the use case is not acceptable; 1–2, the use case is possibly acceptable; 3–5, the use case is probably acceptable; and 6–7, the use case is most acceptable.)* 5. Compute the total applicability scores (APS × factor rating), and compute the sum total applicability score for the strategy

2.3 *Fermi Estimate for the Sightseeing/Leisure and UAM Use Cases*

This study conducted Fermi estimate to determine the market size by calculating s_{UAM} (i.e., the number of sales for the UAM use case with weather, public acceptance, regulatory, and fare constraints) and $s_{\text{Sightseeing}}$ (i.e., the number of sales for the Sightseeing/Leisure use case with weather, public acceptance, regulatory, and fare constraints) using formulas (1) to (4). The base demand is that of the wealthy people who are likely to use eVTOL services within a year, and the aforementioned constraints are considered. The definition of variables and constants is shown in Table 2.

$$s_{UAM} = \beta \cdot n_{Households} \cdot R_{Wealth} \cdot R_{Prospect} \tag{1}$$

$$s_{Sightseeing} = \beta \cdot n_{Hotel} \cdot R_{Occupancy} \tag{2}$$

where β and r_{Usage} are shown below.

$$\beta = (Y - d) \cdot \left(1 - r_{Regulation}\right) \cdot r_{Usage} \cdot r_{Utilization} \cdot x_{Fare} \tag{3}$$

$$r_{Usage} = f\left(r_{Utilization}, x_{Fare}\right) \tag{4}$$

Table 2 Definition of variables and constants

Variable/constant	Unit	Explanation
s_{UAM}	PHP/THB	Number of sales (UAM use case)
$s_{\text{Sightseeing}}$	PHP/THB	Number of sales (Sightseeing/Leisure use case)
$n_{\text{Households}}$	–	Number of households in the capital
n_{Hotel}	–	Number of 5-star hotel rooms
R_{Wealth}	%	Percentage of population classified under the wealthiest socioeconomic status (1.2% for the Philippines; 2% for Thailand)
$R_{Prospect}$	%	Percentage of people that may avail eVTOL services (5%)
$R_{Occupancy}$	%	Occupancy rate (80%)
$r_{Regulation}$	%	Percentage affected by regulation constraints
Y	day	Year (365 days)
d	day	Number of rainy days
r_{Usage}	%	Usage rate
$r_{Utilization}$	%	Utilization rate (60%, 80%, 95%)
x_{Fare}	PHP/THB	Fare

Note PHP: Philippine Peso; THB: Thai Baht; UAM: urban air mobility

For the UAM use case, the number of wealthy households in the capital is computed using $n_{Households}$ (the number of households in the capital) and R_{Wealth} (the percentage of population classified under the wealthiest socioeconomic status) assumed as below. In the Philippines, the Philippine Institute of Development Studies compiled the distribution of residents by income clusters, showing the percentages of people who would be considered rich, upper class, and upper middle class within urban areas based on data from the National Family and Income Expenditure Survey conducted by the Philippine National Statistics Office in 2015 [14]. R_{Wealth} is assumed to be 1.2% for the Philippines with only the wealthiest households considered for the purpose of this study. In Thailand, the Thai Marketing Research Society has compiled the percentage of those within similar classes [15]. R_{Wealth} is assumed to be 2% for Thailand, with only the percentages of those under the A+ used in this study.

For the Sightseeing/Leisure use case, n_{Hotel} (the number of five-star hotel rooms) is computed for Cebu in the Philippines and Chiang Mai in Thailand; the number of 5-star hotel rooms was listed and totaled. $R_{Occupancy}$ (occupancy rate) considers n_{Hotel} and is assumed to be 80% based on the interview survey of this study.

$R_{Prospect}$ (the percentage of people that may avail eVTOL services) is assumed to be 5% based on the interviews to INAEC Aviation Corp. (the Philippines [PH]), Japanese Chamber of Commerce in Manila (PH), Tourism Authority of Thailand (Thailand [TH]), and Thai Aviation Services (TH) for the UAM use case, and AirTaxi PH (PH), Advance Aviation Corp. (TH), Tourism Authority of Thailand (TH), and Veranda High Resort (TH) for the Sightseeing/Leisure use case.

The demand for weather constraint is subtracted by subtracting d (the number of rainy days), as this study assumed that eVTOLs may not be able to operate during rainy days. The following information was obtained for the number of rainy days in each location:

- Philippines (nationwide): 143 days [16]
- Thailand:
  - Bangkok (Central region): 128 days [17]
  - Chiang Mai (Northern region): 121 days [18].

Then, the demand with public constraints is multiplied, considering $r_{Regulation}$ (the percentage affected by regulation constraints). The following assumptions for the percentages and their breakdowns for each use case are determined based on our interview information (Table 3).

This study assumes $r_{Utilization}$ (utilization rate), r_{Usage} (usage rate), and x_{Fare} (fare) (Fig. 2). $r_{Utilization}$ is the annual percentage of eVTOL flights booked that have flown successfully without cancellations (excluding those caused by inclement weather), and aids in telling the reliability of the service's availability. r_{Usage} is the percentage of people within the country that will avail the service given the utilization rate and fare. r_{Usage} is calculated using formula (4) using $r_{Utilization}$ listed in the row and x_{Fare} in the column in the matrices . Three percentages are assumed for $r_{Utilization}$: 60%,

Table 3 Percentage assumptions for public constraints

Philippines—UAM: 30%	**Thailand—UAM: 99%**
• *5% inconvenience from 1-week period for attaining flight permit* • *5% noise and visual disturbance* • *10% inconvenience from dealing with bureaucratic rules and processes* • *10% building/property landing approval*	• *99% government restriction of airspace*
Philippines—Sightseeing: 27%	**Thailand—Sightseeing: 20%**
• *7% inconvenience from 10-day period for attaining flight permit* • *5% noise and visual disturbance* • *10% landing approval* • *5% airport no-fly zone restriction*	• *5% inconvenience from 1-week period for attaining flight permit* • *5% noise and visual disturbance* • *10% landing approval*

Philippines

Fare (PHP)	Utilization rate	60%	80%	95%
30,000		1%	5%	7%
15,000		5%	10%	25%
5,000		10%	25%	45%

Thailand

Fare (THB)	Utilization rate	60%	80%	95%
30,000		1%	10%	20%
15,000		5%	20%	30%
5,000		15%	30%	60%

Fig. 2 Usage rate assumption matrices

which is the current usage rate for helicopter operations as noted by our interviewees (Advanced Aviation Corp. and AirTaxi PH), a midway point of 80%, and an assumed optimum of 95%. These three percentages, along with the three assumed fares for each country, aid in painting nine separate scenarios of service availability depicted as usage rates.

The estimated percentages of the usage rates are more conservative for the Philippines than those for Thailand because of the following considerable factors. (1) The Philippines has a smaller percentage of wealthy population. (2) Safety is of utmost concern for consumers as more interviewees from the Philippines, especially non-regulation-related interviewees, have stressed that in our interviews.

3 Results

3.1 *EMS Use Case*

3.1.1 Comparison of Current Aeromedical Transport Services Between Japan and Thailand

The comparison of the current aeromedical transport services in Japan and those in Thailand is shown in Table 4. In Japan, municipal fire departments are responsible for emergency transport based on the Fire and Disaster Management Organization Act. For aeromedical transport services, DHs are mainly and commonly used with 29,055 missions dispatched (primary missions for treatment in the field: 19,938 cases; secondary missions for inter-facility transport: 5,000 cases) in FY2018 [19]. Moreover, 45 of 47 prefectures operate DHs dedicated to medical transport as of May 2021 [20]. If the helicopter could not be dispatched because of aircraft unavailability

Table 4 Comparison of the current aeromedical transport services in Japan and Thailand

	Japan	Thailand
Responsibility of transporting emergency patients	Municipal fire departments	–
Aeromedical transport services	1. Civil aeromedical service (Free) – Prefectural aeromedical service # *doctor heli* # Firefighting helicopter service – Private aeromedical service # Fixed-wing aircraft service 2. Exceptional case in emergency (Free) # Helicopters of Police, Japan Coast Guard, Self-Defense Force	1. Private aeromedical service (Paid) # Hospital-based # Non-hospital-based 2. Public aeromedical service (Free) # Thai Sky Doctor Service # Military aeromedical service

or limitation of the aircraft performance, the prefectural government's firefighting helicopter is dispatched.

However, in Thailand, the central or municipal governments are not responsible for transporting emergency patients. Moreover, 80%–90% of emergency patients still rely on friends, relatives, noncertified persons, or bystanders as of 2016 [21]. HEMS has been provided in Thailand since 2007 in the private sector and since 2010 in the public sector; however, the services are not as common as those in Japan with 302 private missions in 2011 [22], and the average of 41 requested public cases per year between 2010 and 2015 [21]. The private HEMS is a paid service called "BDMS Sky ICU," provided by Bangkok Dusit Medical Service. The public HEMS is a free service called "Thai Sky Doctor Service." The service was initiated to improve the accessibility of health services for emergency patients in rural and remote areas of Thailand. For both services, not only helicopters but also fixed wing aircraft are used, unlike Japan where helicopters are mainly used.

3.1.2 Challenges for Private/Public HEMS and Requirements for eVTOLs

The main current challenges using helicopters and main requirements for two-seater eVTOLs for private HEMS in Thailand were as follows, based on our interview with the personnel of Samitivej Sukhumvit Hospital.

- The main challenge is that private HEMS helicopters, unlike public HEMS helicopters, cannot land anywhere, especially in Bangkok. This restriction allows hospitals to fly 2–3 times of 10–12 missions per month in the city.
- Doctor dispatch by a compact eVTOL would be useful considering the short range required, where only 10 km is needed to beat traffic jams within the city.

- Noise is mainly not a challenge with high-rise buildings; however, it is a difficulty with ground landings. For eVTOLs, there is a need for diplomacy with local residents, considering that frequency will determine the magnitude of the challenge.

Next, the main current challenges using helicopters and main requirements for two-seater eVTOLs for public HEMS in Thailand were as follows, based on our interview with the doctor of NIEM.

- No transport is provided in case the aircraft is unavailable, since there is no dedicated aircraft for public air ambulance service operated under MOU with aircraft available only.
- Government's support with high cost for helicopters. High cost is also a challenge for private HEMS.
- Inability to operate at night, which is also the case for private HEMS.

- Doctor dispatch by a low-cost eVTOL would be useful if it can fly 100 km (100 km/h), though current HEMS helicopters often fly 600 km to reach patients in rural areas.
- Exterior noise must be as silent as helicopters. In-flight noise must be as silent as a car to ensure conversations onboard.

3.2 Sightseeing/Leisure and UAM Use Cases

The perceptions of multiple stakeholders on challenges surrounding the implementation of eVTOLs within the specified locations were evaluated using market- and operation-related factors, regulation-related factors, short-term specification-related factors, and long-term specification-related factors using QSPM (Table 5).

3.2.1 Market- and Operation-Related Factors

The following were considered when assuming the scores for the locations:

- Market size: Operators in Manila and Chiang Mai have noted that their clientele for the respective use cases comprises a small percentage of their targeted populations.
- Noise and visual disturbance: There are minimal concerns regarding noise, the same for Chiang Mai, where there are concerns regarding noise and visual pollution disrupting resort ambience.
- Availability of infrastructure: Manila has numerous helipads located within major financial districts.

The results suggest that the UAM use case in Bangkok is the most optimal application of eVTOLs from the market and operation perspectives as their perceived market size, compared with those in other locations and their respective use cases, raises the score.

3.2.2 Regulation-Related Factors

The following were considered when assuming the scores for the locations:

- Gaining flight/landing approval: Processes take more than 1 week for most locations. Bangkok has the lowest applicability because of military-controlled airspace.
- Attitudes toward eVTOL: The Civil Aviation Authority of the Philippines is more open toward negotiating eVTOL regulations and testing in provincial areas, using current regulations for aerial recreation aircraft as the basis.

The results suggest that the Sightseeing/Leisure use case in Cebu is the most optimal application of eVTOLs considering regulatory limitations and concerns,

Table 5 Quantitative Strategic Planning Matrix (QSPM)

Factors		The Philippines				Thailand				Highest score	
	Weight	Manila: UAM		Cebu: Sightseeing/Leisure		Bangkok: UAM		Chiang Mai: Sightseeing/Leisure			
		APS	TAPS	APS	TAPS	APS	TAPS	APS	TAPS	APS	TAPS
Perception of market size	0.3	3	0.9	5	1.5	5	1.5	3	0.9	7	2.1
Ease of acquiring skilled HR	0.1	5	0.5	5	0.5	4	0.4	4	0.4	7	0.7
Noise and visual disturbance	0.3	6	1.8	5	1.5	5	1.5	4	1.2	7	2.1
Availability of infrastructure	0.1	6	0.6	4	0.4	5	0.5	4	0.4	7	0.7
Operation cost	0.2	3	0.6	4	0.8	5	1	4	0.8	7	1.4
Sum weight (market- and operation-related)	1	**4.4**		**4.7**		**4.9**		**3.7**		**7**	
Ease in gaining flight/landing approval	0.15	6	0.9	5	0.75	1	0.15	6	0.9	7	1.05
Government's attitude to novel technology	0.4	3	1.2	4	1.6	2	0.8	2	0.8	7	2.8
Government's attitude to negotiating new regulations for eVTOLs	0.25	2	0.5	6	1.5	3	0.75	3	0.75	7	1.75
Ease of gaining political backing	0.2	4	0.8	5	1	3	0.6	3	0.6	7	1.4
Sum weight (regulation-related)	1	**3.4**		**4.85**		**2.3**		**3.05**		**7**	

(continued)

Table 5 (continued)

Factors		The Philippines				Thailand				Highest score	
	Weight	Manila: UAM		Cebu: Sightseeing/Leisure		Bangkok: UAM		Chiang Mai: Sightseeing/Leisure			
		APS	TAPS	APS	TAPS	APS	TAPS	APS	TAPS	APS	TAPS
Range (<10 km)	0.5	4	2	1	0.5	5	2.5	0	0	7	3.5
Flight time (20 min)	0.4	3	1.2	4	1.6	4	1.6	1	0.4	7	2.8
Size (4.3 × 3 × 2 m)	0.1	7	0.7	7	0.7	6	0.6	6	0.6	7	0.7
Sum weight (specification-related (short term))	1	**3.9**		**2.8**		**4.7**		**1**		**7**	
Range (30–100 km)	0.5	7	3.5	7	3.5	7	3.5	7	3.5	7	3.5
Flight altitude (150–500 m)	0.4	5	2	5	2	5	2	7	2.8	7	2.8
Noise (1/2 compared with heli)	0.1	7	0.7	5	0.5	7	0.7	5	0.5	7	0.7
Sum weight (specification-related (long term))	1	**6.2**		**6**		**6.2**		**6.8**		**7**	
Sum total applicability score		**17.9**		**18.35**		**17.3**		**14.55**		**28**	

(*Note*) APS: applicability scores; TAPS: total applicability scores as the Civil Aviation Authority of the Philippines has recognized that the environment within provincial areas in the Philippines is forgiving in terms of testing and experimentation.

3.2.3 Specification-Related Factors (Short Term)

The following were considered when assuming the scores for the locations:

- Range and flight time: Shorter ranges are more easily applicable to city-based locations. However, sightseeing in Chiang Mai requires flight time for at least 20 min to access major sightseeing spots.

The results suggest that the UAM use case in Bangkok is the most optimal application of eVTOLs considering the chosen short-term specifications, as the environment within the city allows eVTOLs to work within limitations of these specifications.

3.2.4 Specification-Related Factors (Long Term)

The following were considered when assuming the scores for the locations:

- Flight altitude: Chiang Mai is forgiving with how low pilots can fly in the area. Helicopters can go as low as 150 m for sightseeing flights.
- Noise: Helicopter operators complain of noise in island-sightseeing locations, which may pose a general concern for operation in those locations.

The results suggest that the Sightseeing/Leisure use case in Chiang Mai is the most optimal application of eVTOLs considering the chosen long-term specifications.

Considering the summary of information acquired from the interviews, the overall evaluation was finally made. Overall, the Sightseeing/Leisure use case in Cebu appears to be the most applicable use case and location for eVTOLs among the use cases and locations examined in this study.

3.3 *Fermi Estimate for the Sightseeing/Leisure and UAM Use Cases*

The number of sales generated for each location and use case is shown in JPY in Table 6. The conversion rate was based on Morningstar currency data on July 2, 2020 (1 PHP = 2.16 JPY; 1 THB = 3.46 JPY).

Considering the estimation, Bangkok has the highest market potential based on the estimated demand of the four locations. With all assumptions considered, having a utilization rate of 95% at a fare of 30,000 THB would generate the most sales of the four locations.

Table 6 Comparison of sales for each use case

Usage rate (PH) (%)	Usage rate (TH) (%)	Utilization rate (%)	Fare (PHP/THB)	UAM Manila sales (JPY)	Sightseeing /Leisure Cebu sales (JPY)	UAM Bangkok sales (JPY)	Sightseeing/Leisure Chiang Mai sales (JPY)
1	1	60	30,000	7,910,136	12,223,872	36,397,678	2,179,800
5	10	80	30,000	52,734,240	81,492,480	485,302,368	29,064,000
7	20	95	30,000	87,670,674	135,481,248	1,152,593,124	69,027,000
5	5	60	15,000	19,775,340	30,559,680	90,994,194	5,449,500
10	20	80	15,000	52,734,240	81,492,480	485,302,368	29,064,000
25	30	95	15,000	156,554,777	241,930,800	864,444,843	51,770,250
10	15	60	5,000	13,183,560	20,373,120	90,994,194	5,449,500
25	30	80	5,000	43,945,200	67,910,400	242,651,184	14,532,000
45	60	95	5,000	93,932,867	145,158,480	576,296,562	34,513,500

(*Note*) PH: the Philippines; TH: Thailand; JPY: Japanese Yen

4 Discussion

For the EMS use case in Thailand, in terms of one-way range, 10 km and 100 km are required for private and public EMS, respectively, based on our interview results. However, considering the specifications currently proposed for two-seater multirotor eVTOLs, one-way range is likely to be limited to approximately 27–35 km for approximately 100-km/h cruise speed, with 27 km and 102 km/h proposed for Volocopter 2X [23] and 35 km and 100–130 km/h for EHang 216 [24]. Thus, two-seater multirotor eVTOLs could be applied to private EMS, but unlikely public EMS, in Thailand considering the requirements from doctors.

For the Sightseeing/Leisure and UAM use cases, considering the evaluation of stakeholders and the demand estimation for locations from both countries, the Philippines may be a better initial testing ground for eVTOL implementation; however, Thailand may be better for implementation in the long term as it has a larger market demand and higher profitability than the Philippines in both use cases considered.

Three factors are largely considered when implementing eVTOLs within the Philippines and Thailand: the attitudes of regulatory bodies, fare, and percentage of wealthy population/tourists in the area. The first factor affects the availability of eVTOL services within the region. The latter two factors mainly affect the existing market demand and profitability of the use cases.

In the Philippines, the country has higher applicability scores because of its major advantage in how authorities are more open to the discussion of eVTOLs, especially when it comes to the Sightseeing/Leisure use case. The eVTOL technology can be tested through entertainment/aerial recreation use cases because of the lower regulations for the use case. The attitude of regulatory bodies toward specific use cases and locations for operation affect the availability of services provided. This can be observed in the existing case for helicopters and in challenges for eVTOLs. Regulators and operators from both countries have hesitated when testing out UAM because of safety concerns from the density of cities. However, this can be circumvented through the Sightseeing/Leisure use case.

In Thailand, it has a greater advantage in its higher demand from its wealthier population. However, unless safety can be proven with eVTOL technology, the demand could still be undermined by the strictness of its aviation authorities.

5 Conclusion

This study aims to select and evaluate the promising use cases of multirotor eVTOLs in Philippine and Thai markets. As a result of evaluation and market estimate of high-potential eVTOL use cases, this study showed the following. For the EMS use case, HEMS hospitals in Thailand need two-seater eVTOLs to dispatch only a doctor to the field as in Japan; however, the feasibility of this use case is lower than that of other use cases considering the specifications required by doctors. Moreover, the

Sightseeing/Leisure use case in Cebu is best for eVTOL implementation, considering the perceptions of stakeholder, whereas the UAM use case in Bangkok is better for sales based on demand estimation. Comparing the two countries, the Philippines may serve as better initial testing grounds for eVTOL implementation; however, Thailand may be better for implementation in the long term as it has a larger market demand and higher profitability than the Philippines in both Sightseeing/Leisure and UAM use cases.

However, these findings remain limited. Future work can improve with the following points:

- Analysis on medical and cost benefits of applying two-seater eVTOLs to EMS in Thailand.
- Expansion of location scope. Stakeholders from other tourist or metropolitan areas in the Philippines and Thailand would help provide a better idea of the situation. Moreover, a similar research should be conducted in other Southeast Asian countries.
- Other constraints to be considered in the estimation. The estimates should be refined and made more accurate through the inclusion of other constraints, such as vertiports, and other regulatory constraints.

Acknowledgements This work was supported by SkyDrive, Inc. and Keio University Doctoral Student Grant-in-Aid Program.

References

1. Electric VTOL News. https://evtol.news/aircraft. Cited Aug 25, 2021
2. SkyDrive. https://en.skydrive2020.com. Cited Aug 25, 2021
3. Airbus (2017) Global Helicopter Forecast. https://www.airbus.com/content/dam/corporate-topics/publications/backgrounders/Global-Helicopter-Forecast-2017.pdf. Cited Aug 25, 2021
4. Booz Allen Hamilton (2018) Urban air mobility (UAM) market study. https://ntrs.nasa.gov/api/citations/20190000519/downloads/20190000519.pdf. Cited Aug 25, 2021
5. McKinsey & Company (2019) Urban air mobility (UAM) market study. https://ntrs.nasa.gov/api/citations/20190001190/downloads/20190001190.pdf. Cited Aug 25, 2021
6. Balac M, Rothfeld RL, Hörl S (2019) The prospects of on-demand urban air mobility in Zurich, Switzerland. In: Proceedings of the 2019 IEEE intelligent transportation systems conference (ITSC)
7. Xu Z, Zhou Z (2019) Energy-based conceptual design of battery-powered eVTOL aircraft for air ambulance services in China. In: Proceedings of the Asia Pacific international symposium on aerospace technology (APISAT)
8. Vascik PD, John Hansman R (2018) Scaling constraints for urban air mobility operations: air traffic control, ground infrastructure, and noise. In: Proceedings of the 2018 aviation technology, integration, and operations conference
9. Eker U, Ahmed SS, Fountas G, Anastasopoulos PC (2019) An exploratory investigation of public perceptions towards safety and security from the future use of flying cars in the United States. Anal Methods Accid Res 23:54–66

10. Samitivej Sukhumvit Hospital. https://www.samitivejhospitals.com/center/detail/samitivej-trauma-center. Cited Aug 25, 2021
11. National Institute for Emergency Medicine (NIEM). https://www.niems.go.th/1/SubWebsite/?id=1096. Cited Aug 25, 2021
12. Nakano M (2019) System design of flying cars: air mobility revolution through technology and services, Tokyo, Japan
13. David FR (1986) The strategic planning matrix—a quantitative approach. Long Range Plann 19:102–107
14. Albert JRG, Santos AGF, Vizmanos JFV (2018) Profile and determinants of the middle-income class in the Philippines. https://www.econstor.eu/bitstream/10419/211040/1/166320005X.pdf. Cited Aug 25, 2021
15. Poomontre J, Setthawong P (2016) An updated prediction system for the TMRS standardized socio-economic status (SES) classification of Bangkok and Metropolitan subjects. In: Proceedings of the 1st international conference on information technology, 2016
16. Weather Atlas. https://www.weather-atlas.com/en/philippines/manila-climate. Cited Aug 25, 2021
17. Weather Atlas. https://www.weather-th.com/en/thailand/bangkok-climate. Cited Aug 25, 2021
18. Thai Meteorological Department. https://www.tmd.go.th/en/climate.php?FileID=7. Cited Aug 25, 2021
19. Japanese Society for Aeromedical Services (2019) Data on doctor helicopters in Japan, FY2018. J Jpn Soc Aeromed Serv 20:54–55
20. HEM-Net. https://hemnet.jp/know-base. Cited Aug 25, 2021
21. Pangma A, Taneepanichskul S (2017) The outcome of emergency patient transported by public air ambulance service in Thailand. J Health Res 31
22. Suriyachaisawat J, Surakarn E (2013) The characteristics of aeromedical transport missions at Bangkok Hospital, Thailand. Bangkok Med J 6
23. Electric VTOL News. Volocopter 2X (defunct). https://evtol.news/volocopter-2x/. Cited Aug 25, 2021
24. Electric VTOL News. EHang 216. https://evtol.news/ehang-216/. Cited Aug 25, 2021

Numerical Investigation of Droplet Impact on the Surface by Multiphase Lattice Boltzmann Flux Solver

Qingyong Bian, Chang Shu, Ning Zhao, Chengxiang Zhu, and Chunling Zhu

Abstract The dynamic behaviors of the micro-sized water droplet collision onto the wings of the aircraft are essential to the flight safety. The details on the small droplet in the airflow in contact with the aircraft wing surface play a quite important role in the ice accretion process. In this paper, multiphase lattice Boltzmann flux solver coupled with phase field method is applied to simulate the water droplet impact onto the solid hydrophilic/hydrophobic surface to further understand the interactions between droplet and surface at mesoscopic level. The reliability and accuracy of the numerical method is validated by the comparison with experimental data and computational results in other literatures, which shows that the solver is capable of predicting the droplet dynamic behaviors. Then, the effects of different physical parameters such as impact velocity, droplet diameter, surface contact angle and impact inclination angle, are systematically studied. The computational results reveal that when the collision is normal to the surface, the water droplet may experience spreading phase, recoiling phase as well as rebounding phase and finally shows the adhesion state or detachment from the surface. The higher velocity and larger diameter contribute to spread the droplet wider and jump higher during the droplet impact process. And a shorter physical time is taken to reach the spreading factor maximum for higher velocity while it is opposite for the droplet with a lager diameter. Moreover, the whole evolutionary process of smaller-sized droplet is accelerated and smaller diameter as well as higher contact angle of the surface advances the droplet detachment from the hydrophobic surface. It is also found that the surface with higher contact angle impedes the droplet spreading and removes the temporal lag of its performance in lifting up the upper end of droplet during recoiling phase and rebounding phase, which is distinct to the results of higher velocity and larger diameter. Besides this,

Q. Bian · N. Zhao · C. Zhu · C. Zhu (✉)
College of Aerospace Engineering,
Nanjing University of Aeronautics and Astronautics, Nanjing 210016, Jiangsu, China
e-mail: clzhu@nuaa.edu.cn

State Key Laboratory of Mechanics and Control of Mechanical Structures, Nanjing University of Aeronautics and Astronautics, Nanjing 210016, Jiangsu, China

C. Shu
Department of Mechanical Engineering, National University of Singapore, Singapore 119260, Singapore

S. Lee et al. (eds.), *The Proceedings of the 2021 Asia-Pacific International Symposium on Aerospace Technology (APISAT 2021), Volume 1*, Lecture Notes in Electrical Engineering 912, https://doi.org/10.1007/978-981-19-2689-1_52

droplet impact with an inclination angle causes reduction on the spreading factor maximum and jump height after detachment from the surface due to the decrease on the normal velocity of the droplet. And the increase of the tangential velocity accounts for the longer contact time with the surface for the droplet, and causes the difference of the spreading factors in spreading directions, which forms an oval contact area on the surface until the droplet detaches. The analysis and quantitative comparison of the temporal morphology evolutions of the micro-sized droplet in this paper help to reveal the interaction mechanism between the different-sized droplets and surfaces with different properties, which can be considered specially in the numerical prediction of the aircraft icing.

Keywords Lattice Boltzmann method · Large density ratio · Droplet impact · Rebound and adhesion

1 Introduction

When the aircraft passes through the clouds, the supercooled micro-scale water droplets suspended in the air will impact onto the wing surfaces of the aircraft and they may show various collision dynamic behaviors [1]. During this process, freezing could be triggered and changes the geometric shape of the wing. Thus, the aerodynamic characteristics would deviate from the original design [2]. An accurate description of the droplet behaviors would help to predict the iceshapes and evaluate the deterioration of the aerodynamic performance as well as provide theoretical foundations for the anti-icing techniques to ensure the flight safety.

On the experimental research, some works have been carried out [3–7]. Different-sized droplets at various velocities impact on the smooth substrates with contact angles ranging from 6° to 107° were performed by Dong et al. [8]. Liang et al. [9] investigated droplet impact on the wetted inclined surfaces to illustrate the spreading and splashing processes. The numerical simulation is able to provide more detailed information [10–14]. The droplets with diameters of millimeters at low velocities were simulated by volume of fluid method where different contact angle models were used [15]. Lee and Liu [16] proposed a lattice Boltzmann method (LBM) to model contact line dynamics with Reynolds number up to 685 and Weber number up to 103. Besides this, much attention has been paid to the surface chemical inhomogeneity which accounts for the contact angle. A lattice Boltzmann solver with phase field method was employed to analyze the role of the surface wettability distribution with a gradient [17] and interactions of two droplets collision onto the dry solid wall were further investigated [18].

Despite the studies performed by the previous researches, more detailed descriptions of the interplay among the droplet, air and surface should be given. LBM is an effective as well as efficient tool which has been extensively employed in multiphase flows [19–24] and the improved multiphase lattice Boltzmann flux solver (MLBFS) developed by Wang et al. [13, 14] performs well in many challenging complex flows. The aim of this paper is to study the effects of the impact factors on the droplet dynamic behaviors in wider ranges with MLBFS to help optimize the practicability in engineering applications.

The rest of the paper is organized as follows. The numerical method for flow field and interface capturing is introduced in Sect. 2. And numerical validation is implemented with quantitative comparisons in Sect. 3. The simulation results and discussions on the effect of impact velocity, droplet size, surface contact angle and impact inclination angle are presented in Sect. 4. Finally, the conclusions are drawn in Sect. 5.

2 Numerical Method

For computation of the flow field and distinguishing different fluids, the improved multiphase lattice Boltzmann flux solver [14] and the phase field method are applied. The vectorized variables to be solved could be written as $\left(p, \rho u c_s^2, \rho v c_s^2, \rho w c_s^2, C\right)^T$ while the governing equations are given in the finite-volume unified form:

$$\frac{\partial}{\partial t}\begin{pmatrix} p \\ \rho u c_s^2 \\ \rho v c_s^2 \\ \rho w c_s^2 \\ C \end{pmatrix} = -\frac{1}{\Delta V_i}\sum_k \begin{pmatrix} n_x \sum\limits_{\alpha=0}^{N} e_{\alpha x} f_\alpha^{eq} + n_y \sum\limits_{\alpha=0}^{N} e_{\alpha y} f_\alpha^{eq} + n_z \sum\limits_{\alpha=0}^{N} e_{\alpha z} f_\alpha^{eq} \\ n_x \sum\limits_{\alpha=0}^{N} e_{\alpha x} e_{\alpha x} f_\alpha^{\wedge} + n_y \sum\limits_{\alpha=0}^{N} e_{\alpha x} e_{\alpha y} f_\alpha^{\wedge} + n_z \sum\limits_{\alpha=0}^{N} e_{\alpha x} e_{\alpha z} f_\alpha^{\wedge} \\ n_x \sum\limits_{\alpha=0}^{N} e_{\alpha y} e_{\alpha x} f_\alpha^{\wedge} + n_y \sum\limits_{\alpha=0}^{N} e_{\alpha y} e_{\alpha y} f_\alpha^{\wedge} + n_z \sum\limits_{\alpha=0}^{N} e_{\alpha y} e_{\alpha z} f_\alpha^{\wedge} \\ n_x \sum\limits_{\alpha=0}^{N} e_{\alpha z} e_{\alpha x} f_\alpha^{\wedge} + n_y \sum\limits_{\alpha=0}^{N} e_{\alpha z} e_{\alpha y} f_\alpha^{\wedge} + n_z \sum\limits_{\alpha=0}^{N} e_{\alpha z} e_{\alpha z} f_\alpha^{\wedge} \\ n_x u C + n_y v C + n_z w C \end{pmatrix}_k \Delta S_k + \begin{pmatrix} \boldsymbol{u}\cdot\nabla\left(\rho c_s^2\right) \\ F_{sx} c_s^2 \\ F_{sy} c_s^2 \\ F_{sz} c_s^2 \\ M\nabla^2 \mu_C \end{pmatrix} \tag{1}$$

where ρ is the fluid density, $\boldsymbol{u}$ is the velocity, p is the pressure, $\boldsymbol{F_s}$ is the surface tension force, c_s represents the speed of sound which equals to $1/\sqrt{3}$ in the simulation, $\boldsymbol{e_\alpha}$ indicates the particle velocity in the α direction and $f_\alpha^{\wedge}$ is to evaluate the flux between the adjacent control volume. C indicates the volume fraction and M is mobility which is a constant 0.01 in the following simulations while μ_C represents the chemical potential. And ΔV_i denotes the volume of local cell, ΔS_k is the area of the kth surface enclosing Ω_i and $\boldsymbol{n}$ is the unit outer vector normal to the surface.

In order to maintain the numerical stability in the updating of variable C, the convective term is discretized by finite difference scheme of 5th-order WENO algorithm [25], and the temporal evolution technique adopted in the current work is 3rd-order TVD Runge–Kutta scheme [26].

3 Numerical Validation

A single droplet collision on the hydrophobic flat surface is simulated and the result is compared to the experimental data [8] to test the accuracy of the numerical method. Figure 1 describes the problem setup and the factors affecting the dynamic behaviors of droplet.

U is the impact velocity of the droplet, D is the initial diameter, ρ_L and μ_L are liquid density and dynamic viscosity while σ is the surface tension between the liquid and gas phase. In consistent with the experiment [8], the corresponding numerical parameters in the present case are: $U = 0.02, D = 75, \rho_L = 1.0, Re = 689$, the density ratio $\rho_L/\rho_G = 841.8$, the dynamic viscosity ratio $\mu_L/\mu_G = 51$, $We = 103$ and the static contact angle (CA) of the flat surface θ is 107° while the impact angle between the impact velocity and vertical direction of the surface α is 0°. In this validation case, one-fourth of the droplet is computed where a computational domain size of 151 × 151 × 151 is employed and the open boundary conditions are implemented for lateral boundaries as well as the top boundary while the bottom wall is treated by applying nonslip boundary condition.

The spreading factor D^* and dimensionless height H^* are scaled by droplet initial diameter. The definition of D^* is specifically referred to the root diameter of the droplet in contact with the solid surface in this paper while another analogous indicator $Dmax^*$ defined by the instant maximal spreading diameter [12] is introduced in the next section. Figure 2 depicts the evolutions of predicted D^* and H^* where experimental data [8] and numerical results obtained in previous studies [12, 14] are also displayed.

It is evident that the predicted D^* and H^* match well with the experimental results. Although a slight discrepancy appears in the late stage of the dimensionless height evolution, in general, the computational result is satisfactory and the numerical method is accurate and reliable.

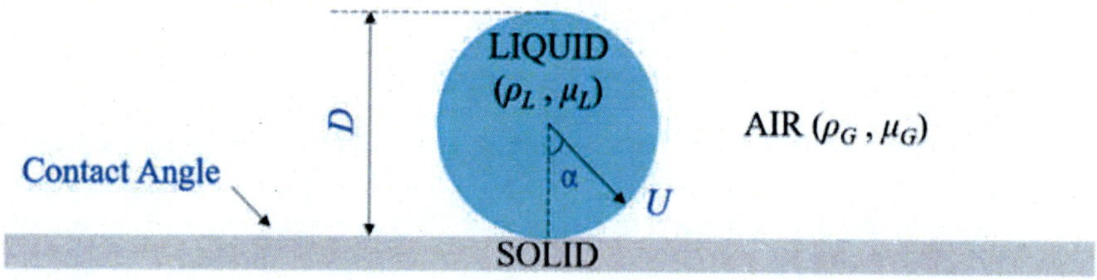

Fig. 1 Schematic representation of the droplet collision and impact factors

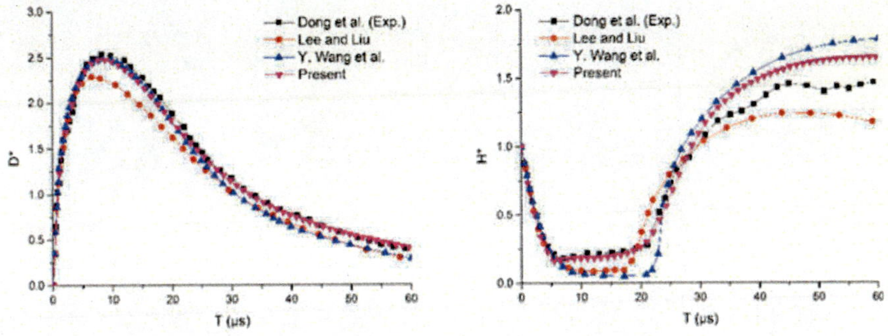

(a) Spreading factor (b) Dimensionless height

Fig. 2 Comparison of spreading factor and dimensionless height of the droplet to reference data

4 Results and Discussion

In this study, the single droplet impinging onto the solid surface is numerically investigated, which is affected by the velocity U and diameter D of the droplet, contact angle θ and the droplet impact inclination angle α. In the following, the effects of these factors will be illustrated. The physical properties of the fluids are specified as follows: $\rho_L = 1000\ \text{kg/m}^3$, $\rho_L/\rho_G = 841.8$, $\mu_L = 0.000894$ Pa·s, $\mu_L/\mu_G = 51$, and $\sigma = 0.073$ N/m. For the case of $\alpha = 0°$, the computational domain only takes one-fourth of the whole physical domain while half of the whole domain is used as the computational domain for $\alpha \neq 0°$.

4.1 *Effect of Impact Velocity*

In this subsection, droplets with distinctive velocities are considered. The droplet size D, the contact angle θ and impact angle α are fixed as: $D = 50\ \mu\text{m}$, $\theta = 120°$ and $\alpha = 0°$ while the velocity U varies from 5 to 20 m/s with an increment of 5 m/s. The evolutions of the droplet shape are presented in Fig. 3.

It can be seen that all the droplets experience the phase of spreading, recoiling, rebounding and relaxation. In the initial stage of the impact, in which the inertial force firstly works, the droplet is squeezed to a flat shape in different level based on the impact momentum brought to the surface. As shown at $t = 18\ \mu\text{s}$, the dominance has already been altered to the surface tension, which competes with viscous force as well as inertial force, and manages to pull the rim of the droplet back after reaching the maximum spreading factor and proceeds to lift up the crater in the center of the droplet. In the later stage, droplet transits from recoiling phase to rebounding phase as depicted at $t = 38\ \mu\text{s}$ and during rebounding motion, the inertial force and surface tension take control in the upper and lower part of the droplet, respectively. The droplet fully rebounds to the air early or late as observed at $t = 68\ \mu\text{s}$. After the

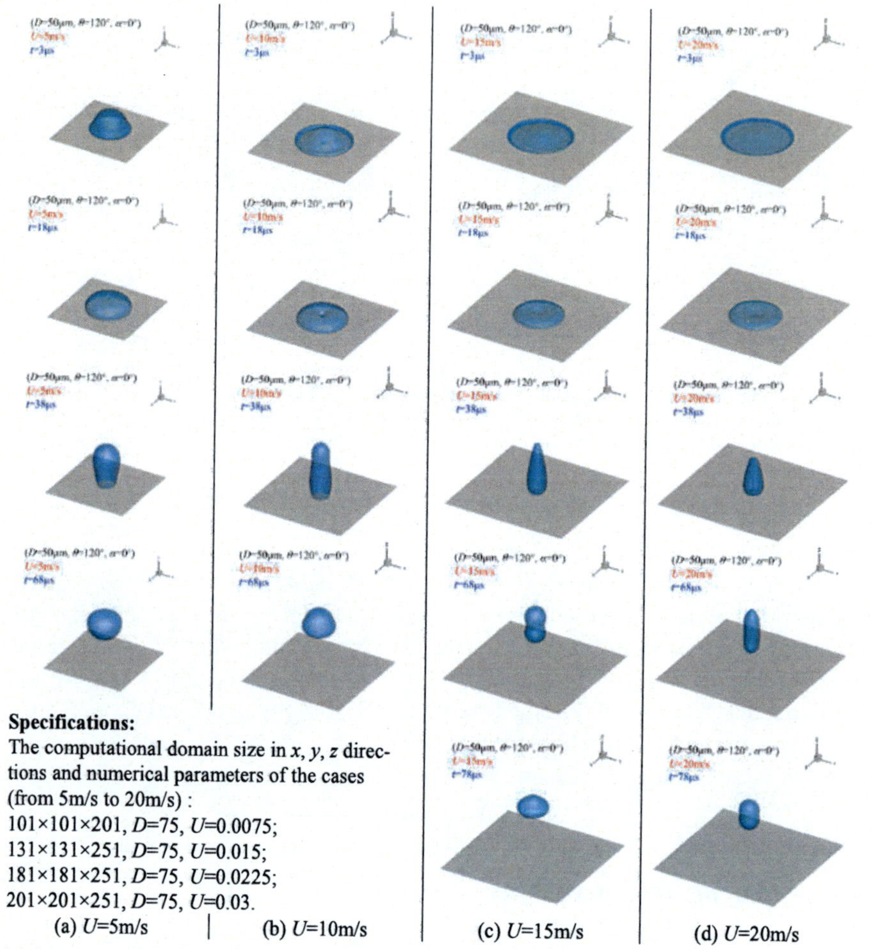

Fig. 3 Temporal evolutions of droplet collision onto the flat hydrophobic surface with different velocities

surface tension hauls up the lower part of the droplet and usually outperforms, the droplet is crushed and then relaxes itself with the dissipation of the energy.

The spreading factor D^* and dimensionless height H^* for each case are plotted in Fig. 4a, b. It should be illustrated that $Dmax^*$, indicating the instant maximal diameter of the droplet in the spreading phase, likely not identical to D^*, is also displayed in Fig. 4a while $Hlow^*$, indicating the lower position of the droplet, is shown in Fig. 4b as well. From Fig. 4a, it is found that the maximums of D^* and $Dmax^*$ increase with the impact velocity while the time taken to reach the maximum decreases. This implies that in spreading phase, it takes shorter time for droplet with larger velocity to spread wider. In the rebounding phase, as observed in Fig. 4b, the droplet with larger velocity does not jump higher all the time, but reaches higher position later

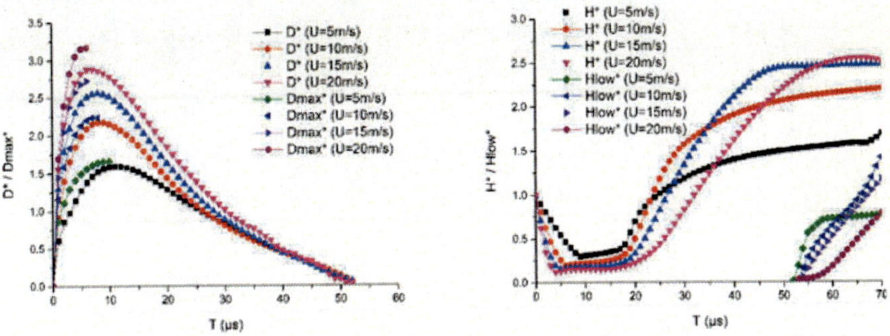

(a) Spreading factor / Maximal spreading factor

(b) Upper and lower positions of the droplet

Fig. 4 Comparison of spreading factor and droplet position when impact velocity varies

and the lower part of the droplet with lower velocity lifts up faster after the droplet just detaches.

4.2 Effect of Droplet Diameter

Next, we focus on the effect of the droplet diameter and 20 μm, 50 μm as well as 100 μm are considered. The velocity U, the contact angle θ and inclination angle α are fixed as: $U = 10$ m/s, $\theta = 120°$ and $\alpha = 0°$ and the corresponding numerical parameters in simulations for 20 μm-sized and 100 μm-sized droplets are: $D = 50$, $U = 0.0125$ and $D = 75$, $U = 0.015$ while a computational domain size of 131 × 131 × 201 and 181 × 181 × 251 are applied, respectively. The simulated morphological evolutions of the droplets are shown in Figs. 5 and 6.

With the numerical result of 50 μm-sized droplet shown in Fig. 3b, it is obvious that three droplets form thin-disk shapes with varying degrees in spreading phase and they all jump out of the surface. The duration time they take and their shapes are of great difference. For the larger droplet, it takes a longer time to jump with a slenderer morphology. As seen from Fig. 3b, the droplet shows nearly a cylinder shape at $t = 38$ μs while the droplet of 20 μm diameter evolves to a stout cylinder at $t = 9$ μs as displayed in Fig. 5 and the 100 μm droplet jumps out of the surface with a sharper protrusion tip at $t = 98$ μs and forms a dumbbell-like shape at $t = 164$ μs as indicated in Fig. 6. Besides this, the elongation caused by the surface tension which recollects inertial energy incurs severer squeeze for larger droplet and the relaxation phase continues much longer as observed in Fig. 6.

In Fig. 7a, b, the spreading factor D^*, $Dmax^*$ and dimensionless height H^* as well as lower position of the droplet $Hlow^*$ are plotted. Because the overall time for different-sized droplets differs even up to one order of magnitude, the non-dimensional time T^* is introduced and defined as: $T^* = Ut/D$. From Fig. 7a, the

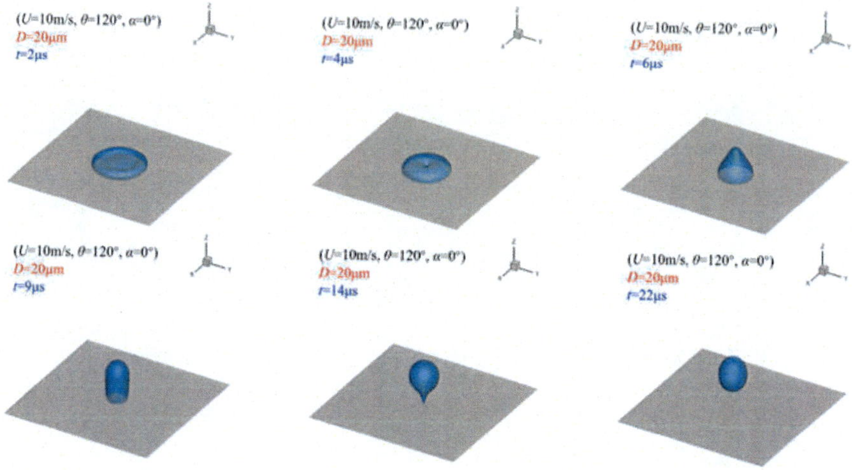

Fig. 5 Temporal evolution of 20 μm-sized droplet collision onto the flat hydrophobic surface

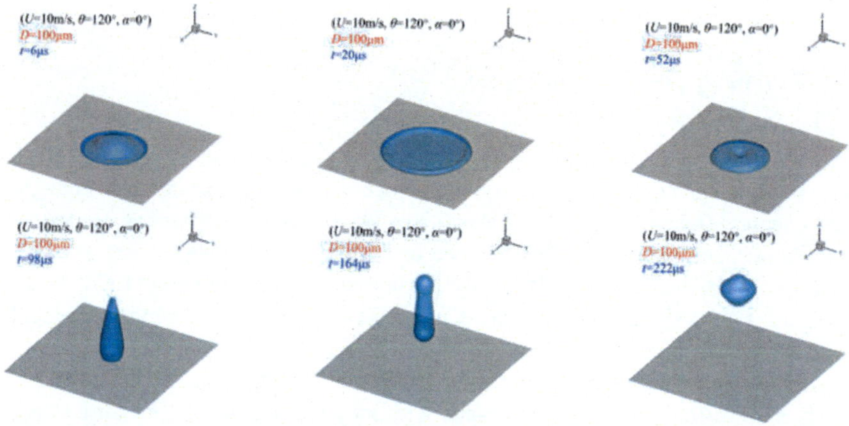

Fig. 6 Temporal evolution of 100 μm-sized droplet collision onto the flat hydrophobic surface

droplet with a larger diameter could spread much wider than small-sized droplet and it takes longer time to reach its maximum spreading factor. The spreading factors mount to approximately 1.67, 2.17 and 2.65 at $T^* = 1.5$, 1.6 and 2.2 corresponding to $t = 3$ μs, 8 μs and 22 μs, respectively. Additionally, the larger-sized droplet jumps fully out of the surface at a much later time. This is also confirmed by $Hlow^*$ in Fig. 7b. It should be illustrated that at the same impact velocity, the droplet with a larger diameter is squeezed to be a flatter disk and thus, it requires a longer time to jump up to a higher position. Furthermore, the whole evolution process of small-sized droplet seems to accelerate as compared to larger-sized droplets, which is obviously shown in Fig. 7b where the 20 μm-sized droplet has already entered the relaxation phase while larger droplets are still in rebounding phase or recoiling phase.

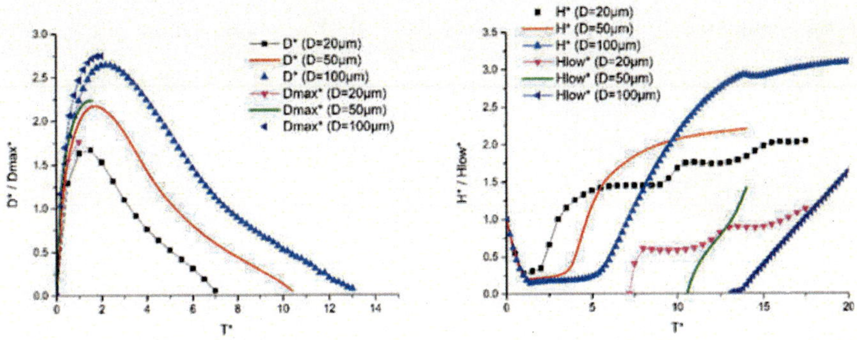

(a) Spreading factor / Maximal spreading factor

(b) Upper and lower positions of the droplet

Fig. 7 Comparison of different-sized droplet spreading factor and position

4.3 *Effect of Contact Angle*

In this subsection, the effect of the surface contact angle is illustrated. The cases of $\theta = 60°, 90°, 120°$ and $135°$ are simulated with other parameters fixed as: $U = 10$ m/s, $D = 50\ \mu$m and $\alpha = 0°$ which correspond to $U = 0.015$, $D = 75$ in computation and the computational domain size of $151 \times 151 \times 151$ is employed for $\theta = 60°$ and $90°$ cases while that of $131 \times 131 \times 251$ is adopted for $\theta = 135°$ case.

The obtained numerical results of the droplet interface are described in Figs. 8, 9 and 10, respectively. In Fig. 8, because the contact angle of the surface is small, the

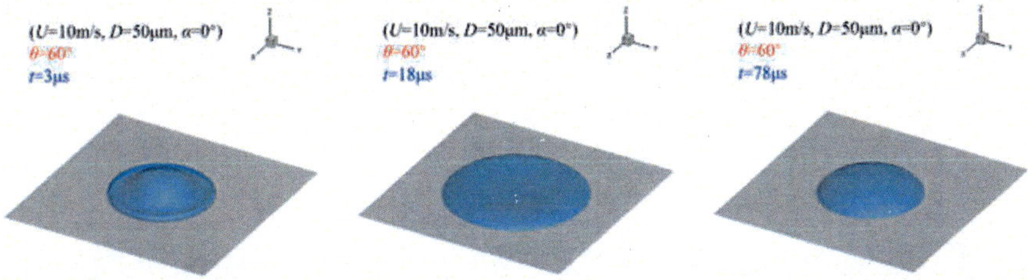

Fig. 8 Temporal evolution of droplet collision onto the flat hydrophilic surface ($\theta = 60°$)

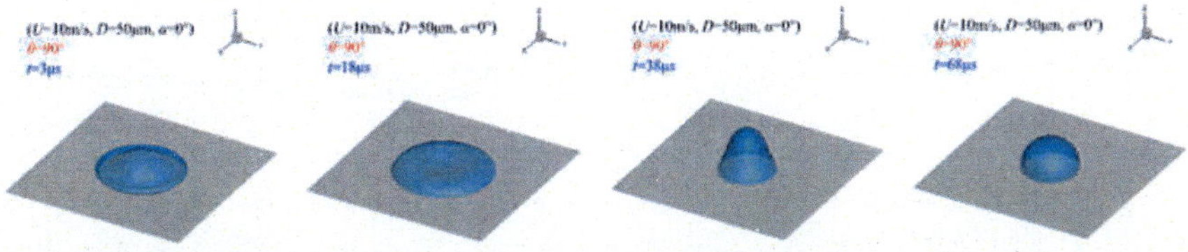

Fig. 9 Temporal evolution of droplet collision onto the flat neutral surface ($\theta = 90°$)

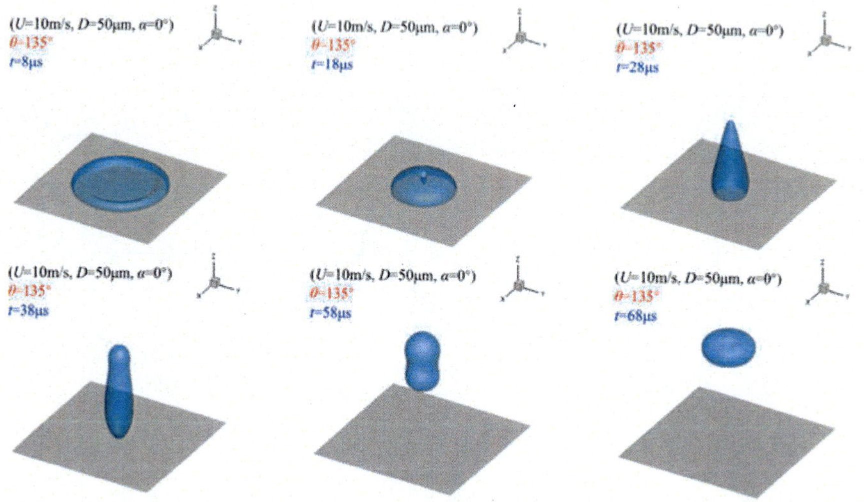

Fig. 10 Temporal evolution of droplet collision onto the flat hydrophobic surface ($\theta = 135°$)

droplet only experiences the spreading phase and recoiling phase, during which the droplet retracts slowly and tends to the equilibrium state. When the surface contact angle increases to 90°, the upper part of droplet seems to lift up higher at $t = 38$ μs but descends later as observed at $t = 68$ μs in Fig. 9. However, when the surface contact angle varies to hydrophobic range, as shown in Figs. 3b and 10, it is seen that droplets are able to fully jump out of the surface in slim shapes and for the surface with contact angle up to 135°, the droplet detaches earlier and forms a thinner disk-like morphology.

Figure 11a, b shows the evolutions of D^*, $Dmax^*$, H^*, and $Hlow^*$ for cases studied. The data of droplet impact on 120° contact angle surface is also displayed as a

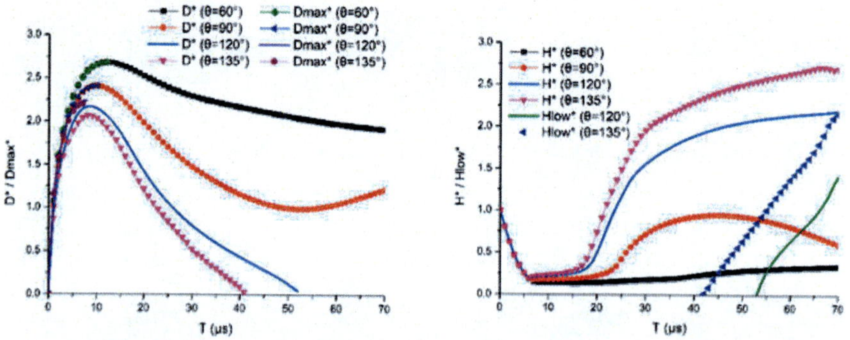

(a) Spreading factor / Maximal spreading factor

(b) Upper and lower positions of the droplet

Fig. 11 Comparison of droplet spreading factor and position when surface contact angle varies reference. The spreading factor history in Fig. 11a illustrates that droplet collision on a hydrophilic surface could spread much wider compared to that on a hydrophobic surface. And with the increase of the contact angle, the droplet spends less time to reach the spreading factor maximum. Nevertheless, the time gap is diminishing as the spreading factors peak almost simultaneously for the cases of 120° and 135° contact angles. It is also noteworthy that the curves of $Dmax^*$ nearly overlap with those of D^* for $\theta = 60°, 90°$ because the root of the droplet rim is able to catch up the spreading diameter which may be separated from the surface by air for hydrophobic case. Moreover, in the recoiling phase of the droplet impact on the neutral surface, re-spreading phenomenon appears which is also corroborated by the dimensionless height curve in Fig. 11b. Besides this, it is revealed in Fig. 11b that for the hydrophobic surface, the droplet is squeezed to be a thicker disk and it rebounds higher, which is evidently different from the previous discussions on the effect of velocity and diameter where there are temporal lags in lifting up the droplet.

4.4 *Effect of Impact Angle*

The above discussions focus on the normal collision of the droplet on the surface. In order to illustrate the flow pattern of the oblique droplet collision, the case of $\alpha = 45°$ is examined with other impact conditions fixed as: $U = 10$ m/s, $D = 50$ μm, and $\theta = 120°$. In the simulation, a computational domain size of $105 \times 450 \times 211$ is adopted with numerical parameters set to be: $U = 0.015$ and $D = 75$. The morphological evolution of the droplet is demonstrated in Fig. 12.

It is noted that the droplet collision with an inclination angle also experiences the spreading phase, recoiling phase and rebounding phase like normal collision of

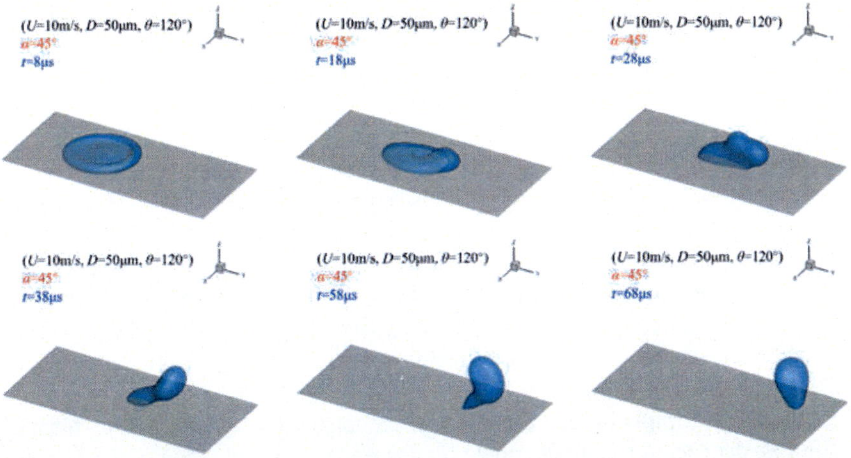

Fig. 12 Temporal evolution of droplet collision onto the hydrophobic flat surface when $\alpha = 45°$

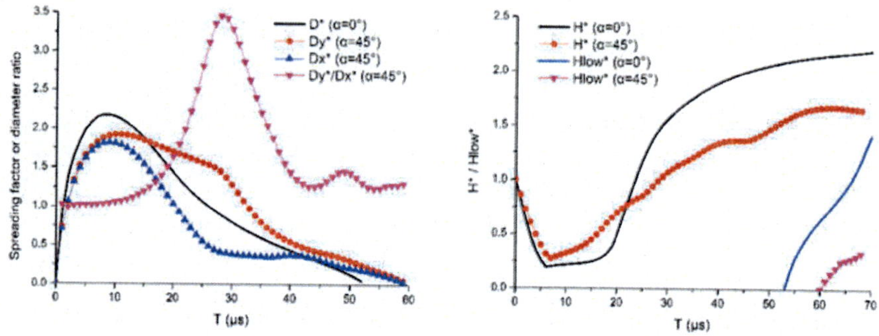

(a) Spreading factor along x and y direction and their ratio

(b) Upper and lower positions of the droplet

Fig. 13 Comparison of droplet spreading factor and position when impact inclination angle varies

the droplet. Due to the tangential velocity, the mass distribution of the droplet is not symmetric about xz plane and initial velocity pushes the mass onwards so that a thick rim is formed as shown at $t = 8$ μs. The surface tension in recoiling phase lifts up the crater in the droplet center. Under the influence of the surface hydrophobicity, two upper ends of the droplet appear as observed at $t = 28$ μs which gradually merge into a single end at $t = 38$ μs. During this period, the leading edge in contact to the surface moves backwards which diminishes the contact area. Subsequently, the leading edge proceeds to move onwards while the trailing edge moves faster and catches up the leading edge as shown at $t = 58$ μs. The tangential component of the velocity leads to forward motion of the droplet, but the property of the surface hinders the movement tendency. Because of the surface hydrophobicity and relatively high impact velocity, the droplet is able to detach from the surface as observed at $t = 68$ μs in Fig. 12.

The quantitative comparison of the spreading factors defined along x and y directions Dx^*, Dy^* with D^* of normal impact as well as the ratio Dy^*/Dx^* are shown in Fig. 13a, and the evolutions of H^* and $Hlow^*$ for $\alpha = 0°$ and 45° are plotted in Fig. 13b. It seems that Dx^*, Dy^* and D^* reach their respective maximums almost at the same time, although both of the maximal spreading factors in x and y directions for the oblique collision are a bit smaller than maximum of D^* for normal collision. A possible explanation for this might be that the maximum spreading factor is determined by the normal component of the impact velocity which could be transferred to spreading velocities in opposite directions. Moreover, the spreading in y direction is always wider than that in x direction because the tangential velocity impedes the droplet retraction. Another finding worth attention is Dy^*/Dx^* peaks at $t = 28$ μs, and after that the effect of tangential velocity abates, despite there appears another peak later which is caused by the faster secondary recoiling of the droplet in x direction than retraction in y direction. And the droplet detaches from the surface in an oval contact area, implying that the effect of the tangential velocity still performs. In addition, the decrease of the normal velocity leads to the reduction of the droplet jump height in later stage and delays the detachment from the surface compared to normal collision as shown in Fig. 13b.

5 Conclusion

In this study, the improved MLBFS is applied to investigate the micron-sized droplet collision onto the surface and the influences of four parameters are discussed, including impact velocity U (from 5 to 20 m/s), droplet diameter D (from 20 to 100 μm), surface contact angle θ (from 60° to 135°) and impact inclination angle α (0° and 45°). By analyzing the temporal morphology evolutions of the droplet and quantitative comparison of the spreading factors, dimensionless heights and other quantity indicators, the obtained conclusions are summarized below.

(1) When the collision is normal to the surface, the increase of the impact velocity causes the droplet spreading wider in a shorter time. And the lower part of the droplet with a smaller velocity detaches from the surface faster but the jump height of the upper end is lower in later stage.
(2) The larger droplet takes a longer time to spread wider and jumps out of the surface later albeit the upper end could jump higher. And the entire evolution process of smaller droplet is expedited.
(3) The spreading factor maximum decreases with the increase of surface contact angle whereas upper and lower ends of the droplet rebound higher. Different from higher velocity and larger diameter, there is no temporal lag in lifting up both droplet ends for higher contact angle.
(4) For the oblique droplet impact, the tangential component of the velocity reduces the spreading factor maximum and highest dimensionless height while the detachment is also deferred.

This work focuses on the droplet collision without considering the effect of temperature, which is expected to be explored in the near future.

Acknowledgments This work is supported by the National Natural Science Foundation of China (NSFC) (Grant No. 11832012) and the Research Fund of State Key Laboratory of Mechanics and Control of Mechanical Structures (Nanjing University of Aeronautics and Astronautics) (Grant No. MCMS-I-0120G01).

References

1. García-Magariño A, Sor S, Velazquez A (2015) Experimental characterization of water droplet deformation and breakup in the vicinity of a moving airfoil. Aerosp Sci Technol 45:490–500
2. Cao Y, Tan W, Wu Z (2018) Aircraft icing: an ongoing threat to aviation safety. Aerosp Sci Technol 75:353–385

3. Scheller BL, Bousfield DW (1995) Newtonian drop impact with a solid surface. AIChE J 41(6):1357–1367
4. Rioboo R, Marengo M, Tropea C (2002) Time evolution of liquid drop impact onto solid, dry surfaces. Exp Fluids 32:112–124
5. Rioboo R, Marengo M, Tropea C (2002) Outcomes from a drop impact on solid surfaces. Atom Spray 11:155–165
6. Graham PJ, Farhangi MM, Dolatabadi A (2012) Dynamics of droplet coalescence in response to increasing hydrophobicity. Phys Fluids 24:112105
7. Zhang C, Liu H (2016) Effect of drop size on the impact thermodynamics for supercooled large droplet in aircraft icing. Phys Fluids 28(6):062107
8. Dong H, Carr WW, Bucknall DG et al (2007) Temporally-resolved inkjet drop impaction on surfaces. AIChE J 53:2606–2616
9. Liang G, Guo Y, Yong Y et al (2013) Spreading and splashing during a single drop impact on an inclined wetted surface. Acta Mech 224(12):2993–3004
10. Hirt CW, Nichols BD (1981) Volume of fluid (VOF) methods for the dynamics of free boundaries. J Comput Phys 39:201–225
11. Sussman M, Smereka P, Osher S (1994) A level set approach for computing solutions to incompressible two-phase flow. J Comput Phys 114:146–159
12. Lee T, Liu L (2010) Lattice Boltzmann simulations of micron-scale drop impact on dry surfaces. J Comput Phys 229:8045–8063
13. Wang Y, Shu C, Huang HB et al (2015) Multiphase lattice Boltzmann flux solver for incompressible multiphase flows with large density ratio. J Comput Phys 280:404–423
14. Wang Y, Shu C, Yang LM (2015) An improved multiphase lattice Boltzmann flux solver for three-dimensional flows with large density ratio and high Reynolds number. J Comput Phys 302:41–58
15. Lunkad SF, Buwa VV, Nigam KDP (2007) Numerical simulations of drop impact and spreading on horizontal and inclined surface. Chem Eng Sci 62:7214–7224
16. Lee T, Liu L (2010) Lattice Boltzmann simulations of micron-scale drop impact on dry surfaces. J Comput Phys 229:8045–8063
17. Raman KA, Jaiman RK, Lee TS et al (2016) Lattice Boltzmann simulations of droplet impact onto surfaces with varying wettabilities. Int J Heat Mass Transf 95:336–354
18. Raman KA (2018) Dynamics of simultaneously impinging drops on a dry surface: Role of inhomogeneous wettability and impact shape. J. Coll. Interf. Sci. 516:232–247
19. Inamuro T, Yokoyama T, Tanaka K et al (2016) An improved lattice Boltzmann method for incompressible two-phase flows with large density differences. Comput Fluids 137:55–69
20. Chen Z, Shu C, Tan D et al (2018) Simplified multiphase lattice Boltzmann method for simulating multiphase flows with large density ratios and complex interfaces. Phys Rev E 98:063314
21. Inamuro T, Echizen T, Horai F (2018) Validation of an improved lattice Boltzmann method for incompressible two-phase flows. Comput Fluids 175:83–90
22. Yuan HZ, Wang Y, Shu C (2017) An adaptive mesh refinement-multiphase lattice Boltzmann flux solver for simulation of complex binary fluid flows. Phys Fluids 29:123604
23. Wang Y, Shu C, Yang LM et al (2018) On the re-initialization of fluid interfaces in diffuse interface method. Comput Fluids 166:209–217
24. Li Y, Niu XD, Wang Y et al (2019) An interfacial lattice Boltzmann flux solver for simulation of multiphase flows at large density ratio. Int. J. Multiphase Flow 116:100–112
25. Liu XD, Osher S, Chan T (1994) Weighted essentially non-oscillatory schemes. J Comput Phys 115:200–212
26. Shu CW, Osher S (1988) Efficient implementation of essentially non-oscillatory shock-capturing schemes. J Comput Phys 77(2):439–471

Computation of Hypersonic Transitional Flows Over Cones Using Gas-Kinetic Scheme Coupled with the Turbulence Model

Chengrui Li, Hualin Liu, Zhongzheng Jiang, and Weifang Chen

Abstract Accurate prediction of the laminar-turbulent transition with the aerodynamic force and heat in hypersonic boundary flows is essential for the design of a hypersonic vehicle. However, subject to the complex boundary layer mechanism of the angle of attack (AOA) effects, the numerical prediction with high accuracy of hypersonic transitional flows over a cone at a small AOA is still a challenging job in current computational fluid dynamic (CFD) modeling. In this paper, our objective is to improve the prediction accuracy of the traditional turbulent model in the boundary layer flow problems. We employ a gas-kinetic scheme (GKS) strongly coupled with the turbulent kinetic energy equation in the shear stress transport (SST) turbulence model. Langtry-Menter's two-equation transition model is also implemented under the GKS framework. The turbulent relaxation time obtained from the turbulence model is also implemented into the kinetic equation of the BGK model as an enhanced particle collision time related to turbulent fluctuation. We further validate this method for several classic cases of the hypersonic transitional flows over cones with various free-stream conditions at zero AOA. We also compare the numerical solutions by the current GKS coupled with the turbulent kinetic energy equation and traditional Reynolds-Averaged Navier–Stokes (RANS) method. The numerical results confirm the consistency of the transition onset locations predicted by the proposed method with the experiments. The new coupled method also significantly improves numerical accuracy. The heat flux is also in good agreement with the experimental data. The hypersonic flows past a sharp cone are also computed at various small angles of attack ranging from 2° to 4°, where the numerical results agree with the existing experimental data. Using our results we also analyze the impact of AOA on a sharp cone.

Keywords Hypersonic transitional flows · Gas-kinetic scheme · Turbulence model · AOA effect · Langtry-Menter two-equation transition model

C. Li · H. Liu · Z. Jiang (✉) · W. Chen
School of Aeronautics and Astronautics, Zhejiang University, Hangzhou 310027, China
e-mail: jzhongzh@zju.edu.cn

S. Lee et al. (eds.), *The Proceedings of the 2021 Asia-Pacific International Symposium on Aerospace Technology (APISAT 2021), Volume 1*, Lecture Notes in Electrical Engineering 912, https://doi.org/10.1007/978-981-19-2689-1_53

1 Introduction

Laminar–turbulent transition in hypersonic boundary layers has a significant impact on heat transfer, skin friction, and separation. Hence it is essential in the development of hypersonic flight vehicles. The convenience and accuracy of the hypersonic transition prediction, therefore, are greatly important in the design of high-speed aerodynamic shapes and aerothermal protection systems. For a hypersonic vehicle, the three-dimensional flows over its surface mainly consist of cone boundary layer flow and sweep back wing boundary layer flow.

Many research works have been done on investigating the supersonic/hypersonic boundary layer transition flows over cones at zero or small AOAs. Such investigations include mechanical studies, numerical simulations, and experimental measurements. For instance, the experiments performed by James F. Muir and Amado A. Trujillo study the boundary layer transition on an 8-degree half-angle cone at the angle of attack varying from 0 to 8° [1]. Further, Kenneth F. Stetson investigated the effects of tip bluntness and angle of attack on boundary layer transition by obtaining wind tunnel data for the cones and biconic configurations [2]. An experimental study has been also conducted by M.S. Holden which investigated the effects of asymmetric boundary layer transition on the aerothermal characteristics of slender blunted cones at the AOA in hypersonic flow [3]. A detailed review of the experimental measurements is also presented by Schneider [4].

Besides the experimental studies, direct numerical simulation (DNS) is also a reliable method to obtain accurate transition results. For instance, Xinliang Li [5] performed direct numerical simulations of the boundary layer transition over a 5° half-cone-angle blunt cone with free-stream Mach number 6 and 1° angle of attack. The whole boundary layer flow over the cone is also simulated by the DNS. The mechanism of the delayed transition is further studied by using joint frequency spectrum analysis and linear stability theory (LST). Furthermore, Su et al. [6] developed a modified e^N method and calculated the transition locations of the hypersonic boundary layer over a blunt cone with free-stream Mach number 6 and showed that results are in agreement with the DNS data. The DNS is not fit for engineering because of its expensive computational consumption. The same applies to the e^N method as it is difficult to be adapted to the 3D shape.

To address the above issues, the transition models based on Reynolds-averaged Navier–Stokes (RANS) are widely applied to engineering transition prediction mainly due to their reasonable computational costs. For the high-speed flows, Wang and Fu [7] also proposed a new k-ω-γ transition/turbulence model considering the modes of instability. Their model was validated in the hypersonic transition flow past straight cones at zero AOA. Also, Zhou et al. [8] improved the prediction accuracy of the k-ω-γ transition model by adding the crossflow-induced elements. The obtained prediction of the transition onset for the leeward side of the sharp cone in their model also agrees with the experimental results for AOA values smaller than 4°. Qiao et al. also proposed a fully local transition closure model for hypersonic boundary layers. The numerical results of a circular cone at an angle of attack of

3° for their model showed that the transition onset location predicted by the transition model was consistent with the experimental results [9]. Xiang et al. further modified the γ-Re_θ transition model by appending cross-flow strength and surface roughness based on the Chant2.0 computing platform. It is shown in [10] that the extended transition model is suitable for the prediction of cross-flow transition in the 3D hypersonic boundary layer over a cone at several small AOAs. Some of the researchers also studied extending the traditional transition/turbulence model by appending cross-flow effects, different instability modes effects, and surface roughness effects. Nevertheless, investigations of improving the accuracy of the transition model considering the microscopic molecular motion have not been investigated before in the literature.

In this paper, we validate a new computational method for the hypersonic transitional flow simulations which is based on the k-ω SST turbulence model [11] and Langtry-Menter γ-Re_θ transition model [12]. The new method [13] is coupled with the GKS method, in which the turbulent kinetic energy equation with conservation equations is employed in the formulation of the GKS scheme. In this method, the internal energy term of the turbulent kinetic energy equation is also added to the Maxwell velocity distribution function. In our new calculation method, we further consider the energy transport route from the turbulent fluctuating field to the mean-field. We also simulate four Hypersonic transition flows over cones at zero AOA using the proposed coupled method. The hypersonic flows past a cone are also computed at several small angles of attack ranging from 0° to 4°. The numerical results are compared with the existing experimental data.

2 Theory

The numerical method in this paper is adopted from [13] which is briefly presented in the following.

2.1 *Gas-Kinetic Scheme Coupled with Turbulent Kinetic Energy*

The Bhatnagar-Gross-Krook (BGK) model governing equation for the current method is

$$\frac{\partial f}{\partial t} + u_i \frac{\partial f}{\partial x_i} = \frac{g - f}{\tau + \tau_t} \tag{1}$$

where g represents the extended Maxwellian distribution function that is expressed as

$$g = \rho \left(\frac{\lambda}{\pi} \right)^{\frac{K+3}{2}} e^{-\lambda[(u_i - U_i)^2 + \xi^2]} \left(\frac{\lambda_t}{\pi} \right)^{\frac{K_t}{2}} e^{-\lambda_t \xi_t^2} \tag{2}$$

f represents the gas distribution function, τ and τ_t represents the collision time for laminar, and turbulent flow, respectively. In the above, (U, V, W) represents the macroscopic velocity and (u, v, w) represents the microscopic velocity and

$$\lambda = \frac{m}{2kT}, \lambda_t = \frac{m}{2kT_t} \tag{3}$$

where m is the molecular mass, k is the Boltzmann constant, T is the temperature, and T_t is the temperature related to the turbulent kinetic energy. For the most common diatomic molecules, the degree of freedom of the internal energy, K, is 2. The turbulent flow K_t represents the degrees of freedom of the internal energy which corresponds to the turbulent kinetic energy. In this paper, we set $K_t = 3$.

The macroscopic physical quantities are obtained by integrating the collision term ψ_α with the distribution function f as

$$Q = \begin{pmatrix} \rho \\ \rho U \\ \rho V \\ \rho W \\ \rho E \\ \rho E_t \end{pmatrix} = \int \psi_\alpha f d\Xi, \alpha = 1, 2, 3, 4, 5, 6 \tag{4}$$

The collision term, ψ_α, and phase space, Ξ, are:

$$\psi_\alpha = \left(1, u, v, w, \frac{1}{2}(u^2 + v^2 + w^2 + \xi^2 + \xi_t^2), \frac{1}{2}\xi_t^2 \right) \tag{5}$$

$$d\Xi = dudvdwd\xi d\xi_t \tag{6}$$

The compatibility condition is written as

$$\int \psi_\alpha \frac{g - f}{\tau + \tau_t} d\Xi = (0, 0, 0, 0, 0, S_{\rho E_t}), \alpha = 1 - 6, \tag{7}$$

where $S_{\rho E_t}$ is the source term for turbulent kinetic energy which can be obtained from the k-ω SST turbulence model. The laminar flow collision time, τ, and turbulence collision time, τ_t, are obtained by the laminar flow viscosity coefficient, and turbulence viscosity coefficient, respectively as

$$\tau = \frac{\mu}{p}, \tau_t = \frac{\mu_t}{p} \tag{8}$$

In (8), the laminar viscosity μ is calculated by Sutherland's law and turbulent viscosity μ_t is obtained by the turbulence model. Finally, the total viscosity coefficient is divided into two parts corresponding to laminar and turbulent flows:

$$\mu = \mu_l + \mu_t \tag{9}$$

The corresponding total heat conduction coefficient is, therefore,

$$\kappa = \kappa_l + \kappa_t \tag{10}$$

where κ_t is the turbulent heat conduction coefficient.

2.2 *K-ω SST γ-Re$_\theta$ Turbulence/Transition Model*

The turbulence and transition models employed in the current coupled method are the k-ω SST turbulence and γ-Re$_\theta$ transition models. The k-ω SST model is described by the two transport equations in the following

$$\frac{\partial(\rho k)}{\partial t} + \frac{\partial}{\partial x_j}\left[\rho u_j k - (\mu + \sigma_k \mu_t)\frac{\partial k}{\partial x_j}\right] = P - \beta^* \rho \omega k \tag{11}$$

$$\frac{\partial(\rho \omega)}{\partial t} + \frac{\partial}{\partial x_j}\left[\rho u_j \omega - (\mu + \sigma_\omega \mu_t)\frac{\partial \omega}{\partial x_j}\right] = \frac{\gamma}{\nu_t} P - \beta \rho \omega^2 + 2(1 - F_1)\frac{\rho \sigma_{\omega 2}}{\omega}\frac{\partial k \partial \omega}{\partial x_j \partial x_j} \tag{12}$$

where the production term, P, in Eq. (11) and Eq. (12) is defined as:

$$P = \min(\mu_t \Omega^2 - \frac{2}{3}\rho k \delta_{ij}\frac{\partial u_i}{\partial x_j}, 10\beta^* \rho \omega k) \tag{13}$$

The turbulent Eddy viscosity is also

$$\nu_t = \frac{a_1 k}{\max(a_1 \omega, S F_2)} \tag{14}$$

and the blending function F_1 in Eq. (12) and F_2 Eq. (14) is

$$F_1 = \tanh\left\{\min[\max(\frac{\sqrt{k}}{0.09\omega y}, \frac{500\mu}{\rho \omega y^2}), \frac{4\rho \sigma_{\omega 2} k}{N y^2}]\right\}^4 \tag{15}$$

$$F_2 = \tanh[\max(\frac{2\sqrt{k}}{0.09\omega y}, \frac{500\mu}{\rho\omega y^2})]^2 \tag{16}$$

The γ-Re$_\theta$ transition model increases the intermittency factor transport equation, Eq. (17), and the transport equation for transition momentum thickness Reynolds number, Eq. (18), based on the k-ω SST turbulence model.

$$\frac{\partial(\rho\gamma)}{\partial t} + \frac{\partial}{\partial x_j}\left[\rho u_j \gamma - (\mu + \frac{\mu_t}{\sigma_\gamma})\frac{\partial \gamma}{\partial x_j}\right] = P_\gamma + E_\gamma \tag{17}$$

$$\frac{\partial(\rho \widetilde{Re}_{\theta t})}{\partial t} + \frac{\partial}{\partial x_j}\left[\rho u_j \widetilde{Re}_{\theta t} - \sigma_{\theta t}(\mu + \mu_t)\frac{\partial \widetilde{Re}_{\theta t}}{\partial x_j}\right] = P_{\theta t} \tag{18}$$

The production term, P_γ, and destruction term, E_γ, in Eq. (17) are also given as

$$P_\gamma = F_{length} c_{a1} \rho S (\gamma F_{onset})^{0.5}(1 - c_{e1}\gamma) \tag{19}$$

$$E_\gamma = c_{a2}\gamma\rho\Omega F_{turb}(1 - c_{e2}\gamma) \tag{20}$$

where F_{length} is the transition region length control function, and F_{onset} is designed to trigger the production of intermittency. The production term of the transition momentum thickness Reynolds number in Eq. (18) is also defined as

$$P_{\theta t} = c_{\theta t}\frac{\rho}{t}(Re_{\theta t} - \widetilde{Re}_{\theta t})(1.0 - F_{\theta t}) \tag{21}$$

The transition momentum thickness Reynolds number outside the boundary layer is then obtained from the freestream turbulence intensity and local pressure gradient parameters.

To combine the γ-Re$_\theta$ transition model with the SST turbulence model, the intermittency factor γ_{eff} is introduced to Eq. (11) as the following

$$\frac{\partial(\rho k)}{\partial t} + \frac{\partial}{\partial x_j}\left[\rho u_j k - (\mu + \sigma_k \mu_t)\frac{\partial k}{\partial x_j}\right] = \widetilde{P}_k - \widetilde{D}_k \tag{22}$$

where $\widetilde{P}_k = \gamma_{eff} P$, $\widetilde{D}_k = \min[\max(\gamma_{eff}, 0.1), 1.0]\beta^* \rho\omega k$, and P, $\beta^*\rho\omega k$ are the production and destruction terms of the original turbulence model.

3 Results and Discussions

In this section, we present an extensive set of validation test cases of hypersonic transition cones at zero AOA. In our experiments, the range of the flow velocity is from Mach 6 to Mach 10. The nose radiuses are also 5.0 and 6.35 mm. We further conduct numerical simulations on the hypersonic transition flows over the cones at small AOAs and nalyse the impact of the AOA. All the numerical results are also compared to the experimental data for cross-validation.

3.1 Hypersonic Transition on Blunt Cones at Zero AOA

Here we consider four hypersonic transition cases with different conditions and various nose radiuses. These cases are referred to as Case A, Case B, Case C, and Case D. The experiments for these cases are performed in the CUBRC LENS I hypervelocity shock tunnel [14]. The geometry of Cases A and B includes a 7-degree half-angle, 2.35 m cone with a nose radius of 5.0 mm. The geometry of Case C and Case D also includes a cone with a nose radius of 6.35 mm. The experimental results are also obtained from [14, 15].

The computational grid of the half-cone model includes 186, 100, and 31 points in the streamwise, wall-normal, and circumferential directions. The distance of the first grid point from the wall is 1×10^{-6} m through grid independence study. The test conditions for these four cases are presented in Table 1. As it is seen these four cases are all tested at zero AOA. The two main hypersonic flow velocities are performed including Mach 7 for Case A and Case B, and then Mach 10 for Case C and Case D.

Figure 1 shows the heat flux predictions of the four cone cases in the GKS coupled method and traditional RANS method compared with the experimental results. The colored symbols represent the experimental data which are measured by the instrumentation on the cone model consisting of four rays of thin-film heat transfer gauges located at 0°, 90°, 180°, and 270° around the model. The blue and green solid lines also represent the numerical results simulated by GKS coupled with the turbulent kinetic energy method, and the RANS method, respectively

It is seen that for Case A and Case B, the current coupled method and traditional RANS method both efficiently predict the transition onset location. Nevertheless,

Table 1 The test conditions for the four considered cases

	H_0 (MJ/kg)	Ma_∞	T_∞ (K)	ρ_∞ (kg/m^3)	Re_∞ (1/m)	T_{wall} (K)	R_n (mm)
Case A	2.62	7.15	231.1	0.07060	9.92e+6	301.0	5.0
Case B	2.09	6.58	214.0	0.12568	1.67e+7	301.3	5.0
Case C	4.49	9.91	206.71	0.0121	2.52e+6	295.61	6.35
Case D	4.56	9.93	209.33	0.0160	3.33e+6	299.00	6.35

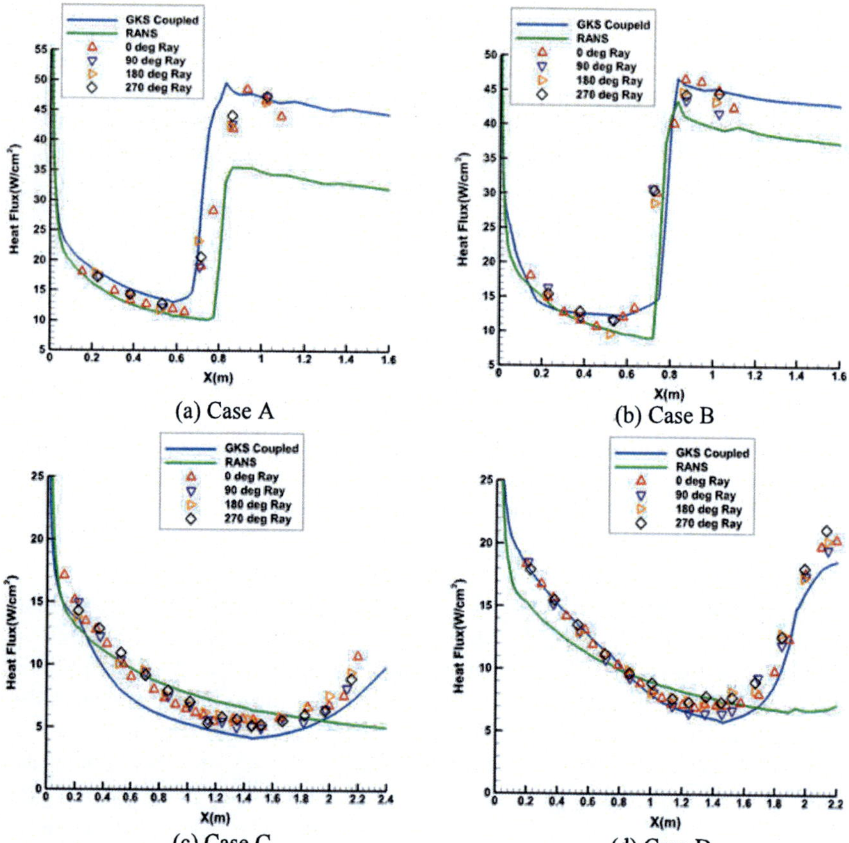

(a) Case A (b) Case B

(c) Case C (d) Case D

Fig. 1 Comparison of the heat flux predictions and the experimental results for the cones at zero AOA

the new coupled method shows better prediction accuracy than that of the RANS method in the turbulent flow zone. In Case C and Case D where the free-stream Mach number is 10 which is much faster than the former cases, the coupled method can predict the transition onset location. It is also seen that the numerical results agree well with the experiments effectively capturing the distribution of the heat flux and predicting the curve trend. On the contrary, the RANS method is no longer efficient in these two higher-speed cases and its prediction trends run counter to the real situation. The results above suggest that the current method is more efficient for predicting the three-dimensional hypersonic transition flow. In such cases, the microscopic molecular motion is considered where the turbulent kinetic energy is added into the velocity distribution function.

3.2 *Hypersonic Transition on a Sharp Cone at Small AOAs*

To validate the performance of the GKS coupled turbulent kinetic energy method in the transition flow prediction, here we consider a sharp cone at different angles of attack in the Φ 1 m hypersonic wind tunnel at the CARDC [16]. The test model is a 7 half-angle cone with a nose radius of 0.05 mm. The length of the cone is equal to 800 mm and the freestream parameters are $Ma_\infty = 6$, and $Re_\infty = 1 \times 10^7$/m. The angles of attack are 0°, 2°, and 4°. Due to the experimental model, the experimental data are only valid for x between 165, and 760 mm. The experimental results and details of the experiment are adopted from [16]. The half model grid with 213 × 130 × 32 cells in streamwise, normal, and circumferential direction is then used in the computation through the grid independence study.

Figure 2 presents the comparisons of the surface Stanton number predicted by the current coupled method (shown in solid lines) and the experimental results (shown by symbols) along the windward and leeward centerline. In Fig. 2 different colors present different AOAs. The experimental results are the instantaneous temperature rise data measured on the cone surface. The temperature rise can be approximated to be proportional to the Stanton number and the transition onset location is obtained from the distribution of temperature rise. Therefore, we directly compare the calculated Stanton number with the temperature rise measured in the experiment and ignore their different units.

The two Y axes are set for these two variables of which the right-side axis represents the temperature rise and the left-side denotes the Stanton number. It is seen in Fig. 2 that the predicted numerical results are consistent with the experimental data and transition onset location can be identified when Stanton number begins to depart from the laminar value. It is also evident that along the windward centerline by increasing AOA, the Stanton number increases gradually and the transition onset location moves continuously downstream. On the leeward side, however, increasing

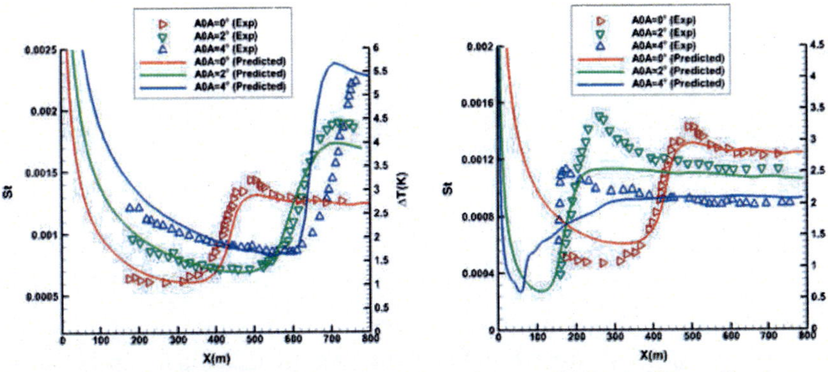

(a) Windward centerline (b) Leeward centerline

Fig. 2 Comparison of the transition onset for the flow over sharp cones at different AOAs

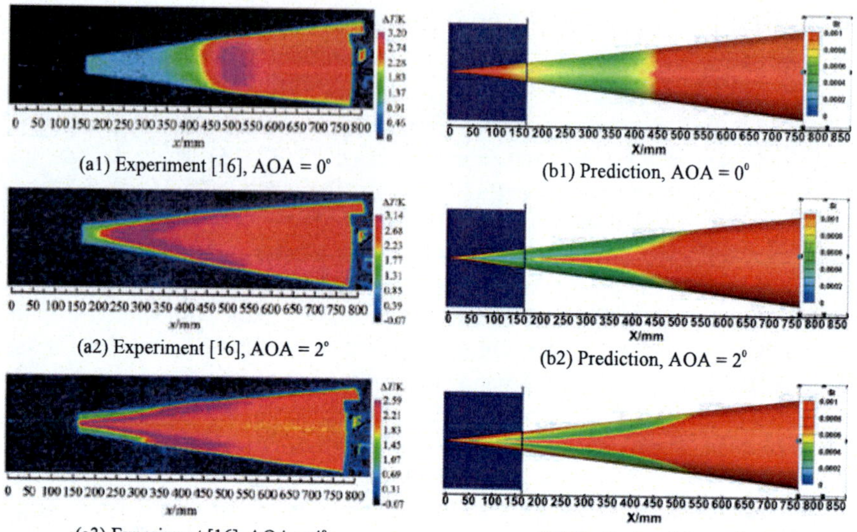

(a1) Experiment [16], AOA = 0°
(b1) Prediction, AOA = 0°
(a2) Experiment [16], AOA = 2°
(b2) Prediction, AOA = 2°
(a3) Experiment [16], AOA = 4°
(b3) Prediction, AOA = 4°

Fig. 3 The transition distributions on the leeward side of the sharp cone at different AOAs

the AOA results in decreasing the Stanton number and the location of the transition onset is gradually moved upstream.

In Fig. 3, we present the transition onset distribution on the leeward side of the sharp cone for the computational and experimental results [16] at different AOAs. The experimental images of temperature rise contour at different AOAs are shown in Fig. 3(a1)–(a3) and the Stanton number results computed from the GKS coupled method are shown in Fig. 3(b1)–(b2). Since the effect measured zone is x = 165–760 mm [16], the zone before x = 165 mm is invisible in the experimental images and a dark blue background is added to cover the unmeasured zone to enable comparisons. It is seen that the transition onset distribution is axisymmetric at zero AOA. By increasing the AOA, the transition onset on the leeward side moves upstream in both computational results and experimental images. Although the three-dimensional numerical results are not completely matched with the experimental results in some areas, the prediction trends and rules are in agreement with the experiment.

4 Conclusion

We introduced a GKS coupled turbulent kinetic energy method and validated its performance on hypersonic cones at different small angles of attack. Compared with the traditional macroscopic turbulence/transition approaches, the coupled method behaves with multi-scale effects in space and time. This is because the macroscopic turbulent fluctuation and microscopic molecular motion are both taken into consideration within the same framework of the velocity distribution function. The coupled method is more accurate than that of the traditional RANS method in predicting the hypersonic blunt cones at zero AOA. The introduced method also provides surprisingly good performance in predicting three-dimensional hypersonic transition flow even in cases where the Mach number reaches up to 10. The hypersonic transition flows over a sharp cone at different small AOAs are also simulated by the coupled method. The predicted results are in good agreement with the experimental results and the conclusion of small AOA effects is verified by the current numerical simulation. It is seen that by increasing AOA the transition is leeside-forward and windside-aft. It can be concluded that the current GKS method coupled with turbulent kinetic energy can be effectively used to obtain the hypersonic turbulent and transitional flows with reasonable accuracy and good compatibility with the modern parallel CFD codes. Hence it can be used for simulating hypersonic transitional flows over a vehicle in various engineering applications.

Acknowledgements This work was supported by National Numerical Wind Tunnel Project (NNW2019ZT3-A08).

References

1. Muir J, Trujillo A (1972 January) Experimental investigation of the effects of nose bluntness, free-stream unit Reynolds number, and angle of attack on cone boundary layer transition at a Mach number of 6. In: 10th aerospace sciences meeting, p 216
2. Stetson K (1979 January) Effect of bluntness and angle of attack on boundary layer transition on cones and biconic configurations. In: 17th aerospace sciences meeting, p 269
3. Holden M (1985 January) Experimental studies of the effects of asymmetric transition on the aerothermal characteristics of hypersonic blunted slender cones. In: 23rd aerospace sciences meeting, p 325
4. Schneider SP (2004) Hypersonic laminar–turbulent transition on circular cones and scramjet forebodies. Prog Aerosp Sci 40(1–2):1–50
5. Li X, Fu D, Ma Y (2010) Direct numerical simulation of hypersonic boundary layer transition over a blunt cone with a small angle of attack. Phys Fluids 22(2):025105
6. Su C, Zhou H (2009) Transition prediction of a hypersonic boundary layer over a cone at small angle of attack—with the improvement of e^N method. Sci China Ser G Phys Mech Astron 52(1):115–123
7. Wang L, Fu S (2009) Modelling flow transition in a hypersonic boundary layer with Reynolds-averaged Navier-Stokes approach. Sci China, Ser G 52(5):768–774
8. Zhou L, Li R, Hao Z, Zaripov DI, Yan C (2017) Improved k-ω-γ model for crossflow-induced transition prediction in hypersonic flow. Int J Heat Mass Transf 115:115–130
9. Qiao L, Xu J, Bai J, Zhang Y (2021) Fully local transition closure model for hypersonic boundary layers considering crossflow effects. AIAA J 59(5):1692–1706
10. Xiang X, Chen J, Yuan X (2021) Study on C-γ-Reθ model for hypersonic three-dimensional boundary layer transition prediction. Acta Aeronautica et Astronautica Sinica. https://kns.cnki.net/kcms/detail/11.1929.v.20210510.1355.014.html. Accessed 20th August 2021
11. Menter FR (1994) Two-equation eddy-viscosity turbulence models for engineering applications. AIAA J 32(8):1598–1605

12. Langtry RB, Menter FR, Likki SR, Suzen YB, Huang PG, Völker S (2006) A correlation-based transition model using local variables—part II: test cases and industrial applications
13. Liu H, Cao G, Chen W, Agarwal RK, Zhao W (2021) Gas-Kinetic scheme coupled with turbulent kinetic energy equation for computing hypersonic turbulent and transitional flows. Int J Comput Fluid Dyn, 1–12
14. Wadhams T, MacLean M, Holden M, Mundy E (2007) Pre-flight ground testing of the full-scale FRESH FX-1 at fully duplicated flight conditions. In: 37th AIAA fluid dynamics conference and exhibit, p 4488
15. Johnson H, Alba C, Candler G, MacLean M, Wadhams T, Holden M (2007) Boundary layer stability analysis to support the HIFiRE transition experiment. In: 45th AIAA aerospace sciences meeting and exhibit, p 311
16. Chen J, Ling G, Zhang C, Xie F, Xu X, Zhang Y (2020) Infrared thermography experiments of hypersonic boundary layer transition on a 7° half-angle sharp cone. J Exp Fluid Mech 34(1):60–66

Effect of Solidity on Efficiency for High-Advance-Ratio Propeller

Koju Hiraki and Hiroki Kai

Abstract The use of a propeller is an option for powered flight on Mars. Powered flight in the Martian atmosphere is quite peculiar in that the propeller operates with low speed under low Reynolds number conditions. The improvement of the performance of the propeller for Martian flight is investigated, focusing on the blade solidity. First using the blade element momentum theory, the effect of the increase in solidity is investigated. Then, in order to verify this solidity effect experimentally, wind tunnel tests are carried out. Two propellers with different solidity are specially designed and tested. Through the measurement of thrust and torque, the effect of the solidity is discussed in terms of efficiency.

Keyword Propeller · Efficiency · Advance ratio · Low reynolds number

1 Introduction

In 2021, NASA achieved the first flight of the Mars helicopter Ingenuity. The Martian atmosphere has a density and gravity that are about 71 and 2.64 times smaller than those on Earth, respectively, which makes vehicle design a daunting task. When a propeller operates in a low-density atmosphere, it performs with poor efficiency. This is mainly due to the degradation of airfoil performance in such an environment. The challenge of achieving efficient flight in a low-density atmosphere is further aggravated by a lower speed of sound. Thus, a key to Martian flight is to achieve higher efficiency of a propeller with lower rotational speed. In order to verify the performance of a designed propeller experimentally, it is necessary to match both the Reynolds number and the advance ratio. Thus, the blades are designed on the condition that is easy to be simulated in a low-speed tunnel. The target Reynolds number of the propeller blade is 30,000, which is equal to that in the Martian flight. The target advance ratio is greater than 1, which means that the blade-tip speed is

K. Hiraki (✉) · H. Kai
Kyushu Institute of Technology, 1-1 Sensui-cho, Tobata-ku, Kitakyushu, Fukuoka 804-8550, Japan
e-mail: hiraki.koju735@mail.kyutech.jp

S. Lee et al. (eds.), *The Proceedings of the 2021 Asia-Pacific International Symposium on Aerospace Technology (APISAT 2021), Volume 1*, Lecture Notes in Electrical Engineering 912, https://doi.org/10.1007/978-981-19-2689-1_54

lower than the translational speed. This combination of the Reynolds number and the advance ratio is quite peculiar to Martian flight.

2 Design Conditions

A Martian vehicle with a deployable wing is considered in the present study. The flexible wing is stowed in a container and will be deployed by gas injection. This vehicle is separated from an entry probe at an altitude of 10 km. Without any propulsion system it rapidly descends to the Martian surface. In order to make an observation flight for a long time, the vehicle should be equipped with some propulsion system. Like a powered paraglider, this vehicle can have a propeller as propulsion system. As the vehicle touches down on the Martian ground eventually after a long-duration flight, the minimum required thrust is equal to the drag it experiences in the atmosphere, which strongly depends on the aerodynamic performance of the deployable wing. In Table 1 the design conditions are summarized.

The assumed thrust is 4.66 N for a cruise condition, in which the vehicle attains level flight. However, one has to account for some disturbances like wind, a change in the density of the atmosphere, insufficient deployment of the wing etc. Thus, the target thrust is set to 5.4 N, which includes a margin of 16 percent of the nominal thrust for horizontal flight. It would be beneficial to use a large-radius propeller to generate this thrust, however, a large-radius propeller is not suitable to be stowed in the entry probe. As the diameter of the entry probe is limited, the largest radius of the propeller should not exceed 0.5 m.

In order to avoid the degradation of the propeller performance due to the compressibility of the atmosphere, the tip speed of the propeller should be lower than 140 m/s, which corresponds to Mach 0.6. Figure 1 illustrates how the tip-speed limit imposes restrictions on both the radius and rotational number of the propeller. As the generated thrust should be greater than 4.66 N, the combination of the radius and the rotational number of the propeller is determined by the thrust coefficients. Assuming a range of the thrust coefficient between 0.01 to 0.1, three curves are shown in Fig. 1. It is necessary to find a combination of radius and rotational number below the tip-speed-limit line. When a moderate thrust coefficient of $C_t = 0.05$ is taken, the available range of the rotational number is between 2000 and 3000 rpm, under the constraint of the radius being below 0.5 m.

Table 1 Design conditions of propeller for Martian flight

Flight speed	50 m/s
Thrust	4.66 ~5.4 N
Blade radius	<0.5 m
Blade tip speed	<140 m/s

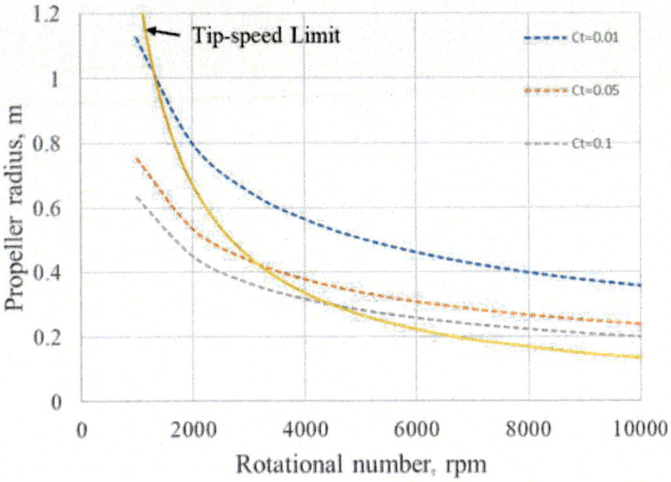

Fig. 1 Possible propeller radius/rotational number combinations for given tip-speed limit

3 Blade Design by Blade Element Momentum Theory

An appropriate propeller can be designed based on the design conditions, as described below. The blade element momentum theory (BEMT) is used here. In the theory the thrust can be calculated, once the radius R, rotational number n and solidity σ of the blade are given. The blade solidity is defined by

$$\sigma = \frac{Bc}{\pi R}$$

where B and c are the number of blades and the chord blade length, respectively.

The induced velocity into the propeller λ can be written as

$$\lambda = \sqrt{\left(\frac{\sigma C_{la}}{16F} - \frac{V}{2\pi nR}\right)^2 + \frac{\sigma C_{la}}{8F}\theta r} - \left(\frac{\sigma C_{la}}{16F} - \frac{V}{4\pi nR}\right) \tag{2}$$

$$F = \frac{2}{\pi}\cos^{-1}\left[e^{\frac{B}{2}\left(1-\frac{r}{\lambda}\right)}\right] \tag{3}$$

where C_{la}, θ, r, F and V are the lift line slope, pitch angle, radial position of the element of the blade, correction factor and flight speed, respectively. The details of the derivation can be found in [1]. Finally, the thrust coefficient can be obtained by

$$C_T = 4\int_0^1 \lambda\left(\lambda - \frac{V}{2\pi nR}\right) r\,dr \tag{4}$$

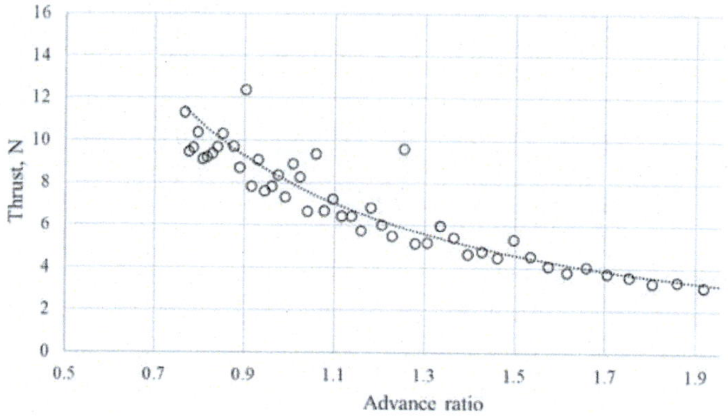

Fig. 2 The generated thrust as a function of advance ratio for Type-B combination

Varying the radius, rotational number and solidity, the generated thrust is evaluated based on the theory. After a number of attempts, the following four combinations were found as candidates which can meet the design conditions.

- Type A: $R = 0.5, \sigma = 0.16$
- Type B: $R = 0.489, \sigma = 0.25$
- Type C: $R = 0.42, \sigma = 0.30$
- Type D: $R = 0.49, \sigma = 0.35$

In each type, the rotational number should be specified to evaluate the thrust. In Fig. 2 this process is shown for Type B, as an example. When the rotational number of the propeller is varied, the advance ratio also changes. The advance ratio is defined by

$$J = \frac{V}{nD}$$

where V and D are flight speed and diameter of the blade, respectively. Figure 2 presents the relationship between the advance ratio and the thrust. The generated thrust becomes larger, as the advance ratio decreases. The dotted line is the fitted curve on the points, and with this curve the advance ratio can be determined to achieve the target thrust, 5.4 N. The advance ratio is found as 1.33, which can be converted into a rotational number of $n = 2304$ rpm. With similar procedures, the rotational number of the propeller can be determined for all four types listed above.

4 Experimental Verification

The design process in the previous section does not take into account the Reynolds-number effect. In reality, the propeller works under low Reynolds number condition, and this significantly degrades the propeller performance. Therefore, an experimental verification is needed for further design. However, it is not easily possible to simulate experimentally the actual Martian environment. Thus, in the present study a low-speed wind tunnel facility is used alternatively. It has a square test section of 0.5 m, and its wind speed can be varied up to 30 m/s.

4.1 Blade Redesign for Wind Tunnel Test

There are three non-dimensional parameters, advance ratio J, Reynolds number R_e and solidity σ. When all three parameters in the experiment match those for Martian flight, the propeller performance can be appropriately evaluated. However, it is quite difficult to match all three parameters in the experiment using a low-speed wind tunnel. Thus, only two of them are selected to match, the advance ratio and Reynolds number. The rest, solidity, is left unmatched to those listed in Table 2.

As Type D operates in lowest rotational speed, it is thought unsuitable for the present tests in the wind tunnel. Thus, the experimental comparison between Type A and B is made, excluding Type D. The Reynolds numbers can be determined once the number of blades is decided. If two blades are selected for both Type A and B, the Reynolds number for Type B is 1.6 times larger than that for Type A. In such case the difference in Reynolds number may become dominant in the propeller performance. It is more important to see an effect of solidity in the propeller performance in the present study. Thus, two blades are selected for Type A, and three blades are selected for Type B.

Here, Xrotor is used to design the experimental propellers for Type A and B. In Table 3 the designed blades for Type A and B are summarized. The advance ratio and

Table 2 Appropriate combination of design parameters to generate 5.4 N in thrust

Type	A	B	C	D
Solidity	0.16	0.25	0.30	0.35
Rotation speed [rpm]	2680	2304	3071	2073
Radius [m]	0.50	0.489	0.420	0.490
Flight speed [m/s]	50	50	50	50
Thrust [N]	5.4	5.4	5.4	5.4
Advance ratio	1.119	1.331	1.1629	1.477
Power [W]	352	338	350	326
Efficiency	0.768	0.799	0.770	0.828

Table 3 Designed blades for the wind tunnel tests

Number of blades	2	3
Section shape	SD7037	SD7037
Radius [m]	0.08	0.078
Flight speed [m/s]	12	12
Rotation number [rpm]	4020	3470
Thrust [N]	0.137	0.137
Advance ratio	1.12	1.33
Power [W]	2.13	2.15
Solidity	0.073	0.114
Reynolds number	17,499	15,400
Efficiency	0.773	0.766

Reynolds number match those for Martian flight. However, the solidities are 0.073 for Type A and 0.114 for Type B, respectively. These values are different to those for Martian flight, however, the ratio of solidity (0.073 to 0.114) is equal to that of Martian flight (0.16–0.25).

The designed blades are fabricated by an optical 3D printer. With this printer the whole shape of the propeller including the hub can be generated as a single piece. It enables the propellers to revolve rapidly without disintegration. In the experiment the rotational number is kept constant. By changing the wind speed, the advance ratio is varied. The generated thrust and torque are measured by independent settings of the instruments. The thrust is measured by a loadcell with the rated capacity of 50 N. The torque is measured by a torquemeter with the rated capacity of 1 Nm. The power can be calculated from the measured torque, and the propeller efficiency can be obtained from the power and thrust.

4.2 *Performance in Wind Tunnel Test*

As described, Type-A propeller has two blades, and Type-B has three. The measured thrust for both propellers is presented in Fig. 3. The dotted line indicates the target value of the thrust. The vertical solid lines in blue and orange show the design points in advance ratio for Type A and Type B, respectively. For both propellers, the generated thrust values at the design points are beyond the target thrust. Thus, it is shown that both propellers meet the requirement.

The measured torque is also presented in Fig. 4 for both propellers. In general, the measured torque with the Type-B propeller is larger than that with Type A. The measured torque shows a monotonous decrease with the increase of advance ratio for both types of propellers. The vertical solid lines in blue and orange show the design points in advance ratio for Type A and Type B, respectively. The measured torque at the design point for Type B is mostly identical to that for Type A. The increase

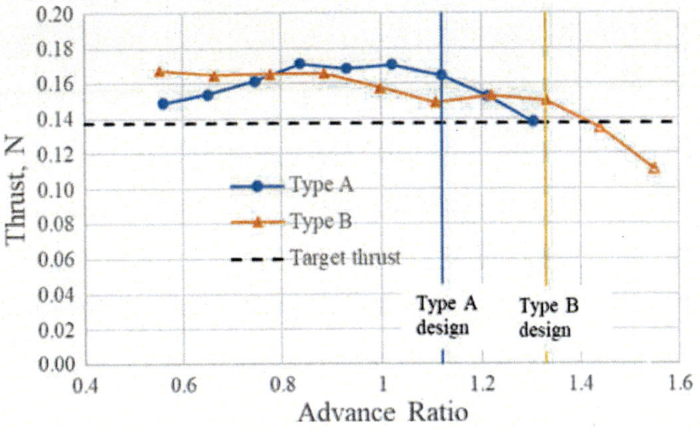

Fig. 3 Measured thrust of Type A and B propellers in the wind tunnel tests

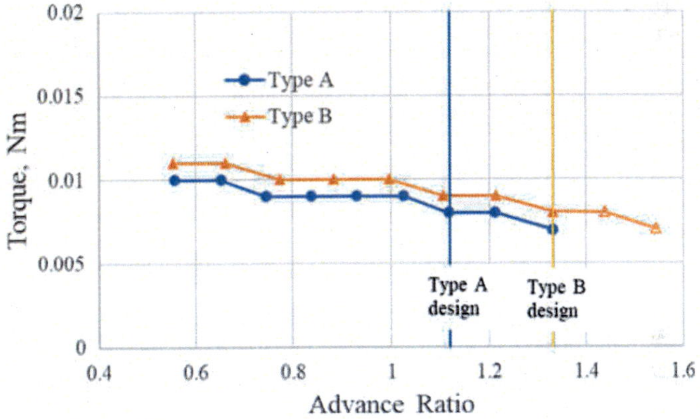

Fig. 4 Comparison of measured torque for Type-A and B propellers in the wind tunnel tests

in the solidity of the blade shifts the design point to larger advance ratios, and then, the required torque to drive Type B propeller decreases. As a result, similar torque is required to drive both propellers.

From the measured thrust and torque, the propeller efficiency can be obtained. The obtained propeller efficiency for both propellers is presented in Fig. 5. Both propellers show the same tendency, namely an increase of efficiency with increasing advance ratio. Again, the vertical solid lines in blue and orange show the design points in advance ratio for Type A and Type B, respectively. When the efficiencies are compared at the design points between Type A and B, the efficiency for Type A is slightly higher than that for Type B.

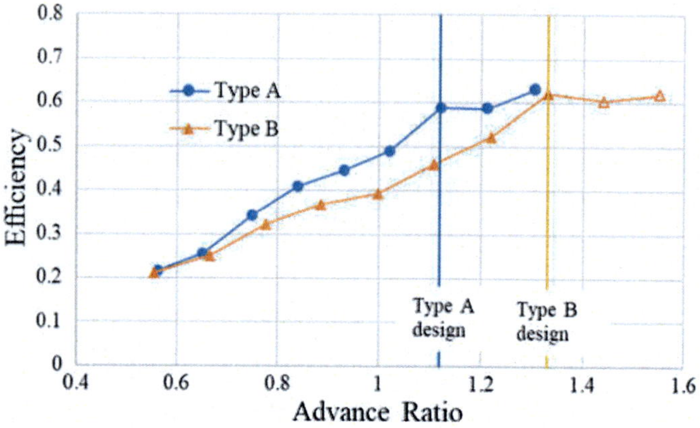

Fig. 5 Comparison of propeller efficiency for Type-A and B propellers in the wind tunnel tests

As shown by the experimental verification, both propellers clear the target thrust, and they require nearly the same torque to drive. The attained efficiency for Type B is slightly better than that for Type A. This improvement is related to the increase in solidity. It indicates the possibility of improvement by further increase in solidity.

5 Conclusion

The improvement of the performance of a propeller for Martian flight was investigated, focusing on the blade solidity. In the design using the blade element momentum theory, the increase in solidity gives better efficiency at lower blade-tip speeds. In order to verify this solidity effect experimentally, wind tunnel tests were carried out. Two propellers with different solidity were tested. Both propellers achieved the target thrust. They required nearly the same torque to drive. The attained efficiency for the propeller with larger solidity was slightly higher than that for the propeller with smaller solidity, so the improvement is seen as a result of the increase in the solidity of the blade.

References

1. Gessow A, Myers G (1985) Aerodynamics of the helicopter, 8th edn. Frederick Ungar Publishing, New York, NY

Aerodynamic Study of Cone-Derived Waverider as Supersonic Transport

Nomin Buyanbaatar, Yuhei Ishikawa, and Wataru Yamazaki

Abstract In this research, a new type of supersonic transport (SST) configuration is investigated that can contribute to improve aerodynamic as well as noise (sonic boom) performances, by generating supplemental lift force from generated shock waves. Inspired by the previous study of a cone-derived waverider conceptual design that achieved a significant increase in of lift-to-drag ratio (L/D), a concave study is examined on the waverider (supersonic) configuration. The main purpose of this study is to increase L/D value as well as to reduce the sonic boom strength by designing the lower side of the waverider configuration. The effect of the concave shape added on the bottom side of the body is examined in this study. The numerical simulation approaches are used to evaluate the objective performance functions of L/D and sonic boom strength. As a result, the effect of the concave shape added on the lower surface of the body influences the sonic boom strength. Various types of concave models are examined and important design knowledge for the waverider configuration is extracted. Strong shock waves on the lower side of the body cause high-pressure regions to generate higher lift force, and its efficiency depends on the concave position and/or size.

Keywords Supersonic transport · Computational fluid dynamics · High lift fuselage · Waverider · Sonic boom

1 Introduction

In recent years, computational fluid dynamics (CFD) has been widely used in the macro-field of study to model fluid flow problems. One of the application fields is aerodynamic design optimization as well as multidisciplinary studies. The aerodynamic design of an efficient fuselage for SST is one of the most challenging and time-consuming tasks. Even though the SST of Concorde had been retired from its commercial flight in 2003, new SST design concepts for solving economical as well as

N. Buyanbaatar (✉) · Y. Ishikawa · W. Yamazaki
Department of Mechanical Engineering, Nagaoka University of Technology, Nagaoka, Japan
e-mail: s175064@stn.nagaokaut.ac.jp

N. Buyanbaatar
Department of Mechanical Engineering and Transportation, Mongolian University of Science and Technology, Ulaanbaatar, Mongolia

S. Lee et al. (eds.), *The Proceedings of the 2021 Asia-Pacific International Symposium on Aerospace Technology (APISAT 2021), Volume 1*, Lecture Notes in Electrical Engineering 912, https://doi.org/10.1007/978-981-19-2689-1_55

environmental issues are being investigated worldwide. To develop a new type of SST configuration, it is necessary to control the noise relating to the sonic boom, which was a critical problem of the Concorde, in addition to securing a sufficient L/D. In our previous studies, twin-body/biplane-wing SST configurations [1], bottom flat fuselage configurations [2], and waverider configurations [3] have been investigated with the assumption of 200 passengers, whose target performance values were set for the maximum pressure rise of the sonic boom waveform P_{max} (under 24.0[Pa]) and the lift-to-drag ratio L/D (over 8.0[-]). We focus on the waverider configuration in this research due to its promising aerodynamic performance compared to the other configurations. In this research, a concave shape is added on the lower surface of the waverider configuration to acquire design knowledge to increase the L/D value and to reduce P_{max}.

2 Modeling Method

2.1 *Definition of the Waverider Model*

Waverider configurations can provide lift force from high-pressure of shock waves generated at its lower surface. In this research, waverider configurations are defined based on a cone-type approach [4]. The analysis model is shown in Figs. 1 and 2. This type of waverider generates shock waves from its leading edge. The lower surface (compression surface) has a higher pressure because of shock waves. In contrast, the upper surface (freestream surface) has no pressure variation because this surface is parallel to the uniform flow.

In the procedure to define the waverider configurations, firstly, the uniform supersonic flow field (freestream Mach number for design M') which has an imaginary cone with a tip angle of δ is considered. Then, the conical shock waves generated from the imaginary cone are also considered. In this configuration, the leading edge

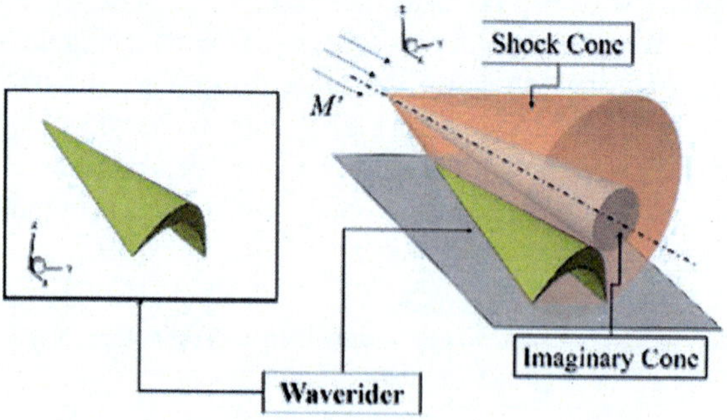

Fig. 1 Waverider configuration based on cone type approach

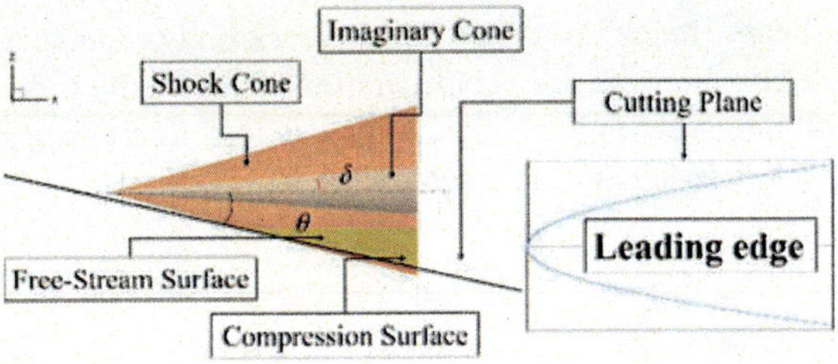

Fig. 2 Definition of waverider model

is defined as contacting the shock cone. A cutting plane is defined to form the leading edge, and its angle is defined as θ.

In an actual flow field which has no imaginary cone, shock waves are generated from the leading edge. With this definition, the generated shock waves concentrate on the lower side of the body and the lift force is then generated.

2.2 *Definition of Combined Fuselage Configuration*

A combined fuselage configuration is created using the waverider configuration and the Sears-Haack body [5]: The lower-half of the fuselage is the waverider; the

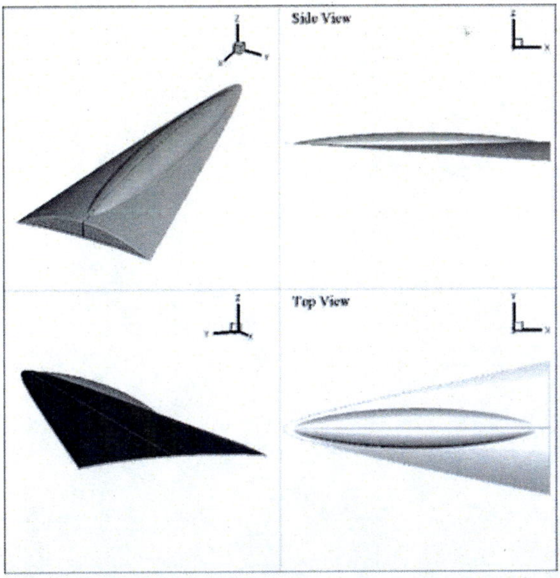

Fig. 3 Combined fuselage configuration as baseline model

upper-half is the Sears-Haack body. The Sears-Haack body is added for making the passenger seat area on the waverider configuration. The length of the fuselage is set to 62[m] and with the target volume for 200 passengers. The analysis model of the combined fuselage configuration is shown in Fig. 3.

3 Numerical Methods

3.1 CFD Methods

Three-dimensional supersonic flows were analyzed using an unstructured grid CFD solver of Tohoku University Aerodynamic Simulation (TAS)—code [6]. In this research, inviscid simulations were mainly performed. The compressible Euler equations are discretized using a finite-volume cell-vertex scheme. In this study, the flow conditions were set to the freestream Mach number M of 1.7. For the evaluation of L/D, the drag coefficient of a waverider model is calculated as the sum of the inviscid pressure drag (obtained from inviscid CFD) and skin friction drag coefficients. The skin friction drag is estimated from a simple algebraic model based on the wetted area of objects [1]. Regarding the further evaluation process, Reynolds-averaged Navier–Stokes (RANS) simulations with the Spalart–Allmaras turbulence model [7] are also performed to consider the viscous effect. In the RANS simulations, prism elements are generated around the body to capture the boundary layers (highlighted in red box), as shown in Fig. 4. The mixed element unstructured grid is generated by MEGG3D [8, 9]. The typical numbers of grid points are 2.65 million

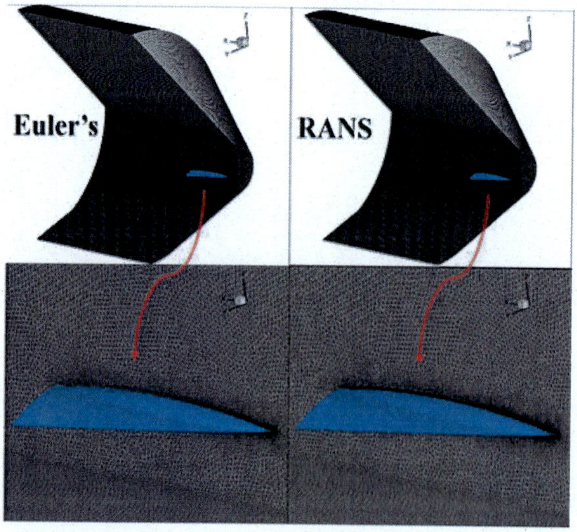

Fig. 4 Computational grids for Euler and RANS simulations

for inviscid simulations and 3.72 million for RANS simulations, respectively. In the inviscid simulations, a special boundary condition was used for the base plane of the waverider configuration for stable simulations, whereas a conventional non-slip (viscous) boundary condition was used for all object walls in the RANS simulations.

3.2 *Sonic Boom Analysis Methods*

The sonic boom waveforms on the ground are predicted by a nonlinear acoustic propagation solver Xnoise, which was developed by the Japan Aerospace Exploration Agency (JAXA) [10, 11]. The augmented Burgers equation is numerically solved using an operator splitting method, which takes into an account the effects of nonlinearity, geometrical spreading, stratification of the atmosphere, thermo-viscous absorption, and molecular relaxation. In this approach, the initial (input) pressure distributions are extracted from the CFD solutions on the lower side of the SST configurations. Then, the propagation of the pressure distribution to the ground is solved by the augmented Burgers equation. Finally, the maximum pressure rise P_{max} and A-weighted sound exposure level L_{AE} are evaluated from the waveform on the ground. In this analysis, the fuselage length and cruise altitude are respectively set to 62[m] and 18,000[m], that are given from the conditions of the Concorde.

4 Results and Discussions

4.1 *Results of Baseline Configuration*

We have designed/evaluated various waverider models before choosing a baseline configuration for this study. In this study, the tip angle of the imaginary cone is fixed to $\delta = 1[°]$ and the freestream Mach number for the design is set to $M' = 5$. The length of the waverider is set to 62[m]. The pressure distributions around the baseline configuration are shown in Fig. 5 at the freestream Mach number M of 1.7 with a fixed angle of attack (α) of 0.227°.

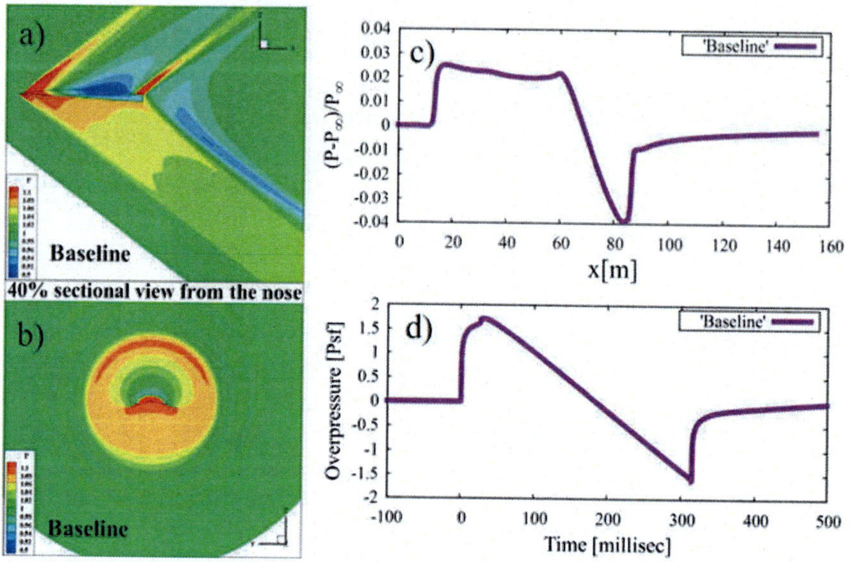

Fig. 5 Simulation results of baseline waverider model, **a**, **b** pressure distributions around the baseline model, **c** near-field pressure distribution used as input waveform for sonic boom analysis, **d** pressure waveform on ground

4.2 *Effect of Concave Adding on Lower Surface*

A concave shape is attached on the bottom side of the waverider to investigate its aerodynamic effect. The concave shape is created by engraving its lower surface as shown in Fig. 6. As for a comparison reason, the same concave model is added at several different positions on the lower surface from 25 to 40% chordwise location, as shown in Fig. 7.

The simulation results at $M = 1.7$ and $\alpha = 0.227°$ are summarized in Table 1. We can observe the reduction of sonic boom strength by adding the concave shapes, while the aerodynamic performance (L/D) is reduced. Regarding the ConcD design, it has the largest L/D and lowest P_{max} among the concave designs due to the reduction of high-pressure regions at the lower back side of the body (indicated by the arrow in

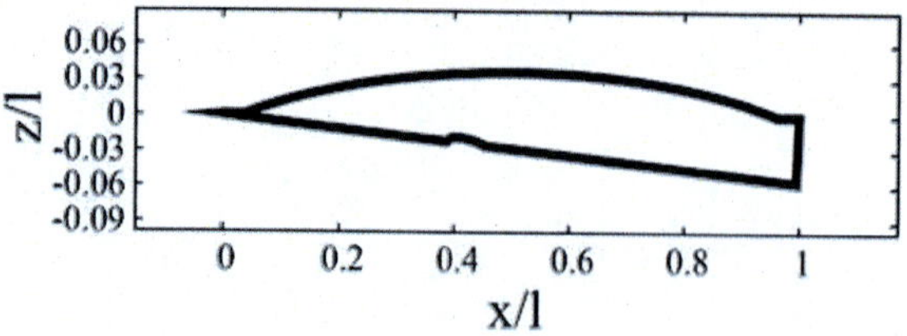

Fig. 6 Waverider model with concave shape at symmetrical plane, adding on 40% location

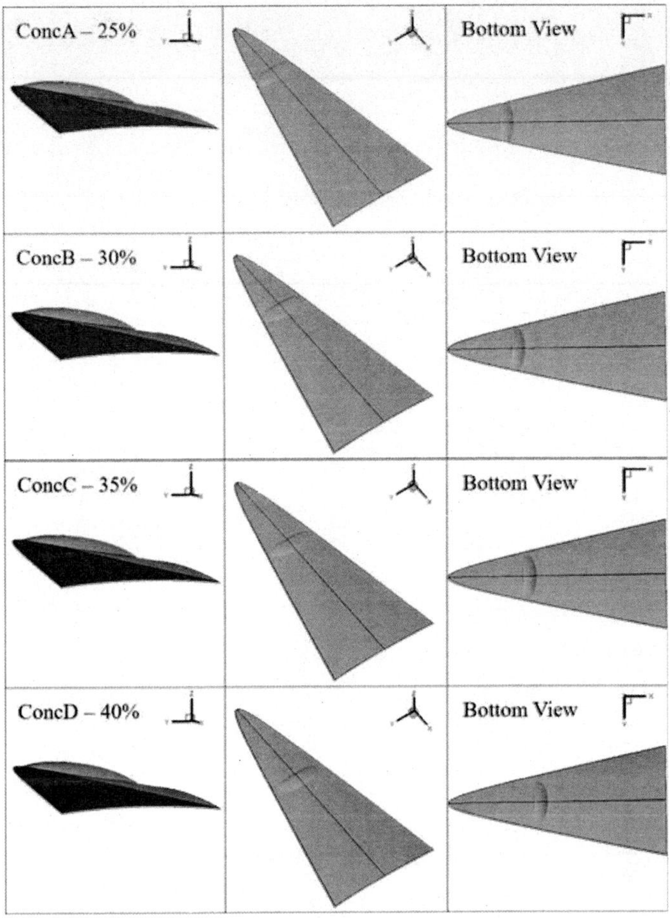

Fig. 7 Waverider models with concave shape, adding on 25, 30, 35, and 40% location

Table 1 Aerodynamic / sonic boom performance of waverider with concave shapes

Configurations	C_L	C_{DP}	C_{Df}	C_D	L/D	P_{max}, [Pa]	L_{AE}, [dB]
Baseline	0.13829	0.01004	0.00973	0.01977	6.99	82.30	86.34
ConcA	0.13795	0.01277	0.00854	0.02131	6.47	82.77	85.65
ConcB	0.13789	0.01266	0.00854	0.02121	6.50	82.31	85.14
ConcC	0.13827	0.01278	0.00854	0.02133	6.48	81.57	85.55
ConcD	0.13858	0.01270	0.00854	0.02125	6.52	80.25	85.88

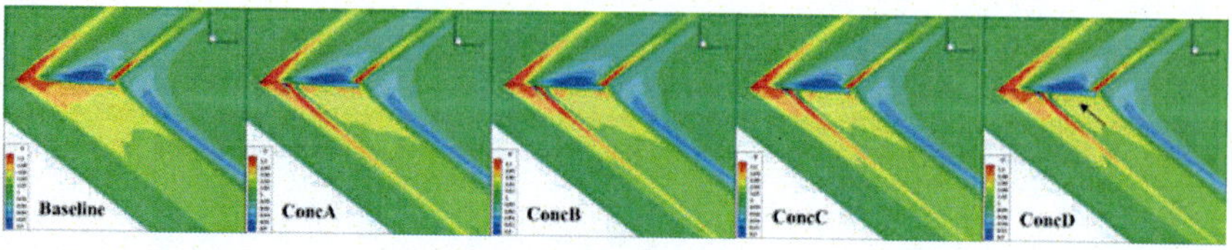

Fig. 8 Comparison of pressure distributions at symmetrical plane with different concave locations

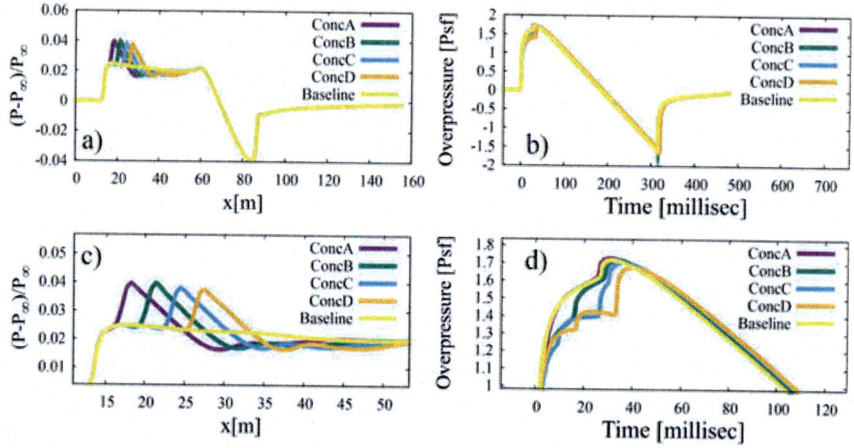

Fig. 9 Comparison of pressure distributions around different concave locations of models, **a** near-field pressure distributions, **b** distributions on ground, **c** and **d** zoomed-in figures of **a** and **b**

Fig. 8). The significant pressure reduction that observed in ConcD plays an important role in the concave model in this study. In Fig. 8, the pressure distributions at the symmetrical plane are compared. The concave designs have a low-pressure (blue) regions at the upstream side of the concave area. In Fig. 9, the pressure distributions at the near-field a), c) and ground b), d) are compared. The effect of adding concave shapes can be observed in the near-field distributions, which results in the reduction of sonic boom strength. Three divided peaks are observed at ConcD design in the distribution on the ground which is given from the effect of the concave shape.

In Fig. 9, the pressure distributions at the lower side of SST in (a), (c), the tendency of representative designs are almost same. Then the three divided peaks are observed at ConcD design in the distribution on the ground which is given from the effect of the concave shape.

By using the extracted design knowledge of the ConcD design, we investigated more designs with respect to the concave shapes/sizes. After examining 45 various concave shapes, representative designs were obtained as ConcD_S14, ConcD_S26, and ConcD_S27 as shown in Fig. 10, whose performance values are compared in Table 2. In Fig. 11, the pressure distributions at the symmetrical plane are shown which reveals that the concave shape generates expansion waves which decrease the

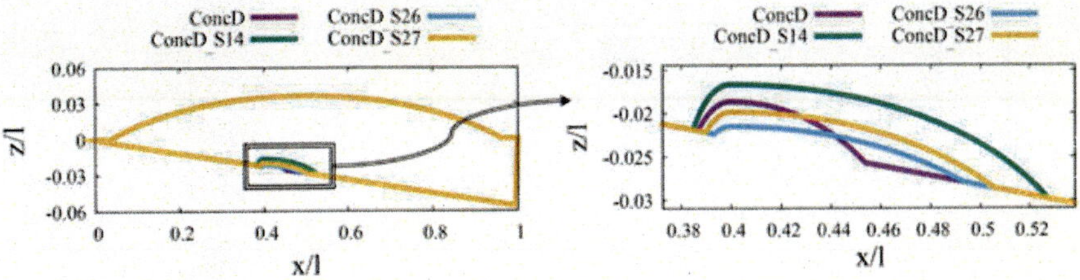

Fig. 10 Comparison of waverider model with various concave shapes at symmetrical plane, adding on 40% location

Table 2 Aerodynamic/sonic boom performance of representative concave designs of waverider

Configurations	C_L	C_{DP}	C_{Df}	C_D	L/D	P_{max}, [Pa]	L_{AE}, [dB]
Baseline	0.13829	0.01004	0.00973	0.01977	6.99	82.30	86.34
ConcD	0.13858	0.01270	0.00854	0.02125	6.52	80.25	85.88
ConcD_S14	0.14128	0.01627	0.00854	0.02482	5.69	69.22	86.59
ConcD_S26	0.13872	0.01041	0.00854	0.01896	7.32	81.77	86.00
ConcD_S27	0.13930	0.01158	0.00854	0.02012	6.92	79.39	86.11

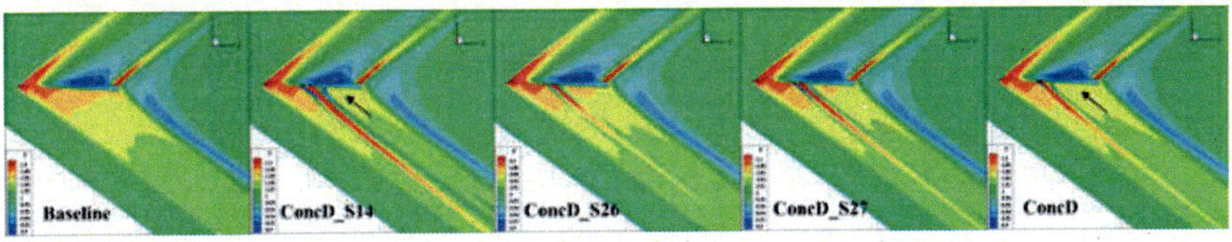

Fig. 11 Comparison of pressure distributions at symmetrical plane with various concave shapes

high-pressure regions (indicated by the arrow) in ConcD_S14. Similar but smaller expansion waves are generated also in the ConcD_26 and ConcD_27 designs. For all representative designs, the pressure distributions on the upper side of the body are almost identical and the concave shape is effective in controlling the flow around the lower side of the body. In Fig. 12, the near and far-field pressure distributions are compared. The three divided peaks are observed in the pressure waveforms on the ground at the design ConcD_27, similar to the ConcD design. In Fig. 13, the pressure distributions on the bottom surface are visualized. The front edge of the concave region affects the generation of low-pressure regions (visualized in blue) at the concave and next to the concave area (green pressure region of Fig. 13 at ConcD_S14), which has a positive impact on the reduction of sonic boom strength. A second low-pressure region is obtained at approximately 60% chordwise location in ConcD_S14, as observed from Fig. 13. A larger concave shape is likely to generate more low-pressure regions on the lower surface. The change in the concave front edge (ConcD_S26) provides more opportunities to improve aerodynamic performance by making a flatter concave.

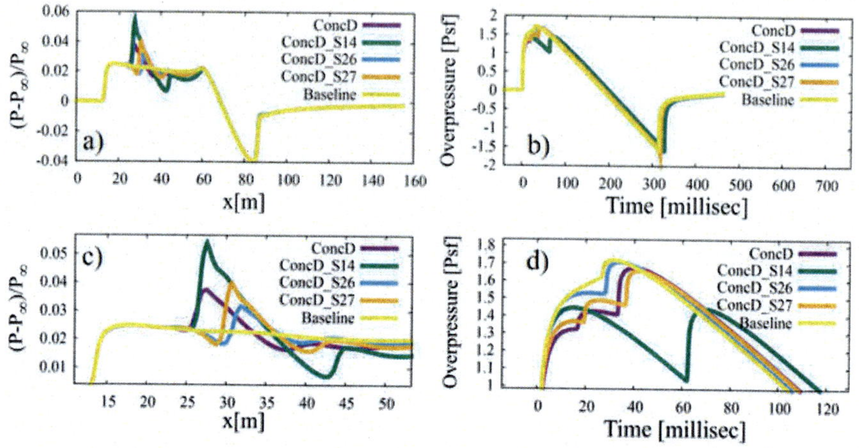

Fig. 12 Comparison of pressure distributions around various concave shape models **a** near-field pressure distributions, **b** distributions on ground, **c**, and **d** zoomed-in figures of **a** and **b**

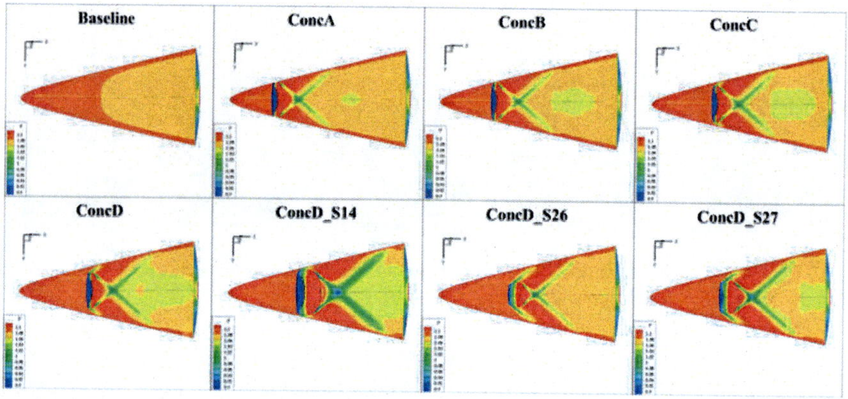

Fig. 13 Pressure distributions on the bottom surface of the representative designs

4.3 Validity of CFD Models

The validity of the CFD/sonic boom analysis method with inviscid simulation is investigated using the RANS method at the Reynolds number of 3.2 million. The aerodynamic as well as sonic boom performances are compared in Table 3, and it can be confirmed that the variation in sonic boom performances of the P_{max} and L_{AE} values

Table 3 Aerodynamic/sonic boom performance of representative concave designs of waverider, by Euler and RANS simulation methods

Configurations	C_L	C_{DP}	C_{Df}	C_D	L/D	P_{max}, [Pa]	L_{AE}, [dB]
Baseline EULER	0.13829	0.01004	0.01664	0.02668	5.18	82.30	86.34
Baseline RANS	0.13588	0.06539	0.02199	0.08738	1.55	88.07	85.88
ConcD EULER	0.13858	0.01270	0.01665	0.02935	4.72	80.25	85.88
ConcD RANS	0.12277	0.06658	0.02425	0.09083	1.35	86.49	88.14
ConcD_S27 EULER	0.13930	0.01158	0.01665	0.02823	4.93	79.39	86.11
ConcD_S27 RANS	0.13430	0.06655	0.02479	0.09135	1.47	85.24	88.29

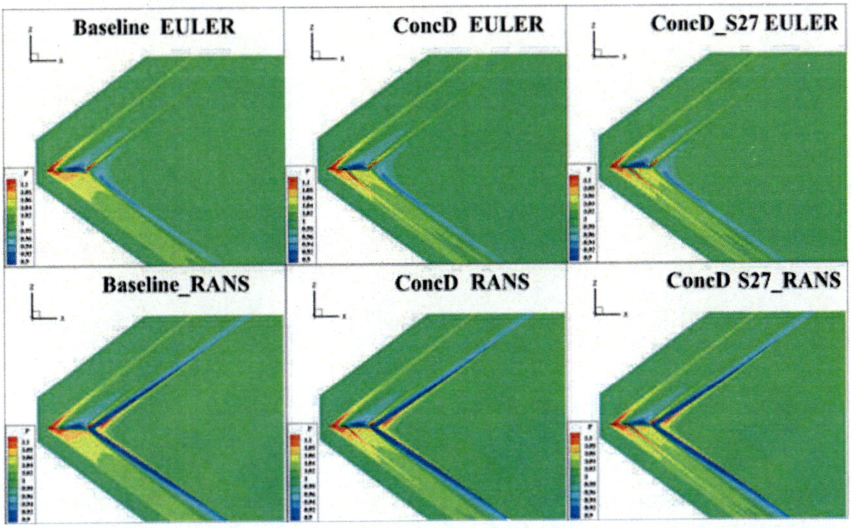

Fig. 14 Comparison of pressure distributions of representative concave designs, evaluated by Euler and RANS simulations

are qualitatively same. The pressure contours around the waverider configurations are visualized in Fig. 14. The flow field pressure distributions are qualitatively same, except for the base plane area. The pressure distributions at the near-field and ground are compared in Fig. 15. Overall the simulation results were qualitatively same, except the regions around the base plane.

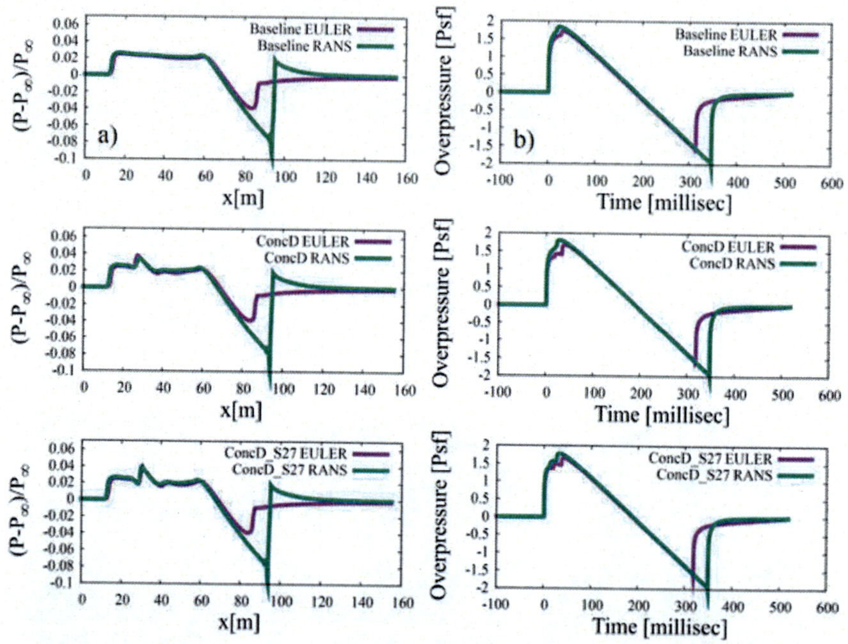

Fig. 15 Comparison of pressure distributions around the representative designs **a**, and pressure waveforms on the ground **b**

5 Conclusion

In this research, the combined waverider model with the Sears Haack body has been studied with its concave designs, and it showed that the attachment of the concave at the bottom side of the body is promising for further SST developments. First, we examined the efficient concave location at the bottom side of the waverider. The ConcD provided the most promising result, and then its concave sizing was also investigated. Consequently, obtained representative models were capable to generate more low-pressure regions at the lower side of the body than the baseline configuration which results in creating the expansion waves. The larger concave shapes have a positive impact for improving the sonic boom performances, while the flat and smaller concave shapes have the highest aerodynamic performance. The stronger/greater expansion waves had a positive impact on sonic boom performances. The main motivation of the addition of the concave shape is the successful interaction of compression/expansion waves at the lower side of the body, which could improve the performance of the sonic boom strength. Thus, it can be concluded that expansion waves are important for generating low-pressure regions next to the concave areas. For strengthening the performance of the introduced models, its aerodynamic design optimization is required in the future study.

Acknowledgements The authors are grateful to Japan Aerospace Exploration Agency for providing us the software of Xnoise.

References

1. Ban N, Yamazaki W, Kusunose K (2018) Low-boom/low drag design optimization of innovative supersonic transport configuration. Journal of Aircraft, doi 10(2514/1):C034171
2. Kamiyama K, Yamazaki W (2019) Aerodynamic performance augmentation of supersonic transport by fuselage shape. In: Sixteenth International Conference on Flow Dynamics, Sendai, Japan
3. Ishikawa Y, Kamiyama K, Yamazaki W (2020) An efficient fuselage concept using waverider configuration for supersonic transport. In: Seventeenth International Conference on Flow Dynamics, Sendai, Japan
4. Kim BS, Rasmussen ML, Jischke MC (1981) Optimization of waverider configurations generated from axisymmetric conical flows. J Spacecraft Rock. https://doi.org/10.2514/3.25630
5. Sears WR (1947) On projectiles of minimum wave drag. Q Appl Math 4:361–366
6. Yamazaki W, Matsushima K, Nakahashi K (2008) Drag prediction, decomposition and visualization ins unstructured mesh CFD solver of TAS-code. Int J Numer Meth Fluids. https://doi.org/10.1002/fld.1643
7. Spalart P R, Allmaras S R (1992) A one-equation turbulence model for aerodynamic flows. In: Journal of 30th aerospace sciences meeting and exhibit. https://doi.org/10.2514/6.1992-439
8. Ito Y, Nakahashi K (2002) Improvements in the reliability and quality of unstructured hybrid mesh generation. Int J Numer Meth Fluids. https://doi.org/10.1002/fld.669
9. Ito Y, Murayama M, Koike S, Yamamoto K, Nakakita K, Kusunose K (2019) Computational investigations of vertical stabilizer with vortex generator and dorsal fin. J Aircraft. https://doi.org/10.2514/1.C035301
10. Kanamori M, Takahashi T, Makino Y, Naka Y, Ishikawa H (2018) Comparison of simulated sonic boom in stratified atmosphere with flight test measurements. AIAA J. https://doi.org/10.2514/1.J056155
11. Yamamoto M, Hashimoto A, Aoyama T, Sakai T (2015) A unified approach to an augmented Burgers equation for the propagation of sonic booms. J Acoust Soc Am. https://doi.org/10.1121/1.4916833

Boundary-Layer Transition Control by Plasma Crossflow Reduction in Swept Wing

Yuto Miwa, Ikuya Yoshimi, Takashi Matsuno, and Dongyoun Kwak

Abstract The boundary-layer transition caused by the crossflow instability in a swept wing is controlled by a plasma actuator (PA). In this study, the control concept is to suppress the crossflow generated at leading edge of the wing by using PAs. The PAs are installed at the leading edge so that the body force can be applied in the direction that cancels the crossflow. An attempt was made to delay the boundary-layer transition by driving a PA. In a wind tunnel experiment, we investigate the transition position and velocity in the boundary layer at U = 20 m/s. The experimental results showed that the transition position delayed when PA was driven. From the spectral analysis of the velocity in the boundary layer, we found that the boundary-layer transition process changed from crossflow instability mode when PA was driven.

Keywords Boundary layer · Transition · Flow control · Plasma actuator

1 Introduction

The importance of improving fuel efficiency has increased further as the demand for air transportation has grown, for both economic and environmental reasons. Since most fuel is consumed during cruise flights, reducing cruise drag is imperative to reducing fuel consumption. The skin friction drag accounts for more than 50% of the total drag during cruising [1]; thus, minimizing viscous drag is crucial to reduce fuel consumption.

A method to reduce skin friction drag is keeping the boundary layer flow around the aircraft laminar. Since the skin friction drag increases significantly as the boundary-layer flow transitions from laminar to turbulent, the skin friction drag can be reduced by maintaining a laminar boundary-layer [2, 3]. Turbulent transitions are caused by

Y. Miwa · I. Yoshimi · T. Matsuno (✉)
Tottori University, Tottori 680-8552, Japan
e-mail: matsuno@tottori-u.ac.jp

D. Kwak
Japan Aerospace Exploration Agency, Mitaka, Tokyo 181-0015, Japan

S. Lee et al. (eds.), *The Proceedings of the 2021 Asia-Pacific International Symposium on Aerospace Technology (APISAT 2021), Volume 1*, Lecture Notes in Electrical Engineering 912, https://doi.org/10.1007/978-981-19-2689-1_56

instability in the flow. The process of turbulent transition depends on the onset and growth of the instability of flow field.

In a swept wing, used for the main wing of subsonic commercial aircraft, the main cause of turbulent transition is crossflow instability (CFI). The sweepback leading edge of the wing causes a gradient in the pressure over the wing surface in the wing spanwise direction. This pressure gradient induces a flow in the wing spanwise direction at the leading edge of the wing, thereby causing the main flow to be curved. At the leading edge of the wing, a three-dimensional boundary-layer velocity profile with inflection points is formed by the curved main flow. CFI is caused by the instability of the inflection point, and wave growth by the CFI induces crossflow vortices downstream [4]. At the downstream, the Kelvin–Helmholtz instability in the shear layer of the crossflow vortex becomes a cause of secondary instability generating waves at specific wavelengths, which grow and collapse before the transition to turbulence. The CFI mitigation is imperative to suppress the turbulent transition in the swept wing.

For the significance of reducing the abovementioned skin friction drag, several studies have been conducted on the turbulent transition control. One method is reducing aerodynamic drag by passively controlling the boundary layer using small depressions called dimples [5]. In the study of the swept wing, active control methods, such as suction of crossflow vortices [6], have been employed to delay the transition. Other methods that use discrete roughness elements (DRE) to generate arbitrary vortices to suppress the growth of crossflow vortices have also been proposed [7].

For the active flow control method of the boundary layer, plasma actuators (PAs) have attracted research attention. A PA comprises two electrodes and a dielectric layer that is put between the electrodes (as will be shown later). Applying AC voltage between the electrodes creates electric discharges, which ionize the surrounding gas. Then, the generated ions generate an induced jet by the body force subjected to the electric field [8, 9].

An essential feature of PAs is their very small invasion to the flow field. Since only the exposed electrodes are in contact with the flow, they have an advantage in controlling phenomena sensitive to the surface roughness of the wing, such as boundary-layer transition. Nevertheless, the CFI is significantly affected, even by minute surface roughness. For example, from the results of the abovementioned DRE study [10], a roughness of 0.012 mm in height was sufficient to grow flow instability in wind tunnel tests at a sweepback angle of 45° and a Reynolds number of $Re_c = 2.4 \times 10^6$. The thickness of the electrodes of the PA must also be carefully considered because the stability of the flow becomes weaker as the uniform flow velocity increases.

This study aims to reduce the CFI and delay the turbulent transition by mitigating the crossflow component in the boundary layer using a PA. In this study, the boundary-layer transition characteristics and the growth of CFI were investigated by wind tunnel experiments using a flat plate model with a sweptback angle of 60°. Then, the effect of controlling the boundary-layer transition was investigated with PAs installed to mitigate the crossflow at a leading edge of the model.

2 Experimental Method and Conditions

2.1 *Overview of Experiment and Flow Control Concept*

First, we investigated the effect of the disturbance from an exposed electrode of PA on the flow field. The transition positions were compared under the condition that the PA was installed and NOT driven, and the condition that nothing was placed on the model ("nude" condition). The nude condition was assumed to be a natural flow field with minimal external disturbance to the boundary layer.

Next, the transition characteristics were examined while varying the angle of attack (AOA) to determine if the turbulent transition was due to the CFI. In a flow field where CFI is occurring, the crossflow causes an inflection point in the boundary-layer velocity distribution at the leading edge of the airfoil. A vortex is generated starting from this inflection point [11]. In this study, we measured the transition position at various AOAs to identify the laminar region and searched for the AOA under which crossflow vortices can be observed in the laminar region.

The boundary-layer transition control test driven by PAs was performed under the above conditions. From our previous study [12], the crossflow was generated at the leading edge of the wing toward the wing tip. Figures 1 and 2 show the results of numerical analysis of a flat plate wing model for a uniform flow velocity of 20 m/s. The spanwise velocity profile of the boundary layer showed that the flow inside the boundary layer was toward the wing root from the leading edge to x/c = 0.01, then toward the wing tip after x/c = 0.049. The result of the oil-flow test (Fig. 3) confirmed that the flow to the wing tip was generated in the wind tunnel test. From these results, it was confirmed that the wing tip direction flow was generated, except for near the leading edge.

We attempted to suppress the crossflow component by introducing a volume force in the wing root direction using a PA to cancel out the crossflow. A PA was installed

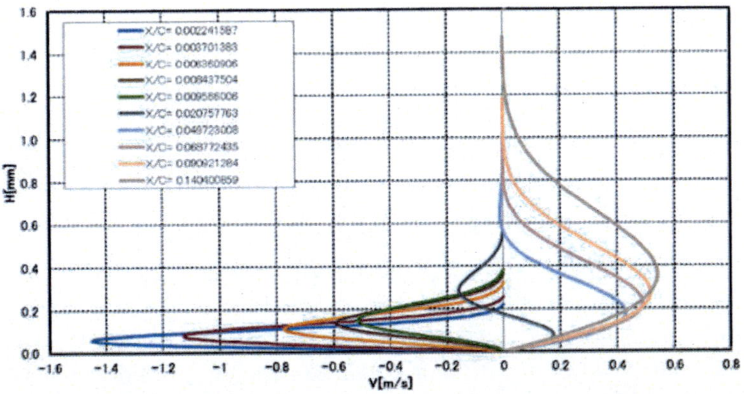

Fig. 1 Spanwise velocity profile

Fig. 2 Oil-flow visualization of surface flow by numerical simulation

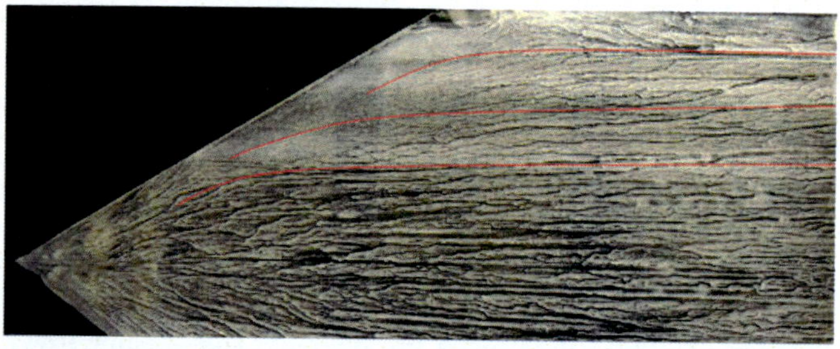

Fig. 3 Oil-flow streamline

parallel to the leading edge of the wing to generate the body force toward the wing root direction.

2.2 *Experimental Apparatus and Method*

The wind tunnel experiments were conducted using a low-speed wind tunnel of Tottori University. The measurement section was 600 × 600 mm^2 and had a floor plate of 1800 mm in the flow direction. The maximum uniform flow velocity was 30 m/s. The level of turbulence was <0.7% of the freestream velocity.

Figure 4 shows the wing model used for this study. The model is a 20 mm thick acrylic flat plate with a 60° leading-edge sweepback angle (Fig. 4b). The leading-edge shape has a 1:10 elliptical distribution in the cross-section perpendicular to the leading edge (Fig. 4c). At the trailing edge, the bottom side is cut off at an angle (Fig. 4d).

The PA used in this study is a single-dielectric barrier discharge type. The schematic configuration is shown in Fig. 5. Figure 6 shows the configuration of the dielectric layers, which comprise three layers of 0.18 mm polytetrafluoroethylene (PTFE) and one layer of 0.055 mm polyimide. The thickness of the dielectric layers is 0.595 mm. The exposed electrode width is 10 mm, and the discharge length

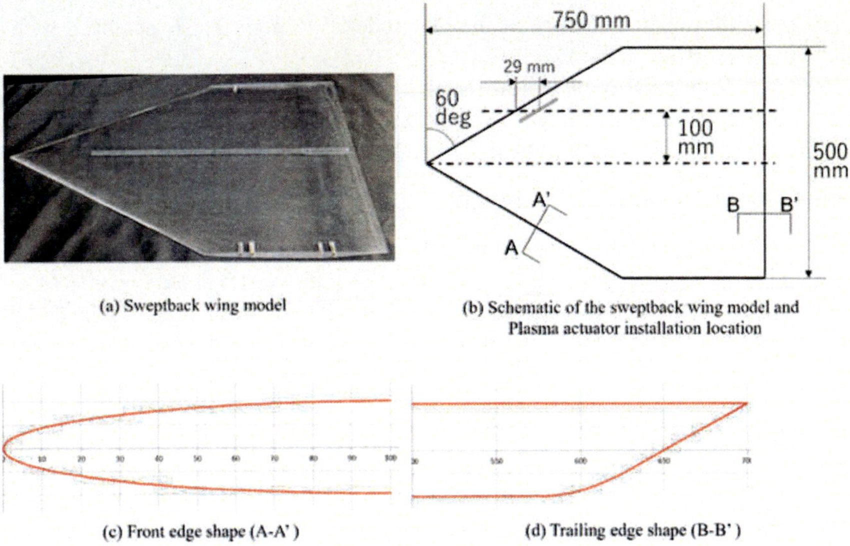

(a) Sweptback wing model

(b) Schematic of the sweptback wing model and Plasma actuator installation location

(c) Front edge shape (A-A')

(d) Trailing edge shape (B-B')

Fig. 4 The flat plate model

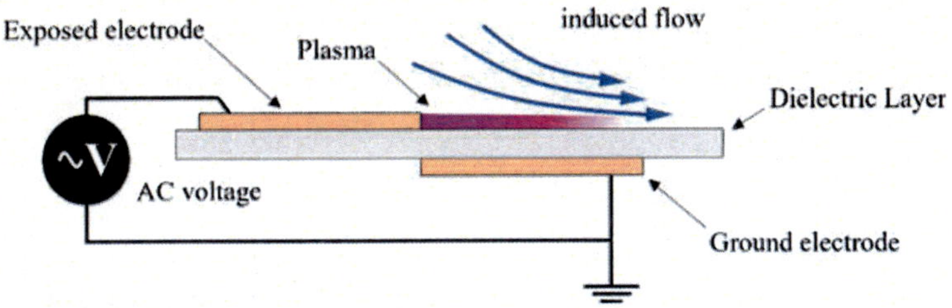

Fig. 5 Schematic configuration of a single-dielectric barrier discharge PA

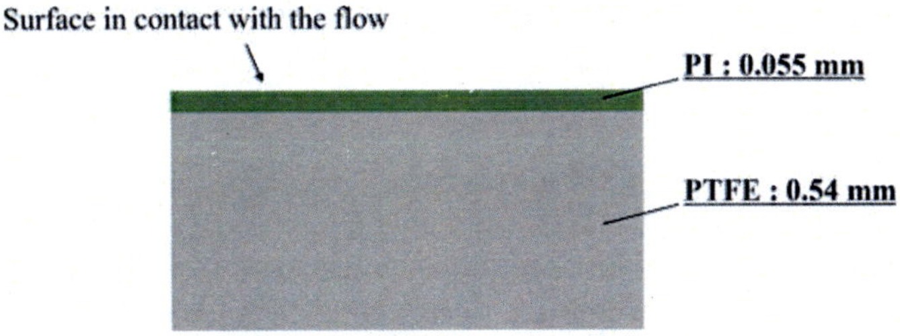

Fig. 6 Configuration of the dielectric layer

is 100 mm. The mean thickness of the electrode is <6 mm, that is achieved with conductive nanosilver painting.

For the driving conditions of PA, the AC voltage applied to the electrodes was generated as a sinusoidal wave with a 10 kHz frequency and a 15.2 kV_{pp} voltage by amplifying the signal from the signal generator using a series of transformers.

Total Pressure Measurement to Identify Transition Position

A Preston tube measurement was employed for transition position identification. The thickness of the Preston tube was 0.35 mm. The tube was pressed to contact the model wall to measure the total pressure at the same height in the boundary layer. In this measurement, a three-axis traversing system was used to move the Preston tube. The total pressure distribution on the wall was measured by traversing in the wing chord direction. The transition position of turbulence was identified in the following manner: when the boundary layer transitioned from laminar to turbulent, the velocity increased, even at the same height, due to the increased momentum transport caused by turbulence, which resulted in a dynamic pressure increase. Therefore, the turbulent transition position could be identified once the dynamic pressure, calculated from measured total pressure and static pressure, increases was observed. For the experiments, 42 measurements were taken with 10 mm intervals in the downstream direction from the $x/c = 0.1$–0.8.

Total Pressure Measurement to Investigate Crossflow Vortices

Preston tubes were also used for vortex measurements. The Preston tube used in these measurements had the same specifications described above. Similar to the measurement method for the transition identification, the total pressure distribution at the wall surface in the wing spanwise direction was measured by traversing the Preston tube in the wing spanwise direction using a three-axis traversing system. When CFI was developed, vortices were generated with a specific wavenumber in the spanwise direction from the crossflow component inflection point. The vortex identification was done by measuring the dynamic pressure variation at the wall, that is the velocity variation at the wall caused by the longitudinal vortex. The vortex identification and measurement of wavenumber were done by measuring the velocity fluctuations at the wall.

Boundary-Layer Velocity Measurement

A hot-wire anemometer was used to measure the boundary-layer velocity distribution. By traversing the hot-wire probe away from the wall, the velocity distribution in the height direction was measured (Fig. 7). Figure 8 shows an example of boundary-layer velocity distribution. The vertical axis of Fig. 8 is normalized height by the boundary-layer thickness (δ_{99}), whereas the horizontal axis is normalized streamwise local velocity by uniform flow velocity (U_∞). By comparing the measured velocity distribution with the Blasius solution of the flat laminar boundary layer and the 1/7th power law of the turbulent boundary layer, the laminar/turbulent flow was determined. The power spectrum of the velocity was analyzed from the hot-wire anemometer data and then compared with Kolmogorov's −5/3 power law to determine whether the flow was laminar or turbulent.

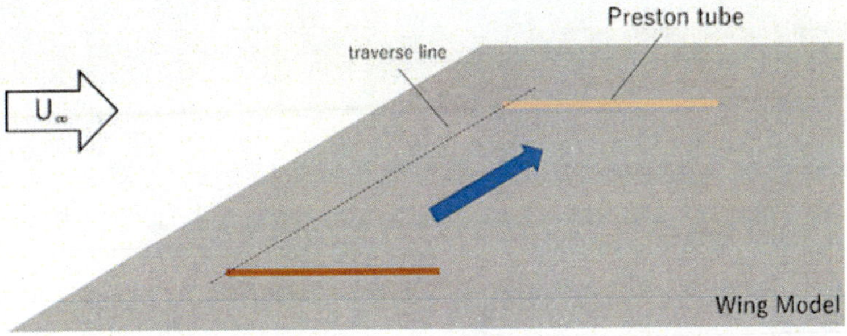

Fig. 7 Spanwise total pressure measurement near the wall using Preston tube

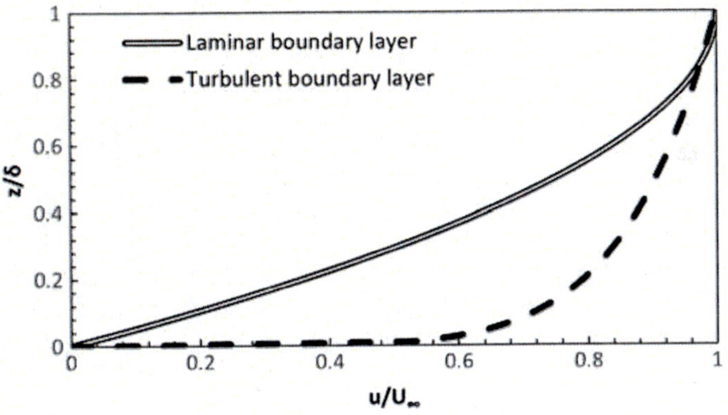

Fig. 8 Boundary-layer velocity distribution

3 Results and Discussion

3.1 Characteristics of the Boundary-Layer Transition

First, we investigated the effect of the disturbance from the exposed electrode of PA on the transition characteristics. Figure 9 shows the chordwise dynamic pressure distribution of the model at $U_\infty = 20$ and 25 m/s and AOA = 2° measured by Preston tube. Two configurations as mentioned above, "nude" and the PA installed, were compared. Figures 9a and b show that the transition positions for these configurations were almost the same, which confirmed that the thickness of the electrode did not affect the boundary layer.

In the experiments, the position where the dynamic pressure began to rise rapidly is defined as the transition start position, and the position where the dynamic pressure changed from a rapid upward trend to a gradual downward trend is defined as the transition completion position. The transition position was defined as the midpoint between the transition start and completion positions.

Fig. 9 Investigation of the effect of electrode roughness. **a** Transition position, U = 20 m/s. **b** Transition position, U = 25 m/s

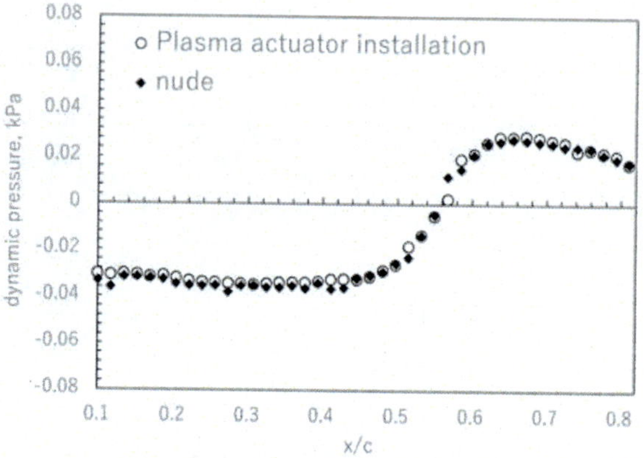

(a) Transition position, U = 20 m/s

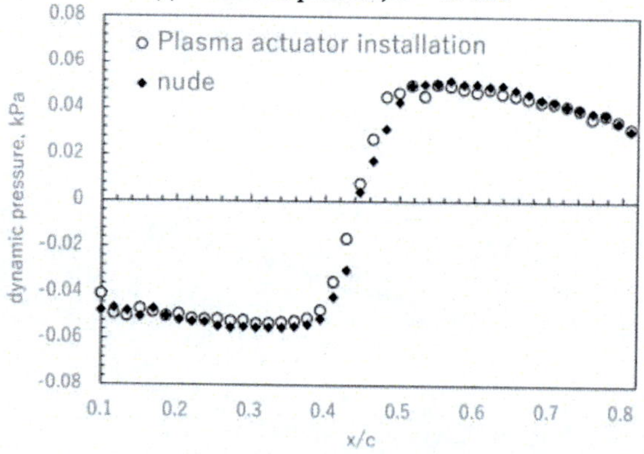

(b) Transition position, U = 25 m/s

Next, we investigated the transition characteristics for several AOAs. A series of wind tunnel experiments was conducted at $U_\infty = 20$ m/s. Figure 10a shows the results of the transition position measurement when the AOA was set to 2.0°, 2.3°, and 3.0°. When the AOA was 2.0° and 2.3°, the transition positions were at x/c = 0.46 and 0.22, respectively. For AOA = 2.0°, the area upstream of x/c = 0.46 was a laminar region, whereas, at AOA = 2.3°, the area upstream of x/c = 0.22 was a laminar region. When AOA = 3.0°, the dynamic pressure increased from the leading edge, so the region was in the transition zone from the leading edge, and the laminar region could not be identified.

The dynamic pressure distribution in the spanwise direction near the wall at the above AOAs is shown in Fig. 10b. These experiments were to verify if vortices occurred in the laminar region. The measurement position was set to x/c = 0.13, which was the laminar region, both for AOA = 2.0° and 2.3°. As shown in Fig. 10b, a periodic structure of about 8 mm was observed in the total pressure distribution at AOA = 2.3°, indicating a vortex generation. Meanwhile, at AOA = 2.0°, no periodic change in dynamic pressure was observed. These results strongly suggested that

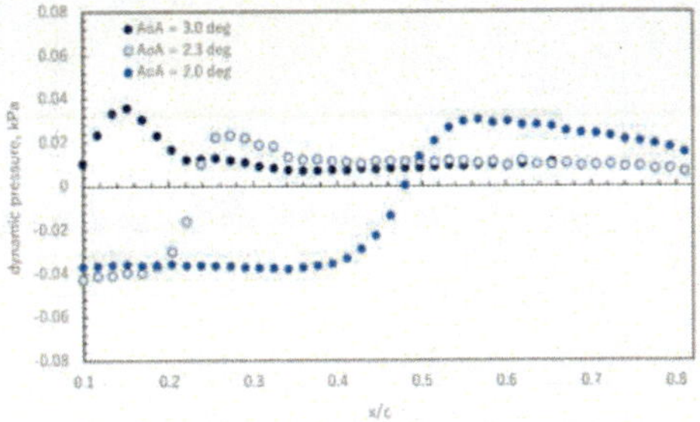

(a) Dynamic pressure distribution on the wall surface in the chord direction

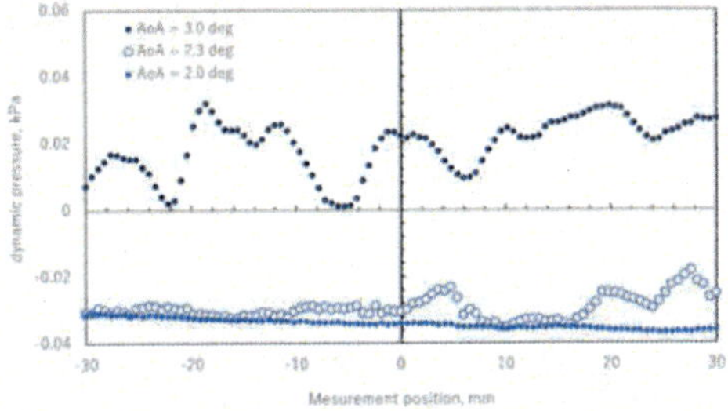

(b) Dynamic pressure distribution on the wall surface in the span direction at x/c = 0.13

Fig. 10 Investigation of transition characteristics for various angles of attack. **a** Dynamic pressure distribution on the wall surface in the chord direction. **b** Dynamic pressure distribution on the wall surface in the span direction at x/c = 0.13

the turbulent transition at AOA = 2.3° was dominated by the CFI. Notably, these results did not indicate that the CFI did not appear at AOA = 2.0°. Possibly at the downstream, the longitudinal vortex series may be observed, even at AOA = 2.0°.

3.2 Controlling Boundary-Layer Transition

The transition position was measured when the PA induced the body force. Figure 11 shows the chordwise dynamic pressure distribution of the model at $U_\infty = 20$ m/s and AOA = 2.3° measured by Preston tube. Two cases, PA is operated ("PA on") and PA is not operated ("PA off") were compared. The differences in the transition position were investigated by comparing their dynamic pressure distribution. As mentioned above, a high-voltage AC sinusoidal waveform with a 10 kHz frequency and a 10 kV_{pp} voltage was applied during the operation. Figure 11 shows that the transition occurred at x/c = 0.22 for PA off. Meanwhile, for PA on, the transition position was successfully delayed to x/c = 0.46.

The velocity distributions of the boundary layer at several chordwise locations with the PA operation are shown in Fig. 12. Measurements were performed using a hot-wire anemometer. The driving conditions of the PA were identical to the case in Fig. 11. Figure 12a and b show that all cases present velocity profiles similar to the

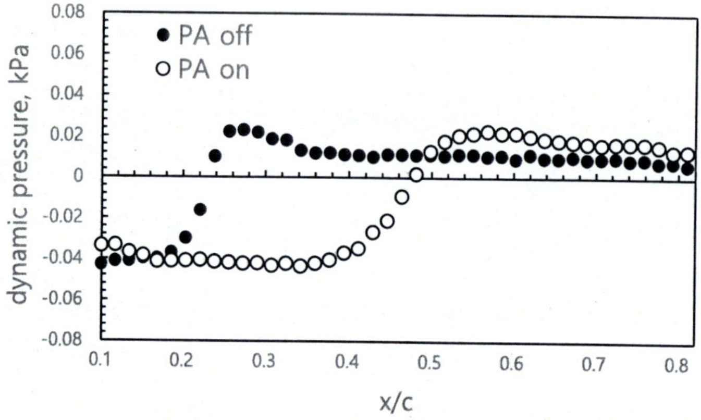

Fig. 11 Comparison of transition positions driven by PA

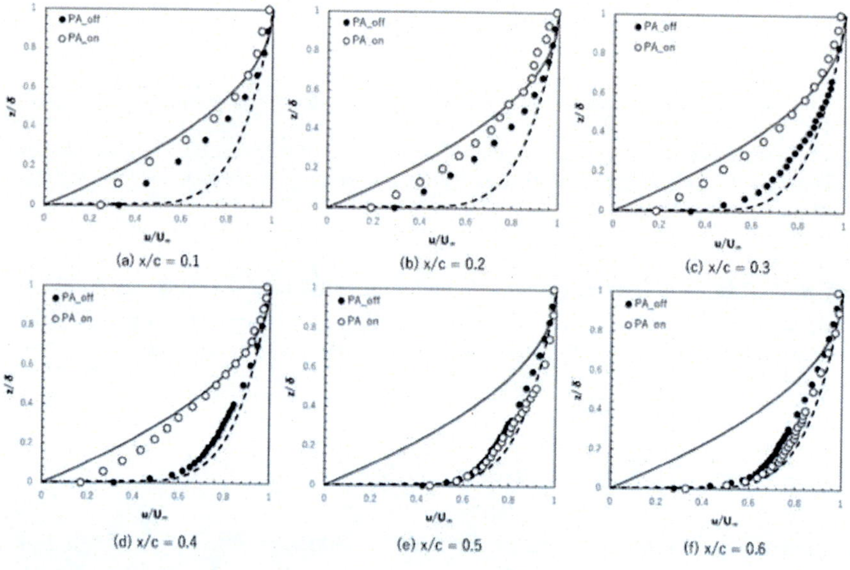

Fig. 12 Dimensionless boundary-layer velocity distribution

Blasius solution. We determined that the flow was laminar in both conditions for x/c = 0.1 and 0.2 from the plots. At x/c = 0.3 (Fig. 12c), there was a large difference in the velocity profile between the PA on and PA off cases. For PA on, the velocity profile was similar to the Blasius boundary layer, the same as in Fig. 12a and b. Meanwhile, for PA off, the velocity profile approached the 1/7th power law curve. At x/c = 0.4 (Fig. 12b), the profile was similar to the result for x/c = 0.3. Finally, in Fig. 12e and f, the velocity profiles for both cases became closer to the 1/7th power law curve. These results suggested that the laminar boundary layer could be maintained up to x/c = 0.4 for PA on, whereas the laminar region shrunk to x/c < 0.3 for PA off. These results agreed well with the measurement of the transition position by the Preston tube. The transition delay by the PA was also confirmed by the measurement using the hot-wire anemometer.

Next, the variation in the power spectrum of velocity in the mainstream direction with flow direction is shown in Fig. 13. By the spectral analysis of the hot-wire anemometer data, the frequency component of the flow direction velocity fluctuation was expressed as energy. The power spectrum was averaged from nine data points measured in 0.1 mm increments from the wall position (z = 0 mm) to 0.9 mm height at each chord position. The vertical and horizontal axes are both logarithmic. The vertical axis can be interpreted as the amount of energy of the turbulent eddies or vortices, whereas the horizontal axis is the size of the eddies, i.e., the higher the frequency, the smaller the scale of the vortex. Kolmogorov's −5/3 power law is also shown on the plot.

For PA off, a peak at approximately 1 kHz appeared at x/c = 0.13, and the energy level increased at x/c = 0.16 (Fig. 13a and b). Further downstream, at x/c = 0.20, the energy at this peak significantly developed (Fig. 13c). At x/c = 0.30, the peak suddenly disappeared and became a continuous spectral distribution. This distribution shows Kolmogorov's −5/3 power law, suggesting that the boundary layer was turbulent there.

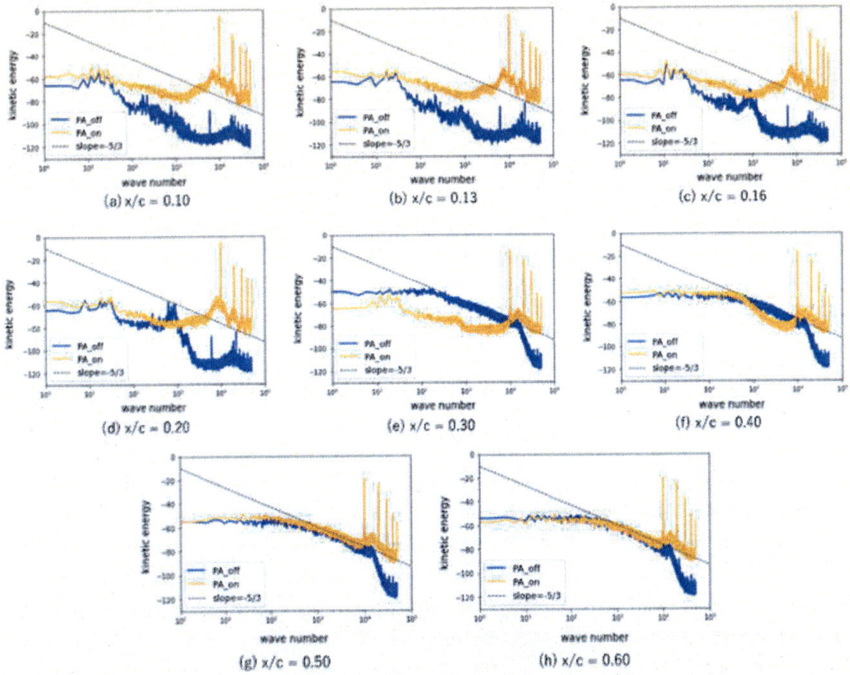

Fig. 13 Power spectra taken at various streamwise stations

The energy peak found in Fig. 13a–c shows similar characteristics as the findings of Kawakami et al. [13], which investigated the development of secondary instability in crossflow vortices. They reported that the energy peak was amplified and the turbulent transition occurred immediately after the large increase in the peak energy level, which indicated the physical mechanism of the transition; high-frequency secondary unstable modes were generated from the shear layer of the crossflow vortices, generated by the alternation of low-speed and high-speed regions due to the lifting of low-speed flow mass by the crossflow vortices. This unstable mode significantly influenced the final stage of the turbulent process. From the comparison of these transition processes, the energy peak seen in Fig. 13a–c was determined due to secondary instability on the crossflow vortices.

For PA on, the spectral distribution showed that the fluctuation level was relatively higher for $x/c < 0.16$. The distribution gradually showed a turbulent profile, and finally, Kolmogorov's power law at $x/c = 0.50$. The change in the spectral distribution also confirmed that the effectiveness of PA on the transition delay in this experiment. Compared to PA off, no similar peak in the energy distribution was observed until the transition was completed, which implied that the transition mode for PA on was completely altered from CFI to another, e.g., Tollmien–Schlichting wave. The existence of sharp peaks above 10 kHz at all x/c positions was due to the electric noise from discharge on the PA because they are consistent with the PA-driving frequency.

4 Conclusion

To reduce the CFI and delay the turbulent transition, by mitigating the crossflow component in a boundary layer using a PA, the boundary-layer transition characteristics and PA effect on the boundary-layer transition were investigated. By wind tunnel experiments using a flat plate model with a sweptback angle of 60° and their analysis, the following findings were obtained.

The exposed electrode formed by nanosilver particles generated a minute disturbance, which is lower than the natural flow. At $U_\infty = 20$ m/s, the boundary layer on the sweptback flat plate model showed transition by the CFI. A fluctuation of about 8 mm period was observed in the dynamic pressure distribution at AOA = 2.3°, indicating the turbulent transition was dominated by the CFI. From the spectral analysis of the velocity fluctuation in the boundary layer, we found the energy peak and their development that supported the transition process governed by secondary instability on the crossflow vortices.

By applying the PA to the flow field where the CFI was the dominant mode of the transition, the turbulent transition position was successfully delayed to the downstream. The trend of the power spectrum distribution showed the difference in the transition process when PA was operated.

Acknowledgements A part of this work was supported by JSPS KAKENHI Grant Number 20KK0104.

References

1. Group on International Aviation and Climate Change (2009) GIACC 3rd Meeting Working Paper GIACC/3–WP/6
2. Ueda Y, Ishikawa H, Yoshida K (2004) Three dimensional boundary layer transition analysis in supersonic flow using a Navier–Stokes code. In: Proceedings of 24th congress of the international counsil of the aeronautical science 2004-2.8.2
3. Chenyshev SL, Kiselev APH, Kuryachii AP (2011) Laminar flow control research at TsAGI: past and present. Prog Aerosp Sci 47:69–185
4. Saric WS, Reed HL, White EB (2003) Stability and transition of three-dimensional boundary layers. Ann Rev Fluid Mech 35:413–440
5. Aoki K, Muto K, Okanaga H (2010) Mechanism of drag reduction by dimple structures on a sphere. Trans JSME (B) 76(770):1473–1480
6. Friederich T, Kloker MJ (2012) Control of the secondary cross-flow instability using localized suction. J Fluid Mech 706:470–495
7. Carpenter A, Saric WS, Reed HL (2008) Laminar flow control on a swept wing with distributed roughness. In: 26th AIAA Applied Aerodynamics Conference 2008-7335
8. Thomas FO, Corke TC, Iqbal M, Kozlov A, Schatzman D (2009) Optimization of dielectric barrier discharge plasma actuators for active aerodynamic flow control. AIAA J 47(9):2169–2178
9. Enloe CL, McLaughlin TE, VanDyken RD, Kachner KD, Jumper EJ, Corke TC (2004) Mechanisms and responses of a single dielectric barrier plasma actuator: plasma morphology. AIAA J 42(3):589–594
10. Hunt LE, Saric WS (2011) Boundary-layer receptivity of three-dimensional roughness arrays on a swept-wing. In: 41th AIAA fluid dynamics conference and exhibit 2011-3881
11. Kohama Y, Saric WS, Hoos JA (1992) Laminar-turbulent transition on swept wing. Trans JSME (B) 58(554):107–113
12. Matsuno T, Esaki H, Tanaka K, Aono H, Kwak D (2020) Transition control for a swept wing by crossflow stabilization using plasma actuators. AIAA Scitech 2020 Forum 2020-0826
13. Kawakami M, Kohama Y, Okutsu M (1999) Stability characteristics of stationary crossflow vortices in three-dimensional boundary-layer. AIAA Paper 99-0811

Numerical Investigations of Ground Effect of a Quadcopter

Koichi Yonezawa, Kazuki Akiba, Hao Liu, Hideaki Sugawara, Yasutada Tanabe, Hiroshi Tokutake, and Shigeru Sunada

Abstract The hovering in ground effect (IGE) of a quadcopter is investigated numerically in terms of flow patterns and aerodynamic characteristics with two different rotor configurations with two different rotor-to-rotor distances. A model of variable-pitch-controlled rotors revealed complicated outwash flow patterns around the quadcopter, showing a pronounced directivity, particularly when the rotor-to-rotor distance was shortened. A narrow and intense outwash (jet) region formed between the rotors. A similar outwash pattern was also found in the diagonal plane. Our results further show that the rotor configuration effect on the aerodynamic performance of the quadcopter can vary significantly with changes in altitude from the ground. Moreover, the rotor configuration with a short rotor-to-rotor distance displayed a relationship between aerodynamic characteristics and altitude that was similar to that of a single rotor.

Keywords Quadcopter · Aerodynamics · Numerical simulation · Ground effect

K. Yonezawa (✉)
Central Research Institute of Electrical Power Industry, Abiko, Japan
e-mail: koichi-y@criepi.denken.or.jp

K. Yonezawa · H. Liu
Center for Aerial Intelligent Vehicles, Chiba University, Chiba, Japan

K. Akiba · H. Liu
Graduate School of Engineering, Chiba University, Chiba, Japan

H. Sugawara · Y. Tanabe
Japan Aerospace Exploration Agency, Tokyo, Japan

H. Tokutake
Institute of Science and Engineering, Kanazawa University, Kanazawa, Japan

S. Sunada
Graduate School of Engineering, Nagoya University, Nagoya, Japan

S. Lee et al. (eds.), *The Proceedings of the 2021 Asia-Pacific International Symposium on Aerospace Technology (APISAT 2021), Volume 1*, Lecture Notes in Electrical Engineering 912, https://doi.org/10.1007/978-981-19-2689-1_57

1 Introduction

The development and applications of electric-powered aircraft are widely spreading. In most electric-powered aircraft, such as multicopters and e-VTOL, multiple propellers are used as thrusters. To enhance the efficiency, safety, and environmental compatibility of such aircraft, it is important to understand the aerodynamic characteristics around the rotors while considering the aerodynamic interactions between rotors and between rotors and surrounding structures. Li et al. reported the influences of the rotor configuration of a quadcopter, such as differences in distance and height or the tilt angles between rotors, on the aerodynamic characteristics [1]. That study found that a long distance and a certain height difference with a zero-tilt angle were appropriate. Kohno et al. carried out an experimental investigation of the aerodynamic characteristics of a quadcopter in ground effect (IGE) and clarified that the distance between the rotors significantly affected the ground effect [2]. The total thrust increased monotonically as the rotor height decreased, which is well-known ground effect of a single rotor. When the rotor gap was >0.3 times the rotor radius, on the other hand, the relationship between thrust and altitude was not monotonic unlike case with a single rotor. Flow patterns were also observed, and flow recirculation was found as a result of the interaction between the rotors and the ground plane [3]. Some of the authors of the present paper previously reported a numerical study of the ground effect on a quad rotor [4, 5]. In those studies, a variable-pitch-controlled quadcopter was modelled, and the ground effect was examined under constant thrust condition by trimming the rotor pitch angle. The results showed that the required power of the quadcopter did not change monotonically due to the altitude change. The outwash pattern of the quadcopter showed directivity. Although this pattern resembles that of a single rotor in the cross section through the diagonal rotor center, high-velocity outwash occurred along the gap between neighbouring rotors. The influence of the central fuselage was also examined, and the results showed that the lift force increased but the required power did not, due to the presence of the fuselage at low altitude. Upwash took place at the center of the quadcopter as a ground effect, and the aerodynamic lift force acted on the fuselage but the recirculation flow around the quadcopter was not affected significantly.

In the present study, the influence of the distance between rotors is discussed as a continuation of the work in our previous studies. The geometries of the rotor blades are identical to those used in the previous studies. The fuselage is not considered. Flow patterns of the outwash and upwash are compared, as is the required power change with changes in altitude.

2 Quadcopter Modelling

The examined quadcopter with variable-pitch-control mechanisms, which was developed by the authors, is shown in Fig. 1 [6]. The rotational frequency of the rotors is

Fig. 1 Quadcopter with variable-pitch-control mechanism

fixed at 90 Hz, and the thrust is varied by changing the collective pitch of the rotor. The thrust can be changed within 1/10 of the time lag of rotor-speed control [7].

The rotors were evenly spaced, and the configuration was determined using diagonal distance, L, and the gap, d, between neighbouring rotor blade tips, as shown in Fig. 2. In the present study, two kinds of rotor configurations were examined, as shown in Table 1. Configuration 2 was examined in previous studies [4, 5]. The coordinate system is also determined as shown in Fig. 2. The heights of the ground and the rotors are determined as 0 and h along the z-axis, respectively. The design parameters of the examined quadcopter are shown in Table 2.

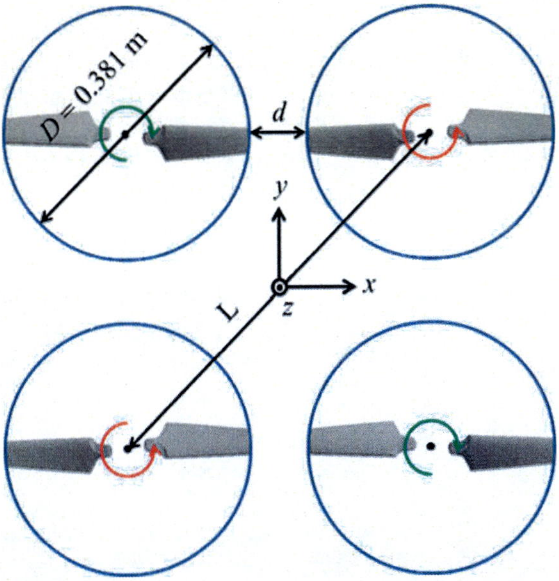

Fig. 2 Layout of rotors

Table 1 Dimensions of rotor configurations

	L/D	d/D
Configuration 1	1.41	0
Configuration 2	1.76	0.243

Table 2 Quadcopter specifications

Number of rotors	4
Number of blades	2
Rotor rotational frequency	90 Hz
Rotor diameter	0.381 m
Blade root chord length	0.0665 m
Blade tip chord length	0.0402 m
Blade airfoil	OAF117
Blade twist	−21° (Linear in-blade span)

3 Numerical Methods

Numerical simulations were carried out using a rotorcraft CFD code, rFlow3D, which was developed at JAXA [8]. The numerical schemes used in rFlow3D are shown in Table 3.

A three-dimensional moving overlapping grid technique is used in rFlow3D as shown in Fig. 3, in which the blade grids move in an outer background grid. Eight blocks of the blades, in which 121 × 143 × 61 points of grid nodes in blade-span × chord × normal directions, are used in the computational domain, as are a block of the background field, with 449 × 449 × 333 points of grid nodes in $X \times Y \times Z$ (altitude) directions. The bottom of the background grid was treated as a nonslip wall to simulate the ground. The same grid set was used for all cases.

Table 3 Numerical methods used in rFlow3D

	Background grid	Blade grid
Governing equations	3-D compressible Navier–Stokes equations	
Spatial discretization method	Cell-vertex FVM	Cell-centred FVM
Time integration	4-stage Runge–Kutta method	Dual-time-stepping LU-SGS
Numerical flux reconstruction	Modified SLAU with 4th-order compact MUSCL TVD	
Viscous flux	2nd-order central difference method	
Turbulence model	Not used	

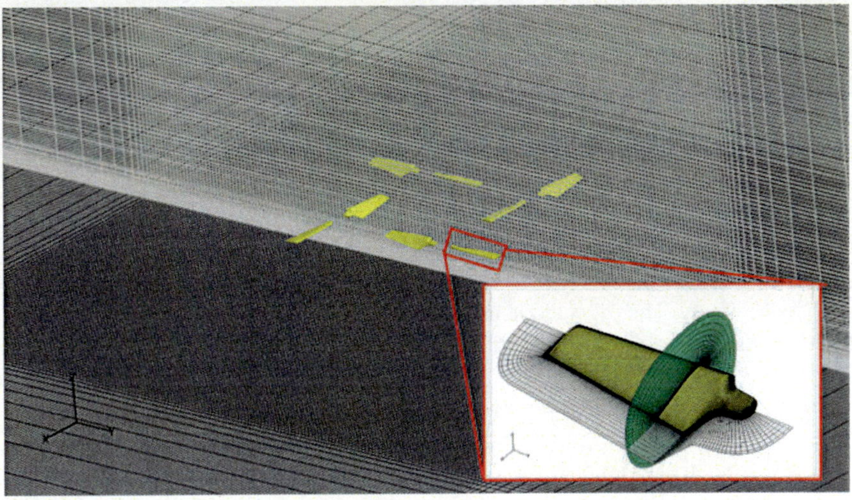

Fig. 3 Overlap grid system

These numerical methods were validated carefully in the previous studies [4, 5].

To discuss aerodynamic characteristics, aerodynamic thrust and torque are normalized as follows:

$$C_T = \frac{\text{Thrust}}{\rho\pi R^2 V_{tip}^2} \quad (1)$$

$$C_Q = \frac{\text{Torque}}{\rho\pi R^3 V_{tip}^2} \quad (2)$$

where, ρ, R, and V_{tip} are fluid density, rotor radius, and blade tip speed, respectively. The normalized required power of the rotor is identical to C_Q. The averaged thrust and torque of the four rotors are used to evaluate these values. In addition, flow velocity is normalized using the ideal induced velocity evaluated from the thrust as follows:

$$v_i = \sqrt{\frac{\text{Thrust}}{2\rho\pi R^2}} \quad (3)$$

4 Results and Discussion

Before discussing the ground effect, we describe the numerical simulations without the ground below the quadcopter (out of ground effect, OGE) that we carried out for the two kinds of configurations. In these simulations, C_T was fixed at 0.0106. The resultant values of the required power, C_Q, were 0.00128 and 0.00124 for configurations 1 and 2, respectively. This means that the aerodynamic loss increases when the distance between rotors decreases, in agreement with a previous report [1]. The velocity distributions around the rotors are shown in Fig. 4. Axial velocity distribution shows that aerodynamic interaction takes place in configuration 1 because the outline of the rotational disks of the rotors contact each other. Mixing of the rotor wakes was also enhanced in the downstream region, as can be seen at the bottom of Fig. 4. The averaged pitch angles of the rotors in configurations 1 and 2 are 9.48 and 9.17°, respectively. The pitch angle of configuration 1 is larger than that of configuration 2 to ensure the designated thrust in the aerodynamic interactions; this results in an increase in the required power.

The change in the required power to ensure the designated thrust, $C_T = 0.0106$, was evaluated to discuss the ground effect, as can be seen in Fig. 5. Here, h is the height from the ground to the rotors. A calculated result for the single rotor

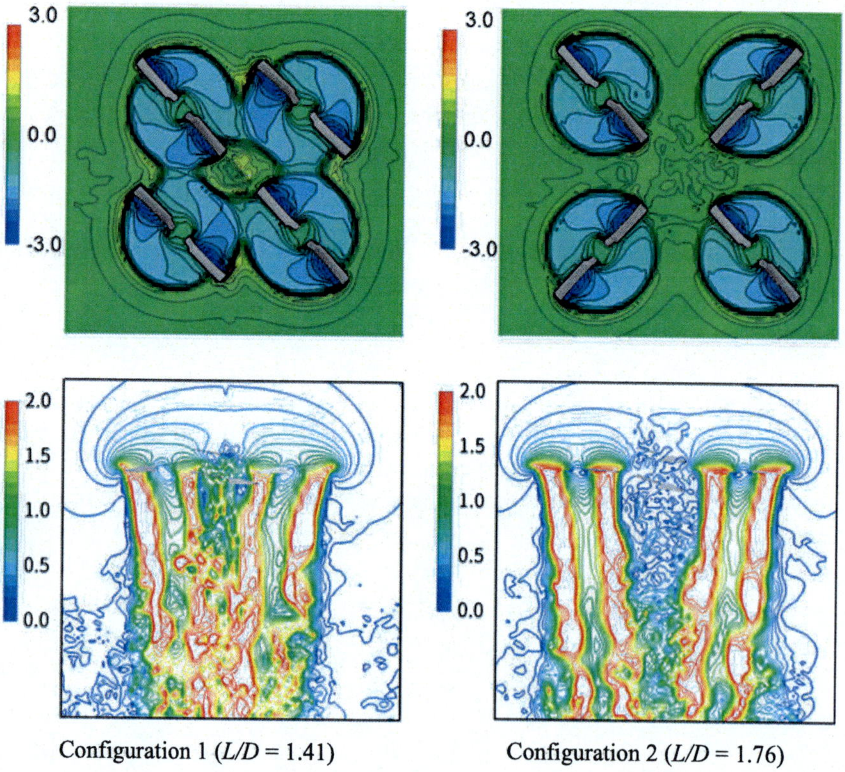

Fig. 4 Instantaneous normalized axial velocity fields (w/v_i, top) in the cross section along the rotational disk of rotors and velocity magnitude fields (V/v_i) in the cross section along the diagonal rotor centers (OGE)

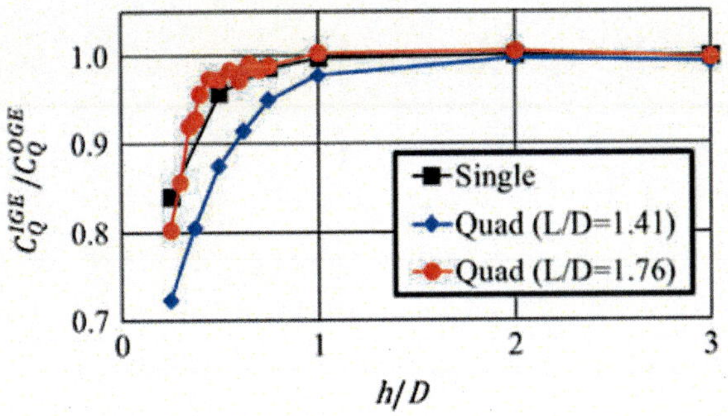

Fig. 5 Required power versus altitude of rotors

case is also plotted in Fig. 5. The results of the single rotor and of the quadcopter with configuration 2 are almost identical in the ground effect (IGE) at $h/D < 1$, where the required power decreases as the altitude decreases. On the other hand, the result for the quadcopter with configuration 1 ($L/D = 1.41$) shows the ground effect at $h/D < 2$.

The normalized velocity field with rotors at various altitudes are shown in Fig. 6. The rotor wakes impinge on the ground, and outwash flow patterns take place along the ground surface at $h/D = 2.0$ in Fig. 6a. When the altitude decreases at $h/D = 1.0$, an upwash takes place at the center of the four rotors, as can be seen in Fig. 6b. The previous reports pointed out that the occurrence of upwash led to the recirculation of the rotor-induced flows. The flow rate of the upwash increases as the altitude decreases further, as can be seen in Fig. 6c. The upwash regions around the rotor rotational disks are visualized in Fig. 7. The upwash regions take place between neighbouring rotors as well as at the center of the four rotors in Fig. 7a. On the other hand, the upwash region takes place only at the center of the four rotors in Fig. 7b. These results suggested that the upwash is enhanced by getting closer to each rotor of the quadcopter, and this leads to the increase in the recirculation and the reduction of required power.

The outwash flow patterns are visualized in Fig. 8. These figures present the time-mean distributions of the velocity during three rotations of the rotors. The rotor altitude is $h/D = 0.5$ and the flow patterns are visualized at $z/D = 0.25$, which is between the ground and the rotor rotational disks. White arrows show the rotational directions of the rotors. These figures show that the outwash has obvious directivity. Four outwash jets take place in parallel to the *x*- and *y*- axes. It is also observed that the outwash jets are stronger in the *y*-direction than in the *x*-direction for both configurations 1 and 2.

The velocity distribution of the outwash jets is visualized in the side-view cross section between the neighbouring rotors in Fig. 9. The velocity profiles along the lines presented in Fig. 9 are also plotted in Fig. 10. The velocity profile of the single rotor case is also presented in Fig. 10. Figure 9 shows that the outwash jet is as high as the rotor altitude for both cases. The jet velocity increases as the distance between

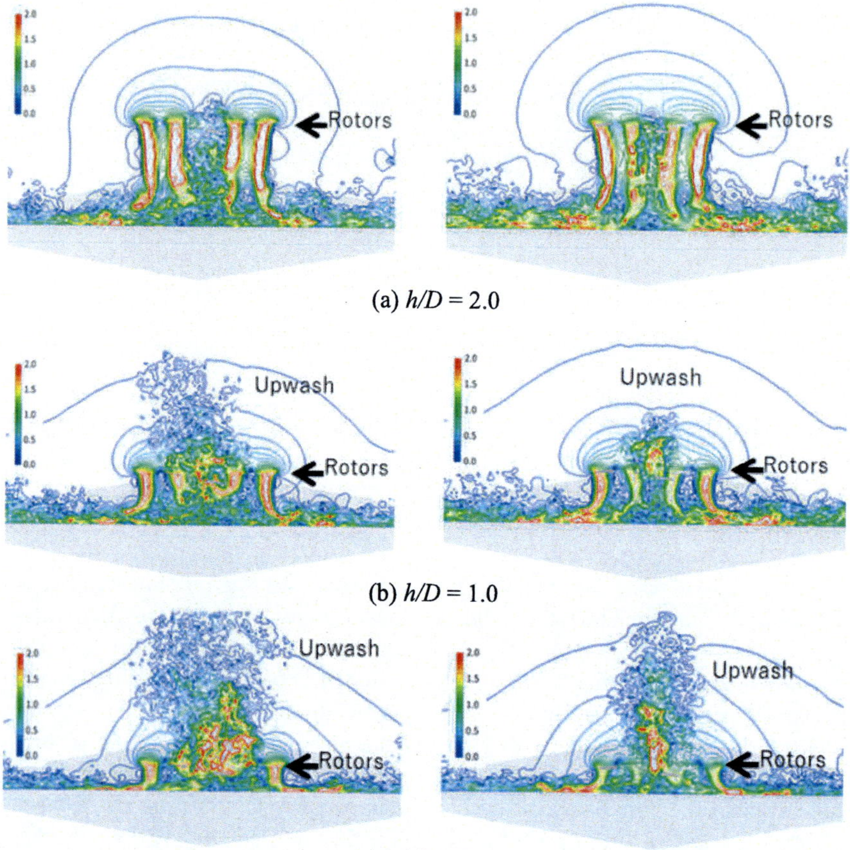

(a) $h/D = 2.0$

(b) $h/D = 1.0$

(c) $h/D = 0.5$

Fig. 6 Instantaneous normalized velocity magnitude fields (V/v_i) in cross section along diagonal rotor centers with rotos at various altitudes from the ground (Left, configuration 1 ($L/D = 1.41$), right: configuration 2 ($L/D = 1.76$))

rotors decreases, as can be seen in Figs. 9 and 10. Compared with the single-rotor case, the results for the quadcopter with both configurations show distinct velocity profiles.

The velocity distribution of the outwash is visualized in the side-view cross section along the diagonal rotor centers in Fig. 11. The velocity profiles along the lines presented in Fig. 11 are also plotted in Fig. 12. The outwash is restricted in the vicinity of the ground surface, as can be seen in Fig. 11. In Fig. 12, all results, including those for the single rotor case, show almost identical values and tendencies at each location. This suggests that the influence of aerodynamic interaction between rotors does not appear in the diagonal cross section and that the ground effect can be evaluated using the conventional procedures for single rotor helicopters.

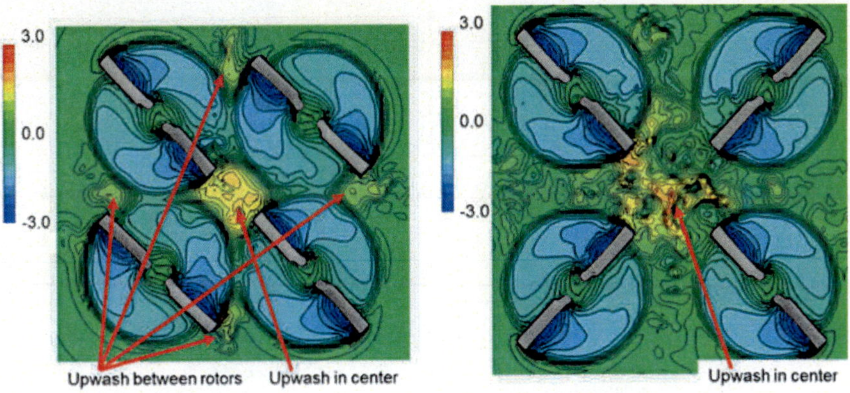

(a) Configuration 1 ($L/D = 1.41$) (b) Configuration 2 ($L/D = 1.76$)

Fig. 7 Instantaneous normalized axial velocity fields (w/v_i) in the cross section along the rotational disk at $h/D = 1.0$

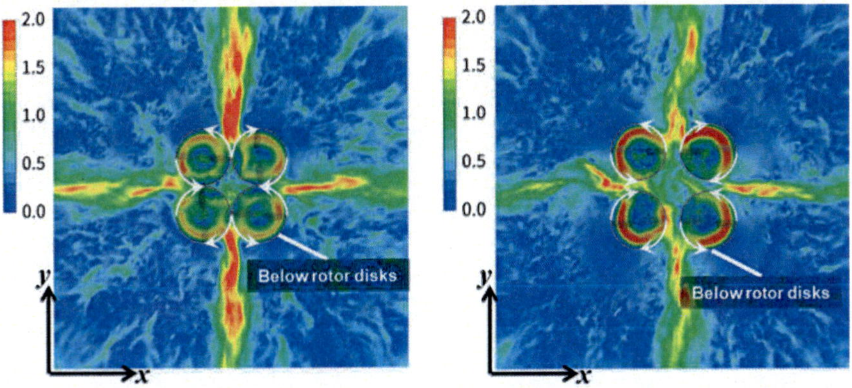

(a) Configuration 1 ($L/D = 1.41$) (b) Configuration 2 ($L/D = 1.76$)

Fig. 8 Time mean velocity magnitude fields in the cross section along $z/D = 0.25$ when the rotor altitude is $h/D = 0.5$

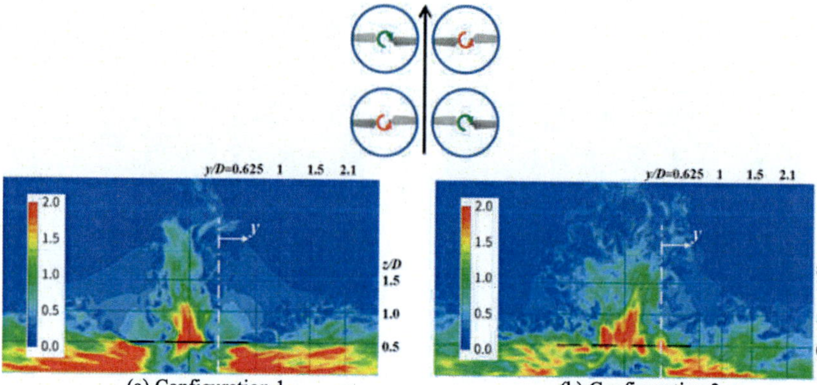

Fig. 9 Time-mean velocity magnitude distribution in the cross section between neighboring rotors when the rotor altitude is $h/D = 0.5$

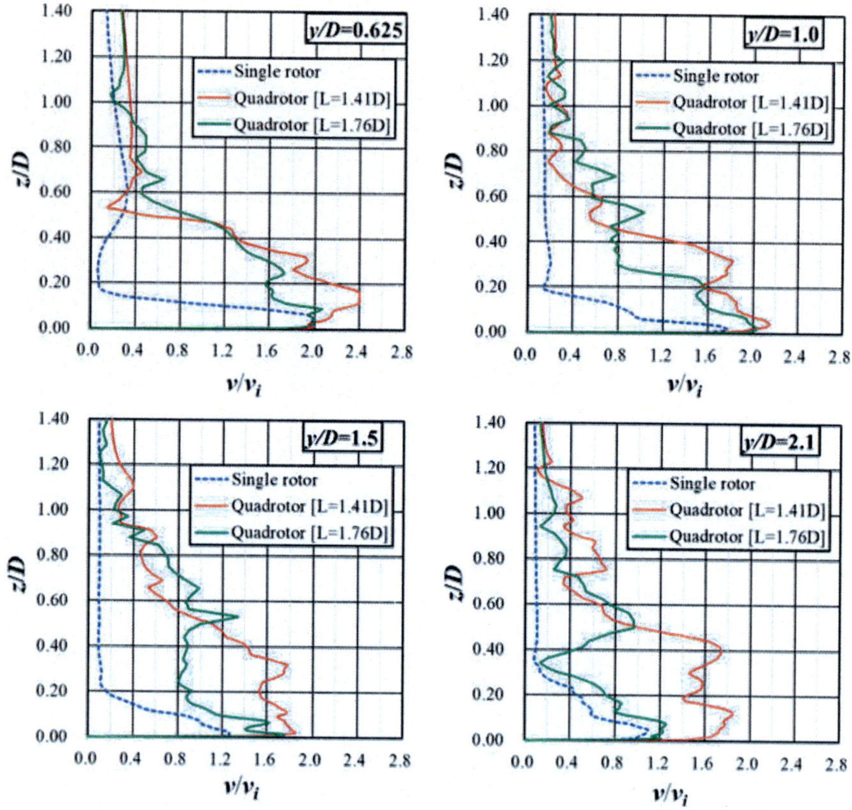

Fig. 10 Outwash velocity profiles in the cross section between neighboring rotors when the rotor altitude is $h/D = 0.5$

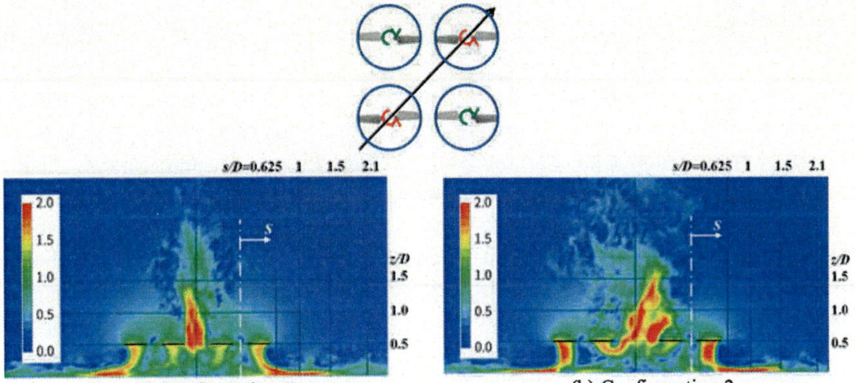

(a) Configuration 1 (b) Configuration 2

Fig. 11 Timemean velocity magnitude distribution in the cross section along the diagonal rotor centres when the rotor altitude is $h/D = 0.5$

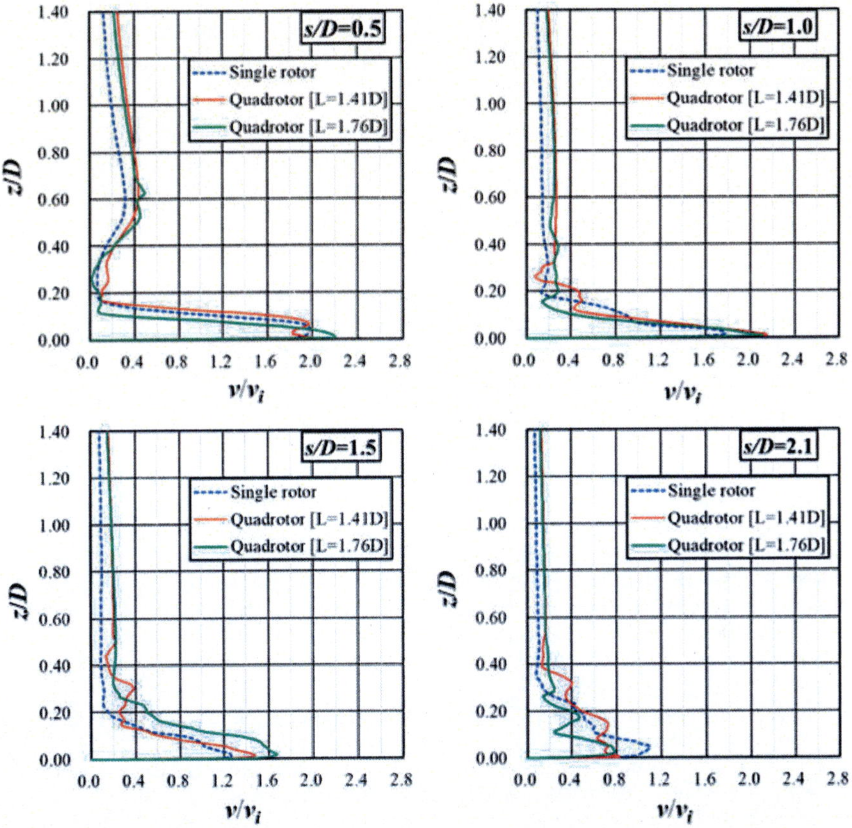

Fig. 12 Outwash velocity profiles in the cross section along the diagonal rotor centres when the rotor altitude is $h/D = 0.5$

5 Conclusion

The hovering in ground effect (IGE) of a quadcopter was examined numerically using two rotor configurations with different rotor-to-rotor distances. It was confirmed that the rotor-to-rotor distance has a significantly greater effect on the aerodynamic characteristics than the ground effect. The quadcopter turns out to be more efficient aerodynamically when the rotor-to-rotor distance is reduced, which also leads to a reduction in the required power, in the case of IGE. Flow visualization demonstrated that the upwash in the center of the quadrotor configuration as well as the recirculation flow are obviously enhanced due to the aerodynamic interaction between rotors, which can be dominated significantly by the rotor-to-rotor distance. As a result, the outwash jet streams turn out to be highly strengthened when the rotor-to-rotor distances are reduced, whereas the surrounding flow structures away from the outwash jet streams, e.g., in the diagonal cross section of the quadcopter, resemble those of the single rotor case.

Acknowledgements This study was financially supported by JSPS KAKENHI Grant No. JP19H02344 as part of research on the "Enhancement of safety of manned and unmanned multi-rotor aircraft and assessment of influences to surroundings."

References

1. Li Y, Yonezawa K, Xu R, Liu H (2021) A biomimetic rotor-configuration design for optimal aerodynamic performance in quadrotor drone. J Bionic Eng 18:824–839. https://doi.org/10.1007/s42235-021-0069-0
2. Kohno M, Otsuka H, Kiribayashi S, Nagatani K (2017) Investigation on relationship between rotors axis length and ground effect on a small quadrotor UAV performance. In: The robotics and mechatronics conference 2017 in Fukushima, 1P2-F02 (in Japanese)
3. Kohno M (2017) Visualization of the upstream flow of rotors and analysis on thrust of a small quadrotor UAV in ground effect. In: 55th aircraft symposium, Shimane, JSASS-2017-5022 (in Japanese)
4. Tanabe Y, Sugawara H, Sunada S, Yonezawa K, Tokutake H (2021) Quadrotor drone hovering in ground effect. J Robot Mechatron 33(2):339–347. https://doi.org/10.20965/jrm.2021.0339
5. Tanabe Y, Sugawara H, Sunada S, Yonezawa K, Tokutake H (2021) Numerical investigation of quad rotors in ground effect. In: 47th European Rotorcraft Forum, Paper #57
6. Yonezawa K, Matsumoto H, Sugiyama K, Tanabe Y, Tokutake H, Sunada S (2019) Aerodynamic characteristics of a quad-rotor-drone with ducted rotors. In: 8th Asian/Australian Rotorcraft Forum
7. Sunada S, Tanabe Y, Yonezawa K, Tokutake H, Umezaki S, Yamaguchi K, Sugawara H (2021) A study on improvement of performance of a multiple rotor drone. J Robot Soc Jpn 39(4):357–362
8. Tanabe Y, Saito S, Sugawara H (2012) Construction and validation of an analysis tool chain for rotorcraft active noise reduction. In: 38th European Rotorcraft Forum

A Review of Recognition in Military Airworthiness Regulatory Frameworks

Zawar A. Nawaz Bhatti and Nicholas S. Bardell

Abstract Military aviation does not have a globally standardised airworthiness framework like that maintained by ICAO for civil aviation. As a result, the distinct and unique nature of each State's military airworthiness regulatory framework makes safety assurance difficult during joint operations, collaborations in aircraft development or during procurement of foreign aviation assets. The process of recognition of military airworthiness frameworks can, to some extent, alleviate many of these difficulties and help to build trust, allow resource sharing, enable mutual learning, provide economic benefits, and enhance safety assurance between collaborating Military Airworthiness Authorities. A review of the various military aviation forums, alliances, and collaborations is presented here in which recognition activities are described and evaluated. Attention is then focused on the most successful recognition process currently in use worldwide, namely the European Military Airworthiness Document–Recognition (EMAD-R). The main reason for its success is that it incorporates much of the globally accepted ICAO oversight and airworthiness requirements in its Military Authorities Recognition Question-set (MARQ), thus providing the necessary confidence for recognition between collaborating Military Airworthiness Authorities. This work concludes by examining the EMAD-R process and showing it can effectively be used for recognition of disparate airworthiness frameworks even outside the European paradigm.

Keywords Airworthiness · Military aviation · Recognition · Regulatory framework · Safety assurance · Oversight · Military airworthiness authority

1 Introduction

Commercial aviation has evolved over the years from a high-risk mode of transport to the safest that is available to the public [59]. This enviable state of affairs is generally attributed to the standardisation and oversight provided by the International

---

Z. A. Nawaz Bhatti · N. S. Bardell (✉)
School of Engineering, RMIT University, Melbourne, VIC 3000, Australia
e-mail: nick.bardell@rmit.edu.au

S. Lee et al. (eds.), *The Proceedings of the 2021 Asia-Pacific International Symposium on Aerospace Technology (APISAT 2021), Volume 1*, Lecture Notes in Electrical Engineering 912, https://doi.org/10.1007/978-981-19-2689-1_58

Civil Aviation Organisation (ICAO), assuring safe operations globally [53]. Military aviation does not come under the jurisdiction of the 1944 Chicago Convention that laid the foundations of ICAO and the resultant commercial aviation framework. The main task of military aviation is to preserve the aerial defence of a State and ensure combat capability [50]. Although military airworthiness authorities are influenced by their respective civil regulatory frameworks, the distinctive characteristics of military aviation and diversity in its modes of operations create several differences from civil aviation [30]. Military airworthiness frameworks are formalised explicitly according to the capability, safety understanding and requirements of each State, making them similar but unique [47]. Military airworthiness authorities (MAA) obtain their regulatory powers through various policies at the government level. There are instances where each State service has an independent airworthiness system; for example, the United States (US) Army, Navy and Airforce have distinct frameworks while working under the US Department of Defense (DoD). The uniqueness of MAAs means there is little likelihood that any two airworthiness frameworks will be identical or closely aligned [47].

Many countries worldwide do not possess a sovereign capability in military aircraft design or production and rely on friendly States to supply their needs. Many of the States that do retain some level of sovereign capability are finding the costs of developing new military aircraft programs prohibitive, and find it advantageous to collaborate on joint ventures with other States to benefit from each other's expertise and experience [31], not to mention risk sharing and cost reductions. However, joint ventures, purchase of foreign aircraft, modifications and service-life support from other countries can prove problematic since these activities entail dependency upon the standards, processes, and qualification systems of foreign military airworthiness authorities [41]. The solution to such issues of safety assurance between disparate military airworthiness frameworks is through the incorporation of recognition activities.

1.1 Aims

This paper aims to address three important areas concerning recognition amongst military airworthiness regulatory authorities. Firstly, an in-depth review of recognition is presented in which the motivation for accomplishing recognition, and the benefits of obtaining recognition, are examined and discussed. Having thus established the *raison d'être* for recognition, attention is then focused on the current challenges that military airworthiness authorities face gaining recognition; potential solutions are presented in the form of the European EMAD-R process. Finally, the question of whether it is possible to construct an MAA recognition process that can be applied at a *global* level is discussed and answered in the affirmative, citing evidence from recent recognition activities between various non-European countries that nonetheless adopted and used the EMAD-R process.

2 Introduction

2.1 Recognition Definition

The Oxford Learner's Dictionary defines recognition as "*the act of accepting that something exists, is true or is official*". The UK Military Aviation Authority (MAA) describes recognition as the process that assesses other military regulators for the subsequent possibility to utilise their products and services in one's own airworthiness system [57]. Lawson [34], while providing an update on the US services' recognition assessment, described it as a "*cooperative activity between participating nations*" to gain insight into the regulatory framework of others to get more confidence in making informed airworthiness decisions. The review of these definitions and descriptions creates a holistic view of recognition as a systematic process that thoroughly assesses a regulatory framework for comparison and gap analysis to further accept activities.

2.2 Recognition in Civil Aviation

ICAO provides standardisation in civil aviation through its global standards and recommended practices which are accepted by all member States for implementation. ICAO has defined eight safety tools, known as critical elements (CE), that collectively assure an effective safety oversight system. See below. The ICAO Safety Oversight manual (ICAO Document 9734) outlines the requirements for each critical element [27].

1. Primary Aviation Legislation
2. Specific Operating Regulations
3. State Civil Aviation System and Safety Oversight Functions
4. Technical Personnel Qualification and Training
5. Technical Guidance, Tools and the Provision of Safety–Critical Information
6. Licensing, Certification, Authorization and Approval Obligations
7. Surveillance Obligations
8. Resolution of Safety Concerns

Safety and airworthiness are subsequently assured through the oversight and audit of contracting States [12]. ICAO determines the capabilities and effectiveness of the regulatory frameworks of its member States through the Universal Safety Oversight Audit Programme (USOAP); eight audit areas are assessed individually to ensure the State has effectively and consistently implemented the critical elements of the safety oversight system described above. The USOAP assesses these eight audit areas through a Continuous Monitoring Approach (CMA) to determine the system's maturity and level of safety assurance [28]. The performance of States in each area in the audit is available globally to review and compare with others, as shown in Fig. 1.

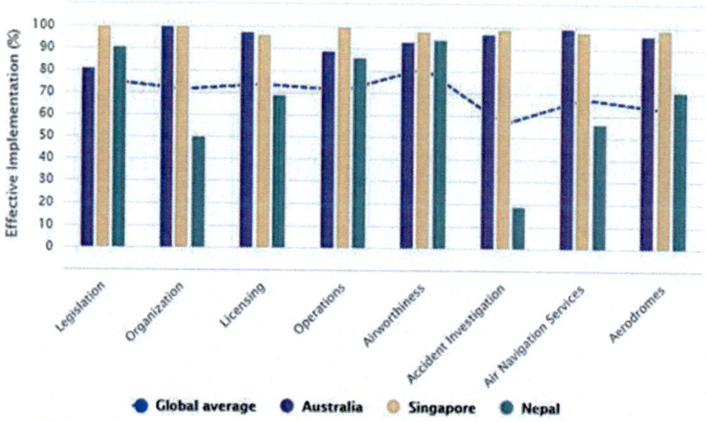

Fig. 1 USOAP performance of selected aviation authorities in each audit area [25]

The global standard of assessment provides confidence to civil aviation authorities to recognise others through bilateral agreements. For example, the performance of the aviation authorities of Australia and Singapore is above the global average in each area of USOAP (Fig. 1), which provides them with the confidence to accept selected airworthiness instruments and artefacts of the other authority. As a result, Australia and Singapore have current bilateral agreements for airworthiness certification [6] and maintenance [5]. In contrast, both countries would remain cautious about entering into similar agreements with the Nepalese Aviation Authority, given its performance in four audit areas is well below USOAP global standards.

2.3 *Recognition in Military Aviation*

Military aviation does not have a global oversight governing body to assure standardization; different States maintain different standards and processes for ensuring military airworthiness. Therefore, it is possible that a product or certification approval by one military airworthiness authority may be below the acceptable risk levels maintained by another [57]. Aviation regulators try to ensure that the regulated community they oversee maintains the desired levels of safety. However, assets procurement outside the area of control of the regulator makes it difficult for the regulator to maintain the necessary oversight. For example, the Australian Defence Force (ADF) relies on products and services from several foreign States and their industries. Figure 2 illustrates the interaction of the ADF with various States, with the width of the colour band depicting the level of interaction and supply of military hardware and services.

De Luis [13] highlights that for the ADF, it may not be practical or cost-effective to oversee the processes of design, production, and certification of each product and maintain the same oversight for in-service support. The concept of recognition

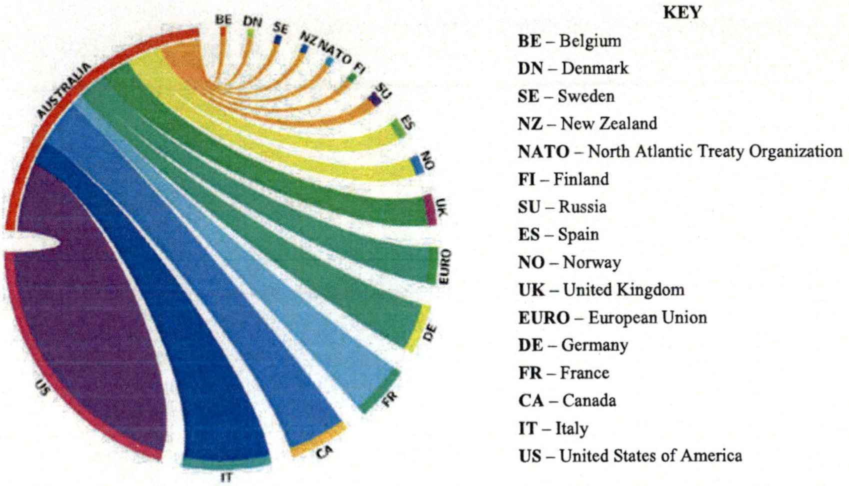

Fig. 2 Graphic illustration of ADF reliance on others for services and equipment. Image reproduced with kind permission from Purton [41]

supports these actions. The UK MAA emphasises the importance of recognition in military airworthiness by highlighting that it supports the judgements and decisions of accepting other regulatory frameworks through a systematic and coordinated approach while identifying the areas of difference or ones requiring additional checks [34]. However, it is also cautioned by the UK [56] that the process does not allow complete access to the recognised airworthiness system and requires informed decisions for utilisation of acceptable processes and systems only.

De Luis [13] has identified that MAAs can utilise the recognition processes in three different ways:

1. **Unilateral Recognition**. Recognition of one authority by another, usually when a one-way delivery of aeronautical products is planned. The Australian Defence Aviation and Safety Authority (DASA) recognises the US Naval airworthiness authority (NAVAIR) for the certification process of military aviation assets such as the F/A-18 aircraft [14].
2. **Bilateral Recognition**. Two authorities mutually recognise each other for the interchange of services, information, or products. The Military Aviation Authority of the Republic of Korea (DAPA) and the US DoD have a bilateral (mutual) recognition agreement [51].
3. **Multilateral Recognition**. The recognition agreements between several authorities on the interchange of services, information, or products. Multilateral recognition is valuable for the design and production of joint program products. The MAAs of Germany, Italy, Spain, and the UK made a multilateral recognition agreement for certain design changes required by the Eurofighter Typhoon programme in 2016 [48].

2.4 *Efforts for Development of a Recognition Process*

The details of efforts, in the available literature, for developing and pursuing recognition processes are now explained in Sects. 2.4.1–2.4.5.

2.4.1 European Defence Agency (EDA)

The difficulties faced during certification and subsequent utilisation of equipment by the European Union (EU) member States highlighted the necessity of efforts for a common airworthiness platform [46]. The Eurofighter Typhoon incurred excessive cost overruns and development delays, some of which were due to the variation in design and certification standards of the collaborating nations (Germany, Italy, Spain, and the UK) [23]. As a result, the Military Airworthiness Authorities (MAWA) forum was established in 2008 by the European Defence Agency (EDA) to generate a framework of common airworthiness requirements and create harmonisation in the airworthiness processes [8]. The forum achieved its goal through the launch of the European Military Airworthiness Requirements (EMARs) [20]. A significant part of the EMARs was derived purposefully from the regulatory framework of its corresponding civilian counterpart, the European Aviation Safety Authority (EASA). The EMARs also allowed the development of a recognition process through the MAWA document EMAD-R, which accomplished recognition by examining the Military Authorities Recognition Question-set (MARQ).

2.4.2 North Atlantic Treaty Organisation (NATO)

The North Atlantic Treaty Organisation (NATO) has limited aviation assets and therefore utilises the aircraft of its member States or leases/charters aircraft from non-members for supporting operations. The aerial missions organised by NATO become its responsibility whether the aircraft used are owned or leased by NATO [15]. Even though NATO is responsible for the conduct and safety of its aerial operations, the organisation did not have an integrated airworthiness system. In addition to these issues, there were certain instances where the responsibility of continuing airworthiness remained vague. In the case of NATO owned E-3A Sentries, registered in Luxembourg, the country does not have any MAA to oversee the continuing airworthiness aspects of this fleet [16]. The absence of an airworthiness system was immediately felt after the fatal crash of a NATO chartered aircraft (UM Airlines Flight 4230) in Turkey in 2003 while transporting 62 Spanish soldiers [15]. This accident occurred just four years after a similar fatal accident in Kosovo and hence precipitated the establishment of the Airworthiness Working Group in June 2010; the subsequent issuance of a NATO Airworthiness Policy (NAWP) followed in June 2013 [39]. The policy led to establishing a NATO Airworthiness Executive (NAE) to determine, ensure, and oversee the airworthiness and safety requirements of assets operated under NATO. The recognition process developed by NAE finalised the use of MARQ (EMAD-R), building on its use of the EDA MAWA forum and the Air Force Interoperability Council (AFIC) to assess the airworthiness authority of the charter State [17].

2.4.3 US Department of Defense—Harmonising Defence Service

The three defence services of the United States (US)—the Army, Navy and Air Force—have historically been independent of each other and competitive. Each has its own autonomous airworthiness system in place and has individual certification and qualification processes. Independence provides each service with the necessary freedom to pursue tasks according to their priorities and particular needs in military projects. However, this independence creates additional barriers for external agencies to meet the desired standards of each system. Eventually, the United States Department of Defense (DoD) finalised an agreement with the three services to harmonise their regulatory frameworks [46]. The agreement has led to the issuance of the DoD Airworthiness Policy [58]. The policy requires all three services to accept each other's certification process and to work in the future on the generation of a common airworthiness framework [34]. The US DoD and its three services also utilise the MARQ for recognition activities.

2.4.4 Air Force Interoperability Council (AFIC)

The Air Force Interoperability Council (AFIC) was initially formed as the Air Standardization Coordination Committee in 1948 to maintain aviation standardisation between Canada, the UK, and the US. The committee evolved over time and included Australia and New Zealand as its members [2]. The council has several working groups that maintain standardisation in various operational domains through the issuance of air standards to which all members adhere for interoperability. Sensing the requirement of maintaining safety through assured airworthiness during joint operations and assets utilisation, AFIC formulated an Airworthiness Working Group in 2011 to generate a robust but pragmatic recognition process for the AFIC nations. The group gained impetus from work already done by the MAWA forum through the MARQ. AFIC uses the MARQ as the baseline for the initiation of recognition between member States [43].

2.4.5 Academic Contributions

Presently, due to the national interests and confidential nature of each matter involving military aviation, there is limited research carried out on military airworthiness regulatory frameworks employed by various countries [46]. However, one novel process has been established by Purton et al. [47], the Product-Behaviour-Process (PBP)

Bow-Tie model, that could effectively be utilised for comparing two disparate regulatory frameworks without depending on the EMARs. The method is based around the principles of using the Bow-Tie model for representing technical airworthiness while comparing the independence of technical integrity assurance in a regulatory framework. Test points are generated around the steps of a certification process for each phase of the aircraft life cycle (Design-Production-Maintenance). Each test point is rated from 1 to 5 depending upon the level of independence (1-No check to 5-External regulator checks). The results of the test points of one authority are compared with another. The method uses Iris charts for graphical representation of the comparison, as shown in Fig. 3. The method was subsequently utilised to compare the regulatory frameworks of Australian (ADF) and New Zealand Defence Force (NZDF) [44] and comparison of the three United States services with the ADF [45]. The comparison was further enhanced to include the military technical airworthiness systems of the remaining AFIC nations, Canada, and the UK [42]. The proposed method is useful for recognising non-EU nations, benefitting from a pictorial display for a simple visual assessment of the differences.

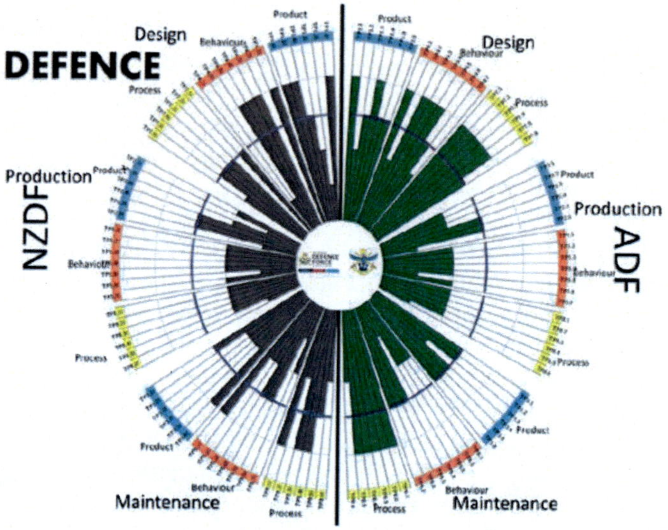

Fig. 3 Iris Chart depiction comparing the ADF and the NZDF regulatory frameworks. Image reproduced with kind permission from Purton et al. [44]

3 Methodology

Literature-based research is a powerful qualitative tool since it utilises the integration of findings and views of various researchers from several sources to form a comprehensive inference [52]. As such, this forms the main basis for the work presented herein, noting the following limitations:

1. There is a dearth of public research available on military airworthiness [47], and therefore this work uses an inductive process to explore the available publications and other applicable sources. The motivation behind the increased interest in recognition and the benefits of recognition in military airworthiness is extracted from the available research to provide succinct information for military regulators and organisations.
2. This research uses secondary data extracted from available literature and will use it to explore the details of the EMAD-R recognition process. An analysis of the MARQ (EMAD-R) will be carried out for a detailed understanding of its utilisation in non-EMAR based airworthiness framework recognitions.
3. The methodology will encompass content analysis of a variety of online publications, including airworthiness documents, policies, regulations, official briefings, journal articles and related books.

4 Increased Motivation for Recognition

States have realised that assuring the safety of personnel and resources even during military conflicts is an essential duty of care within the military covenant. Maintaining acceptable levels of safety through regulatory oversight is not only necessary due to the legal barriers of Work, Health and Safety legislation but also supports attaining the desired level of operational efficiency. However, activities beyond the control and oversight authority of the respective MAA make it challenging to assure safety in all domains. The need to achieve safety assurance of aviation assets and personnel in such conditions motivates recognition activities.

4.1 Foreign Military Procurements

Most States around the world do not design or manufacture military aircraft or aeronautical products. These MAAs make intelligent assessments based upon experience and available knowledge, complemented by additional checks that may incur significant economic effort. The verification process can be adversely affected due to limited knowledge of the airworthiness system of the State of design [57]. Moreover, pressing defence and national security requirements frequently outweigh the requirement of detailed certification of the aircraft. Even though rational assessments are performed to assure the aircraft meets acceptable safety standards, there will always be some doubts and uncertainties associated with the product. The process of recognition provides confidence in the airworthiness systems of other States, along with the necessary information to plan additional checks (if required) to achieve the desired airworthiness outcomes.

4.2 Interoperability

States and countries have historically aligned with their strategic friends to support each other during wars, conflicts and even during national disasters. The alliances necessitate joint training and operations during peace times, such as Exercises like *Pitch Black* and *Red Flag* [40]. Resources of each member of the alliance, including aviation assets and personnel, are often utilised by others, initiating the system of interoperability. Even though the mutual sharing of resources remains beneficial economically and strategically, safety aspects remain paramount, especially during peacetime operations. The uniqueness of military aviation airworthiness of each State and a reluctance to share military details limits safety knowledge and lowers the awareness of risks during interoperability. The need to assure the safety of personnel and assets during interoperability makes recognition activities desirable.

4.3 Military Alliance Organisations

The interoperability between two or more strategic allies is a limited and less dynamic situation, with the utilisation of each other's resources only during joint operations. However, the conditions become highly complicated within global alliance operations where aircraft may even be chartered or leased from a State that is not even part of the alliance [16]. NATO faces such issues regularly wherein a large group of countries work together utilising NATO aircraft, aircraft from one of the member States, or aircraft chartered/leased from non-member States. The non-availability of an indigenous regulatory authority and operating aircraft of other States with diverse safety oversight systems makes the airworthiness assurance capacity of NATO extremely difficult. The airworthiness issues, coupled with the legal matters faced due to losses in accidents persuades alliances, such as NATO, to work towards achieving assurance through recognition of relevant MAAs [15].

4.4 Aircraft Development Collaborations

Collaborative aircraft development and certification projects provide economic benefits as member States share costs and resources to efficiently complete the process.

The collaboration also provides each member State access to the diverse pool of knowledge and experience of other members. However, the uniqueness of military airworthiness frameworks of each State can hinder the effectiveness of any activity in the collaboration. Each State relies on its unique qualification and certification process to assure the airworthiness of the product, hence generating an increased workload [20]. For example, some minor design changes in the Eurofighter Typhoon, a joint project between the UK, Germany, Italy, and Spain, took 18 months to be accepted as each member State had to independently satisfy the safe accomplishment of the task [49]. Recognition of MAAs of member States simplifies the entire process, providing an opportunity to achieve the desired financial and strategic benefits from collaboration.

5 Recognition Benefits

The continued persistence of States to enhance safety assurance in military aviation has been the primary motivator behind the increased interest in recognition activities in the last two decades. However, the motivation behind recognition cannot be sustained if no tangible benefits can be accrued from the process. Table 1 highlights the various project areas where recognition of MAAs within the EDA proved beneficial.

5.1 Mutual Learning Process

Sehoon [51] highlights that the recognition process provides opportunities to learn from the processes and procedures of other authorities. It creates a scenario for a State to review its own airworthiness framework and identify weaknesses in the system while processing the self-assessments. In addition, it also provides an opportunity to learn from other airworthiness frameworks and adapt their strengths [10]. Masterton [36] signifies that safety assurance can be achieved through a diverse range of methodologies. The recognition process gives insight into these methodologies that other authorities have incorporated to attain similar goals. It is an opportunity to learn global best practice and integrate it into one's own system.

5.2 Resource Sharing

States utilise the resources of others during joint/collaborative projects and operations with the realisation achieved through recognition activities that adequate safety standards have been attained. Designs, certification processes and even the design/production facilities of other States can be assessed as safe for utilisation

Table 1 Beneficial experiences of military recognition activities (all images have been reproduced from the Creative Commons databank)

Project	MAAs involved	Benefit areas	Source
 A400M	Germany & Spain	Retrofit	[22]
 Eurofighter Typhoon (EF2000)	Germany, Italy, Spain, and the UK	Modifications	
 A400M	Germany, France	Technical Training	

through the recognition process. Recognition also provides the opportunity for MAAs to enhance their knowledge through mutual training and sharing experiences [9]. Airworthiness instruments, artefacts and organisational approvals can be utilised between recognised MAAs depending upon the underlying caveats [7].

5.3 *Access to Global Supply Chain and MRO Facilities*

The global market has numerous suppliers of aeronautical parts and products duly certified by their respective regulatory authorities. Even if not restricted due to strategic ties between States, access to each remains difficult if they are not independently certified by the buyer MAA. Recognition provides the opportunity to access the various global sources for spares support while assuring the desired safety standards [21]. Recognition of an MAA can further provide evidence of the reliability of the respective foreign maintenance, repair, and overhaul (MRO) facilities for subsequent utilisation. Recognition between the Australian MAA and the US Navy regulator [14] has provided opportunities to both parties to utilise MRO resources of each other. Subsequently, RUAG Australia is now providing MRO facilities for US Navy F/A-18 aircraft in the Asia Pacific region [1]. A safety assured global supply chain complemented by the availability of certified maintenance support in different world regions makes the continuing airworthiness tasks simpler and more efficient [38].

5.4 *Economic Gains*

In an age of shrinking defence budgets, the concept of shared resources, including equipment, personnel, and knowledge, achieved through recognition, becomes a source of economic efficiency. Conversely, new collaboration entails multiple qualifications and certifications of an aeronautical product to meet the requirements of member States' MAAs, making the complete process not only exceptionally time-consuming but also economically expensive. Defence aviation industries highly recommend the use of harmonised regulatory frameworks by different MAAs to avoid such issues. However, accepting the defence and sovereignty issues involved, it is advocated that a recognition process is utilised to avoid repetitive certification work for a product planned to be used by two or more States [36].

5.5 *Enhanced Trust and Respect*

Grintz [22] explains that the detailed review of each other's regulatory framework and oversight systems, while maintaining transparency, creates trust and respect. Moreover, the process itself gives the required evidence to States about the quality of the airworthiness framework and its implementation in other States, subsequently giving confidence in utilising their artefacts and procedures. Childs [7] highlights that field commanders require confidence in the military aviation assets of other States during joint operations to utilise the assets effectively. The required confidence and trust are acquired through the availability of recognition. The complete process enhances the cooperation between States and constructs positive relationships [51].

6 Military Airworthiness Recognition Process

The motivation behind recognition in military airworthiness has increased significantly in recent times, with benefits through collaborations being realised [22]. The most effective and widely implemented recognition process for military airworthiness is the EMAD-R process, developed under the MAWA forum [22]. The focus of Sect. 6 is hence devoted to examining the EMAD-R process and understanding what makes it so versatile on the world stage.

6.1 EMAD-R Process Overview

The EMAD-R is a structured process that highlights the regulatory capability of any MAA. The process is based around the concept that the most effective way through which the acceptability of any other system can be assessed is through the means of "*look and compare*" [21]. The process uses this methodology to identify similarities to gain confidence and highlight any differences for further risk analysis.

6.1.1 Guidance from ICAO Air Safety Goals

A significant portion of the EMARs has a similar structure to the ICAO-based EASA framework. The links with the ICAO framework provide the EMARs with a unique opportunity to implement and benefit from a globally proven civil aviation framework that is now applied to a military environment. Similarly, the EMAD-R process also builds upon the ideology of the ICAO airworthiness systems and retains a strong link to the ICAO safety oversight practices. ICAO delegates the responsibilities of States to maintain airworthiness through the ICAO Airworthiness Manual Doc 9760 [29]. Moreover, ICAO guides its member States through the Safety Oversight Manual (ICAO Doc 9734) to achieve the desired levels of effectiveness of their oversight systems [27]. Both manuals collectively achieve the air safety goals of the aviation industry for technical airworthiness. EMAD-R aims to review and compare the regulatory and oversight process of two or more MAAs through the MARQ air safety goals derived from the requirements outlined within ICAO Safety Oversight and Airworthiness manuals [21].

6.1.2 Recognition Need

The first and most vital part of any recognition process is the need or purpose to achieve the recognition, such as collaboration in projects, the requirement of interoperability, use of organisational approvals or certification artefacts from other MAAs.

The value of recognition and subsequent efforts to achieve it depends on the importance of the need to achieve it. Depending upon the need for recognition, the process provides four types of recognition options that the members can pursue [21]:

1. **Mutual Recognition**. All parties plan on utilising each other's assurance systems, such as organisational approvals or certification processes. This recognition is beneficial during joint operations or in collaboration projects and may be bilateral or multilateral. The UK MAA and French MAA (DSAE) signed a mutual recognition in 2013 for continuing airworthiness (EMAR Part 145) of the Airbus A400M aircraft [19]
2. **Internal Recognition**. A unilateral recognition where one or more MAAs recognise another MAA to utilise its airworthiness outputs. The process is usually beneficial when aircraft belonging to another MAA are planned to be procured or operated during joint operations. The Finnish MAA recognises the UK MAA for maintenance and design approvals of Eurofighter Typhoon (EF2000) aircraft [56].
3. **External Recognition**. A mutual or one-way recognition activity involving one or more MAAs that have not incorporated the principles of EMARs. The recognition by the UK MAA of the US Army Airworthiness Authority (AMRDEC-AED) is an external recognition [56]. Most recognitions outside the EDA utilise an external recognition process.
4. **Transitive Recognition**. An MAA utilises a prior recognition between two MAAs to recognise one or both parties.

6.2 Process Details

The recognition process consists of three distinct phases [21]:

1. **Initial Recognition**. The initial recognition consists of the central planning activities, assessments, and finalisation of certifications.
2. **Recognition Review**. The recognition review commences after a predefined time has elapsed since the initial recognition, and an extension in the recognition is deemed valuable to all parties. Under EMAD-R, a recognition remains valid for four years. A significant change in a State's regulatory system affecting the recognition or the expiry of the validity period necessitates a recognition review with the issuance of a new certificate.
3. **Realisation of Benefits**. The realisation of benefits identifies and quantifies the actual benefits accrued from recognition activity during airworthiness assurance activities.

However, this work focuses only on initial recognition because it forms the backbone of the entire process, and the remaining two phases develop from its effective implementation.

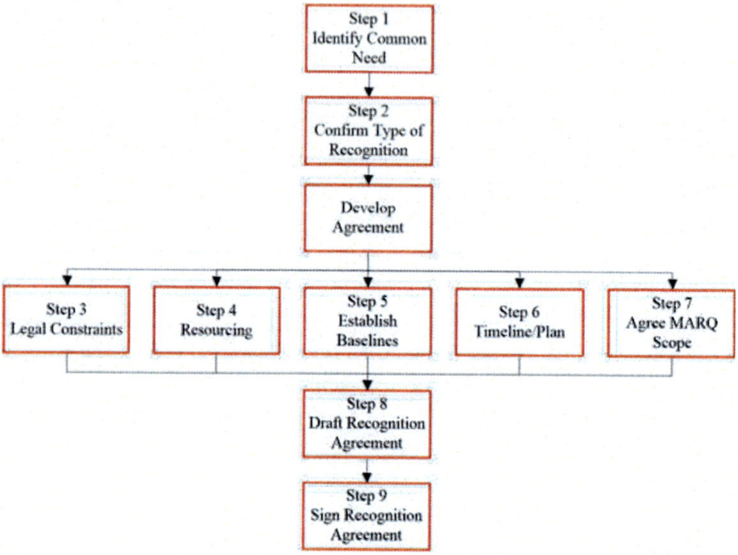

Fig. 4 EMAD-R initial recognition baseline steps. Image based on EDA [21]

6.2.1 Planning and Recognition Agreement

The initial recognition phase consists of 16 steps, where the starting nine steps formulate the foundation for the recognition process. The identification of recognition need, scope, timeline, plan and resources are finalised in these nine steps. Although the steps do not cover the actual comparison and assessment component of the process, these steps create the core strength of the entire activity. This phase is culminated by the signing of the recognition agreement. Figure 4 details the sequential flow of the initial 9 steps.

6.2.2 Self-assessment, MARQ Review and Visit

The remaining 7 steps of the process consist of the actual assessment and analysis stage of the process while dealing with the MARQ. It is prudent to emphasise that recognition is not a plain process that generates a quantitative output from simple yes and no responses. There is no definitive answer to any question within the MARQ, and the strength of a reply can only be substantiated by the detailed justifications and supporting material provided [19]. The process requires an in-depth self-assessment by each party to give detailed responses to the applicable question sets. Comprehensive responses with supporting artefacts provide a better understanding of the recognition partner. The MARQ assessment review is complemented by an onsite visit. After clarification of all issues, each MAA generates a recognition report to state its findings and justify decisions. The last activity is the signing of recognition certificates. Figure 5 details the sequential steps from 10 to 16.

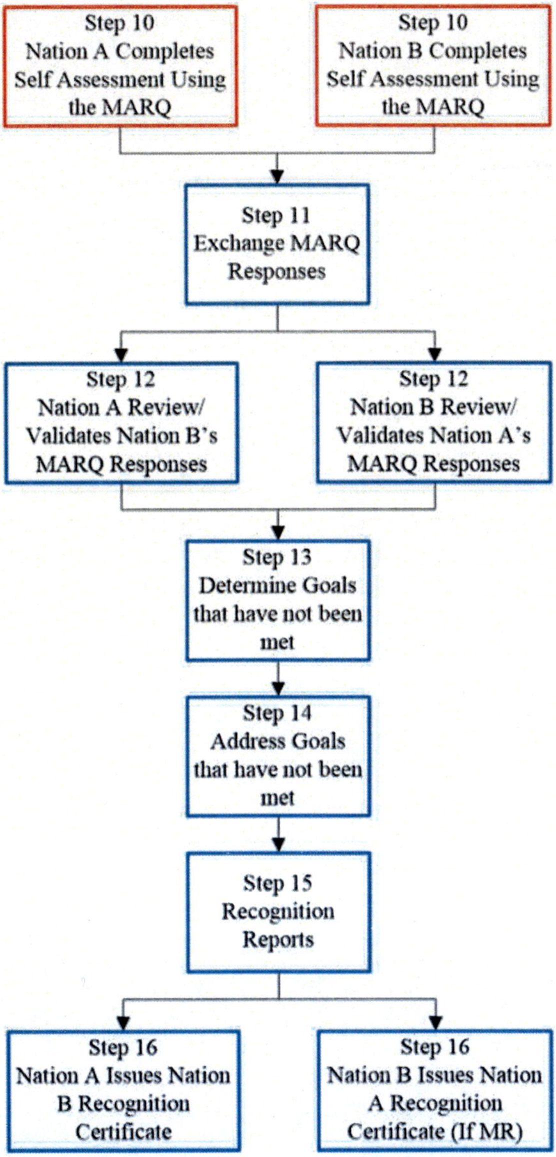

Fig. 5 EMAD-R initial recognition assessment & finalisation steps. Image based on EDA [21]

6.3 Review of MARQ

The entire assessment element of a recognition activity relies on processing the MARQ. The MARQ aims to assess the oversight capabilities of the MAA and its ability to assure safety under its jurisdiction. The details of the exact processes and procedures through which an MAA achieves these results are not significant.

6.3.1 MARQ Goals

The MARQ is divided into four parts, each relating to one specific aspect of airworthiness:

- Sub-section A—Airworthiness Authority (1.1)
- Sub-section B—Airworthiness Inspection (1.2)
- Sub-section C—Production Oversight (1.3)
- Sub-section D—Aircraft Certification (1.4)

The four parts achieve two distinct layers of air safety goal sets: organisational and assurance. The organisational layer provides an assessment of the fundamental oversight capability of any MAA and consists of sub-section A only. The assurance layer aims to evaluate the capability of the MAA to assure safety in the three phases of airworthiness and consists of sub-sections B, C and D. The goal layers with the distribution of MARQ sub-sections are illustrated in Fig. 6.

The Organisational Goals Layer is mandatory for any recognition activity. However, selecting one or more sub-sections from within the assurance layer depends upon the recognition requirement by the involved MAAs. The maintenance support for common platforms may only require MARQ assessment involving sub-section A and B (the organisational layer and airworthiness inspection in the assurance layer). In contrast, the procurement of foreign aviation products, along with a requirement of in-service support, may need assessment of all sub-sections (the organisational layer and the complete assurance goal layer).

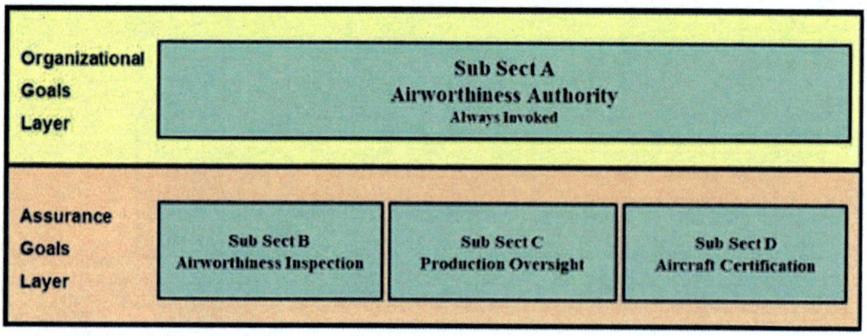

Fig. 6 MARQ goal layers. Image based on EDA [21]

6.3.2 MARQ Breakdown

As discussed earlier, in Sect. 6.1.1, the air safety goals of MARQ were derived from the ICAO Safety Oversight and Airworthiness manuals. The MARQ encompasses the requirements stated in these manuals that are necessary for an active regulatory oversight system. Sub-section A (Airworthiness Authority), which assesses the oversight capability of an MAA, consists of the evaluation of the eight CEs that ICAO has defined as the safety tools necessary for effective oversight (see Sect. 2.2). The eight CEs are supported by an additional element to assess risk management. Similarly, sub-sections B, C and D are also formulated through the regulatory requirements stipulated in the ICAO airworthiness and safety oversight manuals [21]. However, MARQ only covers the ICAO requirements that could equally apply to military aviation. Figure 7 highlights the constituent elements of sub-section A (Airworthiness

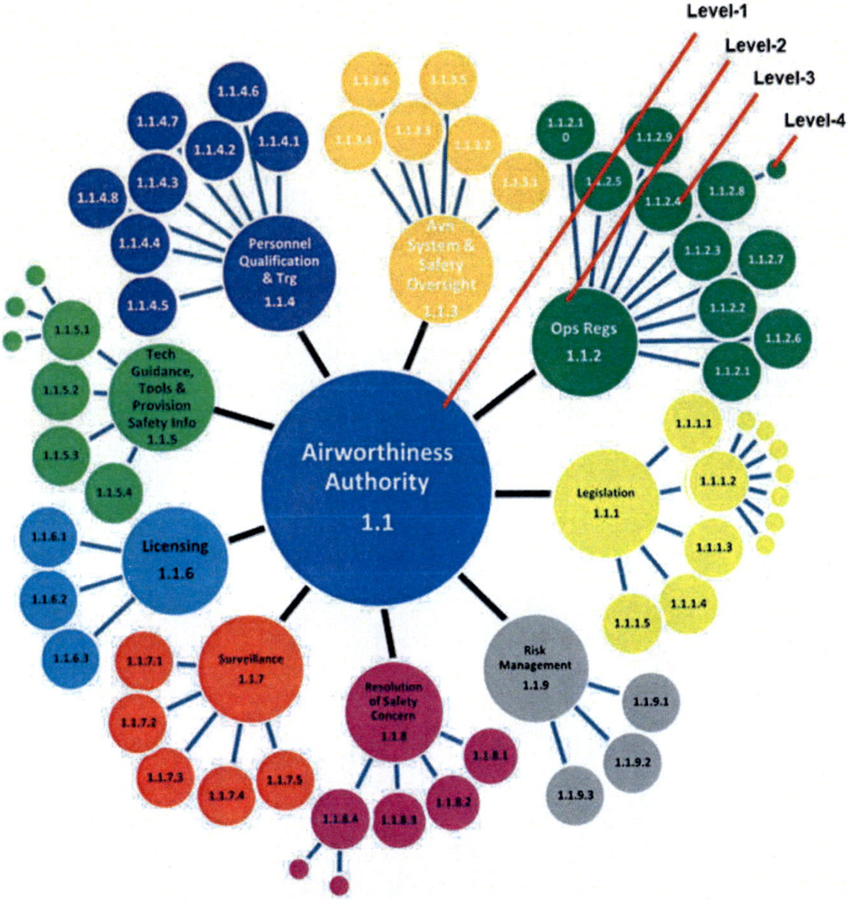

Fig. 7 Hierarchical grouping of sub-section A elements (Airworthiness Authority) (*Source* Author)

Authority) assessed to ascertain the oversight capability of an MAA. The details for the remaining MARQ sub-sections are included in Appendix 1.

Each MARQ sub-section works on a building block approach in four levels, wherein the lower and more detailed levels combine to establish the assessment of the higher levels. Figure 7 illustrates the MARQ building block concept using a spoke diagram, where the assurance assessment of the central 'Airworthiness Authority' (level-1) is achieved through the verification of the nine elements (level-2) and collective support of the respective third and fourth layers (level-3 & 4).

Although the MARQ may include details of specific system processes, the selection and construction of questions are based upon assessing the airworthiness assurance of an MAA and not the methodology used to achieve it [21]. Each MARQ sub-section consists of the key level-2 elements that can collectively highlight the capability of the MAA for the respective sub-section. The significance of a question in MARQ cannot be qualitatively analysed independently as each question is equally essential for recognition assessment. However, the quantitative strength of questions that collectively support a higher level can be used to identify the importance of each level-2 element. Figure 8 lists the level-2 elements for Sub-Section A: Airworthiness Authority and visually depicts their quantitative strength using a sunburst chart. Although the spoke diagram in Fig. 7 effectively illustrates the hierarchy of the MARQ sub-section, the sunburst chart shown in Fig. 8 provides a visual comparison based upon the number of questions within each level-2 element (a combination of level-3 and 4 questions). Each ring layer of the sunburst chart relates to one level in the hierarchy, with decreasing hierarchy moving radially outwards. The size of each segment in a ring is based upon the number of questions in its lower levels.

Figure 8 reveals that each level-2 element of sub-section A is further supported through questions in lower levels, with relatively more focus on legislation, operational regulations and personnel qualification and training (These items are shown highlighted in bold in Fig. 8). However, as shown by the spoke and sunburst charts for sub-sections B, C, and D in Appendix 1, the relative distribution of questions is

Fig. 8 Quantitative comparison of key elements—sub-section A (numerically focused areas in bold) (*Source* Author)

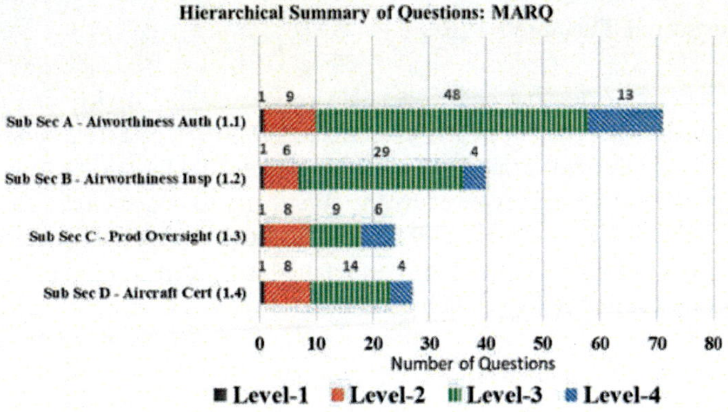

Fig. 9 Hierarchical summary of questions (*Source* Author)

not even. Over half of the questions within sub-sections B & C focus upon oversight and organisational approvals only; sub-section D, which assesses aircraft certification, focuses more on questions pertaining to type certification activities and design issues. The airworthiness activity being pursued in each sub-section influences the selection of the type of Level-2 elements and the number of questions required to assess it.

The MARQ is detailed and extensive in the task that it has been created to achieve. A quantitative summary of the questions shows 162 questions spread across the four levels of each of the four sub-sections. The comprehensive responses to these questions provide an efficient approach to gain confidence in the assurance activities of an MAA. Figure 9 provides a summary of questions at each level for the four sub-sections. The pictorial representation highlights the fact that level-3 questions form a major part of MARQ (100 questions from a total of 162) and extensively support the respective key elements of level-2. Level-4 questions are relatively less in number and provide appropriate assistance where further elaborations are required. The recognition process has a higher emphasis on the airworthiness oversight capabilities of the MAA, and therefore the questions to assess sub-section A (Airworthiness Authority) are significantly more numerous and detailed than others.

6.3.3 Assessment Through MARQ

The requirement of extensive and in-detailed responses for MARQ can only be achieved through honest and open self-assessments, followed by active communication [19]. Limited information attracts higher clarification requests [18]. The MARQ responses for each MAA are reviewed by the other party to understand their airworthiness oversight and assurance capability. Clarifications are sought for any queries until a clear understanding is achieved. The detailed MARQ responses and finalisation of all queries, followed by validation visits, create the confidence for recognition of another MAA.

6.4 Implementation of EMAD-R and Global Utilisation

The pilot project for assessing the efficacy of the EMAD-R process within the EDA was labelled the *Pathfinder* and involved the recognition of continuing airworthiness (EMAR Part 145—Approved Maintenance Organisations) of the Airbus A400M aircraft. This was achieved between the UK MAA and French MAA (DSAE) in 2013 [19]. The first trial activity of a non-EMAR based MAA was mutual recognition (Airworthiness Authority and Aircraft Certification) between the UK MAA and US Army in 2013 using the external recognition process [19]. Subsequently, EMAD-R is successfully being employed as the designated recognition process for all EDA States.

Purton et al. [44] suggested that there might be challenges in the global use of EMAD-R, as its primary design supports EMAR harmonised military airworthiness frameworks only. However, while EDA members were pursuing EMAD-R for recognition activities, the EMAR implementation status review of the EDA members by Kouskouridis [33] highlighted that most members had still not fully incorporated the EMARs (Table 2).

Some States, such as Hungary and France, had the legacy system running alongside the EMARs [18]. Belgium had already carried out recognitions while still waiting for legislative approval to implement EMARs [18]. This partial and varied implementation of EMARs across the EDA members, while EMAD-R was successfully

Table 2 EMAR implementation status in the EU [33]

EMAR part	Adopt (%)	Full compliance (%)	Partial adoption (%)	Partial compliance (%)
21	11	26	0	11
M	4	4	0	11
145	19	41	0	4
147	15	26	0	11
66	7	15	0	4

Table 3 Recognition status of some non-European MAAs utilising EMAD-R

Countries	Recognition Status	Source
Australia	Australia (DASA) has recognised MAAs of France, Germany, Italy, Netherlands, Spain, UK, USAF, & US Navy. Recognition of Canada, US Army, Singapore and New Zealand MAAs is in process	[11]
Canada	Canada (TAA) recognises MAAs of Spain and US Services. Recognition is in process for MAAs of Australia, Italy and New Zealand	[54]
Republic of Korea	US DoD (Mutual Recognition). The recognition is with DoD and encompasses all three branches of the US services	[51]

being utilised, supports the viewpoint that the EMAD-R can also be utilised for other frameworks.

Additionally, a process tailored to any single framework would be too intricate and challenging during implementation for recognition activities with other disparate frameworks. Therefore, it is vital that a process that has common and globally standardized oversight systems as its foundation is utilised within the military systems. As discussed in Sect. 6.1.1, EMAD-R retains a strong link to the ICAO airworthiness system by incorporating its safety oversight principles. Moreover, even though military airworthiness systems are disparate and unique to each State, their influence from their respective civil regulatory frameworks makes the employment of EMAD-R as a recognition process as straightforward as possible. Consequently, following the development of the process and its proven efficacy in the trial recognition between the UK MAA and the US Army, EMAD-R has found application with several non-EMAR based authorities. Table 3 shows a selection of the recognition activity status of certain non-European States utilising EMAD-R. Furthermore, EMAD-R is also the prime recognition process adapted by military alliances such as NATO and AFIC.

7 Conclusion

Military airworthiness systems have evolved significantly over time while learning from past lessons and experiences. However, the non-standardisation of military airworthiness systems makes it difficult to assure the safety of foreign-produced aviation products or operating foreign aircraft during joint ventures. The process of recognition simplifies safety assurance processes when interacting with other systems. This paper has investigated the concept of recognition in military airworthiness while exploring the increased motivation of MAAs to engage in recognition activities in recent times. Recognition becomes essential while assuring airworthiness of foreign procured aviation products or during collaborative developments. Recognition equally remains vital to assure safety when strategic ties of States necessitate interoperability or operations within military alliance organisations. Although the desire to gain confidence in safety assurance necessitates recognition, it also provides opportunities for resource sharing, mutual learning, and economic benefits while generating trust and respect between MAAs. However, it is essential to understand that the recognition process is as strong as the need to establish it. Childs [7] succinctly summarises that recognition is not a "*science project*" for merely acquiring a certificate to place on the wall. Extensive preparatory work to ascertain the need to achieve recognition and an evaluation of the expected benefits is necessary for any recognition activity. The literature search presented herein has revealed various efforts by different MAAs and military organisations for the development of a recognition process along with novel academic contributions by Purton et al. [44].

The EMAD-R process was developed for recognition activities supporting the EDA members to enhance safety assurance during interactions, collaborations, and joint operations. The process provides an opportunity for the MAAs to compare other systems and make informed decisions on utilising their airworthiness instruments and approvals. The EMAD-R retains a strong link to the globally proven ICAO framework and incorporates the ICAO safety oversight and airworthiness requirements within its assessments. An overview of the process highlights that harmonising State regulatory frameworks towards EMARs would make the process simpler and avoid misunderstandings. However, an in-depth understanding of the MARQ assessment process and implementation of EMAD-R in various recognition activities outside the EMAR frameworks supports the effective use of this process for global utilization and demonstrates quite convincingly that the EMAD-R process can effectively be used for recognition of disparate airworthiness frameworks outside the European paradigm. This is a novel finding, substantiated by countries like Australia, Canada, and the Republic of Korea, who are all currently using the EMAD-R process to develop recognition of other States' military airworthiness regulatory frameworks. Supplementary processes, such as that developed by Purton et al. [44], can provide support and further strengthen the recognition assessment activity of EMAD-R.

Acknowledgements Guidance was sought from the Australian Defence Aviation Safety Authority on the recognition activities in military airworthiness for clarity concerning EMAD-R implementation.

Appendix 1: Spoke Diagrams and Sunburst Charts of MARQ Sub-sections

See Figs. 10, 11, 12, 13, 14 and 15.

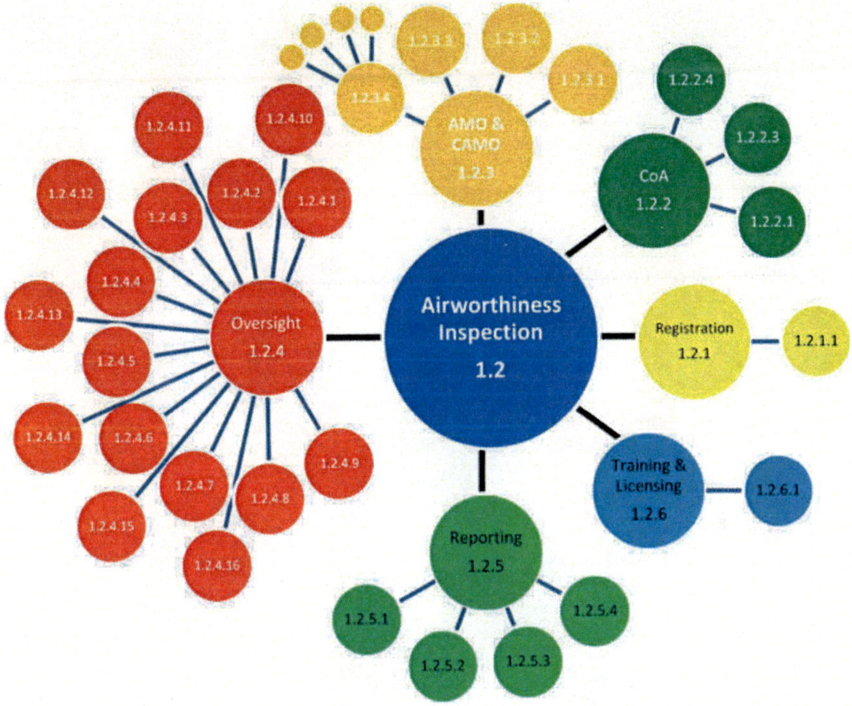

Fig. 10 Hierarchical grouping of sub-section B elements (airworthiness inspection) (*Source* Author)

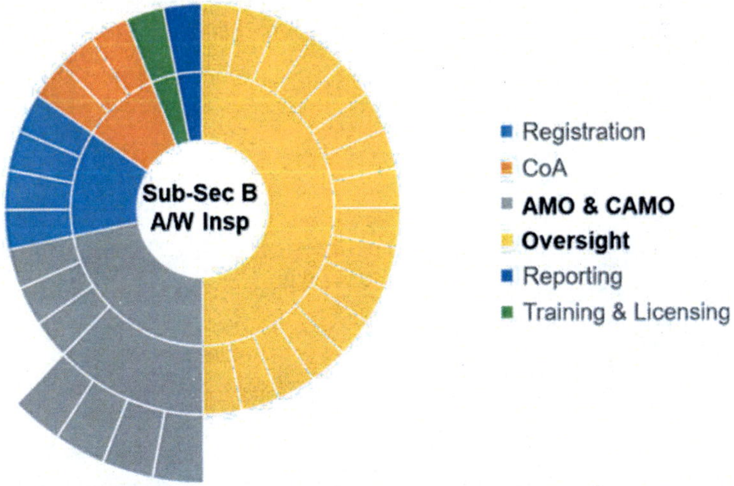

Fig. 11 Quantitative comparison of key elements—sub-section B (numerically focused areas in bold) (*Source* Author)

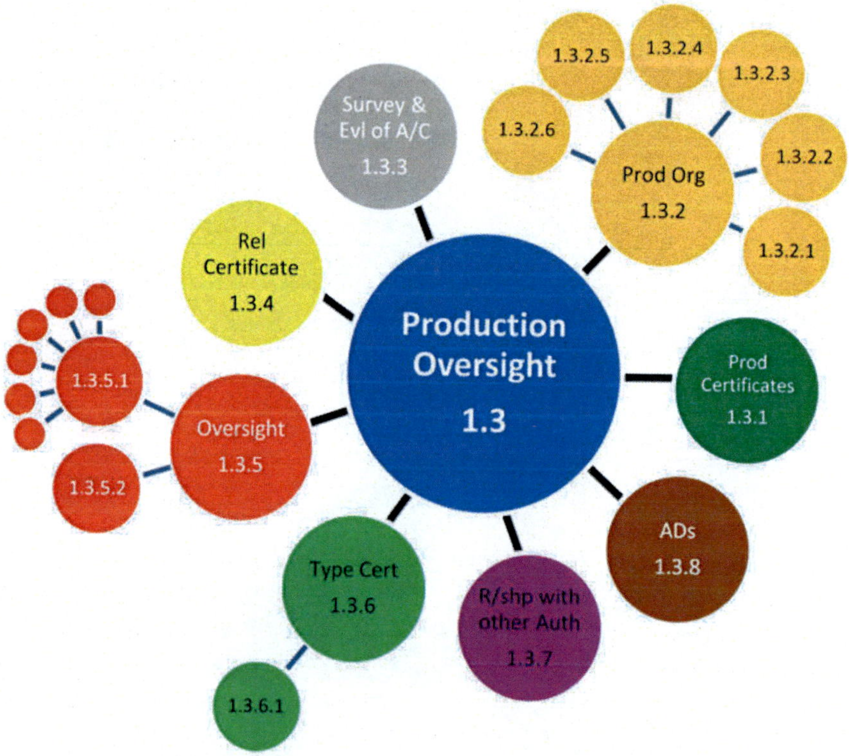

Fig. 12 Hierarchical grouping of sub-section C elements (production oversight) (*Source* Author)

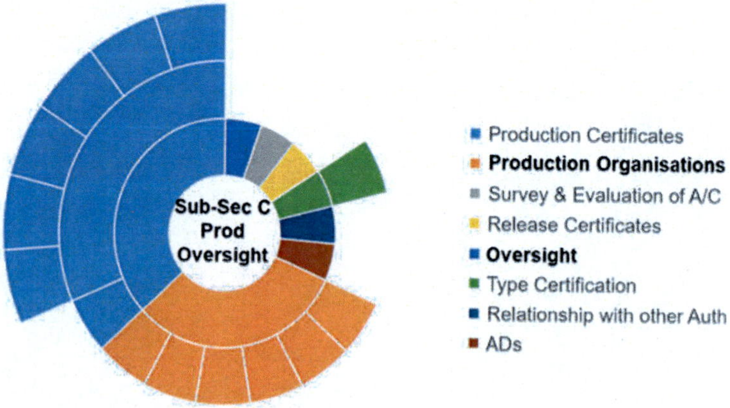

Fig. 13 Quantitative comparison of key elements—sub-section C (numerically focused areas in bold) (*Source* Author)

Fig. 14 Hierarchical grouping of sub-section D elements (aircraft certification) (*Source* Author)

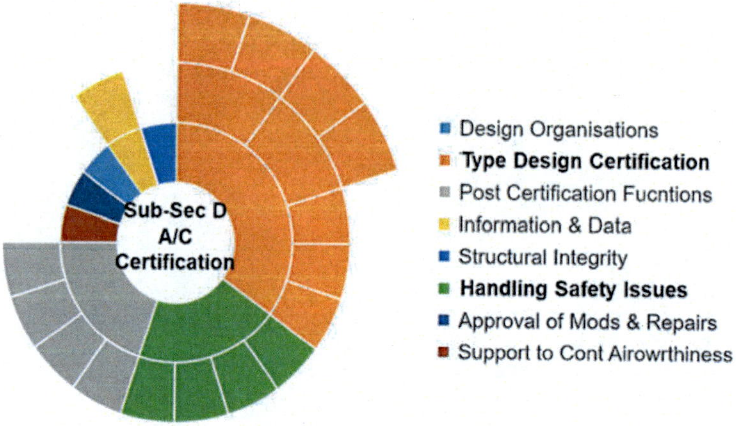

Fig. 15 Quantitative comparison of key elements—sub-section D (numerically focused areas in bold) (*Source* Author)

References

1. ADM (2020) RUAG awarded first US Hornet MRO contracts. Austr Def Mag. https://www.australiandefence.com.au/. Accessed 30 Sept 2020
2. AFIC (2020) Air Force Interoperability Council (AFIC). https://community.apan.org/wg/afic/. Accessed 3 Aug 2020
3. Australian Government Department of Defence (2011) DI(G) OPS 02–2 Defence Aviation Safety Program, 2nd edn. Department of Defence, Canberra
4. Bourgault P (2016) A400M: ensuring continued airworthiness [Powerpoint presentation]. In: EDA Military Airworthiness Conference, Lisbon
5. CAAS & CASA (2013) Technical arrangement on aviation maintenance between the Civil Aviation Authority of Singapore and the Civil Aviation Safety Authority. https://www.casa.gov.au. Accessed 28 Sept 2020
6. CAAS & CASA (2014) Technical arrangement for airworthiness certification between the Civil Aviation Authority of Singapore and the Civil Aviation Safety Authority. https://www.casa.gov.au. Accessed 28 Sept 2020
7. Childs D (2017) Recognition—What next? In: EDA military airworthiness conference, Athens
8. Clark M (2015) Harmonising airworthiness. Aerospace, February, pp 36–40
9. Coperet S (2016) Airworthiness in State's Aviation France: 10 years of experience [Powerpoint presentation]. In: International military airowrthiness regulation conference, Melbourne
10. Cripps D (2016) Collaboration in Airworthiness: The United States National Airworthiness Council [Powerpoint presentation]. In: International military airworthiness regulation conference, Melbourne
11. DASA (2021) Defence Aviation Safety Authority. https://www.defence.gov.au/DASP/DASR-Regulations/Recognition.asp. Accessed 5 Aug 2021
12. De Florio F (2016) Airworthiness—an introduction to aircraft certification and operations, 3rd edn. Elsevier Ltd., Oxford
13. De Luis C (2016) 10 ways to better aviation regulation: assuring safety of defence aviation, 2nd edn. Defence Publishing Service, Department of Defence Australia, Canberra (ACT)
14. Defence Aviation Safety Authority (2018) Airworthiness recognition: United States Navy Military Airworthiness Authority. DASA, Canberra
15. Duriez R (2012) NATO AwWG, ASIC, DoD and others [Powerpoint presentation]. In: Military airworthiness authorities conference, Larnaca

16. Duriez R (2013) Synergies between MAWA forum and NATO AwWG for recognition of MAAs [Powerpoint presentation]. In: Military airworthiness authorities forum conference, Aix-en Provence
17. Duriez R (2017) NATO recognition process [Powerpoint presentation]. In: EDA military airworthiness conference, Athens
18. Duriez R (2019) NATO recognition of airworthiness authorities: return of experience [Powerpoint presentation]. In: EDA military airworthiness conference, Vilnius
19. Duriez R, Toy M (2013) FR DSAE and UK MAA mutual recognition [Powerpoint presentation]. In: Military airworthiness authorities forum conference, Aix-en-Province
20. European Defence Agency (2015) Military Airworthiness Authorities (MAWA) forum. https://eda.europa.eu/docs/default-source/documents/mawa-frequently-asked-questions-ed-1-1.pdf. Accessed 14 Sept 2020
21. European Defence Agency (2016) European military airworthiness document: EMAD-R recognition process, 2.0 edn. EDA, Ixelles
22. Grintz M (2017) Recognition/Ideas for further development. In: EDA military airworthiness conference, Athens
23. Heinrich MN (2015) The eurofighter typhoon programme: economic and industrial implications of collaborative defence manufacturing. Def Stud 15(4):341–360
24. Hinz C (2016) Collaboration benefits in the area of mutual recognition [Powerpoint presentation]. In: EDA military airworthiness conference, Lisbon
25. ICAO (2020) International Civil Aviation Organization. https://www.icao.int/safety/Pages/USOAP-Results.aspx. Accessed 13 Aug 2020
26. Intenational Civil Aviation Organization (2010) Safety evolution: moving forward with CMA. ICAO J 65(4):24–25
27. Internation Civil Aviation Organisation (2017) Safety oversight manual: Part A—The establishment and management of a state safety oversight system (Doc 9734), 3rd edn. ICAO, Montreal
28. International Civil Aviation Organization (2011) Universal safety oversight audit programme continuous monitoring manual doc 9735, 3rd edn. ICAO, Montreal
29. International Civil Aviation Organization (2013) Airworthiness manual doc 9760, 3rd edn. ICAO, Montreal
30. Jilian G, Kangming B, Lintong J (2011) Research on airworthiness management system about military aircraft development. Procedia Eng 17:375–381
31. Jones SG (2006) The rise of a European defense. Polit Sci Quar 121(2):241–267
32. Kitson I (2016) As military as necessary as civilian as practicable. QinetiQ Pty Ltd., Canberra
33. Kouskouridis S (2017) Quantifying EMAR implementation [Powerpoint presentation]. In: EDA military airworthiness conference, Athens
34. Lawson K (2014) US services' military airworthiness authority recognition assessments: process & lessons learned [Powerpoint presentation].In: EDA military airworthiness conference, Rome
35. Li G (2009) Higher education research methodology-literature method. Int Educ Stud 2(4):179–181
36. Masterton C (2012) A vision for European military airworthiness harmonisation [Powerpoint presentation]. In: Military airworthiness authorities conference, Larnaca
37. Mawdsley J (2013) The A400M project: from flagship project to warning for European Defence Cooperation. Def Stud 13(1):14–32
38. Moclair T (2016) Safety of flight. Austr Aviat, January/February, 42–45
39. North Atlantic Treaty Organization (2016) NATO policy for civil/military aircraft operating in support of NATO or NATO-led missions and operations. https://www.nato.int/. Accessed 27 July 2020
40. Parliament of Australia (2006) Australia's defence relations with the United States. https://www.aph.gov.au/. Accessed 28 Sept 2020
41. Purton L (2015a) The changing face of military airworthiness interactions. In: 16th Australian international aerospace congress, Melbourne, pp 431–436

42. Purton L (2015b) Development of an airworthiness framework assessment tool; novel and visual. In: 16th Australian international aerospace congress, Melbourne
43. Purton L (2016) Translating airworthiness frameworks: the complexity of recognition. In: International military airworthiness regulation conference, Melbourne
44. Purton L, Clothier R, Kourousis K (2014) Assessment of technical airworthiness in military aviation: implementation and further advancement of the Bow-Tie model. Procedia Eng 80:529–544
45. Purton L, Clothier R, Kourousis K, Massey K (2014b) The PBP Bow-Tie framework for the systematic representation and comparison of military aviation regulatory frameworks. Aeronaut J 118
46. Purton L, Kourousis K (2014) Military airworthiness management frameworks: a critical review. Procedia Eng 80:545–564
47. Purton L, Kourousis KI, Clothier R, Massey K (2014) Recognition of national military airworthiness authorities: a streamlined assessment process. Int J Aeronaut Space Sci 15(1):54–62
48. Robinson P (2016) Benefits of mutual recognition in support of the Typhoon Programme [Powerpoint presentation]. In: EDA military airworthiness conference, Lisbon
49. Roman JM (2019) The collaboration between civil and military airworthiness authorities in Europe: a field in continous evolution [Powerpoint presentation]. In: Aircraft airworthiness & sustainment conference, Brisbane
50. Schmidt N, Monaghan J, Keirnan J, Vivian B (1995) The use of airworthiness codes for military applications by the Royal Australian Airforce. Melbourne, Barton, A.C.T: Institution of Engineers, Australia, pp 305–310
51. Sehoon C (2016) Mutual recognition between ROK-US Airworthiness Authority. In: EDA military airworthiness conference, Lisbon
52. Snyder H (2019) Literature review as a research methodology: an overview and guidelines. J Bus Res 104:333–339
53. Spence TB, Fanjoy RO, Lu C-T, Schreckengast SW (2015) International standardization compliance in aviation. J Air Transp Manag 49:1–8
54. Technical Airworthiness Authority (2020) Recognition of Airworthiness Authorities (TAA Advisory Number 2016-04). TAA, DND/CAF Canada, Ottawa
55. Thian CV (2015) Civil and military airworthiness challenges in Asia. Aviation 19(2):78–82
56. UK MAA (2020) Guidance—MAA recognition. https://www.gov.uk/government/publications/maa-recognition. Accessed 2 Aug 2020
57. UK Military Aviation Authority (2015) Recognition of other military airworthiness regulators: MAA/RN/2015/08 (D Tech). UK MAA, Bristol
58. US Department of Defence (2013) DOD Airworthiness Policy: DoD Directive 5030.61. US DoD, Washington
59. Vasigh B, Fleming K, Tacker T (2016) Introduction to air transport economics; From theory to application, 2nd edn. Routledge, New York

Structures and Materials

Material Property Measurement of 3D Printed Carbon Fiber Composite Using a Digital Image Correlation Method

Feng Quan, Rui Hai Xin, and Nam Seo Goo

Abstract Three-dimensional (3D) printing of carbon fiber composite is a promising manufacturing process for aerospace and other industry fields because it is fast and easy to implement and requires less labor. Knowledge of the material properties of the 3D printed composite is important for applications to aerospace and other industrial structures. In this study, a digital image correlation (DIC) technique is employed to measure the material properties of 3D printed composites. This is a non-contact measurement method, and able to measure the full-field displacement and strain of the structure. This study provides a suitable testing method for determining the material properties of 3D printed continuous fiber composite materials according to ASTM standards using the DIC method.

Keywords 3D printed composite · Material property · Digital image correlation method · ASTM

Nomenclature

Symbol	Meaning
BD	Bending Test
C	Composite
E	Elastic Modulus
G_{12}	Shear Modulus
IP	In-Plane Shear Test
O	Onyx
S	Ultimate Strength
TE	Tensile Test

(continued)

F. Quan · R. H. Xin · N. S. Goo (✉)
Department of Mechanical and Aerospace Engineering, Konkuk University, Seoul, Republic of Korea
e-mail: nsgoo@konkuk.ac.kr

S. Lee et al. (eds.), *The Proceedings of the 2021 Asia-Pacific International Symposium on Aerospace Technology (APISAT 2021), Volume 1*, Lecture Notes in Electrical Engineering 912, https://doi.org/10.1007/978-981-19-2689-1_59

(continued)

Symbol	Meaning
cf	Carbon Fiber
ε,σ	Strain,Stress
v_{12}	Poisson's Ratio

1 Introduction

In recent years, with the continuous expansion of the applications of carbon fiber reinforced polymer (CFRP) materials in aerospace engineering and other industries, carbon fiber materials manufactured by 3D printing technology have received much attentions due to their advantages of easy operation and less labor. Their mechanical properties are of great interest. However, due to the lack of a complete understanding of their performance, the applications of CFRP products, especially the applications of new materials, are still limited. The main purpose of this research was to determine the various mechanical properties of CFRP materials manufactured by 3D printing technology. Another purpose is to evaluate the feasibility of a new digital technology called digital image correlation (DIC) method to determine the mechanical properties of CFRP materials.

The biggest problem of the material property testing is the measurement of strain. In several material property test, simple and high-resolution measurement is possible by using contact measurement methods such as strain gauge [1] and extensometer. Various non-contact measurement methods with capacitive displacement sensors [2], optical measurement [2–5] or digital microscopes [6] are used for small-scale specimens. Among these methods, the DIC method [7–10], which uses a digital image of a deformed specimen captured by a camera, have been widely used to measure the displacement or strain of a specimen, recently. This method is a non-contact measurement method that enables high-precision measurement and high test precision of the specimen.

In this study, we followed ASTM standards to fabricate testing specimens and measured the mechanical properties of 3D printed composite specimens using the DIC method.

2 Material Preparation

The test specimens for the material property measurement test were fabricated by the 3D printing method at KAIST (Korea) according to ASTM-D3039 [11], ASTM-D3518 [12], and ASTM-D790 [13] standard. We used a universal testing machine

(UTM) to create applied loads in various experiments, the heating and cooling chamber to implement temperature control. The GOM-Aramis DIC system was used to monitor and obtain the strain of specimens. In addition, surface of the testing specimen was sprayed with black and white paints to achieve the random speckle pattern which is required for the DIC method measurement, as shown in Fig. 1a. The 3D printed composite specimens are consisted of two onyx layers and one carbon fiber layer as shown in Fig. 1b.

In this study, we used the codes in Table 1 to represent five experiments:

1. We use a ± 45°onyx tensile test specimen (TE-O-45) to obtain the Young's modulus E^{o} and ultimate strength $S^{o}$of onyx material.
2. We use 0° carbon fiber and onyx composite tensile test specimen (TE-C-0) to obtain Young's modulus E_1^c, the ultimate strength S_1^c, and Poisson's ratio ν_{12} of the 3D printed carbon fiber composite material.
3. We use 90°carbon fiber and onyx composite tensile test specimen (TE-C-90) to obtain Young's modulus E_2^c and the ultimate strength S_2^c of the 3D printed carbon fiber composite material.
4. We use ±45° carbon fiber and onyx composite shear test specimen (IP-C-45) to obtain the shear modulus G_{12}^c, and ultimate strength S_{12}^c of the 3D printed carbon fiber composite material.
5. We use 0° carbon fiber and onyx composite bending test specimen (BD-C-0) to obtain the flexural modulus E_f^c, and flexural strength S_f^c of the 3D printed carbon fiber composite material.

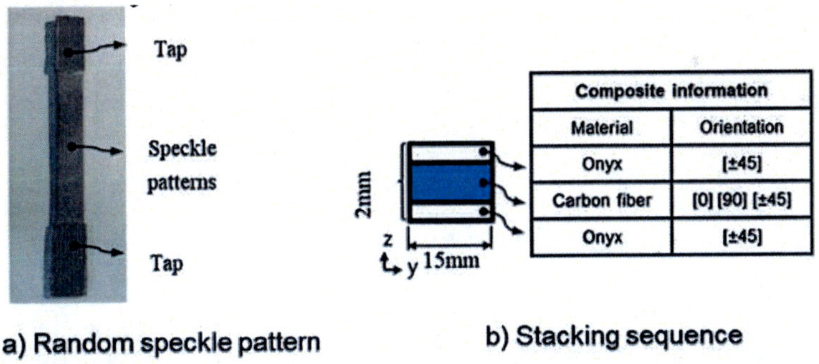

Fig. 1 A 3D printed composite specimen

Table 1 Description of specimen codes

Type	Tensile test	In-plane shear test	Bending test	3D printed composite	Onyx	[0°]	[90°]	[±45°]
Code	TE	IP	BD	C	O	0	90	45

3 Experiments

In this study, we conducted several experiments to measure the mechanical properties of the 3D printed composite material. The mechanical properties of the 3D printed composite were determined through stress–strains curves. The experimental setup for the tensile test and the in-plane shear test is shown in Fig. 2a. Two FLIR cameras with 35 mm lenses were used for the 3D DIC system. Three specimens for each test were used to ensure experimental repeatability. Specimens were mounted in the fixture of the UTM. A LabVIEW program was created to automatically obtain deformed images. The tensile load was obtained from the load cell of the UTM. Finally, deformed images of specimens were analyzed through ARAMIS software to measure the strain of the specimen.

The 3D DIC system was calibrated in advance of the experiment. The calibration procedure for the 3D DIC system is described in the ARAMIS Software User Manual [14]. During the calibration procedure, the calibration panel was moved, tilted, and rotated in multiple directions to ensure the multi-angle measurement accuracy of a 3D DIC system. The instruction manual of the ARAMIS software states that the calibration deviation should be smaller than 0.04 pixels to guarantee accuracy.

The schematic diagram of the experiment is shown in Fig. 2b. The specimen was mounted on the grip of the UTM and tensioned at a rate of 2 mm/min until it was broken; the deformed images of the specimen were obtained through the DIC system. An external light source from two halogen lamps was used to minimize the error in the experimental results and increase the accuracy. Two cameras were simultaneously controlled using a LabVIEW program, and images were automatically acquired at intervals of 1 s. After the experiment was finished, the deformed images were analyzed through ARAMIS software.

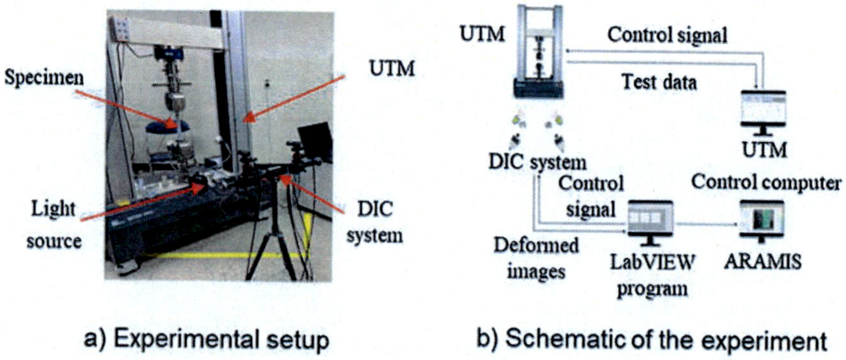

Fig. 2 Experimental setup and schematic of tensile and in-plane shear tests

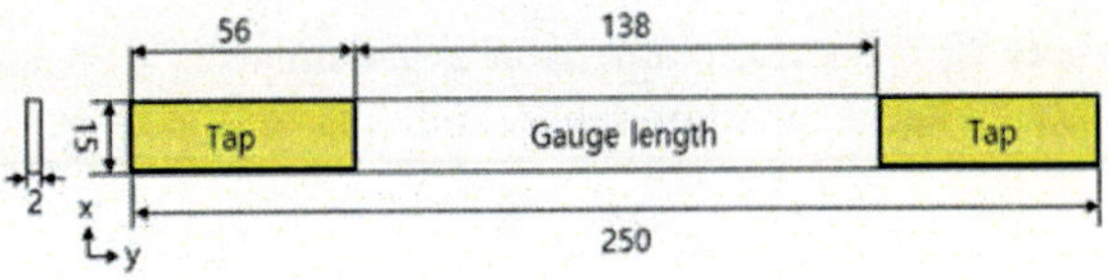

Fig. 3 Specimen schematic of TE-C-0, unit: mm

3.1 *Tensile Test of 3D Printed Composite [TE-C-(0&90)] and Onyx (TE-O-45)*

The specimen schematic of TE-C-0 is shown in Fig. 3. The tensile modulus of the specimen was obtained through Eq. (1). The tensile strain of the specimen was obtained through the DIC system, and the stress was obtained through the load cell in the UTM.

$$E = \frac{\Delta\sigma}{\Delta\varepsilon} \tag{1}$$

where E is Young's modulus, $\Delta\sigma$ is the stress difference between two strain points, $\Delta\varepsilon$ is the strain difference between two sampling points.

Equation (2) was employed to obtain the Poisson's ratio of the specimen.

$$\nu_{12} = -\frac{\Delta\varepsilon_y}{\Delta\varepsilon_x} \tag{2}$$

where ν_{12} is Poisson's ratio, $\Delta\varepsilon_x$ is the longitudinal strain, and $\Delta\varepsilon_y$ is the transverse strain.

3.2 *Shear Test of 3D Printed Composite (IP-C-45)*

The shear strain can be determined through Eq. (3) using the longitudinal strain and transverse strain obtained by the DIC system, and the shear stress was determined through Eq. (4) using the maximum load obtained from the load cell of UTM. Finally, shear modulus (G) was determined through Eq.(5).

$$\gamma_{12} = \varepsilon_x - \varepsilon_y \tag{3}$$

where γ_{12} is the shear strain, ε_x is the longitudinal strain, and ε_y is the transverse strain.

$$\tau_{12}^{m} = \frac{P^{m}}{2A} \tag{4}$$

where τ_{12}^{m} is the maximum plane shear stress, P^{m} is the maximum load within the 5% engineering shear strain range, and A is the cross-sectional area of the specimen.

$$G = \frac{\tau_{12}^{m}}{\gamma_{12}} \tag{5}$$

3.3 Bending Test of 3D Printed Composite (BD-C-0)

Three-point bending test was performed using a standard specimen according to ASTM D790 standard to obtain the flexural modulus and ultimate flexural strength of the specimen. The experimental setup for the three-point bending test is shown in Fig. 4a.

The schematic diagram of the experiment is shown in Fig. 4b. The specimen was mounted on the supports of the UTM and the loading nose was driven at a speed of 1.18 mm/min. The driving speed of the loading nose was obtained by Eq. (6).

$$R = \frac{ZL^2}{6d} \tag{6}$$

where R is the driving speed of the loading nose, L is the length of the support span, d is the thickness of the specimen, and Z is the strain of the outer fiber ($Z = 0.01$). Under the experiment condition, the specimen was bent until it fractured, and the deformed images of the specimen were obtained through the DIC system. Finally, the deformed images were analyzed through ARAMIS software.

When a uniform elastic material receives a force at the center between two supporting points, the maximum flexural stress occurs at the center of the outer surface of the specimen. The flexural stress is calculated by Eq. (7), and the flexural strain is calculated by Eq. (8). Finally, flexural modulus (E_{fc}) was determined

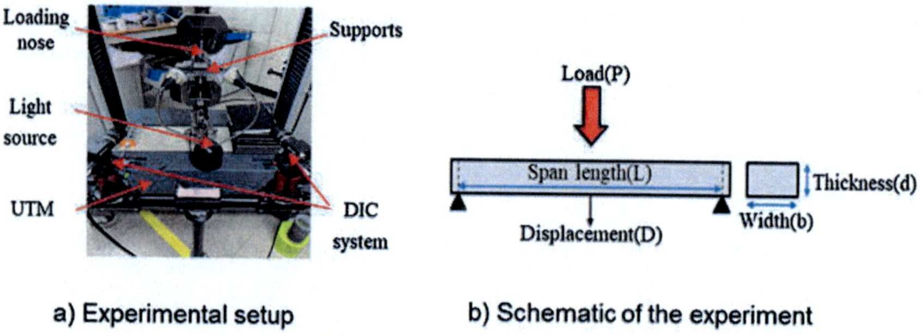

Fig. 4 The experimental setup and schematic of bending test

by Eq. (9).

$$\sigma_f = \frac{3PL}{2bd^2} \tag{7}$$

where σ_f is the bending stress, P is the applied load, L is the support span, d is the thickness of the specimen, and b is the width of the specimen.

$$\varepsilon_f = \frac{6Dd}{L^2} \tag{8}$$

where ε_f is the strain at the center of the outer edge of the specimen, D is loading arm displacement, L is the length of the support span.

$$E_f^c = \frac{\sigma_f}{\varepsilon_f} \tag{9}$$

4 Results

4.1 *Mechanical Properties of Onyx*

The material property of onyx(matrix material) must be clarified in advance to determine the material property of the carbon fiber(reinforced material). Young's modulus and the ultimate strength of the onyx material calculated experimentally through the material tests according to ASTM standard. The stress–strain curves of the tensile modulus test of onyx material are shown in Fig. 5.

The experimental results of the tensile test of the onyx material were summarized in Table 2. The mean value of tensile modulus is 1.08 GPa and the average of ultimate strength is 29.3 MPa.

4.2 *Mechanical Properties of 3D Printed Composite*

4.2.1 Young's Modulus and Poisson's Ratio of 3D Printed Composite

Young's modulus in the longitudinal direction, the ultimate strength, and Poisson's ratio of a TE-C-0 specimens for the tensile test of a 3D printed composite specimen were measured experimentally. The stress–strain curves of the longitudinal tensile modulus of the 3D printed composite experiment are shown in Fig. 6.

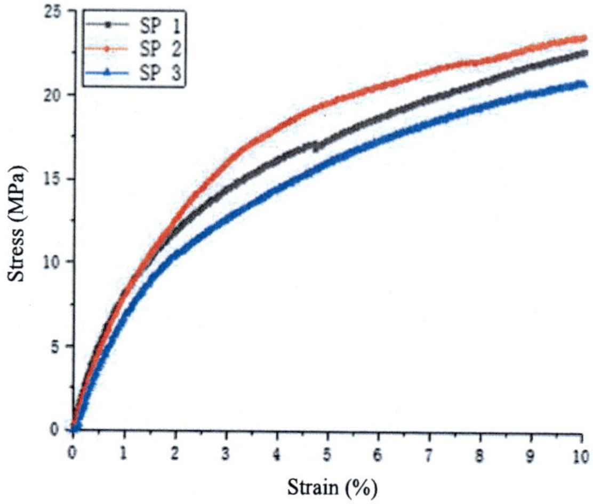

Fig. 5 Stress–strain curve of standard specimens for acquiring the tensile modulus and ultimate strength of onyx

Table 2 Test results of tensile specimens for acquiring tensile modulus and ultimate strength of onyx

Sample No.	E^o (GPa)	S^o_{12} (MPa)
1	1.06	N/A
2	1.1	30.69
3	1.09	27.91
Average	1.08	29.3

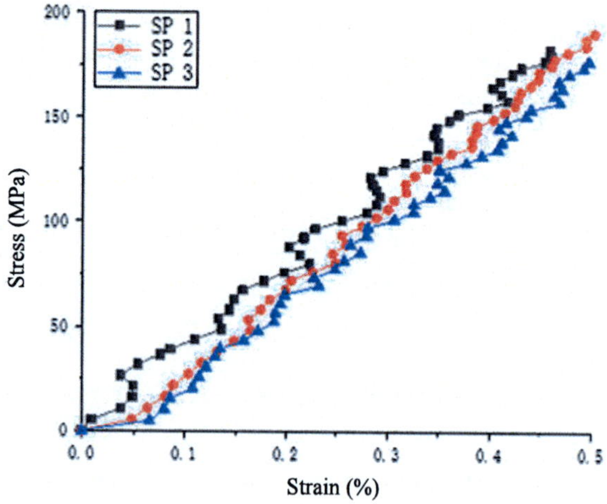

Fig. 6 Stress–strain curve of tensile test results for obtaining the longitudinal tensile modulus of the 3D printed composite

The test results of the TE-C-0 specimen were summarized in Table 3. The mean value of the longitudinal tensile modulus of the composite material is 40.86 GPa, the mean value of the ultimate strength is 450.7 MPa, and Poisson's ratio is 0.34. The tensile modulus of carbon fiber E_1^{cf} was determined through the rule of mixture

Table 3 Tensile test results for obtaining the mechanical properties of the 3D printed carbon fiber composite in longitudinal direction

Sample no.	E_1^c (GPa)	S_1^c (MPa)	ν_{12}	E_1^{cf} (Use R.O.M)
1	41.81	486.2	0.35	–
2	40.64	441.9	0.34	–
3	40	408.2	0.32	–
Average	42	466.6	0.33	80.64

(R.O.M) (Eq. (10)).

$$E_1^{cf} = E^o + \frac{1}{V_{cf}}\left(E_1^c - E^o\right) \tag{10}$$

where V_{cf} is volume fraction.

We determined the Young's modulus and ultimate strength of a TE-C-90 specimensexperimentally. The stress–strain curves of transverse modulus of 3D printed composite are shown in Fig. 7.

The experimental results of the (TE-C-90) specimens were summarized in Table 4. The mean value of the transverse tensile modulus of the composite material was 2.88 GPa and the mean value of the ultimate strength was 15.05 MPa.The tensile modulus of carbon fiber E_2^{cf} was determined through rule of mixture (Eq. (11)).

$$E_2^{cf} = E^o + \frac{1}{V_{cf}}\left(E_2^c - E^o\right) \tag{11}$$

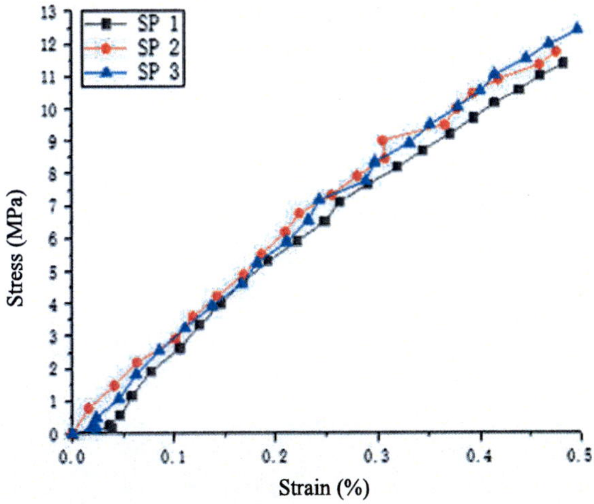

Fig. 7 Stress–strain curve of tensile test results for the transverse modulus of 3D printed composite

Table 4 Tensile test results for obtaining the mechanical properties of the 3D printed composite

Sample no.	E_2^c (GPa)	S_2^c (MPa)	E_2^{cf} (Use R.O.M)
1	2.93	13.52	–
2	2.83	15.85	–
3	2.88	15.79	–
Average	2.88	15.05	3.78

4.2.2 Shear Modulus of 3D Printed Composite

Shear modulus and ultimate shear strength of the IP-C-45 specimens for the shear test of the 3D printed composite specimen were measured experimentally. The shear stress-shear strain curves of the 3D printed composite are shown in Fig. 8.

The test results of the IP-C-45 specimen were summarized in Table 5. The mean value of the shear modulus of the composite is 1.57 GPa and the mean value of ultimate strength is 52.62 MPa. The shear modulus of carbon fiber G_{12}^{cf} was determined through the rule of mixture as shown in Eqs. (12) and (13).

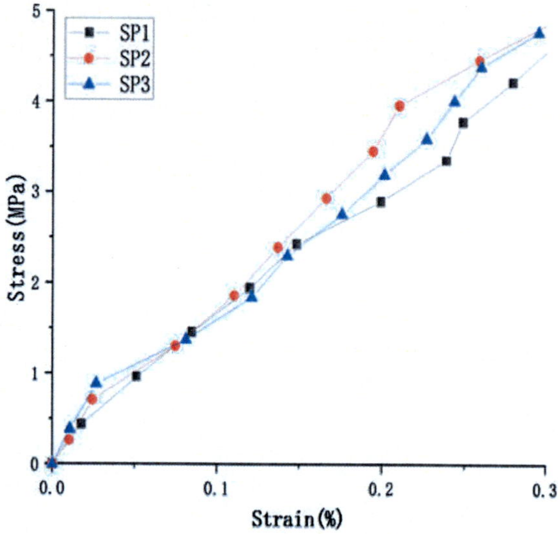

Fig. 8 Shear stress–shear strain curve of shear test results for obtaining the shear properties of the 3D printed composite

Table 5 Shear test result for obtaining shear properties of the 3D printed carbon fiber composite

Sample no.	G_{12}^c (GPa)	S_{12}^c (MPa)	G_{12}^{cf} (Use R.O.M)
1	1.51	N/A	–
2	1.6	48.28	–
3	1.59	56.96	–
Average	1.57	52.62	1.96

$$\sigma_{12}^{cf} = \frac{\varepsilon_c}{V_{cf}}\left(E_{12}^{c} - V_o E^o\right) \tag{12}$$

$$G_{12}^{cf} = \frac{1}{2}\frac{\Delta\sigma_{12}^{cf}}{\Delta\gamma_{12}} \tag{13}$$

4.2.3 Flexural Modulus of 3D Printed Composite

Flexural modulus E_f^c and flexural strength S_f^c of the BD-C-0 specimens for the bending test of the 3D printed composite were measured experimentally. The flexural stress-flexural strain curves are shown in Fig. 9.

The experimental results of the BD-C-0 specimens for the bending test are summarized in Table 6. The mean value of the flexural modulus of composite material is 36.1 GPa and the mean value of the flexural strength is 551.25 MPa.

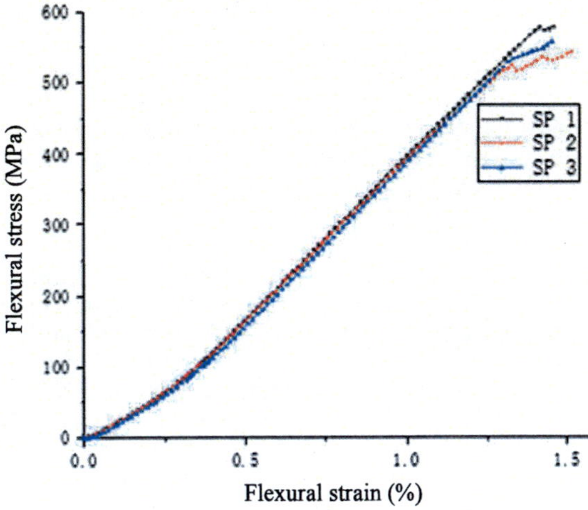

Fig. 9 Flexural stress–flexural strain curve of the bending test results for obtaining flexural properties of the 3D printed composite

Table 6 Bending test results for obtaining the flexural properties of 3D printed carbon fiber composite

Test no.	E_f^c (GPa)	S_f^c (MPa)
Test 1	36.4	590.82
Test 2	35.9	491.35
Test 3	36	571.59
Average	36.1	551.25

5 Conclusion

In this study, we successfully measured the material properties of the 3D printed continuous carbon fiber composite using a 3D DIC method. In the experiments, the composite specimens were fabricated by a 3D printing method according to the ASTM standards for various mechanical property tests. Full-field deformation and strain of the specimens were measured through the DIC system. The data obtained from each experiment had similar values, which confirmed experimental repeatability. The results demonstrated the applicability of the DIC method in measuring the material property of the3D printed composite material.

Acknowledgements This research was carried out through the Unmanned Mobile Source Technology Development Project with the support of the National Research Foundation and the Unmanned Mobile Source Technology Development Project with funding from the Ministry of Science and ICT. (NRF-2020M3C1C1A01084220)

References

1. Wang Y, Zhu S (2021) Effect of stringer design on the stability of stiffened composite panel under in-plane shear: an experimental study. Int J Aeronaut Space Sci 22(3):590–601
2. Kim C-Y, Sharpe WN (2010) Development of a fatigue testing system for micro-specimens. In: Proceedings of the Korean society of mechanical engineers A., vol 34, no 9, pp 1201–1207
3. Sharpe WN, Yuan B, Edwards RL (1997) A new technique for measuring the mechanical properties of thin films. J Microelectromech Syst 6(3):193–199
4. Lee SJ, Han SW, Kim JH, Lee HJ (2007) Measurement of mechanical properties for thin film using visual image tracing method. Solid State Phenom 124–126:1701–1704
5. Lee S, Kim SH, Lee HJ et al (2018) Density acquisition and aero-optics measurement from BOS images for a hot jet. Int J Aeronaut Space Sci 19:563–574
6. Kim M, Park J (2021) Stiffness prediction of triaxial braided composites accounting for manufacturing parameters. Int J Aeronaut Space Sci 22:602–612
7. Zhao T, Le VT, Goo NS (2020) Global-local deformation measurement of stress concentration structures using a multi-digital image correlation system. J Mech Sci Technol 34:1655–1665
8. Le VT, Goo NS (2020) Dynamic characteristics and damage detection of a metallic thermal protection system panel using a three-dimensional point tracking method and a modal assurance criterion. Sensors 20(24):7185
9. Le VT, Ha NS, Jin T, Goo NS, Kim JY (2016) Thermal interaction of a circular plate-ring structure using digital image correlation technique and infrared heating system. J Mech Sci Technol 30(9):4363–4372
10. Ha NS, Le VT, Goo NS, Kim JY (2017) Thermal strain measurement of austin stainless steel (ss304) during a heating-cooling process. Int J Aeronaut Space Sci 18(2):206–214
11. ASTM D3039/D3039M-17 (2017) Standard test method for tensile properties of polymer matrix composite materials. ASTM International, West Conshohocken, PA. www.astm.org
12. ASTM D3518/D3518M-18 (2018) Standard test method for in-plane shear response of polymer matrix composite materials by tensile test of a ±45° laminate. ASTM International, West Conshohocken, PA. www.astm.org

13. ASTM D790-17 (2017) Standard test methods for flexural properties of unreinforced and reinforced plastics and electrical insulating materials. ASTM International, West Conshohocken, PA. www.astm.org
14. Techniques GOM (2005) ARAMIS@v5. 4 user manual. GOM mbH

Research on Helicopter with Emergency Floating System Impact Load Characteristics During Ditching Based on ALE Method

Zhijie Feng, Aokun An, Huajin Lei, and Chuanyi Ke

Abstract This paper based on the ALE algorithm, the helicopter with emergency floating system load characteristics during ditching are numerically simulated and analyzed. Provide support for the design of emergency floating system, gasbag installation point and personnel evacuation channel. First of all the ditching model of the helicopter and the floats is established, and the accuracy of the simulation model and the calculation method is verified by comparing the results of the scale model and the physical testing, verified the accuracy and effectiveness of the simulation model, modeling method and calculation method. Then by using this method establish the full-size model. Based on this method, the influence of different roll angles, horizontal velocities and vertical velocities on the helicopter impact load during ditching is studied. This paper provides support for the design of emergency floating system, gasbag installation point and personnel evacuation channel. It not only overcomes the difficulty that the full-size prototype cannot carry out physical experiments, but also can effectively shorten the development cycle and save development funds.

Keywords Helicopter ditching · Scale model · Full-size model · Impact load · Emergency floating system

1 Foreword

In order to ensure the safety of the helicopter during ditching, it is essential to research the emergency floatation system at the stage of product design. The ditching process of the helicopter is a very complicated fluid–solid coupling phenomenon. At present, the helicopter impact load, influence of attitude changes and float impact load during ditching are mainly researched through test and simulation.

The research on test and simulation technology of helicopter ditching has been performed abroad. A ditching test on the helicopter is conducted by Groenen et al. [1] to obtain data such as the pressure and the acceleration of the helicopter. The

Z. Feng (✉) · A. An · H. Lei · C. Ke
General Technology R&D Department, AVIC Aerospace Life-Support Industries, LTD, Xiangyang, People's Republic of China
e-mail: zhijie_feng@sina.com

S. Lee et al. (eds.), *The Proceedings of the 2021 Asia-Pacific International Symposium on Aerospace Technology (APISAT 2021), Volume 1*, Lecture Notes in Electrical Engineering 912, https://doi.org/10.1007/978-981-19-2689-1_60

helicopter scale model is adopted by Muller et al. [2] to conduct the ditching test and measure the helicopter impact load, considering the 2/3 lift generated by the rotor. Test and simulation on helicopter ditching are conducted by Alcedo et al. [3] to obtain the test data and simulation data of the pressure distribution and the impact load of the helicopter. A scale test on the ditching and stability of the helicopter floatation system is conducted by Ye et al. [4] to analyze the influence of the location and pressure of the floats on the ditching performance. A ditching test on the helicopter scale model is conducted by Wang et al. [5] of China helicopter design and research institute, and the test results can match those of Moscow institute of Russia by comparison. The ditching progress of the helicopter is analyzed by Li [6] with the numerical simulation method to obtain the law of the helicopter impact load at the different pitch angles and yaw angles.

At present, the scale prototype is used for ditching test and relevant tests at home and abroad. Although data such as ditching impact load and attitude influence can be obtained, problems including long cycle, high cost and high risk exist. In the previous numerical simulation of the emergency floatation system, the realization of simulation methods and the peak of the helicopter impact load have been fully researched, but the helicopter attitude change during ditching has not been tested and the related factors have not been systematically researched.

In this paper, the simulated attitude change result of the scale model is compared with the test data, and the ditching ofthe full size helicopter at different vertical velocities, horizontal velocities, pitch angles, yaw angles and roll angles is systematically simulated, and then the horizontal acceleration and the vertical acceleration of the helicopter and the impact load of the helicopter and floats are obtained, for providing the reference for the design of the emergency floatation system.

2 Control Equation

In this paper, the fluid–solid coupling characteristics of the helicopter ditching are researched by using ALE method. The general transport equation is integrated in a finite volume to obtain:

$$\frac{\partial}{\partial t}\int_V \rho\, f\, dV + \int_A \boldsymbol{n} \bullet (\rho\, \boldsymbol{\mu f})dA = \int_A \boldsymbol{n} \bullet (\boldsymbol{\Gamma}\, grad\ f)\, dA + \int_V S\, dV \tag{1}$$

When the outer surface A of the finite control volume V moves at the velocity of $\underline{\boldsymbol{u}}$, the integral expression of the transport equation is:

$$\frac{\partial}{\partial t}\int_V \rho f\, dV + \int_A n \bullet (\rho(\mu - \overline{\mu}) f) dA = \int_A n \bullet (\Gamma\, grad\, f)\, dA + \int_V S\, dV \tag{2}$$

When the velocity $\boldsymbol{u}$ of the finite control volume is equal to zero, namely Eulerian method, the mesh is fixed in space. When $\overline{\mu}$ is equal to zero, that is Lagrangian method, and when $\overline{\mu}$ is not equal to zero, that is either one of the above methods (ALE).

3 Verification of Simulation Model and Scale Model

3.1 Simulation Model

In order to analyze the fluid–solid coupling dynamic characteristics of the helicopter emergency ditching system, the calculation area of helicopter ditching is established, as shown in Fig. 1. The length of the calculation area is 76 m, and the distance between the helicopter tail and the rear end of the calculation area is 5 m. The width of the calculation area is 26 m. The upper area is the air area, the height of which is 7 m. The lower area is the water area, the height of which is 10 m. At the beginning, both the helicopter and the floats are above the water surface, and the lowest point of the floats is at a distance of 0.01 m from the water surface. The emergency floatation system mainly consists of the helicopter, front floats and rear floats. Keep the original main geometric shape, remove the smaller geometric surface, and delete the landing gear and rotors. The helicopter is 15 m long, 2.5 m wide and 3 m high. The helicopter direction is along X-axis, and the gravity direction is Z axis.

A multi-substance ALE algorithm is used for the element, which allows many kinds of substances to coexist in a fluid area. The fluid–solid coupling between the structure and water is simulated by defining the contact. The solid unit made of Euler empty material is used for water and air, and the pressure gradient in the water is generated by gravity. The membrane unit made of fabric material is used for the floats and the connecting strap, and the Lagrange shell unit made of rigid body material is used for the helicopter, all of which apply a lift of 2/3 gravity on the helicopter

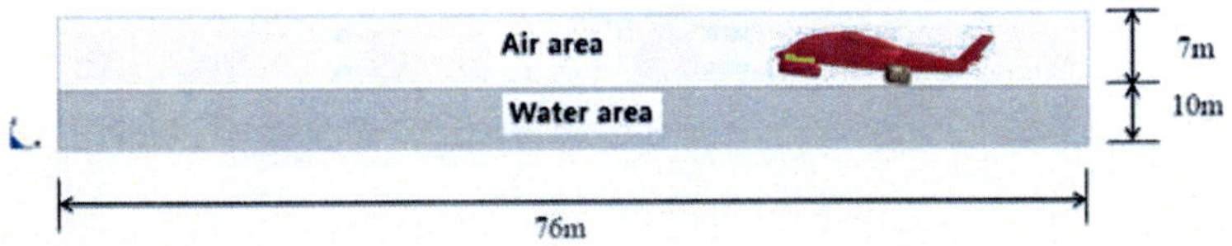

Fig. 1 Calculation area of helicopter ditching

center of mass. The control volume method is used to simulate the floats. The contact between the floats and the helicopter, the connecting strap and floats, and connecting strap and the helicopter are defined by automatic contact. In order to reduce the influence of the fluid area size, the outer boundary of the air area and water area is unreflected.

3.2 *Model Verification*

The helicopter ditching simulation is a complicated process greatly affected by parameters. In order to verify the validity and accuracy of the established model, simulation method and parameter setup, the impact ditching process of the 1:5 scale model of the helicopter is simulated and compared with the test data. The working condition parameters of the selected scale verification model are shown in Table 1, and the initial attitude and position of the model are shown in Fig. 2.The positive and negative definitions of attitude angles are as follows: (1) For the pilot, the pitch angle

Table 1 Working condition of scale verification model

Forward velocity (m/s)	Vertical velocity (m/s)	Yaw angle (°)	Pitch angle (°)	Roll angle (°)
6.9	0.9	+15	+10	+10

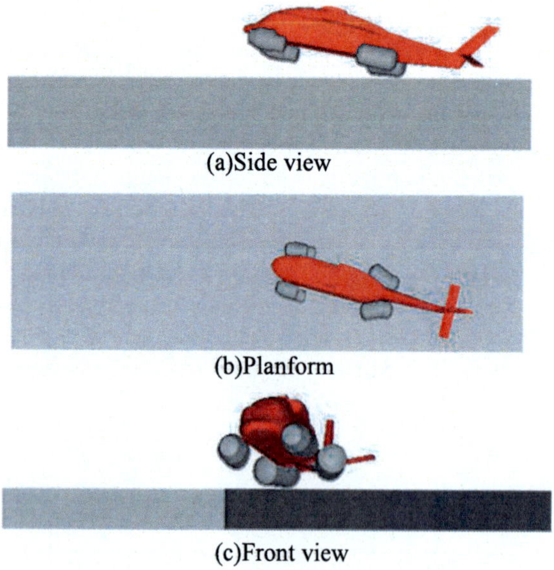

Fig. 2 Helicopter initial attitude and position of scale verification model

is positive when the nose of the helicopter lifts, and negative when the nose of the helicopter descends. (2) The yaw angle is positive when the nose of the helicopter is tilted to the right, and negative when the nose of the helicopter is tilted to the left. (3) The roll angle is positive when the helicopter is tilted to the right, and negative when the helicopter is tilted to the left.

The vertical (z-direction) acceleration at the center of gravity of the helicopter is selected to compare the simulation value and the test value, as shown in Fig. 3, The test process is shown in Fig. 4. It shows that the change rules of the acceleration curves

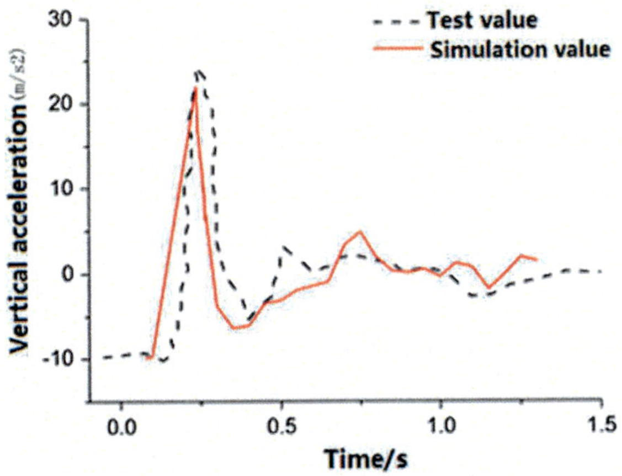

Fig. 3 Comparison of z-direction acceleration history of simulation value and test value

Fig. 4 The test process

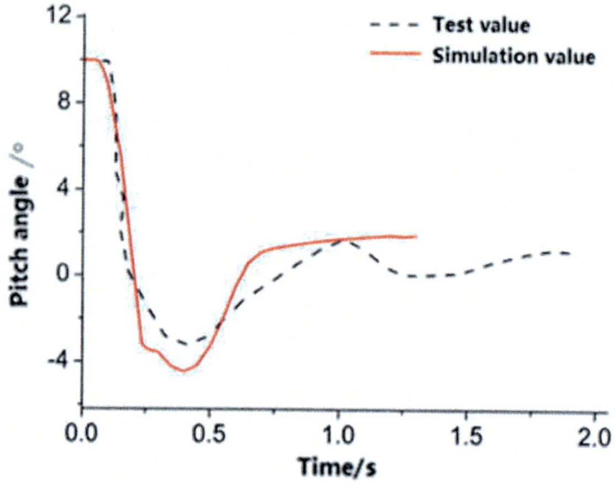

Fig. 5 Comparison of pitch angle attitude change between simulation and test

of the simulation value and the test value are consistent. Before the helicopter and the floats approach the water surface, the z-direction acceleration of the helicopter is the gravity acceleration. In the process of pitching, the acceleration increases in the opposite direction, then decreases, and finally fluctuates around zero to keep in the phase of force balance. The peak acceleration obtained by simulation is 24.5 m/s^2, and the peak acceleration measured by test is 23.4 m/s^2, with a difference of 4.5% between them.

In order to compare the helicopter attitude changes during ditching, the change of the angle between the connection line on the simulation model and that on the water in the corresponding direction is selected and analyzed to obtain the change curves of the pitch angle and roll angle of the helicopter and compare them with the test curves.

The pitch angle change during pitching of test model is compared with that of simulation model in Fig. 5. It shows that the pitch change rule obtained by simulation is consistent with that obtained by test. During ditching, the pitch angle first decreases and then increases, which means that the helicopter first lowers its nose and then raises its nose. Finally, the pitch angle fluctuates in the horizontal position to tend to balance. The minimum pitch angles of the simulation value and the test value are −4.5° and −3.2° respectively, and the difference of the pitch angle change amplitude during pitching is 8.9%.

The roll angle change history during pitching of the test model is compared with that of the simulation model in Fig. 6. It shows that the roll angle change rule obtained by simulation is consistent with that obtained by test. During ditching, the roll angle first decreases and then fluctuates in the horizontal position to tend to balance. The minimum pitch angles of the test value and the simulation value are −8.1° and −9.6° respectively, and the difference of the roll angle change amplitude during pitching is 7.6%.

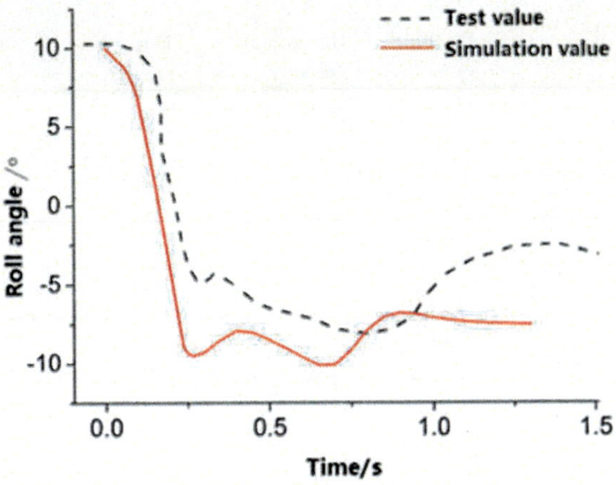

Fig. 6 Comparison of roll angle attitude change between simulation and test

By comparing the simulation value history and the test value history on the vertical load, the pitch angle and the roll angle of the helicopter scale verification model during pitching, it shows that the simulation value is consistent with the test value, with an error within 10%, which verifies the reliability and accuracy of the simulation model and parameters.

4 Analysis on Full-Size Ditching Working Condition

4.1 Full-Size Simulation Working Condition

The impact load of the full-size helicopter during ditching is analyzed systematically by using the grid scale and calculation method of the above scale verification model. The relevant research and analysis show that maintaining a certain pitch angle facilitates the ditching of the helicopter, and the yaw angle has little effect on the impact load. Therefore, the pitch angle of the full-size helicopter in all working conditions is 6°, without yaw angle. In order to analyze the relationship between the helicopter impact load during ditching and the horizontal velocity, vertical velocity and the roll angle of the helicopter, design three groups of working conditions. The effect of the roll angle of the helicopter is considered in the working condition of Table 2, the effect of the horizontal velocity of the helicopter is considered in the working condition of Table 3, and the effect of the vertical velocity of the helicopter is considered in the working condition of Table 4.

Table 2 First working condition (roll angle change)

Working condition	Horizontal velocity (m/s)	Vertical velocity (m/s)	Roll angle (°)
G0	15.5	2	0
G1	15.5	2	5
G2	15.5	2	10
G3	15.5	2	15

Table 3 Second working condition (horizontal velocity change)

Working condition	Horizontal velocity (m/s)	Vertical velocity (m/s)	Roll angle (°)
G4	5.5	2	0
G5	10.5	2	0
G0	15.5	2	0
G6	20.5	2	0

Table 4 Third working condition (vertical velocity change)

Working condition	Horizontal velocity (m/s)	Vertical velocity (m/s)	Roll angle (°)
G7	15.5	1	0
G8	15.5	1.5	0
G0	15.5	2	0
G9	15.5	2.5	0

4.2 Attitude Change During Ditching

Figure 7 shows the helicopter attitudes at typical moments during ditching in G0 working condition. Figures 8 and 9 show the horizontal velocity and vertical velocity change curves of the center of mass of the helicopter respectively. The three Fig. s show that, at the beginning, the helicopter keeps moving forward at its initial horizontal velocity and moves downward to accelerate under the action of gravity. At about 0.3 s, the rear floats of the helicopter contact the water surface, and the water resistance gradually decreases the horizontal velocity of the helicopter. With the increase of the interaction between water and the helicopter (including floats),

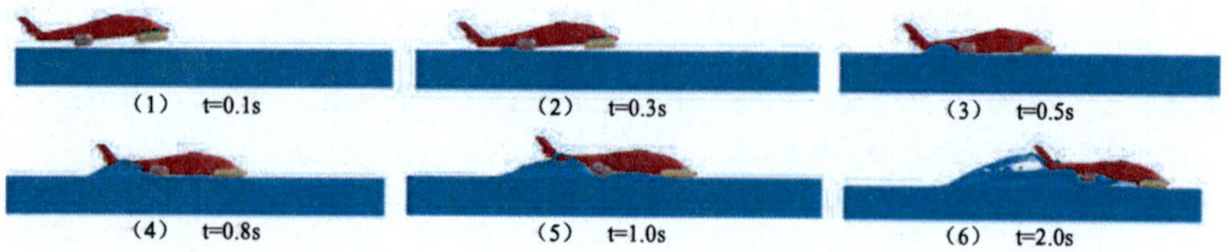

Fig. 7 Helicopter attitudes at typical moments during ditching (G0 working condition)

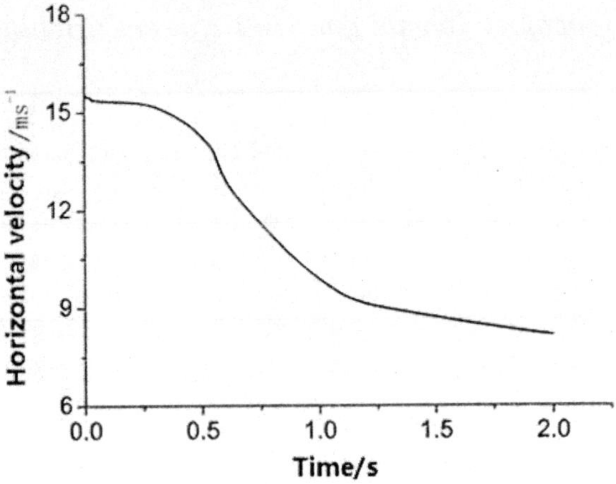

Fig. 8 Horizontal velocity change curve of center of mass of helicopter (G0 working condition)

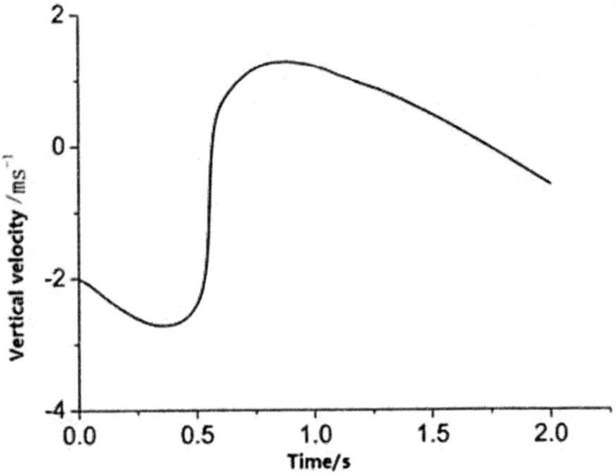

Fig. 9 Vertical velocity change curve of center of mass of helicopter (G0 working condition)

the helicopter moves downward and slows down, and under the action of water, the helicopter rotates. At about 0.5 s, the helicopter begins to move upward and accelerate. At about 1 s, as the helicopter moves upward, the action of water applied on floats and the helicopter decreases, and the helicopter moves downward again under the action of gravity.

4.3 Law of Helicopter Impact Load in Different Ditching States

In order to analyze the relationship between the helicopter impact load during ditching and the horizontal velocity, vertical velocity and the roll angle of the helicopter, horizontal and vertical acceleration peaks of center of mass of the helicopter are selected as the analysis index of impact load. Figures 10 and 11 show the horizontal acceleration change curve and the vertical acceleration change curve during ditching respectively. It can be seen that, during ditching, the acceleration of the center of mass first increases and then decreases with the interaction between the helicopter

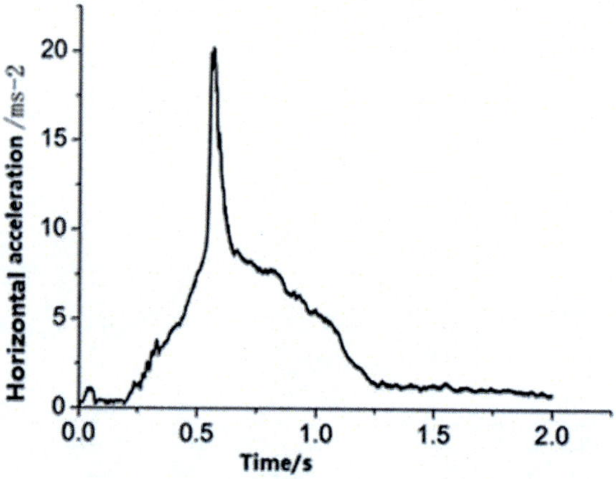

Fig. 10 Horizontal acceleration change curve of center of mass of helicopter (G0 working condition)

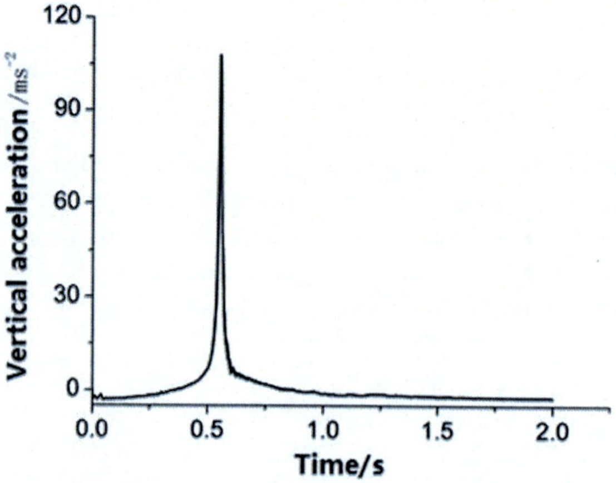

Fig. 11 Vertical acceleration change curve of center of mass of helicopter (G0 working condition)

(including floats) and water. The horizontal acceleration peak is 20.2 m/s^2, and the vertical acceleration peak is 107.9 m/s^2.

Figures 12 and 13 show the horizontal acceleration peak and the vertical acceleration peak of the center of mass of the helicopter at different roll angles. It can be seen that, within a certain range of the roll angle, both the horizontal acceleration and the vertical acceleration of the center of mass decrease with the increase of the roll angle, that is, the impact load of the helicopter decreases. It is because that the helicopter enters the water at a certain roll angle, and the floats are below the helicopter and buffers the impact load of the helicopter, so that the acceleration peaks of the helicopter decrease to a certain extent.

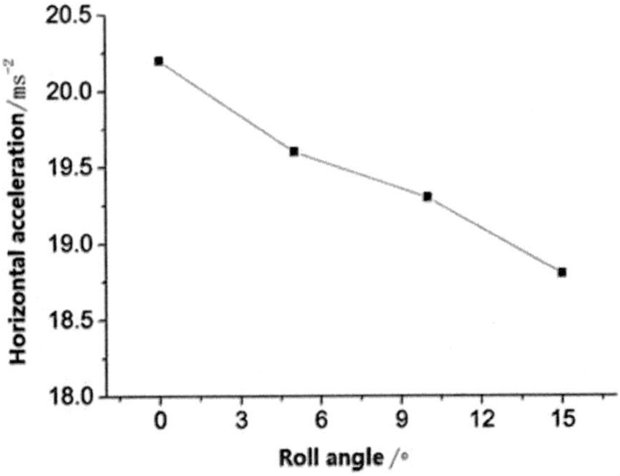

Fig. 12 Horizontal acceleration peak of center of mass of helicopter at different roll angles

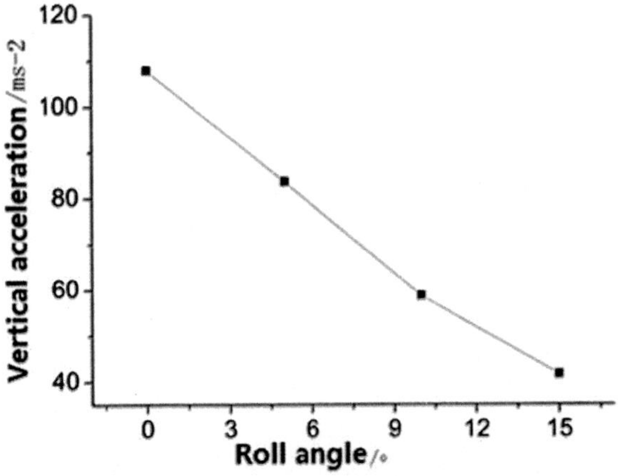

Fig. 13 Vertical acceleration peak of center of mass of helicopter at different roll angles

Figure 14 and Fig. 15 show the horizontal acceleration peak and the vertical acceleration peak of the center of mass of the helicopter at different horizontal velocities. It can be seen that, both the horizontal acceleration and vertical acceleration of the center of mass increase with the increase of the horizontal velocity, that is, the impact load of the helicopter increases. It is because that with the increase of the horizontal velocity, the "wave resistance" between the helicopter (including floats) and water increases, and because the helicopter has a certain angle of attack, the vertical impact load increases accordingly.

Figures 16 and 17 show the horizontal acceleration peak and vertical acceleration peak of the center of mass of the helicopter at different vertical velocities. It can be

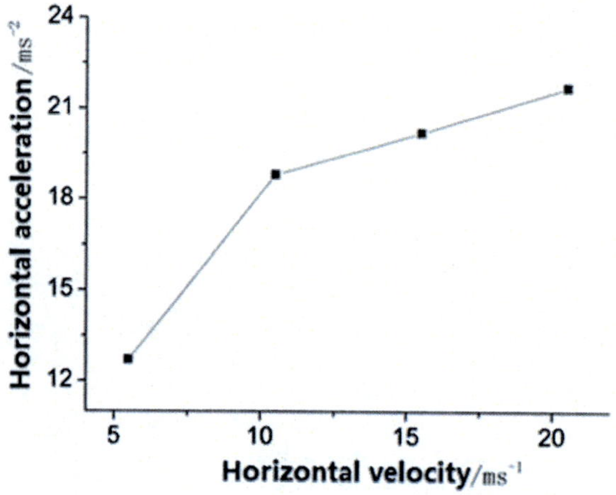

Fig. 14 Horizontal acceleration peak of center of mass of helicopter at different horizontal velocities

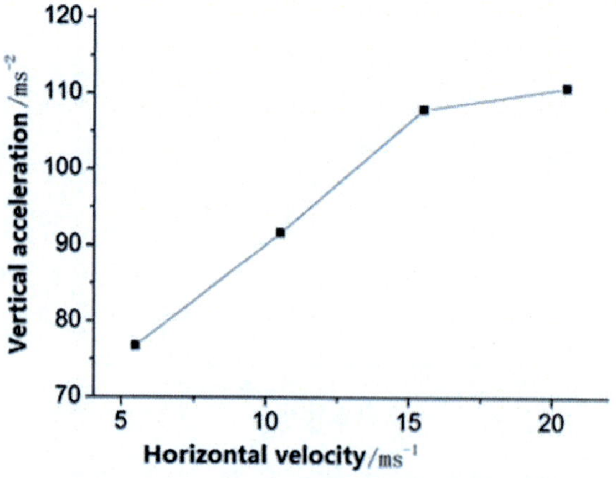

Fig. 15 Vertical acceleration peak of center of mass of helicopter at different horizontal velocities

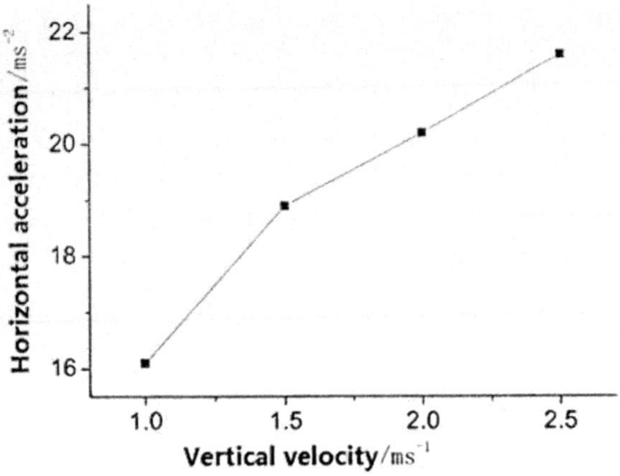

Fig. 16 Horizontal acceleration peak of center of mass of helicopter at different vertical velocities

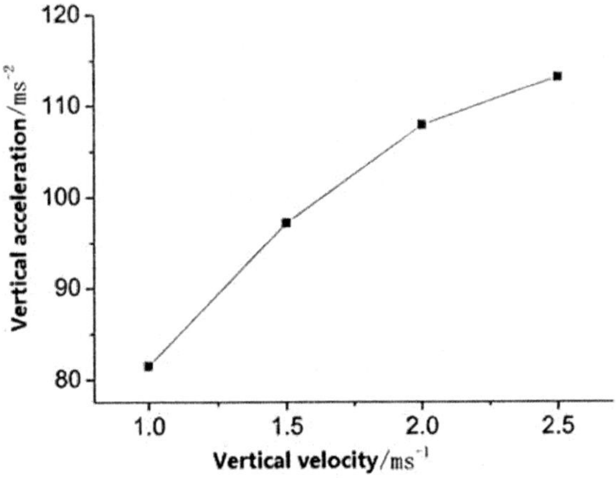

Fig. 17 Vertical acceleration peak of center of mass of helicopter at different vertical velocities

seen that, with the increase of the vertical velocity, the horizontal acceleration of the center of mass increases and the vertical acceleration increases more obviously, that is, the impact load of the helicopter increases. It is because that with the increase of the vertical velocity, the "flap" applied on water by the helicopter and floats increases, and the vertical impact load increases accordingly.

5 Conclusion

(1) ALE method is adopted to compare the simulation value history and the test value history of the three helicopter scale verification model indexes during ditching, including the vertical overload, the pitch angle and the roll angle. The reliability and accuracy of the simulation model and parameters are verified.
(2) The influence of different roll angles on the impact load of the helicopter is studied. At a certain roll angle, the floats play a buffering role in the impact load of the helicopter, namely the impact load decreases with the increase of the roll angle.
(3) The influence of different horizontal and vertical velocities on the impact load of the helicopter is studied. With the increase of the velocity, the "wave resistance" and "flap" applied on water by the helicopter and floats increase, and the impact load increases accordingly.

References

1. Groenen PHL et al (2010) Hydrodynamics and fluid-structure interaction by coupled SPH-FE method. J Hydr Res 48(1):61-73
2. Muller M et al (1996) Survey and analysis of rotorcraft floation system [R]. DOT/FAA/AR-95/53
3. Alcedo AM (1979) Design and test of float landing gear system for helicopters. In: Regional Lichten award paper presented at the american helicopter society southwest regional meeting, Fort Worth, TX
4. Ye S (2018) The research of the test platform and its data acquisition system for helicopter with airbag landing on water
5. Wang Z et al (2017) Test research on helicopter ditching load. J Nanjing Univ Aeron Astron 49(2):258–263
6. LI M (2008) The subscale test and simulation on water impacting characteristic for emergency flotation system. Nanjing Univ Aeron Astron

Study on shapes of Double Cylindrical Structure for Wing Twist Morphing

Hiroaki Tanaka and Yusuke Arai

Abstract We previously developed a twisted morphing wing system based on a double-cylinder structure that consists of outer and inner parts. Here, trade-off studies on the shape of the outer part of the double-cylinder structure are carried out. The open section of the outer part has four beams. The equivalent bending stiffness, equivalent torsional stiffness, and buckling force are calculated for various shapes of the open section using previously derived formulas, and their relationships are obtained. The effects of the shape parameters, namely beam height, width, aspect ratio, and length, on the equivalent bending stiffness, torsional stiffness, and buckling force are investigated, and the optimal cross-sectional shape of the outer part for wing twist morphing is determined. The open section should have low torsional stiffness to achieve wing twist morphing, high bending stiffness, and a large buckling force to bear the aerodynamic load. The results indicate that the smallest equivalent torsional stiffness is achieved by using a cross-sectional shape with a large aspect ratio. Moderately small torsional stiffness and moderately large bending stiffness are achieved with a small aspect ratio. Therefore, the shape of the open section can be designed by considering the trade-off between bending and torsional stiffness. Furthermore, the open section length should be set to avoid buckling of the beam and realize a large twist angle.

Keywords Morphing wing · Wing twist · Double-cylinder structure · Trade-off study

H. Tanaka (✉)
Department of Aerospace Engineering, National Defense Academy of Japan, 1-10-20, Yokosuka 239-8686, Japan
e-mail: tanakah@nda.ac.jp

Y. Arai
Japan Air Self-Defense Force, Kagamihara, Japan

S. Lee et al. (eds.), *The Proceedings of the 2021 Asia-Pacific International Symposium on Aerospace Technology (APISAT 2021), Volume 1*, Lecture Notes in Electrical Engineering 912, https://doi.org/10.1007/978-981-19-2689-1_61

1 Introduction

Morphing wing systems, which have been extensively studied [1, 7, 8], are a promising technology for future aircraft. There are several kinds of morphing, including chord-wise bending, span-wise bending, and twist morphing. Wing twist morphing improves aerodynamic behaviour without the need for large platform modifications and can achieve multiple tasks simultaneously [1]. The Wright brothers used a wing twist morphing system in the Wright Flyer for roll control [7]. Vos et al. investigated wing twist based on warping displacement [9]. Jenett et al. developed a digital morphing wing, which can achieve wing twist morphing, and demonstrated its effectiveness [6]. Cramer et al. developed a twist morphing wing system that uses a modular, lattice-based, ultralight material [5]. The authors previously developed a twist morphing wing system based on a double-cylinder structure and demonstrated its feasibility and effectiveness through numerical simulations and experiments [3, 4].

A schematic diagram of the double-cylinder structure is shown in Fig. 1. Figure 2 shows the application of the double-cylinder structure to a morphing wing. The double-cylinder structure, which can achieve an arbitrary twist distribution, is applied to a spar of the twist morphing structure in the system. The double-cylinder structure

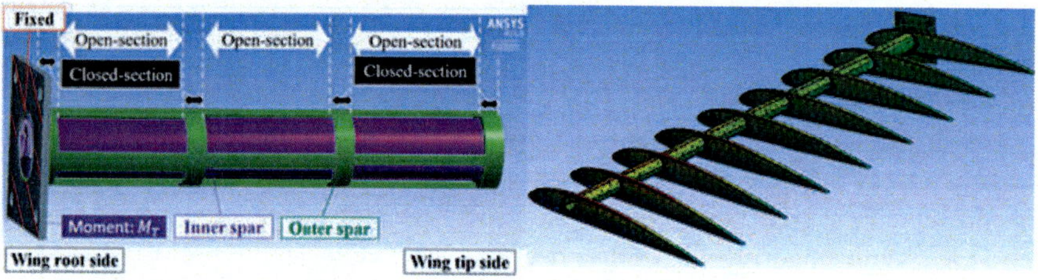

Fig. 1 Schematic diagram of double-cylinder structure [3]

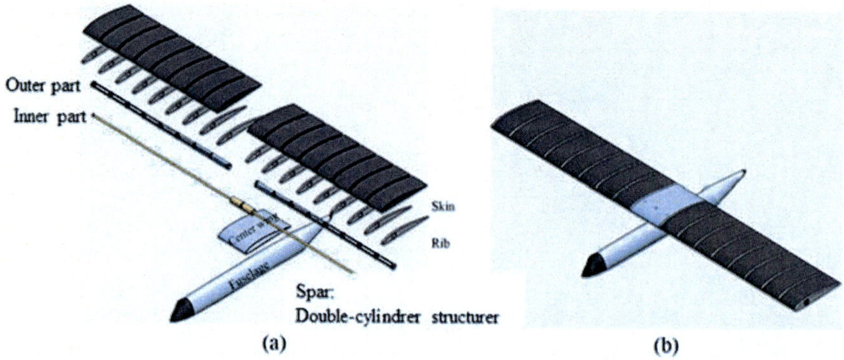

Fig. 2 Application of double-cylinder structure to morphing wing. **a** Exploded view, **b** Assembly diagram

consists of outer and inner parts. For these parts, one set of ends is connected at the wingtip side; the other set of ends is connected to the airplane body or a motor that gives a torsional moment at the wing root side. The inner part transmits a torsional moment from the motor to the outer part. The outer part has open sections and closed sections. The open sections consist of beams designed for achieving the target twist angles. The ribs of the morphing wing are attached to the closed sections of the outer part. The double-cylinder structure has to be twistable and bear the bending force during flight. Therefore, the outer cylinder must have a shape that allows it to have high bending stiffness and low torsional stiffness. Furthermore, some beams are subjected to compression forces and must thus be designed to prevent buckling. We previously derived theoretical formulas for twisting, bending, and buckling, and demonstrated their validity via a comparison with numerical simulation results [2]. However, we did not optimize the outer part shape. Here, trade-off studies on the shape of the outer part of the double-cylinder structure are performed. The bending stiffness, torsional stiffness, and buckling force are calculated for various shapes of the open section using previously derived formulas, and their relationships are obtained and used to optimize the shape for wing twist morphing.

2 Basic Concept of Open Section Design

The double-cylinder structure has to be twistable for wing twist morphing and must bear the bending force during flight. Consequently, the structure should have high bending stiffness and low torsional stiffness. Furthermore, some beams of the upper side of a wing are subjected to compression forces due to the aerodynamic load and thus must be designed to prevent buckling. We previously derived formulas for bending stiffness, torsional stiffness, and buckling force of the open section of the double-cylinder structure [2]. In the double-cylinder structure, beams that make up the outer part have an arc-shaped cross-section, as shown in Fig. 1. The outer part need not be arc-shaped. Therefore, we changed the cross-sectional shape of beams of the outer part to a rectangle to increase design freedom.

In the double-cylinder structure, the open section of the outer part has four beams. A schematic diagram of one segment of the open section of the outer part is shown in Fig. 3. The closed sections at both sides of the open section are also shown. The force due to the aerodynamic load is applied to the closed section of the outer part, which connects to the ribs of the wing. This force deforms the open section. Here, b and h denote the width and height of each beam, respectively. d denotes the distance from a beam to the center of the cross-section. l and D denote the length of the open section and that of the closed section, respectively. A is the cross-sectional area of a beam. E and G denote Young's modulus and the shear modulus, respectively.

The equivalent bending stiffness of the open section, EI_{eq}, was derived by considering the bending and torsion of each beam as follows:

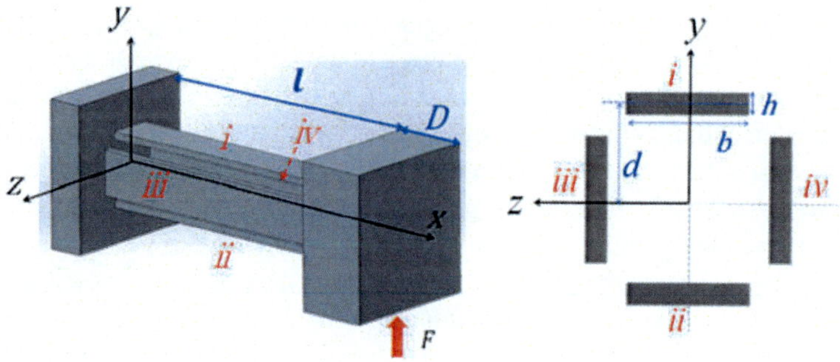

Fig. 3 Schematic diagrams of outer cylinder segment

$$EI_{eq} = \frac{E(2l+D)^3(I_1+I_2)\left(I_1+I_2+Ad^2\right)}{l^2\left\{(I_1+I_2)(4l+3D)+lAd^2\right\}}, \tag{1}$$

where I_1 and I_2 are the moments of inertia of the area of each beam. They are obtained as follows:

$$I_1 = \frac{bh^3}{12},\quad I_2 = \frac{hb^3}{12} \tag{2}$$

The equivalent torsional stiffness of the outer part, GJ_{eq}, is obtained using Eqs. (3) or (4) depending on the size relation between the width and height of a given beam.

$$GJ_{eq} = 48EI_2\left(\frac{d}{l}\right)^2 + 4G\frac{b^3h}{3}\left(1 - \frac{192b}{\pi^5h}\tanh\frac{\pi h}{2b}\right) \quad (h \geq b), \tag{3}$$

$$GJ_{eq} = 48EI_2\left(\frac{d}{l}\right)^2 + 4G\frac{h^3b}{3}\left(1 - \frac{192h}{\pi^5b}\tanh\frac{\pi b}{2h}\right) \quad (h < b). \tag{4}$$

The force due to the aerodynamic load is applied to the closed section on which a rib is attached during flight, as indicated in Fig. 3. Each beam must have sufficient buckling strength. The force at which one of the beams will buckle, F_{cr}, was derived as follows:

$$F_{cr} = \min\left\{\frac{16\pi^2EI_1\left(I_1+I_2+Ad^2\right)}{Adl^2(l+D)}, \frac{16\pi^2EI_2\left(I_1+I_2+Ad^2\right)}{Adl^2(l+D)}\right\} \tag{5}$$

Hereafter, this force is referred to as the buckling force. At the buckling force, beam *i* of the open section will buckle. Accordingly, the buckling force should be large enough to avoid buckling during flight.

3 Trade-Off Studies on Outer Part Shape of Double-Cylinder Structure

The effects of shape parameters, namely the beam height, width, and length, on the equivalent bending stiffness, torsional stiffness, and buckling force were investigated using previously derived formulas. Here, we introduce the aspect ratio of a beam, defined as the ratio of the beam width to the beam height.

$$AR = b/h. \tag{6}$$

The equivalent bending stiffness, torsional stiffness, and buckling force were calculated for various sectional shapes of the beams. The size ranges of the open section shape are shown in Table 1. The material parameters used in this study are summarized in Table 2. These parameters were determined under the assumption that the double-cylinder structure will be used in small unmanned aerial vehicles. The material is acrylonitrile butadiene styrene (ABS). The length shown in Table 1 is that of one segment of the double cylinder; the total length of the structure is set to the required value for a given application.

The relationships among the equivalent bending stiffness, equivalent torsional stiffness, and buckling load for various open section lengths (represented by colors) are shown in Figs. 4 and 5. High bending stiffness, high buckling force, and low torsional stiffness are required for the double-cylinder structure's outer part. The lower-right regions of these figures thus indicate the optimal design. These figures show that a longer open section length leads to low torsional stiffness but also dramatically decreases the buckling force, which may lead to wing damage during flight. Therefore, the open section length should be shorter than the length at which the

Table 1 Parameters of outer part shape

Parameter	Range
Height, *h*	1–6 mm in 0.1 mm increments
Width, *b*	1 to 18 mm in 0.1 mm increments
Distance from beam to center of cross-section, *d*	4 mm +h/2
Length of open section, *l*	30–200 mm in 10 mm increments
Length of closed section, *D*	10 mm

Table 2 Parameters of open section material (ABS)

Parameter	Value
Young's modulus, *E*	1.67 GPa
Shear modulus, *G*	0.56 GPa

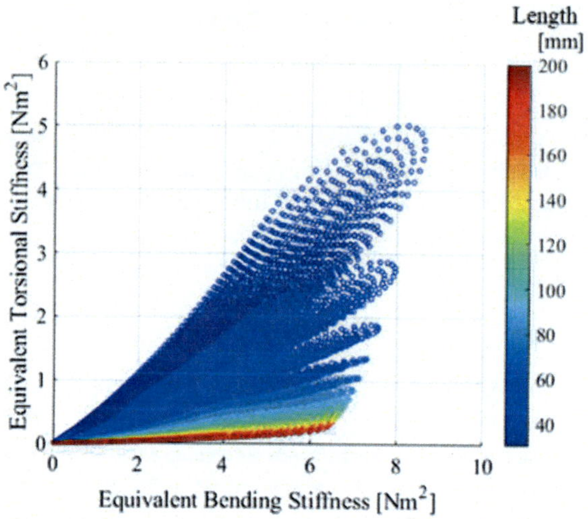

Fig. 4 Relationship between bending stiffness and torsional stiffness for various open section lengths

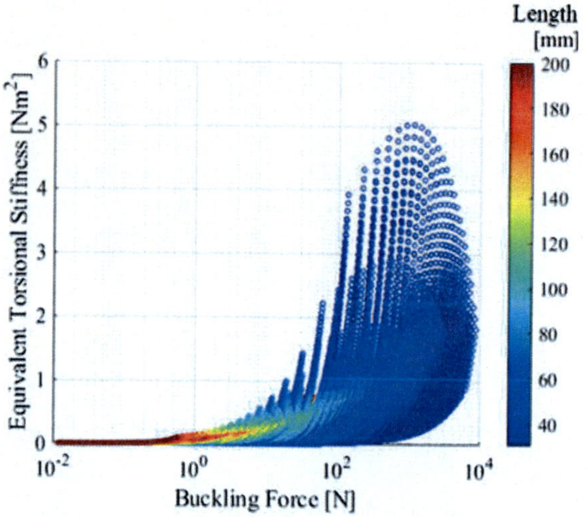

Fig. 5 Relationship between buckling force and torsional stiffness for various open section lengths

buckling force becomes larger than that due to aerodynamic load multiplied by a safety factor.

The effects of the aspect ratio were next investigated. To focus on the effects of the aspect ratio, the area of the beams of the open section was set to 10 mm^2. The beam height used in this investigation is shown in Table 1. The beam width was set to a value that made the cross-sectional area of the beam equal to 10 mm^2. The

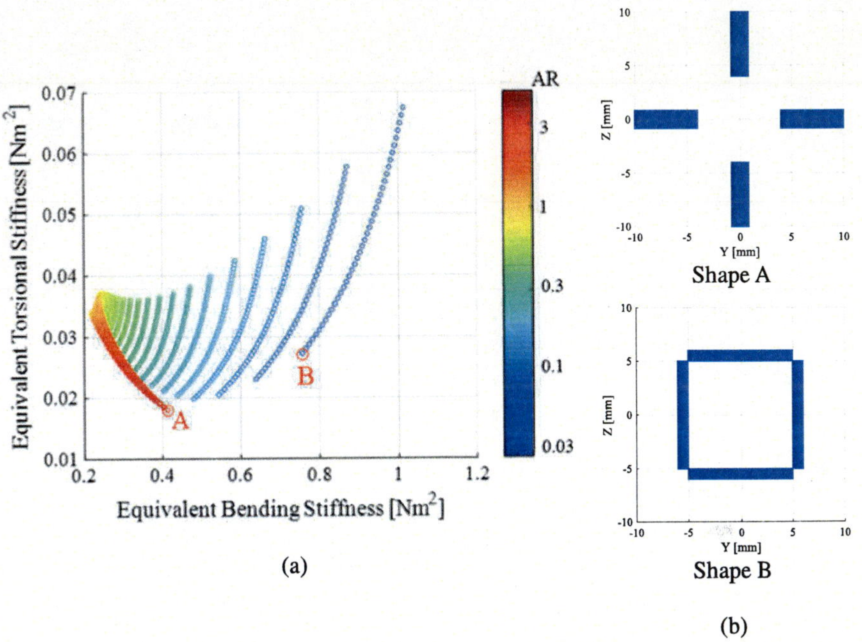

Fig. 6 **a** Relationship between bending stiffness and torsional stiffness for various aspect ratios. **b** Typical cross-sectional shapes of open section ($l = 100$)

relationships among the equivalent bending stiffness, equivalent torsional stiffness, and buckling load for various aspect ratios are shown in Figs. 6, 7 8 and 9. Figures 6 and 7 show the results for an open section length of 100 mm and Figs. 8 and 9 show those for an open section length of 50 mm. The typical cross-sectional shapes are also shown. The largest buckling force is achieved by using a cross-sectional shape with an aspect ratio of 1 (shapes C and F). However, these shapes make the equivalent bending stiffness smaller than that for other shapes. The smallest equivalent torsional stiffness is achieved by using a cross-sectional shape with a large aspect ratio (shapes A and D). Moderately small torsional stiffness and moderately large bending stiffness are achieved with a small aspect ratio (shapes B and E). These results indicate that there is a trade-off among the equivalent bending stiffness, torsional stiffness, and buckling force. However, a large buckling force can be obtained by using a short open section. Therefore, the shape of the open section can be designed by considering the trade-off between bending and torsional stiffness. The open section length can then be determined to satisfy the buckling force requirement.

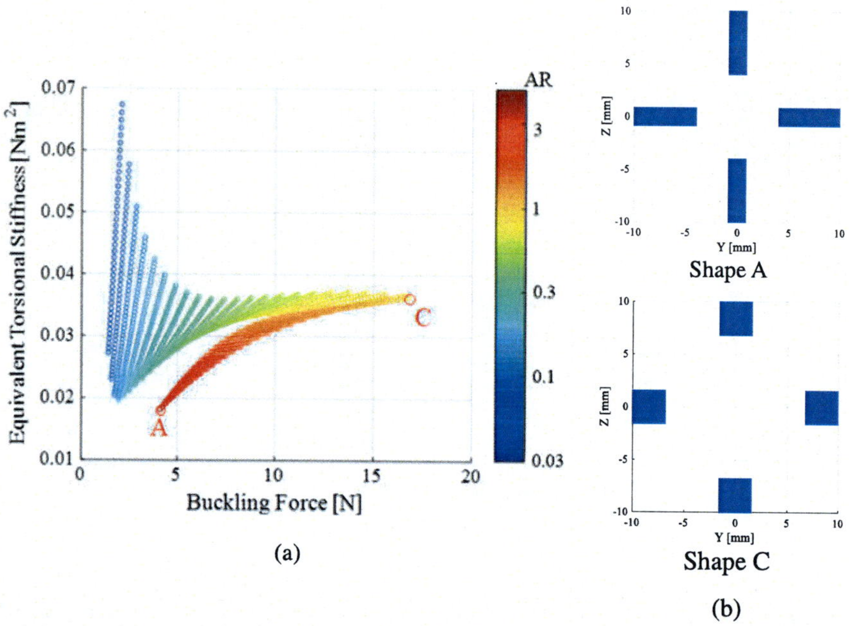

Fig. 7 **a** Relationship between buckling force and torsional stiffness for various aspect ratios. **b** Typical cross-sectional shapes of open section (l = 100)

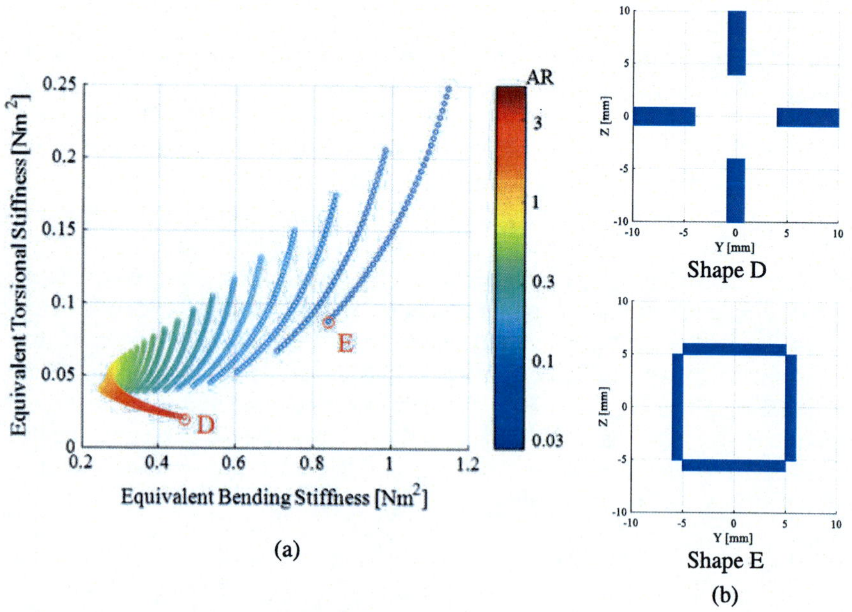

Fig. 8 **a** Relationship between bending stiffness and torsional stiffness for various aspect ratios. **b** Typical cross-sectional shapes of open section (l = 50)

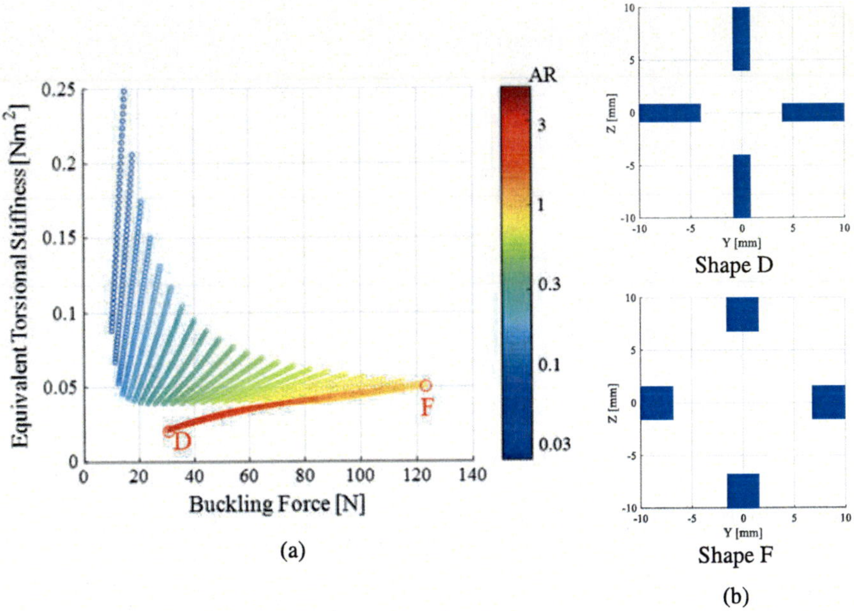

Fig. 9 **a** Relationship between buckling force and torsional stiffness for various aspect ratios. **b** Typical cross-sectional shapes of open section ($l = 50$)

4 Conclusion

The effects of the shape parameters for the open section of the outer part of a double-cylinder structure, namely the beam height, width, and length, on the equivalent bending stiffness, torsional stiffness, and buckling force were investigated. The equivalent bending stiffness, equivalent torsional stiffness, and buckling force were calculated using previously derived formulas for various shapes of the open section. The results indicate that the smallest equivalent torsional stiffness is achieved by using a cross-sectional shape with a large aspect ratio, and that a moderately small torsional stiffness and a moderately large bending stiffness are achieved with a small aspect ratio. Therefore, the shape of the open section can be designed by considering the trade-off between the bending and torsional stiffness. A large buckling force can be obtained by using a short open section. Therefore, the open section length should be used to avoid buckling of the beam and to realize a large twist angle.

Acknowledgements This work was supported by a Japan Society for the Promotion of Science (JSPS) KAKENHI grant (grant number 21H01535).

References

1. Ajaj RM, Parancheerivilakkathil MS, Amoozgar M, Friswell MI, Cantwell WJ (2021) Recent developments in the aeroelasticity of morphing aircraft. Progr Aerosp Sci 120:100682. https://doi.org/10.1016/j.paerosci.2020.100682
2. Arai A, Tanaka H (2021) Study on cross-sectional shape of outer cylinder for twist type morphing blade using double cylinder structure. T JSME 87:21–00078 (in Japanese). https://doi.org/10.1299/transjsme.21-00078
3. Aso A, Perrey M, Tanaka H (2016) Experimental study on wing twist-morphing structure using a double-tube cylinder. T JSASS Aerosp Technol Japan APISAT a1–a6. https://doi.org/10.2322/tastj.15.a1
4. Aso A, Tanaka H, Mizoguchi M (2019) Application of double cylindrical tube having arbitrary twist distribution to twistable morphing wing and experimental demonstration. T JSME 85:19–00004 (in Japanese). https://doi.org/10.1299/transjsme.19-00004
5. Cramer NB, Cellucci DW, Formoso OB, Gregg CE, Jenett BE, Kim JH, Lendraitis M, Swei SS, Trinh GT, Trinh KV, Cheung KC (2019) Elastic shape morphing of ultralight structures by programmable assembly. Smart Mater Struct 28:055006. https://doi.org/10.1088/1361-665X/ab0ea2
6. Jenett B, Calisch S, Cellucci D, Cramer N, Gershenfeld N, Swei S, Cheung KC (2017) Digital morphing wing: active wing shaping concept using composite lattice-based cellular structures. Soft Rob 4:33–48. https://doi.org/10.1089/soro.2016.0032
7. Sofla AYN, Meguid SA, Tan KT, Yeo WK (2010) Shape morphing of aircraft wing: Status and challenges. Mater Des 31:1284–1292. https://doi.org/10.1016/j.matdes.2009.09.011
8. Tsushima N, Tamayama M (2019) Recent researches on morphing aircraft technologies in Japan and other countries. Bulletin JSME Mech Eng Rev 6:19–00197. https://doi.org/10.1299/mer.19-00197
9. Vos R, Gürdal Z, Abdalla M (2010) Mechanism for warp-controlled twist of a morphing wing. J Aircr 47:450–457. https://doi.org/10.2514/1.39328

Multi-objective Optimization for Resilient Operation of Adaptive Morphing Flap Divided into Span Directions

Nozomu Kogiso and Kento Konishi

Abstract This study proposes a multi-objective optimization method to enable the recovery of lifting force using the remaining flaps after failure of one flap, such that the flap angle returns to zero. Recoverability is regarded as a resilient operation in this study. Therefore, a multi-objective optimization problem is formulated to minimize the bending moment at the wing root from the structural point of view, to minimize the sum of the flap angle differences for remaining flaps corresponding to the minimum effort of the operation and the difference of the lifting force from the viewpoint of flight mechanics in terms of the flap angles of the remaining flaps, where the lifting force distribution is obtained by vortex lattice method. The validity of the proposed resilient operation is demonstrated through numerical examples, and the efficiency of the adaptive wing divided flaps is evaluated. Additionally, the lifting force distribution and the flap angle distribution along the span direction are shown for different flap failures.

Keywords Resilient operation · Multi-objective optimization · Morphing flap · Smart structure · Vortex lattice method

1 Introduction

In the field of aerospace engineering, research has been widely conducted on smart structural systems such as morphing wings that can adaptively deform its geometrical shape to improve aerodynamic performance [1]. The morphing technologies are classified into three categories: planform morphing, airfoil morphing and out-of-plane morphing. In case of airfoil morphing, a corrugated structure consisting of fiber-reinforced composite material was proposed [2]. Optimization methods such as shape

N. Kogiso (✉) · K. Konishi
Department of Aerospace Engineering, Osaka Metropolitan University, Sakai, Osaka 599-8531, Japan
e-mail: kogiso@omu.ac.jp

K. Konishi
e-mail: konishi.kento.q8@dc.tohoku.ac.jp

S. Lee et al. (eds.), *The Proceedings of the 2021 Asia-Pacific International Symposium on Aerospace Technology (APISAT 2021), Volume 1*, Lecture Notes in Electrical Engineering 912, https://doi.org/10.1007/978-981-19-2689-1_62

optimization of morphing airfoil have been applied to the designs of morphing wings to improve the aerodynamic performance [3]. Additonally, topology optimization has been applied to realize the ideal shape with proposed multi-layered structural configuration [4].

Conversely, such a smart structural system is prone to significant damage, if malfunction of the built-in sensor and actuator occurs. A conventional technique to overcome this problem is to reduce the failure probability of sensors and actuators. However, the resultant increase in cost and mass will reduce the feasibility of the system.

Recently, resilience engineering has become increasingly popular because it considers the sustainability of social system. It ensures that the recovery of a lost function of a system subjected to unexpected events occurrs within minimum time [5]. This was originally proposed in the field of social engineering, which considers the vulnerability of our social system beyond our expectations, such as the Great East Japan Earthquake. Therefore, the importance of a system safety framework that considers unexpected situations was recognized.

This study introduces the concept of resilience engineering to a morphing wing to sustain flight safety, even if a part of the morphing function is lost. As the first trial, this study proposes "resilient operation" that enables the recovery of a lost function using the survived functions of an adaptive wing proposed by Tashiro et al. [6, 7]. The adaptive wing has multiple slats and flaps divided along the span direction that can be driven independently to adjust the lift distribution and the bending moment at the wing root. In this study, the divided flap is replaced by divided morphing flaps and the resilient operation is applied to the adaptive morphing wing to ensure that the function of the failed flap is replaced by the survived flaps based on the concept of Safety-II [8]. The adaptive wing requires the same lifting force as that of during normal operation. In addition, the flight operation should reduce the bending moment at the wing root and the magnitude of change of the flap angles of the devices as much as possible. Therefore, the design problem of the resilient operation is formulated by using the multi-objective optimization problem.

The paper is organized as follows. Section 2 briefly describes the concept of the resilience engineering. Section 3 explains the aerodynamic analysis followed by introduction of the adaptive morphing wing. Then, Sect. 4 explains the resilient operation with proposal of the formulation using the multi-objective optimization method. Furthermore, numerical examples demonstrate the efficiency of the proposed method in Sect. 5. Finally, conclusions are remarked in Sect. 6.

2 Resilience Engineering

Resilience is a combination of features such as adaptability, robustness, recoverability, and sustainability. It is the ability of a system to sustain functions under an unexpected change in the environment. However, its typical meaning depends on the field [5]. The resilience triangle shown in Fig. 1 is widely used in the field of

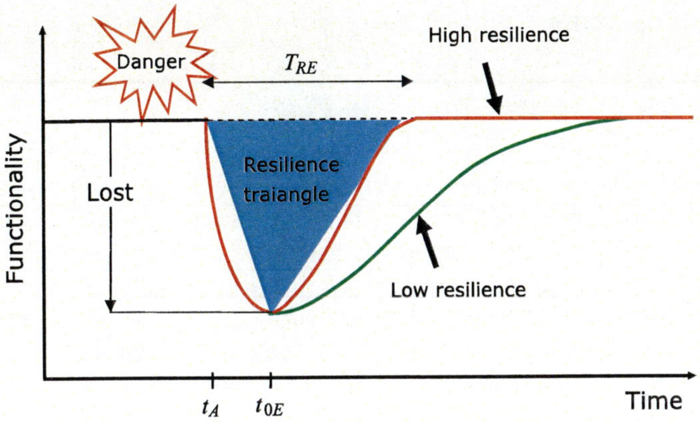

Fig. 1 Resilience triangle

disaster prevention. The triangle signifies that quick recoverability is an important property of resilience. This figure shows the recoverability of system performance after a damage event occurs and the triangle surrounded by damaged and recovered functionality in terms of time indicates the importance of quick recoverability for a resilient system.

In the field of aerospace engineering, resilience engineering has been applied for thruster allocation design of the attitude control system in HTV (H-II transfer vehicle) developed in JAXA, (Japan Aerospace Exploration Agency) [9]. Under the conventional design, the reaction control system (RCS) thrusters were allocated to three translation and three rotational directions independently with redundancy. Introducing the concept of resilience engineering, the number of thrusters and the total mass of an HTV can be reduced by alternating the combination of the survived RCS thrusters when a thruster fails, without assigning redundant ones to each thruster. The resilient operation in this case can flexibly respond to the changes in the environment using the survived functions.

A civil aircraft can perform "NO FLAP" landing without emergency when a flap fails. However, this requires additional operations as weight reduction by fuel consumption flight (current twin-engine aircraft does not install a dumping of fuel). This leads to delayed arrival and delayed departure of the next flight. Moreover, hard braking might be required to reduce the landing runway distance against the faster approach speed. This hard braking will require additional maintenance, which required additional cost and time. Although "NO FLAP" landing is a sufficient safe operation and it is not included in the incident, it increases the operational cost.

3 Morphing Wing

3.1 *Adaptive Wing with Morphing Flap*

A morphing wing is expected to be lightweight and it should improve the aerodynamic performance by adaptively deforming its geometrical shape. This study proposes an adaptive wing with eight divided morphing flaps along the span direction as shown in Fig. 2 with reference to the original adaptive wing proposed by Tashiro et al. [6, 7]. The dimensions are determined with reference to the future passenger aircraft named TRA2022 proposed by JAXA [10]. The length of the half-span is 17.05 m, the chord lengths at the wing root and the tip are 5.522 m and 1.657 m, respectively, and the chord length is linearly varied from the root to the tip. The swept angle at the 25% chord is set to zero degree for simplicity. Eight morphing flaps with a chord length of 1.9 m is allocated after 25% of the chord length, except for wing root which has a chord length of 1.95 m.

The morphing shape of the flap proposed by one of the authors [3] shown in Fig. 3 is adopted in this study. The intermediate flap angle shown in this figure is linearly approximated the shape.

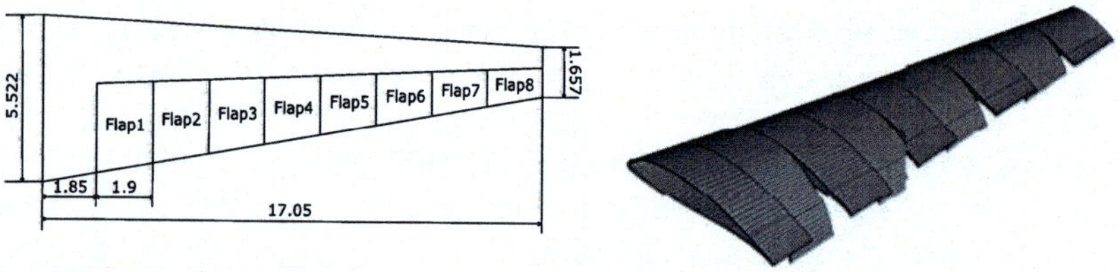

Fig. 2 Adaptive morphing wing with divided morphing flaps, (left) Wing planform, (right) Three-dimensional shape

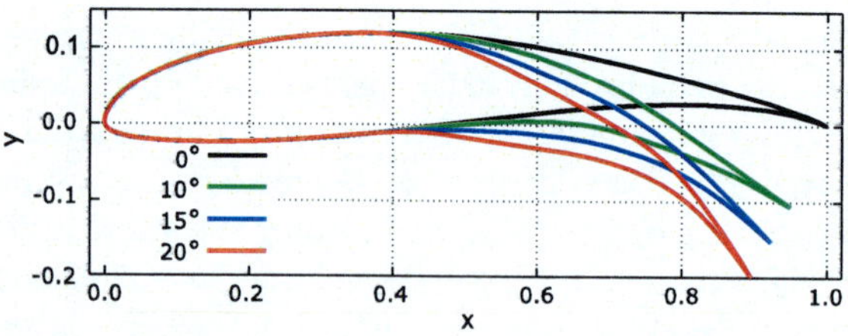

Fig. 3 Ideal morphing shape of morphing flap

3.2 *Aerodynamic Analysis, Lift Distribution and Bending Moment*

The aerodynamic performance is evaluated using free software, Athena Vortex Lattice (AVL) [11, 12]. It adopts the vortex lattice method in non-viscous and uncompressed potential flow. The method divides the wing surface into finite number of lattices and horseshoe vortices are allocated to the lattice. Subsequently, a system of linear equations equal to the number of lattices is derived from the boundary conditions such that the sum of the uniform flow and the induced flow from the horseshoe vortices flows along the wing surface at each lattice in terms of the strength of the horseshow vortices. The lifting force can be evaluated through Kutta-Jukowski theorem.

The analysis model used in this research is shown at the upper left in Fig. 4, where the number of lattices is set to 20 in the chord direction, 10 in span directions for each morphing wing and 5 for the fixed wing at the wing root. AVL software obtains the lift coefficient and the drag coefficient of the entire wing, and the lift coefficient distribution along the span direction and the downwash angle. This is shown at the lower right side in Fig. 4.

The bending moment at the wing root for the span length b can be evaluated from the lifting force $L(y)$ at the point with distance y from the root as follows.

$$M_{root} = \int_0^{b/2} L(y)ydy \tag{1}$$

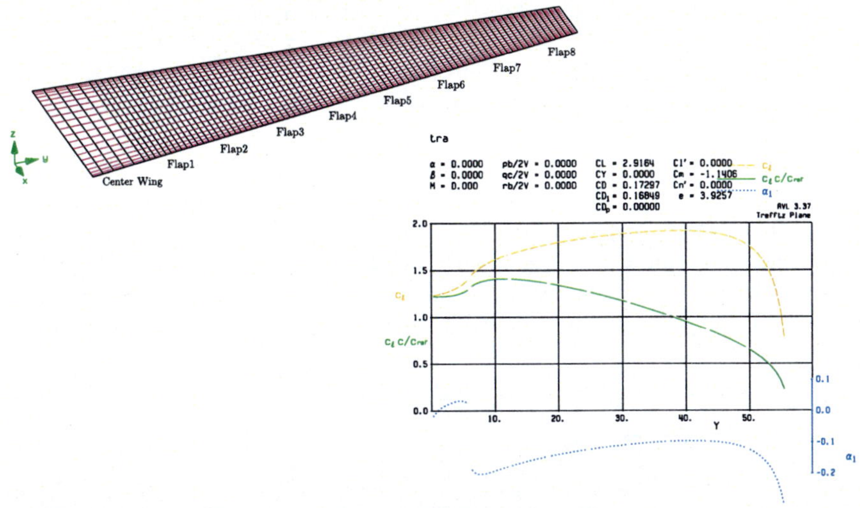

Fig. 4 Aerodynamic analysis model and output image of AVL

4 Resilient Operation of Morphing Wing and Formulation as Multi-Objective Optimization Problem

Resilient operation that does not lead to late arrival and additional maintenance cost is expected, even if a flap mechanism fails. Therefore, this research proposes resilient operation that enables the recovery of lifting force equivalent to that of during landing after failure of one flap (out of eight flaps) by changing the flap angles of the survived flaps. In addition to the recovery of lifting force, the operation considers reducing the bending moment at the wing root from the viewpoint of structural engineering and the sum of changes in the flap angles that reduces the energy required for morphing.

This resilient operation can be formulated using the multi-objective optimization problem. The objective functions are described as follows.

$$\text{Minimize}: \; f_1(\boldsymbol{x}) = M_{root} \tag{2}$$

$$\text{Minimize}: \; f_2(\boldsymbol{x}) = \sum_{i=1}^{8} |x_i - x_{0_i}| \tag{3}$$

$$\text{Minimize}: \; f_3(\boldsymbol{x}) = |L - L_0| \tag{4}$$

where M_{root} is the bending moment at the wing root, L and L_0 are the lifting force after the resilient operation and before the failure, and x_i and x_{0_i} are the morphing angle of i-th flap after the resilient operation as design variables and before failure of some flap. Lower and upper bounds of the flap angle, the design variable is set as $0^\circ \leq x_i \leq 20^\circ (i = 1, \cdots, 8)$ to avoid the stall. The flap angle of the failed flap is back to 0°.

Although the absolute value of the difference in the lifting force between the normal and the resilient operations can be formulated as an equal constraint, this study formulates as an objective function. Therefore, it is easier to determine the Pareto solutions and it is easier to investigate the trade-off relationships.

This study adopts NSGA-II (Non-dominated Sorting Genetic Algorithms-II) [13], which is a multi-objective genetic algorithm for numerical implementation. The flap angles treated as design variables take only integer values for $0^\circ \leq x_i \leq 20^\circ$. The population size and the number of generations are set as 100 and 150, respectively, and the crossover and the mutation probabilities are set as 0.9 and 0.05, respectively.

5 Numerical Examples

The validity of the proposed resilient operation is demonstrated by obtaining the Pareto solutions that correspond to flap angle distribution that has the same lifting force as that before failure of single flap between Flaps 1 and 8. Additinally, the flap angle of the failed flap is set to 0°, where the flap angles of the original condition are set to 10° for all flaps.

Pareto set distribution in case of failure in Flap 1 is shown in Fig. 5, where Fig. 5(a) indicates the three-dimensional objective function space. Figure 5(b) shows the trade-off relationship between the bending moment at the wing root and the difference between the original and the recovered conditions. Figure 5(c) shows the trade-off relationship between the total magnitude of change in flap angle and the bending moment at the wing root. The other cases of failures of other flaps have the same tendency, though the Pareto solution distributions are not shown here.

Although the feasible solutions are limited to that lies close to $L - L_0 = 0$, it is found that the trade-off relationship between $L - L_0$ and M_{root} by treating the objective function as Eq. (4).

Pareto solutions in cases of failure of Flaps 1, 2, 3, 4, 5 and 7 are listed in Table 1, where the resilient operation is limited such that the relative difference between the recovered and the original lifting force is less than 1%, i.e., $|L - L_0|/L_0 \leq 0.01$. The objective functions are arranged acording to the total magnitude of change of

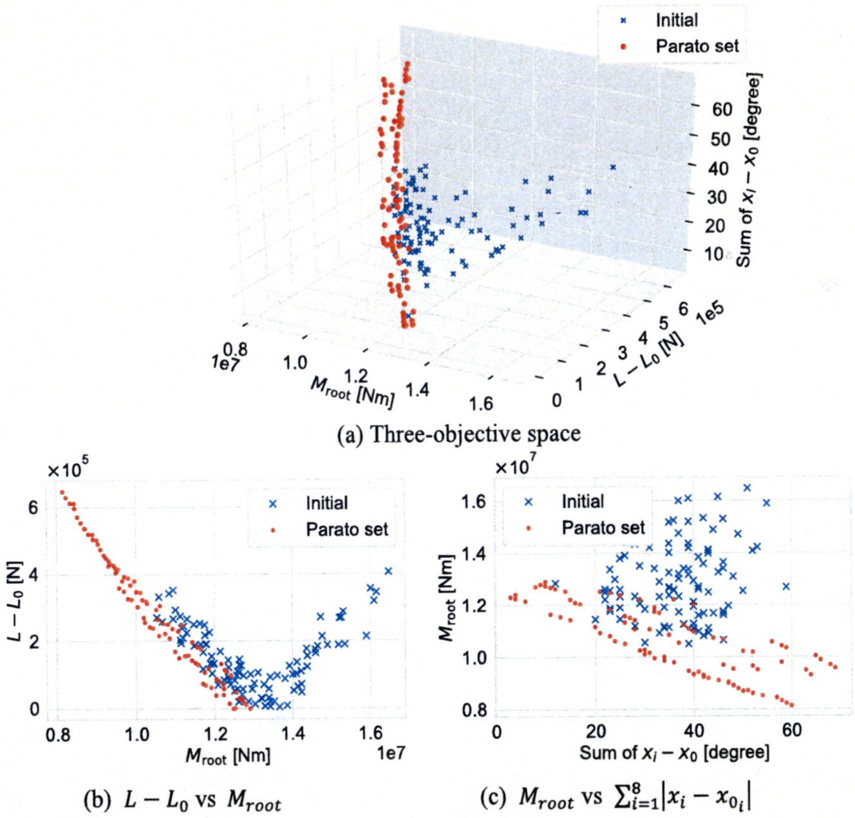

(a) Three-objective space

(b) $L - L_0$ vs M_{root}

(c) M_{root} vs $\sum_{i=1}^{8}|x_i - x_{0_i}|$

Fig. 5 Initial designs and Pareto solutions in case of failure in Flap 1

Table 1 Pareto results for resilient operation within 1% of $|L - L_0|/L_0$

(a) In case of failure of Flap 1

	Flap angle (degree)								Total flap angles (degree)	Decrease rate in M_{root} (%)	$\frac{\vert L-L_0\vert}{L_0}$
	x_1	x_2	x_3	x_4	x_5	x_6	x_7	x_8			
Case 1	0	10	16	10	11	10	10	8	9	1.14	-6.93×10^{-3}
Case 2	0	10	16	10	11	11	10	8	10	0.340	-3.90×10^{-3}
Case 3	0	10	20	10	9	11	9	0	23	3.63	-8.22×10^{-3}

(b) In case of failure of Flap 2

	Flap angle (degree)								Total flap angles (degree)	Decrease rate in M_{root} (%)	$\frac{\vert L-L_0\vert}{L_0}$
	x_1	x_2	x_3	x_4	x_5	x_6	x_7	x_8			
Case 1	17	0	9	10	10	10	10	9	9	2.38	$+3.32 \times 10^{-3}$
Case 2	17	0	9	8	10	10	10	9	11	3.14	-5.47×10^{-3}
Case 3	20	0	11	19	9	9	0	1	50	10.1	-8.17×10^{-4}

(c) In case of failure of Flap 3

	Flap angle (degree)								Total flap angles (degree)	Decrease rate in M_{root} (%)	$\frac{\vert L-L_0\vert}{L_0}$
	x_1	x_2	x_3	x_4	x_5	x_6	x_7	x_8			
Case 1	16	10	0	10	11	10	10	9	8	2.90	$+3.14 \times 10^{-3}$
Case 2	16	10	0	10	11	10	8	9	10	3.81	-3.27×10^{-3}
Case 3	16	10	0	10	11	10	8	8	11	4.24	-5.94×10^{-3}
Case 4	19	10	0	10	11	10	8	1	21	6.33	$+3.69 \times 10^{-3}$
Case 5	19	10	0	10	11	10	8	0	22	7.19	-1.98×10^{-3}
Case 6	19	10	0	10	11	0	0	0	49	12.20	$+1.17 \times 10^{-5}$

(d) In case of failure of Flap 4

	Flap angle (degree)								Total flap angles (degree)	Decrease rate in M_{root} (%)	$\frac{\vert L-L_0\vert}{L_0}$
	x_1	x_2	x_3	x_4	x_5	x_6	x_7	x_8			
Case 1	10	15	10	0	11	10	12	8	10	2.95	-2.70×10^{-3}
Case 2	9	15	11	0	11	10	12	8	12	2.55	$+4.02 \times 10^{-4}$
Case 3	19	10	10	0	11	10	1	8	21	7.86	-4.60×10^{-3}
Case 4	19	10	11	0	11	10	0	8	23	8.17	-3.66×10^{-3}
Case 5	19	15	8	0	9	10	0	1	36	10.4	$+6.67 \times 10^{-5}$

(e) In case of failure of Flap 5

	Flap angle (degree)								Total flap angles (degree)	Decrease rate in M_{root} (%)	$\frac{\vert L-L_0\vert}{L_0}$
	x_1	x_2	x_3	x_4	x_5	x_6	x_7	x_8			
Case 1	10	14	10	11	0	10	10	10	5	4.03	-6.28×10^{-3}
Case 2	10	14	10	11	0	10	10	11	6	3.26	-1.56×10^{-3}
Case 3	9	14	10	11	0	10	10	10	7	3.65	-1.50×10^{-3}
Case 4	10	15	10	11	0	10	8	10	8	4.28	-3.22×10^{-3}

(continued)

Table 1 (continued)

(e) In case of failure of Flap 5

	Flap angle (degree)								Total flap angles (degree)	Decrease rate in M_{root} (%)	$\frac{\lvert L-L_0\rvert}{L_0}$
	x_1	x_2	x_3	x_4	x_5	x_6	x_7	x_8			
Case 5	9	15	10	11	0	10	9	11	9	3.33	$+8.41 \times 10^{-5}$
Case 6	20	14	10	0	0	10	1	10	33	9.90	-1.01×10^{-3}

(f) In case of failure of Flap 7

	Flap angle (degree)								Total flap angles (degree)	Decrease rate in M_{root} (%)	$\frac{\lvert L-L_0\rvert}{L_0}$
	x_1	x_2	x_3	x_4	x_5	x_6	x_7	x_8			
Case 1	9	15	9	11	10	8	0	10	10	4.60	$+1.44 \times 10^{-3}$
Case 2	11	15	9	11	10	5	0	10	13	5.32	$+4.36 \times 10^{-3}$
Case 3	3	19	9	11	10	5	0	10	23	5.04	-1.38×10^{-4}

flap angles, reduction rate of the bending moment at the wing root and the difference in the lifting force from the left to the right side. As shown in Fig. 5(c), the bending moment at the wing root becomes smaller, as the total magnitude of change of the flap angle is larger. However, the Pareto solutions within $|L - L_0|/L_0 \leq 0.01$ are not sufficient and only few solutions can be obtained. Further improvement of searching efficiency is required.

Finally, the lifting force distribution after recovery is shown in Fig. 6 for several cases of failures of Flaps 1, 2, 3, 4, 5 and 7. Although the lifting force at the failed flap position is reduced, the lifting force is recovered at the other flap positions corresponding to the difference in the flap angles. In addition, it is observed that a smaller bending moment at the wing root is achieved by reducing the flap angles closed to the wing root and increasing that closed to the wing tip. Conversely, in case that the total magnitude of change of the flap angle is small, the change in the bending moment is also small that corresponds to the small change on the lifting force distribution along the span direction.

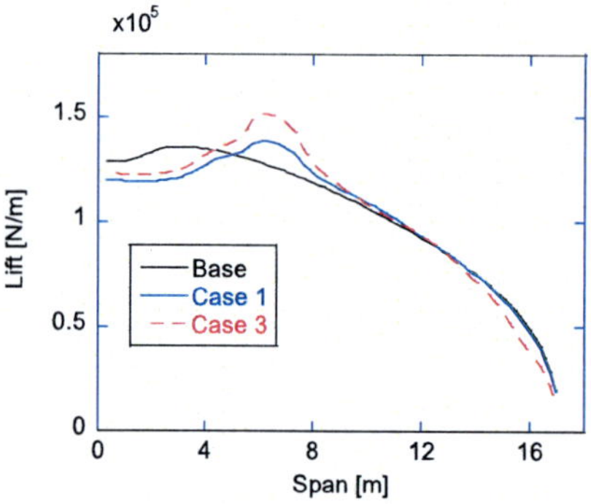

(a) In case of failure of Flap 1

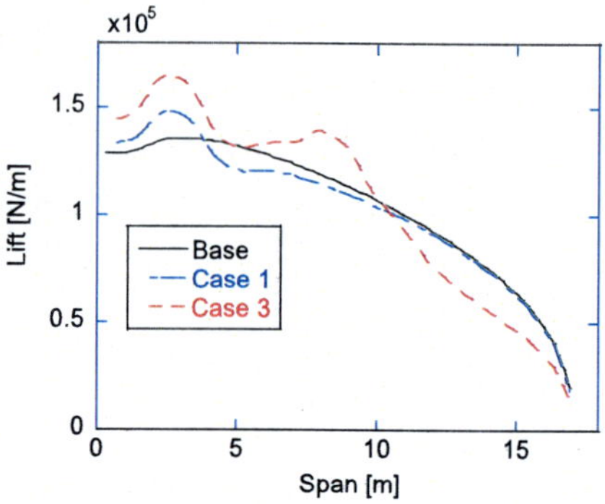

(b) In case of failure of Flap 2

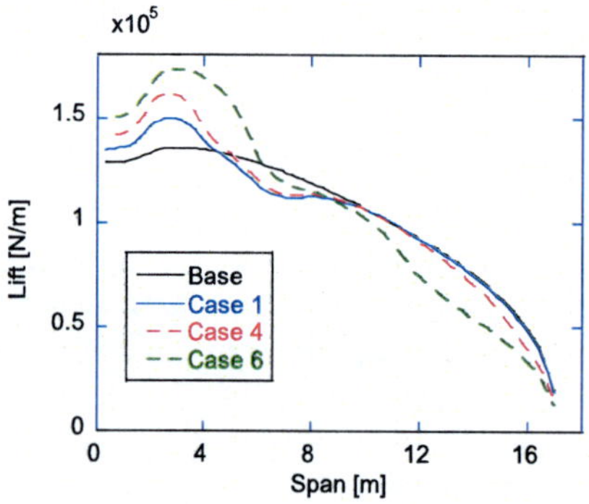

(c) In case of failure of Flap 3

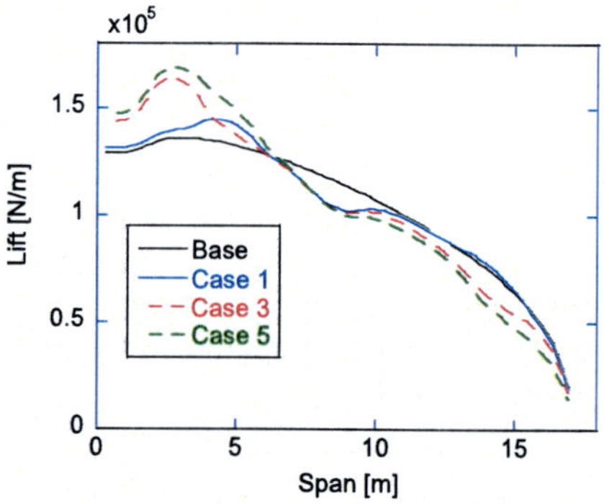

(d) In case of failure of Flap 4

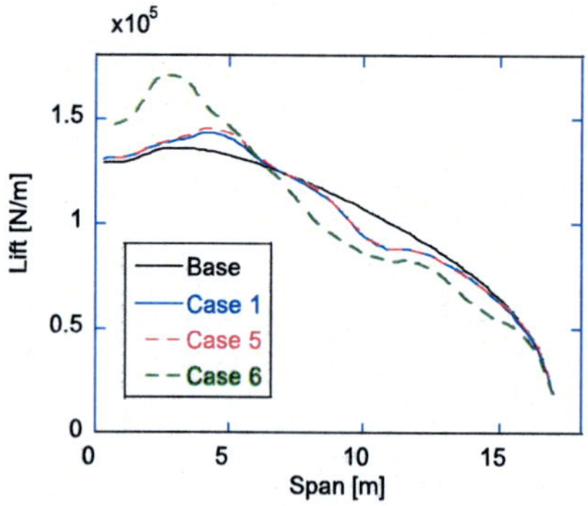

(e) In case of failure of Flap 5

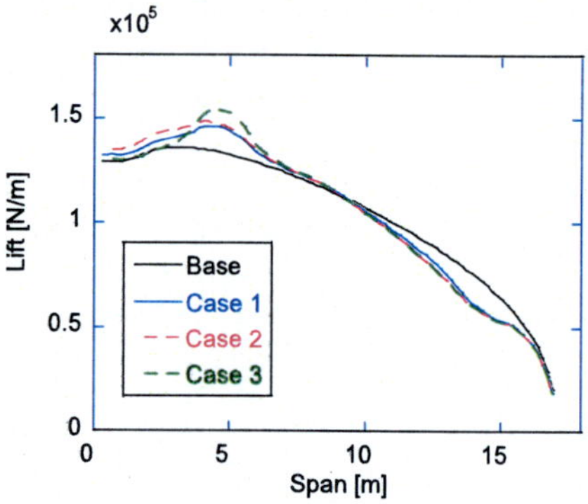

(f) In case of failure of Flap 7

Fig. 6 Lifting force distributions of some Pareto solutions

6 Conclusion

This research proposed the resilient operation of the adaptive wing with morphing flaps divided along the span direction based on the concept of the resilience engineering. The total lifting force is recovered by adjusting the flap angles of the survived divided flaps after failure of single flap such that the flap angle moves back to during the flight. This resilient operation problem is formulated as a multi-objective optimization problem that minimizes the absolute value of the lifting force between the original and recovered condition, minimizes the total magnitude of change of the flap angles, and minimizes the bending moment at the wing root for efficient operation. NSGA-II, which is a multi-objective genetic algorithm method is adopted for implementation. The validity of the proposed resilient operation is demonstrated through numerical examples.

Further improvement of the searching efficiency is a future problem. In addition, the resilient operation problem is extended to the case when the lifting force balance of right and left wing is lost. In addition, this research assumes the aerodynamic analysis for simplicity that the interference of the neighbouring flaps with different flap angles is ignored. Since this will deteriorate the aerodynamic performance, further detail analysis is required.

Acknowledgements This research was partially supported by JSPS KAKENHI Grant Numbers JP18K18812 and JP21H01535.

References

1. Tamayama M (2009) Survey on morphing activities. Nagare 28(4):277–284 (in Japanese)
2. Yokozeki T, Sugiura A, Hirano Y (2014) Development of variable camber morphing airfoil using corrugated structure. J Aircraft 51:1023–1029
3. Nakamura G, Uehara K, Kogiso N, Yokozeki T (2017) Optimum morphing shape design for morphing wing with corrugated structure using RBF network. In: 12th world congress of structural and multidisciplinary optimization, p 214
4. Kambayashi K, Kogiso N, Yamada T, Izui K, Tamayama NS, M, (2020) Multiobjective topology optimization for multi-layered morphing flap considering multiple flight conditions. Trans JSASS 63(3):90–100
5. Furuta K eds (2017) Introduction on resilience engineering. Nikka-Giren, (in Japanese)
6. Tashiro Y, Tsushima N, Yokozeki T, Tamayama M (2018) Investigation of efficient optimization method for load control using adaptive wing structures. In: Proceedings of 60th JSASS/JSME/JAXA structure conference, 2B12 (in Japanese)
7. Tashiro Y, Yokozeki T, Tamayama M (2019) Optimum load control of wing lift distribution using morphing wing structure. In: Proceedings 61st JSASS/JSME/JAXA/SURCAS structure conference, 2A02 (in Japanese)
8. Kitamura M (2014) New concept of safety pursued and implemented by resilience engineering. IEICE Fund Rev 8(2):84–95 (in Japanese)
9. Nomoto H, Ueda S, Takata S, Kasai T, Fukatsu T, Kobayashi R, Nogami M, Wakabayashi Y (2015) Resilient redundancy: Design study of the new HTV (H-II transfer vehicle). Springer, Space Safety is No Accident, pp 529–536

10. Nomura T (2013) Conceptual design of future passenger aircraft aimed at reducing fuel consumption. JAXA-RR-13–007, (in Japanese)
11. http://web.mit.edu/drela/Public/web/avl/. Accessed 15 Jun 2021
12. Budziak K (2015) Aerodynamic analysis with athena vortex lattice (AVL). Hamburg University of Applied Sciences
13. Deb K, Pratap A, Agarwal S, Meyarivan T (2002) A fast and elitist multiobjective genetic algorithm: NSGA-II. IEEE Trans Evol Comp 6(2):182–197

On the Evaluation and Design Methods of Structural Recoverability

Yuting He

Abstract Usually, maintainability, which is often expressed by means of *MTTR* (Mean Time to Repair), is used to characterize whether it's easy to repair when the equipment structures have some common damages during the normal service period. But, it's hard for equipment structures to avoid accidental and unexpected damages in service, such as accident damages, battle damages, etc. And then recoverability can be used to express whether it's easy to repair and recover when equipment structures suffer some accidental and unexpected damages. After expounding the basic concept of structural recoverability, the significance of structural recoverability to maintaining structural operational integrity, which is categorized into structural static operational integrity and structural dynamic operational integrity, is discussed. Then, parameters $R_c(t)$ (Recovery Degree), *MTTRC* (Mean Time to Recovery), *MRCC* (Mean Recovery Cost), and *MWTTRC* (Mean Work Time to Recovery) are proposed to characterize and measure structural recoverability. Furthermore, the normal evaluation, assessment, and design methods of structural recoverability are put forward primarily. Finally, a brief example of recoverability evaluation for accidentally damaged vehicle front bumper structures of three brands of the same model class is shown through analyzing *MTTRC* and *MRCC*, and followed by the recoverability design discussion on an aircraft monolithic backswept wing structure constructed by ribs, stringers and skins. It is also stated here that the concept, measurement, evaluation, assessment, and design methods of structural recoverability are suitable for other structures, such as building structures, bridge structures, and other infrastructure structures, etc.

Keywords Structure · Recoverability · Evaluation · Assessment · Design · Structural operational integrity

Y. He (✉)
Aeronautics Engineering College, Air Force Engineering University, Xi'an, China
e-mail: heyut666@126.com

S. Lee et al. (eds.), *The Proceedings of the 2021 Asia-Pacific International Symposium on Aerospace Technology (APISAT 2021), Volume 1*, Lecture Notes in Electrical Engineering 912, https://doi.org/10.1007/978-981-19-2689-1_63

1 Introduction

In the design of equipment structures, just like aircraft structures, helicopter structures, space vehicle structures and so on, reliability, safety, maintainability, supportability and testability have become important design parameters [1–5]. In fact, these parameters reflect the special characteristics of the structure in normal service, which is of great significance to ensure the integrity of the structure under normal service conditions. Moreover, crashworthiness [6] was proposed to reflect the crashworthiness of the structure for modern aircraft structures. For military aircraft structure, survivability [7] was also proposed to reflect the ability of the structure not to be found by the enemy and to resist the attack of the enemy's weapons under operational conditions.

However, in operational processes of equipment structure, in addition to normal conventional damages caused by the accumulation of service cycles, it is inevitable for structures to have some accidental or unconventional damages, such as accident damages, battle damages, bird-strike damages, hailstone-induced damages, typhoon-induced damages, tool-drop damages, foreign body impact damages and so on. These accidental damages usually come from accidental and uncertain events, and it is difficult to predict the damaged sites and extents because of their uncertainty. After these accidental events, it is necessary to repair the damaged equipment as well as structure to recover it to the original perfect status or one certain state for implementation of missions or functions. For example, in the 1973 Middle East War, the ratio of battle-killed aircraft to battle-damaged aircraft of the Israeli Air Force increased from 1:3 at the beginning to 1:7 later. If the Israeli Air Force couldn't had the ability to repair their damaged aircraft, they would have lost their operational effectiveness completely on the eighth day of the war. In the repair of damaged aircraft, more than 90% of the repairing work is the repair of the structures [8].

Usually, the ability of a structure to be repaired to a sound state (or a specified state) after some common damages, such as fatigue damages, corrosion damages, abrasion damages, and so on, can be measured often using *MTTR* (Mean Time to Repair). It is obvious that the repair of accidental structural damages is totally different from the normal maintenance activities of the structures during the normal service period, including repair tools, repair methods, repair time limit, repair requirements, and so on. The ability of a structure to recover to a sound state (or a specified state) after accidental damage is an inherent property of the structure, which is determined in the design and manufacture of the structure. This ability of structure can be characterized by means of structural recoverability.

2 Basic Concept of Structural Recoverability

In 2007, Dr. Lowell Tonnessen of the United States put repairability, together with susceptibility, vulnerability, and fire protection, into the research category of operational survivability in "the historical prospect of live-fire test and evaluation" [9]. The main research connotation of repairability is battle damage evaluation and repair. In 2015, the Test and Evaluation Office of the U.S. Department of defense proposed in the "Integrated Survivability Assessment-Guidance" [10] that recoverability, together with susceptibility, vulnerability, and fire protection, should be included in the scope of survivability assessment. The connotation includes two parts: battle damage assessment and repair, and mission capability recovery. Authors have given a strict definition of aircraft structural recoverability in Refs. [11, 12]. Moreover, it can be generalized to define the recoverability of all equipment structures. Equipment structural recoverability is the ability of a structure to recover to sound status or one certain required status in the specified mission from the status with accidental damages or unexpected damages, such as accidental damage status, battle damage status, bird-strike damage status, hailstone-induced damage status, tool-drop damage status, foreign body impact damage status, etc., within the specified time and under specified conditions, by means of the specified repair methods and procedures. Structural recoverability is the inherent property of a structure, which is formed to the structure through the processes of structural design and solidified to the structure through the processes of structural manufacturing.

It should be pointed out that the structural recoverability and maintainability are the inherent properties of a structure gained in the design and manufacturing processes, so there are some common requirements for structure design, such as accessibility design, interchangeability design, mistakes preventability design, etc. However, there are obvious differences between recoverability and maintainability. Maintainability is aimed at natural faults of structures, including faults caused by reliability and quality problems as well as human errors. The focus is on "maintain and repair". Generally, there are expected fault (or damage) sites, fault modes, fault extents, and standard maintenance or troubleshooting schemes. Hence, the maintainability parameters and requirements of the structures are usually for the normal use of the structures. Recoverability is aimed at accidental or unconventional damages, including accidental damages (such as damages caused by collision and crush, etc.) and battle damages (such as weapon-induced damages and secondary damages, etc.). The emphasis here is on "repair and recovery", and more emphasis is put on the repair measures for unforeseen or unconventional damages. The site and extent of these accidental damages are often unpredictable. When the structure with good maintainability in normal service has accidental damages or unconventional damages, it may be difficult to repair or irreparable due to different damage sites and original structure constructions. That is to say, in this case, usually the recoverability is poorer than maintainability for structures.

Some structural design methods can improve the maintainability and recoverability at the same time, such as openness or accessibility design, which can make it convenient for maintenance in ordinary service and repair in case of accidental damages. Some structural design methods can improve the maintainability, but reduce the recoverability. For example, the application of the integral structure construction in the structural design can significantly improve the maintainability and reliability of a structure, because the key critical parts needed to be inspected carefully in operational processes are significantly reduced compared with the combined structure constructions. And if there are cracks, corrosion, and other damages in the critical parts, it is also convenient for inspection and repair. Even if small cracks, corrosion, and other damages are found, it is also easy to use grinding and other methods to quickly repair and get rid of them. However, the integral structure is not as easy to repair as the combined structure in case of serious accidental or battle damages, which reduces its recoverability. For another example, in designs of various pipeline structures, to improve the maintainability and reliability, the integral pipeline design can be adopted to minimize the pipeline joints. Because the more joints there are, the more likely leakage failures will occur. However, once the integral pipeline is accidentally damaged (for example, battle damage occurs in a certain part of the middle section.), it will be very difficult to repair. When the conditions do not meet the requirements, it can't even be recovered, so the recoverability of the pipeline system is reduced. Ref. [13] makes a serious analysis of the difference between the equipment battle damage repair and the usual maintenance work, and believes that they are basically two things, or even "have nothing in common". All these show that whether the structure is easy to repair after failure in normal service is not the same as whether it is easy to repair in case of accidental or unconventional damages, and both types of damages can't be measured and characterized by means of the same index parameters. It can be seen that the maintainability and recoverability of a structure are not the same things. They are two different inherent quality properties of structures.

At the same time, the structural recoverability is different from the structural battle damage first-aid repair, which refers to the work that the user or maintenance personnel can use emergency means and local materials to make the damaged structures quickly put into combat again when the structures are damaged on the battlefield. However, usually the structural battle damage first-aid repair can't recover all the functions of the structures, and it is required to recover the partial functions for the equipment to perform the tasks at that time, which is closely related to the logistics support conditions. Structural recoverability is aimed at all accidental damages as well as unconventional damages, and battle damage repair is only a case of its research. The purpose of recoverability is to recover the original sound state firstly. Only when it is impossible to recover the original sound state, can it be recovered to a specified functional state. The normal repair method is the first recovery method, while the first-aid repair method is only used in an emergency. Moreover, when the recoverability is studied as an inherent quality characteristic of a structure, its logistics support conditions are specified. For example, for a structure with battle damages, the emergency repair on the battlefield belongs to the battle damage first-aid repair, and most of the repairs sent to the rear base are normal repairs to recover the original state of the structure. The accidental damage repair of equipment structure in non-wartime is not the battle damage repair, and it is generally to recover its original state. At this time, whether it is the battle damage repair or the normal repair, it is closely related to the recoverability. It can be seen that the connotation of structural recoverability is broader than that of battle damage repair, and it is also a more general inherent quality characteristic of a structure.

It should also be noted that the structural recoverability is not the same as whether the structure can be repaired in case of accidental or unconventional damages. The recoverability characterizes the difficulty of repairing the structure when the same level of accidental or unconventional damage occurs. For example, for two aircraft wingtips with the same levels of deformation damages caused by collision, the recoverability here is directly related to whether the wing is easy to recover or not. Whether the structure can be repaired in case of accidental or unconventional damage is related to recoverability, but it mainly depends on the damage extent of the structure. If the damage extent is too large, it will directly lead to irreparability, which is mainly related to economical efficiency.

3 The Significance of Structural Recoverability

The recoverability is of great significance to maintaining the structural operational integrity of equipment. The term integrity has become an important concept to characterize the quality characteristics of structures [14–19]. On this basis, with regard to performing and completing the functions and missions, the concept of military aircraft structural operational integrity is illustrated in Refs. [11, 12]. And aircraft operational integrity is discussed in Ref. [20]. Therefore, it can be generalized that for all aircraft, including military aircraft and civil aircraft, the definition of the aircraft structural operational integrity (ASOI) in service is the attribute which exists when a structure is sound and unimpaired while providing the desired levels of structural durability, supportability, safety, performance, survivability, recoverability and so on in operational processes. Generally speaking, as an attribute of a structure, it includes static attribute and dynamic attribute, showing the status and ability of a structure separately. So, in Ref. [21], aircraft structural operational integrity (ASOI) is categorized into aircraft structural static operational integrity (ASSOI) and aircraft structural dynamic operational integrity (ASDOI). Aircraft structural static operational integrity is defined as the status which exists when an aircraft structure is sound and unimpaired in service while the aircraft structural dynamic operational integrity is defined as the ability which exists when an aircraft structure is sound and unimpaired while providing the desired levels of structural durability, supportability, safety, performance, survivability, recoverability and so on in operational processes. It can be seen that the aircraft structural static operational integrity concerns the status of aircraft structure at the exact moment while the aircraft structural dynamic operational integrity concerns the ability of aircraft structure to keep sound and unimpaired during one exact service period. To be generalized, the equipment structural operational integrity (ESOI) can be described here similarly and categorized into equipment structural static operational integrity (ESSOI) and equipment structural dynamic operational integrity (ESDOI). And it can be seen that, in equipment structural dynamic operational integrity, the deficiency of any index factors of structural durability, supportability, safety, performance, survivability, and recoverability will lead to the overall low level of the equipment structural operational integrity. That is to say, the structural operational integrity conforms to the "cask effect", which means that all the factors mentioned above must be considered comprehensively and coordinately in the whole life cycle of the equipment structure.

From the discussion above, it can be seen that if the recoverability of the structure is poor, it will directly lead to poor structural operational integrity, especially for military equipment structures. It also means that once the structure has accidental damages or unconventional damages (such as accident damages, battle damages, etc.), it can't be recovered effectively in time, which will directly affect its ability to continue to use. For military equipment, if the equipment structural integrity cannot be maintained effectively in operational processes, it will directly affect the maintaining of the operational effectiveness of the troops. For civil equipment, if the equipment cannot be put into operation again in time, it directly affects the economical benefits of the company.

4 Assessment Method of Structural Recoverability

4.1 Measurement and Characterization of Structural Recoverability

The structural recoverability of equipment can be described by structural recoverability degree, which is a probability index and is mainly reflected in the recovery time of equipment structures, represented as $R_c(t)$. The meaning of the recoverability degree $R_c(t)$ can be described as: with the accidental or unconventional damages (such as occasional damages, battle damages, etc.) of equipment structures, the probability of the structures to be repaired within the specified time, under the specified conditions in accordance with the specified procedures and methods, while resuming the structural abilities to complete the specified missions or functions. The other meaning of $R_c(t)$ is the proportion of equipment structures subjected to accidental or unconventional damages, resuming the ability status to complete the specified missions or functions. $R_c(t)$ can be expressed as follows,

$$R_c(t) = P(T \le t) = \frac{q_t - q_{n,t}}{q_t} \tag{1}$$

where T is the recovery time of equipment structure under the specified conditions, procedures, and methods; t is the specified time; q_t and $q_{n,t}$ refer to the total number of equipment structures that have been subjected to accidental or unconventional damages and the number of equipment structures that cannot be restored to the ability to perform the specified missions or functions during the specified period of time.

The recoverability parameters of equipment structures can be also expressed by recovery time parameters, as well as recovery cost parameter and recovery work time parameters. These parameters can be used to quantify the recoverability of equipment structure, compare the recoverability of similar equipment structures or compare the recoverability of different design schemes of the same equipment structure.

The recovery time parameter of structure can be expressed by the mean recovery time *MTTRC* (Mean Time to Recovery, recorded as T_{mr}), which is the average time taken by a required group of persons with the required technical level to repair the equipment structures with a certain level accidental or unconventional damages to restore specified functions using specified procedures and resources (other equipment, tools, spare parts, manpower, logistics, etc.). It can be expressed as follows,

$$T_{\mathrm{mr}} = E(Y) = \int_0^\infty tm(t)\mathrm{d}t \tag{2}$$

where Y is the recovery time, and $m(t)$ is the recovery time density function, which is the probability of completing the repair work in time $\Delta(t)$.

The recovery cost parameter of structures can be characterized by the accidental or unconventional damage recovery cost *MRCC* (Mean Recovery Cost, recorded as C_{mr}). *MRCC* is the mean amount of resources used by a required group of persons with a specified technical level to repair the same model equipment structures with a certain extent of accidental or unconventional damages to the specified functional status by using the prescribed procedures and methods. Hence, *MRCC* can be expressed as follows,

$$C_{\mathrm{mr}} = E(X) = \int_0^\infty cn(c)\mathrm{d}c \tag{3}$$

where X is the recovery cost, $n(c)$ is the recovery cost density function, which is the probability of completing the repair work with cost $\Delta(c)$.

Recovery work time of structures can be characterized by the accidental or unconventional damage recovery work time *MWTTRC* (Mean Work Time to Recovery, recorded as T_{mwr}), which is mean work time used by a required group of persons with specified technical level to repair the same model equipment structures with a certain extent of accidental or unconventional damages to the specified functional status by using the prescribed procedures and methods. Here, *MWTTRC* can be expressed as follows,

$$T_{\mathrm{mwr}} = E(Z) = \int_0^{\infty} t w(t) \mathrm{d}t \tag{4}$$

where Z is the recovery work time, and $w(t)$ is the recovery work time density function, which is the probability of completing the repair work in work time $\Delta(t)$.

Obviously, structural recoverability is an inherent quality characteristic of equipment structure, which is mainly improved by the design and manufacturing of equipment structure. For the structures with high recoverability, in the case of accidental damages or unconventional damages (such as occasional damages, battle damages, etc.), the maintenance personnel can repair and recover the equipment structures conveniently and quickly under limited conditions. Especially in the battlefield environment, high recoverability is of great significance to improve the re-deployment capabilities of military combat equipment and maintain the scale of the combat equipment fleet, which will directly affect the sustainment of operational effectiveness and the victory of the combat.

When repairing damaged structures, due to different damage levels and repair resource conditions, the damaged structural repair parameters representing structural recoverability will be completely different. According to the severity of the damages, the damage extents of the structures can be divided into some levels, for example, serious damage extent (or irreparable damage extent), relative serious damage extent, medium damage extent, minor damage extent, and slight damage extent [8, 22]. Of course, damage extents can also be classified in other ways. Therefore, when describing the requirements of structural recoverability parameters, it is necessary to describe the damage extents of structures and repair resource conditions.

For example, 5 damage levels are used to show the damage severity: serious damage, relative serious damage, medium damage, minor damage, and slight damage. When discussing the recoverability of a certain structural member, it can be stated as: under the condition of deport, the recovery time for relative serious damage of structure is not more than 5 h; or, under the condition of deport, the recovery cost for relative serious damage of structure is not more than 5 thousand US Dollars. Moreover, under the condition of deport, the recovery time for relative serious damage of structure is not more than 100 work hours, etc. Under the two-level maintenance system of aircraft, the conditions of repair resources of aircraft can be generally considered as two conditions: field level and deport (or manufacturer) level. Therefore, the recoverability requirements of structure can also be described as follows: under the two-level maintenance system, the recovery time for the relative serious damage of the same model of the structure shall not be more than 5 h; or, the recovery cost shall not be more than 5 thousand US Dollars; or, the recovery work time shall not be more than 100 work hours. And these statements are some typical requirements of structural recoverability.

4.2 Evaluation Method of Structural Recoverability

As an inherent property of structures, recoverability can not only be designed but also be evaluated and verified by certain means. To confirm whether the structures can meet the established quantitative and qualitative requirements is the overall objective of the recoverability assessment. The evaluation method is not a single one, but a combination of analysis, inspection, test, and demonstration. The recoverability assessment runs through the whole life cycle of the structure and can involve all levels. The design defects of the structures can be found as early as possible and the improved designs can be carried out through the prediction and evaluation of recoverability. The prediction and evaluation methods of structural recoverability are similar to the prediction and evaluation methods of structural maintainability, but it is aimed at the situation when the structures are repaired due to accidental or unconventional damages and the restoring of structural functions. The main evaluation methods of structural recoverability are as follows:

(1) Probability simulation method. Based on the basic recovery work of the structural system, the parameters of recovery time are derived step by step through the comprehensive process of time distribution. The time distribution parameters (mean/logarithm mean, standard deviation/logarithm standard deviation) of basic recovery work of structural system can be specified by relevant standards. Detailed information on the repairable elements of the structural system is required for such a predicted assessment.

(2) Functional levels analysis method. The decomposition of the structural system function level is carried out from top to bottom according to the structural system. Generally, it starts from the structural system, and decomposes to the level where accidental or unconventional damage can be located, damaged components can be replaced or repaired in situ, and structural system functions can be restored or adjusted. When this method is applied, the analyses of structural connection and joint relationship become the most basic and effective work.

(3) Sampling and scoring method. It samples enough replaceable or in-situ repaired components from the structural system, scores the recovery work according to the checklist, and estimates the recovery time by using the relevant empirical formulas. Only the empirical formulas for calculating the recovery time of the structural system need sufficient and enough test data supporting.

(4) Operational function prediction method. This method does not provide any ready-made data, but only provides a program. It is required to evaluate the recovery time of structural system on the basis of historical experience, subjective evaluation, expert judgment, and other data supporting.

(5) Time accumulation method. This method is based on historical experience or ready-made data, and determines the recovery time or work hours required for each recovery item, each recovery work, each recovery activity, and even each basic recovery work one by one, according to the structural system design or

design scheme as well as accidental or unconventional damage repair conditions, then accumulates the recovery time or work hours and calculates the average values of them, and finally estimates and evaluates the recoverability parameters of the structural system.

(6) Component comparing method. Among the components of a structural system, a basic component with recovery time known can always be found as a benchmark. By comparing with the benchmark component, the recovery time of each component can be estimated, and then the recovery time of the structural system can be determined. The basic parameter predicted by means of this method is the average recovery time.

At the same time, it should be pointed out that the structural recoverability assessment should be carried out according to the actual structural damage level and repair resource conditions. Whether the conclusion of structural recoverability assessment is accurate and credible or not is based on the structural recoverability test results. In the case of a lack of test data, the predicted assessment results can be used as the parameters of structural recoverability instead. If conditions permit, the conclusions of structural recoverability should be drawn through Research & Develop tests and evaluations. Some special tests such as interchangeability can be combined with maintainability and supportability tests. For tests of accidental damages or unconventional damages, such as accidental damages, battle damages, bird strike damages, etc., they can be combined with survivability tests, crash tests, etc. Moreover, special verification tests (such as live fire test, etc.) can also be carried out for accidental damages or unconventional damages. After the completion of the relevant tests, the damaged structures can be repaired immediately, and the recoverability test results can be carried out. According to the results of the structural recoverability tests, and with the modification of previous evaluation results, we can get a more accurate assessment conclusion of structural recoverability. Of course, the true level of structural recoverability can only be obtained through operational test and assessment of the actual operational processes of equipment structures.

5 Design Method of Structural Recoverability

The recoverability design of structure is to take into account the inevitable occurrence of accidental or unconventional damages in service. For example, the aircraft structures encounter bird strike and lightning strike in flight, and the military equipment structures suffer battle damages in the combat environment, etc. Since these damages are always inevitable, it is necessary to consider how to recover structures with these accidental or unconventional damages easily, quickly, and economically. Especially for military equipment, most of the workload of battle damage repair is structure repair, and the recovery of equipment is directly related to the victory or failure on the battlefield. For civil equipment, it is about economic benefits and market share. Therefore, the recoverability level of equipment structures must be guaranteed by effective design methods. In order to improve the recoverability of structures, the following main design methods are usually employed.

(1) Modular design. The application of the modular design concept to structure design will enable the structure to recover its specified functional state only by replacing the damaged modules in case of accidental or unconventional damages. The replacement of damaged modules is often faster, more convenient, and more economical compared to conventional repairs. The recoverability of the structure system will be improved significantly through modular design.
(2) Interchangeability design. The same parts of different types of equipment structures can be replaced with each other physically and functionally through the interchangeability design. When the structure system is damaged by accidents or weapon attacks, the damaged parts can be replaced and repaired in time to achieve rapid recovery of the damaged equipment.
(3) Standardized design. The standardized design and manufacturing of equipment structural parts will be very beneficial to the supply and storage of parts, so that the equipment is more convenient to repair in case of accidental or unconventional damages. For example, for the battle damage repairs of combat equipment structures in battlefield environments, we can use the standardized parts to repair the damaged parts quickly, which greatly improves the structural recoverability of equipment.
(4) Accessibility design. In the case of accidental or unconventional damages, if the accessibility of the damaged structural parts is good, it is easy to reach the damaged parts without disassembling other parts or with simple disassembling actions, and it also has enough operational spaces. All of these will make the recovery work of damaged structure system more convenient and quicker.
(5) Topological optimization design. Taking the structural recoverability as one of the design objectives, the topology optimization design of structures can make the structures satisfactorily recoverable with more excellent quality performances while meeting some special requirements.
(6) Repair process design. When the equipment is damaged accidentally or unconventionally, the position and extent of the damages cannot be predicted accurately in advance, and the environments and places where the structures need to be recovered may be strange and crude. Military equipment grounding to a strange environment after the equipment structures are damaged in battle. All of these require that reducing the skill level of repairmen should be taken into account when carrying out structural recoverability design, and the demand for recovery tools and additional equipment should be as simple as possible so as to make it easy to support recovery events. That is to say, the recovery processes should be designed as simple as possible to be easy to operate, with reducing complex operation steps, repair processes requirements, and using replacement repair or simple repair methods. For example, in the design of quick-lock connections, conventional bolt connections instead of welding connection, pulling riveting processes instead of conventional riveting

processes, solid lubrication processes instead of liquid lubrication processes, non-sealing processes instead of sealing processes, normal temperature environment processes instead of high temperature and high humidity or constant temperature and humidity environment processes are all effective methods of repair process design. In the case of battles, we can also use such technological methods as switching, cutting or bypassing, dismantling, replacing, reconstructing, on-site repairing, manufacturing, and matching, etc.

(7) Self-repair design. In the process of recoverability design of equipment structures, if the technologies are feasible, self-repair design technology should be adopted as far as possible. With the development of technologies, the self-repair ability of structure must be improved day by day. If the self-repair structure design is adopted, when the structure encounters accidental damages or unconventional damages (such as accident damages, battle damages, bird strike damages, hailstone-induced damages, typhoon-induced damages, tool-drop damages, foreign body impact damages, etc.), the structure can be quickly self-repaired and recover the damaged parts, or recover the functions of the damaged parts so that the damaged structure system can recover its mission abilities quickly. It can be seen that the self-repair technology must be one of the key technologies for equipment structures in the future.

Of course, there are other methods of structural recoverability design. Through the design and manufacture, the structural recoverability can be solidified, maintained, and improved. At the same time, it can be seen that it may affect the changes of other characteristics of the structures, or even reduce the overall characteristics of the structures. For example, the employment of combined structures instead of integral structures can significantly improve the structural recoverability, but reduce the reliability and maintainability of the structures, and may also increase the weight of the structures at the same time. These also show that the structure design is a process of comprehensive trade-off design, and the final goal is to get the structures with the best comprehensive performances, or the structures with the best operational integrity, instead of only pursuing the best one performance of the structures at the expense of other performances.

6 Cases Analyses

6.1 A Brief Example of Structural Recoverability Assessment

In this paper, three brands of vehicle equipment of the same model with the same functions and the same price range are selected to briefly evaluate the structural recoverability. In this case, structural damages of the front bumpers of the vehicles caused by accidental collision damages need to be repaired for the three brands of vehicle equipment. These three brands of equipment are represented by K, P, and

Table 1 Recovery time and cost of damaged front bumpers of three brands of vehicles

Time/cost Company	K		P		V	
	Time/hour	Cost /CNY	Time/hour	Cost /CNY	Time/hour	Cost /CNY
Company 1	36	900	48	1300	48	12,000
Company 2	40	1050	48	1200	60	14,000
Company 3	36	1000	40	1000	72	14,000
Company 4	36	850	40	1100	60	13,000
Company 5	36	900	48	1250	60	12,000
Company 6	40	1000	40	1150	48	11,000
Average value	37.3	950	44	1166.7	58	12,666.7

V whose front bumper structures damaged by accidental collision need to be recovered. According to the market situation in 2019, the actual recovery time and recovery cost from six qualified maintenance companies are randomly selected for statistical analyses, and the results are shown in Table 1. Although the sample size of random sampling statistics here is not large enough, and the statistics also have some contingency, it is still a case example of recoverability analyses. Of course, one or more companies can make statistical analyses on the recovery time and actual cost of three brands of vehicle equipment. When the sample size is large enough, the conclusions of these methods should be consistent. In fact, this example is a comparative case of structural recoverability given by the operational test and evaluation method of equipment.

From the analysis results in Table 1, although the functions and prices of the three vehicles of the same model are basically the same, and their front bumper structures with accidental collision damages need to be repaired (replaced with new ones), the average recovery time *MTTRC* and average recovery cost *MRCC* are different. For Brand-K vehicle equipment, the average recovery time (*MTTRC*) is 37.3 h, and the average recovery cost (*MRCC*) is 950 CNY, which is the best. Therefore, it can be seen that in the case of three brands of vehicle equipment with the same-class accidental damages on the front bumpers, Brand-K vehicle equipment has the best structural recoverability.

6.2 A Brief Example of Structural Recoverability Design

The typical aircraft monolithic backswept wing structure as shown in Fig. 1a is selected to analyze the structural recoverability design briefly. The wing consists of typical wing ribs, spars, stringers, and skins. Under normal service conditions, the force-bearing and transmission characteristics of the wing are very good. However, if accidental damage occurs in one part of the wing, a collision that causes the wing tip more seriously damaged, as shown in Fig. 1b. The damaged wing needs

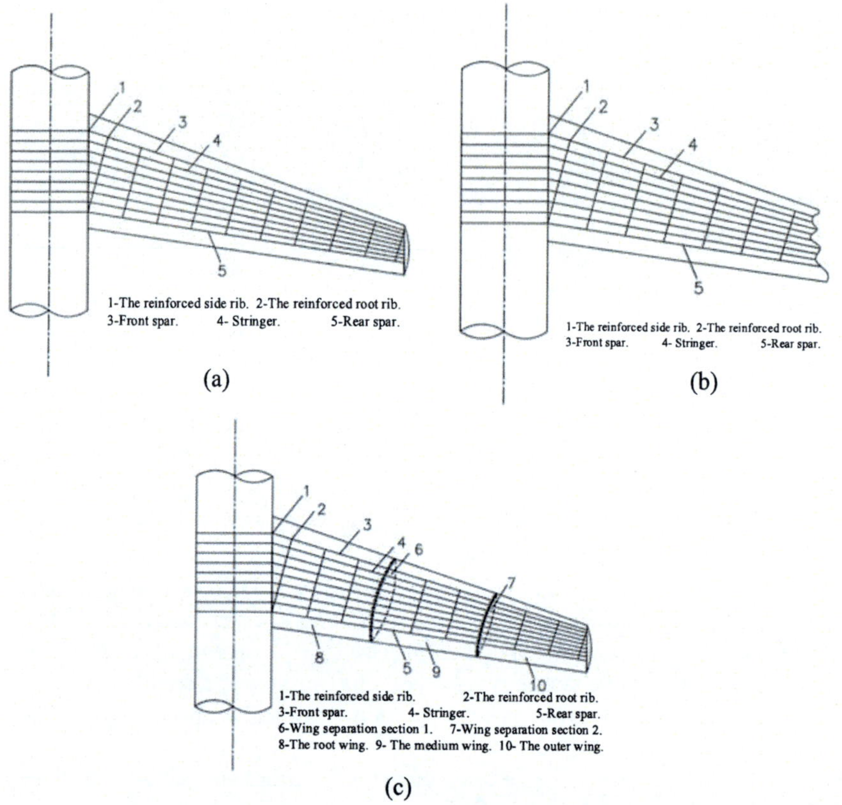

Fig. 1 Example of recoverability design for a monolithic backswept wing structure constructed by ribs, stringers and skins: **a** Typical aircraft monolithic backswept wing structure, **b** Aircraft monolithic backswept wing structure with damaged wing tip, **c** Recoverability-designed aircraft monolithic backswept wing structure

to be recovered in accordance with the following main steps: first, the damaged stringers and skin structures are cut away, then the damaged stringers are removed and replaced by new compartments with reliable connections, and finally, the new skins are covered. This process shows that the repair process is complex with the different types of repairing tools required and the high skill level of the repair personnel. So, this damaged wing will need a long repair time and a high repair cost to recover.

If the wing is designed with a recoverability idea, the wing structure can be designed as a combination of several structural segments, shown in Fig. 1c. If the same accidental damage occurs to the wing, only the damaged wing tip segment needs to be replaced directly with another new segment. Then both recovery time and recovery efficiency will be greatly improved. Especially in the case of battle damages of military aircraft structures, the recoverability design idea will enable combat-damaged aircraft to be quickly repaired and restored to the status of carrying out combat missions and providing direct support for combat victory. Even for civil aircraft structures, the improvement of recoverability can significantly improve the operational economic efficiency of aircraft.

From this design example, it shows that the design of recoverability, as an important inherent quality characteristic of the equipment structure system, will greatly improve the equipment recovery capabilities in the event of uncertain, accidental, or unconventional damages. It is of great significance for the damaged equipment structures to be put into service again and to sustain the equipment fleet scale.

7 Conclusions

In order to solve the problem of whether the structure system is easy to repair when it suffers from accidental or unconventional damages, the following preliminary work is carried out:

(1) The basic concepts of structural recoverability and structural operational integrity are described;
(2) From the point of view of maintaining the structural operational integrity of equipment, the significance of structural recoverability is discussed.
(3) The measurement and characterization methods of structural recoverability are proposed. Some basic evaluation, assessment, and design methods of structural recoverability are provided here.
(4) A case of structural recoverability evaluation is presented when the front bumper structures of a model of vehicles are damaged seriously. Meanwhile, a brief example of the structural recoverability design of a typical aircraft monolithic backswept wing structure system is cited.

To conclude, the concept of structural recoverability, the methods of structural recoverability design, measurement and assessment, as well as the concept of structural operational integrity are also applicable to other structures, such as building structures, bridge structures, infrastructure structures, and so on, so there still leaves a lot of work to do in these aspects.

Acknowledgements The author is grateful for the grant from the National Science and Technology Major Special Funding of China (J2019-I-0016-0015) and the National Natural Science Foundation of China (52175155).

References

1. He Y et al (2014) Safety engineering of aircraft[M]. National Defense Industry Press, Beijing, pp 20–32

2. He Y et al (2017) Theory and technology of aircraft structural life control[M]. National Defense Industry Press, Beijing
3. The General Armaments Department of PLA (2017) Quality management system requirements[S]. The People's Republic of China military standard, GJB 9001C-2017
4. Zhang H (2015) Review on quality characteristic of equipment. J Dual Technol Products 09(2):252–253
5. The General Armaments Department of PLA (2005) Terminology of reliability and maintainability[S]. The People's Republic of China military standard, GJB451.A–2005
6. Zhang H (2016) Design guidelines of general-purpose plane anti-the drop and crash[M]. Aviation Industry Press, Beijing, pp 3–9
7. Li S, Chang F, He Y et al (2016) Survivability analysis and evaluation of military aircraft in operational processes[M]. National Defense Industry Press, Beijing, pp 23–42
8. Zhang J (2001) Aircraft battle injury emergency repair engineering[M]. Aviation Industry Press, Beijing, pp 12–31
9. Tonnessen L (2007) An historic perspective on live fire test and evaluation (LFT&E). J Aircraft Surv Live Fire Test Eval (3):8–10
10. DoT&E Program Office, Integrated Survivability As-sessment-Guidance [EB/OL] (2015). http://www.dote.osd.mil/docs/TempGuide3/Integrated_Survivability_Assesment_Guidance_3.0.pdf
11. He Y, Zhang T, Ma B (2019) Basic connotation and evaluation of military aircraft structural operational integrity. J Air Force Eng Univ (Natural Science Edition) 9(10):433–439
12. He Y, Zhang T, Ma B et al (2019) Structural integrity control technology based on structural damage monitoring[C]. In: Poland:36th conference and 30th symposium of the international committee on aeronautical fatigue and structural integrity
13. Chen X (2003) Modern maintenance theory[M]. National Defense Industry Press, Beijing, pp 87–98
14. USA Department of Defense (2016) Aircraft structural integrity program (ASIP)[S]. Department of defense standard MIL-STD-1530D(USFA)
15. The General Armaments Department of PLA (2012) Structural integrity program of military aircraft[S]. The People's Republic of China military standard, GJB775.A–2012
16. Wang L (1988) Aircraft equipment integrity. J ACTA Aeron ET Astron SINICA 9(10):433–439
17. USA Department of Defense (2013) Mechanical equipment and subsystems integrity program [S]. Department of defense standard MIL-STD-1798C(USFA)
18. USA Department of Defense (1984) Engine structural integrity program (ENSIP)[S]. Department of defense standard MIL-STD-1783 (USFA)
19. USA Department of Defense (2011) Avionic integrity program (AVIP) [S]. Department of defense standard MIL-STD-1796A(USFA)
20. He Y (2021) On aircraft operational integrity[C]. In: China:32nd congress of the international council on the aeronautical sciences
21. He Y (2021) Aircraft structural operational integrity—the comprehensive quality characteristic of aircraft structure[C]. In: China:2021 International Symposium on Structural Integrity
22. Yao W (2016) Criteria and methods for safety flight assessment of battle injured aircraft[M]. National Defense Industry Press, Beijing, pp 45–67

Simulation and Analysis of Automatic Passenger Door Actuation System

Weijuan Zheng, Yi Wu, Wenjing Zhi, and Dongping Liu

Abstract Aiming at the requirement of automatic operation of passenger door of a civil aircraft, a set of mechanism and electric actuation system suitable for all-electric automatic intelligent operation of passenger door is designed. The system can complete the automatic operation of unlocking, opening pressurization prevention door, unlatching, lifting, opening and the electric operation of girt bar mechanism, and has the function of manual operation all the mechanism. In the system design and refines the top-down design method adopted in analysis of systems engineering, and at the same time in the mechanism design of organization performance analysis Real-time adjustment and optimization, to ensure that automation involved institutions include lock mechanism, latch mechanism, pressurization prevention door mechanism, lifting mechanism, open mechanism, gust lock, flight lock mechanism and girt bar mechanism to meet the design performance requirements, the performance of latch mechanism, lifting mechanism, opening mechanism and girt bar mechanism is analysed emphatically. It can be used as the basis to choose the parameters of the downstream transmission train and the power source. It can better design the downstream drive line and choose the suitable power source. According to the two-dimensional diagram of the mechanism of CATIA, a two-dimensional linear frame mechanism model was established. On the basis of considering the size of the entity, the dynamic simulation model was established by using the dynamic simulation analysis software LMS Virtual Lab Motion, and the simulation analysis was carried out by adding constraint joints and mechanical model. According to the Hertz contact force model, the contact model is added by using solid contour, and the empirical data is used to make the model closer to the actual contact, so that the simulation analysis results have a certain reliability. The design and simulation analysis and optimization of the parallel method makes it possible to scientifically analyse and verify the performance of the mechanism before the physical manufacturing and molding, realizes the coordination of design and analysis, exposes the development risk in advance, and reduces the cost and time of the physical verification. After the model design, the wire frame simulation analysis method can be simply and quickly changed to carry out a new round of simulation calculation. There is no need to

W. Zheng (✉) · Y. Wu · W. Zhi · D. Liu
The Aeronautical Facility Research Institute, AVIC Qing'an Group., Co., Ltd, Xi'an, China
e-mail: zhengwj003@avic.com

S. Lee et al. (eds.), *The Proceedings of the 2021 Asia-Pacific International Symposium on Aerospace Technology (APISAT 2021), Volume 1*, Lecture Notes in Electrical Engineering 912, https://doi.org/10.1007/978-981-19-2689-1_64

rebuild the model and change the entity to complete the simulation. Fast multi-round iterative calculation greatly shortens the development cycle. In this paper, it is found that the difference between the two results is small by using MATLA kinematics simulation analysis and local processing calculation, which can be used as the basis for the load design of the passenger door actuation system.

Keywords Passenger door · Actuation system · LMS · Simulation · Kinetics

1 Introduction

The passenger door is used for boarding and deplane members, and is the normal channel for boarding and deplaning crew, passenger, maintenance and service personnel [1]. It is required that the passenger door can be opened normally or in an emergency when boarding the plane and landing on the ground. It is required that the passenger door can ensure the safety of members when flying in the air and reduce the probability of passenger door opening accidentally in the air to extremely unlikely. The mechanism of the passenger door must ensure the completion of the function of the passenger door, and the operating model of the passenger door is to ensure that the door can complete the drive of the mechanism movement according to the requirements. In order to improve the passenger comfort, also increase the size of the airplane, passenger door increase is a result of the increase of exit, on the large aircraft, the passenger door for type A basic door, in order to satisfy the security requirements, setting up mechanisms complex, making weight of the door bigger, door automatic needs more and more strong. In this paper, based on the design requirements of automatic cabin door, relevant design is carried out.

2 About Passenger Door

In order to ensure the safety of the passenger door in flight and the realization of the operation function on the ground, the passenger door usually includes: locking mechanism, lifting mechanism, opening mechanism, pressurization prevention mechanism, girt bar mechanism, gust mechanism and flight lock mechanism and the door structure to ensure the reliable support of the mechanism.

Figure 1 is the schematic diagram of A320 passenger door. It can be seen from the diagram that this door contains most of the institutions. Each type of the cabin door adopts four specific institutions and technologies to meet the requirements of current airworthiness. It can be seen from the figure that the passenger door contains a handle mechanism. The unlocking movement of the door is completed manually, the opening mechanism and girt bar mechanism are also completed manually, and the emergency opening completed by pneumatic mode.

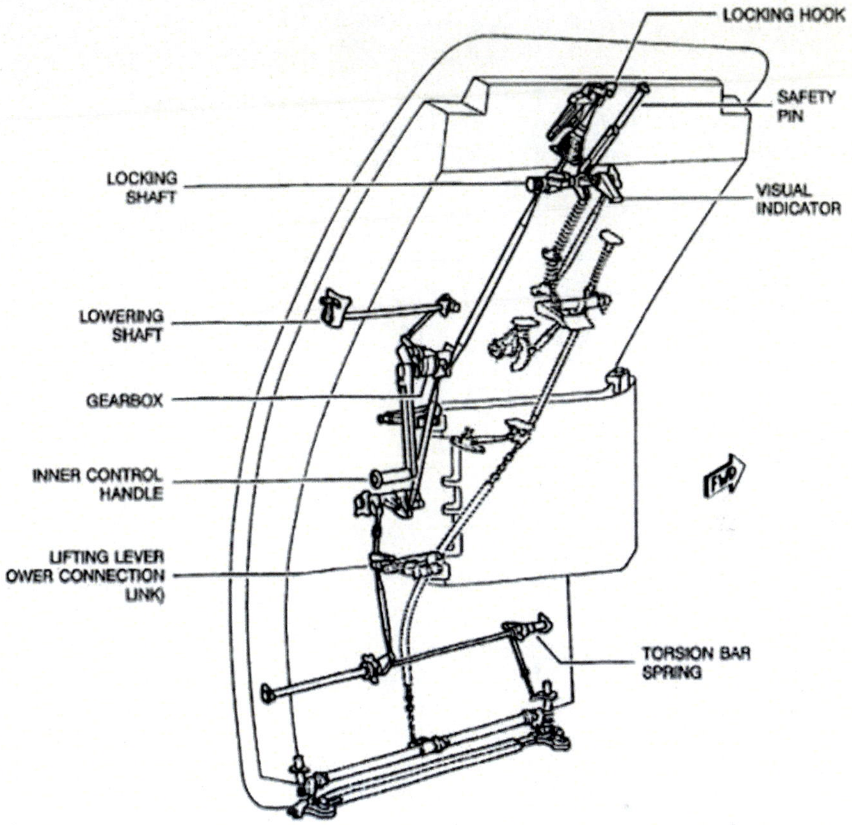

Fig. 1 A320 passenger door mechanism

Based on the idea of automatic passenger door, it is the key to realize the electric drive of the passenger door mechanism. There is a sequential relationship between the movements of the mechanism and the mechanism can not be driven by a single driver. The logic between the drives needs to be sorted out clearly and controlled.

3 Load Analysis of Passenger Door Mechanism

Door is a set of structural, institutional and instructions, such as the warning system in the integration of complex components, in order to guarantee the door to meet the design requirements, reduce the risk of developing, modern commercial door design to follow the double "v" pattern design flow of system engineering [2], the decomposition confirmed at the design stage to follow a top-down design process, reanalysis and so on. In the integration validation phase, the bottom-up design validation process is followed.

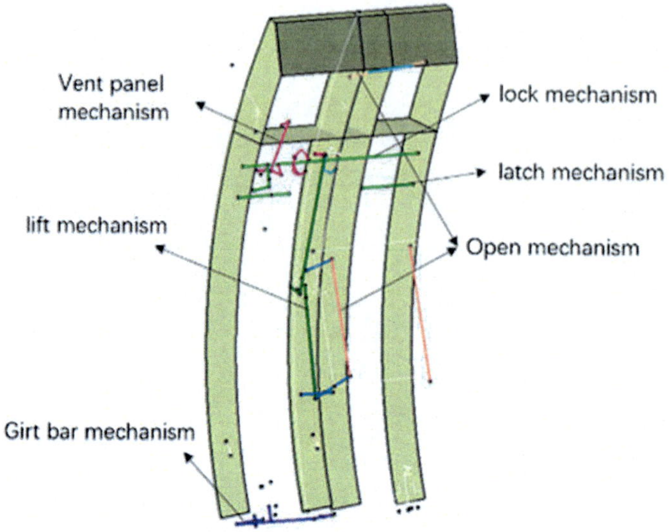

Fig. 2 The wire frame of passenger door mechanism

3.1 *The Design of Passenger Mechanism*

Based on the above requirements and some constraints of the airworthiness regulations, the initial design of the passenger door can be started. As shown in Fig. 2, the wire frame containing the basic door structure and mechanism has been formed. The wire frame of door lock mechanism, pressurization prevention mechanism, lifting mechanism, opening mechanism, girt bar mechanism has been formed, and the motion principle has been verified by the CATIA sketch.

The latch mechanism can lock the cabin door, the lifting mechanism can make the cabin door be lifted to prepare for the opening, the opening mechanism can realize the opening of the boarding door, and the girt bar mechanism can realize the control of the slide package. All the mechanisms compromise the driving system. The automatic operation of the passenger door can be realized by adding an electric drive to these mechanisms.

In order to evaluate the space position of the door mechanism on the door structure, the wire frame mechanism simply materialize, as Fig. 3 shown.

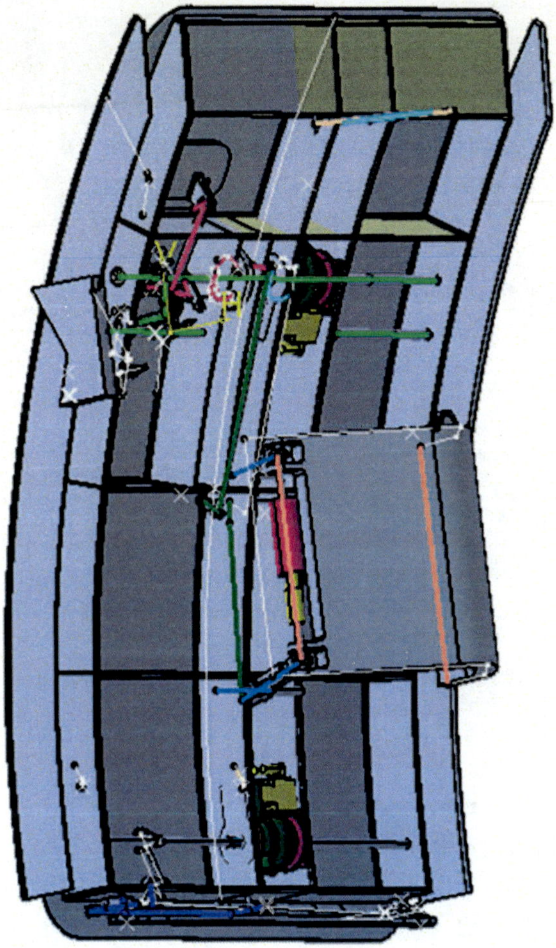

Fig. 3 Preliminary physical view of door

3.2 *Dynamic Simulation*

After the preliminary design and evaluation of door mechanism by CATIA, the model is needed to import to LMS Virtual Lab motion to simulate, using virtual prototype instead of physical prototype, so as to shorten the development cycle, reduce cost and improve efficiency [4]. The dynamic software is compatible to CATIA, the interface and the operation are the same as CATIA.

After the model enters the mechanism design platform of Motion, the model needs to be processed. The constraints are added in LMS based on coordinates. In fact, dynamic analysis and simulation can be carried out after two-dimension wire frame of mechanism comes out, which does not require high precision of model (Fig. 4).

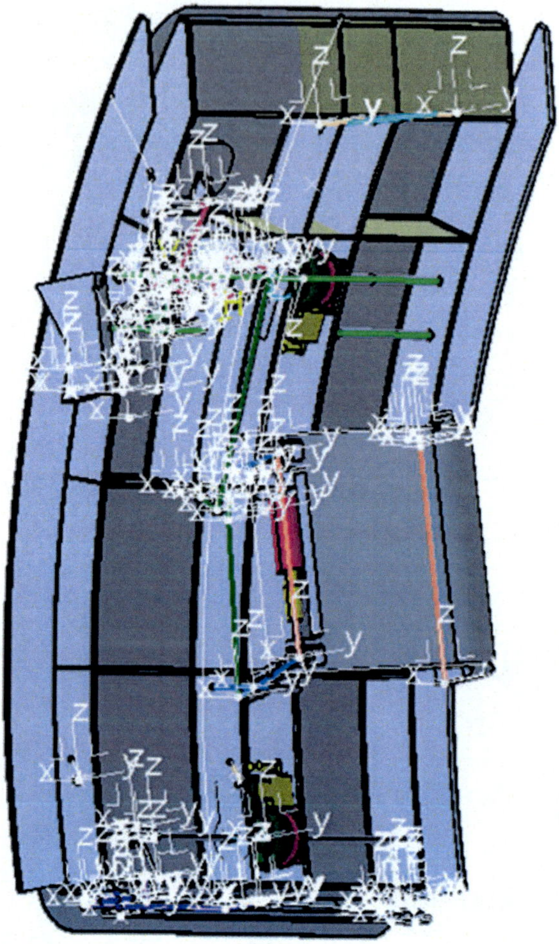

Fig. 4 Model preprocess

Based on the coordinates of the linkage system of the mechanism, motion pairs, such as rotation pairs and movement pair are added Hertz contact model is adopted where the mechanism involves the transfer of large sum of driving force in physical collision. Contact models have 6 types, there are point to point contact, sphere to extruded surface, sphere to revolved surface contact, extruded surface to revolved surface contact, revolved surface to revolved surface contact and CAD contact. There are 7 contact constraints in the model in Fig. 5, which are respectively located at the lock mechanism, latch roller, flight lock drive mechanism, gust mechanism, lock cam, pressurization prevention cam and girt bar mechanism. Based on long-term design and simulation experience, it can be obtained as shown below.

1. Point to point contact, the force is too big;

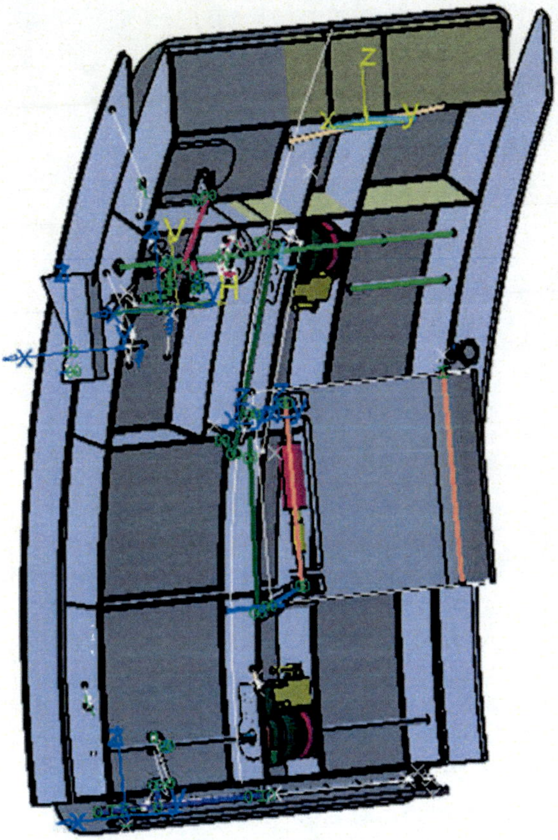

Fig. 5 The dynamic simulation model of passenger door mechanism

2. Extruded surface to revolved surface contact, the contact force has big vibration;
3. Sphere to extruded surface, the force is too big;
4. Adding the mass and inertia, sphere to extruded surface, the force is reliable [3].

So all the contact constraints should be the fourth constraints.

To simplify the model, all parts are treated as rigid bodies, but this does not interfere with the elastic parameters assigned when material is assigned. All the parts are as steel, its mass can be transmitted to LMS models.

Defining two kinds load cases to the model: the gravity force only and landing gear breaking force. It includes single and combination of forward landing gear breaking, left gear breaking, right gear breaking.

After adding the constraints and drive force, the model needs spring force in some place, the final model is as Figs. 5 and 6.

First uses a wire frame view institutions design whether meets the requirements, based on door needs interlock of the mechanism, the need to focus on the institutions

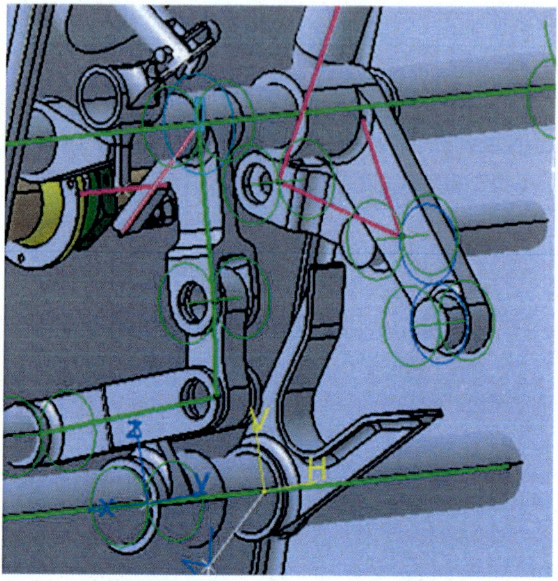

Fig. 6 Contact model

successively movement order, as shown in Fig. 7, the kinetic basically follows the opening pressurization prevention first, unlock the door, the door open and the order, from the point of the simulation results, the design of the basic meet the requirements.

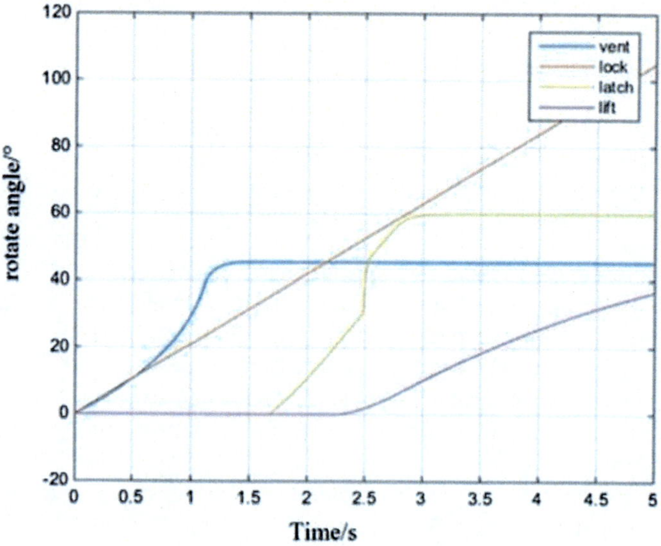

Fig. 7 The sequence of mechanism movement

Table 1 Data compare

	Sketch		Simulation	
	Lock shaft (°)	The end mechanism (°/mm)	Lock shaft (°)	The end mechanism (°/mm)
Pressurization prevention	36.5	44.92	29.4	45
Lock	31.29		31.29	
Latch	35.07–63	60	35.07–63	60
Lift	44.7–76	30 mm	48.3–105	46.5 mm

Compare the simulation result and the sketch design object, the result is as data in Table1.

It can be seen from the table that there is no difference between the design and the simulation of lock mechanism and latch mechanism, mainly in that their design and implementation are consistent. The theoretical design of the pressurization prevention mechanism and the lifting mechanism are quite different from the actual simulation. The main reason is that the two mechanisms adopt time-sharing cam mechanism, and the sliding phenomenon of the cam curve can be predicted in the theoretical design of entity, which leads to difference between the two mechanisms.

With the help of dynamic analysis software, the door lock actuator, open actuator and girt bar actuator based on working conditions are analyzed, and the load value is obtained. Several landing gear breaking conditions were analyzed for emergency opening conditions of the passenger door, and the following results were obtained (Fig. 8).

In the process of analysis, by changing different contact modes and adjusting parameters, sphere surface to extrude surface is finally selected for impact force transfer analysis.

It can change the sphere radius easily by change the number in Fig. 9. The simulation software in the measure of the sphere of the collision is used to set the size as a physical boundary, with the combination of surface or guide to add the impact force calculation, which can be a simple interaction to determine the best of the size of the size of the sphere, saved using the entity beam time, reduce the repeated.

The load of the driving end analyzed above is taken as the basis for selecting the driving source, which largely depends on the accuracy of the data.

As there are many and complex mechanisms mentioned above, there is no time and unnecessary to verify every kind of mechanism. Plane four-bar mechanism is the most widely used mechanism on the passenger door, and the transmission of mechanism force depends on a number of four-bar series. This paper taking the most typical four-bar mechanism as the checking mechanism to evaluate the rationality of the power credit.

Taking typical planar four-bar mechanism as the object, two methods are used to analyzed system load and motion parameters (Fig. 10).

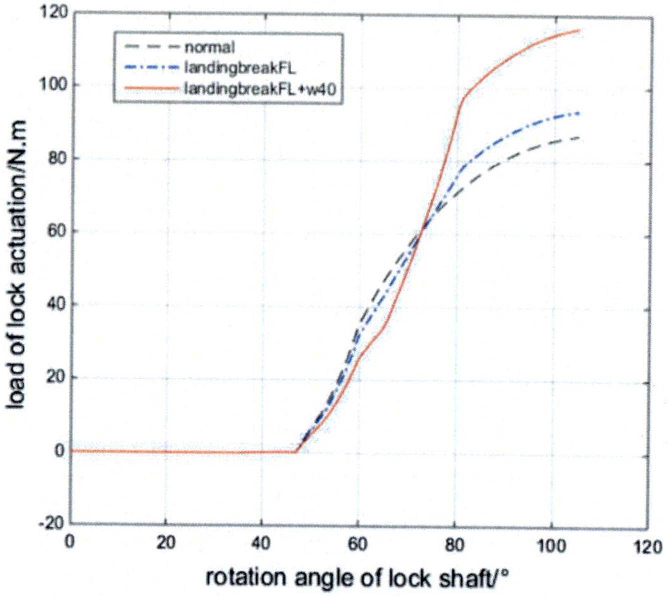

Fig. 8 The lock actuator load of front and right landing gear break

Simulation model method and MATLAB mathematical model method were used to analyze respectively, and to get the difference of the angular velocity of 13 under constant speed (Fig. 11).

As known is Fig. 12, the error is between −2.5%−1.5%。

Simulation model is based on the mathematical model of simulation model, by contrast, the simulation and the result of dynamic difference within the acceptable range, compared with the balance equation established by analyzing the stress, and build a simulation model and the results seem to be simple and convenient, more capacity. It is easy to be realized in engineering and can be used as the load evaluation in the early stage of subsequent scheme design.

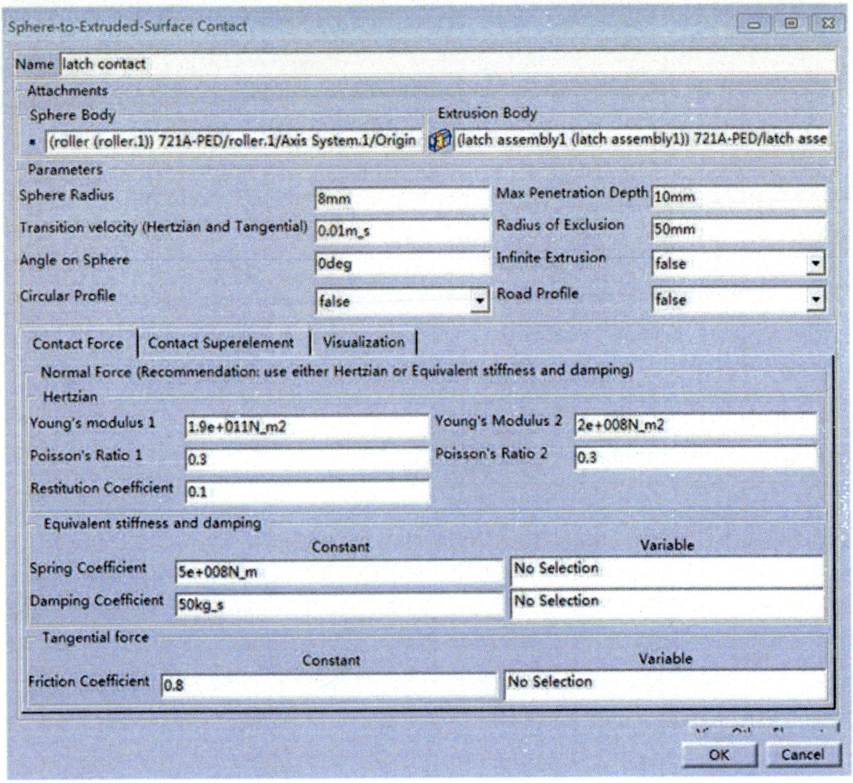

Fig. 9 Contact parameter set

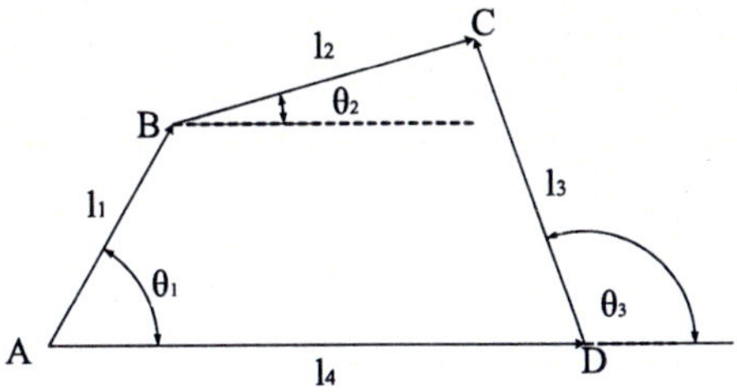

Fig. 10 Typical planar four-bar mechanism

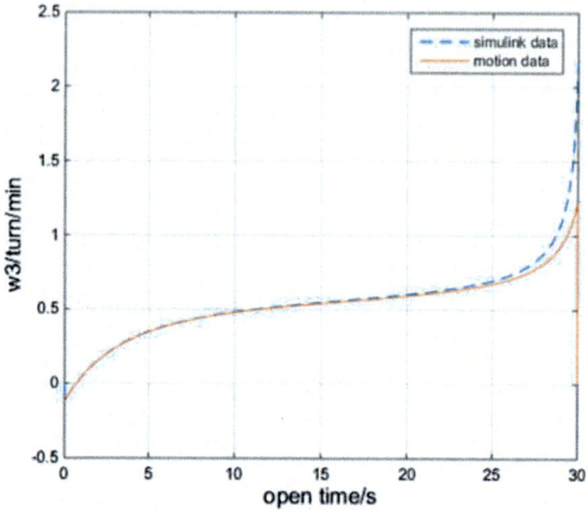

Fig. 11 The difference of w3

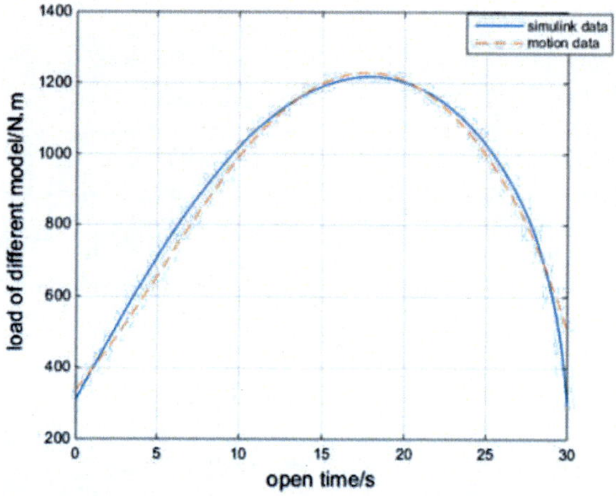

Fig. 12 The difference of load in point A

4 Conclusion

Based on the analysis of the general gate mechanism, based on the demand of automatic door, the paper designed the door mechanism and electrical actuation system suitable for automatic control, and analyzed the actuation demand by system engineering method to capture the design points. Dynamic simulation analysis was carried out based on 2-dimension CATIA sketch design, and the constraints of the solid collision were loaded according to the reality, and the most realistic constraint was found. The load of the system was obtained through dynamic simulation analysis, and compared with the simulink results, confirming that the dynamic simulation results can be used as the selection basis on the driving source of the door actuation system.

References

1. Xionghua Y, Junfeng D, Yunwen F (2017) The doors design of transport airplane[M]. National defense industry press, Beijing
2. INCOSE (2015) Systems engineering handbook 4th edn. USA
3. Gu CH, Tu Q, Tong MB (2018) Dynamic analysis for unlock and lifting mechanism of passenger door based on LMS. Aeron Comput Tech 46:74–77
4. Wang W, Gao M (2014) Dynamics simulation analysis of the airplane door lock mechanism. J Aeron Sci Technol 25(12):30–34

Research on the Aerodynamic Failure Load for Civil Aircraft

Linlin Tan

Abstract This paper introduces the failure list selecting methods based on the system function safety evaluation report, and introduces the flight loads calculation methods of the time at the failure occurrence and continuous flight after the failure based on the selected failure list. Taking a large civil airplane as an example, it proposes the principle of failure list selecting, and creates the concept of control surfaces jam biased Normally Encountered Position rather then the extreme deflection. Finally it quantitatively analyzes failure load such as elevator jam, aileron jam, rudder jam, stabilizer jam and flap actuator disengagement. During the procedure it compares with the conventional flight load. The analysis of elevator jam show that it would create an asymmetrical moment at the tail due to asymmetric deflection of the left and right elevators. But the moment is not covered by normal load, which constitutes the critical case of the rear fuselage. And also the flap actuator disengagement would be the serious situation in load design. So both of them should be carried out. This paper comes up with an effective, reliable failure load calculation and analysis method.

Keywords Civil airplane failure list · Failure load · Elevator jam · Asymmetrical moment

1 Introduction

The load caused by the system failure would exceed the normal flight load by improper design of load. The consequence is that the structure gains weight and even the system and structure is destroyed, and then the plane crashes. So it is very important for the design of failure load, but the failure load of civil aircraft is a difficult point in flight load, especially the analysis of the cause of system failure, the impact on the flight envelope, the maneuvering flight simulation of the time at the failure occurrence and continuous flight, unconventional pressure distribution in

L. Tan (✉)
Shanghai Aircraft Design and Research Institute, Shanghai 201210, China
e-mail: linglingtanken@163.com

S. Lee et al. (eds.), *The Proceedings of the 2021 Asia-Pacific International Symposium on Aerospace Technology (APISAT 2021), Volume 1*, Lecture Notes in Electrical Engineering 912, https://doi.org/10.1007/978-981-19-2689-1_65

case of failure. In summary, it is necessary to research the method of failure load to ensure the reliability and integrity of the load design [1].

2 Failure List Selecting Methods

Aircraft system for special conditions include flight control system、automatic flight control system/stabilization system、load mitigation system、flutter management system and fuel management system. In this paper the failure load is taken out only for flight control systems and automatic flight control systems/automatic flight control system and fuel management system, because only these systems is installed on the civil aircraft [2].

The following failure condition are selected according to filter reports based on failure list of aircraft specific conditions SCA-2.

3 Design Principles

According to the requirement of special condition of SCA-2 [3],The flight load of the following failure situation would be considered (Fig. 1).

(1) The time at the failure occurrence: it is necessary to establish the actual situation from the 1 g leveling state, including the pilot's corrective measures, to determine the load at the time at failure occurrence or immediately after the failure. These loads are multiplied by the appropriate safety factor to obtain the ultimate load for static strength verification. The safety factor is defined as shown below.

(2) Continuous flight in the state of failure: when the system is in this failure state, the configuration is adjusted, meanwhile the aircraft is controlled with flight

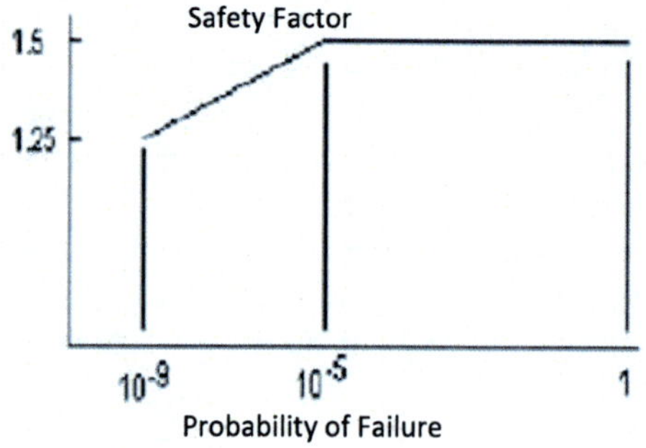

Fig. 1 Safety factor at the failure occurrence

restrictions in the range of up to VC/MC or prescribed continue flight speed, the load in the following cases should be determined.

(a) Limited symmetric maneuvering conditions specified in CCAR 25.331 and CCAR 25.345;
(b) Limited gust and turbulence conditions specified in CCAR 25.341 and CCAR 25.345;
(c) Limited rolling conditions specified in CCAR 25.349/limited asymmetric maneuvering conditions specified in CCAR 25.367 and CCAR 25.427(b) (c);
(d) Limited yaw maneuver conditions specified in CCAR 25.351.
(e))The safety factor depends on the probability of the failure state, the safety factor as shown.

4 Design Method

4.1 Jam and the Hardover Position Determination

For the jam of number 1–4 for Table 1, the jamming position need to be determined to calculate failure load. The first method is that the control surface is assumed to jam at ultimate position, but the probability of this condition is extremely improbable to occur. The second method is that jamming in the normal use of the operating position considering aircraft take-off, cruise, normal turning, descent and landing is take account in. That is mean the jamming position is determined by normally encountered position (called NEP) of control surface inside the normal flight envelope. It is so reasonable that the failure conditions from 1 to 4 of the table are calculated according this method.

For the spoiler's failure cases of number 5–6, the position of hardover is determined according to the maximum capacity of the actuator.

Table 1 Failure case load calculation

Serial number	Fail condition	Probability of Occurrence
1	Jam of one aileron	< ~ 1.0E-5
2	Jam of one elevator	< ~ 1.0E-5
3	Loss of pitch trim of horizontal stabilizer	< ~ 1.89E-6
4	Jam of rudder	< ~ 1.03E-8
5	One multi-function spoiler hardover	< ~ 1.0E-5
6	Uncommanded motion of both ground spoilers	< ~ 1.0E-7

4.2 The Load Calculation Method of the Time at the Failure Occurrence

For the failure cases of number 1–4 of Table 1, after the NEP is defined, it is necessary to determine the pilot's control mode for the trim aircraft, and then the airplane is trimmed to 1G level. During aircraft trim, the load is calculated, which is determined to be the load of the time at the failure occurrence.

For the spoiler hard-over of number 5,the position of the spoiler is determined according to the maximum capability of actuator, and the airplane is trimmed to 1G level, and then the load is calculated, which is determined to be the load of the time at the failure occurrence.

For the uncommanded motion of both ground spoilers, the load is determined according to the maximum capability of actuator.

4.3 The Load Calculation Method of the Continuous Flight in the State of Failure

For the failure cases of number 1–4, according to the special condition of SCA-2, it is calculated up to the speed of VC including all the maneuver conditions of longitudinal conditions, lateral conditions, rolling, after that critical conditions are selected, and then the loads of all components is calculated.

For the failure case of number 5–6, the spoiler on is little effect on the aerodynamic characteristics of the whole plane, and the speed is just to VC(the cruise speed). The load factor is only 1.0 from curve in Fig. 2, so that the load all components of the failure case would not exceed the normal load. In summary,the load of the failure cases is not used to be calculated, only the spoiler's own load needs to be considered.

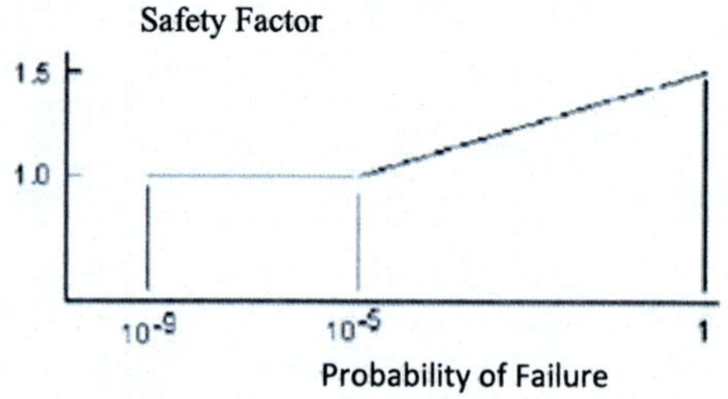

Fig. 2 Load safety factor for continuous flight after failure

5 Analysis and Comparison of the Elevator Jam

Taking the elevator as an example, the failure load analysis of the elevator jam is given including the load of the time at the failure occurrence and the load of the continuous flight.

5.1 *Simulation Calculation Method*

The scene at the time of elevator jamming is simulated as that the initial state of the aircraft is 1 g level flight. The pilot controls the side lever to make the aircraft pitch maneuver. When the aircraft reaches stable pitch, the sides tick returns to the middle, and in the process of returning the elevator is jammed. According to AC25.671, the load factor is 0.8–1.3 g during pitch maneuver. After jamming occurs, the pilot manipulates the elevator on the other side to correct the deviation of the aircraft to 1 g trim state according to the change of aircraft response attitude. The side rod displacement and deflection of elevators is as shown in the Figs. 3 and 4 below.

When the failure case occurs, the aircraft continues to fly with jam of one elevator, and the maneuver of the aircraft needs to be controlled by deflection of the other elevator. But the asymmetry of the elevators will cause a scissors difference, which would lead to an unsymmetrical load. In this case, the effect of this asymmetric load on the strength of tail and the rear fuselage must be checked. Balanced maneuver of the airplane between 0.25 and 1.75 g with high lift devices fully retracted and in

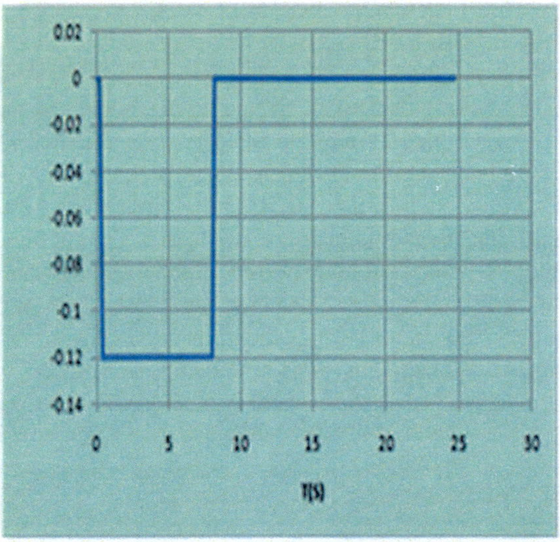

Fig. 3 Side rod displacement

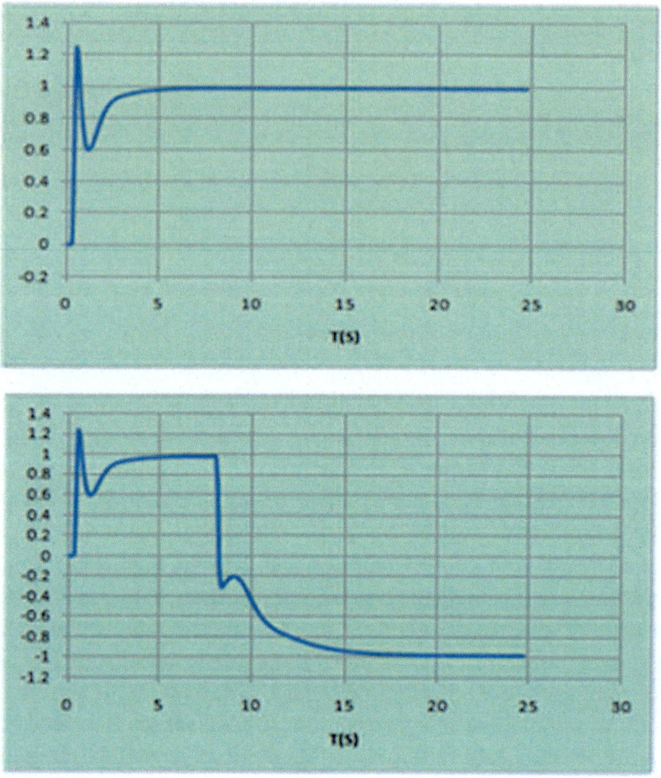

Fig. 4 Left and right elevators

enroute configurations, and between 0.6 and 1.4 g with high lift devices extended [4]. For yaw maneuver, rolling maneuver, the load of the failure condition would not be the critical of the whole plane. Analyzed three reasons, the first point is that the elevator jamming mainly affects the local aerodynamic characteristics of the h-tail, which has little effect on the whole aircraft, and it has little effect for the lateral and rolling maneuvering. The second point is that the load of the continuous flight just consider the speed of vc, which is below normal flight envelope. The third point is that the failure probability of the jam of one elevator is just $10^{\wedge -9}$.According to the special conditions, the safety factor is 1.0, which is less than the normal safety factor of 1.5. Based on the above factors, in terms of the ultimate load, the yaw maneuver and the rolling maneuver load when the elevator is jammed will not exceed the normal load.

5.2 Analysis and Comparison

The total load and distributed load are calculated after screening out the critical load at time of occurrence of and the continuous flight after the jam. The net load at time of occurrence and the normal load envelope are given below. It can be seen from the Figs. 5, 6, 7 and 8 that the net load of wing and the h-tail could not exceed the normal load, and will not constitute a serious situation.

For the continuous flight after the elevator is jammed, the load factor only needs to consider 0.25–1.75 g with high lift devices fully retracted, and 0.6–1.4 g with high lift devices extended, so the load of the horizontal and the wing will not exceed the normal load with the 2.5 g load factor.

For the h-tail, when one elevator is jammed, it is necessary to ensure that the aircraft can continue to fly safely and land, relying on the deflection of the other elevator. The asymmetrical movement of the left and right elevators produces a scissors difference, and an asymmetric load is generated at the h-tail, commonly known as the scissors differential load. Various different pitch maneuvers are considered in the load calculation, from which the serious situation of the flat tail load is selected as the basic symmetric load. According to the left and right elevator scissors difference angle data, the h-tail additional load due to the scissors difference is calculated. The loads are equal in magnitude and opposite in direction, acting on the left and right flat tails, creating an asymmetrical moment at the tail. Three methods are given below to calculate the asymmetry scissors differential load [5].

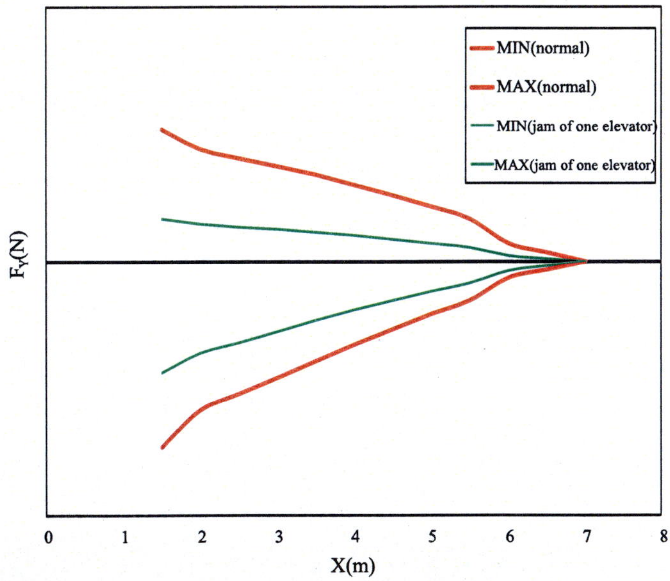

Fig. 5 h-tail shear curve

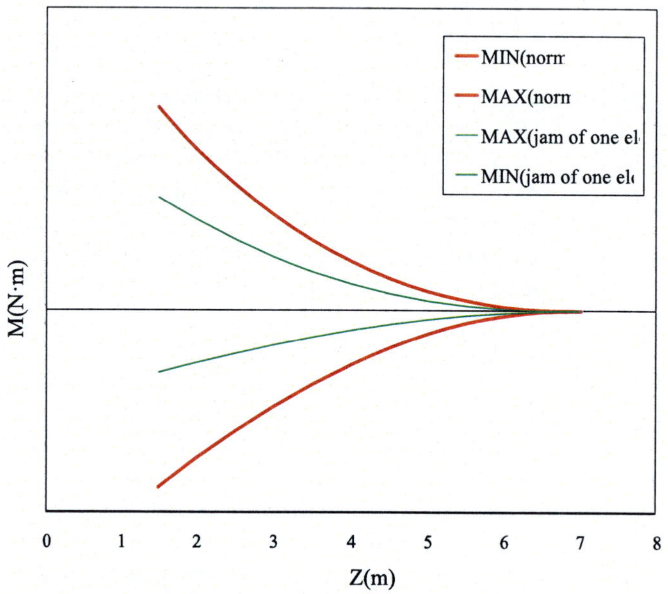

Fig. 6 h-tail bending moment curve

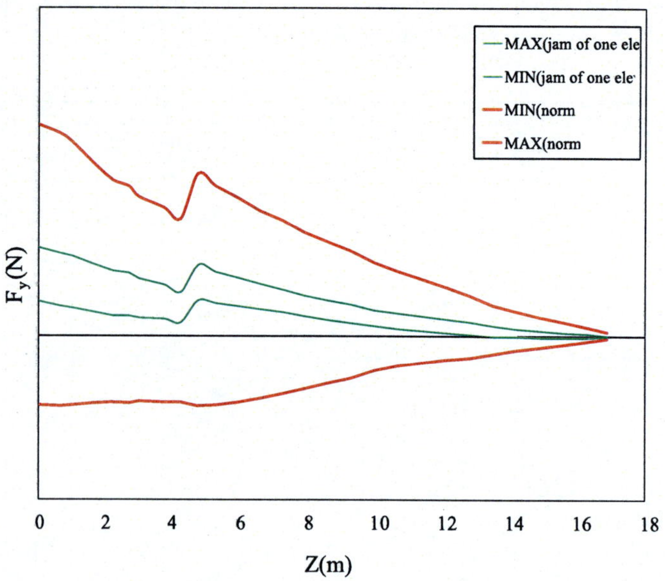

Fig. 7 Shear curve of wing

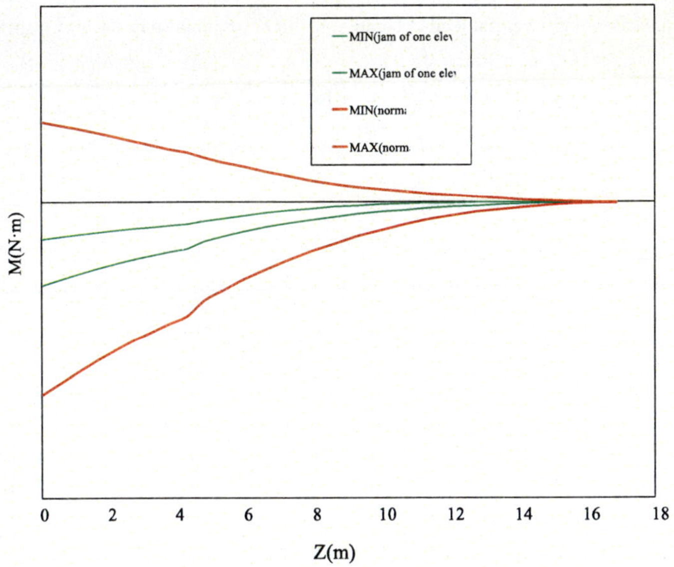

Fig. 8 Bending moment curve of wing

(1) One elevator is jammed in the NEP position, and the other elevator is deflected to meet the minimum requirements for operation. The range of load factor is 0.8–1.3 g.
(2) One elevator is jammed in the NEP position, and the other elevator is not limited, which can be deflected to the extreme deflection according to the actual;
(3) One elevator is jammed in the NEP position, and the other elevator is deflected to meet the requirements of AC 25.671 (Tables 2 and 3).

Method 2 has no limitation on the other side of the elevator after the jamming, and it is considered that the bias can be too conservative. The method 1 considers that the

Table 2 h-tail scissors differential load at continuous flight of one elevator jamming (Method 1 & 2)

Condition 1	The load of h-tail (N)			Asymmetrical moment (Nm)
	FX	FY	FZ	MX
Based load	1103	−55,207	5772	–
	1103	−55,207	−5772	
Method 1	495	−23,514	2460	**−168,263**
	−495	23,514	2460	
Method 2	1046	−45,617	4772	**−324,340**
	−1046	45,617	4772	

Table 3 h-tail scissors differential load at continuous flight of jam of one elevator (Method 3)

Condition 1	The load of h-tail (N)			Unbalanced moment (Nm)
	FX	FY	FZ	MX
Based load		−10,696		–
		− 0,696		
Method 3		−27,618		−193,692
		27,618		

other side of the deflection only needs to meet the minimum maneuvering requirement. Method 3 uses advisory circular AC25.671 "The loads considered as ultimate should be derived from the following conditions at speeds up to the maximum speed allowed for the jammed position or for the failure condition: (1) Balanced maneuver of the airplane between 0.25 and 1.75 g with high lift devices fully retracted and in enroute configurations, and between 0.6 and 1.4 g with high lift devices extended". Method 3 satisfies the failure load calculation requirement, and finally uses this method to calculate the elevator jamming load for strength check. Method 3 satisfies the fault load calculation requirement, and finally uses this method to calculate the elevator jamming load for strength check. Due to the elevator jam, a rolling torque is superimposed on the rear fuselage, the rear fuselage load is severe. Therefore, it is necessary to design the fuselage according to the elevator jamming of failure condition, the rear fuselage structure must be strengthened.

6 Conclusion

For modern aircraft, advanced control technology is adopted, and the load of failure condition analysis is required. According to the requirements of the special condition SC A2, the failure conditions such as the elevator jam, the stability surface jam, the rudder jam, the aileron jam and the spoiler hard over are calculated. In this paper, the analysis method and comparison of the elevator jam are given. The results show that the elevator jamming constitutes a serious load on the rear fuselage and requires structural reinforcement. In summary, the load of failure condition design is indispensable and must be incorporated into the load design. The screening of the fault list and the selection of the jamming position of the NEP are essential, affecting the load analysis and structural design. the load analysis and structural design.

References

1. China Civil Aviation Regulations Part 25: Airworthiness Standards for Transport Aircraft
2. Aircraft Design Manual (9)
3. Problem Summary SC-A2 Structure and System Interaction

4. FAR JAR 25.671 FCHWG - ARAC Report.pdf
5. Xie S (2001) Hand book of aircraft design:9th fascicle[M]. The Publisher of Aeronautic Industry, Beijing

The Vibration Transfer Path Analysis Based on One Large Passenger Aircraft

Li Yixuan, Li Kaixiang, and Liu Jijun

Abstract The transfer path analysis (TPA) method, which combines the testing and calculating, is a very useful and accurate tool for the structural vibration analysis, and has a wide application in the automotive field. However, the aircrafts have a very complicated loading condition, so the TPA method according to the aircrafts should be studied further. Nowadays, a lot of structural transfer path analysis cases are based on the commercial software, such as LMS and Head. Not mastering the core analysis technology of TPA technology is adverse to the further research in the aviation field, so it is necessary to carry out the foundational TPA technology. This article based on one large passenger aircraft, the vibration transfer paths from the engine and the landing gear to the four seat rail points in the cabin were identified. The vibration contribution and the sensitivity analyses of each path was also analysed. All the above were achieved through basic programming.

Keywords The Transfer Path Analysis (TPA) · Original program · The vibration contribution · The sensitivity analyses

1 Introduction

The vibration of a modern large passenger aircraft is often composed of multiple incentives, after reaching the target position through different transmission paths. To comprehensively consider the situation of each incentive and transmission path, the transmission path analysis (TPA) is an effective method [1–3]. The vibration transfer path method describes the vibration characteristics of the system in the frequency domain, combining experiments and computational simulation, providing a fast and accurate tool for the rapid diagnosis of the vibration problems.

Through the transmission path analysis, the proportion of the excitation energy in different transmission path is determined, and the transmission paths leading to the cabin vibration are found. By controlling the main paths, including the excitation

---

L. Yixuan (✉) · L. Kaixiang · L. Jijun
AVIC Aircraft Strength Research Institute, Xi'an, China
e-mail: liyixuan0430@126.com

S. Lee et al. (eds.), *The Proceedings of the 2021 Asia-Pacific International Symposium on Aerospace Technology (APISAT 2021), Volume 1*, Lecture Notes in Electrical Engineering 912, https://doi.org/10.1007/978-981-19-2689-1_66

strength, path vibration sensitivity and other parameters, in a reasonable range, to control the vibration of the cabin structure within the predetermined target value. At present, TPA technology has a preliminary application in automobile vibration and noise transfer path recognition, but it is difficult to clearly and correctly clarify the contribution amount of multi-dimensional path. The research on the study of vibration source, transfer path and structure interaction is still in the initial stage, mainly focusing on experimental methods and energy transfer methods. Because the redundant incentive sources, the source and structure coupling, and coupling between sources and structure, the transmission path research of the complex aviation structure is still in the initial stage. Especially for the transmission path, most is still in a qualitative research stage, so the research is relatively difficult, and need to carry out a series of simulation analysis and test verification work.

Based on the finite element model of a large passenger plane, this paper defines two input incentives, one at the engine and the other at the landing gear compartment. The output is the vibration response at the four guide rails in the passenger cabin section. The input and output directions are defined as vertical vibration. Based on this, the vibration transmission path analysis method is studied and the vibration transmission path model is established after the vibration contribution and path sensitivity of each path are verified.

2 Theoretical Base of Vibration TPA

For a system, the transmission characteristics usually constitute three parts: system input, system output and transmission paths that link input and output. When the system is stimulated, the system response would happen. The system transmission function is the transfer characteristics describing the corresponding relationship between such incentive and response [4]. In the engineer field, a complex system is generally simultaneously motivated by multiple excitation sources, each transmitting along the respective transmission paths to multiple different response locations and superimpose with each other to form the system outputs. If the transfer characteristics of the system can be obtained, we can not only better study the system itself in order to discover the problems of the system, but also predict the response of the system under different motivations, but also improve the transmission characteristics of the system.

Transmission path analysis is an effective method to study such system characteristics. In the transmission path analysis method, the output of the system is superimposed by the input of the system propagating along the respective transmission path to the output position, and the schematic diagram of the system transmission path analysis model is as shown in Fig. 1.

The linear system shown in the schematic of the transfer path analysis model can be expressed as Formula (1) in the frequency domain.

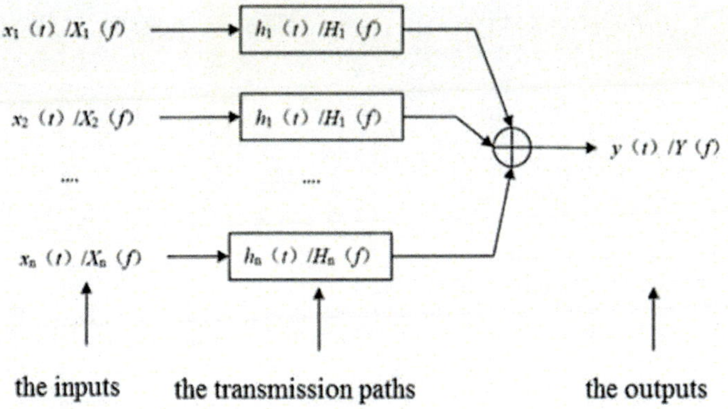

Fig. 1 Schematic diagram of the transfer path analysis model

$$Y = X_1H_1 + X_2H_2 + \cdots + X_nH_n = \begin{bmatrix} X_1 & X_2 & \cdots & X_n \end{bmatrix} \begin{bmatrix} H_1 \\ H_2 \\ \ldots \\ H_n \end{bmatrix} \tag{1}$$

In the formula, Y represents system output (vibration acceleration, etc.) signal; X_i represents the input (force, acceleration, etc.) signal of the system; H_i represents the input to output transfer function of the system (vibration transfer function, etc.).

When the system has multiple outputs, you can extend Formulas (1)–(2) to represent a multi-input, multi-output system (MIMO).

$$\begin{bmatrix} Y_1 & Y_2 & \cdots & Y_m \end{bmatrix} = \begin{bmatrix} X_1 & X_2 & \cdots & X_n \end{bmatrix} \begin{bmatrix} H_{1,1} & H_{1,2} & \cdots & H_{1,m} \\ H_{2,1} & H_{2,2} & \cdots & H_{2,1} \\ \vdots & \vdots & \ddots & \vdots \\ H_{n,1} & H_{n,1} & \cdots & H_{n,m} \end{bmatrix} \tag{2}$$

In the formula, the Y_j represents the jth output (vibration acceleration, etc.) signal; X_i represents the ith input (force, acceleration, etc.) signal; $H_{i,j}$ represents the transfer function between the ith input to the jth output of the system.

Write the Formula (2) in a matrix form of:

$$Y = XH \tag{3}$$

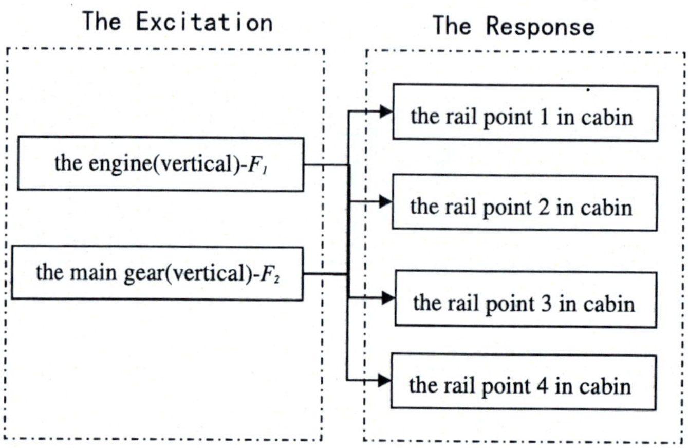

Fig. 2 Schematic diagram of the vibration transmission path of the cabin

3 Establishment of the Vibration TPA Model

According to the analysis goal, a simplified finite element model of a large plane is established, and the cabin section is a partial detailed model.

The vibration TPA model in the paper includes the vibration incentives, the vibration responses and the vibration transmission paths. The vertical direction responses of the four guide rails in the passenger cabin are selected. The incentives and the transmission paths of the aircraft in the flight are very complex and in the modelling process, the model is simplified and there are two incentive sources, one is at the landing gear and the other is at the engine, only considering the vertical direction.

Based on the above assumptions, the vibration transmission path model is established, which is referred in Fig. 2 [5].

4 Identification of the vibration TPA Model

4.1 Verification of the Vibration TPA Model

In this paper, it mainly studies the vibration level at 0–300 Hz, and only the vertical direction is considered. The vibration transmission path model is established based on Formula (2) and programmed in MATLAB software. The transfer function corresponding to each path, that is, the frequency-response function of each path from the excitation force to the acceleration response is calculated according to the established TPA model. Then the contribution amount of each path to the vertical vibration of the cabin floor is estimated according to the basic principle of the transmission path

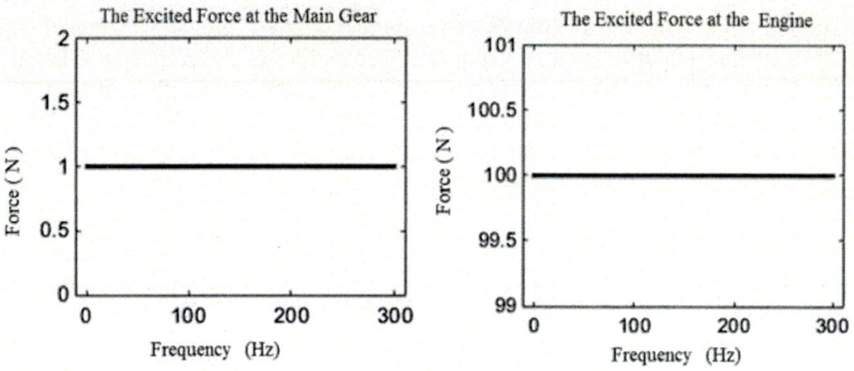

Fig. 3 Incentive force at the landing gear compartment and the engine

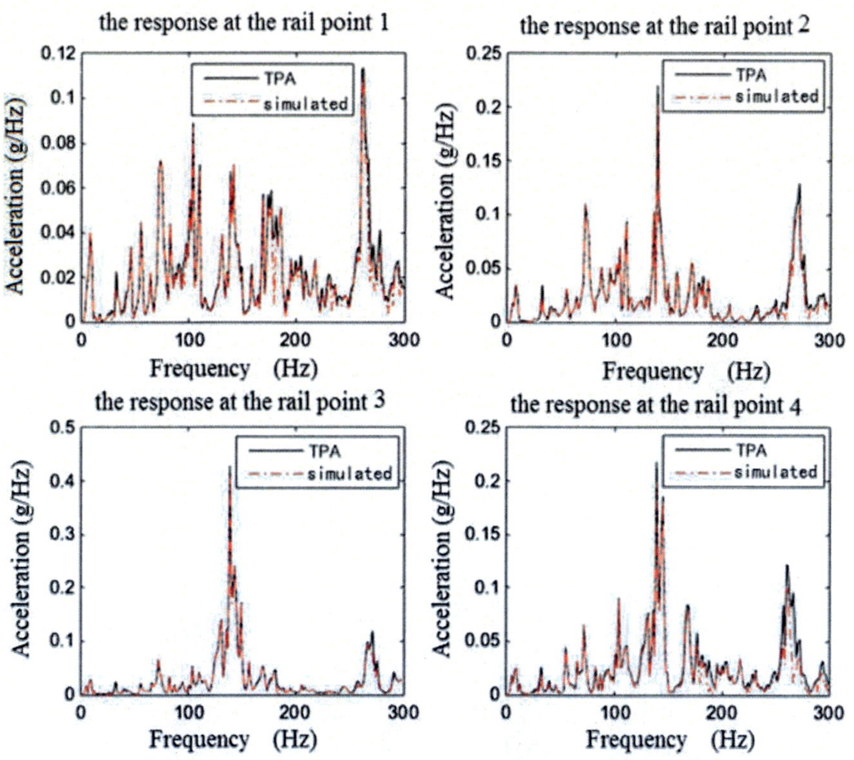

Fig. 4 Comparison of cabin floor vibration response in TPA with simulated value

analysis. Two incentives at the engine and the landing gear compartment are defined in Fig. 3.

The vibration response spectrum of the cabin floor based on the TPA model is calculated, which is shown in Fig. 4. Compared with the response spectrum calculated in the TPA model, the frequency domain distribution and amplitude is basically consistent with the results of the simulation software, and the error is within 5%. Therefore, it shows that the transmission path analysis model established here is reasonable, and the analysis based on this model has a certain reliability.

4.2 Analysis of the vibration response contribution

Figure 5 shows the comparison of the vertical vibration response at the cabin floor in the whole analysis band. It is learned from the figure that the vertical vibration response of the four rails on the floor is roughly at the same level, and the vertical vibration response at the rail 3 is slightly larger than the other three rails, which needs to pay attention.

The main frequency band of the aircraft structure is 0–300 Hz. At the frequency of 137 Hz, the vertical vibration response of the cabin floor guide rail 1, 2, 3 is large (see Fig. 6), especially the guide rail 2, at 137 Hz with a peak of 0.23 g. Furthermore, the vertical vibration response of the four guide rails is also large near the frequency 260 Hz.

It is known from Fig. 7 that the vibration response at the guide rail 2 is large at some frequency values, which need to study the contribution diagram of the two paths (Path-1, Path-2) to the total response from the engine to the cabin guide rail 2.

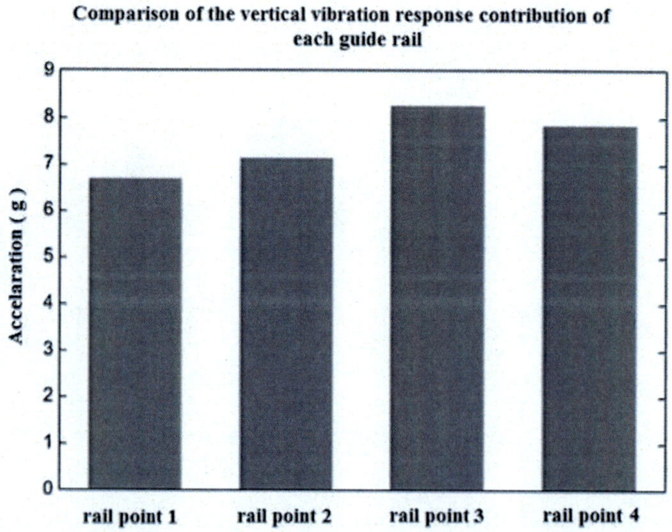

Fig. 5 Comparison of the vertical vibration response of each guide rail in the aircraft cabin

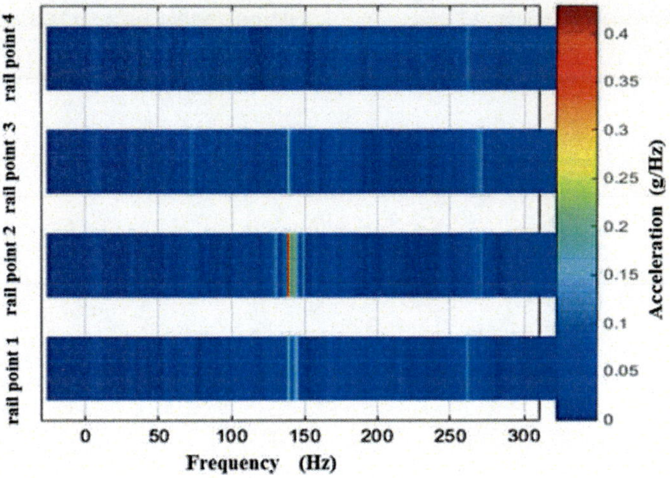

Fig. 6 Cloud map of the vertical vibration response of each guide rail in the aircraft cabin

Fig. 7 Schematic diagram of the effective contribution

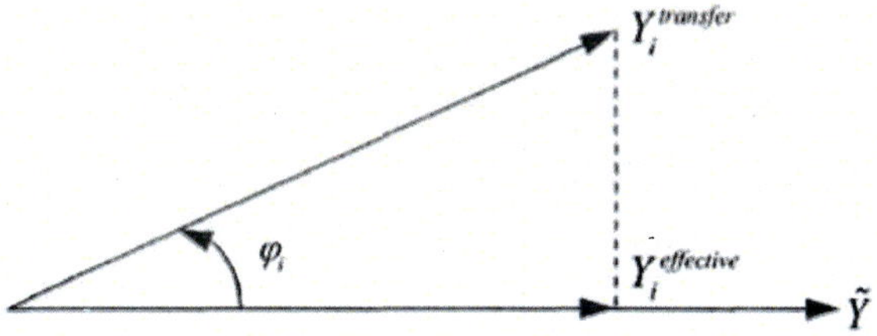

The vibration contribution at the guide rail 2 is sorted by two vibration transfer paths at 137, 262 Hz, as shown in Table 1.

The vibration contribution ranking in Table 1 is only based on the amplitude of vibration contribution, in fact the superposition of vibration is not simply amplitude subtraction. Since the vibration acceleration is a vector, the vector superposition of amplitude and phase should be considered, it is not enough to consider the amplitude only. In order to further study the vibration contribution of each path, the concept of effective vibration contribution is introduced. The vibration contribution amount of the respective Path-1, Path-2 is defined as $Y_1{}^{transfer}$ and $Y_2{}^{transfer}$. The total contribution is defined as $\tilde{Y}$.The effective vibration contribution of the ith path is defined as $Y_i{}^{effective}$ $(i = 1, 2, 3, 4)$, which is the projection of $Y_i{}^{transfer}$ for $\tilde{Y}$, as shown in Fig. 7.

Table 1 Each path vibration contribution amplitude order at 1 137 and 262 Hz

Frequency (Hz)	The sorted results
137	Path-2 > Path-1
262	Path-1 > Path-2

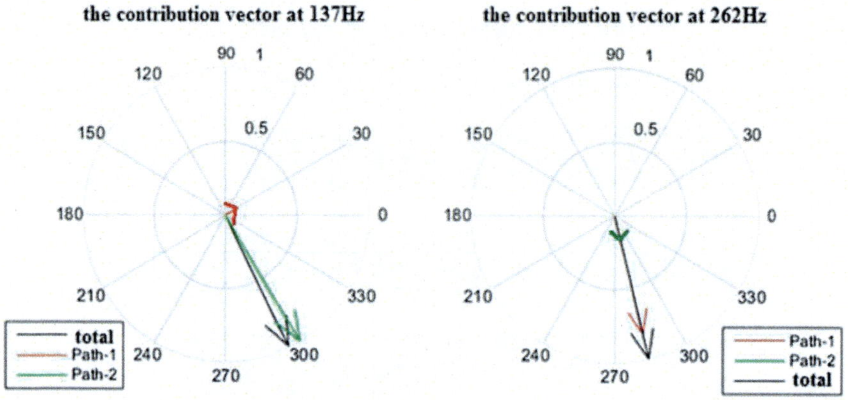

Fig. 8 Vibration response at guide rail 2 at 137 and 262 Hz

In Fig. 7, φ_i (i =1, 2, 3, 4) is the phase difference between $Y_i^{transfer}$ and $\tilde{Y}$, the effective vibration contribution of the ith path is:

$$Y_i^{effective} = \left|Y_i^{transfer}\right| cos\varphi_i (i = 1, 2, 3, 4) \tag{4}$$

It is seen from Fig. 8 that at the frequency of 137 Hz, the phase between the Path-2 vibration contribution and the total contribution are not much different, while the vibration contribution of the Path-1 is basically vertical to the total contribution vector diagram, that is, the total contribution is basically entirely contributed by the Path-2. At 262 Hz, the vibration contribution is in the same direction as the total contribution, and the Path-1 is greater.

4.3 *Transmission Path Sensitivity Analysis*

According to the transmission path analysis, both of the excitation and vibration path sensitivity will cause greater contribution, in the paper, the excitation force is artificial, so only the path sensitivity is compared. Figure 9 is the trend of each path sensitivity spectrum, the path sensitivity amplitude of h21 and h31, that is, the excitation force from the landing gear compartment to the rail 2 and rail 3, is larger than other transmission paths.

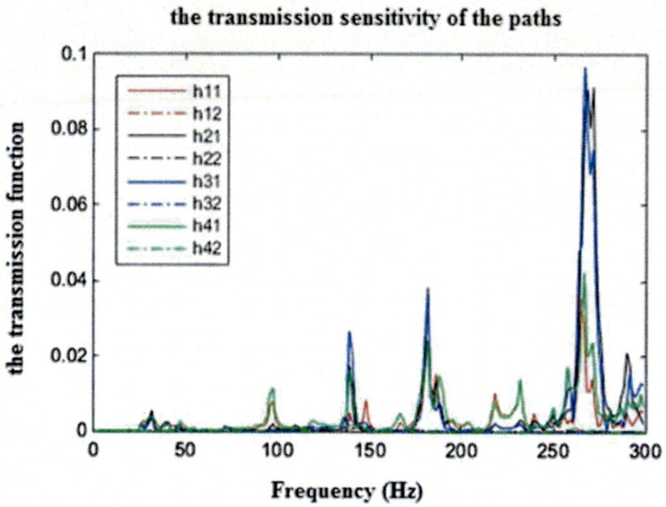

Fig. 9 The transmission sensitivity of the main contribution path

5 Conclusion

Based on MATLAB transmission path analysis method, a large passenger aircraft vibration transmission path analysis model, which defines the engine excitation and landing gear excitation as the excitation source, is established and verified by the finite element calculation. Based on the established model, the vibration contribution and sensitivity of the acceleration response at the four guide rails in the large passenger plane cabin are analyzed, which lays the foundation for the later analysis of the aircraft vibration transmission path under complex load.

References

1. Dubbaka KR, Zweng FJ, Haq SU (2003) Application of noise path target setting using the technique of transfer path analysis. SAE Technical Papers: SAE International
2. Gregor K (2003) Panel noise contribution analysis: an experimental method for determining the noise contributions of panels to an interior noise. SAE Technical Papers: SAE International
3. Dongming L, Dang X, Qing L et al (2007) Application of transmission path analysis technology in vehicle noise and vibration. Noise Vibr Control Manuf (4):73–77
4. Xiaoke P (2014) Study on excavator seat vibration based on working condition transmission path analysis. Shandong University, Jinan
5. Mule ZH (2011) Identification and improvement of steering disc under idle condition based on transmission path analysis. Automot Technol (4):15–21

Manufacturability and Deformation Performances of CFRP Twist Morphing Wing Structure with Applying the Electrodeposition Resin Molding Method

Kazuaki Katagiri, Daekwi Kim, Choong Sik Park, Sonomi Kawakita, Masato Tamayama, Koki Kayano, Shinya Honda, Katsuhiko Sasaki, and Makoto Yamazaki

Abstract To achieve efficient flight by minimizing energy consumption, a morphing wing that allows large and smooth deformation similar to a wing of a migratory bird is required. Previously, the authors have developed a migratory-sized unmanned aerial vehicle (UAV) with a twist morphing wing. The twist morphing wing was manufactured using a carbon fiber reinforced plastics (CFRP) with applying the electrodeposition resin molding (ERM) method, which was developed by authors. In the ERM method, resin impregnation proceeds in a liquid by electrophoresis. Thus, manufacturing process is very efficient because neither pressurization nor evacuation are necessary. On the other hand, although the molding of the twist morphing wing is efficient, an optimal design is important for the twist morphing wing to lower the torsional rigidity to reduce the driving force of the morphing and increase the bending rigidity to support aerodynamic loads. Therefore, in this study, the spar of

K. Katagiri (✉) · S. Kawakita
Research Division of Machining and Molding, Osaka Research Institute of Industrial Science and Technology, Osaka, Japan
e-mail: katagirika@tri-osaka.jp

D. Kim
Research and Development Department, Tokyo Metropolitan Industrial Technology Research Institute, Tokyo, Japan

C. S. Park
Research Division of Electronic and Mechanical Systems, Osaka Research Institute of Industrial Science and Technology, Osaka, Japan

M. Tamayama
Aeronautical Technology Directorate, Japan Aerospace Exploration Agency, Tokyo, Japan

K. Kayano · K. Sasaki
Graduate School of Engineering, Hokkaido University, Sapporo, Japan

S. Honda
Faculty of Engineering, Hokkaido University, Sapporo, Japan

M. Yamazaki
Rec Control, Co., Ltd., Osaka, Japan

S. Lee et al. (eds.), *The Proceedings of the 2021 Asia-Pacific International Symposium on Aerospace Technology (APISAT 2021), Volume 1*, Lecture Notes in Electrical Engineering 912, https://doi.org/10.1007/978-981-19-2689-1_67

the twist morphing wing was designed to be arranged in a triangle. Furthermore, the twist morphing wing structures were fabricated by the ERM method, and by a 3D printer for benchmarking. So, torsion and bending deformation properties of the twist morphing wing were confirmed under the aerodynamic loading condition, and validated by finite element analysis. As a result, it was confirmed that the manufacturing efficiency of the ERM method was superior to that of the 3D printer. Additionally, the torsional rigidity of the twisted morphing wing was almost the same, regardless of the manufacturing method. Furthermore, the bending rigidity of the specimen manufactured by the ERM method was higher than that of the 3D printer.

Keywords Electrodeposition · CFRP · Morphing wing structure · UAV

1 Introduction

In recent years, the number of drones or UAVs has been increasing [1]. Particularly, the drones with four rotors are used in many fields from children's toys to military applications. One of the important factors for UAVs is the battery performance, namely, they need high-output/large-capacity and lightweight [2]. However, for further extensions of flight distance and time, sole reliance on advances in battery performance should be avoided because it is more appropriate to efficiently fly using wings, instead of by driving rotors similar to a helicopter. In nature, many birds fly efficiently with their wings. Particularly, migratory birds have excellent flight efficiency. Historically, numerous studies on the flight of birds have been conducted and reported [3]. The morphing deformation of wings is the key to the excellent flight that allows birds to take off and land without runways, and to ascend or descend, freely. Hence, the development of morphing wings is expected to be indispensable [4–6]. Currently, taking Boeing 787 as an example, the application of CFRPs to aircraft is remarkable. However, although various structures such as corrugations and lattices were proposed, a CFRP morphing wing has not been practically applied [7, 8]. The design and manufacturing of a CFRP morphing wing are difficult because of the requirements of the flexibility, strength, and lightness, simultaneously.

Previously, the authors have developed a migratory-sized UAV with a twist morphing wing. The twist morphing wing was manufactured with a CFRP using the ERM method, which was developed by the authors [9]. In the ERM method, resin impregnation proceeds in a liquid. Therefore, manufacturing process is very efficient because neither pressurization nor evacuation are necessary. By immersing the carbon fibers in an electrodeposition solution containing a polymer with epoxy group and applying current, the resin is precipitated and impregnated between the carbon fibers. And, thermally cured, a CFRP with sufficient strength was fabricated. By setting a carbon fiber bundle in the electrodeposition solution, even a closed lattice CFRP structure was efficiently fabricated [10]. However, although the molding of the twist morphing wing was efficient, an optimal design is important for the twist morphing wing structure to lower the torsional rigidity as much as possible to reduce the driving force of the morphing and increase the bending rigidity to support aerodynamic loads.

In this study, to optimize the torsional and bending rigidity of the above twist morphing wing structure, the spars were designed to be arranged in a triangle. Furthermore, the specimens of the twist morphing wing were fabricated using the ERM method and a 3D printer for CFRPs, based on our previous research [11]. The specimen manufactured by the 3D printer was used for benchmarking to examine the manufacturing efficiency and deformation properties. Then, torsional properties of the twist morphing wings manufactured by using a 3D printer and the ERM method were confirmed. Additionally, the bending deformation properties were compared under the aerodynamic loading conditions, and were validated by the finite element analysis.

2 Method

2.1 Design of Skeletal Structure of the Twist Morphing Wing

In our previous research, a small UAV weighing 3 kg with a twist morphing wing was prototyped [11]. A brushless motor (HSL81501, Align Co. Ltd, Japan) that drives morphing was installed in the wing. Further, design of the twist morphing wing is required to reduce its torsional rigidity for a smooth transition to twist morphing while simultaneously maintaining flexural rigidity to sustain aerodynamic loads. Thus, the finite element analysis was conducted, and it was found that the delta arrangement of the spars is the most optimized shape to achieve the above.

Therefore, in this study, the skeletal structure of the twist morphing wing was redesigned and manufactured. The wing outline geometry of the twist morphing wing was assumed to be NACA 2415. The upward and downward morphing deformations are shown in Fig. 1 by broken and dotted lines, respectively. The spars were arranged in a triangular shape considering the bending rigidity and the twist morphing deformation. For installation in a UAV, one-third of each conventional wing was replaced by the morphing wing, and the surface of the wing was covered by a flexible rubber film.

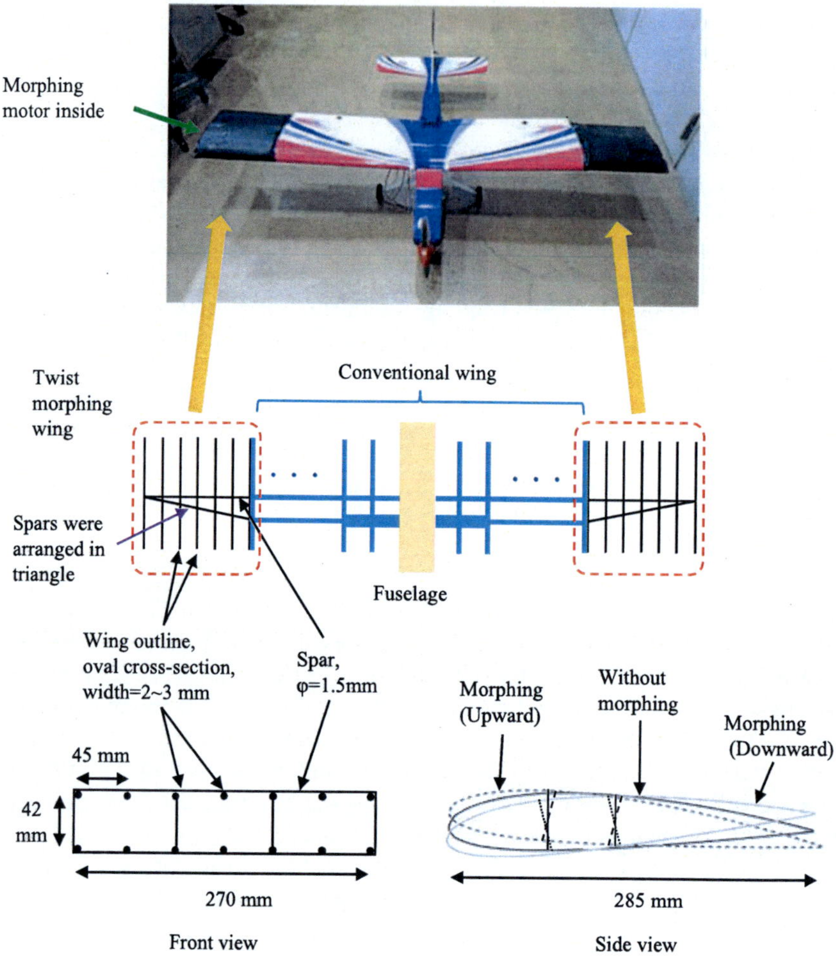

Fig. 1 Design of skeletal structure of the twist morphing wing

2.2 *Manufacturing Methods*

Two skeletal structures of the twist morphing wing, Specimen #1 and Specimen #2, were manufactured using a 3D printer for CFRPs (Mark Two, Mark Forged, USA) and the ERM method, respectively. In both methods, the wing outline and the spars were bonded by epoxy adhesives.

In the 3D printer for a benchmarking, a thermoplastic resin (nylon) was impregnated between continuous carbon fiber by heating the nozzle [12, 13]. Resin-impregnated carbon fibers were ejected from the nozzle to draw two-dimensional contours of the wing outline and the spars, respectively. Finally, the outer circumference of the CFRP (carbon fiber /thermoplastic resin) was wrapped in ONYX (mixture

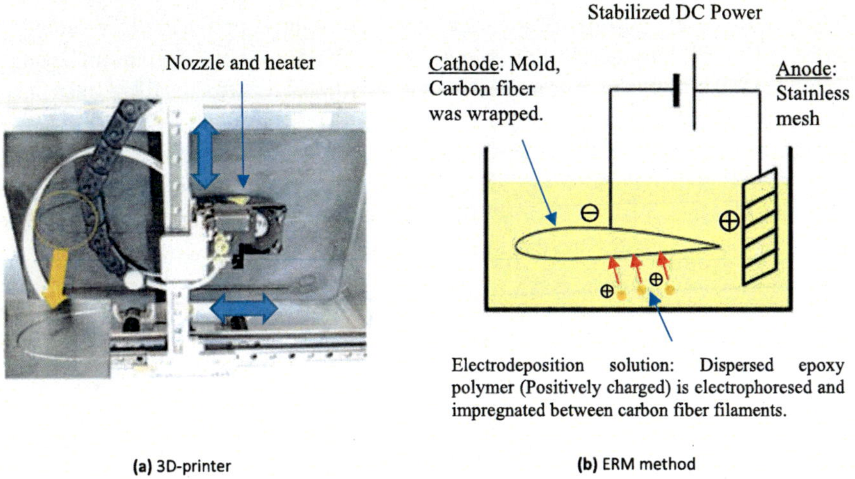

Fig. 2 Manufacturing methods of skeletal structure of the twist morphing wing. **a** 3D-printer. **b** ERM method

of short carbon fibers and nylon). The lamination thickness per layer was 0.125 mm, and 12 layers were laminated up to 1.5 mm.

In the ERM method, a carbon fiber bundle was fixed to the jig to mold the wing outlines and the spars. The jig with the carbon fiber bundle was immersed in an electrodeposition solution containing an epoxy resin [10, 11]. A schematic of the method is shown in Fig. 2. A cationic electrodeposition solution was used (Elecoat AR, Shimizu, Osaka, Japan). By applying the optimal current based on our previous research [14], the epoxy resin was impregnated between the carbon fiber filaments. As shown in Fig. 3, the energization condition is initially 0.1 A × 2 min, and subsequently

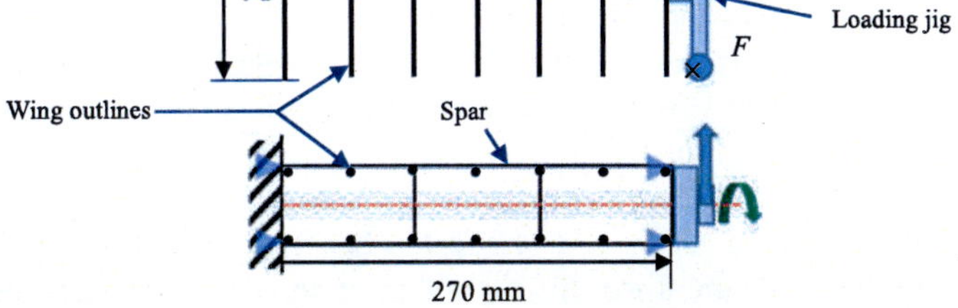

Fig. 3 Loading conditions of the torsion test for the skeletal structures of the morphing wing

0.2 A × 2 min. Following this, a 0.4 A current was constantly applied until the voltage reached to 125 V. Subsequently, this voltage (125 V) held constant. The current began to decrease, and it was turned off when it reached below 0.1 A. After the ERM, thermal curing was conducting (180 °C × 1 h, under vacuum).

Additionally, to confirm the elastic modulus, test pieces (length 75 mm, $\varphi =$ 1.5 mm) were prepared using the 3D printer and the ERM method. Tensile tests were conducted by using a desktop tension and compression testing machine (A&D MCT-2150, Japan), and the loading speed was 10 mm/min.

2.3 Torsion Test

For the developed UAV, the torque for the twist morphing was applied using the brushless motor inside the wing. To confirm the relationship between the torque and angle of the skeletal structures of the twist morphing wing, torsion test was carried out. As shown in Fig. 3, a torsion torque is applied around the elastic axis of the twist morphing wing using a loading jig. The jig was connected to a desktop tension and compression testing machine. By applying tension and compression loads, the torque could be controlled. The loading speed was 10 mm/min. The procedures were as follows. Prior to the torsion test, a balance test was conducted, which confirmed that the position of the elastic axis was approximately in the middle of the front and backward spars. First, a tension load was applied, twisting up to 27°. Subsequently, unloading was applied to downward. The direction of the load was reversed when a maximum of −21° twist was realized. Finally, the wing was twisted to a maximum of 27° again and then returned to the no load condition.

2.4 Bending Tests

Bending tests of the skeletal structures of the morphing wing were conducted to investigate the entire rigidity during a horizontal flight. On specimens #1 and #2, two types of loading conditions were applied to confirm the strength and deformation properties as follows.

First, based on our previous study [11], a concentrated load (F_a) was applied to the wing tip of the elastic axis to investigate the strength against the bending moment generated at the wing root on the main wing side, this is shown in Fig. 4a. The loading speed was 10 mm/s. Displacement of the wing tip was measured, and the load was applied up to a displacement of 25 mm.

Second, an aerodynamic load, corresponding to the distributed load during a horizontal flight, was applied discretely, and the bending deformation properties were investigated. During the flight, the lift acting on the wing was assumed to be parabolic, and discretized into two forces: f_1 and f_2; this is shown in Fig. 4b [15, 16]. The force (F_b) was loaded using the principle of leverage as follows. Namely,

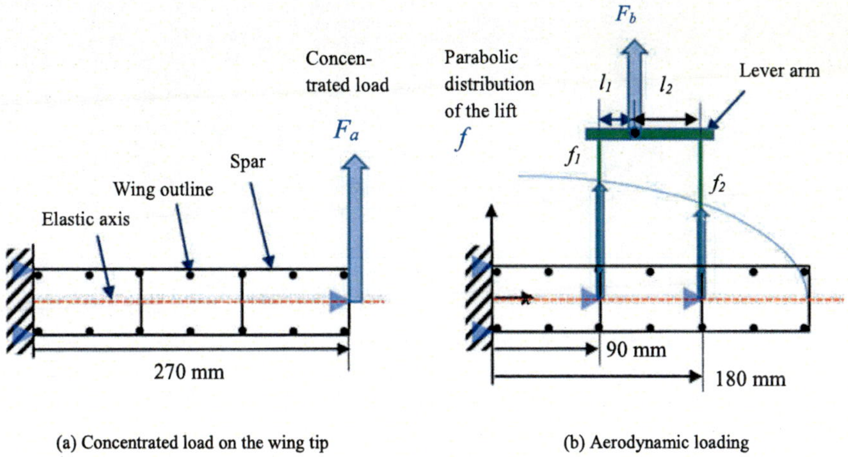

Fig. 4 Loading conditions o bending test for skeletal structures of the twist morphing wing. **a** Concentrated load on the wing tip. **b** Aerodynamic loading

the lift is given by using Eq. (1),

$$f = ax^2 + b \tag{1}$$

$$F_b = f_1 + f_2 \tag{2}$$

where f is the lift, x is the position coordinate, and a and b are constants. From the boundary conditions, $(x, f) = (90, f_1)$, $(180, f_2)$, and $(270, 0)$, a and b were determined as $a = -2.4 \times 10^{-5} \times F_b$, and $b = 7.3 \times 10^4 \times F_b$. Therefore, $f_2/f_1 = 0.63$, and the lever arm ratio $(R = l_2/l_1)$ is determined as 1.6. Actually, the load was applied to a screw that was fixed to the elastic axis by a wire. To confirm the effect of the fluctuation of the distributed load on the bending deformation, cases of $R = 1$ and 2 were also tested. The loading speed was 10 mm/s. The displacement of the fulcrum of the lever was measured, and the load was applied up to a displacement of 10 mm.

2.5 *Finite Element Analysis*

The finite element analysis was conducted to validate the bending tests of the skeletal structures of the morphing wing manufactured by the ERM method. Using the FEM analysis software ANSYS 19.0 (ANSYS Inc., USA), the finite element models shown in Fig. 5 were created, based on Figs. 1 and 4. The element type is a 3-node beam element. The sectional shape of a spar is a solid circle (φ = 1.5 mm), and the wing

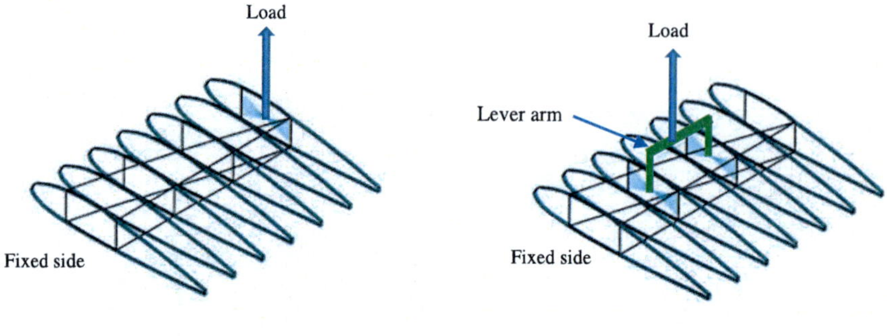

(a) Concentrated load on wing tip (b) Aerodynamic loading

Fig. 5 Finite element models. **a** Concentrated load on wing tip. **b** Aerodynamic loading

outline is a rectangle (W = 4 mm, H = 1 mm). The longitudinal elastic modulus (E_L) of the CFRP members was determined by the tensile test in Sect. 3.1. Two types of boundary conditions were considered; a concentrated load at the wing tip and an aerodynamic load as shown in Figs. 5a and b, respectively. The loads were applied at the point where the elastic axis penetrates the surface consisting of the wing outline. The loading direction was set to be vertically upward and free to the horizontal direction. Calculating the displacements, the relations between the load and the displacement of the specimens were simulated.

3 Results and Discussions

3.1 Skeletal Structures of the Twist Morphing Wing

The skeletal structures of the twist morphing wing fabricated as designed by using the 3D printer (specimen #1) and the ERM method (specimen #2) are shown in Fig. 6a and b, respectively. In case of the 3D printer, total manufacturing time was approximately 4–5 h. In case of the ERM method, the relation between current and voltage was shown in Fig. 7. The resistance was small before 300 s. But, it suddenly increased after 300 s. This is because that resin could be slowly precipitated, and be sufficiently impregnated between carbon fiber. Total manufacturing time was 1.5–2 h including thermal curing. It was confirmed that the manufacturing efficiency of the ERM method was superior to that of the 3D printer. Additionally, from the tensile test, the elastic modulus of the CFRP manufactured by the 3D printer and ERM method were approximately 23.6 GPa and 38.7 GPa, respectively.

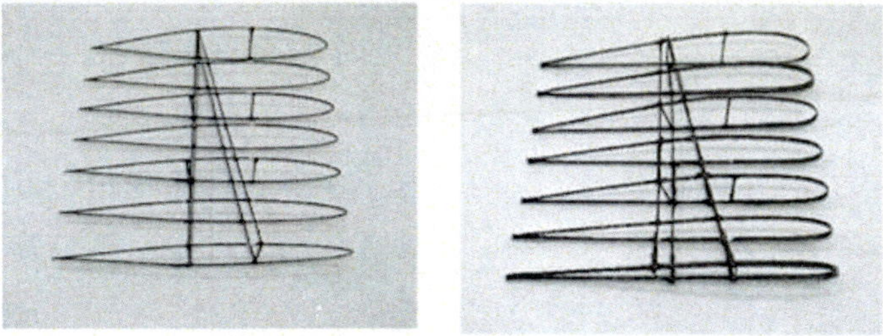

Fig. 6 Skeletal structures of the morphing wing. **a** Specimen #1. **b** Specimen #2

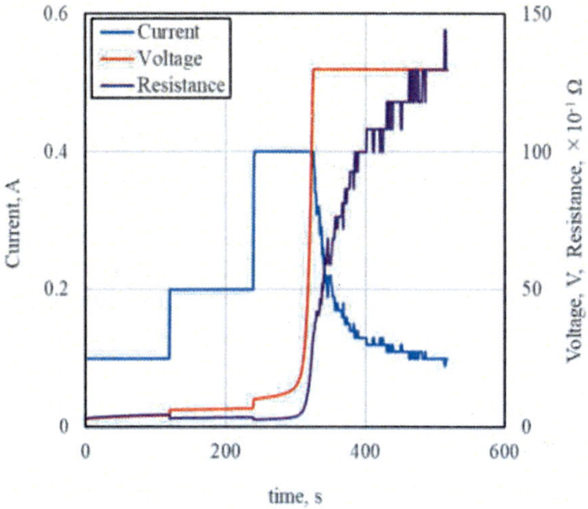

Fig. 7 Relationship between time, current, voltage and resistance under the ERM

3.2 *Torsional Properties*

Relationships between the torque and angle of the skeletal structures of the twist morphing wing are shown in Fig. 8. The deformation properties of specimen # 1 and specimen # 2 were almost the same, that is, the effects of the manufacturing method were seemed to be small. The torque was within the output range of the motor to be installed in the wing, therefore, twist morphing might be possible. However, since torsional morphing is driven by a motor, a non-linearity makes smooth deformation difficult. Nonlinear deformation properties were observed for both specimens, and was considerable for the twist morphing wing manufactured by 3D printer. For the specimen manufactured by the ERM method, upward deformation (deformation angle is positive) was linear although hysteresis was large. It was thought that the hysteresis could be due to local buckling.

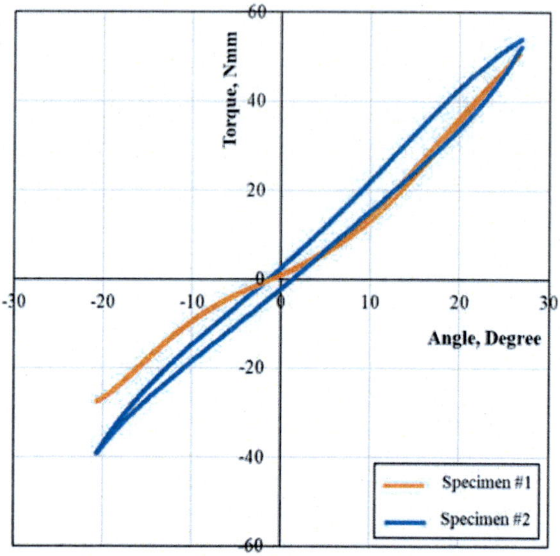

Fig. 8 Relationships between torque and torsional angle

3.3 Bending Deformation Under the Concentrated Load on Wing Tip

The relationships between the concentrated load and the displacement of the wing tip are shown in Fig. 9. The relationship for specimen #1 was nonlinear, whereas, that of specimen # 2 was almost linear. The finite element simulation was coincident with the experimental tendencies. The rigidity of Specimen #1 was lower than that of specimen #2. That is, the skeletal structure of the twist morphing wing with applying the ERM method was stiffer than that made by using a 3D printer. Substantially, the material properties of the component were affected. The maximum deformations of Specimens # 1 and # 2 are shown in Fig. 9. As indicated by the red line in the photograph of Fig. 9, little deformation was seen on the left side of specimen #1, but, it was deformed in the center and on the right side. However, Specimen #2 was almost uniformly deformed.

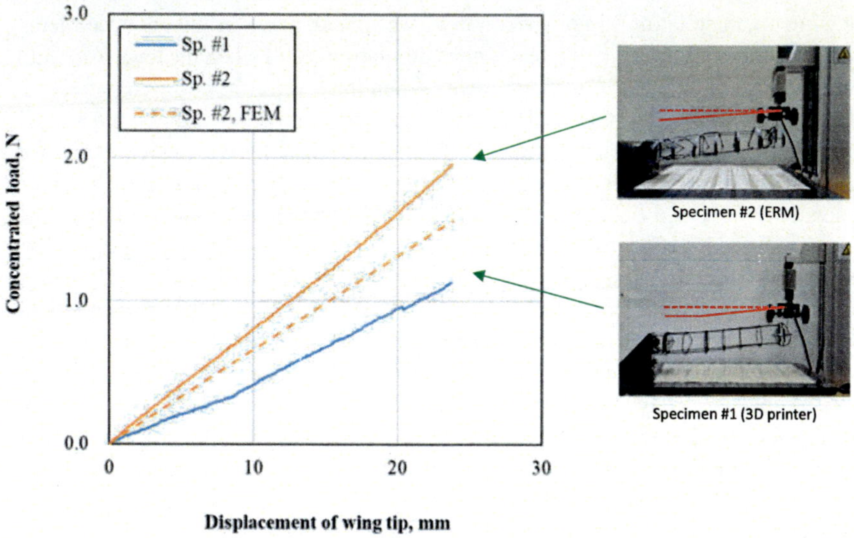

Fig. 9 Relationships between concentrated load and displacement of the wing tip, and deformations at the maximum loading

3.4 *Bending Deformation Under Aerodynamic Load*

Relationships between the aerodynamic load and displacement of the fulcrum of the lever are shown in Fig. 10. Unlike the case that the concentrated load was applied to

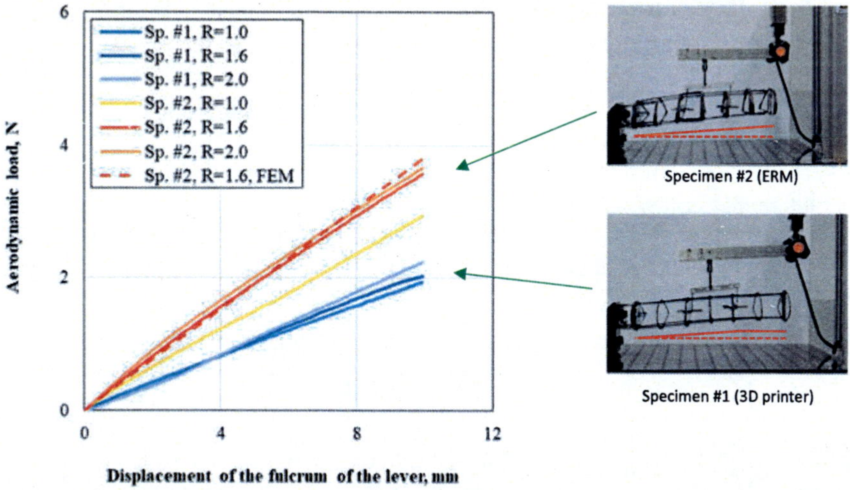

Fig. 10 Relationships between aerodynamic load and displacement of the fulcrum of the lever, and deformations at the maximum loading

the wing tip, the relationship of specimen #1 was linear, however, that of specimen # 2 was slightly nonlinear. The finite element simulation result when the lever arm ratio = 1.6 coincided well with the experiment. When the lever arm ratio was increased, the rigidity was also increased. The rigidity of specimen #1 manufactured by 3D printer was lower than that of specimen #2 manufactured by the ERM method for all lever arm ratios. When the lever arm ratio = 1.6, the maximum deformations of specimens # 1 and # 2 are shown in Fig. 10. As indicated by the red line in the photograph of Fig. 10, deformation of specimen #1 was seen in the left side, and the deformation from the center to the right side is small and almost horizontal. However, the deformations of specimen #2 were uniform. This might be due to the effects of material anisotropy and differences in geometry details.

Considering the span of the skeletal structure, the displacement of the fulcrum of the lever approximately corresponded to the half of that of the wing tip. Therefore, in case that the displacement of the fulcrum of the lever was 10 mm, the displacement of the wing tip could be assumed to be approximately 20 mm, therefore, the aerodynamic load might be almost twice of the concentrated load when the deformation geometry was same.

4 Conclusions

In this study, the manufacturability and deformation performances of a twist morphing wing structure with applying the ERM method were investigated. The findings obtained were as follows:

(1) The twist morphing wing structure could be fabricated by applying the ERM method. As a benchmarking, 3D printer was also adopted. Manufacturing efficiency of the ERM method was higher than that of the 3D printer.
(2) The spar of the twist morphing wing were designed to be arranged in a triangle to reduce torsional rigidity to drive morphing deformation by the motor, while increase bending rigidity to support aerodynamic loads.
(3) The torsional properties of the twist morphing wing structures manufactured by the 3D printer and ERM method were almost the same. Furthermore, the bending rigidity of the twist morphing wing manufactured by the ERM method was higher than that of the twist morphing wing manufactured by the 3D printer.

Acknowledgements This work was supported by Japan Society for the Promotion of Science KAKENHI Grant Numbers JP20H02039, and Izumi city (Osaka prefecture, Japan) research fund.

References

1. Nehme CE, Crandall JW, Cummings ML (2007) An operator function taxonomy for unmanned aerial vehicle missions. In: 12th international command and control research and technology symposium, pp 19–21
2. Dörfler S, Walus S, Locke J, Fotouhi A, Auger DJ, Shateri N et al (2021) Recent progress and emerging application areas for Lithium-Sulfur battery technology. Energ Technol 9(1):2000694
3. Lilienthal O (1889).Der Vogelflug als Grundlage der Fliegekunst
4. Li D, Zhao S, Da Ronch A, Xiang J, Drofelnik J, Li Y et al (2018) A review of modelling and analysis of morphing wings. Prog Aerosp Sci 100:46–62
5. Sun J et al (2016) Morphing aircraft based on smart materials and structures: a state-of-the-art review. J Intell Mater Syst Struct 27(17):2289–2312
6. Tsushima N, Tamayama M (2019) Recent researches on morphing aircraft technologies in Japan and other countries. Mech Eng Rev 6(2):19–00197
7. Yokozeki T, Takeda SI, Ogasawara T, Ishikawa T (2006) Mechanical properties of corrugated composites for candidate materials of flexible wing structures. Compos A Appl Sci Manuf 37:1578–1586
8. Jenett B et al (2017) Digital morphing wing: active wing shaping concept using composite lattice-based cellular structures. Soft Rob 4(1):33–48
9. Katagiri K, Sasaki K, Honda S, Nakashima H, Yamaguchi S et al (2017) CFRP manufacturing method using electrodeposition Resin molding for curvilinear fiber arrangements. Compos A 102:108–116
10. Katagiri K, Park CS et al (2020) Fabrication and deformation test of UAV morphing wing structure by CFRP applying electrodeposition resin impregnation method (in Japanese). In: Proceedings of 58th aircraft symposium, 3E07
11. Katagiri K, Yamaguchi S et al (2020) Fabrication of the twist morphing wing for the UAV by CFRP with applying the electrodeposition resin molding method. AIAA Scitech Forum 2020:0883
12. Dickson AN, Ross KA, Dowling DP (2018) Additive manufacturing of woven carbon fibre polymer composites. Compos Struct 206:637–643
13. Fasel U, Keidel D, Baumann L, Cavolina G, Eichenhofer M, Ermanni P (2020) Composite additive manufacturing of morphing aerospace structures. Manuf Lett 23:85–88
14. Sakamoto H (2005) Environment friendly electrolyzed activate deposition material with a high electric insulation (in Japanese). J Oleo Sci 5(10):489–496
15. Anderson D (2016) Fundamentals of aerodynamics, 5th ed. (translated by Yamaguchi H, Kashitani M)
16. Peery D (2017) Aircraft structures (translated by Taki T)

Doubly Coupled Antagonistic Shape Memory Alloy Actuator for Morphing Wing

Shingo Suzuki, Atsuhiko Senba, Tadashige Ikeda, Masato Tamayama, and Hitoshi Arizono

Abstract Shape memory alloy (SMA) wires have received considerable attention for use in actuators in morphing wings. This research presents an extension of our previous study on an antagonistic SMA wire actuator to a doubly coupled antagonistic SMA wire actuator. The actuator is composed of one main antagonistic system and two sub-systems. The experimental results demonstrated the potential of the proposed system to drastically reduce the wire length compared to a single antagonistic system.

Keywords Shape memory alloy · Morphing · Actuator · Wing

1 Introduction

In recent years, various environmental problems have developed in various areas, such as the automotive and aerospace industries. Owing to the increase in air traffic, the aircraft industry has focused on reducing its CO_2 emissions. In line with this, the improvement of flight operations and use of alternative aviation fuels instead of fossil fuels, such as plant-derived biofuels, have garnered attention.

To improve the fuel efficiency of aircraft, morphing wings, which differ from conventional flaps, are being actively studied in Japan and overseas [1]. This technology reduces the maintenance costs and inspection time by reducing the number of parts, and achieves an adjustable flap angle to reduce drag, thereby maximizing fuel efficiency. In addition, shape memory alloys (SMAs) are expected to reduce

S. Suzuki · A. Senba (✉)
Department of Vehicle and Mechanical Engineering, Meijo University, 1-501 Shiogamaguchi, Tenpaku-ku, Nagoya 468-8502, Japan
e-mail: senba@meijo-u.ac.jp

T. Ikeda
Department of Astronautics and Aeronautics, Chubu University, 1200 Matsumoto-cho, Kasugai, Aichi, Kasugai 487-8501, Japan

M. Tamayama · H. Arizono
Aeronautical Technology Directorate, Japan Aerospace Exploration Agency, 6-13-1 Ohsawa, Mitaka, Tokyo 181-0015, Japan

S. Lee et al. (eds.), *The Proceedings of the 2021 Asia-Pacific International Symposium on Aerospace Technology (APISAT 2021), Volume 1*, Lecture Notes in Electrical Engineering 912, https://doi.org/10.1007/978-981-19-2689-1_68

the weight of the actuator compared to conventional hydraulic actuator mechanisms because of their superior force generation and deformation per unit mass [2, 3].

Kojima et al. [4] investigated a simple antagonistic SMA wire actuator in a wing model by extending the SMA actuator constructed of Ti–Ni. In a wind tunnel test, the actuator achieved a target flap angle of $\pm 10°$ or more. [4] Ikeda et al. [5] demonstrated the energy saving of the antagonistic SMA wire actuator system using the hysteresis behavior of SMA wires. The total length of the SMA wire was found to be too long to achieve the required flap angle, thereby needing improvement for its compact integration into the wing.

Therefore, from the viewpoint of the compactness and lightness required to incorporate an actuator into a morphing wing, the driving mechanism of the original antagonistic SMA wires was revised; a doubly coupled antagonistic SMA actuator concept was proposed. This paper presents the experimental validation of the proposed concept.

2 Doubly Coupled Antagonistic SMA Wires

2.1 Single Antagonistic SMA Actuator

In this study, the SMA wire has a unidirectional and irreversible shape memory effect. After the wire is deformed at a temperature below the transformation temperature, the wire returns to its original shape when it is unloaded and subsequently heat is applied to it. Therefore, to use SMA wires as a two-way actuator, two pre-strained wires were set in antagonistic configurations, as shown in Fig. 1.

When SMA 1 is heated, the slider moves to the left, and SMA 2 is pulled in the low-temperature state. When the heating of SMA 1 is stopped, the elastic force of SMA 2 attempts to return to its original state; however, this is inhibited by the residual strain applied to SMA 2. A similar phenomenon occurs with the heating of SMA 2. Thus, antagonistic SMA wires can be used as two-way actuators.

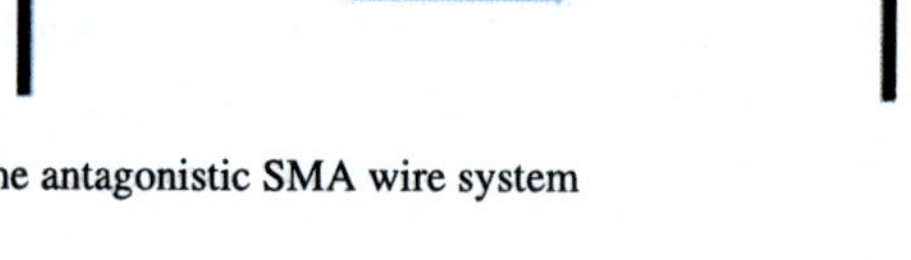

Fig. 1 Schematic of the antagonistic SMA wire system

2.2 Doubly Coupled Antagonistic SMA Actuator

In a single antagonistic SMA actuator, the driving force at the slider between two SMA wires can be calculated as: (contraction force of the wire in the hot state) − (pulling force of the wire in the cold state) = (driving force to move the slider). Based on this formula, the force of a single antagonistic SMA always decreases with the pulling force of the wire in the cold state. In contrast, the pulling force on the wire in the cold state can be reduced by fixing a sub-slider of the antagonistic actuator to the main slider, which is connected to the flap, instead of fixing both ends of the wire as in a single antagonistic type, as shown in Fig. 2a. Another sub-slider connected to the heated wire side of the main slider could also work like a rigid wall by heating the wire on the same side, thereby improving the stroke of

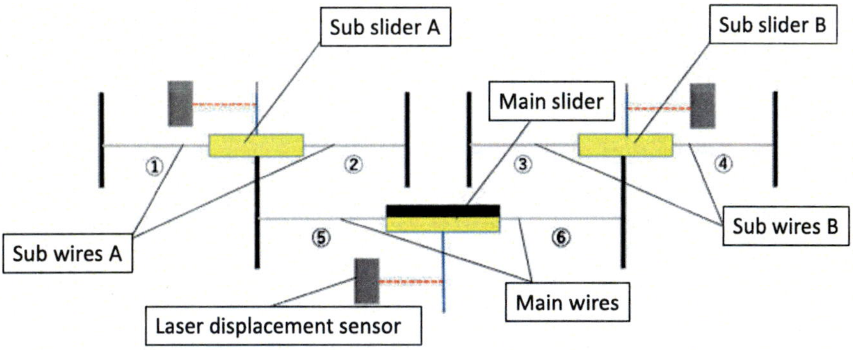

a)

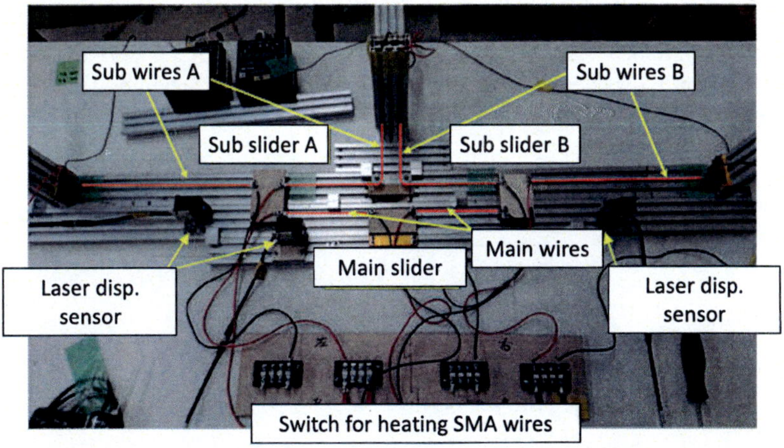

b)

Fig. 2 Doubly coupled antagonistic SMA wire system: **a** schematic of the system and **b** experimental setup the system. Consequently, the doubly coupled SMA actuator needs additional SMA wires as sub-systems. Nonetheless, the system reduces the maximum continuous SMA wire compared to a single antagonistic actuator by dividing the wire into main and sub-wires.

In Fig. 2a, the system includes one main slider and two sub-sliders. If the movement of the two sub-sliders is restricted, the system becomes a single antagonistic SMA wire. For the doubly coupled antagonistic system, the wires on the left side of each slider (①, ③, and ⑤ in Fig. 2a) are heated to drive the main slider to the left. Similarly, the wires on the right side of each slider (②, ④, and ⑥ in Fig. 2a) are heated to drive the main slider to the right. Owing to the hysteresis property, the slider does not return to its neutral position (zero point) before the next cycle after the current heating is completed. Therefore, the stroke at each cycle was determined as the difference between the slider positions before and after wire heating.

2.3 Experimental Methods

This study used an SMA wire with a diameter of 0.3 mm and transformation temperature $A_f = 85$ °C, manufactured by Yoshimi Corporation, was used. As shown in the schematic diagram in Fig. 2b), we constructed an experimental apparatus for the doubly coupled antagonistic actuator model. We measured the stroke of the slider without any external load to verify the validity of the proposed system. The length of the wire installed in the main system was 130 mm, and the length of the wire in the sub-part was varied, as shown in Table 1. Case (i) represents single antagonistic SMA wires. The other three cases were doubly coupled antagonistic wires. The displacement of each slider was measured using laser displacement sensors, as shown in Fig. 2.

The current applied was set to 1.1 A to increase the SMA temperature up to 110 °C, and heating was carried out for 30 s. To converge the actuation stroke, the wires were driven for 40 cycles (left → right → left → ...), as shown in Fig. 3.

Table 1 Length of the main and sub-wires for the four experimental cases

Case	Length of main wires (mm)	Length of sub-wires (mm)
(i)	130	NA
(ii)	130	130
(iii)	130	260
(iv)	130	390

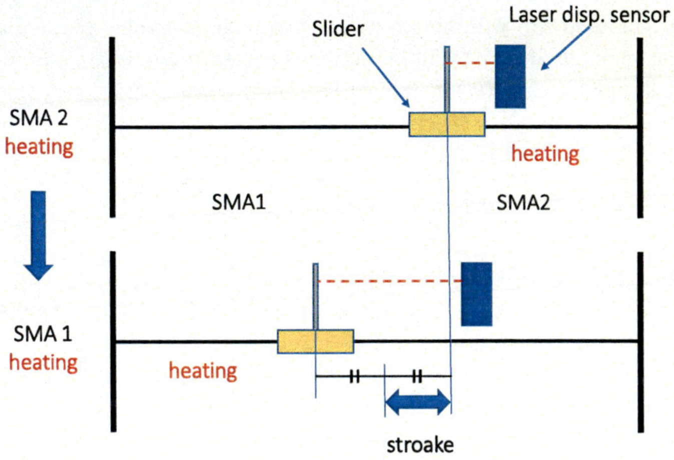

Fig. 3 Schematic of the stroke of the doubly coupled antagonistic SMA wire system

3 Results and Discussion

3.1 Experimental Results Without External Force

The experimental results are shown in Fig. 4. At the 40th cycle, the strokes were obtained to be 1.8, 2.8, and 5.4 mm with 130, 260, and 390 mm sub-wires, respectively. Meanwhile, the single antagonistic 130-mm wires obtained a stroke of 1.1 mm. Although there was no proportional relationship between the different sub-wire

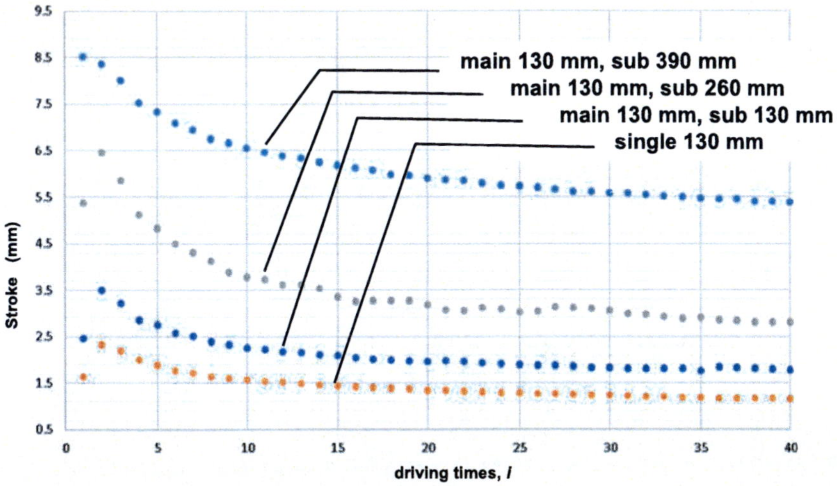

Fig. 4 Measured stroke versus driving time

lengths, the increased sub-wire length resulted in a larger stroke. If each wire for the single antagonistic SMA was 390 mm, the expected stroke is 3.3 mm, which is less than the stroke for Case (iv) in Table 1 (5.4 mm).

3.2 Experimental Results with External Force

Finally, to design an actuator to drive the flap, the relationship between the stroke and driving force is discussed. An experiment with external load was conducted based on the results obtained without load.

As shown in Fig. 5, linear springs were installed at both ends of the main slider. The springs were extended 10 mm from its natural length considering the range of the initial tension and maximum load of the spring. The spring constants for the experiments were $k = 1.408$ and 2.676 N/mm, which were selected to simulate the flexible flap used in previous experiments [6]. For comparison, three experimental patterns were measured based on the no-load drive results: 130 mm main and 260 mm sub, and 390 mm main and 130 mm sub doubly antagonistic SMA wire system; and 130 mm single antagonistic SMA wire system.

The experimental results are shown in Figs. 6 and 7. In Fig. 6, at the 40th cycle, the strokes were obtained as 2.5, 3.5, and 1 mm with 260 and 390 mm sub-wires, and single antagonistic 130 mm wires, respectively. In Fig. 7, at the 40th cycle, the strokes were obtained as 1.8, 2.1, and 130 mm with 260 and 390 mm sub-wires, and single antagonistic 130 mm wires, respectively.

From these results, the validity of the doubly coupled antagonistic SMA wires was confirmed. The total number of SMA wires for each case was not equal, but the main wire length was set to 130 mm for all cases. Future work will evaluate the actuator performance with constant total length of SMA wires.

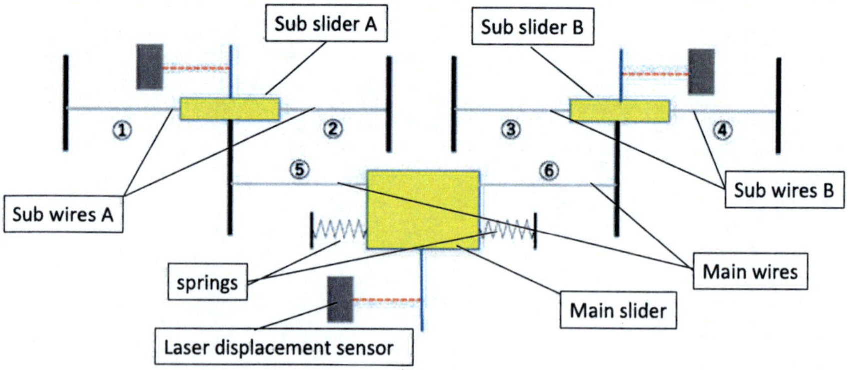

Fig. 5 Doubly coupled antagonistic SMA wire system with linear springs as a morphing flap

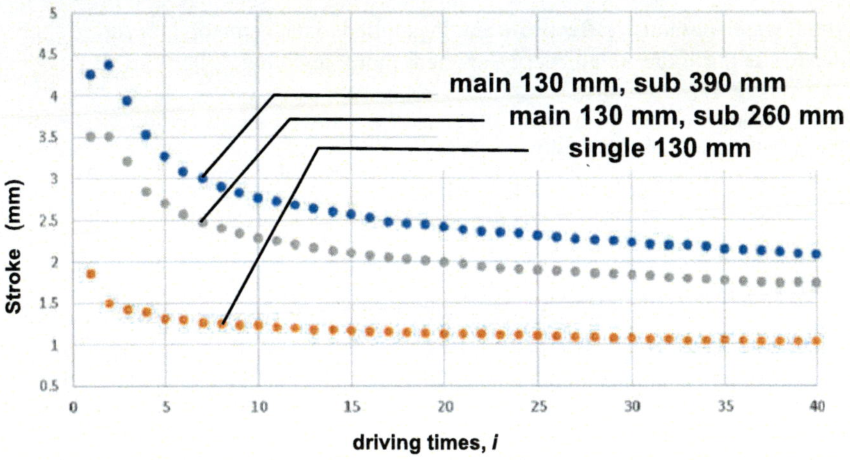

Fig. 6 Measured stroke versus driving times with linear spring #1 (k = 1.408 N/mm)

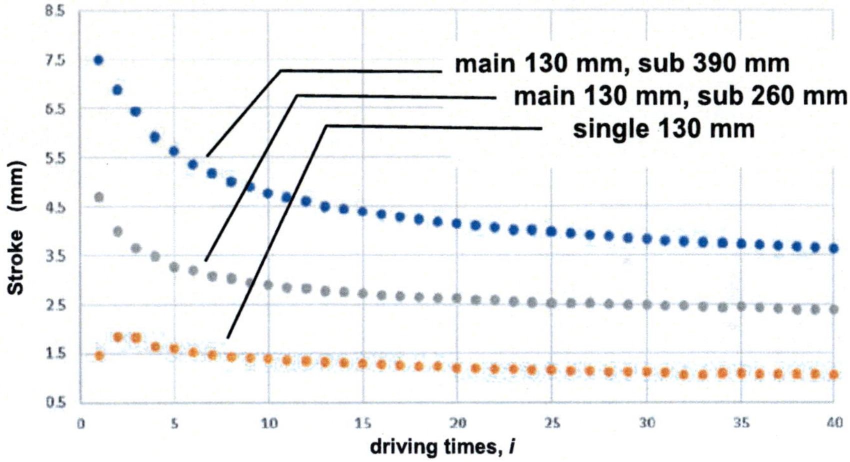

Fig. 7 Measured stroke versus driving times with linear spring #2 (k = 2.676 N/mm)

4 Conclusions

This research presents an extension of our previous study of an antagonistic SMA wire actuator to a doubly coupled antagonistic SMA wire actuator. The actuator has one main antagonistic system and two other sub-systems. The experimental results demonstrated that the proposed system worked well for different parameter settings, thereby clearly indicating the reduction of the SMA wire length. However, we have not yet investigated the comparison of the total length of SMA wires used in the new system, which obviously affects the energy applied to the actuator. Our future study will address this issue for an energy-efficient morphing wing design.

Acknowledgements This work was partially conducted under the contract of cooperative research at Meijo University, Chubu University, and JAXA.

References

1. Barbarino S, Bilgen O, Ajaj RM, Friswell MI, Inman DJ (2011) A review of morphing aircraft. J Intell Mater Syst Struct 22(9):823–877
2. Barbarino S et al (2014) A review on shape memory alloys with applications to morphing aircraft. Smart Mater Struct 23(6):063001
3. Kang WR, Kim EH, Jeong MS, Lee I, Ahn SM (2012) Morphing wing mechanism using an SMA wire actuator. Int J Aeronaut Space Sci 13(1):58–63
4. Kojima T, Ikeda T, Senba A, Tamayama M, Arizono H (2017) Wind tunnel test of morphing flap driven by shape memory alloy wires. Trans Jpn Soc Aeronaut Space Sci Aerosp Technol Jpn15(APISAT-2016):a75–a82
5. Ikeda T, Sawamura K, Senba A, Tamayama M (2018) Zero-electric-power position retaining utilizing hysteresis on an antagonistic shape memory alloy system. Trans JSME 84(866): 18-00101 (in Japanese)
6. Koga H, Senba A, Ikeda T, Kojima T, Tamayama M, Arizono H (2019) Development of morphing wing driven by antagonistic shape memory alloy wires. In: APISAT 2019: Asia Pacific international symposium on aerospace technology, pp 481–488. Gold Coast, Australia: Engineers Australia

Dynamical Analysis of Rotor Loads in Helicopter Circling Flight Condition

Xu Zhou, Xiayang Zhang, Bo Wang, and Qijun Zhao

Abstract In order to study the dynamic loads of a helicopter rotor during circling flight, the aerodynamic and structural dynamic models of the rotor are developed. In terms of aerodynamics, the free wake method based on relaxation iteration is adopted, wherein the wake model is distorted for circulating flight. With respect to structural dynamics, the Rayleigh–Ritz method is employed in mode superposition manner. The dynamic loads and properties of the rotor blade are calculated and analyzed. The accuracies of the modeling methods are verified by a numerical analysis and experimental comparison. Thereafter the influences of circling radii are emphasized. The distortion of wake changes the distribution of induced velocity and the aerodynamic load increases with a decrease of the circling radius. However, the dynamic loads share the same variation trend as the circulating radius.

Keywords Rotor · Free wake method · Mode superposition method · Helicopter circling flight · Dynamical loads

1 Introduction

As an important moving part, the main rotor, whose aerodynamic and structural performance play important roles in flight, has a strong influence on the maneuverability of the whole helicopter. Both aerodynamic and structural loads have significant differences from those in steady flight when the helicopter is circling, which is an important flight maneuver. Some potential physical phenomena should be considered when the aeroelastic characteristics in circling are studied [1].

When a helicopter is maneuvering, it is difficult to predict the aerodynamic load with high precision and to analyze the structural response of the blades. Battey [1] loosely coupled a hybrid wake rotorcraft CFD method to a comprehensive multibody dynamic model, and then calculated the aerodynamic loads in pull-up maneuvering and dive turn maneuvering. Xu [12] expanded the Pitt-Peters inflow model

X. Zhou · X. Zhang · B. Wang · Q. Zhao (✉)
National Key Laboratory of Rotorcraft Aeromechanics, Nanjing University of Aeronautics and Astronautics, Nanjing 210016, China
e-mail: zhaoqijun@nuaa.edu.cn

S. Lee et al. (eds.), *The Proceedings of the 2021 Asia-Pacific International Symposium on Aerospace Technology (APISAT 2021), Volume 1*, Lecture Notes in Electrical Engineering 912, https://doi.org/10.1007/978-981-19-2689-1_69

and Peters-He wake model. Afterwards, he analyzed the influence of wake distortion on the inflow of the rotor disc. This method was employed to calculate the aerodynamic loads of the rotor in pitching motion. In addition, the regularity of the inflow distribution was summarized. According to the established free wake method, Lv [7] analyzed the transient aerodynamic loads when the collective pitch overshoots, and explained the effect of an opposite vortex.

Dynamic characteristics analysis and vibration load prediction are important parts of a rotor performance analysis. Many researchers [4, 5, 9, 11] at home and abroad have established the dynamic equation of the blade. Among them, Ferhatoglu et al. [3] analyzed the nonlinear vibration of the beam according to the modal superposition method by using different vibration modes. The blade has infinite degrees of freedom. However, the aerodynamic load, whose frequency is an integer multiple of the rotating velocity, can be decomposed into different orders. As the order increases, the amplitude of the aerodynamic force decreases. Therefore, the response under low-order aerodynamic excitation is mainly considered during the analysis of the blade vibration. When the aerodynamic load changes, especially in cases where its frequency is close to the natural frequency of the blades, the structural response will change significantly.

The free wake method based on relaxation iteration [2, 6, 8, 10, 13, 16] is used to solve the flow field of the rotor and to calculate the aerodynamic performance of the rotor. The modal superposition method [14, 15], combined with the Rayleigh–Ritz method, is used to solve the blade vibration, which can transform the blade to a model with limited degrees of freedom.

2 Model Establishment and Solution Method

2.1 *Aerodynamic Model*

Wake is a major structure in the flow field. When the blades are rotating, the tip vortex quickly appears, which can influence the blade flow field, and in turn affect the aerodynamic load. The intensity of the vortex, the main research object of the free wake method, is represented by the vorticity equation of incompressible viscous flow.

$$\frac{\partial \boldsymbol{\omega}}{\partial t} + (\boldsymbol{u} \cdot \nabla)\boldsymbol{\omega} = \nu \cdot \nabla^2 \boldsymbol{\omega} \tag{1}$$

Combined with the definition of vortex, the passion equation can be obtained as follows.

$$\nabla^2 \mathbf{u} = -\nabla \times \boldsymbol{\omega} \tag{2}$$

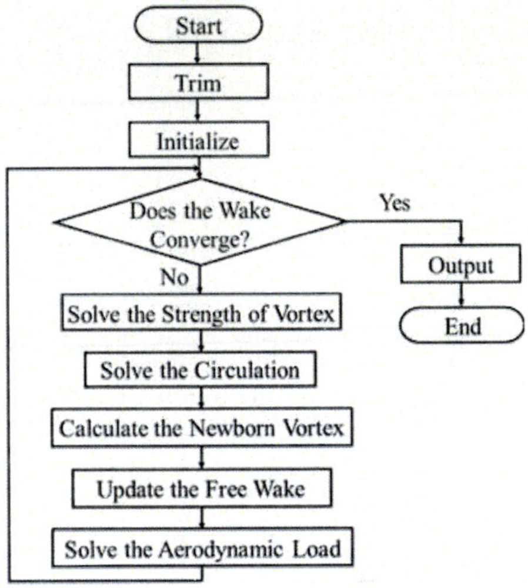

Fig. 1 Flow chart for solving the free wake

Induced velocity at any position in the whole flow field is solved by the Bio-Safar formula.

The Weissinger-L model, combined with the Beddoes model and Prandtl-Glauer compressibility correction factor, is employed to solve the aerodynamic load. With the movement of the rotor, the blades shed newborn vortices continually. After obtaining the distribution of the attached vortex on the blade in the spanwise position, the intensity of the newborn vortex can be solved according to the conservation of mass.

$$\gamma_{\omega} = -\frac{d\Gamma_b}{dt} + \nu_b \nabla \cdot \Gamma_b \tag{3}$$

where γ_{ω} represents the circulation of the newborn vortex; ν_b represents the relative velocity of air; and Γ_b represents the circulation at the blade profile. The positions of the newborn vortex are decided by control points on the blade surface.

A flow chart for solving the free wake is shown in Fig. 1.

2.2 *Modeling of Blade Dynamics*

The blade element is selected for a force analysis. The force in the flapping direction is shown in Fig. 2.

f consists of aerodynamic load and inertial force on the blade element. According to the balance of force and moment, the following equation can be obtained.

$$\left(EI_z y''\right)'' - \left(Ty'\right)' + m\ddot{y} = f_{aero} \tag{4}$$

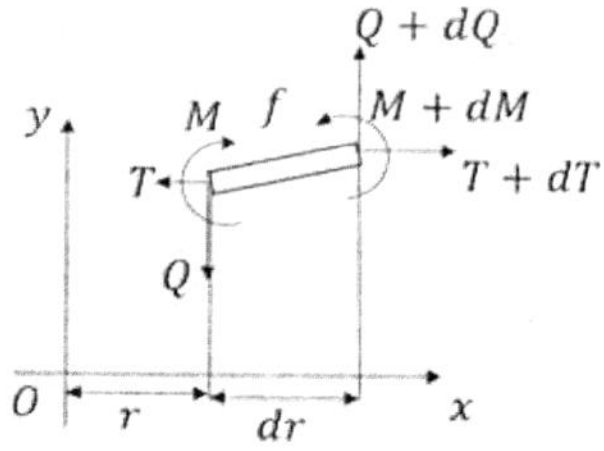

Fig. 2 Force analysis of blade element

The modal superposition method is used to solve the dynamic equations. The calculation result is as the following equation.

$$y(x,t) = \sum_{i=1}^{n} \gamma_i(x) q_i(t) \tag{5}$$

where $\gamma_i(x)$ is the ith natural vibration shape of the blade and $q_i(t)$ is the response corresponding to the ith-order modal shape. The approximate inherent characteristics of the blade are determined by using the general displacement function $\widetilde{\gamma}$ and the Rayleigh–Ritz method. The expressions of the mass matrix and stiffness matrix are as follows.

$$\begin{aligned} M_{ij} &= \int_e^R m\tilde{\gamma}_i\tilde{\gamma}_j dx \\ K_{ij} &= \int_e^R \left(EI_z\tilde{\gamma}_j''\right)''\tilde{\gamma}_i dx + \int_e^R \left(T\tilde{\gamma}_j'\right)'\tilde{\gamma}_i dx \end{aligned} \tag{6}$$

where e is the offset of the flapping hinge.

The displacement function should satisfy the boundary conditions of the hub. The actual rotation angle at the blade root is small, and therefore the boundary condition can be linearized. By solving the eigenvalues of the corresponding general displacement function, the eigenvalues are approximately equal to the natural frequencies of the blade. In addition, the natural vibration mode is as given below.

$$\gamma(x) = \Phi\widetilde{\gamma}(x) \tag{7}$$

where Φ is the matrix composed of eigenvectors.

The modal mass and modal stiffness can be obtained by repeating the same method. Therefore, a system with infinite degrees of freedom can be converted into a decoupling system under different modal coordinates. The expression of the model load is as follows.

$$F_i = \int_e^R \gamma_i f_{aero} dx \tag{8}$$

The thrust of the rotor has a strong periodicity and can be expanded by Fourier transform.

$$f_{aero}(x,t) = f_0(x) + \sum_{i=1}^{n} [f_{ic}(x)\cos(i\Omega t) + f_{is}(x)\sin(i\Omega t)], (i = 1, 2, 3 \cdots) \tag{9}$$

where Ω is the rotating speed of the rotor.

The response equation under different load excitations is as given below.

$$A_i = \frac{F_i}{K_{pi}(1 - \overline{\omega}_i^2)} \tag{10}$$

where K_{pi} is the stiffness of the ith model in the matrix of the model stiffness and $\overline{\omega}_i = \omega/\omega_i$ is the ratio of excitation force frequency to natural frequency. The response of the blade is obtained by adding the response of each order aerodynamic load. The structural load distribution of the blade is solved by the force integration method. A flow chart of the dynamic solution is shown in Fig. 3.

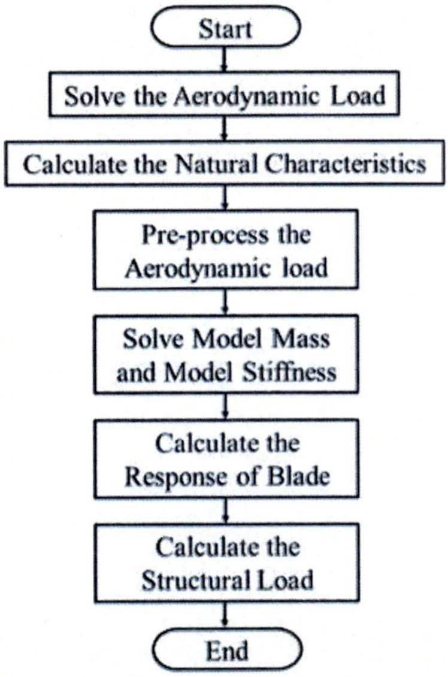

Fig. 3 Flow chart of dynamic solution

3 Numerical Analysis

3.1 Example Verification

On the basis of the mathematical method established in the previous section, the aerodynamic loads of UH-60 are calculated. Figure 4 depicts the normal force distribution at r/R = 0.775 when $\mu = 0.368$.

Compared with the rigid CFD method, the free wake method not only calculates the vortex structure in the flow field, but also introduces the rigid flapping motion of the rotor. It can be seen from the calculation results in the figure that the free wake method takes into consideration the influence of blade flapping motion on the profile angle of attack.

The aerodynamic force obtained by the free wake method is the excitation of the blade vibration, and its accuracy has a strong influence on the structural load of the blade. It is shown in Fig. 4 that there is a significant difference between the experimental results and the computation in the forward side of the rotor. Therefore, when the aerodynamic load is expanded by Fourier transform, the amplitude of the corresponding frequency will be different, resulting in a large error in the structural load. As a result, the inherent mode with large response in the process of blade motion should be selected as the main mode of rotor vibration. Moreover, after analyzing the order of load's frequency, the order that has larger aerodynamic amplitude should be selected as the main excitation force of rotor vibration.

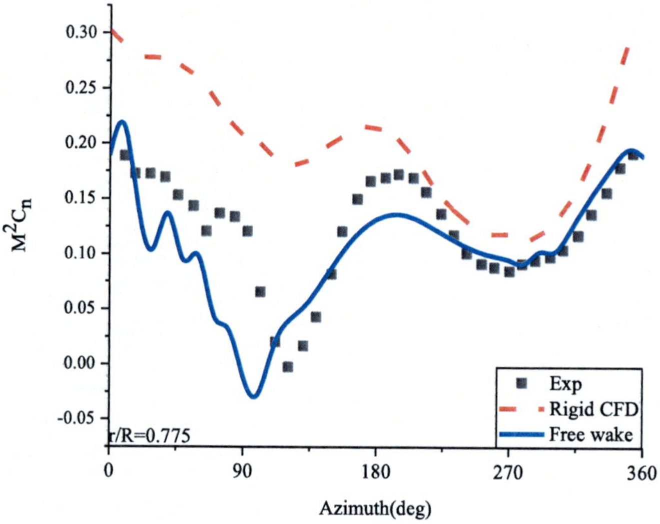

Fig. 4 The normal force distribution at r/R = 0.775

The inherent characteristics of the UH-60 blade are solved by using shape functions that satisfy the boundary conditions. The calculated inherent characteristics of the blade are presented in Table 1.

As seen in Table 1, the natural frequency calculation results of the four modes in the flapping direction have a certain estimation accuracy and small errors, which can be used for subsequent calculation of the blade structural load.

Natural modes of the calculated flapping direction are shown in Fig. 5.

The error that the natural vibration shape of the first three flapping modes is slight compared with the results calculated by the CSD method, but the natural mode of the fourth flapping mode has a larger error. The reason is that the linearization method is used in the root boundary condition processing, which affects the natural vibration shape of all blade sections. It is difficult for higher-order modes to match the actual natural vibration shapes. Therefore, the modal superposition method has poor fitting accuracy for high-order inherent characteristics. However, the aerodynamic force has a small amplitude at high-order frequencies. As a result, the response of high-order modes can be ignored.

Table 1 Natural frequency comparison of UH-60 helicopter blades

Flapping frequency	1/rev	2/rev	3/rev	4/rev
Experimental value/Ω	1.04	2.87	4.46	7.81
Computation/Ω	1.0397	2.7800	4.6228	7.8732
Error	−0.0288%	−3.1359%	3.6502%	0.8092%

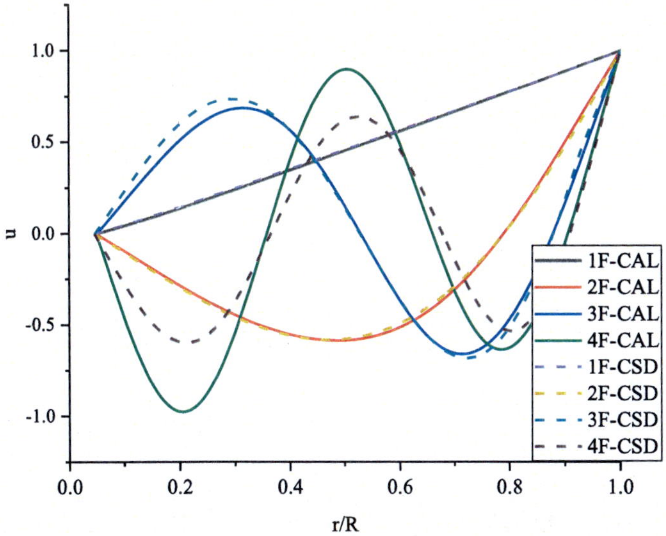

Fig. 5 Natural modes of UH-60 helicopters blades

In the analysis of the dynamic bending moment, a specific type of helicopter is selected for the example verification of the calculation. The parameters are given in Table 2.

According to the following formula, when $\mu = 0.27$, the dynamic bending moment of each section is calculated.

$$M_d = 0.5 * (M_{max} - M_{min}) \tag{11}$$

where M_{max} and M_{min} are the maximum and minimum bending moment in the circumferential direction. A comparison between the dynamic bending moment and experimental values is shown in Fig. 6.

The calculated results of this method are in good agreement with the experimental values. For the blade with an articulated hub configuration, the rigid body mode response is more obvious during the movement, and consequently the rotation angle of the root changes obviously. Due to the elastic constraint at the position of the flapping hinge, the value of the dynamic bending moment at the root position is

Table 2 The parameters of a specific helicopter

Number of blades	The radius of blade/m	Rotating speed/rpm	Chord length/m	Twist/°	The shape of blade tip
3	5.35	380	0.35	−12	Parabolic sweep

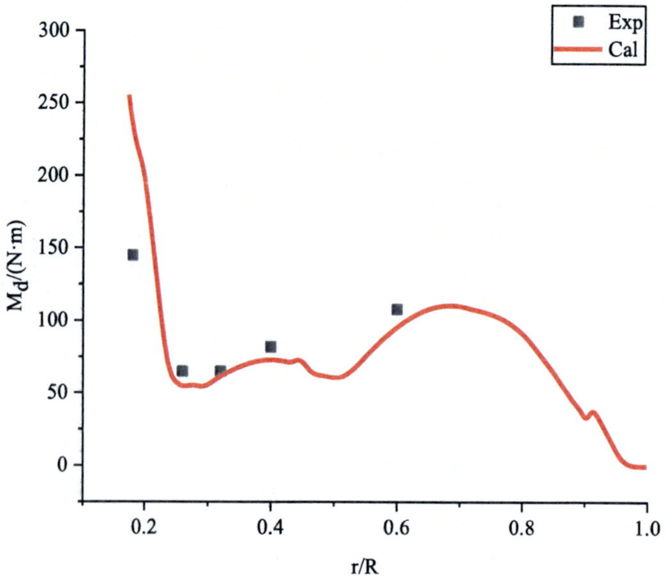

Fig. 6 Comparison of dynamic bending moment and experimental values

Table 3 Natural frequency of the blades

The order	1/rev	2/rev	3/rev	4/rec
The flapping frequency/Ω	1.024	2.718	4.907	8.064

relatively large. The aerodynamic load on the inside of the blade close to the root of the blade is small, as well as the vibration response and the dynamic bending moment. As the spanwise position gradually moves outward, the aerodynamic load gradually increases and the value of the dynamic bending moment also slowly rises. Since the tip of the rotor is free, the bending moment is always zero. Therefore, the distribution trend of the dynamic moment shown in the figure is formed.

3.2 *The Influence of Circling Radius on the Dynamic Load*

Before calculating the rotor response and structural loads, the inherent characteristics of the blades should be identified and they are shown in Table 3.

Natural vibration modes are presented in Fig. 7.

Figure 8 shows the flapping dynamic bending moment under different circling radii when $\mu = 0.1$. Figure 9 presents the circumferential distribution of the flapping bending moment at r/R = 0.9.

Under a large circling radius, the change of aerodynamic load with the radius is not obvious, leading to an inconspicuous change of the dynamic bending moment

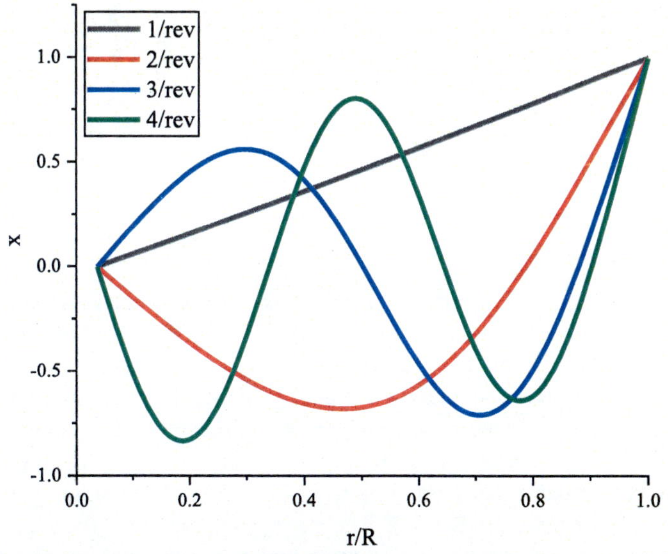

Fig. 7 Natural vibration modes in flapping direction

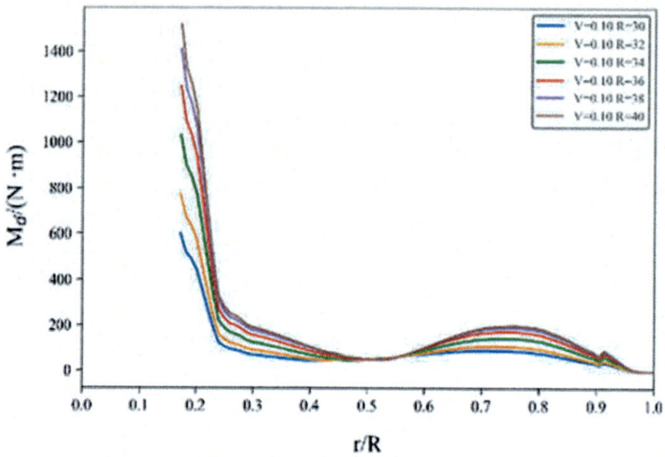

Fig. 8 Dynamic bending moment distribution under different circling radii

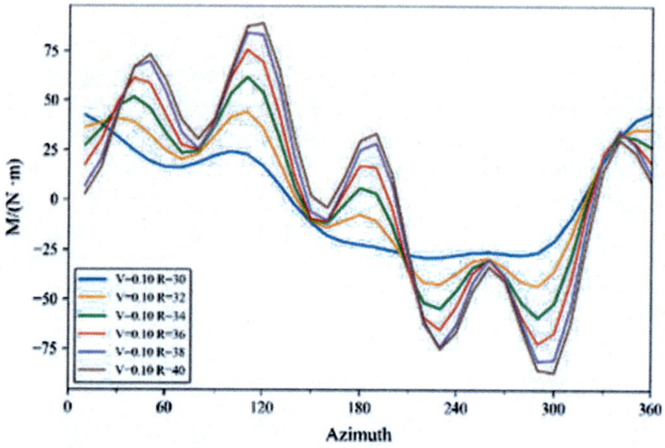

Fig. 9 Circumferential distribution of bending moments under different circling radii

distribution on the blade. When the circling radius decreases to a certain extent, the dynamic bending moment on the blade falls, and the gap between adjacent curves becomes increasingly larger. The distribution of the bending moment on the blade in the circumferential position tends to be flatter with a decrease of the spiral radius.

In order to analyze the reason for this, the aerodynamic load is calculated and the amplitude of each order is extracted. Figure 10 shows a comparison of normal force coefficients under different circling radii. Figure 11 shows aerodynamic load components of each order at r/R = 0.9. The static displacement of the blade caused by the steady component of the aerodynamic load is not the cause of the reciprocating motion of the blade. Since the amplitude of higher order aerodynamic load

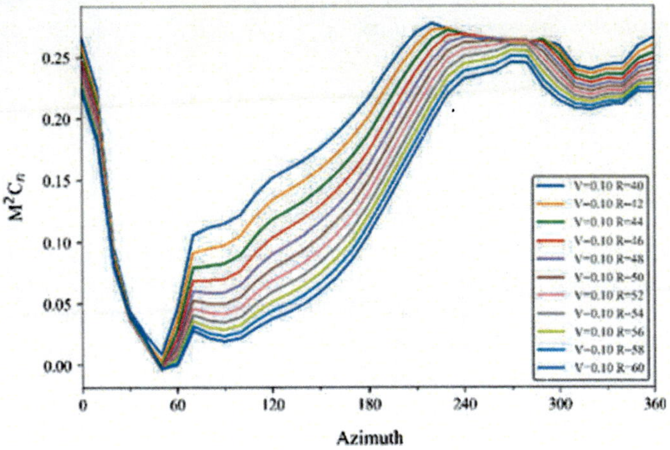

Fig. 10 Comparison of normal force coefficients under different circling radii

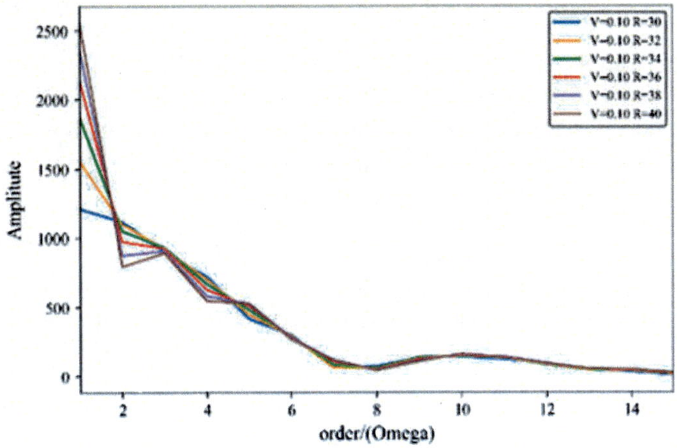

Fig. 11 Aerodynamic load components of each order

is low, the first fifteen order aerodynamic forces are selected as the main excitation load for the analysis. The first-order load is the largest part of the aerodynamic load. With a decrease of the circling radius, the amplitude of the first-order load declines significantly, which leads to a decrease of the bending moment response in a period of the blade. This is the main reason for the decrease of the dynamic bending moment. Although the other harmonic loads have changed, the amplitude changes little. Furthermore, the response phases caused by harmonic excitation forces of different orders are different, forming the bending moment distribution in the figure. The influence of wake distortion on the induced velocity at different phases of the rotor is also changed. In the aerodynamic load distribution, the phases of extreme points are gradually approaching in a period, which also shows the decrease of the first-order aerodynamic load.

4 Conclusion

In this paper, a free wake solution method based on the relaxation iteration method is established. Additionally, the accuracy of the algorithm is verified by the experimental values of the steady flight state. On this basis, considering the distortion of the wake in the flow field, the algorithm is extended to helicopter circling flight. In order to solve the rotor structural response, the modal superposition method is introduced to calculate the blade bending moment. Finally, the following conclusions are obtained.

1. In circling flight, the wake position transforms with a different circling radius, consequently bringing about a change of the induced velocity distribution in the flow field, together with the aerodynamic load and harmonic components of the thrust.
2. When the circling radius decreases, the distortion of the wake changes the amplitude of the aerodynamic load at different orders. Therefore, the circumferential distribution of the bending moment on the blade gradually becomes stable, and the dynamic bending moment decreases.

Acknowledgements This work is supported by the National Key Laboratory Foundation of China (no. 61422200101), the National Natural Science Foundation of China (nos. 11872211, 12032012 and 12102186), the Natural Science Foundation of Jiangsu Province (no. BK20200433), as well as the Priority Academic Program Development of Jiangsu Higher Education Institutions.

References

1. Battey LS (2018) A hybrid navier stokes/vortex particle wake methodology for modeling helicopter rotors in forward flight and maneuvers. Georgia Institute of Technology
2. Deng S, Jiang C, Wang Y et al (2018) Helicopter flight dynamics simulation with continues-time unsteady vortex lattice-free wake and multibody dynamics. In: Asia-Pacific international symposium on aerospace technology
3. Ferhatoglu E, Cigeroglu E, Nevzat zguven H (2018) A new model superposition method for nonlinear vibration analysis of structure using hybrid mode shapes. Mech Syst Sig Process 107(July):317–342
4. Hodges D (1974) Nonlinear equations of motion for the elastic bending and torsion of twisted nonuniform rotor blades. NASA TN D-7818
5. Houbolt B (1958) Differential equations of motion for combined flapwise bending, chordwise bending, and torsion of twisted nonuniform rotor blades. NASA Report 1346
6. Mckillip RM Jr, Wachspress DA, Tischler MB, Juhasz O (2017) A free wake linear inflow model extraction procedure for rotorcraft analysis

7. Lv WL (2013) Research on the stability, numerical method of rotor wake systems and the aerodynamic characteristics of ABC helicopters. Nanjing University of Aeronautics and Astronautics
8. Rand O, Khromov, V (2017) Free-wake–based dynamic inflow model for hover, forward, and maneuvering flight. J Am Helicopter Soc
9. Wang JY, Zhao QJ, Ma L et al (2015) High-precision prediction of unsteady aeroelastic load in helicopter rotor blade-vortex interaction state. J Aerospace Power 30(05):1267–1274
10. Wei P (2012) Research on viscous vortex numerical algorithm and its hybrid method for rotor unsteady flow field. Nanjing University of Aeronautics and Astronautics
11. Weng C (2014) The prediction and analysis of rotor vibration loads based on aeroelastic characteristics in the helicopter steady level flight condition. Nanjing University of Aeronautics and Astronautics
12. Xu J (2007) Research on rotor unsteady aerodynamics in helicopter large maneuver flight. Nanjing University of Aeronautics and Astronautics
13. Xue ZP, Li M, Jia HG et al (2015) Dynamic modeling of cantilever/quasi-cantilever elastic components based on modal method. Opt Precis Eng 23(8):2250–2257
14. Zhang XY, Zhang K, Zhao QJ et al (2020) Application analysis of Durbin method in solving dynamic response of damping beam. J Beijing Univ Aeronaut Astronaut 1–13
15. Zhang XY, Zhu M, Wu Z (2018) Analysis of vibration response of timoshenko beam based on K-V damping model. J Beijing Univ Aeronaut Astronaut 44(3):500–507
16. Zhao QJ, Xu GH (2006) Hybrid method of rotor flow field simulation based on Navier-Stokes equation/free wake/full potential equation. Acta Aerodynamica Sinica 24(1):15–21

Study on Laser Shock Peening Technology for Hole Position of Aviation Parts

Zhi Yang, Xiangyu Ding, and Sijie Ma

Abstract In engineering applications, the opening position is a typical stress concentration structure on aircraft components, which is prone to fatigue cracks under alternating loads. In order to study the effect of different power density and laser shock peening trajectory on the residual stress at the opening position of AL7050-T7451 alloy and improve the fatigue life of aviation parts. The results show that when multi-point impact peening is adopted, the peening effect is better when the power density is 3 GW/cm^2, and the Mises average stress value is 178 MPa. When single-point impact peening is adopted, the Mises stress values under different power densities have little difference, and the power density has little effect on the impact peening effect. The single-point impact peening with small power density is the best, about 3 GW/cm^2. By comparing multi-point and single point, when the laser power density is reduced, the peening effect can be improved by changing the process. The research results have certain reference significance for the opening position of laser shock peened aviation parts, and lay the foundation for improving the fatigue life of laser shock peened hole structure.

Keywords Laser shock peening · Al7050–T7451 alloy · Numerical simulation · Residual stress

1 Introduction

Al7050-T7451 aluminum alloy is widely used in aviation due to its low density, good mechanical properties and high corrosion resistance [1]. In engineering applications, the fastener hole is a typical stress concentration structural detail on aircraft components, and it is easy to produce fatigue cracks under alternating loads [2]. Laser shock peening is a new surface peening technology, which has the characteristics of high pressure, high energy and ultrahigh strain rate. It can introduce deeper

Z. Yang · X. Ding (✉) · S. Ma
School of Aircraft Engineering, Nanchang Hangkong University, Nanchang 360100, China
e-mail: dingxiangyu@nchu.edu.cn

X. Ding
Fundamentals Department, Air Force Engineering University, Xi'an 710038, China

S. Lee et al. (eds.), *The Proceedings of the 2021 Asia-Pacific International Symposium on Aerospace Technology (APISAT 2021), Volume 1*, Lecture Notes in Electrical Engineering 912, https://doi.org/10.1007/978-981-19-2689-1_70

residual compressive stress layer and produce dislocation slip and grain refinement in the surface layer [3], so as to improve the mechanical properties of the material and improve its fatigue life [4].

At present, a large number of studies have been conducted on aluminum alloy hole parts peened by laser shock in China and abroad. Zhang Xingquan [5] discussed the distribution of residual stress field induced by laser spot at different center distance by using experimental and numerical simulation methods, studied the influence of different overlap ratio on the distribution of residual stress, and the distribution characteristics of surface residual stress under two different loading sequences. Ren [6] studied the effect of residual stress on fatigue behavior and crack propagation. The influence of compressive stress on the fatigue behavior and crack propagation of holes was revealed. Yang [7] characterized the fatigue crack initiation life and fatigue crack growth rate of aluminum alloy with different notch shapes by experimental research, and compared with untreated materials. Ren [8] studied the effect of laser shock peening on fatigue crack initiation and propagation of 7050-T7451 aluminum alloy. Ivetic [9] and other experiments have studied the effect of operation sequence on the effectiveness of laser shock processing in improving the fatigue performance of perforated aluminum specimens. However, domestic and foreign researchers rarely carry out the influence of laser power and single point, multi-point impact peening process on the distribution of residual stress. Therefore, this paper mainly analyzes the residual stress distribution of Al7050-T7451 aluminum alloy hole under different laser power and single point and multi-point impact peening by numerical simulation.

In this paper, Al7050-T7451 aluminum alloy was taken as the research object, and its peening treatment was carried out. The finite element model was used to numerically simulate and analyze the laser shock peening hole parts. The effects of laser power and single-point and multi-point impact peening on residual stress distribution were investigated. The research results have certain reference significance for laser power selection of impact peened Al7050-T7451 aluminum alloy hole parts.

2 Materials and Methods

2.1 Material Properties

Al7050-T7451 aluminum alloy is often used on aircraft due to its high strength and machinability. The chemical composition of Al7050-T7451 aluminum alloy is: Si $\leq$ 0.12%, Fe $\leq$ 0.15%, 2.0% $\leq$ Cu $\leq$ 2.6%, Mn $\leq$ 0.10%, 1.9% $\leq$ Mg $\leq$ 2.6%, Cr $\leq$ 0.04%, 0.08% $\leq$ Zr $\leq$ 0.15%, 5.7% $\leq$ Zn $\leq$ 6.7%, Ti $\leq$ 0.06%, the other is Al [8]. Its mechanical properties are shown in Table 1.

Table 1 Mechanical property of Al7050–T7451alloy

Material	Density kg/m^3	Elastic modulus GPa	Poisson ratio
Al7050-T7451	2820	70	0.33

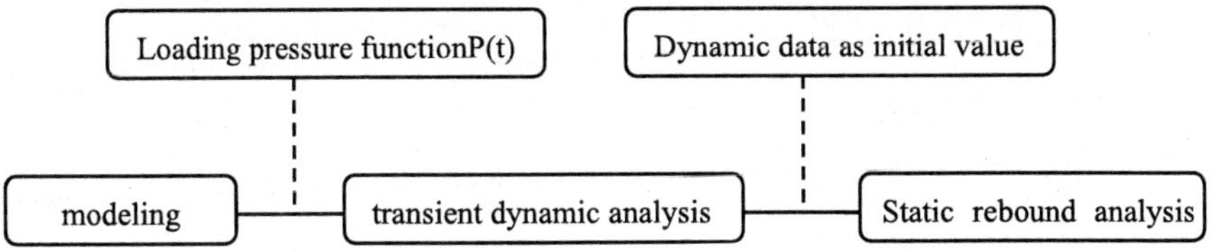

Fig. 1 Simulation process of LSP

2.2 *Experimental Method*

This paper carries out numerical simulation, uses Python language to write scripts, and then uses subroutine loading to realize custom loading. The simulation flow chart is shown in Fig. 1. In the numerical simulation of laser shock peening, although the dynamic calculation of shock wave is an important part of the simulation of laser shock peening, during the propagation of shock wave, the stress wave will reflect and interact within the material, so the response of shock wave on the material will last for a long time to reach stability. So only static analysis can get stable residual stress field. Therefore, two analysis steps are set up: the first step is to establish the dynamic process of shock wave propagation in the material and its interaction; the second step is to establish the static rebound process of residual compressive stress field formed in the material after shock wave. Among them, the first step adopts explicit dynamic analysis, and the second step requires implicit static analysis in the process of material stress rebound and residual stress field stability.

Different laser power density were selected as 3 GW/cm^2, 5 GW/cm^2, 7 GW/cm^2. Single-point and multi-point laser shock peening simulation were carried out under different laser power.

3 Numerical Model

3.1 *Modeling*

The model was established by Rhino software, and the impact peening model was imported into the simulation software. The model size was 1/4 of 30 mm × 30 mm × 5 mm. The diameter of the hole is 7 mm and the chamfer is 1 mm × 45°. Using YS0805-R200A laser shock peening equipment of Xi ' an Tianruida Company, laser

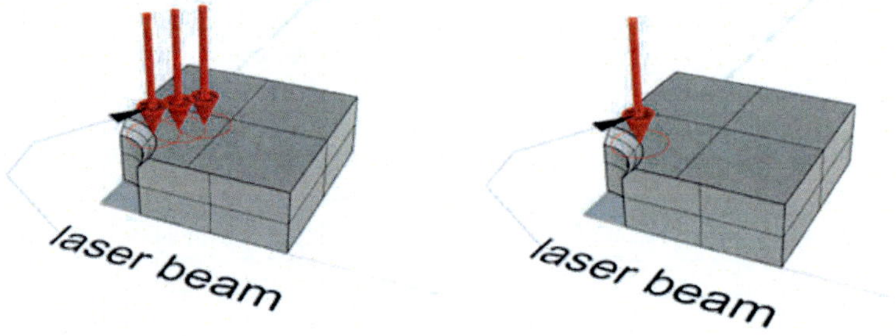

Fig. 2 Shock paths

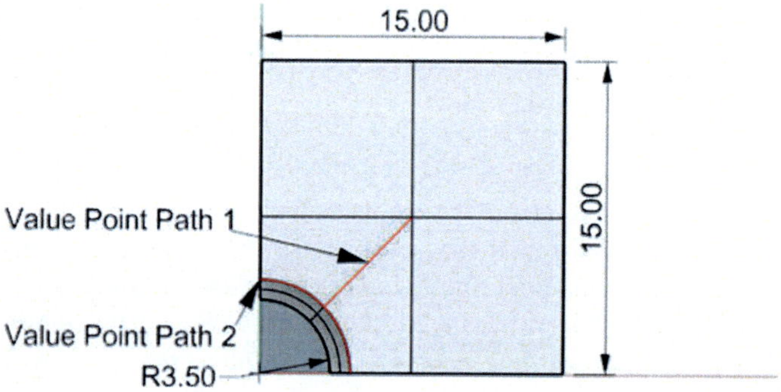

Fig. 3 Model size and path selection

energy E = 8 J, pulse width 18 ~ 20 ns, spot diameter is calculated according to power density. The numerical model is meshed by C3D8R grid. The boundary conditions on both sides of the outer side of the model are set to a fixed constraint, and the symmetrical boundary conditions are adopted on both sides of the hole. Shock trajectory as shown in Fig. 2, spot overlap rate is 50%. The residual stress extraction path is shown in Fig. 3.

3.2 Laser Shock Model

The peak pressure estimation formula of laser shock wave is:

$$P(t) = 0.01\sqrt{\frac{a}{2a+3}}\sqrt{Z}\sqrt{\lambda I_0} \tag{1}$$

where α is the interaction coefficient between plasma and target, λ is the energy absorption rate, Z is the equivalent acoustic impedance,

Table 2 Simulation experiment parameters

Power density (GW/cm^2)	3	5	7
Spot diameter (mm)	4	3	2.7
Peak pressure (GPa)	2.18	2.81	3.33

$$Z = \frac{2Z_w Z_t}{Z_w + Z_t} \tag{2}$$

$\mathbf{Z_w}$ $\mathbf{Z_t}$ is the acoustic impedance of the water constraint layer and the target, respectively. The constraint layer used in this paper is water, and the absorption layer is aluminum foil. α and λ are 0.25 and 0.7 [10], respectively.

$$Z_w = 0.165 \times 10^6 g \cdot cm^{-2} \bullet s^{-1} \tag{3}$$

$$Z_t = 3.950 \times 10^6 g \cdot cm^{-2} \bullet s^{-1} \tag{4}$$

$\mathbf{I_0}$ is the average power density of laser.

The calculation formula of spot diameter is:

$$S = \frac{E}{t \bullet I_0} = \frac{\pi}{4} \cdot d^2 \tag{5}$$

where **E** – energy, **t** – pulse width, $\mathbf{I_0}$ – laser power density.

Thus, the peak laser shock pressure and spot diameter at the corresponding laser power density can be calculated. The corresponding simulation experiment parameters are obtained, as shown in Table 2.

The half-width-height time length of the laser shock pressure loading curve is 2 – 3 times of the laser pulse width, so the half-width-height time length of the laser shock pressure loading curve is set to be about 40 ns, as shown in Fig. 4.

Fig. 4 Pressure versus time curve

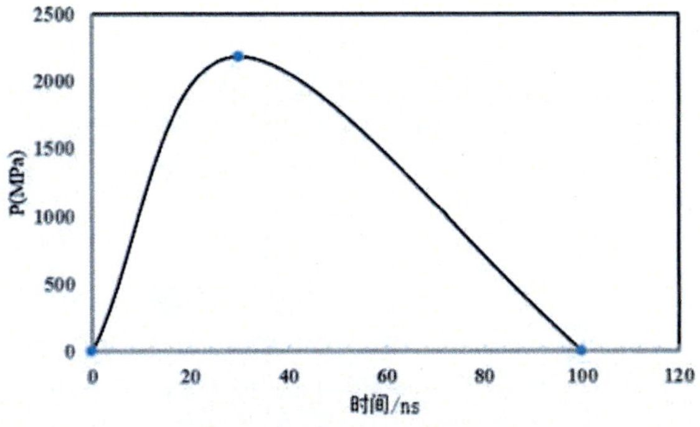

Table 3 Constitutive model parameter of J-C

Material	A/MPa	B/MPa	n	m	C
Al7050-T7451	435.7	534.624	0.504	0.97	0.019

3.3 *Material Constitutive Model*

The strain rate is as high as above in the process of laser shock peening, so this paper uses the J-C constitutive model to analyze the dynamic response process. The J-C constitutive model parameters of Al7050-T7451 material are shown in Table 3. where A is the yield strength of the material, B is the working hardening modulus, n is the hardening coefficient, C reflects the strain rate hardening effect of the material, T_0 is room temperature, T_m is the melting point of the material, and m reflects the temperature softening effect of the material. The melting point is 833 K and room temperature is 298 K.

4 Result and Discussion

4.1 *Multipoint Impact Peening at Different Power Density*

For the laser average power densities of 3 GW/cm^2, 5 GW/cm^2 and 7 GW/cm^2, the laser shock peening simulation is carried out according to the above. As shown in Fig. 5, the Mises stress distribution on the stress extraction path 1 is shown. It can be seen that the front end is close to the end of the hole wall, and the stress values at the laser power densities of 5 GW/cm^2 and 7 GW/cm^2 are better, and the rising edge rate is larger. However, it can be seen that the maximum value of the front end is not

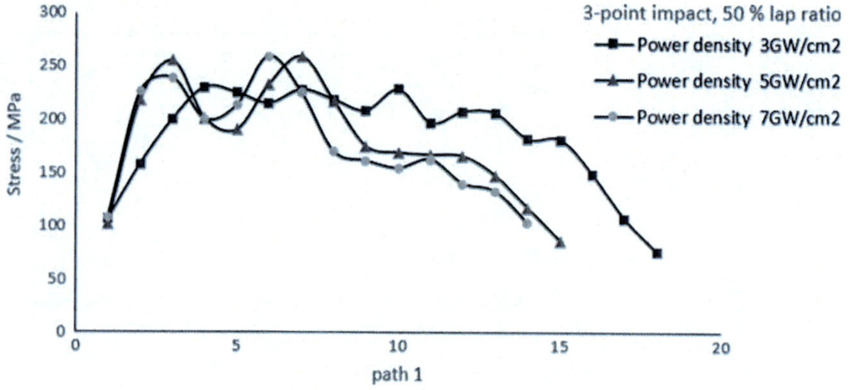

Fig. 5 Stress distribution under path 1 extraction

under the action of laser power 7 GW/cm^2, but under the action of 5 GW/cm^2. It can be obtained that the larger the power density is, the better the peening effect will be. Because of the center effect, the peening effect of the center of the impact point is not very good, so the overlap way to enhance the peening effect of the center can be seen under 5 GW/cm^2, 7 GW/cm^2 curve fluctuation is very obvious. The curve under the action of 3 GW/cm^2 is more gentle and the peening effect is better. Because of its larger impact radius, the peening range is wider at the same impact point. Figure 6 is the average Mises stress histogram under path 1, and it can be seen that the Mises stress value is larger under the power density of 3 GW/cm^2.

Figure 7 below is the Mises stress distribution under path 2. It can be seen that the stress distribution curve trend along the hole wall under three different laser densities is roughly the same. The Mises stress values of 5 and 7 GW/cm^2 are almost the same. Compared with 3 GW/cm^2, the stress value is larger and the peening effect is better.

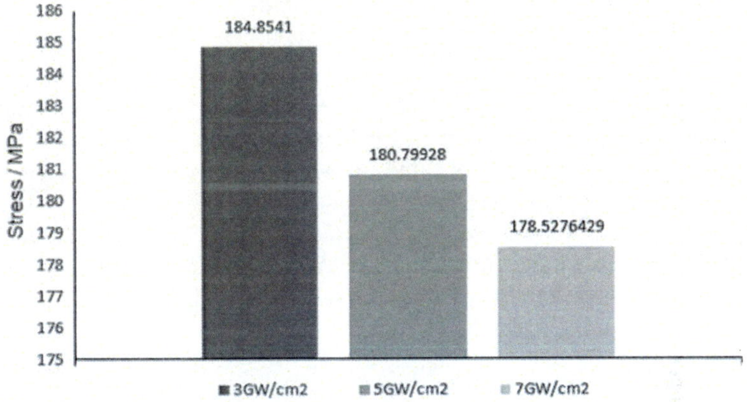

Fig. 6 Average stress under path 1

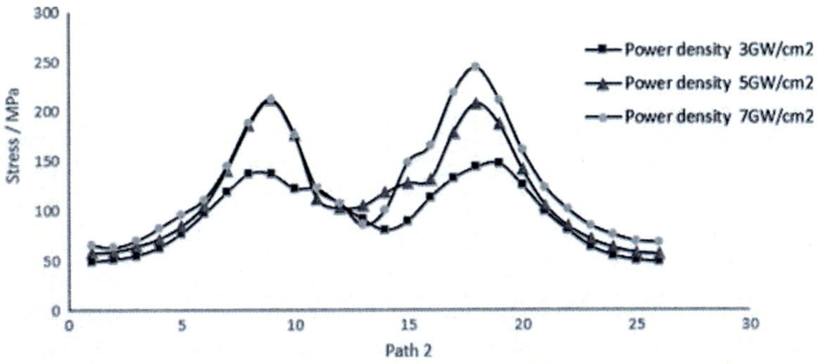

Fig. 7 Stress distribution under extraction path 2

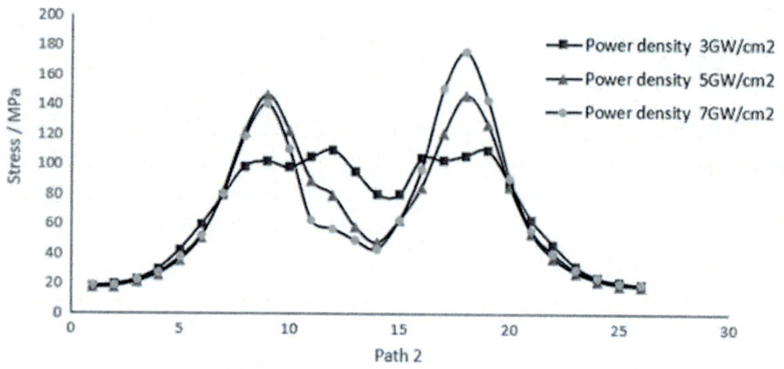

Fig. 8 Stress distribution along path 2

4.2 *Single Point Impact Peening Under Different Power Densities*

Figure 8 shows the Mises stress distribution along path 2 under single point impact peening under different laser powers. It can be seen that the curve trend is steeper under the action of power density of 5 and 7 GW/cm^2, and the peak and valley lines are more obvious. In contrast, the curve under the action of 3 GW/cm^2 is smoother. Although the Mises stress value under the action of 5 and 7 GW/cm^2 around the peak is greater than that under the action of 3 GW/cm^2, it is opposite at the peak and valley. The Mises stress value under the action of 3 GW/cm^2 is better. Overall, when the power density is 3GW/cm^2, the Mises stress distribution is more stable, and its value is not too large for the power density of 5 and 7GW/cm^2, the peening effect is better.

4.3 *Comparison of Single Point and Multipoint Laser Shock Peening*

For single and multi-point laser shock peening, the average Mises stress value is taken on path 2, as shown in Fig. 9. It can be seen from the figure that the average Mises stress of multi-point impact peening is generally larger than that of single-point impact peening, and the difference is the largest when the power density is 7 GW/cm^2. Simply looking at the Mises stress value of single point impact peening, there is little difference under different laser power density, and almost no effect. The Mises stress value of multi-point impact peening increases with the increase of power density.

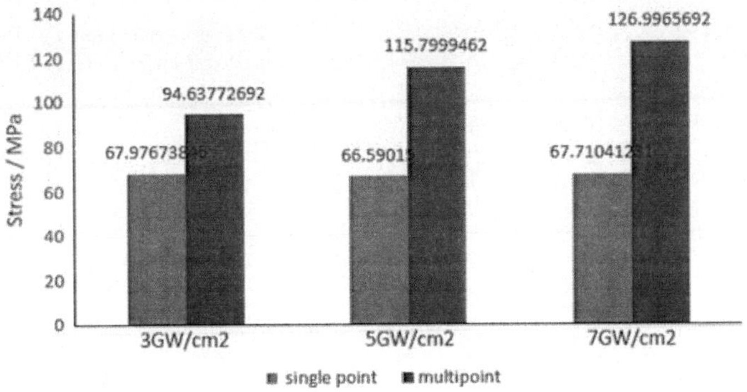

Fig. 9 Stress histogram

5 Conclusion

When the multi-point impact peening is adopted according to the peening path shown in this paper, the peening effect is better when the power density is 3 GW/cm^2, and the average Mises stress value along path 1 is 178 MPa. The average Mises stress under other power densities is large, but the stress distribution is not stable and the effect is poor. If only the peening effect near the wall is considered, the effect is better under the action of large power density. When single-point impact peening is adopted, the Mises stress values along path 2 under different power densities are not significantly different. The power density has little effect on the impact peening effect, and high power density may cause damage to the microstructure of Al7050-T7451 aluminum alloy specimen. Therefore, the single point impact peening with low power density is the best, about 3 GW/cm^2. By comparing multi-point and single point, multi-point impact peening can effectively improve the peening effect.

References

1. Trdan U, Grum J, Li Y (2015) Investigation of corrosion behaviour of aluminium alloy subjected to laser shock peening without a protective coating. Adv Mater Sci Eng
2. Zhang H, Tang Y, Yu C, Zhang Y, Guo D, Wu H (1996) Effects of laser shock processing on the fatigue life of fastener holes. Chinese J Lasers 1996(12):1112-1116
3. Montross CS, Wei T, Ye L et al. (2002) Laser shock processing and its effects on microstructure and properties of metal alloys: a review. Int J Fatig 24(10)
4. Peipei H, Chunyi Q, Qingyang J et al (2021) Effect of laser shock processing on residual stress and mechanical properties of 7050–T7451 aluminium alloy. Heat Treat Met 46(02):190–195
5. Zhang X, Li H, Huang Z et al. (2015) Numerical simulation of residual stress induced in 7075 aluminum alloy by repeated high-power laser pulses. Chinese J Lasers 42(12):80–88
6. Ren XD, Zhan QB, Yang HM et al. (2013) The effects of residual stress on fatigue behavior and crack propagation from laser shock processing-worked hole. Mater Design 44
7. Yang J-M, Her YC, Han N et al. (2001) Laser shock peening on fatigue behavior of 2024-T3 Al alloy with fastener holes and stopholes. Mater Sci Eng A 298(1)

8. Ren XD, Zhang YK, Yongzhuo HF et al. (2011) Effect of laser shock processing on the fatigue crack initiation and propagation of 7050-T7451 aluminum alloy. Mater Sci Eng A 528(6)
9. Ivetic G, Meneghin I, Troiani E et al. (2012) Fatigue in laser shock peened open-hole thin aluminium specimens. Mater Sci Eng A 534
10. Shuai S, Wu J, Che Z et al., Residual stresses of Al7050–T7451 alloy fastener holes with laser shock peening. Aeronaut Manufact Technol 1–8

Calculation Method and Experimental Verification of Optical Fiber Limit Value of Aeroengine Rotor Blade

Qing Du, Jian Zhang, Yixiong Liu, and Fayong Wu

Abstract In this paper, the implementation and validation of the proposed methods for optical fiber limit value calculation were investigated. The intersection distance method (IDM) and finite element drawing method (FEDM) were proposed and compared with the common similar triangle method (STM). The advantages and disadvantages of the three methods were analyzed from the aspects of accuracy, limitation and computing resources. The simulation results were verified by the fan rotor blade dynamic test. By comparing the calculated amplitude obtained by dynamic stress inversion with the measured amplitude, it is found that the calculated results of the three methods are in good agreement with the measured amplitude and could be used to estimate the amplitude under low stress condition. In torsional mode, the errors between the amplitude calculated by the STM, IDM, FEDM and the measured amplitude are 1.8%, 1.3% and 3.1% respectively. In high order mode, the amplitude difference obtained by each calculation method is up to 12%, and the errors between the amplitude by the FEDM and the measured amplitude is the lowest, reaching 0.3%. Furthermore, the errors of different calculation methods under stress higher than 200 MPa were analyzed. In conclusion, the results indicate that FEDM is the most accurate method for any vibration modes, and STM and IDM are also applicable in simplified calculation of low-order vibration modes of blades.

Keywords Tip-timing · Vibration monitoring · Amplitude · Rotor blade · Optimal fiber sensor

1 Introduction

With an increasing aerodynamic load of engines, the shape of blades becoming more and more complex (sweep and wide chord, etc.), and the improvement of the engine performance also leads more abundant resonance problems to the blades.

Q. Du (✉) · J. Zhang · Y. Liu · F. Wu
AECC Shenyang Engine Research Institute, Shenyang, China
e-mail: tju_duqing@126.com

S. Lee et al. (eds.), *The Proceedings of the 2021 Asia-Pacific International Symposium on Aerospace Technology (APISAT 2021), Volume 1*, Lecture Notes in Electrical Engineering 912, https://doi.org/10.1007/978-981-19-2689-1_71

Therefore, dynamic monitoring of blade vibration parameters, extraction of characteristic information, early identification and troubleshooting are urgent problems to be solved [1]. In the process of dynamic stress monitoring of engine rotor blades, strain gauge method and non-contact vibration testing method based on tip timing are widely used, and the latter is more suitable for the requirements of a high speed, high precision and comprehensive real-time monitoring, which has become a hot research field home and abroad [2, 3].

Based on the tip timing measurement technology, scholars have carried out extensive research on the problem of blade vibration. Giovanni Rigosi applied an improved two-parameter method to the timing measurement of engine tips, and the blade amplitude as well as frequency obtained through the improved two-parameter method was basically consistent with that through the strain gauge method [4]. Changbo He monitored the amplitude of each blade of a centrifugal compressor through the tip timing measurement technology, and warned the blade fault by judging whether the amplitude of one blade was too large compared with that of others under the condition of forced vibrations [5]. Meiru Liu applied the tip timing measurement technique to the blade modification of a certain type of compressor rotor, through which the blade resonance frequency and amplitude were obtained, and inversely calculated the dynamic stress at key points, thus realizing the prediction of a high cycle fatigue life of the blades [6]. In this paper, the tip timing measurement technology is applied to the amplitude monitoring of engine fan rotor blades through an optical fiber sensor, which is based on the fact that the high cycle fatigue life of the blades meets the requirements. The optical fiber limit value of the optical fiber monitoring position of the blades under different vibration modes is given.

2 Influencing Factors in the Calculation of Optical Fiber Limit Value

2.1 Static Deformation

In the process of dynamic measurement, the engine rotor blades are subjected to the joint action of centrifugal force, aerodynamic load and temperature load, and there is a certain static deformation before the blades resonate.

On the one hand, the axial component of the static deformation will lead to a certain axial offset between the actual monitoring position and the cold marking position of optical fiber. For the vibration modes of tip torsional vibration and edge vibration, different optical fiber monitoring positions may have a great impact on the analysis results. Therefore, the axial components of engine static deformation in different states should be considered, and the limit value should be calculated on the basis of an accurate optical fiber monitoring position.

On the other hand, the circumferential component of the static deformation will lead the blades to form an initial amplitude without resonance. Some scholars have studied the above influencing factors, e.g. Yue Lin [7] calibrated the angle value of the blades arriving at the optical fiber monitoring position online in the process of frequency scanning, and carried out a linear fitting calculation in the background of the system, so as to get the theoretical angle of the blades at any speed. Through the above calibration method, the influence of the circumferential component of static deformation can be eliminated from optical fiber monitoring, which does not need to be considered in the calculation of optical fiber limit value.

2.2 Vibration Order

For a certain order resonance of the engine rotor blades, the tip position corresponding to the optical fiber measuring point usually has an obvious displacement in both circumferential and axial direction, which should be comprehensively considered in the amplitude calculation.

Some scholars such as Fajie Duan have considered the above geometric factors while comparing the tip timing measurement data with the finite element analysis results, during which the blade installation angle is taken into account, so that the test results are projected to the circumferential direction, which is consistent with the finite element calculation results [8]. In essence, it is assumed through this method that the blades translate along the normal direction of the tip, and good results are obtained in the measurement of the first-order vibration mode of the tested blades. However, under the same maximum vibration stress of the blades, the deformation of other vibration modes at the tip of the blades is obviously smaller than that under first bending mode. According to the results of Campbell diagram analysis and previous tests, the corresponding optical fiber limit values for possible resonance modes need to be given respectively to ensure the safety of the engine test.

3 Calculation Methods of Optical Fiber Limit Value

From the view point of a geometric analysis, the following three calculation methods of optical fiber limit value are introduced and put forward in this section, and a comparative analysis is made.

3.1 Common Similar Triangle Method (STM)

At present, in the calculating process of the limit value of the tip optical fiber, the vibration direction of the tip is usually simplified to be perpendicular to its tangential direction, while the torsion effect of the blades is ignored. The top view of the blade tip before and after blade deformation is shown in Fig. 1a. The flow chart of STM is

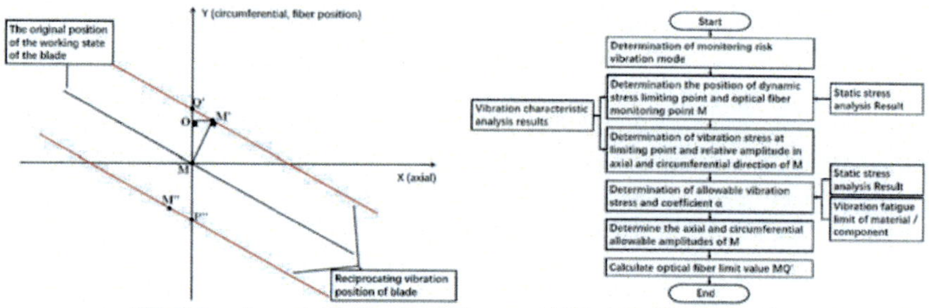

a) Tip Deformation under First Bending Mode   b) Calculation Flow Chart of STM

Fig. 1 Calculation of optical fiber limit value by STM

shown in Fig. 1b. Through the calculated circumferential and axial displacement of the blade tip and the proportional relationship among similar triangles, the optical fiber limit value corresponding to the measuring point can be calculated.

$$MQ^{'} = \frac{(\alpha D_{My})^2 + (\alpha D_{Mz})^2}{\alpha D_{My}} \tag{1}$$

Among them: α is the conversion coefficient of the corresponding vibration mode and limit point, $\alpha = \sigma_{\text{allowable}}/\sigma_{\text{relative vibration}}$, D_{Mz} and D_{My} are the axial and circumferential relative vibration displacements of M point, respectively.

The calculation through this method is simple and the train of thought is clear. For the first ideal bending mode, the vibration direction of the blades is perpendicular to the chord direction of the tip, and the above calculation method is reasonable and reliable. However, in the actual work of the blades, torsional, compound and other modes often appear, even a common bending mode is usually accompanied by corner vibration, which have a great impact on the calculation of optical fiber limit value.

3.2 *Intersection Distance Method (IDM)*

In order to consider the influence of torsional deformation on the tip amplitude, an IDM is proposed to solve the optical fiber limit value of the blades.

Considering the combined bending and torsion deformation of the blades, the local segment of the blade tip is simplified to a straight-line segment, and the tip top view before and after blade deformation is shown in Fig. 2a. Affected by the installation angle of the blades, the maximum amplitude output in the optical fiber monitoring is MQ′, so an analysis is carried out according to the original position MQ and vibration position M′Q′. The flow chart of the IDM is shown in Fig. 2b. By extracting the coordinates of M′ and N′ from the finite element results, and in conjunction with the circumferential Y-axis coordinates, the optical fiber limit value corresponding to the measuring points can be calculated:

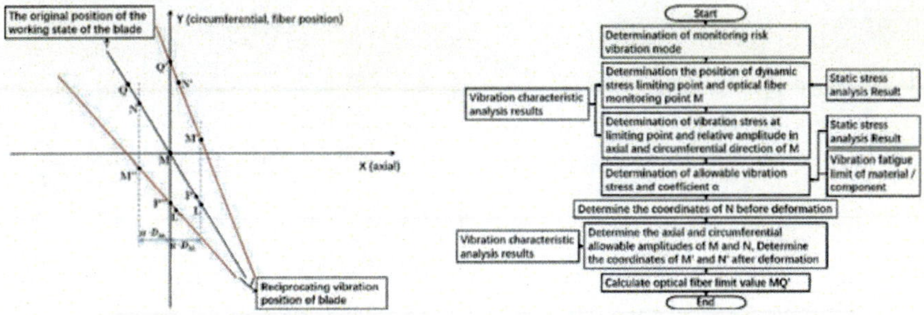

a) Tip Deformation under Combined Bending and Torsion Mode b) Calculation Flow Chart of IDM

Fig. 2 Calculation of optical fiber limit value by IDM

$$MQ^{'} = \frac{-\alpha D_{Mz}(y_N + \alpha D_{Ny} - \alpha D_{My})}{x_N + \alpha D_{Nz} - \alpha D_{Mz}} + \alpha D_{My} \quad (2)$$

Among them: α is consistent with the definition in Sect. 3.1; X_N and Y_N are the axial and circumferential coordinates of N point relative to M point in the finite element model, as shown in Fig. 2a; D_{Mz}, D_{My}, D_{Nz} and D_{Ny} are the axial and circumferential relative vibration displacements of M point and N point respectively.

This method can be used to effectively consider the influence of torsional deformation on the amplitude of blade tip. However, in high-order modes, there is often an obvious bending deformation at the leading and trailing edge position of the blade tip (optical fiber monitoring position), and simplifying the monitoring position to a straight line may cause errors.

3.3 Finite Element Drawing Method (FEDM)

In order to consider the influence of the tip bending deformation on the amplitude of high-order corner vibration, a FEDM is proposed to solve the optical fiber limit value of the blades.

With the help of the finite element tool, the tip nodes are defined as groups and combined with the relative vibration displacement of each node as well as conversion coefficient α, the top view of the blades before and after deformation is drawn and shown in Fig. 3a. The principle of the FEDM is similar to that of the IDM. The optical fiber limit value is calculated by drawing a profile of the blade tip before and after deformation and obtaining their intersection point at the optical fiber monitoring position. The difference is that through FEDM, it does not need to determine the coordinates of the auxiliary points M and N in advance. Instead, the tip amplitude is analyzed as a whole, as is shown in Fig. 3b.

Because this method can be used to effectively consider the bending deformation of the blade tip itself, it has a good adaptability to the sharp curl of the blade tip in the high-order vibration mode of the blades.

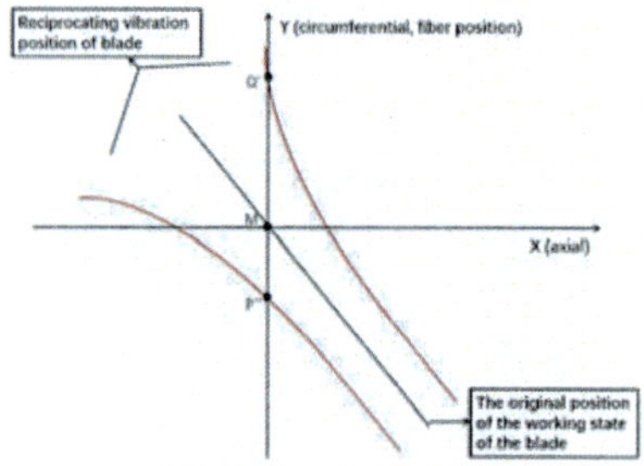

a) Tip Deformation under Corner Vibration

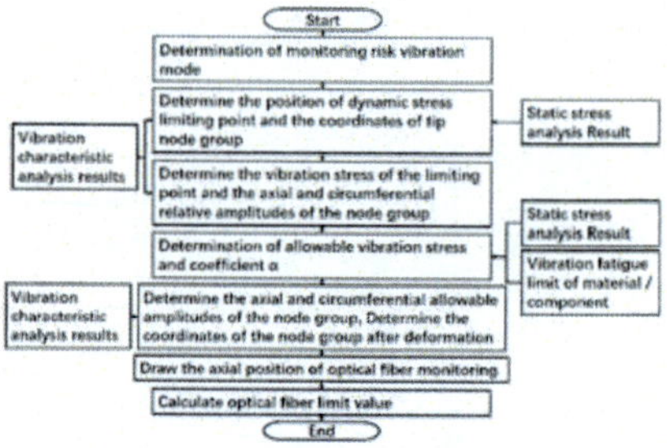

b) Calculation Flow Chart of FEDM

Fig. 3 Calculation of optical fiber limit value by FEDM

Table 1 Comparison of calculation methods of optical fiber limit value

Methods	STM	IDM	FEDM
Applicability	Ideal first bending vibration	Combined bending and torsional vibration	High-order angular vibration
The limitation	Torsional mode and high-order corner vibrations can not be considered	High-order angular vibration can not be considered	——
The amount of calculation	The amplitude of one node is extracted and calculated by the method of similar triangles	The amplitudes of two nodes are extracted and the intercept is calculated by connecting lines	The amplitudes of all nodes are extracted and calculated by drawing contour lines

3.4 Comparison of Calculation Methods of Optical Fiber Limit Value

The advantages and disadvantages of the above three calculation methods are compared and summarized, as is shown in Table 1.

4 Experimental Verification

4.1 Test Scheme

In order to verify the accuracy of the above optical fiber limit value calculation methods, a dynamic test of the whole blade disk of the fan rotor was carried out, and an optical fiber monitoring device was set up at the leading edge of the first rotor blade to collect the amplitude. According to the analysis of the previous dynamic test results, the third-order and fifth-order resonance easy occurred, which was used as

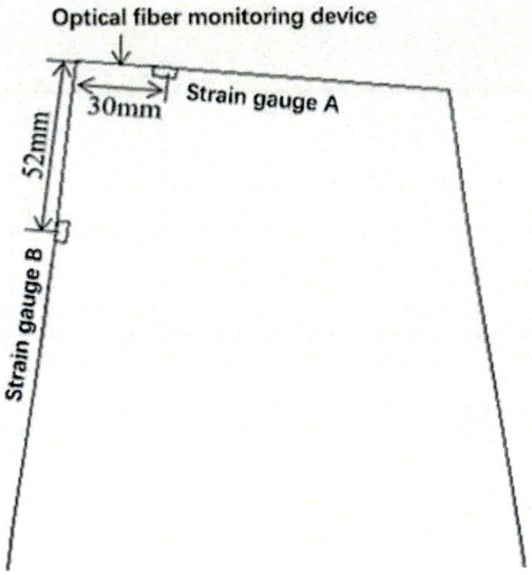

Fig. 4 Strain gauge setting position and optical fiber monitoring position

the monitoring dangerous vibration mode, and a strain gauge was set at the position where the vibration stress of the resonance mode was larger. The strain gauge setting position and the optical fiber monitoring position are shown in Fig. 4. Under a cold condition, the optical fiber monitoring device is 7.4 mm from the leading edge of the blade tip.

4.2 Test Results and Comparative Analysis

The results of dynamic test and optical fiber test on the first-stage rotor blades are shown in Table 2. According to the resonance frequency and the relationship between the dynamic stress ratio of A and B, the resonance mode of the blades is identified, as is shown in Table 2, and the relative deformation distribution of the tip under the corresponding mode is shown in Fig. 5.

Table 2 Dynamic test and optical fiber test results of first-stage rotor blade

Incentive factors	Resonance speed (r/min)	Resonance frequency (Hz)	Resonance mode	Stress (MPa)	Optical fiber monitoring value (mm)
19E	2384	755	One torsional mode, third order	15.49(A)	1.053
	3123	989	Corner vibration, fifth order	19.12(B)	0.375

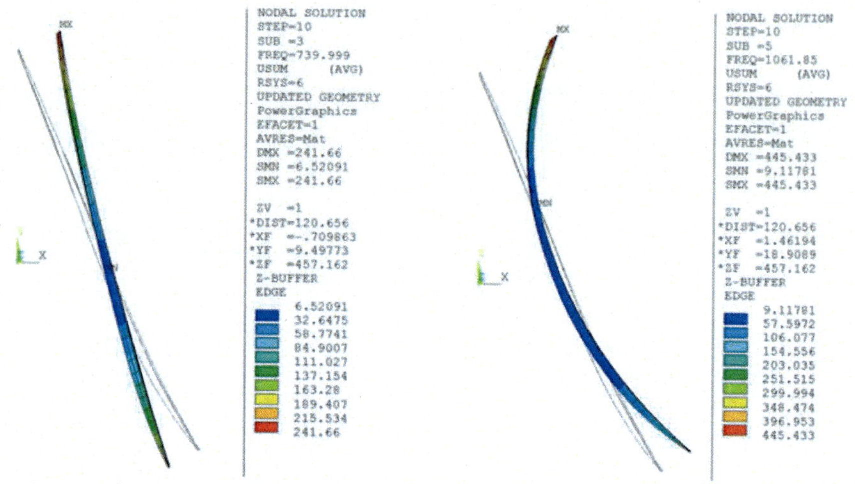

One Torsional Mode, Third Order — Corner Vibration, Fifth Order

Fig. 5 Relative deformation distribution of blade tip under resonance mode

According to the resonance speed in the test, and combined with the results of static stress analysis, the axial displacement at the optical fiber monitoring position under the corresponding working condition is obtained and considered in the analysis. According to the dynamic stress measurement results in Table 2, the amplitudes are calculated through the above three methods and compared with the measured optical fiber amplitudes, as is shown in Table 3.

It can be seen from the table that the amplitude results obtained through each calculation method are in good agreement with the measured optical fiber amplitude, which can be used to calculate the amplitude under a lower stress. At the same time, it can be found that under the first torsional mode, the difference among the amplitude obtained through each calculation method is less than 2%, and the error between the IDM and the measured optical fiber amplitude is the smallest, which is 1.3%; in the corner vibration, the maximum difference among the amplitude obtained through

Table 3 Comparison of measured and calculated amplitudes obtained by different methods

	Conversion coefficient α	Calculating method	Calculated amplitude	Measured amplitude	Error (%)
One torsional mode, third order	2.20×10^{-3}	STM	1.034	1.053	1.8
		IDM	1.039		1.3
		FEDM	1.020		3.1
Corner vibration, fifth order	6.20×10^{-4}	STM	0.351	0.375	6.4
		IDM	0.394		5.1
		FEDM	0.376		0.3

each calculation method is 12%, and the error between the FEDM and the measured amplitude is the smallest, which is 0.3%.

4.3 Deviation Analysis of Different Calculation Methods Under High a Stress

In the calculating process of the optical fiber limit value under a certain vibration mode, the maximum stress point of the blades should reach the material fatigue limit (considering safety reserve). In order to analyze the deviation of different calculation methods under a high stress, with the above blades as an example, the amplitude of optical fiber monitoring position under different dynamic stress levels is calculated, as is shown in Fig. 6.

The following rules can be seen from the figure:

1. In the same vibration mode, the amplitude obtained by the STM has a linear relationship with the stress value of the measuring point, and the coefficient is related to the vibration displacement of the measuring point and allowable vibration stress; the IDM and the FEDM have a non-linear relationship with the stress of the measuring point because the torsional deformation of blade and tip is taken into account.
2. Because the normal vibration and the torsional deformation of the blade play a positive superposition role in the amplitude analysis of the leading edge of the tip, the amplitude calculation results of the IDM and the FEDM are larger than that of the STM, and the difference is gradually obvious as the increasing of blade torsion angle. For example, for first torsional mode, when the stress at point An is 20 MPa, the difference between the IDM and the STM is 0.9%,

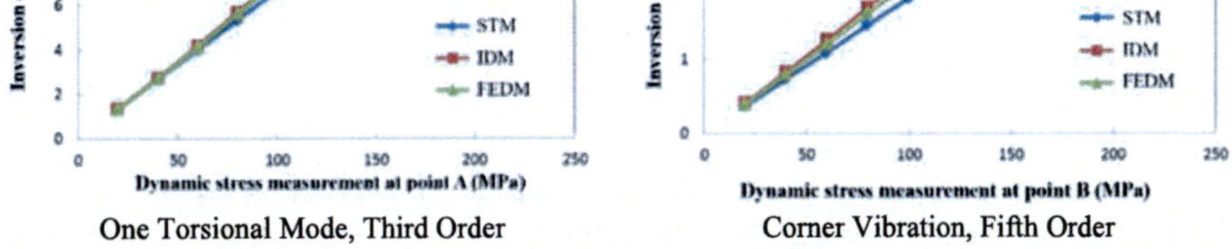

Fig. 6 Inversion of leading-edge amplitude of blade tip under different dynamic stresses

while when the stress is 220 MPa, the difference between the two methods is 23.1%.

3. In the corner vibration, the torsional crimp deformation of the leading edge of the tip leads to a certain error in the IDM, which increases approximately linearly with the increase of the stress of the measuring point. For example, in first torsional mode, the blade body is torsional and the tip has no obvious crimp deformation, and there is no significant difference between the IDM and the FEDM, while in the fifth-order mode, the tip is obviously crimped and a concave function is formed in the normal direction of the tip, so that the amplitude obtained by the IDM is always larger than that obtained by the FEDM, and the difference is about 5.0%.

To sum up, the results indicate that FEDM is the most accurate method for any vibration modes, and STM and IDM are also applicable in simplified calculation of low-order vibration modes of blades.

5 Summary

1. It is emphasized that the influence of blade static deformation and blade resonance mode should be taken into account in the calculation of blade optical fiber limit value.
2. Aiming at the problem of calculating the limit value of optical fiber, the IDM and FEDM are proposed for the first time, and compared with the commonly used STM, and the above three methods are compared and analyzed. The advantages and disadvantages of the three methods are analyzed from three aspects: applicability, limitation and amount of calculation.
3. The dynamic test is carried out for the rotor blade of a fan, and the calculated optical fiber amplitude is compared with the measured optical fiber amplitude. It is found that the calculated results of the three methods are in good agreement with the measured amplitudes. Furthermore, the differences of different methods under high stress are analyzed, and it is pointed out that FEDM is the most accurate method for any vibration modes, and STM and IDM are also applicable in simplified calculation of low-order vibration modes of blades.

References

1. Fan BN, Zhang YB, Wang HD, Xu BS. Research status for dynamic monitoring impellers of blades based on blade tip-timing. J Vibrat Shock (5)
2. Fang ZQ, Duan FJ, Zhang YG, Ye SH (2008) Research on vibration measurement for high-speed rotating blades based on optical fiber sensor. Transd Microsyst Technol (2)
3. Wang P (2013) Application review of blade tip timing method for foreign areo-engine blade vibration measurement. Aeronaut Sci Technol (6)

4. Rigosi G, Battiato G, Berruti TM (2017) Synchronous vibration parameters identification by tip timing measurements. Mech Res Commun 79(1):7–14
5. He CB (2020) Research on on-line condition monitoring and fault warning methods for centrifugal compressor blade. Dalian University of Technology
6. Liu MR, Zhu J, Liang EB, et al. (2019) Vibration measurement on compressor rotor blades of aero-engine based on tip-timing. J Aerosp Power (9)
7. Liu H (2018) Investigation on rotating blade vibration monitoring and analysis method. Nanjing University of Aeronautics and Astronautics
8. Duan FJ, Li G, Ye DC et al (2016) Method for inversing dynamic stress of blade based on tip-timing. Nanotechnol Precis Eng 14(3):161–166

Analysis of Laser Positioning Milling Process for Aircraft Cabin Door Opening

Shengxiong Li

Abstract One of a main characteristics of aircraft manufacturing is to solve a lot of interchange and coordination problems in process preparation and production. Traditionally, analog quantity transfer method is adopted, which can not meet the manufacturing requirements of "low cost, short circle and high quality", based on the three dimensional design model and the advanced measurement device of aircraft, it is an effective way to establish and adopt the interchange and coordination model of digital quantity transfer to meet the needs of modern aircraft development, with the rapid development of computer and numerical control technology, digital technology is gradually applied to the field of aviation industry, the level of aircraft design and manufacturing technology has been greatly improved. In terms of assembly technology, digital technology is used in some parts, but analog quantity transmission is still the main coordination method of many aircraft manufacturing enterprises, and special tooling assembly is still used in assembly. Hoped that all aircraft manufacturing enterprises can achieve new breakthroughs in digital assembly technology. The application of digital technology will bring great changes to the manufacturing and inspection process. Under traditional conditions, after assembly, the products can be measured and inspected by analog equipment such as tooling and standard measuring tools. Under the conditions of digital manufacturing, due to the application of direct installation technology such as numerical control technology and laser tracker, the necessary tools such as frame and measuring tools are reduced. In this paper, the problem of interchange and coordination in milling the opening of rear cargo door frame of a cargo aircraft studied. Referring to the traditional aircraft interchange and coordination theory, and combined with the characteristics of aircraft digital manufacturing technology. The key technology of aircraft manufacturing interchange and coordination based on digital quantity transfer is studied. The laser aided positioning and milling technology for the opening of rear cargo door frame of a cargo aircraft is based on rear cargo door frame product digital model and Drawings, and with the help laser tracker, the positioning, marking and milling of frame door opening of the

S. Li (✉)
International Cooperation Technology Office of Civil Aircraft and Subcontracting Project Department, AVIC Xi'an Aircraft Industry Group Co, Ltd, 1 Xifei Avenue, Yanliang District, Xi'an 710089, China
e-mail: 1403850015@qq.com

S. Lee et al. (eds.), *The Proceedings of the 2021 Asia-Pacific International Symposium on Aerospace Technology (APISAT 2021), Volume 1*, Lecture Notes in Electrical Engineering 912, https://doi.org/10.1007/978-981-19-2689-1_72

Fig. 1 Illustration of door frame opening of rear cargo compartment door fuselage

aircraft fuselage door frame are carried out. The advantages of this technology way are: it can save a set of tooling for milling the door frame opening, and no matter how to modify the drawing/digital model size of the door frame opening, this technology way can complete the milling of the door frame with the fastest speed, and adapt to and meet the latest size requirements of the drawing/digital model (because the use of laser tracker does not need to spend the time and cost of designing, modifying and manufacturing tooling).

Keywords Assembly positioning · Positioning datum · Laser positioning · Laser measuring · Cut and milling

1 Product Overview

For a cargo aircraft, the door frame opening of the rear cargo compartment door of the fuselage of a cargo aircraft is located in the Station 17319–18149 of section 16 of the fuselage, the rear cargo compartment door frame is located at located in the Station 17319–18149 of section 16 of the fuselage, open an opening on the floor beam—the door frame of the rear cargo compartment door, which is the same size as the door leaf of the rear cargo compartment door, and the shape is consistent with the shape of the aircraft skin. The length and width dimension is 1830 mm(length)x 780 mm(width), as shown in Fig. 1.

2 Processing Technology and Procedure

2.1 Processing Technology

The laser assisted milling process for the door frame opening of the rear cargo compartment of a cargo aircraft is a process method for positioning, marking and milling the door frame opening of the aircraft fuselage with the help of a laser tracker based on the product digital model and drawings of the door frame opening, the advantage of this process method is that it can save a set of tooling for milling the door frame opening, and no matter how the drawing/digital mode size of the door frame opening is changed, this method can complete the door frame milling at the fastest speed, and adapt to and meet the latest size requirements of the drawing/digital mode (because the laser tracker does not need to spend design time and cost of modifying and manufacturing tooling).

2.2 Processing Procedure

(1) Firstly, use a fixture for positioning and making hole, take the upper plane of floor beam, the door frame plates of Station 17,319 and Station 18,149 and the out surface of the skin as the positioning datum to position the fixture, to obtain the 4 positioning holes on the door frame opening and the 2 hinge positions at the door lintel above the door frame opening (as shown in Figs. 2, 3, 4).

Then, at the hinge position of the rear cargo door, drill assembly holes on the fuselage skin according to the holes on the hinge and locate the hinge, at the same time, determine that the 4 positioning holes on the door frame opening are 4 φ 6.35 mm reference points (holes) for establishing laser measurement system, positioning and marking during milling (as shown in Fig. 5).

(2) and then, the laser tracker was used to established the laser measurement system based on 4 φ 6.35 mm reference points determined by positioning and hole making fixture and the axis of the hinge hole of the rear cargo compartment door (as shown in Fig. 6);

(3) Compare the established laser measurement system with the data of door position of rear cargo compartment door on the skin digital model, determine the outline of rear cargo compartment door frame, and then take an appropriate number of points around the door frame of the digital model (the more points, the higher the accuracy) (as shown in Fig. 7).

(4) and then, use the laser tracker target to take an appropriate number of points corresponding to the digital model of the door frame around physical door frame of the fuselage (this number of points are taken from the net edge of contour of the final door frame) (as shown in Fig. 8).

Measure the data, input it into the measurement system, and compare it with door frame data of digital model, if the data are consistent, remove the tracker to make

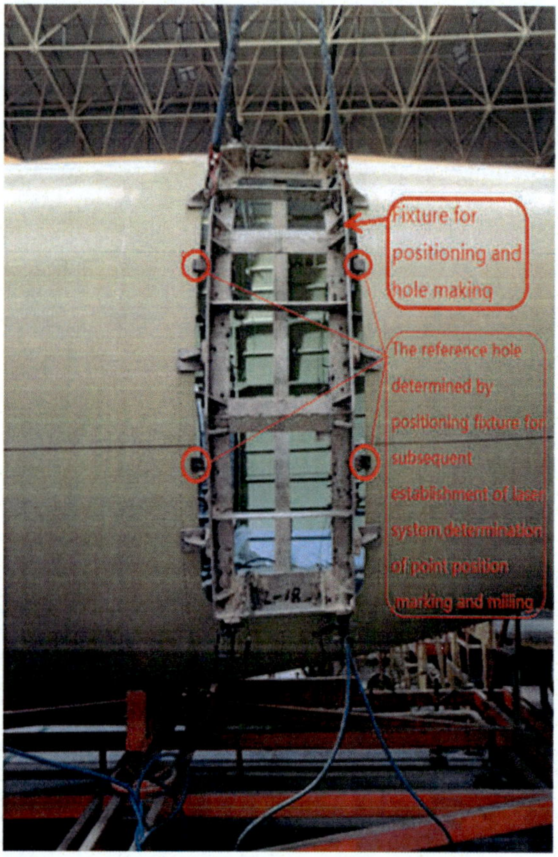

Fig. 2 Illustration of how to use the positioning fixture

the final contour position of the door frame 2 mm smaller than each side of the theoretical door frame opening (for the margin left for the next level assembly of the door frame and door leaf), connect the points on door frame with a continuous and smooth line with a pencil, and finally, cut along the draw line with a pneumatic milling tool/milling cutter (as shown in Fig. 9), and grind and deburr slightly, the resulting opening is the final opening size of the rear cargo compartment door frame.

And then, use the laser tracker to take the same number and position points on the door frame opening of the physical door, measure the door frame opening again. And compare with the position data of the previous points, the results are consistent. After the assembly verification of the rear cargo door of more than 10 sorties, the method of laser positioning, marking and milling the door opening fully meets the drawings/digital model requirements and the assembly requirements.

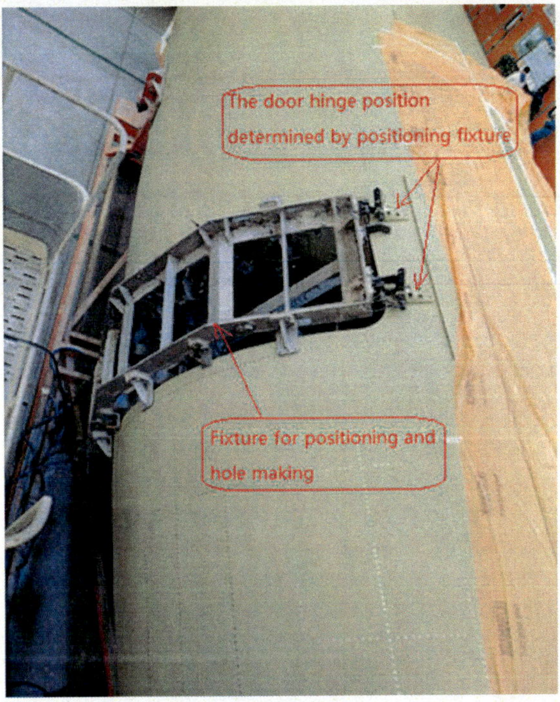

Fig. 3 Illustration of how to use the positioning fixture

3 Problem Analysis and Solution of Milling Opening of Rear Cargo Door Frame

After milling, the skin opening of the rear cargo door frame is basically the final assembly level of the aircraft skin, and the errors in many previous processes will accumulate here, whether the rear cargo door frame opening meets the drawing requirements, and whether it meets the gap and mismatch requirements between any aircraft physical door leaf and the door frame (to ensure the interchangeability of rear cargo door frame opening), how to reasonably use the positioning fixture and laser tracker will have a great impact on the final gap and mismatch difference, and will eventually affect the water tightness and air tightness requirements of the aircraft. If the water tightness and air tightness requirements of the aircraft are not good, it will cause great potential safety hazards to the aircraft in flight, for example, in high-altitude flight, the air in the cabin will be thin and the oxygen content in the air will be reduced, which will threaten the life safety of passengers and crew, therefore, it is particularly urgent to solve the problem of out of tolerance of gap and mismatch difference of cargo door frame behind the cargo aircraft. In view of this, the on-site construction personnel are required to use the laser tracker to measure the points

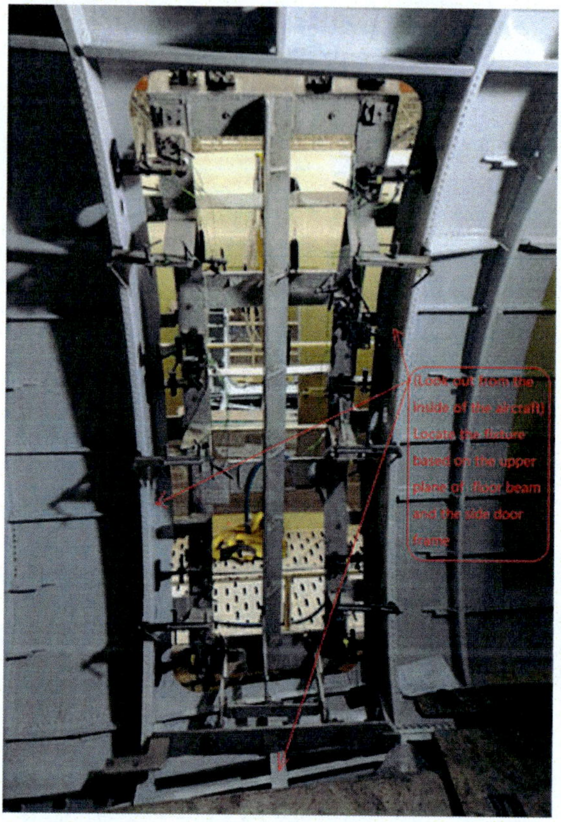

Fig. 4 Illustration of how to use the positioning fixture

on the door frame opening 2 to 3 times, if the data are completely consistent, the subsequent marking and milling can be carried out.

4 Find Out the Cause of the Problem and Corrective Action

In order to prevent out of tolerance of gap and mismatch of rear cargo door frame opening, after continuously tracking the construction process of door frame opening cutting, and repeatedly studying the positioning fixture and laser tracker when cutting the door frame opening of the cargo aircraft, it was found that there were some problems when the constructors took points and placed laser target on the door frame, as well as some deficiencies in the cutting methods of the constructors, and the remedial and correction work was carried out immediately.

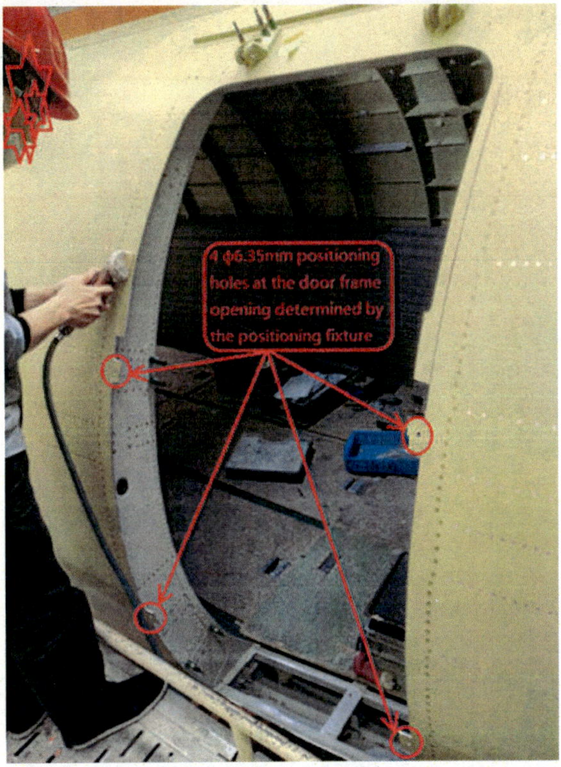

Fig. 5 4 positioning holes at door frame opening determined by positioning fixture

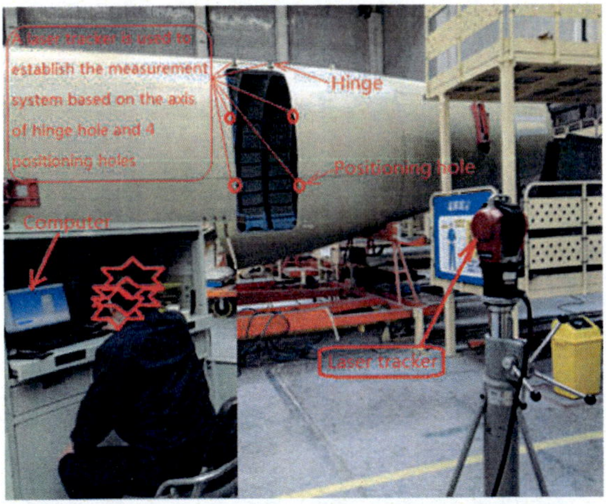

Fig. 6 The laser measurement system was established on 4 φ 6.35 mm reference points and the axis of the hinge hole

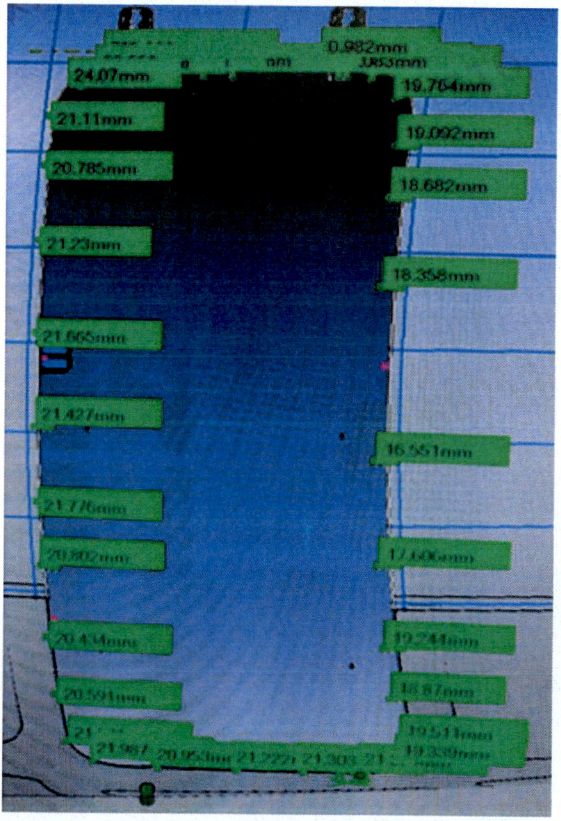

Fig. 7 Illustration of door frame opening and the collected points on the digital model

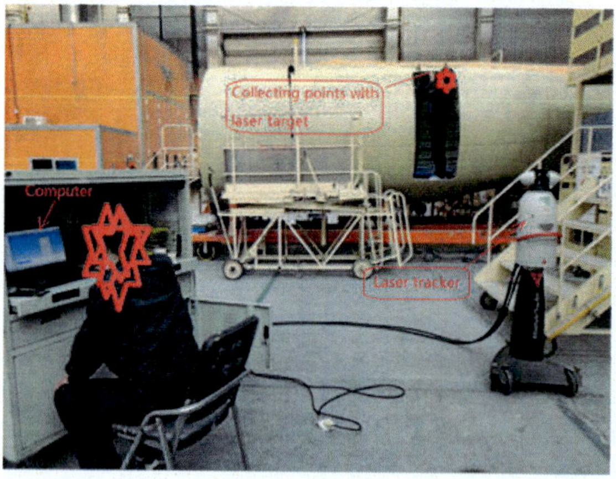

Fig. 8 Take points around the door frame with laser tracker

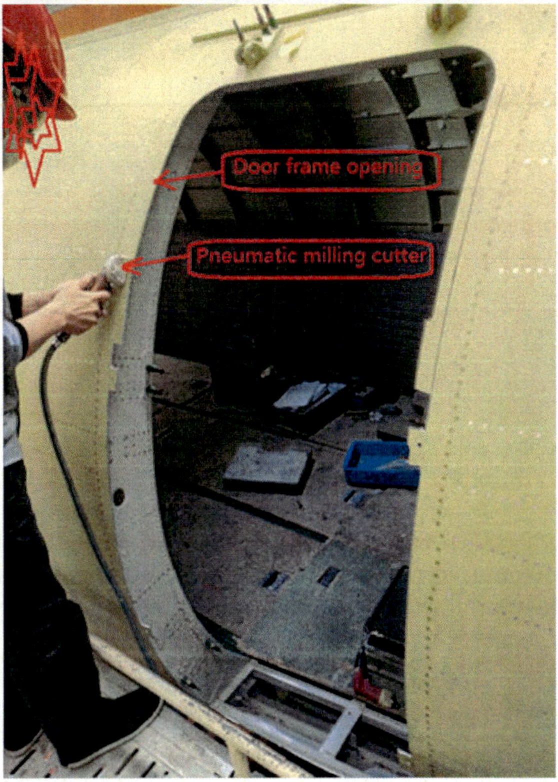

Fig. 9 Milling door frame opening with a pneumatic milling cutter

4.1 Deviation of Collecting Points Around Door Frame Opening and Its Treatment

When collecting measuring points on the door frame opening of the rear cargo compartment door, they must correspond to the number and position of points on the door frame digital model one by one, and must be consistent; when placing the laser target, be sure to place the target at the determined position of each point, so that the measurement results can be consistent with the actual values, after measuring the collecting points on the door frame opening for many times, and comparing with the position data of the previous points, the results are consistent, and then the next marking and trimming operation can be carried out.

4.2 *Methods of Cutting by Construction Personnel*

After emphasizing the collect method of points on the door frame opening, and tracking the milling process of the door frame opening of the rear cargo compartment of the three sorties, it is found that the qualified rate of the door frame opening is 100%, which fully meets the drawing/digital model requirements, and the qualified rate of the door leaf assembly of cargo aircraft is also 100%.

However, still feel that there are some defects and risks, it is suggested that when the construction personnel use the pneumatic cutting saw/milling cutter to mill the door frame opening, should not mill it in place at one time, reserve 0.5 mm allowance, and then grind and trim it to the size required by the drawing/digital model, so as to avoid milling to the negative difference (lack of material) state due to improper operation when milling to the final size at one time, on the contrary, it has become an unnecessary risk. After continuously tracking the construction process of door frame opening cutting of 3 sorties, and finalize when using the risk prevention method determined above, a margin of 0.5 mm should be reserved when milling the door frame opening for the first time, and finally repaired and polished to the final size, so that the gap and mismatch of the door frame opening can more easily meet the requirements of the drawing/digital model, and avoid the risk that negative difference (lack of material) cannot be repaired.

5 Conclusion

The milling of the door frame opening of the rear cargo compartment door of the cargo aircraft seems simple, but in fact, it is greatly affected by the accumulated errors of many previous assembly process. The errors of many previous assembly process will accumulate here, so that the aircraft skin shape of each sortie is different from that of other sorties, which is more or less different. Therefore, the construction personnel need to be certified and carefully complete the use requirements of all the above tooling, milling allowance should be carefully reserved and carefully polished, in this way, the impact of aircraft assembly cumulative error can be minimized, and the gap and mismatch range required for the opening of cargo door frame can be met as far as possible, so as to meet the water tightness and air tightness requirements of the aircraft, ensure the safety of the aircraft, reduce the potential safety hazards of the aircraft in flight, create a comfortable and safe flight environment for passengers and crew members.

It is very important link for the smooth progress of this project to study the opening positioning, milling and trimming technology and coordination methods of the rear cargo compartment door frame of cargo aircraft, which is of great significance to improve the aircraft assembly level. At the same time, some assembly and coordination methods and experiences adopted in the assembly can be popularized and applied.

Reference

1. Aircraft product Drawings and digital Models,-----Leonardo Aerostructures

Numerical Simulation of TC17 Titanium Alloy Thin Blade Strengthened by Laser Shock Processing

Sijie Ma, Xiangyu Ding, Zhi Yang, Jiayu Hao, Wei Xiao, and Longgang Fan

Abstract Laser shock processing (LSP) is a new surface strengthening technology, it can produces relatively large residual compressive stress on the metal surface and a certain depth, and refines the grain on the metal surface, so as to strengthen the metal. In this paper, the stress distribution and deformation of thin blade edge of TC17 titanium alloy under multi-point impact were studied by numerical simulation. By controlling the action distance of plastic wave, the process parameters of single-sided laser shock strengthening of thin blade edge were designed. The results show that the solution time of dynamic shock is 10000 ns according to the energy curve. By changing the boundary constraints of thin blade, such as bottom constraint, cantilever constraint and both ends constraint, to simulate the actual processing situation, the distribution of residual stress in bottom constraint is the most uniform, the increase of residual stress in the processing area is more obvious, and the material deformation is the least. Based on the commonly used processing parameters of 3 mm laser spot and 20 ns pulse width, 30% overlap ratio is adopted to avoid uneven impact or excessive local stress. The residual stress is about 104, 247 and 450 MPa when 2, 3 and 5 J energy are used respectively; the blade edge deformation is controlled at about 4 μm. In this paper, the actual multi-point machining is simulated, which provides a theoretical reference for the selection of the actual machining parameters of thin blade.

Keyword Laser shock processing · TC17 titanium alloy · Residual stress · Material deformation · Numerical simulation

S. Ma · X. Ding (✉) · Z. Yang · J. Hao · W. Xiao · L. Fan
School of Aircraft Engineering, Nanchang Hangkong University, Nanchang 360100, China
e-mail: dingxiangyu@nchu.edu.cn

X. Ding
Fundamentals Department, Air Force Engineering University, Xi'an 710038, China

S. Lee et al. (eds.), *The Proceedings of the 2021 Asia-Pacific International Symposium on Aerospace Technology (APISAT 2021), Volume 1*, Lecture Notes in Electrical Engineering 912, https://doi.org/10.1007/978-981-19-2689-1_73

1 Introduction

TC17 material is a type of dual phase titanium alloy, which has many advantages, such as high strength, good fracture toughness, high hardenability and wide forging temperature range, it is widely used in the manufacturing field of aero-engine blades, rotor shafts and other aviation components [1]. The working environment of aero engine is relatively bad. In order to ensure the service life of the engine, very high requirements are put forward for the performance of its key components, especially fatigue performance [2]. In the aerospace field, high-performance thin-walled parts are mainly used to reduce the weight of aircraft [3]. Advanced aero engines are required to develop towards integration and lightweight. Titanium alloy is widely used in compressor blades, with thinner and thinner design thickness and low strength reserve. High cycle fatigue fracture is easy to occur under the coupling action of aerodynamic, centrifugal force and vibration, which seriously threatens flight safety. It is a weak link in the service process of aviation equipment. Because the maximum working stress and crack initiation are generally located on the material surface, surface strengthening technology has become an important means to solve the fatigue fracture of parts. Laser shock strengthening technology can use the force effect of shock wave to form thick residual compressive stress on the surface and cause the change of material microstructure without changing the design structure of parts and the performance of material matrix, so as to significantly improve the fatigue performance of materials [4].

However, when laser strikes thin-walled parts, the propagation law of shock wave is complex, and it is difficult to choose the appropriate processing technology. Numerical simulation analysis is an important means for the research of laser shock strengthening process. Since Braisted and Brockman [5] first used ABAQUS platform to analyze the mechanical model and residual stress field of laser shock strengthening materials in 1999, many scholars domestic and abroad have used numerical simulation technology to make a more comprehensive analysis on the application of laser shock strengthening technology to thin-walled parts. Ivetic et al. [6] studied the residual stress in the depth direction of aluminum alloy perforated thin-walled parts strengthened by laser shock. Zhang et al. [7] simulated and studied the single point laser shock strengthening of titanium alloys with different thickness. Bhamare et al. [8] explored the relationship between machining parameters and residual stress distribution of Ti-6Al-2Sn-4Zr-2Mo based on the numerical method of three-dimensional nonlinear finite element LSP simulation. The above research is of great significance to understand the residual stress and distribution law of laser shock strengthening on thin-walled parts. However, the current research mainly focuses on the stress distribution of single point impact, and there are few studies on the approximation between multi-point lapping and actual processing.

Laser shock strengthening may lead to macro deformation of blades and affect the accuracy and aerodynamic performance of parts. Therefore, in order to better apply this technology, different processing parameters need to be studied according to different situations. Complex blade simulation takes more modeling time and calculation time. In this paper, a regular constant section blade simulator is established to simulate different conditions, which provides a theoretical reference for the actual machining process of TC17 thin blade.

2 Brief Introduction of LSP

Laser shock processing (LSP) is a new surface strengthening technology, which has significant technical advantages. Its basic principle is to irradiate the metal surface with short pulse (tens of nanoseconds) and high peak power density (>10^9 w/cm^2), which is an absorption protective layer coated on the metal surface to absorb laser energy and produce explosive vaporization, resulting in high temperature (>10^7 k) and high pressure (>1 GPa) plasma; The plasma is constrained by the constraint layer to form a high-pressure shock wave and propagate to the interior of the material. The force effect of the shock wave is used to produce plastic deformation on the surface of the material, change the microstructure of the surface material, and leave residual compressive stress on the deeper thickness, so as to improve the fatigue resistance, wear resistance and stress corrosion resistance of metal materials [9].

Laser shock processing produces relatively large residual compressive stress on the metal surface and a certain depth, and refines the grain on the metal surface, so as to strengthen the metal material. Residual compressive stress can increase the micro crack closing force, curb the surface crack propagation, block the crack propagation, and prolong the fatigue life. The residual tensile stress is not conducive to the improvement of fatigue life. The grain refinement on the surface of metal materials and the increase of dislocation density increase the resistance of dislocation movement, increase the resistance of fatigue crack formation and delay the time, so as to prolong its fatigue life. It inhibits the generation and propagation of cracks, and it has certain thermal stability.

3 Numerical Simulations

Laser shock strengthening is a nanosecond instantaneous dynamic process. Under the existing experimental conditions, it is difficult to accurately obtain the propagation and reflection of shock wave in materials [10]. Therefore, in order to simplify the calculation and speed up the calculation, the blade simulator with equal thickness is used for simulation analysis. Firstly, the analysis and calculation are carried out in ABAQUS/Explicit module. The dynamic analysis time is set to 5000 ns, which is much greater than the loading time of laser shock wave. Then, the ABAQUS/Explicit transient response results are imported into the ABAQUS/Standard module for static rebound calculation, and the stress distribution and deformation of the material are obtained.

3.1 Finite Element Model

In order to simplify the calculation and speed up the calculation speed, the blade simulator with equal thickness is used for simulation analysis. The blade root size of blade simulator is 25 × 20 × 4 mm, blade size is 25 × 90 × 1.5 mm. Figure 2 is a model of blade simulator with constant cross section. Since the main damage form of blade is external object damage, impact strengthening is carried out on some areas of the edge, and the strengthening area is shown in the figure. In order to shorten the calculation time, the grid at the blade root is coarse, and the size is 1 × 1 × 0.6 mm, the side length of the grid in the thickness direction of the blade part is 0.15 mm, the side length of the grid in the impact area is 0.15 mm, and that in other areas is 0.4 mm. The unit used for grid division is C3D8R, and the number of grids in the whole model is 341562. Figure 3 is the impact path, with a 30% overlap rate, forming a serpentine impact from the blade edge to the middle and from the free end to the blade root. This is because according to Yang's [11] research, it is found that the stress distribution of 30% overlap is more uniform than that of 50% overlap, and the stress fluctuates greatly at 50% overlap.

TC17 titanium alloy was used as the research object. TC17 is a rich β Stable element α + β Phase titanium alloy. Table 2 shows the chemical composition of TC17 alloy. In the process of laser shock strengthening, higher strain rate will cause the change of mechanical properties of materials, so it is far from enough to describe the physical properties of materials by static mechanical constitutive equation. According to engineering experience, experimental results and physical phenomena, a large number of constitutive models have been proposed to describe the dynamic response process of materials, such as Power-Law (PL) model, Steinberg–Guinan (SG) model, Zerilli-Armstrong (ZA) model, Johnson–Cook (JC) model and so on. Among them, the Johnson cook constitutive model comprehensively considers the factors such as strain, strain rate and temperature. Its constitutive formula is shown in Formula 1, and the parameters of the constitutive model are shown in Table 3.

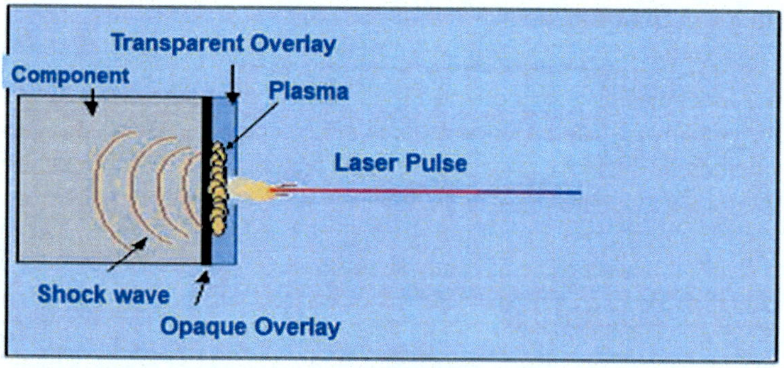

Fig. 1 Schematic diagram of laser shock processing

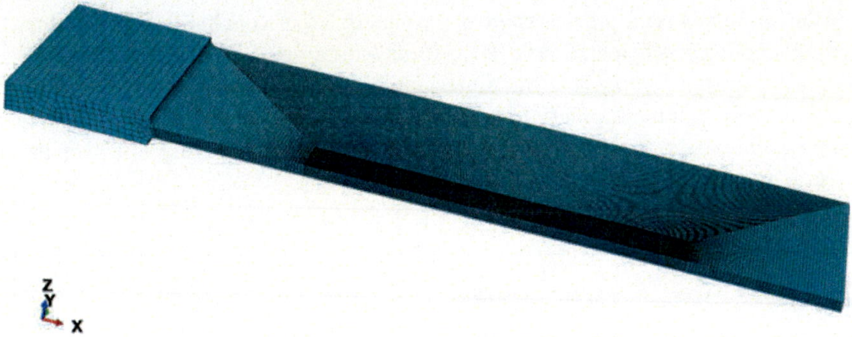

Fig. 2 Simulation model and mesh of constant section blade

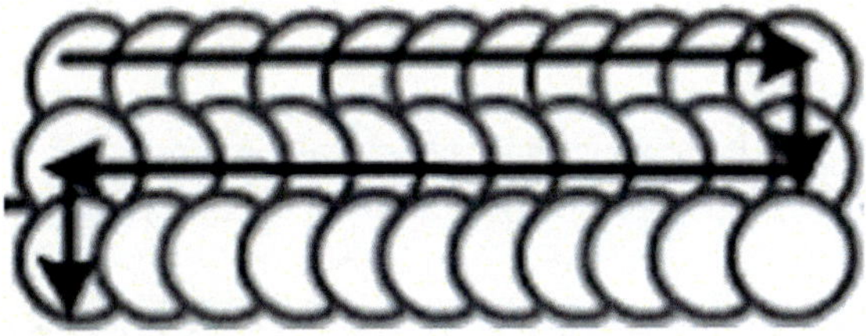

Fig. 3 Schematic diagram of impact trajectory

Table 1 The interaction between laser and matter with different power density

Power density W/cm^2	10^3–10^4	10^4–10^6	10^6–10^8	10^9–10^{10}
Phenomenon	Heating	Melt	Vaporization	Plasma

Table 2 Chemical composition of TC17 titanium alloy (%, mass fraction)

Al	Sn	Zr	Mo	Cr	Fe	Impurity	Ti
4.5 ~ 5.5	1.6 ~ 2.4	1.6 ~ 2.4	3.5 ~ 4.5	3.5 ~ 4.5	0.3	≤ 0.15	allowance

$$\sigma = (A + B\varepsilon^n)\left[1 + C\ln(\dot{\varepsilon}^*)\right]\left[1 - (T^*)^m\right] \quad (1)$$

Table 3 Parameters of Johnson cook constitutive model for TC17 titanium alloy

Parameter	Density ρ/(kg•m^{-3})	Elastic modulus E/MPa	Poisson's ratio μ	Melting temperature T/°C	A (MPa)	B (MPa)	C	n	m
Value	4680	115,800	0.29	1675	1100	590	0.0152	0.41	0.833

In the formula, A is the yield strength of the material; B is work hardening modulus; n is the hardening coefficient; C reflects the strain rate hardening effect of the material; ε Is plastic strain;$\dot{\varepsilon}^* = \dot{\varepsilon}/\dot{\varepsilon}_0$ is the dimensionless plastic strain rate; $T^* = (T - T_0)/(T_m - T_0)$ is dimensionless temperature; T_0 is room temperature; T_m is the melting point of the material; m reflects the temperature softening effect of the material.

3.2 Shock Wave Pressure Model

The simulation parameters are based on the actual equipment parameters. In the numerical simulation of laser shock, the loading condition of shock wave is the key factor to form the residual stress field. In the 1990s, Fabbro [12] proposed a calculation model of shock wave pressure, which can better reflect the variation law of detonation wave peak pressure with energy. It is the most commonly used shock wave pressure model. The pressure intensity calculation formula of laser-induced plasma shock wave is shown in formula (2)–(4).

$$P_0 = 0.01 \bullet \sqrt{\frac{\alpha}{2\alpha + 3}} \bullet \sqrt{Z(g \bullet cm^{-2} \bullet s^{-1}) \bullet A \bullet I(GW \bullet cm^{-2})} \tag{2}$$

$$\frac{2}{Z} = \frac{1}{Z_1} + \frac{1}{Z_2} \tag{3}$$

$$I_0 = \frac{4E}{\pi d^2 \tau} \tag{4}$$

In the formula α Is the ratio coefficient of plasma thermal energy to internal energy, usually α value is 0.2 ~ 0.5; Z is the converted acoustic impedance of the target and the constraint layer; Z_1 and Z_2 are the acoustic impedance of black tape and water restraint layer respectively; I_0 is the laser power density (GW•cm^{-2}); E is laser energy (J); d is the laser spot diameter (mm);τ Is the laser pulse width (s).

The laser energy of the laser presents a near Gaussian distribution in space, so the spatial distribution of laser shock wave pressure in the simulation is also Gaussian distribution. The variation formula of shock wave pressure with spatial coordinates is shown in formula (5).

$$P(x, y, t) = P_0(t)\exp[-\frac{(x^2 + y^2)}{2R^2}] \tag{5}$$

In the formula, (x, y) is the coordinate of any point based on the positive center of the impact area, $P_0(t)$ is the peak pressure value varying with time, and R is the radius of the impact area.

For the temporal distribution of shock wave, some scholars have shown that the pulse width of shock pressure wave is about 2 ~ 3 times that of laser pulse [13]. The laser pulse width in this paper is 20 ns, and the established shock wave time distribution model is shown in Fig. 4b.

The laser-induced plasma shock wave will change between plastic wave and elastic wave in the process of propagation, and its effect on the material can be divided into two processes: first, when the shock wave pressure exceeds the Hugoniot elastic limit of the material, the shock wave causes irreversible high strain rate plastic deformation of the material in the form of plastic wave, and the resulting plastic deformation is the basis of the formation of residual stress field; Second, when the shock wave pressure decreases with propagation, when it is lower than the Hugoniot elastic limit, the shock wave rebounds elastically in the material in the form of elastic wave until the energy dissipation is zero. This is the most important factor for setting parameters.

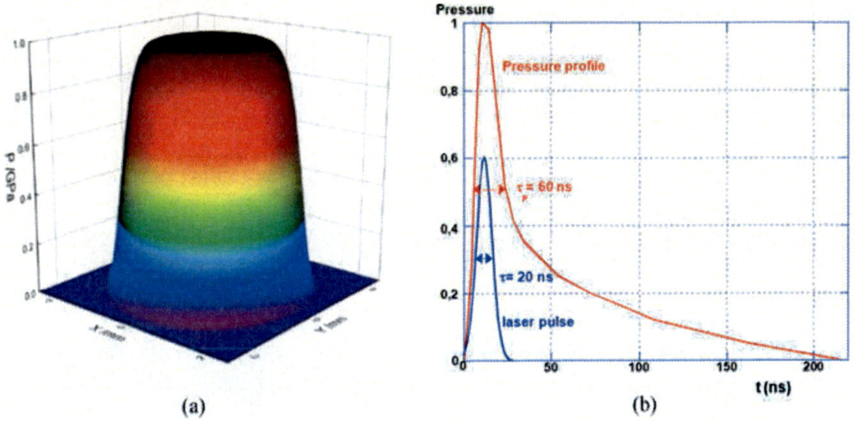

Fig. 4 Schematic diagram of spatial and temporal distribution of laser shock wave

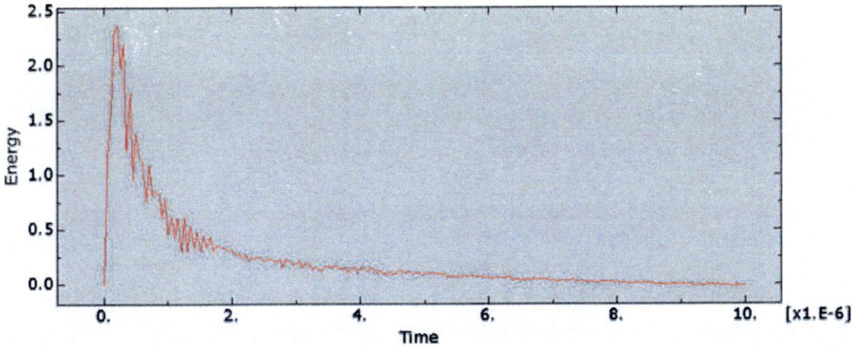

Fig. 5 Overall energy change curve

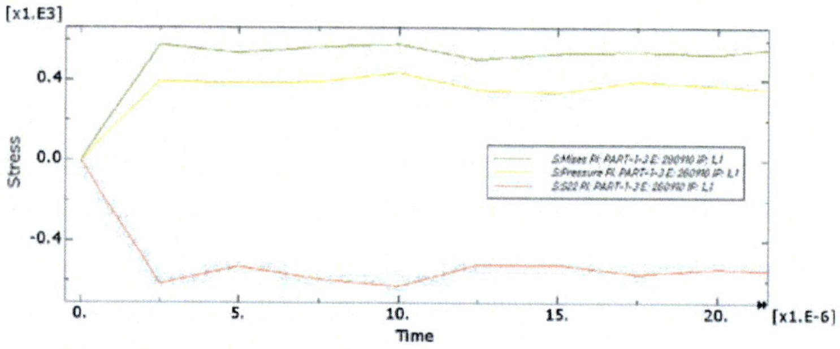

Fig. 6 Curve of various stresses at impact point with time

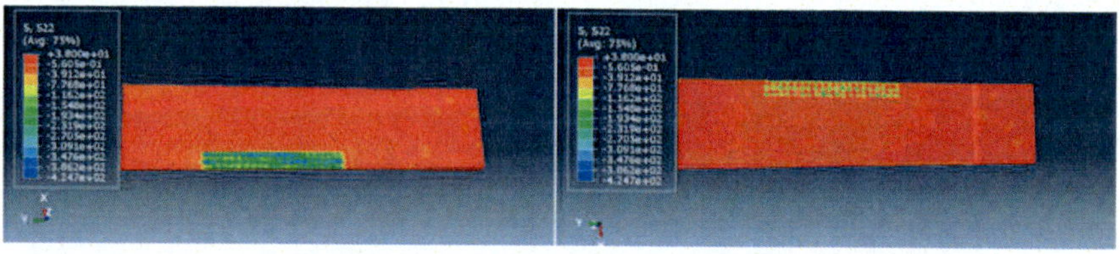

Fig. 7 Stress distribution on both sides under bottom constraint

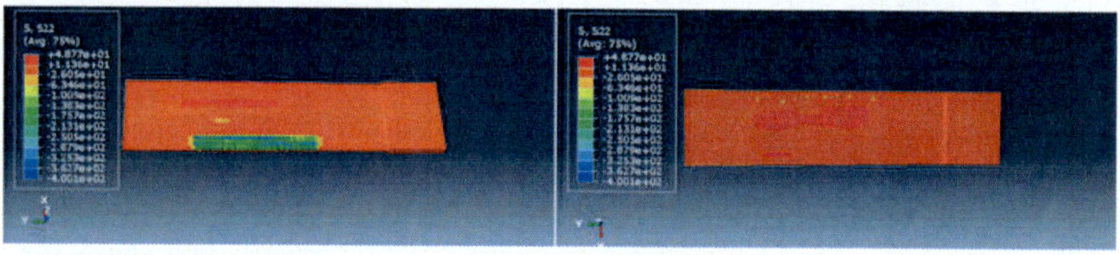

Fig. 8 Stress distribution on both sides under both ends constraint

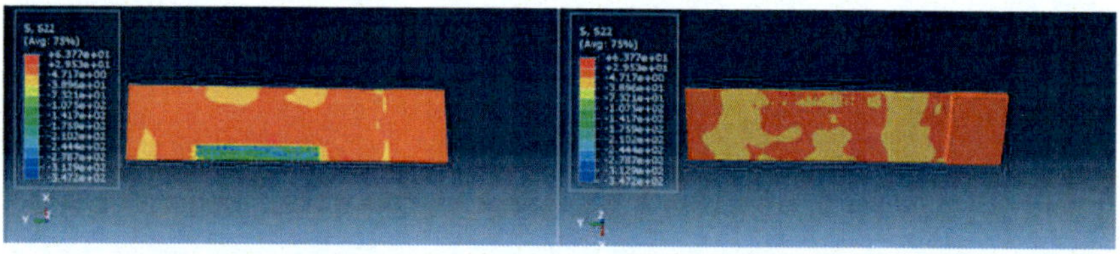

Fig. 9 Stress distribution on both sides under cantilever constraint

The peak pressure must at least exceed the Hugoniot elastic limit to form effective plastic deformation, achieve the purpose of strengthening.

The solution formula of Hugoniot elastic limit of material is:

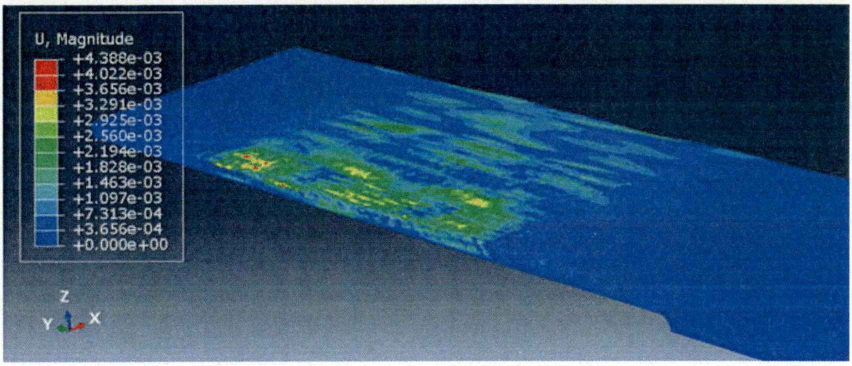

Fig. 10 Deformation under bottom constraint

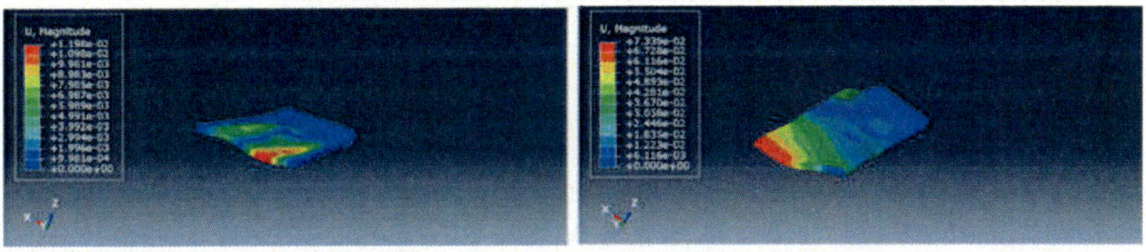

Fig. 11 Deformation under both end constraints and cantilever constraints

$$\sigma_{HEL} = (\frac{K}{2G} + \frac{2}{3})\sigma_{0.2} \tag{6}$$

In the formula, σ_{HEL} is the Hugoniot elastic limit, K is the bulk modulus and G is the shear modulus. It can be obtained that the Hugoniot elastic limit of TC17 titanium alloy is 1.836 GPa. The pressure loading in the simulation process is realized by editing the load application subroutine VDLOAD in FORTRAN language.

4 Results and Analysis

4.1 Determination of Dynamic Solution Time

In the numerical simulation of laser shock strengthening using ABAQUS/Explicit solver, the solution time has a great impact on the accuracy and calculation efficiency of the final residual stress distribution results. If the solution time is too short, the elastic rebound may not end and the stress distribution may not be completely stable; if the solution time is too long, it can ensure the completion of the elastic rebound process, but it will lead to low computational efficiency, especially in the simulation of multi-point impact of large-scale model. This paper simulates the actual machining, and there are many machining points. Therefore, it is particularly important to determine a reasonable dynamic solution time.

The finite element model shown in Fig. 2 is established and solved dynamically by ABAQUS/Explicit solver. Using 5 J energy for simulation, make the solution time sufficient to meet the needs of other conditions. Whether the elastic rebound tends to be completed can be clearly judged by the change trend of energy in the impact process. The overall energy curve and various stress data at the center of the impact point are extracted, and the horizontal stress and the variation trend of the total stress with time are shown in the figure. It can be concluded that the energy and stress tend to be stable after 8000 ns, so choosing 10000 ns as the dynamic solution time is a reasonable choice.

4.2 Restraint Mode on Residual Stress Distribution and Deformation

Different boundary conditions will greatly affect the propagation of shock wave, and finally affect the residual stress distribution and plastic deformation of parts after laser shock strengthening. Different constraint methods will also change the difficulty of deformation during machining, resulting in the difference of machining deformation. Improper constraint will also increase the macro deformation of parts. Reasonable restraint or certain pre-stress before processing will help to accurately control the shape parameters of parts and reduce deformation.

In order to study the effects of different restraint methods on blade residual stress and deformation, under the condition of maintaining the same laser parameters (laser energy 3 J, peak pressure 3.072 GPa, pulse width 20 ns, spot diameter 3 mm, 30%overlap rates), three methods of cantilever restraint, end restraint and bottom beam restraint were used for laser shock respectively. When the cantilever is restrained, fix the end of the blade root; when both ends are constrained, the blade root and free end are fixed and suspended in the middle; the bottom constraint is to make the whole lower surface contact with the fixture.

It can be seen from the figure that in the impact area, after the three impact methods are processed, the residual compressive stress on the upper surface fluctuates between 100 and 200 MPa, with little difference. However, from the distribution of residual stress on the bottom surface, it can be seen that the influence depth of bottom constraints is the largest, the effects of both end constraints and cantilever constraints are less than bottom constraints, and the distribution of residual stress under bottom constraints is more uniform.

In order to more intuitively reflect the deformation of the blade under the three constraints, the deformation proportion of the cloud diagram is enlarged by 100 times. The deformation under the constraint of the bottom surface shown in the figure is only reflected on the blade surface, and the scale is small, while the constraints on both sides and cantilever will produce the overall deformation and torsion of the blade.

In conclusion, the use of bottom constraint can not only ensure the strengthening effect, but also better reduce the machining deformation.

4.3 *Influence of Different Energy Parameters on Impact Effect*

Based on the bottom constraint, the appropriate energy parameters are studied. The laser energy of 2, 3 and 5 J are selected for multi-point impact strengthening under other conditions. As shown in Fig. 12, the stress distribution from top to bottom is 2, 3 and 5 J energy respectively. The surface path of the blade processing area is selected to draw the residual stress curve, which roughly reflects the magnitude of the residual compressive stress of the blade under different energy. The average value of the residual compressive stress under 2 J energy is about 104 MPa, the average value of the residual compressive stress under 3 J energy is about 247 MPa, and the average value of the residual compressive stress under 5 J energy is 450 MPa. Due to the limited blade size, there will be rebound of shock wave in the boundary area. In addition, there are residual stress holes [14] in the impact process, resulting in slightly uneven distribution of stress.

The residual stress distribution in the blade depth direction is shown in Fig. 13, under different energies, the residual stress decreases in the depth direction, but the residual stress on the surface and back is the largest under 5 J energy, the strengthening effect is the best.

The deformation curves under three different energies are shown in Fig. 14. For this simulation, the three parameters meet the requirements, will not cause excessive deformation, and are only controlled in the range of 4 μm. Therefore, to sum up, the energy parameter of 5 J is more appropriate under the bottom constraint, which can not only strengthen the blade, but also avoid large deformation.

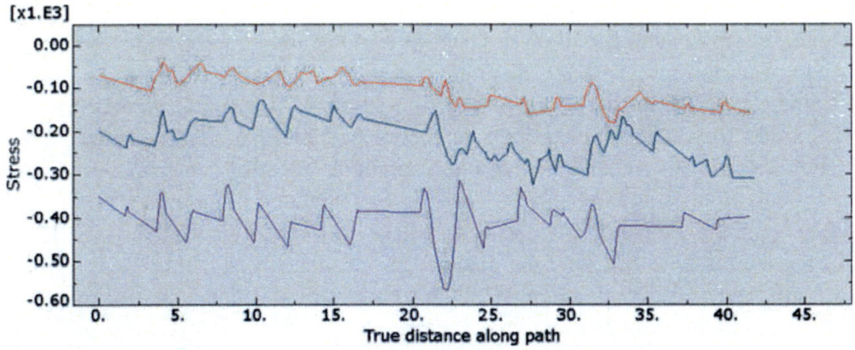

Fig. 12 Surface stress distribution curve under different energy

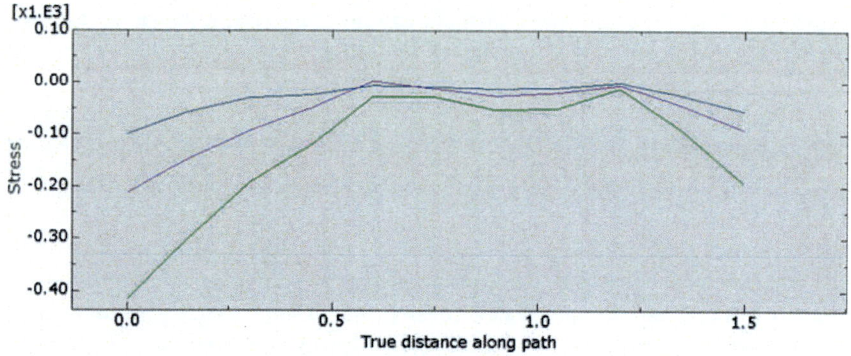

Fig. 13 Residual stress distribution curve in depth direction under different energy

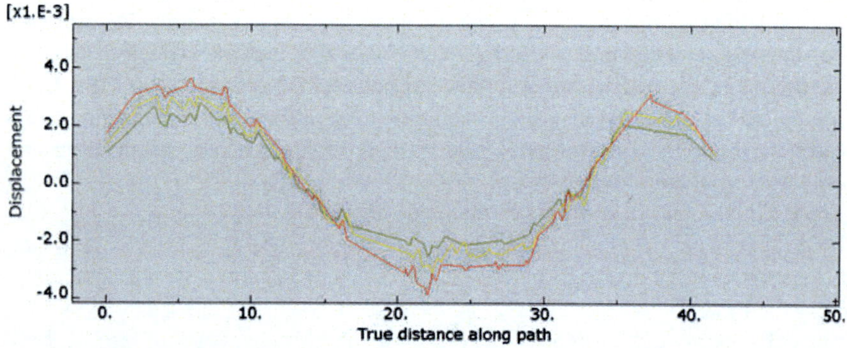

Fig. 14 Deformation curve under different energy

5 Conclusion

A constant section blade model is established to simulate the impact of actual machining on the blade. Based on the change trend of total energy and stress, the dynamic solution time is 10000 ns. There are great differences in the shape variables produced by the impact strengthening simulation blade under different restraint modes. Under the cantilever restraint state, the impact deformation is the largest, the two end restraint is the second, the bottom restraint deformation is the smallest, and the stress distribution is the most uniform. In this simulation, the impact with 5 J laser energy, 3 mm diameter spot and 20 ns pulse width can produce about 450 MPa residual compressive stress, and the blade edge deformation is controlled at about 4 μm. This way of using numerical simulation to control machining parameters has reference significance for practical machining.

References

1. Brown B, Aaron M (2001) The politics of nature. In: Smith J (ed) The rise of modern genomics, 3rd edn. Wiley, New York
2. Dod J (1999) Effective substances. In: The dictionary of substances and their effects. Royal Society of Chemistry. Available via DIALOG. http://www.rsc.org/dose/title of subordinate document. Cited 15 Jan 1999
3. Slifka MK, Whitton JL (2000) Clinical implications of dysregulated cytokine production. J Mol Med. https://doi.org/10.1007/s001090000086
4. Smith J, Jones M Jr, Houghton L et al (1999) Future of health insurance. N Engl J Med 965:325–329
5. Xue-de WANG, Lei YANG, Xin ZHOU et al (2012) Residual stress and microstructure of laser shock peened layer of titanium alloy. Mater Mech Eng 36(4):77–79
6. Hong-chao QIAO, Ji-bin ZHAO, Yan-feng YU (2013) Analysis on laser processing parameters and processing effects of laser peening of TC4 titanium alloy. Laser Optoelectron Progr 50(4):139–144
7. Wang XY, Wang YF, Jiang H, et al. (2014) Laser cladding forming of round thin-walled parts with slope angle. Chinese J Lasers 41(1):0103006
8. Qipeng LI, Weifeng HE, Chonglou TONG et al (2008) Study of laser shock processing used in aero engine blades. Aviat Precis Manufact Technol 44(4):37–39
9. Braisted W, Brockman R (1999) Finite element simulation of laser shock peening. Int J Fatigue 21(7):719–724
10. Ivetic G, Meneghin I, Troiani E, et al. (2012) Fatigue in laser shock peened open-hole thin aluminium specimens. Mater Sci Eng A 534:573–579
11. Zhang XQ, Li H, Duan SW et al (2015) Modeling of residual stress field induced in Ti-6Al-4V alloy plate by two sided laser shock processing. Surf Coat Technol 280:163–173
12. Bhamare S, Ramakrishnan G, Mannava SR et al. (2013) Simulation-based optimization of laser shock peening process for improved bending fatigue life of Ti-6Al-2Sn-4Zr-2Mo alloy. Surface Coat Technol 232:464–474
13. Yinghong LI (2013) Laser shock peening theory and technology. Science Press, Beijing
14. Shabani-Nooshabadi M, Ghoreishi SM, Behpour M (2009) Electropolymerized polyaniline coatings on aluminum alloy 3004 and their corrosion protection performance. Electrochim Acta 54:6989–6995
15. Fenghuai Y (2019) Numerical simulation study on laser shock of TC4 titanium alloy blade simulator. Guangdong University of technology
16. Fabbro R, Fournier J, Ballard P et al (1990) Physical study of laser-produced plasma in confined geometry. J Appl Phys 68:775
17. Peyre P, Berthe L, Scherpereel X et al (1998) Experimental study of laser-driven shock waves in stainless steels. J Appl Phys 84(11):5985–5992
18. Yongxiang H (2008) Numerical modeling of laser shock treatment process and Study on shock effect. Shanghai Jiaotong University

Aerodynamic Study for Wing Shapes Based on a Stuffed Bird

Y. Tanahashi, T. Kato, M. Amano, M. Sagata, N. Kishimoto, and T. Ikeda

Abstract A fundamental aerodynamic study for the effect of wing shape, based on a stuffed bird, is presented. Birds fly efficiently and optimally under various conditions by changing the shape of the wing quickly during flight. Aerodynamic forces on bird models with several flow conditions generated by changing the shape of the wing are examined with wind tunnel experiments and numerical analyses to understand how the aerodynamic characteristics change with the wing shape of the stuffed bird using similar 3D printer models. The goal is to explore the optimum wing shapes, and flow control devices according to flight conditions and flight methods of delaying flow separation on the wing surface at the maximum lift and controlling stall. The results show that the aerodynamic characteristics of the bird models were confirmed through the experiment and the analysis. The lift coefficients of the bird models are relatively close in both data obtained in the wind tunnel test and the numerical analysis, except for the stall angles. It is considered that the stall angles caused errors due to angle of attack adjustment, surface roughness of the similar 3D printer models to the stuffed bird, and the method of data reduction of the experiments and the analyses. The stall angles of the bird models occur at angles of attack of 20 to 25 degrees, which is higher than those obtained with conventional wings. From the flow visualization patterns of the experiments and the analyses, the angle of attack where the stall occurred due to the separation of the wing clarified that the shape of the leading edge of the wing effectively prevented the separation on the bird wing, where a simple shaped slat delayed separation and controlled the stall moderately.

Y. Tanahashi (✉) · M. Amano · M. Sagata · T. Ikeda
Department of Astronautics and Aeronautics, Chubu University, 1200 Matsumoto-cho, Kasugai, Aichi 487-8501, Japan
e-mail: tanahashi@isc.chubu.ac.jp

T. Kato
Department of Mechanical Engineering, Chubu University, Kasugai, Aichi, Japan

N. Kishimoto
Department of Mechanical Engineering, Setsunan University, Neyagawa, Osaka, Japan

S. Lee et al. (eds.), *The Proceedings of the 2021 Asia-Pacific International Symposium on Aerospace Technology (APISAT 2021), Volume 1*, Lecture Notes in Electrical Engineering 912, https://doi.org/10.1007/978-981-19-2689-1_74

Keywords Stuffed bird · Wing shape · Aerodynamic characteristics · Flow separation

1 Introduction

Birds fly efficiently and optimally under various conditions by changing the shape of the wing quickly during flight. A lot of research has been conducted on morphing wings that continuously change shape like bird wings, leading to fuel reduction by reducing drag. While learning from the flight of birds, the development of AI technology and communication technology, which have made remarkable progress in recent years, has attracted new attention in the field of unmanned aerial vehicles for research to optimize flight characteristics by wing deformation.

In this research, the basic aerodynamic characteristics of birds are investigated, which is the basis of airplane development, by wind tunnel tests and numerical analyses, examining the flight performance of birds, and exploring application methods for unmanned aerial vehicles. The goal is to explore the optimum wing shapes, flow control devices according to flight conditions and flight methods for delaying flow separation on the wing surface at the maximum lift and controlling stall.

2 Bird Model Descriptions

2.1 Bird Wing

The structure of a bird's wing is similar to that of a human arm in some respects, as shown in Fig. 1. A bird's wing is divided into three parts: the tertiary feather (upper arm), the secondary feather (forearm), and the primary feather (fingers). These parts are used to change the wing into the optimal shape under various flight conditions to obtain optimal gliding performance. First, folding the upper arms causes forward and backward motion, which changes the wing area of the forearms and changes the overall shape of the wing. In addition, by changing the angle of the forearm during flapping, the leading edge of the wing is turned downward during downward

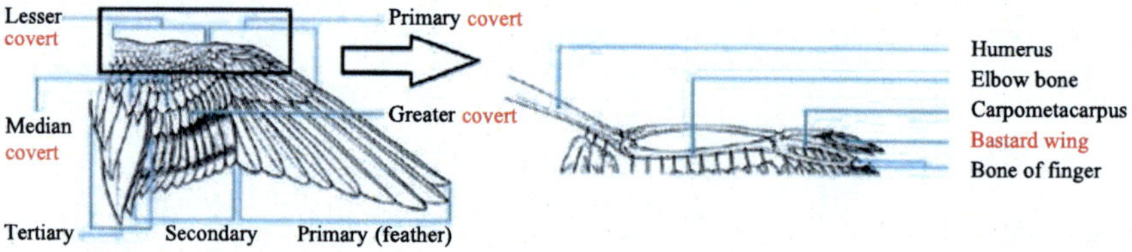

Fig. 1 Structure of bird wing [1]

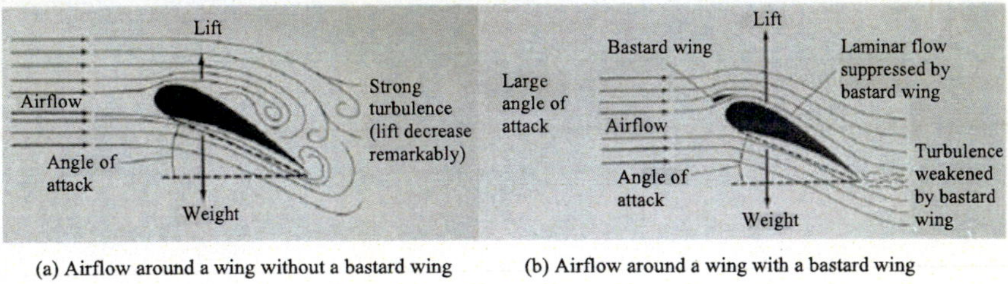

Fig. 2 Effect of bastard wing [2]

flapping, and the leading edge is turned upward during upward flapping, thereby adjusting the angle between the secondary feather and the primary feather [1].

Each feather has a different function, which is important for flying. Roughly 10 pieces of long and narrow primary feathers located at the tip of the bird wing flap to generate propulsion. About 6–10 pieces of short and thick secondary feathers located near the center of the wing generate lift. The tertiary feathers located near the root of the wing on the body side reduce drag and increase lift by filling the gap between the wing and body during flight and regulating the airflow between the wing and the upper surface of the body. Other wing feathers, such as the covert and bastard wing near the wing bone, are used during takeoff and landing to maintain lift by delaying wing surface separation and regulating airflow. The bastard wing is the three or so feathers on the leading edge of the primary feather on the wing, and corresponds to the first finger of the hand. As shown in Fig. 2, the bastard wing creates a gap between the wing feathers and the leading edge of the wing, which prevents a large separation of the airflow over the wing surface and delays stalling when the angle of attack of the wing increases at low speeds, such as just before landing. In addition, the 12 tail feathers located on the tail of the bird are used to maintain flight stability, change direction, and reduce flight speed. These feathers allow the bird to achieve optimal gliding performance [2].

2.2 *Measurement of Stuffed Bird and Setting of Deformed Wing*

The starling stuffed bird used in the experiment is shown in Fig. 3a. The starling has a total length of about 24 cm, a wingspan of about 40 cm, and a weight of about 70 to 90 g. They fly in groups of 100–10,000 at a cruising speed of about 11.1 m/s, flapping their wings, gliding, and sailing. The number of feathers and the flat shape of the starlings were measured visually and with a scale. Then, as shown in Fig. 3b, the wing cross-sectional shape was measured using a wire, seven sections on each side. The planar shape image of the stuffed bird was measured using the painting software of a personal computer.

(a) Measurement of Wing 1 (b) Wing section of wire (c) Wing 2 based on Wing 1 (d) Wing 3 based on Wing

Fig. 3 Setting of deformed wings based on Wing 1

Specifically, for image measurements, a point was placed at the desired location and the pixel between the two points was calculated, and the distance between the two points was calculated as 10 mm = 31.7 pixels. The number of feathers on the wing was visually measured: 9 for the primary feather, 6 for the secondary feather, 3 for the tertiary feather, and 12 for the tail feather. While the tail feathers were horizontal, the main wing feathers had a stronger camber in the order of, the primary feather < the secondary feather < the tertiary feather. Two types of wing planform shapes were predicted for the case of wingspan shortening, as shown in Fig. 3c and d, and the wings were set up by mapping the wings according to the movement of the outer wing feathers.

2.3 Design of the Bird Model

For the design of the bird model using 3D CAD, the measurement results of the aforementioned stuffed bird were read into 3D CAD SolidWorks to create solid data, and the model was divided into three parts (body, left wing, and right wing) considering the limitation of the 3D printer's modeling size (310 × 295 × 320) mm, the work efficiency of the wind tunnel experiment, and the accuracy of the experiment. The body was designed to have a common shape. Therefore, it was designed as a simple shape for convenience in setting up and handling the model. Figures 4 and 5

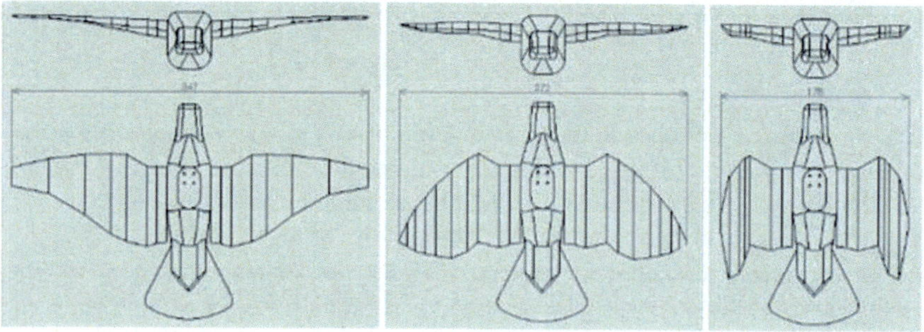

(a) Wing 1 (b) Wing 2 (c) Wing 3

Fig. 4 Drawings of Wing 1 to Wing 3

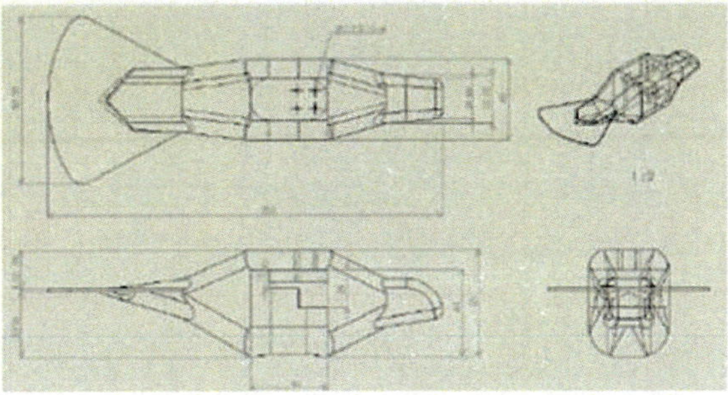

Fig. 5 Drawing of the body model

show the assembly drawings of Wings 1 to 3 and the three-dimensional view of the body, respectively.

2.4 *Modeling of the Bird Model Using a 3D Printer*

We used Simplify3D, a slicer software that converts the design data into G-Code for 3D printers, to set the stacking speed and materials, and fabricated the bird model with Moment2 3D printer. There are several types of 3D printers, such as fused deposition modeling, stereolithography, and powder lamination method, but fused deposition modeling is the most commonly used method to create 3D objects by melting solid materials such as ABS resin and PLA resin at a high temperature of around 260 °C, and then layering them one by one while injecting them from a nozzle. The disadvantage of fused deposition modeling is that it often leaves lamination marks on the object and requires a support material. For this reason, in this study, we used the fused deposition modeling to form the model using PLA resin; the conformed bird models of Wings 1 to 3 formed by the 3D printer are shown in Fig. 6. This also shows a comparison of the starling stuffed bird and Wing 1 of the 3D printed bird model. The wing area of the modeled wing is $S_1 = 0.0173\ m^2$, $S_2 = 0.0164\ m^2$, and $S_3 = 0.0119\ m^2$, respectively.

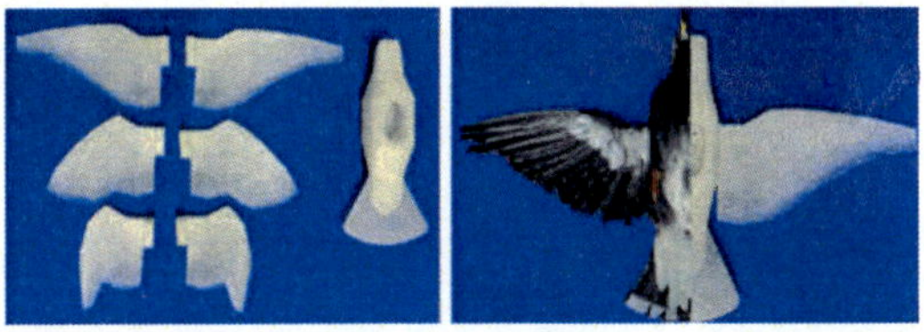

Fig. 6 3D printed wings and a body compared with the original stuffed bird

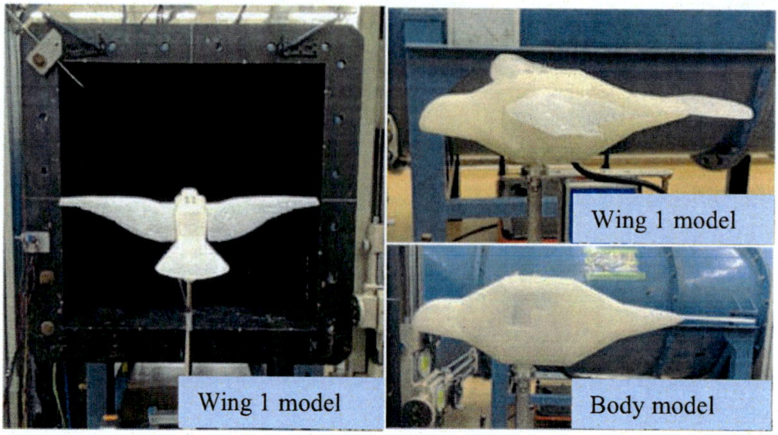

Fig. 7 Setting of 3D printed Wing 1 model and the body model

3 Wind Tunnel Experiment

3.1 Overview of the Wind Tunnel Facility

A Goettingen (circular flow) type low-speed wind tunnel (measurement section dimensions: 0.5 m × 0.5 m × 0.9 m long, maximum wind speed: approx. 60 m/s) at Chubu university was used. The 6-component force sensor is a compact and lightweight 20 g Leptrino sensor, model CFS018CA101U (main dimensions: φ18 mm × 23 mm in length × 26 mm in height, capacity: Fx, Fy: ± 50 N, Fz: ± 100 N, Mx, My: ± 0.5 Nm, Mz: ± 0.5 Nm) [3]. The sensors were calibrated by varying the load on the forces in the three axes and the moment around the three axes, and the six-component force interference correction was performed. Photographs of the wind tunnel installation of the whole Wing 1 model and the body model are shown in Fig. 7.

3.2 Overview of the Wind Tunnel Experiment

For the 6-component force measurement of the bird model, measurements were taken for each wing and body at wind speeds of 5, 10, and 15 m/s, respectively, and angles of attack from −15 to 30° in increments of 5°. The data were averaged over 30 s at a sampling rate of 100 Hz. As the angle of attack of the bird model increases, it is thought that the measured values are affected by the flow into the gap of the 6-component force sensor at the bottom of the body, so in this study, the measured values of the wing were obtained by subtracting the data of the body configuration from the data of the all-plane configuration of each bird model. The drag and lift

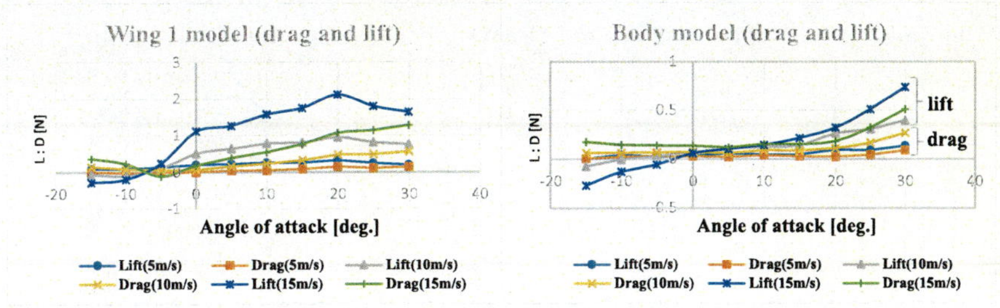

Fig. 8 Measured forces of drag and lift with Wing 1 model and the body model

forces of the full-plane configuration (Wing 1) and the body configuration are shown in Fig. 8 as an overlay of the output values at each air speed. For the processing of the wind tunnel experiment data, the mean and standard deviation of the data output from the six component force sensors were summarized for each flow velocity for each bird model and body. The error bars of the standard deviation are omitted from the figure because they are extremely small in the graph.

3.3 *Wind Tunnel Test Results and Discussion*

The angle-of-attack variation of the drag and lift coefficients for each bird model configuration is shown in Figs. 9 and 10. In the lift coefficient of each wing, which was obtained by subtracting the body configuration from the total configuration of each bird model, it was confirmed that stalling occurred around 20° angle of attack.

The results of the wind tunnel tests on each wing showed that the aerodynamic characteristics of the drag and lift coefficients at wind speeds of 10 and 15 m/s were relatively close to each other. On the other hand, the coefficients at wind speed of 5 m/s showed a large change with respect to the angle of attack. This is thought to be due to the fact that, the detected value of the 6-component force sensor was small and that the fluctuation was large due to the slow flow speed. On comparing the drag

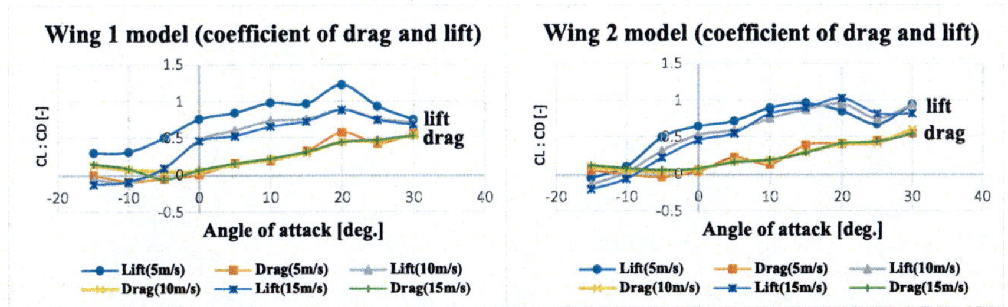

Fig. 9 Measured force coefficients of drag and lift with Wing 1 model and Wing 2 model

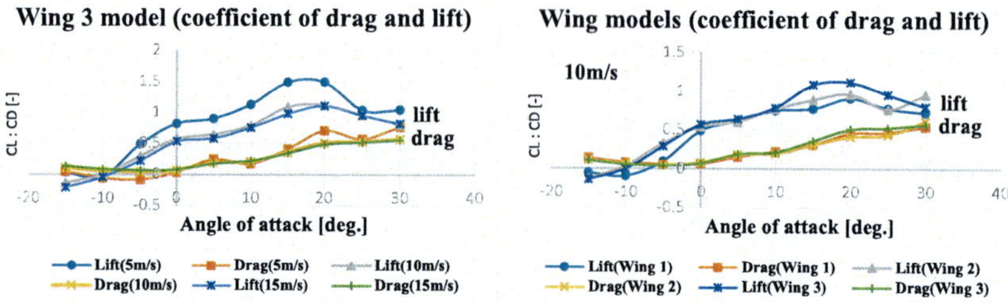

Fig. 10 Measured force coefficients of drag and lift with Wing 3 model and comparison at airspeed 10 m/s

and lift coefficients of each wing with respect to the change in wing shape, the lift coefficient of the wing with the smaller wing area became larger, which means that a relatively large lift force was generated with respect to the wing area due to the 3D effect of the wing aspect ratio, stall angle, chord length, camber, etc.

4 3D Numerical Analysis

4.1 Overview of 3D Numerical Analysis

The 3D analysis was performed with SolidWorks Flow Simulation, a fluid simulation tool using the finite volume method, based on the 3D data of the bird model used in the wind tunnel experiment. Figure 11 shows the shape of each airfoil analyzed in

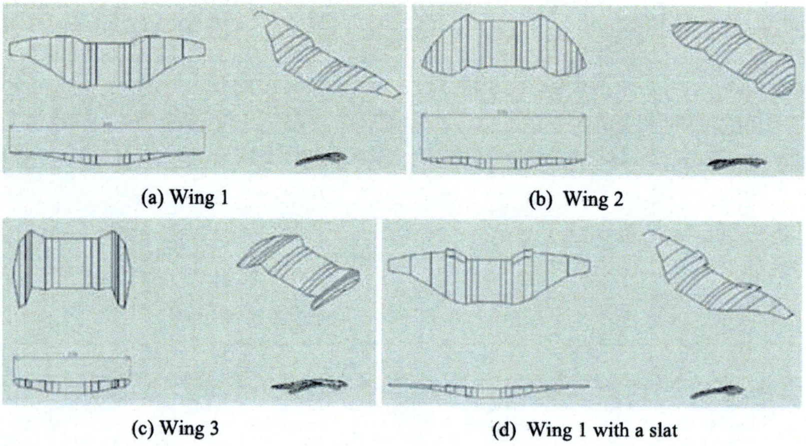

Fig. 11 Drawings of Wing 1 to Wing 3 and Wing 1 with a slat

3D, as well as the shape of the wing with the addition of the functional slats, which slow down the separation and prevent stalling, to Wing 1.

4.2 Results of 3D Analysis and Discussion

In the 3D analysis, three types of wings were analyzed in increments of 5° from 15 to 30° angle of attack, as in the wind tunnel experiment. The analytical results of the drag coefficient and lift coefficient of each wing are shown in Fig. 12.

When the results of the 3D analysis are compared with the results of the wind tunnel experiment, it can be seen that the drag forces show good agreement between the 3D analysis and the experiment. The 3D analysis also shows that the values of drag and lift decrease as the wing area decreases, as planned when the wing shape was designed.

As for the stall, the results of the wind tunnel experiment showed that the stall occurred at around 20° angle of attack, while the 3D analysis showed that the stall occurred at around 25° angle of attack. The reason for this is that the wind tunnel test results are the values obtained by subtracting the body-only configuration from the bird model configuration, while the numerical analysis shows the wing-only configuration; so, the experimental values include the interference components of the wing and body. In addition, the roughness of the wing surface and the accuracy of the shape near the leading edge of the wing produced by the 3D printer, the adjustment of the angle of attack during the experiment, and the analysis conditions are considered.

In this study, the slats designed to simulate the stall suppression of small wing feathers were added to the leading edge of the wing, which slightly reduced the value of the maximum lift coefficient but tended to mitigate the change in the lift coefficient with angle of attack; however, it did not delay the stall. Therefore, in the design of slats, it is necessary to conduct case studies on the shape, relative position (vertical and longitudinal gap) and angle with the leading edge of the main wing, as well as

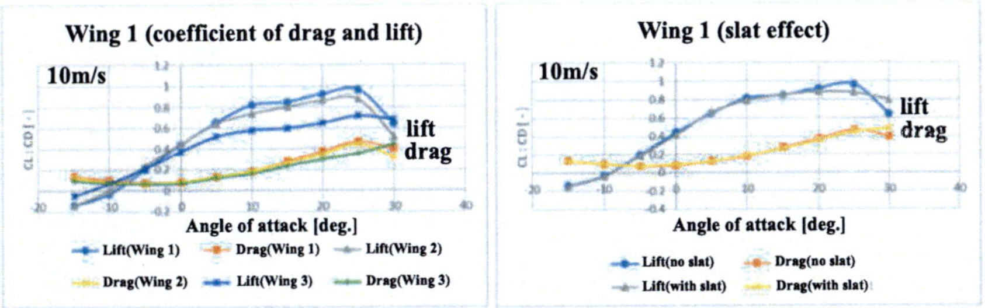

Fig. 12 Analyzed force coefficients of drag and lift with Wing 1 to Wing 3 and Wing 1 compared to that with a slat

the analysis method, and compare and examine the aerodynamic characteristics with those of the stuffed wing.

5 Flow Visualization

5.1 Flow Visualization Method

In order to study the drag and lift characteristics, flow visualization was conducted. The flow field near the wing was observed in a wind tunnel experiment using a simple tuft method. For a more detailed study, SolidWorks Flow Simulation was used to visualize flow path lines as shown in Fig. 13.

5.2 Flow Visualization Results and Discussion

In the flow visualization of Wing 1, airflow in the vicinity of the wing was first visualized on a 2D airfoil with the cross-section as shown in Fig. 13a, and after confirming the angle of attack and position of the flow separation, the flow field of the 3D wing was visualized as shown in Fig. 13b–d. Changes in the airflow near the wing and the separation can be seen with changes in the angle of attack.

Figure 14 shows the visualization of the flow field of the 2D airfoil and 3D wing with Wing 1 slats at the angle of attack near the point of attachment of the slats (point 2) where separation occurs, which is a simplified model of a small wing feather that suppresses separation.

From the visualization of the flow of Wing 1, it can be inferred that the angle of attack at which separation occurs differs depending on the thickness of the leading edge of the 2D wing, the wing chord length, and the size of the wing camber. At the angle of attack of 30°, where the separation was observed at each point of wing 1, the lift decreased, and the stall was observed. In the comparison between the wing with and without slats at the same angle of attack, a slight change in the flow on the upper surface of the wing was observed, and the change in the lift coefficient

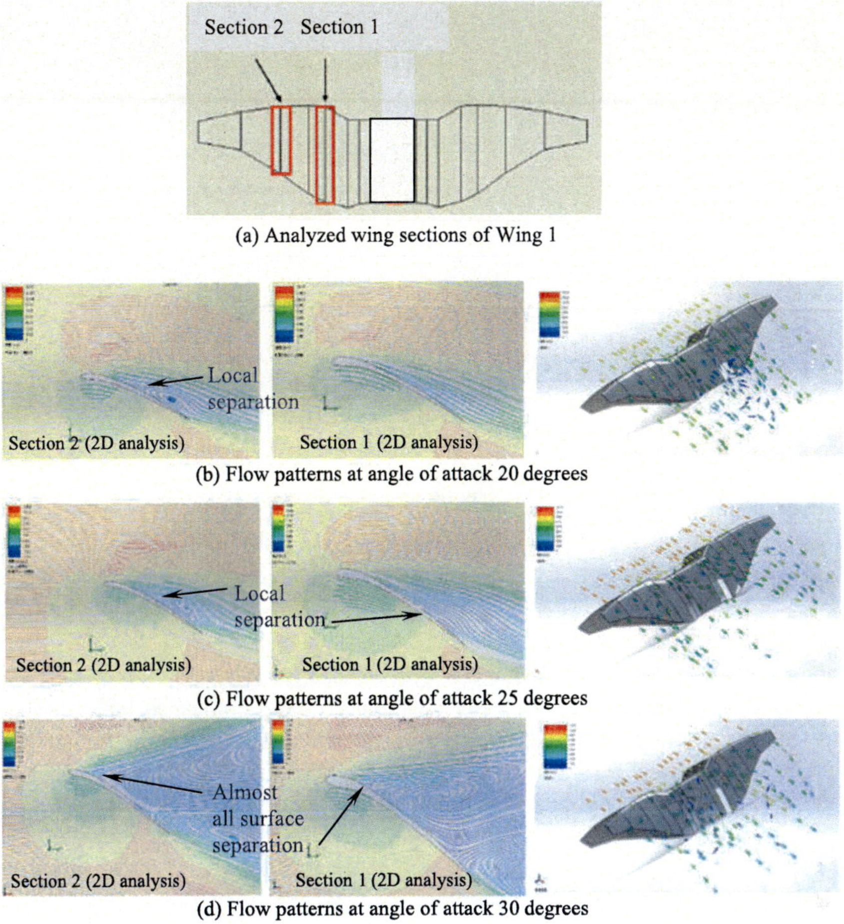

Fig. 13 Analyzed flow patterns with Wing 1

due to the angle of attack was mitigated, but the effect of delaying the separation and suppressing the stall, which was the original purpose, was not fully obtained. However, when the locations where separation tends to occur and the structure of the bird's wing were compared from the visualization of the flow, the locations where separation tends to occur were located in the covert and bastard wing that prevent large separation in the bird's wing, and a relationship between separation and the bird's function could be found.

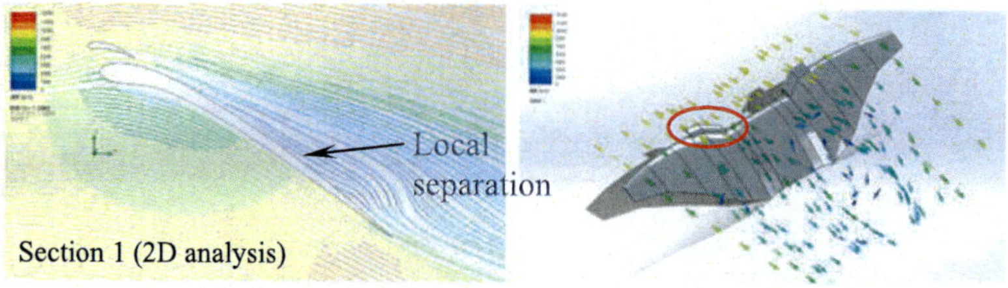

(a) Flow patterns at angle of attack 25 degrees

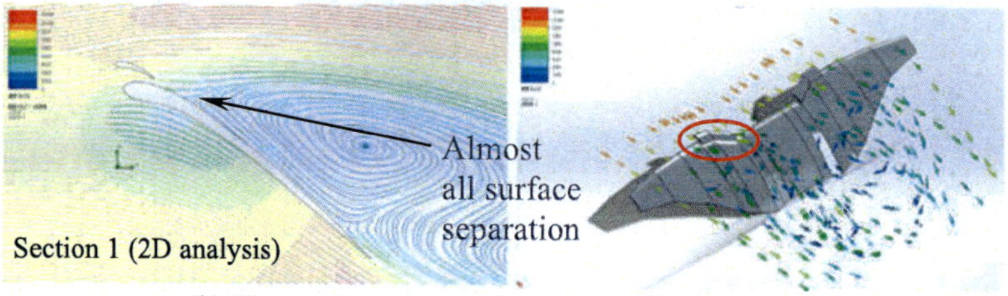

(b) Flow patterns at angle of attack 30 degrees

Fig. 14 Analyzed flow patterns with Wing 1 with a slat

6 Conclusion

The aerodynamic characteristics of the bird models were confirmed through the wind tunnel experiment and the 2D and 3D flow field analyses around the wings. The following findings were obtained from the experimental and analytical results of aerodynamic characteristics and airflow visualization. As a result of comparing the results of the wind tunnel experiment on the bird model with those of the 2D and 3D analyses, the values are relatively close, except for the slightly different stall angle of attack in the wind tunnel experiments.

The reason for the difference in the stall angle of attack is that the wind tunnel test results were obtained by subtracting the body-only configuration from the bird model configuration, while the numerical analysis results were obtained by using the wing-only configuration; so, the experimental values include the aerodynamic interference components of the wing and the relatively large body. In addition, the roughness of the wing surface and the accuracy of the shape near the leading edge of the wing fabricated by the 3D printer, the adjustment of the angle of attack during the experiment, and the analysis conditions are considered.

In the bird model, stalling occurred at around 20–25° angle of attack. From the visualization of the flow, it was found that the angle of attack and the position of the wing where the stall occurred due to flow separation were related to the covert and the bastard wing on the leading edge of the bird wing that prevent the flow separation. The results of the analysis of the slat, in which the bastard wing was modeled in a simplified manner, showed the effect of mitigating the change in the lift coefficient near the stall angle of attack, but the effect of delaying the stall was not fully obtained.

As the next theme, we will examine the pitching moment characteristics and measure the aerodynamic characteristics of stuffed birds and evaluate the deformation effects of feather bending and torsion while attempting to observe actual bird flight conditions. This will lead to improved flight stability and stalling characteristics. The model of variable wing and slats can be constructed. In addition to the angle of attack and speed of the wing, the angle of sideslip can be changed as a parameter, and the stability and control of flight can be examined from the six-component forces generated in the bird model. In the future, the development of a variable wing like a bird wing could be applied to a bird-shaped flying robot that is effective in combating bird and animal damage.

References

1. The Encyclopedia of Japan (1994) Shogakukan
2. Gill FB (2009) Translated by Yamashina Institute of Ornithology, Ornithology, Shinki-Sha
3. Suguru M, Yoshiyuki N (2018) The fundamentals and mechanism of six axis force sensor. J Jpn Soc Precis Eng 84(4)

Investigation of Temperature Generalization for PCM-Based Cooling Plate Used for Remotely Located Electronics

Bo Wu, Bin Wang, Liang Zhao, and Junjie Han

Abstract In the present study, investigation of temperature generalization is performed to characterize the complex heat transfer process within a PCM -based cooling plate used for electronics remotely located in aircraft. Firstly, a simplified physical model is abstracted from the PCM-based cooling plate serving in actual applications where the cooling plate undergoes a complex heat transfer process with thermal conduction, thermal storage, thermal convection, and thermal radiation coupling with each other. Dimensionless numbers are defined to generalize the physical quantities associated with transient heat transfer with phase change. The comprehensive heat transfer process is then numerically simulated with the entropy-porous model to predict the melting process of PCM. Transient dimensionless temperature curves are presented in terms of dimensionless numbers including Rayleigh (*Ra*), Fourier (*Fo*), and Stefan (*Ste*) numbers. Transient heat fluxes of natural convection and thermal radiation during the whole heat transfer process are characterized by regression analysis. Finally, a modified product of dimensionless numbers is derived based on the results of regression analysis. It is found that the temperature could be generalized well with the modified product of dimensionless numbers. The results could provide guidelines for generalizing the performance of PCM-based cooling plates utilized in remotely located electronics and could be served as a quick design tool.

Keywords Aircraft · Temperature generalization · Phase change material · Cooling plate · Remotely located electronic

1 Introduction

Phase change material (PCM) -based cooling plate is one kind of efficient thermal management subassembly for remotely located electronics used in aircraft [1, 2]. It is very suitable for resolving short-period thermal load within narrow space as

B. Wu (✉) · B. Wang · L. Zhao · J. Han
AVIC Xi'an Aeronautics Computing Technique Research Institute, 710068 Xi'an, Shaanxi, PR China
e-mail: wub003@avic.com

S. Lee et al. (eds.), *The Proceedings of the 2021 Asia-Pacific International Symposium on Aerospace Technology (APISAT 2021), Volume 1*, Lecture Notes in Electrical Engineering 912, https://doi.org/10.1007/978-981-19-2689-1_75

remotely located electronics have. In actual applications, such cooling plate undergoes a complex heat transfer process with thermal conduction, thermal storage, thermal convection, and thermal radiation coupling with each other. There is a serious lack of a general description of this heat transfer process.

It is a general method to use dimensionless numbers to characterize the inherent rules underlying such heat transfer process. Shatikian et al. [3, 4] studied the processes of melting of a PCM-based heat sink with both a constant temperature boundary and a constant heat flux boundary. They proposed corresponding correlations based on numerical parameters studies. Furthermore, Saha and Dutta [5] found that the correlation of Nusselt number (*Nu*) with a single Rayleigh number (*Ra*) is not applicable for all aspect ratios of PCM-based heat sink. Thus they identified an appropriate characteristic length scale for each type of heat sinks. Arshad et al. [6] also reported modified heat transfer correlations combining two kinds of Rayleigh numbers to accounting for natural convection within the PCM better. Also, Motahar et al. [7] proposed an artificial neural network (ANN) based on feed-forward back -propagation multilayered perceptron to predict the Nusselt number. Analytical correlations have been developed to quickly estimate the melting rates by Sharifi et al. [8], and the predictions of the correlations showed a good agreement with those of the detailed model.

In brief, previous studies have clarified the heat transfer rule of PCM-based heat sink well. However, the situation with both natural convection and thermal radiation existing at the top face has not been taken enough consideration. The underlying general rules of this heat transfer phenomenon remains unclear. The method of ANN could achieve high accuracy of predicting for internal datasets but might lose fidelity for external cases. And it doesn't give explicit mathematical equations to characterize the rules. Thus, a semi-theoretical and semi-empirical method is utilized in the present work to generalize the temperature of PCM-based cooling plate having a complex heat transfer process with thermal conduction, thermal storage, thermal convection, and thermal radiation coupling with each other. A two-dimensional PCM-based cooling plate with constant heat flux at the bottom face is involved in this study and the effect of heat transfer at the top face is emphasized.

2 Description of the Physical Problem and Mathematical Formulation

The physics of the problem is introduced in this chapter, and dimensionless numbers are defined to characterize the inherent rules of the melting phenomena. Also, the numerical method of how we solve the heat transfer process with phase change is described in detail.

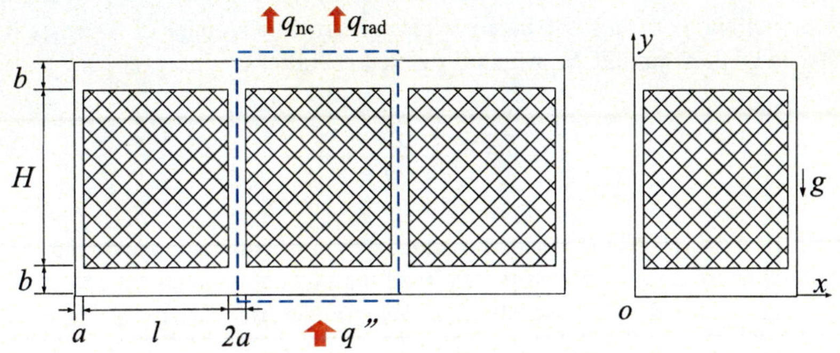

(a) (b)

Fig. 1 **a** Schematically diagram of the typical unit of the PCM-based cooling plate, **b** the representative domain

2.1 Physics of the Problem

A typical unit of the two-dimensional PCM-based cooling plate used for remotely located electronics is schematically illustrated in Fig. 1a. The PCM is embedded between the fins. The bottom face is supplied by a constant heat flux, whereas the heat transfers into the ambient by natural convection and thermal radiation at the top face. The geometry sizes (*a, b, H, l*) of the PCM-based cooling plate are the same as those in the author's previous study [9]. The heat flux at the bottom face is uniformly distributed, and it is varied as 1784 W/m^2, 3123 W/m^2 and 4462 W/m^2. The heat fluxes of natural convection and thermal radiation q_{nc} and q_{rad} are also defined as the same as those in the author's previous study [9]. RT80 is employed as the PCM material in the present study. The heat sink is made of aluminum 6061. The thermal properties of RT80 and aluminum are introduced in the literature [10].

A representative domain (as shown in Fig. 1b) is chosen for numerical simulation to reduce the computational cost. The coordinate system is shown in Fig. 1b, the value of g is equal to 9.81 m/s^2.

2.2 Non-Dimensional Equations

The Fourier number (*Fo*) is utilized to characterize the dimensionless time, which is defined as follows:

$$Fo = \frac{\alpha t}{l^2}, \quad (1)$$

where α refers to the thermal diffusivity, t refers to the time.

The Stefan number (*Ste*) is utilized to characterize the amount of superheating experienced by the liquid PCM, which is defined as follows:

$$Ste = \frac{q'' c_p l}{\lambda_p L_p}, \tag{2}$$

where c_p, λ_p, and L_p refers to the specific heat, the thermal conductivity and the latent heat of PCM, respectively, and l is the width of the PCM zone as shown in Fig. 1.

The Rayleigh number (*Ra*) is utilized to account for the effect of natural convection in molten PCM, which is defined as follows:

$$Ra = \frac{g \beta_p \rho_p^2 c_p q'' H^4}{\lambda_p^2 \mu_p}, \tag{3}$$

where β_p, ρ_p, and μ_p refers to the thermal expansivity, the density and the viscosity of PCM, respectively, and H is the height of the PCM zone as shown in Fig. 1.

A new parameter Z is defined to characterize the dimensionless temperature, which is given as follows:

$$Z = \frac{T - T_i}{T_m - T_i}, \tag{4}$$

where T refers to the base temperature of the PCM-based cooling plate. T_i refers to the initial temperature of the computation, which is equal to 343.15 K. And T_m refers to the melting temperature of PCM, which is equal to 353.15 K.

Another parameter w is defined to transfer the heat flux supplied to the bottom face to a dimensionless one, which is given as follows:

$$w = \frac{q''}{q_o}, \tag{5}$$

where q_o is equal to 1784 W/m^2.

2.3 Numerical Simulation

The entropy-porous method is applied to simulate the transient melting process of the PCM-based cooling plate's representative domain. The flow of liquid PCM when melting is assumed to be laminar. Thermophysical properties of PCM are constant in each phase, except that the density is assumed to match the Boussinesq hypothesis. The thermal resistance is omitted between the metal and the PCM. The government equations of above-described heat transfer phenomena are the same as those in literature [11]. The bottom face has a boundary with constant heat flux, while the top face has a boundary with varied heat flux as described above. The left and right sides of the representative domain are symmetric boundaries. As a contrast, the cases with the adiabatic conditions at the top face are also calculated.

3 Results and Discussions

Dimensionless temperature curves are presented in terms of two different products of dimensionless numbers for cases with and without the natural convection and thermal radiation at the top face. Then the characteristics of the heat flux of natural convection and thermal radiation are examined by regression analyses. Finally, a modified product of dimensionless numbers accounting for the natural convection and thermal radiation at the top face is achieved to better generalize the temperature.

3.1 Dimensionless Temperature

According to the studies of Shatikian et al. [3, 4], the transient melting fraction and Nusselt number curves could converge as function of the product of the Fourier, the Stefan, and the Rayleigh numbers. Thus, it is believed that the temperature could also be generalized by such product of dimensionless numbers. To verify this hypothesis, the dimensionless transient temperature curves are illustrated in Fig. 2 as function of products of dimensionless numbers under different power consumptions.

The product of the Fourier and the Stefan numbers, $FoSte$, serves as a fundamental parameter for generalizing heat transfer process with phase change. Firstly, the dimensionless temperature Z derived from cases without natural convection and thermal radiation at the top face is presented as function of $FoSte$ in Fig. 2a under three different heat flux consumptions ($w = 1,1.75$ and 2). Obviously, the temperature curves don't converge, which is expected, because the product, $FoSte$, doesn't account for the effect of motion in the liquid phase of the PCM [3, 4]. Thus, the Rayleigh number, accounting for the motion of the liquid PCM, is employed to form another product, $FoSteRa^{0.27}$. As shown in Fig. 2b, the dimensionless temperature curves converge well as function of the product at this time.

Then, the transient temperature curves derived from cases with natural convection and thermal radiation at the top face are examined. The dimensionless temperature curves are presented as function of $FoSte$ and $FoSteRa^{0.27}$, respectively, as shown in Fig. 2c and d. It is found that the temperature curves don't converge in terms of either product of dimensionless numbers. Furthermore, the temperature curves trend to be horizontal after the melting process has finished, especially under a low heat flux. It is because that a considerable amount of heat is transferred into the ambient through natural convection and thermal radiation at the top face, especially after the

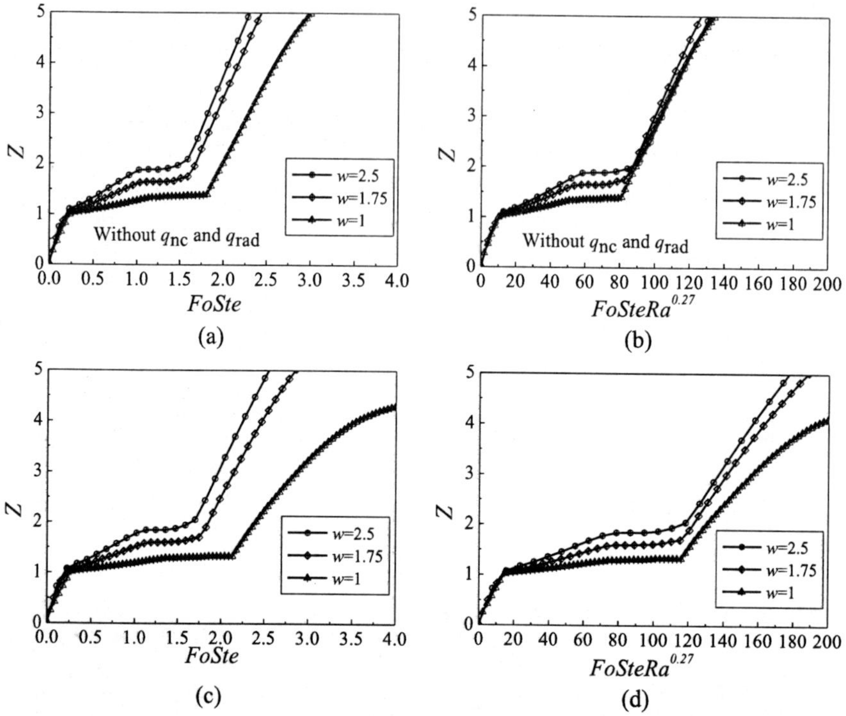

Fig. 2 Dimensionless temperature curves as function of two different products of dimensionless numbers under the conditions with and without natural convection and thermal radiation at the top face. **a** As function of *FoSte*, without q_{nc} and q_{rad}, **b** As function of $FoSteRa^{0.27}$, without q_{nc} and q_{rad}, **c** As function of *FoSte*, with q_{nc} and q_{rad}, **d** As function of $FoSteRa^{0.27}$, with q_{nc} and q_{rad}

melting process has finished. Hence, it is necessary to take the effect of the natural convection and thermal radiation at the top face into consideration for generalizing the temperature.

3.2 Characteristics of the Heat Flux of Natural Convection and Thermal Radiation

To quantify the effect of the natural convection and thermal radiation, two parameters, natural convection ratio (η_{nc}) and thermal radiation ratio (η_{rad}), are defined as follows:

$$\eta_{nc} = \frac{q_{nc}}{q''}, \tag{6}$$

$$\eta_{rad} = \frac{q_{rad}}{q''}, \tag{7}$$

The transient natural convection ratio curves under $w = 1$ and 2.5 are illustrated as function of Fo in Fig. 3a and b. It is found that the natural convection ratio increases following the pattern of the base temperature, and it is smaller totally under a higher w. Notably, both curves of the natural convection ratio are similar to linear ones. Thus, linear regressions are performed to get a deeper insight. As we can see from the results of linear regression in Fig. 3a and b, the Person's coefficient (r) and goodness of fit (R^2) indicate that good fits are achieved. Similarly, the thermal radiation ratio curves under $w = 1$ and 2.5 are illustrated as function of Fo in Fig. 3c and d. The results of regression indicate the same rule. This phenomenon could be approximately described by mathematics equation as follows:

$$q_{\text{nc}} + q_{\text{rad}} = q''(\eta_{\text{nc}} + \eta_{\text{rad}}) = q''(K_1 Fo + K_2), \quad (8)$$

where K_1 and K_2 are unknown parameters.

As we can see from the four regression equations, they are similar but not identical, especially the intercepts. The intercepts are smaller under higher power consumption no matter in terms of the natural convection or thermal radiation, and vice versa. Thus, the parameters, K_1 and K_2 in Eq. (8) may not be constant under different

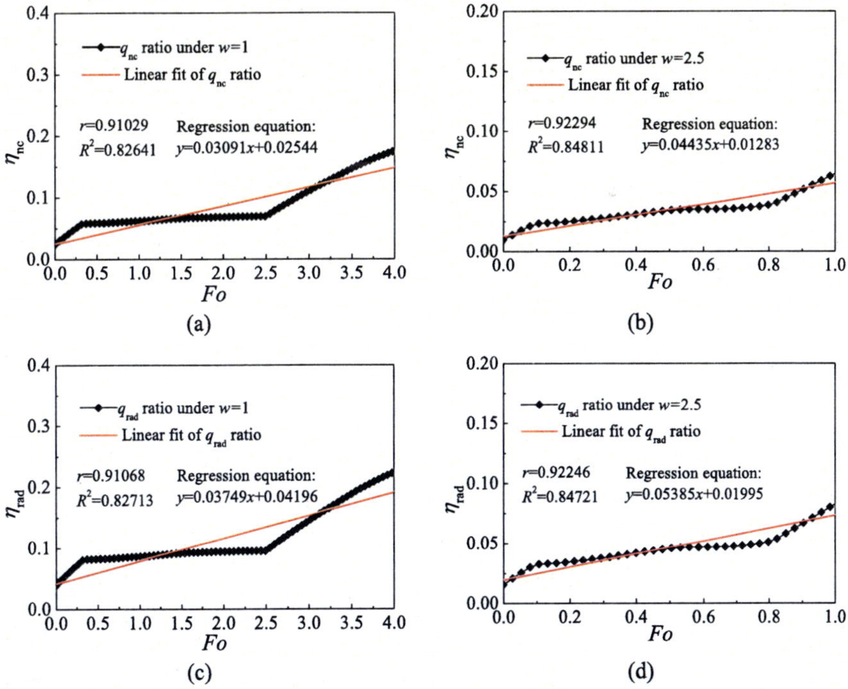

Fig. 3 Transient natural convection ratio and thermal radiation ratio curves under two different power consumptions and related linear regression results. **a** q_{nc} ratio, $w = 1$, **b** q_{nc} ratio, $w = 2.5$, **c** q_{rad} ratio, $w = 1$, **d** q_{rad} ratio, $w = 2.5$

power consumptions. But for simplicity, in the present study, these two parameters are assumed to be constant. They could be approximately quantified by the four slopes and four intercepts as follows:

$$K_1 = \frac{S_1 + S_2 + S_3 + S_4}{4} = 0.04165, \tag{9}$$

$$K_2 = \frac{I_1 + I_2 + I_3 + I_4}{4} = 0.0250, \tag{10}$$

where S_1, S_2, S_3, $S_4$and I_1, I_2, I_3, I_4, are the four slopes and four intercepts in the four regression equations, respectively.

3.3 *Modification of the Product of the Dimensionless Numbers*

The reason why the temperature curves don't converge is that the combination of $FoSteRa^{0.27}$ doesn't account for the natural convection and thermal radiation at the top face. In cases without the natural convection and thermal radiation at the top face, the heat flux *q*" is all absorbed by the PCM-based cooling plate, whereas in cases with them, only part of the heat is absorbed by the cooling plate. Thus, the *q*" included in related dimensionless numbers should be modified, so that it could reflect the real heat flux that is absorbed by the cooling plate. According to the energy conservation principle and Eq. (8), the real heat flux q_{real} could be expressed as:

$$\begin{aligned} q_{\text{real}} &= q'' - (\eta_{\text{nc}} + \eta_{\text{rad}}) = q'' - q''(K_1 Fo + K_2) = q''(1 - K_1 Fo - K_2) \\ &= q''(K_3 - K_1 Fo) \end{aligned} \tag{11}$$

where K_3 is equal to 1-K_2.

Accordingly, modified Stefan number (Ste^*) and Rayleigh number (Ra^*) are defined as:

$$Ste^* = \frac{q_{\text{real}} c_{\text{p}} l}{\lambda_{\text{p}} L_{\text{p}}} = \frac{q''(K_3 - K_1 Fo) c_{\text{p}} l}{\lambda_{\text{p}} L_{\text{p}}} = (K_3 - K_1 Fo) Ste, \tag{12}$$

$$Ra^* = \frac{g \beta_{\text{p}} \rho_{\text{p}}^2 c_{\text{p}} q_{\text{real}} H^4}{\lambda_{\text{p}}^2 \mu_{\text{p}}} = \frac{(K_3 - K_1 Fo) g \beta_{\text{p}} \rho_{\text{p}}^2 c_{\text{p}} q'' H^4}{\lambda_{\text{p}}^2 \mu_{\text{p}}} = (K_3 - K_1 Fo) Ra, \tag{13}$$

Similarly, the previous combination of $FoSteRa^{0.27}$ could be modified as $FoSte^*Ra^{*0.27}$, it is given as:

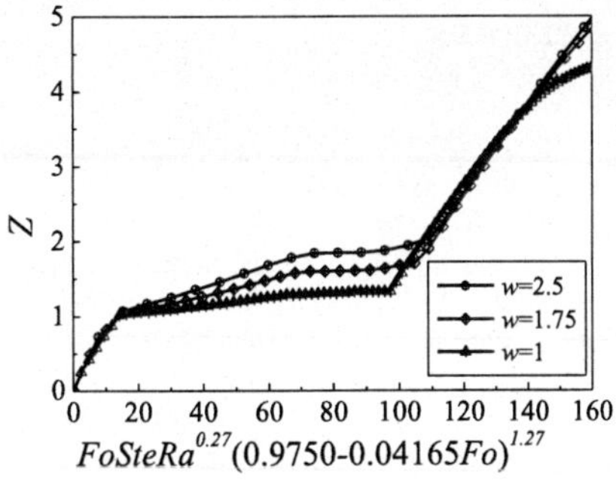

Fig. 4 Dimensionless temperature curves as function of the modified product of dimensionless numbers under the conditions with natural convection and thermal radiation at the top face

$$FoSte^{*}Ra^{*0.27} = FoSteRa^{0.27}(K_3 - K_1Fo)^{1.27}, \tag{14}$$

It is believed that the temperature curves could converge as function of the modified product of dimensionless numbers, $FoSte^{*}Ra^{*0.27}$. The temperature curves are illustrated again in terms of the new modified product, as shown in Fig. 4. It is found that the temperature curves converge well at this time. In another word, the temperature of PCM-based cooling plate could be generalized with the modified product of dimensionless numbers. Thus, a quick method to predict the temperature evolution is achieved without numerical simulations, which has an important significance in the engineering design of PCM-based cooling plates used in remotely located electronics.

4 Conclusions

Based on the numerical results, regression analyses, and dimensionless analyses, the following conclusions are drawn:

1. For the PCM-based cooling plate with natural convection and thermal radiation at the top face, the temperature curves couldn't converge as function of the traditional product of dimensionless numbers.
2. The transient ratios of natural convection and thermal radiation at the top face could be approximately considered as increasing in a linear manner as indicated by corresponding regression analyses. At the same time, an equation is obtained, which could quantify the heat of natural convection and thermal radiation.
3. A modified product of dimensionless numbers is derived from a comprehensive, semi-theoretical and semi-empirical analysis, and it performs well in generalizing the temperature of PCM-based cooling plate considered in the present study.

The authors initial the study of temperature generalization of PCM-based cooling plate with thermal convection and radiation, and give explicit mathematics description firstly in that field. It could serve as a useful tool in engineering design.

References

1. Pal D, Joshi YK (1997) Application of phase change materials to thermal control of electronic modules: a computational study. J Electron Packag 119(1):40–50
2. Fossett AJ, Maguire MT, Kudirka AA (1998) Avionics passive cooling with micro-encapsulated phase change materials. J Electron Packag 120(3):238–242
3. Shatikian V, Ziskind G, Letan R (2005) Numerical investigation of a PCM-based heat sink with internal fins. Int J Heat Mass Transf 48(17):3689–3706
4. Shatikian V, Ziskind G, Letan R (2008) Numerical investigation of a PCM-based heat sink with internal fins: constant heat flux. Int J Heat Mass Transf 51(5–6):1488–1493
5. Saha SK, Dutta P (2010) Heat transfer correlations for PCM-based heat sinks with plate fins. Appl Therm Eng 30(16):2485–2491
6. Arshad A, Jabbal M, Sardari PT, Bashir MA, Yan Y (2020) Transient simulation of finned heat sinks embedded with PCM for electronics cooling. Therm Sci Eng Prog 18:100520
7. Motahar S, Jahangiri M (2020) Transient heat transfer analysis of a phase change material heat sink using experimental data and artificial neural network. Appl Therm Eng 167:114817
8. Sharifi N, Bergman TL, Faghri A (2011) Enhancement of PCM melting in enclosures with horizontally-finned internal surfaces. Int J Heat Mass Transf 54(19–20):4182–4192
9. Wu B, Zhang FH, Tian F, Cu QY (2021) Operating time prediction of heat storage device in avionics module. Chinses J Eng Des 28(01):48–55
10. Hosseinizadeh SF, Tan FL, Moosania SM (2011) Experimental and numerical studies on performance of PCM-based heat sink with different configurations of internal fins. Appl Therm Eng 31(17–18):3827–4383
11. Wu B, Li P, Zhang FH, Tian F (2021) A novel phase change material-based heat sink with an orthotropic plate to enhance the temperature field uniformity for avionics. J Mech Sci Technol 35(5):2237–2246

Structural Design and Analysis of Large-Load Fire-Fighting Multi-rotor UAV

Zhan-ke Li, Hai-yang Han, Yi Wu, Hai-bo Wei, and Liang-yang Zhang

Abstract This paper designs a large-load multi-rotor fire-fighting UAV. Use CATIA three-dimensional drawing software for auxiliary design to control the size and weight of the drone within a reasonable range. After obtaining the three-dimensional model of the fire-fighting drone, HyperWorks finite element analysis software was used to perform finite element analysis on the airframe structure composed of carbon fiber composite materials and aluminum alloy. The results showed that the strength and rigidity of the airframe structure meet the design and use requirements.

Keywords Large-load · Fire-fighting UAV · Structural design · Finite element analysis

1 Introduction

With the continuous development of society, China's urbanization process is rapid, and the number of high-rise buildings continues to increase. In the past 10 years, there have been 31,000 high-rise building fires in China. In 2019 alone, there were a total of 6,974 fires in high-rise buildings, and the fires in high-rise buildings showed an increasing trend. Firefighters belong to high-risk occupations, and it is difficult to protect their personal safety when dealing with larger fires. If drones are used to replace part of human actions, the casualties of firefighters can be greatly reduced.

The MC6-1200 six-rotor fire-fighting drone developed by Shenzhen Corbett Airlines can view the drone's aerial images in real time and provide high-definition images of the fire scene accident scene to the fire-fighting headquarters, but it does not carry fire-fighting equipment and does not have the ability to extinguish fire [1]. Chen Bin and Qin Zhengnan have designed an eight-rotor fire-fighting early warning UAV. The UAV has communication, video system and ground station. The UAV can carry fire extinguishing bombs with a load of 5 kg and a very limited fire-fighting range [2]. Wu Ruibo, Cheng Xinyuan, etc. analyzed the application of multi-rotor

Z. Li (✉) · H. Han · Y. Wu · H. Wei · L. Zhang
School of Aeronautics, Northwestern Polytechnical University, Xi'an, China
e-mail: lzk@nwpu.edu.cn

S. Lee et al. (eds.), *The Proceedings of the 2021 Asia-Pacific International Symposium on Aerospace Technology (APISAT 2021), Volume 1*, Lecture Notes in Electrical Engineering 912, https://doi.org/10.1007/978-981-19-2689-1_76

Fig. 1 Drone model

UAV in urban fire-fighting and rescue, and proposed a systematic application method of fire UAV [3].

The current shortcomings of former fire-fighting drones are small and cannot meet the actual fire-fighting needs. There is an urgent need for a flexible, low-cost, high-efficiency fire-fighting equipment to fill the gap. This paper intends to design a large-load multi-rotor fire-fighting UAV, which can carry fire-extinguishing bombs for bombing operations, and is a new type of fire-fighting equipment that can replace firefighters in high-risk operations near the fire site. It can take off and land vertically, is not restricted by the site, is flexible and easy to operate, and can also realize the integration of "observation and extinction". It has a good extinguishing effect for high-rise building fires, forest fires, etc., and has great application prospect.

In this paper, a large-load multi-rotor fire-fighting UAV is designed. At the same time of structural design, the finite element analysis of key stressed components is carried out by combining the theoretical calculation methods of material mechanics and composite structural mechanics. The analysis results prove that the design fully meets the index requirements. The digital model of the drone is shown in Fig. 1.

Table 1 Epoxy carbon UD prepreg material properties

Young's modulus(MPa)		Shear modulus(MPa)		Poisson's ratio		Maximum stretch stress(MPa)		Maximum compressive stress(MPa)		Maximum shear stress(MPa)		Density(g/cm^3)
Ex	1.21×10^5	Gxy	4700	μxy	0.27	x	2231	x	1082	x	60	1.49
Ey	8600	Gyz	3100	μyz	0.4	y	29	y	100	y	32	
Ez	8600	Gxz	4700	μxz	0.27	z	29	z	100	z	60	

2 Structural Design

2.1 *Structural Material Selection*

The UAV structure design mainly refers to the relevant theories and design methods of the manned aircraft structure design to design UAV that meet the overall design indicators proposed in advance. The problems that need to be solved include the UAV's mission characteristics, carrying loads, and flight performance. And the matching problem between structural strength requirements [4].

The selection of structural materials is an important part of the UAV structural design. This paper selects Epoxy Carbon UD Prepreg as the main material of the body. The specific parameters are shown in Table 1; 7075 aluminum alloy material is selected for the connection with relatively complicated shape.

2.2 *UAV Wheelbase Design*

The multi-rotor UAV G40 × 13.1CF carbon fiber propeller designed in this paper is used as the power source, and the wheelbase of the flying platform should be designed around the size of the propeller.

The diameter R of the G40 × 13.1CF carbon fiber propeller is 1016 mm. According to design experience and Fig.2, in order to reduce the aerodynamic interference between adjacent rotors, the distance between adjacent propellers is selected:

$$L = 2.1r \sim 3r \tag{1}$$

In this paper, the value of L is 1250 mm, which conforms to the range of formula (1), and the wheelbase D of the drone is calculated:

$$D = 2L = 2500\,\text{mm} \tag{2}$$

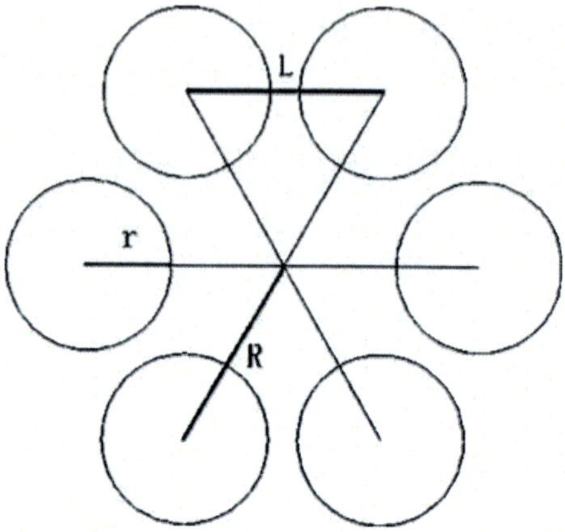

Fig. 2 The layout of the UAV rotor

2.3 Center Main Bearing Box Design

The central main load-bearing box is an important part of the UAV flight platform structure. In addition to fixing and supporting other components, it is also the force foundation of the entire UAV and needs to bear the load from the arm and landing gear connected to it. At the same time bear the gravity and inertial force of the power battery and flight control components loaded inside the box [3].

The length of the main load-bearing box is 974 mm, the width is 616 mm, the thickness of the side plate of the load-bearing box is set to 2.4 mm, the ply angle is [45/0/-45/90/45] 3 s, and the thickness of the single-layer board The thickness of the bottom plate of the load-bearing box is set to 3.6 mm, the laying angle is [45/0/-45/90/45/0] 3 s, the thickness of the single-layer board is 0.2 mm, and the total laying is 18 floors. The three-dimensional view of the main load-bearing box is shown in Fig. 3, and the two-dimensional top view with the upper cover removed is shown in Fig. 4.

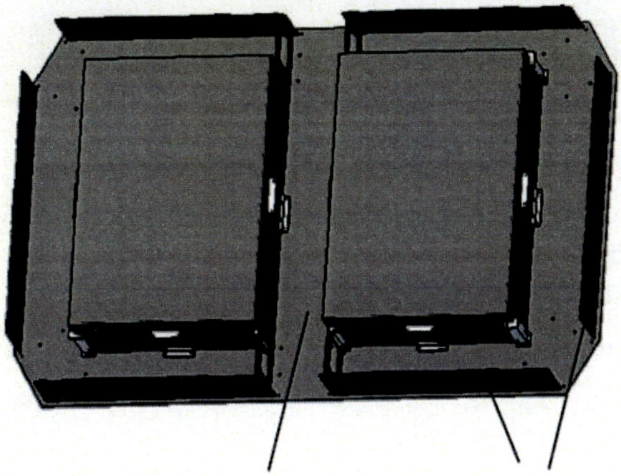

Fig. 3 Three-dimensional drawing of the bearing box

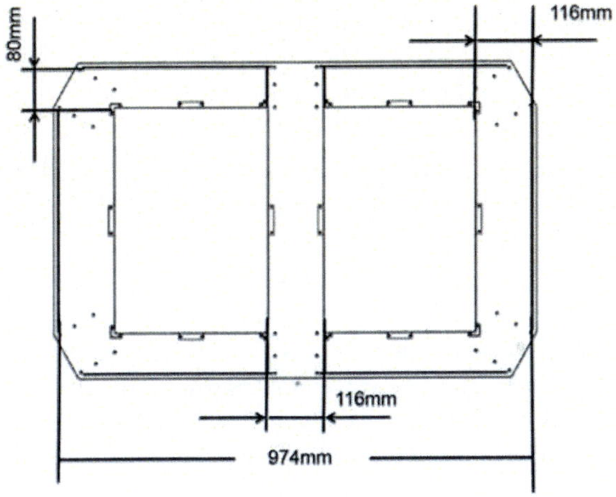

Fig. 4 Two-dimensional top view of the load-bearing box

2.4 *Arm Design*

The UAV designed in this paper needs to provide a pulling force of 14.58 kg for a single motor when hovering without wind, and two motors are installed on a single arm, so the task load of a single arm is 29.16 kg. According to the UAV, the wheelbase and the size of the center main bearing box are designed with two different length arms. The thickness of the arm is set to 2 mm. The dimensions of the carbon fiber arm are shown in Figs. 5 and 6. The square part on the right is aluminum alloy motor base, the left side is the aluminum alloy joint connecting the machine arm and the center bearing box, and the machine arm is a pluggable structure. The layering angle

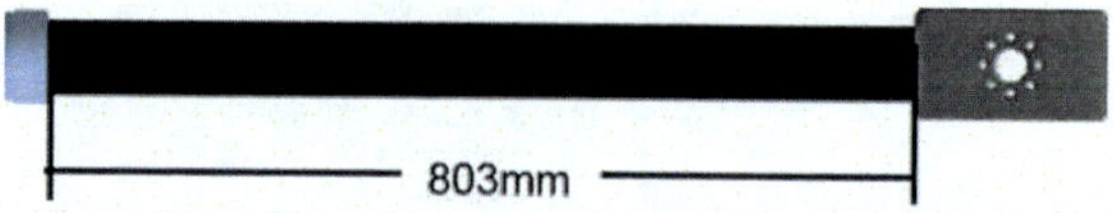

Fig. 5 UAV arm model 1

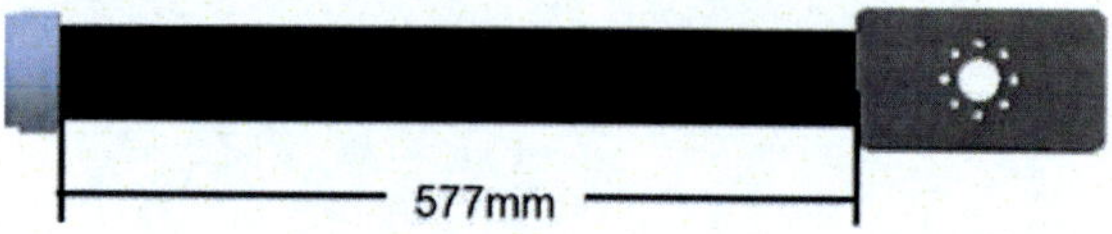

Fig. 6 Machine arm model 2

of the arm is set to [45/0/-45/0/45]s, the thickness of the single-layer carbon fiber board is 0.2 mm, and there are ten layers in total. The arm model 1 and arm model 2 use the same Laying angle and layer thickness.

In material mechanics, the deflection value of the farthest cantilever beam:

$$\omega_B = -FL^3/(3EI) \tag{3}$$

where F is the concentrated force on the free end of the cantilever, L is the length of the cantilever, E is the Young's modulus of the cantilever material, and I is the moment of inertia of the cantilever beam to the Z axis. Since the arm is a composite material laminated tube, its Young's modulus needs to be estimated based on the composite material mechanics.

In composite material mechanics, it can be considered that unidirectional composite materials can be regarded as a plane stress state, that is, $\sigma_3 = \tau_{23} = \tau_{31} = 0$. And the deformation of the single-layer plate is a small deformation, which conforms to the law of linear elasticity. The constitutive equation between stress and strain along the main axis of the material is [5]:

$$\begin{bmatrix} \varepsilon_x \\ \varepsilon_y \\ \gamma_{xy} \end{bmatrix} = \begin{bmatrix} \overline{S}_{11} & \overline{S}_{12} & \overline{S}_{16} \\ & \overline{S}_{22} & \overline{S}_{26} \\ sym & & \overline{S}_{66} \end{bmatrix} \begin{bmatrix} \sigma_x \\ \sigma_y \\ \tau_{xy} \end{bmatrix} \tag{4}$$

In formula (4):

$$\begin{cases} \overline{S_{11}} = m^4 S_{11} + m^2 n^2 (2S_{12} + S_{66}) + n^4 S_{22} \\ \overline{S_{12}} = m^2 n^2 (S_{11} + S_{22} - S_{66}) + S_{12}\left(m^4 + n^4\right) \\ \overline{S_{22}} = n^4 S_{11} + m^2 n^2 (2S_{12} + S_{66}) + m^4 S_{22} \\ \overline{S_{16}} = 2m^3 n (S_{11} - S_{12}) + 2mn^3 (S_{12} - S_{22}) - mn\left(m^2 - n^2\right) S_{66} \\ \overline{S_{26}} = 2mn^3 (S_{11} - S_{12}) + 2m^3 n (S_{12} - S_{22}) + mn\left(m^2 - n^2\right) S_{66} \\ \overline{S_{66}} = 4m^2 n^2 (S_{11} - S_{12}) - 4m^2 n^2 (S_{12} - S_{22}) + \left(m^2 - n^2\right)^2 S_{66} \end{cases} \tag{5}$$

Single-layer board is affected by σ_x n the x direction, Other stress components are 0, according to formula (4):

$$\begin{cases} \varepsilon_x = \overline{S_{11}}\sigma_x \\ \varepsilon_y = \overline{S_{12}}\sigma_x \\ \gamma_{xy} = \overline{S_{16}}\sigma_x \end{cases} \tag{6}$$

Hooke's Law, $\sigma_x = E_x\varepsilon_x$, The combined formula (6) is:

$$\frac{1}{E_x} = \frac{\varepsilon_x}{\sigma_x} = \overline{S_{11}} \tag{7}$$

Substituting formula (5) into formula (7), we get:

$$\frac{1}{E_x} = S_{11}m^4 + S_{22}n^4 + m^2n^2(2S_{12} + S_{66}) = \frac{m^4}{E_1} + \frac{n^4}{E_2} + m^2n^2\left(\frac{1}{G_{12}} - \frac{2\mu_{12}}{E_1}\right) \tag{8}$$

The composite material arm in this article uses 0° and ±45° layup, where Ex = E1 = 1.21 × 105 Mpa for 0° layup, when the layup angle is ±45°, $m^2 = n^2 = 1/2$, $E_1 = 1.21 \times 10^5$ Mpa, $E_2 = 8600$ Mpa, $G_{12} = 4700$ Mpa, $\mu_{12} = 0.27$ are substituted into formula (8) gives an E_x of ±45° layering of 1.2 × 104 Mpa.

According to the estimation method of the in-plane elasticity of the laminate:

$$Laninate\ Modulus = \sum(SLayer\ Modulus \times Thinkness)/\sum(SLayer\ Thinkness) \tag{9}$$

Then the E_x of the composite material arm is:

$$E_x = 1.21 \times 10^{11} \times 0.2 \times 4/2 + 1.20 \times 10^{10} \times 0.2 \times 6/2 = 5.56 \times 10^4\ \text{Mpa} \tag{10}$$

Substituting $E_x = 5.56 \times 104$ Mpa, $F = 578N$ into formula (3), get:

$$\omega_B = 578\text{L}^3/\left(3\text{I} \times 5.56 \times 10^4\ \text{Mpa}\right) \tag{11}$$

Since there are two types of arm length, for arm model 1, the cantilever length L1 is 753 mm, and the cantilever length L2 of arm model 2 is 527 mm. Under the same load, the end deformation of the two arms is different, but the amount of deformation is not There are too many differences, so as not to affect the balance of the machine body. Based on this principle, the diameter D of the machine arm is designed. The section moment of inertia of the machine arm $I = \pi/64(D^4 - d^4)$. Use MATLAB to draw the diameter of two types of machine arms on the same coordinate axis. The change curve between D and ω_B is shown in Fig. 7.

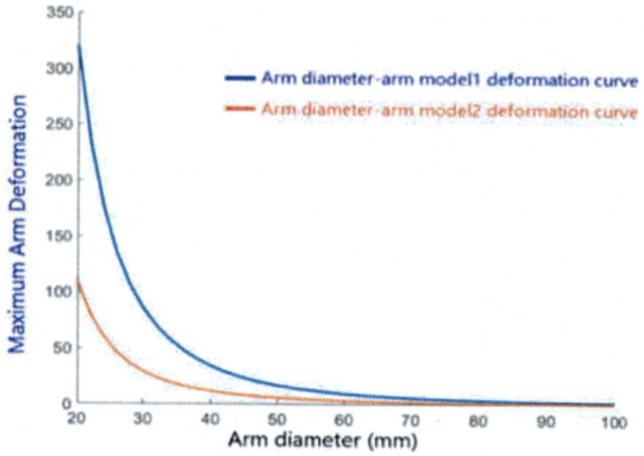

Fig. 7 Curve of maximum deformation of the arm

From Fig. 7, when the arm diameter is greater than 60 mm, the maximum deformation of the two types of arms is small, and the difference in the amount of deformation is also small. The arms with a diameter of 60–70 mm have basically met the requirements for use. Continue to increase the diameter of the arm. In order to carry out a detailed diameter selection, the deformation amount of the arm with a diameter of 60–70 mm is listed in Table 2.

According to Table 2, when the diameter is greater than 65 mm, the relative deformation of arm model 1 is less than 1%, and the difference between the deformation of the two models is less than 5 mm, and the final diameter of the arm is selected as 68 mm.

Table 2 Deformation of the arm with a diameter of 60–70 mm

Arm diameter (mm)	Arm 1 deformation (mm)	Arm 2 deformation (mm)	Deformation difference (mm)	Relative deformation of model 1	Relative deformation of model 2
60	9.643	3.306	6.337	0.0128	0.0063
61	9.162	3.141	6.021	0.0121	0.0059
62	8.711	2.986	5.725	0.0116	0.0056
63	8.290	2.842	5.448	0.0110	0.0053
64	7.896	2.707	5.189	0.0105	0.0051
65	7.526	2.580	4.946	0.0100	0.0048
66	7.179	2.461	4.718	0.0095	0.0046
67	6.853	2.349	4.504	0.0091	0.0045
68	6.546	2.244	4.302	0.0087	0.0042
69	6.258	2.145	4.113	0.0083	0.0040
70	5.986	2.052	3.934	0.0079	0.0038

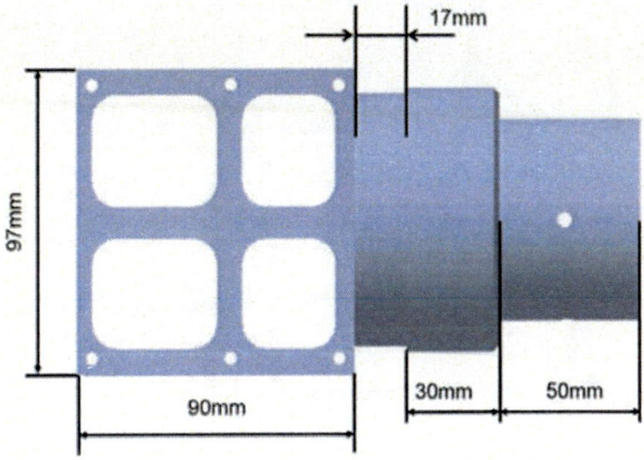

Fig. 8 The top view of the arm connecting piece

Fig. 9 The front view of the arm connecting piece

2.5 *Arm Connection Design*

The arm connector must be connected to the center main load-bearing box and the arm at the same time. The arm connecting piece is made of 7075 aluminum alloy, and the central load-bearing box connecting block is designed as a hollow structure, which is convenient for wires to pass through and can also reduce the weight of the structure (Figs. 8 and 9).

2.6 *Landing Gear Design*

The main function of the landing gear is to support the parking, take-off and landing of the UAV, and to protect the mission equipment below. The structural design of the landing gear is shown in Fig. 10. The landing gear is made of carbon fiber tubes

Fig. 10 Schematic diagram of the landing gear structure

and aluminum alloy joints. Assembled, the structure is relatively simple, and the disassembly and assembly are convenient.

3 Structural Finite Element Analysis

3.1 Finite Element Method Software

The finite element structure analysis software used in this article is Altair HyperWorks2018, which contains a wealth of CAE modeling and finite element analysis product modules. This article mainly uses HyperMesh pre-processing module, OptiStruct solver, and HyperView post-processing visualization module.

3.2 Model Simplification and Meshing

In order to reduce the workload of meshing and software calculation, the model can be simplified appropriately.

The UAV structure designed in this paper is mainly composed of aluminum alloy and carbon fiber composite materials. The plate and pole structure be made of carbon fiber, and the grid is divided into a quadrilateral grid with a size of 5 $\times$ 5 mm; the connection structure is aluminum alloy and is divided into a tetrahedral grid with a side length of 3 mm. The total number of grids in the whole UAV is 1738366. After the Quality Index panel test, the QI value of the 2D grid is 0.02, which meets the calculation requirements. The fuselage grid is shown in Fig. 11, and the detailed diagram is shown in Fig. 12.

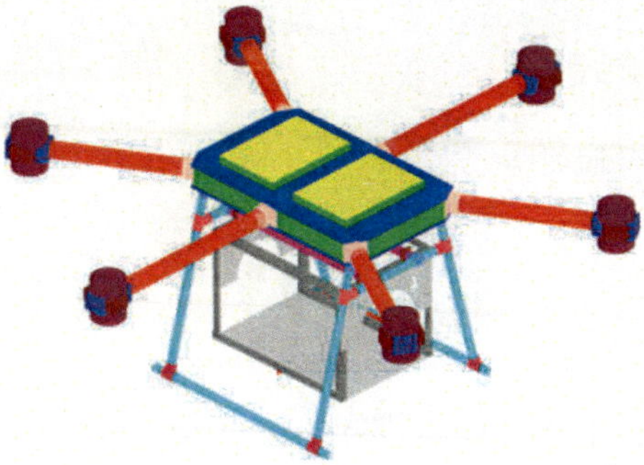

Fig. 11 Full machine grid

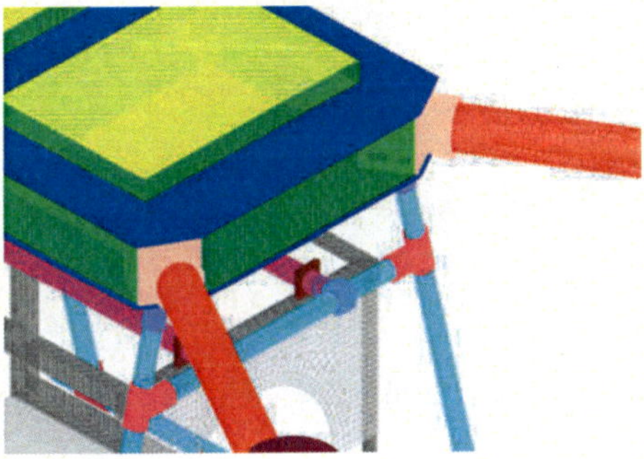

Fig. 12 Details of the fuselage grid

3.3 Load Application

After the Hypermesh software body structure gives the material properties, gravity will not automatically appear. Therefore, it is necessary to design the load application scheme according to the force characteristics of the body structure. The lift provided by the 12 propellers is applied to the lower end of the motor in the form of distributed force. The size of the lift can be set according to the calculation needs; the weight of the 705.6 N battery is added to the bottom of the central main force box in the form of distributed force; the mass of the 509.6 N fire extinguishing bomb applied in the form of distributed force on the elongated rectangular area in the middle of the base plate of the hanger. The load application scheme of the UAV designed in this paper is shown in Fig. 13.

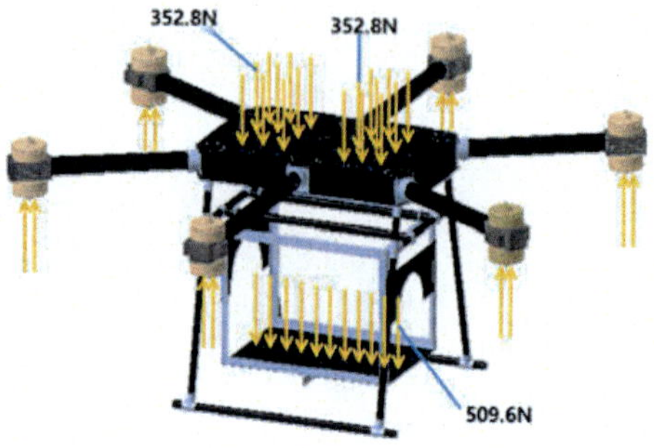

Fig. 13 Load application scheme

3.4 Model Constraints and Load Application Form

After determining the load application plan, impose constraints on the model. First, set the inertia release constraint for the whole machine. For the connection of each part of the machine body, the RBE2 unit is used for multi-point restraint. Figure 14 is a schematic diagram of the connection between the body and the arm joint, and Fig. 15 is a schematic diagram of the connection between the machine arm and the arm joint.

According to the load application plan of Sect. 3.3, the simplified mass point CONM2 of the battery is applied to the RBE3 flexible unit with an area equal to the actual area of the battery; the simplified mass point CONM2 of the fire extinguishing bomb is applied to the carbon fiber board RBE2 unit; The lift generated by the propeller is applied to the motor under the arm with the "force" command, as shown in Fig. 16.

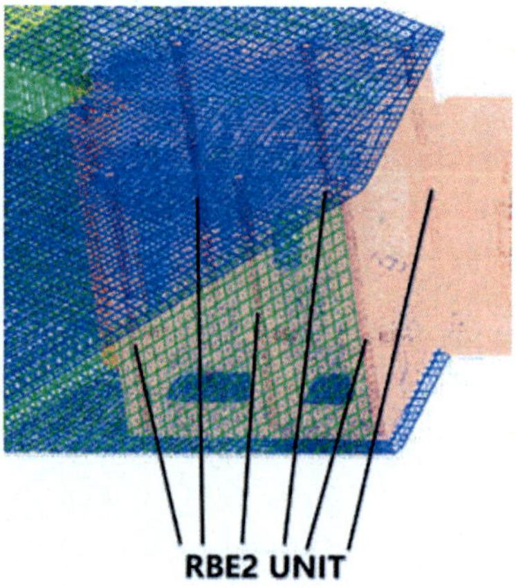

Fig. 14 Fuselage-arm joint connection

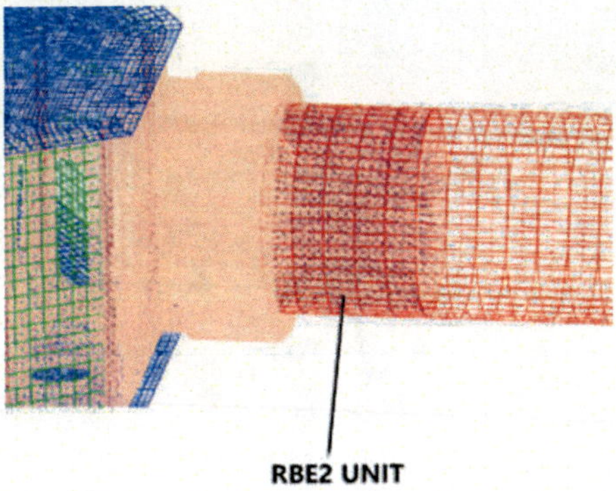

Fig. 15 UAV arm-machine arm joint connection

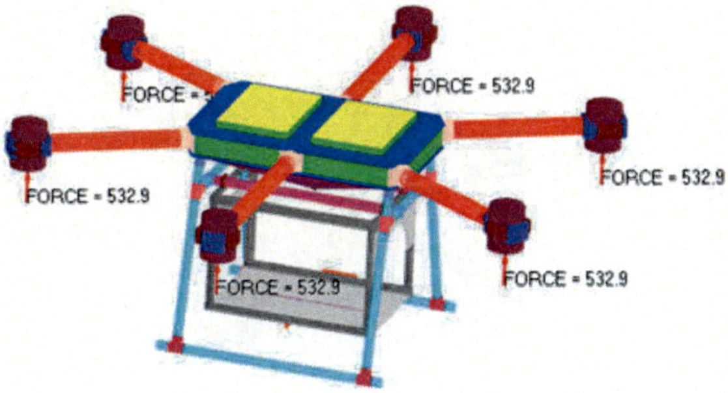

Fig. 16 Schematic diagram of propeller pulling force application

3.5 *Inertial Release Finite Element Analysis*

Calculated according to the maximum lift of the body. According to the test results of the power system, the maximum lift generated by the two propellers installed on a single arm is 578 N. When setting the lift, the weight of the two motors, ESCs, and propellers must be subtracted:

$$T = 578 - 2 \times (1.74 + 0.279 + 0.237) \times 10 = 532.88\,\text{N} \tag{12}$$

(1) Total Deformation Finite Element Calculation Results

The structural deformation diagram of the whole UAV after the load is applied is shown in Fig. 17. The maximum deformation along the positive z-axis is 9.834 mm. The deformation diagram of the six arms is shown separately as shown in Fig. 18. The maximum deformation of the model 1 arm is 7.224 mm and the relative deformation is 0.89%, and the maximum deformation of the model 2 arm is 3.184 mm and the relative deformation is 0.55%. The deformation of the two arms is very small and will not affect the lift of

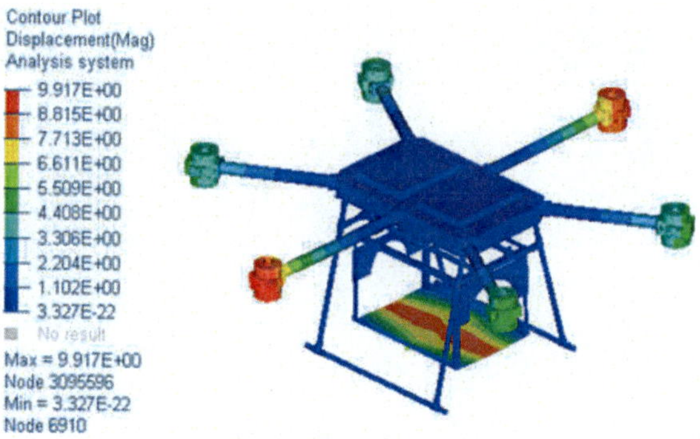

Fig. 17 Deformation diagram of structure direction

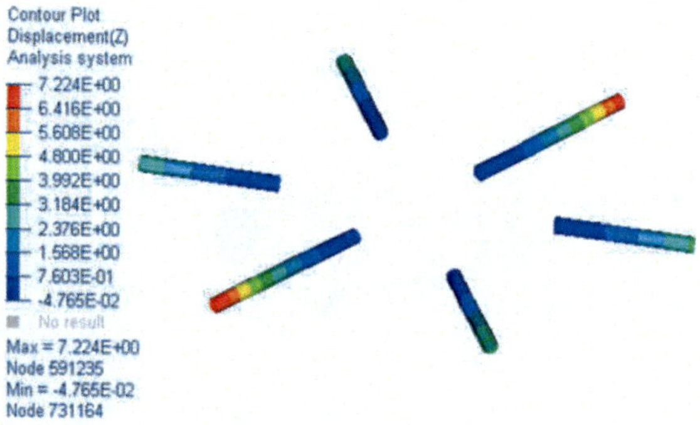

Fig. 18 Machine arm deformation diagram

the propeller; the deformation diagram of the upper and lower plates of the main bearing box of the fuselage is shown in Fig. 19, and the deformation diagram of the landing gear is shown in Fig. 20, and the main bearing box

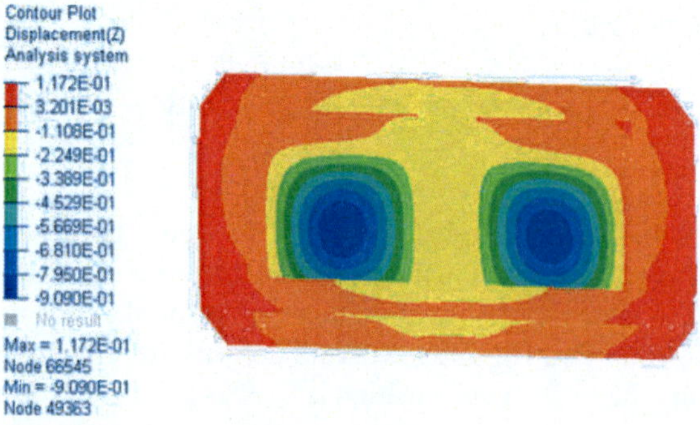

Fig. 19 Deformation of plates of the main bearing box

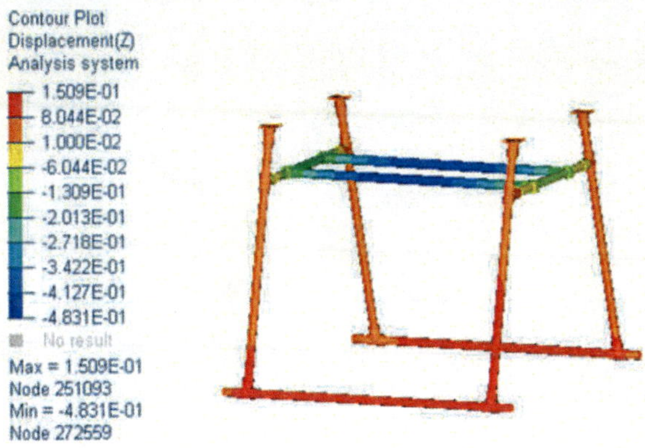

Fig. 20 Deformation of the landing gear

is positive along the Z axis. The maximum deformation along the Z axis is 0.1172 mm, the maximum deformation along the negative Z axis is 0.909 mm, and the maximum deformation along the positive Z axis of the landing gear is 0.1509 mm. The analysis results show that the structural deformation of the UAV does not affect the normal use of the airframe.

(2) Structural Stress Display of the Whole UAV

From Fig. 21, the maximum stress of the body structure is 351.7 Mpa, the tensile strength of the carbon fiber board is 2231 Mpa, the compressive strength is 1082 Mpa, and the yield strength of the aluminum alloy is 455 Mpa, all of which do not exceed the strength limit of the material.

The stress of the six arms is shown separately as shown in Fig. 22, the stress of the main load-bearing box is shown in Fig. 23, and the stress of the landing gear is shown in Fig. 24.

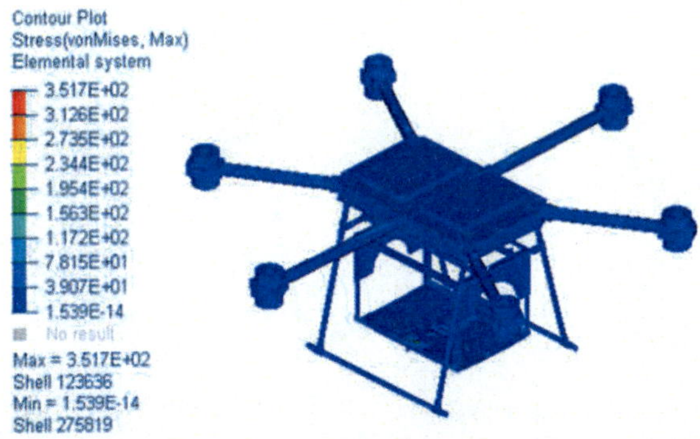

Fig. 21 Stress diagram of the whole UAV structure

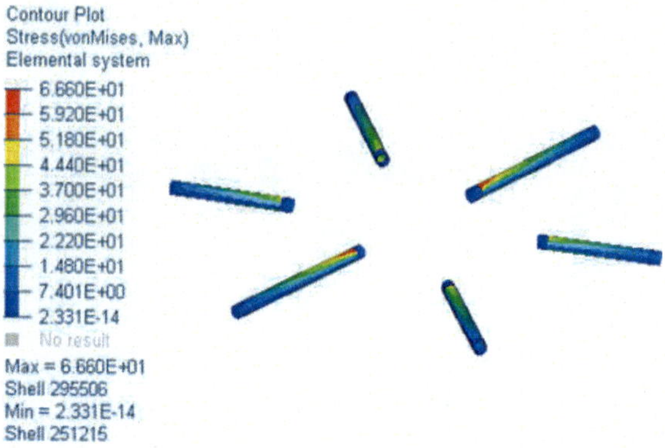

Fig. 22 Stress diagram of the UAV arm

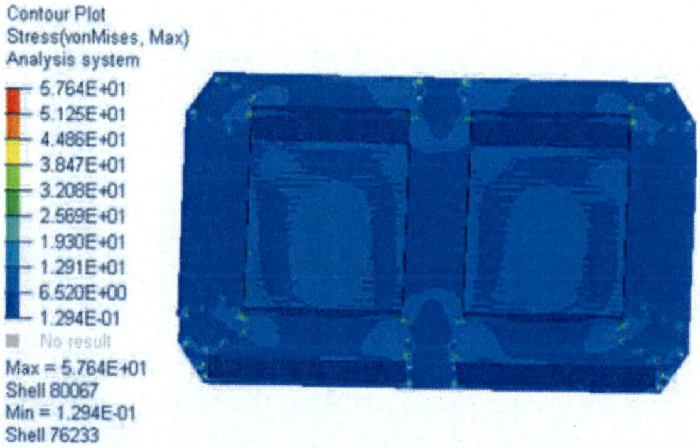

Fig. 23 Stress diagram of the main bearing box

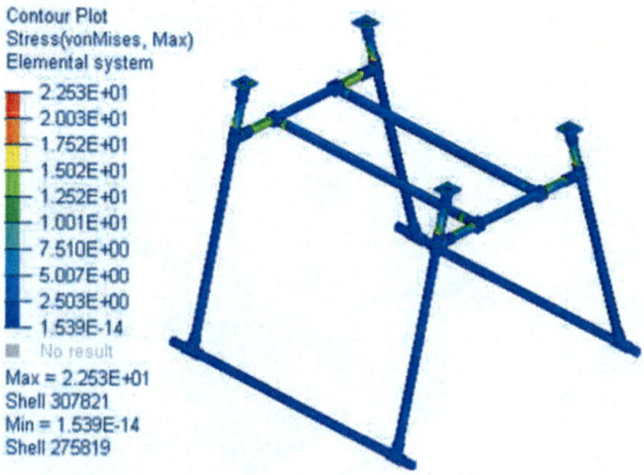

Fig. 24 Stress diagram of the landing gear

Through the stress diagram of the whole UAV structure and the stress diagrams of each component, it can be seen that when the lift of the propeller reaches the maximum of the unmanned aerial vehicle with load under the restraint of inertia release, the structural stress is within a safe range, and the structural strength meets the requirements of use.

4 Conclusion

According to the structural design index of the large-load fire-fighting drone, combined with the general principles of composite material drone structure design, the detailed design of the large-load fire-fighting unmanned structure is carried out, and the detailed dimensions of the key stressed components of the airframe structure Design and strength verification, integrate the size of each part of the UAV, and draw a full CATIA three-dimensional model.

Use Hyperworks finite element analysis software to carry out the finite element analysis of the body structure of the large-load fire-fighting UAV. First, the finite element model of the body structure is constructed according to the geometric and material characteristics of the body structure, and the model simplification, mesh division, and application are completed. Constraint and load work; then, according to the working characteristics of the UAV, the structure finite element analysis of the body structure with the fire extinguishing bomb load is carried out, and the stress and deformation of the body structure under the maximum lift condition are analyzed. After inspection, the strength of the body structure and the rigidity can meet the requirements of use.

References

1. Introduction to MC6–1200 six-rotor fire-fighting police drone. Fire Technol Product Inf (08):25 (2016)
2. Bin C, Zhengnan Q (2019) Design and Implementation of octo-rotor fire fighting early warning UAV. J Jiamusi Univ
3. Wu R, Xingyuan C, Zheng W, Ji W (2019) Discussion on the application of multi-rotor drones in urban firefighting and rescue. Modern Manuf Technol Equip 06:147–149
4. Changqing F, Xiaoqin F (2019) UAV composite material structure design and manufacturing technology. Xi'an: Northwestern Polytechnical University Press, 28, 55, 78, 89–90
5. Jianqiao C (2016) Mechanics of composite materials. Huazhong University of Science and Technology Press, Wuhan, pp 20–29

Research on Bearing Fault Recognition Based on PSO-MCKD and 1D-CNN

Yinling Wang and Xianming Yin

Abstract This paper proposes a bearing fault feature extraction and recognition method based on particle swarm optimization optimization Maximum Correlated Kurtosis Deconvolution (MCKD) and one-dimensional convolutional neural network to solve the non-stationarity of rolling bearing fault signals, Non-linear and complex characteristics, as well as the problems of noise interference and unclear fault characteristics in the process of fault identification. First, the multi-channel signals of the rolling bearing is analyzed, in order to select the signal containing the impact component as the fault feature. Next, the signal containing the fault feature is filtered through MCKD, where the best parameters of MCKD are obtained by improving the particle swarm algorithm to achieve the feature enhancement of the main signal. Finally, a one-dimensional convolutional neural network (One Dimensional Convolutional Neural Network, 1D-CNN) is used to model the characteristic signals under different damage conditions in order to obtain the fault recognition model of the rolling bearing.The experimental results show that the method can effectively extract the main characteristic signals of the faulty bearing, and realize the accurate identification of the bearing fault in the noisy environment.

Keyword Rolling bearings · Particle swarm optimization algorithm · Maximum correlation kurtosis deconvolution · One-dimensional convolutional neural network

1 Introduction

The working environment of rolling bearings is complex, and they will be affected by non-linear factors such as load, friction, and impact in during the detection process, as a result, the fault signal collected by the acceleration sensor contains a lot of background noise. Therefore, it is a hot and difficult point in the study of rolling bearing fault detection for accurately detect the fault signal by analyzing the nonlinear and non-stationary random signal generated during the damage process.

Y. Wang (✉) · X. Yin
Engineering Technology Center, Southwest University of Science and Technology, Mianyang, China
e-mail: wangyinling@swust.edu.cn

S. Lee et al. (eds.), *The Proceedings of the 2021 Asia-Pacific International Symposium on Aerospace Technology (APISAT 2021), Volume 1*, Lecture Notes in Electrical Engineering 912, https://doi.org/10.1007/978-981-19-2689-1_77

In bearing fault diagnosis, the damage signal is regarded as a periodic shock signal, and its convolution with the system response is the output signal measured by the sensor. If the influence of noise is not considered, an inverse filter can be found, and the original fault signal can be recovered by deconvolution through the observation signal output by the sensor. In 1978, Wiggin [1] combined the idea of minimum entropy convolution filtering, and proposed the Minimum Entropy Deconvolution (MED) method, which uses iterative selection of finite pulse filters to maximize kurtosis during the deconvolution process, this method is also called maximum kurtosis deconvolution. The MED method can only process a single impact signal, not only is not suitable for rotating machinery, but also false components will be extracted when the filter is improper selected. In 2012, McDonald [2] proposed the maximum correlation kurtosis deconvolution algorithm, using Correlation Kurtosis (CK) as the optimal index, and calculating the optimal filter through iteration to obtain the original fault Signal. Since the MCKD algorithm was proposed, many scholars have studied it and applied it to many fields of fault diagnosis. A fault feature extraction method based on local feature scale decomposition (LCD) and maximum correlation kurtosis deconvolution (MCKD) was proposed by Su [3]. A fault diagnosis method for rolling bearings based on autocorrelation analysis and maximum correlation kurtosis deconvolution algorithm was proposed by Xiaoyan [4]. Shangkun et al. [5] proposed an early fault identification method for rolling bearings based on the combination of Teager energy operator and MCKD.

In recent years, deep learning algorithms have shown unique advantages and potential in feature extraction and pattern recognition, with the rapid development of deep learning algorithms. Deep learning is good at abstracting feature representations from raw data, and is especially suitable for extracting features from huge amounts of data. A motor fault diagnosis method based on sparse autoencoder was proposed by Liyang [6], which uses the envelope spectrum of the motor fault signal as input to realize effective classification of the fault signal; Heng [7] aimed at the rolling bearing vibration signal with strong non-stationarity and easy to be disturbed by strong background noise, the one-dimensional signal was transformed into two-dimensional time–frequency signal through short-time Fourier transform, and then realized by convolutional neural network End-to-end failure mode identification. Yuxuan [8] classified and diagnosed the operating state of the bearing through a long and short-term memory network (LSTM).

In order to realize the accurate identification of bearing fault characteristic signals under different damage conditions, this article first analyzes the collected multi-channel fault signals to study their characteristics under different damage characteristics; secondly, the MCKD algorithm optimized by PSO is used to filter the fault information to obtain the original shock signal; finally, the improved one-dimensional convolutional neural The network (1 Dimensional Convolutional Neural Network, 1D CNN), through the feature information training under different damage situations, obtains the bearing damage recognition model.

2 PSO Optimizes Maximum Correlation Kurtosis Deconvolution

The rolling bearing fault signal $\boldsymbol{x(n)}$ collected by the sensor can be expressed as:

$$x(n) = h(n) * y(n) + e(n) \tag{1}$$

where $h(n)$ is the impulse response of the system, $y(n)$ is the system fault signal, and $e(n)$ is the noise. Without considering the noise, find the optimal filter $f = [f_1, f_2, ..., f_L]^T$, and deconvolve the signal collected $x(n)$, in order to restore the fault signal $y(n)$, which is:

$$y(n) = f(n) * x(n) \tag{2}$$

The calculation formula of the correlation kurtosis is:

$$CK_M(T) = \frac{\sum_{n=1}^{N}\left(\prod_{m=0}^{M} y_{n-mT}\right)^2}{\left(\sum_{n=1}^{N} y_n^2\right)^{M+1}} \tag{3}$$

where $y_n = \sum_{k=1}^{L} f_k x_{n-k+1}$, when $n \neq 1, 2..., N$, there are $y_n = 0, x_n = 0$. Where N is the sampling length, and M is the number of digits of the period offset, and T is the period of the convoluted pulse signal, which is determined by the sampling signal frequency f_s and the characteristic frequency f_r of the fault signal, and satisfies the relational expression $T = f_s/f_r$. And in order to obtain the impact component of the convolved signal, the filter parameters need to be selected to maximize the correlation kurtosis value, which is:

$$MCK_M(T) = \max_{\vec{f}} CK_M(T) = \max_{\vec{f}} \frac{\sum_{n=1}^{N}\left(\prod_{m=0}^{M} x_{n-mT}\right)^2}{\left(\sum_{n=1}^{N} \left(x_n^2\right)^2\right)^{M+1}} \tag{4}$$

where $f = [f_1, f_2, ..., f_L]^T$, and L is the filter length.

Derivation of the objective function, such that:

$$\frac{d}{df_k} CK_M(T) = 0, k = 1, 2, 3, ...L \tag{5}$$

Finally get the filter, that is:

$$\vec{f} = \frac{\|\vec{y}\|^2}{2\|\vec{\beta}\|^2}\left(X_0X_0^T\right)^{-1}\sum_{m=0}^{M} X_{mT}\vec{\alpha}_m \quad (6)$$

where

$$X_r = \begin{bmatrix} x_{1-r} & x_{2-r} & x_{3-r} & \ldots & x_{N-r} \\ 0 & x_{1-r} & x_{2-r} & \ldots & x_{N-1-r} \\ 0 & 0 & x_{1-r} & \ldots & x_{N-2-r} \\ \vdots & \vdots & \vdots & \ddots & \vdots \\ 0 & 0 & 0 & \ldots & x_{N-L+1-r} \end{bmatrix}_{L\times N} \quad (7)$$

$$r = [0, T, 2T, ..., mT]$$

$$\vec{\alpha}_m = \begin{bmatrix} y_{1-mT}^{-1}\left(y_1^2 y_{1-T}^2 \cdots y_{1-MT}^2\right) \\ y_{2-mT}^{-1}\left(y_2^2 y_{2-T}^2 \cdots y_{2-MT}^2\right) \\ \vdots \\ y_{N-mT}^{-1}\left(y_N^2 y_{N-T}^2 \cdots y_{N-MT}^2\right) \end{bmatrix} \quad (8)$$

$$\vec{\beta} = \begin{bmatrix} y_1 y_{1-T} \cdots y_{1-MT} \\ y_2 y_{2-T} \cdots y_{2-MT} \\ \vdots \\ y_N y_{N-T} \cdots y_{N-MT} \end{bmatrix} \quad (9)$$

The MCKD algorithm first loads the original signal $x(n)$, selects the appropriate filter length L, and the appropriate deconvolution signal period T, and the appropriate displacement offset M. Secondly, calculate X_{mT}, X_0^T and $\left(X_0X_0^T\right)^{-1}$ according to the signal $x(n)$, and initialize the filter $\vec{f}$. Thirdly, calculate the filtered output fault signal $y(n)$, and calculate $\vec{\alpha}_m$ and $\vec{\beta}$ in accordance with $y(n)$, and then bring it into the Eq. (6) to update the filter $\vec{f}$, and circulate in turn until the difference before and after filtering is less than the given error, the program ends, and the filtered fault signal $y(n)$ is obtained.

When using MCKD to perform deconvolution filtering on $x(n)$, the period T, the offset M, the filter length L, and the number of iterations N need to be set. Therefore, this paper takes the maximum kurtosis as the objective function, and optimize the above by the particle swarm algorithm, in order to obtain the optimal parameters.

The PSO algorithm was first proposed by American electrical engineer Eberhart and social psychologist Kennedy in 1995 based on an intelligent optimization algorithm based on flock of birds foraging. The algorithm regards each individual as a particle, and the flock of birds can be regarded as a particle swarm. Assuming that there is a particle swarm composed of *m* birds in a d-dimensional target search space, the position of the *i*-th particle (*i* = *1,2,...m*) in the d-dimensional space can be expressed as $x_i = [x_{i1}, x_{i2}, ..., x_{id}]$, where each the position of each particle is a potential solution. The fitness value can be calculated by substituting x_i into the objective function, and the quality of this particle can be measured according to the size of the fitness value. The best position experienced by an individual particle is recorded as $p_i = [p_{i1}, p_{i2}, ..., p_{id}]$, and the best position experienced by the entire particle swarm can be recorded as $p_{gbest} = [p_{g1}, p_{g2}, ..., p_{gd}]$. The velocity component of the particle position can be expressed as $v_i = [v_{i1}, v_{i2}, ..., v_{id}]$. The speed and position used by the particle swarm algorithm are updated as follows:

$$v_{id} = \omega v_{id} + c_1 r_1 (p_{id} - x_{id}) + c_2 r_2 (p_{gd} - x_{id}) \tag{10}$$

$$x_{id} = x_{id} + \alpha v_{id} \tag{11}$$

where ω is a non-negative number, called the inertia factor. c_1 and c_2 are non-negative numbers and are called acceleration constants. r_1 and r_2 are random numbers in the interval [0, 1]. α is called the constraint factor, which is used to control the weight of the speed. The termination condition of the particle swarm algorithm can be set to the maximum number of iterations, or it can be set to meet a certain goal to terminate.

3 One-Dimensional Convolutional Neural Network

3.1 Convolutional Neural Network

Since the 2012 ImageNet competition, Alexnet has achieved breakthrough development and became the first application to apply deep neural networks, making Convolutional Neural Networks (CNN) beginning to be accepted and widely developed, becoming the most used deep learning method currently, has developed rapidly in the fields of machine vision, natural language processing, speech recognition, and semantic segmentation. CNN is mainly developed by imitating the characteristics of local perception and hierarchical processing in the work of neural networks in the biological cerebral cortex. A typical CNN mainly includes an input layer, a convolutional layer, a pooling layer, a fully connected layer, and an output layer.

For the input layer, CNN mainly processes two-dimensional image data. For black and white images or grayscale images, the input is single-channel two-dimensional data. But for color images, it is composed of the three primary colors of R, G, and B, so its input should be three-channel. For 1D CNN, the input is a row or column vector.

The convolution layer performs convolution operations on the input data. For one-dimensional data, the convolution kernel is a one-dimensional matrix. Each element that composes the convolution kernel corresponds to a weight coefficient and a deviation. Each neuron in the convolution layer is connected to multiple neurons in a region close to the previous layer. The size of the region depends on the convolution. For a one-dimensional convolutional neural network, the calculation of convolution can be expressed as:

$$Y_j^{l+1} = f\left(\sum\nolimits_{i \in M_j}\left(Y_i^l \otimes \omega_{ij}^{l+1}\right) + b_j^{l+1}\right) \tag{12}$$

where Y_j^{l+1} is the j-th output of the $l + 1$ layer, M_j is the j-th convolution area of the l layer, ω_{ij}^{l+1} is the corresponding convolution kernel, b_j^{l+1} is the offset, and f is the activation function.

Perform a nonlinear transformation on the output in each convolution through the activation function, commonly used activation functions include Sigmoid function, Tanh function and ReLU function, etc. Because the Sigmoid function will cause the gradient to disappear when the number of convolutional layers increases, in order to prevent this deficiency, the ReLU activation function is usually used in two-dimensional image processing. The Tanh function is also known as the hyperbolic tangent function, with a value range of $[-1, 1]$. It has a better effect when the characteristics are significantly different, and it will continue to expand the characteristic effect during the cycle. According to the characteristics of the one-dimensional time-varying fault characteristic signal, Tanh is finally used as the activation function, which has incomparable advantages over other functions.

The pooling layer is also called the down-sampling layer. The main function of this layer is to scale and map the data of the upper layer, and use the overall characteristics of a certain perceptual domain of the upper layer as the output of neurons, which can speed up the calculation of the network. It also contributes to the overfitting of the network. Common pooling operations include maximum pooling and average pooling, the article uses maximum pooling to extract features.

The function of the fully connected layer is to classify, each node of which is connected to all nodes of the upper layer, so the parameters of the fully connected layer are the most. The fully connected layer finally realizes the classification through the Softmax classifier, and finally realizes the mapping from the abstract feature to the sample label space.

3.2 1D-CNN Bearing Fault Recognition Model

First, analyze the one-dimensional data collected by the sensor. Since the sampling period is 12 kHz and the bearing speed per minute is between 1720 and 1797, the data length for training is set to 400, which ensures that each training sample will contain one cycle. Each fault data contains approximately 120 k sampling points, and each 400 sampling points intercepts the training data as a feature set, so that each fault combined plan is divided into approximately 300 samples.

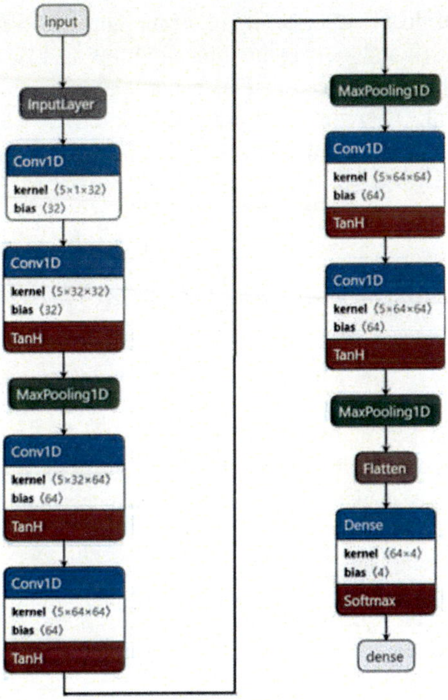

Fig. 1 1D- CNN model structure

Next, design the network structure, the input data vector length of the 1D-CNN structure is 400, and the network contains 3 sets of convolution and pooling layers. The first layer contains two convolutions and one pooling, and each convolutional layer has 32 convolution kernels of size 5, the pooling window size is 5, and the maximum pooling method is adopted. The structure of the second and third layers is exactly the same as that of the first layer, except that each convolution layer has 64 convolution kernels of size 5, and the convolution uses the Tanh function as the activation function. Then convert the pooled data into one-dimensional data, and form a fully connected layer, and finally output the probability distribution of the four categories of the sample. Here, the activation function selects Softmax. The final model is shown in Fig. 1.

4 Test Verification

Experimental design: EDM is performed on the inner and outer rings and rolling elements of the bearing to form a single-point damage with a diameter of 0.007 to 0.040 inches to simulate bearing damage. The acceleration sensor is used to record the fault signal data of different loads and different damage positions, and the data sampling frequency is 12 and 48 kHz, and the default is 12 kHz. The experimental data adopts the CWRU failure data set. This paper mainly studies the failure of the outer ring of the bearing. Therefore, the four cases for the drive end rolling element are studied: no load, 0.007 inch load is 3 horsepower, 0.014 inch load is 2 horsepower, and 0.014 inch no load.

4.1 Frequency Domain Analysis of Bearing Fault Signal

First study the signal characteristics of no damage and no load on the outer ring. It contains two sets of data at the fan end and the drive end, the sampling frequency is 12 kHz. Figure 2 shows the time-domain waveform and Fourier transform. From the figure, it can be seen that the waveforms of the fan end and the drive end are similar, and the Fourier transform is also similar, and the frequencies at 1.1 K and 2.1 kHz have maximum values.

Envelope spectrum analysis of the impact signal can extract the fault frequency of the signal. Therefore, the fan end and the drive end of the normal signal are decomposed by EMD, and then the envelope spectrum analysis is performed on the first two components of each signal. The result is shown in Fig. 3.

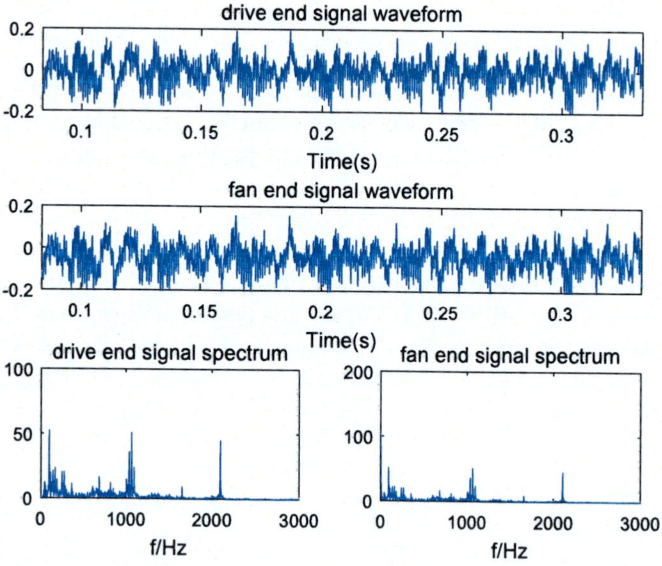

Fig. 2 Normal signal waveform and frequency spectrum

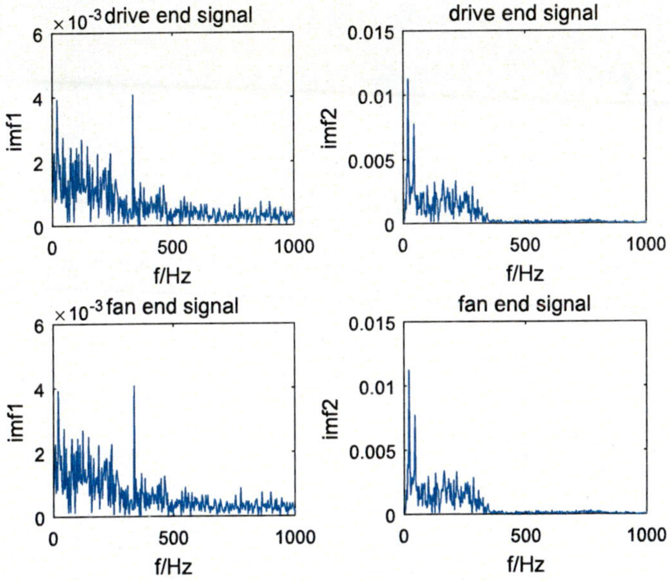

Fig. 3 Envelope spectrum of the first two sections of normal signal

It can be seen from Fig. 3 that the envelope spectrum of the signal does not have a large amplitude at the multiplier, which also proves that the normal signal does not contain impact components, and there is no fault signal.

Next, analyze the characteristics of the fault signal. Choose a single point damage with an outer ring of 0.007 inches and a load of 3 horsepower, including three sets of data for the fan end, the drive end and the base. Figure 4 shows part of the time-domain waveform of the signal, and Fig. 5 shows the frequency-domain waveform. Figure 6 shows the envelope spectrum of each end signal. From the above figure, it can be clearly seen that the signal frequency is at 81 Hz and its multiplier, and the amplitude of the frequency has a maximum value, which also proves the impact component in the signal.

From the above analysis, it can be seen that the fan end, base and drive end in the fault signal data set can detect the impact component of the drive end, and their time domain, frequency domain, and envelope spectrum characteristics are similar. Therefore, the next step is to analyze this The similarity of the three signals. Perform correlation analysis on these three sets of data, and the calculation formula for correlation analysis is:

$$r = \frac{\sum_{i=1}^{n}\left(X_i - \overline{X}\right)\left(Y_i - \overline{Y}\right)}{\sqrt{\sum_{i=1}^{n}\left(X_i - \overline{X}\right)^2}\sqrt{\sum_{i=1}^{n}\left(Y_i - \overline{Y}\right)^2}} \tag{13}$$

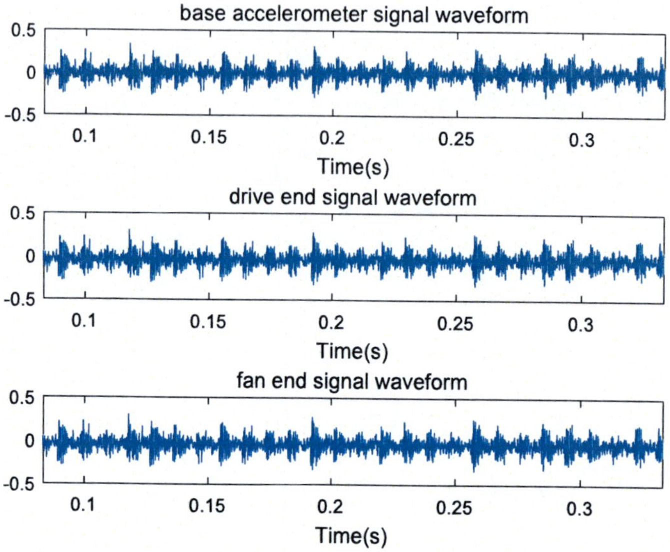

Fig. 4 Time-domain waveform of damage signal

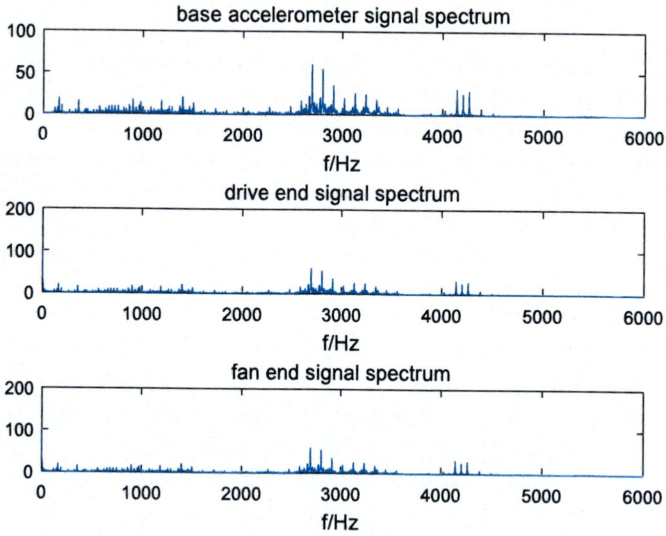

Fig. 5 Frequency domain characteristics of damage signal

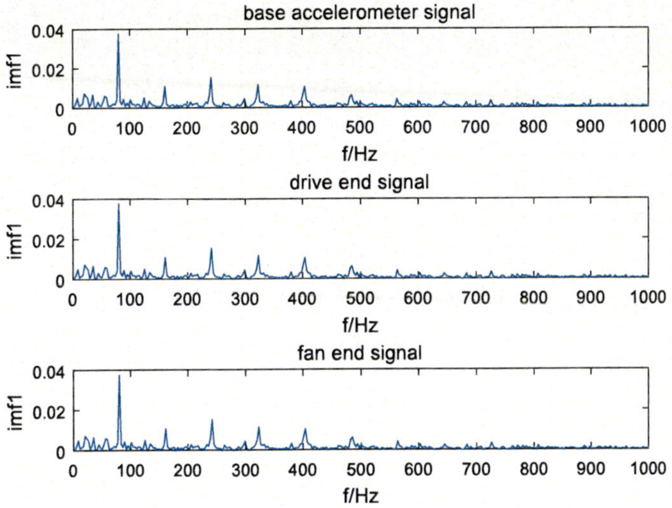

Fig. 6 Damage signal envelope spectrum

In formula (13), if *r* is close to 1, it means that the two signals are very similar, and 0 means that the signals are completely different. Calculating the similarity between the signals, the similarity of the three signals is 1, indicating that the three signals are very similar, so in the subsequent processing, the characteristics of the drive end will be the research object, which will greatly reduce the amount of subsequent data processing. Also reduces the computational complexity. Calculating the similarity between the signals, the similarity of the three signals is 1, indicating that the three signals are very similar, so in the subsequent processing, the characteristics of the drive end will be the research object, which will not only greatly reduce the amount of subsequent data processing, but also reduces the computational complexity.

4.2 *Demising Analysis of Bearing Fault Signal*

Because the original signal contains a lot of noise, it needs to be denoised in order to enhance the characteristics of the impact signal. When the signal is deconvolved through MCKD, the period T, the offset M, the filter length L, and the number of iterations N need to be set. Such as: the calculated value of the period T is not necessarily optimal, and there is usually a deviation from the actual. The filter length L directly affects the size of the entire calculation and storage data, and the offset M is generally between 1 and 7. The number of iterations may not be too small The optimal solution greatly affects the calculation efficiency, so 30 is generally selected.

In this paper, the maximum kurtosis is the objective function, and the particle swarm optimization (PSO) is used to optimize the period T, the offset M, and the

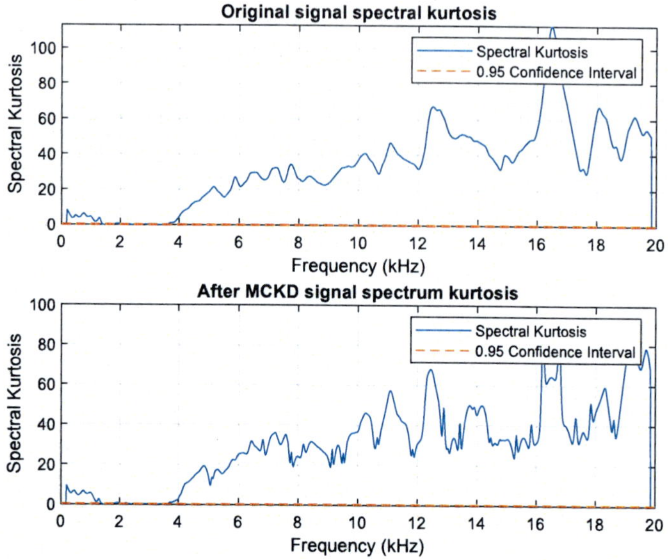

Fig. 7 Signal spectral kurtosis before and after MCKD processing

filter length L, and select the optimal combination. It is calculated that when the filter size L = 89, the number of iterations N = 30, the deconvolution period is T = 32, and the offset M = 4, the optimal solution is obtained. The signal spectrum kurtosis before and after MCKD processing is shown in Fig. 7.

4.3 *Bearing Fault Identification Based on 1D-CNN*

Through the filtered data, the identification model of the damaged bearing is established, and the network structure of the model adopts 1D-CNN. Next, the model needs to be trained. First, these samples are divided into training set, validation set and test set, and the division ratio is 7:2:1. The parameter update in the training process adopts the stochastic gradient descent strategy, the learning rate is set to 0.01, the learning rate decay value after each update is 1e-6, the momentum parameter is set to 0.9, and the Nesterov momentum is enabled. In the training process, the loss function is set as the cross-entropy loss function, and Adam is selected as the optimizer. Set the batch size to 4, epoch = 25, and train the model. Figure 8 shows the training loss and accuracy, Fig. 9 shows the loss and correctness of the validation set. After 25 rounds of training, the accuracy is 1, and the loss is reduced to 0.0241. This also proves that the method is effective.

Finally, the fault is predicted through the obtained fault identification model, and the test data set is brought into the model. The loss is in the range of 0.0364 to 0.0092, and the calculation accuracy is 1, which can correctly identify the fault type.

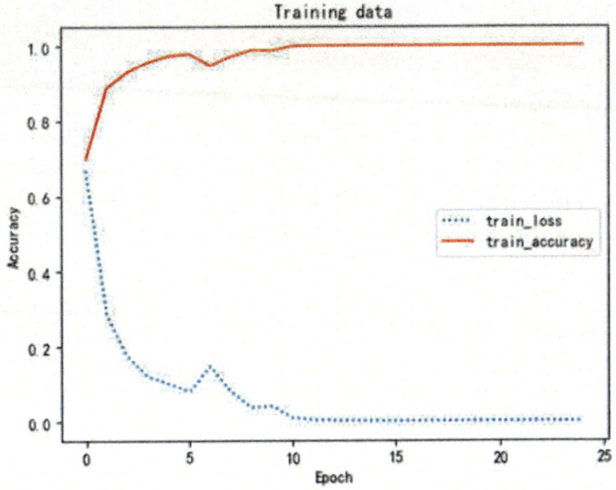

Fig. 8 The loss and accuracy of the training process vary with Epoch

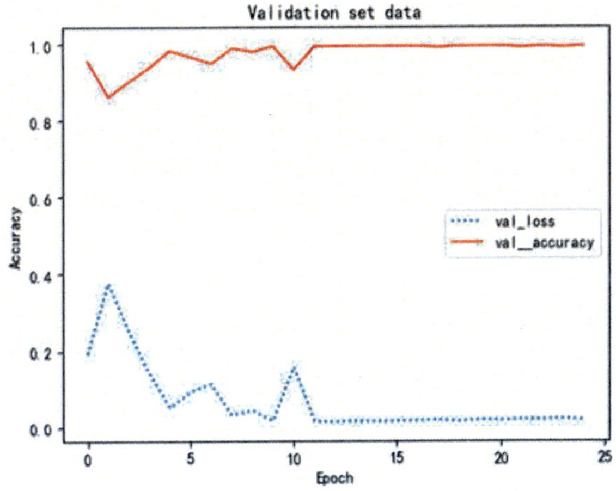

Fig. 9 The verification process loss and accuracy rate vary with Epoch

5 Conclusion

This paper proposes a bearing fault diagnosis algorithm based on particle swarm optimization to optimize the combination of MCKD and 1D-CNN, and verify the effectiveness of the method through experiments. The main research results of this paper are as follows.

(1) Using MCKD to extract the impact component of the signal, the purpose is to eliminate the background noise of the bearing and highlight the impact component of the damage. Because, the MCKD algorithm is greatly affected by the

selection of parameters, and it is not easy to obtain the maximum kurtosis problem, so the particle swarm optimization method is used to obtain the optimal solution.

(2) The signal is analyzed in the time domain, frequency domain, and envelope spectrum, and the similarity characteristics of the signals at different measurement points of the same damage are obtained, and then the correlation analysis results of the signals are carried out to realize the selection of similar data and reduce the subsequent processing.

(3) Design a 1D-CNN network structure, establish a fault bearing identification model, and realize 100% accurate fault identification. This method can be applied to fault detection to improve the efficiency and accuracy of bearing fault detection.

Acknowledgements This paper is supported by Doctoral Fund of Southwest University of Science and Technology (No.21zx7137).

References

1. Wiggins RA (1978) Minimum entropy deconvolution. Geoexploration 16(1):21–35
2. McDonald GL, Zhao Q, Zuo MJ (2012) Maximum correlated kurtosis deconvolution and application on gear tooth chip fault detection. Mech Syst Signal Process 33:237–255
3. Su L, Hairun H, Li K et al (2019) A method for extracting fault features of rolling bearings based on LCD-MCKD. J Huazhong Univ Sci Technol 47(9):19–24
4. Xiaoyan Z, Yongjie W (2019) Early fault diagnosis of rolling bearing based on autocorrelation analysis and MCKD. Vibrat Shock 38(24):183–188
5. Shangkun L, Guiji T, He Y (2016) Teager energy operator combined with MCKD early fault identification of rolling bearings. Vibrat Shock 35(15):98–102
6. Liyang W, Chong D, Xin W et al (2019) A motor fault diagnosis method based on sparse auto encoder. Noise Vibrat Control (5):197–202,249
7. Li H, Hydro Z, Xianrong Q, Yuantao S (2018) Bearing fault diagnosis method based on short-time Fourier transform and convolutional neural network. Vibrat Shock 37(19):124–131
8. Yuxuan Z, Li Q, Bingru Y et al (2019) End-to-end method of bearing fault diagnosis based on LSTM. Noise Vibrat Control 39(6):187–193
9. Xin W, Dongxing M, Li X (2021) Research on motor fault diagnosis based on acoustic signal and one-dimensional convolutional neural network. Noise Vibrat Control 41(02):125–129
10. Zhuang Y, Jianbo Y (2020) Gearbox fault diagnosis method based on multi-channel one-dimensional convolutional neural network feature learning. Vibra Shock 39(20):55–66
11. Haiyan (2019) Rolling bearing fault feature extraction method based on exponential distribution parameters. Automat Technol Applic 38(3):27–30, 42
12. Pu W, Li T, Xuejin G et al (2019) A composite fault feature extraction method for rolling bearings based on LMD and MSEE. Bearings 3:63–69
13. Gilles J (2013) Empirical wavelet transform. IEEE Trans Signal Process 61(16):3999–4010
14. Mohanty GKK, Raju KS (2014) Bearing fault analysis using variational mode decomposition. Int Conf Indust Inf Syst 4(2):1–6

Low-Speed Wind Tunnel Testing of a Passive Camber Morphing Airfoil Using a 3D-Printed Compliant Mechanism

Shoko Kai, Shogo Takazawa, Shuji Ochi, Taro Imamura, Tomohiro Yokozeki, and Kenichi Rinoie

Abstract Experimental investigations into the aerodynamics of the passive trailing-edge morphing airfoil model are presented. While most morphing airfoils are controlled using actuators, the passive morphing airfoil deforms due to the balance between the airfoil surface's static pressure and the model flexibility. In our previous study, the designed morphing airfoil model had an interesting characteristic. As lift force increased, the camber of the airfoil increased and a higher lift was achieved. Meanwhile, the internal structure of the model was complex; thus, 24% thickness airfoil (NACA0024) had to be used, and maintaining a model-manufacturing accuracy was difficult. Considering adaptation to aircraft, it is necessary to study thinner airfoils in terms of aerodynamic performance and to manufacture models more accurately. Therefore, a new passive morphing airfoil model was manufactured based on NACA0012. The internal structure was simplified to realize thickness reduction and high model-manufacturing accuracy. The deformation and aerodynamic characteristics were measured through low-speed wind tunnel experiments. Two deformation characteristics, (1) decreasing of the effective angle-of-attack (AOA) and (2) increasing of the camber, were observed. In addition, the deformation gradually increased as the AOA increased. The maximum lift coefficient increased by 13% compared with the rigid airfoil. Further, the lift coefficient of the passive airfoil tailed off slower than the rigid airfoil at a stall angle.

Keywords Morphing wing airfoil · Low reynolds number flow · Wind tunnel testing · Passive · Adaptable structure

Funded by the Key R&D Projects of Shaanxi Province

S. Takazawa · S. Ochi · T. Yokozeki · K. Rinoie
Department of Aeronautics and Astronautics, The University of Tokyo, Tokyo, Japan

S. Kai (✉) · T. Imamura (✉)
Department of Aeronautics and Astronautics, The University of Tokyo, 7-3-1 Hongo Bunkyo-ku, Tokyo, Japan
e-mail: imamura@g.ecc.u-tokyo.ac.jp

S. Lee et al. (eds.), *The Proceedings of the 2021 Asia-Pacific International Symposium on Aerospace Technology (APISAT 2021), Volume 1*, Lecture Notes in Electrical Engineering 912, https://doi.org/10.1007/978-981-19-2689-1_78

1 Introduction

Morphing airfoil technologies are beneficial for improving fuel efficiency, yielding environmental protection [1]. Morphing airfoil technologies can be broadly classified into two: active morphing airfoils controlled using actuators and passive morphing airfoils that deform due to the balance between the airfoil surface's static pressure and the model flexibility. Most studies are devoted to active morphing airfoils. Woods et al. [2] proposed a fish-bone-active-camber morphing structure, whose lift-to-drag ratio improved 20–25% in the maximum compared with a flapped airfoil. Sugiuchi et al. [3] demonstrated that a morphing wing using a corrugated structure was superior to a conventional hinged wing in terms of lift and rudder effectiveness.

However, passive morphing airfoils could be more eco-friendly than active ones because actuators are not required. Taguchi et al. [4] demonstrated that lift will increase when using passive morphing airfoils, whose trailing edge ($0.3 \leq x/c \leq 1.0$) was manufactured using soft materials (Fig. 1). In general, when a beam is placed in a free stream, it bends along the free stream. However, the internal structure shown in Fig. 1 has an interesting characteristic. The flexible trailing-edge part bends against the free stream Fig. 1b. Flexible materials are used for the upper and lower skins, which are connected by a trailing edge. Between the skins, five links are attached to the surface by hinges that can rotate freely. Then, wind tunnel experiments were conducted; it was revealed that the deformation was created by the balance between the airfoil surface's static pressure and the model flexibility, which increases the maximum lift coefficient, unlike the rigid airfoil. However, three challenges have come to light through this research. The first point is the trailing edge rises due to the deformation, which reduces the lift Fig. 1b. The second point is the limitation of the wing thickness. Because of the complexity of the internal structure, a thick airfoil (NACA0024, 24% thickness) was selected. Considering adaptation to actual aircraft, it is necessary to study thinner airfoils in terms of aerodynamic performance. The third point is the difficulty in maintaining the model-manufacturing accuracy. Multiple components need to be assembled manually to complete the model, which leads to manufacturing errors.

This study investigates the aerodynamic characteristics of a thin trailing-edge morphing airfoil. A new passive morphing airfoil model was manufactured based on the NACA0012. The wing thickness reduced and model-manufacturing accuracy

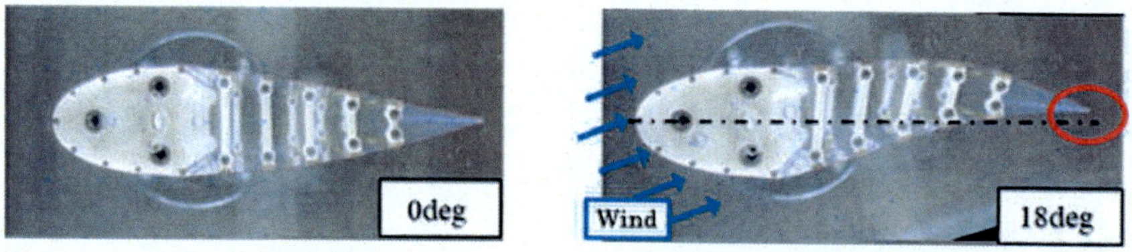

(a) Before deformation (b) After deformation

Fig. 1 Deformation of passive morphing airfoil [4]

improved by simplifying the internal structure. Aerodynamic measurements were performed through low-speed wind tunnel experiments.

2 Wind Tunnel Experiment

The setup of the wind tunnel experiment is explained. The geometry and instrumentation of the wind tunnel test used will be described first, followed by the test parameters in this study. Then, manufacturing methods of airfoil models are explained, with particular attention to the details of the passive morphing airfoil model. Finally, the methodology of deformation analysis will be described.

2.1 *Experimental Equipment*

Wind tunnel experiments were performed in a low-speed blowdown type wind tunnel at the Rinoie-Imamura laboratory, the University of Tokyo (Fig. 2). The test section was an open type, and the outlet was a 600 × 600-mm square. As shown in Fig. 3, the upper and lower plates were placed parallel to the floor 200 mm apart for 2D experiments. The model was installed perpendicular to the plates to prevent the model from being deformed by its weight.

The angle of attack (AOA) was controlled with a rotary stage (ARS-936-HP), which was placed under the lower plate, and aerodynamic forces were measured using a 6-axis force sensor (SFS080F300M5R0U6). In addition, the deformations of the airfoil were measured using a camera installed above the upper plate.

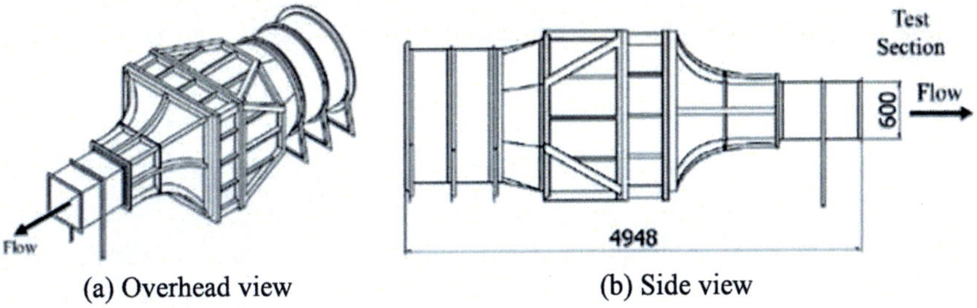

Fig. 2 The wind tunnel overview

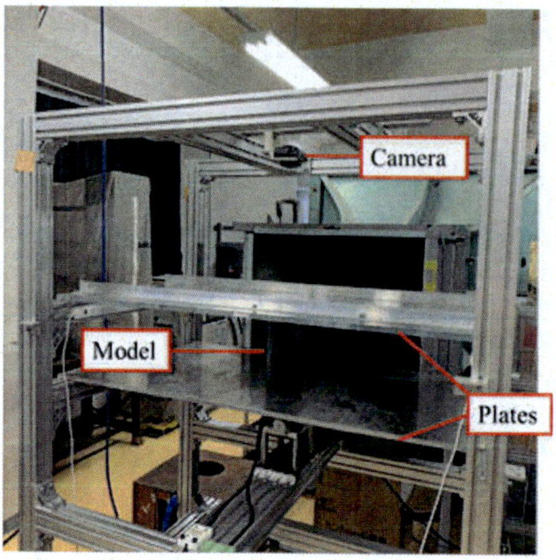

Fig. 3 The wind tunnel experimental setup

Table 1 Experimental parameters

Wind speed	10 m/s
Reynolds number (based on chord length)	1.3×10^5
AOA	$-30° \leq \alpha \leq 30°$ (1° interval)

2.2 *Experimental Parameters*

The tunnel wind speed was set to 10 m/s, and the resulting chord-based Reynolds number was 1.3×10^5 (Table 1). The AOA was varied from −30° to 30° at 1° interval, where the AOA is defined using the chord line of undeformed airfoil geometry. The clockwise rotation viewed above the upper plate is defined as the positive direction.

2.3 *Wind Tunnel Model Construction*

To evaluate the passive morphing airfoil, two models were manufactured: passive and rigid airfoil models. Both airfoils were based on the NACA0012 airfoil. Table 2

Table 2 Geometric parameters of wind tunnel models

Referenced airfoil	NACA0012
Chord length	200 mm
Span length	200 mm
Aspect ratio	1.0

shows an overview of the geometric parameters. The models have a 200-mm span length, a 200-mm chord length, and an aspect ratio of 1.

2.3.1 Passive Morphing Airfoil Model

Figure 4a shows the manufactured passive morphing airfoil model. The model comprises a rigid Sect. ($0 \leq x/c \leq 0.3$) and a morphing Sect. ($0.3 \leq x/c \leq 1.0$).

A rigid section comprises a leading-edge part and two metal round bars (ϕ10). A leading-edge part was 3D-printed from resin (White V4) using a Stereolithography (SLA) 3D printer (Form3) [5]. It had a little step at $x/c = 25\%$ chord so that the

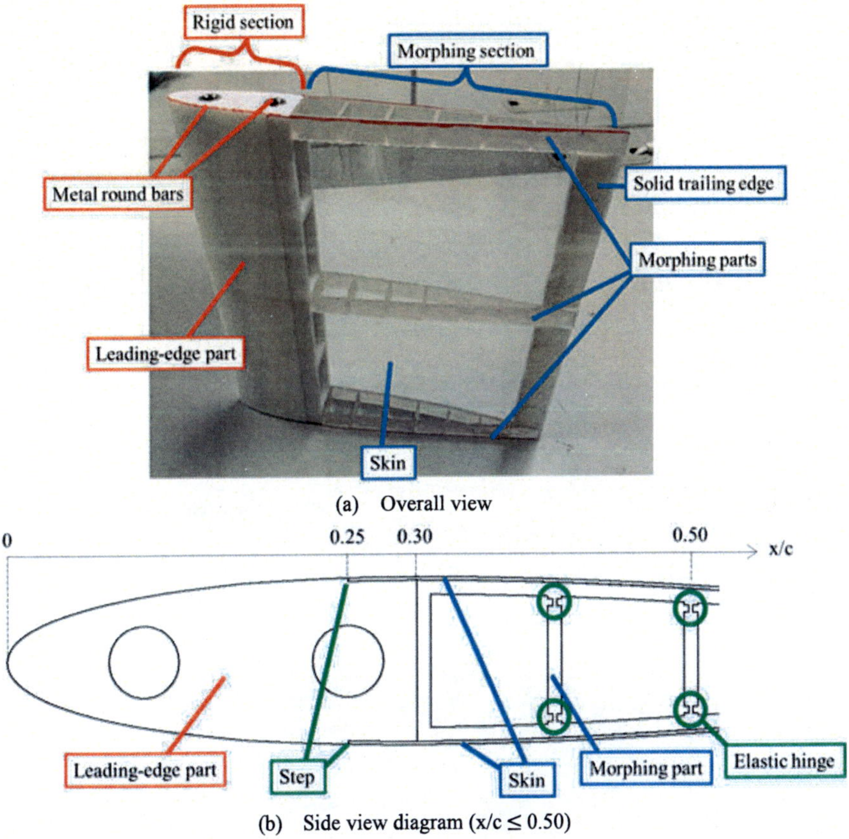

Fig. 4 Passive morphing airfoil model

surface contour would be continuous after the skin was placed Fig. 4b. The two-round bars had female threads on both sides and were used to connect to the circular metal plate which was attached to the force sensor.

A morphing section comprises morphing parts, solid trailing edges, and skin. The key to simplifying the structure of the passive morphing airfoil is the morphing mechanism. In the previous study, five links connecting upper and lower surfaces with hinges were used to realize the morphing mechanism (Fig. 1). On the other hand, 3D printed flexible parts called morphing parts were installed in this study. They were 3D-printed integrally from a material called "Flexible 80A" using an SLA 3D printer (Form2) [6]. Resin (Flexible 80A) is an elastomeric resin designed to print stiff flexible parts that simulate the flexibility of rubber [7]. Hinges in the previous study are replaced by elastic hinges Fig. 4b. By installing morphing parts instead of five links and rotation mechanisms (Fig. 1), manufacturing and structure became simple, yielding two superior points. The first point was that a model using a NACA0012 airfoil could be manufactured, which had a 50% thinner wing thickness than the previous model [4]. The second point was the improvement in the model-manufacturing accuracy. This model assembling procedure will be the basis for future studies on various internal structures. Notably, the edges of the airfoil were colored in red to measure the deformation of the airfoil cross-section, as described in Sect. 2.4.

2.3.2 Rigid Airfoil Model

Figure 5 shows the manufactured rigid airfoil model, which was 3D-printed integrally from Acrylonitrile butadiene styrene using a Fused Filament Fabrication 3D printer (Raise 3D Pro2) [8].

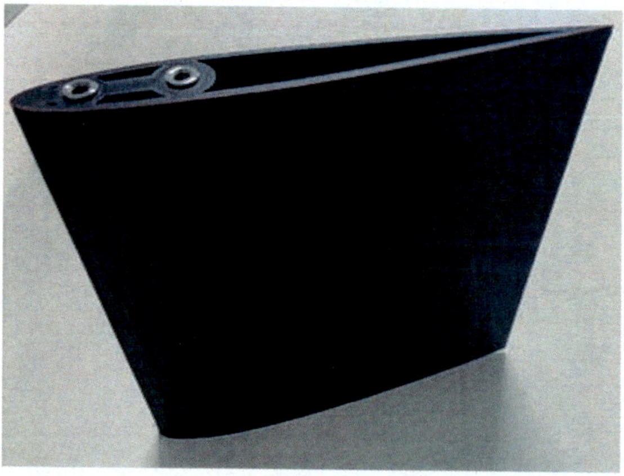

Fig. 5 Rigid airfoil model

(a) Image data (b) Representative points (c) Airfoil deformation data

Fig. 6 Deformation analysis of the passive morphing airfoil

2.4 *Deformation Analysis*

The deformations of the passive morphing airfoil model were measured using a camera located above the upper plate Fig. 3. To analyze deformations in detail, the image data Fig. 6a was processed as described below. Pixels of a red line were extracted by RGB value, and representative points were selected with K-means clustering Fig. 6b. Then, cubic spline interpolation was applied to create airfoil deformation data Fig. 6c. These geometry data were used for numerical analysis described in Sect. 3.

3 Results

The results of the wind tunnel experiment are presented. First, deformation characteristics will be discussed and compared with the previous model [4]. Second, the dependency of the deformation characteristics in terms of the AOA will be discussed. Finally, aerodynamic characteristics will be compared between the passive and rigid models.

3.1 *Deformation Characteristics* ($\alpha = 15°$)

The deformation result at $\alpha = 15°$ is shown in Fig. 7. The static pressure distribution on the airfoil surface raised the trailing edge, which would reduce the lift force because of the decreasing effective AOA. Meanwhile, static pressure increased the

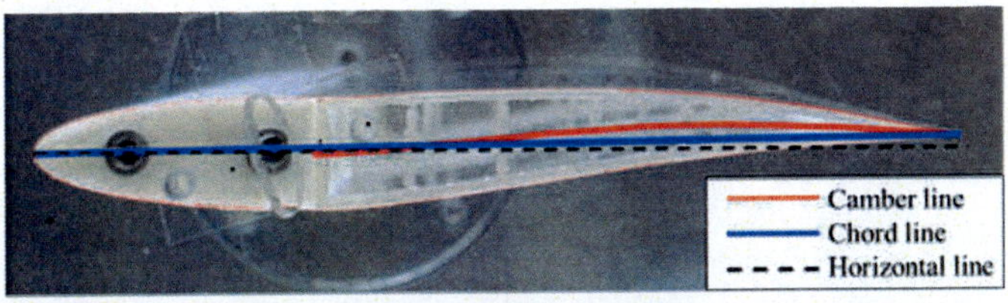

Fig. 7 Deformation of the passive morphing airfoil using a NACA0012 airfoil

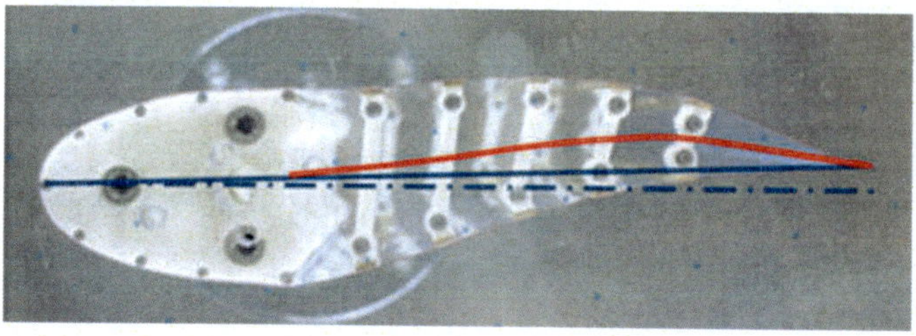

Fig. 8 Deformation of the passive morphing airfoil using a NACA0024 airfoil [4]

camber, which would increase the lift force. The deformation characteristics were the same as the previous airfoil model based on NACA0024 (Fig. 8) [4]. The deformation characteristics observed by the previous model (NACA0024) were realized on the passive morphing wing model based on a NACA0012 airfoil. In Sect. 3.3, the lift force results will be discussed along with these deformation characteristics.

3.2 The AOA Dependencies of the Deformation Characteristics

To investigate the deformations at each AOA, the deformation results were analyzed (Sect. 2.4) (Fig. 9). The deformations were compared with models rotated so that rigid sections matched, and only the morphing sections are shown. The deformations gradually increased as the AOA increased.

To evaluate the deformations quantitatively, two indicators, "decreasing of the effective AOA" and "increasing of the camber," were defined (Fig. 10). There was a negative correlation between the "decreasing of the effective AOA" and AOA Fig. 11a. Meanwhile, there was a positive correlation between the "increasing of the camber" and AOA Fig. 11b.

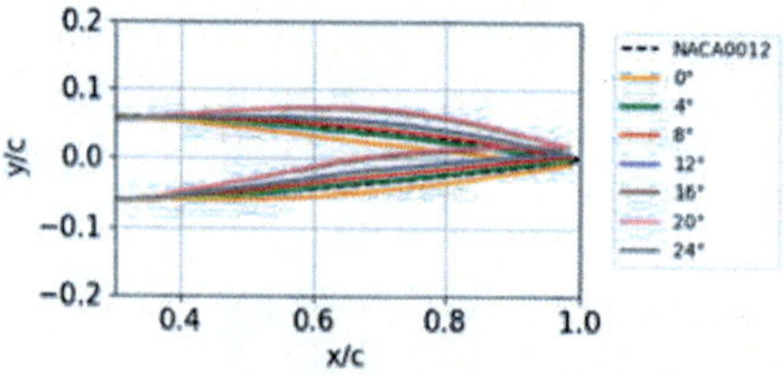

(a) Positive AOA

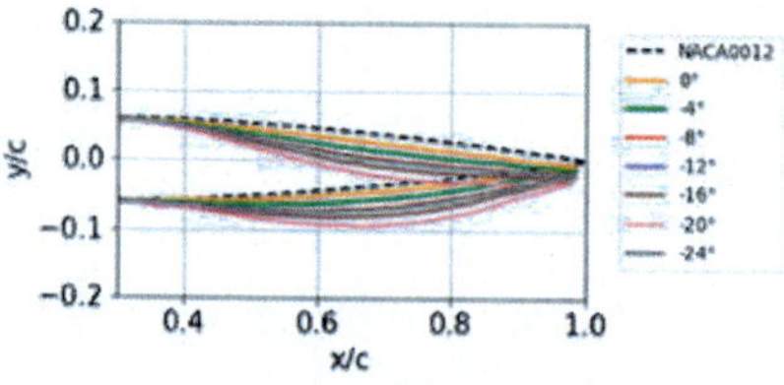

(b) Negative AOA

Fig. 9 Deformation of the passive morphing airfoil model

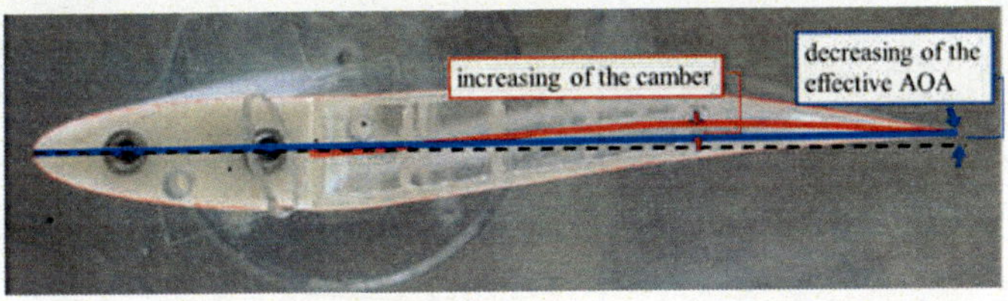

Fig. 10 Definition of "decreasing of the effective AOA" and "increasing of the camber"

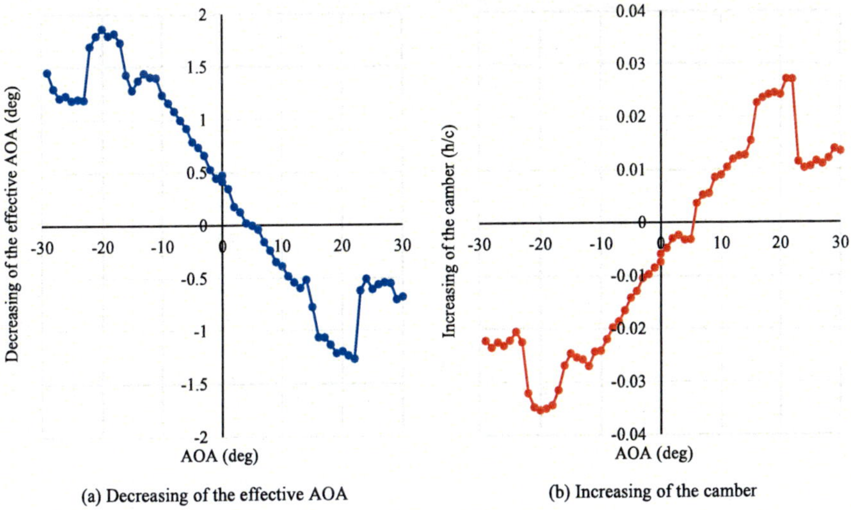

Fig. 11 The AOA dependencies of the deformation characteristics

3.3 *Aerodynamic Characteristics*

Figure 12 shows the lift coefficient C_l against AOA curves for the rigid and passive airfoils. On each figure, two curves are plotted, one is the results at positive AOAs and the other is the results at negative AOAs displaced point-symmetrically to the origin. Since NACA0012 airfoil is a symmetric airfoil, the measured lift coefficient for the rigid airfoil was point symmetric to the origin Fig. 12a. In addition, the curve for the passive airfoil was almost point symmetric to the origin since the passive airfoil was based on NACA0012 airfoil, and the internal structure was symmetric to the chord line Fig. 12b. Therefore, the averaged lift coefficient of positive and negative AOAs was calculated for the comparison of the two models (Fig. 13). The lift curve slope did not significantly change between the two models. However, the linear ranges of the lift curves were different. The linear range of the rigid airfoil was only below 5 °, whereas that of the passive airfoil was below 10 °. Owing to this difference, a larger lift force was observed when the AOA was in the range of 8 °–16 °. In Sect. 3.1, we discussed the deformation characteristics of the passive airfoil.

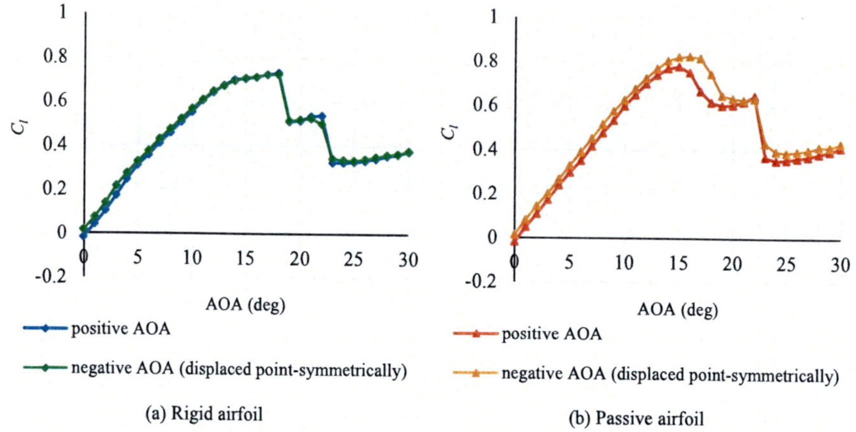

Fig. 12 The lift coefficient comparison between the positive AOA and the negative AOA

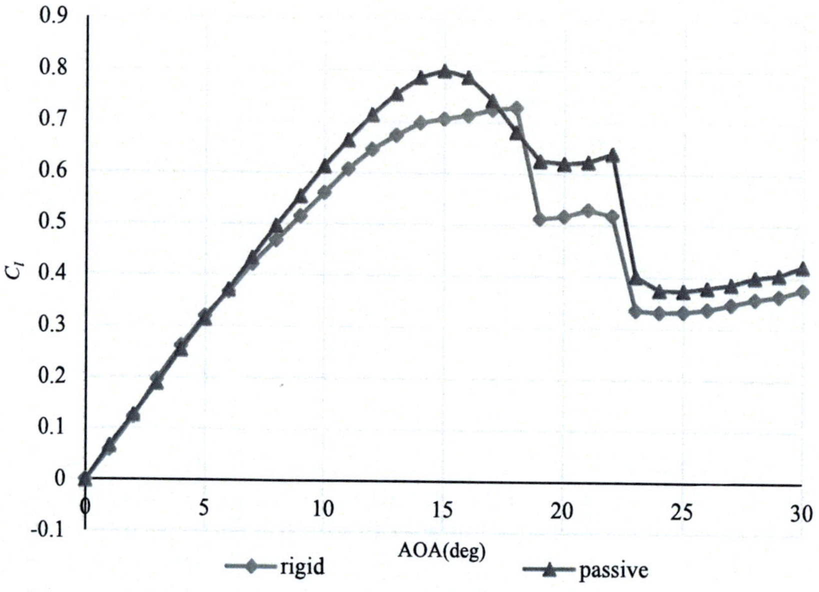

Fig. 13 The lift coefficient comparison between the passive and rigid airfoils

The current force measurement results indicate that the effect of "increasing of the camber" worked better than "decreasing of the effective AOA."

Table 3 compares the lift curve slopes, maximum lift coefficients, and stall angles for a passive airfoil with the rigid airfoil. The lift curve slopes were only about 58% of the theoretical value (2π/rad). The main reason could be the low aspect ratio of the model. A maximum lift coefficient increased by 13%, whereas a stall angle decreased

Table 3 Comparison of lift curve slope, maximum lift coefficient, and stall angle

Airfoil	Rigid	Passive
Lift curve slope	0.064	0.063
Maximum lift coefficient	0.728	0.800
Stall angle	18 °	15 °

by 3 ° compared with the rigid airfoil. Further, while C_l of rigid airfoil dropped at 19 °, C_l of the passive airfoil tailed off slowly above 16 °.

From these observations, we can conclude that the passive airfoil was superior in terms of lift coefficient and maximum lift coefficient, and the lift of the passive airfoil tailed off slower than the rigid airfoil at the stall angle.

4 Conclusion

A passive morphing airfoil model based on a NACA0012 airfoil, having a simple internal structure, was manufactured. Owing to the 3D-printed morphing parts, this model had two superior points. (1) Wing thickness became 50% thinner than the previous model. (2) Model-manufacturing accuracy was improved. This model assembling procedure will be the basis for future studies on various internal structures.

Subsequently, experimentally measured characteristics were discussed. The deformation characteristics of the passive airfoil, which were (1) decreasing of the effective AOA and (2) increasing of the camber, were observed. These characteristics were comparable to those of the previous model. The deformation analysis revealed that the deformations gradually increased as the AOA increased. Next, the lift coefficient against AOA was measured. The passive airfoil was superior to the rigid airfoil above 8°. Further, the maximum lift coefficient increased by 13% compared with the rigid airfoil. These results indicate that the effect of "increasing of the camber" worked better than "decreasing of the effective AOA." In addition, the lift coefficient of the passive airfoil tailed off slower than the rigid airfoil at the stall angle.

Acknowledgements This work was supported by Grant-in-Aid for Scientific Research(B), JSPS KAKENHI Grant Numbers 21H01527.

References

1. Tamayama M (2009) Survey on morphing activities. Nagare J Japanese Soc Fluid Dyn 28:277–284 (in Japanese)
2. Sugiura A, Yokozeki T, Hirano Y (2014) Development of corrugated morphing wing and its driving demonstration in the wind tunnel test. J Aircr 51(3):1023–1029

3. Woods BK, Bilgen O, Friswell MI (2014) Wind tunnel testing of the fish bone active camber morphing concept. J Intell Mater Syst Struct 25(7):772–785
4. Taguchi K, Fukunishi K, Takazawa S, Sunada Y, Imamura T, Rinoie K, Yokozeki T (2020) Experimental study about the deformation and aerodynamic characteristics of the passive morphing airfoil. Trans Japan Soc Aeronaut Space Sci 63(1):18–23
5. Formlabs Inc (2021) Form 3: industrial-quality desktop SLA 3D Printer. Accessed 11 August 2021. https://formlabs.com/3d-printers/form-3/
6. Formlabs Inc (2021) Form 2: affordable desktop SLA 3D Printer. Accessed 11 August 2021. https://formlabs.com/3d-printers/form-2/
7. Formlabs Inc (2021) Resin family: flexible and elastic. Accessed 11 August 2021. https://formlabs.com/materials/flexible-elastic/
8. Raise 3D Technologies, Inc (2021) Best Dual Extruder 3D Printer - Pro2. Accessed 11 August 2021. https://www.raise3d.com/products/pro2-3d-printer/

Surrogate Model-Based Parametric Structural Design of a Composite Tiltrotor Blade

Fan Sun, Chen Jiang, Yang Shen, and Haowen Wang

Abstract Tiltrotor aircraft combine the vertical take-off and landing capability of a helicopter with the high cruising speed of a fixed-wing airplane. During flight, the aircraft switches between helicopter and airplane modes. Meanwhile, structural vibration and deformation are affected by changes in the aerodynamic loads, collective pitch, and rotation speed of blades. Therefore, it is crucial to consider dynamic characteristics in different modes. This paper presents a surrogate-based structural design framework of a tiltrotor blade to speed up calculations while maintaining accuracy. The shape generation method, rapid finite element method (FEM) model generation, and simplified boundary conditions are implemented to compute the blade modal frequencies, mass and inertia. A B-spline is used to construct the shape of the blade with the distribution of airfoils, twist angles, chord length, swept angles, and dihedral angles. These geometric features are realized by approximation, rotation, translation, and scaling of the control points. The spanwise distribution of the C-beam area and the width and length of the skins are selected to parameterize and establish the FEM model. The blind kriging model is applied to develop a surrogate model. The precision of this surrogate model is evaluated and compared to the kriging model and backpropagation neural network (BPNN) model based on the blade of an eight-ton-weight aircraft. A blade with a linear spanwise distribution of the C-beam area is designed using these three surrogate models trained by 13 samples. The results reveal that the error of the blind kriging model concerning the first four modal frequencies is less than 0.5% and lower than that of the other two models. A gimbal hub is modeled using a combination of a simply supported and a clamped boundary condition at the root. The two boundary conditions and two working conditions are unified into one working condition since a linear relationship is found. Moreover, the effects of the segmented linearly distributed cross-sectional area of the C-beam and the geometric parameters of the skins are investigated. The tip and root of the C-beam have the most significant influence on the low-order modal frequency. The parameters of the skins have opposite effects on the 1st and 2nd flap frequencies and the 1st torsion frequency. This framework provides a complete design process on the basis of geometric features and design parameters in the structure design, reducing

F. Sun · C. Jiang (✉) · Y. Shen · H. Wang
School of Aerospace Engineering, Tsinghua University, Beijing 100084, PR China
e-mail: jc2017@mail.tsinghua.edu.cn

S. Lee et al. (eds.), *The Proceedings of the 2021 Asia-Pacific International Symposium on Aerospace Technology (APISAT 2021), Volume 1*, Lecture Notes in Electrical Engineering 912, https://doi.org/10.1007/978-981-19-2689-1_79

the design variables. The use of the surrogate model and simplified working conditions reduce the consumption of time in the design. Conclusions about the tiltrotor blade have reference significance in the design stage.

Keywords Tiltrotor · Surrogate model · Structural design

1 Introduction

Tiltrotor aircraft are a kind of aircraft that combine the vertical or short take-off and landing capability of helicopters with a high cruising speed. By virtue of this unique advantage, they have been studied and investigated since their advent in the 1960s. The first practical tiltrotor demonstrator, named XV-15, made its first flight in 1977 [1]. Later in 1989, V-22 was the first tiltrotor in service and made its first flight [2]. In 2017, the first flight of a new version of tiltrotor V-280 was achieved for the United States Army's Future Vertical Lift (FVL) program [3].

Tiltrotor blades must be designed to operate in helicopter and airplane modes due to variations in the rotation speed and collective pitch of the proprotor. Moreover, it is necessary to introduce complex boundary conditions because of the special hub type. Multiple working conditions and special boundary conditions make the structure design more complicated [4]. Since traditional metal proprotors cannot meet the specifications, composite rotors have attracted more attention due to their long fatigue life with good maintainability and light weight than traditional rotors.

There are two main methods for optimizing the blade structure: the cross-section characteristics, for instance, stiffness distribution and mass distribution, and the geometric parameters, such as skin, girders, Z beams, lead bars, and other structural components [5, 6]. The former method is carried out by modeling the blade by the beam model with centrifugal stress [7]. However, more work is needed to obtain the actual blade design scheme in this method. The latter method is carried out using two-dimensional finite element analysis of each cross-section [8, 9] or three-dimensional finite element analysis of the blade [10]. For multivariable optimization, a surrogate model is applied to improve computational efficiency [11–13].

B-splines represent a class of control point based curves and surfaces. Due to the advantages of variable continuity, intuitiveness, and local control, they have been widely applied in shape design [14–16]. In this paper, parametric modeling for composite blades and a three-dimensional finite element analysis are used on the basis of the shape generated by B-splines. Different approaches to obtain surrogate models are compared. Simplification of the working conditions and boundary conditions are illustrated and is beneficial for quick computation. Furthermore, a sensitivity analysis is conducted to provide specific insight into the effects of the C-beam cross-sectional area and skins.

2 Numerical Methodology

2.1 Blade Shape Generation

B-splines are widely used for producing smooth curves based on basis functions and discrete control points $\boldsymbol{P}_i \in \mathbb{R}^3$, which can be defined as:

$$\boldsymbol{C}(u) = \sum_{i=0}^{n-1} N_{i,p}(u)\boldsymbol{P}_i, u \in [r_0, r_l] \tag{1}$$

$$N_{i,0}(u) = \begin{cases} 1, r_i \leq u \leq r_{i+1} \\ 0, \text{otherwise} \end{cases}$$
$$N_{i,p}(u) = \frac{u-u_i}{u_{i+p}-u_i} N_{i,p-1}(u) + \frac{u_{i+p+1}-u}{u_{i+p+1}-u_{i+1}} N_{i+1,p-1}(u) \tag{2}$$

where u_i are called knots and $U = \{u_0, \ldots, u_m\}$ is a nondecreasing sequence called the knot vector. The ith basis function of p-degree or p + 1 order is denoted as $N_{i,p}(u)$.

Each airfoil is represented by an upper and a lower curve. The cosine distribution of the control points is applied and defined as:

$$P_0 = (0, 0),\ P_i = \left(\frac{1}{2}\left[1 - \cos\left(\frac{\pi(i-1)}{n}\right)\right], a_i\right),\ P_{n-1} = (1, z_{te}) \tag{3}$$

where a_i is the design variable and z_{te} is the width of the trailing edge.

The approximation method of an airfoil curve is carried out in three steps:

1. The x coordinates of the new data points are calculated according to the control points;
2. The z coordinates of the new data points are obtained by spline interpolation;
3. Least squares curve approximation is applied.

In a study by Kulfan [17], the geometric errors of the solution airfoils are defined as

$$z^{error} = \begin{cases} 4 \times 10^{-4} if x/c < 0.2 \\ 8 \times 10^{-4} if x/c > 0.2 \end{cases} \tag{4}$$

where z^{error} are the differences between the approximated shape functions and airfoil data points.

Assume that $C_k(u) = \sum_{i=0}^{n} N_{i,p}(u)\boldsymbol{P}_{i,k}, k = 0, \ldots, K$ are K upper or lower curves of the airfoil and are defined on the same knot vector U and same degree p. Then, for the spanwise direction, a degree q, parameters $\{v_k\}$, and knot vector V are

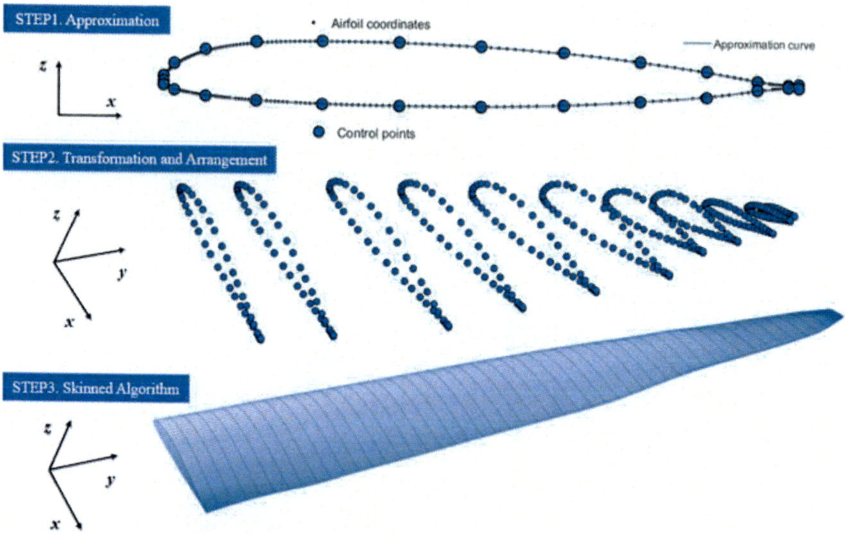

Fig. 1 Shape generation of a blade. A set of control points after transformation are arranged along the span to generate the 3D blade via the skinned algorithm

computed. Next, $\boldsymbol{Q}_{i,j}$ are yielded by n + 1 curve interpolations across the control points of the section curves.

The skinned B-spline surface [18] is obtained as

$$S(u, v) = \sum_{i=0}^{n} \sum_{j=0}^{m} N_{i,p}(u) N_{j,q}(v) \boldsymbol{Q}_{i,j} \tag{5}$$

First, an approximation is applied to change the airfoil coordinates into a series of control points. The upper and lower curves of an airfoil are B-splines of degree 3 with 14 control points. Then, the B-spline curve is transformed with the transformation of all control points, including rotation, translation, and scaling. Therefore, the generated blade can have different twist angles, chord lengths, swept angles, and dihedral angles by this method, improving the flexibility of the design. Next, the skinned algorithm is applied to generate the shape (Fig. 1).

2.2 *Parametric Modeling for Composite Blades*

Blades mainly carry centrifugal force and lift and are generally composed of the skin, main beam, foam, etc. The counterweight, Z-beam, and trailing-edge strip are optional. The functions of these components are shown in Table 1.

Table 1 Blade cross section components

Part name	Major function	Material
C-beam	Force bearing, tensile stiffness, flapping stiffness	Carbon fiber prepreg
Skins	Torsional stiffness	Carbon fiber prepreg
Z-beam	Torsional stiffness	Carbon fiber prepreg
Trailing edge	Lagging stiffness	Carbon fiber prepreg,
Foam	Shape maintenance	Polymethacrylimide

The material of the C-beam, Z-beam, and skins is a kind of carbon fiber prepreg, while the material of the foam is a kind of polymethacrylimide. These two materials are used in actual production, and their properties are shown in Tables 2 and 3.

Since the grid has a significant impact on the calculation accuracy of the finite element, it is appropriate to use a structured grid. Conventional shell elements and continuum shell elements are the two most common forms of skin models. The conventional shell elements discretize the skin by defining the geometry along the surface. High computational efficiency is achieved due to one layer of mesh in this model with relatively compromised accuracy. The continuum shell elements model each layer of the skin separately with low calculation efficiency but high accuracy. This paper uses conventional shell elements in the optimization process and continuum shell elements in the analysis process, considering the different simulation requirements.

The C-beam, foam, and other components inside the blade can be modeled by a hexahedral grid while avoiding distortion as much as possible. Due to the low stiffness of the foam, a sparse mesh can be used in the model. The design parameters of each component are listed in Fig. 2. In the following section, the C-beam and skins are analyzed, while the parameters of other components without bolds are predefined. Furthermore, the area of the C-beam (s_{cbeam}) is used intuitively to replace d_1.

In this paper, the numerical analysis is carried out using MATLAB with FEM software as the solver. The establishment of the surrogate model and optimization are conducted using MATLAB.

According to related research, a gimbal hub can be modeled as a combination of a simply supported and a clamped boundary condition [4].

Table 2 Carbon fiber prepreg

E_1(MPa)	E_2(MPa)	ν_{12}	G_{12}(MPa)	G_{13}(MPa)	G_{13}(MPa)	$\rho(kg/m^3)$
145,000	8250	0.25	4100	4100	2750	1570

Table 3 Polymethacrylimide

E (MPa)	ν	$\rho(kg/m^3)$
74	0.3	52

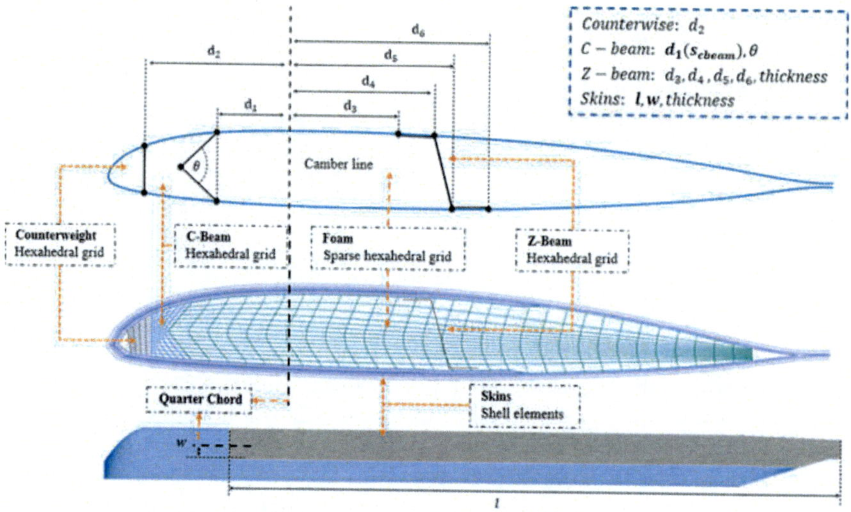

Fig. 2 Schematic diagram of finite element meshing for a blade

The assumptions are listed as follows:

1. The edgewise stiffness is rigid;
2. Each blade is in the same periodic aerodynamic environment;
3. The rotor is composed of three blades, and the phase angles are $-\frac{2\pi}{3}$, 0, $\frac{2\pi}{3}$.

In rotational coordinates, φ_m is the rotation angle, and the moment of each blade can be written as:

$$\begin{array}{c} M_*^{(1)} = M_{*_0} + \sum_{n=1}^{\infty} M_{*_c}\cos(\varphi_m) + M_{*_s}\sin(\varphi_m) \\ M_*^{(2)} = M_{*_0} + \sum_{n=1}^{\infty} M_{*_c}\cos\left(\varphi_m - \frac{2\pi}{3}\right) + M_{*_s}\sin\left(\varphi_m - \frac{2\pi}{3}\right) \\ M_*^{(3)} = M_{*_0} + \sum_{n=1}^{\infty} M_{*_c}\cos\left(\varphi_m + \frac{2\pi}{3}\right) + M_{*_s}\sin\left(\varphi_m + \frac{2\pi}{3}\right) \\ * = flapping, lagging, torsional \end{array} \tag{6}$$

The flapping moment of the proprotor M_{flap} can be written as:

$$\begin{array}{c} M_{flap} = M_{flap}^{(1)} - \frac{1}{2}M_{flap}^{(2)} - \frac{1}{2}M_{flap}^{(3)} \\ M_{flap} = \sum_{n=1}^{\infty} [1 - \cos(\frac{2n\pi}{3})][M_{flap_c}\cos(\varphi_m) + M_{flap_s}\sin(\varphi_m)] \end{array} \tag{7}$$

Similarly, the lagging moment $M_{lagging}$ and torsional moment $M_{torsional}$ are:

$$\begin{array}{c} M_* = M_*^{(1)} + M_*^{(2)} + M_*^{(3)} \\ M_* = 3M_{*_0} + \sum_{n=1}^{\infty} [1 + \cos(\frac{2n\pi}{3})][M_{*_C}\cos(\varphi_m) + M_{*_s}\sin(\varphi_m)] \\ * = lagging, torsional \end{array} \tag{8}$$

Two coefficients $[1 - cos(\frac{2n\pi}{3}), 1 + cos(\frac{2n\pi}{3})]$ directly influence the type of boundary condition and can be valued as [0, 2] or [1.5, 0.5]. For [0, 2], the hub can be regarded as the clamped boundary condition, while the hub approximates the simply-supported boundary condition for the other situation.

2.3 Surrogate Model

A surrogate model is used to construct an approximative mapping from inputs to outputs of the simulation model while being computationally favorable. The key point is to generate sample points and build the approximate model.

The commonly used sampling methods include complete sampling, random sampling, Latin hypercube sampling (LHS), Sobol sampling, Halton sampling, etc. [19] A reasonable distribution of sample points supports better prediction accuracy, even if a small number of sample points is used [20]. In this paper, LHS is used because of its small quantity of computation and uniform sampling.

The surrogate models are obtained based on the samples. The frequently used approaches are the polynomial model, radial basis function (RBF), kriging model, support vector regression (SVR), and neural network regression (NNR) [20]. In this paper, the kriging model, blind kriging model, and backpropagation neural network (BPNN) are used to fit the first six frequencies of the blade, mass, chordwise center of gravity, and inertia.

The ordinary kriging model [21–23] can be formulated as:

$$Y(\boldsymbol{x}) = \mu_0 + Z(\boldsymbol{x}) \tag{9}$$

where μ_0 is the known trend and $Z(x)$ is a random error with a mean of zero. The covariance function is denoted as:

$$\mathrm{cov}(Y(\boldsymbol{x}+\boldsymbol{h}), Y(\boldsymbol{x})) = \sigma^2\psi(\boldsymbol{h}) \tag{10}$$

Let $\boldsymbol{y} = (y_1, y_2, \ldots, y_n)^T$ be the function values at $\{\boldsymbol{x}_1, \boldsymbol{x}_2, \ldots, \boldsymbol{x}_n\}$. The best linear unbiased predictor is written as:

$$\widehat{y}(\boldsymbol{x}) = \widehat{\mu}_0 + \boldsymbol{\psi}(\boldsymbol{x})^T\boldsymbol{\Psi}^{-1}(\boldsymbol{y} - \widehat{\mu}_0\mathbf{1}) \tag{11}$$

where $\mathbf{1}$ is a vector of size $n \times 1$ with a value of 1; $\boldsymbol{\psi}(\boldsymbol{x})^T = (\psi(\boldsymbol{x}-\boldsymbol{x}_1), \psi(\boldsymbol{x}-\boldsymbol{x}_2), \ldots, \psi(\boldsymbol{x}-\boldsymbol{x}_n))$, where $\boldsymbol{\Psi}$ is an $n \times n$ matrix with elements $\psi(\boldsymbol{x}_i - \boldsymbol{x}_j)$; and $\widehat{\mu}_0 = \mathbf{1}^T\boldsymbol{\Psi}^{-1}\boldsymbol{y}/\mathbf{1}^T\boldsymbol{\Psi}^{-1}\mathbf{1}$.

Specifying a trend or a value for the mean when it is unknown may lead to inaccuracy in the prediction. In the blind kriging model, the functions used to describe the trend are unknown and identified through data-analytic procedures [24]. The model is given by:

$$Y(x) = v(x)^T \mu_m + Z(x) \tag{12}$$

where $v(x)^T = (1, v_1, \ldots, v_m)$, $\mu_m = (\mu_0, \mu_1, \ldots, \mu_m)^T$, and m are unknown. The predictor is written as:

$$\begin{aligned} \hat{y}(x) &= v(x)^T \hat{\mu}_m + \psi(x)^T \Psi^{-1}(y - V_m \hat{\mu}_m) \\ V_m &= (v(x_1), v(x_2), \cdots, v(x_n))^T \\ \hat{\mu}_m &= \left(V_m^T \Psi^{-1} V_m\right)^{-1} V_m^T \Psi^{-1} y \end{aligned} \tag{13}$$

where V_m is an $n \times (m + 1)$ matrix. $v(x)$ can be identified by choosing from a set of candidate functions.

BPNN [25] is a multilayer feedforward neural network and is often used to address nonlinear problems. A typical BPNN usually consists of the input layer, output layer, and several hidden layers. The input layer indicates the input information, and the output layer shows the learning result. Each connection corresponds to a weight value, and supervised algorithms train the network. The weight is updated according to the gradient descent algorithm, while the error between the actual output and the expected result is sequentially propagated to the input layer.

2.4 Aircraft Aerodynamics Model

Referring to existing tiltrotor aircraft, for instance, XV-15, V-22, and V280, the aircraft model consists of two proprotors, a wing, a fuselage, and a V-shaped tail. The schematic diagram is shown in Fig. 3a.

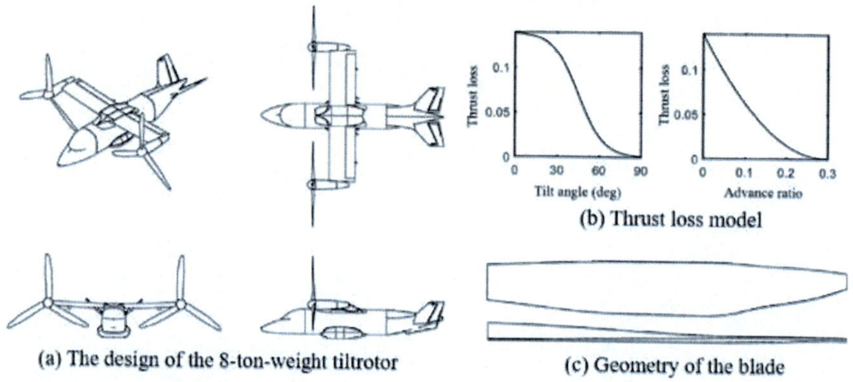

Fig. 3 Schematic diagram

Table 4 Working condition

	$\theta_0/°$	C_T	T/kg	C_P	P/kW	FM	Ω/rpm
Hovering	17.45	0.0135	4891	0.0014572	1184	0.7600	520
Cruising	49.97	0.0015	426	0.0016381	921	0.7556	460

The blade element momentum theory is applied in the proprotor model [26]. The lift coefficient.

Cl and drag coefficient Cd of the blade airfoil segments are obtained by lookup tables. The tables are generated by using CFD software and XFOIL [27]. The aerodynamic coefficient of the wing and the drag coefficient of the fuselage are calculated using the panel method. The interference of the wing and the proprotor is modeled as a function of the tilt angle and advance ratio in Fig. 3b, according to the study by Felker et al. [28]. The geometry of the blade is shown in Fig. 3c.

The parameters of this tiltrotor model under the two conditions are listed in Table 4. It is shown that the differences in collective pitch and thrust are great under the two working conditions.

where θ_0 is the collective pitch; T is the thrust; and C_T is obtained by $T/\rho A(\Omega R)^2$ while A is the disc area, R is the radius and ρ is the air density. Similarly, the power coefficient C_P is acquired by $P/\rho A(\Omega R)^3$ and P is the power. FM, namely, figure of merit, is defined as Tv_i/P, where v_i is the induced velocity. Usually, FM is used to evaluate the efficiency of the rotor.

3 Result Analysis

3.1 *Comparison of Surrogate Models*

In this paper, the length of the inner six skins, the width of the inner six skins, and the distribution of the cross-sectional area of the C-beam are chosen and marked as the design parameters in Table 4, while others are fixed in the design.

The components are designed stepwise, and the C-beam is the first due to its greatest impact on the low order modal frequency of the blade. The cross-sectional area of the C-beam changes linearly from the root to the tip. $S_{root}(mm^2)$ denotes the area of the root of the C beam, and $S_{tip}(mm^2)$ denotes the area of the tip. The bound of S_{tip} is [100, 500], and it is [2500, 4500] for S_{root}.

The characteristics of the blade to be considered in the design are listed in Table 5. The parameters of the blade are listed in Table 6.

Through Latin hypercube sampling and corner sampling shown in Fig. 4a, 13 sample points are selected as the training set, and ($S_{root} = 4000$, $S_{tip} = 300$) is selected as the test set.

Table 5 Characteristics

F1	F2	F3	F4	
1st flap frequency	1st lag frequency	2nd flap frequency	1st torsion frequency	
F5	F6	cg	M	I_ρ
3rd flap frequency	4th flap frequency	Center of gravity	Mass	Inertia

Table 6 Parameters

Skins								
Layer	1	2	3	4	5	6	7	8
Thickness/mm	0.15	0.15	0.15	0.15	0.15	0.15	0.15	0.15
Ply angle/°	±45	0	0	0	0	0	0	0
Width	Whole blade		B1	B1	B2	B2	B3	B3
Length	Whole blade		A1	A1	A2	A2	A3	A3
C-beam								
Ply angle/°	0							
Sectional area/mm^2	$(S_{root}, S_{tip}) or x_1 \sim x_9$							
Z-beam								
Ply angle/°	±45							
Sectional area/mm^2	Fixed							

In Fig. 4b, it is shown that the blind kriging model has higher accuracy in all characteristics, and the error is less than 0.5% in the first four modal frequencies, better than the other two surrogate models. The kriging model performs poorly in estimating the center of gravity, while the BPNN has a larger error in the estimation of mass and inertia. Comprehensively considering the performance of the three methods, blind kriging is chosen and used in the following analysis.

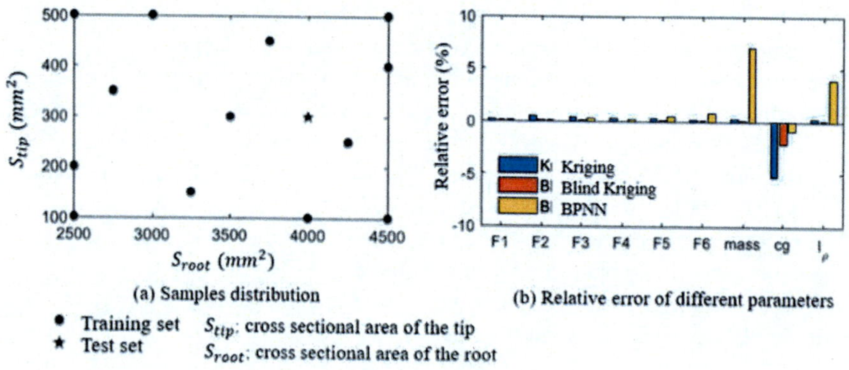

Fig. 4 Sampling and comparison of different surrogate models

3.2 Simplified Boundary Conditions and Working Conditions

The modal frequencies of the blade under two working conditions and two boundary conditions are obtained using the blind kriging model. The distribution trend of each order frequency under different working conditions and boundary conditions is similar. Thus, F2 is taken as an example to illustrate the relationship.

In Fig. 5, $Freq_{hover}$ is the F2 frequency of the blade when hovering, $Freq_{cruise}$ is the F2 frequency when cruising, $Freq_C$ is the F2 frequency under the clamped boundary condition, and $Freq_S$ is the F2 frequency under the simply supported boundary condition. Figure 5a, b shows that the frequency varies with S_{root} and S_{tip} under the clamped boundary condition, and Fig. 5c shows the linear relationship between $Freq_{hover}$ and $Freq_{cruise}$. Figure 5d, e shows that the frequency varies with S_{root} and S_{tip} under the simply supported boundary condition, and Fig. 5f shows the linear relationship between $Freq_{hover}$ and $Freq_{cruise}$. Moreover, the linear relationship between the two boundary conditions when hovering and cruising is also presented in Fig. 5g.

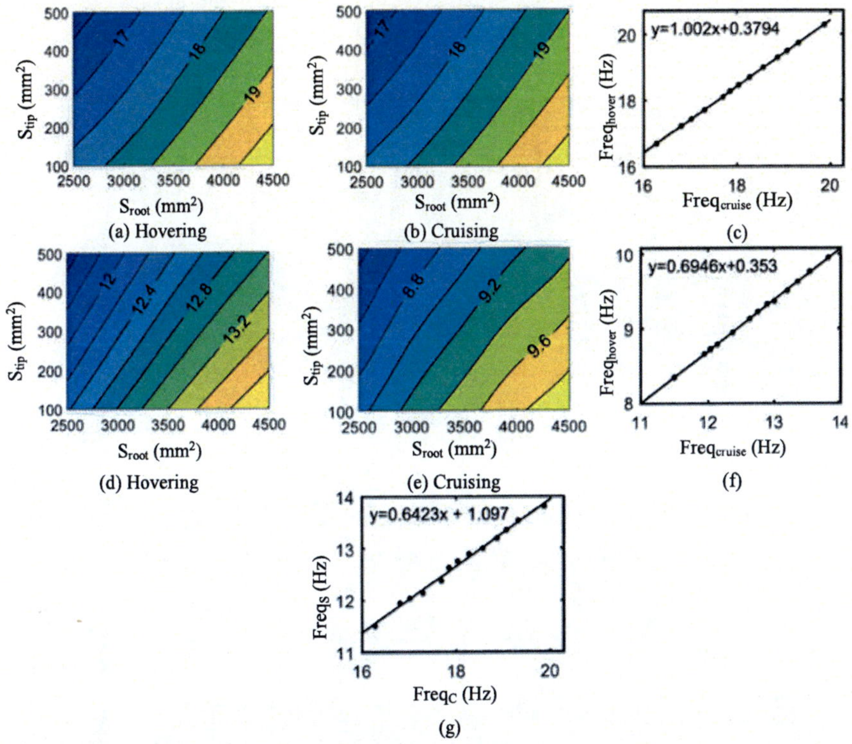

Fig. 5 F2 distribution and linear relationship

From the above results, the two boundary conditions and two aerodynamic loads are unified into one working condition. This simplification can reduce the calculation time and improve the efficiency.

3.3 *Effects of Parameters*

The influence of the cross-sectional area distribution of the C-beam and skin geometry on the blade frequency is analyzed. The sensitivity analysis is conducted and shown in Fig. 6.

For the cross-sectional area distribution of the C beam, nine cross sections are selected at intervals of 0.1R from 0.2R, denoted as $x_1 \sim x_9$. Each cross-sectional area floats by 5% based on the previous linear distribution in Sect. 3.1. The lower bound of the parameters is [−215, −188.875, −162.75, −136.625, −110.5, −84.375, −58.25, −32.125, −6], and the upper bound is [215, 188.875, 162.75, 136.625, 110.5, 84.375, 58.25, 32.125, 6]. Latin hypercube sampling is used to obtain 42 sample points. The length and width of the skins are selected as optimization variables, with a total of 6, denoted as A1, A2, A3, B1, B2, and B3. The basic values are [4200,

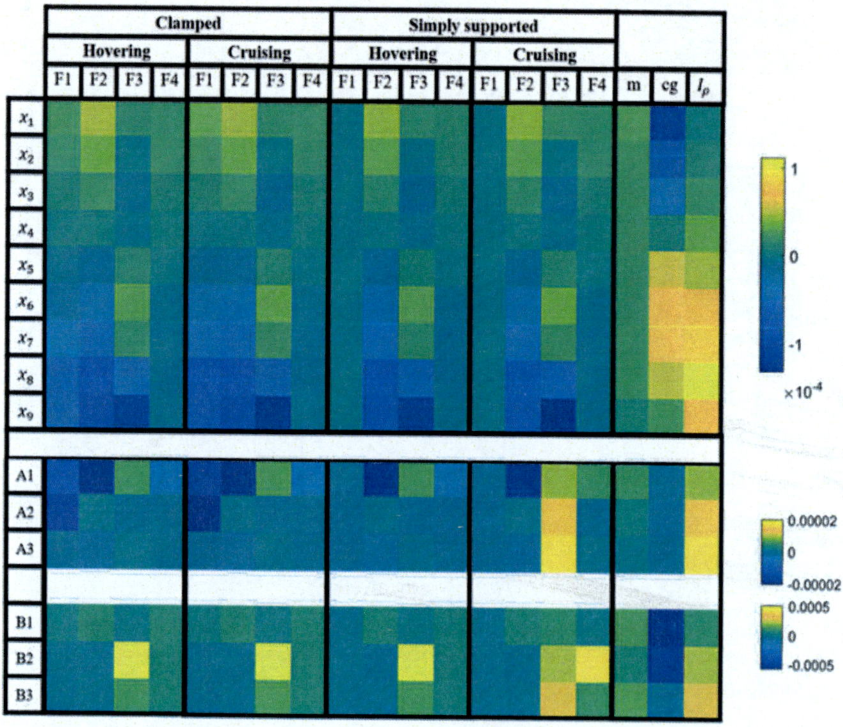

Fig. 6 Sensitivity analysis

4200, 4200, 60, 0, −20], the lower bound is [840, 840, 840, −20, −20, −20], and the upper bound is [4200, 4200, 4200, 60, 60, 60], where the quarter chord is defined as 0. Latin hypercube sampling and corner sampling are used to obtain 21 sample points. The blind kriging algorithm is used to build two surrogate models for these two kinds of samples. The parameters are listed in Table 6.

The C-beam at the tip has the most significant influence on the 1st flap frequency (F1), 1st lag frequency (F2), and 2nd flap frequency (F3), whereas the cross-sectional area of the C-beam at the root is secondary. However, the cross-sectional area of the C-beam has little effect on the 1st torsion frequency (F4). The 1st flap frequency (F1), 1st lag frequency (F2), and 2nd flap frequency (F3) are positively correlated with the root cross-sectional area and negatively correlated with the blade tip cross-sectional area. Therefore, to increase the low-order frequency of the blade, the most effective method is to reduce the cross-sectional area at the tip. The cross-sectional area also has a significant impact on the center of gravity and inertia, with the cross-sections at 0.8R and 0.9R having the most significant impact.

The effects of the width and length of the skins reveal that lengths mainly influence the 1st flap frequency (F1) and the 1st lag frequency (F2), while widths impact the 2nd flap frequency (F3) and 1st torsion frequency (F4). Moreover, the width of the skins has a more significant influence on the center of gravity.

4 Summary and Conclusion

This paper presents a surrogate-based structural design framework for tiltrotor aircraft. The differences in rotation speed and collective pitch when hovering and cruising must be considered in the design and optimization. B-splines are used to generate the shape of the blade with inputs of airfoils, twist angles, chord length, and swept angles. Based on the shape, a parametric modeling method and rapid grid division are applied. The FEM solver is then used to calculate the characteristics of samples, which are utilized for training and testing the surrogate model. The results show that:

(1) The blind kriging model is verified to be more accurate with an error of less than 0.5% when estimating the first four modal frequencies, mass, center of gravity, and inertia;
(2) In the design of the cross-sectional area of the C-beam, a linear relationship is found between hovering, cruising, clamped boundary conditions, and simply supported boundary conditions, which can improve the design efficiency;
(3) The cross-sectional area of the C beam affects the 1st flap frequency, 1st lag frequency, and 2nd flap frequency, while the geometric parameters of the skins affect the 1st flap frequency, 1st lag frequency, and 1st torsion frequency.

References

1. Maisel MD (2000) The history of the XV-15 tilt rotor research aircraft: from concept to flight: national aeronautics and space administration, office of policy and plans~…
2. Bolkcom C (2004) V-22 osprey tilt-rotor aircraft. Library Of Congress Washington Dc Congressional Research Service
3. Bell V-280 valor achieves first flight (2017) https://news.bellflight.com/en-US/162449-bell-v-280-valor-achieves-first-flight
4. YU Guorui JL, Jun H (2017) Research and development of tilt-rotor structure dynamic design. Helicopter Technique (1):1–7
5. Celi R (1999) Recent applications of design optimization to rotorcraft-a survey. J Aircraft 36(1):176–189
6. Sobieszczanski-Sobieski J, Haftka RT (1997) Multidisciplinary aerospace design optimization: survey of recent developments. Struct Optimization 14(1):1–23
7. Eun W, Sim J, Lee S, Shin S (2018) Further improvements in the SNUF blade design by numerical design optimization framework. J Aerospace Eng. 31(1):4017081
8. Cesnik CE, Hodges DH (1997) VABS: a new concept for composite rotor blade cross-sectional modeling. J AM Helicopter Soc 42(1):27–38
9. Yang Jianling ZL, Shaohua Z (2013) Composite helicopter rotor blade optimization design based on parametric module definition. Acta Aeronautica ET Astronautica Sinica 34(3):554–65
10. Skiba K (2019) Designing and FEM simulation of the helicopter rotor and hub. IOP Publishing, p 12003
11. Glaz B, Friedmann PP, Liu L (2008) Surrogate based optimization of helicopter rotor blades for vibration reduction in forward flight. Struct Multidiscip O 35(4):341–363
12. Glaz B, Friedmann PP, Liu L (2009) Helicopter vibration reduction throughout the entire flight envelope using surrogate-based optimization. J Am Helicopter Soc 54(1):12007
13. Glaz B, Goel T, Liu L, Friedmann PP, Haftka RT (2009) Multiple-surrogate approach to helicopter rotor blade vibration reduction. AIAA J 47(1):271–282
14. P E Rez-Arribas F, P E Rez-Fern A Ndez RUL (2018) A B-spline design model for propeller blades. Adv Eng Softw 118:35–44
15. Jin X, Sun G (2012) Aerodynamic optimization for parameterized wing based on improved particle-swarm-optimization algorithm. Chinese Quart Mech 3
16. Zhang X, Jesudasan R, M U Ller (2019) J Adjoint-based aerodynamic optimisation of wing shape using Non-Uniform Rational B-splines. Evolutionary and deterministic methods for design optimization and control with applications to industrial and societal problems. Springer, p 143–58
17. Kulfan B, Bussoletti J (2006) Fundamental parameteric geometry representations for aircraft component shapes, p. 6948
18. Piegl L, Tiller W (1996) The NURBS book. Springer Science & Business Media
19. Forrester A, Sobester A, Keane A (2008) Engineering design via surrogate modelling: a practical guide. John Wiley & Sons
20. Yuepeng BU, Wenping S, Zhonghua H, Zhang Y, Zhang L (2020) Aerodynamic/aeroacoustic variable-fidelity optimization of helicopter rotor based on hierarchical kriging model. Chinese J Aeronaut 33(2):476–492
21. Currin C, Mitchell T, Morris M, Ylvisaker D (1991) Bayesian prediction of deterministic functions, with applications to the design and analysis of computer experiments. J Am Stat Assoc 86(416):953–963
22. Wackernagel H (2003) Multivariate geostatistics: an introduction with applications. Springer Science & Business Media
23. Welch WJ, Buck RJ, Sacks J, Wynn HP, Mitchell TJ, Morris MD (1992) Screening, predicting, and computer experiments. Technometrics 34(1):15–25
24. Joseph VR, Hung Y, Sudjianto A (2008) Blind kriging: a new method for developing metamodels

25. Rumelhart DE, Hinton GE, Williams RJ (1986) Learning representations by back-propagating errors. Nature 323(6088):533–536
26. Leishman GJ (2006) Principles of helicopter aerodynamics. Cambridge university press
27. Drela M (1989) XFOIL: an analysis and design system for low Reynolds number airfoils. Springer, Low Reynolds number aerodynamics, pp 1–12
28. Felker FF (1988) A review of tilt rotor download research